Advances in Intelligent Systems and Computing

Volume 843

Series editor

Janusz Kacprzyk, Polish Academy of Sciences, Warsaw, Poland
e-mail: kacprzyk@ibspan.waw.pl

The series "Advances in Intelligent Systems and Computing" contains publications on theory, applications, and design methods of Intelligent Systems and Intelligent Computing. Virtually all disciplines such as engineering, natural sciences, computer and information science, ICT, economics, business, e-commerce, environment, healthcare, life science are covered. The list of topics spans all the areas of modern intelligent systems and computing such as: computational intelligence, soft computing including neural networks, fuzzy systems, evolutionary computing and the fusion of these paradigms, social intelligence, ambient intelligence, computational neuroscience, artificial life, virtual worlds and society, cognitive science and systems, Perception and Vision, DNA and immune based systems, self-organizing and adaptive systems, e-Learning and teaching, human-centered and human-centric computing, recommender systems, intelligent control, robotics and mechatronics including human-machine teaming, knowledge-based paradigms, learning paradigms, machine ethics, intelligent data analysis, knowledge management, intelligent agents, intelligent decision making and support, intelligent network security, trust management, interactive entertainment, Web intelligence and multimedia.

The publications within "Advances in Intelligent Systems and Computing" are primarily proceedings of important conferences, symposia and congresses. They cover significant recent developments in the field, both of a foundational and applicable character. An important characteristic feature of the series is the short publication time and world-wide distribution. This permits a rapid and broad dissemination of research results.

More information about this series at http://www.springer.com/series/11156

Faisal Saeed · Nadhmi Gazem
Fathey Mohammed · Abdelsalam Busalim
Editors

Recent Trends in Data Science and Soft Computing

Proceedings of the 3rd International Conference of Reliable Information and Communication Technology (IRICT 2018)

IRICT 2018

Springer

Editors
Faisal Saeed
College of Computer Science and Engineering
Taibah University
Medina, Saudi Arabia

Fathey Mohammed
Information Technology Department, Faculty of Engineering and Information Technology
Taiz University
Taiz, Yemen

Nadhmi Gazem
Information Systems Department, Faculty of Computing
Universiti Teknologi Malaysia
Skudai, Malaysia

Abdelsalam Busalim
Information Systems Department, Faculty of Computing
Universiti Teknologi Malaysia
Skudai, Malaysia

ISSN 2194-5357 ISSN 2194-5365 (electronic)
Advances in Intelligent Systems and Computing
ISBN 978-3-319-99006-4 ISBN 978-3-319-99007-1 (eBook)
https://doi.org/10.1007/978-3-319-99007-1

Library of Congress Control Number: 2018950928

This Springer imprint is published by the registered company Springer Nature Switzerland AG
The registered company address is: Gewerbestrasse 11, 6330 Cham, Switzerland

Preface

It is an honor and a great pleasure to welcome all of you to the 3rd International Conference of Reliable Information and Communication Technology 2018 (IRICT 2018) held at Hotel Bangi-Putrajaya, Malaysia, on July 23–24, 2018.

The conference is organized by the Yemeni Scientists Research Group (YSRG), Information Service Systems and Innovation Research Group (ISSIRG) in Universiti Teknologi Malaysia (UTM), Faculty of Computer and Mathematical Sciences in Universiti Teknologi Mara (UiTM), Information Systems Department in Universiti Teknologi PETRONAS (UTP), Information Systems Department and Data Science Research Group in Taibah University (Kingdom of Saudi Arabia), and School of Science and Technology in Nottingham Trent (UK).

IRICT 2018 is a forum for the presentation of technological advances and research results in the field of ICT. The conference aims to bring together leading researchers, engineers, and scientists in the domain of interest from around the world.

IRICT 2018 attracted a total of 170 submissions from 27 countries including Australia, Bangladesh, Canada, China, India, Indonesia, Iraq, Malaysia, Mali, Morocco, Nigeria, Oman, Pakistan, Palestine, Saudi Arabia, Singapore, Somalia, Spain, Sri Lanka, Sudan, Taiwan, Thailand, Tunisia, Turkey, United Arab Emirates, UK, and Yemen.

These submissions underwent a rigorous double-blind peer review process. Of those 170 submissions, 103 submissions (66%) have been selected to be included in this book.

The book presents several hot research topics which include data science and big data analytics, artificial intelligence: machine learning and optimization techniques; business intelligence for Industrial Revolution 4.0; Internet of things (IoT): issues, techniques, and applications; intelligent communication systems; advances in information security; computational vision: detection and simulation techniques; advances in health informatics; information systems: modeling the adoption of innovations; social media services and applications; advances in information technology for education and recent trends on software engineering.

Many thanks go to the keynote speakers for sharing their knowledge and expertise with us and to all the authors who have spent the time and effort to contribute significantly to this event. And we would like to thank the organizing committee for their great efforts in ensuring the successful implementation of the conference. In particular, we would like to thank the technical committee for their thorough and timely reviewing of the papers; Prof. Dr. Janusz Kacprzyk, AISC series editor; and Dr. Thomas Ditzinger, Ms. Almas Schimel, and Ms. Parvathidevi Krishnan from Springer.

Finally, we thank all the participants of IRICT 2018 and hope to see you all again in the next IRICT.

Organization

IRICT 2018 Organizing Committee

Honorary Co-chairs

Rose Alinda Alias	Association for Information Systems—Malaysian Chapter, Head of the Information Service Systems and Innovation Research Group (ISSIRG) in Universiti Teknologi Malaysia
Anne James-Taylor	Computing and Technology, School of Science and Technology, Nottingham Trent University, UK
Ahmad Hawalah	Information Systems Department, Deanship of Information Technology, Taibah University, Kingdom of Saudi Arabia

International Advisory Board

Abdul Samad Haji Ismail	Universiti Teknologi Malaysia, Malaysia
Ahmed Yassin Al-Dubai	Edinburgh Napier University, UK
Ali Bastawissy	Cairo University, Egypt
Ali Selamat	Universiti Teknologi Malaysia, Malaysia
Ayoub AL-Hamadi	Otto-von-Guericke University, Germany
Eldon Y. Li	National Chengchi University, Taiwan
Kamal Zuhairi Zamil	Universiti Malaysia Pahang, Malaysia
Kamalrulnizam Abu Bakar	Universiti Teknologi Malaysia, Malaysia

Mohamed M. S. Nasser	Qatar University, Qatar
Srikanta Patnaik	SOA University, Bhubaneswar, India

Conference General Chair

Faisal Saeed	Yemeni Scientists Research Group (YSRG), Data Science Research Group in Taibah University, Kingdom of Saudi Arabia

Program Committee Co-chairs

Nadhmi Gazem	Universiti Teknologi Malaysia, Malaysia
Fathey Mohammed	Taiz University, Yemen

Publicity Committee

Hakim Qaid Abdullah Abdulrab (Chair)	Universiti Teknologi Malaysia
Abdulalem Ali	Universiti Teknologi Malaysia
Mohammed A. Al-Sharafi	Universiti Malaysia Pahang
Abdullah Aysh Dahawi	Universiti Teknologi Malaysia
Muaadh Shaif Mukred	Universiti Kebangsaan Malaysia
Mohammed Abdulrahman Ebrahim Abdo Nashtan	University of Malaya
Eissa Mohammed Mohsen Alshari	Universiti Putra Malaysia
Ali Ahmed Ali Salem	Universiti Tun Hussein Onn Malaysia
Adnan Haidar Yusef Sa'd	Universiti Science Malaysia
Abdulrahman Farah Gubran Ahmed Alfareh	UniMAP
Mohammed Yahya Qaid Algumani	Universiti Teknikal Malaysia Melaka
Yahya Aysh Qasem Dahawi	Universiti Malaysia Pahang
Mohammed Abdulkaliq Abduullah Al-Mahfadi	University International Islamic Malaysia
Taha Hussain Qasem Dahawi	Multimedia University
Ali Saleh Amer Maaodhah	Universiti Teknologi Malaysia, KL
Abdullah Abdurahman Mohamed Ahmed	Universiti Utara Malaysia

Technical Committee Chair

Tawfik Al-Hadhrami	Nottingham Trent University, UK

Publications Committee

Abdel Salam Alzwawi (Chair)	Universiti Teknologi Malaysia
Ali Balaid	Universiti Teknologi Malaysia
Abdulaziz Al-Nahari	Universiti Teknologi Malaysia
Abdulalem Ali	Universiti Teknologi Malaysia
Mohammed Yahya Mohammed Al-Fasih	Universiti Teknologi Malaysia
Mohammed Sultan Ahmed Mohammed	Universiti Teknologi Malaysia
Maged Naeser	Universiti Teknologi Malaysia

IT Committee

Fuad Abdeljalil Al-shamiri (Chair)	Universiti Teknologi Malaysia
Bander Ali Saleh Al-rimy	Universiti Teknologi Malaysia
Amer Alsaket	Universiti Putra Malaysia

Logistic Committee

Abdullah Sharaf Saleh Al-Hammadi	Universiti Teknologi Malaysia
Majed Ali Obeid Mohamed	Universiti Teknologi Malaysia

Treasure Committee

Abdullah Aysh Dahawi (Chair)	Universiti Teknologi Malaysia
Hamzah Gamal Abdo Allozy	Universiti Teknologi Malaysia

Registration Committee

Sameer Hasan Albakri (Chair)	Universiti Teknologi Malaysia

International Technical Committee

Ab Razak Che Hussin	Universiti Teknologi Malaysia, Malaysia
Abdulbasit Darem	Mysore University, India
Abdulghani Ali Ahmed	Universiti Malaysia Pahang, Malaysia
Abdulrahman Alsewari	Universiti Malaysia Pahang, Malaysia
Abdulrazak Alhababi	UNIMAS, Malaysia
Ahmed Al-Saman	Universiti Teknologi Malaysia, Malaysia
Ahmed Hamza	King Abdulaziz University, KSA
Ahmed Ibrahim Alzahrani	King Saud University, KSA
Aiman Abusamra	Islamic University of Gaza, Palestine
Alaa El-Halees	Islamic University of Gaza, Palestine
Ali Al-Awadhi	Lincoln University College, Malaysia
Ali Balaid	Universiti Teknologi Malaysia, Malaysia
Ali Jwaid	Nottingham Trent University, UK
Aliza Sarlan	Universiti Teknologi PETRONAS, Malaysia
Ameer Tawfik	KMIT University, Australia
Ammar Abu-Hudrouss	Islamic University of Gaza, Palestine
Ashraf Osman	Alzaiem Alazhari University, Sudan
Asma Ahmed Alhashmi	Mysore University, India
Azlan Ismail	Universiti Teknologi Mara, Malaysia
Azlin Ahmed	Universiti Teknologi Mara, Malaysia
Azlinah Hj Mohamed	Universiti Teknologi Mara, Malaysia
Beverley Cook	Nottingham Trent University, UK
Dhanapal Dhurai Dominic	Universiti Teknologi PETRONAS, Malaysia
Ezzaldeen Edwan	Palestine Technical College, Palestine
Ghahida Mutasher	University of Technology, Iraq
Faisal Saeed	Taibah University, Kingdom of Saudi Arabia
Fathey Mohammed	Taiz University, Yemen
Funminiyi Olajide	Nottingham Trent University, UK
Georgina Cosma	Nottingham Trent University, UK
Haryani Haron	Universiti Teknologi Mara, Malaysia
Hussein Abu Al-Rejal	University Utara Malaysia, Malaysia
Ihab Zakout	Al Azhar University, Gaza, Palestine
Iyad Alagha	Islamic University of Gaza, Palestine

Low Tang Jung	Universiti Teknologi PETRONAS, Malaysia
Manzoor Ahmed Hasmani	Universiti Teknologi PETRONAS, Malaysia
Marina Yusoff	Universiti Teknologi Mara, Malaysia
Mohammed Alhanjouri	Islamic University of Gaza, Palestine
Mohammed Alshargabi	Najran University, KSA
Mohammad Mikki	Islamic University of Gaza, Palestine
Mohd Nordin Zakaria	Universiti Teknologi PETRONAS, Malaysia
Mohd Soperi	Universiti Teknologi PETRONAS, Malaysia
Motaz Saad	Islamic University of Gaza, Palestine
Murad Rassam	Taiz University, Yemen
Muthukkaruppan Annamalai	Universiti Teknologi Mara, Malaysia
Nadhmi Gazem	Universiti Teknologi Malaysia, Malaysia
Nasrin Makbol	Universiti Science Malaysia, Malaysia
Noorminshah Iahad	Universiti Teknologi Malaysia, Malaysia
Nordin Abu Bakar	Universiti Teknologi Mara, Malaysia
Norhaslinda Kamaruddin	Universiti Teknologi Mara, Malaysia
Norjansalika Janom	Universiti Teknologi Mara, Malaysia
Nursuriati Jamil	Universiti Teknologi Mara, Malaysia
Osamah Ibrahim Khalaf	Al-Nahrain University, Iraq
Qais Alnuzaili	Universiti Teknologi Malaysia, Malaysia
Redhwan Q Shaddad	Taiz University, Yemen
Rohiza Ahmad	Universiti Teknologi PETRONAS, Malaysia
Sameer Albakri	Universiti Teknologi Malaysia, Malaysia
Sharifah Aliman	Universiti Teknologi Mara, Malaysia
Stephen Clark	Nottingham Trent University, UK
Taha Hussein	Universiti Malaysia Pahang, Malaysia
Tawfik Al-Hadhrami	Nottingham Trent University, UK
Thabit Sabbah	Al-Quds Open University, Palestine
Wan Fatimah Wan Ahmad	Universiti Teknologi PETRONAS, Malaysia
Wesam Ashour	Islamic University of Gaza, Palestine
Yahya M. Al-dheleai	Universiti Teknologi Malaysia, Malaysia
Yap Bee Wah	Universiti Teknologi Mara, Malaysia
Yogan J. Kumar	Universiti Teknikal Malaysia Melaka, Malaysia

Contents

Intelligent Communication Systems

Advances in Information Security

Data Science and Big Data Analytics

Deep Belief Network for Molecular Feature Selection in Ligand-Based Virtual Screening

Maged Nasser[1(✉)], Naomie Salim[1], Hentabli Hamza[1], and Faisal Saeed[2]

[1] Faculty of Computing, Universiti Teknologi Malaysia, 81310 Johor Bahru, Johor, Malaysia
maged.m.nasser@gmail.com, naomie@utm.my, hamza@yahoo.fr

[2] College of Computer Science and Engineering, Taibah University, Medina, Saudi Arabia
fsaeed@taibahu.edu.sa

Abstract. It became obvious that it is required to reduce the high-dimensional of data in many data mining researches and applications. Virtual Screening (VS) is a set of computational methods that aim to score, rank and/or filter a set of chemical structures using one or more computational procedures to ensure those molecules with the largest prior probabilities of activity. 2D fingerprint descriptors are used to represent molecule features, most of these features are important and has ability to improve the molecules similarity and the others are not important and taking more computational time without any effect on the similarity score. Deep belief networks is one of the deep learning methods used to select the important features to reduce the high dimensionality by using stack of Restricted Boltzmann Machine and fine tune to enhance weights and reduce the reconstruct feature error. Thus, the features that have more reconstruct error are removed and only features with less constrict error will be used. The experimental results showed the enhancements on VS results using the proposed methods.

Keywords: Deep learning · Deep belief networks
Restricted Boltzmann Machine · Feature selection

1 Introduction

Computer aided drug design has been used to aid the process of drug discovery and design so that the expenditure and cost to discover and develop new drugs can be optimized. In a computer-aided drug discovery program, virtual screening methods have been used to select compounds for testing or design of combinatorial libraries. To lessen these costs, replacing the old traditional-hand crafted has been necessitated. Therefore, pharmaceutical companies needed to look for new technologies for testing new chemical entities. Examples of these technologies include high throughput screening (HTS), virtual screening, and combinatorial chemistry (CC) [1].

F. Saeed et al. (Eds.): IRICT 2018, AISC 843, pp. 3–14, 2019.
https://doi.org/10.1007/978-3-319-99007-1_1

Virtual Screening (VS) is one of the common computational methods used, for search offing about small molecules libraries, in drug discovery. VS is frequently used in protein, in order to identify structures that are most likely to bind to a drug target [2]. It is not like high-throughput screening (HTS), where compound must exist physically, the virtual screening is done using computational techniques which easily performs screening and searching extremely large parts of the chemical space and an enormous amount of molecules in a short time with less cost, and this is one of the VS advantages compared to HTS.

Majority of the similarity-based virtual screening techniques that are currently in use handle vast quantities of data that possess redundant and/or irrelevant fragments or features. The fingerprint of the current molecule is made up of multiple features. Furthermore, since their importance levels are not the same, the recall of similarity measure may be enhanced by removing some features [3]. Data that has irrelevant and redundant features can misguide the results of the virtual screening and clustering and make the results difficult to be explained [4]. Real world applications issues of high complexity and poor learning performance are brought by large dimensional datasets comprising redundant, irrelevant and relevant features [5]. Adoption of feature selection is among the most effective ways for overcoming these issues [6]. Feature selection provides most useful or relevant features contributing to the model construction in a more effective and efficient way. Feature selection is not only beneficial for the simplification of the learning model and minimizing the training time but also for performance classification. Consequently, selecting relevant and useful features for the model is not easy because of the huge search space and complex interactions between features which usually occurs in two, three or more ways. Subsequently, an irrelevant feature can happen to be useful for the classification or learning performance while interacting with other features. Relevant features may, on the other hand, become redundant when interacting with other features.

Deep learning makes it possible for computational models that are made up of multiple processing layers to study data representations with several abstraction levels. Representation learning refers to group of methods that makes it possible for raw data to be fed to a machine and for that machine to automatically discover what representations are necessary to detect or classify certain information [7]. Deep-learning methods are a type of representation-learning methods that possess multiple representation levels that are obtained by creating models that are both simple and non-linear and that are capable of transforming the representation from one level (starting with the raw input) into one that is at a higher level and a slightly more abstract one [8]. When enough transformations are composed, functions that are very complex can be learned. In deep learning techniques, some researches agreed on better performance of deep architecture with various hidden layers over shallow architecture. Fortunately, Deep belief networks (DBN) model was suggested to solve the difficulty in training for deep architecture by bringing two-step framework, pretraining and fine-tuning [9]. In this work, DBN has been represented and used in ligand based virtual screening to select the only important features that will be used to enhance the performance of similarity methods.

2 Related Works

Many of the similarity approaches assume that molecular features/fragments that do not relate to biological activity carry the same weight as the important ones. Generally, the chemists may consider some fragments as being more important than others through the chemist structure diagrams, such as functional groups. For this reason, the researchers have investigated the weight for each fragment in chemical structure compounds by giving more weight to those fragments that are more important. Therefore, a match between two molecules on highly weighted features would make a greater contribution to the overall similarity than a match on a less important features [10, 11]. Abdo et al. [12] have investigated various weighting functions and introduced a new fragment weighting scheme for the Bayesian Inference Network in Ligand-Based Virtual Screening. Ahmed et al. [13] have developed fragment reweighting technique using reweighting factors and relevance feedback to improve the retrieval recall of Bayesian inference network.

According to Vogt [3], features selection can enhance the recall of similarity measure and allow important fragments to be given more weight while ignoring the unimportant fragments. Many types of currently used fingerprints are highly complex, containing many fragments or features, and therefore many bit positions, often more than 1000 features or fragments [14–16]. Some studies have enhanced the Bayesian Inference Network (BIN) by using only a subset of selected molecular features [17], by removing unimportant fragment using the minifingerprint idea, where the fundamental concept behind minifingerprinting is to reduce features and keep the percentage of compounds of biological activity [18, 19].

DBN has been applied to different data analysis, e.g., computer vision [20], speech and language processing [21] and emotion recognition [22]. In addition, DBN has been successfully used to investigate feature abstraction and reconstruction of images [23–26]. Also, DBN was successfully implemented for multi-level feature selection to select the most discriminative genes that can be used to enhance the sample classification accuracy [27], and implemented for Feature Selection for Remote Sensing Scene Classification [26].

In Remote Sensing Scene Classification, Zou et al. [26] suggested DBN for remote sensing feature selection which devises the problem for feature selection as a problem for feature reconstruction in a DBN; the standard data set having 2800 remote sensing pictures of seven categories was developed for evaluation. The results from the research validated the efficacy of the proposed technique.

DBNs have been utilised to generate audio-visual attributes for emotion sorting, even in an unsupervised situation [22]. The comparison of the categorisation performances among the baseline and the suggested DBN models showed that it is crucial to retain complex non-linear attribute relationships (by deep learning methods) in emotion classification undertakings. DBN applied deep belief network to network intrusion detection, and the combination of BP neural network model, by comparing the results with other methods. The results showed that the method in the intrusion detection feature selection has certain advantages [28] such as, reducing the miss rate and the dimension of data, to accelerate the speed of data processing and providing a method for the quasi real-time detection of intrusion detection system.

3 Materials and Methods

3.1 Deep Belief Network (DBN) Architecture

The basic concept behind feature selection is selecting a subset of all variables by getting rid of a large amount of features that possess little predictive and discriminative information [29, 30]. The current molecule fingerprint is made up of several features, though not all of them have the same level of importance. Furthermore, getting rid of some features may improve the similarity measure's recall [3]. Not all the features in a dataset are important; some are irrelevant while some are redundant. Data that possesses several irrelevant features may lead to clustering results that are misguided and hard to explain [4]. Dimensionality can be reduced in two ways: feature selection and feature transformation. Feature transformation lessens the dimensions by implementing a nonlinear or linear function to the original features while feature selection chooses a subset from the original features. Rather than transformation, feature selection can be performed to preserve the features' original meaning.

In deep neural networks, some researches agreed on better performance of deep architecture with various hidden layers over shallow architecture. Fortunately, the DBN model was suggested to solve the difficulty in training for deep architecture by bringing two-step framework, pretraining and fine-tuning [9].

3.1.1 Restricted Boltzmann Machines (RBM)

A general architecture of the pretraining in the DBN can be illustrated in Fig. 1. This stage is a stack of Restricted Boltzmann Machines (RBMs) and the output from $(i-1)^{th}$ RBM hidden layer in the stack is the input to the visible layer of the ith RBM [31], in which the output of the lower-level RBM can be treated as the input of the higher-level RBM. Therefore, each RBM can be trained individually beginning with the 1st RBM in the stack. Once the training is finished for the first RBM in the stack, an abstract output representation, also known as features, of the input has been formed at the hidden layer. These features are passed on to the 2nd RBM in the stack and the 2nd RBM is trained. This procedure is repeated until the last RBM in the stack has been trained. The RBM is one of the undirected graphical models with two layers [32–34]. The first layer in RBM is called visible units which contains the observed variables while the second layer is known as hidden units and consists of latent variables. Inheriting from Boltzmann Machine [35], RBM is a particular form of log-linear Markov Random Field (MRF) [36, 37]. The difference between RBM and general Boltzmann Machine is that RBM only contains inter-layer connections among visible and hidden units. It is as equivalent to that intra-layer connections which are connections among visible-visible, hidden-hidden units are excluded. Figure 1 shows the stack of Restricted Boltzmann Machines architecture.

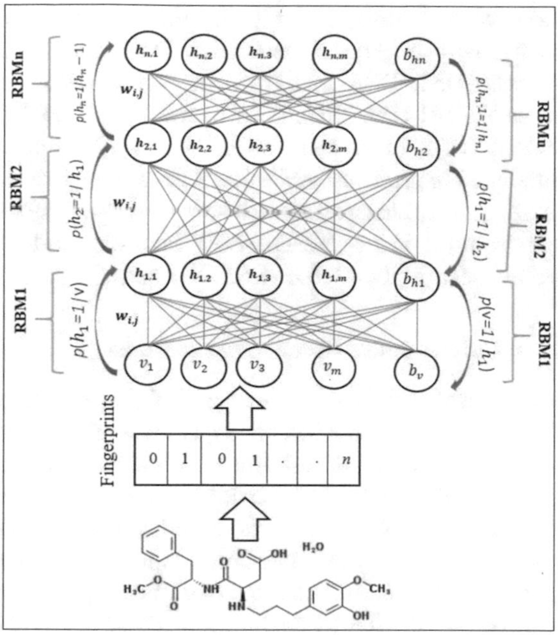

Fig. 1. DBN architecture

3.1.2 Reconstruction Weight Fine-Tuning

After training all the RBMs and saving the weight for all RBMs, it is considered that the DBN pretraining has been done. DBN performs a backpropagation to fine-tune the weights. A reconstruction error can be calculated for each input. The variation of features would commonly make the difference of reconstruction errors. Intuitively, a feature with a lower reconstruction error is more re-constructible. To get the reconstruct feature error, we compare the input features molecule with the reconstruct features molecule that have been obtained by the backpropagation. The fine tune of DBN that has been used to get the reconstruct feature weight. Pretraining and fine tune should be trained for all molecules in the dataset to get the reconstruct features for all molecules in the dataset. The more re-constructible features, the more prone to hold the feature intrinsic. As feature intrinsic is vital in protein-ligand binding, the features that are more re-constructible are the discriminative features.

3.2 Dataset

The MDDR database is still one of the most popular databases in chemo-informatic, which contains over 102,516 chemical compounds with several hundred distinct activities, some of these activities are related to therapeutic areas such as Antihypertensive, and some of which are related to specific enzymes such as Renin inhibitors.

The molecules that have been turned into Pipeline Pilot ECPC_4 (folded to size 1024 bits with their connectivity fingerprints extended) [38]. The screening experiments utilised three data sets (DS1, DS2, and DS3) that were selected from the MDDR database (signified as MDDR-DS1, MDDR-DS2, and MDDR-DS3). There are eleven activity classes in MDDR-DS1, with some of the classes taking part in actives that are considered structurally homogeneous while the others are parts of actives that are considered structurally heterogeneous (i.e. with structural diversity). Ten homogeneous activity classes are included in the MDDR-DS2 data set, while ten heterogeneous activity classes are included in the MDDR-DS3 data set.

3.3 Using DBN to Calculate the Reconstruct Feature

The DBN has been trained with both stages (pre-training and fine tune).In the first stage, three RBMs are used and each RBM has input layer and output layer. The size of the input layer in the first RBM is 1024 as same size of the dataset vector, which refers to the whole features of molecule and the size of the output layer from the first RBM is 2000. The output from the first RBM becomes input layer to the second RBM and the size of the output layer in second RBM is 1500, and the output from the second RBM becomes input layer to the third RBM and the size of the output layer in the third RBM is 1000, using the epoch = 10, Gibbs steps = 10, batch size = 10 and learning rate = 0.001. In this study, only three RBMs were used and not many iterations were performed because the size of dataset is huge and using more RBM will take long training time. After training the first stage, the backpropagation is used to fine the fine tune.

After training the DBN, the output is a new matrix that has the same size of the input dataset (102516 * 1024), but this new matrix represents the new reconstruct feature weight for all molecules in the input dataset, as shown in Fig. 2.

3.4 Reconstruct Feature Error

The reconstruction error of feature v_i is calculated as $e_i = \|v_{i_{re}} - v_i\|$, $v_{i_{re}}$ is the reconstruction feature weight corresponding to v_i [26]. The output from this section is reconstruct feature error matrix that has the same size of the reconstruct feature weight (102516 * 1024). This new matrix represented the new reconstruct feature error for all molecules in the input dataset as it shows in Fig. 2.

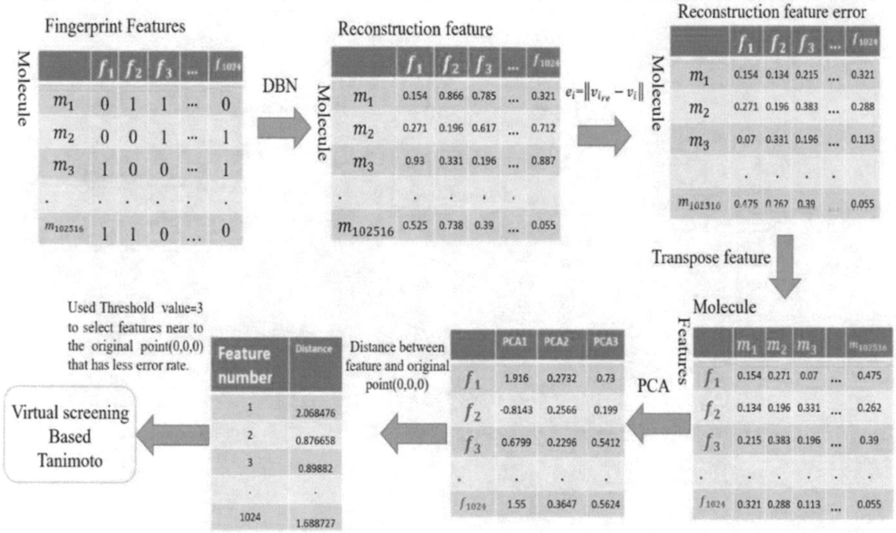

Fig. 2. Ligand-based virtual screening using Deep Belief Network (DBN)

3.5 Principal Component Analysis (PCA)

PCA is one the fast algorithms that have been implemented for non-linear dimension reduction and introduced to reduce the dimensionality of the feature vectors and gain more effective features [39]. Before implementing PCA, the reconstruct feature error matrix is transformed to become (1024 * 102516) and each vector presents only one feature for all molecules in the dataset. After using PCA, three values only (PCA1, PCA2 and PCA3) for each feature are obtained, as shown in Fig. 2. These values are used to draw 3D coordinates to show which features are near to the original point (0, 0, 0) that have less error rate. After that, the distance between all features and original point is calculated by using this formula: $\mathrm{D} = \sqrt{(x_i - x_j)^2 + (y_i - y_j)^2 + (z_i - z_j)^2}$ and because the distance is to the original point (0, 0, 0) so $x_j = y_j = z_j = 0$, which becomes $D = \sqrt{(x_i)^2 + (y_i)^2 + (z_i)^2}$.

The output is shown in Fig. 3 presents thatthe features that have high distance value from the original point (0, 0, 0), also have high error rate. This study uses threshold value = 3 to cut only the features with error rate less than the threshold and only these features are used for virtual screening.

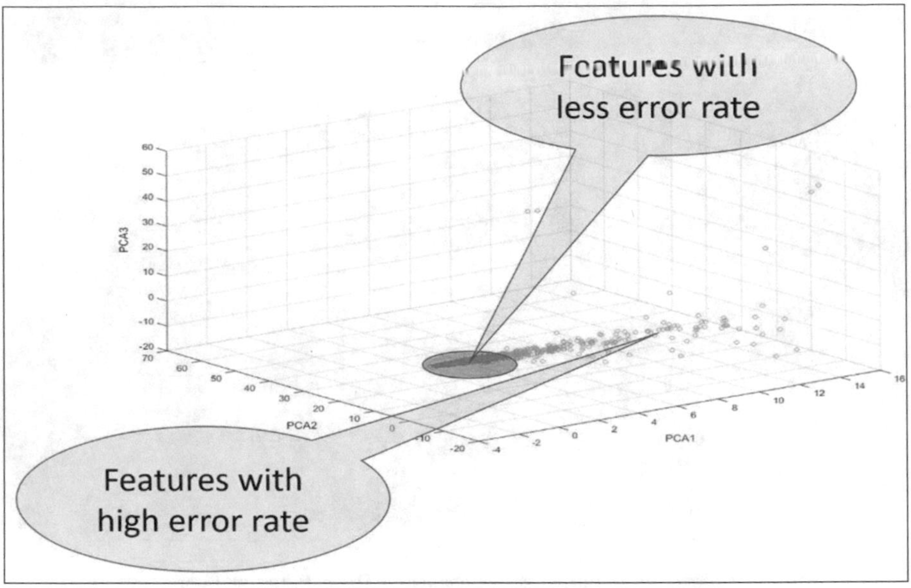

Fig. 3. PCA for reconstruction features error rate

4 Results and Discussion

Most of the currently used similarity methods for virtual screening approaches deal with a huge of data with redundant and irrelevant features. In this study, Deep Belief Networks is implemented for ligand based virtual screening to be a new feature selection method that enhances the similarity searching result compared with the base line similarity searching method.

Simulated virtual screening experiments with MDL Drug Data Report (MDDR) data sets showed that the proposed method provides different ways of enhancing the effectiveness of ligand-based virtual screening searches, especially for high diverse data sets.

This work used three MDDR datasets - MDDR-DS1, MDDR-DS2, MDDR-DS3- and the results are presented in Tables 1, 2 and 3 respectively, using cut offs at 1% and 5%. In these tables, the results of Deep Belief Networks method compared with benchmark TAN method are reported to show the performance enhancement.

In all the tables results, each row presents the recall for top 1% and 5% of the activity class, and the best recall rates in each row are shaded. The Mean rows in the tables correspond to the mean when averaged over all activity classes (the best average is bolded).

The results shown in Tables 1, 2 and 3 emphasize that the DBN can be used for the feature selection to enhance the similarity searching. The results obtained by DBN is the best compared with TAN method for the three MDDR datasets.

Table 1. Retrieval results of top 1% and 5% for MDDR-DS1 dataset.

	TAN		DBN	
Activity Index	1%	5%	1%	5%
31420	69.69	83.49	70.21	86.55
71523	25.94	48.92	27.61	56.03
37110	9.63	21.01	23.51	46.5
31432	35.82	74.29	40.63	79.99
42731	17.77	29.68	17.67	26.89
06233	13.87	27.68	14.63	26.25
06245	6.51	16.54	7.43	17.54
07701	8.63	24.09	10.56	26.18
06235	9.71	20.06	11.48	23.97
78374	13.69	20.51	12.8	19.69
78331	7.17	16.2	6.97	12.97
Mean	19.85727	34.77	**22.13636**	**38.41455**

Table 2. Retrieval results of top 1% and 5% for MDDR-DS2 dataset

	TAN		DBN	
Activity Index	1%	5%	1%	5%
07707	61.84	70.39	71.85	74.44
07708	47.03	56.58	96.73	99.36
31420	65.1	88.19	74.58	94.19
42710	81.27	88.09	81.27	90.72
64100	80.31	93.75	86.94	99.5
64200	53.84	77.68	63.16	95.63
64220	38.64	52.19	68.16	90.33
64500	30.56	44.8	73.49	89.21
64350	80.18	91.71	82.06	90.39
75755	87.56	94.82	97.67	98.64
Mean	62.633	75.82	**79.37**	**92.181**

Table 3. Retrieval results of top 1% and 5% for MDDR-DS3 dataset

	TAN		DBN	
Activity Index	1%	5%	1%	5%
09249	12.12	24.17	17.21	28.78
12455	6.57	10.29	8.98	13.75
12464	8.17	15.22	12.72	21.76
31281	16.95	29.62	21.62	35.43
43210	6.27	16.07	10.86	18.71
71522	3.75	12.37	7.49	13.24
75721	17.32	25.21	25.39	35.81
78331	6.31	15.01	7.67	13.86
78348	10.15	24.67	7.14	20.93
78351	9.84	11.71	16.42	18.15
Mean	9.745	18.434	**13.55**	**22.042**

5 Conclusion

This study focuses on using the deep learning to investigate the enhancement of the similarity searching in virtual screening. One technique of deep learning that is called Deep Belief Network was used for molecular features selection to select only the important features that should be used for the similarity searching. The MDDR dataset was used in this study and the experimental results showed that the proposed method enhanced the cost effectiveness of ligand based virtual screening in chemical databases. The results showed that the DBN has been implemented successfully to enhance the performances of similarity searching, particularly when the molecules are structurally high diverse.

Acknowledgments. This work is supported by the Ministry of Higher Education (MOHE) and the Research Management Centre (RMC) at the Universiti Teknologi Malaysia (UTM) under the Research University Grant Category (VOT Q.J130000.2528.16H74 and R.J130000.7828.4F985).

References

1. Lionta, E., et al.: Structure-based virtual screening for drug discovery: principles, applications and recent advances. Curr. Top. Med. Chem. **14**(16), 1923–1938 (2014)
2. Rollinger, J.M., Stuppner, H., Langer, T.: Virtual screening for the discovery of bioactive natural products. Natural Compounds as Drugs, vol. 1, pp. 211–249. Springer, Berlin (2008)
3. Vogt, M., Wassermann, A.M., Bajorath, J.: Application of information—theoretic concepts in chemoinformatics. Information **1**(2), 60–73 (2010)
4. Liu, H., Motoda, H.: Computational Methods of Feature Selection. CRC Press, London (2007)
5. Xue, B., et al.: A survey on evolutionary computation approaches to feature selection. IEEE Trans. Evol. Comput. **20**(4), 606–626 (2016)

6. Unler, A., Murat, A., Chinnam, R.B.: mr2PSO: a maximum relevance minimum redundancy feature selection method based on swarm intelligence for support vector machine classification. Inf. Sci. **181**(20), 4625–4641 (2011)
7. Pradipta Lie, F.A., Go, T.H.: Reconfiguration control with collision avoidance framework for unmanned aerial vehicles in three-dimensional space. J. Aerosp. Eng. **26**(3), 637–645 (2011)
8. Liu, N., et al.: Sparse representation based image super-resolution on the KNN based dictionaries. Laser Technol. Opt. (2018). https://doi.org/10.1016/j.optlastec.2018.01.043
9. Hinton, G.E., Osindero, S., Teh, Y.-W.: A fast learning algorithm for deep belief nets. Neural Comput. **18**(7), 1527–1554 (2006)
10. Klinger, S., Austin, J.: Weighted superstructures for chemical similarity searching. In: Proceedings of the 9th Joint Conference on Information Sciences (2006)
11. Arif, S.M., Holliday, J.D., Willett, P.: The Use of Weighted 2D Fingerprints in Similarity-Based Virtual Screening. Elsevier, Amsterdam (2016)
12. Abdo, A., Salim, N.: New fragment weighting scheme for the bayesian inference network in ligand-based virtual screening. J. Chem. Inf. Model. **51**(1), 25–32 (2010)
13. Ahmed, A., Abdo, A., Salim, N.: Ligand-based Virtual screening using Bayesian inference network and reweighted fragments. Sci. World J. (2012). https://doi.org/10.1100/2012/410914
14. Unity. Tripos Inc
15. Matter, H., Pötter, T.: Comparing 3D pharmacophore triplets and 2D fingerprints for selecting diverse compound subsets. J. Chem. Inf. Comput. Sci. **39**(6), 1211–1225 (1999)
16. James, C., Weininger, D., Delany, J.: Daylight Theory Manual. Daylight Chemical Information Systems, Inc., Irvine (1995)
17. Ahmed, A., Salim, N., Abdo, A.: Fragment reweighting in ligand-based virtual screening. Adv. Sci. Lett. **19**(9), 2782–2786 (2013)
18. Xue, L., et al.: Mini-fingerprints detect similar activity of receptor ligands previously recognized only by three-dimensional pharmacophore-based methods. J. Chem. Inf. Comput. Sci. **41**(2), 394–401 (2001)
19. Xue, L., et al.: Profile scaling increases the similarity search performance of molecular fingerprints containing numerical descriptors and structural keys. J. Chem. Inf. Comput. Sci. **43**(4), 1218–1225 (2003)
20. Krizhevsky, A., Sutskever, I., Hinton, G.E.: Imagenet classification with deep convolutional neural networks. In: Advances in Neural Information Processing Systems (2012)
21. Mohamed, A.-R., Dahl, G.E., Hinton, G.: Acoustic modeling using deep belief networks. IEEE Trans. Audio Speech Lang. Process. **20**(1), 14–22 (2012)
22. Kim, Y., Lee, H., Provost, E.M.: Deep learning for robust feature generation in audiovisual emotion recognition. In: 2013 IEEE International Conference on Acoustics, Speech and Signal Processing (ICASSP). IEEE (2013)
23. Peng, Z., et al.: Deep boosting: joint feature selection and analysis dictionary learning in hierarchy. Neurocomputing **178**, 36–45 (2016)
24. Semwal, V.B., Mondal, K., Nandi, G.C.: Robust and accurate feature selection for humanoid push recovery and classification: deep learning approach. Neural Comput. Appl. **28**(3), 565–574 (2017)
25. Suk, H.-I., et al.: Deep sparse multi-task learning for feature selection in Alzheimer's disease diagnosis. Brain Struct. Funct. **221**(5), 2569–2587 (2016)
26. Zou, Q., et al.: Deep learning based feature selection for remote sensing scene classification. IEEE Geosci. Remote Sens. Lett. **12**(11), 2321–2325 (2015)

27. Ibrahim, R., et al.: Multi-level gene/MiRNA feature selection using deep belief nets and active learning. In: 2014 36th Annual International Conference of the IEEE Engineering in Medicine and Biology Society. IEEE (2014)
28. Baoyi Wang, S.S., Zhang, S.: Research on feature selection method of intrusion detection based on deep belief network. In: Proceedings of the 2015 3rd International Conference on Machinery, Materials and Information Technology Applications (2015)
29. Blum, A.L., Langley, P.: Selection of relevant features and examples in machine learning. Artif. Intell. **97**(1), 245–271 (1997)
30. Beltrán, N.H., et al.: Feature selection algorithms using Chilean wine chromatograms as examples. J. Food Eng. **67**(4), 483–490 (2005)
31. Hinton, G.E., Salakhutdinov, R.R.: Reducing the dimensionality of data with neural networks. Science **313**(5786), 504–507 (2006)
32. Freund, Y., Haussler, D.: Unsupervised learning of distributions on binary vectors using two layer networks. In: Advances in Neural Information Processing Systems (1992)
33. Hinton, G.E.: Training products of experts by minimizing contrastive divergence. Training **14**(8), 1771–1800 (2006)
34. Smolensky, P.: Information processing in dynamical systems: foundations of harmony theory. Parallel Distributed Processing, Explorations in the Microstructure of Cognition, vol. 1, p. 18. MIT Press, Cambridge (1986)
35. Ackley, D.H., Hinton, G.E., Sejnowski, T.J.: A learning algorithm for Boltzmann machines. Cogn. Sci. **9**(1), 147–169 (1985)
36. Darroch, J.N., Lauritzen, S.L., Speed, T.P.: Markov fields and log-linear interaction models for contingency tables. Ann. Stat. **8**, 522–539 (1980)
37. Lauritzen, S.L.: Graphical Models, vol. 17. Clarendon Press, Oxford (1996)
38. Pipeline Pilot Software: SciTegic Accelrys Inc. http://www.accelrys.com/. San Diego Accelrys Inc. (2008)
39. Yuan, C., Sun, X., Lv, R.: Fingerprint liveness detection based on multi-scale LPQ and PCA. China Commun. **13**(7), 60–65 (2016)

Energy-Efficient Resource Allocation Technique Using Flower Pollination Algorithm for Cloud Datacenters

Mohammed Joda Usman[1,3](✉), Abdul Samad Ismail[1], Abdulsalam Yau Gital[2], Ahmed Aliyu[1,3], and Tahir Abubakar[3]

[1] Faculty of Engineering, School of Computing, Universiti Teknologi Malaysia, 81310 Skudai Johor Bahru, Malaysia
umjoda@gmail.com, ahmedaliyu8513@gmail.com, absamad@utm.my

[2] Faculty of Science, Department of Mathematics, Abubakar Tafawa Balewa University, Bauchi 81027, Bauchi State, Nigeria
asgital@yahoo.com

[3] Faculty of Science, Department of Mathematics, Bauchi State University, Gadau 81007, Bauchi State, Nigeria
tahir_abubakr@yahoo.com,

Abstract. Cloud Computing is modernizing how Computing resources are created and disbursed over the Internet on a model of pay-per-use basis. The wider acceptance of Cloud Computing give rise to the formation of datacenters. Presently these datacenters consumed a lot of energy due to high demand of resources by users and inefficient resource allocation technique. Therefore, resource allocation technique that is energy-efficient are needed to minimize datacenters energy consumption. This paper proposes Energy-Efficient Flower Pollination Algorithm (EE-FPA) for optimal resource allocation of datacenter Virtual Machines (VMs) and also resource under-utilization. We presented the system framework that supports allocation of multiple VMs instances on a Physical Machine (PM) known as a server which has the potential to increase the energy efficiency as well resource utilization in Cloud datacenter. The proposed technique uses Processor, Storage and Memory as major resource component of PM to allocate a set of VMs, such that the capacity of PM will satisfy the resource requirement of all VMs operating on it. The experiment was conducted on Multi-RecCloudSim using Planet workload. The results indicate that the proposed technique energy consumption outperform the benchmarking techniques which include GAPA, and OEMACS with 91% and 94.5% energy consumption while EE-FPA is around 65%. On average 35% of energy has been saved using EE-FPA and resource utilization has been improved.

Keywords: Cloud Computing · Datacenter · Resource allocation
Energy consumption · Flower Pollination Algorithm

F. Saeed et al. (Eds.): IRICT 2018, AISC 843, pp. 15–29, 2019.
https://doi.org/10.1007/978-3-319-99007-1_2

1 Introduction

Cloud Computing as the name implies is a worldwide name employed for providing physical and logical computing resources and services over the Internet using virtualization technology. The technology allows single Physical Machine (PM) to host multiple Virtual Machines (VMs) which then imitates or pretend to be server. This approach has been considered to be the best resource management technique in Cloud environment. The model of operating and deployment of Cloud Computing services allows companies, industries, and organization to acquire computing resources based on the pay-per-use model. The Cloud service consumers will only pay for the services they have used like other utility such as electricity or water [1]. The wider acceptance of Cloud Computing give rise to the formation of datacenters that render Cloud services. These Cloud datacenters offered these services in several forms and at a distinct point as shown in Fig. 1.

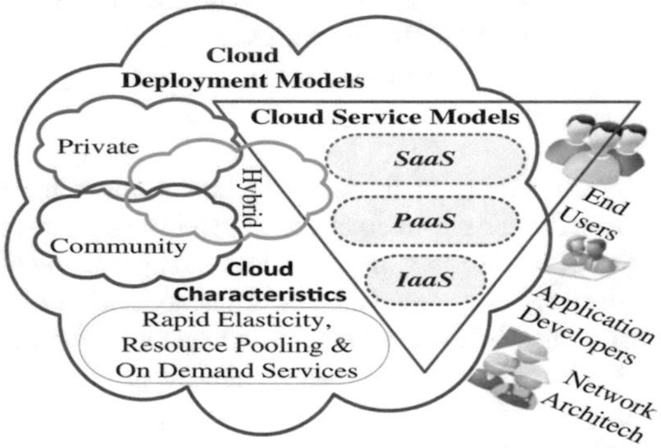

Fig. 1. Component of cloud computing

The main services that Cloud offers are categorized into Infrastructure as a Service (IaaS), Platform as a Service (PaaS) and Software as a Service (SaaS) [1, 2]. Among the advantages of employing Cloud Computing includes; pay-per-use, prompt on-demand service provision, rapid elasticity, and lower cost of ownership. The perspective of resource allocation (RA) has a significant effect on the datacenter resource management. Cloud datacenters are usually deployed as public, private, community and or hybrid. In these models, useful resource utilization is pointed with the aid of Cloud service provider's tools used for service deployment. This can be achieved when appropriate datacenter resources are selected to run for actual application requirement in IaaS. Once the required resources are determined, VMs instances of these resources are allocated in order to execute user's request. In spite of the benefit derived from using this technology, Cloud datacenter consumed a lot of energy. This high energy consumption has attracted many researchers due to the growing demand for computing resource by the business organizations and other scientific research applications. It has been observed that, the datacenter energy consumption is proportional to the utilization of various Computing resources and are considered to be among the largest consumers of electricity in the world [3]. In datacenter, utilization level of resource with their corresponding energy used are not trivial. This is due to the fact that, less utilized resources waste energy than resources which are totally utilized. Various scheduling strategies and techniques that take energy consumption into consideration by exploring Nature-Inspired algorithms have been designed [4–6] to avoid if not minimize under-utilization of resource, which is liable for arousing the excessive energy intake by the Cloud datacenters [7]. Furthermore, [8–11] have also proposed resource scheduling technique using metaheuristics algorithms to scale down the datacenter energy consumption, resource utilization that include other related service parameters. Therefore, datacenter energy consumption remains a challenge for resource allocation and management by the Cloud providers.

Furthermore, the approach used to overcome this hassle is to decrease the inherent capacity of the Cloud datacenters by exploring different virtualization technique [1]. Virtualization has been considered as one of the supporting technology of Cloud computing that permits Cloud managers to commence VM placement on PM. By this, the demand for physical resources will be minimized, thus resource utilization is maximized. Alternatively, reducing energy consumption will be realized by turning off or switching PMs that are in the idle state to low-power mode state (i.e. sleep, hibernation). This approach also reduces the idle power consumption in the datacenters but has become complicated and difficult to manage due to the in efficiency in allocating VMs on servers. However, the method enhances the condition, but it is usually not sufficient because of the Cloud environment dynamisms.

In addition, a large number of research has been performed in this area of resource management and energy efficiency [1, 12, 13]. However, the researchers have focused on only one dimensional resource which is CPU of PM and the computing requirement of VMs. Therefore, the main idea of this study is design and develop a new technique for efficient resource utilization in datacentre based on Cloud provider perspective. This is because of the fact that, effective resource allocation technique and management for IaaS helps in reducing datacenter consumption of energy and resource underutilization. This paper presents a new technique for energy-efficient resource allocation using Flower Pollination Algorithm (FPA) to optimize datacenter resource allocation while reducing energy consumption. The allocation technique uses the FPA local and global search approach for the initialization of VM deployment on PM, to expand the exploration space for the most excellent allocation in managing the datacenter resources. The technique has been proved to be efficient by comparing it with the best technique in the literature and most used ones in the domain of resource allocation strategy.

2 Related Work

Efficient allocation of virtual resources is an essential feature of Cloud datacenter infrastructure. The advantage of virtualization in Cloud datacenter is to improved efficiency regarding utilization and availability of PM resource components [15]. Li et al. [16] proposed a method referred to as EnaCloud that assist active placement of VMs that are currently being utilize in the datacenter. Bin-packing approach has been employed in this technique in order to allocate users request efficiently by considering energy consumption issue due to the applications that are being executed. Energy-aware heuristic algorithm has been considered to efficaciously schedule the running VMs applications that ensures virtualization of resources [17, 18]. In these technique, the VMs usually captured running applications for live VM migration implementation to minimizing energy consumption. However, they only focus on energy without considering resource utilization.

Phan et al. [10] presented a Green Monster framework and designed Evolutionary Multi-objective Optimization Algorithm (GM-EMOA) to move services dynamically across a federation of geographically dispersed datacenters. Shu et al. [14] present a resource scheduling inspired by Clonal Selection Algorithm (ICSA) to cut down the energy consumption of datacenter IaaS. ICSA has a major influence on local optima avoidance within a range of a given feasible solution. The approach uses less running time as well. Once a new request arrives, the system runs the algorithm to adjust the overall resource allocation. Clearly, ICSA proofs to be more effective in optimizing energy consumption and resource allocation compared with existing algorithms. However, it uses only few physical machines in its implementation and thus, may not support large datacenters.

Deore et al. [8] proposed resource scheduling technique to efficiently schedule the datacenter resources in order to reduce energy consumption. The proposed technique achieved minimum migration cost, clone, pause, and resume concepts on VMs. Therefore, the scheme consumes less energy during the test in Cloud environment using Virtual Box. Likewise, Ghribi et al. [15] proposed energy-aware resource placement and migration technique to migrate VM from over utilize PMs to under-utilize ones. Thereby making the datacenter to be energy-efficient. The authors combined optimal resource allocation algorithm by exploring VM placement that is dependent on VM migration. However, the method suffers from inefficient allocation of resources due to the slow convergence and imbalance amid the local and global search of the algorithm.

A green datacenter resource allocation that uses Genetic Algorithm (GA) has been developed to efficiently minimize energy consumption and resource wastage by the datacenters [16]. The model used in the technique combined VM placement and transport network using Integer Linear Programming (ILP). The results shows that the GA has the ability and flexibility to handle the datacenter resource allocation. However, the technique takes into consideration of only CPU as a resource component. Sharma and Reddy [6] proposed another hybridization of GA with Dynamic Voltage Frequency Scaling (DVFS) to reduce the energy consumed by the datacenter and at the same time improve resource utilization with faster convergence for making near optimal resource allocation. The method map VMs to PMs randomly by mapping VMs to PM by the coded chromosomes with their offspring's using the fitness evolution to fit in VMs on PMs with small number of running application. Moganarangan et al. [9] proposed a hybridized algorithm for minimizing datacenter energy consumption with makespan. The algorithm bring together ACO and Cuckoo Search Algorithms (CSA) to effectively reduce the high energy consumed by the datacenter. The jobs are processed based on arrival from 1 to n jobs. After applying transition rules, jobs that arrive are processed. However, HA focuses only on the energy consumption of the CPU neglecting other components that also consume energy of the datacenter. Therefore, there is need for energy-efficient resource utilization based on Cloud service provider's perspective in order improve on datacenter resource management. Based on the current literature, we are the first to use FPA in optimizing resource allocation of Cloud datacenter infrastructure.

3 Energy-Efficient Flower Pollination Resource Allocation

This section presents discussion on the proposed energy efficient allocation technique of datacenter resources. The technique encompass the system framework, Flower Pollination Algorithm (FPA) with their assumption and corresponding energy and resource utilization models are described.

3.1 Proposed System Architecture

The system architecture of resource allocation in Cloud environment assigned available resources that are accessible to various usage that is generally considered as plan for provisioning in datacenters environment. There a number of approaches for solving resource allocation problems. For example resource can be allocated using manual approach, dynamic and or combination of the two (static/dynamic). It can also be used to reduce energy wastage by the datacenter together with resource under-utilization. Figure 2 illustrates the architecture of the proposed energy-efficient resource allocation. It depict how the proposed technique handled user request and forward the request to the Cloud broker and ultimately to Cloud datacenter.

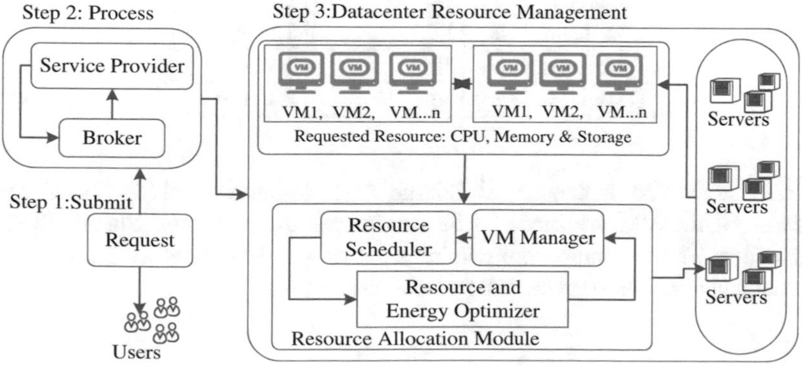

Fig. 2. The proposed system architecture

The user requests are first presented to Cloud provider, then the Cloud broker process the request based on resource availability and users requirement. The broker will also forward the request to the datacenter resource management which includes VM manager and resource scheduler in order to make a decision on placement of the request in the datacenter for execution. Furthermore, the aforementioned resource management modules will find out whether the requirement sends by the user such as on-demand service, availability and allocation will be supported by the VM. The user request attractiveness is determined by the Cloud resource manager based on system readiness. The proposed technique considered the aforementioned hassle using FPA to Address the issue of resource utilization and energy consumption. In the following section we described the user request model, resource mapping and energy models that are used in realizing energy-efficient resource allocation of datacenter environment.

The User Request Model. The Cloud users make a request to the Cloud datacenter through the broker or Cloud provider for their various application needs. The user request a set of resources known as VMs. Each of the requested resource (VMs) has its required components to perform its task. We denote user request as UR which is the subject. The users send one or many request at a time which are UR $(A_i, \ldots, A_n)$ for $i = 1, 2, 3, \ldots, n$ which are executed based on a First-Come-First-Serve (FCFS). their order of arrival using queuing theory of first come first served (FCFS). A_i is the components of the VMs. The resource components of VMs include α_s^i as Processor, β_s^i as Memory and γ_s^i as Storage. The corresponding i and s represent the number of resources and their measuring capacity respectively. Table 1 introduces the four samples of universal plan of VM instances allowed by Amazon EC2. By examining CPU, memory, and storage as type 1, type 2, and type 3 resources, respectively. For example, request (10; 0; 0; 5; 2) describes a user demanding 10 C3.medium VM

Table 1. Categorization of VM instance offered by Amazon EC2.

Name	CPU	Memory (GB)	Storage (GB)
C3.medium	2	3.75	2 × 8
C3.large	4	7.5	2 × 16
C3.x large	8	15	2 × 40
C3.2xlarge	16	30	2 × 80
C3.4xlarge	32	60	2 × 160

0 C3.xlarge VM instance, C3.2xlarge VM instances and 2 C3.4xlarge VM instances. All the VM are made to be single-core to satisfy the PlanetLab data set specifications. The simulation consists of a set of 125 PMs conforming a datacenter. Mathematically we can represent the whole request as:

$$Ai \subset UR$$

$$\alpha_s^1, \beta_s^1, \gamma_s^1 \subset Ai$$

$$\alpha_s^1, \beta_s^1, \gamma_s^1 \subset Ai \subset UR \Rightarrow \alpha_s^1, \beta_s^1, \gamma_s^1 \subset UR$$

Therefore, when the user send request of only one resource it will be expressed as in Eqs. 1 and 2.

$$UR^1 = A_i \tag{1}$$

where $i = 1$, and UR^1 is when the resource required is only one.

$$= Ai = \left(\alpha_s^1 + \beta_s^1 + \gamma_s^1\right) \tag{2}$$

Whereas, if the user request is more than one, then the request will be expressed as in Eqs. 3 and 4.

$$= Ai = A_1 + A_2 + A_3 + \cdots\cdots\cdots A_n \tag{3}$$

$$= Ai = \left(\alpha_s^1 + \beta_s^1 + \gamma_s^1\right) + \left(\alpha_s^2 + \beta_s^2 + \gamma_s^2\right) + \cdots\cdots\cdots\cdots \left(\left(\alpha_s^n + \beta_s^n + \gamma_s^n\right)\right)$$

$$UR^n = \sum\nolimits_{i=1}^{n} \left(\alpha_s^i\right) + \sum\nolimits_{i=1}^{n} \left(\beta_s^i\right) + \sum\nolimits_{i=1}^{n} \left(\gamma_s^i\right) \tag{4}$$

Energy-Efficient Flower Pollination Technique. The FPA is a new Nature-Inspired algorithm inspired by the analogy of biological process of flower pollination. A detail of FPA optimization can be found in [28]. The algorithm followed the pollination which has been idealized based on four rules:

1. Biotic and cross-pollination is the process of moving pollen by the pollinators using Levy flights strategy. This method is considered as the global pollination process.
2. The term abiotic and or self-pollination describes the local pollination method.
3. The mean of flower constancy is regarded as probability of reproduction system of the flower that is directly associated to other distinct flowers.
4. Switching probability is employed to control between exploration and exploitation that are commonly known as local and global search. It is define as p [0, 1]. Figure 3 shows the algorithm pseudocode.

The aforementioned rules are mathematically expressed in the following paragraph. The performance and effectiveness of FPA are verified using some widely used benchmark problems. The results support its applicability in solving optimization tasks [17]. The synopsis found from FPA evolutionary line indicate that the algorithm has the flexibility, capability and proficiency that is suitable in solving several kinds of problems in various NP-hard states. Previous researchers also implement this algorithm to solve the NP-hard problems either in continuous or discrete search spaces, and found to outperform the compared metaheuristics algorithms [18–20]. In FPA, the pollination takes place in two classes except for the fittest element, switching probability, and the levy flight. One of these classes is called the global solution and or global search. In this class, every flower receives single pollen and individual flower drops one pollen gamete. Consequently a solution X_i is equal to a flower and a pollen gamete. Pollinators searches the whole solution space to find the position of the optimum solution. Therefore, global optimization conforms to the biotic method and or cross-pollinators to move the pollen to different location obeying levy flight law. Levy flight is also efficient compared to the Brownian law in exploring unknown large search space, and this can be express as in Eq. 5.

Algorithm: *Flower Pollination Algorithm*
Require: *Set of population of n flowers/pollen gametes with random solutions*
*Find the best solution g * in the initial population*
Ensure: *Define a switch probability* $P = 0.5\,\frac{(Max_{iteration} - t)}{Max_{iteration}}$

Input: PM list, VM, set of parameters
Output: VM allocation
Execute: Objective *Max* $R^{A} = (\mathrm{RU}^{\mathrm{DC}})\left(DC_{U}^{Energy}\right)$
Initialize a population of n flowers/pollen gametes with random solutions
Find the best solution $\boldsymbol{g}_{*}$ *in the initial population*
Define a switch probability P
While (t < $MaxGeneration$)
  For $i = 1{:}n$ $(all\ n\ flowers\ in\ the\ population)$
    If $rand < p,$
  $Draw\ a\ (d - dimentional)\ step\ vector\ L\ which\ obeys\ a\ Levy\ distribution$
    $Global\ pollination\ via\ X_i^t = X_i^t + L\,(\boldsymbol{g}_{*} - X_i^t)$
    else
      $Draw\ \in from\ a\ uniform\ distribution\ in\ [0,1]$
      $Randomly\ choose\ j\ and\ k\ among\ all\ the\ solutions$
    $Do\ local\ pollination\ via\ X_{i+1,G} = X_{i,G} + \emptyset\left(X_{n-best_iG}\right) + \quad \vartheta\,(X_{p,G} - X_{q,G}$
    $where\ \emptyset = \vartheta = \in;$
  end if
    Evaluate new solutions
    If new solutions are better, **Then**
Update them in the population
  end for
    Find the current best solution $\boldsymbol{g}_{*}$
End while

Fig. 3. Energy-efficient Flower Pollination Algorithm pseudocode

$$s \tag{5}$$

Here, X_i^t denotes the *ith* pollen (solution) at iteration *t*, $\boldsymbol{g}_{*}$ denotes the current best solution obtained among all existing solutions during the iteration. L is a parametric value indicating movement and scope of the Levy diversion represented as Eq. (6).

$$L \sim \frac{(\beta + 1) * \sin\left(\frac{\pi\beta}{2}\right)}{\pi} * \frac{1}{s^{1} + \beta}, \quad (s \gg s0 > 0) \tag{6}$$

Here β the conventional gamma function, and the Levy division is efficient for large steps > 0.

The second class which is the local search is the composition of flowers after the getting the global optimum to find a more near optimal solution in the neighborhood. FPA requires local pollinators for exploitation as they can adequately exploit the search region of the algorithm as shown in Fig. 3. This position usually finds an improved solution from the current solution of the objective function. Mathematically it is expressed as Eq. (7).

$$X_i^t = X_i^t + \in \left(X_j^t - X_k^t\right) \tag{7}$$

where, X_j^t and X_k^t are the pollens of identical flowers types that fits to distinct flowers. $\in$ is taken from the uniform distribution in [0, 1] which become a random walk. The switch probability P is then used to switch between the global and the local pollination as presented in Eq. (8).

$$P = 0.5\frac{(Max_{iteration} - t)}{Max_{iteration}} \tag{8}$$

Modelling of Energy-Efficient Resource Allocation. Given a set of $VM = \{vm_i \backslash i = 1, 2, \ldots, n\}$ to be allocated on a set of $PM = \{PM_j \backslash j = 1, 2, \ldots, m\}$. Each VM is denoted as a d-dimensional vector of require resources [21], i.e. $VM_i = \left(A_{i,1}, A_{i,2} \ldots, A_{i,d}\right)$. Similarly, each PM is denoted as a d-dimensional vector size of resources [22], i.e. $PM_j = \left(B_{j,1}, B_{j,2} \ldots, B_{j,d}\right)$. The various PM_j resources considered in this study are 3 including processor (CPU), physical memory (RAM), and storage hence, the dimension $d = 3$ [23]. Furthermore, the vm allocation has life-cycle with begin time and end time, i.e., each VM_i is initialize at a static time (S_{t_i}) with execution time (E_{t_i}), therefore the overall time expended during the allocation of vm_i is mathematically represented as $S_{t_i} + E_{t_i}$.

Where $A_{i,s}$ is resource capacity $(\alpha_s^1, \beta_s^1, \gamma_s^1)$ requested by the $VM_i(1, 2, \ldots, n)$ and $B_{i,s}$ resource capacity $(\alpha_s^1, \beta_s^1, \gamma_s^1)$ of the $PM_j(1, 2, \ldots, m)$.

The resource allocation problem has the following (hard) Requirements:

1. Requirements 1: Resource most be compatible with the request
2. Requirements 2: ∀VMs request is $\leq$ the PMs' total resource capacity.
3. Requirements 3: ∀VMs each VM is run by a PM at any given time.
4. Requirements 4: Assume that a_j (t) is the set of VMs that are assigned to a PM.

$\sum_{i=1}^{n} UR^{n}$ of these assigned VMs is $\leq$ the PMs' total resource capacity.

$$For\,all\,capacity\,(\forall s) = 1, \ldots, d : \sum\nolimits_{vm_{i \in a_j}(t)} A_{i,s} \leq B_{i,s} \tag{9}$$

A feasible resource allocation R^A shows a positive allocation of all VMs to PMs, i.e. $\forall i \in \{1, 2, \ldots, n\}, \exists j \in \{1, 2, \ldots, m\}$, mapped ($VM_i$, PM_j) holds when VM_i is assigned to physical machine PM_j. Therefore, the objective function of R^A is to maximizes resource utilization (RU^{DC}) and energy efficiency $\left(DC_U^{Energy}\right)$ of the Cloud datacenter.

Firstly, maximization of RU of PM_j is considered as denoted in Eq. (10).

$$\mathrm{RU}_j^d = \frac{\sum_i^n = \rho_{ij} * \mathrm{VM}_i^d}{\mathrm{PM}_j^d} \forall \mathrm{d} \in \left\{\alpha_s^1, \beta_s^1, \gamma_s^1\right\} \tag{10}$$

To achieve the total RU of the datacenter, the individual PM_j resources are integrated and formulated as thus:

$$RU^{DC} = \int_{t1}^{t2} \frac{\sum_{j=1}^{m} U_j^{\alpha_s^1} + \sum_{j=1}^{m} U_j^{\beta_s^1} + \sum_{j=1}^{m} U_j^{\gamma_s^1}}{|d| \sum_{j=1}^{m} \rho_j^i} \partial t \tag{11}$$

$$\begin{cases} 1 & \text{if VM}_\text{i}\text{ assigned to PM}_\text{j} \\ 0 & \text{else} \end{cases}.$$

We assume that every host PM_j can run any VM_i and the energy consumption model $P_j(t)$ of the host PM_j has a linear relationship with resource utilization [24]. [21] uses same model with different resource energy consumption. The energy consumption model is given as follows:

$$TEC_j^{\alpha} = \int_{t1}^{t2} P(EU(t))dt \quad PM_j \in P \tag{12}$$

where $EU(t)$ is the immediate PM utilization at the given time t and $P(EU(t))$ is the power consumption associated with $EU(t)$.

$$Maximize \sum_{j=1}^{m} = EU(t)_j \tag{13}$$

$$TEC_j^{\alpha} = \sum_{j=1}^{m} = \mathrm{EU(t)_j} \tag{14}$$

where $EU(t)_j$ with $j = 1, 2, \ldots, m$ is total energy consumption of the PM_j. $i \in \{1, 2, \ldots, n\}, j \in \{1, 2, \ldots, m\}, t \in [0; T]$.

The overall energy consumption for the datacenter can be expressed as presented in Eq. (15).

$$DC_U^{Energy} = \int_{t1}^{t2} P(EU(t))(\alpha)dt + EU(\beta) * P_{max} + EU(\gamma) * P_{max} \tag{15}$$

4 Experimental Settings and Evaluation

The proposed technique has been implemented using the Multi-RecCloudsim version 3.0.3 with IntelliJ IDEA java development environment [21]. The simulation tool is design for the study energy-efficiency in Cloud datacenter environment. The system software specification used during the simulation of the algorithm are Intel CoreTM i7 processor, 2 GHz processor speed, 1 TB storage drive with 8 GB memory. The results are based on real workload of PlanetLab [25]. During the simulations, each VM is randomly assigned a workload trace based on the user request. For uniformity, the PMs are examined to be homogeneous during the simulation, though heterogeneous configuration can also be simulated. Table 1 illustrates the experiment parameter settings. The datacenter have 125 servers to serve the user request concurrently as shown in Table 2. The table also include other parameter setting of the datacenter.

Table 2. Parameter settings for the experiment

S/NO	Parameter setting				
	Techniques	No of hosts	No of VM	No of users	VMware
1	EE-FPA	125	285	25	Xen
2	GAPA	125	285	25	Xen
3	OEMACS	125	285	25	Xen

5 Results and Discussion

The EEE-FPA is first verified using the benchmarking function proposed by [26] which are important in obtaining the optimization algorithm real operation in terms of user request execution. The table below shows the functions used in order to compare the EEE-FPA with FPA and ACO. The various algorithms are examined in different settings, i.e., switching the iteration to different numbers (50, 100, and 200). It is outstanding that the algorithms have been run 20 times for each of the above mentioned requirements, and the concluding result has been collected from the average of 20 times. The proposed technique EE-FPA can transmit information within the search space, therefore it converges very fast against the FPA and ACO. The convergence of the proposed EE-FPA scheme is presented and contrasted with the OEMACS with the result achieved by experimenting. In this analysis, the VM requests are 1 to 120 and P = 0.5. The time needed to place the VMs over the PM is estimated for both algorithms and is depicted in Fig. 4. The proposed technique outperforms FPA and OEMACS. OEMACS presents guaranteed of convergence. However, the convergence time is problematic due to the distribution of random decisions.

The experiment result shows that our proposed EE-FPA technique takes less than 3 min to solve an uncertain allocation problem. This is due to the EE-FPA readiness and flexibility in searching near optimal solution. Therefore, we can say that our technique is appropriate and more suitable for large-scale datacenters due to faster convergence.

The user request ranges from 1–100 at a time and the EE-FPA is applied at the datacenter on the arrival of new request from the user. To show the effectiveness of our proposed technique the (RU^{DC}) of (PMs) and DC_U^{Energy} of datacenter are computed using Eqs. 11 and 15 respectively. The used of Minimax at the initialization of our technique produced a global optimum solutions for allocating VMs to PMs which minimizes the datacenter energy consumption. The results of the proposed EE-FPA technique are evaluated with the existing GAPA [4] and OEMACS [27] techniques on the basis of resource utilization and datacenter total energy consumption as shown in Figs. 5 and 6 below.

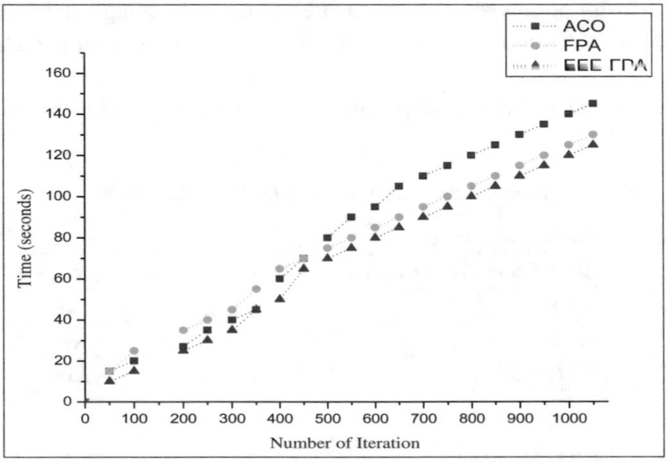

Fig. 4. Convergence and computational performance

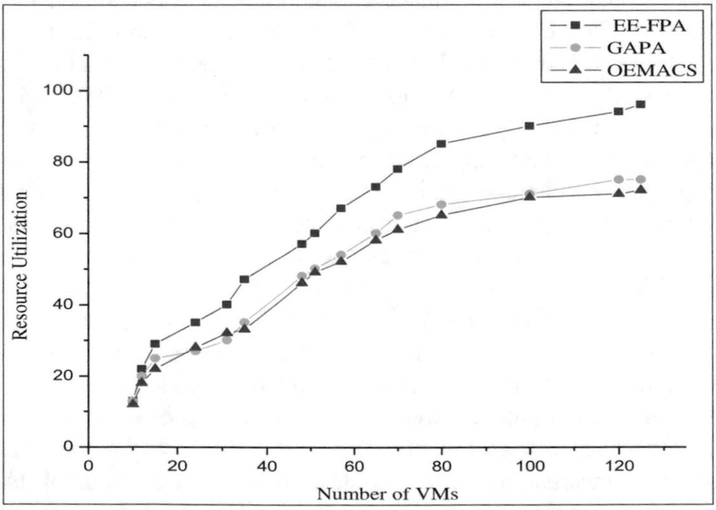

Fig. 5. Resource utilization

The energy consumed by Cloud datacenter escalates when the VM request changes. For example, there is an increase in the total energy consumption whenever there is an increase in the VM request by users. The energy consumption increases due to the efficiency of our proposed technique that allocated many VMs to PMs. Therefore, EE-FPA outperforms GAPA and OEMACS regarding resource utilization and energy efficiency.

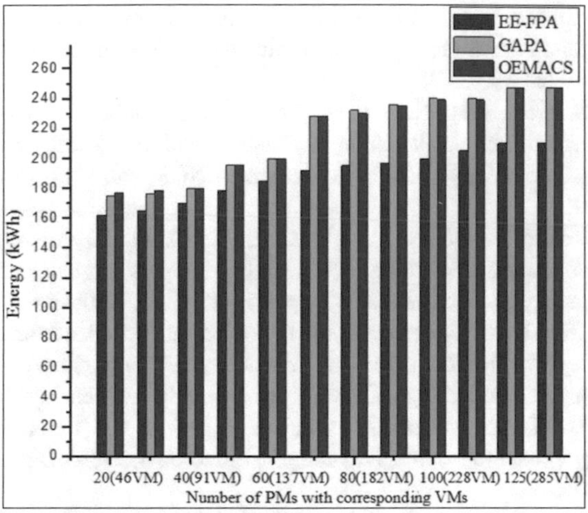

Fig. 6. Energy consumption

6 Conclusion

This paper proposes Energy-Efficient Flower Pollination Algorithm (EE-FPA) for near optimal resource allocation in datacenter VM resources that minimizes energy consumption of datacenter and also resource under-utilization. We presented the system framework that supports allocation of multiple VMs instances on a Physical Machine (PM) known as a server which has the potential to increase the energy efficiency as well resource utilization in Cloud datacenter. The proposed technique uses Processor, Storage and Memory as major resource component of PM to allocate a set of VMs, such that the capacity of PM will satisfy the resource requirement of all VMs operating on it. The framework of the proposed resource utilization and energy consumption model and Algorithm's pseudo code were elaborated. The experiment was conducted on Multi-RecCloudSim with IntelliJ IDEA development tool has been used as the testbed to simulate our algorithm using Planet workload. The results indicate that the proposed technique energy consumption outperform the benchmarking techniques which include GAPA, and OEMACS with 91% and 94.5% energy consumption while EE-FPA is around 65%. On average 35% of energy has been saved using EE-FPA and resource utilization has been improved.

References

1. Beloglazov, A., Abawajy, J., Buyya, R.: Energy-aware resource allocation heuristics for efficient management of data centers for cloud computing. Future Gener. Comput. Syst. **28** (5), 755–768 (2012)
2. Foster, I., et al.: Cloud computing and grid computing 360-degree compared. In: Proceedings of Grid Computing Environments Workshop (GCE), pp. 1–10. IEEE, Austin (2008)

3. Buyya, R., Beloglazov, A., Abawajy, J.: Energy-efficient management of data center resources for cloud computing: a vision, architectural elements, and open challenges. arXiv preprint arXiv:1006.0308 (2010)
4. Quang-Hung, N., Nien, P.D., Nam, N.H., Tuong, N.H., Thoai, N.: A genetic algorithm for power-aware virtual machine allocation in private cloud. In: Proceeding of Information and Communication Technology-EurAsia Conference. LNCS, vol. 7804, pp. 183–191. Springer, Heidelberg (2013)
5. Rodero, I., Jaramillo, J., Quiroz, A., Parashar, M., Guim, F., Poole, S.: Energy-efficient application-aware online provisioning for virtualized clouds and data centers. In: Proceeding of Green Computing Conference (GCC), pp. 31–45. IEEE, Chicago (2010)
6. Sharma, N.K., Reddy, G.R.M.: Novel energy efficient virtual machine allocation at data center using Genetic algorithm. In: 3rd International Conference on Signal Processing, Communication and Networking (ICSCN), pp. 1–6. IEEE, Chennai (2015)
7. Usman, M.J., Ismail, A.S., Chizari, H., Gital, A.Y., Aliyu, A.: A conceptual framework for realizing energy efficient resource allocation in cloud data centre. Indian J. Sci. Technol. **9** (46), 73–82 (2016)
8. Deore, S., Patil, A., Bhargava, R.: Energy-efficient scheduling scheme for virtual machines in cloud computing. Int. J. Comput. Appl. **56**(10), 79–86 (2012)
9. Moganarangan, N., Babukarthik, R.G., Bhuvaneswari, S., Basha, M.S., Dhavachelvan, P.: A novel algorithm for reducing energy-consumption in cloud computing environment: web service computing approach. J. King Saud Univ.-Comput. Inf. Sci. **28**(1), 55–67 (2016)
10. Phan, D.H., Suzuki, J., Carroll, R., Balasubramaniam, S., Donnelly, W., Botvich, D.L.: Evolutionary multiobjective optimization for green clouds. In: Proceedings of the 14th Annual Conference Companion on Genetic and Evolutionary Computation, pp. 19–26. ACM, Philadelphia (2012)
11. Shu, W., Wang, W., Wang, Y.: A novel energy-efficient resource allocation algorithm based on immune clonal optimization for green cloud computing. EURASIP J. Wirel. Commun. Netw. **14**(1), 1–9 (2014)
12. Jennings, B., Stadler, R.: Resource management in clouds: survey and research challenges. J. Netw. Syst. Manag. **15**(7), 1–53 (2014)
13. Madni, S.H.H., Latiff, M.S.A., Coulibaly, Y.: Resource scheduling for infrastructure as a service (IaaS) in cloud computing: challenges and opportunities. J. Netw. Comput. Appl. **2** (68), 173–200 (2016)
14. Shu, W., Wang, W., Wang, Y.: A novel energy-efficient resource allocation algorithm based on immune clonal optimization for green cloud computing. EURASIP J. Wirel. Commun. Netw. **2**(41), 1–9 (2014)
15. Ghribi, C., Hadji, M., Zeghlache, D.: Energy efficient VM scheduling for cloud data centers: exact allocation and migration algorithms. In: International Symposium of Cluster, Cloud and Grid Computing (CCGrid), pp. 671–678. ACM/IEEE, Delft (2013)
16. Rocha, L.A., Cardozo, E.: A hybrid optimization model for green cloud computing. In: Proceedings of the 7th International Conference on Utility and Cloud Computing, pp. 671–678. ACM/IEEE, London (2014)
17. Yang, X.-S., Karamanoglu, M., He, X.: Multi-objective flower algorithm for optimization. Proc. Comput. Sci. **5**(18), 861–868 (2013)
18. Abdelaziz, A., Ali, E., Elazim, S.A.: Flower pollination algorithm and loss sensitivity factors for optimal sizing and placement of capacitors in radial distribution systems. Int. J. Electr. Power Energy Syst. **7**(8), 207–214 (2016)
19. Abdel-Raouf, O., Abdel-Baset, M.: A new hybrid flower pollination algorithm for solving constrained global optimization problems. Int. J. Appl. Oper. Res. Open Access J. **4**(2), 1–13 (2014)

20. Babu, M., Jaisiva, S.: Optimal reactive power flow by flower pollination algorithm. Asian J. Appl. Sci. Technol. **1**(3), 137–141 (2017)
21. Lin, W., et al.: Multi-resource scheduling and power simulation for cloud computing. Inf. Sci. **3**(9), 168–186 (2017)
22. Xiong, A.-P., Xu, C.-X.: Energy efficient multiresource allocation of virtual machine based on PSO in cloud data center. Math. Prob. Eng. **2**(14), 23–31 (2014)
23. Mishra, M., Sahoo, A.: On theory of VM placement: anomalies in existing methodologies and their mitigation using a novel vector based approach. In: Proceeding of Cloud Computing (CLOUD) Conference, pp. 275–282. IEEE, Washington, D.C. (2011)
24. Beloglazov, A.: Energy-efficient management of virtual machines in data centers for cloud computing. University of Melbourne, Department of Computing and Information Systems (2013)
25. Park, K., Pai, V.S.: CoMon: a mostly-scalable monitoring system for PlanetLab. ACM SIGOPS Oper. Syst. Rev. **4**(1), 65–74 (2006)
26. Jamil, M., Yang, X.-S.: A literature survey of benchmark functions for global optimisation problems. Int. J. Math. Model. Numer. Optim. **4**(2), 150–194 (2013)
27. Liu, X.-F., Zhan, Z.H., Deng, J.D., Li, Y., Gu, T., Zhang, J.: An energy efficient ant colony system for virtual machine placement in cloud computing. IEEE Trans. Evolut. Comput. **22** (1), 113–128 (2016)
28. Yang, X.S.: Flower pollination algorithm for global optimization. In: Durand-Lose, J., Jonoska, N. (eds.) Unconventional Computation and Natural Computation, UCNC 2012. LNCS, vol. 7445, pp. 240–249. Springer, Heidelberg (2012)

Schema Proposition Model for NoSQL Applications

Abdullahi Abubakar Imam[1,2,3], Shuib Basri[1,2](✉), Rohiza Ahmad[1,2], and María T. González-Aparicio[4]

[1] Computer and Information Sciences Department, Universiti Teknologi PETRONAS, Bandar Seri Iskandar, 32610 Seri Iskandar, Perak, Malaysia
aiabubakar3@gmail.com, {shuib_basri,rohiza_ahmad}@utp.edu.my
[2] SQ2E Research Cluster, Universiti Teknologi PETRONAS, Bandar Seri Iskandar, 32610 Seri Iskandar, Perak, Malaysia
[3] Ahmadu Bello University, Zaria, Nigeria
[4] Computing Department, University of Oviedo, Gijon, Spain
maytega@uniovi.es

Abstract. This paper proposes a schema proposition model for physical and logical modeling of NoSQL (Not Only SQL) document-store databases. Unlike the traditional databases which are rigid in design, the NoSQL databases are flexible, scalable, and low-latent making them to house big data efficiently. However, issues such as dataset complexity coupled with programmers' low competency are making database modeling and designing very challenging, especially in balancing the significant parameters like availability, consistency and scalability. The proposed model aims to address this balancing by abstractly proposing schemas to programmers with respect to user defined system requirements, Create, Read, Update and Delete (CRUD) operations, etc. Exploratory and experimental approaches were adopted in this research. System requirement formulas were introduced to compute the expected performance. A mini system was developed to validate the proposed model by comparing the manually generated schema to the schema generated through the proposed model. Results obtained indicated a significant improvement in the schema generated by the proposed model.

Keywords: NoSQL databases · Database modeling · Big data Schema proposition · Document-store databases

1 Introduction

In database technologies and big data research, new concerns been debated increasingly are the storage capabilities of the non-conventional applications. Traditional databases predominantly occupied all aspect of data storage for decades. They provide engines to centrally control data, eliminate inconsistencies and control redundancy [1]. The emergence of big data with its untraditional characteristics has necessitated the invention of flexible, scalable and latent databases called NoSQL [2–4]. The term

F. Saeed et al. (Eds.): IRICT 2018, AISC 843, pp. 30–39, 2019.
https://doi.org/10.1007/978-3-319-99007-1_3

NoSQL comes from "Not only SQL", which means NoSQL are not introduced to replace SQL or traditional databases, but to fill the gap created due to the continuous expansion in data size, complexity, variety and variability [5].

NoSQL databases have since been widely accepted and adopted by many big organizations such as Google, Amazon, Facebook among others [6]. However, the dynamism nature of today's data makes it difficult for programmers to model NoSQL databases appropriately for reasons such as low competency, traditional modeling mindset and complex datasets [7–9]. These have attracted the attention of both the industry and academia to search for viable solutions to help NoSQL data modelers. Nonetheless, some of these solutions are vendor specific [6], while others focused on a different scope of research [10].

In our earlier contributions to the same research scope, we conducted two different studies which were industrially motivated, theoretically grounded, and empirically validated. The first paper proposed new cardinality notations and styles [11], while the second paper presented modeling guidelines for NoSQL document-store databases [12]. The outcome of these studies laid down a solid foundation for the current study in the aspects of semantic mapping of entities and relationship prioritization. They also contributed in understanding and prioritizing system requirements in relations to user requirements.

In this paper, we proposed a schema proposition model for NoSQL document-store databases. This model aims to simplify further the NoSQL modeling process. As mentioned earlier, our first solution to the aforementioned problem was to establish cardinality standards for modeling NoSQL databases. Second, we produced, categorized and prioritized NoSQL modeling guidelines to guide programmers on how NoSQL databases can be best modeled. These solutions require considerably high level of expertise thus motivated the invention of the proposed model. Programmers need not have to write schemas from scratch. They only have to feed some parameters into the proposed model, and the rest is handled by the model. The main contributions of this paper are outlined as follows:

- Automatic schema proposition model for NoSQL document store databases. This model accepts parameters like system requirements (availability, consistency or scalability), CRUD operations, entities and their expected number of records etc. to produce a modeled schema at its output.
- New ways to calculate cluster or document availability. Cluster availability is categorized as Series, Parallel and Partial.
- An algorithm which can be directly translated into any programming language for the proposed model.

With such model, programmers can supply parameters with associated data to produce an initial schema based on solid theoretical and empirical foundations. It is important to note that, the proposed model aims to propose schema which can be used at the initial stage of database design only. This means the schema generated by the proposed model require expert review and enhancement before the commencement of system development. It's not meant to be 100% copy-and-paste functional schema.

This paper is organized as follows: Sect. 2 discusses the related literature with respect to NoSQL data modeling. Section 3 presents the architecture, algorithms and components of the proposed model. Section 4 shows the results of the proposed model with comparisons. Section 5 concludes the paper with future focus.

2 Literature Review

NoSQL document-store databases are highly flexible [13]. They are based on a flexible model that allows schemas to be written and managed by the client side application developers [5, 14]. However, this may lead to incorrect or inappropriate schema design especially in modeling the relationships between datasets and entities [5, 6]. In efforts to address these challenges, several researches were conducted.

In the work of [5], conceptual modeling was proposed using Formal Concept Analysis (FCA). This was proposed to assist developers model the document based databases. It adopted three (3) types of relationships from relational databases which are (i) one-to-one ➔ 1:1, (ii) one-to-many ➔ 1:M, and (iii) many-to-many ➔ M:M relationships. These relationships were directly inherited from relational database and applied onto document-store databases. This method reveals the effectiveness of the aforesaid relationships when applied to document-store databases. However, the types of data stored in document-stores are much more complex and bulkier than the one stored in relational databases. Therefore require more detailed and deeper cardinality breakdowns. Also, foreign keys are not directly supported in document-store databases. In addition, other contributing factors to document-store modeling such as embedding and referencing are not considered in this research despite their importance to NoSQL database modeling practice.

On the other hand, a study of patterns and techniques was conducted by Arora and Aggarwal [15] This study aims to model data using different NoSQL database categories. However, this approach does not have a modeling engine for schema propositions. While in [6, 16], data modeling was proposed for MongoDB using JSON and UML Diagram class. These approaches were developed for MongoDB only which is considered vendor specific.

On the contrary, optimizing storage efficiency and retrieval performance for geographic data was given less attention in recent year despite its great importance. This is why techniques such as Modeling Technique for Geographic Applications [17] (OMT-G), GMOD [18], data modeling (GISER) [19] and object-oriented analysis (GeoOOA) [20] for geographic IS and OMTEXT was proposed to minimize data inconsistency in NoSQL databases.

In a similar concept, some contributors, such as technical experts from JSON [9] and mongoDB [6] explained some ways to achieve relatively good data modeling relationships. These approaches are however sort of proprietary, focusing on the functionalities of the database in which they set to promote. There is a need to have a generalized approach which can be adopted by at least one category of NoSQL databases [5, 7, 21].

As can be noticed, these approaches focus more on spatial data, which is in most cases, managed by graph databases or are concerned with fundamental principles only, leaving out the automated process of modeling. Generating schemas automatically not only simplify modeling process but also makes it more accurate and secure.

3 Proposed Model

The proposed model is presented in this section. The model aims to minimize data modeling hardship faced by programmers with these parameters such as availability, consistency and scalability. The model is presented in three sections, namely, model architecture, algorithms and model components.

3.1 Model Architecture

The architecture of the proposed model is represented in Fig. 1. The rectangles used in the architecture represent phases or layers, arrow-line to indicate flow direction, curly brackets to signify document schema. Other symbols are labeled as shown in the diagram.

The architecture consists of four interconnected phases, with presentation layer (Client side) being the first. The user is expected to select from the selectable parameters, and supply values to the non-selectable parameters which require system peculiar inputs like entity name. Starting from phase 3, the model takes over and proceeds with semantic mappings of the entities based on the new generation cardinalities [11] and modeling guidelines [12]. The following algorithms explain the logic.

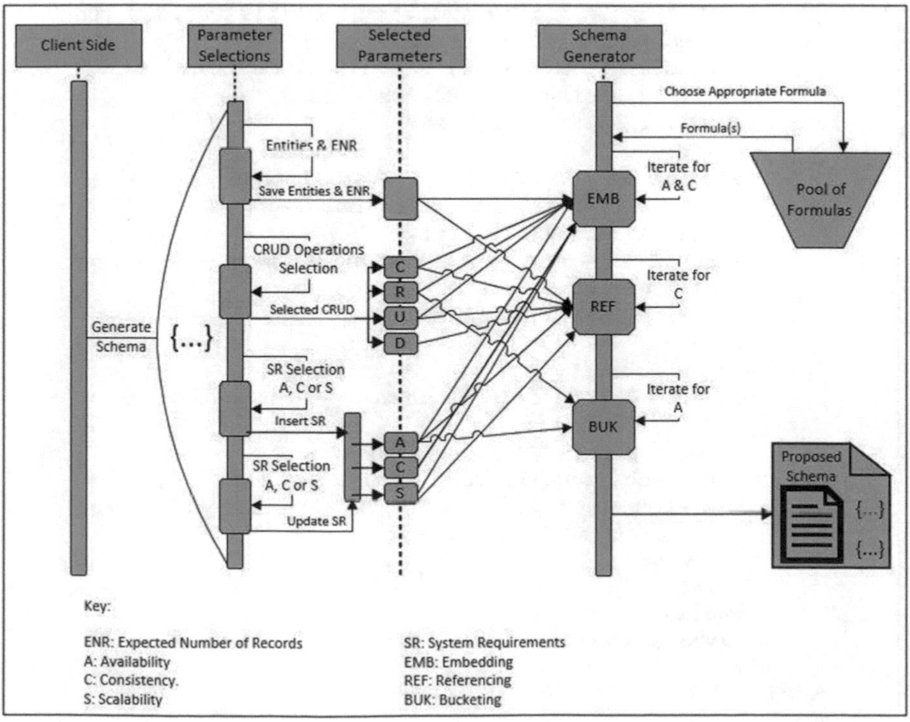

Fig. 1. Schema Proposition Model Architecture

3.2 Schema Generator Algorithms

This section presents the step-by-step procedure which the proposed model follows in order to generate a desired schema.

```
Algorithm Schema Generator.
Input:  List of entities E and Expected number of records
        per entity ENR.
        CRUD Operations Cr (C, R, U, D).
        System Requirements SR(A=Availability,C L,M,H= Con-
                                    sistency S=Security).
Output: Schema for NoSQL document-store databases.
Definitions: C L,M,H(Low, Medium and High Consistency)

begin
variables (E(E + ENR), CRUD, SR, i)
if E.size != Ø: then
  SR ← get preferred requirement
  Cr ← get selected CRUD operation Ø
  while i < E.size do
    ENR ← get ENR of an entity
      for each item in E do
        if SR = C_H and (Cr = C or U) then
          if item(ENR)[i]>0 and item(ENR)[i]<1000,000 then
            embedding is preferred;
          else if item(ENR)[i] > 100,000,000 then
            referencing is preferred; but…
            call RollBack(series);
          end;
        else if SR = C_L and (Cr = C or U) then
          referencing is preferred; but…
          call RollBack(parallel);
        else if SR = C_m and (Cr = C or U) then
          referencing is preferred; but…
          call RollBack(partial);
        end;
        //Availability section (retrieval)
        if SR = A and (Cr = R) then
          pg ← newPg[i];
          Tpg = ROUND(ENR / RecordsPerPage);
          foreach(range(1, Tpg) as pg)
            if(pg=1 or pg=Tpg or(pg>=newPg and pg<=newPg))
              pg ← newPg (+ or - 1);
            end;
          end;
        end;
        //var a ← ENR + ROUND((ENR / 100) * 20) /
      end;
  end;
else
  return null;
end;
```

```
Algorithm RollBack(method).
Input:  Series, Parallel, Partial.
Output: Cluster availability.

Definition: C = cluster and x is the cluster counter

begin
if method = series then
  Av = False
  for (i = 1; i <= 3; i++)
    Check availability of C1, C2, C3 ... Cn
    if availability is = 100% (Cx1Cx2...Cxn) then
      Av = True;
      return Av; set i = 4; exit();
    else
    repeat
  end;
else if method = parallel then
  Av = False
  for (i = 1; i <= 3; i++)
    Check availability of C1, C2, C3 ... Cn
    if availability is = 50% (1-(1-Cx)^2) then
      Av = True;
      return Av; set i = 4; exit();
    else
    repeat
  end;
else if method = partial then
  Av = False
  for (i = 1; i <= 3; i++)
    Check availability of C1, C2, C3 ... Cn
    if availability is = 75% (Formula as in Eq(3)) then
      Av = True;
      return Av; set i = 4; exit();
    else
    repeat
  end;

  end;
```

Next, we present the components that make up the mapping module which includes selections storage and formulas.

3.3 Model Components

The components for the semantic mapping module are presented here, starting with selections storage structure then followed by formulas.

Selections Storage. This storage is used to save user selected parameters and values supplied to the parameters. These records are vital for the computation, semantic and systematic mappings. The storage structure is described as follows:

- projectID: this attribute serves as a unique identifier of a particular project. The model is designed to handle multiple modeling processes for one user, so, a unique Id is required to differentiate between projects that belong to a single user.
- UserID: in case the model repository is centralized, this ID will be used to differentiate between users, so that successfully generated schemas cannot be interchanged between users.
- Entities: this attribute holds the names of all entities.
- ENR: this stands for expected number of records, which are used to calculate the possibility of either embedding or referencing a particular document.
- CRUD: this attribute house the selected CRUD operations. The model will iterate based on the saved selections such as *C* for create.
- SR: this stands for system requirements. The user selects a priority requirement such as availability, consistency or scalability.
- Relations: this attribute saves the relationship between entities.
- Active: this indicates if an entity is related to other entities or not.
- Flag: this attribute saves the computational status of all entities, where the values 0 and 1 represent not attended and completed respectively.

Formulas. We introduced some mathematical formulas to provide accurate systematic mappings, especially when checking cluster or document availability. These formulas are presented as follows.

$$C_s = A_x A_y \tag{1}$$

Equation 1 is the formula for highest level of consistency. This means all nodes (availability A of cluster $_x$ and cluster $_y$) must be available for an operation to take place, otherwise roll back function is called to retry or terminate the operation if the number of trials are more than 3 times or any number defined by the user.

$$C_L = 1 - (1 - A_x)^2 \tag{2}$$

If the choice of consistency is low C_L, then parallel operation is allowed. Which means if cluster $_x$ fails to report, cluster $_y$ can take over and cluster $_x$ is updated upon its next availability.

$$C_{m_{c,n}} = \sum_{i=0}^{n} \frac{c!}{i! \times (c-i)!} \times C^{(c-i)} \times (1-A)i \tag{3}$$

On the other hand, if the system requirement is consistency C and the level is medium $_m$, then Eq. 3 is used to calculate the availability of each cluster $_{C..N}$. For instance, if a system has $_C$ number of clusters, under this formula, the system is considered to be available when at least $_{C-N}$ components are available, which means not more than $_N$ clusters can fail.

In the following section, we present and discuss the results of the proposed model as well as the results of the traditionally generated schemas.

4 Results and Discussion

In this section, a comparison is made between the schema generated by the proposed model and the schema produced using conventional method. One parameter (medium $_m$ consistency) was considered for this experiment. Also, C of the CRUD operations are selected to apply and test the consistency requirement using create and update queries.

By referring to Fig. 2, it can be noticed that, when data size is around 5000 to 100,000 records, the difference between the schema generated by the proposed model and the schema produced using the traditional method is insignificant with regards to the speed of data creation. However, as data size increases, the gap begins to be substantially high. The proposed model maintained its speed of data creation from around 4 s to 15 s across different sizes of records, while the traditionally generated schema performs very low, up to nearly 60 s as data size increases exponentially (10, 20 and 50 million records).

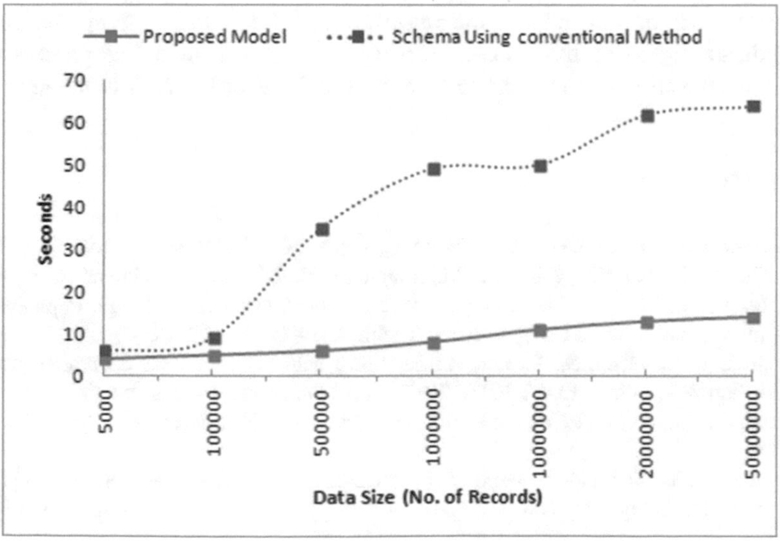

Fig. 2. Rate of record creations *C*: a proposed model and conventional method schemas comparison

This significant difference can be attributed to appropriate use of embedding and referencing when modeling NoSQL document-stores. For example, if entity E_x is related to E_y and E_z, embedding should not be applied since E_y can function in the absence of E_z as defined by C_m formula (Eq. 3). In this case referencing would be the best choice here, since it allows documents to be separated and referenced using ID.

With this significant improvement in NoSQL automated modeling, we can conclude that, despite the existence of several NoSQL modeling techniques, the model that does not require much technical skills is need in practice not only to alleviate modeling difficulties but also to accurately relate entities and propose desired schema.

5 Conclusion and Future Focus

In this paper, we presented a schema proposition model for modeling NoSQL document-store databases. The model consists of a conceptual architecture, algorithms, repository and mathematical formulas which are used to determine cluster availability. The proposed model aims to minimize difficulties faced by database modelers and erroneous implementation. To assess its performance, two separate schemas were generated. The first schema was generated using the traditional method while the proposed model produced the second schema. Both the schemas were implemented using MongoDB database. Records were added in batches to each database simultaneously. Results show a significant decrease in time taken to save records into the database that was generated using the proposed model. In addition to this performance increment, the modeling time is considerably reduced, and modeling process is simplified. This advancement may continue to be the focus of research in the near future.

References

1. Garcia-Molina, H., Ullman, J.D., Widom, J., Özsu, M., Valduriez, P., Connolly, T., Begg, C., Elmasri, R., Navathe, S.B., Lin, M., Tsuchiya, M., Member, S., Mariani, M.P., Sharma, M., Singh, G., Virk, R.: Database systems: a practical approach to design, implementation, and management. Int. J. Comput. Appl. Technol. **49**(4), 90–107 (2010)
2. Chouder, M.L., Rizzi, S., Chalal, R.: Enabling self-service BI on document stores. In: Workshop Proceedings EDBT/ICDT 2017 Joint Conference Venice, Italy (2017)
3. Atzeni, P., Bugiotti, F., Rossi, L.: Uniform access to NoSQL systems. Inf. Syst. **43**, 117–133 (2014)
4. Enríquez, J.G., Domínguez-Mayo, F.J., Escalona, M.J., Ross, M., Staples, G.: Entity reconciliation in big data sources: a systematic mapping study. Expert Syst. Appl. **80**, 14–27 (2017)
5. Varga, V., Jánosi, K.T., Kálmán, B.: Conceptual design of document NoSQL database with formal concept analysis. Acta Polytech. Hungarica **13**(2), 229–248 (2016)
6. William, Z.: 6 Rules of Thumb for MongoDB Schema Design. MongoDB, (2014). [Online]. https://www.mongodb.com/blog/post/6-rules-of-thumb-for-mongodb-schema-design-part-1. Accessed 23 Jan 2017

7. Atzeni, P.: Data modelling in the NoSQL world: a contradiction? In: International Conference on Computer Systems and Technology – CompSysTech 2016, June, pp. 23–24 (2016)
8. April, R.: NoSQL Technologies: Embrace NoSQL as a relational Guy – Column Family Store. DBCouncil (2016). https://dbcouncil.net/category/nosql-technologies/. Accessed 21 Apr 2017
9. CrawCuor, R., Makogon, D.: Modeling Data in Document Databases. United States: Developer Experience and Document DB (2016)
10. Mior, M.J., Salem, K., Aboulnaga, A., Liu, R: NoSE: schema design for NoSQL applications. In: IEEE Transactions on Knowledge and Data Engineering From 2016 IEEE 32nd International Conference on Data Engineering, ICDE 2016, vol. 4347, pp. 181–192 (2016)
11. Imam, A.A., Basri, S., Ahmad, R., Abdulaziz, N., González-aparicio, M.T.: New cardinality notations and styles for modeling NoSQL document-stores databases. In: IEEE Region 10 Conference (TENCON), Penang, Malaysia (2017)
12. Imam, A.A., Basri, S., Ahmad, R., Aziz, N., Gonzlez-aparicio, M.T., Watada, J., Member, S.: Data modeling guidelines for NoSQL document-store databases. Computer Science Education. SI on Advancing Theory About the Novice Programmer. Taylor & Francis (2018). Under Review
13. Han, J., Haihong, E., Le, G., Du, J.: Survey on NoSQL database. In: Proceedings - 2011 6th International Conference on Pervasive Computing Applications, ICPCA 2011, pp. 363–366 (2011)
14. Alhaj, T.A., Taha, M.M., Alim, F.M.: Synchronization wireless algorithm based on message digest (SWAMD) for mobile device database. In: 2013 International Conference on Computer, Electrical and Electronic Engineering Synchronization, pp. 259–262 (2013)
15. Arora, R., Aggarwal, R.: Modeling and querying data in mongodb. In: International Journal of Scientific and Engineering Research (IJSER 2013), vol. 4, pp. 141–144 (2013)
16. Kaur, K., Rani, K.: Modeling and querying data in NoSQL databases. In: IEEE International Conference on Big Data, pp. 1–7 (2013)
17. Borges, K.A., Davis, C.A., Laender, A.H.: Omt-g: an objectoriented data model for geographic applications. Geoinformatica **5**(3), 221–260 (2001)
18. De Oliveira, J.L., Pires, F., Medeiros, C.B.: An environment for modeling and design of geographic applications. Geoinformatica **1**(1), 29–58 (1997)
19. Shekhar, S., Coyle, M., Goyal, B., Liu, D.-R., Sarkar, S.: Data models in geographic information systems. Commun. ACM **40**(4), 103–111 (1997)
20. Kosters, G., Pagel, B.-U., Six, H.-W.: Gis-application development with geoooa. Int. J. Geogr. Inf. Sci. **11**(4), 307–335 (1997)
21. Mior, M.J.: Automated schema design for NoSQL databases. In: Proceedings of 2014 SIGMOD PhD Symposium – SIGMOD 2014, Ph.D. Symposium, pp. 41–45 (2014)

Data Mining Techniques for Disease Risk Prediction Model: A Systematic Literature Review

Wan Muhamad Taufik Wan Ahmad(✉),
Nur Laila Ab Ghani, and Sulfeeza Mohd Drus

College of Computer Science and Information Technology, Universiti Tenaga Nasional, Jalan IKRAM-UNITEN, 43000 Kajang, Selangor, Malaysia
taufikwanahmad@gmail.com,
{Laila, Sulfeeza}@uniten.edu.my

Abstract. Risk prediction model estimates event occurrence based on related data. Conventional statistical metrics that utilized primary data generates simple descriptive analysis that often provide insufficient knowledge for decision making. In contrast, data mining techniques that have the capability to find hidden pattern from the secondary data in large databases and create prediction for de- sired output has become a popular approach to develop any risk prediction model. In healthcare particularly, data mining techniques can be applied in disease risk prediction model to provide reliable prediction on the possibility of acquiring the disease based on individual's clinical and non-clinical data. Due to the increased use of data mining in healthcare, this study aims at identifying the data mining techniques and algorithms that are commonly implemented in studies related to various disease risk prediction model as well as finding the accuracy of the algorithms. The accuracy evaluation consists of various method, but this paper is focusing on overall accuracy which is measured by the total number of correctly predicted output over the total number of prediction. A systematic literature review approach that search across five databases found 170 articles, of which 7 articles were selected in the final process. This review found that most prediction model used classification technique, with a focus on decision tree, neural network, support vector machines, and Naïve Bayes algorithms where heart-related disease is commonly studied. Further research can apply similar algorithms to develop risk prediction model for other types of diseases, such as infectious disease prediction.

Keywords: Disease risk prediction · Data mining · Data mining techniques
Data mining algorithm

1 Introduction

Risk prediction can be described as a study to predict the occurrence of events such as disaster, accident, disease, and even to detect fault of sub-system in real-time [1]. The many advantages of risk prediction include early precaution for safety purpose and reducing cost of production in an organization. For many years, various sectors have

F. Saeed et al. (Eds.): IRICT 2018, AISC 843, pp. 40–46, 2019.
https://doi.org/10.1007/978-3-319-99007-1_4

been implementing risk prediction to assist in decision making. In healthcare, risk prediction usually involves prediction model that is used to predict the occurrence of disease in an individual. For instance, a doctor could detect an individual who has high possibility to have heart attack based on prediction model that has been developed by analyzing patient's information, medical diagnostic and images [2]. In addition to that, having dependable disease risk prediction model seems to be increasingly important as World Health Organization (WHO), stated that around 40 million deaths per year is reportedly occurred due to non-communicable diseases such as cardiovascular disease and diabetes which equivalent to 70 percent of all deaths worldwide [3]. Hence, having this knowledge will act as decision support system that assist health practitioners with their clinical decision-making tasks.

One of the approach in developing disease risk prediction model is through data mining. Data mining refers to the process of extracting useful information from the collection of data and present them into understandable structure, where the process involves two fields which are computer science and statistic [4]. Previously, people used traditional statistical analysis to analyze the data. Statistical analysis is mostly concerned on primary data analysis whereby the data are collected with a predefined set of questions in mind, while data mining gives greater attention of analyzing secondary data from large databases [5].

Data mining consist of several techniques that can be applied on different dataset. In developing prediction model, techniques that are commonly used are classification and regression [6]. Classification is used to assign each item in a set of data into one of predefined set of target class [7]. For example, in healthcare application, new target class can be set, namely as "high risk" or "low risk", where the type of patient that will fall under which class will be decided by classification technique, that learn pattern from the provided attributes [8]. Contrary to regression technique, the purpose is still the same, which is to predict the unknown class of the items, but the difference is in terms of the nature of attributes involved. Regression is used when the attributes of targeted class is continuous like numbers and figures, while classification is used when the attributes are categorical [9]. For example, regression is used in business domain, to predict the number of sales [10].

In this study, a systematic review of existing literatures was performed mainly to identify the data mining techniques that have been applied in disease risk prediction model. Various type of diseases, type of data mining techniques and algorithms, and prediction accuracy are presented in this review. Physicians are concerned on medical decision that they made especially when there is a doubt in diagnosis, thus improving data mining algorithms in term of accuracy is crucial in helping them. Review on accuracy performance will give some insight on selection of algorithm for their medical case. The paper aims to answer following research questions:

RQ1. How can we categorize the diseases?

RQ2. What are the data mining techniques used for disease risk prediction and their accuracy?

This paper is organized as follows: Sect. 2 discusses on the systematic review methodology. Section 3 presents the findings, Sect. 4 analyze the findings, before the concluding remarks in Sect. 5.

2 Methods

Title search was performed focusing on keywords: (disease AND risk AND prediction AND model). The process of searching was conducted using five databases: Google Scholar, Scopus, ACM, IEEExplore and PubMed. All of the retrieved articles were reviewed whereby duplicate articles were eliminated followed by selection based on abstract and finally based on full-text reading. The main inclusion criterion in this study are:

- Any study that implemented or proposed data mining technique in developing risk prediction model of any type of disease.
- Any study that discuss the overall accuracy of the selected algorithms.

Any study would be excluded if the topic of discussion is unrelated to disease risk prediction. For example, studies that involved disease outbreak prediction, crop disease, death prediction or general reviews were excluded from the literature review. Furthermore, studies that does not discussed the accuracy of selected data mining techniques were also excluded. For this purpose, articles published from 2008-2018 are taken into consideration during the search process. Figure 1 below shows the systematic literature review flow diagram.

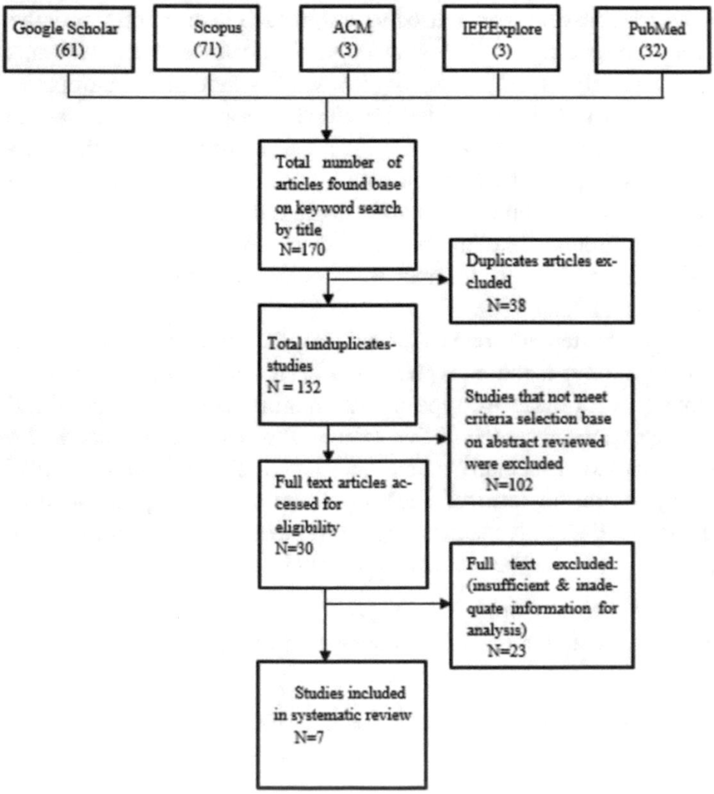

Fig. 1. Systematic literature review flow diagram

3 Results

From the five selected databases, a total of 170 articles were found using selected keywords based on title searching, of which 61 articles are from Google Scholar, 71 from Scopus, 3 from ACM, 3 from IEEExplore and finally 32 from PubMed database. After that, duplicate articles were checked and removed leaving only 132 of articles. The screening process continued by excluding articles that did not meet selection criteria based on abstract review resulting in 30 selected articles. Finally, 7 selected articles were used for this review paper after 23 articles were excluded because the articles did not contain sufficient discussion on the accuracy of the selected algorithms. The articles are summarized in Table 1 with their respective year, types of publications and number of citations to date.

Table 1. Summary of articles.

Source	Year	Type	Citation[a]
[11]	2015	Journal	19
[12]	2017	Journal	2
[13]	2015	Conference	2
[14]	2014	Conference	0
[15]	2014	Journal	0
[16]	2014	Journal	0
[17]	2016	Conference	2

[a]Taken from Google Scholar as at 14th April 2018

4 Analysis and Discussion

Based on Table 2, heart-related disease such as congenital heart disease and coronary heart disease were found to be the most studied disease using data mining techniques from the year 2008 until recently. Congenital is a heart disease related to malformation on the blood vessel around the heart, while coronary problem is related to the plaque form on the coronary arteries. Other types of diseases include diabetes, hypertension, and thyroid. There were also studies that investigates the correlation between different types of disease such as the possibility of having heart disease with the precedence case of diabetes or hypertension.

The data mining technique that is used for disease risk prediction model is classification, because the model produced categorical output. The most commonly used algorithms are artificial neural network (ANN), decision tree (DT), support vector machine (SVM), logistic regression (LR), fuzzy logic (FL), and Naïve Bayes (NB), where ANN produce the most accurate result.

Table 2. Data mining (DM) techniques and algorithms for various diseases.

Source	Disease type	Disease	DM techniques	DM algorithms	Accuracy (%)
[11]	Heart-related	Coronary heart disease	Classification	ANN	62.78
				FL	69.51
				DT	53.36
				SVM	67.71
				LR	63.23
[12]	Heart-related	Congenital heart disease	Classification	ANN	86.00
[13]	Heart-related	Heart disease	Classification	ANN	100.00
[14]	Diabetes	Type II diabetes	Classification	SVM	65.90
[15]	Diabetes and cardiovascular	Cardiovascular disease in Type II diabetes patient	Classification	SVM	93.80
[16]	Hypertension & coronary heart disease	Coronary heart disease in elderly hypertensive patient	Classification	RBF neural network	87.30
[17]	Thyroid	Thyroid	Classification	NB	95.38

As can be observed in Table 2, artificial neural network algorithm is able to predict different type of heart disease which are coronary, congenital, and general heart diseases with high overall accuracy result. However, for study in [11], fuzzy logic algorithm achieved higher accuracy than artificial neural network, support vector machine, decision tree and logistic regression when predicting coronary heart disease. One of the factor that may influenced this result is the ability of fuzzy logic algorithm itself that can remove uncertainty of data which cannot be done properly by other algorithms. The uncertainty problem usually emerged from the continuous data. Hence, both artificial neural network and fuzzy logic could be considered when predicting heart-related disease.

For disease prediction related to diabetes and cardiovascular, support vector machine could be chosen as one of the algorithms, but further experiment need to be done manipulating another dataset that may have different inputs.

Thyroid disease is a disease caused by over or under-functioning of human's thyroid gland that control body metabolism. Naïve Bayes works well with predicting thyroid disease in a person with 95.38% accuracy. However, depending on single algorithm could lead to different results when using different parameter of attributes. To solve this, another experiment needs to be done on thyroid disease using different algorithm or dataset to confirm its most suitable algorithm.

Table 3 below provides suggestion of algorithms based on previous researches, whereby few diseases like diabetes, cardiovascular and thyroid are suggested to have another experiment manipulating various dataset of particular disease in order to have better performance and strengthen the confidence on selected algorithm.

Table 3. Categorization of disease and performance of algorithm.

Disease type	Algorithms	Suggested algorithm for other researcher
Heart-related	Artificial neural network	Artificial neural network and fuzzy logic
	Decision tree	
	Fuzzy logic	
	Support vector machine	
Diabetes	Support vector machine	Support vector machine and further experiments needed
Cardiovascular	Support vector machine	Support vector machine and further experiments needed
Thyroid	Naïve Bayes	Naïve Bayes and further experiments needed

5 Conclusion

This paper is focusing on reviewing data mining techniques and algorithm for developing disease risk prediction model. The performance of data mining algorithm was evaluated based on its accuracy. It is found that, for heart disease, artificial neural network is frequently used for prediction and achieved highest accuracy for general heart disease prediction. For other diseases, lack of experiment using different type of inputs and algorithms, provide the need for further research in order to have more reliable prediction on those diseases. The application of the classification algorithms also can be further investigated in other types of diseases.

Acknowledgement. This work was supported by the Ministry of Higher Education, Malaysia under FRGS/1/2016/ICT04/UNITEN/03/1.

References

1. Kim, Y.-K., Jeong, C.-S.: Risk prediction system based on risk prediction model with complex event processing: risk prediction in real time on complex event processing engine. In: 2014 IEEE Fourth International Conference on Big Data and Cloud Computing, pp. 711–715 (2014)
2. Takci, H.: Improvement of heart attack prediction by the feature selection methods. Turk. J. Electr. Eng. Comput. Sci. **26**, 1–10 (2018)
3. WHO: Non-communicable diseases (2017). http://www.who.int/news-room/fact-sheets/detail/noncommunicable-diseases. Accessed 28 May 2018

4. Tomar, D., Agarwal, S.: A survey on data mining approaches for healthcare. Int. J. Bio-Sci. Bio-Technol. **5**(5), 241–266 (2013)
5. Srivastava, J., Srivastava, A.K.: Understanding linkage between data mining and statistics. IJETMAS **3**(10), 4–12 (2015)
6. Patel, S., Patel, H.: Survey of data mining techniques used in healthcare domain. Int. J. Inf. Sci. Technol. **62**(1), 1–8 (2016)
7. Kesavaraj, G., Sukumaran, S.: A study on classification techniques in data mining. In: 2013 Fourth International Conference on Computing, Communications and Networking Technologies (ICCCNT), pp. 1–7 (2013)
8. Leopord, H., Kipruto Cheruiyot, W., Kimani, S.: A survey and analysis on classification and regression data mining techniques for diseases outbreak prediction in datasets. Int. J. Eng. Sci. **5**, 1–11 (2016)
9. Sagar, P.: Analysis of prediction techniques based on classification and regression. Int. J. Comput. Appl. **163**(7), 975–8887 (2017)
10. Sharma, A., Kaur, B.: A research review on comparative analysis of data mining tools, techniques and parameters. Int. J. Adv. Res. Comput. Sci (2017). https://doi.org/10.26483/ijarcs.v8i7.4255
11. Kim, J., Lee, J., Lee, Y.: Data-mining-based coronary heart disease risk prediction model using fuzzy logic and decision tree. Healthc. Inform. Res. **21**(3), 167 (2015)
12. Li, H., et al.: An artificial neural network prediction model of congenital heart disease based on risk factors. Medicine (Baltimore) **96**(6), e6090 (2017)
13. Yazdani, A., Ramakrishnan, K.: Performance evaluation of artificial neural network models for the prediction of the risk of heart disease. In: Ibrahim, F., Usman, J., Mohktar, M., Ahmad, M. (eds.) International Conference for Innovation in Biomedical Engineering and Life Sciences. ICIBEL 2015. IFMBE Proceedings, vol. 56, pp. 179–182. Springer, Singapore (2016)
14. Pak, M., Shin, M.: Developing disease risk prediction model based on environmental factors. In: The 18th IEEE International Symposium on Consumer Electronics (ISCE 2014), pp. 1–2 (2014)
15. Radha, P., Prof, A.: Hybrid prediction model for the risk of cardiovascular disease in type-2 diabetic patients. Int. J. Adv. Res. Comput. Sci. Manag. Stud. **2**(10), 2321–7782 (2014)
16. Wang, X., Guo, S., Han, L.: GW28-e0440 the risk prediction model of coronary heart disease for elderly hypertensive patients. J. Am. Coll. Cardiol. **70**(16), C72 (2017)
17. Dash, S., Das, M.N., Mishra, B.K.: Implementation of an optimized classification model for prediction of hypothyroid disease risks. In: 2016 International Conference on Inventive Computation Technologies (ICICT), pp. 1–4 (2016)

Quality Features for Summarizing Text Forum Threads by Selecting Quality Replies

Akram Osman[1(✉)], Naomie Salim[1], Faisal Saeed[2], and Ibrahim Abdelhamid[3]

[1] Faculty of Computing, Universiti Teknologi Malaysia, 81310 Johor Bahru, Johor, Malaysia
oyoakram2@live.utm.my, naomie@utm.my

[2] College of Computer Science and Engineering, University of Taibah, Medina, Saudi Arabia
fsaeed@taibahu.edu.sa

[3] Faculty of Computer Studies, International University of Africa, Khartoum, Sudan
ibrahim_almahy@iua.edu.sd

Abstract. Text Forum Threads contain a huge volume of user-generated content derived from the discussion and information exchange among users who have similar interests. Often, some of the replies in a thread are completely off-topic which changes the discussion's direction. This phenomenon impacts negatively on the user's desire to continue with the discussion hence, a user might be interested in reading a few selected replies that provide a brief summary of the discussion topic. This paper aims at selecting quality replies about a topic raised in the initial-post which provide a short summary. We present a detailed analysis of the text forum threads structure based on a set of quality features for the forum summarization. Moreover, crowdsourcing platforms were used for judging the quality of the replies. Therefore, we have performed a text forum threads summarization based on replies weights and human judgment. TripAdvisor dataset has been used, therefore, the system summary helps the traveler in planning a journey. The experimental results conducted showed that the proposed approach can improve the performance of the text forum threads summarization based on forum quality features and crowdsourcing.

Keywords: Crowdsourcing services · Quality dimensions
Initial-post replies pairs · Text forum threads
Text forum threads summarization · Text summarization

1 Introduction

The form of content presented on the Internet has changed in the last few years. From the early 1990s onwards, the majority of web participant were consumers of traditionally published content created by a small number of publishers. Since the early 2000s, more users participated in generating content rather than being simply consumers. This transformation can be seen from the use of web forums, blogs, sites for social bookmarking, sharing communities for multimedia as well as social networking

F. Saeed et al. (Eds.): IRICT 2018, AISC 843, pp. 47–56, 2019.
https://doi.org/10.1007/978-3-319-99007-1_5

platforms such as Twitter and Facebook. These offer a combination of all social media, in addition, emphasizing the relationship among the participants of the community [1]. Text forum threads can be defined as an online community or a web-based application that is used for a wide range of users with common interests to gather and discuss problems and shared information. Text forum threads are also characterized as collaborative tools with large volumes of information which is accessible to the general community as a source of consultation in a virtual environment across the globe [2]. A text forum thread is formed by an aggregation of replies created by its users over a long period of time. The replies are organized into threads, each of which has a specific topic which is initiated by the first post in the thread [3]. The volume of information contained in a forum thread can grow exponentially with the popularity of the forum. For any user who is interested in the topic discussed in a given thread, it is difficult to go through all the replies and collect vital information especially in a long thread [4].

Text forum threads have become an important source for search engines for retrieving information that are relevant to users' queries. Many forum search facilities are able to help in finding threads with interesting topics. Although the main post (inquiry) is clearly represented in the initial-post (first post) of a topic, in order to find the actual responses or most relevant reactions to the inquiry, the thread needs to be read manually. Unfortunately, users have to face a growing amount of redundant thread information when browsing text forums [5]. Generally, people often prefer to read a few replies that appear at the beginning of the thread which reflects only the views of some contributors to the incomplete conversation [6]. Thus, with an increase in size of threads, the need to summarize each thread to help human understanding is becoming more and more urgent [7–9]. We argue that most of the initial-post replies pairs in the text forum threads consider inquiries and have more than one correct response (reply) and are therefore again more suited to the discussion threads being summarized. We can conclude from the above that it would be greatly beneficial to the users if each thread could present some replies as a brief summary of the topic discussed.

The rest of this paper is organized as follows: Sect. 2 presents related works of text forum thread summarization, Sect. 3 presents our Materials and methods and Sect. 4 provides the result and discussion. Finally, Sect. 5 concludes the paper.

2 Related Work

In the recent past, some researches have been directed at the summarization of text forum threads [9–14]. The most similar work to this study, in the context of discussion forum summarization, is the work in [15]. The authors dealt with whole replies in a thread instead of sentences. This enabled them to choose some replies after filtering irrelevant replies. The authors therefore selected the highly related replies to form a summary. Similarly, in [8], textual features and dialog act information of individual messages are used to select a subset of replies to unveil an extractive summary for thread. In [8, 15], the authors are argued that sentence selection from the reply would lead to loss of coherence in the thread summary.

According to [11, 16], the first work in threads summarization on specific context of discussion groups was carried out by Farrell et al. [17]. In this work, the authors selected salient sentences extracted from individual replies into variable-length summaries. In [18], a summarizing blog post was conducted, the authors extracted representative sentences from a blog post that best represent the topics discussed among its comments. In [11], the authors examined the use of cross-document structural theory in summarizing the thread content by exploiting semantic links between the replies. In [19], the authors defined a set of features to capture relevance, text quality and subjectivity their usefulness in choosing concise sentences to form a summary in an opinion forum. [6] considered each reply in a thread as a mixture of topics. The authors therefore employed modelling each topic dependencies during the asynchronous discussion. In order to give a brief statement for each thread involving multiple dynamic topics, the authors accomplished this task by extracting the most significant sentences in a thread. In [20], crowdsourcing services are used for text summarization in the tourist domain. The authors aimed at retrieving relevant information about a place or an object that is pictured in order to provide a short summary. In this paper, we proposed an approach for summarizing text forum threads by selecting quality initial-post replies pairs from text forum threads based on a set of forum quality features and a crowdsourcing service that determine quality replies.

3 Materials and Methods

This section presents the proposed model for selecting quality replies for text forum thread summarization. Figure 1 illustrates the structure of Text Thread Summarization Model. The details are described in the subsequent sections.

3.1 Data Description

For the experiments conducted in this paper, we used a dataset from the TripAdvisor online forum for New York City (NYC) [21]. The data consists of threads crawled from a discussion forum,[1] where each initial-post replies pairs form a thread. Raw texts should be preprocessed in order to achieve a suitable representation of them; so that they can be efficiently utilized in the experiments. Each thread is parsed by using traditional information retrieval techniques, including tokenization, stop words removal, and stemming to extract terms for indexing [22]. Each feature is normalized by given a value between 0 and 1.

3.2 Definition of Quality Replies

In this section, we describe how classifying user replies according to their response to the initial-post in the thread can be useful in text forum threads summarization. We hired a crowdsourcing platform[2] for judging the quality of the initial-post replies pairs. This platform was used for assigning class labels to every reply posted in response to

[1] https://www.tripadvisor.com.my/ShowForum-g28953-i4-New_York.html.

[2] www.crowdflower.com/.

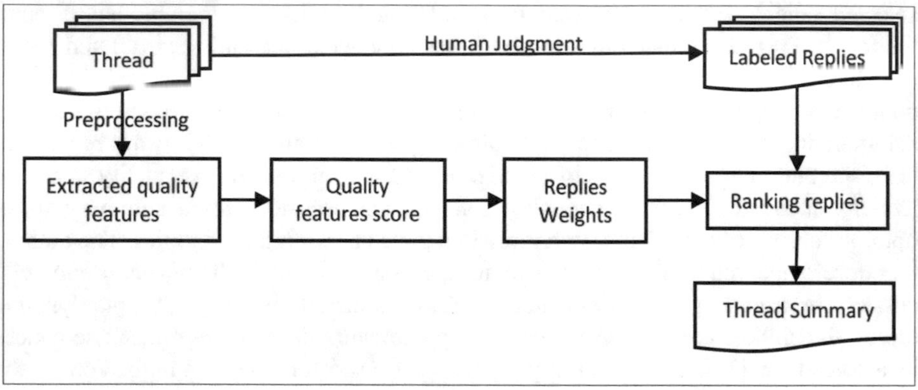

Fig. 1. Structure of text thread summarization model.

the initial-post. Humans were asked to consider a reply relevant if the information in the reply was about the given initial-post. Based on human judgments, each reply to the initial-post in the thread was classified as either non-quality, low-quality or high-quality". Table 1 shows an example of the thread[3] with initial-post and its replies with class labels. These can be represented by nominal values respectively. These class labels show the following information.

- High-quality Replies are totally relevant to the initial-post, informative, trustful and objective.
- Low-quality Replies are partially relevant to the initial-post and somehow informative.
- Non-quality Replies are irrelevant to the initial-post and uninformative.

From the dataset, 100 threads were randomly selected consisting of 816 replies. A majority of human decisions achieved high quality for 342 (42%) replies out of 816 and low quality for 303 (37%) and non-quality for 171 (21%) replies. Moreover, for creating data for the summarization task, we obtained gold summaries for each thread from the sampled 100 threads [8]. These hand-written summaries were then used to identify the most relevant replies in a discussion thread. The annotators were requested to keep the length of summaries roughly between 10% and 25% of the original text length.

3.3 Set of Forum Quality Features

Regarding the nature of this work, that is, summarizing text forum threads by filtering irrelevant replies to initial-post in a thread, this paper deals with initial-post replies pairs without any labeled human judgement. We relied on a set of forum quality features for

[3] https://www.tripadvisor.com.my/ShowTopic-g60763-i5-k3128263-How_to_get_from_JFK_to_New_Rochelle-New_York_City_New_York.html.

Table 1. An example of the initial-post replies pairs with class labels.

Topic Title: How to get from JFK to New Rochelle	
Initial post: *Can someone please advise the best way to get from JFK to New Rochelle using public transport? I know that by Taxi it is $80–85 USD which is quite expensive.*	
High-quality reply	Relpy 1: *New Rochelle is on the New Haven line of the Metro-North Railroad:* http://www.mta.info/mnr/html/mnrmap.htm *(It's the red line.)* *You can take a bus from JFK to Grand Central Station, where you can board the Metro-North train:* www.nyairportservice.com/page.php?id=48
Low-quality reply	Reply 2: *What are you going to do when you get to New Rochelle. Since it is not in the city you will need transportation of some sort to get around. In that case you may want to rent a car at Jfk and drive. The one day cost would be equal to the taxi fare and you will be in control.*
Non-quality reply	Reply 3: *One more thing, would I have to be going up and down a lot of stairs? I was thinking about my luggage...*

ranking the replies in the threads. We chose to follow the automatic unsupervised statistical techniques. We focus on quality dimensions as the most common features that are used to enhance thread retrieval and information retrieval [23–27]. We chose 12 quality features for the text forum threads. The text forum thread is regarded as the text document and the task is to select the replies that are highly relevant to the initial-posts in the threads. The replies are carried out as our selection unit such as the sentence or paragraph when dealing with ordinary text documents. The forum quality features that were used to select a quality replies that include the relevance of the replies, authors activity, replies structure, replies types, sentiment replies, and amount of data in replies are as follows in detail:

1. Cosine similarity between this reply and the initial-post *(f1)*.
   This feature computes the cosine value between the reply and initial-post to measure their similarity.
2. Reply words overlapping with the initial-post *(f2)*.
   This feature computes the overlap cosine value between the reply and initial-post to measure their overlapping.
3. Centroid of a reply to other replies in the thread. *(f3)*.
   This feature is obtained by computing the cosine similarity score between the reply vector and the vector obtained as the centroid of all the reply vectors of the thread. Similarity with centroid measures the relatedness of each reply with the initial-post. A reply with a higher similarity score with the thread centroid vector indicates that the reply better represents the basic idea in the initial-post.
4. The overlap with the previous replies *(f4)*.

This feature uses the overlap cosine value between initial-post and their replies in a thread. The importance of the replies in a thread is determined according to their overlap of occurrence when calculating the cosine of the thread. The greater the overlap is the most important.

5. Does this user create the initial-post? *(f5).*
   This feature verifies the owner of the reply. Reply that is created by the author of a thread should be relevant to the initial-post.
6. The position of a reply in the thread *(f6).*
   This feature determines the sequence of replies within thread ordered by ascending creation time. The reply position is defined as the position of the reply in the thread divided by the total numbers of replies in the thread.
7. Does this reply contain positive feedback words? *(f7).*
   This feature helps to determine the appreciative words in the replies. For instance, the reply followed by a "Thank you" or "This worked!". It reflects user satisfaction with the previous reply in their response to initial-post.
8. Does this reply contain negative feedback words? *(f8).*
   This feature helps to determine the refusal words in the replies. For instance, the reply followed by a "do not" or "does not!". It reflects user displeasure with the previous reply in their response to the initial-post.
9. Number of replies generated by the user in the thread *(f9).*
   This feature assesses the reply-author's activities in the thread. A highly active author has more experience. The interaction of the reply's author with the initial-post author increases the trust of the responding author.
10. Number of words in a reply *(f10).*
    This feature measures how much information was written in the reply. The higher the number the more meaningful the reply would be and more relevant to the initial-post.
11. Does this reply contain W-H question words? *(f11).*
    This feature helps to check the type of questions in the replies. A reply that contains any of 5WH-Q words reflects the inquiries in the initial-post for example, what, where, when, why, who and how. This reply should be relevant to the initial-post.
12. Does this reply contain question marks (?)? *(f12).*
    The feature helps to determine if a reply that contained a question mark (?) reflects the inquiries in the initial-post. This reply should be relevant to the initial-post.

Finally, the reply-weight can be computed by scoring all above quality features as shown in the following Eq. 1.

$$\mathrm{Wgt(Re)} = \sum_{\mathrm{f}=1}^{12} f_s\_\mathrm{Score} \tag{1}$$

where Wgt(Re) is the weight of the reply, *fs*_Score is the score of the quality features, f is the number of quality features.

4 Results and Discussion

We defined the quality of replies and labeled all replies as high-quality, low-quality or non-quality replies as described in Sect. 3.2. Also, all forum quality features score for each reply in the thread are calculated using the replies weights that as explained in Sect. 3.3. Table 2 provides the distribution of the Top K ranked replies based on a forum quality features score with human judgment in case K = 10%, K = 20% and K = 30% of the original 816 replies. Moreover, the replies are then ranked according to their weights by the model. The replies weights with their corresponding human judgment are matched. We note that the majority of the top K replies weights are similar to human judgment. Using the weight for each reply, the replies can be ranked and the top K replies selected to be part of the threads summary.

Figure 2 shows the performance using a forum quality features score with human judgment for creating threads summary. By looking at the top K ranked replies, we found that most of the top K ranked replies belong to the high-quality reply class (62.20%, 57.05% and 51.43%) in three K cases respectively. However, we also found some replies that belonged to low-quality reply class (28.05%, 32.52% and 34.29%) and non-quality replies (9.75%, 10.43% and 14.29%) which are part of the top K replies and in three cases respectively. This distribution indicates that top K replies had characteristics like high-quality replies. However, they provided information that is partly useful or did not provide information that was very useful to the traveler respectively. Finally, the top K initial-post replies pairs were selected as a thread summary. To illustrate this, Table 3 shows an example of the ranking forum quality features scores, comparing them to ones previously judged by crowdsourcing for selecting top k replies.

As shown in the results above, we considered a weight of each reply in the thread by calculating the score for 12 quality features. The Replies were then ranked and divided into three equal periods depending on their weights. Furthermore, replies that had higher weights and were defined as high-quality replies. The replies that had average weights were known as low-quality replies. Finally, the replies that had lower weights were considered as irrelevant replies (non-quality replies). Therefore, the task of the proposed approach was to choose the high-quality replies which achieved the same evaluation through their weights and human judgment and were then selected as a final summary of the thread.

Table 2. Top K ranked replies based on forum quality features and human judgment.

Thread summary	Top K ranked replies			Total of selected replies
	High-quality replies	Low-quality replies	Non-quality replies	
K = 10%	51	23	8	82
K = 20%	93	53	17	163
K = 30%	126	84	35	245

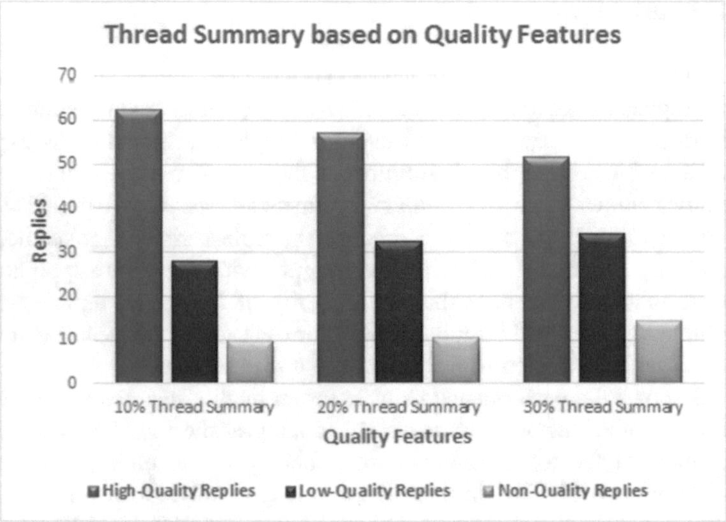

Fig. 2. Selected top K ranked replies as threads summary.

Table 3. Distribution of the top replies weights in case K = 20% as sampled threads summary.

Thread ID	High-quality reply	Low-quality reply	Non-quality reply
3056374	1	0	1
3060698	2	0	0
3064008	1	0	1
3064362	1	1	0
3064470	1	0	0

5 Conclusion and Future Work

This paper addresses the problem of generating an automatic summary for long threads in discussion forums on the internet. We approached it by finding the highest quality replies for each thread which are considered acceptance responses to the initial-post. Through the proposed approach, we aimed to exploit forum quality features and crowdsourcing services. A set of forum quality features were studied on the threads' structure for selecting quality replies which included the relevance of the replies, authors activity, replies structure, replies types, sentiment replies, and amount of data in replies. The process of calculating the quality features score gives weight to each reply in a thread. By using crowdsourcing services, replies are divided into the three quality classes. Replies weights were then compared against all replies classes selected by crowdsourcing services. The top 10% replies weights that were similar to human judgment as follows 62.20% high-quality, 28.05%, low-quality, 9.75% non-quality reply. The high-quality replies which were given the same evaluation through their weights and human judgment were selected as a summary of the thread.

The experimental result showed outstanding performance. In future work, classifying the replies into different forum discussion context could be done so that better selection of replies can be achieved and appropriate corresponding summary can be provided to the discussion forum.

Acknowledgment. This work is supported by the Ministry of Higher Education (MOHE) and the Research Management Centre (RMC) at the Universiti Teknologi Malaysia (UTM) under the Research University Grant Category (VOT Q.J130000.2528.16H74).

References

1. Agichtein, E., et al.: Finding high-quality content in social media. In: Proceedings of the 2008 International Conference on Web Search and Data Mining. ACM (2008)
2. Martínez Carod, N., et al.: Búsqueda de estrategias para la clasificación del contenido en foros técnicos de discusión. In: XIX Workshop de Investigadores en Ciencias de la Computación (WICC 2017, ITBA, Buenos Aires) (2017)
3. Fan, W.: Effective Search in Online Knowledge Communities: A Genetic Algorithm Approach. Virginia Polytechnic Institute and State University, Virginia (2009)
4. Krishnamani, J., Zhao, Y., Sunderraman, R.: Forum summarization using topic models and content-metadata sensitive clustering. In: 2013 IEEE/WIC/ACM International Joint Conferences on Web Intelligence (WI) and Intelligent Agent Technologies (IAT). IEEE (2013)
5. Tigelaar, A.S.: Automatic discussion summarization: a study of Internet fora (2008)
6. Ren, Z., et al.: Summarizing web forum threads based on a latent topic propagation process. In: Proceedings of the 20th ACM International Conference on Information and Knowledge Management. ACM (2011)
7. Almahy, I., Salim, N.: Web discussion summarization: study review. In: Proceedings of the First International Conference on Advanced Data and Information Engineering (DaEng-2013). Springer (2014)
8. Bhatia, S., Biyani, P., Mitra, P.: Summarizing online forum discussions-can dialog acts of individual messages help? In: EMNLP (2014)
9. Grozin, V.A., Gusarova, N.F., Dobrenko, N.V.: Feature selection for language independent text forum summarization. In: International Conference on Knowledge Engineering and the Semantic Web. Springer (2015)
10. Verberne, S., et al.: Automatic summarization of domain-specific forum threads: collecting reference data. In: Proceedings of the 2017 Conference on Conference Human Information Interaction and Retrieval. ACM (2017)
11. Almahy, I., et al.: Discussion summarization based on Crossdocument relation using model selection technique. In: Advances in Neural Networks, Fuzzy Systems and Artificial Intelligence, pp. 218–229 (2014)
12. Kabadjov, M.A., et al.: The OnForumS corpus from the Shared Task on Online Forum Summarisation at MultiLing 2015. In: LREC (2016)
13. Buraya, K., et al.: Mining of relevant and informative posts from text forums
14. Grozin, V., Buraya, K., Gusarova, N.: Comparison of text forum summarization depending on query type for text forums. In: Advances in Machine Learning and Signal Processing, pp. 269–279. Springer (2016)

15. Altantawy, M., Rafea, A., Aly, S.: Summarizing online discussions by filtering posts. In: IEEE International Conference on Information Reuse and Integration, IRI 2009. IEEE (2009)
16. Tigelaar, A.S., Den Akker, R.O., Hiemstra, D.: Automatic summarisation of discussion fora. Nat. Lang. Eng. **16**(2), 161–192 (2010)
17. Farrell, R., Fairweather, P.G., Snyder, K.: Summarization of discussion groups. In: Proceedings of the Tenth International Conference on Information and Knowledge Management. ACM (2001)
18. Hu, M., Sun, A., Lim, E.-P.: Comments-oriented blog summarization by sentence extraction. In: Proceedings of the Sixteenth ACM Conference on Information and Knowledge Management. ACM (2007)
19. Ying, D., Jiang., J.: Towards Opinion Summarization from Online Forums. ACL, Stroudsburg (2015)
20. Lloret, E., Plaza, L., Aker, A.: Analyzing the capabilities of crowdsourcing services for text summarization. Lang. Resour. Eval. **47**(2), 337–369 (2013)
21. Bhatia, S., Biyani, P., Mitra, P.: Identifying the role of individual user messages in an online discussion and its use in thread retrieval. J. Assoc. Inf. Sci. Technol. **67**(2), 276–288 (2016)
22. Manning, C.D., Raghavan, E.-P., Schütze, H.: Introduction to Information Retrieval, vol. 1. Cambridge University Press, Cambridge (2008)
23. Osman, A., Salim, N., Saeed, F.: Quality-based text web forum summarization-a review. Int. J. Soft Comput. **12**(1), 31–44 (2017)
24. Hoogeveen, D., et al.: Web forum retrieval and text analytics: a survey. Foundations and trends®. Inf. Retriev. **12**(1), 1–163 (2018)
25. Chai, K.E.K.: A Machine Learning-Based Approach for Automated Quality Assessment of User Generated Content in Web Forums. Curtin University, Digital Ecosystems and Business Intelligence Institute, Perth (2011)
26. Albaham, A.T., Salim, N., Adekunle, O.I.: Leveraging post level quality indicators in online forum thread retrieval. In: Proceedings of the First International Conference on Advanced Data and Information Engineering (DaEng-2013). Springer (2014)
27. Biyani, P., et al.: Online thread retrieval using thread structure and query subjectivity. Google Patents (2016)

Modelling and Forecasting Indoor Illumination Time Series Data from Light Pipe System

Waddah Waheeb[1(✉)], Rozaida Ghazali[1], Lokman Hakim Ismail[2], and Aslila Abd Kadir[3]

[1] Faculty of Computer Science and Information Technology, Universiti Tun Hussein Onn Malaysia, 86400 Parit Raja, Batu Pahat, Johor, Malaysia
waddah.waheeb@gmail.com, rozaida@uthm.edu.my

[2] Department of Design Engineering and Architecture, Faculty of Civil and Environmental Engineering, Universiti Tun Hussein Onn Malaysia, 86400 Parit Raja, Batu Pahat, Johor, Malaysia
lokman@uthm.edu.my

[3] Department of Civil Engineering, Centre for Diploma Studies, Universiti Tun Hussein Onn Malaysia, 86400 Parit Raja, Batu Pahat, Johor, Malaysia
aslila@uthm.edu.my

Abstract. Light pipe system allows the transmission of natural light from outside to inside of a room. It is needed where daylight cannot reach due to some reasons such as building design. Modeling and forecasting the indoor illumination amounts obtained using such system could help to estimate the needed artificial lighting. Thus, it provides occupants with a high satisfaction of visual comfort along with low energy consumption for lighting. In this research work, the indoor illumination data was obtained using a system under Malaysia climate condition. Four methods are used for modeling and forecasting tasks namely seasonal naïve, seasonal autoregressive integrated moving-average, seasonal and trend decomposition using Loess (STL), and TBATS. Results showed that the forecasts of the decomposed time series using STL method outperforms other methods in terms of the widely used forecasting metrics such as root mean squared error, mean absolute error, and mean absolute percentage error. Therefore, it can be used to forecast the indoor illumination, and as a result the needed artificial light amounts per day can be provided for the room on time.

Keywords: Time series · Forecasting · Green building
Sustainable daylighting system · Light pipes system
Seasonal and trend decomposition · STL

1 Introduction

Daylighting is the practice of using natural light to provide illumination in building during day time. In contrast, artificial lighting refers to any light source that is produced by electrical means.

Daylighting can reduce the need for artificial lighting in buildings especially during daytime. Utilizing daylighting system in buildings has many advantages and benefits, ranging from the aesthetic to physiological and economic [1]. Spaces illuminated by

F. Saeed et al. (Eds.): IRICT 2018, AISC 843, pp. 57–64, 2019.
https://doi.org/10.1007/978-3-319-99007-1_6

daylight will provide occupants with a high satisfaction of visual with low energy consumption for lighting [2]. Moreover, reducing the dependency on artificial light by using natural light to illuminate indoor area is one of the principle in green building index [1].

Light pipe system can be used to direct daylight into the interior of a building. Numerous studies have been reported on the performance of light pipe system as natural daylighting system [3–7]. Recently, an evaluation for the daylight amount offered by a light pipe system used under the Malaysian climate conditions was conducted in [1]. Indoor and outdoor illuminances of a windowless experiment room, in which the light pipe system was installed, were measured. It was found that the average internal illuminance levels offered by the system met the MS 1525:2007 recommendation for application in Malaysian buildings.

Therefore, the indoor illumination data measured by that system on different periods was considered in this research work for modeling and forecasting tasks. Modeling this data from that system could help to estimate the needed artificial lighting on a specified time where the system is installed. Thus, a supplementary artificial lighting can be provided on time to ensure a high satisfaction of visual comfort for the occupants along with low energy consumption for lighting.

The initial analysis for this data shows that this data has a seasonal pattern. Therefore, four methods applied by researchers to deal with seasonal time series are used in this research work namely; the seasonal naïve, seasonal autoregressive integrated moving-average, seasonal and trend decomposition using Loess, and TBATS. These methods have been used by researchers for runoff forecasting [8] traffic flow forecasting [9], airborne pollen concentrations modelling [10], and wind speed analysis [11].

This paper consists of four sections. Section 2 describes the indoor illumination time series data and methodology to model and forecast this data using the different methods mentioned above. Section 3 consists of results and discussion. And finally, Sect. 4 concludes the paper.

2 Materials and Methods

The following subsections describe the source of the used indoor illumination, the steps for preparing the data for modeling, the methods used for modeling and forecasting tasks, and the metrics in evaluating the forecasting performance of the four methods.

2.1 Data Source

The data used in this research work is collected based on a light pipe system designed as reported in [1]. The chosen area was built at 5 3′N latitude, 100 3′E longitude in Batu Pahat, Johor, Malaysia. In this research work, illuminance level was selected from 8:00 am to 5:00 pm starting from 5 February 2017 to 20 February 2017. In order to measure the internal illuminance levels for each minute, Extech Digital Light Meter 401025 was used. All internal illuminance was taken at work plane height 800 mm from the floor level.

2.2 Data Preparation

The data was prepared at 5-min intervals. Since there are 9 h (i.e., 8 am–5 pm) and in each hour there are 12 points; thus the total number of points in each day are 108 points. The missing data was filled in by the average of its 5-min interval. Figure 1 shows the data for the 16 days.

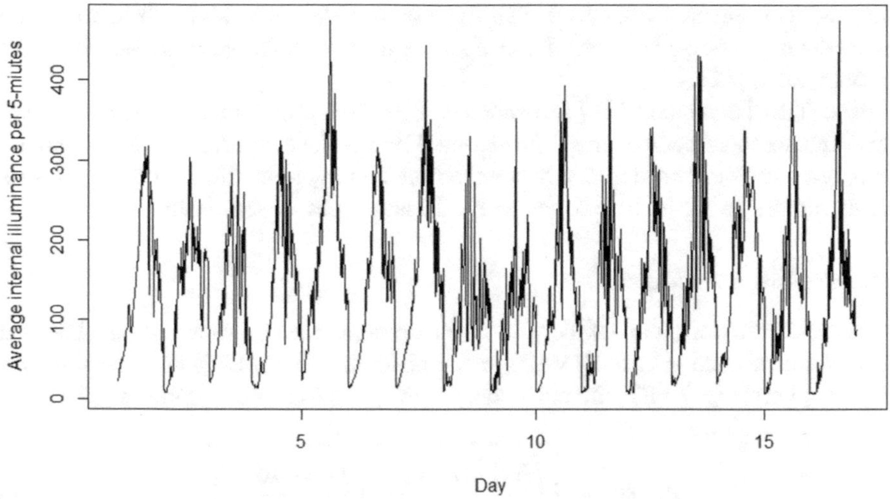

Fig. 1. Indoor illumination time series from 5 February 2017 to 20 February 2017.

Training data was chosen to be from 5 February 2017 to 19 February 2017 (1620 points), while the data on 20 February 2017 (108 points) was used only to validate the models. It basically means that after training the models, the final models will use their forecasts as inputs in forecasting for the next points, and not the original points from the validation (i.e., recursive multi-step forecasting).

2.3 Methods

It can be seen from the Fig. 1 that there is a seasonal pattern. For that, four methods were used in this work namely seasonal naïve, seasonal ARIMA, seasonal and trend decomposition using Loess, and TBATS.

A simple seasonal forecasting method is the seasonal naïve (snaïve). The forecasts produced by snaïve equals to the last observed values from the last season [12]. In our case, it returns the values from the last day in the training data.

The second method is the seasonal autoregressive integrated moving-average (SARIMA) which is formed by including seasonal terms in the ARIMA model. It is written as follows: ARIMA (p, d, q) $(P, D, Q)_m$, where (p, d, q) is the non-seasonal part of the method, (P, D, Q) is the seasonal part of the method, and m is the number of periods per season [12]. The parameters p and P are the order of the autoregressive (AR) component for both seasonal and non-seasonal parts, while q and Q are the order of moving-average (MA) component for both parts. The parameters d and D are the degree of differencing involved for both parts. Differencing is a way to make a time series stationary.

Seasonal and trend decomposition using Loess (STL) [13] is a method aimed at decomposing time series into seasonal, trend and irregular components. STL has several advantages such as the seasonal component is allowed to change over time, the rate of seasonal change and the smoothness of the trend-cycle can be controlled by the user, and its robustness to outliers [12]. Even though decomposition is primarily useful for analyzing time series data, it can also be used in forecasting as we did in this research work [12].

The TBATS method [14] incorporates **B**ox-Cox transformation, ARMA errors, **t**rend and seasonal components. The seasonal components are represented in terms of **t**rigonometric Fourier series, and hence the first T (Trigonometric) in TBATS. TBATS has the ability to model time series with different types of seasonality.

2.4 Performance Metrics

In this research work, three widely used performance metrics were chosen. These are the root mean squared error (RMSE), mean absolute error (MAE) and mean absolute percentage error (MAPE). The equations for these metrics are as follows:

$$\text{RMSE} = \sqrt{\left(\frac{\sum_{i=1}^{T}(Actual_i - Forecast_i)^2}{T}\right)} \tag{1}$$

$$\text{MAE} = \frac{1}{T}\sum_{i=1}^{T}|Actual_i - Forecast_i| \tag{2}$$

$$\text{MAPE} = \frac{100}{T}\sum_{i=1}^{T}\left|\frac{Actual_i - Forecast_i}{Actual_i}\right| \tag{3}$$

where T is the number of test samples.

3 Results and Discussion

The four methods are available in R language [15–17]. Therefore, after data preparation step, training data set was used to train and estimate methods' parameters. Following that, the next day was forecasted. Then, the forecasts were compared with the real data.

For snaïve method, there is no training step since it returns the values from the last day in the training data as a forecast for the next day.

With regards SARIMA, a seasonal difference was taken. However, it appears to be non-stationary, and so an additional first difference was taken. Our aim is to find an appropriate ARIMA model based on the plots shown in Fig. 2. The significant spike at lag 1 and 2 in the autocorrelation function (ACF) suggests a non-seasonal MA(2) component, and the significant spike at lag 108 in the ACF suggests a seasonal MA(1) component. Furthermore, there are significant spikes in the partial autocorrelation (PACF) and thus they suggest seasonal and non-seasonal AR components. Therefore, we begin with ARIMA (0, 1, 2) $(0, 1, 1)_{108}$, then several seasonal ARIMA models were fitted. The best forecast is produced by ARIMA (0, 1, 2) $(0, 1, 1)_{108}$. It is good to note that the available tests to estimate the required number of seasonal and non-seasonal differences suggest that the time series is stationary so no further difference is needed. However, the best fitted model is with the used first seasonal and non-seasonal difference.

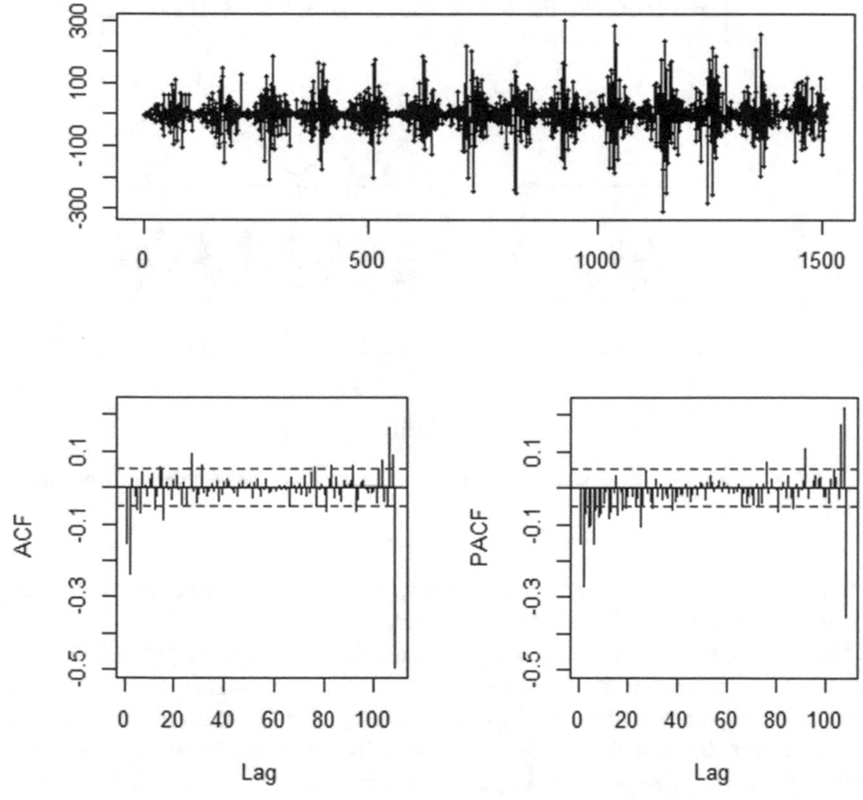

Fig. 2. Double differenced indoor illumination time series.

In the case of the STL method, different span of the loess window for trend extraction were investigated. The best model is shown in Fig. 3 that reveals the three components obtained from this method. Furthermore, it is clear that the seasonal pattern is repeated with a fixed period of time and the trend moves around between 130 and 170 lux.

Finally, with respect to TBATS method, the best fitted model is TBATS (0.292, {3,2}, –, {<108, 6>}). This translates that the Lambda for BoxCox transformation is 0.292, {3, 2} means three autoregressive components and two moving-average

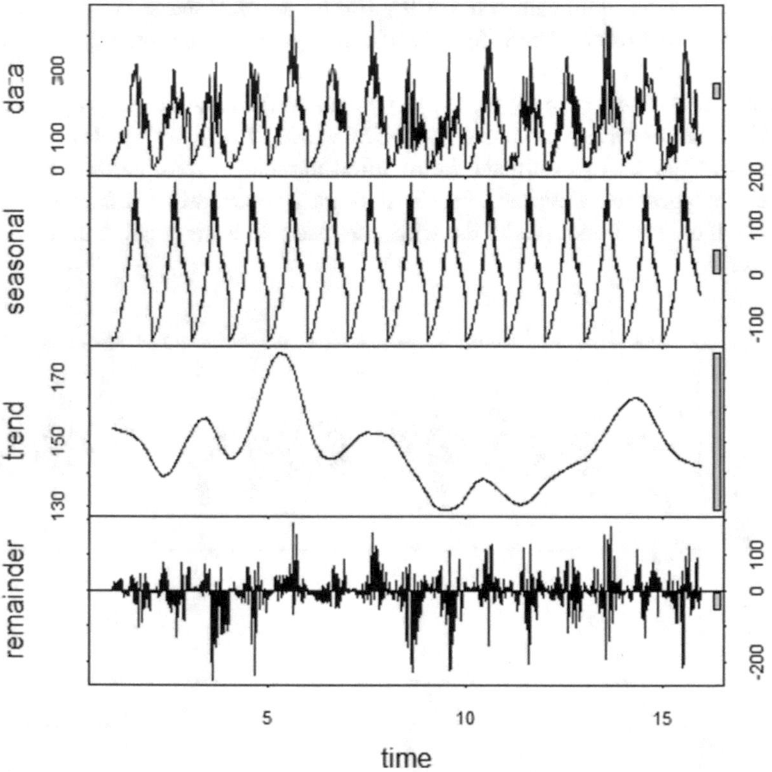

Fig. 3. The illumination time series (top) and its three components obtained from the STL method.

components, – denotes no damping parameter, <108, 6> means one seasonal of length 108 and TBATS fits it using six Fourier terms.

The forecasting performances using the four methods are tabulated in Table 1. It is clear that the forecasts of the decomposed time series using STL method outperforms all the methods in terms of all performance metrics. The forecasts of the decomposed time series can be obtained easily by adding the forecasts from the seasonal and seasonally adjusted components (i.e., trend and irregular components). The former is achieved using seasonal naïve method while the latter using the naïve method. The benefit of decomposing the time series into three components and to combine the forecasts from the seasonal and seasonally adjusted components leads to better results.

The forecasts obtained using STL + naïve and snaïve is shown in Fig. 4. Clearly, this method can be considered to forecast the indoor illumination for the room. As a result, the needed artificial light amounts per day can be provided for the room on time. Thus, ensuring a high satisfaction of visual comfort for the occupants plus consuming low energy.

Table 1. Best forecasting results using the four methods. Best value is in boldface

Method	RMSE	MAE	MAPE
snaïve	64.95083	42.6272	49.45309
SARIMA	48.29141	32.98855	36.28694
STL + naïve and snaïve	**43.92559**	**29.40703**	**32.00246**
TBATS	50.93877	40.1169	76.29157

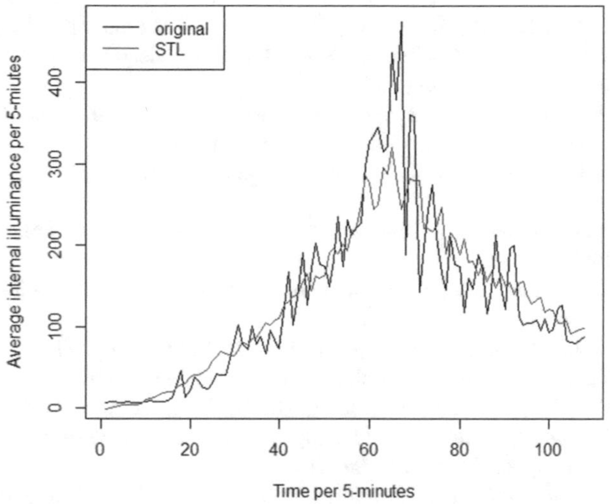

Fig. 4. The best forecasting results obtained from the STL + naïve and snaïve.

4 Conclusions and Future Works

This paper investigated modeling and forecasting the indoor illumination amounts obtained using light pipe system. The aim is to design a model that can forecast the indoor illumination amounts in order to estimate the needed artificial lighting for the room thus providing occupants a high satisfaction of visual comfort along with low energy consumption for lighting. Four methods were used namely seasonal naïve, seasonal ARIMA, STL, and TBATS. Results showed that the forecasts of the decomposed time series using STL method is superior as compared to the other methods in terms of forecasting metrics. Therefore, it can be concluded that the procedures used in modeling and forecasting using STL method is the most suitable for the problem at hand. The future works will be exploring machine learning methods to model and forecast this data. Furthermore, outdoor illumination data and weather information can be considered as factors to improve the forecasting accuracy.

Acknowledgments. The authors would like to thank Universiti Tun Hussein Onn Malaysia and the Office for Research, Innovation, Commercialization and Consultancy Management (ORICC) for funding this research under the Postgraduate Research Grant (GPPS), VOT # U612, and the Ministry of Higher Education (MOHE) Malaysia under the Fundamental Research Grant Scheme (FRGS), Vote No. 1641.

References

1. Kadir, A.A., Ismail, L.H., Kasim, N., Kaamin, M.: Potential of light pipes system in malaysian climate. In: IOP Conference Series: Materials Science and Engineering, vol. 160, no. 1, p. 012071. IOP Publishing (2016)
2. Reinhart, C.F., Wienold, J.: The daylighting dashboard-a simulation-based design analysis for daylit spaces. Build. Environ. **46**(2), 386–396 (2011)
3. Oakley, G., Riffat, S.B., Shao, L.: Daylight performance of lightpipes. Sol. Energy **69**(2), 89–98 (2000)
4. Carter, D.J.: Developments in tubular daylight guidance systems. Build. Res. Inf. **32**(3), 220–234 (2004)
5. Marwaee, M.A., Carter, D.J.: A field study of tubular daylight guidance installations. Light. Res. Technol. **38**(3), 241–258 (2006)
6. Mohelnikova, J.: Tubular light guide evaluation. Build. Environ. **44**(10), 2193–2200 (2009)
7. Shin, J.Y., Yun, G.Y., Kim, J.T.: Evaluation of daylighting effectiveness and energy saving potentials of light-pipe systems in buildings. Indoor Built Environ. **21**(1), 129–136 (2012)
8. Valipour, M.: Long-term runoff study using SARIMA and ARIMA models in the United States. Meteorol. Appl. **22**(3), 592–598 (2015)
9. Kumar, S.V., Vanajakshi, L.: Short-term traffic flow prediction using seasonal ARIMA model with limited input data. Eur. Transp. Res. Rev. **7**(3), 21 (2015)
10. Rojo, J., Rivero, R., Romero-Morte, J., Fernández-González, F., Pérez-Badia, R.: Modeling pollen time series using seasonal-trend decomposition procedure based on LOESS smoothing. Int. J. Biometeorol. **61**(2), 335–348 (2017)
11. Condeixa, L.D., Bastos, L.D.S.L., Oliveira, F.L.C., Barbosa, S.D.: Wind speed time series analysis using TBATS decomposition and moving blocks bootstrap. Int. J. Energy Stat. **5** (02), 1750010 (2017)
12. Hyndman, R.J., Athanasopoulos, G.: Forecasting: Principles and Practice, 2nd edn. OTexts, Melbourne (2016)
13. Cleveland, R.B., Cleveland, W.S., Terpenning, I.: STL: a seasonal-trend decomposition procedure based on loess. J. Off. Stat. **6**(1), 3 (1990)
14. De Livera, A.M., Hyndman, R.J., Snyder, R.D.: Forecasting time series with complex seasonal patterns using exponential smoothing. J. Am. Stat. Assoc. **106**(496), 1513–1527 (2011)
15. R Core Team: R: a language and environment for statistical computing (2013)
16. Hyndman, R.J., Khandakar, Y.: Automatic time series forecasting: the forecast package for R. J. Stat. Softw. **26**(3), 1–22 (2008)
17. Hyndman, R.J.: Forecast: forecasting functions for time series and linear models. R package version 8.2 (2017). http://pkg.robjhyndman.com/forecast

A Model for Addressing Quality Issues in Big Data

Grace Amina Onyeabor[1,2](✉) and Azman Ta'a[1]

[1] School of Computing, College of Arts and Sciences, Universiti Utara Malaysia, 06010 Sintok, Kedah, Malaysia
grace_amina@ahsgs.uum.edu.my, azman@uum.edu.my
[2] Federal University of Technology, Minna, Nigeria
grace.onyeabor@futminna.edu.ng

Abstract. Big Data (BD) is everywhere and quite a lot of benefits have been derived from its usage by different organizations. Notwithstanding, there are still numerous technical and research challenges that must be tackled to comprehend and gain its full potential. The major challenges of BD are not just its processing, storage and analytics, there are also challenges associated with it that run across the BD value chain such as the data collection phase, integration and the enforcement of quality. This paper propose a DQ transformation model to evaluate BD quality from the data collection phase through to the visualization phase involving both data-driven and process-driven quality evaluation by assessing the quality of data itself first then assessing the process quality. This is still an ongoing research and hopefully will be experimented using specific Data Quality Dimensions (DQDs) like completeness, consistency, accuracy and timeliness with process quality dimensions such as Throughput, response time, latency with their corresponding metrics.

Keywords: Big Data · Data Quality · Data Quality Dimensions

1 Introduction

Going by the wide acceptance and usefulness of BD by several organizations, it comes with some challenges. These challenges are recognized in [1] namely: DQ, adequate characterization of data, interpreting correctly the results, visualization of data, real time data view versus retrospective view and determination of the importance of projects and results. The challenge that is fiddlier is DQ, which is, ensuring quality in the BD. Realizing high quality of data has always been a significant element of data management in business organizations due to the fact that it could help the organizations in articulating or enhancing their business strategies for better decision making. Organizations have suffered a lot from the effects of poor DQ such as wrong decision making, increase in production cost and the ability to satisfy their various customers [2]. Besides, with the vast amount of BD with unknown quality [3], the speed at which they are being exchange and in the various formats and structure they are created, DQ is far from being fetch. Business organizations such as online retailers are gradually taking advantage of data to increase discernibility into expenditures. The recent

F. Saeed et al. (Eds.): IRICT 2018, AISC 843, pp. 65–73, 2019.
https://doi.org/10.1007/978-3-319-99007-1_7

research works such as [4–8] proposed some kinds of good initiatives for the integration of DQ in BD, but, most of the initiatives did not evaluate DQ across the BD value chain (from data collection to the visualization phase). Only a few initiatives proposed all-inclusive solutions to guarantee BD quality and there's still requirements for that in organizations such as the online retail. It is vital to ensure the construction and implementation of DQ across the BD value chain. Therefore, this paper propose DQ model to implement quality of data across the BD value chain.

2 Big Data and Data Quality

According to [9], BD is a used to describe massive data sets that are of diverse format created at a very high speed, the management of which is near impossible by using traditional database management systems. According to [10], BD does not only concern the large volume of data but it also includes the ability to search, process, analyze and present meaningful information obtain from huge, varied and rapidly moving datasets. These three attributes lead to the foundational definition of BD regarding volume, variety, and velocity. Data are created from an extensive range of sources such as social media, the internet, databases, websites, sensors, and so on. But before these data are stored, processing and cleansing with the help of numerous analytical algorithms are performed on them [11]. However, because of the nature of BD, oftentimes, organizations encounter issues and challenges. These concerns and challenges need to be looked into to make proper business decisions prospectively. The researchers in [12] identified the challenges and the most risky of them all is DQ.

It has not been easy to define the concept of DQ [13]. Notwithstanding, the definition has been given by various researchers since the days of the traditional data. DQ is defined as fitness for use and being able to meet the set purpose by the user [14, 15]. The Authors on recent researches in DQ such as [4, 6, 16–18] have proposed some ideas relating to DQ. Some of them even delivered all-inclusive solutions guaranteeing quality across the BD value chain which is a good significant progress in this research area. Nonetheless, there's still room for improvement of DQ. According to [19], it was pointed out that improvement of DQ involves approaches of data-driven and process-driven. Data-driven handles the data the way it is, making use of methods and actions like cleansing to enhance the quality of the data while Process-driven strategy tries to detect originating poor DQ sources then redesigns the way the data is produced. DQ problems exist, right before the introduction of BD in the field. Because of these joint issues, the processes of BD cleaning and sifting are phases to be implemented before the analyses of data with quality that is unknown. [20] pointed out that DQ problems are more pronounced when dealing with data from multiple data sources. This problem obviously multiplies the data cleansing needs. Also, the huge amount of data sets that comes in at an unprecedented speed creates an overhead on the cleansing processes [17]. With the magnitude of data generated, the velocity at which the data arrives, and the huge variety of data, the quality of these data has left so much to be desired which cost organizations billion dollars yearly [21, 22].

3 Data Quality Requirements for Big Data

This section provides the description of DQ requirements for carrying out effective and efficient evaluation and assessment of DQ across BD value chain. These requirements include Data Quality Dimensions (DQDs) and their associated metrics. Data process quality and their metrics as introduced by [6].

3.1 Data Quality Dimensions

DQ is measured using multiple dimensions. A DQD is a characteristics or information part use for data requirements. DQD affords us the way to measure and manage DQ [6, 16, 23–25]. DQDs sometimes in the literature are referred to as characteristics, or attributes [26]. There are many DQDs in the literature but they are usually categorized into intrinsic and contextual. Intrinsic DQDs deals with the schema of the data and contextual deals with the values of the data [27, 28]. The commonly used DQDs for BD are the intrinsic DQDs which includes: Completeness DQD, Consistency DQD, Accuracy DQD and Timeliness DQD [29, 30].

In addition, for the evaluation of data process-driven quality, [7] introduced some DQDs which are associated with BD quality across the chain of BD. Some of these DQDs are namely: 1. Latency: which measures the delay in receiving the results of the processed data. 2. Response time: refereeing to the maximum time it takes to process the data records. 3. Throughput: refers to the number of processed records over a period of time. 4. Capacity: means the maximum number the processes run concurrently and to add the fifth one – Scalability: This has to do with the hardware- showing how much the hardware can scale the processing of the data.

3.2 Data Quality Metrics

DQ metrics measures specific properties of a DQD. This evaluates the extent of the presence of the DQD within the dataset. All DQDs are associated with one or more than one metric [7]. Table 1 shows different DQDs with their associated quality metrics.

Table 1. Data quality dimensions and metrics

Data quality dimension	Description	Metrics functions
Completeness	Null or missing values in the data set	$Comp = \left(\frac{Nnv}{N}\right)$
Consistency	Consistency in the format and structure of data values	$Cons = \left(\frac{Nvc}{N}\right)$
Accuracy	Error free data	$Acc = \left(\frac{Ncv}{N}\right)$
Timeliness	Currency and volatility of the data	TMa = (1 − CMa / VMb)

3.3 Metadata

Metadata is referred to as data about data [31] It provides information that are pertinent to data like the quality or provenance of the data. Using metadata is a stress-free and faster way of data features extraction and process. The description of metadata is represented by making use of vocabularies which are of definite models and standard [32]. According to [7], Java Script Object Notation (JSON) [31], is a standard of metadata that is used to signify large dataset into a structure built property graph models.

4 Related Research Works

Studies have been carried out to address DQ assessment and evaluation on other organizational data. The researchers in [20] pointed out there's increment in DQ issues when dealing with data from numerous sources. Furthermore, the authors in [33] proposed a classification model and categorized the DQ issues during data preprocessing according to (i) errors correction, (ii) unstructured data to structured conversion and (iii) integrating data from various sources of data. With the inception of BD, [34] looked at data provenance as a pertinent data source to enable the evaluation of its quality. Data provenance refers to a concept being used for databases that are distributed with business and scientific data for the evaluation of the DQ. Tracing of data from its gathering through any transformation until the data visualization can be done by provenance data. The authors in [35] did something like that by proposing a multiple layer framework for data provenance collection. Another proposal is by [36] providing data cleaning tools for BD. The researchers called it NADEEF and this idea was extended in [30] handling the quality issues of data streaming. Furthermore, [37] proposed data semantics to guarantee the consistency of DQ dimension for BD. The replication of data with consistency below an efficient network bandwidth optimization was achieved. Literature also include extensive explanations on the identification and discussions on DQ issues and challenges [38, 39]. In fact, [38] proposed a more inclusive process for DQ assessment for the evaluation of BD quality after identifying the main challenges of DQ in BD. Nevertheless, [40] distinguished objective DQ assessment from subjective DQ assessment for identifying quality inconsistencies and suggest actions for enhancement.

None of the authors from the above research work conducted addressed BD quality across the BD value chain. Only a few studies in the literature attempted to do that. For example, [41] proposed a framework for the integration of different DQ areas and find out the assessment of DQ dimensions. Also, [42] proposed a cross layer approach the assessment of DQ. The approach was applied to trustworthiness evaluation of sensory data. Furthermore, in the efforts to provide an all-inclusive DQ assessment and evaluation, a recommendation system for repairing data was proposed by [43]. This scheme is to handle degradation in DQ. And the authors in [44] proposed a framework for the evaluation and management of social media DQ all through the BD value chain. The framework was extended and evaluated in [44] by implementing a reference architecture for DQ management in social media data. So far so good, there's still room for more research work in the literature to address product DQ issue of DQ in BD.

To address the above challenges relating with BD quality across the BD value chain. This paper is proposing a model for quality evaluation and assessment of BD from its collection until visualization enforcing DQ at all phases.

5 Proposed Research Model

Figure 1 below gives the description of the key processes constituting the BD transformation model for assessing DQ. The processes consists of about six phases namely: BD collection, the pre and post BD quality evaluation, BD processing and analytics evaluation and the finally the quality evaluation of the BD visualization phase. The phases will work together for the achievement of a comprehensive assessment of quality across the chain. The dataset that will be used for this study will be online retail product data from different heterogeneous sources online.

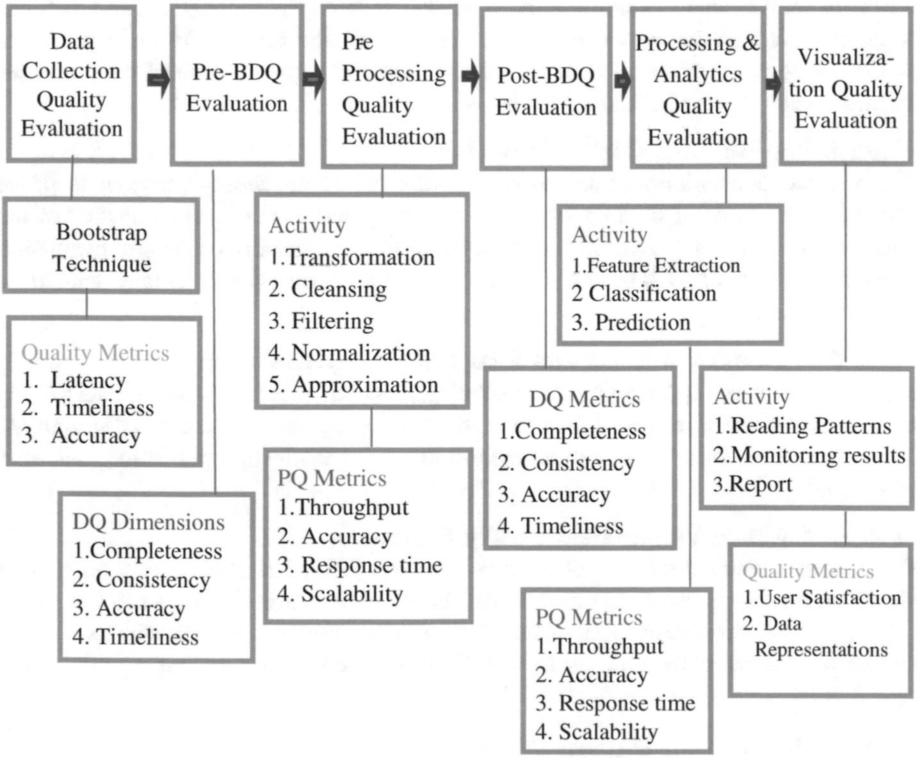

Fig. 1. A model for assessing quality issues in Big Data value chain

Phase 1: Data Collection
This is the phase where the data will be collected from the sources and stored on a temporary storage location. Due to the volume of the BD, Bootstrap sampling technique will be used to sample the data. The DQ metrics expected to be measured here are timeliness, latency and accuracy of the process of the data extraction and/or collection.

Phase 2: Pre-Big Data Quality Evaluation
This evaluation is going to be performed on the collected data itself (data-driven quality evaluation) before data preprocessing. The essence of this is to have a knowledge of the percentage of DQ to be able to choose the appropriate preprocessing activity that would be applicable. The DQDs that will be measured here will be completeness, consistency, accuracy using their metrics.

Phase 3: Data Pre-Processing Quality Evaluation
In this phase, the result of the previous phase will determine which of the preprocessing activity will be selected and performed on the dataset. Owing to the diversity of the data sources, the data might have been collected with diverse quality levels containing for example, redundancies or noise. Also, some data analysis applications and methods could involve requirements that are stringent or specific on DQ. Hopefully, in this phase, we could improve on throughput, accuracy and response time of the processes and the reliability and scalability of the systems.

Phase 4: Post-Big Data Quality Evaluation
The post DQ evaluation will be carried out after the preprocessing activities to prove that the percentages of the DQ has improved compared to the results obtained in the phase before the preprocessing. It is also going to be a data-driven quality evaluation measuring the same DQDs, that is, completeness, consistency, accuracy with their associated metrics.

Phase 5: Big Data Processing and Analytics Quality Evaluation
This phase involves applying machine learning procedures and methods of data mining leading to a required results of DQ. Here, the same preprocessing metrics (throughput, accuracy and response time) will be measured for the evaluation processing and analytics quality.

Phase 6: Big Data Visualization Quality Evaluation
To view and validate data result of data from the BD value chain is the process of visualization. The presentation of data will be done here using diverse sorts of views such as graphs, monitoring results summary, reading patterns. DQDs metrics such as quality of data representation, user satisfaction will be used for evaluating this phase.

6 Conclusion and Future Work

There is an urgent need for researcher in both industry and academia to give more attention to evaluation and assessment of BD quality because there are only few initiatives so far that addressed this vital aspect of BD. Thus, this paper proposed a model to tackle this problem. The model will handle the evaluation and assessment of

BD from the data collection phase all through to visualization phase using both data-driven and process driven quality. This is research in progress and the future work will be the implementation of the proposed model.

References

1. Tee, J.: The Server Side (2013). http://www.theserverside.com/feature/Handling-the-four-Vof-big-data-volume-velocity-varietyand-Veracity
2. Levitin, V., Redman, T.C.: Data as a resource: properties, implications, and prescriptions. Sloan Manag. Rev. **40**, 89–101 (1998)
3. Izham Jaya, M., Sidi, F., Ishak, I., Suriani Affendey, L.I.L.L.Y., Jabar, M.A.: A review of data quality research in achieving high data quality within organization. J. Theor. Appl. Inform. Technol. **95**(12), 2647–2657 (2017)
4. Hu, H., Wen, Y., Chua, T.-S., Li, X.: Toward scalable systems for Big Data analytics: a technology tutorial. IEEE Access **2**, 652–687 (2014)
5. Idi, F., Shariat Panahy, P. H., Affendey, L.S., Jabar, M.A.H., Ibrahim, H., Mustapha, A.: Data quality: a survey of data quality dimensions. In: 2012 International Conference on Information Retrieval Knowledge Management (CAMP), pp. 300–304 2(012)
6. Glowalla, P., Balazy, P., Basten, D., Sunyaev, A.: Process-driven data quality management-an application of the combined conceptual life cycle model. In: 47th Hawaii International Conference on System Sciences (HICSS), pp. 4700–4709 (2014)
7. Serhani, M.A., El Kassabi, H.T., Taleb, I., Nujum, A.: An hybrid approach to quality, evaluation across Big Data value chain. In: IEEE International Congress on Big Data (BigData Congress), pp. 418–425. IEEE. (2016)
8. Pääkkönen, P., Pakkala, D.: Reference architecture and classification of technologies, products and services for Big Data systems. Big Data Res. (2015). https://doi.org/10.1016/j.bdr.2015.01.001
9. Maier, M., Serebrenik, A., Vanderfeesten, I.T.P.: Towards a Big Data Reference Architecture. University of Eindhoven, Eindhoven (2013)
10. Malik, P.: Governing Big Data: Principles and Practices. IBM J. Res. Dev. **57**, 1–13 (2013)
11. Soares, S.: Big Data Governance: An Emerging Imperative. MC Press, Boise (2012)
12. Feldman M.: The Big Data challenge: intelligent tiered storage at scale. White Paper (2013)
13. Strong, Y.W., Lee, Y.E., Wang, R.Y.: Data quality in context. Commun. ACM **40**(5), 103–110 (1997)
14. Wang, R.Y.: A product perspective on total data quality management. Commun. ACM **41** (2), 58–65 (1998)
15. Fürber, C., Hepp, M.: Using SPARQL and SPIN for data quality management on the semantic web. In: International Conference on Business Information Systems pp. 35–46 Springer, Berlin, Heidelberg (2010)
16. Sidi, F., Shariat Panahy, P.H., Affendey, L.S., Jabar, M.A., Ibrahim, H., Mustapha, A.: Data quality: a survey of data quality dimensions. In: International Conference on Information Retrieval Knowledge Management (CAMP) (2012)
17. Taleb, I., Dssouli, R., Serhani, M.A.: Big Data pre-processing: a quality framework. In: 2015 IEEE International Congress on Big Data (BigData Congress), pp. 191–198. IEEE (2015)
18. Hazen, B.T., Boone, C.A., Ezell, J.D., Jones-Farmer, L.A.: Data quality for data science, predictive analytics, and Big Data in supply chain management: an introduction to the problem and suggestions for research and applications. Int. J. Prod. Econ. **154**, 72–80 (2014)

19. Loshin, D.: Big Data Analytics: From Strategic Planning to Enterprise Integration with Tools, Techniques, NoSql, and Graph. Elsevier, Amsterdam (2013)
20. Rahm, E., Do, H.H.: Data cleaning: problems and current approaches. IEEE Data Eng. Bull. **23**(4), 3–13 (2000)
21. Eckerson, W.W.: Data Quality and the Bottom Line: Achieving Business Success Through a Commitment to High-Quality Data. Data Warehousing Institute, Chatsworth (2002)
22. Fan, W., Geerts, F.: Foundations of Data Quality Management. Morgan & Claypool, San Rafael (2012)
23. Batini, C., Cappiello, C., Francalanci, C., Maurino, A.: Methodologies for data quality assessment and improvement. ACM Comput. Surv. **41**(3), 1–52 (2009)
24. McGilvray, D.: Executing Data Quality Projects: Ten Steps to Quality Data and Trusted Information. Morgan Kaufmann, Burlington (2008)
25. Jayawardene, V., Sadiq, S., Indulska, M.: An analysis of data quality dimensions, pp. 1–32 (2015)
26. Loshin, D.: The Practitioner's Guide to Data Quality Improvement. Elsevier, Amsterdam Morgan Kaufmann OMG Press (2011)
27. Batini, C., Cappiello, C., Francalanc, C., Maurino, A.: Methodologies for data quality assessment and improvement. ACM Comput. Surv. (CSUR) **41**(3), 16 (2009)
28. Taleb, I., Dssouli, R., Serhani, M.A.: Big Data pre-processing: a quality framework. In: IEE International Congress on Big Data (2015)
29. Saha, B., Srivastava, D.: Data quality: the other face of Big Data. In: IEEE 30th International Conference on Data Engineering (ICDE), pp. 1294–1297 (2014)
30. Tang, N.: Big Data cleaning. In: Chen, L., Jia, Y., Sellis, T., Liu, G. (eds.) Web Technologies and Applications, pp. 13–24. Springer, Berlin (2014)
31. Introducing JSON. http://www.json.org/
32. Understanding Metadata. NISO Press, Bethesda, MD, USA, (2004)
33. Oliveira, P., Rodrigues F., Henriques, P.R.: A formal definition of data quality problems. In: IQ (2005)
34. Glavic, B.: Big Data Provenance: Challenges and Implications for Benchmarking. In: Specifying Big Data Benchmarks, pp. 72–80 Springer, Berlin Heidelberg (2014)
35. Cheah, Y-W., Canon, R., Plale, B., Ramakrishnan, L.: Milieu: lightweight and configurable Big Data provenance for science. In: 2013 IEEE International Congress on Big Data (BigData Congress) pp. 46–53 (2013)
36. Ebaid, A., Elmagarmid, A., Ilyas, I.F., Ouzzani, M., Quiane-Ruiz, J.-A., Tang, N., Yin, S.: NADEEF: a generalized data cleaning system. Proc. VLDB Endow. **6**(12), 1218–1221 (2013)
37. Recuero, A.G., Esteves, S., Veiga, L.: Towards quality-of-service driven consistency for Big Data management. Int. J. Big Data Intell. **1**(1/2), 74 (2014)
38. Juddoo, S.: Overview of data quality challenges in the context of Big Data. In: International Conference on Computing, Communication and Security (ICCCS), pp. 1–9 (2015)
39. Rao, D., Gudivada, V.N., Raghavan, V.V.: Data quality issues in Big Data. In: IEEE International Conference on Big Data (Big Data) (2015)
40. Pipino, L.L., Lee, Y.W., Wang, R.Y.: Data quality assessment. Commun. ACM **45**(4), 211–218 (2002)

41. Cheah, Y.-W., Canon, R., Plale, B., Ramakrishnan, L.: Milieu: lightweight and configurable Big Data
42. Monga, M., Sicari, S.: Assessing data quality by a cross-layer approach. In: IEEE International Conference on Ultra Modern Telecommunications & Workshops (ICUMT 2009) (2009)
43. Ding, X., Wang, H., Zhang, D., Li, J., Gao, H.: A fair data market system with data quality evaluation and repairing recommendation. In: Web Technologies and Applications, pp. 855–858 (2015)
44. Immonen, A., Pääkkönen, P., Ovaska, E.: Evaluating the quality of social media data in Big Data architecture. In: IEEE Access, vol. 3, pp. 2028–2043 (2015)

Climbing Harmony Search Algorithm for Nurse Rostering Problems

Mohammed Hadwan[1(✉)], Masri Ayob[2], Mohammed Al-Hagery[3], and Bassam Naji Al-Tamimi[4]

[1] Department of Information Technology, Qassim University, Qassim-Buraidah 51452, Saudi Arabia
m.hadwan@qu.edu.sa

[2] Data Mining and Optimisation Research Group (DMO), Centre for Artificial Intelligent (CAIT), Fakulti Teknologi dan Sains Maklumat, Universiti Kebangsaan Malaysia, UKM, 43600, Bangi, Selangor, Malaysia

[3] Department of Computer Science, College of Computer, Qassim University, Qassim-Buraidah 51452, Saudi Arabia

[4] Department of Information Systems, College of Computer Science and Engineering, Taibah University, Medina, Saudi Arabia

Abstract. Nurse Rostering Problem (NRP) is a well-known NP-Hard combinatorial optimization problem. The fact is that coping real-world constraints in allocating the shift duties fairly among the available nurses is still a hard task to accomplish. The problem becomes more serious due to the shortage of nurses. Thus, this work aims to tackle this problem by hybridizing an Enhanced Harmony Search Algorithm (EHSA) with the standard Hill climbing (HC). This hybridization may help to strike the balance between exploration and exploitation in the searching process. The proposed algorithm is called Climbing Harmony Search Algorithm (CHSA) where it applied to solve a real-world NRP dataset, which arises at the Medical Center of Universiti Kebangsaan. The results show that CHSA performs much better than EHSA alone and Basic Harmony Search algorithm (BHSA) in all instances in terms of obtained penalty values (PVs), desirable patterns (DPs) and computational time as well.

Keywords: Nurse Rostering Problem (NRP) · Harmony search algorithm
Hill climbing algorithm · Metaheuristic algorithms · Optimization

1 Introduction

One of the life aspects that rely heavily on timetabling is shift distribution among personnel in healthcare organizations. Within the area of healthcare, there is a worldwide shortage of nursing personnel; a phenomenon known as the "crisis in nursing" [1]. One of the solutions to this problem is to use automated tools to construct a fair duty roster for staff nurses [2]. Currently, metaheuristic algorithms showed a very promising solution to this NP-hard problem. Harmony search algorithm (HSA) is a musical based metaheuristic algorithm which was proposed by Geem et al. in 2001 [3] to solve the difficult combinatorial optimization problems. HSA has been applied successfully to solve many optimization problems; one of them is the nurse rostering

F. Saeed et al. (Eds.): IRICT 2018, AISC 843, pp. 74–83, 2019.
https://doi.org/10.1007/978-3-319-99007-1_8

problem [4]. The main purpose of this work is to study the efficiency of the standard HC algorithm when it hybridized within the framework of Enhanced Harmony Search Algorithm (EHSA) [5] and to have a balance between exploration (controlled by the EHSA) and exploitation (controlled by Hill Climbing (HC)). The proposed algorithm is applied to solve NRP at one of the public hospitals in Malaysia where the number of nurses exceeds 1400 nurses working around the clock. For more information about UKMMC please refer to [6, 7]. To evaluate the performance of CHSA, the obtained results will be compared with other algorithms that were used to solve the same dataset namely, basic harmony search algorithm (BHSA) [6] and enhanced harmony search algorithm (EHSA) [5].

The remainder of this research is organized as follows: In Sect. 2 reviews the most important studies related to the work in this research. Section 3 provides brief background information on the Universiti Kebangsaan Malaysia Medical Center. The CHSA algorithm is explained in Sect. 4. Results of using CHSA on UKMMC dataset are discussed in Sect. 5. Finally, Sect. 6 concludes the paper.

2 Background

In spite of the great improvement and progress in the area of metaheuristic methods to solve timetabling problems, the intention paid more to concentrate on solving benchmark problems and competitive advances. This perhaps has drifted away the researchers from solving real-world problems [8]. Metaheuristics are iterative improvement methods that apply a set of operators to explore the search space. HSA is a population-based metaheuristic algorithm with local search-based components [3]. HSA has been successfully applied to solve many difficult optimization problems, nurse rostering problem is one of these problem as in [7, 9–11]. As far as the literature of HSA is concerned, it is noticed that the fully random mechanism is the method used to initialize the harmony memory (HM), and that is one of the reasons behind the occurrence of slow convergence. Researchers have not paid much attention to the initialization step of the HM (refer to [12]). More research was directed towards improving the improvisation step, however. It is also noticed that there is still room to investigate the performance of local search algorithms in different stages of HSA framework.

In [7], the first implementation of Basic HSA is reported to solve NRP at UKMMC where the proposed algorithm called BHSA. Also, the basic Genetic Algorithm (BGA) is implemented as well to solve NRP at UKMMC [7]. It was found that BHSA overcomes BGA in the entire tested instances. This is due to the better convergence mechanism of BHSA compared to BGA but slow convergence was noticed. Thus in [5], we built the HM randomly and using fixed parameter values for Harmony Memory Considering Rate (HMCR) and Pitch Adjustment Rate (PAR).

A semi-cyclic shift pattern approach in [13] proposed to build HM and dynamic parameters for HMCR and PAR is introduced in [5] to overcome the problems of BHSA. The proposed algorithm called Enhanced harmony search algorithm (EHSA). In addition, It is very obvious that local search algorithms are better in exploiting the solution search space while the population search algorithms are good in the exploring of the search space [14]. The hybridization of the strength points of these two methods will definitely result in promising and successful metaheuristic approaches capable of solving the complex combinatorial optimization problems [15].

Many researchers such as [10, 16–18] supported the research trend of hybridizing local search and population search algorithms. Hill Climbing (HC) algorithm is one of the well-known local search methods that attempt to find improved solutions. The improving process starts by exploring the neighbor solutions of the current state by changing a single element of the solution [15]. The effectiveness and usefulness of using the HC inside population based algorithms is highlighted by many researchers such as [19, 20]. Due to the complexity nature of the time tabling problems, the HC is normally used as an aid component inside other metaheuristics frameworks.

In order to apply the HC, at least one solution must be existed. To enhance the exploitation mechanism of population based algorithms, the HC is hybridized with different population based algorithms such as the GA which results in a new algorithm called memetic algorithm. Özan in [21] used a memetic algorithm (GA + HC) to solve NRP where the exploitation mechanism of the HC makes it a good choice due to its fast performance in obtaining the local optima. A hybridization between a greedy local search and a harmony search algorithm (HHSA) is proposed by [11] to solve the NRP using a modified idea of shift pattern approach that proposed by [22]. Where the shift patterns that obey subset of constraints are generated in advance and a repairing procedure used to maintain the coverage followed by improving the roster's quality by using a greedy local search [11]. Jin et al. in [10] proposed a hybrid metaheuristic approach to solve NRP where they combined HSA and artificial immune systems in order to make a balance between local and global searches and to prevent slow the occurrence of slow convergence and prematurity.

3 Universiti Kebangsaan Malaysia Medical Center (UKMMC)

The Universiti Kebangsaan Malaysia Medical Center (UKMMC) is one of the biggest educational hospitals in Malaysia. It is owned by National University Malaysia (UKM). UKMMC operates 24 h a day to provide its services for the public. The hospital has a workforce exceeding 1400 nurses attending more than 900 beds for 24 h a day and 7 days a week. For the set of hard and soft constraints and the penalty values for each soft constraint in addition to the mathematical formulation for NRP of UKMMC refer to [6, 13].

3.1 UKMMC Constraints

The constraints can be classified into hard constraints and soft constraints. Hard constraints represent the constraints that must be satisfied for any generated rosters. Hard constraints are those constraints that must be fulfilled at all times. Whilst the soft constraints represent the set of constraint that we would like to satisfy in order to improve the quality of the generated roster. A penalty value is given for each soft constraint that cannot be satisfied in the generated roster.

Based on the gathered data from the interviews, observations and questionnaires at UKMMC, the problem we are dealing with has the following constraints:

UKMMC Hard Constraints

(H1). The required coverage for all shifts must be satisfied. Understaffing is unacceptable.
(H2). At most, a shift per day for each nurse.
(H3). At least one senior nurse must be allocated for every shift.
(H4). An isolated working day is not allowed. For example, a day off followed by working-day then followed by a day off.
(H5). For each two weeks, the maximum working days are twelve days and the minimum is ten working days.
(H6). No more than four consecutive working days to be allocated for each nurse.
(H7). For any four consecutive night shifts, it must be followed by two days off.

UKMMC Soft Constraints

(S1). Attempts to give working days and days off equally to all nurses during the roster's period.
(S2). Attempts, at least, to allocate a day off to in the weekend (Saturday or Sunday) during the roster's period.
(S3). Attempts to give patterns of four consecutive morning shifts then a day off.
(S4). Attempts to give patterns of four consecutive evening shifts then a day off.
(S5). Attempts to give a day off or evening shift after the 2 days off that follows the patterns of night shift (NNNNxxE) or (NNNNxxx).

After several discussions with the administration of nurses at UKMMC, we found that (SC1) and (SC2) are more important than the others. Each soft constraint is associated with a particular penalty value. Normally, the higher penalty value indicates the main importance of satisfying the constraints. Refer to Table 1 for the penalty values (PVs) associated for each soft constraint.

3.2 Objective Function

The objective function for this research presented in Eq. (1), aims to minimize total penalty value that occurs from violating the soft constraints during the duty roster construction.

$$\begin{aligned}\min f = {} & W_1 \sum_i (d1_i^+ + d1_i^-) + W_2 \sum_i d2_i^+ + W_3 \sum_i \sum_d d3_{id}^- \\ & + W_4 \sum_i \sum_d d4_{id}^- + W_5 \sum_{p \in \{1,2,3,4,5,6\}} d5_p^+ \end{aligned} \tag{1}$$

Table 1. The penalty values of soft constraints

Weights	Soft constraints	Penalty value
w1	Equal workload: Setting equal working days and off days for all nurses	100
w2	A day off: For each nurse, give at least one day off a week (particularly, on the weekend)	100
w3	Morning shift: The number of consecutive morning shift should be four, followed by one day off	10
w4	Evening shift: The number of consecutive evening shift should be four, followed by one day off	10
w5	Gives either a day off or an evening shift after the night shift pattern	1

3.3 Coverage Demand

The required set of nurses for each shift per day called the coverage demand, which is the most important constraint that must be satisfied. The minimum required number of nurses and the combinations of wanted skills for each shift and day (Refer to Table 2 for more details about the coverage demand) [5, 9]. We deal with coverage demand as a hard constraint that cannot be violated at any time.

Table 2. The minimum coverage demand for UKMMC dataset [5, 9]

			Weekdays (Monday–Friday)			Weekend (Saturday–Sunday)		
Instance	Total nurses	Senior nurses	Morning	Evening	Night	Morning	Evening	Night
CICU	11	8	3	3	2	2	2	2
SGY5	18	11	4	4	3	4	4	2
MD1	19	12	4	4	3	4	4	2
NICU	49	29	10	10	10	8	8	7
N50	50	31	10	10	10	10	10	10
ED	57	32	13	13	10	11	11	8
GICU	73	38	16	16	15	15	15	14
ICU	79	41	17	17	16	16	16	15

4 Climbing Harmony Search Algorithm (CHSA)

In this section, the focus is paid to show how a sequential hybridizes between EHSA and HC. The mechanism of the basic HC is very simple and fast which makes it stuck very easily in the local optima. The HC starts with an initial point x and in each iteration, it moves from the current solution to the neighboring solution $x^{new} \in N(x)$. If the solution x^{new} is better than x, the new solution is accepted and becomes x^{new}. Otherwise, it rejects it and searches in the other neighboring solutions until it gets stuck in the local optima. Figure 1 shows the pseudo code of the standard HC.

```
Procedure HILL CLIMBING;
x= Initial_solution ()
while not stopping criterion is met do
    Generate neighbor solution
    Move to x^new where x^new ∈ N(x)
    if f(x^new )<f(x) then
        x = x^new
    end if
end while
```

Fig. 1. Pseudo code of basic hill climbing algorithm

The HC is used after the improvisation step of the EHSA is terminated to improve the best harmony obtained so far. This harmony is allocated in the HM [0]. Since the HC improves only the best harmony obtained so far, the stopping criterion is when there is no improvement (stuck in local optima). The pseudo code of hybridizing CHSA is presented in Fig. 2.

```
Procedure CHSA;
Initialize (HM)
while (No termination criterion is met) do
        x^new = Φ
        improvise new harmony (x^new)
        if f(x^new) better than HMS-1 then
            update and sort (HM)
        endif
end while
improve f(x^new) in HM[0] using HC   /* Algorithm HC in Fig 1*/
```

Fig. 2. Procedure of climbing harmony search algorithm (CHSA)

5 Results of Using CHSA on UKMMC Dataset

The stopping criteria of the CHSA are as follows: (i) reaching maximum number of iterations = 1000 iterations (determined based on preliminary test), (ii) reaching Penalty Value (PV) = 0 and (iii) exceeding computational time = 600 s. Table 3 shows the computational results of using CHSA on UKMMC benchmark. The experiments were conducted on windows vista Intel 1.73 GHz and 32-bit laptop with RAM of 2-GB.

Table 3. Results of using CHSA on UKMMC dataset

Instance	Best (PV)	Average (PV)	Worst (PV)	Median	Stddv.	DPs	Time (s)
CICU	70.1	76	80.1	76.6	2.941	11	45.7
SGY5	69.1	71.3	75	70.6	2.152	19	62
MD1	77.8	80.9	85.4	80.6	2.423	20	67.8
NICU	83	88.6	97.8	88.1	4.736	53	110.1
N50	93.7	97.7	112.7	95.9	5.818	56	110
ED	97.6	109.3	139.8	106.6	12.712	59	208.9
GICU	122.7	141.3	165.7	139.7	13.355	75	243.2
ICU	233.8	249.7	275.7	249.7	13.126	86	378.5

5.1 Discussion of Using CHSA on UKMMC Dataset

The results of the 20 runs are shown in Table 3. For the conducted experiments, we used the obtained penalty value (PV) and number of desirable patterns (DPs) as main evaluation criteria. According to the results, the CHSA achieves slightly better than the EHSA alone. However, the decreased amount of the PVs compared to the EHSA is not that much remarkable and the execution time slightly increases, as well. Due to the weakness of the acceptance criteria of HC (accept only improving solutions), the algorithm gets stuck very fast in the local optima.

5.2 A Comparison Between CHSA, EHSA and BHSA

In this section, the CHSA is compared to EHSA and BHSA. The results are presented in Table 4. The average of penalty values (PV), the standard deviation (Stddv.) of the Penalty Values, and the number of desirable patterns (DPs) together with the execution time of each algorithm for each instance being used as the evaluation criteria in this comparison.

Results show that the CHSA is better than the EHSA and the BHSA. This is due to the lower PVs and higher number of generated DPs that the CHSA obtains for all instances. The best results in each instance has been shown in bold, for PV, Stddv. and execution time, the lowest number is the better. Whereas the highest numbers are the better in case of the number of DPs. The results presented above show that the CHSA is better than the EHSA and BHSA in all the evaluation criteria for all of UKMMC datasets. This is apparent in Table 4 where the CHSA is able to produce rosters with a better quality for all the instances of UKMMC dataset but in slightly more computational time compared to EHSA and less than BHSA. Using the HC in performing the random selection out of the HM causes two main problems: (i) premature convergence owing to that some of the new solution vectors selected out of the HM are already enhanced by the HC, and (ii) extra computational time is consumed in each random selection for constructing one entire solution (harmony) and in enhancing it using the HC.

Table 4. Comparison results: BHSA, EHSA and CHSA on UKMMC dataset

Instance	Criteria	BHSA [7]	EHSA [5]	CHSA
CICU	PV	341.1	80.75	**76**
	Stddv.	40.781	**2.781**	2.941
	DPs	5	10	**11**
	Time (s)	105.1	**35.1**	45.7
SGY5	PV	301.4	72.85	**71.3**
	Stddv.	60.662	2.663	**2.152**
	DPs	8	18	**19**
	Time (s)	115.3	**55.6**	62
MD1	PV	406.6	86.85	**80.9**
	Stddv.	60.626	2.624	**2.423**
	DPs	9	**20**	**20**
	Time (s)	127	**57.9**	67.8
NICU	PV	981.35	95.55	**88.6**
	Stddv.	150.188	8.178	**4.736**
	DPs	18	52	**53**
	Time (s)	168.8	**98**	110.1
N50	PV	1075.65	135.65	**97.7**
	Stddv.	180.689	10.784	**5.818**
	DPs	21	55	**56**
	Time (s)	175	**105.4**	110
ED	PV	1194.4	151.75	**109.3**
	Stddv.	180.123	**11.231**	12.712
	DPs	24	58	**59**
	Time (s)	265.9	**195.7**	208.9
GICU	PV	1607.1	179.65	**141.3**
	Stddv.	120.212	**11.721**	13.355
	DPs	27	**75**	**75**
	Time (s)	295.2	**225**	243.2
ICU	PV	1773.15	280.45	**249.7**
	Stddv.	170.822	**12.151**	13.126
	DPs	29	85	**86**
	Time (s)	395.7	**355.8**	378.5

6 Conclusion

In this paper, a novel hybrid HSA-based scheduling algorithm for nurse rostering problem is proposed. The algorithm is applied to solve a NRP at UKMMC Malaysia. The algorithm is utilized to search the solution search space for the optimal shift patterns to construct the duty roster under the provided constraints. The evolution of constructed rosters is driven by lowest penalty value (PV) and the maximum number of provided desirable shift patterns (DPs).

The performance of the CHSA evaluated using a real-world NRP from UKMMC. The experimental results show the superiority of the CHSA algorithm in generating good duty roster that copes with all of the hard and soft constraints. The performance of the CHSA algorithm is compared to the performance of the Basic Harmony Search Algorithm (BHSA) and enhanced harmony search algorithm (EHSA). The results showed that CHSA yields better solutions for all instances. In the future, the performance of the EHSA algorithm can be improved by using advanced local search algorithms such as simulated annealing, Tabu search and great deluge to search for better exploitation.

Acknowledgments. This work was supported by Universiti Kebangsaan Malaysia grant Dana Impak Perdana (DIP-2014-039).

References

1. WHO: World Health Statistics 2009 (2009). http://www.who.int/whosis/whostat/EN_WHS09_Full.pdf
2. Mihaylov, M., et al.: Facilitating the transition from manual to automated nurse rostering. Health Syst. **5**(2), 120–131 (2016)
3. Geem, Z.W., Kim, J.H., Loganathan, G.V.: Original harmony search, a new heuristic optimization algorithm: harmony search. J. Simul. **76**(2), 60–68 (2001)
4. Masri, A., Mohammed, H., Hafiz, M.S.: The harmony search algorithms in solving combinatorial optimization problems. Res. J. Appl. Sci. **8**, 191–198 (2013)
5. Masri, A., et al.: Enhanced harmony search for nurse rostering problem. J. Appl. Sci. **8**, 846–853 (2013)
6. Hadwan, M., Ayob, M.B.: An exploration study of nurse rostering practice at Hospital Universiti Kebangsaan Malaysia. In: 2nd Conference on Data Mining and Optimization, DMO 2009, pp. 100–107. IEEE, Bangi (2009)
7. Mohammed, H., et al.: A harmony search algorithm for nurse rostering problems. Inf. Sci. **233**, 126–140 (2013)
8. McCollum, B.: University timetabling: bridging the gap between research and practice. In: Proceedings of the 5th International Conference on the Practice and Theory of Automated Timetabling. Springer (2006)
9. Yagmur, E.C., Sarucan, A.: Nurse scheduling with opposition-based parallel harmony search algorithm. J. Intell. Syst. (2017). https://doi.org/10.1515/jisys-2017-0150
10. Jin, S.H., et al.: Hybrid and cooperative strategies using harmony search and artificial immune systems for solving the nurse rostering problem. Sustainability **9**, 1090 (2017)
11. Nie, Y., Wang, B., Zhang, X.: Hybrid Harmony Search Algorithm for Nurse Rostering Problem. Springer, Berlin (2016)
12. Alia, O.M., Mandava, R.: The variants of the harmony search algorithm: an overview. Artif. Intell. Rev. **36**, 49–68 (2011)
13. Hadwan, M., Ayob, M.: A semi-cyclic shift patterns approach for nurse rostering problems. In: 3rd Conference on Data Mining and Optimization (DMO 2011) IEEE (2011)
14. Blum, C., et al.: Hybrid Metaheuristics: An Emerging Approach to Optimization. Studies in Computational Intelligence, vol. 114, p. 290. Springer, Heidelberg (2008)
15. Blum, C., Roli, A.: Metaheuristics in combinatorial optimization: overview and conceptual comparison. ACM Comput. Surv. **35**(3), 268–308 (2003)

16. Talbi, E.G.: Metaheuristics: from Design to Implementation. Wiley, Hoboken (2009)
17. Qu, R., et al.: A survey of search methodologies and automated system development for examination timetabling. J. Sched. **12**(1), 55–89 (2009)
18. Blum, C., et al.: Hybrid metaheuristics in combinatorial optimization: a survey. Appl. Soft Comput. **11**(6), 4135–4151 (2011)
19. Radcliffe, N.J., Patrick, D.S.: Formal memetic algorithms. In: Evolutionary Computing: AISB Workshop. Springer (1994)
20. Özcan, E.: Memetic Algorithms for Nurse Rostering. Springer, Berlin (2005)
21. Özcan, E.: Memes, Self-generation and nurse rostering. In: Practice and Theory of Automated Timetabling VI. Springer, Berlin (2007)
22. Hadwan, M., Ayob, M.: A constructive shift patterns approach with simulated annealing for nurse rostering problems. In: International Symposium in Information Technology (ITSim 2010), IEEE, Kuala Lumpur

Analyzing Traffic Accident and Casualty Trend Using Data Visualization

Abdullah Sakib, Saiful Adli Ismail(✉), Haslina Sarkan, Azri Azmi, and Othman Mohd Yusop

Advanced Informatics School, Level 5, Menara Razak, University Teknologi Malaysia, Jalan Sultan Yahya Petra, Kuala Lumpur, Malaysia
sakib7910@gmail.com,
{saifuladli,haslinams,azriazmi}@utm.my

Abstract. Motor vehicle is the backbone of the modern transportation system worldwide. However, the excessive number of motor vehicles tend to cause traffic accidents leading to numerous casualties. Analyzing existing works on this area, this study has identified prime reasons behind traffic accidents and casualties. They include driving over the speed limit, Age of drivers and pedestrians, environmental condition, location and road types. It has also reviewed and identified several data visualization methods and visualization techniques that have been proposed by many researchers. The objective of this research endeavour is to identify the factors behind traffic accidents, determine the techniques that are used to visualize data, develop a dashboard using data visualization tools to visualize traffic accident trend and to evaluate the functionality of the dashboard which is developed on United Kingdom's (UK) traffic accident dataset from 2014 to 2016. Upon performing data cleaning, pre-processing and filtering, the raw data has been converted into cleaned, filtered and processed data to create a coherent and properly linked data model. Then, using the Power BI visualization tool, various interactive visualizations have been produced that illustrated several significant trends in accident and casualties. The visualization trend revealed that between 2014 to 2016, majority of the accidents in the UK occurred in the urban area, in the single carriageway, on the dry road surface, under the daylight with fine weather, and when the speed limit was below 30 mph. This research may assist UK's traffic management authority to identify the underlying factors behind the traffic accident and to detect the traffic accident and casualty trend in order to take necessary steps to reduce casualties in traffic accidents.

Keywords: Traffic accident · Data visualization · Data analysis

1 Introduction

The overarching goal of this research study is to control and reduce the overall traffic accidents and casualty rate in UK's highways. To do so, visual analytics technique will be used to develop a dashboard using Power BI tool which will help the UK's traffic management authority to comprehend the contemporary trend of traffic accidents. Visual analytics can lead to revealing many hidden and unnoticed information,

F. Saeed et al. (Eds.): IRICT 2018, AISC 843, pp. 84–94, 2019.
https://doi.org/10.1007/978-3-319-99007-1_9

identifying hazardous road locations, determining the type of vehicle involved in accidents, finding out risky weather conditions, understanding age's and gender's role in accidents, identifying accident frequency in a different period of time and much more. These pieces of information will help the authority to improve decision making process and to take effective action to initiate a vital countermeasure in order control traffic accidents and improve road safety.

1.1 Background of the Problem

Modern transportation system heavily depends on the motor vehicles. The motor vehicles enable passengers to move from one place to another within a very short span of time. However, an excessive number of the automobiles are also the reason of numerous problems and one of the major complications is that they lead to traffic accident or traffic fatality. Accidents differ in their geographic location, tending to occur more frequently at certain sites which are designated as "black spots" [1]. Many researchers have identified speed as the most responsible factor behind most of the traffic accidents [2–4]. Tulu et al.'s [2] study reveal that vehicles' speed is highly correlated to casualty and death. Even though speed limit has been imposed in most of the countries, 30–50% drivers outpace the posted or given speed limit, jeopardizing the road safety regulation [5–7]. Various environmental factor escalates the possibility of traffic accidents [8–11]. Bergel-Hayat et al. [11] finds in their study that, rain causes the most weather-related accidents in the road. Besides, snow and fog are also responsible for a traffic accidents [9, 10]. Other than these, drivers' characteristics, particularly age of the drivers and pedestrians have been identified as some major factors of road accidents [12–14]. Clarke et al. [15] have observed in his study that driver aged up to and below 20 years are nearly 12 times more likely to have caused a fatal accident compared to people over 20 years of age. Researchers have also identified that traffic accident and casualty vary on different locations and the types of roads. In another fascinating piece of research, Vedagiri and Kadali [16] have noticed that injuries in vehicle crashes were greater in the areas with higher population density compared to the areas where population density is low. Moreover, [6, 17, 18] have identified road inter-sections with traffic controls are most vulnerable spot for traffic casualties. Day, time and light conditions are found to be somehow interrelated in causing traffic accidents. Zhang et al., [19], have discovered in their findings that show an unusual increase in fatal injuries in the road during the morning hours which is normally rushing hours for workers. Moreover, evening period (6 p.m.–12 a.m.) is found to be highly risky time for traffic casualties, which is studied and acknowledged by many researchers [20, 21].

2 Data Visualization Techniques

There are plenty of data visualization techniques available to visualize the data digitally. To choose the proper visualization techniques in accordance with the the contest is the tricky part. Inappropriate selection of visualization chart may lead to ambiguous or misleading information to the audience. Abela [22] has introduced a technique that assists us to choose the right type of visualization for data presentation. He divided all data visualization tool into four main categories. They are; Comparison, Composition, Distribution and Relationship.

First of all, under Comparison category, visualization tools are compared with one or more value sets, and they can easily show the low and high values in the data sets. Column chart, horizontal bar chart, circular area chart, line chart, scatter plot are utilized in order to create a Comparison chart [22, 23]. In in aim to visualize how individual parts make up the whole of something Composition category should be chosen to show use of Composition in pie charts, stacked bar, stacked column, area and waterfall charts [22, 23]. Distribution category helps to understand the outliers, the normal tendency, and the range of information in your values. The following charts are suitable to show distribution: scatter plot, line chart, column chart, bar chart and histogram [22, 24]. Finally, Relationship category is suitable to show how one variable is related to one or numerous other different variables. It illustrates how something positively effects, has no effect, and/or negatively affects another variable. Scatter plot, bubble, line charts are effective to represent Relationship category [22–24].

3 Related Work

Many researchers have used integrated GIS and spatial analysis, and visual analytics to identify the pattern of the traffic casualties and hazardous road segments.

A study conducted by Yu et al. [25] shows that the researchers picked two spatial analysis methods and four traditional methods in the search for a hot-spot location. An exploratory data analysis technique named Comap was implemented to exhibit the spatial-temporal interaction effect in vehicle crashes. Denham et al. [26] have developed a system called GeoTAIS using EBM and GIS, which can represent data visually. Thus, by applying numerous modern and non-biased statistical methods, they identify wildlife crash locations and hot-spot.

A Power BI report was posted by Dales [27] on Power BI Community blog where he explores and reviews road traffic accidents on the M25 motorway around London on an Open Data dataset from the UK's Department of Transport from 2012 to 2014. The first page of the report explores accidents over time and he uses the map to drill into specific accidents (see Fig. 1).

Another dashboard report developed by O'Donnell [28] using Tableau visual that illustrates vehicle crash data in New York City from July 2012 to July 2015. The first page of this report depicts the relationship between location and type of accident (see Fig. 2).

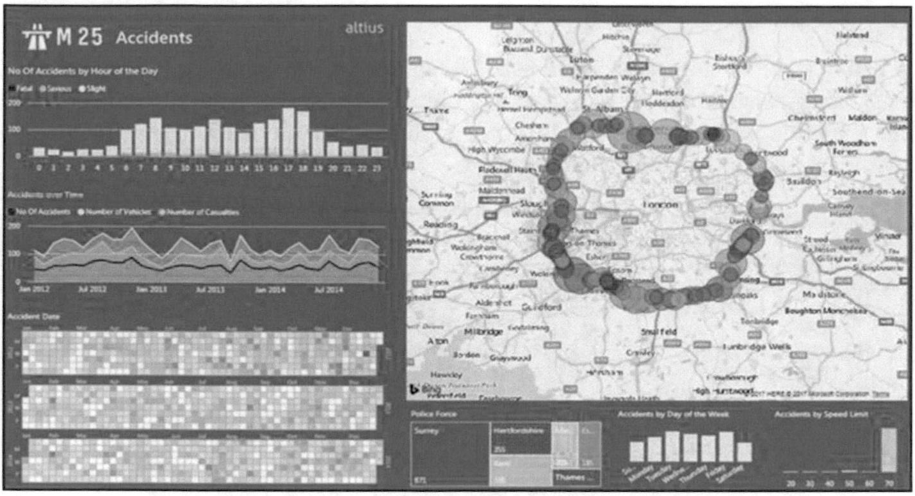

Fig. 1. Power BI dashboard on accident over time and map location (Source: Dales [27])

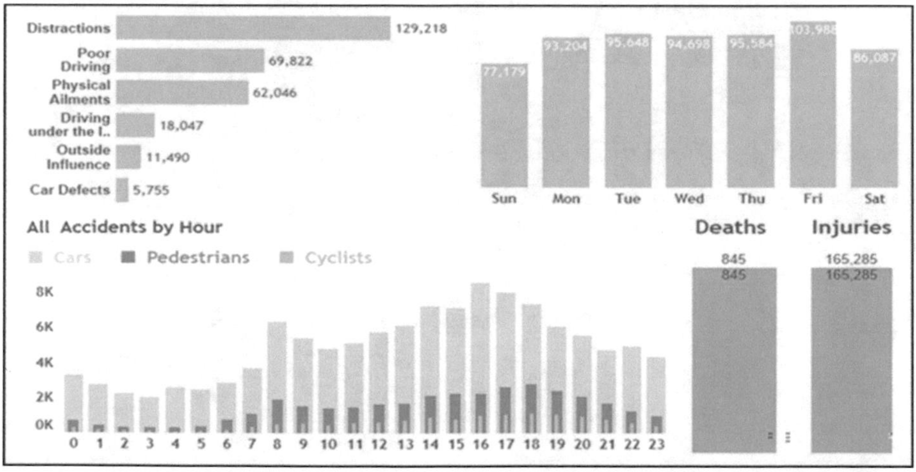

Fig. 2. Tableau report on relationship between location and type of accident (Source: O'Donnell [28])

4 Methodology

Research design explains all necessary steps that are required to conduct this study and also addresses the research questions and objectives in a logical and meaningful way. To analyse the traffic accident trend by developing the dashboard, a proper research design has been constructed (see Fig. 3). This research design consists of six phases. Each of these phases is explained below.

4.1 Initiation and Planning

Initiating and Planning phase is followed by three activities: preparing project proposal, formulating research questions and setting objectives, and finalizing research proposal. The expected outputs produced by these activities are problem statement, research questions, objectives and research method.

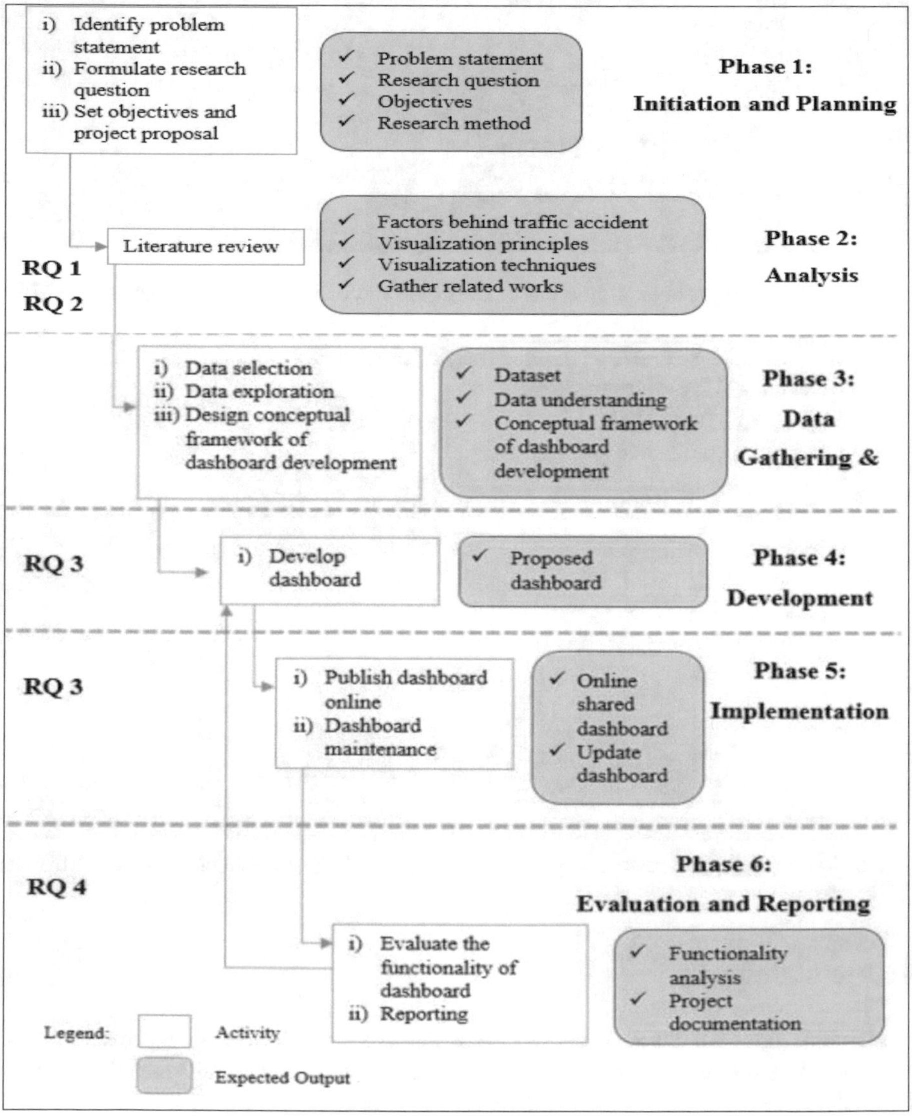

Fig. 3. Research design

4.2 Analysis

Analysis intends to identify the cause of traffic accident and determine relevant techniques to visualize the data.

4.3 Data Gathering and Design

In Data Gathering and Design phase, three activities are performed, which are data selection, data exploration and design conceptual framework of the dashboard. The selected dataset, data understanding, and conceptual framework of dashboard development are the outputs of the prior mentioned activities.

4.4 Development

Development phase is accomplished by developing the dashboard followed by prototyping the method to visualize the accident trend using Power BI tool.

4.5 Implementation

Implementation phase performs two key activities that include publishing dashboard online and maintaining dashboard after the publication. The expected outputs of these activities are sharing the dashboard online and updating it timely.

4.6 Evaluating and Reporting

Evaluating and Reporting phase is the last step of this research activity where evaluation of the functionality of the dashboard and reporting takes place.

5 Design and Implementation

This dashboard has been developed by performing three steps. It starts with data gathering, followed by data cleaning, pre-processing and filtering, and it ends with developing the dashboard.

5.1 Data Gathering

In data gathering step, the dataset has been collected from UK Government's open data website (https://data.gov.uk/dataset/). For the purpose of this project, three years' dataset – starting from 2014 to 2016 – has been used. Each year's dataset comprises of four files which are Accidents, Casualties, Vehicles and Vehicle by Make, and Model. Each of these datasets is in comma separated values file or in CSV format.

5.2 Data Cleaning, Preprocessing and Filtering

This step handles pre-processing, cleaning, filtering and integrating the retrieved data. Several steps are performed in order to prepare the data for developing the dashboard. At first, the three years' data is loaded altogether from the stored folder to Power BI. Then, the loaded gets pre-processed, cleaned and combined the data into one single file. After completing the data pre-processing and cleaning, dataset is loaded into data model to establish the relationship among themselves. Moreover, lookup table is also related to this data model to present meaningful information.

5.3 Develop the Dashboard

This dashboard has been developed, using visualization tool which has been identified in the literature review and the proposed prototyping method mentioned in methodology section. Crucial information is identified and illustrated through visualization.

6 Findings and Analysis

The dashboard has been developed to analyse and visualize traffic accidents and casualty trend. In this section the detailed analysis of each visualizations is explained.

Figure 4 shows the overall accident statistics. The dashboard is comprised with three bar charts which represents the last three years' overall accident statistics in the UK. The first bar chart illustrates the number of accidents by an hour of the day. It shows that after the sunrise from early morning until the sunset in the evening, the accident count increases gradually. Particularly, in the morning at 8:00 pm and in the afternoon from 3:00 pm to 5:00 pm, the number increases sharply. Weekly accident

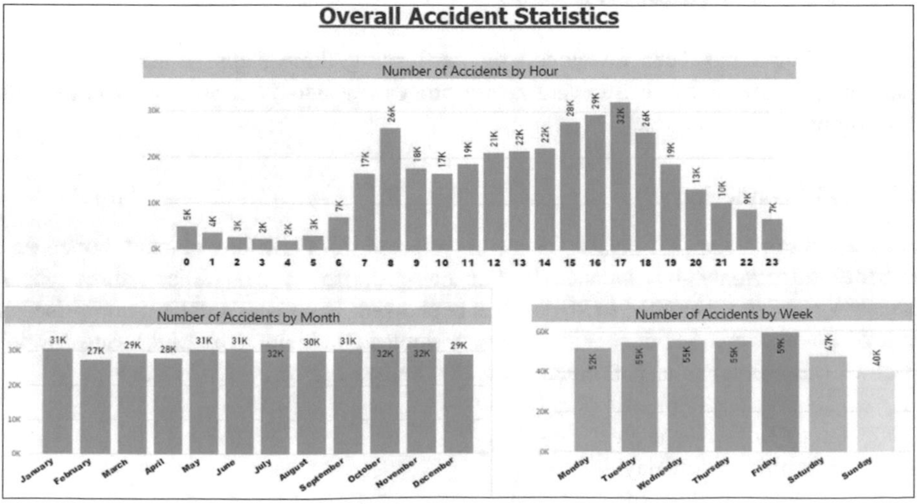

Fig. 4. Overall accident statistics

graph shows that, at beginning of the week, accident number is comparatively low but gradually the number increases throughout the week and it reaches at a pick point on Friday. On the contrary, during the weekend, casualty occurrence decreases, because perhaps less number of people travel for work during the weekend. Whereas the bar chart depicting the number of accident by month shows an upward trend with some fluctuation throughout the year. However, in the month of July, October and November, the highest number of road accidents are recorded.

Variation in accidents between urban and rural area has been highlighted in Fig. 5. The stalked bar chart reveals that, under relatively slow speed limit, which is 30 mph, most of the accident occurred and drivers aged between 26 and 35 mostly committed these accidents. In terms of accident severity, slight type of accident in urban area outnumbered the rest of the categories. The horizontal stacked bar chart represents the total number of accident number by road type and location. As it is presented, Single and dual carriageway and dual carriageway roads are the most accident-friendly zones compared to other types of roads. Moreover, accident rates in the different speed limit in terms of the urban and rural area show that most of the accidents happened in the urban areas at the speed limit 30 mph. However, from this bar chart, it is distinctively noticeable that as the speed limit increases, the accident number increase in the rural area but decrease in the urban area.

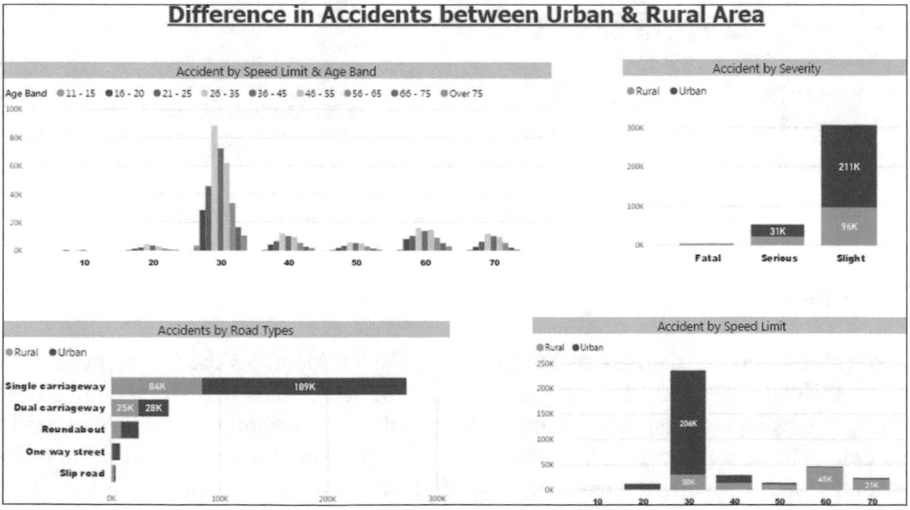

Fig. 5. Difference in accidents between urban and rural areas

Several weather-related factors have been identified as key factors in traffic casualties. The pie chart demonstrates the light condition during accident which shows that 74% of total accident happened in daylight condition, where only 20% of accident occurred in mild or low daylight. Surprisingly, little less than 5% accident are counted under no light or total darkness. Considering the road surface condition, the donut chart affirms more than 73% of total accident are counted on dry road surface even though it is considered as drive-friendly condition whereas slightly higher than 26% accident occurred in less drive-friendly condition or in the wet road surface. In terms of weather, the bar chart acknowledges the fine and no high wind condition are responsible for the highest number of accidents, mounting up to 30000 incidents. Surprisingly, relatively more hostile weather such as heavy rain, high wind and even the combination of rain and high wind experienced less number of accidents. Moreover, in many cases, on-going roadworks also cause many traffic accidents whereas mud and defective road surface are less hazardous (Fig. 6).

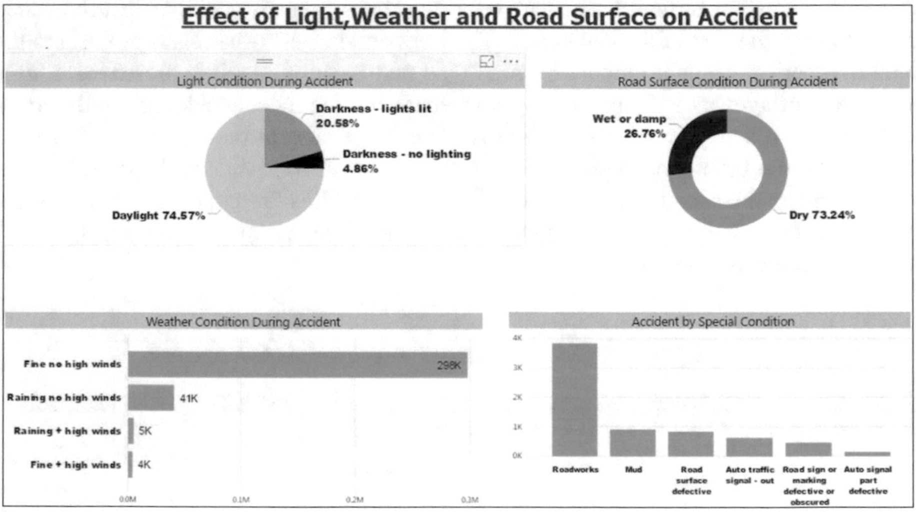

Fig. 6. Effect of light, weather and road surface on accident

7 Conclusion

This research focuses on three main objectives. First, it identifies the factors behind the traffic accidents. Second, the study determines the techniques used to visualize data. Third, it develops a dashboard using data visualization techniques to analyse traffic accident and casualty trend. The study reveals five key factors that lead to traffic accidents and casualties. They are; Speed, Environmental Condition, Age Effect, Vehicle Attributes and Accident Location. Moreover, Road Types, Time, Day and Light Conditions have been identified as most responsible factors for traffic accidents. The study has implemented Abela's [22] data visualization technique to develop the dashboard. Finally, the study has analysed several traffic accidents and casualties from this dashboard.

Acknowledgments. The authors would like to thanks Advanced Informatic School (AIS), Universiti Teknologi Malaysia (UTM) for the support in publishing the findings. This work was financially supported through research grant of RUG UTM under vote number 14H76

References

1. Triguero, I., Figueredo, G.P., Mesgarpour, M., Garibaldi, J.M., John, R.I.: Vehicle incident hot spots identification: an approach for big data. In: 2017 IEEE Trustcom/BigDataSE/ICESS, pp. 901–908 (2017)
2. Tulu, G.S., Washington, S., Haque, M.M., King, M.J.: Injury severity of pedestrians involved in road traffic crashes in Addis Ababa, Ethiopia. J. Transp. Saf. Secur. **9**(sup1), 47–66 (2017)
3. Ram, T., Chand, K.: Effect of drivers' risk perception and perception of driving tasks on road safety attitude. Transp. Res. Part F Traffic Psychol. Behav. **42**, 162–176 (2016)
4. Masuri, M.G., Isa, K.A.M., Tahir, M.P.M.: Children, youth and road environment: road traffic accident. Asian J. Environ. Behav. Stud. **2**(4), 13–20 (2017)
5. Guo, M., Wei, W., Liao, G., Chu, F.: The impact of personality on driving safety among Chinese high-speed railway drivers. Accid. Anal. Prev. **92**, 9–14 (2016)
6. Liao, Y., Li, S.E., Wang, W., Wang, Y., Li, G., Cheng, B.: Detection of driver cognitive distraction: a comparison study of stop-controlled intersection and speed-limited highway. IEEE Trans. Intell. Transp. Syst. **17**(6), 1628–1637 (2016)
7. Goniewicz, K., Goniewicz, M., Pawłowski, W., Fiedor, P.: Road accident rates: strategies and programmes for improving road traffic safety. Eur. J. Trauma Emerg. Surg. **42**(4), 433–438 (2016)
8. Omranian, E., Sharif, H., Dessouky, S., Weissmann, J.: Exploring rainfall impacts on the crash risk on Texas roadways: a crash-based matched-pairs analysis approach. Accid. Anal. Prev. **117**, 10–20 (2018)
9. Lin, L., Wang, Q., Sadek, A.W.: A combined M5p tree and hazard-based duration model for predicting urban freeway traffic accident durations. Accid. Anal. Prev. **91**, 114–126 (2016)
10. Xi, J., Zhao, Z., Li, W., Wang, Q.: A traffic accident causation analysis method based on AHP-Apriori. Procedia Eng. **137**, 680–687 (2016)
11. Bergel-Hayat, R., Debbarh, M., Antoniou, C., Yannis, G.: Explaining the road accident risk: weather effects. Accid. Anal. Prev. **60**, 456–465 (2013)
12. Mohamed, M.G., Saunier, N., Miranda-Moreno, L.F., Ukkusuri, S.V.: A clustering regression approach: a comprehensive injury severity analysis of pedestrian–vehicle crashes in New York, US and Montreal, Canada. Saf. Sci. **54**, 27–37 (2013)
13. Ferenchak, N.N.: Pedestrian age and gender in relation to crossing behavior at midblock crossings in India. J. Traffic Transp. Eng. Engl. Ed. **3**(4), 345–351 (2016)
14. Zhang, W., Wang, K., Wang, L., Feng, Z., Du, Y.: Exploring factors affecting pedestrians' red-light running behaviors at intersections in China. Accid. Anal. Prev. **96**, 71–78 (2016)
15. Clarke, D.D., Ward, P., Bartle, C., Truman, W.: Killer crashes: fatal road traffic accidents in the UK. Accid. Anal. Prev. **42**(2), 764–770 (2010)
16. Vedagiri, P., Kadali, B.R.: Evaluation of pedestrian–vehicle conflict severity at unprotected midblock crosswalks in India. Transp. Res. Rec. J. Transp. Res. Board **2581**, 48–56 (2016)
17. Shirazi, M.S., Morris, B.T.: Looking at intersections: a survey of intersection monitoring, behavior and safety analysis of recent studies. IEEE Trans. Intell. Transp. Syst. **18**(1), 4–24 (2017)
18. Huang, H., Zhou, H., Wang, J., Chang, F., Ma, M.: A multivariate spatial model of crash frequency by transportation modes for urban intersections. Anal. Methods Accid. Res. **14**, 10–21 (2017)
19. Zhang, G., Yau, K.K.W., Zhang, X., Li, Y.: Traffic accidents involving fatigue driving and their extent of casualties. Accid. Anal. Prev. **87**, 34–42 (2016)

20. Green, C.P., Heywood, J.S., Navarro, M.: Traffic accidents and the London congestion charge. J. Public Econ. **133**, 11–22 (2016)
21. Smith, A.C.: Spring forward at your own risk: daylight saving time and fatal vehicle crashes. Am. Econ. J. Appl. Econ. **8**(2), 65–91 (2016)
22. Abela, A.: Choosing a good chart. In: The Extreme Presentation(tm) Method, 06 Sept 2006
23. Gulbis, J.: Data visualization – how to pick the right chart type? In: eaziBI (2016)
24. Oetting, J.: Data visualization 101: how to choose the right chart or graph for your data. In: HubSpot, 31 Aug 2016
25. Yu, H., Liu, P., Chen, J., Wang, H.: Comparative analysis of the spatial analysis methods for hotspot identification. Accid. Anal. Prev. **66**, 80–88 (2014)
26. Denham, B., Eguakun, G., Quaye, K.: GeoTAIS: an application of spatial analysis for traffic safety improvements on provincial highways in Saskatchewan (2011)
27. Dales, J.: M25 road traffic accidents, 02 Feb 2016. http://community.powerbi.com/t5/Best-Report-Contest/M25-Road-Traffic-Accidents/m-p/17210#M14. Accessed 20 Dec 2017
28. O'Donnel, M.: Exploring NYC Vehicle Crash Data in Tableau. InterWorks, Inc., 26 Aug 2015

Towards Real-Time Customer Satisfaction Prediction Model for Mobile Internet Networks

Ayisat W. Yusuf-Asaju[1,2](✉), Zulkhairi B. Dahalin[2], and Azman Ta'a[2]

[1] University of Ilorin, Ilorin PMB 1515, Nigeria
ayisatwuraola@gmail.com
[2] Universiti Utara Malaysia, 06010 Sintok, Malaysia
zuldahlim@gmail.com, azman@uum.edu.my

Abstract. Satisfying the customers' service requirements and expectation, especially customer satisfaction had been one of the major challenges faced by the mobile network operators in most telecommunication organizations. This article implemented an analytical customer satisfaction prediction model by employing the mobile internet traffic datasets collected in real-time through the drive test measurement. To this end, the implementation phase has employed machine learning algorithms in the Microsoft Machine Learning R client Server. The results show that previous user's traffic datasets can be used to predict customer satisfaction and identify the root cause of poor customer experience before the complete deterioration of the service performance, which could lead to larger percentage of customer dissatisfaction. The mobile network operators can also use the proposed model to overcome the drawbacks of the conventional subjective method of analysing customer satisfaction.

Keywords: Telecommunication · Quality of experience
Mobile network operators · Traffic datasets · Machine learning algorithms
Mean opinion score

1 Introduction

In mobile telecommunication (telecoms) organizations, satisfaction is an individual judgement after the customer experience with a service. Accordingly, the competitive nature of the telecoms market, have made the Mobile Network Operators (MNOs) focus on both the technical aspect of the services provided to the customers, customer expectations and experiences with the aim of satisfying the customers. Generally, quality of experience (QoE) is used to analyse customer perception based on the network performance. Previous studies have demonstrated QoE as a measure of customer perception in relation to the technical parameters representing the quality of service performance, context, expectations and others factors that can influence the customer perception to determine the diversity of satisfied and dissatisfied customers using a specific service [1]. It should be noted that customer expectation is an ideal standard presented as a service level agreement (SLA), which is an agreement between the customers and the MNOs about the service characteristics provided by the MNOs

F. Saeed et al. (Eds.): IRICT 2018, AISC 843, pp. 95–104, 2019.
https://doi.org/10.1007/978-3-319-99007-1_10

[2], usually monitored by the telecoms regulators. In line with the QoE description, both the expectation and other factors are to be used to analyse the overall perceived QoE for a specific service to determine the mean opinion score (MOS).

The reliable and standard method of measuring the perceived QoE (i.e., MOS) is through the subjective method. However, the subjective method is time consuming, lacks usability in real time, lacks repeatability in real time, and in most cases, the study is usually conducted after the service consumption, which might be too late to take actions to prevent customer dissatisfaction [3]. Thus, there is a need for an analytical forecasting model for telecoms network service operation that can predict subjective measurement of the perceived QoE from the objective measurement in real time or near real time.

MOS is often used to estimate the perceived QoE through the prediction of the subjective measurement from the objective measurement [3, 4]. However, recent studies have demonstrated MOS was not sufficient enough to quantify the diversity of satisfied and dissatified customers, because MOS often eliminates the diversity of the customer assessment while quantifying the customer satisfaction [5]. Accordingly, this article improved and implemented an analytical customer satisfaction (ACSAT) prediction model proposed in the study of Yusuf-Asaju et al. [6] to demonstrate the effectiveness of predicting customer satisfaction with the mobile network realtime dataset. The remainder of this article discusses customer satisfaction issues in telecommunications, mobile internet QoE, ACSAT prediction model, the research methodology, analysis, discussion, and conclusion.

2 Literature Review

2.1 Customer Satisfaction in Telecommunications

Previous literatures have provided several thoughtful interpretations on customer satisfaction. Within this context, previous literature has demonstrated performance-specific expectation and expectancy disconfirmation play significant role in customer satisfaction analysis [7]. However, in different cases, disconfirmation measure has the greatest impact on satisfaction. As a result, Xiao and Boutaba [7] hold the view that there is a substantial basis that can be used to derive an analytical forecasting model for the telecoms network services through the use of theories, models of perceived service quality and customer satisfaction adopted from the marketing and economics literatures. Consequently, the authors proposed an analytical modelling approach termed Customer satisfaction (CSAT) model. The CSAT model consists of three service utilities (Network Quality of Service (QoS), Network Availability, and Customer care), expectation update, perceived utilities, disconfirmation, customer satisfaction, market dynamics, customer behavior and profitability and deployed through a mathematical model.

A recent study enhanced the CSAT model by defining the service utilities elements as four QoE dimension (Network coverage, service availability, QoS support and Price) of customers satisfaction [8]. Service content and security was included to the service utility in the CSAT model [8]. The study holds the view that QoS is closely related to perceive QoE and customer satisfaction model is based on QoS criteria and parameters.

Therefore, this study strongly believed that CSAT model can be modified and improved through the QoE components, QoE measurements, and QoE estimations.

2.2 Mobile Internet Quality of Experience

Qualinet in [1] describes QoE as the "*degree of delight or annoyance of the user of an application or service. It results from the fulfilment of his or her expectations with respect to the utility and/or enjoyment of the application or service in the light of the user's personality and current state.*" Perceived QoE measurements are classified into subjective and objective measurement [3]. The subjective measurements are based on customer perception of the services delivered to the customers, while objective measurement is the means of estimating subjective quality solely from the measurement obtained from the network traffic [3, 4]. The MOS is an opinion score on five-point category-judgement scales and it is mostly used in many applications such as audio, video, and data to estimate the perceived QoE [4]. The MOS is illustrated by allocating values to the scores such as Excellent = 5; Good = 4; Fair = 3; Poor = 2; Bad = 1 [5].

According to Le Callet et al. [1], QoE influence factor is "*any characteristic of a user, system, service, application, or context whose actual state or setting may have influence on the Quality of Experience for the user*". This description implies the QoE Influence factors can be classified into system, context and human influence factors. System QoE Influence Factor comprises of the properties and characteristics of the technical quality of the application [1]. This technical quality consists of the QoS parameters (such as throughput, delay, jitter, and loss). Several studies have analysed perceived QoE by employing the delay, jitter, and loss [4, 9]. However, limited studies have used throughput for the analysis of the perceived QoE. According to Battisti et al. [10], throughput is the actual data transmission speed experienced by the internet users.

Context QoE Influence Factor is any situational property that describes the users' environment [1]. Generally, previous studies often combine context factors with human and system factors without any specific structure or categorization [11]. In mobile network scenario, context factors in this study represents the physical components. Physical component describes the characteristics of location and space [11]. Therefore, the physical component variables used for the implementation of the ACSAT prediction model was location (longitude and latitude) and time of the day.

Human QoE Influence Factor describes any characteristics of users including demographic, socio-economic background, socio-cultural background, physical and mental constitution, and emotional state [1, 11]. Although, human QoE influence factor is an important dimension, human factor has been limitedly studied in most empirical studies, due to the difficulties in assessing the way it influences QoE [11]. Moreso, user demographic is difficult to ascertain in real-time, since the MNO often analysed the user experience based on contextual factors. Therefore, this study considers user expectation as SLA. In addition, this study considers the use of region, which describes the population densities of users in an area. This is an example of socio-cultural. Both expectation and region are used as a variable of the human influence factors for the implementation of the ACSAT prediction model.

2.3 Analytical Customer Satisfaction Prediction Model

The ACSAT prediction model comprises of five different assumptions. The first assumptions is to employ network performance as the underlying component of ACSAT prediction model. Network performance is a principal determinant of customer satisfaction. The second assumption is the use of customer traffic data consisting of the mobile internet usage dataset. This assumption was as a result of the limited use of large scale database comprising of customer behaviour to estimate the percieved QoE [12]. The third assumption is to incorporate the QoE influence factors in the ACSAT prediction model. Therefore, modelling of the perceived QoE, considers all the three perceived QoE dimensions of the influence factors in the ACSAT prediction model, which enables an adequate estimation of the perceived QoE in relation to mobility (such as time and location).

The fourth assumption is the estimation of perceived QoE with respect to expectation as indicated in SLA. Several marketing and perceived QoE literatures reviewed showed that expectation is an influence factor of perceived QoE [1, 12]. Although, the expectation in respect to SLA was used in the CSAT model to model the perceived service quality [7]. Tsiaras and Stiller [13] specifically stated that using expectation as stated in SLA would be adequate for the modelling of perceived QoE. In addition, the CSAT model reversed the expectation to formulate the expectation update [10]. However, Ibarrola et al. [2] stated that web survey is usually built to collect customer requirements and preferences. This paper argues that the arrow has to indicate that expectation can used to estimate the perceived QoE and at the same time the customer behaviour observed while estimating the perceived QoE can be used to update the expectation.

Lastly, the fifth assumption proposed in the ACSAT prediction model is perceived QoE maximization to determine the variation of customers that are satisfied and not satisfied. Therefore, this article used threshold of acceptance in Standard deviation of Opinion Scores (SoS) hypothesis (that is, the rate of perceived quality is either acceptable or not acceptable) [5]. The percentage of Good or bad and percentage of poor or worse (%GoB and %PoW that is: ratio of satisfied and dissatisfied customers) can be defined as $GoB(u) = P_u(U \geq 60)$ and $PoW(u) = P_u(U \leq 45)$. This depicts categories 1–3 on the ACR MOS can be translated into dissatisfied ($PoW(u) = P_u(U \leq 45)$), while categories 4–5 on the ACR MOS can be translated into satisfied ($GoB(u) = P_u(U \geq 60)$) as explained by Hoßfeld et al. [5], where U is the random variable (throughput variable in this case) for quality ratings. All the five assumptions facilitated the process of determining the distribution of satisfied and unsatisfied customers and the results can be used by the MNOs for allocation of network resources [7, 14]. Figure 1 depicts the ACSAT prediction Model.

3 Methodology

Following the underlying assumptions stated in the ACSAT prediction model and description of the QoE influence factor dimension in the ACSAT prediction model, the following research hypothesis were verified:

Hypothesis 1: Time of the day has significant influence perceived QoE.
Hypothesis 2: Location of the users has significant influence on perceived QoE
Hypothesis 3: QoS parameters (Throughput) has significant influence on perceived QoE.
Hypothesis 4: Region of the users has significant influence on perceived QoE.
Hypothesis 5: Expectation (SLA) has significant influence on perceived QoE.

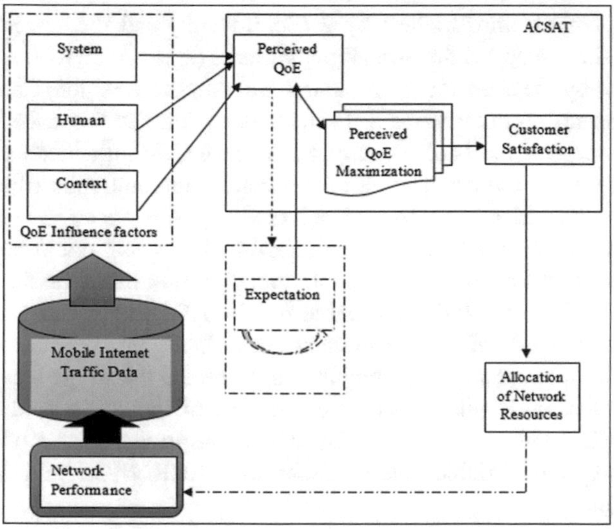

Fig. 1. ACSAT prediction model

3.1 Data Collection

The dataset used in this article was collected in real-time through the drive measurement with the aid of Test Mobile System (TEMS) investigation. TEMS investigation is a software used by the MNOs to collect large amounts of dataset with real-time presentation [15]. TEMS investigation software was used to collect dataset from the real-time measurement of a major MNO in Nigeria at different times of the day for File Transfer Protocol Download (FTP DL) and Hyper-test Transfer Protocol (HTTP) applications because limited studies have evaluated the perceived QoE of these mobile internet applications. The dataset was collected from four different regions (Dense Urban, Urban, Semi-Rural and Rural) in Nigeria. The total observation of the dataset for FTP DL application was 1.8 million while the total dataset of the HTTP application was 1.3 million observations. FTP is a protocol used to facilitate exchange of data between a server and clients(s) and is used as a test due to the specific nature of the data exchange and simulates file download. HTTP simulates general browsing which is the underlying protocol used in WWW and its data exchange.

3.2 Data Preparation

The dataset was collected from the mobile network in real time as discussed in the data collection section. To make the dataset suitable for the modelling of perceived QoE, the dataset passed through a data preparation process. The dataset was cleaned, integrated, and transformed into a form that was suitable for the modelling phase. In addition, the conceptual labels used were based on the population densities of the regions. Region with the highest population density was represented by 4 (Dense Urban), second highest by 3 (Urban), third highest by 2 (Semi-Rural) and the lowest by 1 (Rural). Similarly, this study mapped the throughput variable (aggregated total application layer throughput used by the users on a particular node) into the Absolute Categorical Rating (ACR) MOS scale, represented as a discreet value (that is 5 = excellent. 4 = good, 3 = fair, 2 = poor, and 1 = bad). Equally, SLA of the QoE influence factor dataset that was used to represent the expectation of the mobile internet subscribers was defined into different categorical interval (5 = Excellent, 4 = good, 3 = fair, 2 = poor, 1 = bad) based on the maximum throughput of 42 Mbps that could be achieved on a node as explained by Ericson [16]. The interpretation was based on the maximum and minimum value of the throughput variable for the ACR scale was achieved through probability mass function of the discrete random variable (throughput). Afterwards, the longitude and latitude variable was combined to represent the single location of the user in one single variable. Similarly, the date and time of the day was also combined to represent a single variable time of the day. The descriptive statistics of maximum and minimum throughput variable interpretation into ACR MOS and SLA score are depicted in Table 1.

Table 1. Descriptive statistics of throughput variable interpretation to ACR MOS and SLA score

FTP DL applications		HTTP applications				
TT variable intervals (Mpbs)	ACR MOS score	TT variable intervals (Mpbs)	ACR MOS score	TT variable (Mpbs)	SLA score	Scale interpretation
0–7.2	1	0–6.046	1	0–8.4	1	Poor
7.3–14.3	2	6.047–12.092	2	8.41–16.8	2	Bad
14.4–21.5	3	12.092–18.138	3	16.81–25.2	3	Fair
21.6–28.7	4	18.139–24.184	4	25.21–33.6	4	Good
28.8–35.8	5	24.185–30.23	5	33.61–42.00	5	Excellent

3.3 Data Modelling

This phase involves the modelling of perceived QoE through an abstract representation of data and its relationship within the dataset. The stage adopts the use of regression task in the form of a predictive technique through the machine learning algorithms (such as multiple regression, decision trees, random forest, decision forest and neural network) in Microsoft Machine Learning R Server Platform. Following the deterministic mathematical (DQX) model in [13] that gives QoE formalization as $QoE \coloneqq f(User, Service, Variable)$. A multiple regression was considered to model the relationship between the MOS and the QoE influence factors (variables) through a linear predictor functions, where unknown model parameters are estimated from the data. In this case, the observation of data instances represents the independent variable (that is, the variables in the QoE influence factors dataset) while the predicted variable is the possible values of dependent variables (perceived QoE) which is the outcome. The predicted perceived QoE was evaluated with root mean squared error (RMSE) [17], which is the most used prediction accuracy technique for numerical measures of QoE in previous studies [4, 9].

4 Results and Discussion

The first analysis conducted was multiple regression analysis between the QoE influence factors and the perceived QoE. The results of the correlation and multiple regression for both FTP DL and HTTP applications are presented in the Tables 2 and 3.

Table 2. Correlation and multiple regression result for FTP DL application

Variables	Pearson correlation	Regression estimates	T-value	P-value
QoE ~ Time of the day	−0.036***	0.005	16.71	<2e−16
QoE ~ Location	0.332***	0.019	48.47	<2e−16
QoE ~ Throughput	0.908***	0.417	744.17	<2e−16
QoE ~ Region	0.010***	−0.291	−38.27	<2e−16
QoE ~ SLA	0.903***	0.555	685.97	<2e−16

***Significant ($p < 0.000$)

Following the results presented in Tables 2 and 3, it can be concluded that all the proposed hypotheses were significant ($p < 0.000$) with R^2 of 0.871 and 0.869 for FTP DL and HTTP applications respectively. The results depict all the stated hypotheses have significant influence on the overall perceived QoE. Specifically, for both applications, H1 was accepted and shows a positive significant influence for MOS in both application, this implies that the MOS increases with the time of the day. This is evident among the mobile internet users whereby, the internet services is usually stable during the late hours of the night (off peak time). H2 was accepted for both application and shows a positive significant influence for MOS in both applications. This implies, MOS increases in irrespective of the user's location. In other words, internet users in different locations are bound to experience different mobile internet experience. H3 was accepted for both application, this interprets the MOS increases with the MOS, the higher the throughput, the higher the mobile internet experience. H4 was accepted for both application with a negative significance influence, this indicates the lower the population density (less congested areas) the higher the mobile internet user experience. Lastly, H5, was accepted and shows a positive significance influence which interprets the expectation of the user increases with the mobile internet experience.

Table 3. Correlation and multiple regression result for HTTP application

Variables	Pearson correlation	Regression estimates	T-value	P-value
QoE ~ Time of the day	0.173***	0.013	36.55	<2e−16
QoE ~ Location	0.033***	0.019	36.82	<2e−16
QoE ~ Throughput	0.915***	0.666	994.93	<2e−16
QoE ~ Region	−0.071***	−0.022	−20.50	<2e−16
QoE ~ SLA	0.855***	0.293	270.02	<2e−16

***Significant ($p < 0.000$)

Overall, the regression results show that all the selected QoE influence factors have significant influence on mobile internet users perceived QoE for both FTP DL and HTTP applications.

The second Analysis conducted was the prediction of the perceived QoE with decision trees, random forest, decision forest and neural network. While the actual MOS of FTP DL application was 1.5018 and actual MOS of HTTP application was 1.6170. The best machine learning accuracy was selected based on the predicted MOS that has the closer value to the actual MOS. The RMSE results for both the FTP DL and HTTP applications are presented in the Tables 4 and 5 below.

Table 4. Perceived QoE modelling accuracy result for FTP download applications

Machine learning algorithms	RMSE	MOS
Multiple linear regression	0.274	1.5025
Decision trees	0.120	1.5024
Random forest	0.118	1.5021
Decision forest	0.072	1.5019
Neural network	0.141	1.4937

Table 5. Perceived QoE modelling accuracy result for HTTP applications

Machine learning algorithms	RMSE	MOS
Multiple linear regression	0.310	1.6162
Decision trees	0.116	1.6172
Random forest	0.127	1.6169
Decision forest	0.091	1.6171
Neural network	0.148	1.6013

The third analysis is the prediction of the customer satisfaction model using logistic regression to determine the diversity of the satisfied and dissatisfied customers. In compliance with the SOS hypothesis discussed in Sect. 2.3, the threshold of %GoB = P (MOS $\geq$ 4) and %PoW = P (MOS $\leq$ 2) was used to filter out the satisfied and dissatisfied quality ratings (accepted or not accepted) from the predicted QoE at Pearson correlation MOS ~ TT equals 0.908 (see Table 2) and Decision forest modelling RMSE was 0.072 (see Table 4) of the FTP DL application. Therefore, a correlation was conducted to ascertain the relationship between the predicted MOS and acceptability variable. Afterwards a binary classification prediction modelling was conducted to analyses the percentage of satisfied and dissatisfied users through logistic regression in R platform. The result of the customer satisfaction analysis of FTP DL application shows a correlation of 0.529 between the perceived QoE (MOSp) and acceptability threshold of the SoS hypothesis. And the binary classification prediction indicates logistic regression have the accuracy of 100% (Specificity = Recall = Sensitivity = Precision = F1 measure = 100%). Following the SoS hypothesis benchmark, the correlation of MOSp and acceptability was 0.529 interprets 97.5% is the %PoW was not Dissatisfied and 2.5% is the %GoB was Satisfied.

For the HTTP application, the correlation of MOS ~ TT equals 0.915 (see Table 3) and Decision forest modelling RMSE = 0.091 (see Table 5) of the HTTP application. And the binary classification prediction indicates logistic regression have the accuracy of 100%. Following the SoS hypothesis benchmark, the correlation of MOSp and Acceptability was 0.637 (at accuracy = Specificity = Recall = Sensitivity = Precision = F1 measure = 100%) interprets 95% is the %PoW = Dissatisfied and 5% is the %GoB = Satisfied. Overall, based on the SoS hypothesis, the percentage of dissatisfied users is 95% and percentage of satisfied users is 5% of the HTTP application. Following the presented results, the MNOs can improve allocation of resources in areas where the perceived QoE is below the acceptable threshold.

5 Conclusion

This article presented an ACSAT prediction model to estimate customer satisfaction of two mobile internet applications using the datasets collected in real-time. The empirical evaluation indicated that customer satisfaction is feasible in real-time to overcome the drawbacks that often occurs through the subjective measurement of customer satisfaction analysis. The analysis showed that the time of the day, location, region, throughput, and expectation have a significant influence on the user perceived QoE. Following the dataset that was used in this study, it was observed that high population density with inadequate network resources can be a cause of the poor mobile internet experience. One way the MNOs can rectify this type of issue is to trace the location and region with very poor MOS and provide proactive measures to increase the network resources that would be efficient for areas with high population density. As seen in the results poor QoE lead into a poor customer satisfaction. Therefore, providing a proactive measures to improve the QoE would reduce the poor rate of mobile internet customer satisfaction.

References

1. Le Callet, P., Möller, S., Perkis, A.: Qualinet White Paper on Definitions of Quality of Experience Output version of the Dagstuhl seminar 1218. 12 June 2012. http://www.qualinet.eu/images/stories/whitepaper_v1.1_dagstuhl_output_corrected.pdf
2. Ibarrola, E., Liberal, F., Ferro, A.: Quality of service management for ISPs: a model and implementation methodology based on the ITU-T recommendation E.802 framework. IEEE Communications Magazine, pp. 146–153 (2010)
3. Tsolkas, D., Liotou, E., Passas, N., Merakos, L.: A survey on parametric QoE estimation for popular services. J. Netw. Comput. Appl. **77**(1), 1–17 (2017)
4. Demirbilek, E., Gregoire, J.-C.: Towards reduced reference parametric models for estimating audio-visual quality in multimedia services. In: IEEE International Conference on Communications (ICC), Kuala Lumpur (2016)
5. Hoßfeld, T., Heegaard, P.E., Varela, M., Moller, S.: QoE beyond the MOS: an in-depth look at QoE via better metrics and their relation to MOS. Qual. User Exp. **1**(2), 1–23 (2016)
6. Yusuf-Asaju, A. W., Zulkhairi D., Ta'a, A.: A proposed analytical customer satisfaction prediction model for mobile internet networks. In: Pacific Asia Conference on Information Systems (PACIS) Proceedings, Langkawi, Malaysia (2017)
7. Xiao, V., Boutaba, R.: Assessing network service profitability: modelling from market science perspective. IEEE/ACM Trans. Netw. **15**(6), 1307–1320 (2007)
8. Djogatovic, V. R., Kostic-Ljubisavljevic, A., Stojanovic, M., Mikavica, B.: Quality of experience in telecommunication. In: 8th International Quality Conference, Kragujevac (2014)
9. Aroussi, S., Mellouk, A.: Statistical evaluation for quality of experience prediction based on quality of service parameters. In: 23rd International Conference on Telecommunications (ICT), Thessaloniki (2016)
10. Battisti, F., Carli, M., Paudyal, P.: QoS to QoE mapping model for wired/wireless video communication. In: Euro Med Telco Conference (EMTC) (2014)
11. Reiter, U., Brunnström, K., De Moor, K., Larabi, M.-C., Pereira, M., Pinheiro, A., You, J., Zgank, A.: Factors Influencing Quality of Experience. Quality of Experience: Advanced Concepts, Applications, pp. 55–72. Springer, Cham (2014)
12. Reichl, P., Egger, S., Möller, S., Kilkki, K., Fiedler, M., Hossfeld, T., Christos, T., Asrese, A.: Towards a comprehensive framework for QOE and user behavior modelling. In: Seventh International Workshop on Quality of Multimedia Experience (QoMEX), Pylos-Nestoras (2015)
13. Tsiaras, C., Stiller, B.: A deterministic QoE formalization of user satisfaction demand (DQX). In: Annual IEEE Conference on Local Computer Networks, Canada (2014)
14. Spiess, J., T'Joens, Y., Dragnea, R., Spencer, P., Philippart, L.: Using big data to improve customer experience and business performance. Bell Labs Tech. J. **18**(4), 3–17 (2014). https://doi.org/10.1002/bltj.21642
15. Ascom: Ascom Network Testing (2012). http://www.livingston-products.com/products/pdf/139777_1_en.pdf
16. Ericson: Basic Concepts of HSPA, 02 (2007). http://escrig.perso.enseeiht.fr/HSPA-Concepts.pdf
17. Galit, S., Peter, B.C., Inbal, Y., Nitin, P.R., Kenneth, K.C.J.: Data Mining for Business Analytics, Concepts, Techniques and Application in R. Wiley, Hoboken (2018)

Data Quality Issues in Big Data: A Review

Fathi Ibrahim Salih, Saiful Adli Ismail(✉), Mosaab M. Hamed,
Othman Mohd Yusop, Azri Azmi,
and Nurulhuda Firdaus Mohd Azmi

Advanced Informatics School, Universiti Teknologi Malaysia,
Jalan Sultan Yahya Petra, 54100 Kuala Lumpur, Malaysia
{i.salih,mhsmosaab2}@graduate.utm.my,
{saifuladli,othmanyusop,azrizami,huda}@utm.my

Abstract. Data with good quality has precedence when analyzing and using big data to deduce value from such tremendous volume of data in today's business environments. Decisions and insights derived from poor data has a negative and unpredictable consequences to organizations. At present, due to the lack of comprehensive and intensive research in the field of data quality, especially large data, there is an urgent need to address this issue by researchers to reach the optimal way to estimate and evaluate the quality of large data. Thus, enabling institutions to make rational decisions based on evaluation outputs. In this paper, the current research on the quality of large data was reviewed and summarized by exploring the basic characteristics of large data. The main challenges facing the quality of information were also discussed in the context of large data. Some of the initiatives suggested by the researchers to evaluate the quality of the data have been highlighted. Finally, we believe that the results of these reviews will enhance the conceptual measurements of the large data quality and produce a concrete groundwork for the future by creating an integrated data quality assessment and evaluation models using the suitable algorithms.

Keywords: Big data · Data quality · Quality assessment
Big data quality dimensions · DQ evaluation

1 Introduction

IT environment has changed rapidly following the commencement of the 21st century, for instance, cloud computing, IoT, augmented reality and social networking have been emerged into the market. All the aforementioned technological advancements opened the space for the big data era to be emerged [1]. Nowadays, the measure of worldwide information is developing exponentially. Unit in which data is measured is no longer been gigabyte or terabyte. As indicated by "Advanced Universe" conjectures [2], 40 ZB of information will be produced by 2020.

This high growth of data has been generally accompanied with little attention to the quality of data that these databases contain, being now more than ever true the old phrase at the beginning of the computer days that said, "garbage in garbage out". And while there is abundant literature on the Big Data subject, there are few concrete

F. Saeed et al. (Eds.): IRICT 2018, AISC 843, pp. 105–116, 2019.
https://doi.org/10.1007/978-3-319-99007-1_11

proposals for schemas that directly address the issue of data quality for very large databases.

This paper reviews existing literature on big data to obtain an understanding of the present extravagant publicity on big data and its data quality. Firstly, an overview of big data, its characteristics, and the potential value of such big data utilization in current business practices is presented. Then an attempt to scout the challenges of data quality in big data and conventional data quality and data usage problems all over the literature. Finally, the paper provides an investigation on several ingredient and actions constituting part of data quality management such as dimensions, metrics, assessment schemes being proposed for big datasets as well as the future work to build a special data quality assessment model for big data which is recommended.

2 Overview of Big Data

2.1 The Concept of Big Data

Big data involve the large and varying dynamic data volume created by people, tools, and machinery. To be effectively processed, it needs a new, state-of-the art, and flexible technology to collect, host, and analyze a huge data.

2.2 Features of Big Data in Context of Data Quality

Big data is regarded as organized, semi organized, and unorganized datasets with excessive amount of data, normally it reaches a terabyte and above. Due to this, it is a nightmare task to be seized, stored, controlled, analyzed, managed and presented by traditional hardware, software and database management technologies. Laney [3] is the first one who presented the three main elements that characterize big data: i.e. Volume, Velocity and Variety, known as the (3 V's) of data. Although these main elements have been frequently used to describe big data, extra elements of Veracity and value have been produced to express data integrity. Table 1 below describe the features of big data.

Table 1. Characteristics of big data

Volume: Tremendous amount of data is created in comparison traditional databases.
Variety: Different sources of data. Big data draws from text, audio, images,
Velocity: data generation is extremely fast.
Veracity: Refer to the bias, noise and abnormality in data because data is sourced from different places, as a result you need to test the veracity/quality of the data.
Value: The usefulness of data in its end purpose.

2.3 Creating Big Value from Big Data

Big data produce benefits to organizations in many ways. They can bring such value from analyzing big data for different purposes like understanding the root causes of diminishing revenue in specific product, or analyzing data related to HR to analyze staff turnover reasons. A three-step approach below can avail to determine how to obtain Value from Big Data [4].

The first step is Start with the Right Big Data Store that is aligned with business needs this is typically done by matching the business problem or opportunity with the right technology. The second step would be achieved by building a domain knowledge which involve building the necessary expertise that determine which data, from all the possible sources, are valuable and which are not. Finally, it is vital to Choose the right reporting and analysis tool that enables the right overall big data approach.

2.4 Data Quality Challenges in Big Data

For better understanding data quality issues in big data it is essential to do this using its main characteristics and figure out its relationship with data quality. Currently, data quality in big data is confronted with the following challenges:

Velocity: In a big data era, data is changing very fast and in a very limited time frame, which necessitate higher processing power to keep up with such speed. Due to the fact of rapid changes in data flow in big data, organizations need to acquire/collect these data in a reasonable time before any changes may occur. Not collecting or dealing with data in an exact timing may results in an outdated information's leading to futile or misleading conclusions. Currently, the real processing and analysis program for big data is still in its early stages. One choice to overcome this is by sampling, but this is at the expense of bias.

Variety: In big data era, the diversity of data sources brings too many data with different and complex data structures. This will create challenges in data integration processes. In big data, data will be originated from a variety of data sources with different shapes and sizes and this will affect the data quality. A single data quality metric may not meet the datum of all collected data. so, it is important to utilize multiple data metrics because the process of examining and enhancing data quality of unstructured data is considered more complex when compared to structured data.

Volume: Post industrial revolution, the volume of data generated globally has been doubled exponentially. Due to this tremendous shift of data volume, it is difficult to obtain, clean, migrate, and realize high-quality data in a limited time frame. As a result, the process of converting such tremendous unstructured data to structured data types will take an extensive period of time and more processing power representing a real challenge to the current technologies used.

Value: Cost vs benefit of improving data quality. Basically, it can be thought that data is valuable if it achieves its ultimate purpose. Organizations utilize big data for a variety of purposes and these goals typically determine how data quality is expressed, accounted for, and enhanced. Mainly data quality dependent on what we are wishing to achieve form the data. for example, in some circumstances incomplete data may not affect the value of data in achieving its business goals. Sometimes it is better to data quality improvement because the cost the cost of enhancement exceeds the value of data to the business.

Veracity: What constitutes data quality Honesty is directly related to quality problems. To be honest, data shall be free from errors, biases, inconsistency, trustworthiness and noise. All these problems will directly affect data integrity and accountability. In organizations, there is a diverse needs, goals, and different working processes among data users. This in turn will leads to a diverse perception of what is actually comprises the data quality for each group of data users.

No specific global data quality standard to deal with big data had been developed yet and researches on the data quality of big data has just begun: General data quality standards were evolved in the 1990s, but quality standards were not published in ISO 8000 data until 2011 [5]. Despite the acceptance and participation to deploy this standard reached about 20 countries there is still some ambiguities and issues about it. The standards need to be mature and perfect.

3 Data Quality and Big Data

Data quality doesn't require that data is without errors. Erroneous data is just a single piece of the data quality condition. Most specialists take a more extensive point of view. Larry English says data quality includes "reliably meeting knowledge worker and end-client desires." Others say data quality is the wellness or appropriateness of data to meet business necessities. Regardless, most refer to a few characteristics that altogether describe the quality of data [6]. All together for the expert to decide the extent of the hidden underlying drivers and to design the ways that tools [7] can be utilized to address data quality issues, it is profitable to comprehend these regular data quality attributes [8].

Big data quality analysis is needed to justify the impact that poor data will have on the result of the analysis. Of course, institutions rely on data when developing their short and long-term strategic plans. However, before embarking on these plans and strategies, it is necessary to obtain sufficient assurance about the quality of the large data that will be used in developing these plans and strategies. This is achieved through assessments and analysis of the quality of these data. Here under, a brief description of the essential elements of data quality evaluation of structured data and eventually Big data is provided.

- **Data and Data Types**

As Described by the authors in [7, 8] the data is normally created in data bases using schemas which give a well-organized frame or structure. But with the emergence of social media, complex data types had been created which can be characterized by semi-structured and unstructured data.

- **Data Quality (DQ) Definition**

Authors in [9], summarized data quality from ISO 25012 Standard as "the capability of data to satisfy stated and implied needs when used under specified conditions".

- **Poor Data, Data Quality Issues and Problems**

Too many elements had contributed to the production of a low-quality data: errors resulted from invalid input by individuals, error reading from machine/devices, unstructured data, and lost values. The authors in [9, 10], linked reasons for bad data with data quality dimensions, they concluded that data quality issues could appears as misfielded and contradictory values, missing data, uniqueness constraints and functional dependency violation, those issues are related to Accuracy and completeness of the data.

3.1 Data Quality Dimensions

Data quality dimension is a term used by data management professionals to describe a feature of data that can be measured or assessed against defined standards to determine the quality of data [11]. The dimensions are defined as follows:

- ***Completeness:*** to be complete, the values of all components of a single data are correct. It defines whether or not all the data necessary to fulfil the current and future business objectives are available in the data resource.
- ***Uniqueness:*** Data shall be recorded once.
- ***Timeliness:*** Is a measure the time lag between the data capture, interpretation, and its utilization to achieve a specific business objective. Failure to obtain the data in reasonable timing may affect its usefulness for decision making.
- ***Validity:*** Data are valid if it follows the rules (format, type, range).
- ***Accuracy:*** refer to whether data reflects the dataset. In other words, it is a measure of the degree to which the data correctly describe the theme of the "real world", object or the event described.
- ***Consistency:*** The absence of difference, when comparing two or more representations of a thing against a definition.

Furthermore there is other factors that can affect the effectiveness of data usage such as ease of use, flexibility, confidentiality, and timeliness of data [12, 13].

A number of reasons leads to poor data quality such as the absence of validation protocols [14], sometimes data may be valid but lacks accuracy or correctness [11], others have inconsistency in syntax or formatting. Bad data quality may also be due to ineffective change and configuration management processes within the organization as well as poor system development process i.e. (system design, data conversion errors, validation processes, etc.). According to TDWI Data Quality Survey [6], the typical issues caused by defective data represented in loss of credibility, extra cost and time to refine the data, customer dissatisfaction, and continuous delays in delivering new systems. All of these problems can be illustrated in Fig. 1 below.

3.2 Big Data Quality Dimensions

In existing literature works, there are many arrays of quality dimensions have been considered by different authors. According to [15], authors reviewed how to evaluate or assess the quality of data. They said that part of data quality evaluations could be task

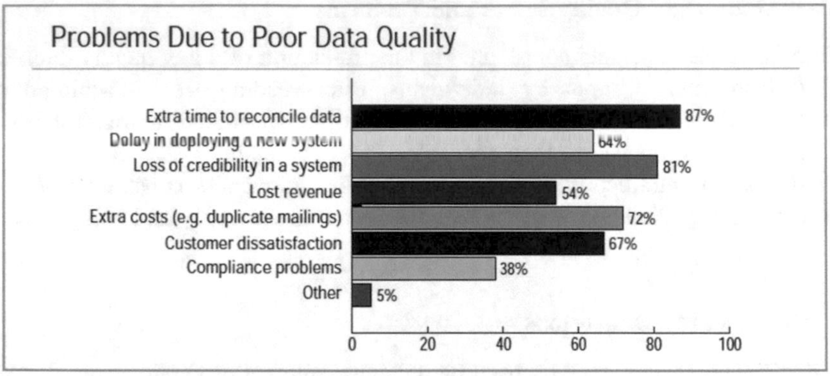

Fig. 1. Causes of data quality problems [6]

independent, hence, not bounded by the context of application. In contrast other are task dependent. Table 2 illustrating the main dimensions they believe were valuable for discussion:

Table 2. The main data quality dimensions [15]

Dimensions	Definitions
Accessibility	Degree of data availability, or ease and speed of retrieval
Volume of data	Degree to which there is a proper size of data to achieve the task
Believability	Degree to which data is true and originated from trustworthy source as well as its value is credible and convincing
Completeness	Degree to which there is no missing or insufficient data
Consistent representation	Degree to which data is displayed/given in the same manner
Ease of manipulation	How easy it is to process and apply data to different tasks
Free-of-error	Degree to which data is honest, accurate, and reliable
Interpretability	Extent to which data is inappropriate languages, symbols, units and the definitions are clear
Objectivity	The degree to which data definitions are clear and free of inappropriate languages or syntax
Relevancy	Degree to which data can be of benefit to solve the issue
Reputation	Degree to which data is trustworthy
Security	The degree of access control provided to maintain security
Timeliness	Data should not be outdated
Understandability	Data provided is easily comprehended by user
Value-added	Data shall be of value

The idea that "each dimension of data quality should cover specific aspects that may fall under the general idea of data quality" was discussed by both Batini and Scannapieca [16]. They concluded that with each dimension, there must be many measurements that could be applied to measure a given dimension, and for each measure, there may be more than one measurement method. Here is a breakdown of what they concluded:

Consistency: Worry about the infringement of specific semantic knowledge across data elements usually reflect integrity constraints

Accessibility: Ease of access to data from everywhere

Believability: Data is true and originated from trustworthy source as well as its value is credible and convincing

Reputation: Considers how trustable is an information source

Objectivity: The source of data is not being characterizes by fairness and free of bias

Value added: Related to whether data is beneficial and can be used attain business goals

Relevancy: How beneficial is the data? And can it be used to solve the prevailing issue?

Understandability: Data should be provided in a manner that it is easily comprehended by user

By a simple comparison between the set of dimensions mentioned above we can realize that there a reasonable level of matching and similarity between what are the standard approaches that could be used in assessing data quality.

When analyzing large data, it is necessary to understand their characteristics in terms of the magnitude of the data size and the speed of change, which necessitate the separation of high-quality data from the weak one during conducting such analysis. However, according to [12], in some cases, improving data may not be important, since the size of the defective data is relatively small compared to the size of the total evidence and has no significant effect on the final outcome, therefore can be neglected. This in turn increases the importance of understanding which dimensions for big data are being disclosed from those that require neglect.

Caballero [13] hypothesized that the basic dimension to be contemplated is consistency. The concept of consistency means the ability of an information system to confirm consistency of databases when moving data across systems and networks. However, many dimensions of traditional data quality have been linked to the sub-bands of consistency as follows (Table 3).

However, the methodology in which data quality dimensions have been assigned to the basic characteristics of large data has been constructed was merely a hypothesis that has not been demonstrated through scientific tests or experiments. Thus, there is an urgent need to research in this area by conducting in-depth researches into the quality of large data to find out which dimensions is more important in the context of big data.

Table 3. Mapped 3V's of big data to the 3C's of data quality [13]

	Velocity	Volume	Variety
Contextual [16] consistency	Consistency, Credibility, Confidentiality	Completeness, Credibility	Accuracy, Consistency, Understandability
Temporal consistency	Consistency, Credibility, Currentness, Availability	Availability	Consistency, Currentness, Compliance
Operational consistency	Completeness, Accessibility, Efficiency, traceability, Availability, Recoverability	Completeness, Accessibility, Efficiency, Availability, Recoverability	Accuracy, Accessibility, Compliance, Efficiency, Precision, Traceability, Availability

4 Quality Criteria of Big Data

The business environment using the data together with the data's feature determines the quality if the data to be considered as qualified (or good quality data) it must meet the requirements and conforms to the relevant uses. Previously, data consumers were the data generators either directly or indirectly, hence ensured quality of the data obtained. But in the era of big data, data consumers are not necessarily data generators which make it very difficult to measure data quality.

A hierarchical data quality standard had been proposed by Cai and Zhu [17], as shown in Fig. 2 authors defined dimensions of data quality which are commonly accepted as big data quality standards and redefined their concepts for the business needs. Each standard had splitted to multiple typical elements associated with it, each element has its indicators of quality. Some detailed quality indicators are given in Table 4.

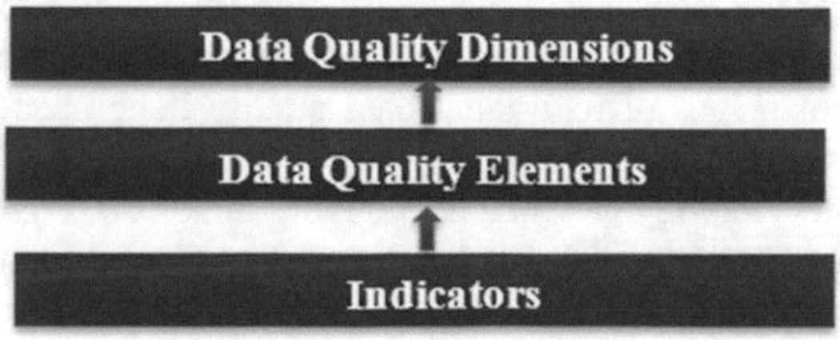

Fig. 2. Data quality framework [17]

Table 4. The hierarchical big data quality assessment framework (partial content).

Dimensions	Ingredient	Key indicators
Availability	(1) Accessibility	• Whether means of access is produced • Data can be easily viewed or acquired
	(2) Timeliness	• Data is available when needed • Whether data is up to date
Usability	Credibility	• Data obtained from trusted source • Regular assurance of data preciseness by professionals • The data is in the range of known or accepted values
Reliability	(1) Accuracy	• Level of preciseness and accuracy of data provided • Consistency between the source and outputs represented • Clear information is presented
	(2) Consistency	• After processing the data, its theory, value ranges and layout still match what they were before treatment • Data is provable and consistent in specific period in time
	(3) Integrity	• Data formatting is not unambiguous and conforms with the used standards • Data conform to structural integrity • The data is consistent with the integrity of the content
	(4) Completeness	• Whether a flaw in a single component will have an effect the data usage for data with multiple components • Whether a flaw in a component will affect data precision and integrity
Relevance	Fitness	• Collected data not fully correspond to the overall object, but they demonstrate one aspect • Information theme provides matches with users' retrieval theme
Presentation quality	Readability	• Data (content, format, etc.) are clear and understandable • It is easy to judge that the data provided meet needs

5 Quality Assessment Process for Big Data

Cai and Zhu [17], also introduce an essential data quality assessment process which based on the big data's attribute to provide a dynamic feedback mechanism, as shown in Fig. 3 below.

In [18], authors shows how to deal with the data quality the Big Data. In their proposal, they map the big data processing pipeline [19] with their proposed quality pipeline, by first presenting a quality processing pipeline, and then concentrating on three key quality characteristics, i.e. consistency, accuracy and confidentiality. Figure 4 illustrates their findings.

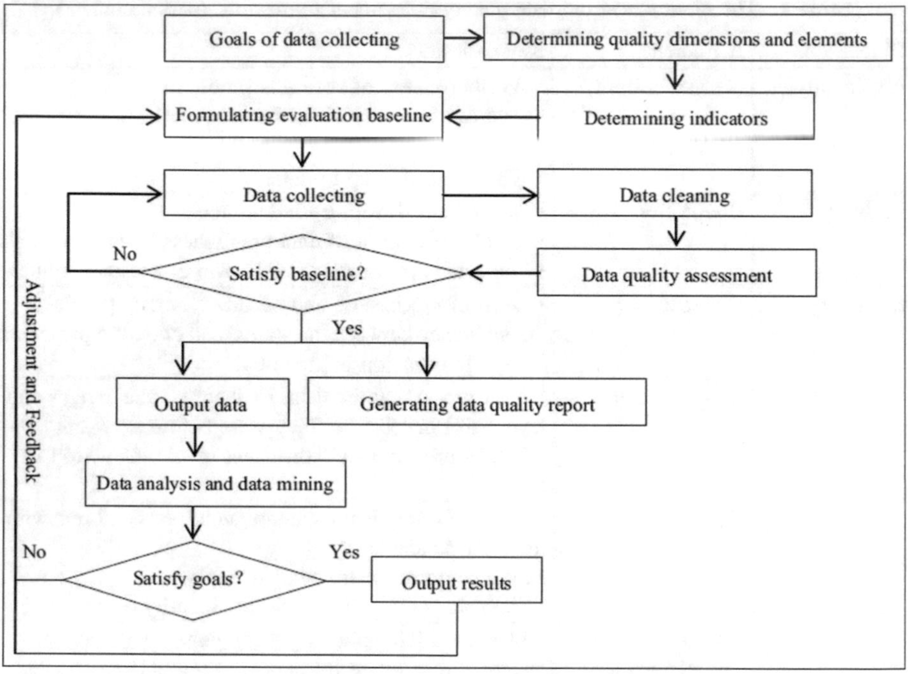

Fig. 3. Quality assessment process for big data [17]

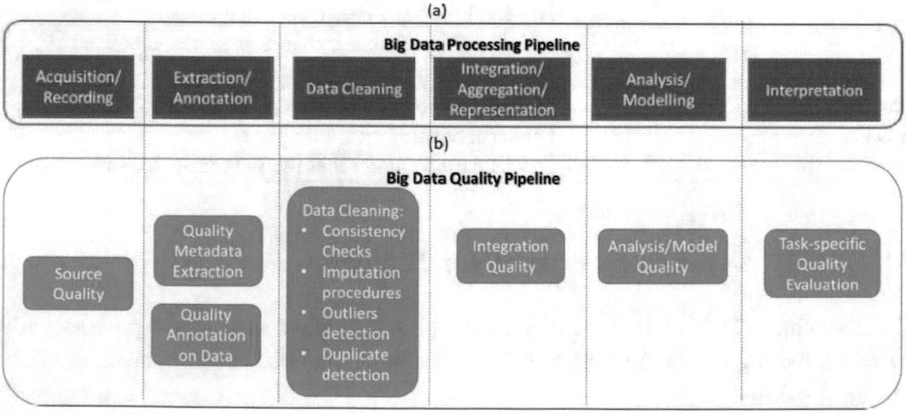

Fig. 4. Part (a) - Big data processing pipeline; Part (b) - Big data quality pipeline [18]

6 Big Data Quality Evaluation Scheme

Taleb et al. [20], applied sampling strategies on large data sets. For a fast quality evaluation, a representative population samples used to reduce the data size. Hence, data attributes, quality and profile can easily be defined. Every data source involved into Big Data generation, profiled and its quality is evaluated before any implication in the lifecycle processes of the Big Data. To reduce resources and computing time authors developed an algorithm depends on Bag of little bootstrap BLB. the concept of it is producing a small subscts of big data set without replacement and for each produced sample another set of samples generated by resampling with replacement. an efficient data quality evaluation is achieved with the aid of BLB sampling.

Figure 5 illustrating the big data quality evaluation scheme, it determines the quality of the data it passes through many modules which consisting of: (a) data sampling, and data profiling, (b) DQD vs attributes selection, (c) data quality Metric selection, (d) samples data quality evaluation.

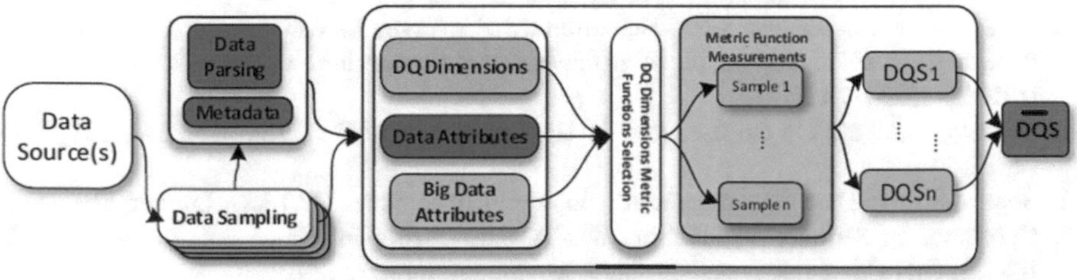

Fig. 5. Big data quality evaluation scheme [20]

7 Conclusion

The big data era leads to concerns on how to ensure the quality of big data and how to analyze and mine information and knowledge hidden behind it. Poor data quality will result in low data utilization efficiency and even bring serious decision-making mistakes. As a future work and based on all the work surveyed above and referring to [20] we suggest including more empirical evaluations and testing of the proposed solutions. Furthermore, an automatic optimizations and discovery of quality should be developed based on the results of data quality dimensions' evaluation. Moreover, a metric model for Big data to be used as a reference to be built in a context of data quality dimensions. In this paper, a review of existing literature on big data quality issues was conducted, proposed frameworks and assessment methodologies to enforce and increase big data quality is highlighted, as well as recommendation of future work for an automatic optimization and discovery of data results discussed previously. Also, this paper recommended a metric model for Big data to be used as a reference to be built in a context of data quality dimensions.

References

1. Meng, X.F., Ci, X.: Big data management: concepts, techniques and challenges. J. Comput. Res Dev. **50**(1), 146–169 (2013)
2. Gantz, J., Reinsel, D.: The digital universe in 2020: big data, bigger digital shadows, and biggest growth in the far east (2012)
3. Douglas, L.: 3D data management: controlling data volume, velocity and variety. Gart. Retriev. **6**(2001), 6 (2001)
4. Brown, B., Chui, M., Manyika, J.: Are you ready for the era of 'big data'. McKinsey Q. **4**(1), 24–35 (2011)
5. Wang, J., Li, H., Wang, Q.: Research on ISO 8000 series standards for data quality. Stand. Sci. **12**, 44–46 (2010)
6. Eckerson, W.W.: Data quality and the bottom line: achieving business success through a commitment to high quality data. The Data Warehousing Institute, pp. 1–36 (2002)
7. Kaisler, S. et al.: Advanced analytics for big data. In: Encyclopedia of Information Science and Technology, 3rd edn. IGI Global, pp. 7584–7593 (2015)
8. Russom, P.: Big data analytics. TDWI Best Pract. Rep. Fourth Quart. **19**(4), 1–34 (2011)
9. Chen, M. et al.: Survey on data quality. In 2012 World Congress on Information and Communication Technologies (WICT). IEEE (2012)
10. Laranjeiro, N., Soydemir, S.N. , Bernardino, J.: A survey on data quality: classifying poor data. In 2015 IEEE 21st Pacific Rim International Symposium on Dependable Computing (PRDC). IEEE (2015)
11. Firmani, D., et al.: On the meaningfulness of "big data quality". Data Sci. Eng. **1**(1), 6–20 (2016)
12. Soares, S.: Big Data Governance: An Emerging Imperative. Mc Press, London (2012)
13. Caballero, I., Serrano, M., Piattini, M.: A data quality in use model for big data. In: International Conference on Conceptual Modeling. Springer (2014)
14. Juddoo, S.: Overview of data quality challenges in the context of big data. In: 2015 International Conference on Computing, Communication and Security (ICCCS). IEEE (2015)
15. Pipino, L.L., Lee, Y.W., Wang, R.Y.: Data quality assessment. Commun. ACM **45**(4), 211–218 (2002)
16. Batini, C., Scannapieco, M.: Data and Information Quality: Dimensions, Principles and Techniques. Springer, Cham (2016)
17. Cai, L., Zhu, Y.: The challenges of data quality and data quality assessment in the big data era. Data Sci. J. **14** (2015)
18. Catarci, T. et al.: My (fair) big data. In: 2017 IEEE International Conference on Big Data (Big Data). IEEE (2017)
19. Bertino, E.: Big data-opportunities and challenges. IEEE, pp. 479–480 (2013)
20. Taleb, I. et al.: Big data quality: a quality dimensions evaluation. In: Ubiquitous Intelligence & Computing, Advanced and Trusted Computing, Scalable Computing and Communications, Cloud and Big Data Computing, Internet of People, and Smart World Congress (UIC/ATC/ScalCom/CBDCom/IoP/SmartWorld), 2016 International IEEE Conferences. IEEE (2016)

Comparative Study of Segmentation and Feature Extraction Method on Finger Movement

Nurazrin Mohd Esa(✉), Azlan Mohd Zain(✉), Mahadi Bahari(✉), and Suhaila Mohd Yusuf(✉)

Faculty of Computing, Universiti Teknologi Malaysia, Skudai, Johor, Malaysia
nurazrin_mohdesa@gmail.com,
{azlanmz, mahadi, suhailamy}@utm.my

Abstract. Myoelectric control prostheses fingers are a popular developing clinical option that offers an amputee person to control their artificial fingers by recognizing the contacting muscle residual informs of electromyography (EMG) signal. Lower performance of recognition system always has been the main problem in producing the efficient prostheses finger. This is due to the inefficiency of segmentation and feature extraction in EMG recognition system. This paper aims to compare the most used overlapping segmentation scheme and time domain feature extraction method in recognition system respectively. A literature review found that a combination of Hudgins and Root Mean Square (RMS) methods is a possible way of improving feature extraction. To proof this hypothesis, an experiment was conducted by using a dataset of ten finger movements that has been pre-processed. The performance measurement considered in this study is the classification accuracy. Based on the classification accuracy results for the three common overlapping segmentation schemes, the smaller the window size with larger increment windows produce better accuracy but it will degrade the computational time. For feature extraction, the proposed Hudgins with RMS feature showed an improvement of average accuracy for ten finger movements by 0.74 and 3 per cent compared to Hudgins and RMS alone. Future study should incorporate more advance classification accuracy to improve the study.

Keywords: Finger movement classification · EMG signal
Segmentation scheme · Feature extraction

1 Introduction

The loss of human lower arms is a noteworthy incapacity that significantly confines the ordinary capacities and connections of people with upper-limb amputation [1]. In Italy, the number of upper limb loss is estimated as 4000 people per year [2] while in a larger country such as the United States of America (USA) there are estimated 340000 people living with this situation [3]. To overcome the communication disability with this present reality, it can reestablished by utilizing myoelectric control [4, 5], where the electromyogram (EMG) signals created by the human muscles are utilized to infer control orders for controlled finger prostheses.

F. Saeed et al. (Eds.): IRICT 2018, AISC 843, pp. 117–127, 2019.
https://doi.org/10.1007/978-3-319-99007-1_12

Normally, an example of an acknowledgment structure is used to characterize the procured EMG signals into one predefined set of lower arm developments [4, 6]. To do this, the control for the finger prostheses system consists of several important stages which are pre-processing, segmentation, feature extraction and classification. Choosing a segmentation method is critical for achieving optimize data separations between each classes [7]. Zawawi et al. [8] stated that by applying the suitable window size will be able to provide accurate information in identifying the characteristics of each class EMG signal represent. However, there are only a few studies that have been conducted in relation to quantitative comparison performance of several used segmentation methods [9]. In fact, there is no standard or well-established segmentation technique that can be used for different case study [10] where it will degrade the computational time. This statement highlights the need for an optimized variable for segmentation methods. Other than determining the need in efficient EMG signal processing, the identification of the best optimized feature extraction methods [7, 11–13] which directly affected the classification accuracy is also important.

In this paper, three common disjoint windowing schemes used by [14–16] which are 200 millisecond (ms) and the increment segment is 50 ms, 100 ms with 25 ms increment segment and 256 ms with 128 ms respectively will be compared. These are the common disjoint windowing scheme used. The goal of this study is to identify the optimize windowing scheme to be employed in finger movement classification. In addition, the study also aims to compare and propose a new combination between the Hudgins feature extraction and root mean square (RMS) feature extraction. To proof this, a support vector machine (SVM) classification technique with tenfold cross validation will be employed to identify the classification accuracy of each experiment. The result of classification accuracy will then be analyzed to identify the optimized windowing the efficiency of the proposed combine feature extraction.

The structure of this paper is as follows: Sect. 2 described the material and methods starting from the EMG data acquisition procedure, the segmentation, feature extraction and classification. Section 3 described result of the experiment and a brief discussion and finally, conclusion will be done in Sect. 4.

2 Related Works

The concerning issues in myoelectric control prostheses fingers is the system performance. An efficient system should be able to efficiently perform the essential movements in terms of both movement classification accuracy and computational time [7]. Based on [7], the myoelectric control system consists of several stages starting from the acquisition of data from the electrode patch on skin in human arms, data segmentation, feature extraction and classification. Figure 1 shows the complete process of myoelectric control of hand prostheses. However, some researcher [17] highlighted that myoelectric control system consists of two main stages which are feature extraction and classification. Which they believes that the segmentation process is mandatory include in feature extraction.

Data segmentation or also known as windowing is a process of dividing up the EMG signal into windows while feature extraction is a process on extracting the representative data from each window. Lastly, classification is a process of classifying the data into a class of data. This step is required in order to extract features from raw, instantaneous sample of EMG. The defect of using raw EMG signal directly is increases the computational time because it contains relatively little information about muscle activity with the large size of dataset. Hence, windowing or smoothing must be performed as wells.

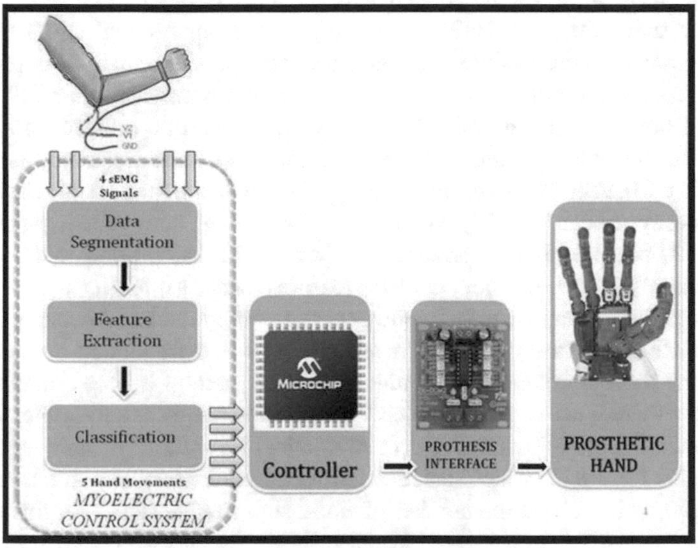

Fig. 1. The process of myoelectric control of hand prostheses [7]

The efficiency of the myoelectric control system required an optimized segmentation process. The main criteria of an optimized windowing scheme should be a good trade-off between the controller delays and classification error [13]. There are two common type of segmentation which are disjoint and overlapping segmentation. In Sujian and Vuskovic [10] and Khusaba et al. [18] they stressed that overlapping segmentation techniques is a proven technique where the controller were four times faster without noticeable degradation in accuracy compared to dis-joint segmentation techniques. Hence, in this study, the overlapping segmentation is implemented. In the overlapped segmentation, window size (WS) and window interval (WI) is the most critical parameter. WS represents the width of the analysis windows while WI variable represents the time interval between the neighboring segments. Generally, the basic guideline of the selection of WS is because identified size of segments cannot be too short which will lead to bias feature estimation, however if it's too long, it will likely failed in real time due to high computational operations. For WI, the length needs to be larger than the processing time but it must be less than WS.

Tang et al. [13] also stated that the WS and processing time should be less than 300 ms in total in order to allow the subject to feel any delay error. For hand prostheses case, the window duration should be acceptable within time delays between user command and prostheses action. The determination of the optimized windowing is also mentioned by Khusaba et al. [18] where they compared three disjointed windowing size 125, 128 and 64 ms. The result shows that 128 and 64 ms segment provides better than 125 ms. However, it is widely known that, the overlapping segmentation has shown better performance compares to disjoint segmentation. Hence, the overlapping windowing scheme will be investigated further in this study.

Another issue in finger movement classification is the feature extraction stage. Feature ex-traction refers to the transformation of the input signal into a set of representative signal features. Feature extraction process was consequently implemented in order to condense the only important EMG parts represent each class [19, 20]. Generally, the features in the analysis of the EMG signal can be divided into three main group. There are time domain, frequency domain, and time-frequency or time-scale representative [3]. Zardoshti-Kermani et al. [20] evaluated a number of features that are now commonly used for EMG classification, including: zero crossing (ZC), autoregressive (AR) coefficient and Variance. Saikia et al. [21] also compared several times domain features and frequency domain features such as RMS, waveform length (WL) and modified mean frequency, power and modified median frequency. The result shows that RMS has a competitive result compared to frequency domain features. However, the most common used feature extraction method is time domain (TD) feature set. This feature set is simple which leads to quick and easy to implement as they can be extracted directly based on the mathematical formula. The most regular feature is the Hudgins feature [5] which includes four combinations of TD - the mean absolute value (MAV), WL, ZC and the number of slope sign change (SSC). In this study, only the time domain group is considered because they are quick and easy to implemented and calculated directly from the EMG time series.

Higher accuracy, less number of computational time and low storage space are the characteristics that measure the efficiency of the recognition system. Nowadays, there are many classifiers that is capable of providing the highest efficiency to the system. Amongst the classifier, SVM fulfills almost all of criteria of the effective classifier. A study by Leon et al. [22] shows that SVM is superior in terms of classification in recognizing multiple hand motion compared to other algorithms. Research conducted by [22] and [6] confirmed the effectiveness of EMG pattern recognition using SVM methods. Hence, in terms of movement recognition, SVM had proven to be a powerful classification tool [23].

3 Material and Method

Figure 2 shows the overall process of this study, in which the myoelectric control system starts from the acquisition of EMG signal from human arms followed by data segmentation. In this study, three widely use overlapping segmentation is implemented and the analysis is based on their classification accuracy of each class movement. Next, a comparison of three feature extraction methods is conducted which involves the

Hudgins method, RMS and the proposed combination of Hudgins and RMS. Then, this extracted data is classified using one versus all Support Vector Machine (OVA-SVM). Later on the result will be analyzed leading to the EMG-based movement classification.

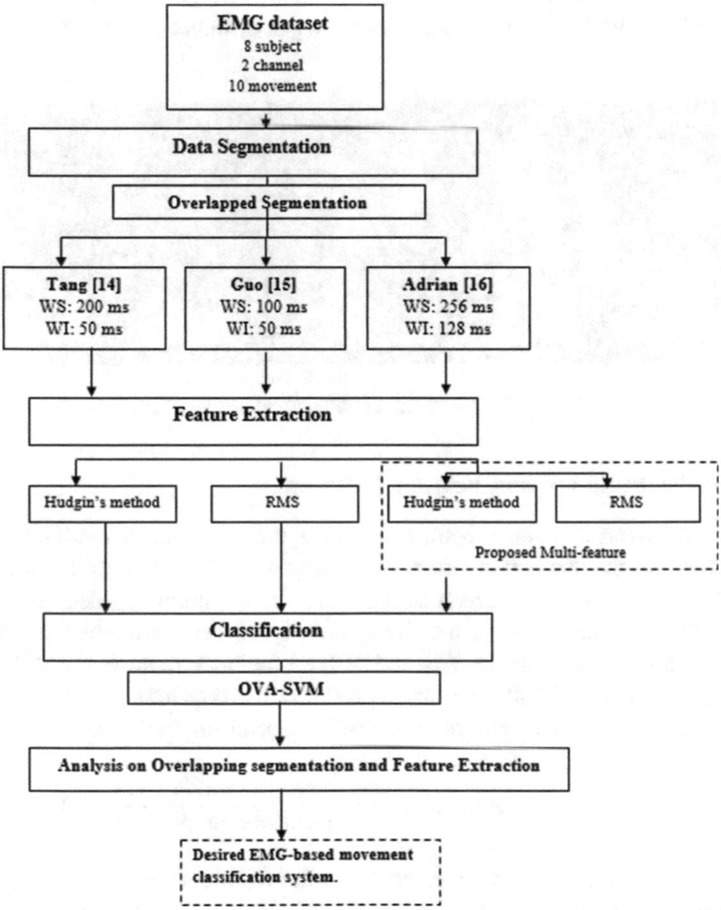

Fig. 2. The overall process of conducting this study

3.1 Data Collection

The data used in this study were acquired from Khusaba et al. [18]. The data was gathered from utilizing the two EMG sensor channel (Delsys DE 2.x series EMG sensors) that was positioned on the wrist of each subject. A total of eight subjects comprising all normally limbed consisting of two (2) females and six (6) males. The subjects are aged between twenty (20) to thirty fives (35) years old. During the data collection, all the subjects are seated with their arm fixed in one position to avoid inference noise in signal collected. After the signal was collected, a Bagnoli Desk-top EMG system is used to process the signal followed by the amplification process using the Delsys Bagnoli-8 amplifier. Then, sampling the signal which was processed at 4000 Hz was done using a 12-bit analog-to-digital converter. The identified ten single and pinch movement are hand close (HC), index (I), little (L), middle (M), thumb (T), ring finger (R) and the pinch movement from a combination of thumb and index (T-I), thumb and little (T-L), thumb and middle (T-M) and thumb and ring fingers (T-R). Figure 3 shows the ten types images of finger movement.

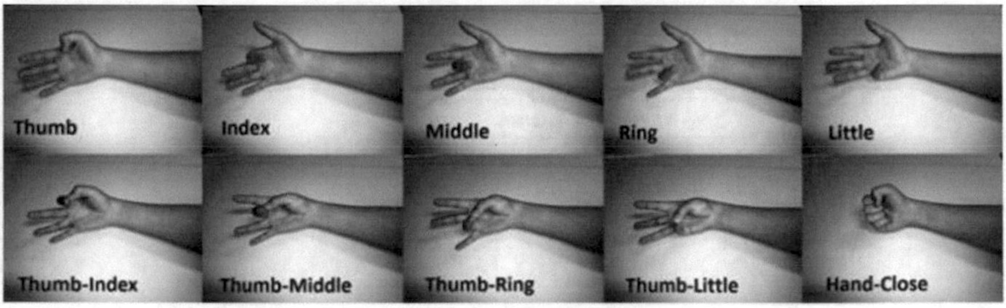

Fig. 3. Types of finger movement studied [18]

3.2 Data Segmentation and Feature Extraction

After the acquisition of data is completed, each motion channel matrix is processed independently. Since the study aims to compare the common used segmentation method by [14–16], the analyzed matrix is split into disjoint segmentation with this three WS and WZ. Number of each training sample for these three methods is different because of the differentiation in WS and WI. Equation 1 represents the formula on calculating the number of training sample produce. The data length for this study is 20, 000 while window size and window interval is depend on methods used.

$$\text{No. of training samples} = \frac{\text{data length} - \text{window size}}{\text{windows interval}} + 1; \quad (1)$$

Once, the signal data has been segmented, the feature were extracted from a time domain feature set which consist of mean absolute value (MAV), (WL), (ZC), (SSC) and RMS. All features were extracted by using myoelectric toolbox provided by Xie et al. [15]. The description of each feature set uses are described briefly in Table 1.

Table 1. Description and mathematical formulation for time domain feature extraction

Feature extraction	Description	Mathematical formulation
MAV	MAV is the average of the absolute value of the EMG signal amplitude in a sample of data	$MAV = \frac{1}{N}\sum_{i=1}^{N} \lvert x_i \rvert$

(*continued*)

Table 1. (*continued*)

Feature extraction	Description	Mathematical formulation
WL	Waveform length (WL) in time domain features represents cumulative length of the EMG waveform over the time segment. It usually uses to identify the complexity of EMG signal sequence [6, 9]. Wavelength (WAVE) is another representative name for this feature	$WL = \sum_{i=1}^{N} \lvert x_{i+1} - x_i \rvert$
SSC	Slope sign change (SSC), is a method of EMG signal represent frequency information [9, 20]. The different with ZC is the value represent cumulative numbers of times that slope of the dataset change sign either negative to positive value or conversely	$SSC = \sum_{i=1}^{N-1} [f(x_i \times x_{i+1}) \cap \lvert x_i - x_{i+1} \rvert \gg threshold];$ $f(x) = \begin{cases} 1, \ if\ x \geq threshold \\ 0, \ otherwise \end{cases}$
ZC	Zero crossing (ZC) in time domain defines as the measure of frequency information of the EMG signal (e.g. [9, 20]). The value represents number of times that the amplitude of EMG signal passes the zero amplitude axes	$ZC = \sum_{i=1}^{N-1} [\mathrm{sgn}(x_i \times x_{i+1}) \cap \lvert x_i - x_{i+1} \rvert \geq threshold];$ $sgn(x) = \begin{cases} 1, & if\ x \geq threshold \\ 0, & otherwise \end{cases}$
RMS	RMS is quite similar t standard deviation where amplitude modulated model related to non-fatiguing and constant force	$RMS = \sqrt{\frac{1}{N}\sum_{i=1}^{N} x_i^2}$

3.3 Classification Using Support Vector Machine (SVM)

SVM is kernel based classification that the theory is from structural risk management (SRM) idea [23]. SVM also well known as kernel based classifier because the kernel function will define the on how to design the hyper-plane. The kernels function of SVM also provides the ability of classify data in non-linear separation without increasing calculation cost significantly. Even, the basic SVM only support binary classification, these kernel function has been extended to support multiclass classification with combining these binary classification ideas. Then, SVM has been expanded on its functionality by producing One-versus-one (OVO) methods and one-versus-all (OVA) to support multiclass classification problem. In this study, the OVA-SVM with 10 fold cross validation is implemented. For each sample, one per ten is used as training sample and the rest as testing using SVM.

4 Result and Discussion

Table 2 shows the comparison of the three segmentation methods based on the classification accuracy. It can be observed that the classification accuracy value calculated for each of ten single and combined pinch finger movements starting from hand close to thumb ring movement were done. SVM with ten-fold cross validation was used to guarantee an exact performance measure for this dataset. The most important parameter in windowing scheme is choosing the length and increment variable value. The increment variable represents the time interval between the neighbour segments. The guideline on choosing the time interval is where the length needs to be larger than the processing time but must be less than the segment length.

Table 2. Comparison of segmentation method based on classification accuracy.

Class/method	Tang [13] WS::200 ms WI: 50 ms	Guo [14] WS: 100 ms WI: 50 ms	Adrian [15] WS: 256 ms WI: 128 ms
No. of training sample	397	399	**155**
Hand close	**0.9667**	**0.9667**	**0.9667**
Index	**0.9**	**0.9**	0.8667
Little	0.9667	0.9	**0.9667**
Middle	**0.9667**	**0.9667**	0.93
Ring	**0.9667**	**0.9667**	0.83
Thumb	**0.9667**	**0.9667**	**0.9667**
Thumb-index	0.9333	**0.9667**	0.8667
Thumb-little	**0.9667**	0.9333	0.8667
Thumb-middle	**1**	**1**	0.9
Thumb-ring	**1**	**1**	0.9667
Average	**0.9634**	0.9569	0.9127

Here the overall classification accuracy of segmentation scheme used by Tang [13] is higher compared to windowing scheme used by Guo [14] and Adrian [15] throughout all movements. However, the fragment used is two times bigger than Adrian [15] windows, where the bigger the fragment used for classification will degrade the computation times. As mentioned by Khusaba et al. [18] for overlapping windowing scheme which produces better performance when comparing the classification accuracy, but if the data is too large it will lead to higher computational cost in the training and testing phase. Table 3 shows the result of time processing for segmentation process. Hence, based on the result below, considering the accuracy and processing time, the Adrian [15] which used the lower time is used in this study.

To identify the optimize feature for finger movement classification an experiment on the comparison of the three feature extraction technique was conducted. The first single robust feature is RMS that shows the best feature that has resulted in group of higher-order time and frequency domain by [15]. The second feature is the most popular combination of time domain feature set which consists of MAV, WL, SSC and ZC [5]. Lastly, is the new combination of RMS and this Hudgins feature sets. Based on the obtained classification accuracy in Fig. 4, the proposed combination of Hudgins feature with RMS improved the classification accuracy by 0.74 and 3% compared to using Hudgins and RMS only respectively. This shows that, this combination can helps to separate each class of finger movement better than the others. Hence, for this study the combination of RMS and Hudgins will be used. Although identifying the optimized features important but there are also other factor that affect the performance such

Table 3. Comparison of the feature extraction method based on average classification accuracy

Method	Time for segmentation (s)
Tang [13]	0.124
Guo [14]	0.127
Adrian [15]	0.94

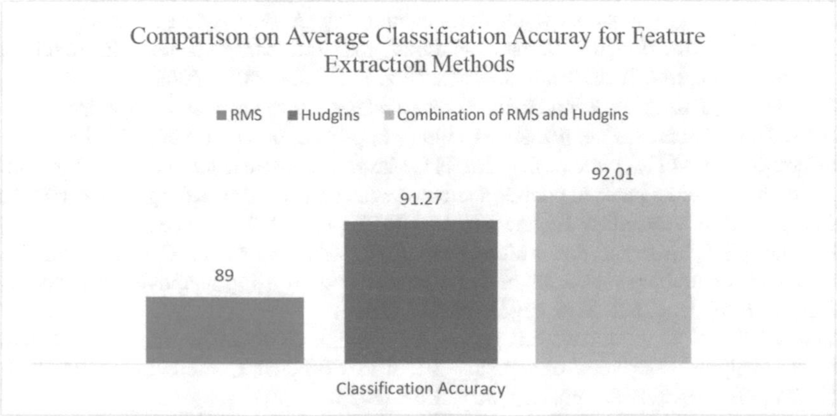

Fig. 4. Graft on comparison on average classification accuracy for feature extraction method

sampling rates of dataset [24] which also contribute to the performance of features. Hence, for the futures, this ‘result will combined with optimized sampling rates to produce better performance.

5 Conclusion

Overall, it can be said that the windowing scheme plays an important part in the classification accuracy of finger movement recognition. The choice of windowing scheme needs to be considered carefully before it’s implemented in the study. Based on the experiment, the bigger the fragment extracted for each case study will lead to better classification accuracy. However, the bigger fragment will degrade the computational tie which is also another important performance of the recognition system. This case is similar to the feature extraction method in which the proposed combination of Hudgins feature with RMS improved the classification accuracy by 0.74 and 3% compared to using Hudgins and RMS only respectively. The classified signal can be efficiently used to design a prosthetic hand to help the people with disabilities which to interact with computer devices and the outside world. Further study can improve this study by incorporating more advance classification accuracy.

Acknowledgements. The author greatly acknowledge the research management center, UTM for financial support through the research university grant scheme (RUG) Vot No Q.J 13000.2528.18H53.

References

1. Kuiken, T.A., Li, G., Lock, B.A., Lipschutz, R.D., Miller, L.A., Stubblefield, K.A., Englehart, K.B.: Targeted muscle reinnervation for real-time myoelectric control of multifunction artificial arms. JAMA **301**(6), 619–628 (2009)
2. Cutti, A.G., Verni, G., Gruppioni, E., Amoresano, A.: Nuove tecnologie e innovazione nelle protesi di arto superiore. Med. Chir. Ortop. **3**, 20–33 (2012)
3. NLLIC Homepage. http://www.amputee-coalition.org. Accessed 23 May 2018
4. Englehart, Kevin, Hudgins, Bernard: A robust, real-time control scheme for multifunction myoelectric control. IEEE Trans. Biomed. Eng. **50**(7), 848–854 (2003)
5. Hudgins, B., Parker, P., Scott, R.N.: A new strategy for multifunction myoelectric control. IEEE Trans. Biomed. Eng. **40**, 82–94 (1993). https://doi.org/10.1109/10.204774
6. Firoozabadi, S.M.P., Oskoei, M.A., Hu, H.: A human–computer interface based on forehead multi-channel bio-signals to control a virtual wheelchair. In: Proceedings of the 14th Iranian Conference on Biomedical Engineering (ICBME), pp. 272–277 (2008)
7. Fariman, H.J., Ahmad, S.A., Marhaban, M.H., Ghasab, M.A., Chappell, P.H.: Hand movements classification for myoelectric control system using adaptive resonance theory. Austral. Phys. Eng. Sci. Med. **39**(1), 85–102 (2016)
8. Zawawi, T.N.S.T., Abdullah, A.R., Shair, E.F., Halim, I., Rawaida, O.: Electromyography signal analysis using spectrogram. In: 2013 IEEE Student Conference on Research and Development (SCOReD), pp. 319–324. IEEE (2013)
9. Reaz, M.B.I., Hussain, M.S., Mohd-Yasin, F.: Techniques of EMG signal analysis: detection, processing, classification and applications. Biol. Proced. Online **8**(1), 11 (2006)

10. Du, S., Vuskovic, M.: Temporal vs. spectral approach to feature extraction from prehensile EMG signals. In: Proceedings of the 2004 IEEE International Conference on Information Reuse and Integration, IRI 2004, pp. 344–350. IEEE (2004)
11. Hargrove, L.J., Englehart, K., Hudgins, B.: A comparison of surface and intramuscular myoelectric signal classification. IEEE Trans. Biomed. Eng. **54**(5), 847–853 (2007)
12. Veer, K., Sharma, T.: A novel feature extraction for robust EMG pattern recognition. J. Med. Eng. Technol. **40**(4), 149–154 (2016)
13. Zecca, M., Micera, S., Carrozza, M.C., Dario, P.: Control of multifunctional prosthetic hands by processing the electromyographic signal. Crit. Rev. Biomed. Eng. **30**(4–6), 459 (2002)
14. Tang, Z., Zhang, K., Sun, S., Gao, Z., Zhang, L., Yang, Z.: An upper-limb power-assist exoskeleton using proportional myoelectric control. Sensors **14**(4), 6677–6694 (2014)
15. Xie, H.-B., Zheng, Y.-P., Guo, J.-Y.: Classification of the mechanomyogram signal using a wavelet packet transform and singular value decomposition for multifunction prosthesis control. Physiol. Meas. **30**(5), 441 (2009)
16. Chan, A.D.C., Green, G.C.: Myoelectric control development toolbox. In: 30th Conference of the Canadian Medical & Biological Engineering Society, Toronto, Canada, M0100 (2007)
17. Pancholi, S., Joshi, A.M.: Portable EMG data acquisition module for upper limb prosthesis application. IEEE Sens. J. (2018)
18. Khushaba, R.N., Takruri, M., Kodagoda, S., Dissanayake, G.: Toward improved control of prosthetic fingers using surface electromyogram (EMG) signals. Expert Syst. Appl. **39**(12), 10731–10738 (2012)
19. Boostani, R., Moradi, M.H.: Evaluation of the forearm EMG signal features for the control of a prosthetic hand. Physiol. Meas. **24**(2), 309 (2009)
20. Zardoshti-Kermani, M., Wheeler, B.C., Badie, K., Hashemi, R.M.: EMG feature evaluation for movement control of upper extremity prostheses. IEEE Trans. Rehabil. Eng. **3**(4), 324–333 (1995)
21. Saikia, A., Mazumdar, S., Sahai, N., Paul, S., Bhatia, D.: Comparative study and feature extraction of the muscle activity patterns in healthy subjects. In: 2016 3rd International Conference on Signal Processing and Integrated Networks (SPIN), pp. 147–151. IEEE (2015)
22. Leon, M., Gutierrez, J.M., Leija, L., Munoz, R:. EMG pattern recognition using support vector machines classifier for myoelectric control purposes. In: 2011 Pan American Health Care Exchanges (PAHCE), pp. 175–178. IEEE (2011)
23. Cristianini, N., Shawe-Taylor, J.: An Introduction to Support Vector Machines and Other Kernel-Based Learning Methods. Cambridge University Press, Cambridge (2000)
24. Phinyomark, A., Scheme, E.: A feature extraction issue for myoelectric control based on wearable EMG sensors. In: 2018 IEEE Sensors Applications Symposium (SAS), pp. 1–6. IEEE (2018)

Realizing the Value of Big Data in Process Monitoring and Control: Current Issues and Opportunities

Saddaf Rubab[1,4(✉)], Syed A. Taqvi[2,3], and Mohd Fadzil Hassan[1]

[1] Department of Computer and Information Sciences, Universiti Teknologi PETRONAS, 32610 Seri Iskandar, Perak Darul Ridzuan, Malaysia
{saddaf_g02754,mfadzil_hassan}@utp.edu.my

[2] Chemical Engineering Department, Universiti Teknologi PETRONAS, 32610 Seri Iskandar, Perak Darul Ridzuan, Malaysia
syed.ali_g03168@utp.edu.my

[3] Department of Chemical Engineering, N.E.D. University of Engineering and Technology, Karachi 75270, Pakistan

[4] National University of Sciences and Technology (NUST), Islamabad, Pakistan

Abstract. With the advancement of Big Data, data analytics have benefited many organizations and industries. The continuous improvement of process industries is a challenging task that requires insights for efficient process monitoring with minimum downtime and safer operations. These industries generate a large amount of data i.e. "Big Data" every second that contains useful information. For better control, different analysis can be executed using statistical methods, data mining or machine learning. This data can be used for daily process operations and decision making. However, the past literature on Big Data have put limited focus on realizing its implication in process monitoring and control. The paper emphasizes on the issues and opportunities faced by process industries in terms of Big Data collection, storage and analysis for process monitoring and control.

Keywords: Advanced process control · Big Data Analytics · Decision making Data mining · Machine learning · Process monitoring and control

1 Introduction

Technology advancements and increase in the complexities of process industries have resulted in challengeable control to attain the desired product quality with minimum downtime [1–3]. Among the most important factors in process industries is the efficient process monitoring to produce the quality product with minimum downtime and maximum profit. This has great prospect in understanding the hidden correlation and statistical dependencies of the process. The data analysis and its utilization using various statistical and machine learning tools can lead towards better decision making. A modern and complex chemical process plant or refinery equipped with an advanced control system generates a large amount of data every single minute. The process industries have combined the science, statistical control, process modeling and

F. Saeed et al. (Eds.): IRICT 2018, AISC 843, pp. 128–138, 2019.
https://doi.org/10.1007/978-3-319-99007-1_13

engineering modeling to grow Advanced Process Control (APC) that is beneficial for process monitoring [1–8]. The rapid growth in Big Data Engineering (BDE) helped to assist the advancement in process control.

For few years, process industries have improved the productivity, downtime, and supply chain visibility by utilizing Big Data. The potential benefits of Big Data for efficient process monitoring, control and predictions are increasing day by day. However, the collection, storage and usage of this data is still a challenge for every upstream and downstream industry [3]. Once the data has been captured from different sensors in process industries, the next task is to store it and utilize for making logical decision for economic and business activities. The data can be obtained and combine from multiple sources to build a self-learning model for the enhancement in decision making process. The Big Data analysis in process industries has numerous advantages by increasing the production quality, efficiency and reducing the cost and time.

This technological advancement develops an interest in fourth Industrial Revolution (IR 4.0), which relates advances in the Information and Communication Technology (ICT). The recent development of soft computing, machine learning, and data mining explains the application of Big Data obtained from process industries [2, 3, 9]. The application of Big Data covers many sectors such as Oil and gas industries, chemical process plants, financial services, supply chain management, manufacturing plants etc. The implication of Big Data can be further extended in different areas such as detection of defective products in manufacturing plants, supply planning, increasing energy efficiency, carbon emission forecasting, design and simulation of the new manufacturing process, and process monitoring [1–8].

The process industries are generating a large amount of data (volume) from different sources such as sensors (variety) and selection of data from decision making (analytics) is also essential in real-time process monitoring (velocity) [2, 10]. Figure 1 shows the flow and utilization of this data for process monitoring, analytics and predictive analytics for efficient monitoring and control to make smarter industries.

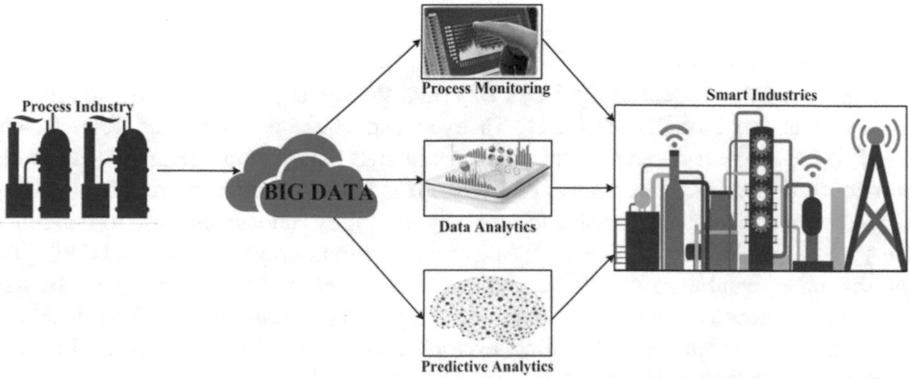

Fig. 1. Application of Big Data in process industries

The rest of the paper is organized as follows. Section 2 discusses Big Data Engineering and Analytics as well as highlight the current techniques/methods adopted in process industries. Section 3 summarizes the challenges and opportunities faced by APC Engineers in utilizing Big Data for process monitoring and control. Finally, Sect. 4 concludes the paper.

2 Big Data Engineering and Analytics

Understanding the details of Big Data require to substantiate between two of its complementary aspects, i.e., Big Data Engineering (BDE) and Big Data Analytics (BDA). Following sections explain BDE and BDA along with the thorough literature review in the process industry.

2.1 Big Data Engineering (BDE)

Big Data Engineering (BDE) make available essentials to support Big Data Analytics (BDA) by providing the storage and processing activities. Hadoop [11], NoSQL, Hive and Berkeley Data Analytics Stack are few examples of BDE technologies commonly employed in processing and storage of data. As the data is growing, the existing Big Data platforms need to fulfill the data processing requirements. Scaling is incorporated into many Big Data platforms to handle this phenomenon. Following are the two groups of Big Data platforms with wide-ranging characteristics to perform scaling:

- Vertical Platforms – installing more number of processors or memory within a server to upgrade the hardware. It scales up a single server by running only one instance of an operating system on upgraded hardware.
- Horizontal Platforms – distributing the processing workload to other servers or scale out by adding more number of servers. Horizontal scaling can also be achieved by running multiple instances of an operating system on a single server. To overcome the limitations researchers has contributed towards developing horizontal scaling tools such as Spark.

2.1.1 Big Data Storage

Storage as a prime aspect of the BDE, Big Data storage is provided by the distributed file systems and NoSQL databases. Tachyon and Hadoop Distributed File System (HDFS) [12, 13] are two commonly used distributed file systems. Tachyon [13] is the component lies at the lowest level of BDAS. It is fault tolerant and provides memory speeds across the cluster. It stores the recently used files and caches them to maximize the speed as well as relying less on the disk/memory for storage workloads. HDFS [12] on the other hand manages large datasets using commodity servers. Due to this potential chances of hardware failures are higher. Therefore, HDFS should exhibit better fault tolerance by distributing or replicating data to other servers. Tachyon provides the following features transcending HDFS performance:

- Run MapReduce and Spark codes without any modifications
- In-memory speed for better performance

Relational databases were considered as de facto of management systems before the introduction of NoSQL [14, 15]. The database systems require higher flexibility, scalability and performance with the recent advancements of distributed applications. Relational databases are incompatible to deal with the emerging needs of such applications, which paved the way for the Not only SQL (NoSQL) systems. NoSQL works with non-relational database systems to store schema-less data.

2.1.2 Big Data Processing

The principal of BDE is to handle the data and computations in parallel as well as in distributed ways. In the recent past, many processing models are introduced such as MapReduce (MR) and Directed Acyclic Graphs (DAG). MR processing model writes analytical data tasks in two functions, i.e., map and reduce. These functions are then submitted to separate processors, mappers and reducers. Mappers are responsible for reading data and then processing it to generate immediate outputs. These outputs are then processed by reducers to provide final results. MR processes the tasks in batch, therefore multiple mappers and reducers can work together on a single MR program. MapReduce is an incredible technology, but there are limitations with it, the serialization as a primary one. Tez [16] uses the concept of a DAG to enforce concurrency and serialization between MapReduce tasks.

2.2 Big Data Analytics (BDA)

The concept of Big Data Analytics (BDA) has been used traditionally in many disciplines such as statistics, data mining, predictive analytics, knowledge discovery, and data science. BDA encompasses processes, methods, and techniques to understand the Big Data. Due to the rapid growth of data, attempts are made to enhance the BDA capabilities to process a large amount of varied data and draw conclusions. Statistical methods deal with the collection, analysis and make decisions from the observed data using the right choice of tools. Massive amount of data collected from process industries can only be useful if it can detect the faults early and help maximize the productivity. Few open-source tools for Big Data statistical analysis are R, SAS Sentiment Analysis and NodeXL [17, 18]. Data mining analyze data to discover the meaningful rules and patterns. Data mining count on the database activities to access, order and group the queries drawn from data provided. Data warehouse supports these databases to collate the data using the Extract, Transform and Load (ETL) process. Frequently used tools for mining Big Data are Weka, RapidMiner, Orange and DataMelt [19, 20]. Machine Learning (ML) classified under Artificial Intelligence (AI), concentrate on computational system learning through data about an explicit task. ML has diverse advantages in process industries such as early detection of faults, predicting the outcomes and estimating the process throughput. Nora, Weka, H_2O.ai and Spark MLbase [10, 21] are few open-source tools available for BDA using machine learning algorithms.

The Big Data analytics plays a key role in the process monitoring, fault detection, fault diagnosis, fault classification, soft sensors, and prediction performance etc. Various statistical, data mining and machine learning techniques have been utilized in process industries for monitoring such as Principal Component Analysis (PCA) Discriminant

Partial Least Square (DPLS), K- nearest neighbor (KNN), Independent Component Analysis (ICA) and Fisher Discriminant Analysis (FDA). For fault detection and classification, Artificial Neural Network (ANN), Bayesian Networks (BN) and Support Vector Machines (SVM) have been used extensively. Table 1 illustrates the brief description of various data analytics methods/techniques applied in process monitoring and control from the year 2010 to 2018 only.

Table 1. Big Data Analytics in process monitoring and control- (2010–2018)

Method/Technique	Objective	Reference
Bayesian	Fault detection in chillers	[9, 22]
Principal component analysis (PCA)	Robust optimization, Outlier detection, fault diagnosis	[23–25]
Independent Component Analysis (ICA)	Process monitoring, fault diagnosis	[26, 27]
Kernel Density Estimation (KDE)	Process monitoring	[4, 6]
K-Means	Fault detection, process monitoring of offshore pipeline	[7, 28, 29]
Self-Organizing Map (SOM)	Process monitoring of industrial wastewater treatment, fault diagnosis	[30–32]
Gaussian	Manufacturing process monitoring, fault diagnosis of complex process	[8, 33]
Support Vector Machine (SVM)	Quality prediction, fault classification	[34, 35]
Principal component regression (PCR)	Process monitoring, prediction	[36, 37]
Partial least squares (PLS) regression	Quality prediction, process monitoring, fault diagnosis	[38–40]
Fisher Discriminant Analysis (FDA)	Fault diagnosis	[41, 42]
Multivariate linear regression	Water quality predictions, process monitoring/profiling	[43, 44]
Artificial neural networks (ANN)	Process and condition monitoring, fault diagnosis	[45–49]
Nearest Neighbors	Fault detection of manufacturing process	[5, 50]
Decision Trees	Fault diagnosis, online process monitoring	[51, 52]
Random Forest	Process quality prediction, process control	[53, 54]
KDE-BN	Process monitoring	[4]
Gaussian-BN	Manufacturing process monitoring	[33]
SVM-PCA	Fault classification	[35]

3 Challenges, Issues and Solutions

3.1 Challenges Faced by APC Engineers

The APC systems are costly and difficult to develop. As mentioned in Table 1, the application of BDA in APC systems is limited to complex and large-scale process plants [4, 23, 54]. Process industries generate a large volume of data however, the data collection and management capabilities are not enhanced appropriately. The quality of the data is also an important factor since a large volume of data does not guarantee a better output [34, 35]. The main issues reported by various researchers as mentioned in Table 1 are the quality and availability of data used for process monitoring because it requires ample time to clean and remove unnecessary features from the noisy data. Conversely, the small amount of data obtained by online or offline historian archives is used for process monitoring [23, 45]. However, to find any historical event in the specific archives for process building requires time and cost. The process historian is essential for data storage and to connect with the real-time process but searching mechanism and annotation ability is not effective. In attempt to solve these issues, many engineers utilized the excel spreadsheets but it has also found to be time consuming and limited to specific functions [36, 37]. Initially, the data collection was limited to important process parameters such as temperature, pressure and flows. Nowadays, in modern oil refinery or gas processing plant collect tens of thousands data from the process measurements, valve positioning, disturbances, manipulated and control variables after every second. This has increased the complexities of data collection and storage which complicates the job of engineers to analyze this flooded data, and to extract the useful. The issues and challenges faced by process industries are associated with BDA and its applications. Several challenges for process industries can be solved by proposing solutions of common attributes offered by Big Data applications. Key challenges are presented in Fig. 2 which can be linked to BDA. Therefore, big data can be helpful in resolving major issues in process plants such as:

1. Control performance measurement
2. Detection, identification and isolation of faults and failures
3. Detection and correction of model plant mismatch
4. Control valve stiction, detection and quantification
5. Identifying the correct control strategy in the process plant

3.2 Big Data Analytics: A Solution for Optimal Control

The complex automation processes have various demanding jobs which play a vital role in the economic operation and process monitoring. These include troubleshooting, identifying the potential failure of the components for process monitoring, fault detection and diagnosis as well as model plant mismatch. The overall performance and efficiency of the process can be enhanced by understanding data features and operational intelligence. Mostly the process engineers require time series data to incorporate process parameters quickly and efficiently. The combination of time-series data and data annotation will help to understand the accurate process behavior and to predict the

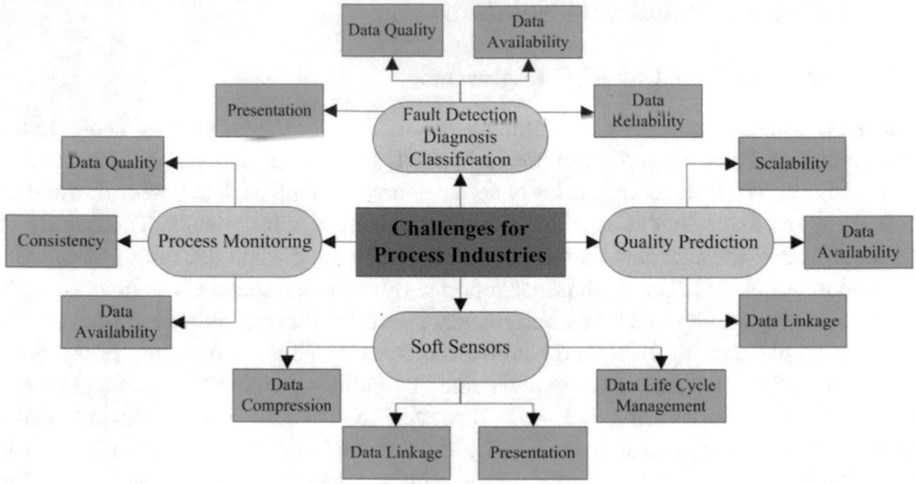

Fig. 2. Challenges for Big Data in process industries

most likely scenario in the continuous and batch operations. The plant data enables the APC engineers to perform predictive analytics which provide the valuable insights of the process.

The use of BDA has potential to interpret and visualize the process data effectively. Hence it is imperative to use BDA to drive process plant economically and target the improvement in a shorter time. Therefore, it is essential to develop a system which is different from the traditional approaches with the following features:

1. A system can be developed based on BDA techniques and deep knowledge of the process which does not only minimize the efforts of data scientist but also lessen the intensive engineering modeling. The data from the plant will help to gain significant knowledge by converting human intelligence into machine intelligence.
2. A model-free, diagnostics and prognostics approach that utilized the available historian data architecture.
3. A plug and play system that supports cost-effective virtualization and cloud security and a have ability to grow into a corporate of Big Data environment.

3.3 Decision Making in Process Industries

The Big Data has an ability to provide significant information and knowledge to make valuable and crucial decision at process plant. The intelligent phase is known as the first phase of decision making in which the data is collected from different sources for the identification of problems and prospects. The next phase is known as design phase which incorporates the planning, data analytics and analyzing. The choice phase is the next phase in which different methods are used for the evaluation of proposed solution from the design phase. The final phase is known as implementation phase in which the proposed solution is implemented. The importance of Big Data has been growing significantly for decision makers in process plant since years. As discussed earlier, the large amount of data obtained from process plants has a potential to provide remarkable benefits to the organization. This time-series data can be properly analyzed to get valuable insights and helps to take decisions in the field of failure detection and diagnosis, production processes, supply chain and development of prognostics systems. Therefore, the industrial sectors are interested in the utilization of Big Data in order to unlock economic value and faster decision making. The huge amount of data can be used for the prediction of unseen patterns and customer intelligence.

4 Conclusion

This paper summarizes the challenges and opportunities faced by APC engineers in process industries. For efficient process monitoring and control, data analytics is the essential tool in all process industries. However, the applications of Big Data in advanced process control, as well as business operations, is growing every day. As mentioned earlier, the process operations contain tons of thousands of data and without effective analysis tool, this data is achieved in history either for maintaining the record or for any emergency analysis. There is a need to use this data for statistical process monitoring, predictive analysis, fault diagnosis and decision-making process.

Acknowledgements. Authors appreciatively acknowledge the international Grant (0153AB-M20) for funding the publication work. The authors also acknowledge the Universiti Teknologi PETRONAS (UTP), Malaysia for providing support.

References

1. Bilal, M., et al.: Big Data in the construction industry: a review of present status, opportunities, and future trends. Adv. Eng. Inf. **30**(3), 500–521 (2016)
2. Konikov, A., Konikov, G.: Big Data is a powerful tool for environmental improvements in the construction business. In: IOP Conference Series: Earth and Environmental Science (2017)
3. Singh, D., Reddy, C.K.: A survey on platforms for big data analytics. J. Big Data **2**(1), 8 (2014)
4. Gonzalez, R., Huang, B., Lau, E.: Process monitoring using kernel density estimation and Bayesian networking with an industrial case study. ISA Trans. **58**, 330–347 (2015)
5. He, Q.P., Wang, J.: Statistics pattern analysis: a new process monitoring framework and its application to semiconductor batch processes. AIChE J. **57**(1), 107–121 (2011)
6. Jiang, Q., Yan, X.: Weighted kernel principal component analysis based on probability density estimation and moving window and its application in nonlinear chemical process monitoring. Chemom. Intell. Lab. Syst. **127**, 121–131 (2013)
7. Tong, C., Palazoglu, A., Yan, X.: An adaptive multimode process monitoring strategy based on mode clustering and mode unfolding. J. Process Control **23**(10), 1497–1507 (2013)
8. Yu, J.: A particle filter driven dynamic Gaussian mixture model approach for complex process monitoring and fault diagnosis. J. Process Control **22**(4), 778–788 (2012)
9. Zhao, Y., Xiao, F., Wang, S.: An intelligent chiller fault detection and diagnosis methodology using Bayesian belief network. Energy Build. **57**, 278–288 (2013)

10. Fowdur, T.P., et al.: Big Data analytics with machine learning tools. In: Dey, N., et al. (ed.) Internet of Things and Big Data Analytics Toward Next-Generation Intelligence, pp. 49–97. Springer International Publishing, Cham (2018)
11. Wright, S.: Path coefficients and path regressions: alternative or complementary concepts? Biometrics **16**(2), 189–202 (1960)
12. White, T.: Hadoop: The Definitive Guide: The Definitive Guide. O'Reilly Media, Newton (2009)
13. Everitt, B.: The Cambridge Dictionary of Statistics/BS Everitt. Cambridge University Press, Cambridge (2002)
14. Jaroslav, P.: NoSQL databases: a step to database scalability in web environment. Int. J. Web Inf. Syst. **9**(1), 69–82 (2013)
15. Leavitt, N.: Will NoSQL databases live up to their promise? Computer **43**(2), 12–14 (2010)
16. Henseler, J., Hubona, G., Ray, P.A.: Using PLS path modeling in new technology research: updated guidelines. Ind. Manag. Data Syst. **116**(1), 2–20 (2016)
17. Smith, M., et al.: NodeXL: a free and open network overview, discovery and exploration add-in for Excel 2007/2010 (2010)
18. Feldman, R.: Techniques and applications for sentiment analysis. Commun. ACM **56**(4), 82–89 (2013)
19. Hofmann, M., Klinkenberg, R.: RapidMiner: Data Mining Use Cases and Business Analytics Applications. CRC Press, Boca Raton (2013)
20. Kukasvadiya, M.S., Divecha, N.H.: Analysis of data using data mining tool orange. Int. J. Eng. Dev. Res. **5**, 1836–1840 (2017)
21. Landset, S., et al.: A survey of open source tools for machine learning with big data in the Hadoop ecosystem. J. Big Data **2**(1), 24 (2015)
22. Verron, S., Li, J., Tiplica, T.: Fault detection and isolation of faults in a multivariate process with Bayesian network. J. Process Control **20**(8), 902–911 (2010)
23. Yao, M., Wang, H.: On-line monitoring of batch processes using generalized additive kernel principal component analysis. J. Process Control **28**, 56–72 (2015)
24. Ning, C., You, F.: Data-driven adaptive nested robust optimization: general modeling framework and efficient computational algorithm for decision making under uncertainty. AIChE J. **63**(9), 3790–3817 (2017)
25. Xu, S., et al.: An improved methodology for outlier detection in dynamic datasets. AIChE J. **61**(2), 419–433 (2015)
26. Cai, L., Tian, X.: A new process monitoring method based on noisy time structure independent component analysis. Chin. J. Chem. Eng. **23**(1), 162–172 (2015)
27. Yu, H., Khan, F., Garaniya, V.: Modified independent component analysis and bayesian network-based two-stage fault diagnosis of process operations. Ind. Eng. Chem. Res. **54**(10), 2724–2742 (2015)
28. Khediri, I.B., Weihs, C., Limam, M.: Kernel k-means clustering based local support vector domain description fault detection of multimodal processes. Expert Syst. Appl. **39**(2), 2166–2171 (2012)
29. Zhao, X., et al.: Active thermometry based DS18B20 temperature sensor network for offshore pipeline scour monitoring using K-means clustering algorithm. Int. J. Distrib. Sens. Netw. **9**(6), 852090 (2013)
30. Liukkonen, M., Laakso, I., Hiltunen, Y.: Advanced monitoring platform for industrial wastewater treatment: Multivariable approach using the self-organizing map. Environ. Model. Softw. **48**, 193–201 (2013)

31. Chen, X., Yan, X.: Using improved self-organizing map for fault diagnosis in chemical industry process. Chem. Eng. Res. Des. **90**(12), 2262–2277 (2012)
32. Voyslavov, T., Tsakovski, S., Simeonov, V.: Surface water quality assessment using self-organizing maps and Hasse diagram technique. Chemom. Intell. Lab. Syst. **118**, 280–286 (2012)
33. Yu, J.: Semiconductor manufacturing process monitoring using gaussian mixture model and bayesian method with local and nonlocal information. IEEE Trans. Semicond. Manuf. **25**(3), 480–493 (2012)
34. Liu, Y., Chen, J.: Integrated soft sensor using just-in-time support vector regression and probabilistic analysis for quality prediction of multi-grade processes. J. Process Control **23**(6), 793–804 (2013)
35. Jing, C., Hou, J.: SVM and PCA based fault classification approaches for complicated industrial process. Neurocomputing **167**, 636–642 (2015)
36. Kolluri, S.S., et al.: Evaluation of multivariate statistical analyses for monitoring and prediction of processes in an seawater reverse osmosis desalination plant. Korean J. Chem. Eng. **32**(8), 1486–1497 (2015)
37. Ebrahimi, M., Gerber, E.L., Rockaway, T.D.: Temporal performance assessment of wastewater treatment plants by using multivariate statistical analysis. J. Environ. Manag. **193**, 234–246 (2017)
38. Peng, K., et al.: Quality-related prediction and monitoring of multi-mode processes using multiple PLS with application to an industrial hot strip mill. Neurocomputing **168**, 1094–1103 (2015)
39. Qin, S.J., Zheng, Y.: Quality-relevant and process-relevant fault monitoring with concurrent projection to latent structures. AIChE J. **59**(2), 496–504 (2013)
40. Zhang, Y., Sun, R., Fan, Y.: Fault diagnosis of nonlinear process based on KCPLS reconstruction. Chemom. Intell. Lab. Syst. **140**, 49–60 (2015)
41. Zhu, Z.-B., Song, Z.-H.: A novel fault diagnosis system using pattern classification on kernel FDA subspace. Expert Syst. Appl. **38**(6), 6895–6905 (2011)
42. Jiang, B., et al.: A combined canonical variate analysis and Fisher discriminant analysis (CVA–FDA) approach for fault diagnosis. Comput. Chem. Eng. **77**, 1–9 (2015)
43. Abyaneh, H.Z.: Evaluation of multivariate linear regression and artificial neural networks in prediction of water quality parameters. J. Environ. Health Sci. Eng. **12**(1), 40 (2014)
44. Amiri, A., et al.: Diagnosis aids in multivariate multiple linear regression profiles monitoring. Commun. Stat. Theory Methods **43**(14), 3057–3079 (2014)
45. Khorasani, A., Yazdi, M.R.S.: Development of a dynamic surface roughness monitoring system based on artificial neural networks (ANN) in milling operation. Int. J. Adv. Manuf. Technol. **93**(1–4), 141–151 (2017)
46. Loboda, I., Robles, M.A.O.: Gas turbine fault diagnosis using probabilistic neural networks. Int. J. Turbo Jet Engines **32**(2), 175–191 (2015)
47. Iliyas, S.A., et al.: RBF neural network inferential sensor for process emission monitoring. Control Eng. Pract. **21**(7), 962–970 (2013)
48. Taqvi, S.A., et al.: Artificial Neural Network for Anomalies Detection in Distillation Column. Springer, Singapore (2017)
49. Taqvi, S.A., Tufa, L.D., Zabiri, H., et al.: Neural Comput. Applic. (2018). https://doi.org/10.1007/s00521-018-3658-z
50. Li, Y., Zhang, X.: Diffusion maps based k-nearest-neighbor rule technique for semiconductor manufacturing process fault detection. Chemom. Intell. Lab. Syst. **136**, 47–57 (2014)

51. Karabadji, N.E.I., et al.: Improved decision tree construction based on attribute selection and data sampling for fault diagnosis in rotating machines. Eng. Appl. Artif. Intell. **35**, 71–83 (2014)
52. Demetgul, M.: Fault diagnosis on production systems with support vector machine and decision trees algorithms. Int. J. Adv. Manuf. Technol. **67**(9–12), 2183–2194 (2013)
53. Ahmad, I., et al.: Gray-box modeling for prediction and control of molten steel temperature in tundish. J. Process Control **24**(4), 375–382 (2014)
54. Xiong, W., et al.: Adaptive soft sensor based on time difference Gaussian process regression with local time-delay reconstruction. Chem. Eng. Res. Des. **117**, 670–680 (2017)

Big Data Analytics in the Malaysian Public Sector: The Determinants of Value Creation

Esmat A. Wahdain[1,2(✉)], Ahmad Suhaimi Baharudin[1], and Mohammad Nazir Ahmad[3]

[1] School of Computer Sciences, Universiti Sains Malaysia (USM), 11800 Gelugor, Penang, Malaysia
esmat_wahdain@yahoo.com, asuhaimi@usm.my

[2] College of Computers and Information Technology, Hadhramout University, Mukalla, Yemen

[3] Institute of Visual Informatics, Universiti Kebangsaan Malaysia (UKM), Bangi, Selangor, Malaysia
mnazir@ukm.edu.my

Abstract. Big data analytics is a crucial pillar of organisational success. Today's organisations need to leverage the troves of big data to achieve competitive advantage over their competitors. This is only possible by extracting valuable meaning from the vast amount of data. Big data analytics is the tool used for this value extraction. This research in progress aims to help Malaysian public organisations leverage the BDA capabilities to improve their performance and provide better services for citizens. The research proposed a modified DELTA model, and after exploratory interviews with MAMPU officials, factors believed to be highly influential in the Malaysian context were added such as knowledge sharing and job turnover.

Keywords: Big data analytics · Malaysia · Public sector · MAMPU

1 Introduction

1.1 Overview

With the emergence of social media, Internet of Things (IoT), and multimedia, the amount of data generated and captured by individuals, businesses, public organisations, industries and scientific research, has increased tremendously in recent years [1, 2]. These data are of different format be it textual data (either structured or unstructured), or multimedia, and it can be captured and shared on many platforms, e.g. social media sites, ubiquitous-sensing mobile devices [3], machine to machine communications and IoT [2]. The International Data Corporation stated that more than 1600 exabytes (1 exabyte equals 1 billion gigabytes) of data were produced in 2015 from the service and manufacturing sectors alone [3]. Also, [4] reported that the world generates about 2.5 quintillion bytes of data daily(1 exabyte equals 1 quintillion) [2]. Data that is created at such an enormous rate is referred to as big data, and it has emerged as a globally recognised trend.

F. Saeed et al. (Eds.): IRICT 2018, AISC 843, pp. 139–150, 2019.
https://doi.org/10.1007/978-3-319-99007-1_14

According to Gartner Inc. for technology research, big data was one of the Top 10 Strategic Technology Trends for 2013, and one of the Top 10 Critical Tech Trends for the Next Five Years [2, 5]. With the proven potential of big data, corporate investments in big data are increasing dramatically from $34 billion in 2013 to $232 billion in 2016 according to Gartner's 2012 predictions. Moreover, a 2014 Gartner survey revealed that 73% of surveyed corporates had either invested or planned to invest in big data in the next two years [6]. With this growth rate of big data adoption, prompt and cost-effective analytics over big data has become a key success factor for many businesses, scientific institutions, and government organisations [7]. Therefore, Big Data Analytics (BDA) has become a hot topic among practitioners and scholars [8]. Some consultative services companies such as McKinsey and the International Institute for Analytics predict that big data analytics will be the primary source of competitive advantage in the coming years [9]. This proven importance of BDA resulted in several research works that have been dedicated to it. However, the topic is still growing and has not yet fully established. As such, a comprehensive understanding of the field, its terminology and classifications are still up for discussion [2].

Additionally, most of the previously published studies investigated BD and BDA in private organisations, which is manifested in the dearth of publications on BD in the public sector [10] and creates the need for more research to be dedicated for the public sector's use of BD and BDA to fill in this gap [11]. Even more, this relatively scarce scholarship on BD in the public sector focuses mainly on the public sector of most developed countries, like [10, 12] in the U.S, [13] in Russia, and [14] in the Dutch public sector.

This study will contribute to the body of literature by investigating a topic that has rarely been discussed in previous studies, namely the adoption of BDA in public sector organisations of a developing country that is in a transition phase from a production economy into a knowledge economy. Furthermore, the study intends to help in the ongoing efforts to implement BDA in different Malaysian public organisations by shedding light on the facilitators and barriers of BDA adoption in Malaysian public organisations.

1.2 Research Problem

Malaysia is one of the Tiger Cub Economies, and "it was one of 13 countries identified by the Commission on Growth and Development in its 2008 Growth Report to have recorded average growth of more than 7% per year for 25 years or more" [15]. To realise transformation towards a developed nation, Malaysia is eager to achieve significant economic and technological progress. In line with that, the Malaysian Administrative Modernisation and Management Planning Unit (MAMPU) was established in 1977 to improve professionalism in public services. Among its major roles is to lead the development of ICT for the public service sector, and to provide consultancy in management organisation and ICT for the public service sector [16, 17]. One of the priorities of the Ninth Malaysia Plan (2006–2010) set by the government of Malaysia (GOM), was to achieve long-term development in higher education to help the transformation from a P-economy (production-based economy) to K-economy (knowledge-based economy) [18, 19].

With this governmental interest in leveraging technology to achieve transformational goals, the ICT industry expanded by 6.8% per year from 2011 to 2015 [20]. In the same period, there were interesting changes within the ICT services subsectors. These changes include - The replacement of traditional computers by smaller ICT devices and gadgets. - The growing amount of interactive multimedia content based on mobile technology. - The increasing popularity of new trends such as cloud computing, social media, Internet of Things and big data analytics [20].

With the massive rise in data generation due to the above mentioned technological changes, big data analytics has become crucial for both public and private organisations to create value for effective decision making, and to improve productivity and innovation [20]. The Government Chief Information Officer also declared big data analytics as one of three key priorities for ICT in the public sector of Malaysia and was incorporated into the Eleventh Malaysia Plan [21]. Moreover, In the 25th MSC Malaysia Implementation Council Meeting (ICM), held on 14 November 2013, two goals were set. The first was to establish a framework for big data in Malaysia, and the second was to initiate four government BDA pilot projects. These projects were planned to cover sentiment analysis, in partnership with the Ministry of Communications and Multimedia; crime prevention with the Home Ministry; infectious disease prevention with the Health Ministry; and price watch with the Ministry of Trade [13, 22].

Despite the foreseen benefits that BDA can bring to the public sector in terms of value creation and improved decision-making, implementing BDA is a challenging project. The 3rd ASEAN CIO Forum 2014, defined the following key challenges that are expected to face any implementation of BDA: -1- Applying governance rules that are inherited from the structured data environments to unstructured/semi-structured data. -2- The real value stems from what to do with the data. -3- How to get everybody talking to each other (social media with traditional data experts) to agree on common rules. -4- Domain expertise and data manipulation skills are keys to creating value.

The evolving nature of BDA research and the shortage of studies that focus on the public sector, along with nascence and immaturity of BDA implementations in the public sector, motivated the researchers to conduct this study. It is expected to contribute positively to both literature and practice of BDA by investigating the factors that hinder the implementation of BDA in the public sector, as well as identifying the facilitators that help overcome the mentioned challenges.

2 Literature Review and Research Model

2.1 Big Data

Despite the nascence of the big data phenomenon, there have been many attempts to define it. [9] referred to big data as "the amount of data just beyond technology's capability to store, manage, and process efficiently". The International Data Corporation (IDC) defined it as "a new generation of technologies and architectures, designed to economically extract value from very large volumes of a wide variety of data, by enabling the high-velocity capture, discovery, and/or analysis."

While [23] defined big data as "the data that has the following characteristics: volume, variety, and velocity", some scholars stated that big data characteristics are not limited to the three Vs that [23] suggested, they extended the definition to include value [24, 25]. [26] proposed the 5 V model to describe big data. The 5 V refers to;

Volume: with the generation and sharing of massive amounts of data, data are produced in the order of zettabytes, and the annual increase in the produced data is estimated by 40% [7, 27].

(a) Velocity: To create the value of the big data, data must be collected and analysed in a timely manner.
(b) Variety: Different types of data, including traditional structured data, semi-structured and unstructured data such as text, web pages, audio, video.
(c) Value: Nowadays cost becomes associated with data, and data itself can be a commodity that can be sold to third parties for revenue.
(d) Veracity: It is important to assure the accuracy of data by eliminating noisy data by using methodologies like data pedigree and sanitisation. This is crucial because the quality of data influence the effectiveness and accuracy of decisions that are made based on that data [7].

To escape the fuzziness about the concept of big data, and about how big the data must be to be considered big data, [14] used a different approach in conceptualising big data, by defining big data by its use characteristics instead of data characteristics.

Despite the differences in defining the concept of big data, there is a consensus that capturing the potential of big data can only be done by employing effective analytic techniques [9]. The following subsection introduces the concept of BDA.

2.2 Big Data Analytics

The high operational and strategic potentials of BDA have made it the new enabler that can bring improved firm efficiency and effectiveness [8]. BDA is defined as "a holistic approach to manage, process and analyse the "5 Vs" data-related dimensions (i.e., volume, variety, velocity, veracity and value) in order to create actionable insights for sustained value delivery, measuring performance and establishing competitive advantages" [28].

BDA is predicted to have a huge impact in different industries such as giant retailing companies are currently leveraging BDA to improve customer experience, minimise the risk of fraud and make timely recommendations [28]. In healthcare, BDA is anticipated to reduce operational expenses and promote quality of life [29]. In manufacturing and operation management, BDA is expected to offer better monitoring of assets and business processes [29].

Wamba et al. defined business analytics as "a set of all the skills, technologies, applications and practices required for continuous iterative exploration and investigation of past business performance to gain insight and drive business planning" [29]. Based on the output of the analytics process, it can be classified into descriptive, diagnostic, predictive, or prescriptive.

(a) **Descriptive analytics**: deals with historical data and tries to describe an observed phenomenon using different measures with the aim of discovering (what happened?)
(b) **Diagnostic analytics:** its purpose is to unravel the root causes of the problem, using exploratory analysis of the available data or by collecting additional data, to answer the question (why did the problem occur?)
(c) **Predictive analytics:** as the name implies, it helps in predicting the potential future outcomes by using statistical techniques and data mining, it answers the question (what might happen in future?)
(d) **Prescriptive analytics:** it goes beyond just describing, interpretation, and prediction to propose (what should be done?) in the future in order to achieve the organisation's objectives, it combines both: the decision alternatives along with the predicted results of each decision [30].

2.3 BD and BDA in the Public Sector

While the BD is being implemented successfully in the private sector and scientific fields, public organisations are lagging behind in BD and BDA adoption. This could explain the scarcity of studies about BD and BDA in the public sector [11]. Moreover, there is a gap between the potentials that BD can offer for government organisations, and between the actual use of BD in those public sector organisations [14].

[14] attributed the scarcity of BD adoption and usage in the Dutch public sector to the uncertainties among decision makers about their organisations' readiness for BD use. They proposed a framework to help organisations to evaluate their readiness to adopt BD.

[13] provided a good overview to understand the application of BDA in the public sector and related issues. The paper emphasised the fact that the success criteria of a project in the public sector are completely different than in the private sector since the success is measured by the commercial benefit, while in public organisations a project is considered successful if it solved significant problems or helped in achieving the social good.

Although the paper focused on the Russian context and identified some specific influential factors such as sanctions and software import substitution, it also identified a group of common challenges that hinder big data diffusion in all countries as follows:

- Costly complete solutions.
- Lack of executive awareness about the importance of data, or about the current insufficiency of data processing.
- Managing risk.
- Lack of methodological foundation.
- Both data integration and data sharing are complex.
- Data security.

2.4 Research Model and Hypotheses

To develop the theoretical framework for this study, we draw on the extant literature on big data analytics, along with the information collected from individual interviews with a MAMPU official who is a member of the steering committee of four BDA pilots projects implemented in the Malaysian public sector.

There are several models proposed to study BDA adoption in different contexts. [8] proposed a Big Data Analytics Capabilities model. The BDAC hypothesises the direct effect of BDAC on firm's performance, as well as the mediating role of Process Oriented Dynamic Capabilities (PODC) on the relationship between BDAC and firm performance. Hence, BDAC is a third order construct expressed in three second-order constructs, namely management capability, BDA infrastructure capability, and personnel capability.

This study contributes to BDA literature by integrating BDAC and PODC in one model to emphasise the importance of the three components of BDAC mentioned above. Also, the findings prove the significance of the proposed mediator (PODC) that organisations that have the flexibility and the ability to acquire the needed capabilities to change the current business processes or develop new ones are more likely to gain and maintain competitive advantage than those organisations that do not have such flexibility.

The primary data of this study was collected quantitatively using a questionnaire that was distributed electronically to 500 Chinese IT managers and business analysts.

Despite the importance of this study, there are some contextual differences between the context covered by the study and the context being investigated by our research. Those differences might affect the applicability of the proposed model to our study context. Firstly, the level of BDA adoption in China is higher than in Malaysia whereby BDA is still in its nascence stage. Most importantly, although the respondents were selected randomly, the results revealed that most of the sample originated from industries such as information and communication (36%) and finance and insurance (13%) and accommodation and food services (6%), while 0% responses were collected from (public administration and defence). This raises questions about the applicability of the model in the public sector organisations, especially when we talk about BDA which is context specific, according to the authors.

In another study, [31] investigated the effect of Big Data Predictive Analytics (BDPA) assimilation on supply chain and organisational performance. BDPA assimilation was conceptualised as a three-phase process (acceptance, routinisation and assimilation). The questionnaire was distributed to 315 Indian firms, and the results proved the significance of the hypothesised links. One limitation of this study according to the authors is that it relied completely on survey-based method, while a mixed research method can offer deeper insights when studying BDPA. Also, the sampled firms were from manufacturing, technology, consultancy, and e-commerce industries, which are significantly different from public sector organisations.

Another comprehensive model was proposed by Seddon et al. [32]. The proposed model was derived from the most cited models from literature. It was predominantly based on three important models, namely the Wixom and Watson (2001) [33] about the factors affecting data warehousing success, and DELTA model proposed by Davenport et al. [34] about the factors that lead to getting business value from business analytics, and finally Clark et al.'s model [35] of the perceived benefits from Management Support Systems (MSS).

Despite the comprehensiveness of this model, it was based on the success stories of analytics software customer organisations collected from vendors' web pages.

Despite the importance of the above-reviewed models, applying a particular model to a specific study domain requires sufficient knowledge about the domain. For that reason, and to get a clearer image about the context being investigated, the author conducted exploratory interviews with one MAMPU official on the steering committee of the four pilot BDA projects. The insights obtained from this exploratory investigation helped to identify the most relevant and applicable model to be adopted in our study.

For the purpose of this research, we adopted the DELTA model as our theoretical foundation with additions proposed based on the preliminary investigation of the study domain. The DELTA model was selected not only because of its comprehensiveness in regard to the factors that influence gaining value from analytics, but also because of its relevance in investigating the effect of those factors in a nascent BDA environment such as the Malaysian public sector. The five factors that constitute DELTA are **D**ata quality: as data is the prerequisite for the analytics process, and it has to be clean and integrated and measure something new and important. **E**nterprise-wide integrated platforms: hence, fiefdoms of data are not allowed, and centralised expertise must be achieved in order for the firm to compete in analytics. **L**eadership: the authors provided several examples of how a leadership that is keen on data can lead to success well-chosen **T**argets. Hence, a major strategic target must be identified beforehand, with one or two minor targets. **A**nalytical people: hence, four groups of analytical people were identified, analytical champions, analytical professionals, analytical semi-professionals and analytical amateurs. Those five factors were consistent with the factors uncovered during the exploratory interviews

Also, the following additions were suggested:

- Knowledge Sharing was added as a predictor variable of the independent variables, particularly the data variable. There is evidence from the literature that knowledge sharing plays an important role in the output of the analytics process, for instance. [13] identified knowledge integration and sharing as one of the commonly faced challenges especially in the public sector. Also, [8] found that unavailability of appropriate data is one of the factors that hinder firms from getting the performance appraisal sought from information sharing (IS) investments.

While [31] found that IS along with connectivity are positively related to BDPA assimilation under the mediation role of Top Management Commitment (TMC), and BDPA assimilation, in turn, is positively related to organisational performance. Most importantly, knowledge sharing was emphasised by the interviewee as one of the prominent problems that led to poor visualisations in some of the implemented BDA modules due to the absence of sufficient data.

- Job Turnover was added as a moderator of the relationship between the people factor and the dependent variable. By looking at HR literature, there is strong evidence that job turnover has a significant effect on firm's performance. For example, [39] investigated 48 months of turnover data collected from major retail chain stores in the U.S and found that there is a correlation between the increase in job turnover and the decrease in-store performance was measured by profit margin and customer service. This is consistent with previous studies that found that there is a negative correlation between job turnover rates and different performance indicators that are related to profitability and customer service [36, 37]. [38] investigated the impact of turnover on operational performance in 114 outpatient centres in the U.S. The results revealed the negative association between turnover and customer service, especially in the organisations wherein a group-oriented culture was missing.

However, none of the previous studies used job turnover as a moderator. We add it to our model based on strong evidence from the case study, whereby job turnover was reported as one of the main obstacles that hinder the implementation phase and post-implementation usage. Hence, projects with low job turnover were completed smoothly, while projects with a high turnover rate flounded. Because of the categorisation role of the moderator, we intend to categorise targeted organisations into organisations with a high turnover rate, and organisations with a low turnover rate (Fig. 1).

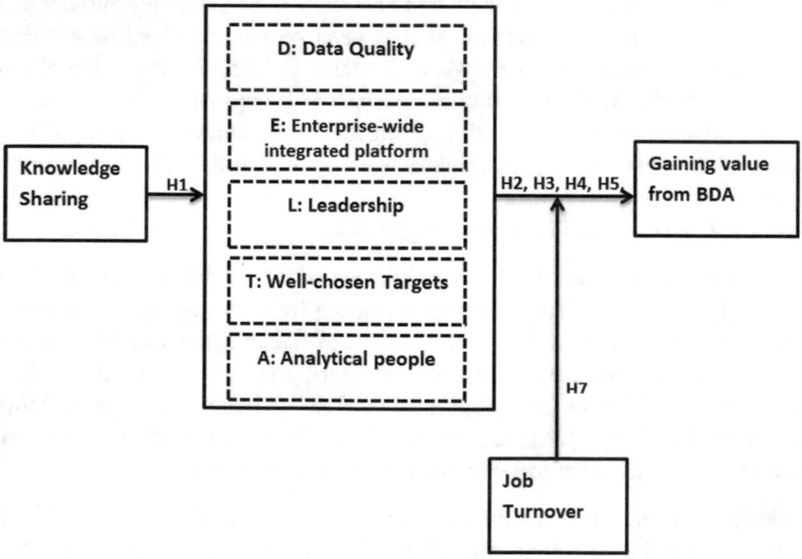

Fig. 1. Research model and hypotheses

The relationships between different factors are hypothesised as follows:

Hypothesis 1: Knowledge sharing has a significant positive effect on the data quality.
Hypothesis 2: Data quality has a significant positive effect on gaining value from analytics.
Hypothesis 3: Enterprise-wide integrated platform has a significant positive effect on gaining value from analytics.
Hypothesis 4: Leadership has a significant positive effect on gaining value from analytics.
Hypothesis 5: Well-chosen targets have a significant positive effect on gaining value from analytics.
Hypothesis 6: Analytical people have a significant positive effect on gaining value from analytics.
Hypothesis 7: The positive relationship between analytical people and gaining value from analytics is stronger in the case of low job turnover rate than in the case of high job turnover rates.

3 Current Research Direction

The current research will be carried out in the Malaysian public sector ministries and agencies that experienced the implementation of BDA projects. The study will combine qualitative and quantitative approaches.

At the early stages of research, exploratory interviews were conducted for a clear image about the influential factors that affect the successful implementation and usage of BDA in the context of the Malaysian public sector. For this reason, we approached MAMPU to conduct interviews because MAMPU is the leading governmental body that drives all BDA projects in the Malaysian public sector. The insights gained along with evidence from the literature were used to develop the conceptual research model.

For the data collection phase, a quantitative approach will be followed. A questionnaire will be developed as an instrument for data collection. Because of the nascent nature of the topic, we expect that not all the constructs' measurement items will be available from the literature. Therefore, some measurement items will be self-constructed. The respondents will be Malaysian public sector employees of agencies that have implemented BDA projects.

The collected response will be analysed by the Structural Equation Modelling method, using the SmartPLS software to test the proposed hypotheses and measure the predictability of the proposed model.

4 Conclusion

This paper introduced a conceptual model that examines the factors that influence gaining value from big data analytics projects in the context of Malaysian public sector organisations. The proposed model was derived from relevant studies from the literature, along with the information collected about the implemented BDA projects through exploratory interviews with MAMPU officials.

The contribution of this study is twofold. First, it contributes to the growing literature of big data analytics by introducing a case study about a context that has rarely been investigated, namely the public sector of a country that is in the transformation phase from a P-economy to K-economy. Moreover, it provides empirical evidence of the significance of the five factors proposed by [34] (DELTA), along with the added factors. Additionally, the study presented the job turnover as a moderator to the relationship between the analytical people and the dependent variable.

Secondly, this study is expected to contribute significantly to the Malaysian public sector. It is expected that the results will shed light on the factors that affect gaining the sought after value from BDA projects. This is very important not only because BDA adoption is still in its early stages in Malaysia, but also because it is still in its infancy [10]. The results of this study should help the decision makers and practitioners to gain a better understanding about the facilitators and barriers of getting better value from the implemented BDA projects in a way that leads to better practices and consequently, better results. This research is a good foundation for future research that investigates the facilitators and barriers of BDA in the public sector of developing countries.

Acknowledgements. The authors would like to thank Dr. Yusminar Binti Yunus, Director of Digital Government Division at the Malaysian Administrative Modernisation and Management Planning Unit (MAMPU) for her continuous help and support in conducting this research.

References

1. Agarwal, R., Dhar, V.: Editorial—big data, data science, and analytics: the opportunity and challenge for IS research. Inf. Syst. Res. **25**(3), 443–448 (2014)
2. Sivarajah, U., Kamal, M.M., Irani, Z., Weerakkody, V.: Critical analysis of Big Data challenges and analytical methods. J. Bus. Res. **70**, 263–286 (2017)
3. Zhong, R.Y., Newman, S.T., Huang, G.Q., Lan, S.: Big Data for supply chain management in the service and manufacturing sectors: challenges, opportunities, and future perspectives. Comput. Ind. Eng. **101**, 572–591 (2016)
4. Dobre, C., Xhafa, F.: Intelligent services for Big Data science. Future Gener. Comput. Syst. **37**, 267–281 (2014)
5. Cearley, D.: Top 10 strategic technology trends for 2013 (2013)
6. Alles, M., Gray, G.L.: Incorporating big data in audits: identifying inhibitors and a research agenda to address those inhibitors. Int. J. Acc. Inf. Syst. **22**, 44–59 (2016)
7. Rodríguez-Mazahua, L., Rodríguez-Enríquez, C.-A., Sánchez-Cervantes, J.L., Cervantes, J., García-Alcaraz, J.L., Alor-Hernández, G.: A general perspective of Big Data: applications, tools, challenges and trends. J. Supercomput. **72**(8), 3073–3113 (2016)

8. Wamba, S.F., Gunasekaran, A., Akter, S., Ren, S.J., Dubey, R., Childe, S.J.: Big data analytics and firm performance: effects of dynamic capabilities. J. Bus. Res. **70**, 356–365 (2017)
9. Manyika, J., Chui, M., Brown, B., Bughin, J., Dobbs, R., Roxburgh, C., Byers, A.: Big data: the next frontier for innovation, competition, and productivity (2011)
10. Desouza, K.C., Jacob, B.: Big Data in the public sector: lessons for practitioners and scholars. Adm. Soc. **49**(7), 1043–1064 (2017)
11. Fredriksson, C., Mubarak, F., Tuohimaa, M., Zhan, M.: Big data in the public sector: a systematic literature review. Scand. J. Public Adm. **21**(3), 22 (2017)
12. Akter, S., Wamba, S.F., Gunasekaran, A., Dubey, R., Childe, S.J.: How to improve firm performance using big data analytics capability and business strategy alignment? Int. J. Prod. Econ. **182**, 113–131 (2016)
13. Anna, K., Nikolay, K.: Survey on Big data analytics in public sector of Russian Federation. Proc. Comput. Sci. **55**, 905–911 (2015)
14. Klievink, B., Romijn, B.-J., Cunningham, S., de Bruijn, H.: Big data in the public sector: uncertainties and readiness. Inf. Syst. Front. **19**(2), 267–283 (2017)
15. The World Bank: Malaysia overview, report (2016). http://www.worldbank.org/en/country/malaysia/overview
16. MAMPU: Roles of MAMPU (2016). http://www.mampu.gov.my/en/corporate-information/role-of-mampu-department
17. Salleh, D.D.M.T.: National integrity plan (2007)
18. Zaleha, M.N.Z., Tan, H.-B., Wong, M.-F.: Education and growth in Malaysian knowledge-based economy. Int. J. Econ. Manag. **1**(1), 141–154 (2006)
19. Regel, O.J.A.H.J.A.J.: Malaysia and the knowledge economy: building a world-class higher education system. Human development sector reports, World Bank Publications (2007)
20. U.S. Embassies abroad: Malaysia—information and communications technology (2017)
21. Big community: Big data and key component of digital economy in the 11th Malaysia plan (2015)
22. Dzazali, D.S.: Public sector big data analytics initiative: Malaysia's Perspective (2014)
23. Zikopoulos, P., Deroos, D., Parasuraman, K., Deutsch, T., Giles, J., Corrigan, D.: Harness the Power of Big Data the IBM Big Data Platform. McGraw-Hill, New York (2012)
24. Gantz, J., Reinsel, D.: Extracting value from chaos state of the universe: an executive summary. IDC iView **1142**, 1–12 (2011)
25. Hashem, I.A.T., Yaqoob, I., Anuar, N.B., Mokhtar, S., Gani, A., Ullah Khan, S.: The rise of 'Big Data' on cloud computing: review and open research issues. Inf. Syst. **47**, 98–115 (2015)
26. Lomotey, R.K., Deters, R.: Towards knowledge discovery in Big Data. In: 2014 IEEE 8th International Symposium on Service Oriented System Engineering, pp. 181–191 (2014)
27. Fan, W., Bifet, A.: Mining big data. ACM SIGKDD Explor. Newsl. **14**(2), 1 (2013)
28. Fosso Wamba, S., Akter, S., Edwards, A., Chopin, G., Gnanzou, D.: How 'Big Data' can make big impact: findings from a systematic review and a longitudinal case study. Int. J. Prod. Econ. **165**, 234–246 (2015)
29. Davenport, T.H., Barth, P., Bean, R.: How 'big data' is different. MIT Sloan Manag. Rev. **54** (1) (2012)
30. Rosebt.com: Descriptive diagnostic predictive prescriptive analytics (2013). http://www.rosebt.com/blog/descriptive-diagnostic-predictive-prescriptive-analytics
31. Gunasekaran, A., Papadopoulos, T., Dubey, R., Wamba, S.F., Childe, S.J., Hazen, B., Akter, S.: Big data and predictive analytics for supply chain and organizational performance. J. Bus. Res. **70**, 308–317 (2017)

32. Seddon, P., Constantinidis, D., Dod, H.: How does business analytics contribute to business value? In: ICIS 2012 Proceedings (2012)
33. Wixom, B.H., Watson, H.J.: An empirical investigation of the factors affecting data warehousing success. MIS Q. **25**(1), 17 (2001)
34. Davenport, R.M.: Analytics at work: smarter decisions, better results. In: Davenport, T.H., Harris, J.G., Morison, R. (eds.) Google Books. Harvard Business Press, Brighton (2010)
35. Clark, C.P.A., Thomas, D., Jones, M.C.: The dynamic structure of management support systems: theory development, research focus, and direction. MIS Q. **31**(1), 579–615 (2007)
36. Koys, D.J.: The effects of employee satisfaction, organizational citizenship behavior, and turnover on organizational effectiveness: a unit-level, longitudinal study. Pers. Psychol. **54** (1), 101–114 (2001)
37. Hurley, R.F., Estelami, H.: An exploratory study of employee turnover indicators as predictors of customer satisfaction. J. Serv. Mark. **21**(3), 186–199 (2007)
38. Mohr, D.C., Young, G.J., Burgess Jr., J.F.: Employee turnover and operational performance: the moderating effect of group-oriented organisational culture. Hum. Resour. Manag. J. **22** (2), 216–233 (2012)
39. Ton, Z., Huckman, R.S.: Managing the impact of employee turnover on performance: the role of process conformance. Organ. Sci. **19**(1), 56–68 (2008)

Comparing the Performance of FCBF, Chi-Square and Relief-F Filter Feature Selection Algorithms in Educational Data Mining

Maryam Zaffar[1(✉)], Manzoor Ahmed Hashmani[1,2], and K. S. Savita[1,2]

[1] Department of Computer and Information Sciences, Universiti Teknologi PETRONAS, 32610 Seri Iskander, Malaysia
maryam.zaffar82@gmail.com, {manzoor.hashmani, savitasugathan}@utp.edu.my

[2] High Performance Cloud Computing Center, Universiti Teknologi PETRONAS, 32610 Seri Iskander, Malaysia

Abstract. Educational Data Mining (EDM) is a very vital area of Data Mining, and it is helpful in predicting the performance of students. This paper is a step towards identifying the factors affecting the academic performance of the students. It is very necessary to increase the quality of dataset as to get better prediction results. There are many feature selection algorithms, however three filter feature selection algorithms FCBF, Chi-Square, and ReliefF are selected due their better performance, and applied on three different student's data sets. The results of three filter feature selection algorithms are evaluated. The result of the paper extracted that Chi-Square and ReliefF perform better than FCBF on a dataset with larger number of features, however the performance of three selected algorithms is found worst on a student dataset with less number of instances. The analysis of the results of algorithms shows that, Home to Institution (School/College) travel time is one of the important feature affecting the performance of the students. Student's previous academic background and Socio-economic factors also appeared to be the important factors for predicting the academic performance of the students.

Keywords: Educational Data Mining (EDM)
Filter feature selection algorithms · FCBF · Chi-Square · ReliefF

1 Introduction

Educational Data Mining (EDM) is the research area concerned with the exploring and creating methods for data from various educational information structures, to understand students, teachers and learning process. The main aim of EDM is to convert the raw student data into use full information to analyze and predict different issues to enhance the academic performance [1].

Students' academic performance is one of the most important area of EDM [2]. Predicting students' performance is very significant, as the main concern are students in

F. Saeed et al. (Eds.): IRICT 2018, AISC 843, pp. 151–160, 2019.
https://doi.org/10.1007/978-3-319-99007-1_15

schools and institutions of all over the world it may affect the financial position, graduations rates and the reputation of educational institutions [3].

The performance of student academic performance prediction model depends upon the selection of features from the student dataset. This can be accomplished by applying different feature selection algorithms on data set. The main aim of the feature selection algorithm is to find out a minimize redundancy and maximize relevance feature subset, while maintaining high accuracy without dropping of some useful information [2, 4].

In previous studies [1], in the field of EDM, prediction models are developed by using e feature selection algorithms. Different Filter Feature selection algorithms are used in the study.

In this study we focus on the process of producing quality data for Predicting the performance of students. The process of improving quality data is very important as it will increase the classification accuracy. Most of research in EDM overlooked the process of obtaining quality data. By realizing the importance of quality data set we have used different filter feature selection algorithms. This is a first research paper in EDM that have taken three different student datasets to evaluate the performance of different feature selection algorithms. There are different types of filter feature selection algorithm, we have selected FCBF, Chi-Square and ReliefF.

The outline of the paper is as follows Sect. 2 describes the methods used in this research, Sect. 3 discusses and describes results of filter feature selection algorithm on three datasets, and the conclusion of the paper is summarized in Sect. 4.

2 Methods

The performance of different filter feature selection algorithms on different student datasets (with different number of features and different number of instances) is analyzed and compared by using three different dataset of students. SVM classification algorithm is used to find the predictions. The main flow of the methodology used in the paper is also explained using Fig. 1. The Fig. 1 displays four main steps of the method used in this paper. In the first step, datasets are collected from different valid sources (details will be explained in Sect. 2.1), and then three filter feature selection algorithms are applied on these three datasets separately (details of filter feature selection algorithms in Sect. 2.2), after applying filter feature selection algorithm on each dataset SVM classification algorithm is applied. In the last step the prediction accuracies are evaluated on each step, to check and analyze the performance of filter feature selection algorithms. Furthermore, the results of each dataset is evaluated to analyze the factors that are considered to the most important in predicting the performance of students.

2.1 Dataset Description

In this paper we have taken three datasets. The description of these datasets is given below.

Dataset1. Dataset1 is collected from UCI Machine Learning repository [2]. The dataset consist of 395 students of higher secondary school in Portuguese, along with 33 attributes. The features include students Demographic, Academic, and Life Style information.

Dataset2. Dataset2 is previously used in [3], it consists of 500 students records and 16 features. The features used in dataset2 are classified into three categories: Demographic, Academic and Behavioral features.

Dataset3. Dataset3 is collected from three different colleges of India by [1]. The dataset consists of 300 students' record along with 21 features.

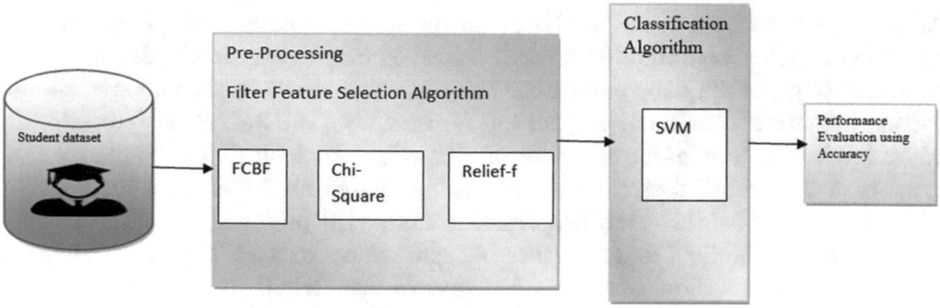

Fig. 1. Methodology

2.2 Filter Feature Selection Algorithms

Feature selection is an important pre-processing technique that is applied in machine learning. The filter feature selection is a type of feature selection that maximizes the evaluation function for getting an optimal feature subset through a search strategy [4]. The filters have three main stages that are feature set generation, measurement and testing by a learning algorithm. The filter feature selection algorithms process quickly, and they calculate the information from the features, so that their results will depend on measured information of the features [5]. The filter feature selection algorithms are preferred because they can perform better with any classification algorithm as they have less computational complexity [6]. In this paper we have taken three filter feature selection algorithms. The description of these three filter algorithms is given below.

Fast Correlation Based Feature Selection (FCBF). FCBF is a multivariate feature selection method that tries to find the best feature subset based on goodness of features [7]. It starts with full set of features and uses symmetrical uncertainty to calculate the dependence of features. Symmetrical uncertainty (SU) is a normalized information theoretic measure which uses the values of entropy and conditional entropy to calculate the dependencies of features FCBF is a correlation based feature subset selection method, which is faster than other subset selection methods [8]. FCBF is being used to rank the features of graduate students in United States universities, in order to identify the factors of high dropout rate and low graduation rate of four-year college students [9]. Another study [10] uses FCBF in pre-processing stage to predict the student interactions in intelligent learning environment. This study suggested that FCBF would be competent on selecting features from the kind of datasets used in EDM.

Chi-Square (Chi^2). The Chi-Square algorithm is a univariant feature selection algorithm that is used as a test of independence. There are two main phases of this algorithm. In the first phase consistency checking is done as the stopping criteria, whereas in phase two, the results of phase one are checked. It continues until there remain no attributes for merging [11]. In EDM Chi^2 algorithm is being used by different studies [12–16] on different datasets of students.

ReliefF. ReliefF algorithm is measured to be one of the most successful algorithms for accessing the quality of features [13]. ReliefF feature selection algorithm is based on random selection. It does not use the values of information gain and entropy. The calculations in this algorithm are based on hit and miss ratio. This algorithm is an extension of relief algorithm to make it robust to incomplete data and improve the reliability of probability approximation. This algorithm uses a user configured k-value for the number of neighbors to search [17]. In study [18] ReliefF feature selection algorithm is used to filter the attributes of high school students dataset to predict slow learners in educational sector. The results of the study reveal that the ReliefF shown 75% accuracy with MLP classifier. Another study [13] proposes a feature selection architecture, uses ReliefF as a relevance filter on student dataset. This study compares different feature selection algorithms and reveals that ReliefF is best among all the existing weighing algorithms.

2.3 Classification Algorithm

Classification is a supervised learning technique and is most commonly used as a data mining technique in predicting the performance of students [19–21]. There are a number of classification algorithms: Decision Tree, Neural Network, Naïve Bayes, Random Forest, Ada Boost and SVM. In this research authors have used SVM as classification algorithm.

SVM (Support Vector Machine). SVM is a supervised machine learning algorithm. It has been applied in many practical problems related to the face recognition, 3D object recognition, text and image classification and in EDM. It has a inimitable benefit of solving small-sample, on-linear, and high dimensional pattern recognition problems [22]. SVM uses a Gaussian function. As a benefit much more complex relationship between the given data points can be captured. SVM is suitable for feature selection problems [23].

3 Results and Discussion

3.1 Results on Dataset1

The results of three selected feature selection algorithms on dataset1 [2] is 87.97% for FCBF, 89.87% for Chi-square, and 88.61% for ReliefF filter feature selection algorithm. According to results the prediction accuracy of Chi-Square feature selection algorithm is better than other two feature selection algorithms. However, the accuracy of FCBF is found lowest on dataset1. Figure 2 clearly presents the difference of accuracy among three feature selection algorithms.

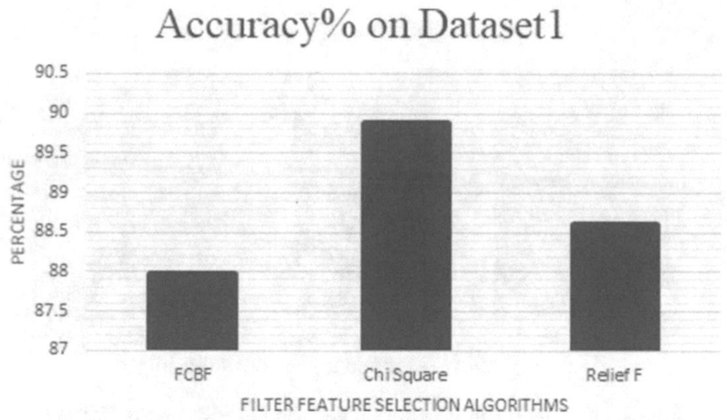

Fig. 2. Results of filter feature selection algorithms on Datset1

3.2 Results on Dataset2

The T results of three selected feature selection algorithms by our study on dataset2 [3] are as follows 72.92% on FCBF, 71.88% on Chi-Square, and 70.83% on ReliefF. FCBF shows highest performance on dataset2. Whereas ReliefF shows the lowest results. Figure 3 presents a clear picture of the performance of feature selection algorithms on dataset2.

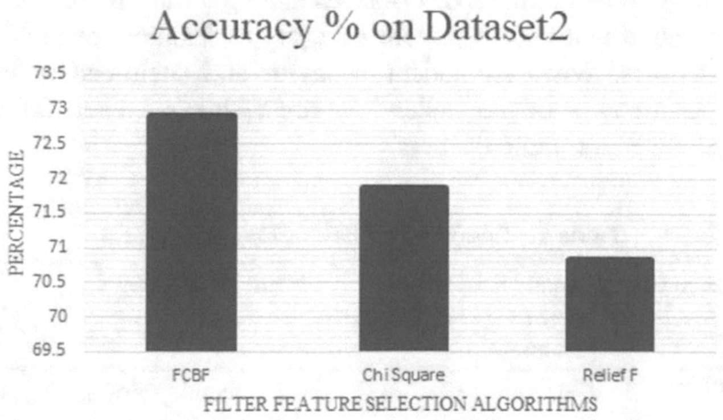

Fig. 3. Results of filter feature selection algorithms on Datset2

3.3 Results on Dataset3

The accuracy of FCBF, Chi-Square and ReliefF on dataset3 [1] are 45%, 45% and 41.67% alternatively. The results on dataset3 are not up to the mark. However, FCBF and Chi-Square shows better performance than ReliefF. Figure 4 presents the performance of three selected feature selection algorithms on dataset3.

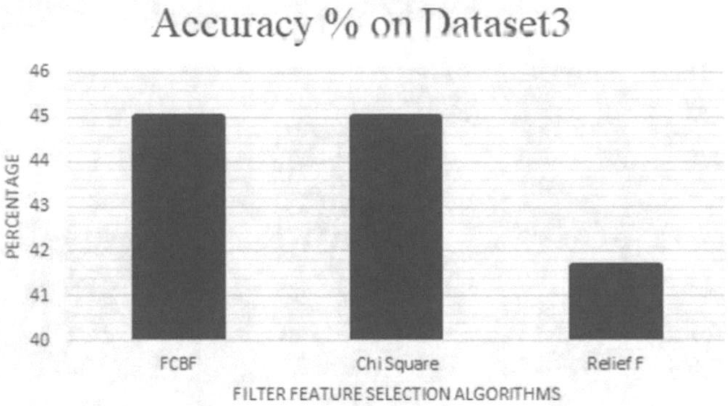

Fig. 4. Results of filter feature selection algorithms on Datset3

3.4 Comparison of Performance of FCBF on Three Datasets

Table 1 presents a detailed analysis of the result of FCBF on three datasets. The datasets 1 and 2 have almost two same categories of features that are Demograic (DF), and academic (AF), whereas dataset1 has life style information (LFI), and dataset2 has behavioral features (BEF) category. Dataset2 has also included the features regarding parent's participation in learning process (PPL). Whereas, the features of third dataset has features regarding demographic, academic and socio-economic information of students. FCBF shows highest accuracy on dataset1. Whereas, FCBF shows lowest on dataset 3, that have lowest number of instances among all 3. The results show that the academic background is a very important category of features for predicting the performance of students. Whereas, student behavior and socio-economic factors also influence the performance of the student. Figure 5 shows a clear difference of performance of the FCBF on three datasets.

Table 1. Results of FCBF on Dataset1, 2 & 3

Datasets	Accuracy	Total no of features	Features extracted	Name of features
Dataset1	87.97	34	2	G2 (second period grade), travel time (home to school travel time)
Dataset2	72.92	16	4	Relation, VisITedResources (BEF), ParentAnsweringSurvey (PPL), StudentAbsenceDays (AF)
Datsaset3	45	21	3	ge (Gender), tnp (Class X %), fmi (Family Monthly Income)

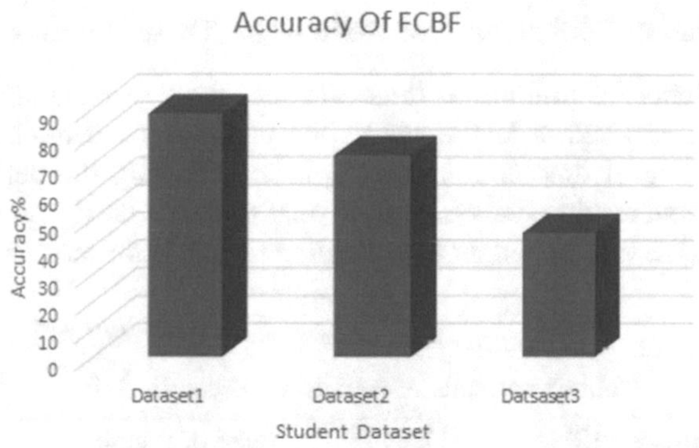

Fig. 5. Performance comparison of FCBF on Datset1, 2 and 3

3.5 Comparison of Performance of Chi-Square on Three Datasets

The Chi-Square feature selection algorithm shows better results on dataset1 having 34 number of features. However, chi-square shows lowest prediction accuracy on dataset3 with 21 features and 300 instances. According to features extracted from dataset1 the travel time (home to school travel time) is one of the important factors influencing the performance of students. The overall results of Chi-square show that educational data miners should not ignore the students' previous academic record while predicting the performance of students. Whereas, students' behavior and socio-economic factors also play vital role (Table 2).

Table 2. Results of Chi-Square on Dataset1, 2 & 3

Datasets	Accuracy	Total no of features	Features extracted	Name of features
Dataset1	89.97	34	11	School, Mjob (mother's job), Guardian (student's guardian), traveltime (home to school travel time), studytime, nursery (attended nursery school?), freetime (free time after school), health (current health status), absences (# of school absences), G1 (first period grade), G2 (second period grade)
Dataset2	71.88	16	11	PlaceofBirth, gender, Topic, Relation, raisedhands, VisITedResources (BEF), AnnouncementsView, Discussion, ParentAnsweringSurvey, ParentschoolSatisfaction, StudentAbsenceDays
Datsaset3	45	21	5	Tnp (Class X %), as (Admission Category), fo (Father Occupation)

3.6 Comparison of Performance of Relief-F on Three Datasets

The results of ReliefF algorithm on three datasets show that the algorithms perform best on dataset 1 and shows lowest prediction accuracy on dataset 3. However, the home to college travel time is found as a factor affecting the performance of the students. The overall results on datasets show that student's previous records and socio-economic factors are very important in predicting the performance of students (Table 3, Figs. 6 and 7).

Table 3. Results of ReliefF on Dataset1, 2 & 3

Datasets	Accuracy	Total no of features	Features extracted	Name of features
Dataset1	88.61	34	11	School, Medu (mother's education), Fjob (father's job), Guardian (student's guardian), studytime, freetime (free time after school), Dalc (workday alcohol consumption), health (current health status), absences (# of school absences), G1 (first period grade), G2 (second period grade)
Dataset2	70.83	16	11	Nationality, PlaceofBirth, StageID, GradeID, SectionID, Topic, raisedhands, VisITedResources, AnnouncementsView, Discussion, StudentAbsenceDays
Datsaset3	41.67	21	11	cst (Caste), tnp (Class x %), twp (Class xii %), esp (End Semester %), fmi (Family Monthly Income), fq (Father Qualification), mq (Mother Qualification), fo (Father Occupation), nf (Number of friends), sh (Study Hours), tt (Home to college Travel Time)

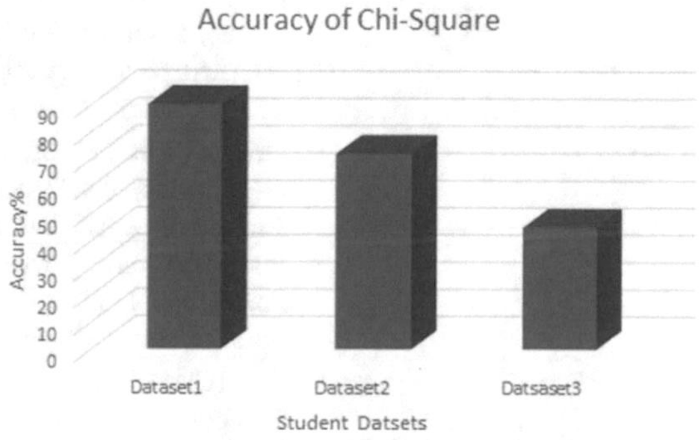

Fig. 6. Performance comparison of Chi-Square on Datset1, 2 and 3

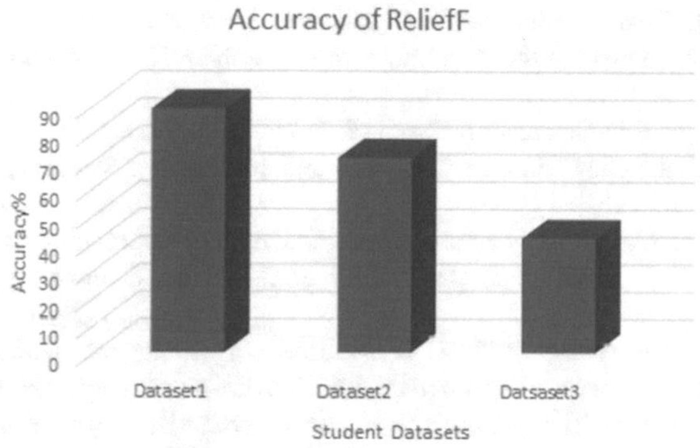

Fig. 7. Performance comparison of ReliefF on Datset1, 2 and 3

4 Conclusion

Predicting the performance of students is very important task for the institutions to improve the quality of education. This paper tries to evaluate to three most prominent filter feature selection algorithms. From the results we conclude that Chi-Square and ReliefF performs well on datasets with highest number of features. Whereas, FCBF performs well on a dataset with highest number of instances. Whereas, all three filter feature selection algorithms show worst results on datasets with less number of instances. There is almost 30% difference of accuracy as compared with other dataset results. In educational sector the datasets might be huge in some cases and in other cases very small. So, in order to solve that issue there is a need of an algorithm which may perform well over small as well as big datasets for predicting the student performance. The results also show that students' previous academic record as well as socio-economic features are most important in predicting the performance of students. Home to school travel time is found to be the most important factor in all results.

References

1. Hussain, S., et al.: Educational data mining and analysis of students' academic performance using WEKA. Indones. J. Electr. Eng. Comput. Sci. **9**(2), 447 (2018)
2. Cortez, P., Silva, A.M.G.: Using data mining to predict secondary school student performance (2008)
3. Amrieh, E.A., Hamtini, T., Aljarah, I.: Mining educational data to predict student's academic performance using ensemble methods. Int. J. Database Theory Appl. **9**(8), 119–136 (2016)
4. Liu, J., et al.: Feature selection based on quality of information. Neurocomputing **225**, 11–22 (2017)
5. Hsu, H.-H., Hsieh, C.-W., Lu, M.-D.: Hybrid feature selection by combining filters and wrappers. Expert Syst. Appl. **38**(7), 8144–8150 (2011)
6. Gnana, D.A.A., Balamurugan, S.A.A., Leavline, E.J.: Literature review on feature selection methods for high-dimensional data. Int. J. Comput. Appl. **136**(1), 9 (2016)

7. Yu, L., Liu, H.: Feature selection for high-dimensional data: a fast correlation-based filter solution. In: Proceedings of the 20th International Conference on Machine Learning (ICML-2003) (2003)
8. Senliol, B., et al.: Fast Correlation Based Filter (FCBF) with a different search strategy. In: 23rd International Symposium on Computer and Information Sciences, ISCIS 2008. IEEE (2008)
9. Gopalakrishnan, A., et al.: A multifaceted data mining approach to understanding what factors lead college students to persist and graduate. In: Computing Conference. IEEE (2017)
10. Mavrikis, M.: Modelling student interactions in intelligent learning environments: constructing bayesian networks from data. Int. J. Artif. Intell. Tools **19**(06), 733–753 (2010)
11. Liu, H., Setiono, R.: Chi^2: feature selection and discretization of numeric attributes. In: Proceedings of the Seventh International Conference on Tools with Artificial Intelligence. IEEE (1995)
12. Ramaswami, M., Bhaskaran, R.: A study on feature selection techniques in educational data mining. arXiv preprint arXiv:0912.3924 (2009)
13. Kavitha, A., Kavitha, R., Gripsy, J.V.: Empirical evaluation of feature selection technique in educational data mining. ARPN J. Sci. Technol. **2**(11), 1103 (2012)
14. Rachburee, N., Punlumjeak, W.: A comparison of feature selection approach between greedy, IG-ratio, Chi-square, and mRMR in educational mining. In: 2015 7th International Conference on Information Technology and Electrical Engineering (ICITEE). IEEE (2015)
15. Zaffar, M., Hashmani, M.A., Savita, K.: Performance analysis of feature selection algorithm for educational data mining. In: 2017 IEEE Conference on Big Data and Analytics (ICBDA). IEEE (2017)
16. Velmurugan, T., Anuradha, C.: Performance evaluation of feature selection algorithms in educational data mining. Perform. Eval. **5**(02), 212 (2016)
17. Kononenko, I., Šimec, E., Robnik-Šikonja, M.: Overcoming the myopia of inductive learning algorithms with RELIEFF. Appl. Intell. **7**(1), 39–55 (1997)
18. Kaur, P., Singh, M., Josan, G.S.: Classification and prediction based data mining algorithms to predict slow learners in education sector. Proc. Comput. Sci. **57**, 500–508 (2015)
19. Dekker, G.W., Pechenizkiy, M., Vleeshouwers, J.M.: Predicting students drop out: a case study. In: International Working Group on Educational Data Mining (2009)
20. Bhardwaj, B.K., Pal, S.: Data Mining: a prediction for performance improvement using classification. arXiv preprint arXiv:1201.3418 (2012)
21. Algur, S.P., Bhat, P., Ayachit, N.H.: Educational data mining: RT and RF classification models for higher education professional courses. Int. J. Inf. Eng. Electron. Bus. **8**(2), 59 (2016)
22. Fradkin, D., Muchnik, I.: Support vector machines for classification. DIMACS Ser. Discrete Math. Theor. Comput. Sci. **70**, 13–20 (2006)
23. Cheng, C.-H., Liu, W.-X.: An appraisal model based on a synthetic feature selection approach for students' academic achievement. Symmetry **9**(11), 282 (2017)

The Study of Co-occurrences Index's Keywords for Malaysian Publications

Nurul Mardhiah Azura Md. Nadzar(✉), Aryati Bakri, and Roliana Ibrahim

Faculty of Computing, Universiti Teknologi Malaysia, 81300 Johor, Malaysia
mardhiahazura@gmail.com, {aryati,roliana}@utm.my

Abstract. Keywords become more crucial in searching any information either in a research database, website or in social media. Another usage of keywords is to identify the pattern for a potential subject area with manipulated elements from the bibliometric information via classification or clustering. There are number of studies were done previously in classification whether in local or international. However, the study regarding clustering was identified as limited to here in Malaysia. Therefore, the intention of this paper is to fill the gap which and identify the pattern for a potential subject area which manipulated one of bibliometric element namely as index's keywords. The approach that was used to obtain the pattern is bibliomining. In our research, we used one promising application namely Vosviewer to achieve our research objective. Data from Scopus databases from 1995 to 2015 with a total of 46,959 documents can be used after data cleaning. We managed to get 4 clusters to represent some theme in our analysis and output. These might be beneficial to management and other researchers to maneuver the research direction and gives an idea which subject area needs to be focused on and which subject area needs new research.

Keywords: Publication · Clustering · Index's keywords · Bibliometic Bibliomining · Text mining

1 Introduction

Keywords. Very simple word with few syllables but gives a huge impact towards something [1]. Either searching in social media platform or searching related articles in database [2]. The ultimate goal is to find suitable and related keywords that can be used for their research [3]. Not to forget, there is research which has been conducted by previous researchers to identify how a novice researcher identifies suitable keywords for the research [4]. Besides, from keywords themselves we can identify the research theme for certain subject areas and identify the research area that becomes a choice for researchers [5]. This emerges when there are number of keywords that keep on appearing in the numbers of publication which we called co-occurrences keywords.

The definition of co-occurrences is a linguistic term that either means concurrence/coincidence or in a more specific sense is the frequent occurrence of two terms from publication [6]. Therefore, they can be considered as important and frequent

F. Saeed et al. (Eds.): IRICT 2018, AISC 843, pp. 161–172, 2019.
https://doi.org/10.1007/978-3-319-99007-1_16

words that researchers used to describe their publication. There are a number of studies from previous researchers using this approach to seek the potential keywords that always occur in their area [7].

Therefore, our intention is to implement this approach in our experiment to identify the subject area from index keywords which one of bibliometric element in publication. Our research environment was related to Malaysian publications due to there being no other studies conducted by previous researchers. Our initial step is to gather and process all author's keywords only. In order to generate the potential subject area from co-occurrences keywords that comes from the index's keyword, we used the application which has been developed by previous researchers namely Vosviewer [8]. Most of the previous studies were done by manipulating and processing the data from Web of Science [9] and fewer studies were conducted in Scopus's data [10]. Therefore this is another reason why we want to process the data using our selected database namely Scopus. This paper is divided into several sections and is presented accordingly.

2 Literature Review

There are many ways to manipulate these words and one of them is text mining or in this case we called it bibliomining [11]. The term bibliomining was coined by a previous researcher who claimed it was the combination of two elements - bibliometric and data mining [11]. In the previous study, the results are more towards pattern of users who are using the library services and the combination of several elements to perform this bibliomining process [12]. The process and result in bibliomining isn't limited to discover the pattern of usage in the library's database only, but it has the potential to be applied in other environments as well. There are numbers of previous studies that manipulate this bibliometric element and one of them is co-occurrences word analysis [13].

Co-occurrences keywords can be identified in several applications and methods [14]. The algorithms were used to calculate the frequency of the keywords as well as the distance between keywords [15]. Among the algorithms which have frequently been studied by previous researchers are term frequency known as TF [16, 17] and others [18].

As we mentioned before, Vosviewer was the selected application to manipulate and process our data. The aim of VOS is to provide a low-dimensional visualization in which objects are located in such a way that the distance between any pair of objects reflects their similarity as accurately as possible [19]. The application is robust enough to handle and process the huge amount of data although it does takes time, but it is still manageable [8].

At the first stage, Vosviewer will calculate all keywords found in the text provided with the publication's information [19]. Next, all keywords were sorted to minimize and reduce the scope of similar text to select the most relevant noun phrases [19]. Then, it will generate the clusters and map the keywords. Final, the visualization of mapping and clustering results. In VosViewer, there is some text mining functionality that provides term map based on the corpus of the document.

3 Methodology

Data were retrieved from selected database from the library of Universiti Teknologi Malaysia from year 1995 to 2015–20 years of publication. Data were downloaded from 21 November 2017 to 23 November 2017. All data were retrieved from Scopus database. Several criteria were used to select related data in the database. One of the criteria was the affiliation country. We chose 'Malaysia' which indicates that all documents that we downloaded from the database were related to Malaysian publications. Another criteria that we chose is related to citation data. Then, we sorted the research result from the most often cited to the least cited. As we know, publication needs time to mature and exposure to other researchers [20], therefore, by sorting the related publications we retrieved the matured and developed publication that is beneficial to our study. The final criteria is the type of publication. Next, our task is data cleaning. Data cleaning is needed especially when integrating assorted data sources and should be addressed together with scheme-related data transformation [21]. A total of 46,959 documents was processed.

3.1 Approach in Vosviewer

VOSviewer constructs a map based on a co-occurrence matrix. The construction of a map is a process that consists of three steps. In the first step, a similarity matrix is calculated based on the co-occurrence matrix. In the second step, a map is constructed by applying the VOS mapping technique to the similarity matrix. And finally, in the third step, the map is translated, rotated, and perhaps also reflected

Step 1: Similarity Index
VOS mapping technique requires a similarity matrix as input. A similarity matrix can be obtained from a co-occurrence matrix by normalizing the latter matrix, that is, by correcting the matrix for differences in the total number of occurrences or co-occurrences of items [22]. VOSviewer, however, does not use one of these similarity measures. Instead, it uses a similarity measure known as the association strength [22, 23]. This similarity measure is sometimes also referred to as the proximity index [15] or as the probabilistic affinity index as shown in 1 [24].

$$Sij = \frac{Cij}{wiwj}. \tag{1}$$

Step 2: Mapping Technique
VOS has the tendency to locate objects close to what we have called their ideal coordinates. The ideal coordinates of an object *i* are defined as a weighted average of the coordinates of all other objects, where the coordinates of objects more similar to object *i* are given higher weight in the calculation of the weighted average [19, 25]. Second, VOS seems to pay more attention to indirect similarities via third objects multidimensional scaling also known as MDS [19, 25]. The idea of the VOS mapping technique is to minimize a weighted sum of the squared Euclidean distances between all pairs of items. The higher the similarity between two items, the higher the weight of

their squared distance in the summation. To avoid trivial maps in which all items have the same location, the constraint is imposed that the average distance between two items must be equal to 1. In mathematical notation, the objective function to be minimized is given by 2 [26]

$$E(xi, \ldots, xn) = \sum\nolimits_{i,j} sij\|xi - xj\|^2 \tag{2}$$

where the vector **x***i* = (*xi*1; *xi*2) denotes the location of item *i* in a two-dimensional map and where $k \cdot k$ denotes the Euclidean norm. Minimization of the objective function is performed subject to the constraint as shows in [27]

$$\frac{2}{n(n-1)} \sum_{i<j} \|xi - xj\| = 1 \tag{3}$$

Step 3: Translation, Rotation, and Reflection
The optimization problem discussed in step 2 does not have a unique globally optimal solution. This is because, if a solution is globally optimal, any translation, rotation, or reflection of the solution must also be globally optimal. It is of course important that VOSviewer produces consistent results. The same co-occurrence matrix should therefore always yield the same map. To accomplish this, it is necessary to transform the solution obtained for the optimization problem discussed in step 2. VOSviewer applies the three transformations to the solution which are translation, rotation and reflection. These three transformations are sufficient to ensure that VOSviewer produces consistent results.

In Vosviewer, there are several stages that we need to choose instead of choosing full counting which means every document and author that was involved in the publication were counted. The idea of full counting is to know the relationship between authors and publication. The next process was to determine the number of occurrences. Normal execution by a previous study is five. It means that if there are terms which occur fewer than five times, their existence will be excluded. Therefore, it depends on the objectives of the study as well as how the interpretation of the data that might happen after the process. In this study, we chose a 5 occurrences keyword to be included in our output which was used in numbers of previous studies [28]. At the final stage, we need to decide whether to choose all words including the general word and word that has in par occurrence or to choose the selected words only. In the system, the suggestion is to take only 60% of the total words to be the output. After all stages are completed, the process behind the system will start to calculate and generate the clusters as the output.

4 Analysis

In this section, we will discuss in depth the analysis obtained from VosViewer. There are three types of output from the application and one of them is a cluster map. A cluster map basically implements the social network map to illustrate the results from the process [29]. It was used to reveal the internal relationship between intra word in the cluster as well as inter-theme words between clusters [29].

4.1 Cluster Analysis

In Fig. 1, there are 4 clusters obtained from our experiment as shown in Fig. 1. Each cluster has its own color and has its own highlighted words which represent the clusters. The clusters consist of a blue cluster, red cluster, yellow cluster and green cluster. Each cluster consists of words that have a relationship with each other, the highlighted word which has a significant value in occurrences as well as the link between each word. As for link between words, the bigger value of link strength, the closer between words with each other. The average number of keywords in each cluster is between 50 to 150 keywords. Each cluster will normally represent one theme or area. In previous studies, all clusters produced from the expert application represented either a research theme, subject area or focus area which they can use as a basis to other research such as to produce a potential subject area or potential research area.

Fig. 1. Cluster map produced from Vosviewer

4.1.1 Dominant Cluster

In our analysis, the most dominant cluster is a red cluster which has the most number of words and the biggest total strength of occurrences for highlight circle. Out of a total of 147 words in the red cluster the word with the most occurrences is 'Chemistry' with a total of 969 occurrences. In Table 1, the top 20 most occurring words in the red cluster is presented.

Table 1. Top 20 most occurrences words in red cluster

	Label	Weight of link	Weight of total link strength	Weight of occurrences	Score of Avg. pub. year	Score of Avg. citations	Score of Avg. norm. citations
Top 5 words	Chemistry	455	11193	969	2011.5851	30.3034	1.1817
	Scanning electron microscopy	332	3970	736	2009.9674	41.5367	1.448
	X ray diffraction	274	3212	538	2011.1152	40.0353	1.525
Bottom 3 words	Industrial waste	179	892	96	2008.6042	91.0208	2.7978
	Dye	163	959	93	2009.3333	96.871	3.0246
	Isotherm	129	1154	93	2009.2796	117.0645	3.4741

In Table 2, the greatest weight value for occurrences with a total of 969 occurrences has the value weight of total link strength with 11193 link strength. Total link strength is the attribute indicating the total strength of the co-authorship links of a given researcher with other researchers. This explains that this word – 'Chemistry' frequently appeared in a number of publications and was always used by number of researchers that collaborate with each other. Nevertheless, it can't conclude that these criteria correlate with each other. In others words, although the number of occurrences has a high value, the numbers of total link strength isn't high if compared between each word in Table 1. This depends on word usage in the publication. Each researcher has their own focus in their research. Still, all words in red clusters were related with each other because there was a connection between each words. Therefore, clusters can represent topic research, research theme and subject area. Sometimes a research topic can be a trend too. It is due to the popularity of the current situation of the environment. Thus, the year of publication is another major factor in the establishment of the clusters.

4.1.2 Average Year of Publication

The average publication year of the documents in which a keyword or a term occurs or the average publication year of the documents published by a source, an author, an organization, or a country. In Table 1, the word 'Chemistry' became prominent in 2011. Whereas the second highest value for occurrence is a word 'scanning electron microscopy' which has an average year of 2009. The differentiation years in Table 1 indicate that although the words were used several years ago, it might be valid to use in current publication. It depends on researcher's creativity to utilize these words to develop new research topic. Table 2 shows the list of words that have been used in recent years. In 2013, there are two words that became so prominent among researchers which are 'drug effects' with a total of 385 occurrences and 'psychology' with a total of 132 occurrences. Among all the words in Table 3, there are words that have low occurrences but have only emerged in recent years such as 'tumor cell line' with a total of 85 occurrences and 'reactive oxygen metabolite' with a total of 80 occurrences.

This might be due to less usage in publications or this area of study is new and has not yet revealed massive discoveries by our researchers. Besides the words less often used, there are words that are situated between these two groups – low usage and frequent usage such as 'nanofluidics – 191 occurrences, 'real time polymerate chain reaction' – 101 occurrences and 'biodiesel production' – 137 occurrences. In conclusion, all words produced by the expert application can be used in various and creative ways in order to develop new and emerging research

Table 2. Analysis for list of words that been used in recent years in all clusters

Label	Weight of link	Weight of total link strength	Weight of occurrences	Score of Avg. pub. year	Score of Avg. citations	Score of Avg. norm. citations
Drug effects	385	5027	310	2013.9323	17.8387	0.9231
Psychology	139	763	132	2013.6667	11.7197	0.5034
Nanofluidics	53	331	191	2012.9162	45.5864	2.0353
Tumor cell line	241	1560	85	2012.4471	28.8235	1.2296
Nanocomposites	185	1141	214	2012.215	47.1682	2.0179
Reactive oxygen metabolite	222	1170	80	2012.175	32.125	1.349
Biodiesel production	109	864	137	2011.8394	67.4526	2.6851
Real time polymerase chain reaction	267	1114	101	2011.8713	28.9802	1.174

4.1.3 Average Citation of Publication

Average citation is also another point that might benefit other researchers. Table 3 is the analysis of top and bottom 5 words that gained the greatest score of average citation in all clusters. Top 5 is the words that gained huge score of average citation and bottom 5 is the word that gained least score of average citation. Although there is a huge difference between the top 5 and bottom 5, nevertheless, it doesn't means that the bottom 5 words are insignificant to use in the current year or environment. It is just to indicate that these words were less used and has less coverage by the researchers. Most of the words included in the bottom 5 were appeared in this recent years; 2010 to 2013. This can indicate that these words have the potential to be discovered and to be used in future research. Therefore, every single word produced by this expert application is beneficial to use in a creative way. Another interesting discovery in Table 4 is in the bottom 5 words, there is a word that gains a high number of weight of total link strength but is low in citation namely 'body mass index'. This might happen when it has a link with another word towards 'body mass index' but still, dos not have a lot of coverage in the publication.

4.1.4 Formation of Subject Area or Research Theme

In clusters shown in Fig. 1, as mentioned before, there are four clusters – red, blue, yellow and green. Each clusters represents a research theme or subject area. In this study, we focus on the formation of potential subject areas by observing and mapping all words in each cluster with another trusted database. It is quite difficult to appoint which cluster represents which subject area. We used a previous database from a previous study as shown in Table 4. In Table 4, all related and major subject areas that possibly can be mapped with our cluster were listed [30]. This acts as a guidance for us to map and assign all clusters. Each listed subject area in Table 4 has their own group of words that can be used for this study as shown in Table 5.

Table 3. Analysis of word for greatest score of average citation in all clusters

Label	Weight of link	Weight of total link strength	Weight of occurrences	Score of Avg. pub. year	Score of Avg. citations	Score of Avg. norm. citations
Water purification	150	952	81	2008.6667	128.3086	3.8258
Practice guideline	200	639	91	2010.3187	119.6264	4.4664
Side effect	167	789	77	2009.7532	119.2078	4.1537
Body mass index	177	1282	125	2011.128	19.648	0.6943
Drug effects	385	5027	310	2013.9323	17.8387	0.9231

Table 4. List of major subject area from previous study to map clusters

No	Subject area	No	Subject area	No	Subject area
1	Biomedical sciences 1	8	Biomedical sciences 2	15	Physical sciences 4
2	Environmental sciences 1	9	Medical sciences 1	16	Physical sciences 5
3	Physical sciences 1	10	Medical sciences 2	17	Physical sciences 6
4	Social and health sciences	11	Food and agricultural sciences	18	Medical sciences 4
5	Mathematics and computer sciences	12	Environmental sciences 2	19	Physical sciences 7
6	Physical sciences 2	13	Physical sciences 3	20	Earth sciences
7	Cognitive sciences	14	Medical sciences 3		

Since the existence of discipline or subject area in the education were quite extensive, it is quite hard to determine the subject area in our experiment. Thus, we used the simple approach to determine the subject area which we already mentioned before.

Our first stage is to identify the major subject areas which has already existed for a long time such as medical, chemistry and others [31] with existing database discovered by previous researchers in Table 4. Consequently, in our experiment, we need to do some another analysis towards words in the existing database which have been grouped according subject area in order to determine which subject area in the existing database really represents a major subject area. Finally, we managed to list the major subject areas by doing a comparison between listed words in existing database [30] with the previous study in discipline in education [32]. After we managed to list the major subject area, only then could we map and assign the cluster accordingly with identifying words which exist in the clusters. Thus, our end results as we described before, we managed to produce four clusters with three subject areas - is Social and Health Sciences for blue cluster, Physical Sciences 1 for red cluster and Biomedical Sciences 1 for yellow cluster. Apparently, green cluster needed to be excluded in this study due to the majority of words representing general words such as 'young adult', 'comparative study', 'methodology', 'physiology', 'prevalence', 'questionnaire', 'cross-sectional studies' - which do not illustrate any related subject area. These words were not in the stop words category but it can be used to give an idea to the researchers which method of study that existed in the research environment. In Table 5 is the list of cluster with the mapped subject area.

Table 5. List of cluster with the mapped subject area

Red cluster (Physical sciences 1)7		Blue cluster (Social and health sciences)		Yellow cluster (Biomedical sciences 1)	
Word in cluster	Weight co-occurrences	Word in cluster	Weight co-occurrences	Word in cluster	Weight co-occurrences
Chemistry	969	Animal experiment	850	Genetics	610
Optimization	755	Metabolism	782	Nucleotide sequence	534
Scanning electron microscopy	736	Plant extract	654	Polymerase chain reaction	460
Algorithms	607	Human cell	628	Isolation and purification	404

From all analysis already presented in this section, there is a lot of knowledge discovered that we can gain from each table. Each of the words produced from the expert application has its own benefits and contribution towards research environment. We can see the analysis in detail – which is either go through each word in every cluster or see the presentation cluster itself. Most previous studies really study in depth and detail each cluster. Some of the previous research even manipulates this cluster to make another new algorithm in order to generate new and comprehensive clusters and words. All in all, whether we just accept the expert application output or manipulate it and produce another clusters, either way is significant to use in order to make sure these words and clusters gives some impact to other researchers and future research. Further discussion regarding this analysis and related discussion will be in the next section.

5 Discussion and Conclusion

With the clusters obtained in this experiment, we allege that our main objective in this study which was to study and identify the potential subject area hidden in index keywords was achieved. The index keywords were manipulated by our selected criteria in order to make sure that we successfully processed the data. As for a conclusion for this experiment, we believed that index keywords has the capabilities to produce the initial idea to form the subject area. Although this is the initial stage to get the overview of subject area for the assessment, we believe that clusters that we earned in this experiment can give a heads up for management to improve the assessment framework. Since this is the first study that has been carried out in Malaysia to determine subject area by manipulating bibliometric element, apparently, there are a number of studies which have already been done by implementing the same approach [13]. Most of the previous research conducted such study to get the idea or state-of-the-art and insight into the conceptual structure of research discipline and to help visualize the division of the field into several interconnected subfields. Bibliometric maps created by co-word analysis can be used by both experts and novices to understand the current state of the art of a scientific field and to predict where future research could lead [33]. Since our main objective is to identify the subject area from index keywords which consist of bibliometric elements in publication, we can claim that our objective was achieved by the clusters produced in this study. This study also can benefit others to get the insight that has the potential to be implemented in Malaysian research assessment [34].

The next recommendation is to identify the relationship between compilations of these subject areas with selected peer review ranking. Our initial hypothesis is there is correlation between the subject areas which form from clusters with peer review ranking in terms of a number of publications and citation analysis. It might looks like a small contribution from our experiment but without it we realize that, among all publications among assessment and publication made by our Malaysian researchers, no studies have been conducted to determine the potential subject area which can be implemented in our local research assessment. Whereas, other local assessment have been carried out in other countries, most of the assessments were done by subject area, therefore, they managed to determine which subject area needs to get full focus and needs the development [35]. Consequently, one of the ways to develop and produce the subject area which needs to be focused on is by manipulating the current publication information such as increase the number of publication or produce variety of research in order to predict for the future.

References

1. John, N.A.: Sharing and Web 2.0: the emergence of a keyword. New Media Soc. **15**(2), 167–182 (2012)
2. Chen, G., et al.: Identifying the research focus of Library and Information Science institutions in China with institution-specific keywords. Scientometrics **103**(2), 707–724 (2015)
3. Davis, M.A.: Title keyword selection and use for optimum document retrieval. Public Access Serv. Q. **2**, 15–22 (1997)
4. Eassom, H.: How to Choose Effective Keywords for Your Article. Discover the Future of Research (2017)
5. Jaewoo, C., Woonsun, K.: Themes and trends in Korean educational technology research: a social network analysis of keywords. Proc. Soc. Behav. Sci. **131**, 171–176 (2014)
6. Sedighi, M.: Application of word co-occurrence analysis method in mapping of the scientific fields (case study: the field of Informetrics). Libra. Rev. **65**(1/2), 52–64 (2016)
7. Muñoz-Leiva, F., et al.: An application of co-word analysis and bibliometric maps for detecting the most highlighting themes in the consumer behaviour research from a longitudinal perspective. Qual. Quant. **46**(4), 1077–1095 (2011)
8. Eck, N.J.V., Waltman, L.: Text mining and visualization using VOSviewer. Centre for Science and Technology Studies, Leiden University, The Netherlands. arXiv preprint arXiv: 1109.2058 (2011)
9. Alfonzo, P.M., Sakraida, T.J., Hastings-Tolsma, M.: Bibliometrics visualizing the impact of nursing research. Online J. Nurs. Inform. **18**(1), 2014 (2014)
10. Bazm, S., Kalantar, S.M., Mirzaei, M.: Bibliometric mapping and clustering analysis of Iranian papers on reproductive medicine in Scopus. Int. J. Reprod. Biomed. (Yazd). **14**(6), 371–382 (2016)
11. Nicholson, S.: The basis for bibliomining: frameworks for bringing together usage-based data mining and bibliometrics through data warehousing in digital library services. Inf. Process. Manag. **42**(3), 785–804 (2006)
12. Azam, I., et al.: Bibliomining on North South University Library Data. In: 2013 Eighth International Conference on Digital Information Management (ICDIM). IEEE (2013)
13. Li, M., Chu, Y.: Explore the research front of a specific research theme based on a novel technique of enhanced co-word analysis. J. Inform. Sci. **43**, 016555151666191 (2016)
14. Leydesdorff, L., Welbers, K.: The semantic mapping of words and co-words in contexts. J. Inform. **5**(3), 469–475 (2011)
15. Peters, H.P.F., van Raan, A.F.J.: Co-word-based science maps of chemical engineering. Part I: representations by direct multidimensional scaling. Res. Policy **22**(1), 23–45 (1993)
16. Wanga, D., et al.: t-Test feature selection approach based on term frequency for text categorization. Pattern Recogn. Lett. **45**, 1–10 (2014)
17. Azam, N., Yao, J.T.: Comparison of term frequency and document frequency based feature selection metrics in text categorization. Expert Syst. Appl. **39**(5), 4760–4768 (2012)
18. Trstenjak, B., Mika, S., Donko, D.: KNN with TF-IDF based framework for text categorization. Proc. Eng. **69**, 1356–1364 (2014)

19. Eck, N.J.V., Waltman, L.: VOS a new method for visualizing similarities between objects. In: Advances in Data Analysis. Springer, Berlin, pp. 299–306 (2007)
20. Lewis, B.R., Templeton, G.F., Luo, X.: A Scientometric investigation into the validity of IS journal quality measures. J. Assoc. Inf. Syst. **8**(12), 619–633 (2007)
21. Rahm, E., Do, H.H.: Data cleaning: problems and current approaches. IEEE Data Eng. Bull. **23**(4), 3–13 (2000)
22. Eck, N.J.V., Waltman, L.: Bibliometric mapping of the computational intelligence field. Int. J. Uncertain. Fuzziness Knowl. Based Syst. **15**(05), 625–645 (2007)
23. Eck, N.J.V., et al.: Visualizing the computational intelligence field [Application Notes]. In: IEEE Computational Intelligence Magazine, vol. 1, no. 4. (2006)
24. Information, M.Z., Bassecoulard, E., Okubo, Y.: Shadows of the past in international cooperation: collaboration profiles of the top five producers of science. Scientometrics **47**(3), 627 (2004)
25. Eck, N.J.V., Waltman, L.: VOSviewer a Computer Program for Bibliometric Mapping (2009)
26. Collinge, W., Yarnold, P., Soltysik, R.: Fibromyalgia symptom reduction by online behavioral self-monitoring, longitudinal single subject analysis and automated delivery of individualized guidance. N. Am. J. Med. Sci. **5**(9), 546 (2013)
27. Papatheodorou, C., et al.: Mining user communities in digital libraries. Inf. Technol. Libr. **22** (4), 152–157 (2003)
28. Zahedi, Z., Eck, N.J.V.: Visualizing readership activity of Mendeley users using VOSviewer. In altmetrics14: Expanding Impacts and Metrics, Workshop at Web Science Conference, vol. 1041819 (2014)
29. Yang, Y., Wu, M., Cui, L.: Integration of three visualization methods based on co-word analysis. Scientometrics **90**(2), 659–673 (2011)
30. Waltman, L., Eck, N.J.V.: A new methodology for constructing a publication-level classification system of science. J. Am. Soc. Inform. Sci. Technol. **63**(12), 2378–2392 (2012)
31. Furlong, J.: Education: An Anatomy of the Discipline: Rescuing the University Project?. Routledge, Abingdon (2013)
32. Giurgiutiu, V., Bayoumi, A.-M.E., Nall, G.: Mechatronics and smart structures emerging engineering disciplines for the third millennium. Mechatronics **12**(2), 169–181 (2002)
33. Viedma-Del-Jesus, M.I., et al.: Sketching the first 45 years of the journal Psychophysiology (1964–2008): a co-word-based analysis. Psychophysiology **48**(8), 1029–1036 (2011)
34. MyRA, e-MyRA (Malaysia Research Assessment) (2014)
35. Calver, M.C., Fontaine, J.B., Linke, T.E.: Publication models in a changing environment: Bibliometric analysis of books and book chapters using publications by Surrey Beatty & Sons. Pac. Consev. Biol. **19**(3–4), 394–408 (2013)

The Impacts of Singular Value Decomposition Algorithm Toward Indonesian Language Text Documents Clustering

Muhammad Ihsan Jambak[1(✉)], Fathey Mohammed[2], Novita Hidayati[1], Rusdi Efendi[1], and Rifkie Primartha[1]

[1] Faculty of Computer Science, Sriwijaya University, Palembang, Indonesia
jambak@ilkom.unsri.ac.id, novitahidayati52@gmail.com, rusdiefendi8@gmail.com, rifkie77@gmail.com
[2] Faculty of Engineering and Information Technology, Taiz University, Taiz, Yemen
fathey.m.ye@gmail.com

Abstract. Data with a high dimension mean that the data has many sets of variables. Conventional clustering algorithms are only able to deal with low-dimensional data conditions, so the clustering of high-dimensional data objects poses a challenge for resolving the solution. Indonesian language documents have specificity due to grammatical differences compared to English documents. Therefore, this research implements a Singular Value Decomposition (SVD) algorithm to reduce these dimensions and analyses its impact on the accuracy of clustering methods such as k-Means and k-Medoids. Results show that combining SVD with both clustering methods increased accuracy by 11 and 10%, respectively. Additionally, processing times were also proven to be faster.

Keywords: Dimension reduction · Singular value decomposition Clustering · k-Means · k-Medoids

1 Introduction

Cluster analysis, also referred to as data segmentation, is essentially a statistical method that performs the grouping or segmentation of a collection of objects. Clustering refers to seeking a partition of data into distinct groups so that the observations within each group are quite similar to each other. Then in a cluster, each data/object one with the other will become more closely related than the data/objects that are in different clusters. To make this concrete, we must define what it means for two or more observations to be similar or different. In other words, clustering looks for homogeneous subgroups among observations. Therefore the clusters formed have high internal homogeneity and high external heterogeneity [1, 2]. From this clustering an object can be described by a set of measurement parameters, or by its relation to other objects. Cluster analysis is also used to form descriptive statistics to ascertain whether the data consist of different sets of subgroups, with each group representing objects with substantially different properties.

F. Saeed et al. (Eds.): IRICT 2018, AISC 843, pp. 173–183, 2019.
https://doi.org/10.1007/978-3-319-99007-1_17

Cluster analysis can be applied in the field of Informatics Science; the development of Big Data technology today increasingly reinforces the need for an accurate clustering method, especially in the development of Data Mining/Text Mining applications. Documents presented in large quantities and types require clustering process to facilitate the search for information about a particular field [3], for example a journal document publication.

Each journal document is a set of keywords where each keyword presents a dimension, thus one document can contain many dimensions or attributes or is said to be high-dimensional data. In most learning algorithms, the complexity depends on the number if input dimensions as well as on the size of the data sample. Meanwhile, conventional clustering algorithms are only able to deal with low-dimensional data conditions; therefore, decreasing the dimension also decreases the complexity of the inference algorithm during testing [4, 5]. Thus, finding clusters of high-dimensional data objects is a challenge because high-dimensional data tend to have noise, sparse, and skew.

The focus of cluster analysis is to compare objects based on sets of variables. The set of cluster variables is a set of variables that represent the characteristics of objects. Cluster solution as a whole depends on variables that are used as a basis for assessing the similarity or dissimilarity of the object, so that the addition or subtraction of relevant variables can affect the substance of the clustering results.

Data with a high dimension mean that the data has many sets of variables. To improve the clustering results in high-dimensional data should be done dimensional reduction [4]. Dimensional reduction is a technique in Text Mining by reducing the dimensions so that clustering processes data with reduced number of features. For textual data, this method is also known as Latent Semantic Analysis (LSA) [6, 7].

Several dimensional reduction studies have been conducted on English documents [8–10]. Indonesian language documents have several grammatical differences compared to English document. In consequence, this research was conducted to determine the effect of dimensional reduction on the accuracy of clustering results in high-dimensional documents in the Indonesian language.

2 Singular Value Decomposition

In a system for decision making, observation data that we believe contain information are taken as inputs and fed to the system. Ideally, we should be able to use whichever features are necessary and discarding the irrelevant [5]. When data dimensions are high, usually only a small number of dimensions are relevant to a particular cluster, but data in irrelevant dimensions can generate a lot of noise and overwhelm the clusters that should be formed. In addition, when the dimension increases, the data cluster usually becomes further away because the data points are generally located in various parts of the spatial dimension. When the data becomes very far, the data points located in various dimensions can be considered to have the same distance. An important distance calculation in clustering will be meaningless.

To solve this problem, reduction can be done by capturing the characteristics of the data by mapping the data set from the original dimension to another relatively low dimension. This mapping produces a component principal which can then be retrieved components or features of a new dimension that have a great influence on the dataset and discard the data that has no effect [4]. Dimensional reduction process applied before clustering process and after words preprocesses and words weighting.

SVD is a form of factor analysis on the matrix, the SVD matrix loading the occurrence frequency of keyword decomposed into three matrix components [10, 11]. The first matrix component (U) describes the line entity as an orthogonal vector matrix; the second matrix component (S) is a diagonal matrix that contains the scalar value of the matrix; and the third component (V) is the column entity's matrix as an orthogonal vector matrix. In this study, the SVD of the TF-IDF matrix will be used in a linear transformation approach (Fig. 1).

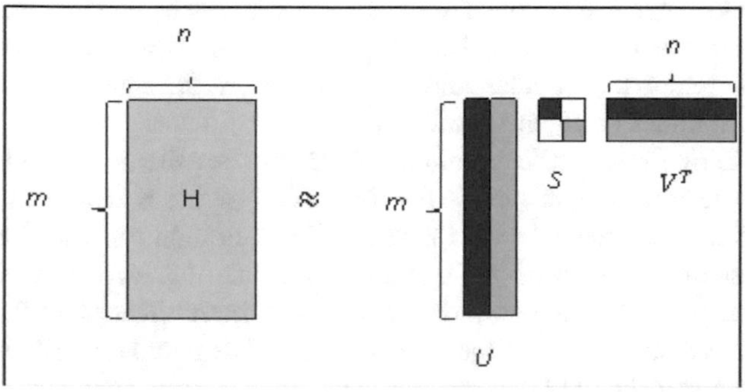

Fig. 1. Singular value decomposition [4]

SVD is basically used to estimate the rank of the matrix. If the H matrix is known by the dimension m × n, where m ≥ n and rank (H) = r, then the singular value decomposition of H is defined by the following equation:

$$H = USV^T \tag{1}$$

where:

H: TF-IDF matrix
S: singular value
U: left singular vector
m: document
V: right singular vector
n: term
T: transpose
r: rank

then:

$$U^T U = V^T V = I_n \tag{2}$$

and meet the conditions:

$$S = diag(\sigma_1, \cdots, \sigma_n) \tag{3}$$

where:

$$\sigma_I > 0 \quad for \quad 1 \leq i \leq r$$
$$\sigma_j = 0 \quad for \quad j \geq r + 1$$

The first column of the U and V matrices defines an orthonormal eigen vector corresponding to the value of the nonzero eigenvector vector of the HH^T and H^TH matrices, respectively. The columns of the U and V matrices contain vectors known as the left and right singular vectors. The singular value of H is the diagonal element of the matrix S, in which the singular value is derived from the square root of the attribute value of the n number of eigen values HH^T.

After obtaining three matrices from the SVD process, the next step in reducing the dimension of the matrix is to reduce the dimension of the S matrix in the form of a diagonal matrix. The smallest scalar matrix value belonging to the S matrix will be omitted so that one column of the U matrix and one V matrix line also disappears according to the singular value of the S matrix. Next, multiplication of the new U and S matrices is performed to produce the new H matrix. The new H matrix will contain the principal component data [12].

The U, S, and V matrix SVD results are then used as inputs for Latent Semantic Analysis (LSA). LSA is a technique of representing a text in the form of a matrix that contains the frequency or number of occurrences of words in a document. Each line in the matrix is the word frequency in each sentence column. LSA requires the SVD process of the resulting matrix with the word line and column of the sentence. SVD is performed as a process of matrix compositing with the intention of being a noise reduction process, so the matrix obtained is a clean matrix of noise dimension. The algorithm combination SVD and clustering methods are as follows [13]: Reduction of dimension matrix of the weighting of all documents using SVD; resulting U, S, and V^T matrix. Multiply the U and S matrices resulting in the H matrix. Perform multiplication between S matrixes. Then, the multiplication matrix is transposed resulting in the J matrix. Perform the clustering process with the matrices H and J that have been reduced as input. Then, use cosine distance to calculate of similarity between documents.

The example of LSA calculations using pre-processing sentences in Indonesian language is as follows:

- Sentence#1: novita yumi putus hengkang usaha
- Sentence#2: yumi putus undur usaha komputer
- Sentence#3: novita yumi jabat wakil direktur komputer undur diri
- Sentence#4: yumi umum facebook tinggal posisi

Each document undergoes the pre-process stages to determine the respective term frequencies (tf-idf) (Table 1).

Table 1. Words weighting results

Term	Sentence#1	Sentence#1	Sentence#1	Sentence#1
novita	1	0	1	0
yumi	1	1	1	1
putus	1	1	0	0
hengkang	1	0	0	0
usaha	1	1	0	0
undur	0	1	1	0
komputer	0	1	1	0
jabat	0	0	1	0
wakil	0	0	1	0
direktur	0	0	1	0
facebook	0	0	0	1
tinggal	0	0	0	1
posisi	0	0	0	1

Next, words' term frequencies are broken down using SVD into U, S, and V^T matrix. The words in the term frequency matrix are represented by the row vectors of U, while the sentence is represented by the columns of the vector V^T (Table 2).

Table 2. Term frequency downsizing into U, S, and V^T matrix

Matrix U			
-0.332	-0.085	-0.102	0.604
-0.561	0.183	0.262	0.019
-0.307	0.381	-0.285	-0.126
-0.144	0.275	-0.162	0.441
-0.307	0.381	-0.285	-0.126
-0.351	-0.253	-0.064	-0.406
-0.351	-0.253	-0.064	-0.406
-0.188	-0.359	0.059	0.162
-0.188	-0.359	0.059	0.162
-0.188	-0.359	0.059	0.162
-0.066	0.162	0.488	-0.017
-0.066	0.162	0.488	-0.017
-0.066	0.162	0.488	-0.017

Matrix S			
3.389	0	0	0
0	2.033	0	0
0	0	1.880	0

Matrix V^T			
Sent#1	**Sent#2**	**Sent#3**	**Sent#4**
-0.487	-0.554	-0.637	-0.224
0.558	0.216	-0.730	0.329
-0.305	-0.232	0.112	0.917
0.599	-0.770	0.220	-0.023

The two vectors are scaled by multiplying singular values of the S matrix. These two matrices become the inputs of the clustering process. Thus, the representation of the words derived from the U × S line vectors are as follows:

$$\text{novita} = \begin{bmatrix} -1.124 \\ -0.172 \\ -0.193 \\ 0.819 \end{bmatrix};\ \text{yumi} = \begin{bmatrix} -1.902 \\ 0.373 \\ 0.493 \\ 0.026 \end{bmatrix};\ \text{putus} = \begin{bmatrix} -1.041 \\ 0.774 \\ -0.536 \\ -0.171 \end{bmatrix};\ \text{hengkang} = \begin{bmatrix} -0.487 \\ 0.558 \\ -0.305 \\ 0.599 \end{bmatrix};\ \text{usaha} = \begin{bmatrix} 1.041 \\ 0.774 \\ -0.536 \\ -0.171 \end{bmatrix}$$

$$\text{undur} = \begin{bmatrix} -1.191 \\ -0.514 \\ -0.120 \\ -0.550 \end{bmatrix};\ \text{komputer} = \begin{bmatrix} -1.191 \\ -0.514 \\ -0.120 \\ -0.550 \end{bmatrix};\ \text{jabat} = \begin{bmatrix} -0.637 \\ -0.730 \\ 0.112 \\ 0.220 \end{bmatrix};\ \text{wakil} = \begin{bmatrix} -0.637 \\ -0.730 \\ 0.112 \\ 0.220 \end{bmatrix};\ \text{direktur} = \begin{bmatrix} -0.637 \\ -0.730 \\ 0.112 \\ 0.220 \end{bmatrix}$$

$$\text{facebook} = \begin{bmatrix} -0.224 \\ 0.329 \\ 0.917 \\ -0.023 \end{bmatrix};\ \text{tinggal} = \begin{bmatrix} -0.224 \\ 0.329 \\ 0.917 \\ -0.023 \end{bmatrix};\ \text{posisi} = \begin{bmatrix} -0.224 \\ 0.329 \\ 0.917 \\ -0.023 \end{bmatrix}$$

and the representation of sentences derived from the SxV^T column vectors are:

$$\text{Sent\#1} = \begin{bmatrix} -1.651 \\ 1.135 \\ -0.573 \\ 0.812 \end{bmatrix};\ \text{Sent\#2} = \begin{bmatrix} -1.878 \\ 0.439 \\ -0.436 \\ -1.044 \end{bmatrix};\ \text{Sent\#3} = \begin{bmatrix} -2.159 \\ -1.485 \\ 0.210 \\ 0.298 \end{bmatrix};$$

$$\text{Sent\#4} = \begin{bmatrix} -0.759 \\ 0.669 \\ 1.725 \\ -0.031 \end{bmatrix}$$

The measurement of similarity value in clustering used is cosine distance, which is a calculation of similarity based on the cosine angle between two vectors, where the value of similarity has a range between 0–1. More closer the value to 1, the more similar the pairs of sentences. Here is an example of calculating the level of similarity between the 1st sentence and the 2nd sentence:

- Sentence#1: novita yumi putus hengkang usaha.
- Sentence#2: yumi putus undur usaha komputer.

$$\text{Sent\#1} = \frac{\begin{bmatrix} -1.124 \\ -0.172 \\ -0.193 \\ 0.819 \end{bmatrix} + \begin{bmatrix} -1.902 \\ 0.373 \\ 0.493 \\ 0.026 \end{bmatrix} + \begin{bmatrix} -1.041 \\ 0.774 \\ -0.536 \\ -0.171 \end{bmatrix} + \begin{bmatrix} -0.487 \\ 0.558 \\ -0.305 \\ 0.599 \end{bmatrix} + \begin{bmatrix} -1.041 \\ 0.774 \\ -0.536 \\ -0.171 \end{bmatrix}}{5} = \begin{bmatrix} -1.119 \\ 0.462 \\ -0.215 \\ 0.220 \end{bmatrix}$$

$$\text{Sent\#2} = \frac{\begin{bmatrix} -1.902 \\ 0.373 \\ 0.493 \\ 0.026 \end{bmatrix} + \begin{bmatrix} -1.041 \\ 0.774 \\ -0.536 \\ -0.171 \end{bmatrix} + \begin{bmatrix} -1.191 \\ -0.514 \\ -0.120 \\ -0.550 \end{bmatrix} + \begin{bmatrix} -1.041 \\ 0.774 \\ -0.536 \\ -0.171 \end{bmatrix} + \begin{bmatrix} -1.191 \\ -0.514 \\ -0.120 \\ -0.550 \end{bmatrix}}{5} = \begin{bmatrix} -1.878 \\ 0.439 \\ -0.436 \\ -1.044 \end{bmatrix}$$

$$|\text{Sent}\#1| = \sqrt{-1.119^2 + 0.462^2 + -0.215^2 + 0.220^2} = 1.249$$
$$|\text{Sent}\#2| = \sqrt{-1.878^2 + 0.439^2 - 0.436^2 - 1.044^2} = 2.236$$

$$\text{Sim}(\text{Sent}\#1, \text{Sent}\#2) = \frac{\text{Sent}\#1.\text{Sent}\#2}{|\text{Sent}\#1||\text{Sent}\#2|} = \frac{\begin{bmatrix} -1.119 \\ 0.462 \\ -0.215 \\ 0.220 \end{bmatrix} \cdot \begin{bmatrix} -1.878 \\ 0.439 \\ -0.436 \\ -1.044 \end{bmatrix}}{1.249 \times 2.236} = \frac{2.168}{2.792}$$
$$= 0.776$$

Thus, the similarity value between 1st sentence and 2nd sentence is 0.776.

3 K-Means Clustering

K-Means is one of the most commonly used centroid-based partitioning clustering methods due to its simplicity. This technique clustering objects into k groups. To perform this clustering, the value of k must be determined first. The process of k-Means is to choose k centers as centroids for use in clustering data. The process of determining the centroids and placement of the data in the cluster is repeated until the centroids value does not change any more. The steps in the k-Means algorithm are as follow [4, 9, 10]:

1. Determine the number of clusters (k) as the number of clusters to be formed;
2. Generate k centroids (starting point) randomly;
3. For each data, locate the nearest cluster center. So, each cluster center has a subset of datasets;
4. For each cluster k, find the center of the cluster area, update the location of each cluster center to the new value of the center area;
5. Repeat steps 3 and 5 to the data in each cluster becomes centralized or completed.

The centroid points will determine success in the k-Means algorithm, the centroid points is often a problem that must be overcome due to its random selection. If the random values of the selected centroid points at the beginning are incorrect, then the cluster results are not maximal.

4 K-Medoid Clustering

K-Medoids is a partition clustering method which is an improvement of the k-means method. Because of grouping objects into k clusters by minimizing absolute error, k-Medoids more resistant to noise compared with the k-Means. In finding k cluster, first conducted random selection of representative data on each cluster. Cluster representation is one of the points chosen to be representative of the cluster that is medoid.

The cluster is built by calculating the proximity it has between the medoid and the non-medoid object. The steps in the k-Medoids algorithm are as follow [4, 14]:

1. Determine the number of clusters (k) as the number of clusters to be formed;
2. Generate the initial medoids randomly;
3. Calculate the similarity medoid with non-medoid using Cosine distance formula;
4. Place the non-medoid objects into the cluster closest to the medoid. For each data, find the closest cluster center so that each cluster center has a subset of datasets;
5. Randomly select one of the non-medoids, called O_{random};
6. Calculate the difference between the total similarity initial medoid (Oj) with the total similarity O_{random}. Where, S = total previous similarity (Oj) - total newest similarity (O_{random});
7. If S < 0 then there is a medoid exchange, change O_{random} to Oj, in other words O_{random} become medoid;
8. Repeat steps 3 to 6 until the data on each cluster becomes centralized or completed.

5 Simulations and Experimental Result

5.1 Data Preprocessing

The testing data included text documents in Indonesian language journals downloaded through the Indonesian Publication Index website (id.portalgaruda.org) stored in the ASCII file (*.txt) format. The data have five natural classes based on the research topic of journals: economic, law, computer, education, and veterinary. Each journal topic consists of 20 documents, for total testing data of 100 documents.

The first phase was text preprocessing, which aims to improve the quality of features and reduce the difficulty in the text mining process. Stages performed on the testing data are case folding, tokenizing, stop words removal, and stemming. The preprocessing results of test data showed that the number of keywords or terms were 8088 words. Furthermore, the terms of the testing data were weighted using the composite weight of term frequency and inverse document frequency (TF-IDF), that is intended to reflect how important a word is to a document in a collection or corpus [15, 16]. The weighting scheme assigned term ***t*** to weight in document ***d***, using the equation:

$$\text{TF - IDF}_{t,d} = \text{TF}_{t,d} \times \text{IDF}_t \tag{4}$$

5.2 K-Means and K-Medoids Clustering

The second phase consists of the clustering processes using the k-Means and k-Medoids methods. The experiment was conducted 10 times on each clustering method, allowing for the cluster center taken at random at the beginning to affect the results of each clustering. The value of k for both k-Means and k-Medoids or the number of clusters formed is 5, in accordance with the number of classes of the journal origin topics. The results of the clustering experiments of using methods obtained the confusion matrix and computation times shown in Table 3.

The confusion matrix contains values for True Positive (TP), False Positive (FP), True Negative (TN), and False Negative (FN). TP shows the number of documents correctly entered into X; FP is the number of documents incorrectly inserted into X; FN is the number of documents incorrectly not included in X; and TN is the number of documents that are not correctly inserted into X. These values will be used to evaluate the clustering results, while the computation times are recorded as shown in seconds.

5.3 Combination of SVD and Clusterings

The third phase was dimension reduction of preprocessed data using SVD algorithm. Then, the reduced dimension data was clustered using both clustering methods. The results of combination of dimension reduction and clustering experiments of both methods obtained the confusion matrix and computation time shown in Table 4.

Table 3. Clusterings result

No.	k-Means					k-Medoids				
	TP	FP	TN	FN	Time	TP	FP	TN	FN	Time
1	637	430	3570	313	269.963	602	520	3479	349	247.201
2	523	721	3119	390	118.681	634	573	3427	316	187.481
3	573	481	3519	377	263.023	554	557	3443	396	211.707
4	658	512	3488	292	244.289	679	568	3432	271	241.417
5	622	421	3579	328	218.688	636	638	3362	314	223.917
6	549	496	3504	401	135.392	478	420	2780	327	270.137
7	577	540	3540	393	134.919	748	595	3405	202	276.616
8	640	409	3591	310	184.010	581	716	3364	389	298.933
9	574	485	3515	376	300.325	670	497	3503	280	299.787
10	621	559	3441	329	244.051	670	807	3193	280	247.201

6 Discussion

Evaluation of clustering results was performed to measure how well the clustering results obtained. To evaluate clustering accuracy, this research used the Stanford's nlp library [17] to obtain Precision, Recall, F-measure, and Rand Index values. Precision is the fraction of the number of true results obtained divided by the sum of all the results obtained. Precision is also a way of measuring the effectiveness of information retrieval systems. Recall is the fraction of the number of true results obtained divided by the sum of all the results that should be obtained. F-measure is a harmonic mean function of Precision and Recall, and the Rand Index presents the clustering accuracy result.

Table 4. Confusion matrix of clustering results with SVD

No.	SVD + k-Means					SVD + k-Medoids				
	TP	FP	TN	FN	Time	TP	FP	TN	FN	Time
1	813	186	3814	137	9.569	788	388	3612	162	29.732
2	828	160	3840	122	12.080	813	211	3789	137	22.838
3	794	198	3802	156	9.766	808	431	3569	142	25.543
4	931	20	3980	19	9.331	811	400	3600	139	24.295
5	810	179	3821	140	11.203	707	250	3750	243	26.076
6	896	60	3940	54	16.514	850	128	3872	100	26.766
7	878	80	3920	72	14.009	886	80	3920	64	25.701
8	758	267	3733	192	12.375	724	409	3591	226	22.810
9	847	394	3606	103	9.239	878	76	3924	72	30.758
10	799	377	3623	151	6.732	736	352	3648	214	22.186

Table 5. Comparison of clusterings accuracy with and without SVD

Evaluation	k-Means		k-Medoids	
	Without SVD	With SVD	Without SVD	With SVD
Precision	0,544013	0,8205432	0,5169140	0,7546735
Recall	0,6300842	0,8793685	0,6658407	0,8422106
F-measure	0,5831531	0,84924459	0,5807763	0,7941583
Rand Index	0,8264532	0,9380394	0,8147600	0,9146666
Time (s)	211,3341	11,0818	262,5183	25,6725

Table 5 shows the average values of Precision, Recall, F-measure, and Rand Index produced by both the k-Means and k-Medoids clustering processes with and without SVD. Rand Index result of clustering and SVD combination shows the success rate of system in clustering with accuracy value equal to 0.93. This is a difference of 11% compared to k-Means accuracy without dimension reduction, which was equal to 0.82. Even though the combination of SVD and k-Medoids the Precision and F-measure values did not reach 0.8, the Rand Index clustering value of 0.91 was 10% greater than the Rand Index value of clustering without dimensional reduction at only 0.81. Furthermore, the clustering execution time that SVD has a significant effect. In each method of clustering, the execution time becomes faster than without dimensional reduction. This is because the clustering method only processes data that has been reduced in terms of dimension.

7 Conclusion

In this research, the Singular Value Decomposition dimension reduction technique was combined with k-Means and k-Medoids algorithms to cluster documents. The results have shown that SVD had an effect on the accuracy results of k-Means and k-Medoids. This was proven to increase clustering accuracy of k-Means by 11% and k-Medoids by 10%, and the processing times were also proven to be significantly faster.

References

1. Hastie, T., Tibshirani, R., Friedman, J.: The elements of statistical learning. In: Data Mining, Inference, and Prediction. Springer, Standford (2008)
2. Wibowo, H.S.: Analisa Statistik II. Modul Pelatihan Statistik. P3M Universitas Indonesia, Depok (2016)
3. Indranandita, A., Susanto, B., Rahmat, A.: Sistem Klasifikasi dan Pencarian Jurnal dengan Menggunakan Metode Naive Bayes dan Vector Space Model. Jurnal Informatika **4**(2) (2011)
4. Han, J., Pei, J., Kamber, M.: Data Mining: Concepts and Techniques. Elsevier, Amsterdam (2011)
5. Alpaydin, E.: Introduction to Machine Learning. MIT Press, Cambridge (2014)
6. Deerwester, S., et al.: Indexing by latent semantic analysis. J. Am. Soc. Inf. Sci. **41**(6), 391 (1990)
7. Dumais, S.T.: Latent semantic analysis. Ann. Rev. Inf. Sci. Technol. **38**(1), 188–230 (2004)
8. Kadhim, A.I., Cheah, Y.-N., Ahamed, N.H.: Text document preprocessing and dimension reduction techniques for text document clustering. In: 2014 4th International Conference on Artificial Intelligence with Applications in Engineering and Technology (ICAIET). IEEE (2014)
9. Madhulatha, T.S.: Comparison between k-means and k-medoids clustering algorithms. In: Advances in Computing and Information Technology, pp. 472–481. Springer, Berlin (2011)
10. Nur'aini, K., et al.: Combination of singular value decomposition and K-means clustering methods for topic detection on Twitter. In: 2015 International Conference on Advanced Computer Science and Information Systems (ICACSIS). IEEE (2015)
11. Leskovec, J., Rajaraman, A., Ullman, J.D.: Mining of Massive Datasets. Cambridge University Press, Cambridge (2014)
12. Burden, R., Faires, J.: Numerical Analysis, 9th edn. Brooks/Cole, Boston (2011)
13. Thomo, A.: Latent semantic analysis (Tutorial). Victoria, Canada, pp. 1–7 (2009)
14. Kaufman, L., Rousseeuw, P.: Clustering by means of medoids. In: Dodge, Y. (ed.) Statistical Data Analysis Based on L1 Norm, pp. 405–416. Elsevier/North-Holland, Amsterdam (1987)
15. Manning, C.D., Raghavan, P., Schütze, H.: Introduction to Information Retrieval, vol. 1. Cambridge University Press, Cambridge (2008)
16. Ullman, J.D.: Mining of Massive Datasets. Cambridge University Press, Cambridge (2011)
17. Manning, C.D., et al.: The Stanford CoreNLP natural language processing toolkit. In: ACL (System Demonstrations) (2014)

Artificial Intelligence: Machine Learning and Optimization Techniques

Pairwise Test Suite Generation Using Adaptive Teaching Learning-Based Optimization Algorithm with Remedial Operator

Fakhrud Din[1,2] and Kamal Z. Zamli[1(✉)]

[1] Faculty of Computer Systems and Software Engineering, Universiti Malaysia Pahang, 26300 Gambang, Kuantan, Pahang, Malaysia
kamalz@ump.edu.my

[2] Department of Computer Science and IT, Faculty of Information Technology, University of Malakand, Chakdara, KPK, Pakistan
fakhruddin@uom.edu.pk

Abstract. Software systems nowadays have large configuration spaces. Pairwise test design technique is found useful by testers to sample only required configuration options of these systems for exploring errors owing to their interactions. Being a NP-complete problem, pairwise test suite generation problem has been addressed using several meta-heuristic algorithms including the Fuzzy Adaptive Teaching Learning-based Optimization (ATLBO) algorithm in the literature. ATLBO is a recent enhanced variant of Teaching Learning-based Optimization (TLBO) algorithm that adaptively applies its search operations using a Mamdani-type fuzzy inference system. Presently, ATLBO enters into stagnation or sometimes converges abnormally after some iterations. To address this issue, this paper proposes ATLBO with a remedial operator so as to further improve its searching capabilities. To evaluate the performance of ATLBO with remedial operator, it is used in a strategy called pATLBO_RO for the pairwise test suite generation problem. Experimental results reveal the strong performance of pATLBO_RO against other meta-heuristic and hyper-heuristic based pairwise test suite generation strategies.

Keywords: Pairwise testing · Adaptive Teaching Learning-Based Optimization Mamdani fuzzy inference system

1 Introduction

Software is indispensable today. It is an integral part of our lives. Virtually everyone uses and/or observes software in computers and in many devices available every-where (our hands, homes, offices, highways, shopping malls, smart buildings, etc.). Today, it is obvious that software affects millions of people and businesses than ever before. Thus, software testing is indispensable too like software itself. As a continuous process or set of processes ("destructive, and even sadistic"), software testing evaluates the under development software in order to raise its quality and reliability [1]. Software testing tries to reduce risks by exploring defects and to ensure that software of interest

F. Saeed et al. (Eds.): IRICT 2018, AISC 843, pp. 187–195, 2019.
https://doi.org/10.1007/978-3-319-99007-1_18

does not deviate from its specifications. Software testing is custodian of software quality and reliability till its retirement [2].

To achieve the goals of testing, software testers wish to test software exhaustively. However, it is infeasible as almost every contemporary software comes with exorbitant number of configuration options or input parameters which may interact with one another. For instance, the Apache server software has 172 input-parameters. Of these, 158 are two-valued, 8 are three-valued, 4 are four-valued, 1 is five-valued and final 1 is six-valued. This results in 1.8×10^{55} unique combinations of the input parameters' values of the server, and is commonly known as the combinatorial explosion problem [3]. Testing these combinations (or interactions) exhaustively is impossible in such cases whereas difficult in other cases owing to time and budget restrictions. Pairwise testing, also known as all-pairs or 2-way, t-way and constrained t-way testing, gained significant research focus [4–14] as it can bypass the combinatorial explosion problem so that to test, with fewer possible test cases, the enormous input combinations of software systems commonly developed nowadays. This paper focuses on only pairwise testing. A pairwise testing strategy generates a set of test cases to cover all interactions (or combinations) of the selected input parameters' values or options for each pair of input parameters or configurations at least once.

Being a NP-complete problem [15], pairwise test suite generation problem has been addressed using several meta-heuristic algorithms such as Genetic Algorithm (GA) [16], Particle Swarm Optimization (PSO) [5], Harmony Search (HS) algorithm [6], Ant Colony Optimization (ACO) [17], Cuckoo Search (CS) [9] and Flower Pollination Algorithm (FPA) [10] in the literature. Recently, a pairwise test suite generation strategy called pATLBO used Fuzzy Adaptive Teaching Learning-based Optimization algorithm (ATLBO) [18]. Although useful, ATLBO often converges abnormally and/or enters into stagnation. To avoid such scenarios, this paper proposed a remedial operator in ATLBO. The operator uses the teacher (i.e., the best candidate solution) to update the population of learners when there is no improvement for certain time and/or when ATLBO converges abnormally. A pairwise test suite generation strategy based on the ATLBO with remedial operator (pATLBO_RO) is proposed and its performance is evaluated against pATLBO, pTLBO and other meta-heuristics and hyper-heuristic based strategies. Experimental results reveal acceptable performance of pATLBO_RO as compared to the referenced pairwise test suite generation strategies. It is found that pATLBO_RO not only produces consistent results but also reduces the test suite generation time as compared to its predecessor method pATLBO.

The rest of the paper is organized as follows. Section 2 describes the pairwise test suite generation problem. Section 3 presents the related works on pairwise strategies based on meta-heuristic algorithms. Section 4 presents the proposed method. Section 5 presents experiments and subsequently describes them. Finally, Sect. 6 concludes this research work.

2 Description of Pairwise Test Suite Generation Problem

To describe the pairwise test suite generation problem, the Notifications part of the Whatsapp messenger is considered here. For simplicity, Fig. 1 shows only few configuration options of the mentioned part. There are 3 parameters or configurations each carrying 2 values. Table 1 shows the test suite consists of only 4 test cases with pairwise or 2-way interaction while there would be 8 test cases if full-strength interaction (3 for this example) was considered. Hence, pairwise testing reduces testing efforts by 50%.

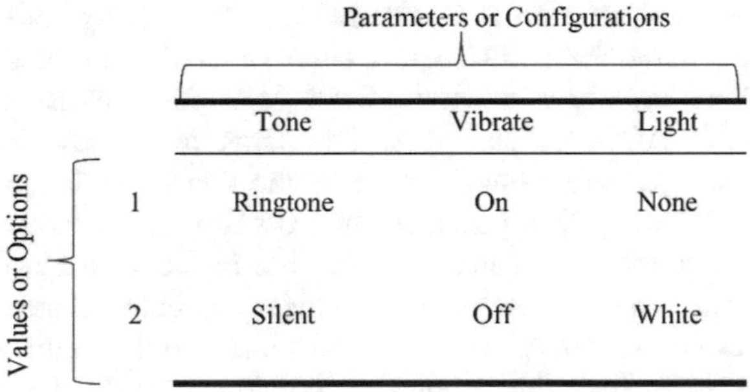

Fig. 1. Whatsapp notifications: 3 configurations/parameters each with 2 options/values.

Table 1. Pairwise test suite for the Whatsapp notifications with 50% reduction in testing efforts

Test Case #	Tone	Vibrate	Light
1	Silent	On	White
2	Silent	Off	None
3	Ringtone	On	None
4	Ringtone	Off	White

Mathematically, pairwise test suites can be represented by covering array (CA) notation with a fixed interaction strength $t = 2$. CA is simply a table that ensures capturing the pairwise interactions among all the required parameters or configurations at least once. The notation CA(N, 2, v^P) mathematically represents a CA. Here, N denotes the number of test cases, 2 denotes the 2-way or pairwise interaction, v denotes values or options each parameter or configuration carries and P denotes the total number of parameters or configurations.

For the running example, the CA notation for the pairwise test suite is CA(4, 2, 2^3). Mixed covering array (MCA) represents a system where parameters carry different number of values. Mathematically it can be represented as MCA(N, 2, $v_{11}^{P} v_{22}^{P} v_{kk}^{P}$). For instance, MCA(N, 2, 2^2 3^4) represents a system which has total 6 parameters where 2 carries 2 values and 4 carries 3 values.

3 Review of Meta-heuristic Based Pairwise Test Suite Generation Strategies

Meta-heuristic algorithms appear to be essential methodologies for researchers to solve hard optimization problems across all engineering and scientific fields [19–21]. These algorithms successfully produce solutions with acceptable optimality for a wide range of complex optimization problems. To solve the pairwise test suite generation problem, a strategy based on a meta-heuristic algorithm generally starts by creating a random population of solutions (test cases). The algorithm then selects the best test case in each iteration based on its maximum coverage of pairwise interactions among the parameters after repeatedly applying one or more search operators to the population for improving the overall fitness. This procedure is repeated till the full coverage of required pairwise interactions. Concerning the pairwise test suite generation problem, several meta-heuristic algorithms have been adopted as the basis for pairwise testing strategies including GA, PSO, ACO, CS and FPA. The literature has also introduced hybrid meta-heuristic based pairwise strategy such as the Pairwise Choice Function based Hyper-Heuristic (PCFHH) [7] to generate optimal pairwise test suites.

GA is adopted as the core searching algorithm in the strategy proposed in [16]. Based on the natural selection processes, the strategy initially creates a random set of test cases called as chromosomes. These chromosomes are then subjected to crossover and mutation processes until the termination criterion is met. The algorithm probabilistically selects and adds best chromosomes to the final set of test cases in each cycle. ACO based strategy is proposed in [17]. The strategy implements the behavior of ant colonies for searching optimal paths between the colony and food sources. The paths with more pheromone amount (i.e., covers more pairwise interactions) represents the best test case. PSO based strategy [5] known as PPSTG represents test cases as particles. Among these particles, the strategy attempts to find optimal particles by iteratively applying its global and local search operations until all required pairwise interactions are covered. HS based strategy [6] called as PHSS uses harmony memory and a number of improvisations in both of its local and global search operations to achieve full pairwise interaction coverage. CS based strategy [9] employs nest generation and their evaluation against existing nests as its two essential operations to generate optimal pairwise test suites. FPA based strategy known as PairFS [10] achieves complete coverage of all interactions with fewer possible pairwise test cases by repeatedly applying its local pollination and global pollination operations on the populations of pollen. Two pairwise strategies (pTLBO and pATLBO) based on TLBO [22] and ATLBO [23] are proposed in [18]. Learners represent test cases in both the strategies. The former applies both the phases whereas the later applies only one phase per iteration to accumulate best test cases. Finally, PCFHH is the first pairwise hyper-heuristic based strategy that adopts the Choice Function as hyper-heuristic for selecting one of the four implemented low-level heuristics to generate optimal pairwise test suites. Some other meta-heuristic and hyper-heuristic based strategies [24–31] for pairwise as well as t-way testing are available in the relevant literature.

4 ATLBO with Remedial Operator

ATLBO [23] is a recent variant of TLBO [22] that adaptively selects its global search (i.e., teacher phase) and local search (i.e., learner phase) operations by using a 3 inputs (the quality measure Q_m, the intensification measure I_m and the diversification measure D_m) and 1 output (Selection) Mamdani-type fuzzy inference system. These linguistic variables are fuzzified using trapezoidal membership functions. The rule base of the system is composed of four fuzzy linguistic rules with max-min fuzzy inference method. Finally, center of gravity (COG) is used as defuzzification method to obtain the single crisp output, the Selection. This value is then used to decide whether to launch either global search or local search rather than executing both as in original TLBO.

The current implementation of ATLBO does not include any method to deal with abnormal convergence. This often results sub optimal solutions owing to no improvement in fitness function for long times. Therefore, a remedial operator is necessary to avoid such conditions during search. The remedial operator monitors both the teacher and learner phases to determine whether the search is improving or not. If there is no improvement in the fitness function evaluation in both phases for long times, the remedial operator updates the population based on the best candidate solution (i.e., the teacher) in the population. Equation 1 shows the corresponding formula for the remedial operator.

$$X_{i,new} = X_{i,current} + X_{i,teacher} \tag{1}$$

where i represents a random i^{th} position value of the current test case updated with the ith position value of the best test case to obtain a new test case. Finally, the whole procedure of the strategy based on ATLBO with remedial operator (pATLBO_RO) for pairwise test suite generation is summarized in Fig. 2.

5 Experiments and Discussion

The proposed pairwise strategy is initially evaluated against pATLBO on four configurable application software (see Table 2). Here, efficiency (i.e., test suite size) and performance (i.e., test suite generation time) of both the implemented strategies are reported. In the second part of evaluation, pATLBO_RO is compared, based on efficiency only, with pATLBO, pTLBO and other pairwise strategies. Tables 2 and 3 present the obtained results. All the best and mean results are reported after running both pATLBO_RO and pATLBO 20 times to ensure statistical significance.

Concerning the mean pairwise test suite sizes in Table 2, pATLBO_RO has improved all the four results obtained with pATLBO previously. Similarly, it has produced the best test suite size for the configuration MCA (N, 2, 2^{13} 4^5). Moreover, the performance of pATLBO_RO is also improved as compared to that of pATLBO with no remedial operator. Based on the results presented in Table 2, it is evident that pATLBO_RO has achieved significant improvement owing to the remedial operator in both performance and efficiency while addressing the pairwise test suite generation problem.

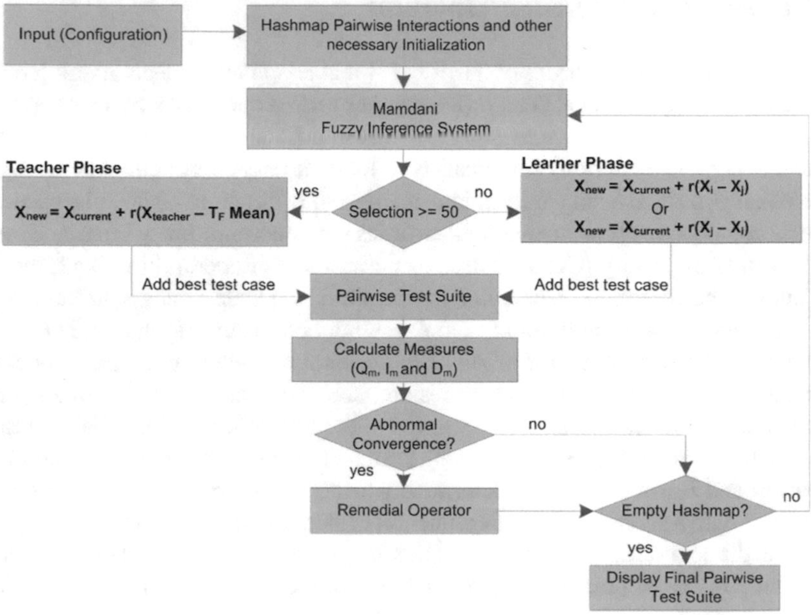

Fig. 2. Pairwise test suite generation strategy (pATLBO_RO).

Table 3 presents comparison of pATLBO_RO with other strategies on configurations with 10 parameters and values varying from 3 to 10. Here, pATLBO_RO improved 6 mean test suite sizes of pATLBO whereas matched the remaining 2. It has produced the best test suite size for the configuration CA (N, 2, 4^{10}). Figure 3 presents the summary of results reported in Table 3 for the best pairwise test suite sizes produced by the strategies.

Figure 3 clearly depicts that pATLBO_RO outperformed all the competing strategies including pATLBO by obtaining best test suite sizes for all the given configurations.

Table 2. Efficiency and performance comparison of pATLBO and pATLBO_RO

Systems	pATLBO				pATLBO_RO			
	Size		Time (sec)		Size		Time (sec)	
	Best	Mean	Best	Mean	Best	Mean	Best	Mean
MCA(N; 2, 2^7 3^1)	8⋆	8.85	7.73	9.01	8⋆	**8.83**	**5.48**	**8.53**
MCA(N; 2, 2^{13} 4^5)	24	25.2	390.07	447.00	23⋆	**24.5**	**389.45**	**443.23**
MCA(N; 2, 2^6 9^1 10^1)	91⋆	94.45	195.57	231.81	91⋆	**93.80**	**94.23**	**195.41**
MCA(N; 2, $2^1 3^4$)	11⋆	11.25	5.71	6.93	11⋆	**11.20**	**5.61**	**6.83**

Entries with ⋆ denote best test suite size whereas bold entries indicate mean best test suite size

Table 3. Comparison of pATLBO_RO against other strategies on CA(N, 2, v^{10}) with p = 10 and with v varying from 3 to 10

V	PCFHH	FairFS	PPSTG	PHSS	pTLBO	pATLBO		pATLBO_RO	
						Best	Mean	Best	Mean
3	16⋆	16⋆	17	17	17	16⋆	17.1	16⋆	**17.0**
4	28	28	29	28	28	28	28.8	27⋆	**28.6**
5	42⋆	42⋆	45	43	42⋆	42⋆	**42.8**	42⋆	**42.8**
6	58⋆	60	62	60	60	58⋆	59.2	58⋆	**59.13**
7	78	79	81	79	77	76⋆	77.6	76⋆	**77.3**
8	100	101	109	105	99	97⋆	98.8	97⋆	**98.4**
8	125	126	139	127	123	121⋆	**122.8**	121⋆	**122.8**
10	154	155	170	155	150	148⋆	150.0	148⋆	**149.9**

Entries with ⋆ denote best test suite size whereas bold entries indicate mean best test suite size

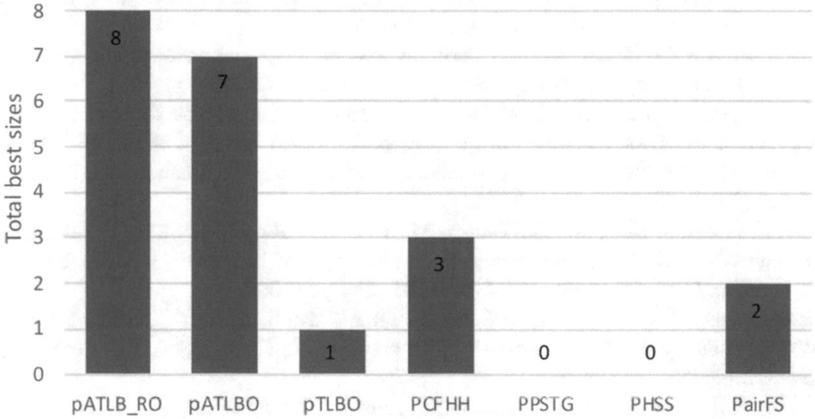

Fig. 3. Total best results obtained by each strategy out of 8 configurable software systems used in Table 3.

6 Conclusion

Pairwise testing is a relatively new test design technique that explores interaction errors in software systems with large configuration options or input parameters. Strategies based on meta-heuristic algorithms have successfully addressed the pairwise test suite generation. In this connection, a pairwise test suite generation strategy based on ATLBO with remedial operator (pATLBO_RO) is proposed in this paper. The remedial operator enabled ATLBO to avoid stagnation and/or abnormal convergence. Experimental results confirmed that both performance and efficiency of ATLBO has improved further after introducing the remedial operator. Owing to the effectiveness of ATLBO with the remedial operator, it will be adopted in future for solving other software engineering optimization problems such as constrained t-way testing, software clustering, test suite redundancy reduction, etc.

Acknowledgments. The work reported in this paper is funded by Fundamental Research Grant from Ministry of Higher Education Malaysia titled: A Reinforcement Learning Sine Cosine based Strategy for Combinatorial Test Suite Generation (grant no: RDU170103). We thank MOHE for the contribution and support. Fakhrud Din is the recipient of the Malaysian International Scholarship from the Ministry of Higher Education, Malaysia.

References

1. Myers, G.J., Sandler, C., Badgett, T.: The Art of Software Testing, 3rd edn. Wiley, Hoboken (2011)
2. Zamli, K.Z., Alkazemi, B.Y.: Combinatorial Testing. UMP Publisher, Pahang (2015)
3. Yilmaz, C., Fouch, S., Cohen, M.B., Porter, A., Demiroz, G., Koc, U.: Moving forward with combinatorial interaction testing. Computer **47**(2), 37–45 (2014)
4. Ahmed, B.S., Zamli, K.Z.: A variable-strength interaction test suites generation strategy using particle swarm optimization. J. Syst. Softw. **84**(12), 2171–2185 (2011)
5. Ahmed, B.S., Zamli, K.Z.: The development of a particle swarm based optimization strategy for pairwise testing. J. Artif. Intell. **4**(2), 156–165 (2011)
6. Alsewari, A.R.A., Zamli, K.Z.: A harmony search based pairwise sampling strategy for combinatorial testing. Int. J. Phys. Sci. **7**(7), 1062–1072 (2012)
7. Din, F., Alsewari, A.R.A., Zamli, K.Z.: A parameter free choice function based hyper-heuristic strategy for pairwise test generation. In: Proceedings of the IEEE International Conference on Software Quality, Reliability and Security Companion, pp. 85–91. IEEE, Prague (2017)
8. Klaib, M., Zamli, K.Z., Isa, N., Younis, M., Abdullah, R.: G2Way a backtracking strategy for pairwise test data generation. In: Proceedings of the 15th Asia-Pacific Software Engineering Conference, pp. 463–470. IEEE, Beijing (2008)
9. Nasser, A.B., Alsariera, Y.A., AlSewari, A.R.A., Zamli, K.Z.: A Cuckoo Search based pairwise strategy for combinatorial testing problem. J. Theor. Appl. Inf. Technol. **82**(1), 154–162 (2015)
10. Nasser, A.B., Alsewari, A.A., Tairan, N.M., Zamli, K.Z.: Pairwise test data generation based on flower pollination algorithm. Malay. J. Comput. Sci. **30**(3), 242–257 (2017)
11. Othman, R.R., Zamli, K.Z.: ITTDG: integrated t-way test data generation strategy for interaction testing. Sci. Res. Essays **6**(17), 3638–3648 (2011)
12. Younis, M.I., Zamli, K.Z., Isa, N.A.M.: MIPOG-modification of the IPOG strategy for t-way software testing. In: Proceedings of the Distributed Frameworks and Applications, pp. 1–6. IEEE, Beijing (2008)
13. Younis, M.I., Zamli, K.Z., Isa, N.A.M.: Algebraic strategy to generate pairwise test set for prime number parameters and variables. In: Proceedings of the International Symposium on Information Technology, pp. 1–4. IEEE, Kuala Lumpur (2008)
14. Zamli, K.Z., Din, F., Ahmed, B.S., Bures, M.: A hybrid q-learning sine-cosine-based strategy for addressing the combinatorial test suite minimization problem. PLoS ONE **13**(5), e0195675 (2018)
15. Ahmed, B.S., Zamli, K.Z., Afzal, W., Bures, M.: Constrained interaction testing: a systematic literature study. IEEE Access. **5**, 25706–25730 (2017)
16. Ghazi, S.A., Ahmed, M.A.: Pair-wise test coverage using genetic algorithms. In: Proceedings of the Congress on Evolutionary Computation, pp. 1420–1424. IEEE, Canberra (2003)

17. Shiba, T., Tsuchiya, T., Kikuno, T.: Using artificial life techniques to generate test cases for combinatorial testing. In: Proceedings of the 28th Annual International Conference on Computer Software and Applications, pp. 72–77. IEEE, Hong Kong (2004)
18. Din, F., Zamli, K.Z.: Fuzzy adaptive teaching learning-based optimization strategy for pairwise testing. In: Proceedings of the 7th IEEE International Conference on System Engineering and Technology, pp. 17–22. IEEE, Shah Alam (2017)
19. Din, F., Zamli, K.Z.: Fuzzy adaptive teaching learning-based optimization strategy for GUI functional test cases generation. In: Proceedings of the 2018 7th international conference on software and computer applications, pp. 92–96. ACM, Kuantan (2018)
20. Ahmed, B.S., Sahib, M.A., Gambardella, L.M., Afzal, W., Zamli, K.Z.: Optimum design of PIλDμ controller for an automatic voltage regulator system using combinatorial test design. PLoS ONE **11**(11), e0166150 (2016)
21. Zamli, K.Z.: A chaotic teaching learning based optimization algorithm for optimizing emergency flood evacuation routing. Adv. Sci. Lett. **22**(10), 2927–2931 (2016)
22. Rao, R.V., Savsani, V.J., Vakharia, D.P.: Teaching-learning-based optimization: a novel method for constrained mechanical design optimization problems. Comput. Aided Des. **43** (3), 303–313 (2011)
23. Zamli, K.Z., Din, F., Baharom, S., Ahmed, B.S.: Fuzzy adaptive teaching learning-based optimization strategy for the problem of generating mixed strength t-way test suites. Eng. Appl. Artif. Intell. **59**, 35–50 (2017)
24. Ahmed, B.S., Gambardella, L.M., Afzal, W., Zamli, K.Z.: Handling constraints in combinatorial interaction testing in the presence of multi objective particle swarm and multithreading. Inf. Softw. Technol. **86**, 20–36 (2017)
25. Ahmed, B.S., Zamli, K.Z., Lim, C.P.: Application of particle swarm optimization to uniform and variable strength covering array construction. Appl. Soft Comput. **12**(4), 1330–1347 (2012)
26. Alsariera, Y.A., Zamli, K.Z.: A BAT-inspired strategy for t-way interaction testing. Adv. Sci. Lett. **21**(7), 2281–2284 (2015)
27. Zamli, K.Z., Alkazemi, B.Y., Kendall, G.: A Tabu Search hyper-heuristic strategy for t-way test suite generation. Appl. Soft Comput. **44**, 57–74 (2016)
28. Zamli, K.Z., Din, F., Kendall, G., Ahmed, B.S.: An experimental study of hyper-heuristic selection and acceptance mechanism for combinatorial t-way test suite generation. Inf. Sci. **399**, 121–153 (2017)
29. Cohen, M.B., Gibbons, P.B., Mugridge, W.B., Colbourn, C.J.: Constructing test suites for interaction testing. In: Proceedings of the 25th International Conference on Software Engineering, pp. 38–48. IEEE, Portland (2003)
30. Alsewari, A.R.A., Zamli, K.Z.: Design and implementation of a harmony-search-based variable-strength t-way testing strategy with constraints support. Inf. Softw. Technol. **54**(6), 553–568 (2012)
31. Nasser, A.B., Zamli, K.Z., Alsewari, A.R.A., Ahmed, B.S.: Hybrid flower pollination algorithm strategies for t-way test suite generation. PLoS ONE **13**(5), e0195187 (2018)

Novel Multi-swarm Approach for Balancing Exploration and Exploitation in Particle Swarm Optimization

Sinan Q. Salih[1,2(✉)], AbdulRahman A. Alsewari[1], Bellal Al-Khateeb[2], and Mohamad Fadli Zolkipli[1]

[1] Faculty of Computer Systems and Software Engineering, Universiti Malaysia Pahang, 26300 Gambang, Pahang, Malaysia
sinan_salih@computer-college.org

[2] Computer Science Department, College of Computer Science and Information Technology, University of Anbar, Ramadi, Iraq

Abstract. Several metaheuristic algorithms and improvements to the existing ones have been presented over the years. Most of these algorithms were inspired either by nature or the behavior of certain swarms, such as birds, ants, bees, or even bats. These algorithms have two major components, which are exploration and exploitation. The interaction of these components can have a significant influence on the efficiency of the metaheuristics. Meanwhile, there are basically no guiding principles on how to strike a balance between these two components. This study, therefore, proposes a new multi-swarm-based balancing mechanism for keeping a balancing between the exploration and exploitation attributes of metaheuristics. The new approach is inspired by the phenomenon of the leadership scenario among a group of people (a group of people being governed by a selected leader(s)). These leaders communicate in a meeting room, and the overall best leader makes the final decision. The simulation aspect of the study considered several benchmark functions and compared the performance of the suggested algorithm to that of the standard PSO (SPSO) in terms of efficiency.

Keywords: Swarm intelligence · Exploration · Exploitation · Metaheuristics Optimization · Computational intelligence

1 Introduction

Over the past 2 decades, nature-inspired metaheuristics have attracted much attention due to their efficiency in establishing accurate solutions to complex industrial and engineering problems, especially the NP-complete problems. Most nature inspired metaheuristics are classified as stochastic techniques. These stochastic algorithms randomly pick a set of solutions and improve them based on the algorithmic mechanism. The solutions are constantly improved until a set stopping criterion is met. Stochastic techniques are classified as random searches but guided to the next iteration by heuristics. In the last few years, many stochastic algorithms have been proposed due to their great success in finding best solutions to science and engineering problems [1–4].

F. Saeed et al. (Eds.): IRICT 2018, AISC 843, pp. 196–206, 2019.
https://doi.org/10.1007/978-3-319-99007-1_19

The Particle Swarm Optimizer (PSO) is one of the most popular algorithms first introduced by Kennedy and Eberhard [5, 6]. The PSO solves optimization problems by emulating the flocking behavior of birds; where each bird is regarded as a solution. The advantage of the PSO when compared to the evolution-based frameworks like the Genetic algorithm, lies in its ease of implementation and in requiring just a few parameters to be adjusted [7, 8]. The PSO has successfully been applied in several instances such as function optimization, fuzzy systems, artificial neural network training, and feature selection [9–17]. It can also be applied in other areas where GA can be employed.

In the original PSO or simple PSO (SPSO), a major difficulty lies in maintaining the balancing between exploration (searching for the global optimum) and exploitation (searching for the local optimum). Although the SPSO can converge quickly in the initial iterations towards an optimum, its problems lie in reaching a near optimal solution. This problem has attracted several attentions in terms of ways to enhance the performance of SPSO, including a proposal for hybrid models [8, 18–20].

In the literature, several multi-swarm attempts have earlier been proposed. The balancing between local search 'exploration' and global search 'exploitation' in PSO using a master-slave approach has been proposed [21]. This approach is a cooperative scheme made up of a superior swarm known as a master swarm and several other inferior swarms known as slave swarms. These slave swarms provide the master swarm with new promising particles (new positions with the best fitness value) as the evolution continues. The state of these new particles is updated by the master swarm with respect to the best position so far discovered by both itself and the slave swarms.

This study proposed a new multi-swarm cooperative scheme for balancing the exploration and exploitation of PSO. The proposed scheme consists of several swarms called 'clans'; each clan has its own leader. The leader of each clan is the best solution in the clan and represents the local best. All the clan leaders meet periodically to select the best among themselves who will represent the global best solution. This best leader has control over all the other clan leaders. The interaction between the selected best leader (the global best) and the individual clan leaders (the local best) has an influence on the balance between their exploratory and local search performances, and maintain a suitable population diversity even when approaching the global best solution, thereby, minimizing the risk of convergence or being trapped to the local sub-optimal.

The remaining part of this study is organized thus: Sect. 2 described the original PSO and its variants, while Sect. 3 described the motivation for the proposed approach and provided the algorithmic pseudo-code as well. In Sect. 4, a description of the benchmark continuous optimization problems for benchmarking the performance of the algorithm was provided, followed by the discussions of the study results. Section 5 provided a brief conclusion of the study.

2 Particle Swarm Optimization (PSO)

The standard version of particle swarm optimization PSO is a well-known optimization algorithm, the swarm is initialized with a random population of solutions. The PSO searches for the best positions by updating its component generations. The generated particles in the PSO (which are the solutions) fly in a D-dimensional search space at a velocity dynamically adjusted based on both their own respective experiences and the experience of their neighbors.

The ith particle in the PSO is denoted in the D-dimensional space as $x_i = (x_{i1}, x_{i2}, x_{i3}, \ldots, x_{iD})$, where $x_{id} \in [LB_d, UB_d]$, $d \in [1, D]$, LB_d, UB_d respectively represents the minimum and maximum limits of the d th dimension. The velocity of particle i is given as $v_i = (v_{i1}, v_{i2}, v_{i3}, \ldots, v_{iD})$, which is maintained at a maximum user-specified velocity V_{max}. The particles, at each time step t, are manipulated based on the following relations:

$$v_i(t+1) = v_i(t) + r_1 c_1 (P_i - x_i(t)) + r_2 c_2 \left(P_g - x_i(t)\right) \quad (1)$$

$$x_i(t+1) = x_i(t) + v_i(t) \quad (2)$$

where r_1 and r_2 represents the random values in the range of 0 and 1. c_1 and c_2 represents the acceleration constants that governs the extent a particle can move within a given iteration. The previous best position of the ith particle is represented by P_i.

Based on the several definitions of P_g, there are 2 variants of the PSO. A global version of PSO is achieved when P_g represents the position of the best particle among the other particles in the same population (also referred to as the as *gbest*). But if P_g is derived from a few number of adjacent particles of a population (called *lbest*), a local version of PSO is achieved. An inertia term w was later introduced by Shi and Eberhard [22] via a modification of Eq. (1) into:

$$v_i(t+1) = w \times v_i(t) + r_1 c_1 (P_i - x_i(t)) + r_2 c_2 \left(P_g - x_i(t)\right) \quad (3)$$

They suggested that a proper balance between global and local explorations can be achieved through a proper selection of w, thus, requiring averagely less iterations to establish an optimal solution. The w, as originally developed, is set using the following equation:

$$w = w_{max} - \frac{w_{max} - w_{min}}{itr_{max}} \times itr \quad (4)$$

where w_{max} represents the initial weight, w_{min} represents the final weights, itr_{max} is the highest number of allowable iterations, and itr represents the present number of iterations.

This version of PSO is in this study, henceforth referred to as a linearly decrease inertia weight method (LPSO).

In addition to LPSO, a random inertia weight factor for dynamic systems tracking has also been suggested [23]. The inertia weight factor in this development is set to randomly change based on the following relation:

$$w = 0.5 - \frac{rand()}{2} \quad (5)$$

where $rand()$ represents a uniformly distributed random number in the range of 0 *and* 1.

The acceleration coefficients were suggested to be maintained at 1.494. This method is henceforth referred to as random weight method (RPSO) in the remaining part of this paper.

3 Multi-swarm PSO Algorithm

The core idea of the multi-swarm is the interaction between several groups while searching for a solution. Many multi-swarm-based schemes have been proposed, each being inspired by a natural behavior. In this paper, a new cooperative multi-swarm scheme inspired by the human social behavior (the interaction between a group of people known as 'Clan' and their leaders) was proposed. The proposed scheme consists of several swarms called clans; each clan consists of several solutions represented by the group members. The best member of each clan is the clan leader and has control over the members of its clan in terms of the time to move and where they are moving to. Figure 1 showed the structure of the individual swarm.

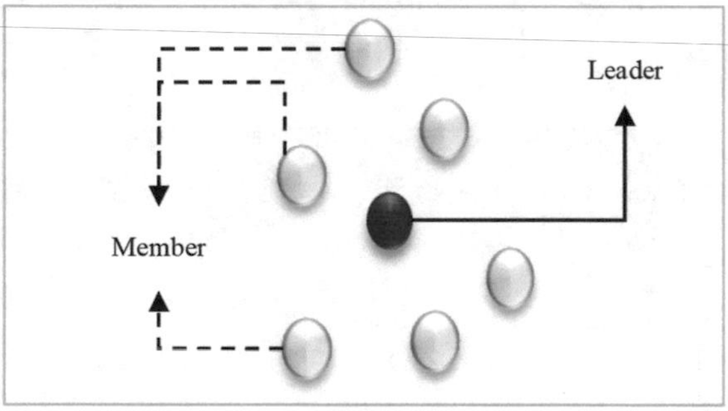

Fig. 1. The structure of the individual swarm

In each generation, the leaders often meet in one room to select an overall best leader who will update the position of the other leaders based on his new-found position. This behavior of knowledge sharing helps to balance the exploration stage with the searching process of the PSO, which represents the exploitation stage. The new multi-swarm approach is called a 'Meeting Room Approach' (MRA). Figure 2 showed the MRA model, where each member in the clan represents a particle in the swarm, and its position and velocity is updated based on the steps of PSO algorithm. Once the new generation of each clan has been set, a new clan leader (the best leader) is elected and sent to the meeting room. The best among the leaders will be selected as the overall best leader (global best) in the meeting room. The newly-selected overall best leader shares his positional information with the other leaders using the following relation:

$$w^{Ln} = \left(\frac{w^{Lg} - w^{Ln}}{Itr}\right) \times rand() \tag{6}$$

$$v_i^{Ln}(t+1) = w^{Ln} \times v_i^{Ln}(t) + rc\left(P_g^L - P_n^L(t)\right) \tag{7}$$

$$x_i^{Ln}(t+1) = x_i^{Ln}(t) + v_i^{Ln}(t) \tag{8}$$

where *Ln* represents the normal leaders, *Lg* represents the overall best leader, x_i^L represents the position of the normal leaders, v_i^{Ln} represents the velocity of the normal leaders, w^{Lg} and w^{Ln} represent the inertia weight of the overall best leader and the normal leaders, respectively.

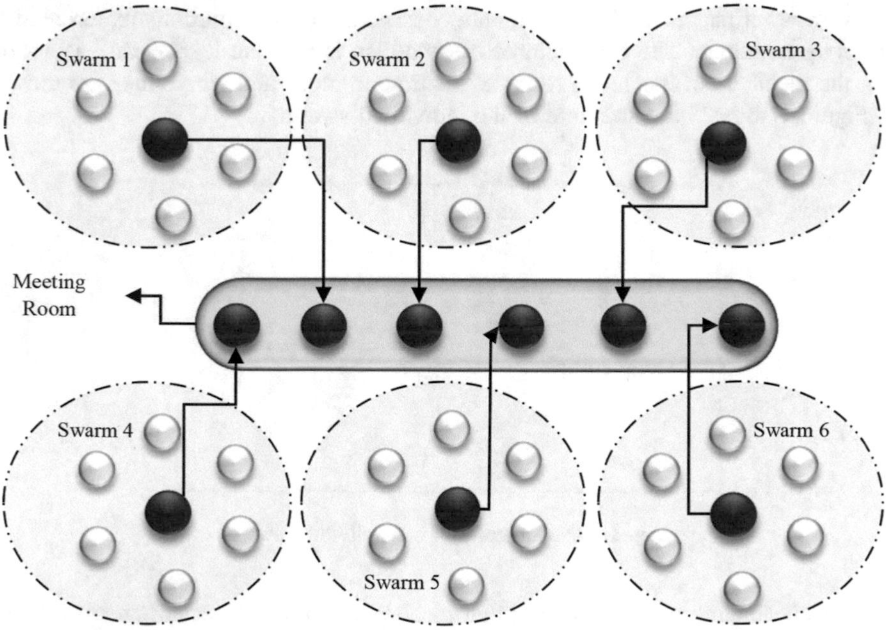

Fig. 2. Meeting room approach

After each generation, a new leader is selected for each swarm because the positions of the members are changed or updated during the meeting. The new equation of the inertia in the meeting room controls the exploration of the search algorithm. The pseudo-code for the MPSO algorithm is listed in Fig. 3.

```
Algorithm MPSO
Input:
      #Swarm, #P, c1, c2, #Dim, #MaxGen
Output:
      Best Solution (Leader)
Procedure:
      Start
            Initialized the Swarms
            Evaluate the fitness of each particle in Swarms
            While (itr ≤ #MaxGen)
                  For each s in Swarm
                        For each member in Swarm s
                              Update the velocity of the member via eq.3
                              Update the position of the member via eq.2
                        Next
                        Select the local best as a Leader_s
                  Next
                  Select the best Leader among all leaders
                  Update the inertia weight of the clan via eq.6
                  Update the velocity of the Leader_s via eq.7
                  Update the position of the Leader_s via eq.8
                  Select the Best Leader ever as the global best.
            Loop
            Return Best Leader
      Stop
```

Fig. 3. MPSO pseudo-code

4 Results and Discussion

This section presents the detailed description of the nonlinear benchmark functions commonly used in the evolutionary literature [24]. Each test function varies in terms of modality (unimodal and multimodal) and the number of dimensions (fixed and dynamic). Table 1 showed the different test functions and their basic characteristics.

The performance of the proposed MPSO was evaluated by benchmarking with two established algorithms (the original PSO (SPSO) [25] and the Master-Slave PSO (MCPSO) [21]). The parameters used for the SPSO have earlier been recommended by [25] with asymmetric initialization method and a linearly decreasing w which was changed from 0.9 to 0.4. Several swarms of SPSO were involved in the MPSO and MCPSO as clans and slaves respectively, to optimize the listed benchmark function. Each of them has the same parameter settings as SPSO1. To investigate the performance of MPSO, different population sizes were employed with different dimensions for each function. The maximum number of iterations was set at 500, which corresponds to 50 dimensions. For each experimental setting, a total of 30 runs were conducted. Table 2 presents the parameters setting for all the algorithms used in this study.

Table 1. Benchmark test functions

Name	Function	Range	Opt.
Sphere unimodal	$f_1 = \sum_{i=1}^{D} X^2$	−100, 100	0
Griewank unimodal	$f_2 = \sum_{i=1}^{D} \frac{x_i^2}{4000} - \prod_{i=1}^{D} \cos\left(\frac{x_i}{\sqrt{i}}\right) + 1$	−600, 600	0
Rastrigin multimodal	$f_3 = \sum_{i=1}^{D} \left(x_i^2 - 10\cos(2\pi x_i) + 10\right)$	−5.12, 5.12	0
Ackley multimodal	$f_4 = -20\exp\left(-0.2\sqrt{\frac{1}{D}\sum_{i=1}^{D} x_i^2 0}\right) - \exp\left(\frac{1}{D}\sum_{i=1}^{D}\cos(2\pi x_i)\right) + 20 + e$	−32, 32	0

Table 2. Parameters Setting

Algorithm	Parameter	Value
SPSO	W	0.9–0.4
	No. of swarms	1
	Swarm size	50
	c_1, c_2	1.5
MCPSO	W	0.9–0.6
	No. of slaves	5
	Swarm size	50
	c_1, c_2, c_3	1.5
MPSO	w^{Ln}	0.8–0.5
	w^{Lg}	0.9–0.7
	c_1, c_2	1.5
	No. of clans	5
	Clan size	10

Table 3 presents the best and mean fitness values of the particle after 30 runs and 4 benchmark functions. Based on the table, MPSO performed better than the other algorithms in almost all the studied cases. A general analysis of the table showed that MPSO had 5 swarms, each consisting of 10 particles, and only 5 particles interacted in the meeting room. It may be concluded that MPSO required less computational complexity, and yet, had a better performance in terms of finding the best solution. Figures 4(a and b) illustrate the sustainability of the MPSO to evolve even when the other algorithms were almost stagnated.

Table 3. Results for benchmark test functions

f_n	Swarm	Algorithm	Best	Mean	S.D
f_1	50	SPSO	2.5457521	2.7647845	0.0784516
		MCPSO	0.9854126	1.0154784	0.0014784
		MPSO	**0.0007845**	**0.0008748**	**0.0000184**
f_2	50	SPSO	0.0884741	0.0964587	0.0078478
		MCPSO	0.0078414	0.0087789	0.0009874
		MPSO	**0.0000897**	**0.0000997**	**0.0000658**
f_3	50	SPSO	21.695847	27.947512	0.0847896
		MCPSO	2.0018977	2.6647845	0.0078487
		MPSO	**0.0004687**	**0.0045214**	**0.0000144**
f_4	50	SPSO	16.4875218	26.110161	0.0238484
		MCPSO	1.99847	2.5869124	0.0084578
		MPSO	**0.0002648**	**0.0017636**	**0.0000584**

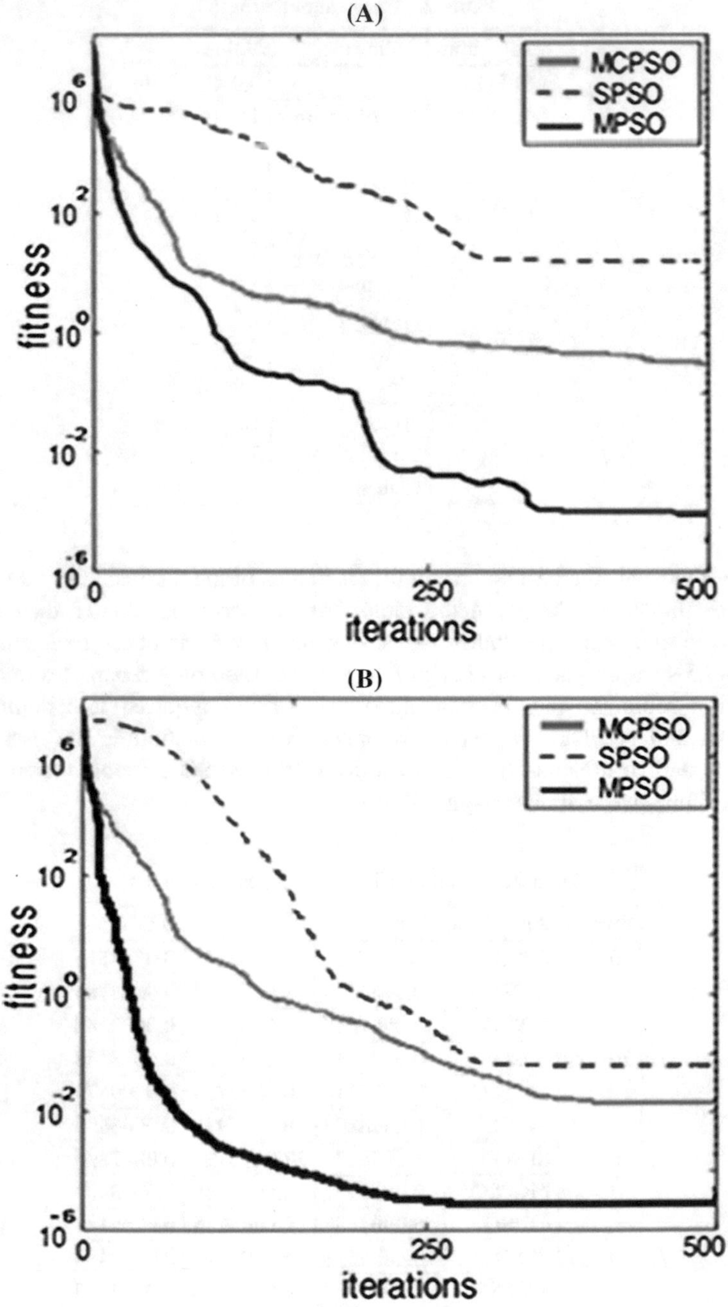

Fig. 4. Convergence curve: (a) Sphere function, (b) Griewank function

5 Conclusion

In this study, a social-inspired mechanism for improving the performance of the PSO was developed. The proposed mechanism simulates the social grouping behavior of human (existing as clans and interacting with their leaders). The proposed algorithm (MPSO) was able to control the balance between exploration and exploitation of PSO. During the simulation stage of the study, 4 benchmark functions were performed using different algorithms. The benchmarking in terms of the performance of the proposed MPSO showed that MPSO had a better performance than SPSO both in the quality and robustness of the solution. In the future works, more emphasis should be laid on applying the proposed MPSO into different swarm metaheuristics such as firefly algorithm, bat algorithm, and grey wolf optimizer.

Acknowledgement. This research is funded by UMP PGRS170338: Analysis System based on Technological YouTube Channels Reviews, and UMP RDU1603119 Grant: Modified Greedy Algorithm Strategy for Combinatorial Testing Problem with Constraints Supports. UMP RDU180367 Grant: Enhance Kidney Algorithm for IOT Combinatorial Testing Problem.

References

1. Slowik, A., Kwasnicka, H.: Nature inspired methods and their industry applications – swarm intelligence algorithms. IEEE Trans. Ind. Inform. **14**, 1004–1015 (2018)
2. Diao, R., Shen, Q.: Nature inspired feature selection meta-heuristics. Artif. Intell. Rev. **44**, 311–340 (2015)
3. Azrag, M.A.K., Kadir, T.A.A., Odili, J.B., Essam, M.H.A.: A global African Buffalo optimization. Int. J. Softw. Eng. Comput. Syst. **3**, 138–145 (2017)
4. Odili, J.B., Kahar, M.N.M., Anwar, S.: African Buffalo optimization: a swarm-intelligence technique. Procedia Comput. Sci. **76**, 443–448 (2015)
5. Eberhart, R., Kennedy, J.: A new optimizer using particle swarm theory. In: Proceedings of the Sixth International Symposium on Micro Machine and Human Science 1995, MHS 1995, pp. 39–43 (1995)
6. Kennedy, J., Eberhart, R.: Particle swarm optimization. In: Proceedings of IEEE International Conference on Neural Networks 1995, vol. 4, pp. 1942–1948 (1995)
7. Naik, B., Nayak, J., Behera, H.S.: A novel FLANN with a hybrid PSO and GA based gradient descent learning for classification. In: Proceedings of the 3rd International Conference on Frontiers of Intelligent Computing: Theory and Applications (FICTA) 2014, pp. 745–754 (2015)
8. Shi, X.H., Liang, Y.C., Lee, H.P., Lu, C., Wang, L.M.: An improved GA and a novel PSO-GA-based hybrid algorithm. Inf. Process. Lett. **93**, 255–261 (2005)
9. Chen, D., Chen, J., Jiang, H., Zou, F., Liu, T.: An improved PSO algorithm based on particle exploration for function optimization and the modeling of chaotic systems. Soft Comput. **19**, 3071–3081 (2015)
10. Palupi Rini, D., Mariyam Shamsuddin, S., Sophiyati Yuhaniz, S.: Particle swarm optimization: technique, system and challenges. Int. J. Comput. Appl. **14**, 19–27 (2011)
11. Mendes, R., Cortez, P., Rocha, M., Neves, J.: Particle swarms for feedforward neural network training. In: Proceedings of the 2002 International Joint Conference on Neural Networks IJCNN02 Cat No. 02CH37290, pp. 1895–1899 (2002)

12. Gudise, V.G., Venayagamoorthy, G.K.: Comparison of particle swarm optimization and backpropagation as training algorithms for neural networks. In: Proceedings of 2003 IEEE Swarm Intelligence Symposium SIS 2003 (Cat. No. 03EX706), vol. 2, pp. 110–117 (2003)
13. Melin, P., Olivas, F., Castillo, O., Valdez, F., Soria, J., Valdez, M.: Optimal design of fuzzy classification systems using PSO with dynamic parameter adaptation through fuzzy logic. Expert Syst. Appl. **40**, 3196–3206 (2013)
14. Niu, Q., Huang, X.: An improved fuzzy C-means clustering algorithm based on PSO. J. Softw. **6**, 873–879 (2011)
15. Xue, B., Zhang, M., Browne, W.N.: Particle swarm optimisation for feature selection in classification: Novel initialisation and updating mechanisms. Appl. Soft Comput. J. **18**, 261–276 (2014)
16. Inbarani, H.H., Bagyamathi, M., Azar, A.T.: A novel hybrid feature selection method based on rough set and improved harmony search. Neural Comput. Appl. **26**, 1859–1880 (2015)
17. Huang, C.L., Dun, J.F.: A distributed PSO-SVM hybrid system with feature selection and parameter optimization. Appl. Soft Comput. J. **8**, 1381–1391 (2008)
18. Mirjalili, S., Hashim, S.Z.M.: A new hybrid PSOGSA algorithm for function optimization. In: Proceedings of ICCIA 2010—2010 International Conference on Computer and Information Application, pp. 374–377 (2010)
19. Premalatha, K., Natarajan, A.M.: Hybrid PSO and GA for global maximization. Int. J. Open Probl. Compt. Math. **2**, 597–608 (2009)
20. Zhang, Y., Wu, L.: A hybrid TS-PSO optimization algorithm. J. Converg. Inf. Technol. **6**, 169–174 (2011)
21. Niu, B., Zhu, Y., He, X., Wu, H.: MCPSO: a multi-swarm cooperative particle swarm optimizer. Appl. Math. Comput. **185**, 1050–1062 (2007)
22. Shi, Y., Eberhart, R.: A modified particle swarm optimizer. In: 1998 IEEE International Conference on Evolutionary Computation Proceedings. IEEE World Congress on Computational Intelligence, pp. 69–73 (1998)
23. Eberhart, R.C., Shi, Y.: Tracking and optimizing dynamic systems with particle swarms. In: Proceedings of IEEE Congress on Evolutionary Computation, pp. 94–97. IEEE, Seoul (2001)
24. Jamil, M., Yang, X.S., Zepernick, H.J.D.: Test functions for global optimization: a comprehensive survey. In: Swarm Intelligence and Bio-Inspired Computation, pp. 193–222 (2013)
25. Shi, Y., Eberhart, R.C.: Empirical study of particle swarm optimization. In: 1999 IEEE International Conference on Evolutionary Computation Proceedings. IEEE World Congress on Computational Intelligence, vol. 3, pp. 1945–1950 (1999)

Computational Analysis of Dynamics in an Agent-Based Model of Cognitive Load and Reading Performance

Hayder M. A. Ghanimi(✉) and Azizi Ab Aziz

Cognitive Artefacts Group, Human-Centred Computing Research Lab,
School of Computing, Universiti Utara Malaysia, 06010 Sintok, Kedah, Malaysia
hayder.alghanami@gmail.com, aziziaziz@uum.edu.my

Abstract. To avoid the development of cognitive load during task-specific actions, technologies like companion robots or intelligent systems may benefit from being aware of the dynamics of related mental performance constructs. As a first step toward the development of such systems, this paper uses an agent-based approach to formalize and simulate cognitive load processes within reading activities, which may involve specific assigned task. The obtained agent-based model is analysed both by mathematical analysis and automated trace evaluation. Based on this description, the proposed agent-based model has exhibited realistic behaviours patterns that adhere to the psychological and cognitive literature. Moreover, it is shown how the agent-based model can be integrated into intelligent systems that monitor and predict cognitive load over time and propose intelligent support actions based on that.

Keywords: Agent-based model · Temporal Trace Language
Cognitive load · Intelligent systems · Verification analysis

1 Introduction

Recent studies have shown a great focus placed upon the development of computational agent-based models as core foundations to understand and analyse physical and mental states of human [1]. These models equipped intelligent systems and robotic artefacts applications with the prime ability to reason in supporting people. The key success of integrating these agent-based models within intelligent systems extremely depends on the correctness of the model as provided by grounding theories in related domain of interest. Therefore, a thorough evaluation process of an agent-based model is essential to ascertain the process and its predictions are accurate and credible significantly [2]. From this spectrum a number of agent models have been developed and evaluated as in [3–5]. Therefore, it inspires this paper to focus on thorough computational analysis (through formal evaluation) of an agent-based model of cognitive load and reading performance that was previously developed in [6]. Two prominent approaches to evaluate the model are implemented namely; *mathematical analysis* and *automated logical analysis* [3, 5, 7, 8]. This paper is organized as follows. First, in Sect. 2, the underlying constructs and a set of factors related to cognitive load and

F. Saeed et al. (Eds.): IRICT 2018, AISC 843, pp. 207–220, 2019.
https://doi.org/10.1007/978-3-319-99007-1_20

reading performance concepts are introduced. Next, the model design and its implementation are shown in Sects. 3 and 4 respectively. The computational analysis of the agent-based model is presented in Sect. 5. Finally Sect. 6 concludes the paper.

2 Background in Cognitive Load and Reading Performance

The process to solve difficult tasks often subjected to a high level of cognitive load that may exceed individuals' cognitive resources and as a result they may not be able to regulate the load imposed by the task (e.g. reading process to solve assignments with certain levels of difficulties). This condition can be explained through some scenarios of monotonous decreases in individual's performance over time [9]. A number of grounding theories was developed to explain the development of a cognitive load level during reading processes through the interplays between related factors and its relationship [9, 11]. The factors are shown in Table 1 and its relationships between those factors are explained in Sect. 3.

Table 1. Nomenclatures for factors of cognitive load and reading performance

Factor	Formalization
Reading task complexity	*Tc*
Time pressure	*Tp*
Task presentation	*Tn*
Physical environment	*Pe*
Personal profile	*Pp*
Experience level	*Ev*
Prior knowledge	*Pk*
Reading norm	*Rn*
Reading demands	*Rd*
Situational aspects	*Sa*
Reading goal	*Rg*
Reading effort	*Rf*
Motivation	*Mv*
Expertise level	*Ev*
Intrinsic load	*Id*
Extraneous load	*Ed*
Germane load	*G)*
Germane resources	*Gr*
Mental load	*Ml*
Mental effort	*Me*
Mental ability	*Ma*
Cognitive exhaustion	*Ce*
Reading engagement	*Rm*

(*continued*)

Table 1. (*continued*)

Factor	Formalization
Critical point	*Ex*
Experienced exhaustion	*Cp*
Recovery effort	*Re*
Short term exhaustion	*Sh*
Accumulative experienced exhaustion	*Ax*
Accumulative exhaustion	*Ae*
Cognitive load	*Cl*
Reading performance	*Rp*
Persistence	*Pr*

It is noteworthy to mention that all the theoretical explanations of identified factors (from related grounded theories) are described in our previous works as in [10, 11].

3 Agent-Based Modeling Approach

This section explains the implementation of *Network-oriented Modeling* approach (based on *Temporal Causal Network*) to construct a computational agent-based model of cognitive and reading performance [1]. Consequently, the identified factors from Sect. 2 are used to conceptualize the causal computational agent model as depicted in Fig. 1.

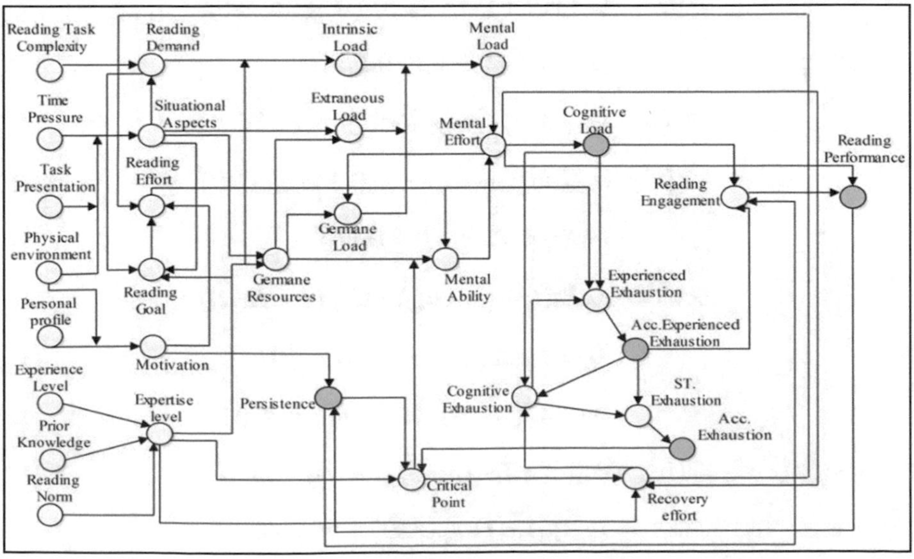

Fig. 1. Global relationships of variables involved in the cognitive load and reading performance

Once the structural relationships in the model have been determined, the formalization process will take place. In our proposed agent-based model, all nodes are designed in a way to have values ranging from 0 (low) to 1 (high). Furthermore, the model was formalized with respect to time using a set of First-Order Differential Equation specifications and can be divided into instantaneous and temporal relationships.

3.1 Instantaneous Relationships

Instantaneous relationships refer to the factors that have direct contribution with respect to time factor. For example, reading demand (*Rd*) is formalized as follows:

$$Rd(t) = \eta_{rd}.Tc(t) + (1 - \eta_{rd.\mathrm{Tc}}).Sa(t) \tag{1}$$

Equation (1) explains that the level of reading demand is determined by the proportional contribution of η_{rd} between basic of reading task complexity (*Tc*) and the level of the load imposed by the situational aspects (*Sa*). Similarly, this formal concept is used to formalize all concepts as in all Eqs. (2)–(19).

$$Gd(t) = \gamma_{\mathrm{Gd}}.Me(t) + (1 - \gamma_{\mathrm{Gd}}).Me(t).(1 - Gr(t)) \tag{2}$$

$$\begin{aligned} Sa(t) = \lambda_{\mathrm{sa}}.[\mathrm{w}_{sa1}.Tp(t) + \mathrm{w}_{sa2}.Pe(t)] + (1 - \lambda_{sa}) \\ .[Tp(t).Pe(t).(1 - Tn(t))] \end{aligned} \tag{3}$$

$$Ev(t) = \zeta_{ev}.(\mathrm{w}_{ev1}.El(t) + \mathrm{w}_{\mathrm{ev2}}.Pk(t)) + (1 - \zeta_{ev}).Rn(t) \tag{4}$$

$$Gr(t) = \omega_{gr}.Ev(t) + \left(1 - \omega_{gr}\right).Ev(t).(1 - Sa(t)) \tag{5}$$

$$Me(t) = (1 - Ma(t)).Ml(t) \tag{6}$$

$$Ml(t) = \mathrm{w}_{ml1}.Id(t) + \mathrm{w}_{ml2}.Ed(t) + \mathrm{w}_{ml3}.Gd(t) \tag{7}$$

$$Ma(t) = \mathrm{w}_{ma1}.Rf(t) + \mathrm{w}_{ma2}.Cp(t) + \mathrm{w}_{ma3}.Gr(t) \tag{8}$$

$$Id(t) = Rd(t).(1 - Ev(t)) \tag{9}$$

$$Ed(t) = \beta_{ed}.Sa(t) + (1 - \beta_{ed}).Sa(t).(1 - Gr(t)) \tag{10}$$

$$Sh(t) = \mu_{st}.Ce(t) + (1 - \mu_{st}).Ax(t) \tag{11}$$

$$Ex(t) = (\mathrm{w}_{ex1}.Cl(t) + \mathrm{w}_{ex2}.Ce(t)).(1 - Rf(t)) \tag{12}$$

$$Re(t) = \mathrm{Pos}((\mathrm{w}_{re1}.Cp(t) + \mathrm{w}_{re2}.Ev(t)) - Me(t)) \tag{13}$$

$$Rm(t) = Pr(t).[1 - (\mathrm{w}_{rm1}.Ax(t) + \mathrm{w}_{rm2}.Cl(t))] \tag{14}$$

$$Ce(t) = (\alpha_{ce}.Cl(t) + (1 - \alpha_{ce}).Ax(t)).(1 - Re(t)) \tag{15}$$

$$Cp(t) = \alpha_{cp}.Ev(t) + \left(1 - \alpha_{cp}\right).Pr(t).Ev(t).(1 - Ae(t)) \tag{16}$$

$$Rg(t) = \zeta_{rg}.\mathrm{Ev}(t) + \left(1 - \zeta_{rg}\right).\left[\mathrm{w}_{rg1}.Rd(t) + \mathrm{w}_{rg2}.(1 - (Sa(t).(1 - Ev(t))))\right] \tag{17}$$

$$Rf(t) = \gamma_{rf}.\left(\mathrm{w}_{rf1}.Mv(t) + \mathrm{w}_{\mathrm{rf2}}.Rg(t)\right) + \left(1 - \gamma_{rf}\right).Re(t) \tag{18}$$

$$Mv(t) = \lambda_{mv}.Pp(t) + (1 - \lambda_{mv}).(1 - Pc(t)) \tag{19}$$

3.2 Temporal Relationships

In an agent-based model, the temporal factors are changeable over time due to the interaction between two or more instantaneous factors and often imposed by the delay or accumulative effects of previous values. For example, in Eq. (20), the prolong exposure towards short-term exhaustion (*Sh*) contributed towards the new level of accumulative exhaustion (*Ae*) in a time interval between t and $t + \Delta t$.

$$Ae(t + \Delta t) = Ae(t) + \beta_{Ae}.[Sh(t) - Ae(t)].Ae(t).(1 - Ae(t)).\Delta t \tag{20}$$

Furthermore, the speed of change over time is proportionally determined by the change parameter β_{Ae}. This concept is used throughout temporal relationships as in (21)–(24).

$$Ax(t + \Delta t) = Ax(t) + \eta_{Ax}.Ex(t).[1 - Ax(t)].\Delta t \tag{21}$$

$$Cl(t + \Delta t) = Cl(t) + \beta_{Cl}.[Me(t) - Cl(t)].Cl(t).(1 - Cl(t)).\Delta t \tag{22}$$

$$\begin{aligned} Pr(t + \Delta t) = {} & Pr(t) + \omega_{Pr}.\left[\left[\mathrm{w}_{\mathrm{pr1}}.Mv(t) + \mathrm{w}_{pr2}.Rp(t)\right] - Pr(\mathrm{t}) - \beta_{dp}\right] \\ & .(1 - Pr(t)).Pr(t).\Delta t \end{aligned} \tag{23}$$

$$Rp(t + \Delta t) = Rp(t) + \eta_{Rp}.[((1 - Me(t)).Rm(t)) - Rp(t)].(1 - Rp(t)).Rp(t).\Delta t \tag{24}$$

In both of instantaneous and temporal equations, several parameters (ranges from 0 to 1) are used. For example, from Eqs. (11) and (18), parameters μ_{st} and γ_{rf} represent the regulation factors, while η_{Rp} (from Eq. (24)) represents the rate change for that particular temporal relationship.

4 Simulation Results

Thereafter, several simulation experiments are conducted by a set of initial variations in different agent exogenous inputs to visualize the patterns of the designed agent-based model. In this paper, different variations for three fictional readers (i.e., in a form of agent) are simulated. These simulated scenarios have taken into account the condition of demanding tasks are assigned to all agents but with different personalized settings. The variations are shown as the following; (1) agent ***A***: (expert reader but less motivated), (2) agent ***B***: (expert reader and highly motivated), and (3) agent ***C***: (non-expert reader and less motivated). For brevity in our discussion, our simulation results only cover intrinsic, extraneous, germane, and cognitive load as depicted in Fig. 2.

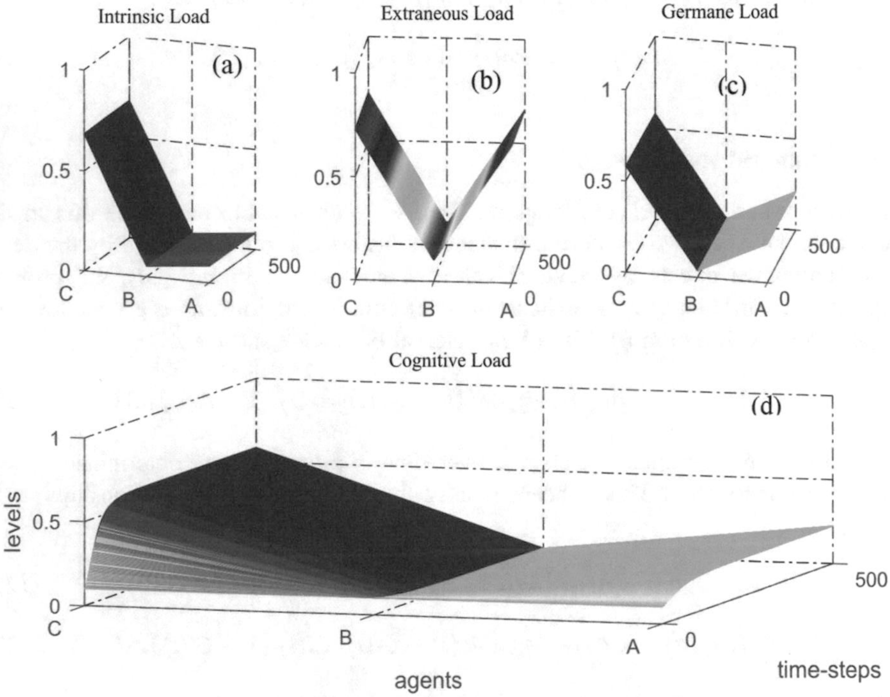

Fig. 2. Simulation results of (a) intrinsic, (b) extraneous, (c) germane, and (d) cognitive load

From Fig. 2, the simulation results show that an agent ***A*** is performing a demanding reading task that requires a high mental effort to be accomplished (e.g. difficult assignments and limited time duration). Nevertheless, the high expertise level has enabled the reader to cope with the demands and curbs the potential progression of cognitive load. However, due to reader's neurotic personality and situated in a non-conducive environment, the observed level of cognitive load is slightly increased. In a long run, this condition can be manifested as a result of high level of extraneous load as seen in Fig. 2(b). Contrary, a high motivation level reduces reader's cognitive load level [12].

In our simulation, similar scenario can be seen within an agent ***B*** (with high expertise level, highly motivated, positive personality (i.e., *openness*) and located within an ambience environment). However, an agent ***C*** shows the contradictory results due to intolerance towards the demands (e.g. *highly difficult task*), not well prepared, discomfort environment, and overwhelming time pressure. Moreover, an agent ***C*** is not motivated to perform the task with inadequate level of knowledge. Detailed explanations about thorough simulation results and settings were used in the experiments can be found in the previous work, as in [6].

5 Formal Verification

In modeling and simulation, correctness and reliability of the models are of the main challenges in developing any cognitive agent models. Within this context, the model correctness is generally understood to mean the extent to which a model implementation conforms to its formal specifications and is free of design errors [13]. The verification process can be defined as can be defined as; "*the process aids of ensuring that the conceptual description and the solution of the model are implemented correctly*". Furthermore, this process is implemented to improve important understanding of the system behaviour, improve computational models, estimate values of parameters, and evaluate (local/global) system performance.

In the light of the above, *mathematical analysis* and *automated logical analysis* are used to obtain an evaluated model of the proposed agent-based model. In mathematical analysis, the equilibrium or stability points are derived to ensure the model is

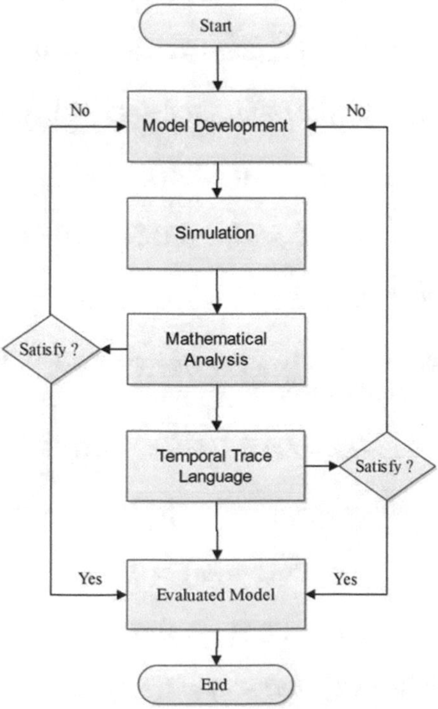

Fig. 3. An agent-based model formal evaluation process

developed as planned. While, automated logical analysis (using Temporal Trace Language) aims to evaluate the validity of the model, which means the generated results of the model adhere to the existing literature. The process of designing and evaluating cognitive agent models are summarized in Fig. 3.

5.1 Mathematical Analysis

The main goal of this approach is to check the theoretical correctness of the model using stable point's analysis of the agent-based model. It is a well-known technique whereas the existence of reasonable equilibria is an indication for the correctness of the model [5]. To analyse the equilibria, both instantaneous and temporal equations are replaced with values for the model variables such that the derivatives are all zero. For instance, the generic differential equation is used to formalize the model can be seen as follows:

$$Y(t+\Delta t) = Y(t) + \beta_y.Y.\langle change\_expression\rangle.\Delta t$$

This differential equation can be re-written into:

$$dY(t)/dt = \beta_y.Y.\langle \text{change\_expression} \rangle$$

Next, the equilibrium points will be generated when all $dY(t)/dt = 0$. The temporal variables of the model written in differential equations as follows:

$$dAe/dt = \beta_{Ae}.(Sh - Ae).Ae.(1 - Ae) \tag{25}$$

$$dPr/dt = \omega_{pr}.\left[\left[\mathrm{w}_{pr1}.Mv + \mathrm{w}_{pr2}.Rp\right] - Pr - \beta_{dp}\right].Pr.(1 - Pr) \tag{26}$$

$$dAx/dt = \eta_{Ax}.Ex.(1 - Ax) \tag{27}$$

$$dCl/dt = \beta_{Cl}.(Me - Cl).(1 - Cl).Cl \tag{28}$$

$$dRp/dt = \eta_{Rp}.[((1 - Me).Rm) - Rp].(1 - Rp).Rp \tag{29}$$

Assuming the parameters ω_{pr}, β_{Ae}, η_{Ax}, β_{Cl}, η_{Rp} are nonzero, from Eqs. 25 to 29, the following cases can be distinguished:

$$(Sh - Ae).Ae.(1 - Ae) = 0,$$

$$\left[\left[\mathrm{w}_{pr1}.Mv + \mathrm{w}_{pr2}.Rp\right] - Pr - \beta_{dp}\right].Pr.(1 - Pr) = 0,$$

$$Ex.(1 - Ax) = 0,$$

$$(Me - Cl).(1 - Cl).Cl = 0,$$

$$[((1 - Me).Rm) - Rp].(1 - Rp).Rp = 0.$$

Assuming $w_{pr1}.Mv + w_{pr2}.Rp = Aa$, and $(1 - Me).Rm = Bb$.

Later these cases can be solved into:

$$
\begin{aligned}
&(Sh = Ae) \vee (Ae = 1) \vee (Ae = 0)\\
&(Pr = 1) \vee (Pr = 0) \vee \left(Pr = Aa - \beta_{dp}\right)\\
&(Cl = Me) \vee (Cl = 1) \vee (Cl = 0)\\
&(Rp = 1) \vee (Rp = 0) \vee (Rp = Bb)\\
&(Ex = 0) \vee (Ax = 1)
\end{aligned}
$$

From here, a first conclusion can be derived as the equilibrium can only occur when $Sh = Ae$, $Ae = 1$, or $Ae = 0$. By combining these three conditions, it can be re-written into a set of relationship in $(A \vee B \vee C) \wedge (D \vee E \vee F)$ expression:

$$
\begin{aligned}
&((Ae = Sh) \vee (Ae = 1) \vee (Ae = 0)) \wedge \left((Pr = 1) \vee (Pr = 0) \vee \left(\mathrm{Pr} = Aa - \beta_{dp}\right)\right)\\
&\quad \wedge ((Cl = Me) \vee (Cl = 1) \vee (Cl = 0)) \wedge ((Rp = 1) \vee (Rp = 0) \vee (Pr = Bb))\\
&\quad \wedge ((Ex = 0) \vee (Ax = 1))
\end{aligned}
$$

This expression can be elaborated using the *law of distributivity* $(A \wedge D) \vee (A \wedge E) \vee, \ldots, \vee (C \wedge F)$ as:

$$
\begin{aligned}
&((Ae = Sh) \wedge (Pr = 1) \wedge (Cl = Me) \wedge (Rp = 1) \wedge (Ex = 0))\\
&\quad \vee ((Ae = Sh) \wedge (Pr = 0) \wedge (Cl = 1) \wedge (Rp = 0) \wedge (Ax = 1))\\
&\quad \vee \ldots \left((Ae = 0) \wedge \left(Pr = Aa - \beta_{dp}\right)\right) \wedge ((Cl = 0) \wedge (Rp = Bb) \wedge (Ax = 1))
\end{aligned}
$$

Therefore, large number of possible equilibrium points are generated (in this case $(2\,(3^4)) = 162$). In addition, mathematical analysis cannot be performed to all equilibrium cases because some of them are not existed in the simulation traces or even in the literatures. For example;

$$
\begin{aligned}
&\left((Ae = 0) \wedge \left(\left(w_{pr1}.Mv + w_{pr2}.Rp\right) - \beta_{dp} = Pr\right)\right)\\
&\quad \wedge ((Cl = 0) \wedge ((Rp = (1 - Me).Rm)) \wedge (Ax = 1))
\end{aligned}
$$

This case shows accumulative exhaustion level is equal to zero ($Ae = 0$) and in the same time the accumulative level of experienced exhaustion is equal to one ($Ax = 1$). Despite the possible combinations in a theoretical construct, this case is not existed in the real-world. Note here, due to the large number of possible combinations, it is hard to provide a complete classification of equilibria. However, for some cases the analysis can be pursued further.

Case #1: $Cl = 1 \wedge Rp = 0 \wedge Ax = 1 \wedge Pr = 0 \wedge Ae = Sh$
For this case (it can be seen in Fig. 4(a–c), by Eq. (11) it follows that

$$Sh = \mu_{st}.Ce + (1 - \mu_{st}) \text{ and } \mu_{st} = 0.5$$

Thus, $Sh = Ce$ which depicts the level of short term exhaustion (Sh) is always determined by the current level of cognitive exhaustion (Ce). In other words, the level of short-term exhaustion is always high when the level of cognitive exhaustion is high and the vice versa. The effect of the stability points in short-term exhaustion and cognitive exhaustion levels can be seen in Fig. 4(c).

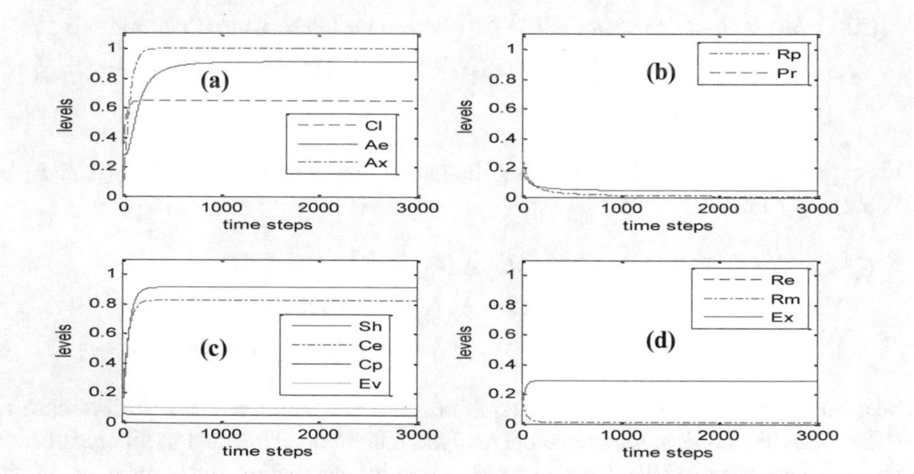

Fig. 4. Simulation results of the stability points in Case #1.

Moreover, by Eq. (12) it follows that:

$$Ex = (w_{ex1} + w_{ex2}.Ce).(1 - Rf), \text{ and } w_{ex2} = 1 \text{ then } Ex = Ce.(1 - Rf)$$

This concept shows recovery effort is negatively contributed to the level of experienced exhaustion. Also, experienced exhaustion is positively correlated with the high cognitive exhaustion level. The simulation result of this analysis can be visualized in Fig. 4 (c and d). Moreover, the Eq. (14) yields: $Rm = 0$ which depicts the level of reading engagement is equal to zero when the reader experiences high level of cognitive load, accumulative experienced exhaustion, and accumulative exhaustion. Figure 4(b) and (d) depicts the effects of the stable points in Rm. Next, by Eq. (15)

$$Ce = (\alpha_{ce} + (1 - \alpha_{ce})).(1 - Re)$$

Assuming α_{ce} is nonzero, then this is equivalent to $Ce = 1 - Re$.

The condition explains the negative correlation between cognitive exhaustion and recovery efforts. The result can be seen in Figs. 3(d) and 4(c).

Finally, Eq. (16) follows that

$$Cp = \alpha_{cp}.Ev \text{ and } \alpha_{cp} \text{ is nonzero } \rightarrow Cp = Ev$$

This condition explains the development of critical point levels is completely depending to the level of expertise a reader has in tackling difficulties of the tasks (as depicted in Fig. 4(c)).

The same concept is also implemented in Cases 2 and 3, and the results were satisfied. The detailed description is as follows:

Case #2: *Cl* = 1
For this case, by Eq. (12) it follows

$$Ex = (w_{ex1} + w_{ex2}.Ce).(1 - Rf)$$

Hence by Eq. (14), it is equivalent to:

$$Rm = Pr.[1 - (w_{rm1}.Ax + w_{rm2})]$$

Moreover, by Eq. (15) it follows $Ce = (\alpha_{ce} + (1 - \alpha_{ce}).Ax).(1 - Re)$.
Assuming $\alpha_{ce} \neq 0$, this equivalent to $Ce = Ax.(1 - Re)$.
Since it can recall that $Ax = 1$, this leads to $Ce = (1 - Re)$.

Case #3: *Ae* = 1 ∧ *Ax* = 1
From Eq. (11) it follows the case is equivalent to

$$Sh = \mu_{st}.Ce + (1 - \mu_{st})$$

Assuming μ_{st} equal to 1, this equals to $Sh = Ce$.

By Eq. (14), it follows

$$Rm = Pr.[1 - (w_{rm1} + w_{rm2}.Cl)]$$

Moreover, the results in Eq. 15, $Ce = (\alpha ce.Cl + (1 - \alpha_{ce})).(1 - Re)$.

Equation (16) is equivalent to $Cp = \alpha_{cp}.Ev$.

5.2 Automated Analysis

To check the model truly yields results adhere to the existing cognitive load literatures, a set of properties have been identified from the literatures. These properties have been specified in a language called Temporal Trace Language (TTL). TTL is built on atoms referring to states of the world, time points, and traces. This relationship can be presented as holds (*state*(γ, *t*), *p*) or *state*(γ, *t*)$| = p$, which means that state property *p* is true in the state of trace γ at time point *t* (details on TTL can be found in [14]). Based on that concept, the dynamic properties can be formulated using a hybrid sorted predicate logic approach, by using quantifiers over time and traces and First-Order Logical connectives (FOPL) such as ¬, ∧, ∨, ⇒, ∀, and ∃. TTL is used by generating a finite state space of a formal model of a system and later verifies a property written in some temporal logic specifications, through an explicit state space search. It can provide an answer in a few minute or even seconds for many models as the search always terminates (due to the finite search space). A number of simulations including the ones described in Sects. 4 and 5 have been used as basis for the automated verification of the identified properties and were confirmed.

VP1: Readers with a high level of persistence tends to reduce cognitive load level.

VP1 ≡ ∀γ:TRACE, ∀t1,t2:TIME, ∀D,B1,B2, R1,R2:REAL, ∀X:AGENT
[state(γ, t1)|= persistence(X, B1) &
state(γ, t2)|= persistence(X, B2) &
state(γ, t1)|= cognitive_load(X,R1) &
state(γ, t2)|= cognitive_load(X,R2) &
t2 > t1 + D & B2 > 0.6 & B2 ≥ B1] ⇒ R2 < R1

The property VP1 is used to check the impact of reader's persistence level on the level of cognitive load. It shows that readers with high level of persistence (having a value greater or equal to 0.6) tend to experience low cognitive load. This result is in line with the literature as in [12].

VP2: Readers with high level of expertise possess high critical power [16].

VP2 ≡ ∀γ:TRACE, ∀t1, t2:TIME, ∀X1, X2, C, D:REAL , ∀A:AGENT
state(γ, t1)|= expertise_level(A, C) &
state(γ, t1)|= critical_power(A, M1) &
state(γ, t2)|= critical_power(A, M2) &
C ≥ 0.8 & t2= t1+D ⇒ M2 ≥ M1

VP3: Non-conducive learning environment increases cognitive load [17].

VP3 ≡ ∀γ:TRACE, ∀t1, t2:TIME, ∀V1, V2, Q, D:REAL , ∀A:AGENT
state(γ, t1)|= ambience_room(A, Q) &
state(γ, t1)|= cognitive_load(A, V1) &
state(γ, t2)|= cognitive_load(A, V2) &
Q < 0.2 & t2= t1+D ⇒ V2 ≥ V1

VP4: Reading performance is high when cognitive load is low.

VP4 ≡ ∀γ:TRACE, ∀t1,t2:TIME, ∀D,V1,V2, R1,R2:REAL, ∀X:AGENT
[state(γ, t1)|= cognitive_load(X, B1) &
state(γ, t2)|= cognitive_load(X, B2) &
state(γ, t1)|= reading_performance(X,R1) &
state(γ, t2)|= reading_performance(X,R2) &
t2 > t1 + D & B1 < 0.2 & B1 ≥ B2] ⇒ R2 ≥ R1

This property used to check the impact of cognitive load on reading performance. It depicts that low level of cognitive load (having a value below 0.2) increases reading performance level [15].

6 Conclusion

Cognitive and psychological models of humans' physical and mental states have been showing a great success in creating intelligent systems able to analyse and perform human-like understanding and base on this understanding a personalize support can be provided. From this standpoint, an agent model of cognitive load and reading performance was developed and formally analysed to ensure the correctness of the model. This paper presented two widely-used approaches to evaluate the cognitive and reading performance model. These approaches are mathematical analysis and automated logical analysis. The results showed that the model is correctly developed and it exhibited realistic behaviour patterns of the intended domain. For the future work, more cases are required to be analysed to consolidate the internal validation of the model. Furthermore, parameter tuning is of much interest to be implemented to confirm the external validation which means individuals experiments to fit with individual differences.

Acknowledgment. This project is funded by UUM Postgraduate Research Scholarship programmes.

References

1. Treur, J.: Network-Oriented Modeling, pp. 463–471. Springer, Berlin (2016)
2. Aziz, A.A., Ahmad, F., ChePa, N., Yusof, S.A.M.: Verification of an agent model for chronic fatigue syndrome. Int. J. Digit. Content Technol. Appl. **7**, 25 (2013)
3. Ojeniyi, A., Aziz, A.A., Yusof, Y.: Verification analysis of an agent based model in behaviour change process. In: International Symposium on Agents, Multi-Agent Systems and Robotics (ISAMSR), pp. 87–92 (2015)
4. Mohammed, H., Aziz, A.A., Pa, N.C., Shabli, A.H.M., Bakar, J.A.A., Alwi, A.: A computational agent model for stress reaction in natural disaster victims. In: The 7th International Conference on Information Science and Applications (ICISA 2016), pp. 817–827. Springer-LNEE (2016)
5. Treur, J.: Verification of temporal-causal network models by mathematical analysis. Vietnam J. Comput. Sci. **3**, 207–221 (2016)
6. Ghanimi, H.M.A., Aziz, A.A., Ahmad, F.: An agent-based model for refined cognitive load and reading performance in reading companion robot. J. Telecommun. Electron. Comput. Eng. **9**(3–5), 55–59 (2016)
7. Bosse, T., Jonker, C.M., Van Der Meij, L., Treur, J.: A temporal trace language for the formal analysis of dynamic properties, pp. 1–15 (2006)
8. Bosse, T., Hoogendoorn, M., Klein, M.C.A., Treur, J.: An ambient agent model for monitoring and analysing dynamics of complex human behaviour. J. Ambient Intell. Smart Environ. **3**, 283–303 (2011)
9. Bosse, T., Both, F., Van Lambalgen, R., Treur, J.: An agent model for a human's functional state and performance. In: Proceedings of the 2008 IEEE/WIC/ACM International Conference on Web Intelligence and Intelligent Agent Technology, vol. 2, pp. 302–307. IEEE Computer Society (2008)
10. Ghanimi, H.M.A., Aziz, A.A., Ahmad, F.: On modeling cognitive load during reading task. Malays. J. Hum. Factors Ergon **1**(1), 55–61 (2016)

11. Ghanimi, H.M.A., Aziz, A.A., Ahmad, F.: An agent-based modeling for a reader's cognitive load and performance. Adv. Sci. Lett. **24**, 966–970 (2016)
12. Schnotz, W., Fries, S., Horz, H.: Motivational aspects of cognitive load theory. In: Wosnitza, M., Karabenick, S.A., Efklides, A., Nenniger, P. (eds.) Contemporary Motivation Research: From Global to Local Perspectives, pp. 69–96. Hogrefe and Huber Publishers, Cambridge (2009)
13. Voogd, J.: Verification and validation for CIPRNet. In: Setola, R., Rosaro, V., Kyriakides, E., Rome, E. (eds.) Managing the Complexity of Critical Infrastructures, pp. 163–193. Springer, Berlin (2016)
14. Bosse, T., Jonker, C.M., Van der Meij, L., Sharpanskykh, A., Treur, J.: Specification and verification of dynamics in agent models. Int. J. Cooper. Inf. Syst. **18**(01), 167–193 (2009)
15. Choi, H.-H., van Merriënboer, J.J.G., Paas, F.: Effects of the physical environment on cognitive load and learning: towards a new model of cognitive load. Educ. Psychol. Rev. **26**, 225–244 (2014)
16. Treur, J.: A virtual human agent model with behaviour based on feeling exhaustion. Appl. Intell. **35**, 469–482 (2011)
17. Cech, P.: Smart classroom study design for analysing the effect of environmental conditions on students' comfort. In: Intelligent Environments 2016: Workshop Proceedings of the 12th International Conference on Intelligent Environments, pp. 14–23 (2016)

Analysis the Arabic Authorship Attribution Using Machine Learning Methods: Application on Islamic Fatwā

Mohammed Al-Sarem[1,2] and Abdel-Hamid Emara[3,4](✉)

[1] Department of Information System, Taibah University, Medina, Kingdom of Saudi Arabia
mohsarem@gmail.com
[2] Department of Computer Science, Saba'a Region University, Mareeb, Yemen
[3] Department of Computer Sciences, Taibah University, Medina, Kingdom of Saudi Arabia
abdemara@gmail.com
[4] Computer Systems and Engineering Department, Al-Azhar University, Cairo, Egypt

Abstract. In context of Arabic, the authorship attribution (AA) problem is not addressed well comparing with other natural languages such English, Chinese and Dutch. This paper addresses the attribution problem in context of Islamic fatwā'. To the best of our knowledge, this is the first study of its kind that addresses this problem in such domain. In term of attribution methods, three machine-learning classifiers namely, the locally weighted learning (LWL) classifier, decision tree C4.5, and Random Forest (RF) are used. The experiment is performed with a selected list of stylomatric features. To extract the most discriminating features, various feature selection techniques are used. The experimental results show that the classifiers have different behaviour respect each feature reduction techniques. Among the used classifiers, the C4.5 method gives the best accuracy.

Keywords: Machine learning classifier · Arabic authorship attribution Stylometric features · PCA · Accuracy

1 Introduction

Authorship attribution AA has a long history. The problem stems directly from the stylometry analysis, where set of features are extracted from a document written by an author. These features are quite enough to compute the probability that an anonymous text document was written by a particular author [1]. Over the past 200 years, studies have yielded several tools and methods [2]. According to [3], AA methods fall into three main classes: unitary invariant, multivariate analysis and machine learning classes. On contrary of the methods of the first two classes, the machine learning methods represent the training textual document as vectors of features that are used to build an attribution model (classifier). The classifier is used later to attribute anonymous documents.

F. Saeed et al. (Eds.): IRICT 2018, AISC 843, pp. 221–229, 2019.
https://doi.org/10.1007/978-3-319-99007-1_21

Although the AA has a long success in various languages such English [4], German [5], Spanish [6] and Chinese [7], number of research in Arabic is still quite limited [8, 9]. Thus, this work aims at drawing researchers' attention to fill this gap. This work addresses the AA problem in context of Islamic fatwa written in Arabic. The fatwā (Arabic: فتوى; plural fatāwā Arabic: فتاوى) as it defined in [10] is an "islamic faith that is a nonbinding but authoritative legal opinion or learned interpretation that the Sheikhul Islam, a qualified jurist or mufti, can give on issues pertaining to the Islamic law". The fatwā's often follows a specific structure. However, we do not make extra attention about that and only deal with it as a regular Arabic text. Indeed, to extract the content of the fatwā', there is a need to implement a specific extractor tool, but in this work we do not address that. Instead of that we focus on the preprocessing phase, features selection, building the attribution model and finally evaluating the model.

This paper is organized as follows: Sect. 2 introduces briefly the necessary background that we stand on. Section 3 describes the research methodology that we follow as well as the conducted experiment and finding results. Finally, Sect. 4 summarizes the whole paper and presents a slight to the future directions of this work.

2 Background

As mentioned earlier, the AA aims at identifying the author of an unseen document based on some extracted features known as "Stylometric features". The researches topics in AA are around the feature selection and attribution techniques.

2.1 Writing Style Features

In the literature, there are several features can be used to quantify the writing style. These features can be categorized into seven main groups [11]: lexical, character, syntactic, semantic, content-specific, structural and language-specific.

Lexical such word frequencies, word length, vocabulary richness, sentence length and word n-grams. To extract lexical features, a text content should be tokenized into sentences, words, numbers, or a punctuation marks.

- *Character*, this type of features require a specific treatment. The text content is considered as sequence of characters. Thus, the extractor tool should differentiate between letters, digits, uppercase and lowercase characters, etc. Furthermore, it should to take in consideration the language characteristics.
- *Syntactic*, this type of features proposes that the author of a text tends to use similar syntactic patterns unconsciously. The most obvious syntactic measure is a part-of-speech POS.
- *Semantic*, features such semantic dependencies, synonyms and functional words.
- *Application-specific*, these features can be: structural, content-specific, or language-specific. In the literature, the work of Zheng et al. [12] and Abbasi and Chen [13] are obvious example of structural features. In contrast of the structural features, the content-specific features are extracted only from the textual data of the same topic

such as in [14]. The language-specific features are specific for a particular language. Thus, to measure them, these features have to be defined manually.

2.2 Typical Authorship Attribution Process

Typically, the AA is considered as a classical supervised learning problem. Figure 1 shows the typical AA process.

In this work, we follow the CRSIP-DM process, so the Fig. 1 is modified as follows:

- at the data preprocessing phase, beside the typical dataset collection, the feature extraction is required process. Practically, this task is time-consuming. The most difficulties, here, is because of absence a suitable extractor tool for Arabic texts. The feature selection and dimension reduction are necessary steps. However, we postponed this task to check first the classifier performance. In case the results are acceptable, the feature selection and dimension reduction can be excluded (the red colour in Fig. 2 denotes that the step is optional).
- at training/testing phase: the classifiers are selected, and the model is built. The text samples are spilt to two subsets. The way that we followed here is to use 10-fold partition.
- to evaluate the constructed model, the 10-fold cross validation is used. The obtained model is evaluated in term of accuracy. In case the accuracy of the classifier is not satisfied, the features are re-check and reduction techniques are applied then the model is re-built.

3 Experiment and Finding Results

3.1 Dataset

Our dataset corpus was collected by extracting the content of fatwā' Dar Al-ifta AL Misriyyah[1]. The original texts were unstructured which make our task more complicated. Figure 3 shows a sample of the fatwā's content that we deal with:

- fatwā' metadata: information such the name of mufti, the fatwā's category and the source of the document.
- fatwā's principles include fatwā's background or the previous fatwā' that the mufti can rely on to issue his opinions.
- fatwā' question summarizes the problem or the question being considered.
- fatwā' answer includes the mufti's response and his interpretation.

Since the fatwā' question can be posed from anyone and the content of the question is not related to the mufti response, we excluded it from the document. Thus, to extract the required text from the whole document, we implemented a special extractor tool.

[1] http://www.dar-alifta.org/Foreign/default.aspx?LangID=2&Home=1.

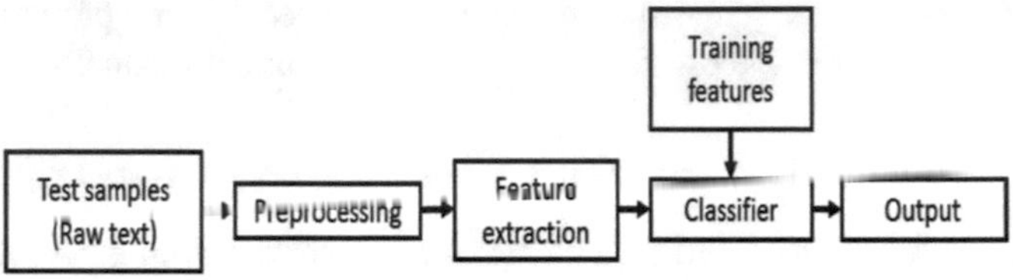

Fig. 1. Typical authorship attribution process 15.

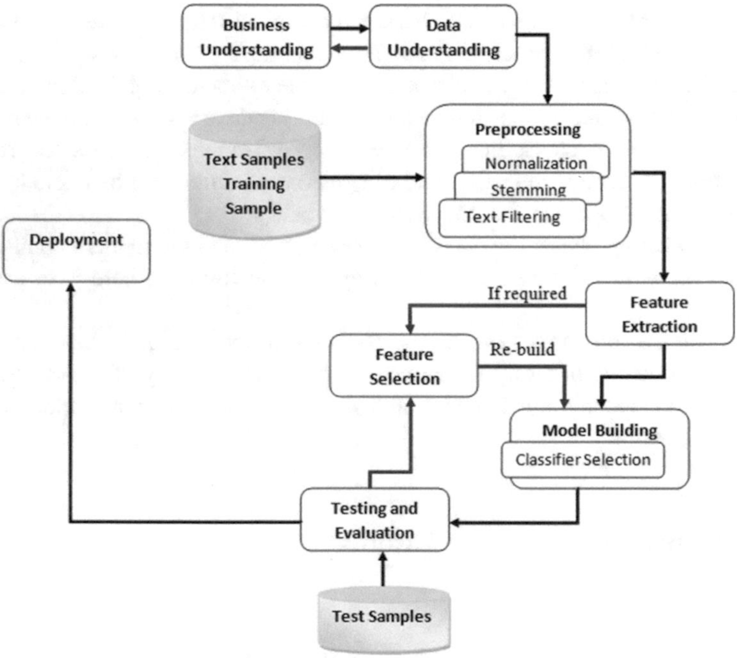

Fig. 2. The AA process based on CRSIP-DM process

3.2 Stylometric Features

Several features can be used to quantify the writing style. These features, as stated earlier, can be categorized into lexical, character, syntactic, semantic, content-specific, structural and language-specific features [11]. Table 1 summarizes the used features in this paper.

3.3 Design of Experiment

Like any supervised learning task, the AA problem requires to build and train the classifier model, to test the model on the set of unseen texts and to evaluate and to improve the model.

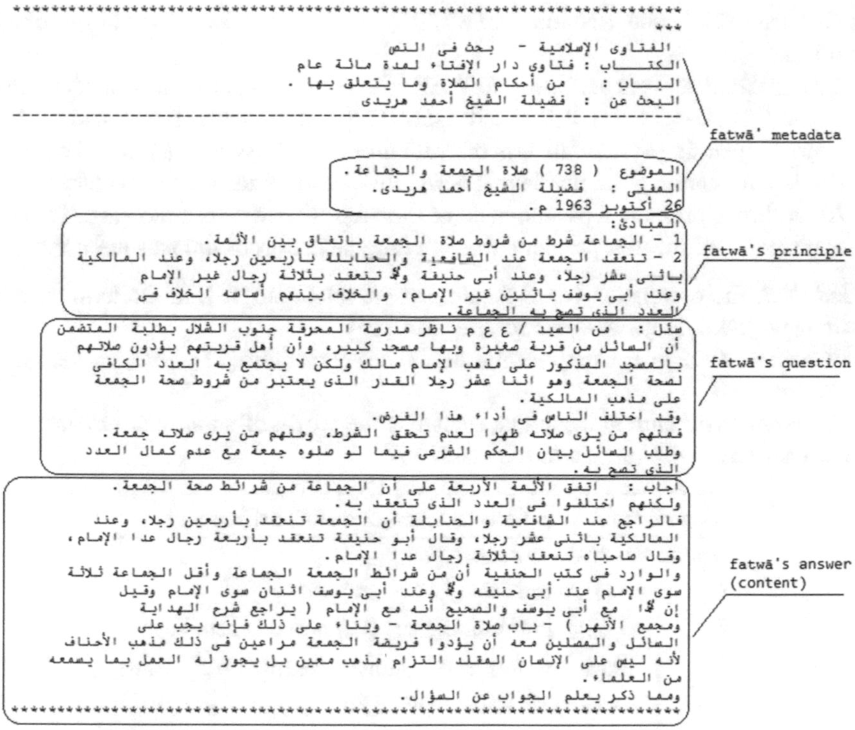
الفتاوى الإسلامية - بحث فى النص
الكتـــــاب : فتاوى دار الإفتاء لمدة مائة عام
البـــاب : من أحكام الصلاة وما يتعلق بها .
البحث عن : فضيلة الشيخ أحمد هريدى

الموضوع (738) صلاة الجمعة والجماعة .
المفتى : فضيلة الشيخ أحمد هريدى.
26 أكتوبر 1963 م .
المبادئ:
1 - الجماعة شرط من شروط صلاة الجمعة باتفاق بين الأئمة .
2 - تنعقد الجمعة عند الشافعية والحنابلة بأربعين رجلا، وعند المالكية باثنى عشر رجلا، وعند أبى حنيفة ومحمد تنعقد بثلاثة رجال غير الإمام وعند أبى يوسف باثنين غير الإمام، والخلاف بينهم أساسه الخلاف فى العدد الذى تصح به الجماعة .
سئل : من السيد / ع ع م ناظر مدرسة المحرقة جنوب الشلال بطلبة المتضمن أن السائل من قرية صغيرة وبها مسجد كبير، وأن أهل قريتهم يؤدون صلاتهم بالمسجد المذكور على مذهب الإمام مالك ولكن لا يجتمع به العدد الكافى لصحة الجمعة وهو اثنا عشر رجلا القدر الذى يعتبر من شروط صحة الجمعة على مذهب المالكية .
وقد اختلف الناس فى أداء هذا الفرض.
فمنهم من يرى صلاته ظهرا لعدم تحقق الشرط، ومنهم من يرى صلاته جمعة .
وطلب السائل بيان الحكم الشرعى فيما لو صلوه جمعة مع عدم كمال العدد الذى تصح به .
أجاب : اتفق الأئمة الأربعة على أن الجماعة من شرائط صحة الجمعة .
ولكنهم اختلفوا فى العدد الذى تنعقد به .
فالراجح عند الشافعية والحنابلة أن الجمعة تنعقد بأربعين رجلا، وعند المالكية باثنى عشر رجلا، وقال أبو حنيفة تنعقد بأربعة رجال عدا الإمام، وقال صاحباه تنعقد بثلاثة رجال عدا الإمام .
والوارد فى كتب الحنفية أن من شرائط الجمعة الجماعة وأقل الجماعة ثلاثة سوى الإمام عند أبى حنيفه ومحمد وعند أبى يوسف اثنان سوى الإمام وقيل إن محمدا مع أبى يوسف والصحيح أنه مع الإمام (يراجع شرح الهداية ومجمع الأنهر) - باب صلاة الجمعة - وبناء على ذلك فإنه يجب على السائل والمصلين معه أن يؤدوا فريضة الجمعة مراعين فى ذلك مذهب الأحناف لأنه ليس على الإنسان المقلد التزام مذهب معين بل يجوز له العمل بما يسمعه من العلماء .
ومما ذكر يعلم الجواب عن السؤال.

Fig. 3. An example of fatwā's structure

Table 1. Stylometric features

Name	Description
#_words	Number of words
#_uniq_words	Number of unique words
vocab_dens	Vocabulary density
#_sent.	Number of sentences
avg_word/sent.	Average words per sentences
#_lines	Number of lines
#_p	Number of paragraphs
avg_word/p	Average words per paragraph
avg_sent/p	Average sentence per paragraph
avg_line/p.	Average lines per paragraph
avg_leng_word	Average word length
Class	Mufti Name

Despite the whole text mining process, our focus is on enhancing the classifier performance. Thus, we applied many feature selection methods to examine whether the classifier performance is enhanced. The locally weighted learning (LWL) classifier, decision tree C4.5, and Random Forest (RF) are selected as authorship attribution classifiers.

All experiments were performed in WEKA[2] 3.6.12 on a personal computer with an Intel Core(TM) i7-4600U CPU @2.70 GHz CPU, a 8-Gbyte RAM and a 64-bit Windows 8 operating system. In term of evaluation, the Cross-validation was employed in 10-folds version and the number of correctly classified instances was taken.

Algorithm 1 presents a pseudo-code of the steps that we are following. To reduce the complexity of the AA problem, three feature selection techniques are used:

- *SubEval*: the correlation-based feature subset selection [15]. In the term of search strategy, the subset size forward selection [16].
- *PCA*: the Principle Components Analysis algorithm [17] with ranker search strategy.
- *GainRatioEval*: current technique evaluates the worth of an attribute by measuring the gain ratio with respect to the class.

Algorithm 1. Experimental Setup, Set of Classifiers (C)

Input: fatwā' documents (D)

1 ***foreach*** *round* $in\ (1,2,\ldots,n)$ ***do***
2   - Split D into training set (D_t) and test set (D_s)
3   - Extract vector of stylomatric features ($\overrightarrow{sf}$) from D_t
4   ***foreach*** classifier $c_i \in C$ ***do***
5   - Train c_i based on the $\overrightarrow{sf}$
6   - Test c_i on unseen document $d_i \in D_s$
7   - Evaluate c_i in term of accuracy
8   ***If*** *accuracy* **not** satisfied ***Then***
9   - Do feature reduction
10   - Go to step 5
11   ***end if***
12   **End**
13 **End**

Output: accuracy of each classifier c_i

3.4 Experimental Findings

As stated earlier, the experiment was conducted in two ways: firstly, we use all stylomatric features to train the classifiers then, the classifier performance is evaluated in term of accuracy. In the next, the feature selection techniques are applied, then some features are excluded.

[2] http://www.cs.waikato.ac.nz/ml/weka/.

(A) *Performance classifier of AA with all features*

As stated earlier, all stylomatric features stated in Table 1 are contributed in examining the classifiers' performance. Table 2 summarizes the experimental findings respect to each of the aforementioned classifiers. The C4.5 method showed the highest accuracy among the other classifiers. The C4.5 achieved 51.55% accuracy. Indeed, this value is still low. However, the result is acceptable especially if we compared this value with that is found in [9]. In contrary of that, applying the Random Forest classifier (RF) leads to decrease the attribution accuracy. In our case, the accuracy of RF method is $\approx$25%. Furthermore, the time to build the model is quite high comparing with the time-consuming by the other methods.

(B) *Performance the Classifiers of AA with selected features*

The goal of this section is to examine if the feature reduction effects positively on the classifier performance and attribution accuracy. Three well-known feature selection techniques are used: SubEval, PCA and GainRatioEval. The experiments showed that:

- no affect on the LWL method when applying the features selection techniques. The only enhancement that we noted is that, despite of the whole stylometric features, few numbers of features can be used to attribute the text.
- the C4.5 classifier is effected differently respect to the feature selection techniques. A little bit improvement was when the SubEval method was applied as well as the time to build the model. On the other hand, applying the PCA and GainRatioEval leads to decrease the accuracy.
- On the opposite of C4.5, applying the PCA and GainRatioEval techniques on RF classifier leads to enhance the accuracy.
- On opposite of the first two experiments, applying the subEval on RF classifier takes a lot of time as well as the accuracy is the worst.

Tables 3, 4 and 5 summarize the experimental findings respect to each of the aforementioned classifiers.

Table 2. Results on all features

Classifier	Acc (%)	Time to build (c)
LWL	40.87%	0.13
Decision tree C4.5	51.55%	0.92
Random forest	24.67%	25.18

Table 3. The effect of feature selection on LWL

Feature selection technique	Number of features	Acc (%)	Time to build (c)
SubEval	5	40.79%	0.23
PCA	4	40.63%	0.12
GainRatioEval	10	40.85%	0.17

Table 4. The effect of feature selection on C4.5

Feature selection technique	Number of features	Acc (%)	Time to build (c)
SubEval	5	51.70%	0.66
PCA	1	45.32%	1.09
GainRatioEval	10	47.27%	1.86

Table 5. The effect of feature selection on RF

Feature selection technique	Number of features	Acc (%)	Time to build (c)
SubEval	5	17.15%	53.19
PCA	6	32.89%	5.2
GainRatioEval	10	40.81%	5.91

4 Conclusions and Future Work

This study addresses the authorship attribution problem in context of Islamic fatwā's. To the best of our knowledge, current task does not address before. As attribution features, a selected list of stylomatric features are used. Indeed, there are a lot of studies discussed the authorship attribution with huge attribution features [3, 5, 9]. However, we limited this experiment on a few lists because of this research, on one hand, is still in the beginning. On the other hand, the absence of Arabic feature extractor, we seek the optimal feature combination that lead to increase the accuracy and save our effort in building the authorship attribution system.

The conducted experiments have examined with the three machine learning classifiers: the locally weighted learning (LWL) classifier, decision tree C4.5, and Random Forest (RF). The experiment showed that the C4.5 is superior to other methods. The accuracy of attribution was achieved 51.55% without feature selection and 51.70% when we applied the feature reduction techniques. In term of feature reduction techniques, the SubEval, PCA and GainRatioEval were applied. It was noted that the attribution methods were affected differentially respect each reduction technique.

In the future, we intend to extend the experiment to cover more machine classifiers, more stylomatric features and more reduction techniques. Also, we are in process of building a free source Arabic attribution tool with a huge embedded feature extractor options.

References

1. Neal, T., Sundararajan, K., Fatima, A., Yan, Y., Xiang, Y., Woodard, D.: Surveying stylometry techniques and applications. ACM Comput. Surv. (CSUR) **50**(6), 86 (2017)
2. Jockers, M.L., Witten, D.M.: A comparative study of machine-learning methods for authorship attribution. Lit. Linguist. Comput. **25**, 215–223 (2010)
3. Altheneyan, A.S., El Bachir Menai, M.: Naıve Bayes classifier for authorship attribution of Arabic texts. J. King Saud Univ. Comput. Inf. Sci. **26**(4), 473–484 (2014)

4. Labbé, D.: Experiments on authorship attribution by intertextual distance in English. J. Quant. Linguist. **14**(1), 33–80 (2017)
5. Savoy, J., Attribution, A.: A comparative study of three text corpora and three languages. J. Quant. Linguist. **19**(2), 132–161 (2012)
6. Oppliger, R.: Automatic authorship attribution based on character n-grams in Swiss German. In: Proceedings of the 13th Conference on Natural Language Processing (KONVENS 2016)
7. Crespo, M., Frías, A.: Stylistic authorship comparison and attribution of Spanish news forum messages based on the Tree Tagger POS Tagger, The Authors. Published by Elsevier, Multimodal Communication in the 21st Century: Professional and Academic Challenges. In: 33rd Conference of the Spanish Association of Applied Linguistics (AESLA), XXXIII AESLA Conference, Madrid, Spain, 16–18 April 2015
8. Shaker, K., Corne, D.: Authorship attribution in Arabic using a hybrid of evolutionary search and linear discriminant analysis. In: 2010 UK Workshop on Computational Intelligence (UKCI), pp. 1–6 (2010). https://doi.org/10.1109/ukci.2010.5625580
9. Al-Ayyoub, M., Jararweh, Y., Rababa'ah, A., Aldwairi, M.: Feature extraction and selection for Arabic tweets authorship authentication. J. Ambient Intell. Hum. Comput. **8**, 383–393 (2017). https://doi.org/10.1007/s12652-017-0452-1
10. https://en.wikipedia.org/wiki/Fatwa. Last visit: 22:48. Accessed 18 Jan 2018
11. Stamatatos, E.: A survey of modern authorship attribution methods. J. Am. Soc. Inf. Sci. Technol. **60**(3), 538–556 (2009). https://doi.org/10.1002/asi.21001
12. Zheng, R., Li, J., Chen, H., Huang, Z.: A framework for authorship identification of online messages: writing-style features and classification techniques. J. Am. Soc. Inf. Sci. Technol. **57**(3), 378–393 (2006)
13. Abbasi, A., Chen, H.: Applying authorship analysis to extremist-group Web forum messages. IEEE Intell. Syst. **20**(5), 67–75 (2005). https://doi.org/10.1109/MIS.2005.81
14. Al-Falahi, A., Ramdani, M., Bellafkih, M., Al-Sarem, M.: Authorship attribution in Arabic poetry' context using Markov Chain classifier. IEEE (2015)
15. Hall, M.A.: Correlation-based feature selection for machine learning. Ph.D. thesis, The University of Waikato (1999)
16. Guetlein, M., Frank, E., Hall, M.: Large scale attribute selection using wrappers. In: Proceedings of IEEE Symposium on Computational Intelligence and Data Mining, pp. 332–339 (2009)
17. Dunteman, G.H.: Principal Components Analysis, vol. 69. Sage, Thousand Oaks (1989)

A Fuzz Logic-Based Cloud Computing Provider's Evaluation and Recommendation Model

Mohd Hilmi Hasan[1(✉)], Norshakirah A. Aziz[1], Emelia Akashah P. Akhir[1], Nur Zarith Natasya Mohd Burhan[1], and Razulaimi Razali[2]

[1] Center for Research in Data Science, Computer and Information Sciences Department, Universiti Teknologi PETRONAS, 32610, Seri Iskandar, Perak, Malaysia
mhilmi_hasan@utp.edu.my

[2] Universiti Teknologi MARA, Bandar Jengka, 26400 Bandar Tun Razak, Pahang, Malaysia

Abstract. The growing number of cloud computing providers has made the selection process difficult and time consuming. This motivates many research to propose cloud computing recommendation system. However, generating an overall rating of a provider based on accumulation of several performance parameters involves uncertainties especially when the computation comprises network-related performance parameters. Hence, this paper proposes a fuzzy logic-based model that evaluates providers based on downlink, uplink and latency parameters, and determines the best provider. The model was constructed based on historical data of the real cloud computing providers implementation. The model was compared against non-fuzzy model, and it showed that the proposed fuzzy logic-based model is better in terms of accuracy. This work implies faster and accurate selection of cloud computing provider.

Keywords: Cloud computing provider's recommendation
Recommendation system · Fuzzy logic recommendation
Fuzzy recommendation system

1 Introduction

For the past several years, cloud computing and networking technologies have been growing rapidly. Cloud computing can be referred as a set of scalable services being hosted online where user can access them via the Internet. Specifically, cloud services are services made available to users on demand via network from data centers operated by cloud computing providers [1]. Cloud service providers have developed a wide range of cloud services to cater for diverse business organizations' information systems. Currently in the market, there are many providers such as Amazon, Google, IBM, Microsoft and others that offer business organizations variety of cloud-based services where users will pay based on service subscription and usage.

F. Saeed et al. (Eds.): IRICT 2018, AISC 843, pp. 230–239, 2019.
https://doi.org/10.1007/978-3-319-99007-1_22

Commonly, there are three cloud service models which are Software as a service (Saas), Platform as a service (Paas) and Infrastructure as a service (Iaas). Those models have different capabilities to be offered to the organization. Cloud computing services can be deployed in three different ways including public, private, and hybrid cloud. With the growth of cloud computing, many organizations started to offer cloud services to various consumers based on their functional and non-functional requirement [2].

The Right Scale 2017 State of Cloud report [3], a survey by the cloud computing service provider company, namely Right Scale, has uncovered the cloud adoption trends. The survey that consisted of 1002 respondents from technical professionals' background across the region of organization showed that many people are currently adopting and planning to adopt cloud computing in their organizations. The company carried out the comparison on the number of users adopting cloud and the results showed that the number has been increasing based on their surveys from 2015 until 2017. This trend leads to a scenario where the number of people adopting cloud services are predicted to be increasing in the future. There are four categories of cloud adoption mentioned on this report namely Cloud Watchers, Beginners, Explorers, and Focused.

One of the reasons why cloud computing can attract many new potential users are due to great service offered over the internet. The organization may consider whether they want to migrate their business information system to cloud thus the organization needs to choose and decide which cloud service provider offer the best service performance. However, choosing the best provider for cloud service is not a simple task due to the huge number of available providers in the list. The selection process consumes so much time because it involves evaluation on cloud service providers one by one with big amount of cloud service providers. This has become one of the motivations for the development of cloud service recommender system. Some previous works related to this system are clustering-based recommendation system [4], review on cloud recommendation [5], and real-time QoWS-based recommendation system [6]. The recommendation system may not only propose the best cloud service provider for each specific QoS, but also for the overall performance. This means that it should produce a result of the recommended provider with the best overall performance based on accumulation of several combined QoS parameters. In this work, we propose a cloud computing provider's evaluation and recommendation model that can propose the best provider based on several QoS parameters. We limited our work so that the model only evaluates three QoS parameters.

Moreover, we argue that the QoS performance of cloud computing services is unpredictable and highly uncertain. The main reasons for this uncertainty are due to cloud computing characteristics of dynamic elasticity, loosely coupling, unstable virtualization performance, and many others [7–9]. Hence, getting exact QoS performance knowledge about any cloud computing provider is a challenging task, if not impossible. Therefore, we proposed a fuzzy logic approach in our model due to the technique's ability to handle uncertainties. In this work, we collected the real QoS performance data of three cloud computing providers. As mentioned earlier, the three parameters are downlink, uplink and latency. The data were clustered so that the performance of the providers can be categorized according to several clusters. Then, the fuzzy inference system (FIS) model was developed based on the clustering outputs. The performance of the model was evaluated based on accuracy and it was compared with a non-fuzzy (crisp) model. The performance measurement showed that our proposed model performed better than the crisp model.

The objectives of this paper are to describe the formulation of the model using fuzzy logic technique, and to present the conducted performance measurements. This paper is structured as follows. The next section will present the related works, Sect. 3 will describe the formulation of the model, Sect. 4 discusses and analyzes the results, and Sect. 5 concludes our work and outlines some future works.

2 Related Works

Garg et al. propose a framework for ranking cloud service providers by using Analytical Hierarchical Process (AHP) [10]. The framework, named as SMICloud, uses all standardized method to measure and compare business service known as Service Measurement Index (SMI) by Cloud Service Measurement Index Consortium (CSMIC). They design the metrics that are reliable to use to measure the cloud service performance. This approach extracts all the qualitative values. Firstly, QoS attributes are sorted hierarchically according to their standard. Secondly, users will assign the weights for each one of the attributes that the users prefer for their cloud selection process. In the third phase, the author finds a solution for values of some attributes that don't have the numerical value by using relative ranking matrix. In the last phase, they combine the relative ranking scheme with the assigned weights earlier.

Another related work is CloudRank, which introduces a framework for ranking prediction by evaluating the service providers that have been chosen for this research [11]. The ranking prediction is based on QoS of cloud providers at the client side. Once a cloud application user sends a request on the ranking prediction from the CloudRank, firstly the system will compute and calculate the similarity of all the current users with the past users that have been using the ranking prediction before. After that, the system will find the similar users to put in the system to apply the ranking prediction. Two ranking prediction algorithms have been projected in this work, which are CloudRank1 and CloudRank2. In this approach, the QoS values of the service provider need to be measured before comparing the service. In solving the problem of incorrect ranking based on the predicted QoS values, both ranking oriented approaches produced to predict the QoS ranking precisely without predicting the equivalent value.

Mamoun and Ibrahim propose a framework for IaaS provider selection system that combines many clouds with a service processing unit [2]. The service processing unit that consists of provision unit, ranking unit and reservation unit, aims to gratify the cloud users based on their QoS requirement needs. They improve the past research work by providing the ranking and reservation unit in their system. The system then will produce a result showing the best IaaS provider based on the preference and numeric weight set by the users on the QoS attributes such as accountability, agility, assurance, cost performance and security.

Most of the past related works were related to evaluating cloud computing providers based on cost, accountability and security. However, our work focused mainly on QoS performance, namely to evaluate the best provider in terms of uplink, downlink and latency. Our work was also unique in terms of the implementation of fuzzy logic approach, which aimed to increase the accuracy the system when uncertainty occur.

Fuzzy logic was introduced by Zadeh in 1965 [12]. It is one of the soft computing techniques used in Artificial Intelligence industries, where it is based on four constructs namely fuzzy sets, membership function, logical operations and if- then rules [13]. Fuzzy logic is a rule based model that uses mathematical to apply the truth values. For instance, fuzzy logic uses continuous range of truth values [0,1] instead of Boolean logic (True or False). This if-then rule has widely been used in many applications. For example, a simple temperature control for air conditioning, controlling washing machine timing, control of ship steering and many more. One of the main strengths of fuzzy logic is that it can deal with uncertainty and vagueness [14]. This is performed by fuzzy logic through its fuzziness in membership function. This means that during evaluation, an input is given a membership degree in the range of [0, 1], instead of 0 or 1 as in crisp technique.

In this work, it was difficult to set the boundaries of clusters that determine whether the performance of any cloud computing provider is good, moderate or poor. This is because the network performance is highly uncertain and it is unrealistic to crisply determine the clusters' boundary values. Therefore, fuzzy logic can handle this issue by providing fuzzy boundaries where each cluster's boundary may overlap each other [15, 16].

3 Formulation of the Model

The tasks involved in formulating the proposed model is shown in Fig. 1. It began with preparation of data sets. Then the data sets were used in clustering validation to identify the optimum number of clusters for each of them. This process was followed by clustering that was carried out to generate the membership functions for our proposed model. Then, the membership functions were used in the construction of fuzzy inference system (FIS). The final stage involved performance measurement activities, which we compared our proposed model with crisp model. The following sub-sections describe each of these tasks.

3.1 Data Sets Preparation

In this research, we used the real QoS data sets of three top cloud computing providers. These data sets were collected from Cloud Harmony website [17]. The literature shows that three main network performance (QoS) measurement parameters are network download speed, network upload speed and network delays (latency) [18]. Hence, we selected downlink, uplink and latency data sets from the data source. The data sets used in this work contained 4500 data points for each of the three QoS parameters. This means that the data set of downlink contained 1500 data points of provider A's downlink, 1500 data points of provider B's downlink and 1500 data points of provider C's downlink. The similar contents applied to uplink and latency data sets. These data sets were used as training data sets, which were utilized to construct the model. Other than that, we also used three testing data sets, one each for downlink, uplink and

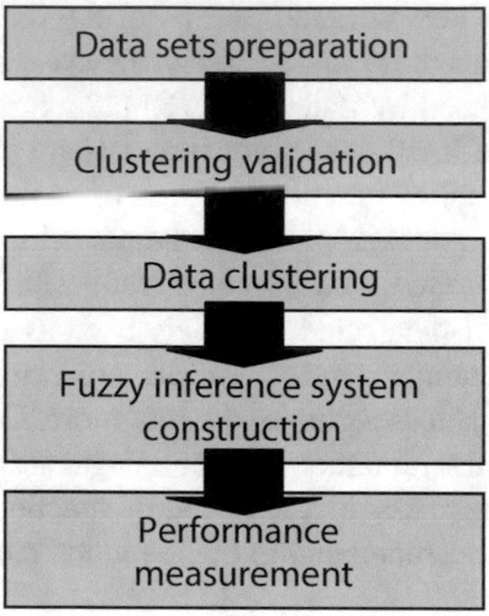

Fig. 1. Research methodology

latency. Each of these data sets consisted of 1500 data points. These testing data sets were used in the performance measurements.

3.2 Clustering Validation

Clustering validation is an important and necessary step in cluster analysis [19]. As the data sets contain big amount of data, the optimum number of cluster for the training data sets need to be identified. In validating and identifying the number of cluster, each training data set was tested with several number of cluster using a clustering validity index. This process showed that our training data sets produced an optimum number of four clusters for downlink and uplink, while for the case of latency the optimum number of clusters was five.

3.3 Data Clustering

An FIS's membership functions are constructed based on either using expert knowledge or automatic development using historical data. The literature show that the former method has disadvantages such as loss of accuracy [20] and may not always available [21]. Hence, in this work we used the latter method through clustering of the training data sets. We chose Fuzzy C-Means (FCM) algorithm for clustering because it provides some advantages such as its results can be used to construct both Mamdani or Sugeno FIS model, and it give the best output for overlapped data [22].

FCM creates several clusters, and assigns a cluster's membership degree to each of data points [23], where each data point becomes member to more than one cluster. These membership degrees of all the data points is stored in a matrix, U, which becomes the output of FCM clustering. Another output of FCM is center of each cluster, c.

3.4 Fuzzy Inference System Construction

The cluster centers, *c*, and matrix of membership degrees, *U*, produced by FCM were used to construct membership function of the model. In this work, we chose Gaussian-typed membership function because its constructs match with the two outputs produced by FCM as mentioned above. Gaussian fuzzy sets are formed based on the function shown in (1) [24, 25]:

$$f(x; w, c) = e^{-(x-c)^\wedge 2/2w^\wedge 2} \quad (1)$$

Hence, the width of the Gaussian membership function, w, from (1) is determined as follows:

$$w = \sum_{1..n} \left(\left(-(X_n - C_i)^2 \right) / (2 * \log(U_n)) \right)^{1/2} / n \quad (2)$$

where n is number of data points.

For downlink and uplink, the four clusters were known as "Bad", "Poor", "Good" and "Excellent", while for latency it consisted of "Bad", "Poor", "Fair", "Good" and "Excellent" clusters. Mamdani-type inference was used for work because it can find the centroid of a two-dimensional function [26]. The defuzzification process efficiency increases when using Mamdani-type of FIS as it simplifies the computation required. The constructed membership functions are shown in Fig. 2.

In this work, the total of eighty number of rules have been implement in the FIS model. The sample of rules is stated below:

1. *If (Downlink is Excellent) and (Uplink is Excellent) and (Latency is Excellent) then (Output is Excellent).*
2. *If (Downlink is Excellent) and (Uplink is Bad) and (Latency is Excellent) then (Output is Good).*
3. *If (Downlink is Good) and (Uplink is Bad) and (Latency is Fair) then (Output is Poor).*
4. *If (Downlink is Poor) and (Uplink is Excellent) and (Latency is Fair) then (Output is Poor).*
5. *If (Downlink is Excellent) and (Uplink is Bad) and (Latency is Excellent) then (Output is Good).*

. .

. .

. .

80. *If (Downlink is Bad) and (Uplink is Bad) and (Latency is Bad), then (Output is Bad).*

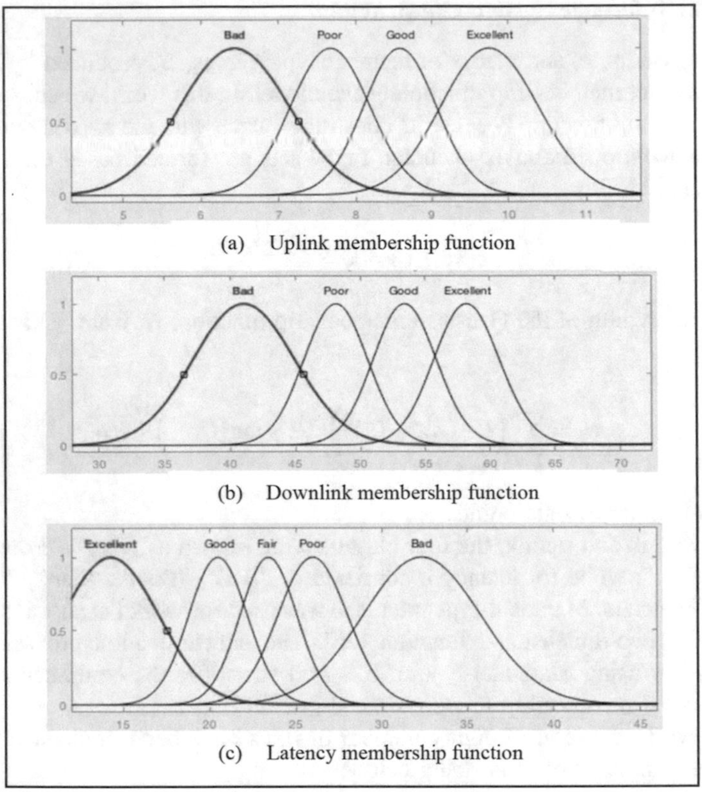

Fig. 2. The constructed membership functions for (a) uplink, (b) downlink and (c) latency

4 Results and Discussion

The FIS model that we constructed was executed with the three testing data sets, namely uplink, downlink and latency data sets. As mentioned in the previous section, each of these data sets comprised 500 data points of provider A, 500 data points of provider B and 500 data points of provider C.

Figure 3 shows the results produced by the model. According to the performance ratings set in the model, an average of 5.0–7.5 was set to "Good" with an average of 7.5–10.0 was rated as "Excellent". These ratings were based on the training done upon the training data sets, as discussed in the previous section. It is therefore shown in Fig. 3 that all providers managed to score "Good" performance but provider B became the best provider by having the highest average performance rating i.e. 7.122. This means that the model will recommend provider B based on the input data sets used in this experiment.

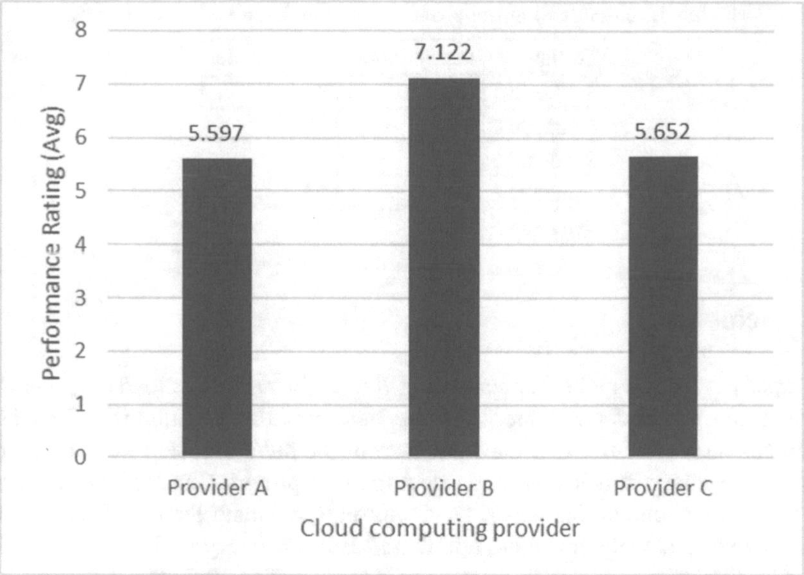

Fig. 3. The performance rating produced by the model

4.1 Performance Measurement

In this work, the proposed model was compared with non-fuzzy (crisp) model in terms of accuracy under the state of uncertainty. We simulated the uncertainty by imposing random errors into the training data sets, which were then known as synthetic data sets [27]. In this work, there were five synthetic data sets used for performance measurement. These synthetic data sets were used to construct five fuzzy synthetic models and five crisp synthetic models. These synthetic models were developed based on the same methodology as shown in Fig. 1, except that the crisp synthetic models were clustered using K-Means algorithm instead of FCM. K-Means algorithm was chosen because its behavior is similar to FCM except that it produces hard clustering results instead of fuzzy clusters [18]. Meanwhile, the original training data sets were used to construct the crisp original model, using K-Means algorithm as mentioned above.

The original and synthetic models were executed with testing data sets, where each of the synthetic models' results were compared with the results of the respective original model. This means that each fuzzy synthetic model's results were compared with the fuzzy original model's results, and each crisp synthetic model's results were compared with the results of their original model. The difference was counted as error. Table 1 shows that fuzzy model outperformed crisp model in terms of number of errors in each of the comparisons between the original model and the synthetic models. This shows that fuzzy model produced better accuracy than the crisp model.

Table 1. Number of errors produced by the fuzzy and crisp models.

Testing	Fuzzy model	Crisp model
Synthetic 1	65	83
Synthetic 2	60	85
Synthetic 3	25	67
Synthetic 4	21	57
Synthetic 5	37	84

5 Conclusion

This paper proposes a model for evaluating the performance of cloud service provider so that it can recommend to the users the best provider to subscribe. This kind of recommendation system is important because of the huge number of cloud computing providers nowadays. It helps users to select the best provider, and reduces the time for them to do selection. In this paper, the focus is to evaluate three QoS parameters of cloud computing namely downlink, uplink and latency. Fuzzy logic is proposed for the implementation of model due to the uncertain nature of these QoS parameters' values. The main objective of this paper has been achieved through the description of the model's formulation. Another objective has also been achieved by proving that the proposed fuzzy model has outperformed non-fuzzy model under the state of uncertainty. For future work, we propose that the model is extended to use type-2 fuzzy logic, which is known to have higher ability than type-1 fuzzy (as proposed in this paper) at handling uncertainties.

References

1. Duan, Q.: Cloud service performance evaluation: status, challenges, and opportunities–a survey from the system modeling perspective. Digit. Commun. Netw. **3**(2), 101–111 (2017)
2. Mamoun, M.H., Ibrahim, E.M.: A proposed framework for ranking and reservation of cloud services. Int. J. Eng. Technol. **4**(9), 536–641 (2014)
3. RightScale 2017 State of the Cloud Report. https://www.rightscale.com/lp/2017-state-of-the-cloud-report
4. Zain, T., Aslam, M., Imran, M.R., Martinez-Enriquez, A.M.: Cloud service recommender system using clustering. In: 2014 11th International Conference on Electrical Engineering, Computing Science and Automatic Control (CCE), Campeche, pp. 1–6 (2014)
5. Aznoli, F., Navimipour, N.J.: Cloud services recommendation. J. Netw. Comput. Appl. **77**, 73–86 (2017)
6. Zhang, M., Ranjan, R., Menzel, M., Nepal, S., Strazdins, P., Jie, W., Wang, L.: An infrastructure service recommendation system for cloud applications with real-time QoS requirement constraints. IEEE Syst. J. **11**(4), 2960–2970 (2017)
7. Tchernykha, A., Schwiegelsohn, U., Alexandrov, V., Talbi, E.: Towards understanding uncertainty in cloud computing resource provisioning. Proc. Comput. Sci. **51**, 1772–1781 (2015)

8. Fard, H.M., Ristov, S., Prodan, R.: Handling the uncertainty in resource performance for executing workflow applications in clouds. In: Proceedings of the 9th International Conference on Utility and Cloud Computing, Shanghai, pp. 89–98 (2016)
9. Minarolli, D., Mazrekaj, A., Freisleben, B.: Tackling uncertainty in long-term predictions for host overload and underload detection in cloud computing. J. Cloud Comput. Adv. Syst. Appl. (2017). https://doi.org/10.1186/s13677-017-0074-3
10. Garg, S.K., Versteeg, S., Buyya, R.: A framework for ranking of cloud computing services. Future Gener. Comput. Syst. **29**(4), 1012–1023 (2013)
11. Zheng, Z., Zhang, Y., Lyu, M.R.: CloudRank: a QoS-driven component ranking framework for cloud computing. In: 29th IEEE Symposium on Reliable Distributed Systems (2010)
12. Zadeh, L.A.: The role of fuzzy logic in the management of uncertainty in expert systems. Fuzzy Sets Syst. **11**(1–3), 199–227 (1983)
13. Supriya, M., Venkataramana, L.J., Sangeeta, K., Patra, G.K.: Estimating trust value for cloud service providers using fuzzy logic. Int. J. Comput. Appl. **48**(19), 28–34 (2012)
14. Wang, S., Liu, Z., Sun, Q., Zou, H., Yang, F.: Towards an accurate evaluation of quality of cloud service in service-oriented cloud computing. J. Intell. Manuf. **25**(2), 283–291 (2014)
15. Klir, G., Yuan, B.: Fuzzy Sets and Fuzzy Logic, vol. 4. Prentice-Hall, Prentice (1995)
16. Performance measurement of Cloud Computing services: https://arxiv.org/abs/1205.1622
17. CloudHarmony: https://cloudharmony.com/
18. Tallamraju, S.R.: Trust metrics for cloud computing (2013)
19. Wang, K., Wang, B., Peng, L.: CVAP: validation for cluster analyses. Data Sci. J. **8**, 88–93 (2009)
20. Guillaume, S.: Designing fuzzy inference systems from data: an interpretability oriented review. IEEE Trans. Fuzzy Syst. **9**, 426–443 (2001)
21. Jang, J.-S.R.: Self-learning fuzzy controllers based on temporal back propagation. IEEE Trans. Neural Netw. **3**, 714–723 (1992)
22. Simhachalam, B., Ganesan, G.: Performance comparison of fuzzy and non-fuzzy classification methods. Egypt. Inf. J. **17**(2), 183–188 (2016)
23. Wang, L., Wang, J.: Feature Weighting fuzzy clustering integrating rough sets and shadowed sets. Int. J. Pattern Recogn. Artif. Intell. **26**(4), 1769–1776 (2012)
24. Castillo, O., Melin, P.: Design of intelligent systems with interval type-2 fuzzy logic. Type-2 Fuzzy Logic: Theory and Applications - Studies in Fuzziness and Soft Computing, vol. 223, pp. 53–76. Springer, Berlin (2008)
25. Tay, K.M., Lim, C.P.: Optimization of Gaussian fuzzy membership functions and evaluation of the monotonicity property of fuzzy inference systems. In: 2011 IEEE International Conference on Fuzzy Systems (2011)
26. Zadeh, L.A.: Outline of a new approach to the analysis of complex systems and decision processes. IEEE Trans. Syst. Man Cybern. **1**, 28–44 (1973)
27. Bolón-Canedo, V., Sánchez-Maroño, N., Alonso-Betanzos, A.: A review of feature selection methods on synthetic data. Knowl. Inf. Syst. **34**(3), 483–519 (2013)

Self-adaptive Population Size Strategy Based on Flower Pollination Algorithm for T-Way Test Suite Generation

Abdullah B. Nasser[1,2](✉) and Kamal Z. Zamli[1](✉)

[1] Faculty of Computer Systems and Software Engineering, Universiti Malaysia Pahang, 26300 Kuantan, Pahang, Malaysia
abdullahnasser83@gmail.com, kamalz@ump.edu.my
[2] Computer Science Department, Systems and Information Technology Center, Hodeidah University, Hodeidah, Yemen

Abstract. The performance of meta-heuristic algorithms is highly dependents on the fine balance between intensification and diversification. Too much intensification may result in the quick loss of diversity and aggressive diversification may lead to inefficient search. Therefore, there is a need for proper parameter controls to balance out between intensification and diversification. The challenge here is to find the best values for the control parameters to achieve acceptable results. Many studies focus on tuning of the control-parameters and ignore the common parameter, that is, the population size. Addressing this issue, this paper proposes self-adaptive population size strategy based on Flower Pollination Algorithm, called saFPA for t-way test suite generation. In the proposed algorithm, the population size of FPA is dynamically varied based on the current need of the search process. Experimental results show that saFPA produces very competitive results as compared to existing strategies.

Keywords: Meta-heuristic · Flower Pollination Algorithm
Self-adaptive population size · T-way testing

1 Introduction

Meta-heuristic algorithms are used to find optimal/near-optimal solutions in reasonable time. It usually used for solving hard optimization problems which cannot be solved by simply adopting exhaustive search.

In the last 40 years, many useful meta-heuristic algorithms have been developed in the literature such as Tabu Search (TS) [1], Simulated Annealing (SA) [2], genetic Algorithm (GA) [3], Anti Colony Algorithm (ACA) [4], Particle Swarm Optimization (PSO) [5], Differential Evolution (DE) [6], Harmony Search (HS) [7], Sine Cosine Algorithm (SCA) [8], Bees Algorithm (BA) [9], Cuckoo Search (CS) [10], Flower Pollination Algorithm (FPA) [11], and Firefly Algorithm(FA) [12], to name a few.

The performance of meta-heuristic algorithms is highly dependents on the fine balance between two components (i.e. intensification to explore the promising neighbouring regions and diversification to ensure that all regions of the search space have been explored). Too much intensification may result in the quick loss of diversity in the

F. Saeed et al. (Eds.): IRICT 2018, AISC 843, pp. 240–248, 2019.
https://doi.org/10.1007/978-3-319-99007-1_23

population, which increases the possibility to make the algorithm being trapped in a local optimum. Therefore, *control-parameters* (such as cognitive parameters (C1 and C2), inertial weight (w) in PSO, mutation rate, and Crossover rate in GA, initial temperature, and cooling rate in SA, switch probability in FPA, and, tabu list size and neighborhood size in TS) are often exploited to ensure effective balance between intensification and diversification. Much recent researchers are focusing on adaptive calibration of these algorithm specific control parameters but less work have been devoted to deal with common control parameter involving the population size.

Taking t-way testing as case studies, many existing meta-heuristics have been proposed in the literature such as GA, ACA, SA, CS, FPA and HS. In our previous work [13], a parameter-free FPA based algorithm is proposed by removing the static probability inherent in FPA. FPA is a flower pollination inspired meta-heuristic algorithm, relies only on one parameter to alter between its intensification and diversification. In the proposed algorithm, the population size of FPA is dynamically varied based on the current need of the search process. Experimental results show that saFPA produces very competitive results as compared to existing strategies.

The rest of this paper is organized as follows: Sect. 2 gives a brief about t-way testing, while Sect. 2 provides an overview of existing strategies. Section 3 presents the design of saFPA. In Sect. 4, the proposed strategy is evaluated against existing strategies. Finally, Sect. 6 concludes the paper and suggests some directions for future works.

2 T-Way Testing

T-way testing, where t indicates the interaction strength, is a sampling mechanism to reduce the number of test cases based on interaction coverage. As illustration, consider the Google Map Service application shown in Fig. 1. Google Map Service allows the user to find directions through driving, public transportation, biking or even walking. The application consists several input components relating to Starting and Destination points, Travel Mode, Avoid highways, Avoid tolls, Avoid ferries, and Distance units.

Each component provides links with associated values as shown in Table 1. To exhaustively test the application requires 120 test cases. However, if only every two pair (2-way testing) of the components are considered, the final test cases can be reduced to 15 test cases. Here, the final test cases ensure that all possibilities of any two components are involved at least one time, which can detect %76 of software failure [14].

On the other hand, considering 3-way testing can reduce the test cases to 35 test cases. It should be noted that as t gets higher, the number of test cases will increase exponentially towards the exhaustive test set.

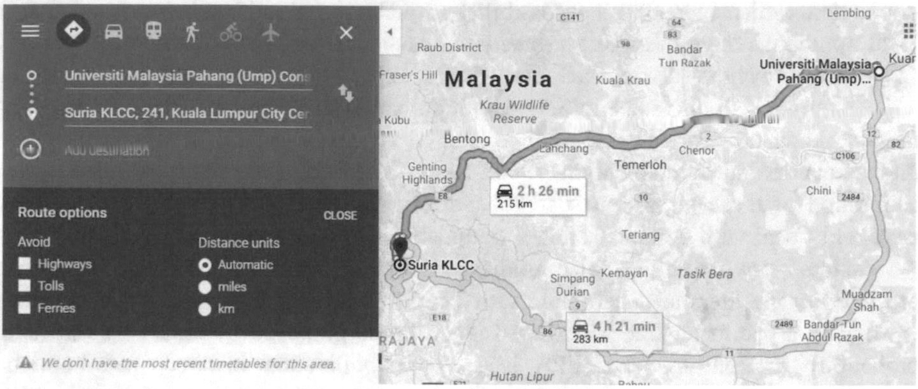

Fig. 1. Google Map application

Table 1. Google map components and values

Starting points	Destination points	Travel mode	Avoid highways	Avoid tolls	Avoid ferries	Distance units
String value	String value	Care	Selected	Selected	Selected	Automatic
		Public transport	Un-selected	Un-selected	Un-selected	Miles
		Walking				Km
		Biking				
		Flight				

3 Related Works

In general, the existing works of t-way testing can be classified into Algebraic and Computational approaches [15]. Algebraic approach often relies on lightweight computations to generate test cases matrix. Strategies of this approach, such as Covering Array (CA), orthogonal Latin squares (OLS), Mixed Covering Array (MCA) and TConfig, [16], arrange the columns and rows using some mathematical functions [17–19].

Computational approach uses a greedy algorithm to construct the test cases, either by adopting one test at a time (OTAT) or one parameter at a time (OPAT) based. OPAT-based strategies, such as IPOG-D [20], IPOG [21], IPOF and IPAD2 [22], first, generate the test cases for the smallest number of components (or parameters), then it horizontally add one parameter each iteration until all parameters are covered. OTAT-based strategies generate *t*-interaction elements of the components which present all possibilities of every t-way parameters. Then OTAT-based strategies, such as TConfig [23], AETG [24], WHITCH [25], and Jenny [26], generate one test case per iteration until all the interaction tuples are covered.

Recently, many meta-heuristic algorithms have been adopted for generating t-way test cases. In general, meta-heuristic-based strategies start with a random set of solutions. These solutions are subjected to a sequence of search operation in an attempt to improve them. In each iteration, the best candidate solution is selected and added to the final test suite. In the literature, many meta-heuristic algorithms have been successfully applied for t-way testing such as TS [27], GA [28], ACA [28], SA [29], PSO [15, 30], HS [31], FPA [32, 33] and CS [34]. Recently, advancement in meta-heuristic algorithms also include hybridization approach such as hybrid-FPA [35], Tabu search hyper-heuristic [36], Fuzzy hyper-heuristic [37], Q-learning sine-cosine [38], and Fuzzy-TLBO [39].

4 Proposed Strategy

This section describes the self-adaptive population size strategy based on FPA, called self-adaptive FPA (saFPA). Standard FPA was adopted for solving Global optimization problems. The comparison between the FPA and other algorithms such as the PSO, and GA showed that FPA was performed better than other algorithms [11]. FPA is a population based algorithm, inspired by the pollination process of flowering plants. The algorithm starts generating a number of population, then applies search process, either global pollination or local pollination, controlled by switch probability $p \in [0, 1]$, on this population. In FPA, the population size is predetermined during tuning process, before the algorithm is used for solving the problem. Hence, for each problem we need to find the best population size that can obtain optimum results.

In saFPA, the population size is determined dynamically based on the problem at hand, therefore the users are not required to find initial value of population size. In saFPA, a dynamic mechanism is used for automatically finding the best population size as follows:

- First, the minimum and maximum population is determined.
- Then, calculate success_rate via the following equation

$$succes\_rate = \frac{sucess}{Old_{pop}} \tag{1}$$

  Where *success* is the total number of pollens that obtain better solution, and Old_{pop} the total number of population.
- Based on success_rate, a new population size is calculated.

  1. If success_rate < 0.5 then increase the population via following equation:

$$New_{pop} = Old_{pop} + round\left(Old_{pop} * (i - succes\_rate)\right) \tag{2}$$

  2. If success_rate > 0.5 then decrease the population via following equation:

$$New_{pop} = Old_{pop} - round\big(Old_{pop} * (i - succes\_rate)\big) \qquad (3)$$

3. If success_rate = 0.5 then maintain the population.

The aforementioned rules can be use as framework in any meta-heuristic algorithm to find the best/suitable population size automatically.

As highlighted earlier, the problem of generating t-way test suite has been adopted as the case studies. The complete pseudocode of saFPA is presented in Fig. 2.

```
Input: (P, v, t):   Set of parameters P, its parameter-value v and interaction strength.
        Pop_min , Pop_max : minimum and maximum population size.
Output:  TS final test cases
Generating Interaction tuples
   Let TS be a set of final test cases;
   Generate initial population of Pop_min pollen randomly
  //Perform search using FPA
  while Interaction tuples are not cover do
     while t < MaxGeneration  or  stop criterion do
        for i = 1 : n (all n pollens in the population)
           if ( rand < pa )
              Global pollination via x_t^{i+1} = x_t^i + L(gbest − x_t^i)
           else
              Randomly choose j and k among all the solutions
              Do local pollination via x_t^{i+1} = x_t^i + ϱ (x_t^j − x_t^k)
           end if
           Evaluate new solutions
           If new solutions are better, update them in the population
            Calculate success_rate
            If success_rate> 0.5 then
                       increase the population Equation (2)
            If success_rate < 0.5 then
                       decrease the population Equation (3)
        end for
        Find the current best pollen gbest
     end while
    Add the best pollen gbest into TS.
     Remove covered interactions tuples.
 end while
```

Fig. 2. saFPA strategy for t-way test cases generation

5 Results and Evaluation

In order to evaluate the performance of the proposed strategy, we compare saFPA with existing strategies involving GA, SA, PSO, HS, CS and HHH.

For experimental setup, the values the minimum population size = 10, minimum population size = 500, and iteration = 500 are used. The experiments have been performed on Intel (R) core i5 CPU, 4 GB of RAM, and Windows 10 Operating System.

Table 2 shows the results obtained from comparison between saFPA and existing strategies and in term of test size. Different system configurations are used as shown in Table 2. x^y indicates that the system configuration has y parameters, each parameter has x values, and t is required interaction strength. Each cell in Table 2 depicts the best test case obtained by association strategy, while the darkened cells indicate the best size obtained among the existing strategies.

Table 2. Comparison of saFPA with existing pairwise strategies

Systems			IPOG	ITCH	Jenny	PICT	TConfig	TVG	GTWay	SA	GA	ACA	PSO	HSS	HHH	CS	FPA	saFPA
No	x^y	t																
S1	3^4	2	12	NA	10	13	10	12	10	9	9	9	9	9	9	9	9	9
S2	3^{13}	2	20	NA	22	20	20	20	19	16	17	17	17	18	NA	NA	18	18
S3	10^{10}	2	176	NA	177	170	170	189	160	NA	157	159	170	155	NA	NA	153	153
S4	5^5	4	908	837	810	773	849	731	625	NA	NA	NA	779	751	746	776	784	779
S5	5^6	4	1239	1074	1072	1092	1128	1027	625	NA	NA	NA	1001	990	967	991	988	988
S6	5^7	4	1349	1248	1279	1320	1384	1216	1125	NA	NA	NA	1209	1186	1151	1200	1164	1165
S7	2^{10}	4	49	58	39	43	45	40	46	NA	NA	NA	34	37	36	28	36	37
S8	3^{10}	4	241	336	221	231	235	228	224	NA	NA	NA	213	211	207	211	211	207
S9	4^{10}	4	707	704	703	742	718	782	621	NA	NA	NA	685	691	668	698	661	657
S10	2^{10}	2	10	6	10	NA	9	10	NA	NA	NA	NA	8	7	8	8	8	8
S11	2^{10}	3	19	18	18	NA	20	17	NA	NA	NA	NA	17	16	16	16	16	16
S12	2^{10}	4	49	58	39	NA	45	41	NA	NA	NA	NA	37	37	36	36	35	37
S13	2^{10}	5	128	NA	87	NA	95	84	NA	NA	NA	NA	82	81	79	79	81	82
S14	2^{10}	6	352	NA	169	NA	183	168	NA	NA	NA	NA	158	158	153	157	158	153

Table 2 shows that the results of saFPA and existing strategies are very competitive. saFPA hits the most minimum results five times (S1, S3, S8, S10 and S14) while, HHH obtains the best six times. In general, we can observe that meta-heuristic based strategies, such as HS, CS, HHH, FPA and saFPA, perform better than other strategies.

In order to evaluate the contribution of this paper, another experiment has been undertaken. In this experiment, FPA and saFPA have been applied on same problem, and number of uncovered tuples are tracked when population size is varied. Figure 3 shows that the population size is changed based on the problem at hand. When the required number of interaction tuples is small, saFPA used only small number of population, however, when the number of interaction tuples is large, saFPA automatically increases its population size. For FPA, population size is fixed.

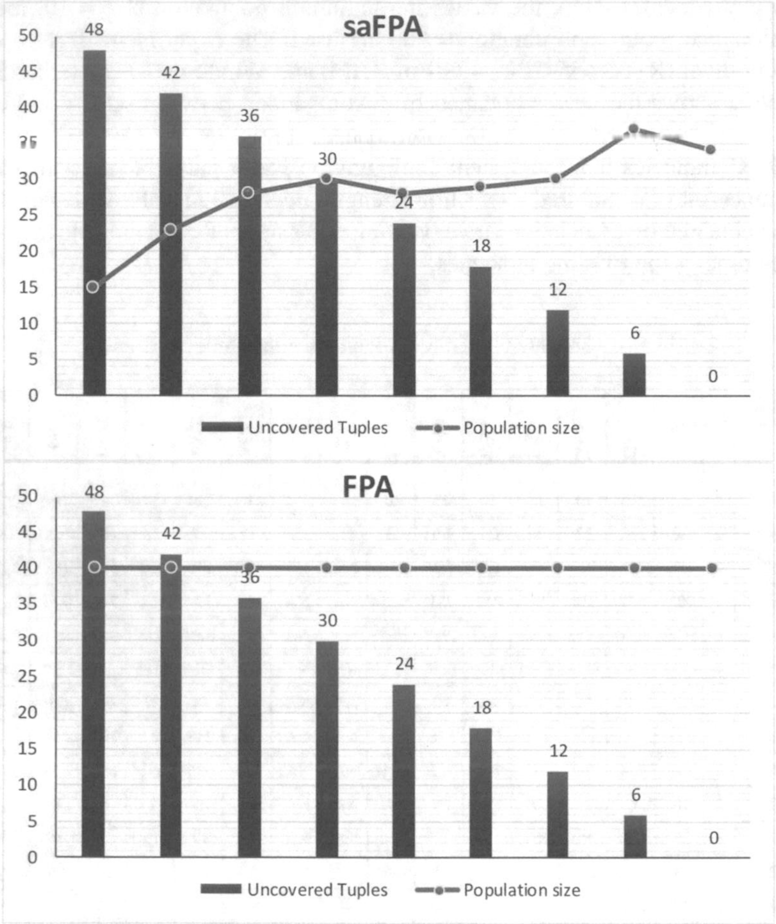

Fig. 3. Uncovered tuples of FPA versus saFPA when population size is varied

6 Conclusion

In this paper, a new strategy based on self-adaptive FPA, called saFPA, have been proposed. saFPA provides a new mechanism for generating the population size dynamically. The self-adaptive population size mechanism proposed in this paper can be used in any meta-heuristic algorithm to find the best/suitable population size automatically. The experimental results show that saFPA produces competitive results comparing with existing strategies. As part of the future work, we are currently investigating for Mamdani fuzzy based approach to improve the performance of saFPA.

Acknowledgements. The work reported in this paper is funded by Fundamental Research Grant from Ministry of Higher Education Malaysia titled: A Reinforcement Learning Sine Cosine based Strategy for Combinatorial Test Suite Generation. We thank MOHE for the contribution and supports, Grant Number: RDU170103.

References

1. Glover, F.: Tabu search-part I. ORSA J. Comput. **1**(3), 190–206 (1989)
2. Kirkpatrick, S.: Optimization by simulated annealing: quantitative studies. J. Stat. Phys. **34** (6), 975–986 (1984)
3. Holland, J.H.: Genetic algorithms. Sci. Am. **267**(1), 66–72 (1992)
4. Dorigo, M., et al.: Ant colony optimization and swarm intelligence. In: 6th International Conference Ants - Theoretical Computer Science and General Issues. Springer, Berlin (2008)
5. Kennedy, J.: Particle swarm optimization. In: Sammut, C., Webb, G.I. (eds.) Encyclopedia of Machine Learning, pp. 760–766. Springer, Berlin (2011)
6. Feoktistov, V.: Differential Evolution. Springer, Berlin (2006)
7. Lee, K.S., Geem, Z.W.: A new meta-heuristic algorithm for continuous engineering optimization: harmony search theory and practice. Comput. Methods Appl. Mech. Eng. **194** (36), 3902–3933 (2005)
8. Mirjalili, S.: SCA: a sine cosine algorithm for solving optimization problems. Knowl.-Based Syst. **96**, 120–133 (2016)
9. Pham, D., et al:. The bees algorithm–a novel tool for complex optimisation. In: International Conference on Intelligent Production Machines and Systems (2011)
10. Yang, X.-S., Deb, S.: Cuckoo search via Lévy flights. In: World Congress on Nature and Biologically Inspired Computing. IEEE (2009)
11. Yang, X.-S.: Flower pollination algorithm for global optimization. In: International Conference on Unconventional Computing and Natural Computation. Springer (2012)
12. Yang, X.-S.: Firefly algorithm, Lévy flights and global optimization. In: Ellis, R., Petridis, M. (eds.) Research and Development in Intelligent Systems, vol. 26, pp. 209–218. Springer, Berlin (2010)
13. Nasser, A.B., Alsewari, A.R.A., Zamli, K.Z.: Parameter free flower algorithm based strategy for pairwise testing. In: 7th International Conference on Software and Computer Applications, Kuantan, Malaysia (2018)
14. Kuhn, D.R., Wallace, D.R., Gallo, J.A.M.: Software fault interactions and implications for software testing. IEEE Trans. Softw. Eng. **30**(6), 418–421 (2004)
15. Chen, X., et al.: Applying particle swarm optimization to pairwise testing. In: 2010 IEEE 34th Annual Computer Software and Applications Conference (COMPSAC). IEEE (2010)
16. Williams, A.W.: Determination of test configurations for pair-wise interaction coverage. In: Testing of Communicating Systems, pp. 59–74. Springer, Berlin (2000)
17. Hartman, A., Raskin, L.: Problems and algorithms for covering arrays. Discrete Math. **284** (1), 149–156 (2004)
18. Mandl, R.: Orthogonal latin squares: an application of experiment design to compiler testing. Commun. ACM **28**(10), 1054–1058 (1985)
19. Bush, K.A.: Orthogonal arrays of index unity. Ann. Math. Stat. **23**(3), 426–434 (1952)
20. Lei, Y., et al.: IPOG/IPOG-D: efficient test generation for multi-way combinatorial testing. Softw. Test. Verif. Reliab. **18**(3), 125–148 (2008)
21. Lei, Y., et al.: IPOG: a general strategy for t-way software testing. In: 14th Annual IEEE International Conference and Workshops on the Engineering of Computer-Based Systems, ECBS 2007. IEEE (2007)

22. Forbes, M., et al.: Refining the in-parameter-order strategy for constructing covering arrays. J. Res. Nat. Inst. Stand. Technol. **113**(5), 287 (2008)
23. Williams, A.: TConfig download page (2008)
24. Cohen, D.M., et al.: The AETG system: an approach to testing based on combinatorial design. IEEE Trans. Softw. Eng. **23**(7), 437–444 (1997)
25. Hartman, A., Klinger, T., Raskin, L.: IBM intelligent test case handler. Discrete Math. **284** (1), 149–156 (2010)
26. Jenkins, B.: Jenny Tool download page (2003). http://www.burtleburtle.net/bob/math. Accessed 16 Dec 2014
27. Nie, C., Leung, H.: A survey of combinatorial testing. ACM Comput. Surv. (CSUR) **43**(2), 11 (2011)
28. Shiba, T., Tsuchiya, T., Kikuno, T.: Using artificial life techniques to generate test cases for combinatorial testing. In: Proceedings of the 28th Annual International Computer Software and Applications Conference, COMPSAC 2004. IEEE (2004)
29. Colbourn, C.J., Cohen, M.B., Turban, R.: A deterministic density algorithm for pairwise interaction coverage. In: IASTED Conference on Software Engineering. Citeseer (2004)
30. Ahmed, B.S., Zamli, K.Z.: A variable strength interaction test suites generation strategy using particle swarm optimization. J. Syst. Softw. **84**(12), 2171–2185 (2011)
31. Alsewari, A.R.A., Zamli, K.Z.: Design and implementation of a harmony-search-based variable-strength tway testing strategy with constraints support. Inf. Softw. Technol. **54**(6), 553–568 (2012)
32. Nasser, A.B., et al.: Pairwise test data generation based on flower pollination algorithm. Malay. J. Comput. Sci. **30**(3), 242–257 (2017)
33. Nasser, A.B., et al.: Assessing optimization based strategies for t-way test suite generation: the case for flower-based strategy. In: 5th IEEE International Conference on Control Systems, Computing and Engineering, Pinang, Malaysia (2015)
34. Ahmed, B.S., Abdulsamad, T.S., Potrus, M.Y.: Achievement of minimized combinatorial test suite for configuration-aware software functional testing using the cuckoo search algorithm. Inf. Softw. Technol. **66**, 13–29 (2015)
35. Nasser, A.B., et al.: Hybrid flower pollination algorithm strategies for t-way test suite generation. PLoS ONE **13**(5), e0195187 (2018)
36. Zamli, K.Z., Alkazemi, B.Y., Kendall, G.: A tabu search hyper-heuristic strategy for t-way test suite generation. Appl. Soft Comput. **44**, 57–74 (2016)
37. Zamli, K.Z., et al.: An experimental study of hyper-heuristic selection and acceptance mechanism for combinatorial t-way test suite generation. Inf. Sci. **399**, 121–153 (2017)
38. Zamli, K.Z., et al.: A hybrid Q-learning sine-cosine-based strategy for addressing the combinatorial test suite minimization problem. PLoS ONE **13**(5), e0195675 (2018)
39. Zamli, K.Z., et al.: Fuzzy adaptive teaching learning-based optimization strategy for the problem of generating mixed strength t-way test suites. Eng. Appl. Artif. Intell. **59**, 35–50 (2017)

Spell Checker for Somali Language Using Knuth-Morris-Pratt String Matching Algorithm

Ali Olow Jimale[1], Wan Mohd Nazmee Wan Zainon[2](✉), and Lul Farah Abdullahi[3]

[1] Faculty of Computing, SIMAD University, Mogadishu, Somalia
colowyare2@simad.edu.so
[2] School of Computer Sciences, Universiti Sains Malaysia, Penang, Malaysia
nazmee@usm.my
[3] Center for Research and Development, SIMAD University, Mogadishu, Somalia
luulf21@simad.edu.so

Abstract. The Somali language is the mother tongue and official language of Somali people in Somalia, a national language in Djibouti. The Somali language is officially written with the Latin alphabet. Before 1990, the Somali language was the main instruction of the educational institutions in Somalia. However, after the collapse of the Somali central government, foreign languages such as Arabic and English have been taking over as the languages of instruction in Somali educational institutions such as schools and universities. It is feared that the language may die out in about 50 years. One reason for this is a lack of computer tools such as word processors with spell checkers which may promote the use of Somali language. The absence of a spell checker in Somali may lead to spelling mistakes in Somali written documents, including social media posts and friendly chats. The focus of this paper is to propose a spell checker for Somali Language using Knuth-Morris-Pratt String Matching Algorithm with Corpus. A spell checker for the Somali language will provide a word processing interface for writing documents that identifies a misspelled word, underlines it, and suggests the proper spelling of the typed word, if any.

Keywords: Somali language · Spell checker · Knuth-Morris-Pratt algorithm Corpus

1 Introduction

The Somali language, the only language spoken throughout Somalia [1], is the official language of the Federal Republic of Somalia. In 1972, the written system of Somali language was implemented. In the golden period of the Somali language, it became the medium of education, replacing English, Italian and Arabic. In 1980s, Somali based textbooks were made available for schools, with some faculties at Somali National University using Somali as a medium of instruction [2].

F. Saeed et al. (Eds.): IRICT 2018, AISC 843, pp. 249–256, 2019.
https://doi.org/10.1007/978-3-319-99007-1_24

However, starting from 1991 onwards, foreign languages such as Arabic and English took over Somali educational institutions such as schools and universities. [3] stated that the use of the Somali language is in danger of extinction. The language will likely die soon if the Somali people do not participate to save and preserve it from possible of extinction.

There are many reasons for the potential death of Somali language. One of the main reasons is the lack of computer tools that can encourage the use of the language, such as a spell checker.

Spell checking is the process of detecting the misspelled words and giving suggestion candidates to correct misspelled words using computer applications commonly known as spell checkers [4]. A spell checker is considered one of the most important Natural Language Processing applications used to increase the success rate of natural language processing applications, including machine translation, word segmentation, information retrieval and natural language understanding [5]. Most spell checker applications are imbedded with text editors. A prominent example of a spell checker is the Microsoft Word processor spell check. Figure 1 shows an example of Microsoft spell checker. This spell checker allows Microsoft Word processing users to check automatically that their documents are free from spelling and grammar errors.

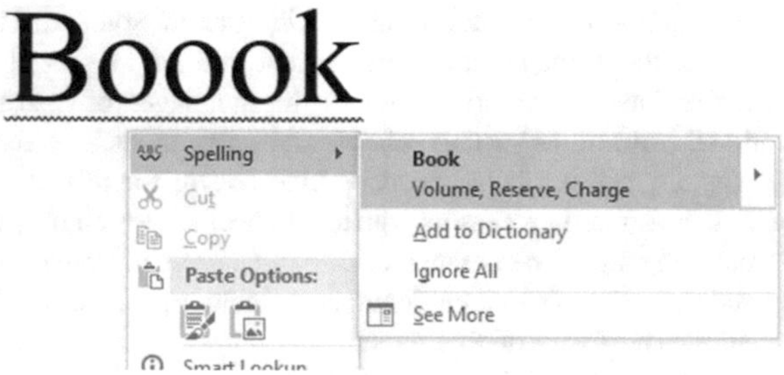

Fig. 1. Example of Microsoft Word spell checking

Spell checkers can be used as a way to promote and maintain the existence of a language. It can help writers to prepare a provisional document in any existing language.

The main advantages of spell checkers are:

1. Checking the accuracy and professionalism of the document.
2. Saving time required for the human to check over the document manually for misspellings.

2 Related Work

This section discusses the literature related to this study. It emphasizes current studies related to spell checking carried by other researchers. Many researchers have tried to solve spelling errors in specific language using different methods and techniques.

An example of spell checking is the work of Mandal and Hossain [6]. These researchers have proposed clustering-based spell-checking technique for Bangla language that can handle both typographical and phonetic errors using Bangla dictionary with clustering algorithm. Their approach was able to reduce both search space and search time. Figure 2 shows the spell correction process of their proposed approach.

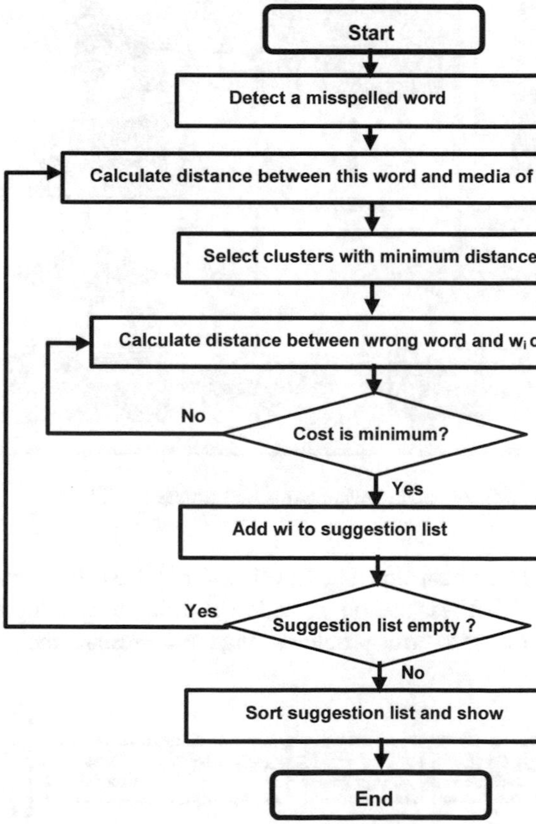

Fig. 2. Clustering-based spell correction process for Bangla language [6]

Another example of a study for solving spelling errors is the work of Basri et al. [7]. They proposed an approach to detect and automatically correct misspelled words in the Malay language without any interaction from the user using lexicon Malay dictionary that contains a list of Malay words. Figure 3 shows their proposed approach.

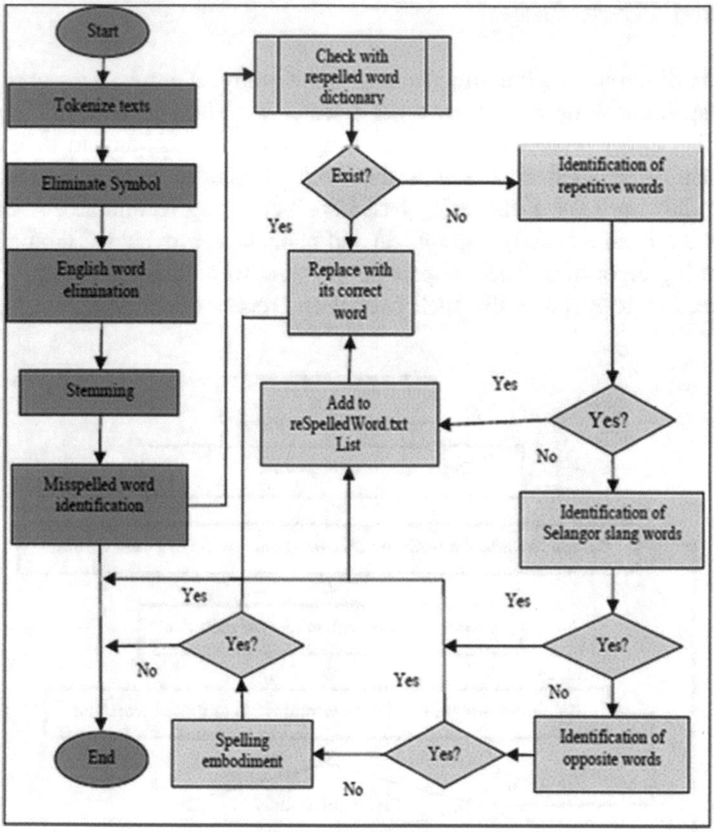

Fig. 3. Malay spell checker proposed approach [7]

In addition to that, a Document Spelling Checker for Bahasa Indonesia was proposed by Kamayani et al. [8]. These authors provide a solution to spelling errors using dictionary. They focused on word error problems. Figure 4 presents their approach.

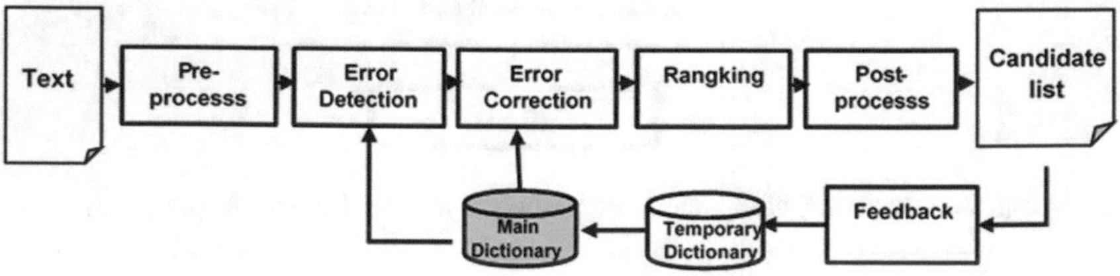

Fig. 4. Document Spelling Checker for Bahasa Indonesia [8]

For Somali language, there have been limited efforts to develop spell checker application. The literature shows the absence of proper word processing application with a spell checker that can detect the spelling errors of Somali language.

3 Proposed Approach

This section discusses the overall techniques used for spell checking and covers in detail the major components of the proposed approach and how these components collectively collaborate each other to detect word errors. Additionally, these techniques offer suggestions to fix the word errors.

The aim of this research is to develop a spell-checking technique that identifies misspelled words of Somali language and gives a proper spelling suggestion. To achieve this objective, a spell-checking approach that can automatically detect misspelled words and give suggestion to do the correction has been proposed. Both the detection and correction of misspelled word can be performed using Knuth-Morris-Pratt String Matching Algorithm with a corpus of Somali words.

The proposed approach of this research takes Somali language as an input text, processes the typed text by the user using Knuth-Morris-Pratt String Matching Algorithm with corpus of Somali words, and generates the detected misspelled word and its correction suggestion as an output.

This consists of three phases: pre-processing, processing, and post processing components. Figure 5 shows the proposed approach. The overall phases of the proposed approach are explained after Fig. 5.

The pre-processing phase is the process of typing words into the proposed approach. The typed words can be misspelled or not. This phase prepares the raw data (Somali language words) to be processed to detect misspelled words and suggest corrections.

Processing phase contains the operations performed to detect and give correction suggestion of the misspelled word. It consists of two sub parts: Knuth-Morris-Pratt String Matching Algorithm and corpus of Somali words. Both subparts collaborate to detect and correct misspelled word.

3.1 Corpus of Somali Words

This component is a document that contains 30,000 Somali words collected from an online dictionary. The dictionary was selected as the corpus of the proposed approach because it is up to date and contains more words than other available Somali dictionaries. The corpus words are used to check the spellings and the occurrence of pattern in the text string. Table 1 shows some of these words.

3.2 Knuth-Morris-Pratt (KMP) Algorithm

KMP is a String Matching algorithm that searches for occurrences of a pattern in a text string. KMP uses information from partial matching of pattern to skip over portions of mismatch and avoid checking characters in text string that we already know matches a

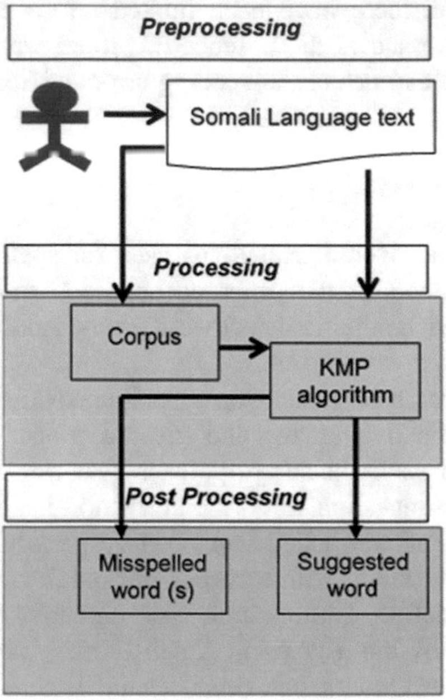

Fig. 5. Proposed approach

Table 1. Example of Somali words

Number	Some of Somali words
1.	*aqoonyahan*
2.	*baafin*
3.	*cabayo*
4.	*daabacaad*
5.	*eeddo*
6.	*falanqayn*
7.	*guuto*
8.	*hoosgunti*
9.	*isir*
10.	*jawaan*
11.	*kalkaaliye*
12.	*laaq*
13.	*macaane*
14.	*nadiifsanaan*
15.	*ogaal*
16.	*qurjujubso*
17.	*raaci*
18.	*shabaq*

prefix of pattern [9]. Assume that our text string, say T, equals 0 0 1 0 0 1 0 0 2 0 0 0 1 0 0 2 0 1 2 2 0 0 and the pattern that we are looking for, say p, is 0 0 1 0 0 2 0 1. Figure 6 shows how the algorithm finds the match of pattern P in the text string T.

```
Iteration        !           !    !
i     0 1 2 3 4 5 5 6 7 8 9 10 10 11 11 12 13 14 15 16 17
j     0 1 2 3 4 5 2 3 4 5 6  7  1  2  1  2  3  4  5  6  7

T     0 0 1 0 0 1 0 0 2 0 0 0 1 0 0 2 0 1 2 2 0 0
P     0 0 1 0 0 2 0 1                               Shift
            0 0 1 0 0 2 0 1                         [5-2 = 3]
                        0 0 1 0 0 2 0 1             [10-1=9]
                          0 0 1 0 0 2 0 1 match     [11-1=10]
```

Fig. 6. KMP example

This algorithm is applied in the proposed approach to generate suggestion lists for misspelled Somali words. KMP was used to speed up the pattern matching with Somali words stored in the Corpus to improve the speed of the proposed approach.

4 Spell Checker for Somali Language

The Spell Checker for Somali Language (HINGAAD HUBIYAHA AF-SOOMAALIGA) is an application designed to detect spelling errors in Somali written document. Figure 7 shows an example the interface of this tool. It contains an editor window which allows the users to type or browse a document. After the user types word(s), the application parses the text and checks if the typed word(s) are correctly spelled or not. The checking is carried out

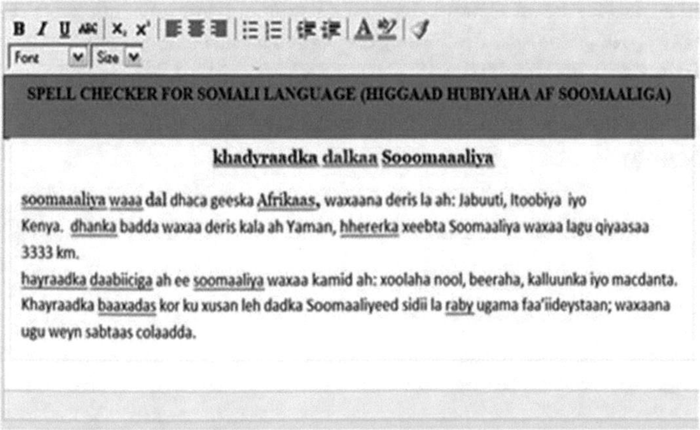

Fig. 7. Spell checker for Somali language

by comparing the words against corpus. If there is no match, it underlines the spelling error and gives suggestion which are similar in structure to the misspelled word.

5 Conclusion and Future Work

In this paper, a spell checker word processing approach for the Somali language was proposed. This method could detect misspelled word(s) and provide suggestion candidates to correct the misspelled words. This study is in the early stage, but the goal is to implement this approach in a functioning text editor computer application to automatically detect misspelled words and suggest the correct words. The proposed application can help the Somali people to ensure the accuracy and professionalism of their typed documents in their native language. This will minimize human spellings errors and reduce the time required to check for spelling errors in a given document.

References

1. Orwin, M.: Somalia: language situation A2. In: Brown, K. (ed.) Encyclopedia of Language and Linguistics, 2nd edn, pp. 509–510. Elsevier, Oxford (2006)
2. Hashi, A.: Developing a model corpus for endangered languages. Graduate thesis, University of Calgary (2014)
3. Patrick, D.: Language endangerment, language rights and indigeneity. Bilingualism: A Social Approach, pp. 111–134. Springer, Berlin (2007)
4. Soleh, M.Y., Purwarianti, A.: A non-word error spell checker for Indonesian using morphologically analyzer and HMM. In: International Conference on Electrical Engineering and Informatics (ICEEI) (2011)
5. Mon, A.M.: Spell checker for Myanmar language. In: International Conference on Information Retrieval and Knowledge Management (CAMP) (2012)
6. Mandal, P., Hossain, B.M.: Clustering-based Bangla spell checker. In: IEEE International Conference on Imaging, Vision and Pattern Recognition (icIVPR) (2017)
7. Basri, S.B., Alfred, R., On, C.K.: Automatic spell checker for Malay blog. In: IEEE International Conference on Control System, Computing and Engineering (ICCSCE) (2012)
8. Kamayani, M., Renainda, R., Simbolon, S.: Application of document spelling checker for Bahasa Indonesia. In: International Conference on Advanced Computer Science and Information System (ICACSIS) (2011)
9. Nonghuloo, M.S., Krishnamurthi, K.: Spell checker for Khasi language. Int. J. Softw. Eng. **7** (1), 1–12 (2017)

Feature Selection Method Based on Grey Wolf Optimization for Coronary Artery Disease Classification

Qasem Al-Tashi[1,2(✉)], Helmi Rais[1], and Said Jadid[1]

[1] Department of Computer and Information Sciences, Universiti Teknologi PETRONAS, 32610 Bandar Seri Iskandar, Perak, Malaysia
qasemacc22@gmail.com, {qasem_17004490, helmim, saidjadid.a}@utp.edu.my

[2] University of Albydha, Al Bayda, Yemen

Abstract. Cardiovascular disease has been declared as one of the deadly illness that affects humans in the Middle and Old ages across the globe. One of the cardiovascular disease known as Coronary artery, has recorded the highest number of motility rates in the recent years. Machine learning tools have been very effective in investigating the causes of such lethal disease which involve analyzing large amount of dataset. Such datasets might contain redundant and irrelevant features which affect the classification accuracy and processing speed. Hence, applying feature selection technique for the elimination of the said redundant and irrelevant features is necessary. In this paper, a novel wrapper feature selection method is proposed to determine the optimal feature subset for diagnosing coronary artery disease. This proposed method consists of two main stages feature selection and classification. In the first stage, Grey Wolf Optimization (GWO) is used to find the best features in the disease identification dataset. In the second stage, the fitness function of GWO is evaluated using Support Vector Machine classifier (SVM). Cleveland Heart disease dataset is used for performance validation of the proposed method. The experimental results showed that, the proposed GWO-SVM outperforms current existing approaches with an achievement of 89.83% in accuracy, 93% in sensitivity and 91% in specificity rates.

Keywords: Feature selection · Grey wolf optimization
Support vector machine

1 Introduction

In recent years, machine learning and data mining has been the most important techniques in medical diagnosis as well as intelligent decision-making processes [1]. In medical industry, data mining techniques are tremendously used in the various diseases prediction models such as heart disease [2], brain tumor [3], skin cancer [4], breast cancer [5] and etc. Heart disease prediction techniques are more essential among other prediction since heart disease is significantly increased in most countries around the globe [6]. Hence, such techniques are used to assist doctors with faster and accurate

F. Saeed et al. (Eds.): IRICT 2018, AISC 843, pp. 257–266, 2019.
https://doi.org/10.1007/978-3-319-99007-1_25

heart disease prediction. Also, the heart disorders are called as Cardio Vascular Diseases (CVD) that describes the blood vessels and diseases of heart [7]. CVD includes coronary heart disease (CHD) or Coronary Artery Disease (CAD), Rheumatic heart disease, Cerebrovascular disease (stroke), hypertension (high blood pressure) and heart failure. Among these various heart disorders, CAD is the commonly occurring heart disorder because of deposition of fatty and cholesterol within the internal wall of the arteries that blocks the required blood flow to the heart [8].

Generally, Coronary Artery Disease (CAD) prediction technique consists of two stages such as feature selection stage and classification stage. Feature selection stage is used to select the feature subsets and then use them for the classification stage [9, 10]. The input datasets of CAD include relevant, irrelevant and redundant features. However, the redundant features as well as irrelevant features create noise to the target class. In addition, these features not only affects the classification performance but also reduce the system response time [11]. Hence, feature selection technique is necessary for removing those features, which in turn reduces the risk of over fitting, improves accuracy and requires less computation [6].

Currently, various feature selection approaches have been proposed. In [12], a feature selection method proposed based on genetic algorithm (GA) algorithm and Bayes Naïve (BN) to obtian the optimal features in the disease identification. This method contains of two stages first generation of a subset of features using GA and the evaluation of the system using BN in the second stage.

In [13], a feature selection method proposed based on Artificial Bee Colony (ABC) algorithm and Support Vector Machine (SVM) to obtain the optimal features in the disease identification. SVM classification is used to evaluate the fitness of ABC.

In [14], to predict the risk level of heart disease, genetic algorithm based fuzzy decision support system has been proposed. The method consists of pre-processing the dataset, features are obtained based on different approaches, and weighted fuzzy rules are produced based on selected attributes using GA and finally fuzzy decision support system (FDSS) used for heart disease prediction.

In [15], the authors proposed a hybrid method for the diagnosis of coronary artery disease. In this method the initial weights of neural network were identified via genetic algorithm. Then, the neural network was learned using training data.

In [16], an automatic fuzzy diagnostic system based a modified dynamic multi-swarm particle swarm optimization (MDMS-PSO) and on genetic algorithm (GA) was proposed for predicting the risk level of heart disease.

The previous studies all of them use the same disease dataset which is Cleveland dataset and were selected for impartial comparison with our proposed method. Particle Swarm Optimization (PSO), ABC, and GA are popular meta-heuristic optimization techniques have been used to search and find relevant and optimal features with some drawbacks such as trapped in local optima [17]. The Grey Wolf Optimizer (GWO) is a new optimization algorithm developed by [18], which mimics the leadership hierarchy of wolves which are well known for their group hunting, the gray wolf optimization method demonstrates much robustness in comparison with PSO and GA optimizers against initialization [19]. Further, this algorithm has a few parameters only and easy to implement, which makes it superior than earlier ones. In this paper a wrapper based feature selection based on gray wolf optimizer to find optimal feature subset for Cleveland dataset is proposed.

For classification, there are several classification techniques have been used for heart disease prediction such as Naive Bayes, linear regression, neural networks, support vector machine, and Fuzzy classifier [1, 20, 21]. Several researches have reported with highly confirmation that the support vector machines (SVM) have greater accurate diagnosis ability [22]. Therefore, SVM will be used for classification stage.

The rest of the paper is organized as follows: a summary of the grey wolf optimization and its mathematical model are provided in the second section. Third Section illustrates the proposed wrapper feature selection method GWO-SVM. Section Fourth, then discusses the experimental setup used with the dataset. In the fifth section experimental results are compared with those of existing approaches. Conclusions are finally drawn in the last Section.

2 Gray Wolf Optimization (GWO)

GWO is inspired by social hierarchy and the hunting approach of grey wolves proposed by [18]. Grey wolves typically prefer to live in a pack of 5–12 individuals and have a strict social hierarchy. As shown in Fig. 1, GWO Consist of four levels as follows:

(1) **Alpha (α)**: male and female are the leaders of a pack of wolves that are responsible for making decisions such as wake-up time, hunting, and sleep place.
(2) **Beta (β):** either male or female wolves, beta probably the best candidate of replacement for alpha. Assisting α in decisions making and suggesting feedbacks are the main roles of β.
(3) **Delta (δ):** The wolves at this level obey α and β wolves and control ω wolves. Delta acts as sentinels, scouts, elders, sentinels, caretakers in the pack, and hunters.
(4) **Omega (ω)**: the wolves at this level are the weakest. Omega (ω) plays a role of scapegoat. Omega (ω) should obey other individuals' orders.

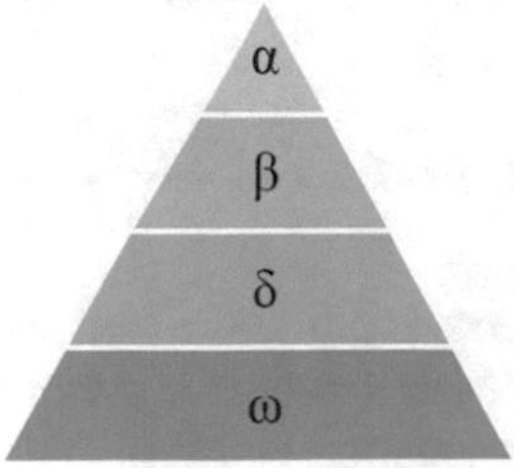

Fig. 1. Grey wolves' social hierarchy represented by [17].

2.1 Mathematical Model of the GWO

In the GWO algorithm, hunting is guided by α, β, and δ, and ω wolves follow them. Mathematically, the grey wolves' encircling behavior can be denoted as:

$$\vec{X}(t+1) = \vec{X}_p(t) + \vec{A}.\vec{D} \tag{1}$$

Where $\vec{X}_p$ is the vector of the prey's positions, $\vec{X}$ is the vector of the grey wolf's positions, (t) is the number of iterations, the coefficient vectors are $\vec{A}$, $\vec{C}$. Where $\vec{D}$ is defined as follows:

$$\vec{D} = \left|\vec{C}.\vec{X}_p(t) - \vec{X}(t)\right| \tag{2}$$

The coefficient vectors $\vec{A}$, $\vec{C}$ can be calculated as following:

$$\vec{A} = 2\vec{a}.\vec{r_1} - \vec{a} \tag{3}$$

$$\vec{C} = 2.\vec{r_2} \tag{4}$$

Where $\vec{a}$ are a vector set decrease over iterations linearly from 2 to 0, $\vec{r_1}$ and $\vec{r_2}$ are random vectors in [0, 1]. The hunting behaviour of grey wolves to be mathematically simulated, alpha (α) is assumed to be the best candidate for the solution, beta (β), and delta (δ) assumed to have more information about the possible position of the prey. Accordingly, three best solutions obtained so far are saved and forces others i.e., (ω) to update their positions according to the best place in the decision space. Such a hunting behaviour can be represented as the following equations:

$$\overrightarrow{D_\alpha} = \left|\overrightarrow{C_1}.\vec{X}_\alpha - \vec{X}\right|, \overrightarrow{D_\beta} = \left|\overrightarrow{C_2}.\vec{X}_\beta - \vec{X}\right|, \overrightarrow{D_\delta} = \left|\overrightarrow{C_3}.\vec{X}_\delta - \vec{X}\right| \tag{5}$$

$$\overrightarrow{X_1} = \overrightarrow{X_\alpha} - A_1.\left(\overrightarrow{D_\alpha}\right), \overrightarrow{X_2} = \overrightarrow{X_\beta} - A_2.\left(\overrightarrow{D_\beta}\right), \overrightarrow{X_3} = \overrightarrow{X_\delta} - A_3.\left(\overrightarrow{D_\delta}\right) \tag{6}$$

$$\vec{X}(t+1) = \frac{\overrightarrow{X_1 + X_2 + X_3}}{3}. \tag{7}$$

Finally, the trade-off between exploration and exploitation is controlled by the updating of the $\vec{a}$ parameter. In each iteration $\vec{a}$ parameter is updated linearly to range from 2 to 0 as according to the equation below:

$$\vec{a} = 2 - t.\frac{2}{max_i ter} \tag{8}$$

Where $max_i\,ter$ indicates the total number of iterations allowed for the optimization and t is the number of iteration.

3 Proposed Method

This study proposed an effective feature selection method, GWO-SVM, to diagnosis a Coronary artery disease. GWO-SVM is consisting of two main phases. In the first phase, GWO proposed by [17] is used to eliminate the redundant and irrelevant features by searching for the best feature in the Cleveland heart dataset. Firstly, GWO produce the initial positions of population, and then update the current positions of population in the discrete searching space. In the second phase, the highly effective SVM classifier is conducted based on the optimal feature subset gained in the first phase. Figure 3 presents a general structure of the proposed GWO-SVM method.

The GWO is used to effectively search the feature space for best feature. The optimal and best feature is the one with high classification accuracy and less number of selected features. The fitness function is used to maximize the classification accuracy also used in GWO to evaluate the selected features is denoted as in Eq. 9:

$$Fitness = \alpha P + \beta \frac{N - L}{L} \tag{9}$$

Where P is the classification accuracy, L is the length of the selected feature, N is the total number of features in the overall dataset, where α and β are the parameters corresponding to the classification accuracy weight and quality of feature selection, $\alpha \in [0, 1]$ and $\beta = 1 - \alpha$. SVM classifier is used for calculating the value (P) in Eq. 9.

The first and primary step for solving a feature selection problem utilizing GWO is to illustrate a feature subset in a solution representation. Figure 2 shows the solution representation. For the proposed feature selection method, we utilized a binary chromosome to illustrate a feature subset. The chromosome's length is represented as d, where d indicates the whole numbers of features. The chromosome's position can take either a '1' or a '0' value. If the value of the ith bit equals one then the feature is selected; otherwise, this feature is not selected (i = 1, 2, …). The number of bits whose values are one are therefore representing the size of a feature subset.

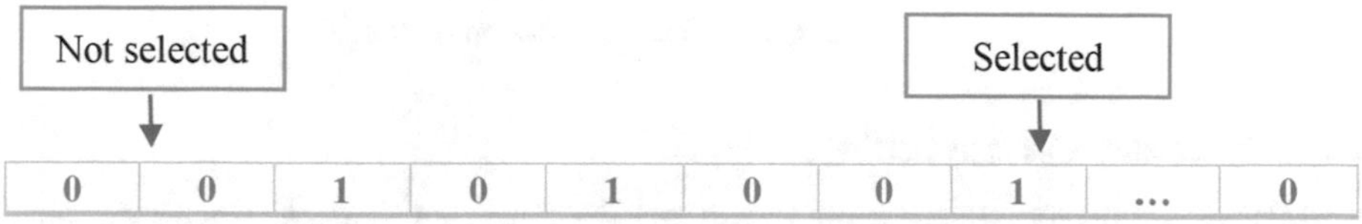

Fig. 2. Solution representation of feature selection

4 Experimental Setup

The proposed method is implemented using MatLab R2018a running on a machine equipped with Intel core i5 processor and a 4 GB RAM capacity and the operating system is Windows 10 Professional 64 bit.

4.1 Dataset Used

Cleveland dataset freely available and can be downloaded from UCI repository [23]. The attributes description of the Cleveland dataset is given in Table 2.

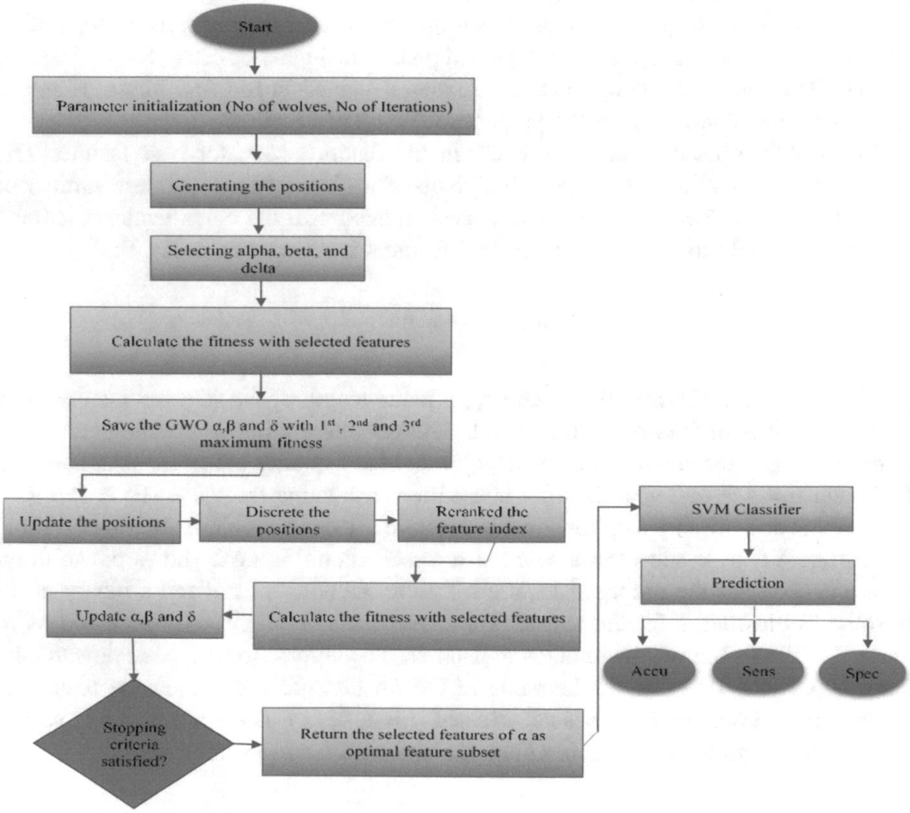

Fig. 3. The proposed feature selection method.

4.2 Parameter Control Setting

Table 1 shows the parameter setting used for the proposed method. Where the iterations, wolves, dimension numbers and search domain are identified. α and β parameters for the fitness function are declared.

Table 1. Parameter setting for the proposed method

Parameter	Numbers
Iterations no.	100
Wolves no.	5
Dimensions no.	14
Search domain	[0 1]
α in fitness function	0.99
β in fitness function	0.01

Table 2. Attributes of Cleveland dataset

No	Attributes	Description
1	Age	Age in year
2	Sex	0 for female and 1 for male
3	Cp	Chest pain type Value 1: typical angina Value 2: atypical angina Value 3: non-anginal pain Value 4: asymptomatic
4	Trestbps	Resting blood sugar in mm Hg on admission to the hospital
5	Chol	Serum cholesterol in mg/dl
6	Fbd	(Fasting blood sugar > 120 mg/dl) (1 = true; 0 = false)
7	Restecg	Resting ECG result
8	Thalach	Maximum heart rate achieved
9	Exang	Exercise induced angina
10	Oldpeak	ST depression induced by exercise relative to rest
11	Slope	Slope or peak exercise ST segment
12	Ca	Number of major vessels colored by fluoroscopy
13	Thal	Defect type
14	num	The predicted attribute

4.3 Performance Evaluation

The performance of the proposed method has been evaluated based on sensitivity, specificity, and accuracy tests, which use the true positive (TP), true negative (TN), false negative (FN), and false positive (FP) terms. These measures are calculated as follows:

$$Sensitivity = \frac{TP}{TP + FN} \times 100\%, \tag{10}$$

$$Specificity = \frac{TN}{FP + TN} \times 100\%, \tag{11}$$

$$Accuracy = \frac{TP + TN}{TP + TN + FP + FN} \times 100\%. \quad (12)$$

5 Result and Discussion

As shown in Table 3, we have conducted two experiments using the same dataset as presented in Sect. 4.1. In the first experiment 10-run time have been conducted and the best result have been chosen, the proposed method GWO-SVM yielded a good result after 10 run time where six features are obtained and achieved accuracy, sensitivity and specificity rates of 87.65%, 88% and 90% respectively. In the second experiment also a 10-run time have been conducted and the best result have been chosen where the number of feature obtained is 7 and the results achieved for accuracy, sensitivity and specificity rates are 89.83%, 93% and 91% respectively. It is found that, Restecg and age features are the major contributing features in cardiovascular disease. Note that, the actual accuracy obtained when all features of Cleveland dataset used as input to SVM is equal to 76.57%.

Table 3. Feature obtained by the proposed method (GWO-SVM)

Experiment	Feature selected	Accuracy	Sensitivity	Specificity
First	Fbs, Thalach, Oldpeak, Slope, Chol, ca	87.65%	88%	90%
Second	Age, Chol, Fbs, Restecg, Thalach, slope, ca	89.83%	93%	91%

A comparison of results between our proposed method and state-of-art methods is shown in Table 4. Studies that used the same datasets (Cleveland dataset) were selected for impartial comparison with our proposed method.

Table 4. Comparison between GWO-SVM and state of art methods.

Study	Method	Accuracy	Sensitivity	Specificity
[13]	ABC-SVM	86.76%	N/A	N/A
[14]	GA-FBSS	80%	84%	75%
[15]	GA-NN	89.4%	88%	91%
Our study	**GWO-SVM**	**89.83%**	**93%**	**91%**

Referring to Table 4. we can see that, the accuracy produced by GWO-SVM outperforms ABC-SVM method proposed by [13]. In addition to accuracy, the ABC-SVM method did not employ the use of sensitivity and specificity evaluation. Moreover, in the second comparison, our method also outperforms the GA-FDSS method proposed by [14], in the areas such as accuracy, sensitivity, and specificity. Finally, we compare our proposed method with the hybrid neural network-Genetic algorithm proposed by [15], different heart datasets have been used such as, Z-Alizadeh Sani dataset, Cleveland, Hungarian and Switzerland. However, the comparison is done only on Cleveland dataset based on which our method shows better performance in terms of accuracy, sensitivity, and specificity.

6 Conclusion

A wrapper feature selection method based on grey wolf optimization and support vector machine has been proposed in this paper. The proposed method has the capability to select relevant features and eliminate redundant and irrelevant features from Cleveland dataset. A comparison was conducted between the proposed method and three competitive counterparts feature selection methods. Experimented results demonstrated that the proposed method performed greatly in terms of accuracy, sensitivity, and specificity and outperformed the state-of-art methods. On the other hand, different classifiers could be used with GWO to further enhance the results. Moreover, some other datasets can be applied in the future to further investigating the robustness of the proposed method.

References

1. Shouman, M., Turner, T., Stocker, R.: Using data mining techniques in heart disease diagnosis and treatment. In: Japan-Egypt Conference on Electronics, Communications and Computers, pp. 173–177 (2012)
2. Dereli, T., Seckiner, S.U., Das, G.S., Gokcen, H., Aydin, M.E.: An exploration of the literature on the use of 'swarm intelligence-based techniques' for public service problems. Eur. J. Ind. Eng. **3**(4), 379 (2009)
3. Arakeri, M.P., Reddy, G.R.M.: Computer-aided diagnosis system for tissue characterization of brain tumor on magnetic resonance images. Signal Image Video Process. **9**(2), 409–425 (2015)
4. Xie, F., Fan, H., Li, Y., Jiang, Z., Meng, R., Bovik, A.: Melanoma classification on dermoscopy images using a neural network ensemble model. IEEE Trans. Med. Imaging **36** (3), 849–858 (2017)
5. Rasti, R., Teshnehlab, M., Phung, S.L.: Breast cancer diagnosis in DCE-MRI using mixture ensemble of convolutional neural networks. Pattern Recognit. **72**, 381–390 (2017)
6. Long, N.C., Meesad, P., Unger, H.: A highly accurate firefly based algorithm for heart disease prediction. Expert Syst. Appl. **42**(21), 8221–8231 (2015)

7. Krishnaiah, V., Narsimha, G., Chandra, N.S.: Heart disease prediction system using data mining technique by fuzzy K-NN approach. In: Emerging ICT for Bridging the Future-Proceedings of the 49th Annual Convention of the Computer Society of India (CSI), vol. 1, pp. 371–384 (2015)
8. Patidar, S., Pachori, R.B., Acharya, U.R.: Automated diagnosis of coronary artery disease using tunable-Q wavelet transform applied on heart rate signals. Knowl. Based Syst. **82**, 1–10 (2015)
9. Khemphila, A., Boonjing, V.: Heart disease classification using neural network and feature selection. In: 2011 21st International Conference on Systems Engineering (ICSEng), pp. 406–409 (2011)
10. Sanz, J.A., Galar, M., Jurio, A., Brugos, A., Pagola, M., Bustince, H.: Medical diagnosis of cardiovascular diseases using an interval-valued fuzzy rule-based classification system. Appl. Soft Comput. **20**, 103–111 (2014)
11. Shilaskar, S., Ghatol, A.: Feature selection for medical diagnosis: evaluation for cardiovascular diseases. Expert Syst. Appl. **40**(10), 4146–4153 (2013)
12. Mokeddem, S., Atmani, B., Mokaddem, M.: Supervised feature selection for diagnosis of coronary artery disease based on genetic algorithm (2013). arXiv Preprint arXiv:1305.6046
13. Subanya, B., Rajalaxmi, R.R.: Feature selection using artificial bee colony for cardiovascular disease classification. In: 2014 International Conference on Electronics and Communication Systems (ICECS) (2014)
14. Paul, A.K., Shill, P.C., Rabin, M.R.I., Akhand, M.A.H.: Genetic algorithm based fuzzy decision support system for the diagnosis of heart disease. In: 2016 5th International Conference on Informatics, Electronics and Vision (ICIEV), pp. 145–150 (2016)
15. Arabasadi, Z., Alizadehsani, R., Roshanzamir, M., Moosaei, H., Yarifard, A.A.: Computer aided decision making for heart disease detection using hybrid neural network-Genetic algorithm. Comput. Methods Programs Biomed. **141**, 19–26 (2017)
16. Paul, A.K., Shill, P.C., Rabin, M.R.I., Murase, K.: Adaptive weighted fuzzy rule-based system for the risk level assessment of heart disease. Appl. Intell. **48**, 1739–1756 (2017)
17. Xue, B., Zhang, M., Browne, W.N., Yao, X.: A survey on evolutionary computation approaches to feature selection. IEEE Trans. Evol. Comput. **20**(4), 606–626 (2016)
18. Mirjalili, S., et al.: Grey Wolf Optimizer. Adv. Eng. Softw. **69**, 46–61 (2014)
19. Emary, E., Zawbaa, H.M., Grosan, C., Hassenian, A.E.: Feature subset selection approach by gray-wolf optimization. In: Afro-European Conference for Industrial Advancement, pp. 1–13 (2015)
20. Srinivas, K., Rao, G.R., Govardhan, A.: Analysis of coronary heart disease and prediction of heart attack in coal mining regions using data mining techniques. In: 2010 5th International Conference on Computer Science and Education (ICCSE), pp. 1344–1349 (2010)
21. Das, R., Turkoglu, I., Sengur, A.: Effective diagnosis of heart disease through neural networks ensembles. Expert Syst. Appl. **36**(4), 7675–7680 (2009)
22. Akay, M.F.: Support vector machines combined with feature selection for breast cancer diagnosis. Expert Syst. Appl. **36**(2), 3240–3247 (2009)
23. Cleveland dataset. http://archive.ics.uci.edu/ml/machine-learning-databases/heart-disease/processed.cleveland.data. Accessed 27 May 2018

Phishing Hybrid Feature-Based Classifier by Using Recursive Features Subset Selection and Machine Learning Algorithms

Hiba Zuhair[1(✉)] and Ali Selamat[2]

[1] College of Information Engineering, Al-Nahrain University, Baghdad, Iraq
hiba.zuhair.pcs2013@gmail.com
[2] Faculty of Computing, Universiti Teknologi Malaysia (UTM), Johor, Malaysia
aselamat@utm.my

Abstract. Machine learning classifiers enriched the anti-phishing schemes with effective phishing classification models. However, they were constrained by their deficiency of inductive factors like learning on big and imbalanced data, deploying rich sets of features, and learning classifiers actively. That resulted in heavyweight phishing classifiers with massive misclassifications in real-time phishing detection. To diminish this deficiency, this paper proposed a new Phishing Hybrid Feature-Based Classifier (PHFBC) which hybridized two machine learning algorithms (Naïve Base) and (Decision Tree) with a statistical criterion of Phish Ratio. In conjunction, a Recursive Feature Subset Selection Algorithm (RFSSA) was also proposed to characterize phishing holistically with a robust selected subset of features. Outcomes of performance assessment via simulations, real-time validation, and comparative analysis demonstrated that PHFBC was highly distinctive among its competitors in terms of classification accuracy and minimal misclassification of novel phishes on the Web.

Keywords: Machine learning · Feature-based classifier
Feature subset selection · Maximal relevance · Minimal redundancy
Active learning

1 Introduction

Although, the web provides a huge communication channel and many services to the users and enterprises, it causes significant digital identity theft and monetary loss annually due to phishing attacks. Phishers evolve their attacks by impersonating the trustworthy web pages in phish web pages to deceive users [1, 2]. Therefore, researchers develop different anti-phishing schemes to tackle phish web pages and mitigate their consequences. Among them are those assisted by machine learning classifiers [1–3]. The developed machine learning classifiers deploy baseline machine learning algorithms such as *Naïve Bayes* (*NB*), *Support Vector Machine* (*SVM*), *Logic Regression* (*LR*), *Sequential Minimum Optimization* (*SMO*), *Random Forest* (*RF*), *C4.5*, and *JRip* [3]. In practice, they outperform their competitors because they rely on various features that distinguish phishing deceptions to give their overall decisions [2–4]. However, they still can be evaded by novel phish web pages due to their

F. Saeed et al. (Eds.): IRICT 2018, AISC 843, pp. 267–277, 2019.
https://doi.org/10.1007/978-3-319-99007-1_26

deficiency of inductive factors which led to partial characterization and inefficient classification in real-time application [3–5]. To hinder their drawbacks, this paper proposes *Phishing Hybrid Feature-Based Classifier (PHFBC)* which is assisted by features subset selection algorithm and hybridizes two most salient machine learning algorithms with a statistical induction criterion. The proposed *PHFBC* demonstrate its effectiveness and supremacy throughout an experiment on benchmarking data sets, a week of real-time practice, and a comparative analysis versus the most salient state-of-the-art classifiers. Next sections gives a bird's eye on PHFBC as follows: Sect. 2 analyzes the related works to address their main problems. Section 3 depicted the work flow of *PHFBC* in details. Section 4 discusses the *PHFBC* performance. Whereas, Sect. 5 presents the conclusions and future insights.

2 Background

Among the prominent phishing classifiers, was $CANTINA^{+}$ that learned *Naïve Bayes* (*NB*), *Support Vector Machine* (*SVM*), and *Logic Regression* (*LR*) with 15 features of web page content and URL to effectively classify phishing in pharming and login web pages. It achieved a *True Positive Rate of* 92% and a *False Positive Rate* of 1.4% [5]. However, it encountered a trade-off in classifying phishing on the up-to date web flows due to the limited number and genericity of the used features. In [6, 7] a phishing classifier with *Support Vector Machine* (*SVM*) algorithm is devoted to classify phish login forms by using 17 features. Despite its effectiveness (*True Positive Rate* and *False Positive Rate* of 99.6% and 0.44% respectively); it performed a heavyweight classification and used external resources for data inquiry. Then, a Chinese e-business phishing website classifier was developed in [8] to characterize phishing with the best ranked set of 15 URL features by using *Chi-Squared* (χ^2). Also, it relied on *Sequential Minimum Optimization* (*SMO*), *Logic Regression* (*LR*), *Naïve Bayes* (*NB*), and *Random Forests* (*RF*) as machine learning algorithms. It achieved an accuracy rate of 95.83% on Chinese e-business web pages execlusively. Later, an ensemble classifier learned *Support Vector Machine* (*SVM*), *Random Forest* (*RF*), *C4.5*, and *JRip* with 12 features was developed in [9]. It was assisted by three methods to select the best features such as *Correlation Features Based Selection* (*CFS*), *Information Gain* (*IG*), and *Chi-Squared* (χ^2). However, it achieved high rate of classification faults (1.44%) with moderate rate of accuracy (94.91%) on big and imbalanced datasets. Other classifier was devoted in [10, 11] to tackle phishing in e-business, login webpages that hosted in English and French. Therefore, it utilized a set of typical and new features to learn *Neural Network* (*NN*) algorithm. It performed well with accuracy rate of (94.07%) but a high rate of misclassifications on big and imbalanced datasets. Then, such classifier was optimized and presented in [12] to learn a set of 212 typical and new features on a big dataset (96,018 samples). However, it achieved significant performance overhead versus high accuracy. Almost revisited classifiers encountered the problems of: (i) evasion by the novel phish web pages continually to cause more damage and illegal gain [1–3]; (ii) limited tolerance of features strongly relevance and redundancy in web pages [1–4, 13–16]; (iii) extensive crawling, processing, and induction for phishing characterization and classification on different web exploits [1–4, 13, 14]; (iv) learning on unreflective datasets versus the vast web page streams of different size, class abundance, web page exploits, and aggregation time [3, 4, 15, 16], (v) inactive learning to adjust induction settings and inadaptability to classify novel phish on the fetched web page streams at any given time [3, 4, 17, 18], and then (v) ambiguous and ineffective phishing real-time detection. Altogether, were attributed to the deficiency of inductive factors that is the main concern to solve and the key contribution in this paper as it will be presented in the next sections.

3 PHFBC

3.1 Features Extraction and Selection

The training set of web pages (dataset) was formulated into a feature space of feature vectors. Each feature vector consisted of the values of 58 various features a they were extracted from web page source code and URLs. Such 58 features included ten *URL* features, 24 *Cross Site Scripting* (*XSS*) features, and 24 *HTML* features as they were proposed in our previously published work [13, 14]. Such features represented the advanced activities and newly explored deceptions of phishers such as imitation of trustworthy web page by embedding dynamic objects and Flash attributes, and injection of suspicious java scripts for malware damage. Then, an optimal selection of features was relied on nominating the most decisive and distinctive features without compromising their minimal exploitation in the input feature space by using a Recursive Features Subset Selection Algorithm (RFSSA) which is assisted by a sub-algorithm FSA and supportive specifics; that was proposed in our previous works [15, 16]. As illustrated in Fig. 1, *FSA* enables RFSSA to boost the mutual information of the targeting class and the mutual dependency among features in the same set of features for the best selection of the most distinctive features (Red_Relv*)* from the input features set (FSet) by using Maximum Relevance and Minimum Redundancy Criterion (mRMR) [17, 18] Then, the candidate (Red_Relv) is projected from (FSet) into (OutSet) to be fed to RFSSA for features subset prioritization so that RFSSA splits (OutSet) into N subsets to validate their goodness (Good Ratio), stability (Stab Ratio), and similarity (SimRatio) [4, 15, 16]. Those supportive specifics are presented in [15, 16] along with their mathematical modelling.

3.2 Phishing Classification

PHFBC boosted the deficiency of inductive factors that Naïve Bayes (NB) and Decision Tree (DT) suffering from. For more decisive classification, it hybridized NB and DT in a synchronized platform and it complemented their deficiency by pruning their induction settings using a statistical measure Phish Indication Ratio (or Phish Ratio), (see Figs. 2 and 3). Then, PHFBC manifested its induction setting iteratively by

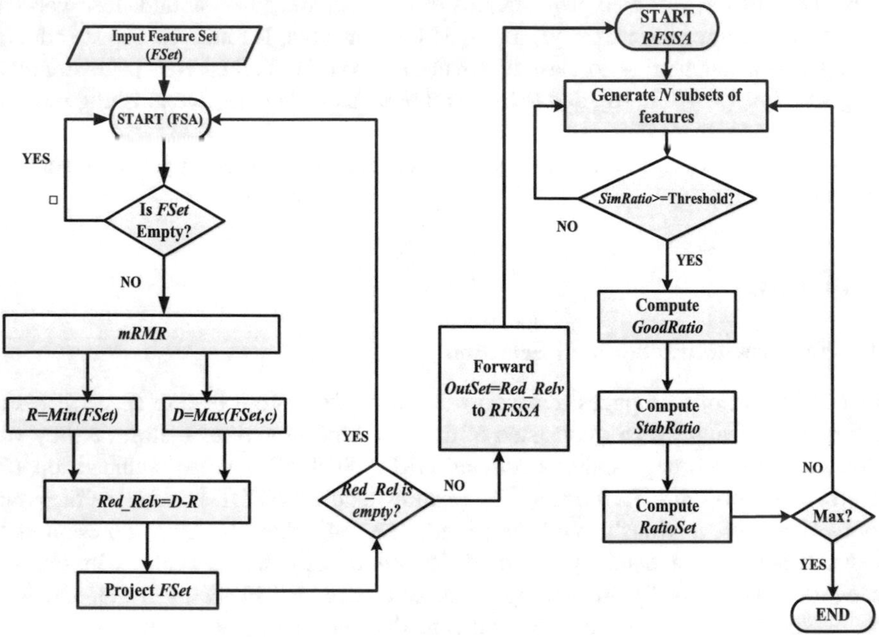

Fig. 1. Flowchart of RFSSA.

mapping the misclassified features in the illegible nodes into their true predictive classes and converged the unclassified feature vectors (overlooked by NB and/or DT) from the remaining feature space. Finally, PHFBC re-learned the feature space progressively and updated its induction settings as along as an unclassified feature vector was inspected. For effective classification, almost revisited phishing classifiers adopted two baseline machine learning algorithms, Naïve Bayesian (NB) and Decision Tree (DT) [5–7, 10–12], see Appendix Table 1. NB pursued Bayes' theorem probabilistically to classify samples on a small training data set to estimate the classifier's parameters correctly and artificially by assuming that all the used features in the training samples were independent from each other [5, 16–18]. Whereas, DT could be sketched in a particular tree structure of nodes, leafs, and branches for mining data statistically and generating the predictive labels effectively. The nodes represented all features in the input feature space (formulated data set), and the leaves denoted the predictive labels of those features while the branches conjunct the inspected features to their relevant predictive labels. Thus, DT set throughout two phases: tree building and tree pruning phases. During tree building phase, the training data set was split recursively to assign each included feature with its relevant predictive labels. In pruning phase, each subtree was pruned by traversing its relevant branches to inspect the minimal training error [5, 16–18]. In spite of their effective classification approach, both NB and DT might perform differently in the case of real-time detection. NB lacked of handling big and imbalanced data, and it fall short in tolerating features heterogeneity and then classifying the duplicate samples belonging to different classes accurately. On the other hand, DT might overlook the unknown features due to the deployment of its default mapping of features and default induction margins via tree building and pruning phases respectively [3, 17, 18].

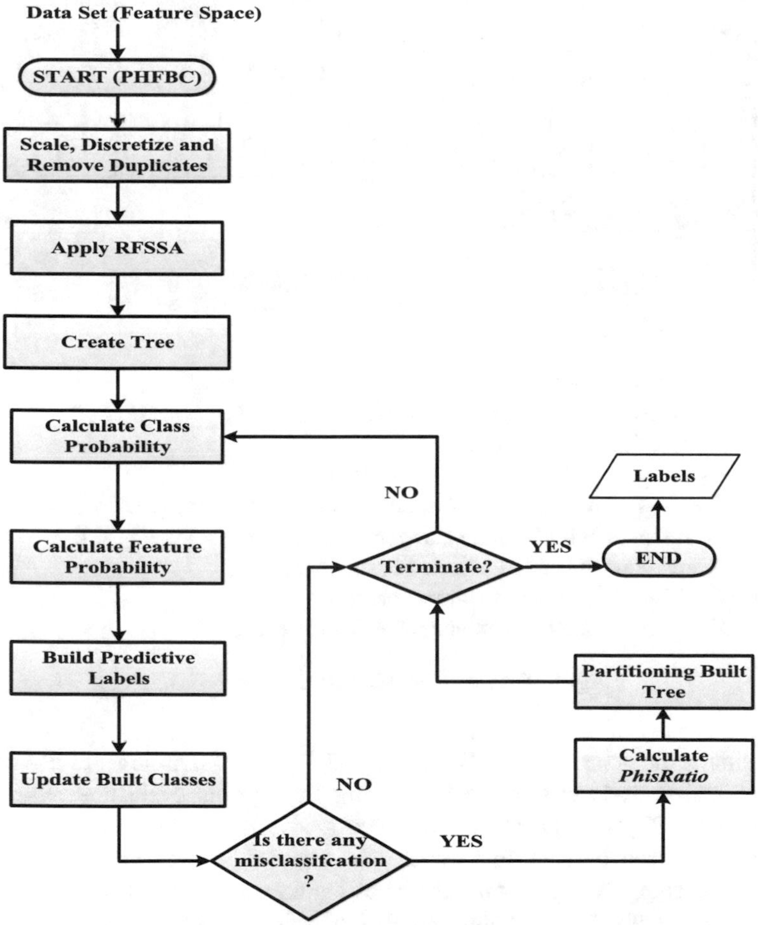

Fig. 2. Workflow of *PHFBC*

4 Performance Assessment

PHFBC's performance was analyzed to justify its distinction among its competitors throughout three strategies: a practice of PHFBC across three benchmarking data sets (see Table 1 in Appendix), a week of real-time validation of PHFBC, and a comparative analysis of PHFBC versus five baseline machine learning-based classifiers.

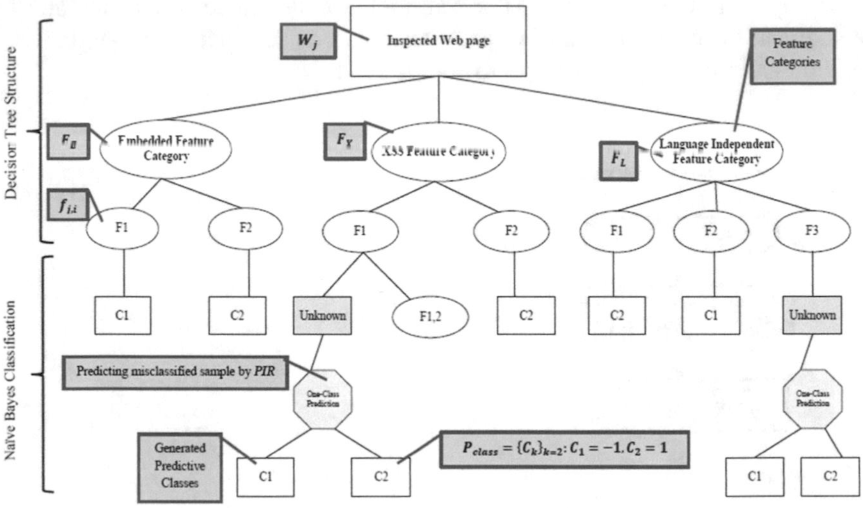

Note: W_j is the examined web page
P_{Class} represents the predictive class such that $P_{class} = \{C_1, C_2\}$;
$C_1 = 1$ denotes phish class;
$C_2 = -1$ denotes non-phish;
$f_{j,i}$ is a given feature in the examined web page;
F_H refers to hybrid feature category such that $F_H = (F_E, F_X, F_L)$;
F_E refers to embedded objects features category;
F_X refers to XSS based features category;
F_L refers to Language independent feature category.

Fig. 3. Classification process through PHFBC.

Related simulation done by using "WEKA 3.5.7-Waikato Environment for Knowledge Analysis" and analysis was pursued by using four typical evaluation metrics: True Positive Rate (TPR) which rationalized the correctly classified phishing over all phish samples, False Positive Rate (FPR) which rationalized the wrongly classified legitimate samples as phishing over the total number of legitimate samples, False Negative Rate (FNR) that rationalized the wrongly labeled phishing samples as legitimate samples overall samples, and AUC that rationalized the plots of TPR versus plots of FPR in the area under the ROC into a scalar value. Experimentally, PHFBC achieved the best rates of TPR (from 0.984 to 0.989), FPR (from 0.051 to 0.066), and FNR (from 0.014 to 0.0156) across the benchmarking data sets. That demonstrated its well performance in the conjunction with RFSSA and 58 hybrid set of features and attributed to the richness and robust compactness of features subset at any settings against the scalable and imbalanced datasets; i.e. RFSSA provided subsets of maximal relevant features to phishing class and minimal redundant features to phish class exploitation and distribution (see Fig. 4 in Appendix).

On the other hand, PHFBC showed progressive and effective phishing classification during 1-week of real-time validation (see Fig. 5 in Appendix). PHFBC reported an ideal classification outcome (AUC was equal to "1") in the last day. That manifested its decisive and effective classification due to the hybridization of two complementary machine learning algorithms (NB and DT) with a statistical criterion (Phish Ratio), update of its default induction margins, and actively learned at every fetched web flow. So far, PHFBC showed its superiority versus five of baseline machine learning algorithms like SMO, SVM, TSVM, NB and DT (see Fig. 6 and Table 1 in Appendix). This was disclosed to: (i) comparable classifiers rendered variations in TPRs and FPRs because they fall short in deploying rich features, tolerating big and imbalanced datasets, and learning inactively (see Fig. 6a and b in Appendix). Whilst, PHFBC achieved the highest TPRs and the least FPRs among its competitors that assured its decisive characterization of different phish exploits and effective classification of novel phishes over the scalable datasets; (ii) the active learning of PHFBC versus inactive learning of the comparable classifiers attained the minimal misclassification cost (FNRs of PHFBC were the closest scores to zero among those of its competitors). Because PHFBC could adjust its initial induction margins by hybridizing NB and DT with Phish Ratio. Then it could adapt to tackle novel phishes on scalable benchmarking datasets.

5 Conclusion and Future Work

This paper addressed inductive factors in phishing machine learning-based classifiers by proposing a Phishing Hybrid Feature-Based Classifier (PHFBC). Conceptually, PHFBC hybridized a Recursive Feature Subset Selection Algorithm (RFSSA) and two complementary machine learning algorithms Naïve Bayes and Decision Tree with a statistical criterion Phish Indication Ratio. Experimentally, PHFBC achieved (97%), (0.7%), (0%), and (98.07%) average rates of TPR, FPR FNR, and AUC respectively. That demonstrated its supremacy as a novel phish-aware and a real-time classifier among its competitors due to its abundance in inductive factors. It employed rich sets of 58 features to characterize Phish web page exploits holistically. It adjusted its margins of induction via the hybridity of statistical and machine learning-based algorithms. It actively learned on big and imbalanced web page streams to adapt novel phishes. Empirical and deductive proofs of its decisive and effective classification, assured that PHFBC can serve as a prospective approach in the future of anti-phishing.

Appendix

See Tables 1, 2 and Figs. 4, 5, and 6.

Table 1. Contributions of this work versus the related works

Merits	Classifier					
	[5]	[6, 7]	[8]	[9]	[12]	This work
Features	15	17	17	15	122	58
Classifier (s)	SVM, LR, BN, DT	SVM	NN, DT	SMO, LR, RF, NB	SVM, RF, C4.5, JRip	PHFBC
Features selection	No	No	Yes	Yes	No	Yes
Active learning	No	No	No	Yes	No	Yes
Data imbalance	No	Yes	No	No	Yes	No
Irrelevance	No	Yes	No	No	Yes	No
Redundancy	Yes	Yes	Yes	Yes	Yes	No
Outcomes	TPR (92%), FPR (1.4%)	TPR (99.6%), FPR (0.42%)	TPR 94.07% FPR: (2.2)	TPR 95.83% FNR: (1.05)	TPR (99%), FPR (0.37%)	TPR 99%, FPR 0.7%

Table 2. Benchmarking datasets

Merits	Data set 1	Data set 2	Data set 3
Size/phishes legitimates	52/36/16	2878/1382/1496	96,018/48009/48009
Data archive	PhishTank	Chinese e-Business	PhishTank/DMOZ
Training split (2/3rd)	34	1918	64012
Testing split (1/3rd)	18	960	32006
Data source	[19]	[8]	[10–12]
Aggregation time	2010	2014	2012–2015
Webpage exploits	Login form/e-Business/Pharming	e-Business	e-Business/homepage/login form
Hosting language	English/French/German	Chinese	English/French/German

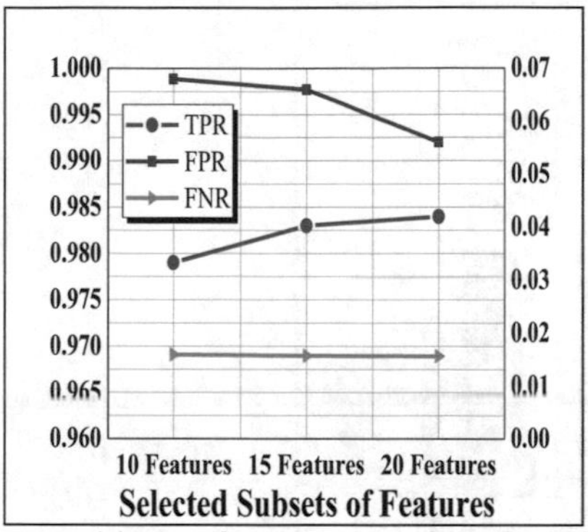

(a) Performance on Data Set1

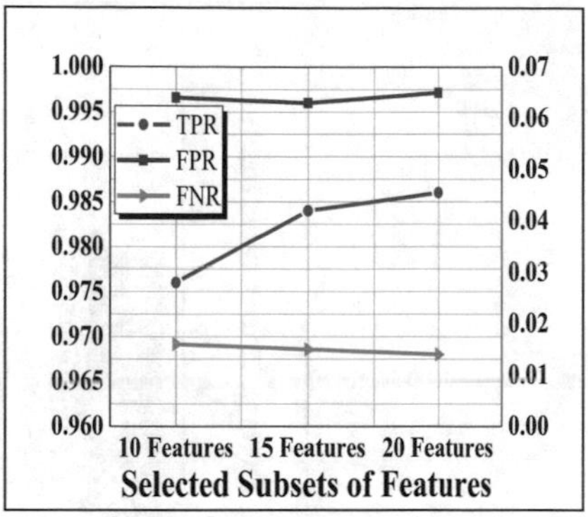

(b) Performance on Data Set2

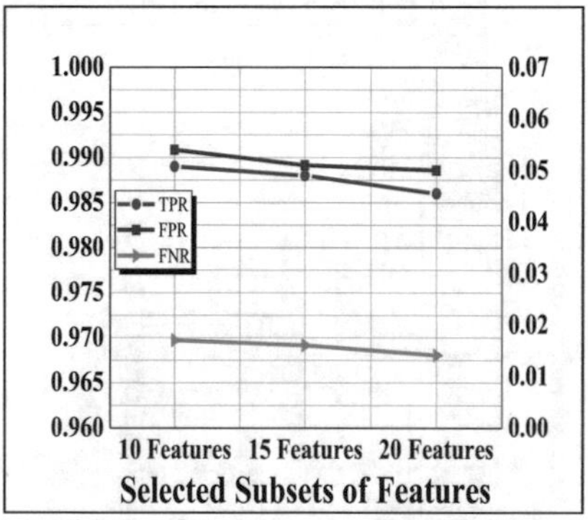

(c) Performance on Data Set3

Fig. 4. Performance assessment of *PHFBC* with respect to features selection

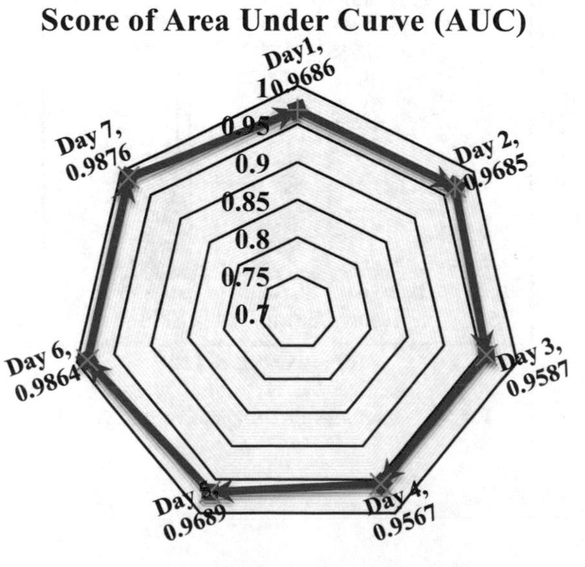

Fig. 5. Real-time validation of *PHFBC*

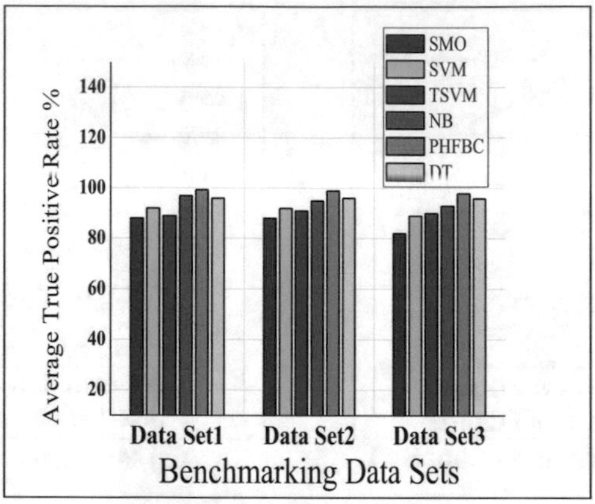

(a) Average of TPR

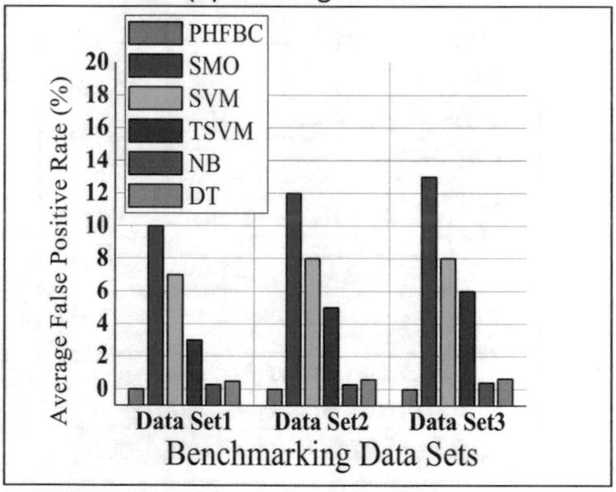

(b) Average of FPR

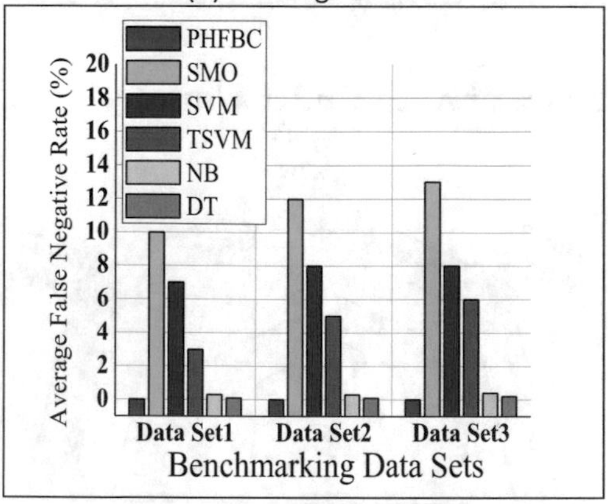

(c) Average of FNR

Fig. 6. Outcomes of comparative analysis

References

1. Zeydan, H.Z., Selamat, A., Salleh, M.: Survey of anti-phishing tools with detection capabilities. In: 14th International Symposium on Biometrics and Security Technologies (ISBAST 2014), pp. 46–54, Kuala Lumpur-Malaysia, August 2014
2. Zeydan, H.Z., Selamat, A., Salleh, M.: Current state of anti-phishing approaches and revealing competencies. J. Theor. Appl. Inf. Technol. **70**(3), 507–515 (2014)
3. Zuhair, H., Selamat, A.: Phishing classification models: issues and perspectives. In: IEEE Conference on Open Systems (ICOS 2017), pp. 26–31, IEEE, Miri-Sarawak (2017)
4. Zuhair, H., Selamat, A., Salleh, M.: Feature Selection for phishing detection: a review of research. Int. J. Intell. Syst. Technol. Appl. **15**(2), 147–162 (2016)
5. Xiang, G.: Towards a phish free world: a cascaded learning framework for phishing detection. Doctoral Dissertation, Carnegie Mellon University, Pittsburgh, PA 15213 (2013)
6. Gowtham, R., Krishnamurthi, I.: A comprehensive and efficacious architecture for detecting phishing webpages. Comput. Secur. **40**, 23–37 (2014)
7. Gowtham, R., Krishnamurthi, I.: PhishTackle-a web services architecture for anti-phishing. Clust. Comput. **17**(3), 1051–1068 (2014)
8. Zhang, D., Yan, Z., Jiang, H., Kim, T.: A domain-feature enhanced classification model for the detection of Chinese phishing e-Business websites. Inf. Manag. **51**(7), 845–853 (2014)
9. Mohammad, R.M., Thabtah, F., McCluskey, L.: Predicting phishing websites based on self-structuring neural network. Neural Comput. Appl. **25**(2), 443–458 (2014)
10. Marchal, S., François, J., State, R., Engel, T.: PhishScore: hacking phishers' minds. In: 10th International Conference on Network and Service Management (CNSM2014), pp. 46–54, IEEE, Rio de Janeiro, 17–21 November 2014
11. Marchal, S., François, J., State, R., Engel, T.: PhishStorm: detecting phishing with streaming analytics. IEEE Trans. Netw. Serv. Manag. **11**(4), 458–471 (2014)
12. Marchal, S.: DNS and semantic analysis for phishing detection. Doctoral Dissertation. University of Luxembourg, 22 June 2015
13. Zuhair, H., Salleh, M., Selamat, A.: New hybrid features for phish website prediction. Int. J. Adv. Softcomput. Appl. **8**(1), 28–43 (2016)
14. Zuhair, H., Salleh, M., Selamat, A.: Hybrid features-based prediction for novel phish website. Jurnal Teknologi **78**(12–13), 95–109 (2016)
15. Zuhair, H., Selamat, A., Salleh, M.: Selection of robust feature subsets for phish webpage prediction using maximum relevance and minimum redundancy criterion. J. Theor. Appl. Inf. Technol. **81**(2), 188–205 (2015)
16. Zuhair, H., Selamat, A., Salleh, M.: The effect of feature selection on phish website detection: an empirical study on robust feature subset selection for effective classification. Int. J. Adv. Comput. Sci. Appl. **6**(10), 221–232 (2016)
17. Vink, J.P., Haan, G.: Comparison of machine learning techniques for target detection. Artif. Intell. Rev. **43**, 125–139 (2015)
18. Kumar, G., Kumar, K., Sachdeva, M.: The use of artificial intelligence based techniques for intrusion detection: a review. Artif. Intell. Rev. **34**(4), 369–387 (2010)
19. Shahriar, H., Zulkernine, M.: Trustworthiness testing of phishing websites: a behavior model-based approach. Future Gener. Comput. Syst. **8**(28), 1258–1271 (2012)

Combined Support Vector Machine and Pattern Matching for Arabic Islamic Hadith Question Classification System

Ali Muttaleb Hasan, Taha H. Rassem(✉), and M. N. Noorhuzaimi@Karimah

Faculty of Computer Systems and Software Engineering, University Malaysia Pahang, 26300 Gambang, Kuantan, Pahang Darul Makmur, Malaysia
alimatlab65@yahoo.com, {tahahussein,nhuzaimi}@ump.edu.my

Abstract. The dimensional phase of Arabic Language, question answering (QA) involves an intrinsic form of question classification (QC) that functions to perform an important task in question answering system (QAS). The purpose of QC is to precisely assign labels to questions that are majorly dependent on the form of answer type. Moreover, classification of user's question is a herculean task based on the tractability that natural language (NL) affords with different forms. The information enshrined in a group of words is not sufficient to effectively classify the question in the quote. Until now, few reports have focused on QC for Arabic Language question answering (QA). The earlier report has employed the technique of handcrafted rules and keyword matching for QC. Nonetheless, these procedures are considered obsolete in terms of applying it to new territories. In this paper, we present a question-answering system combining aims on a combination of model fixed on Support Vector Machine (SVM) and pattern-based Matching techniques for Arabic Language question classification (ALQC). The Islamic Hadith purview on QA in the study was focusing on the effect of a feature set on the performance of SVM for QC. About five patterns were employed in the analysis together with the classification of three types of questions, namely "Who", "Where" and "What". The dataset employed in this study consisted of 200 questions on Arabic Islamic Hadith derived from Sahih Al-Bukhari. The performance generated for the F-measure values for "Who", "Where" and "What" were 88.39%, 87.66% and 87.93% respectively. In this research work, we evaluate the metric performance to combine the SVM with Pattern Matching to get the accuracy answer. The outcome of this answer reflected that the proposed prototype of SVM and pattern-based approach is indispensible from the field of QC in the Arabic language.

Keywords: Question classification system · Question-answering system
Machine learning · Support vector machine · Pattern matching

F. Saeed et al. (Eds.): IRICT 2018, AISC 843, pp. 278–290, 2019.
https://doi.org/10.1007/978-3-319-99007-1_27

1 Introduction

The system of question answering (QA) is simply the method of obtaining a correct response from a given query that is presented in [1]. The researcher in [2] was among the earliest proponents of the QA system. "Lunar" is one type of "Named" systems that is employed to furnish answers for the peculiar purpose of questions that are chemical based subjects. Moreover, it has been suggested by Woods [2] that systems are dependent on the existing knowledge in order to provide the answers. Nonetheless, QA has been justified to tackle this type of issue, where the precise answer would be obtained carefully to answer the user's question. Principally, there are two purposes for QC, which include sitting the answer and selecting the search stratagem. Earlier work had been done and published in text classification. In [3], the QC was conceived specifically for the Quran. The SVM was utilised as a base classifier and utilised many features such as the amount of word and amount of Named Entity Recognition (NER). The NER is a subtask of information extraction that seeks to locate and classify named entities in text into pre-defined categories such as the names of persons, organizations, locations, expressions of times, quantities, monetary values, percentages, etc. example, "Who is killed Musaylimah the liar?" the NER will take the person as NER as which is Msaylimah. "What is the name of the grave where the companions were buried?" the NER will take the location as which Grave is NER. Where the catalogue comprised a set of 230 categorised queries from the Quran dataset. A new multi-label classifier ensemble was formulated by [4], it is popular both in tests as well as through competition. A QC for English QA was design by [5]. Several classifications were used: estimated semantic and semantic groupings, linear SVM in machine learning, support vector machines are unsupervised learning models with associated learning algorithms that analyze data used for classification and regression analysis, in machine learning, kernel methods are a class of algorithms for pattern analysis, whose best known member is the support vector machine (SVM) Gaussian processes, Any linear model can be turned into a non-linear model by applying the kernel trick to the model: replacing its features (predictors) by a kernel function. Given a set of training examples, each marked as belonging to one or the other of two categories, an SVM training algorithm builds a model that assigns new examples to one category or the other, making it a non-probabilistic binary linear classifier. An SVM model is a representation of the examples as points in space, mapped so that the examples of the separate categories are divided by a clear gap that is as wide as possible. New examples are then mapped into that same space and predicted to belong to a category based on which side of the gap they fall, and Lexical level matching (LLM). In order to validate the classification, a large dataset of 50,000 was searched from Wiki and Yahoo answers. In [4], possible answers classification (AC) was made specifically. The categorised answer was classified by machine learning method (MLM) starting at the first level. Structures were obtained from Arabic and English repertoire translated by Google Translate. In addition, several efforts have been done in the field of answering questions from an unstructured text. Firstly, the unstructured text may have information that associates with several domains. It compels specification of an open-domain of QA [5, 6]. Secondly, though the text is reserved to a specific domain, it needs the use of specific registers of words that are customised for a peculiar field of learning, gazetteers or ontologies that provides proper and appropriate semantic correspondences [7]. Thirdly, categorising the question is indispensable in order to establish the question type. The purpose for employing QC is to assign one or more class labels to the question, owing to classification strategy. For instance, the question “What is the name of Surah which begin with the name of a type of fruit”? The purpose of QC is to put together the label of “definition” to this question, as the answer for this type of question is a “definition”. Consequently, the goal of this work is to predict the answering type and QC that are connected to the question as prediction of answer type. This article proposes a combined model based on support vector machine and pattern matching technique for Arabic question classifier (AQC), and Islamic Hadith domain question answering. It is also suggested to depict the intended new methodology for QC in Islamic Hadith-sphere QA using SVM learning and pattern matching classifier that leverage features of questions. Supervision of machine learning model for subjectivity question classification of online posts and used online data set of the online forum was proposed by [9].

2 Methodology

The procedure presents the design of question classifier model for Arabic Islamic Hadith text. However, this classic accomplished text exemplification and features weights. All queries from the row data were exemplified as a row of values for each feature from the predefining set of feature. Afterwards, this output was used as an input for the SVM to make QC and match it on principle of pattern-base. (See Fig. 1) shows the combined QC in Islamic Hadith-domain question answering model.

2.1 Tokenization

The tokenization is the first step of processing task in any natural language. The tokenization aim is to explore words in the sentences. Textual data set at the beginning is only a block of characters. All the following processes in information retrieval required the words of the data set. The main usage of tokenization is to identify the important keywords [10].

2.2 Stop Word Removal

The stop word removal is the processing of the task in any natural language processor, the document is usually found to contain unnecessary words, such as pronouns, prepositions, and conjunctions and others. Therefore, in this regard the process of word removal is common and of significant importance to be considered in Question Classification in Arabic Islamic Hadith-domain question answering. Examples of common stop words use in the Arabic language include: ”“,”من”في,” ”على”,”الى”,”عن corresponding to these words in the English language: ‘from’, ‘about’, ‘to’, ‘on’ and ‘in’ [18].

2.3 Normalization

The normalization step is a primary task before processing any Arabic text for assuring the consistency and predictability of the text. Arabic Natural Language Processing (NLP) for minimizing noise, converting different ways of writing an Arabic word for one form and overcoming the inadequacy of the data [18]. Examples of common normalization, "الفعل المعتل" like لا تنس where the original verb is تنسى.

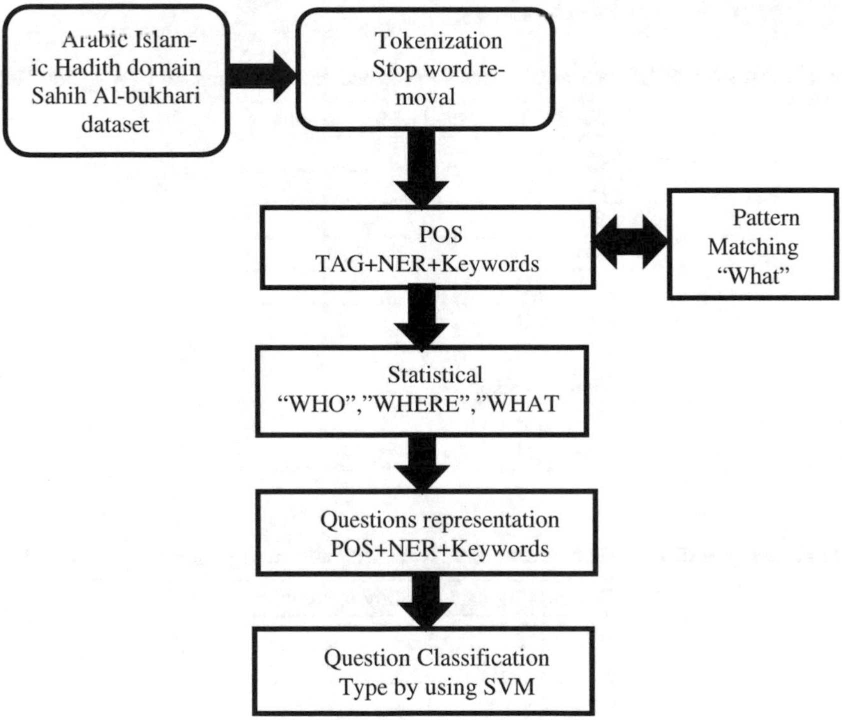

Fig. 1. The QC in Islamic Hadith-domain Q.A architecture

2.4 Part of Speech (POS) Tagging

This section deals with necessary components such as part of speech (POS), which is a component that serves to clarify the capability to arithmetically match word categories to individual words or other symbols such as numerals. Customarily, the traditional eight parts of speech include the following: (1) noun, (2) verb, (3) pronoun, (4) preposition, (5) adverb, (6) conjunction, (7) adjective and (8) article. A tag set is a form and pattern for which certain features are represented. For instance, in the Penn Treebank (PT) tagger set, a singular popular noun is tagged as NN. In this work, the Arabic POS developed by [10] was used. An illustration is given in Tables 1, 2 and 3. POS for Punjabi was used by [10] using SVM to get a good accuracy for an answer.

2.5 Named Entity Recognition (NER)

The NER is a form of a basic method for the detection of queries and suitable response in line with the case of world entities as an individual, position or organisations, referred by name in the dataset. Nonetheless, named entity recognition is crucial in identifying QC tasks. A categorised entity can be a personal, or location with a specific name. Outlining NER in a piece of text such as a question is a data retrieval challenge. For the purpose of this study, NER was used in Islamic Hadith question. Tables 1, 2 and 3 show samples of questions annotated with NER tags.

Table 1. Samples of QC text annotations with named entity recognition tagging for "Who"

Words	Tag named entity recognition
من	O
هو	O
ال	O
صحابي	O
الذي	O
لقب	O
ب	O
جعفر	Person
ال	O
طيار	O

Table 2. Samples of QC text annotations with named entity recognition tagging for "Where"

Words	Tag named entity recognition
أين	O
دفن	Location
ال	O
رسول	O
عليه	O
ال	O
سلام	O

Table 3. Samples of QC text annotations with named entity recognition tagging for "What"

Words	Tag named entity recognition
ما	O
أسم	O
ال	O
مقبرة	Location
التي	O
دفن	O
فيها	O
ال	O
صحابه	O

2.6 Question Representation

Making a question requires all the words pertaining to their part of speech, and named entity recognition are transformed into the element of the vectors. The new features vector is depicted as the significance of the following forms: (1) proper nouns (2) number of noun (3) number of verbs (4) contain person name (5) location name (6) question keywords, and (7) question words. Question words (QW) is taken from a directory of QW, the code word has been obtained on noun. Features that include a number of nouns and a number of verbs in the questions were used as basic features and needed in short text classification and the usage of POS tags as features for QC as reported in other comparable articles. A study by [8] initiated the use of these features. The keyword of features in QC advances the exactness for coarse and fine class classification [8]. Features such as words, POS, question length, proper nouns, named entities (location name, person name) are effectively employed for QC [8, 9].

2.7 Patterns Matching (PM)

Mostly, questions without head word (hw), for instance, the question of "What is Islam?" has no suitable hw for the entity types by placing only noun in the question (Islam) that does not make any provision for classifying this question as "type 3: concept definition". We have as well applied five pattern-rules as given in Table 4, to examine the influence attributed to the classifier. In this section of Patterns Matching (PM) also to use Part of Speech (POS) and Named Entity Recognition (NER) for PM, [10, 17, 18]. Table 4 shows some examples of type of questions for "What". Although, the forms of patterned questions employed started with "What", it also has another connotation referring to "Who" and "Where" as stated in pattern 3 and pattern 4.

Table 4. The developed patterns-rules

Type: Definition pattern 1	If the question beginning with meanings in English is "what" IS/ARE { هوor همfollow in by an optional a, an { لor عل} "What is Islam" { ما هو الاسلام}		الاسلام		هو	ما	
			Noun		Pronoun	Adverb	
			O		O	O	
Type: Definition pattern 2	If the question beginning with what DO/DOES { يفعلor فعل} "(What do we mean by Zakat)" {ماذا تعني الزكاة}		الزكاة	ال	تعني	ماذا	
			Noun	Pre	Verb	Adverb	
			O	O	O	O	
Type: Entity: Location pattern 3	If the question beginning with what followed by one or two words containing a location keyword "(What is the name of the grave where the companions were buried?)" { ما هو أسم المقبرة التي دفن فيها الصحابة}		المقبرة	اسم	هو	ما	
			Noun	Noun	Pro-noun	Adverb	
			Location	O	O	O	
		صحابة	ال	فيها	دفن	التي	
		Noun	Pre	Verb	Noun	Particle	
			O	O	O	O	
Type: Entity: Person pattern 4	If the question beginning with what and followed by many words containing a person keyword "(Who is killed Musaylimah the liar?)" { من هو الذي قتله مسيلمة الكذاب}	الكذاب	مسيلمة	قتله	الذي	هو	من
		Noun	Noun	Verb	Parti-cle	Pro-noun	Ad-verb
		O	Person	O	O	O	O
Type: Entity: Object pattern 5	If the question beginning with what followed by three to four words "(What the first thing build by God Almighty)" { ما أول شيء بناه الله عز وجل}	عز وجل	الله	بناء	شيء	أول	ما
		Noun	Noun	Verb	Noun	Noun	Ad-verb
		O	Person	O	O	O	O

2.8 Question Classification

In this article, the QC process was performed for classification of sample questions. The classification process was similar to the steps as prescribed in the learning process. Moreover, there were three question types which included "Who", "Where" and "What". An arranged query from all sets was utilised as a training data. (See Fig. 2) shows the entered words and the responses obtained from the question classification process (QCP). Training 1 until Training 180 were the given features for a vector of the training dataset (180 questions) from (Who: type 1), (Where: type 2) and (What: type 3). In (see Fig. 2), the feature vector has characterized the values of (1) amount of proper nouns (2) amount of noun (3) amount of verbs (4) location name (5) person name (6) question keywords, and (7) question words. Structures such as a number of noun and verbs in the question were mostly utilised. These structures are commonly used in short text classification, and the use of POS tags as features for QC in other similar works [8, 11] has initiated the approach. The code word structure in QC enhances the efficacy of the coarse and fine-class classification [8]. A study by [8] indicated that this feature would be useful for QC. Features and structures as words,

Question classification input example

من هي اول زوجة للرسول عليه الصلاة والسلام؟
اين حدثت معركة الخندق؟
ماهي اعداد المسلمين الذين قاتلوا الغزو في صلح الحديبية؟

Training
180 Questions (Tr ......Tr180)

T	#No of pronoun	#No of noun	#No of verb	Person	Location	Keywords	Question words
Tr01	0	2	1	0	1	معركة, الخندق	اين
Tr2	1	5	0	1	0	اول, الزوجة, الرسول, السلام	من هي
Tr3	2	2	1	2	1	اعداد, المسلمون, الغزو, صلح, الحديبية	ماهي

ما اول شيء بناه الله عز وجل؟
ما هو اسم الجبل الذي مكث فيه نبي الله محمد (عليه الصلاة والسلام)؟
ما هو الإسلام؟

Testing
20 Questions (Ts ......Ts20)

T	#No of pronoun	#No of noun	#No of verb	Person	Location	Keywords	Question words
Ts1	1	3	1	0	1	بناء	ما
Ts2	0	1	0	0	0	الاسلام	ما هو
Ts3	2	5	1	0	1	الجبل	ما هو

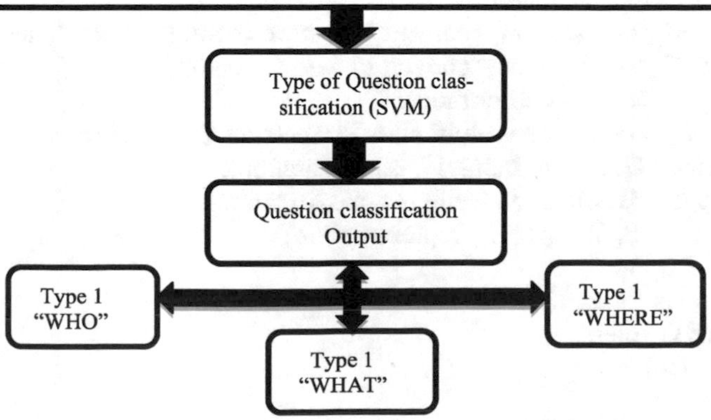

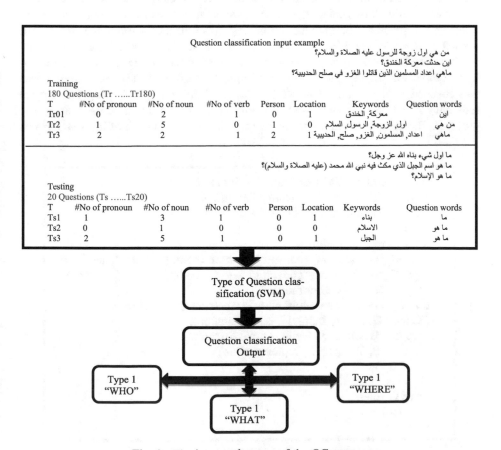

Fig. 2. The input and output of the QC process

POS, question length, proper nouns, named entities (location name, person name) are effectively employed in QC [9]. A question was taken and analysed for feature vector to create the same features for training. As an example, we used 20 questions as a test set of Testing1, Testing2 until Testing20 to reveal classification process. The features or attributes set were extracted from the questions. as shown in (see Fig. 2). However, the tag apportioned to the query was the most ranked answer type. After that, each question passed to the pattern and keywords part of the algorithm. Finally, the evaluation of support vector machine classification and pattern-based matching algorithm was merged. The apportioned tag to each question was "Who: type1", "Where: type2" and "What: type3". (See Fig. 2) shows the training and testing [12, 16, 17].

For the purpose of this study, we used the Rapid Miner tool to route the SVM. The algorithm for SVM kernel methods are a class of algorithms for pattern analysis, whose best known member is the support vector machine (SVM) Gaussian processes, any linear model can be turned into a non-linear model by applying the kernel trick to the model: replacing its features (predictors) by a kernel function. QC was given in (see Fig. 3), the pipeline approach in Rapid Miner for SVM employed in this study comprised extraction of information features and the X-Validation operator, which was a nested operator with two categories under it which included a training sub process and a testing sub-process. The training sub process was utilised for training a model. The trained model was the SVM classifier which was then applied in the testing sub-process. The functional process design was also measured during the trying part. The input dataset was divided into k subsets of equal size. Of the k subsets, a single subset was retained as the testing data set (i.e. input of the testing sub process), and the remaining k − 1 subsets were used as training data set (i.e. input of the training sub process). The cross-validation became repeated k times, with each of the k subset was

```
Algorithm for SVM Question Classification
Input: training Question, testing Questions
Output: SVM model, testing Questions classes list I.
Begin
Step1:// representation
Inputs1_represention = convert to feature vector (training Questions)
Inputs2_represention = convert to feature vector (testing Questions)
Step2:// build classifier model
SVM _ model = new Multi class Support Vector Machine
(Kernel: Gaussian, Inputs1 _ representation);
Step3:// classification using the classifier model
For each (q in Inputs2 _ representation)
    Begin
Q _ class=classify _ question (SVM _ model, q);
L=L+Q _class;
    End
End
```

Fig. 3. The algorithm of the SVM classifier

used exactly once as the testing data. The k results from the k-iterations then can be averaged (or otherwise combined) to produce a single estimation. The factor given as k can be calibrated by using a parameter. The learning process optimised the model to make it fit for training data for the best value.

3 Experimental Setup

The performance of the QC system was measured for the given optimum precision (P), recall (R) and Micro-average (F1). The assessment of the output performance of the initial prototype and evaluation of the competence of the recommended tag set were performed. The seamless system reduced answering period that can be taken with a very minimal window space. Let TP represented the number of questions, which were manually tagged as appropriate and as true positives by the classifier. Let false negative represented a number of questions/queries to be classified as relevant, and classified as insignificant by the classifier (FN).

$$Pr_{micro} = \frac{\sum_{i=1}^{|c|} TP_i}{\sum_{i=1}^{c} TP_i + FP_i} \tag{1}$$

$$Pr_{micro} = \frac{\sum_{i=1}^{|c|} TP_i}{\sum_{i=1}^{|c|} TP_i + FN_i} \tag{2}$$

4 Result and Dissection for Classifier

In this experimental report, a new set of features and structures was studied and their efficiency was scrutinised for QC in Islamic Hadith-domain Q.A. The purpose is to effectively find an enhanced type of classification procedure. Moreover, the objectives of the study are to estimate the expediency and functionalities of various and distinct features with the SVM. In this article, the SVM classification utilised in the testing set employing 10-fold cross-validation. In this section, each question was taken as a feature vector. The efficacy of distinct structures provided in Tables 5, 6 and 7 was studied and estimated to familiarise the effect of these features on the efficiency of the SVM method. Table 5 shows the result of the SVM with regards to the following features (1) number of proper nouns (2) number of noun (3) number of verbs. Table 6 shows the result of the SVM method when the following features are used (4) contains location name (5) contains person name (6) question keywords and (7) question words. Table 7 shows the result of the SVM approach. Tables 5, and 6 show that NER features, hw and question word with enhanced performance and improved the effective applicability of the SVM classification better than that obtained with POS features as shown in Table 5. However, the SVM classification trained under the above listed process features turn out to be the best results among all outcomes of SVM classifier.

Table 5. Implementation of SVM classifier for QC with first three features

	TP	FP	FN	Precision	Recall	F-measure
TYPE1	74	6	19	79.57	92.5	85.55
TYPE2	17	3	7	70.83	85	77.27
TYPE3	79	21	4	95.18	79	86.34
AVG				81.86	85.6	84.06

Table 6. Implementation of SVM classifier for QC with the last four features

	TP	FP	FN	Precision	Recall	F-measure
TYPE1	75	5	15	83.33	93.75	88.23
TYPE2	16	4	5	76.19	80	78.05
TYPE3	84	16	5	94.38	84	88.89
AVG				84.63	85.93	86.07

Table 7. Implementation of SVM classifier for QC with all feature sets

	TP	FP	FN	Precision	Recall	F-measure
TYPE1	73	7	12	85.88	91.25	88.48
TYPE2	18	2	5	78.26	90	83.72
TYPE3	85	15	7	92.39	85	88.54
AVG				85.51	88.75	86.91

5 A Combine SVM Classifier and Pattern-Based Method

This study has applied a bloc concept, which combined Support Vector Machine classifier with a pattern-based matching method for set trials. The major purpose is to effectively generate a corresponding arrangement method, which combines all benefit of pattern-based matching approach and the lead of machine learning method (SVM). In the combination, the first question passed through both SVM module and pattern-based matching module. In addition, Table 8 shows the results as a form of estimation matrices accuracy, recall and F-measure derived from the combined model (SVM classification and pattern-based method).

Table 8. Implementation (precision, recall and F-measure) question type for combined model (SVM and PM) for QC

	TP	FP	FN	Precision	Recall	F-measure
TYPE1	76	4	11	87.36	95	91.01
TYPE2	16	4	3	84.21	80	82.05
TYPE3	88	12	6	93.62	88	90.72
AVG				88.39	87.66	87.93

6 Conclusion

This article has effectively designed a new approach to the QC in Arabic Islamic Hadith-domain QA. The new approach was initiated and produced for combination model of SVM learning classifier and pattern-based approach. The outcome of the experiment revealed that the approach was suitable and appropriately fitted AQC in Islamic Hadith-domain QA. This paper shows the level of efficiency generated when the pattern-based approach was functionalized. These designs were majorly a developer inclined question words, keywords, POS tagging and NER information. The output of the pattern was based on 20 questions. The system output came from the pattern matching algorithms implemented in our system. The manual output was standardized taken manually for familiarity purpose. The evaluation was done by comparing the system output with the manual output. Based on our experiment, 16 out of 20 questions were correctly labelled by the system output.

Acknowledgements. This work is supported by the Universiti Malaysia Pahang (UMP) via Research Grant UMP RDU160349.

References

1. Li, X., Morie, P., Roth, D.: Robust reading: identification and tracing of ambiguous names, DTIC Document (2004)
2. Woods, W.A.: Progress in natural language understanding: an application to lunar geology. In: Proceedings of the June 4–8, 1973, National Computer Conference and Exposition. ACM (1973)
3. Abdelnasser, H., et al.: Al-Bayan: an Arabic question answering system for the Holy Quran. In: ANLP 2014, p. 57 (2014)
4. Hou, Y., et al.: HITSZ-ICRC: exploiting classification approach for answer selection in community question answering. In: Proceedings of the 9th International Workshop on Semantic Evaluation, SemEval (2015)
5. Belinkov, Y., et al.: VectorSLU: a continuous word vector approach to answer selection in community question answering systems. In: Proceedings of the 9th International Workshop on Semantic Evaluation, SemEval (2015)
6. Kumar, N., et al., Leafsnap: a computer vision system for automatic plant species identification. In: Computer Vision–ECCV 2012, pp. 502–516. Springer, Berlin, (2012)
7. Lopez, V., et al.: AquaLog: an ontology-driven question answering system for organizational semantic intranets. Web Semant.: Sci. Serv. Agents World Wide Web **5**(2), 72–105 (2007)
8. Pires, R.M.P.: Query classification and expansion in just. Ask question answering system, Master's thesis, Instituto Superior Tcnico (2012)
9. Li, X., Roth, D.: Learning question classifiers: the role of semantic information. Nat. Lang. Eng. **12**(03), 229–249 (2006)

10. Hammo, B., Abu-Salem, H., Lytinen, S.: QARAB: a question answering system to support the Arabic language. In: Proceedings of the ACL-02 Workshop on Computational Approaches to Semitic Languages. Association for Computational Linguistics (2002)
11. Bomhoff, M., Huibers, T., Van der Vet, P.: User intentions in information retrieval. In: DIR 2005, pp. 47–54 (2005)
12. Metzler, D., Croft, W.B.: Analysis of statistical question classification for fact-based questions. Inf. Retr. **8**(3), 481–504 (2005)

Cloud Service Discovery and Extraction: A Critical Review and Direction for Future Research

Abdullah Ali[1,3](✉), Siti Mariyam Shamsuddin[1], Fathy E. Eassa[2], and Fathey Mohammed[3]

[1] Faculty of Computing, Universiti Teknologi Malaysia, Skudai, Malaysia
abdallah.almenbri@gmail.com

[2] Faculty of Computing and Information Technology, King Abduaziz University, Jaddah, Kingdom of Saudi Arabia

[3] Faculty of Engineering and Information Technology, Taiz University, Taiz, Yemen

Abstract. The advancement of cloud computing has enabled service providers to provide users with diversified cloud services with different attributes and costs. Finding a convenient service that satisfies users' requirements based on both functional and non-functional requirements has become a big challenge. The existing studies on cloud service discovery have addressed this problem and proposed solutions using different techniques. This paper reviews the existing studies on cloud service discovery and covers current approaches, techniques and models used. In addition, the limitations and weaknesses of the proposed solutions are considered. As a result, research issues and gaps for future research are revealed.

Keywords: Cloud service discovery · Approaches · Limitations Review

1 Introduction

Rapid development of information technology has motivated many organizations and enterprises to search for methods that keep operational cost low, increase the scalability, improve performance, and have high efficiency in resource utilization [1]. Distributed systems, parallel computing, grid computing, virtualization, and other technologies have evolved over the years. The inflexibility, high cost and deficiency of scalability of these technologies are ineffective for business requirements. Recently, cloud computing has emerged to meet business requirements by complement these technologies and add new features to resources and application provisioning [2].

Cloud computing is a model that allows users to access to the hosted services (hardware and software) available on the Internet where they are composed and virtualized dynamically upon users' need. There are two players in cloud computing, cloud providers and cloud customers. The cloud providers keep enormous computing resources (cloud services) in their large datacenters and rent these services to customers

F. Saeed et al. (Eds.): IRICT 2018, AISC 843, pp. 291–301, 2019.
https://doi.org/10.1007/978-3-319-99007-1_28

as pay-per-use basis [3]. The majority of popular sites are hosted on a cloud, including social networking, email, document sharing and online gaming. Since cloud services have become very important for the companies, many cloud providers such as Amazon, Google, Rackspace and Microsoft fight to get a foothold on provisioning customers by different cloud services with different performance attributes and costs [4]. The diversity on services description, non-uniformed naming conventions, and the heterogeneous types and features of cloud services make the cloud service discovery a hard problem [5]. The cloud customers have faced a problem of how to find the best cloud service from the growing number of cloud providers that satisfy the Quality of Service (QoS) requirements such as performance, cost and security [6]. Cloud service discovery addresses a problem of returning a suitable service or a set of services that match the specified requirements or objectives. Although various challenges related to cloud computing have been addressed in active research including security, privacy and trust management, cloud service discovery still a rudimentary area of research [7–10].

Several studies have been carried out on cloud services discovery based on different approaches. These approaches differ in the environment of searching form central repository [7, 11–13] or from the Web [14, 15]. In addition, different models were used for discovering such as ontology model. Ontology is used for different purpose such as similarity measuring and reasoning [16], and cloud services representation [5]. Ghazouani and Slimani [17] provided a review on approaches in addressing cloud services discovery. They concentrated on the classification and comparison of previous cloud services discovery approaches by taking into account type of delivery model, the technique adopted, the obtained service representation, and the covered domain. Alkalbani and Hussain [18] presented an overview of the current cloud service discovery trends and challenges. They reviewed and classified the approaches according to service discovery architecture and techniques including approach model/architecture, service type, ontology representation (domain, language, and reasoning), dynamic discovery model, evaluation model, user's preferences techniques, data updates, and public repositories. Moreover, Sun, Dong [19] surveyed cloud service selection approaches, and analysed them from five perspectives: decision making techniques; data representation models; parameters and characteristics of cloud services; contexts, purposes. However, more investigation on the literature may promote the existing techniques of cloud service discovery and reveal new insights for more innovative solutions. This paper presents a critical review of cloud service discovery approaches so that research issues and gaps can be revealed for future works. The next section introduces cloud service discovery concepts. Then, the existing proposed solutions for discovering the required cloud services are presented. Next, a discussion on the reviewed studies and the proposed classification is presented. Finally, the study ends with the conclusion.

2 Cloud Service Discovery

A service has a kind of relation between service providers and the end-users. Consequently, service provisioning is the procedure of empowering the service clients to access the predefined resources and enjoy the provisioned services [20, 21]. Basically, the key interaction between the client and the service provider depends on service provisioning [22]. Similarly, service provisioning plays a significant role for both the cloud provider and the cloud user. The cloud provider should provide the required services as stated in the Service Level Agreement (SLA) including the QoS and pricing. The services must meet the user's requirements, such as on-demand availability, scalability, elasticity, security, and exact billing [6, 23]. One of the service provisioning objectives is providing reasonable comparisons among the offered services. As a result, the users can compare the various cloud service offerings according to their needs and select the services based on several predefined dimensions [24].

The cloud services have attracted much attention in the recent years. Cloud providers offer a diversity of cloud services for customers including processing, storage and other services [25]. It is becoming gradually more difficult for a user to select a suitable service provider that will supply him or her the required services [26]. Service discovery, which is principally related to web services has been intensively researched in the past decade [27]. A great amount of work has been performed on services descriptions and standardisation specifically on the area of service oriented architecture and web services. For service description, the most outstanding language is Web Service Definition Language (WSDL), which describes the service from technical view including features of service, interface operations, binding and others [28]. This description is not sufficient from business point of view for describing the services delivered via the Internet such as the cloud services, which are business services. Business service is concerned with the end to end delivery between a provider and a customer depending on a particular period of time, cost, and SLA [29].

In order to discover suitable cloud services which are able to match specified requirements or objectives, an automated discovery system is necessary to specify the degree of similarity between a user service request and a service provider advertisement. Matching decision can be based on several criteria and parameters related to cloud services such as functional and non-functional requirements. The functional requirements include service tasks to be provided, cost, constraints and others, whereas non-functional requirements include service software applications, software compatibility constraints, hardware policy, service models (IaaS, PaaS, SaaS) and others. The diversity on services description, the heterogeneous types and features of cloud services and non-uniform naming conventions make the cloud service discovery as complex problem [5].

3 Existing Cloud Service Discovery Solutions

There are several techniques, models, algorithms and methods are used to accomplish the process of discovering cloud services. Chen, Bai [30] introduced registry approach for discovering cloud services by extending WSDL to express the specific attributes of cloud services. This description of cloud services is then distributed into a Universal Description Discovery and Integration (UDDI) registry, which enables the automatic discovery from a single location. Furthermore, Ankolekar, Burstein [31] used DARPA Agent Markup Language for Services (DAML-S) to represent the cloud services, semantically. Many researchers developed the relation between cloud providers and consumers by building cloud service broker system [11–13, 32]. These studies introduced a cloud search engine named Cloudle which use agents for searching and checking the similarities between query and cloud services based on the ontology. They designed two ontologies CO-1 and CO-2. CO-1 contained only concepts whereas the CO-2 included a set of cloud concepts as well as individuals of these concepts and the relationship among those individuals. This engine had a database that stored the cloud services provided by cloud service providers. It checked the similarities occurred between cloud services in the database and requested query from users using this ontology. Measuring the similarity is based on three kinds of methods which are data property, concept, and object property. Availability of cloud services depends on the registered services by the providers.

Kanagasabai [33] proposed a novel cloud broker by adopting a semantic service-based approach (Web Ontology Language for Service (OWL-S)) and developed an OWL-based matchmaking system for discovering complex constraint-based services dynamically. However, shared ontology among cloud providers can be used for translating the offered and requested services. Tahamtan, Beheshti [34] designed a business ontology for cloud to help the organisation in searching for suitable cloud services. A framework was designed that provided an integrated unified business service and cloud ontology to assist businesses to search for the suitable cloud service using querying abilities. The purpose of using unified ontology was recording the needed business services by enterprise and built a link between the business functions and provided services in the cloud space.

The SaaS discovery system proposed by Afify, Moawad [35] could establish SaaS service attributes by utilising the SaaS ontology. Service filtering and ranking, service registration as well as service discovery were the three major components included in SaaS system. For matching process, a method was proposed which comprised of ontology matching, semantic-based, and feature similarity matching. However, this system did not support automatic discovering SaaS services and the cloud provider need to subscribe in the system so that it could publish its services. Vasudevan, Haleema [36] suggested a technique for improving the automatic discovery of services by including the semantic representation onto the profile of cloud service. There was different information where a service profile comprised of service name, price, and other features. Each service profile was represented as separate ontology, which later merged together to constitute as a global ontology. Parhi, Pattanayak [9] proposed a framework based on ontology and multi agents to describe and discover cloud services. The benefit of using ontology for cloud services description contributed in increasing the related cloud services requested by customers.

Kang and Sim [32] presented a cloud service discovery system, which publishes the cloud services addresses to make it available through search engine. Then, to extract and retrieve the result based on appropriate information, the semantic filtering is applied. To perform the semantic filtering, the cloud ontology is used for discovering the relationship between required cloud services by advising the matching service. Further, a search engine called Cloudle based on ontology was proposed by Kang and Sim [37]. Various reasoning methods to identify the similarities among cloud services were used. Service type, functions, time and prices were the information needed from the user to be entered through system's interface, where the search engine retrieved the matching service based on the user request using cloud ontology. Finally, the retrieved services were sorted according to the price of time period established by the user. The Cloudle search engine was improved by Sim [38]. The improvement involved a cloud crawler which performed a search on the web for cloud services and stored them in the Cloudle database instead of asking cloud providers to register their services in the Cloudle's database.

Cloud Service Discovery System (CSDS) was proposed by Han and Sim [39]. Agent and ontology were used to discover the cloud services. Google has been used to search for the cloud services on the web. This mechanism was time consuming due to the online search that occurred before matching process between query and retrieved services. The CSDS was improved by Han and Sim [16]. The improvement concentrated on measuring the similarities among services. It used three reasoning methods; similarity, equivalent, and numerical reasoning. However, this system is a time consuming because it based on Google to search for the cloud services before it starts to check the similarities. Furthermore, Chang, Juang [14] proposed an intelligent service discovery framework that integrated mobile agent with ontology and developed a prototype to search for appropriate services in cloud environment. It was assumed that the crawlers knew the structure of clouds data centres, such as the location, schema and others, and had direct access to the data centres, but till now there are no standards for cloud service naming convention and different services description [5]. Also, an inference rule was defined which was executed by Jena inference engine. Results has shown a high degree for precision and low degree for recall, which needs to be verified by applying more evaluation.

A cloud crawler engine was designed by Noor, Sheng [7] which could crawl numerous cloud portals by using cloud ontology. The description of cloud services attributes was saved in repository that built dataset for cloud services. Nevertheless, this dataset showed many shortages for example service name and URL, which represented basic information and also data values were not associated with semantic meaning. As a result, the information of this dataset did not provide benefit to the user or application. Moreover, a platform was proposed by Rodríguez-García, Valencia-García [40] which represented a semantic platform that annotated cloud services represented in natural language descriptions using ontology. Further, Nabeeh, El-Ghareeb [15] proposed a Cloud Service Discovery Framework (CSDF) based on software agents and web services. The framework consists of many agents for achieving different tasks including crawling, cloud services ontology analysing and reasoning, recommendation and other tasks. Cloud services were recommended based on users' profile, ranker and evaluator reports. Nevertheless, the used ontology need to be extended to enhance the crawling process and the framework was not evaluated. Furthermore, Mittal, Joshi [41] developed a prototype system, which automatically extracts the terms definitions and measures for SLA documents written as a text. The extracted terms were saved as Resource Description Framework (RDF) graph to represent the knowledge base. The extraction was performed on SLA terms and did not provide users with ability to compare among SLA of different providers for different cloud services.

Hamza, Aicha-Nabila [42] proposed a mobile agent-based approach for cloud services discovery. An algorithm was proposed to compare the user request with the service description based on keyword search and performed filtering for the retrieved services. The proposed approach based on the keyword matching which may not return the required services, especially if they have different names on several providers. Further, it did not cover all cloud services attributes such as cost. A search engine called CB-Cloudle (centroid-based) was built by [4, 43]. It was based on crawlers in discovering cloud services from different cloud providers and applied k-means clustering algorithm to cluster the cloud services based on similarities and differences between the groups of cloud services. The crawling was performed on most popular cloud providers where each provider had its own crawler.

Wheal and Yang [10] developed a cloud services recommender system called CSRecommender, which was a search engine and recommender system. The CSRecommender system offered recommendation for the user from different cloud services using collaborative and content-based methods and took into account the similar users. The user would visit the CSRecommender website and search for the cloud services; and the ranked cloud services based on ratings done by other users would be returned. The recommender system was applied on a few cloud services. Furthermore, it did not cover all types of cloud services and the accuracy was low. Finally, A central repository for retrieving and storing the SaaS services was proposed by Alkalbani, Shenoy [44] using an open source crawler engine. The repository was applied only for SaaS services. It had some shortcomings regarding to service information, where it provided the service name and URL only. The summary of related works, main ideas and limitations for the existing solutions are listed in Table 1.

4 Discussion

Exploring the related literature shows that the proposed models, frameworks and systems for discovering the cloud services have some limitations. These limitations include time consuming, low performance in discovering Cloud services and lack of standardization for representing Cloud services attributes. Moreover, the process of extracting cloud service attributes represented in different formats and to be represented in a standardized form such ontology was not manipulated. Most common models and technologies used for the cloud services discovery process (searching, comparing and representing) are ontology and agents. However, there is no standardised way to describe and represent the cloud services attributes. The majority of existing works deals with one format of represented cloud services attributes. By scanning of most reputed cloud providers websites, different formats can be used for representing the cloud services attributes such as HTML tables, JSON and text formats. There are few studies which were conducted on extracting the cloud service attributes such as Chang, Juang [14] and Mittal, Joshi [41], whereas the rest only discovered the Cloud services without extracting the cloud service attributes. In addition, few studies represent the retrieved attributes in an ontology form and deal with different models of cloud services (SaaS, PaaS and IaaS). Furthermore, the existing studies do not consider all functional and non-functional requirement for discovering Cloud services.

Table 1. Limitations of existing cloud service discovery solutions

Study	Proposed cloud service discovery solution	Limitations
[39]	A CSDS, which uses agents and ontology to discover the Cloud services over the Internet	It has been evaluated using providers' virtual websites Time consuming Ontology doesn't cover all cloud concepts System was partially implemented
[32]	Cloudle is implemented based on agents and ontology for searching cloud services based on functional, non-functional and technical requirements	It has been judged using providers' virtual websites Cloud providers need to register their services in a central database
[37]	A search engine based on ontology, which uses various reasoning methods to identify the similarities among cloud services. The matching service is retrieved based on the user request and sorted according to the price of time period established by the user	It does not cover all Cloud services attributes Cloud ontology was small Cloud providers need to register their services in a central database
[30]	Cloud services and their specific features are described by extending the WSDL and publishing them on the web service registry (UDDI)	It provides URL and the name for each service only It has issues in coping with the expanding needs, and in presenting the latest updates
[12]	Improved Cloudle which uses two ontologies CO-1 and CO-2. CO-1 contained only concepts and CO-2 contained a set of concepts	The evaluation was done based on virtual websites Cloud providers need to register their services in a central database
[13]	Cloud service broker system based on a cooperative agent-based cloud service discovery protocol using ontology for matching the similarities among Cloud services	The cloud ontology concepts did not take into account upgrading of the ontology to deal with the expanding Cloud services
[16]	A CSDS based on agents and ontology. Different reasoning methods used to determine the similarities among services	Time Consuming The developed cloud ontology covered few concepts It has been evaluated using providers' virtual websites
[14]	An intelligent service discovery framework which integrates mobile agent with ontology for searching appropriate services in cloud environment using inference rules	The developed cloud ontology covered few concepts The framework was partially implemented It needs more evaluation to verify the recall measure
[33]	A cloud broker using semantic service based on OWL for dynamically finding the services which consist of complex multi criteria	Shared ontology among cloud providers is needed for translating the offered and requested services QoS parameters are not included
[42]	A mobile agent is used to compare between user request and cloud service description based on keyword search	Keywords-based search It does not cover all Cloud services attributes

(*continued*)

Table 1. (*continued*)

Study	Proposed cloud service discovery solution	Limitations
[5]	An ontology based on OWL that can represent the functional and non-functional concepts, and the relationship of infrastructure services on the IaaS layer	Applied only to IaaS services
[45]	An improved Cloudle that includes Cloud Crawlers, which searched for cloud services over the web	The evaluation was done based on virtual websites QoS parameters are not included
[34]	A framework that served as a services repository. It provides an integrated, unified business service and cloud ontology	The equalisation between the query and business function is asked by the user The SPARQL is only comprehensive to expert users The providers' data is obtained based on market search
[40]	A semantic platform that annotates cloud services represented in natural language descriptions using ontology	It does not cover all Cloud services attributes The validation was performed on very small services
[7]	A crawler engine to crawl cloud web portals using ontology and store the retrieved information in a local repository	Short of primary services' information for instance, service name and service URL
[35]	A SaaS discovery system that locate SaaS service information based on several parameters using a unified SaaS business ontology	The automatic discovery feature is not supported for SaaS It covers only SaaS services
[36]	Using a single ontology, semantic representations are added to each cloud service profile. All separate ontology was combined to construct a global ontology	It does not cover all Cloud services attributes Cloud service providers need to publish their list of services in the form of RDF
[4, 43]	CB-Cloudle based on crawlers to discover cloud services from different cloud providers. k-means clustering algorithm has been applied to cluster the retrieved Cloud services	It needed to be evaluated
[44]	Distributed methods for crawling only SaaS cloud services based on the Hadoop framework	It only provides URL and the name for each service It covers only SaaS services
[10]	A search engine and recommender system called CSRecommender, which offers recommendations for using collaborative and content-based methods	It was applied on few cloud services Accuracy is low It does not cover all types of Cloud services
[15]	A CSDF based on agents and web services that uses many agents for achieving different tasks related to Cloud services discovering and ranking	No evaluation was done Ontology need to be extended to enhance the crawling process

(*continued*)

Table 1. (*continued*)

Study	Proposed cloud service discovery solution	Limitations
[9]	A prototype based on ontology and agents for describing and discovering Cloud services	Not fully implemented Not evaluated
[41]	A system, which automatically extracts the terms and measures of text file SLA documents. The extracted terms are saved as RDF graph to represent the knowledge base	It covers SAL only Not evaluated No comparison of SLA among different providers

5 Conclusion

The cloud services discovery, the existing cloud services discovery approaches and related works have been investigated. Based on the critical review of the literature of cloud services discovery approaches, there is still a need for models, frameworks and systems that can consider different attributes. The different formats of cloud services attributes representation (tables, JSON, text, etc.) should be considered. In addition, the retrieved cloud services attributes can be represented in an ontology form. Ontology can be used to represent the retrieved attributes so the query of a service can be an ontology-based rather than keyword-based, which will increase the efficiency and effectiveness of the system. Ontology also can be used during the crawling or finding the similarities between retrieved cloud services and queries. In addition, cloud service discovery model can be proposed based on ontology, agents, classification and several extraction methods for extracting Cloud services represented in different formats. Multi agents can be used to achieve different tasks in parallel for overcoming the problem of time consuming and low performance. Furthermore, all cloud service models can be covered. The process of discovering cloud service of different models (SaaS, PaaS, IaaS) and classifying them into their corresponding model can be carried out using classification process. Classifying Cloud services into their models increases efficiency of the process of cloud service attributes extraction which in turn improves the efficiency of the whole process of Cloud services discovery.

References

1. Soundararajan, V., Govil, K.: Challenges in building scalable virtualized datacenter management. SIGOPS Oper. Syst. Rev. **44**(4), 95–102 (2010)
2. Emeakaroha, V.C.: Managing Cloud Service Provisioning and SLA Enforcement via Holistic Monitoring Techniques. Technische Universität Wien, Karlsplatz (2012)
3. Lee, G.: Resource Allocation and Scheduling in Heterogeneous Cloud Environments, in Computer Science. University of California, Berkeley (2012)
4. Gong, S., Sim, K.M.: CB-Cloudle and cloud crawlers. In: 2014 5th IEEE International Conference on Software Engineering and Service Science (ICSESS). IEEE (2014)

5. Zhang, M., et al.: An ontology-based system for Cloud infrastructure services' discovery. In: 2012 8th International Conference on Collaborative Computing: Networking, Applications and Worksharing (CollaborateCom). IEEE (2012)
6. Garg, S.K., Versteeg, S., Buyya, R.: A framework for ranking of cloud computing services. Future Gener. Comput. Syst. **29**(4), 1012–1023 (2013)
7. Noor, T.H., et al.: CSCE: a crawler engine for cloud services discovery on the world wide web. In: 2013 IEEE 20th International Conference on Web Services (ICWS). IEEE (2013)
8. Noor, T.H., et al.: Analysis of web-scale cloud services. IEEE Internet Comput. **18**(4), 55–61 (2014)
9. Parhi, M., Pattanayak, B.K., Patra, M.R. (2015) A Multi-agent-based framework for cloud service description and discovery using ontology. In: Jain, L.C., Patnaik, S., Ichalkaranje, N. (eds.) Proceedings of ICCD 2014 on Intelligent Computing, Communication and Devices, vol. 1, pp. 337–348. Springer, New Delhi (2015)
10. Wheal, J., Yang, Y.: CSRecommender: a cloud service searching and recommendation system. J. Comput. Commun. **3**(06), 65 (2015)
11. Kang, J., Sim, K.M.: A cloud portal with a cloud service search engine. In: International Conference on Information and Intelligent Computing IPCSIT. IACSIT Press, Singapore (2011)
12. Kang, J., Sim, K.M.: Ontology and search engine for Cloud computing system. In: 2011 International Conference on System Science and Engineering (ICSSE). IEEE (2011)
13. Kang, J., Sim, K.M.: Towards agents and ontology for cloud service discovery. In: 2011 International Conference on Cyber-Enabled Distributed Computing and Knowledge Discovery (CyberC). IEEE (2011)
14. Chang, Y.-S., et al.: Integrating intelligent agent and ontology for services discovery on cloud environment. In: 2012 IEEE International Conference on Systems, Man, and Cybernetics (SMC). IEEE (2012)
15. Nabeeh, N.A., El-Ghareeb, H.A, Riad, A.: Integrating software agents and web services in service oriented architecture based cloud services discovery framework. J. Converg. Inf. Technol. (JCIT) **10**(1), 67–79 (2015)
16. Han, T., Sim, K.M.: An agent-based cloud service discovery system that consults a cloud ontology. In: Intelligent Control and Computer Engineering, pp. 203–216. Springer (2011)
17. Ghazouani, S., Slimani, Y.: A survey on cloud service description. J. Netw. Comput. Appl. **91**, 61–74 (2017)
18. Alkalbani, A.M., Hussain, F.K.: A comparative study and future research directions in cloud service discovery. In: 2016 IEEE 11th Conference on Industrial Electronics and Applications (ICIEA), pp. 1049–1056. IEEE, June 2016
19. Sun, L., et al.: Cloud service selection: state-of-the-art and future research directions. J. Netw. Comput. Appl. **45**, 134–150 (2014)
20. Ahmed, E., et al.: Multi-objective optimization model for seamless application execution in mobile cloud computing. In: 2013 5th International Conference on Information and Communication Technologies (ICICT). IEEE (2013)
21. Ahmed, E., Shiraz, M., Gani, A.: Spectrum-aware distributed channel assignment for cognitive radio wireless mesh networks. Malays. J. Comput. Sci. **26**(3), 232–250 (2013)
22. Manvi, S.S., Shyam, G.K.: Resource management for Infrastructure as a Service (IaaS) in cloud computing: a survey. J. Netw. Comput. Appl. **41**, 424–440 (2014)
23. Shamshirband, S., et al.: Co-FAIS: cooperative fuzzy artificial immune system for detecting intrusion in wireless sensor networks. J. Netw. Comput. Appl. **42**, 102–117 (2014)
24. Rahimi, M.R., et al.: MAPCloud: mobile applications on an elastic and scalable 2-tier cloud architecture. In: Proceedings of the 2012 IEEE/ACM Fifth International Conference on Utility and Cloud Computing. IEEE Computer Society (2012)

25. Chang, J.M., Chao, H.C., Chen, J.L., Lai, C.F.: An efficient service discovery system for dual-stack cloud file service. IEEE Syst. J. **6**(4), 584–592 (2012)
26. Zhao, L., et al.: Flexible service selection with user-specific QoS support in service-oriented architecture. J. Netw. Comput. Appl. **35**(3), 962–973 (2012)
27. Segev, A., Sheng, Q.Z.: Bootstrapping ontologies for web services. IEEE Trans. Serv. Comput. **5**(1), 33–44 (2012)
28. Chinnici, R., et al.: Web services description language (WSDL) version 2.0 part 1: Core language. W3C Recommendation 26 (2007)
29. Cardoso, J., et al.: Towards a unified service description language for the internet of services: Requirements and first developments. In: 2010 IEEE International Conference on Services Computing (SCC). IEEE (2010)
30. Chen, F., Bai, X., Liu, B.: Efficient Service Discovery for Cloud Computing Environments. In: Advanced Research on Computer Science and Information Engineering, pp. 443–448. Springer (2011)
31. Ankolekar, A., Burstein, M., Hobbs, J.R., Lassila, O., Martin, D.L., McIlraith, S.A., Narayanan, S., et al.: DAML-S: Semantic markup for web services, 411–430 (2001)
32. Kang, J., Sim, K.M.: Cloudle: a multi-criteria cloud service search engine. In: 2010 IEEE Asia-Pacific Services Computing Conference (APSCC) (2010)
33. Kanagasabai, R.: OWL-S based semantic cloud service broker. In: 2012 IEEE 19th International Conference on Web Services (ICWS). IEEE (2012)
34. Tahamtan, A., et al.: A cloud repository and discovery framework based on a unified business and cloud service ontology. In: 2012 IEEE Eighth World Congress on Services (SERVICES). IEEE (2012)
35. Afify, Y., et al.: A semantic-based software-as-a-service (SAAS) discovery and selection system. In: 2013 8th International Conference on Computer Engineering and Systems (ICCES). IEEE (2013)
36. Vasudevan, M., Haleema, P., Iyengar, N.C.S.: Semantic discovery of cloud service catalog published over resource description framework. Int. J. Grid Distrib. Comput. **7**(6), 211–220 (2014)
37. Kang, J., Sim, K.M.: Cloudle: an agent-based cloud search engine that consults a cloud ontology. In: Proceedings of the International Conference on Cloud Computing and Virtualization. Citeseer (2010)
38. Sim, K.: Agent-based cloud computing. In: IEEE Transactions on Services Computing. IEEE (2012)
39. Han, T., Sim, K.M.: An ontology-enhanced cloud service discovery system. In: Proceedings of the International Multi Conference of Engineers and Computer Scientists, Hong Kong (2010)
40. Rodríguez-García, M.Á., Valencia-García, R., García-Sánchez, F.: Ontology-based annotation and retrieval of services in the cloud. Knowledge-Based Systems. **56**, 15 (2013)
41. Mittal, S., et al.: Automatic extraction of metrics from SLAs for cloud service management. In: 2016 IEEE International Conference on Cloud Engineering (IC2E 2016) (2016)
42. Hamza, S., et al.: A Cloud computing approach based on mobile agents for Web services discovery. In: 2012 Second International Conference on Innovative Computing Technology (INTECH). IEEE (2012)
43. Gong, S., Sim, K.M.: CB-Cloudle: a centroid-based cloud service search engine. In: Proceedings of the International Multi Conference of Engineers and Computer Scientists (2014)
44. Alkalbani, A., et al.: Design and implementation of the hadoop-based crawler for SaaS service discovery. In: 2015 IEEE 29th International Conference on Advanced Information Networking and Applications. IEEE (2015)
45. Sim, K.M.: Agent-based cloud computing. IEEE Trans. Serv. Comput. **5**(4), 564–577 (2012)

Traffic Bottleneck Reconstruction LIDAR Orthoimages: A RANSAC Algorithm Feature Extraction

Md. Nazmus Sakib(✉) and Md. Ashiqur Rahman

World University of Bangladesh, Plot - 13/A, Rd No. 5, Dhaka 1205, Bangladesh
nazmus_sakib70@yahoo.com, rahman.ashiq09@gmail.com

Abstract. This study attempts a solution for autonomous vehicles to avoid immediate collision due to close proximity between cars. Since LIDAR sensors are widely used for capturing images in autonomous car industry, we depict a scope of using RANSAC algorithm and linear regression to reconstruct the orthoimages to escape traffic bottleneck as well as avoid collision. It is found that LIDAR sensors can't suggests much detail in close distance, and cameras don't perform well in conditions with low light or glare images. Dataset is collected from KITTI (Karlsruhe Institute of Technology) containing compressed pixels. Significance resultants focus on error reduction followed by feature extraction simulated with MATLAB. The findings excludes large scale of data size to implement and project in T-way testing for determining strength as well as capability of resultants.

Keywords: Autonomous vehicle · RANSAC algorithm · LIDAR sensor
Orthoimage reconstruction · Linear regression analysis · Traffic bottleneck

1 Introduction

LIDAR sensory data capture information about traffic bottleneck environment and record things that is required to estimate conditions and characteristics. This system applied to many vehicles and helped minimize traffic accident rate. Furthermore, recently a number of auto vehicle manufacturers developed self-driving vehicle technology by active research and advancement. LIDAR sensors generally utilizes redundancy and overlapping capabilities which operates without a driver [1, 2]. RANSAC algorithm is usually used for object determination [3]. The outcome perceptions determined on dataset is gathered by linear regression analysis [4].

1.1 LIDAR Sensors

LIDAR sensors provide both accurate distance and velocity information [2]. Although there has a continuous progress in laser detector technologies over the last few decades, LIDAR maintains constant rotating speed of the prism leading to scan surrounding objects that makes researchers in autonomous car industry a great favor [5].

F. Saeed et al. (Eds.): IRICT 2018, AISC 843, pp. 302–307, 2019.
https://doi.org/10.1007/978-3-319-99007-1_29

1.2 RANSAC Algorithm

RANSAC algorithm can resolve traffic bottleneck problems for autonomous vehicle [5]. Linear regression is also introduced for resultant justification among the changes with RANSAC algorithm to consolidate the resultant output. RANSAC utilizes substantial outcome of possible collision in traffic bottleneck environment [6]. The steps of RANSAC algorithm are described below:

Step 1: Indicate comprehensively minimum range of points to work out model parameters.
Step 2: Resolve parameter model.
Step 3: Correlate dataset points to fit with a predefined acceptance.
Step 4: Predicted fraction of inliers over total consisted dataset points. Exceeds threshold τ, evaluate parameters with complete recognized inliers and discharge.
Step 5: If not, duplicate steps 1 from 4 (maximum N times) [5].

The number of iterations is enough high to ensure probability p (0.99), where minimum one set random model does not including any outlier [2, 5]. Formulated mathematical equation represents probability (u) as selected in the data points, this data points are predictable inlier with the value 1 and $v = 1 - u$. probability of detecting an outlier iteration of multiple points significantly symbolized by m, where formulation is represented as:

$$1 - \mathrm{p} = (1 - \mathrm{u}^{\mathrm{m}})^{\mathrm{N}}$$

And thus with some manipulation,

$$\mathrm{N} = \frac{\mathrm{Log}(1 - p)}{\mathrm{Log}(1 - (1 - V)^{m})}$$

1.3 Orthoimage Reconstruction

Orthoimage reconstruction is observed with feature matching that contains sufficient outlier [8]. LIDAR offers a sensor technology for creating autonomous driving enhancement [7]. For autonomous vehicles a fast and reliable three dimensional monitoring environment is indispensable for managing traffic block situations where orthoimage inputs are compressed [5]. In sudden conditions infrared array of light or ambient sun light is denoted as m caused by high intensity associated to LIDAR sensor [6, 7]. The consistent experimental results derived from RANSAC algorithm for reconstruction of distorted images [7]. This is determined that RANSAC algorithm has ability to excerpt common characteristics from minimum amount of observation enhanced with smallest dataset possible to extract from LIDAR sensor input [4]. RANSAC algorithm requires minimum variety of benchmarks before collision to construct a predictive model and compute solutions under challenging conditions [6] [7].

1.4 Probability Results

The different predictive results from RANSAC algorithm constructs model with parameters calculated from uncontaminated sample to all inliers [9]. This algorithm structure has simple predictive model for image translation. Constant subsets are arbitrarily selected from information and model parameters [9] where size of arbitrary samples is the smallest but sufficient for processing. The amount of distorted images containing condensed pixels are adequate for developing predicted outcome [6]. In this paper we developed that RANSAC algorithm can be used for constructing orthoimage from minimal sample. RANSAC algorithm performs better with small sample size to detect iteration points for reconstructing orthoimage [8].

1.5 Linear Regression Analysis

Linear regression attempts to project matching subsequent results to develop linear regression model. Regression analysis is used for investigating multifactor data. Conducting satisfactory results is a statistical theory of linear regression analysis is necessary to overcome problem that arise when complex traffic bottleneck with real-world data [10]. A statistical factor related with linear regression consists of retrospective study on historical data that includes observational approaches and designed of observational experiments. Linear regression model is an accurate iterative process [7, 8]. Figure 1 represents structural model derived from regression analysis to specify parameterized values for establishing adequacy.

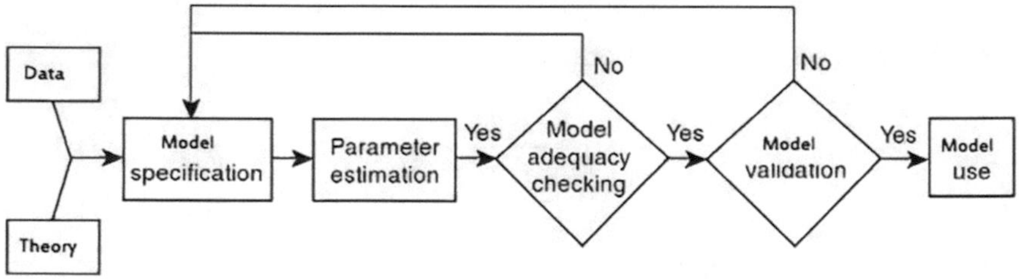

Fig. 1. Regression-model [7].

2 Materials and Methods

Aim of this study concerns with reconstructing orthoimage on dataset collected from KITTI (Karlsruhe Institute of Technology) containing compressed pixel. LIDAR sensory images undergo RANSAC algorithm to extract clear image. Two transformation images gathered from LIDAR sensor were selected with lower resolution values and are subjected to undergo RANSAC algorithm feature extraction between two images with different pixel values. For plotting linear regression a translation is carried out with predicted (P) implied with derived results gained from RANSAC algorithm. Probability of inlier regression analysis is either equal or inliers quantity in data which is divided by number of points.

2.1 Orthoimage Feature Extractions

Observed data points can fit minimum number (n) which is required to adequate model k construction. Maximum iterations determined using RANSAC algorithm threshold value (t) to establish data point that fits derived model. Desired data points (d) essential to assert model fits with data. Best possible results from model parameters are determined to best fit the data (or null if no good model is found). Obtained threshold value is required to determine a data point to establish a model t. Maximized nearest data points associated to data d are projected based on specific requirements of the resultant dataset. Best possible results are based on experimental evaluation to extract orthoimages. The iterations, k can be calculated from a hypothetical result. The assumption made that p stands probability that includes RANSAC algorithm in iteration to chooses from the LIDAR sensor as an input data when it selects n points from model bounds.

2.2 Similar Phenomenon Pattern Matching

Traffic bottleneck is recorded to gain rough value that is homograph from orthoimage pattern matching. Perception n points required for vehicles are selected independently. Probability of all inliers points (n) are considered as matching patterns. The notational presentation of k is the probability that RANSAC never selects which is total of n points consisting of all inliers with same uniformity. Consequently, RANSAC algorithm considers both sides of distorted image. The result assumes data points are selected autonomously. Points (n) are selected once and are replaced again in the identical iteration. Derived value for k phenomenon is rational approach taken as higher limit points selected without replacement. Estimate predicted line based on regression analysis model elucidated in Fig. 1. RANSAC algorithm typically chooses two points among iterations and computes line between the points.

3 Result and Discussion

Findings emphasis on selecting random subset from the dataset by considering iterations probability of trained model extracted from RANSAC algorithm dataset is collected from LIDAR sensor as distorted images. With considered RANSAC algorithm from existing dataset an approximate training model is developed providing a pattern of constructed images.

We demonstrated in Fig. 2 an important finding in the understanding of the subsequent result achieved from RANSAC algorithm with orthoimage extracted from LIDAR sensor. Translations are made from right to left in the Figure that estimate distance differentiation between vehicles. Common strategy used to construct model to be testified consisting RANSAC algorithm. Here we demonstrated results of proposed method, showing RANSAC algorithm that presides error to enable approaches towards developing predictive model. This is especially true for problems of robust geometric environment e.g., image alignment, our method, belongs to family of local methods which are typically fast randomized algorithms, potentially equipped with probabilistic success. RANSAC algorithm provides formulation consisting of iteration quantity n required to reach a probability based on results. Evaluation of constructed method performs extensive tests on both actual as well as synthetic data. No assumptions are made about the data and no (unrealistic) conditions have to be gratified for RANSAC to be successful. Although Fig. 3 originate that RANSAC runs adequately longer even by an order of value than expected.

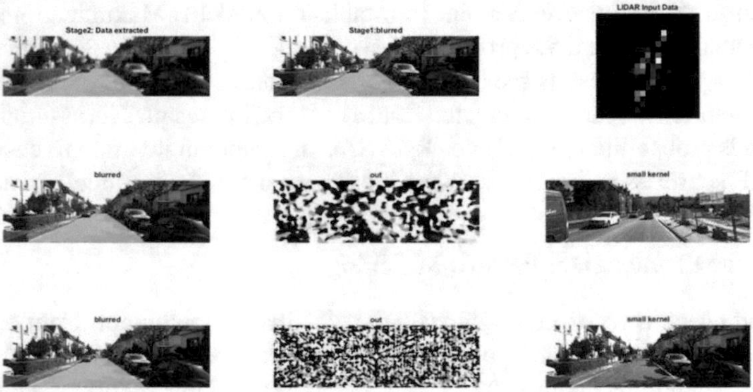

Fig. 2. LIDAR Data Reconstruction

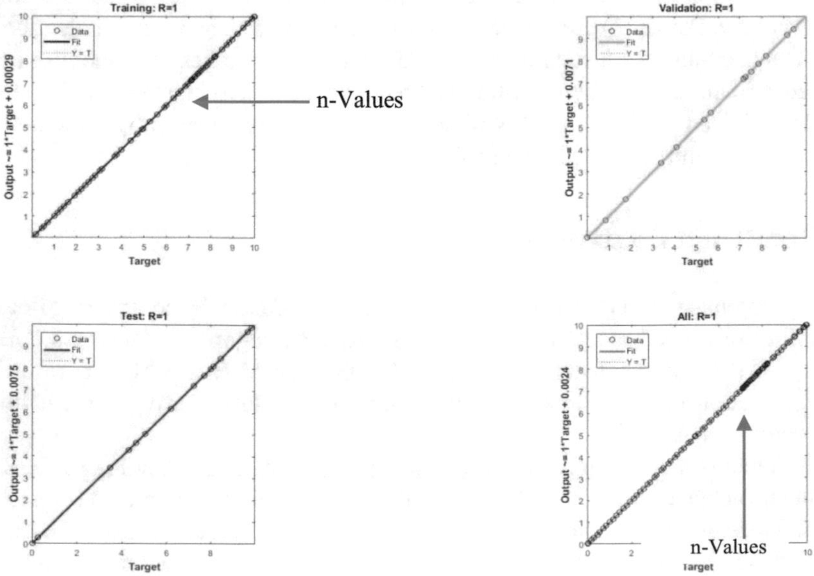

Fig. 3. Regression analysis

Regression analysis result translated in Fig. 3 shows the n-value of linear iteration existing with training data set consist blur orthoimage that is reconstructed to be in same pattern. Points are matched using MATLAB neural networking tool. Challenges and approaches of in-linear regression algorithm consist of value propagation issues.

4 Conclusion

As addressed, this study provides adequate evidence of orthoimage reconstruction procedure to be implemented with the support of LIDAR sensor followed by RANSAC algorithm. Performance and behavior validate decision that is justified using regression analysis. Added transportation with phenomenon would enable to error optimization for image detection. LIDAR outlier dataset reduced from variability in dimension to high reduction of noise in dataset collection. It's almost impossible to search out a real-time dataset excluding outliers. Therefore, it is inevitable to make sure data modeling is developed. LIDAR sensors can't suggest much detail in close distance, and cameras don't perform well in conditions with low light or glare images. The results are very positive; these findings may not translate or testified with various algorithms that are used for orthoimage reconstruction; so, a comparison would require for future classification.

References

1. Sivaraman, S., Trivedi, M.M.: A review of recent developments in vision-based vehicle detection. In: Intelligent Vehicles Symposium (IV). IEEE (2013)
2. Derpanis, K.G.: Overview of the RANSAC algorithm. Image Rochester NY **4**(1), 2–3 (2010)
3. Fischler, M.A., Bolles, R.C.: Random sample consensus: a paradigm for model fitting with applications to image analysis and automated cartography. In: Readings in Computer Vision, pp. 726–740. Elsevier (1987)
4. Van Brummelen, J., et al.: Autonomous vehicle perception: the technology of today and tomorrow. Transp. Res. Part C Emerg. Technol. **89**, 384 (2018)
5. Beer, M., et al.: SPAD-based flash LiDAR sensor with high ambient light rejection for automotive applications. In: Quantum Sensing and Nano Electronics and Photonics XV. International Society for Optics and Photonics (2018)
6. Kidono, K., et al.: Pedestrian recognition using high-definition LIDAR. In: Intelligent Vehicles Symposium (IV). IEEE (2011)
7. Montgomery, D.C., Elizabeth, A.P., Vining, G.G.: Linear regression analysis, ed. F. Edition (2012)
8. Cheng, D., Pang, Y.: Research on sift image recognition algorithm combined with Ransac. Int. J. Adv. Res. Comput. Sci. **9**, 1 (2018)
9. Korman, S., Litman, R.: Latent RANSAC. arXiv preprint arXiv:1802.07045 (2018)
10. Mooi, E., Sarstedt, M., Mooi-Reci, I.: Regression analysis. In: Market Research, pp. 215–263. Springer (2018)

Business Intelligence for Industrial Revolution 4.0

A Fully Adaptive Image Classification Approach for Industrial Revolution 4.0

Syed Muslim Jameel[1(✉)], Manzoor Ahmed Hashmani[1,2], Hitham Alhussain[1,2], and Arif Budiman[3]

[1] Universiti Teknologi PETRONAS, Sri Iskandar, Perak, Malaysia
{muslim_16000370, manzoor.hashmani, seddig.alhussian}@utp.edu.my
[2] High Performance Cloud Computing Center (HPC3), Universiti Teknologi PETRONAS, Sri Iskandar, Perak, Malaysia
[3] University of Indonesia, Jakarta, Indonesia
arif.budiman21@ui.ac.id

Abstract. Industrial Revolution (IR) improves the way we live, work and interact with each other by using state of the art technologies. IR-4.0 describes a future state of industry which is characterized through the digitization of economic and production flows. The nine pillars of IR-4.0 are dependent on Big Data Analytics, Artificial Intelligence, Cloud Computing Technologies and Internet of Things (IoT). Image datasets are most valuable among other types of Big Data. Image Classification Models (ICM) are considered as an appropriate solution for Business Intelligence. However, due to complex image characteristics, one of the most critical issues encountered by the ICM is the Concept Drift (CD). Due to CD, ICM are not able to adapt and result in performance degradation in terms of accuracy. Therefore, ICM need better adaptability to avoid performance degradation during CD. Adaptive Convolutional ELM (ACNNELM) is one of the best existing ICM for handling multiple types of CD. However, ACNNELM does not have sufficient adaptability. This paper proposes a more autonomous adaptability module, based on Meta-Cognitive principles, for ACNNELM to further improve its performance accuracy during CD. The Meta-Cognitive module will dynamically select different CD handling strategies, activation functions, number of neurons and restructure ACNNELM as per changes in the data.

This research contribution will be helpful for improvement in various practical applications areas of Business Intelligence which are relevant to IR-4.0 and TN50 (e.g., Automation Industry, Autonomous Vehicle, Expert Agriculture Systems, Intelligent Education System, and Healthcare etc.).

Keywords: Industrial Revolution 4.0 · Image Classification · Concept Drift Self Adaptability

F. Saeed et al. (Eds.): IRICT 2018, AISC 843, pp. 311–321, 2019.
https://doi.org/10.1007/978-3-319-99007-1_30

Abbreviations

ACNNELM	Adaptive Convolutional Neural Network Extreme Learning Machine
MOSELM	Meta-Cognitive Online Sequential Extreme Learning Machine
OSELM	Online Sequential Extreme Learning Machine
RD	Real Drift
VD	Virtual Drift
HD	Hybrid Drift
N-Ad	Non-Adaptive
S-Ad	Semi-Adaptive
F-Ad	Fully Adaptive
B	Binary
M	Multi
NI	Non Imaging
SI	Synthetic Image
RI	Real Image

1 Introduction

Big Data is participating in the current computing revolution. Industries and organizations are utilizing its insights for Business Intelligence by using Machine Learning. Deep Learning models have been proven to be a better selection than Shallow Learning models. However, Images are one of the complex types of Big Data. Image data is being utilized for predictive analysis, recommendation systems and decision support systems in variety of applications, for example, Automation Industry, Autonomous Vehicle, Healthcare and Education etc. [1]. Especially, Image Classification through Machine Learning models is mostly used in these systems. Image Classification Models (ICM) train from specific image dataset (training data) and later classify different images (testing data). ICM work fine when there is no change in testing image data (feature inputs or outputs). However, due to non-stationary environment, the feature changes may take place in image data over time thereby yielding the phenomenon of CD (CD). Ignoring the proper handling of CD issue is inevitable and degrades the accuracy of results in ICM to such a level that these models are of no further use.

CD issue frequently takes place in the Online Supervise Learning (OSL) environment. In OSL environment, ICM is trained by image dataset to know the predictor features. The ultimate goal is to assign class labels to testing instances [2] (we categorize images according to generic feature space in a unique class and train our ICM to classify different images accordingly). Starting from the hypothesis model, the aim is to maximize this model towards target function, so that ICM can predict the class of unlabelled images correctly. The training mood, being dependent on the environment in which ICM is deployed, can be batch, online or both. In batch learning, the best predictor, is generated after the training of entire image dataset at once. However, in online learning, the best predictor is generated once the model undergoes successive sequential training after the discretization of image data into sizeable chunks. Due to the non-stationary environment of online learning, the statistical properties of images may vary at different time step. For example, a set of class example has legitimate class labels at one time step and different labels at another time step (called as CD issue) which substantially decreases the performance in terms of accuracy in ICM. To avoid this performance degradation, ICM must have sufficient adaptability feature to self-regulate itself according to the dynamic environment.

1.1 Adaptability in Image Classification Models

The concept of adaptability feature is described as the capability of ML models to self-regulate themselves according to environment. Adaptive ML models are more capable to retain their performance accuracy after CD issue. Existing adaptive approaches can be categorized into semi-adaptive and fully adaptive. In semi-adaptive approaches, dynamic changes are limited and directly apply on core classifier (by improving its fundamental structure). However, fully adaptive approach is based on an autonomous module dedicated to incorporate dynamic selection of required features (by classifier) to tune itself as per the environment.

Meta-Cognitive Approach for Ensuring Adaptability. Meta-Cognitive approach was first introduced by experimental psychologists (Thornas 0. Nelson and Louis Narens). This approach describes, how human mind learns new things and makes logical decisions based on learning. This approach, commonly known as "cognition about cognition", tells what to learn, when to learn and how to learn (Fig. 1).

This approach has been used in many Computer Science disciplines (e.g., Machine Learning). However, the practical implementation of this approach is rarely reported in literature. In Machine Learning (ML), Meta-Cognition (refers as self-consciousness of ML models) not only tells it Cognitive part (main classifier) what to learn, when to learn and how to learn but also improves its adaptability.

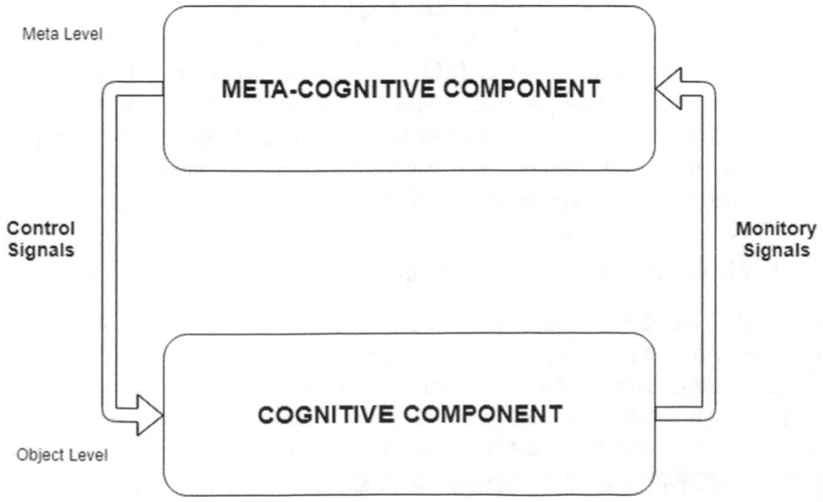

Fig. 1. Thornas 0. Nelson and Louis Narens approach for Meta-Cognition [3].

2 Related Work

Big Data Streams, Online Learning and adaptation to CD have become a few important research topics during the last decade. However, truly autonomous, self-maintaining and adaptive Machine Learning classification models are rarely reported [4]. Many studies focused on CD and adaptation of systems during online sequential environment [5] and urged researchers to further investigate in this direction. After the systematic literature survey in this study, we categorized the existing proposed solution (to handle CD issue in Image Classification Models) as below;

1. Non-Adaptive Shallow Learning Approaches
2. Adaptive Shallow Learning Approaches
3. Non-Adaptive Deep Learning Approaches
4. Adaptive Deep Learning Approaches
5. Adaptive Hybrid Deep Learning Approaches

2.1 Shallow Learning v/s Deep Learning Approaches for CD Handling

Traditional Machine Learning classification models (e.g., Extreme Learning Machine (ELM), Support Vector Machine (SMV), Multi-Layer Perception Neural Network (MLP NN), Hidden Markov Model etc.) handle classification and regression problems efficiently in structured data [6] and are not feasible to handle the large Image dataset [7]. A study argued that finding new means of content-based image retrieval for color image data can be improved using singular value decomposition in Artificial Neural Networks [8]. Deep Learning algorithms are a better selection to handle complex data streams and extract value with more accuracy as compared to conventional approaches.

2.2 Non-adaptive Machine Learning Classification Models Using Single and Ensemble Learning Approach to Handle CD Issue

Classifier Ensemble approach is better than single classifier solution to handle CD [9, 10]. Despite its contribution in different classification areas, it is not sufficient to handle various types of consecutive drifts [11]. The most prominent Machine Learning approaches are single classifier and ensemble learning [12–14]. Various Extreme Learning Machines (ELM) are based on ensemble approach because these approaches provide a solution for CD specific drift case [15]. However, such kind of continuous drift can be handled by means of an adaptive environment [9].

2.3 Adaptive Machine Learning Classification Models

Recently, many adaptive learning approaches and techniques have been introduced [16, 17]. However, these approaches are semi-adaptive in nature and focus on certain types of CD adaptation during online operation. These studies urge to overcome the CD challenge though a more dynamic approach [16]. MOSELM is fully adaptive OSELM (based on additional weighted matrix approach). This model only handles the real CD. However, substantial amount of improvement is still needed in this approach to handle multiple types of CD [13].

2.4 Hybrid Adaptive Machine Learning Classification Models

A recent study proposed an adaptive ML-Model (AOSELM) [4], using single classifier approach based on Online Sequential Extreme Learning Machine (OSELM) and Constructive Sequential Extreme Learning Machine (COSELM), to handle the CD issue of classification problem. This simple solution used a matrix adjustment.. Results were satisfactory for handling Real Drift but not satisfactory to handle Virtual Drift and Hybrid Drift. AOSELM model does not show better results on real dataset. Single classifiers results may not exceed the adaptable ensemble or full batch approach due to their shared weight changes [4]. ACNNELM is a semi-adaptive (based on AOSELM) model that handles all types of CD. ACNNELM combines the Convolutional Neural Network and Extreme Learning Machine. From experimental results, it can be safely stated that ACNNELM is better than all the existing ML models under various Concept Drift conditions.

Table 1. Existing machine learning classification model's scope

Refs.	Technique	Concept Drift			Adaptability			Class		Dataset		
		VD	RD	HD	N-Ad	S-Ad	F-Ad	B	M	NI	SI	RI
[12]	Ensemble integration	✓	×	×	✓	×	×	✓	×	✓	×	×
[18]	AW-SVM	✓	✓	×	×	✓	×	✓	×	✓	×	×
[19]	AUE	✓	×	×	×	✓	×	×	✓	✓	×	×
[20]	WEC-ELM	✓	×	×	×	✓	×	✓	✓	✓	✓	×
[17]	IDS-ELM	✓	×	×	×	✓	×	×	✓	✓	×	×
[16]	Dynamic ELM	✓	✓	×	×	✓	×	✓	✓	✓	×	×
[13]	MOSELM	×	✓	×	×	×	✓	✓	✓	✓	×	✓
[4]	AOSELM	✓	✓	✓	×	✓	×	×	✓	✓	✓	×
[9]	ACNNELM	✓	✓	✓	×	✓	×	×	✓	✓	✓	×

Table 2. Testing accuracy of existing ML models during CD [9].

Dataset	ML model	Testing accuracy
MNIST	OSELM	88.70
Not-MNIST	OSELM	79.02
CIFAR10	OSELM	43.56
MNIST	SVM	89.74
Not-MNIST	SVM	78.05
CIFAR 10	SVM	28.89
MNIST	CNN	90.32
Not-MNIST	CNN	78.18
CIFAR 10	CNN	36.13
MNIST	ACNNELM	94.16
Not-MNIST	ACNNELM	81.18
CIFAR 10	ACNNELM	40.25

2.5 Literature Analysis and Deduction

Traditional Machine Learning Models (e.g., ELM, SVM, and MLPNN) work better for classification of structure data [6]. However, complex structure of image data is proved to be handled better by Deep Learning Classification Models. However, most studies use ELM as a basic classifiers due to its simplicity.

Detecting change in a streaming data has been explored by many researchers in last decade. Many non-adaptive and semi adaptive approaches have been proposed for streaming data classification. However, these classifiers only handle single type of CD but with different frequency (gradual, abrupt or sudden, recurrent and blips). Some work on real and virtual type of CD and categorized them as per their speed and complexity (multi frequencies at same time). But these models work on real dataset streaming. This streaming data is numerical, categorical dataset or mixed datasets. Very few proposed solutions worked on Image Dataset, (e.g., ACNNELM [9] and MOSELM [13]). Unlike MOSELM proposed model (only handles Real Drift), ACNNELM proposed models work on different kinds of CD (Table 1). Through the literature, we can safely state that, the best proposed solution for handling multiple types of CD in Image Classification Models is ACNNELM. However, it needs further improvement in its adapting capability to avoid its performance degradation (Table 2).

3 Fully Adaptive Approach for IR 4.0

In this paper, we propose an autonomous module based on Meta-Cognitive approach which will dynamically modify the existing ACNNELM to handle Concept Drift issue in Image Classification. This proposed model is based on Cognitive (ACNNELM classifier) and Meta-Cognitive part.

3.1 Cognitive Module

Adaptive Convolutional Extreme Learning Machine (ACNNELM) is a multi-parallel hybrid approach based on CNN and ELM. In this study, they studied back propagation and implemented CNN global expansion approach to improve the performance accuracy. To improve the generalization accuracy, they implemented optimal tanh. Stochastic gradient decent method was used to tune kernel weights. In this study, two versions of ACNNELM is proposed to handle CD issue. ACNNELM1 and ACNNELM2 [9].

ACNNELM1: The core of ACNNELM model is AOSELM [4]. ACNNELM1 integrates multiple CNN with a single ELM (Fig. 2).

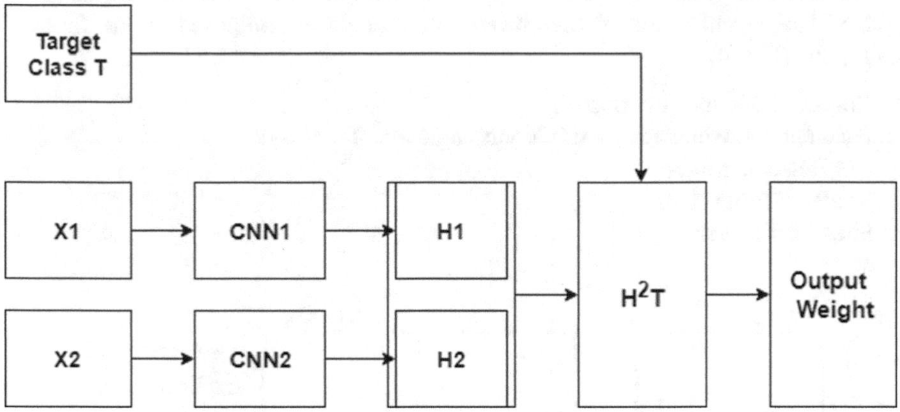

Fig. 2. ACNNELM-1 for CD handling [9]

ACNNELM2: This model integrates multiple parallel CNNELM by using matrices concatenation. As shown in Fig. 3.

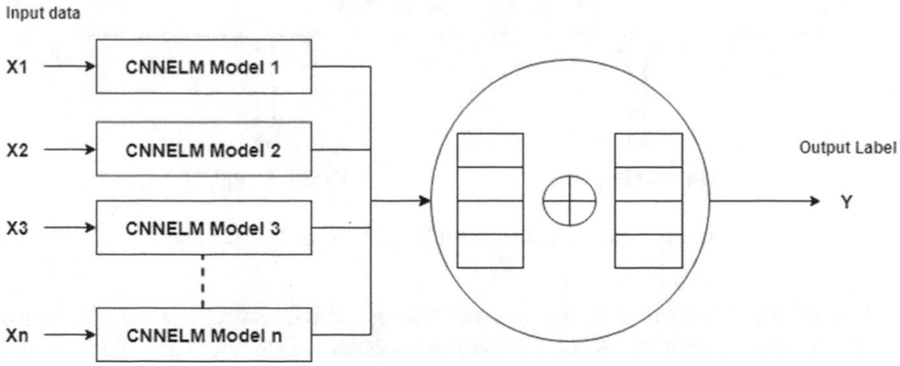

Fig. 3. Multiple CNNELM to boost the performance and CD Handling [9]

3.2 Meta Cognitive Module

In this research paper, we have proposed an autonomous module (Meta-Cognitive module) which dynamically adjusts the core classification module (Cognitive module) of ACNNELM. This autonomous approach increases the level of adaptability in ACNNELM. Architecture of Meta-Cognitive module is composed of the following main parts (Fig. 4):

1. Cognitive Module Interface
2. Dynamic Classifier (Copy of Cognitive Module)
3. CD Detector Module
4. Strategy Controller
5. Strategies Library

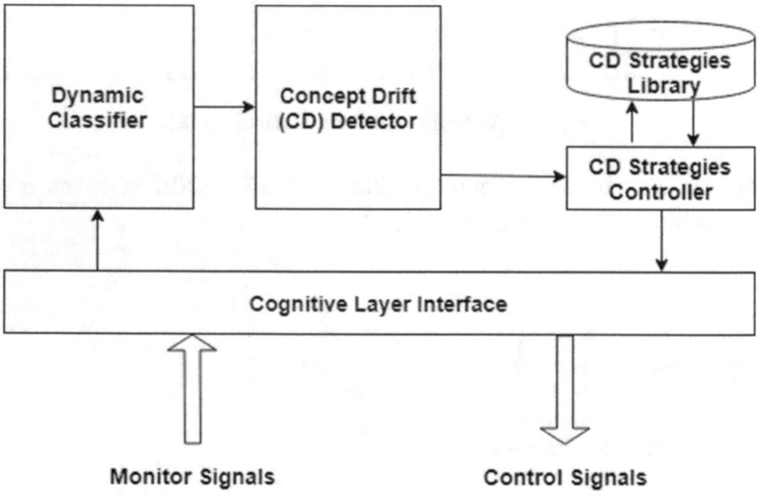

Fig. 4. Meta-Cognitive Module to Improve Adaptability

Cognitive module interface is a gateway through which control and monitor signals transfer between cognitive and Meta-Cognitive modules. Dynamic Classifier is replica of Cognitive module inside the Meta-Cognitive module. Dynamic Classifier is responsible to take the recent sample data and compare it with the previous data sample (identify latest features). CD detector module detects intensity and frequency of CD (Gradual Drift, Sudden Drift and Abrupt Drift), type of CD (Real Drift, Virtual Drift and Hybrid Drift). Strategy controller module is responsible for upgrading the existing strategies according to frequency and type identified by CD Detector. It also sends the appropriate learning strategy to Cognitive module. Strategies library is repository of different CD learning and handling strategies.

How it Works. The new sample data is transferred to Meta-Cognitive module where Dynamic classifier compares existing data sample features (input and output) and new data sample features to identify the difference and sends those details to the CD detector. CD detector analyzes the frequency and type of CD and updates the Strategy Controller to update the existing CD strategies, activation function and number of neurons. Thereafter, Strategy Controller sends the updated strategy to Cognitive module. Finally, the Cognitive module dynamically adapts these recommended changes in it (Figs. 5 and 6).

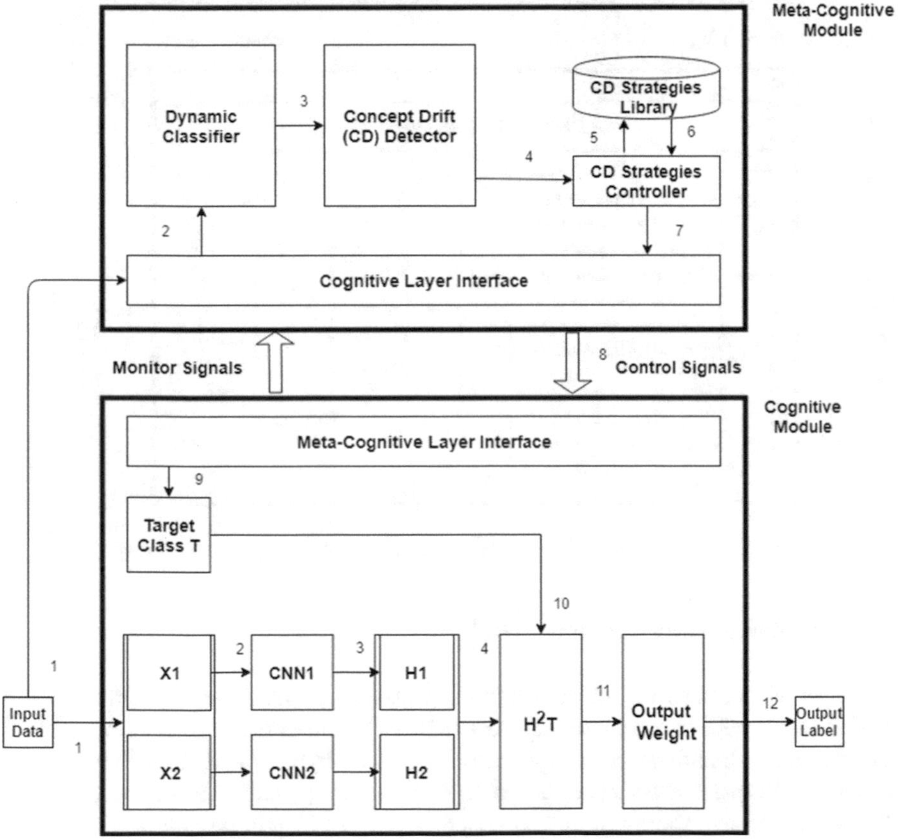

Fig. 5. Meta-cognitive ACNNELM1 model to improve adaptability

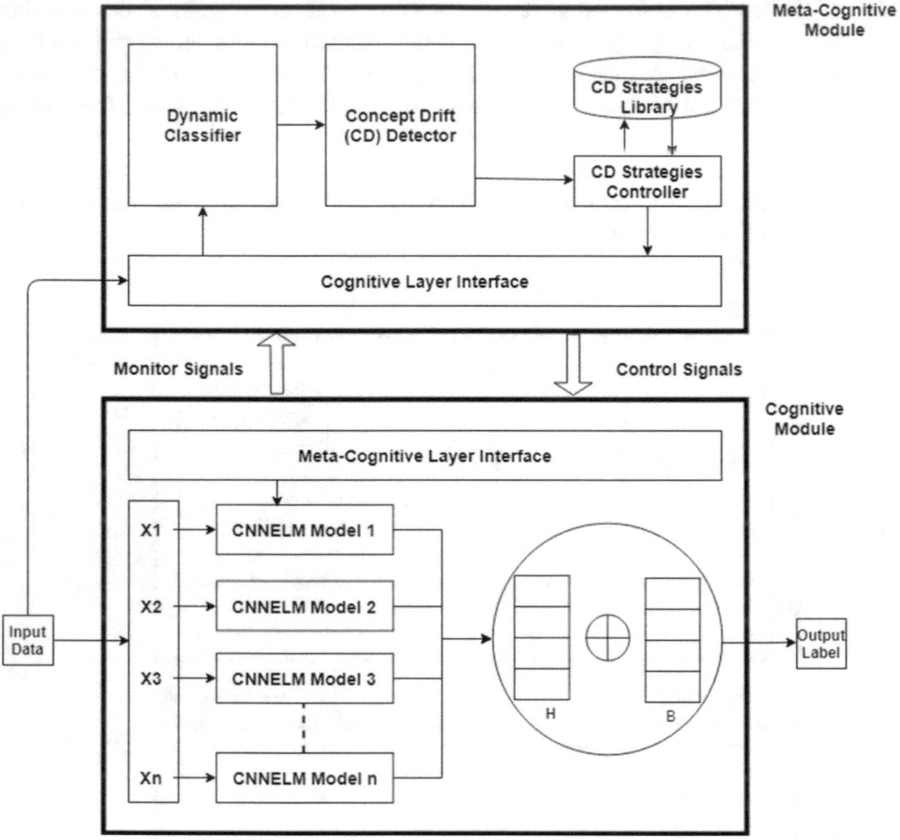

Fig. 6. Meta-Cognitive ACNNELM2 model to improve adaptability

4 Conclusion and Future Work

This paper proposed a fully adaptive approach (based on Meta Cognitive principles) for ACNNELM model. An autonomous part (Meta Cognitive module) is responsible for CD detection, identification of type (real, virtual or hybrid) and frequency (sudden, gradual or abrupt) and selects appropriate CD handling strategy to improve the classification accuracy. Thorough this approach, ACNNELM will be able to dynamically tune itself and adapt to changes. However, some structural changes in ACNNELM are also required along with some special algorithms, to make dynamic changes. ACNNELM-2 is a combination of multiple instances of CNNELM. Therefore, each instance will require different strategies and functions to make the dynamic changes at individual instance level.

In future, we will further investigate the ACNNELM-1 and ACNNELM-2. We will propose detail algorithms to dynamically allocate appropriate weights, tuning parameters, CD strategy, activation function and number of neurons for core classifier. Thereafter, we will implement and evaluate our proposed solution against benchmark dataset.

References

1. Sun, Y., Tang, K., Zhu, Z., Yao, X.: Concept drift adaptation by exploiting historical knowledge. In: IEEE Transactions on Neural Networks and Learning Systems, pp. 1–11 (2018)
2. Kotsiantis, S.B.: Emerging Artificial Intelligence Applications in Computer Engineering, 1st edn. IOS Press, Amsterdam Netherlands (2007)
3. Nelson, T.O., Narens, L.: Meta-memory: a theoretical framework and new findings. Psychol. Learn. Motiv. **26**(1), 125–173 (1990)
4. Budiman, A., Fanany, M.I., Basaruddin, C.: Adaptive online sequential ELM for concept drift tackling. Comput. Intell. Neurosci. **20**, 2016 (2016)
5. Zliobaite, I., et al.: Next challenges for adaptive learning systems. ACM SIGKDD Explor. Newsl. **14**(1), 48 (2012)
6. Huang, G.B., Zhou, H., Ding, X., Zhang, R.: Extreme learning machine for regression and multiclass classification. IEEE Trans. Syst. Man Cybern. **42**(2), 513–529 (2012)
7. Najafabadi, M.M., Villanustre, F., Khoshgoftaar, T.M., Seliya, N., Wald, R., Muharemagic, E.: Deep learning applications and challenges in Big Data analytics. J. Big Data **2**(1), 1 (2015)
8. Hussain S., Hashmani M.: Image retrieval based on color and texture feature using artificial neural network. In: Chowdhry B.S., Shaikh F.K., Hussain D.M.A., Uqaili M.A. (eds.) IMTIC 2012, vol 281. Springer, Berlin
9. Budiman, A., Fanany, M.I., Basaruddin, C.: Adaptive Parallel ELM with Convolutional Features for Big Stream Data. Thesis Dissertation, Faculty of Computer Science, University of Indonesia (2017). https://doi.org/10.13140/rg.2.2.18500.22404
10. Huang, H.K., Chiu, C.F., Kuo, C.H., Wu, Y.C., Chu, N.Y.Y., Chang, P.C.: Mixture of deep CNN-based ensemble model for image retrieval. In: 5th Global Conference on Consumer Electronics, pp. 1–2. IEEE, Kyoto (2016)
11. Kuncheva, L.I.: Classifier ensembles for changing environments. In: Roli F., Kittler J., Windeatt T. (eds.) Multiple Classifier Systems. MCS 2004. LNCS, vol. 3077, pp. 1–15. Springer, Berlin (2004)
12. Tsymbal, A., Pechenizkiy, M., Cunningham, P., Puuronen, S.: Dynamic integration of classifiers for handling Concept Drift. Inf. Fusion Arch. **9**(1), 56–68 (2008)
13. Mirza, B., Lin, Z.: Meta-cognitive online sequential extreme learning machine for imbalanced and concept-drifting data classification. Neural Netw. **80**, 79–94 (2016)
14. Zliobaite, I.: Learning under Concept Drift: An Overview, pp. 1–36. Cornell University Library (2010). arxiv.org/abs/1010.4784
15. van Schaik, A., Tapson, J.: Online and adaptive pseudoinverse solutions for ELM weights. Neurocomputing **149**(Part A), 233–238 (2015)
16. Xu, S., Wang, J.: Dynamic extreme learning machine for data stream classification. Neurocomputing **238**(A), 433–449 (2017)
17. Xu, S., Wang, J.: A fast incremental extreme learning machine algorithm for data streams classification. Expert Syst. Appl. **65**, 332–344 (2016)
18. Krawczyk, B.: Reacting to different types of Concept Drift one-class classifiers. In: 2nd International Conference on Cybernetics, pp. 30–35, IEEE, Gdynia (2015)
19. Brzezinski, D., Stefanowski, J.: Reacting to different types of Concept Drift: the accuracy updated ensemble algorithm. IEEE Trans. Neural Netw. Learn. Syst. **25**(1), 81–94 (2014)
20. Cao, K., Wang, G., Han, D., Ning, J., Zhang, X.: Classification of uncertain data streams based on extreme learning machine. Cogn. Comput. **7**(1), 150–160 (2015)

Success Factors for Business Intelligence Systems Implementation in Higher Education Institutions – A Review

Salamatu Musa(✉), Nazmona Binti Mat Ali(✉), Suraya Binti Miskon(✉), and Mustapha Abubakar Giro(✉)

Faculty of Computing, Universiti Teknologi Malaysia, 81310 Johor Bahru, Johor, Malaysia
salaamat.musa@gmail.com, mustygiro@gmail.com, {nazmona, suraya}@utm.my

Abstract. Considering that organizations are moving towards IT infrastructure, today Business Intelligence (BI) system is the most widely used IT solution. BI is the application of information and specialized tools to support decision making in different organizational environments. Organizations use BI to increase the efficiency of managing information for decision making. Despite the positive effect of BI to organizations, its implementation is frequently plagued with lots of risk, complex processes, drawbacks issues and challenges and the benefit of BI system can only be achieved if the system is successfully implemented. In view of the need and challenge of BI system implementation, this study aims to con-tribute, by proposing the success factors (SFs) of BI implementation for higher education institutions (HEIs). The review period for this study is seven years (2010–2017) and the database used are Science Direct, IEEE, Scopus, Springer link and Google Scholar. This paper serves as a guideline to follow by the HEIs decision makers on the factors that are important to be considered to ensure proper and successful implementation of BI system, this will lead to increase in the chances of successes and reduce the failure rates of BI system implementation in HEIs.

Keywords: Business intelligence · Success factors
Higher education institutions · BI benefit · BI issues

1 Introduction

Business intelligence (BI) has become a significant field of study for practitioners and researchers, indicating the impact and extent of data-related problems that are present in present-day organizations [1]. [2] defined BI as a programmed system that is built to deliver information to the different units of any type of organization.

Presently, BI is deemed as a necessity for all type of organizations because of the rapid growth of data including HEIs [2]. Higher education is considered to be one of the main engines of progress around the world due to its well-known functions of mass tertiary education, academic training and research, and the provision of public service [3, 4].

F. Saeed et al. (Eds.): IRICT 2018, AISC 843, pp. 322–330, 2019.
https://doi.org/10.1007/978-3-319-99007-1_31

Nowadays, to successfully implement an information technology (IT) innovations continues to be both a managerial and a theoretical challenge as several IT implementation projects have a high-risk report and many IT innovations initiated by organizations are either discarded or underused by end users. Even though BI system implementation has a lot of benefits, most of the BI project tends to fail [5, 6]. Various studies confirms that BI systems consumes billions of US dollars annually. Yet, over fifty per-cent of these BI projects are ending up without any benefit [2, 7, 8]. [9] states that the benefits of BI system can only be achieved if it is successfully implemented. So, there is a substantial request to conduct further research to identify more SFs for BI implementation so as to overcome it challenges [5, 10–12].

A number of researchers emphasized that the key reason for failure is the low or lack of deep understanding of the critical success factors that define the success of BI implementation as well as the opportunities and benefits of BI systems [9, 10, 13]. Consequently, there is still inadequate number of successful implemented BI initiatives in higher Education Institutions [2]. There is need to clearly identify the success factors (SFs) that will influence proper and successful BI implementation in HEIs so as to manage all the data efficiently because each data has different requirement and it is very important for the university performance results [1, 14].

Therefore, this study addresses this gap by determining the Success Factors (SFs) of BI system implementation to serve as a guideline to the decision makers of Higher Education Institution (HEIs) on factors that assist in dealing with the challenges of BI systems implementation and by so doing, will lead to increase in the chances of successes and reduce the failure rates of the system implementation.

2 Literature Review

2.1 Business Intelligence (BI)

Researchers defined BI differently [15]. BI can be defined as a system that discovers knowledge from large amount of data using analytical techniques, such as data mining [16]. BI is a set technologies used to convert data into useful information in an organization to improve business performance [17]. Furthermore BI also allow users at any level to have access to data and analyze it according to their needs [18]. [19] indicates that Business Intelligence (BI) involves all the processes of extracting valuable information from large data sets and presenting it to the top management to support them in planning and decision making. These processes are: (1) accessing necessary data; (2) integration of data; (3) cleaning and preparing Data; (4) deciding on machine learning algorithms and techniques; (5) using online analytical processing (OLAP) and data mining tools; (6) performing prescriptive analyses; (7) generating descriptive and predictive reports; and (8) presenting all results in a clear and understandable form.

2.2 Benefits of Business Intelligence in HEIs

The potential benefits of implementing BI systems have been discussed by many academics and IT professionals in different organizational contexts. For instance, with BI system having the capability of combining and analyzing large amount of data, HEIs can get insights of students throughout their studies which can lead to better decision making process and as a result will support the strategic aims and the overall performance [2]. According to [17] BI can enable higher education to timely discover and predict market demands toward improving the graduates' employability and shaping educational goals. With BI application, higher education management and staffs can participate in reality-based decision making which will lead to increase in the overall Performance of the higher education institutions [20].

[21] states that, if BI (known as Educational Intelligence (EI)) system is properly implemented, it can deliver many benefits to HEIs which includes: (1) fast, evidence-based decision making (with readily available and effective analytical data, HEIs personnel can easily act in ways that improve performance and support the overall strategy of the university); (2) present reliable needed information (having all educational data integrated clearly and properly, will allow users to spend less time finding and analysing data to make informed decisions); and (3) EI can maximize the use of student information and data (through the Educational Data Mining (EDM) in EI process, educators achieve a better understanding of the student life cycle, pathways and experience).

Based on [22] Business Intelligence could help higher education institutions to: (1) evaluate the level of success for its programs and also identify recruitment trends; (2) to keep records of the monetary transactions between university and students so as to efficiently manage the finances from tuition fees; (3) provide the appropriate information to government bodies at the right time, in order to enhance their ranks, required funding level and meet compliance requirements; (4) measure the value of the courses offered and know the student enrolment trends to different programs; and (5) provide the decision makers of university with the opportunity to clearly analyse operational data, across functional areas and faculties.

Although, there is massive benefits that a successful BI system can bring to any organization using it, a number of researchers shows that the adoption and usage of BI systems in academic environment is still very low, particularly in HEIs [14, 21].

2.3 Issues of Business Intelligence (BI) Implementation

Many organizations find it difficult to cope with their BI solution as it is complicated and develops continuously [22]. Based on the study conducted by [7] Organizations faces a lot of problems and challenges in implementing or in getting the anticipated results of BI systems and one of the problems is the organizations' readiness for BI systems.

However [23] observed that experts view, is another challenge for BI implementation because experts perceives BI differently. For instance, data mining experts can view BI system as an advance support system with data mining techniques and algorithm while data warehouse experts may regard BI as a supplementary system and statisticians can consider BI as a predictive analysis-based tool. With different views of experts, it becomes hard for organizations to invest in BI.

[24] identified two issues in the universities; the first issue is the increasing amount of educational data in an institution. Making the filtering processes of the important data to become more difficult to manage and handle. Second issue stated by the authors is the cost needed to implement BI system in the university. As a result, it is necessary to discover the factors that facilitate the successful implementation of BI system so that all the resources involve will not be wasted when it is not successful [20, 24].

The key issue of BI implementation, is the organization's low level of knowledge about the opportunities and benefits of BI systems as well as about their critical success factors [13]. Likewise [10] noted that, despite the vibrant BI market and the complexities surrounding the implementation of BI systems, the critical success factors (CSFs) of BI system implementation remain poorly understood.

2.4 Success Factors (SF)

Several studies have addressed the success factors for BI systems using different terms such as: critical success factors (CSFs), implementation success factors (ISFs), success factors (SFs), readiness factors (RFs), key success factors (KSFs) or implementation factors (IFs) [7, 13]. Basically, these terms refer to the factors that influence the success of BI systems implementation [7, 13]. In BI system context, SFs can be viewed as a set of tasks and procedures, which are either adopted or planned, that should be taken to make sure BI system benefits are achieved [25].

A number of researchers have examined the success factors for BI implementations [20, 22]. For instance, Top management support and involvement is found to be a critical success factor in hospital-based business intelligent system (HBIS) development [1]. For hospital-based business intelligence system (HBIS) implementation [16] suggested some elements to be considered such as business processes, organizational culture, governance, resources, policy, organizational environment, people skills and the technology.

A cross-case analysis conducted by [5] reveals that organizational factors are the most important factors that determines BI system implementation success using Engineering asset management organizations. The research findings also point out that those enterprises which had committed management support and sponsorship from the business side and a clear vision and a well-established business case are more likely to have a successful BI system implementation.

[26] states that perceived usefulness and perceived ease of use are considered to be fundamental factors in deciding the adaptation of any innovation. Based on the study conducted by [25] on Vietnamese companies, seven factors were identified as the most important; Committed management support and sponsorship; A clear vision and a well-established business case; Business-centric championship and a balanced team composition; Business-driven and iterative development approach; User-oriented change management; Business-driven, scalable and flexible technical framework; Sustainable data quality and integrity.

Likewise, [25] asserts that SFs are not the same between industries, even in the same sector. Even though there are number of studies that discourse the CSFs for BI systems, there is still need to investigate, examine and validate more CSFs in different sectors [5, 10–12].

3 The Suggested BI Success Factors (SFs) for HEIs

This study adopted the framework of [10] in order to provide a cleverer understanding of the SFs for BI systems as the authors adopted the extended framework of [27] which is the most widely used and agreed CSFs framework for BI implementation success by many researchers, as supported by [8, 13]. Also, it is a recent research with two additional factors (Organizational structure and User Empowerment) and the framework is yet to be validated. Table 1 displays the BI success factors (SFs) for HEIs.

Table 1. Success Factors (SF) of BI implementation for HEIs

Success factors	Authors
Organizational factors	
Top management support	[5, 9, 10, 13, 28, 29]
Clear vision	[5, 9, 10, 13, 28, 29]
Organizational structure	[9, 10]
User empowerment	[10, 12]
Process factors	
Championship and balanced team	[5, 10, 25, 27]
User participation	[10, 28–30]
Technological factors	
System quality	[5, 9, 12, 25, 27, 30]
Information quality	[5, 9, 12, 26, 29, 30]
Service quality	[5, 9, 12, 26, 29, 30]

The success factors (SFs) are grouped into three dimensions, organizational, process and technological dimensions. However, the difference between this study with [10] is the technological factors of this study includes system quality, information quality and service quality as shown in Table 1 which were the quality constructs (BI success) in the framework of [10]. This study included the quality constructs under technological factors as system flexibility and system integration falls under the characteristics of system quality as supported by [5, 9]. The Information Quality and Service Quality are considered to be fundamental factors in deciding the adaptation of any innovation [26]. Furthermore, information quality, system quality and service quality are regarded as important dimensions of BI implementation success [31]. Thus, the BI success factors for this study are described as follows:

(1) **Organizational Factors**

(a) **Top Management Support**

A number of scholars have widely recognized the engagement, commitment and support by top management as an essential success factor for BI implementation [5, 9, 10, 32]. Furthermore, [8] affirmed that top management commitment and support to BI project plays an important role in ensuring the success of BI implementation by allocating resources and by also overcoming the challenges of the new change in the organizational environment.

(b) **Clear vision and well-defined plan**

According to [10] establishing a clear vision and well defined plan is needed during all stages of implementing BI system. It is necessary to have a clear strategic vision and plan in order to align the objective and needs of the organization with BI system and direct the implementation towards achieving the set goals [32].

(c) **Organizational structure**

[10] defined organizational structure is a formal chart that controls the internal communication and relationship, information flow and reporting and allocates authority, activities and tasks to employees within a particular organization. However, [11] confirms that organizational structure is very important in success of BI implementation.

(d) **User Empowerment**

User empowerment is the perception or psychological attitudes of individuals towards there organizational roles and jobs. it involves sufficient training, information and confidence to enable the employees to be responsible for their decisions and actions [10]. Although few studies addressed the influence of BI system implementation, it was observed that user empowerment have a positive impact on BI system success.

(2) **Process Factors**

(a) **Championship and Balanced Team**

In business intelligence context, championship is described as BI project team and manager that have a strong leadership skills and adequate technological and business knowledge in order to have a successful BI implementation [10]. Several studies agrees that having a strong leadership and well balanced team plays a major role in the success of BI implementation.

(b) **User Participation**

User participation refers to the engagement of employees in all the stages of BI implementation from different levels of the organization. Previous studies indicates that user participation is one of the important success factors for BI system implementation [10].

(3) **Technological Factors**

(a) **System Quality**

System quality describes the performance and usability characteristics of the system in terms of integration, customization, accessibility, convenience, and interactivity, flexibility, scalability and response time [27, 31].

(b) **Information Quality**
Information quality describes consistency, accuracy, usefulness, timeliness, completeness, availability and the relevance of the information provided by the system [26].

(c) **Service Quality**
Service quality focus on the quality of support that the system pro-vide to the users in term of empathy, assurance, tangibles, reliability and responsiveness [27]. It also explains the extent to which users believe that utilizing a system will reduce the mental and physical efforts requires to do a task [12].

4 Conclusion

With a successfully implemented BI system, higher education management and staffs can participate in reality-based decision making which will lead to increase in the overall Performance of the higher education institutions. After examining the previous research work, the study come up with nine (9) success factors (SF) which are considered necessary for the implementation of BI system in HEIs. Therefore, this paper serves as a reference for IT managers, BI stakeholders of higher education institutions to use as a guideline to follow when implementing BI system and by doing so, will lead to increase in the chances of success and reduce the failure rate of the system implantation. Furthermore, this study contributes to the literature as it raises the awareness of the success factors of BI system implementation in HEIs. The limitation of this study is that, it was based only on literature review. Further research may focus on investigating and validating the identified success factors within the higher education institution context.

References

1. Kao, H.Y., Yu, M.C., Masud, M., Wu, W.H., Chen, L.J., Wu, Y.C.J.: Design and evaluation of hospital-based business intelligence system (HBIS): a foundation for design science research methodology. Comput. Hum. Behav. **62**, 495–505 (2016). https://doi.org/10.1016/j.chb.2016.04.021
2. Apraxine, D., Stylianou, E.: Business intelligence in a higher educational institution: the case of University of Nicosia. In: 2017 IEEE Global Engineering Education Conference, pp. 1735–1746 (2017). https://doi.org/10.1109/educon.2017.7943085
3. Rodzi, N.A.H.M., Othman, M.S., Yusuf, L.M.: Significance of data integration and ETL in business intelligence framework for higher education. In: Proceedings of the 2015 International Conference on Science in Information Technology Big Data Spectrum for Future Information Economy, ICSITech 2015, pp. 181–186 (2016). https://doi.org/10.1109/icsitech.2015.7407800
4. Sanchez-Puchol, F., Pastor-Collado, J.A., Borrell, B.: Towards an unified information systems reference model for higher education institutions. Procedia Comput. Sci. **121**, 542–553 (2017). https://doi.org/10.1016/j.procs.2017.11.072
5. Yeoh, W., Popovič, A.: Extending the understanding of critical success factors for implementing business intelligence systems. J. Assoc. Inf. Sci. Technol. **67**, 134–147 (2016)

6. Fadhil, A.: Implementation issues affecting the business intelligence adoption in public university. ARPN J. Eng. Appl. Sci. **10**, 18061–18069 (2015)
7. Anjariny, A.H., Zeki, A.M.: The important dimensions for assessing organizations' readiness toward business intelligence systems from the perspective of Malaysian organization. In: Proceedings of the 2013 International Conference on Advanced Computer Science Applications and Technologies, ACSAT 2013, pp. 544–548 (2014). https://doi.org/10.1109/acsat.2013.113
8. Solberg, K., Adamala, S.S., Cidrin, L.: Key success factors in business intelligence. J. Intell. Stud. Bus. **1**, 107–127 (2011)
9. Sangar, A.B., Iahad, N.B.A.: Critical factors that affect the success of business intelligence systems (BIS) implementation in an organization. Int. J. Sci. Technol. Res. **2**, 176–180 (2013)
10. Magaireah, A.I., Sulaiman, H., Ali, N.: Theoretical framework of critical success factors (CSFs) for Business Intelligence (BI) System. In: 2017 8th International Conference on Information Technology, pp. 455–463 (2017). https://doi.org/10.1109/icitech.2017.8080042
11. Anjariny, A.H., Zeki, A.M., Hussin, H.: Assessing organizations readiness toward business intelligence systems: a proposed hypothesized model. In: Proceedings of the 2012 International Conference on Advanced Computer Science Applications and Technologies, ACSAT 2012, pp. 213–218 (2013). https://doi.org/10.1109/acsat.2012.57
12. Kfouri, G., Skyrius, R.: Factors influencing the implementation of business intelligence among small and medium enterprises in Lebanon. Inf. Moksl. **76**, 96–110 (2016)
13. Olszak, C.M., Ziemba, E.: Critical success factors for implementing business intelligence systems in small and medium enterprises on the example of Upper Silesia, Poland. Interdiscip. J. Inf. Knowl. Manag. **7**, 129–150 (2012)
14. Zulkefli, N.A., Miskon, S., Hashim, H., Alias, R.A., Abdullah, N.S., Ahmad, N., Ali, N.M., Maarof, M.A.: A business intelligence framework for higher education institutions. ARPN J. Eng. Appl. Sci. 10 (2015)
15. Bonney, W.: Applicability of Business Intelligence in Electronic Health Record. Procedia Soc. Behav. Sci. **73**, 257–262 (2013). https://doi.org/10.1016/j.sbspro.2013.02.050
16. Brooks, P., El-Gayar, O., Sarnikar, S.: A framework for developing a domain specific business intelligence maturity model: application to healthcare. Int. J. Inf. Manag. **35**, 337–345 (2015). https://doi.org/10.1016/j.ijinfomgt.2015.01.011
17. Elhassan, I., Klett, F.: Bridging higher education and market dynamics in a business intelligence framework. In: Proceedings of the 2015 International Conference on Developments in eSystems Engineering, DeSE 2015, pp. 198–203 (2016). https://doi.org/10.1109/dese.2015.22
18. Akhmetov, B., Izbassova, N., Akhmetov, B.: Developing and customizing university business intelligence cloud. In: Proceedings of 2012 International Conference on Cloud Computing Technologies, Applications and Management, ICCCTAM 2012, pp. 229–233 (2012). https://doi.org/10.1109/iccctam.2012.6488104
19. Silahtaroğlu, G., Alayoglu, N.: Using or not using business intelligence and big data for strategic management: an empirical study based on interviews with executives in various sectors. Procedia Soc. Behav. Sci. **235**, 208–215 (2016). https://doi.org/10.1016/j.sbspro.2016.11.016
20. Kumaran, S.R., Othman, M.S., Yusuf, L.M.: Applying theory of constraints (TOC) in business intelligence of higher education : a case study of postgraduates by research program. In: International Conference on Science Information Technology, pp. 147–151 (2016). https://doi.org/10.1109/icsitech.2015.7407794

21. Gorgan, V.: Requirement analysis for a higher education decision support system. Evidence from a Romanian University. Procedia Soc. Behav. Sci. **197**, 450–455 (2015). https://doi.org/10.1016/j.sbspro.2015.07.165
22. Wanda, P., Stian, S.: The secret of my success: an exploratory study of business intelligence management in the Norwegian industry. Procedia Comput. Sci. **64**, 240–247 (2015). https://doi.org/10.1016/j.procs.2015.08.486
23. Yusof, A.F., Miskon, S., Ahmad, N., Alias, R.A., Hashim, H., Abdullah, N.S., Ali, N.M., Maarof, M.A.: Implementation issues affecting the business intelligence adoption in public university. ARPN J. Eng. Appl. Sci. 10 (2015)
24. Abdul Aziz, A., Jusoh, J.A., Hassan, H., Wan Idris, W.M.R., Md Zulkifli, A.P., Mohamed Yusof, S.A.: A Framework for educational data warehouse (EDW) architecture using business intelligence (Bi) technologies. J. Theor. Appl. Inf. Technol. **69**, 50–56 (2014)
25. Jatana, N., Suri, B., Misra, S., Kumar, P., Choudhury, A.R.: Critical success factors for implementing business intelligence system: empirical study in Vietnam. In: Iccsa 2016, vol. 9790, pp. 585–594 (2016). https://doi.org/10.1007/978-3-319-42092-9
26. Chen, M.: Applying business intelligence in higher education sector: conceptual models and users acceptance (2012)
27. Yeoh, W., Koronios, A.: Business intelligence systems University of South Australia. J. Comput. Inf. Syst. **50**, 23–32 (2010). https://doi.org/10.1109/sisy.2012.6339583
28. Dawson, L., Van Belle, J.-P.: Critical success factors for business intelligence in the South African financial services sector. SA J. Inf. Manag. **15**, 1–12 (2013). https://doi.org/10.4102/sajim.v15i1.545
29. Anjariny, A.H., Zeki, A.M.: Management dimension for assessing organizations' readiness toward business intelligence systems. In: Proceedings of the 3rd International Conference on Advanced Computer Science Applications and Technologies, ACSAT 2014, pp. 21–25 (2014). https://doi.org/10.1109/acsat.2014.11
30. Mudzana, T., Maharaj, M.: Prioritizing the factors influencing the success of business intelligence systems: a Delphi study. Indian J. Sci. Technol. **10**, 1–6 (2017). https://doi.org/10.17485/ijst/2017/v10i25/99981
31. Hawking, P., Sellitto, C.: Business Intelligence (BI) critical success factors. In: ACIS 2010 Proceedings of the 21st Australasian Conference on Information Systems 11 (2010)
32. Mungree, D., Rudra, A., Morien, D.: A framework for understanding the critical success factors of enterprise business intelligence implementation. In: Proceedings of the 19th Americas Conference on Information Systems 1–9 (2013)

A Review on Exploiting Social Media Analytics for the Growth of Tourism

Lim Xiao Yan and Preethi Subramanian(✉)

Asia Pacific University of Technology and Innovation,
57000 Kuala Lumpur, Malaysia
TP048635@mail.apu.edu.my, dr.preethi@apu.edu.my

Abstract. Advances in technology has led to absolutely novel and creative areas for its applications. The popularity of social media has seen the stunning growth of user-generated content that could be potentially useful but yet under-utilized. Analysis of the social media data could produce valuable insights for corporations by performing data analytics on consumer behaviours and predict the market trend. In the recent years, tourism is one of the fastest growing sector in the world and United Arab Emirates (UAE) is deemed as one of the popular travel destination worldwide. This paper focuses to view the growth of tourism sector in UAE, its potential and avenue for growth in the future and most importantly the utilization of technology to contribute to business improvement. It discusses about how social media analytics can improve the competitiveness in tourism and hospitality industry in UAE by discussing the technologies used for social media analytics. It recommends to employ social media analytics for the opportunities it provides towards business improvement as the utilization of analytics can be a turn-around factor to improve business performance.

Keywords: Social media analytics · Tourism · Text analysis
Natural language processing toolkits

1 Introduction

Social Media is a platform that helps the creation and sharing of data, ideas, career interests and other phases of expression via virtual communities and networks. The main source of information is generated from user-generated content in the form of text, pictures and videos [1]. Research found that over 65% of the adults in America are using social networking media and there has been a tremendous increase in the number of users from 10% in 2005 up to 65% in 2015 [2]. The increase in such user-generated content creates opportunities for beneficial analysis, especially in the emerging field of big data analysis. Big Data is the information asset characterized by high volume, velocity and variety of data that requires specific technology and analytical methods for its transformation into value [3]. The big data that can be acquired from social media such as Facebook, Instagram and Twitter have not been utilized to its complete potential. Online booking and review platforms such as TripAdvisor, Booking.com also provides immense data for analysis. Its ability to give insights on customer behaviour by analysing massive and varied amount of data can be deemed as a competitive advantage for many organizations [4].

F. Saeed et al. (Eds.): IRICT 2018, AISC 843, pp. 331–342, 2019.
https://doi.org/10.1007/978-3-319-99007-1_32

On the other hand, Tourism is a major industry that contributes to the economy of the countries. This can be evidenced from the fact that in 2016, it contributed for a direct GDP of $2.3 trillion globally. The GDP contributed by tourism sector is now greater than that of the automotive and chemical manufacturing sector. Due to this growth, it also creates more than 225 million jobs worldwide [5]. Several countries invest heavily in promoting their tourism because of these advantages. United Arab Emirates (UAE) is one of those countries which is seen investing in their tourism sector by building luxury hotels and exotic travel attractions. The GDP contribution by the UAE tourism sector was 12.1% in 2016 and forecasted to increase by 2.9% in 2017. The growth of this sector also supports the country's employment rate by contributing 10.4% of the total employment in the country [6]. Leveraging from the growth of tourism industry and development of Social Media Analytics (SMA), the synchronisation of these can lead to improving the industry standards. Therefore, the research problem for this paper is to look at how SMA can provide competitive advantage to the hospitality and tourism industry together with different methodologies that can be performed on SMA with the objective to obtain insights for business improvement.

The literature search was conducted in Google Scholar with keywords such as smart tourism, tourism in UAE and social media analytics. The search results are limited to last 3 years in order to get the most relevant journal articles that comprises journals from SCOPUS and IEEE database. Other statistic researches were also conducted to obtain facts and figures that supports the objective of the research.

2 Environment and Current Circumstances

This section is divided into 3 parts, starting with tourism and hospitality that describes the concept on Smart Tourism and hospitality future in UAE including the emergence of Halal tourism. The following section on SMA will focus on evolution and limitations of the analysis. The actual case studies of implementation and results of SMA in the hospitality and tourism industry are discussed. Lastly, this section is concluded with the discussion on the methodologies that can be employed to perform SMA.

2.1 Growth of Tourism in UAE

Tourism sector is reaching new heights and along with it has emerged the concept of Smart Tourism. With the integration of smartphones and tourism, it will enable the transformation of data into on-site experience [7]. Smart cities are highly related to smart connected tourism as it will help to bridge the gap between human behaviour and systems. For example, the London Underground App provides the users real time information of the London tube's traffic. When there is any closure or service on any route, the app provides real time feedback to all the parties concerned. Besides that, it can also guide a tourist on the best route to take from one station to another. This will help to improve tourism as it simplifies the transportation aspect and provides convenience to the travellers. Combining this with big data analysis, the businesses can improve service roll-out for peak hour rides, backup plans for accidents or breakdowns. According to Turner and Freiermuth's report, UAE is forecasted to attract 19,131,000 international tourist arrivals in 2017 and it will continue to increase at 5.0% p.a. until 2027 [6]. In 2018, many investments are being done on tourism and hospitality industry in UAE and Saudi Arabia in order to prepare for Expo 2020 Dubai. New projects like UAE Bluewater Island and Warner Brothers Theme Park are also attracting many major hotel groups to invest in this region [8]. From this information, it can be derived that the tourism and hospitality industry in UAE is going to face fierce competition among major business groups. The major hotel groups will need to compete against each other so that they succeed in this market by attracting more customers. In order to survive the competition, the hotels would concentrate immense efforts in their advertising and promotion campaigns.

Furthermore, the growth of the airline industry will also aid in promoting the region's tourism sector as it provides better access for visiting this region. There are 3 major airlines that are currently headquartered in the UAE; Emirates, Etihad and Qatar Air. These airlines will provide ease of access to the country thereby promoting tourism and indirectly contributing to the increase in GDP. For example, most of the Emirates flights from Asia to Europe transit in Abu Dhabi or Dubai. This contributes to additional tourism for the country as the transit time is most probably spent in visiting prominent places. In the recent years, the growth of Low Cost Carriers (LCC) have also contributed to the growth of tourism all over the world as the costs for travel have been minimized [9].

Lastly, with the increase of Muslim travellers in the recent years, halal tourism can be seen as an emerging opportunity. In the Crescent Rating Global Muslim Travel Index, UAE has been ranked at the second place for the Muslim traveller destination mainly due to its air connectivity, travel safety and ease of access of prayer space. Therefore many hoteliers in UAE are creating special packages to attract Muslim travellers [10, 11]. However, there is also a downside of halal tourism; that is the Ramadan period. During the fasting month most of the Muslim will stay into cleanse themselves in preparation for the Islamic Holiday. Therefore during the Ramadan period, the sales will drop drastically as people minimize other activities [12]. Based on the above facts, the review on tourism shows that tourism industry has been expanding rapidly in these few years. This has been supported by the growth on airlines industry, smart technologies, investment on tourism, emergence of Halal tourism and Muslim travellers. UAE is predicted to have a huge inflow of tourist in the future and big data analytics are necessary to help companies to gain competitive advantage over the others. Although, there are some downside where UAE are Islamic countries and currently terrorism issues are still quite serious. That might be a slight hindrance for the travellers in making their decisions. However, the World Travel and Tourism Council has also reported that there will be a continuous increase of tourist to UAE based on their survey results, therefore it might not have much negative impact. With the help of big data analytics, hoteliers can gain insights of customer behaviours and trending activities; these insights will support the hoteliers to design better advertisement campaigns and improve business operations.

2.2 Social Media Analytics in Action

The steady increase in the usage of social media platform has increased the amount of data that can be retrieved and analysed. Current trends and worldwide issues can be identified and therefore many companies are also investing into social media marketing. The definition of Social Media Marketing by Chan and Guillet is that it is basically a transposition of general marketing techniques into social media context [13]. Sharma and Paulsingh created an online sport platform for advertisement and it is focused on a specific target audience [14]. By creating a specific platform to attract a precise target audience, surveys and advertisements can be utilized to gather data on customer preferences. Then the information gathered from the platform can be utilized by the sports company to make better decision as a result of the better understanding of their consumer needs.

Social Media Analytics has been employed by several businesses as a pilot study and the effects have been tested. Vecchio [4] did a research simulation on Smart Tourism destination where the researcher has taken the official hashtags created by the official tourism council for a travel destination for analysis. Using business analytics tool, Keyhole, cluster analysis was performed whereas Buzztrack tool was used for sentiment analysis and social media monitoring. The insights gathered were employed to create a dashboard that enables the tourism council to understand what tourists feel about the attractions. These analyses also aided in the improvement plans as the findings were an accurate representation of customer feedback. The research results were very useful to the local tourism council; however, the limitation of this study was the number of official hashtags selected as a base to represent the whole southern Italian region. Hence, Apulia might not reflect the full traveller opinion on the tourist destination. Another case study has been done on a hotel chain, where the customer reviews were studied and employed the insights to improve the service recovery like solving customer complaints and enhance stay experience [14]. This research selected the techniques of ethnography and netnography. Ethnography is a qualitative content analysis method where later the result can be converted to quantitative data type, whereas netnography is an online research methodology which performs the research based on reviews in TripAdvisor.com. The hotel chain was able to track negative reviews from TripAdvisor and find out the main factor for the complaints and employed them to provide guidance for service improvement. On the other hand, research has also been done on using Twitter message for geo-tagging analysis in San Francisco in relation to smart tourism. The insights generated matched well with the popular tourist destination in San Francisco, therefore it can be replicated to usage in other countries [15]. A different research has also been done on getting insights on tourist behaviour and forecast future demand for better planning by using visual photo context to rank, locate and identify tourist sites [16]. These researches and business case studies have proven the effectiveness of SMA and the methods of utilization to gain insights thereby increasing the competitive advantage for companies.

Nevertheless, there will be some challenges in the implementation of SMA such as the huge data volume. With high volume and velocity of data, it raises the challenges of analytics to a higher level. These are some challenges for SMA as it will be hard to range down to topic, difficulty on getting high quality data and data visualization. Lastly, trend monitoring in Facebook is still under-research which could make it challenging to anticipate and create relevant marketing campaigns [17]. Other than that, there are also other limitations of using social media marketing, that is the lack of feedback control and limitation of target crowd [18]. For instance, industry competitor might use negative post to damage their competitor's brand image and marketing campaign. Furthermore, usage of social media marketing campaigns might not attract much new customers for the company. This is because usually only the current loyal customer will follow your company postings and the chances of the Ads reaching potential new customers might be slim due to the huge number of digital ads available at the moment. In short, the usage of SMA are quite substantial as it helps to provide facts and figures for the management to make important decisions. Multiple case studies have proven that it is widely adopted by the tourism industry and have gain significant benefit from the analysis that are generated. Studying the effect of negative reviews, using geo-tagging and social media visual analysis are some of the ways to gain insights of tourist behaviours. Although there are some limitations on analysing social media data because the posts and comments are not subject to much control and validation, it can be overcome by employing careful data pre-processing techniques as the majority of the user generated content still represents the opinion of the general population.

2.3 Techniques for Business Improvement

There two major techniques to perform social media analytic, which are Sentiment Analysis and Opinion Mining. Sentiment analysis is a series of methods, techniques, and tools about detecting and extracting subjective information, such as opinion and attitudes from language [19]. Opinion mining is defined as a technique which classify text into categories which express positive or negative sentiment [20]. For example, sentiment analysis is commonly performed on reviews posted in the internet. Then from there it can be evaluated whether the reviews are more inclined to positive, neutral or negative opinions. Using sentiment analysis, company can gauge the popularity of their product from the general users.

Mäntylä, Graziotin and Kuutila discovered that most of the research done with sentiment analysis is mainly on keywords such as "social", "online", "media" and "review". The research also reflects the popularity of social media analytics has increased significantly since 2015 where a total of 5699 papers were published in SCOPUS [21]. This is mainly due to the increased usage of online product reviews starting in 2005, thus bringing up the popularity of sentiment analysis. The finding also ties up with the keywords "online" and "review" in the popular research topic in sentiment analysis field. Another research had also been done by Vyas and Uma [22], where they have analysed the majority of the sentiment analysis tools available. The summarization of the different application and extensions of each tool can be viewed from Table 1 [22].

Table 1. Categorization of text analysis tools

Text analysis tool	Applications support	Web source	Extension for sentiment analysis	Related research evidence
SAS text miner	Sentiment analysis, NLP	https://www.sas.com	Nil	[23, 24]
Sysomos	Social media monitoring, text analysis	https://sysomos.com	Media Analysis Platform (MAP)	[25]
Rapid miner	Social media analysis, market search, NLP, text analytics	https://rapidminer.com	Sentiment analysis	[26, 27]
Clarabridge	Social media analysis, sentiment analysis, text analysis	www.clarabridge.com/text-analytics	Nil	
Bitext	Sentiment analysis, concept extraction, text analysis	https://www.bitext.com/	Nil	[28, 29]
Etuma	Social media monitoring, sentiment analysis	https://www.etuma.com/	Nil	
Synapsify	Social media monitoring, text mining	www.gosynapsify.com	Snapify core API	
Medallia	Social media monitoring, text mining	www.medallia.com	Nil	[30, 31]
Abzooba	Social media monitoring, text mining	www.abzooba.com/	XPRESSOInsight	
General sentiment	Sentiment analysis, text mining, social media analytics	www.generalsentiment.com	Nil	
Buzzlogix	Sentiment analysis, social media monitoring	https://buzzlogix.com	Nil	

From Table 1, it can be seen that most of the sentiment analysis tools support social media analytics as well. Hence, it can be derived that the intention of building the tools are mostly for social media analytics usage. Some of these tools selected by popular researches were open source and also paid programs as the major emphasis was on the functionality. In Table 1, column "Related Research Evidence" shows utilization of the respective tools for recent researches conducted under tourism and hospitality area. There are also some tools are offer additional extensions like API and platform for increasing the user usage experience. Rapid Miner tool was selected to perform polarity testing on tweets with Support Vector Machine (SVM), Decision tree and Naive Bayes classification methods. Rapid Miner tool has the advantage of portability and user-friendly interface and therefore it was chosen to conduct the test, where SVM has shown a better accuracy (79.08%) compared to the other classification methods [22].

Opinion mining technique required several pre-processing steps, it is needed for text restructuring and extraction purpose. Tokenization and word segmentation are part of the data pre-processing steps, which needs to be followed by Part of Speech (POS) tagging and parsing. An example for the usage of Opinion Mining would be a review of "The restaurant serves great food". In this example, we can know that the entity is a restaurant and it is a positive sentiment from the word "great". There are several toolkits available for Natural Language processing as listed and explained in Table 2 [32]. But, most of the sentences and words employed in social media analytics could be more complex, have the need for complex pre-processing. Machine Learning plays an important part for Opinion Mining, algorithms such as SVM, Decision tree, Naive Bayes and others play a significant role in the process of designing a tool. Machine learning works by training classifiers to determine the polarity of text [33]. Subsequently the dependency tree-based classification method was proposed to take account of word orders and syntactic relations between words extracted, from that, frequent-word subsequence and dependency subtrees from sentences can prune them to create subtrees for classification.

Table 2. Natural language processing toolkits

Toolkit	Language	Description
NLTK	Python	Natural Language Toolkit (NLTK) is an open source platform for performing NLP tasks including tokenization, stemming, POS tagging, parsing, and semantic reasoning. It provides interfaces for many corpora and lexicons which are useful for opinion mining and sentiment analysis. http://www.nltk.org/
OpenNLP	JAVA	The Apache OpenNLP is a JAVA library for the processing of natural language texts, which supports common tasks including tokenization, sentence segmentation, POS tagging, named entity recognition, parsing, and coreference resolution. https://opennlp.apache.org
CoreNLP	JAVA	Stanford CoreNLP is a framework which supports not only basic NLP task, such as POS tagging, named entity recognition, parsing, coreference resolution, but also advanced sentiment analysis http://stanfordnlp.github.io/CoreNLP/
Gensim	Python	Gensim is an open source library for topic modelling which includes online Latent Semantic Analysis (LSA), Latent Dirichlet Allocation (LDA), Random Projection, Hierarchical Dirichlet Process and word2vec. All implemented algorithms support large scale corpora. LSA and LDA have distributed parallel versions. http://radimrehurek.com/gensim/

(*continued*)

Table 2. (*continued*)

Toolkit	Language	Description
FudanNLP	JAVA	FudanNLP is an open source toolkit for Chinese NLP, which supports word segmentation, POS tagging, named entity recognıtıon, dependency parsıng, coreference resolution and so on. https://code.google.com/archive/p/fudannlp/
LTP	C++/ Python	The Language Technology Platform (LTP) is an open source NLP system for Chinese, including lexical analysis (word segmentation, POS tagging, named entity recognition), syntactic parsing and semantic parsing (word sense disambiguation, semantic role labelling) modules. http://www.ltp-cloud.com/intro/en/
NiuParser	C++	NiuParser is a Chinese Syntactic and Semantic Analysis Toolkit, which supports word segmentation, POS tagging, named entity recognition, constituent parsing, dependency parsing and semantic role labelling. http://www.niuparser.com/index.en.html

Natural Language Processing (NLP) is key component for opinion mining because it contains tokenization for text pre-processing task. Tokenization process will split a sentence into words or phrases and remove stop words such as "a", "the" that does not add value to the analysis. Table 2 presents the current NLP toolkits available, it can be seen that most of the toolkits are open source software as they employ JAVA and Python language to perform. There are also some toolkits which have been specially designed for Chinese NLP due to the strength of the Chinese economy. Thereby, Chinese consumer behaviour analysis has begun to gain popularity in the global market. In general, most of the toolkits can perform word segmentation, POS tagging, named entity recognition and parsing which are the key components in NLP. From the research, it shows NLTK, OpenNLP and Stanford CoreNLP are widely used general NLP toolkits, where it supports most of basic NLP tasks [32].

Case studies have been conducted to test the usage of opinion mining technique in tourism industry. First, a research conducted by Bucur has focused on developing a platform to extract and summarizing opinion from tourism online platform [34]. Web crawlers were employed to extract data from the tourism online platform and then SentiWordNet was employed to evaluate the polarity of the reviews. The samples selected were from hotel reviews in Rome from TripAdvisor. Analysis showed an accuracy of 72% to 76.5% match with their manual categorization. The advantage of this platform is that it does not require a training dataset because it is an unsupervised process thus there will be some resources saving. Although, it shows that the technique is useful and quite accurate, but the research did not provide actual utilization of the technique on companies and how it can help hotels to improve their services [30]. The accuracy of opinion mining techniques can be improved by adding filters to narrow down the subjects [35]. The keywords were specially selected for hotel and restaurant domain to generate better results. In their test, the scenario of object extraction using frequent objects and frequent objects using filters were considered. The results show improvement in accuracy after using filters for both hotel and restaurant domain.

Even after the application of the filters, the accuracy level were at an average of 65%, but it showed 20% improvement from the previous state which was significant. Therefore, it can be concluded that adding filters to select specific domain keywords enhances the accuracy rate of opinion mining technique. From the case studies, it shows opinion mining technique is quite popular in social media analytics for tourism industry. Combining NLP and Machine Learning algorithm, precise analysis can be done to analyse the consumer sentiment on the product and services offered by companies. There are many Open source and paid programs already developed, these tools will provide ease of access for individuals who wish to perform text analysis. Some of the research also prove that adding appropriate filters to the dataset will provide more accurate results and SVM is a better classification method than the others. Yet, all these might have to depend on the nature of the dataset as the compatibility and business requirements could differ between domains and the quality of the data.

3 Conclusion and Recommendations

The review on existing literature proves that tourism in UAE is growing rapidly and it will attract leading business groups to compete in the hospitality industry. The completion of the current projects in UAE will also aid to increase the flow of visitor traffic. Also, with the increase of Muslim travellers worldwide, Halal tourism is also a growing trend that will also bring more traveller to Halal countries like UAE due to the safety, food and facilities that the country can provide to Muslim travellers. One of the ways to improve the competitiveness is by utilizing the data resource that is generated through social media technologies. Social media analytics and technologies provide this ability to analyse customer behaviours and market trends. Companies are investing into this technology because it can help to provide useful insights for decision making process and create products or marketing campaigns that are seemingly close to customer

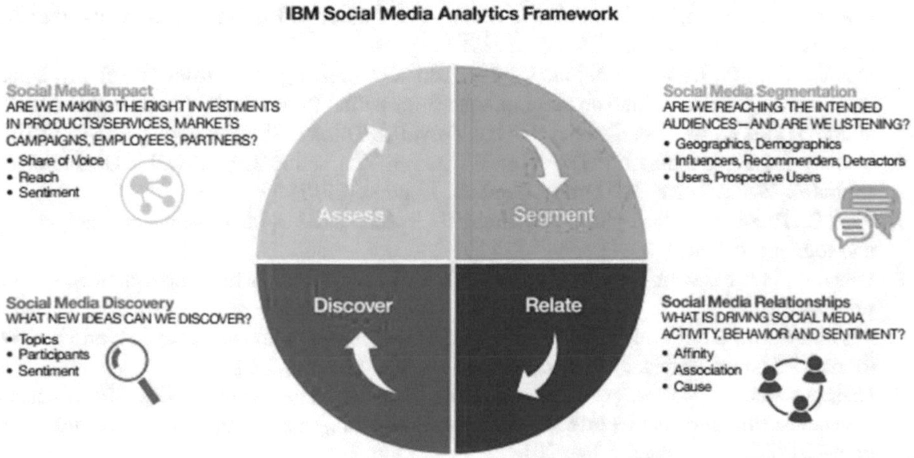

Fig. 1. IBM social media analytics framework [36]

expectations. Along with the increased usage on social media platforms, it is easier to gather data from general consumer worldwide. This platform can also act as a good digital advertisement platform because of its wide reach throughout the world.

SMA benefits on four key areas as shown in Fig. 1: To assess the impact of social media, to perform segmentation of the customers, to relate what is driving social media activity, behaviour and sentiment and finally to discover the topics, participants and sentiments [36]. Careful consideration should be employed on the tools and techniques for the process of social media analytics. Analysis of literature reveals that opinion mining technique is one of the best techniques as it can be combined with NLP and machine learning algorithm. The insights obtained in such a manner would be very useful for companies especially on customer reviews and further analysis. There are many tools available in the market to assist in opinion mining and text analysis, therefore it has lowered down the difficulty to perform analytics on unstructured data. Lastly, this paper reflects that significant improvement can be done in the tourism and hospitality industry with the help of social media analytics. While many researches have been done on evaluating the accuracy of the analysis results, less priority is placed on the actual utilization of the results. Hence, the social media analytics provide great opportunity for business improvement in the Hotel Industry in UAE by the actual utilization of the results of the analysis. Such research can also enhance the operations of the business, sales and image due to faster response and complying with customer expectations. Future research would be to create a pilot study on the utilization of social media analytics in a particular segment as the opportunities for business improvement in the key aspect.

References

1. Obar, J.A., Wildman, S.: Social media definition and the governance challenge. An introduction to the special issue. Telecommun. Policy **39**(9), 745–750 (2015)
2. Perrin, A.: Social Media Usage: 2005–2015. Pew Research Center (2015)
3. Mauro, A.D., Greco, M., Grimaldi, M.: A formal definition of Big Data based on its essential features. Libr. Rev. **65**(3), 122–135 (2016)
4. Del Vecchio, P., Mele, G., Ndou, V., Secundo, G.: Creating value from Social Big Data: Implications for Smart Tourism Destinations. Information Processing & Management (2017)
5. World Travel & Tourism Council: Global Travel & Tourism Benchmarking Report (2017)
6. Turner, R., Freiermuth, E.: Travel and Tourism: Economic Impact 2017 United Arab Emirates. World Travel & Tourism Council, London (2017)
7. Koo, C., Ricci, F., Cobanoglu, C., Okumus, F.: Special issue on smart, connected hospitality and tourism. Inf. Syst. **19**(4), 699–703 (2017)
8. Hoteliermiddleeastcom, News Analysis: UAE and Saudi Arabia's hotel investments, Dubai, pp. 1–4 (2017)
9. Alsumairi, M., HongTsui, K.W.: A case study: the impact of low-cost carriers on inbound tourism of Saudi Arabia. J. Air Transp. Manag. **62**, 129–145 (2017)
10. Crescent Rating: MasterCard-CrescentRating Global Muslim Travel Index 2017. https://www.crescentrating.com/reports/mastercard-crescentrating-global-muslim-travel-index-gmti-2017.html. Accessed 6 Feb 2018

11. Mohsin, A., Ramli, N., Alkhulayfi, B.A.: Halal tourism: emerging opportunities. Tour. Manag. Perspect. **19**, 137–143 (2016)
12. Mujtaba, U.S.: Ramadan: the month of fasting for Muslims, and tourism studies—mapping the unexplored connection. Tour. Manag. Perspect. **19**, 170–177 (2016)
13. Sharma, P., Paulsingh, B.S.: Evolution of social media marketing. Int. J. Adv. Res. Comput. Commun. Eng. **6**(3), 147–151 (2017)
14. Chen, D.W., Tabari, S.: A study of negative customer online reviews and managerial responses on social media—case study of the Marriott Hotel Group in Beijing. J. Consum. Mark. **41**, 53–64 (2017)
15. Brandt, T., Bendler, J., Neumann, D.: Social media analytics and value creation in urban smart tourism ecosystems. Inf. Manag. **54**(6), 703–713 (2017)
16. Miah, S.J., Vu, H.Q., Gammack, J., McGrath, M.: A big data analytics method for tourist behaviour analysis. Inf. Manag. **54**(6), 771–785 (2017)
17. Stieglitz, S., Mirbabaie, M., Ross, B., Neuberger, C.: Social media analytics: challenges in topic discovery, data collection, and data preparation. Int. J. Inf. Manag. **39**, 156–168 (2018)
18. Abrons, R.: The disadvantages of using social networks as marketing tools. http://smallbusiness.chron.com/disadvantages-using-social-networks-marketing-tools-20861.html. Accessed 27 Jan 2018
19. Indurkhya, N., Damerau, F.J.: Handbook of natural language processing, 2nd edn. Taylor and Francis Group, LLC, London (2010)
20. Tsirakis, N., Poulopoulos, V., Tsantilas, P., Varlamis, I.: Large scale opinion mining for social, news and blog data. J. Syst. Softw. **127**, 237–248 (2017)
21. Mäntylä, M.V., Graziotin, D., Kuutila, M.: The evolution of sentiment analysis: a review of research topics, venues, and top cited papers. Comput. Sci. Rev. **27**, 16–32 (2016)
22. Vyas, V., Uma, V.: An extensive study of sentiment analysis tools and binary classification of tweets using rapid miner. Procedia Comput. Sci. **125**, 329–335 (2018)
23. Cajachahua, L., Burga, I.: Sentiments and opinions from Twitter about Peruvian touristic places using correspondence analysis. CEUR Workshop Proc. **2029**, 178–189 (2017)
24. Doğan, Y., Turdu, Y.: Discovering similar cities using text mining: a recommendation application for Turkey. Int. J. Eng. Sci. **1**(12), 8–14 (2017)
25. Minazzi, R.: Social Media Marketing in Tourism and Hospitality. Springer, New York (2015)
26. Sanz-Blaz, S., Buzova, D.: Guided tour influence on cruise tourist experience in a port of call: an eWOM and questionnaire-based approach. Int. J. Tourism Res. **18**(6), 558–566 (2016)
27. Garant, A.: Social Media Competitive Analysis and Text Mining: A Case Study in Digital Marketing in the Hospitality Industry. Aalto University Library, Helsinki (2017)
28. Marcos, E., DeCastro, V., MartínPeña, M.L., Garrido, E.D., Lopez-sanz, M., ManuelVara, J.: Education on service science management and engineering. Exploring services science. In: Lecture Notes in Business Information Processing Book Series, vol. 201, pp. 264–277 (2015)
29. Alcoba, J., Mostajo, S., Paras, R., Mejia, G.C., Ebron, R.A.: Framing Meaningful Experiences Toward a Service Science-Based Tourism Experience Design, pp. 129–140. Springer, Cham (2016)
30. Morgan, J.: Destination ambassadors: examining how hospitality companies value brand ambassadorship from front-line employees: a case study of four seasons hotels and resorts destination ambassadors: examining how hospitality companies value brand ambassadorship. McMaster J. Commun. **11**, 237–262 (2014)
31. Krengel, L.: The financial impact of joining the chain and improving hotel rating: a case study in Russia. Open J. Bus. Manag. **4**(4), 659–674 (2016)

32. Sun, S., Luo, C., Chen, J.: A review of natural language processing techniques for opinion mining systems. Inf. Fus. **36**, 10–25 (2017)
33. Kang, M., Ahn, J., Lee, K.: Opinion mining using ensemble text hidden Markov models for text classification. Expert Syst. Appl. **94**, 218–227 (2018)
34. Bucur, C.: Using opinion mining techniques in tourism. Procedia Econ. Finance **23**, 1666–1673 (2015)
35. Tjahyanto, A., Sisephaputra, B.: The utilization of filter on object-based opinion mining in tourism product reviews. Procedia Comput. Sci. **124**, 38–45 (2017)
36. Sift Analytics Group Pte Ltd, Social Media Analytics. http://www.sift-ag.com/ms-en/products-solutions/predictive-analytics/social-media-analytics. Accessed 17 May 2018

Enterprise Resource Planning and Business Intelligence to Enhance Organizational Performance in Private Sector of KSA: A Preliminary Review

Showaimy Aldossari(✉) and Umi Asma Mukhtar

Faculty of Information Science and Technology, Universiti Kebangsaan Malaysia, 43600 Bangi, Selangor, Malaysia
ceo@tania.sa

Abstract. Enterprise Resource Planning (ERP) and Business Intelligence (BI) employment, regarded as complex, cumbersome and costly, often exceeds the initial estimated resources. It is designed to model and automate basic processes across the organization over a centralized database sheet, or any type of tools that are used by organization. The success of ERP system usage does not require the demolition of the traditional ERP system models, but instead re-platforming and hosting them. Integration of ERP with BI, called ERPBI application, is often viewed as a strategic investment that can provide significant competitive advantage with positive return thus contributing to the company's revenue and growth, which in return enhance the overall performance of private sector organizations. Literature has shown that ERPBI is not supported by any specific model. The absence of models leads to improper use and adoption of ERPBI. The current researches evidence that models for ERPBI adoption are inapplicable or unavailable. Many important factors for successful adoption were overlooked. Thus, the aim of this paper is to identify the significant factors that could influence the behavioural intention toward the adoption of ERPBI in private sector of Kingdom of Saudi Arabia (KSA). In this study, 30 specialists from Private sector of KSA who are involved in ERPBI were interviewed. The findings showed that in order to promote effective ERPBI model all the necessary factors must be included.

Keywords: Enterprise Resource Planning and Business Intelligence Model · Private sector · Saudi Arabia · Adoption · Organizational performance

1 Introduction

In the current business environment, the marketplace is characterized by complexity and environment shift (from place to space) coupled with the economic globalization and information technology development. Consequently, the competition in the market intensifies from day to day [1, 2].

Moreover, the permeating globalization of services and the dynamic development in technology brought on by intelligence and communication technology have led service companies to be competitive in offering new products/services [3].

F. Saeed et al. (Eds.): IRICT 2018, AISC 843, pp. 343–352, 2019.
https://doi.org/10.1007/978-3-319-99007-1_33

Accordingly, majority of organizations have increased their investment in different IT systems [4–7] like ERP [8] and BI [9].

However, majority of ERP systems lack an integrated BI system utilized for accounting and transaction management that covers inventory, material needs or capacity planning. But because the system lacks reporting capabilities, majority of firms do their operation planning and management through spreadsheets [10]. It is therefore more feasible to use ERP in combination with BI, and not individually as the combination of both has been evidenced to improve capabilities of organizations' performance (e.g., [9, 11–15]). A combined ERP-BI, known as ERPBI, is capable of offering additional features, functionality and flexibility via information sharing [10]. The use of ERPBI enables management to understand processes and arrangement of data flows to facilitate access to significant amount of real-time operational data and to generate authentic reports for improving the process of decision-making which will lead to enhancing the overall performance [1, 16].

This paper is organized as follows: the first section is to elaborate the role of ERPBI for Saudi private sector. Thereafter, a related work on factors that affected the behavioral intention to use ERPBI is presented. Methodology that has followed is also discussed followed by results of the proposed model and discussion. The conclusion is the last part of this paper.

2 The Importance of Enterprise Resource Planning and Business Intelligence for Saudi Arabia

In the Saudi Arabian context, however, scholars believe that the existence of models, including the important factors, will play a significant role in promoting ERPBI. Scholars argue that the lack of ERPBI models and ignorant of most of the vital factors are primary reasons hindering ERPBI adoption and use.

In Saudi Arabia, Saudi Vision 2030 primarily aims for the promotion of economic growth through the empowerment of entrepreneurial activities among its citizens (KSA Government Report 2017). In this regard, entrepreneurs (especially online entrepreneurs) have to be provided with guidelines on the adoption of ERPBI. The present study contributes to the Vision 2030 of Saudi Arabia for internet entrepreneurs by recommending the system's adoption to assist in achieving financial objectives and in ultimately building a robust economy.

Moving on to ERPBI, it should be noted that it is not a simple piece of computer software owing to its ability to transform business process in the Kingdom of Saudi Arabia on the basis of different modules. Increasing number of organizations and industries are adopting the system and majority of Saudi organizations have displayed enhanced performance by using ERP and BI in light of their human resource planning, management control and operational control. Currently, researchers are in consensus with regards to the potential promise that ERP holds in the manufacturing industries, with the combination of both systems (ERPBI) successful in bringing about core competency in majority of firms [17, 18].

More importantly, the economy of Saudi Arabia has been developing with the help of enterprise recourse planning that has facilitated the transformation of cognitive process, rules and procedures in organizations. With the dynamic growth of the Middle East software industry, ERP packages are expected to be extensively used in the long-run [17, 19]. Majority of companies claimed that their growth, welfare and dynamism are because of their implementation of ERP system – a system that is distinct for its scope and use. Presently, ERP vendors are generally catering to SMEs for system implementation.

Today's organizations are largely shifting their focus to customers by becoming customer centric and responsive in order to reap the benefits of information systems such as ERPBI. KSA, of the Arab countries, is positioned on top at the world economic forum's networked readiness index of 2017 that measures the ICT performance in economies (Saudi Government Report, 2017). Majority of Saudi SMEs began adopting ERPBI, although use resistance is still prevalent [20, 21].

Moreover, the exposure of many Saudi firms to increasing competitive pressure has led them to adopt modern business processes such as ERPBI solutions in order to provide them with unlimited information access and allow them to obtain competitive advantage [22]. Factors that affect ERP growth in the Saudi market include higher return on investment (ROI), rapid industrialization and easy integration with the legal systems.

The integrated ERPBI system has been considered as a robust system used to handle corporate resource planning and processing of supply chain. ERPBI significantly contributes to integrating and managing enterprise extensive transactional data. Majority of such systems are able to perform the feature successfully but they are lacking data reporting and analytics capabilities. To minimize, if not close such gap in decision making, BI tools can be utilized to keep the information technology innovation path consistent [23–25].

Lastly, as users' at all organizational levels realize the advantages to be reaped from the ERPBI software, its acceptance will increase. In this regard, firms that fail to justify ROI for the implementation of ERPBI are currently in need of adopting ERPBI to enhance the former's usage.

Generally speaking, KSA private sector is in urgent need for model to be used as a guidance for successful ERPBI implementation [26].

3 Factors that Affect the Behavioral Intention to Use ERPBI

Several methods have been adopted in literature to determine the factors of ERPBI implementation, with the topmost listed in Table 1. Each approach is accompanied by strengths and weaknesses and a clarification of the techniques was made by [27].

Khandelwal and Ferguson [28] provide a clarification of these diverse techniques.

In this paper, the studies dedicated to ERPBI were reviewed to determine the general factors utilized by the authors and a total of 71 factors were cited. The factors were then tabulated to determine the top used factors in literature. Based on the frequency table, there are 28 factors of ERPBI use and implementation as seen in Table 2 below. Frequency refers to the number of citations in literature for the factors, but this does not imply that the factors are ordinary and common [38, 39]. The list containing

Table 1. Literature approaches

Research method	Authors
Action-research case studies	[29]
Structured interviewing	[30]
Scenario analysis	[31]
Multivariate analysis	[32]
Literature review	[33]
Group interviewing	[27]
Focus groups	[34]
Delphi technique	[35, 36]
Combination of methods	[28, 37]

the extracted factors is sent to the experts in order to rank the factors with the aim of selecting the factors that impact the integrated ERPBI system.

Table 2. Most cited and frequent factors extracted through intensive literature review

No	Factor	Total	No	Factor	Total
1	Top management support	33	15	Software development	23
2	User involvement	25	16	Partnership	23
3	Perceived usefulness	30	17	External expertise	23
4	Information quality	28	18	System quality	20
5	Effective communication	28	19	Policy	19
6	Clear vision and planning	27	20	Service quality	17
7	Team work and composition	27	21	Consultant selection and relationship	17
8	Legacy system	25	22	Financial support	17
9	Change management	25	23	Competitive pressure	16
10	Vendor support	25	24	Usability	15
11	Government role	24	25	Training	14
12	Data quality and integrity	24	26	Overall satisfaction	13
13	Effective monitoring and control	24	27	Performance	12
14	Risk management	24	28	Reliability	12

The remaining factors in literature were only cited a few times and were thus excluded from the list.

4 Methodology

The first part of the methodology determines the extraction of variables for the purpose of assessing the behavioral intention to use ERPBI among private Saudi sector. After extracting the factors through the previous mentioned steps, the final list was sent to the experts.

Thereafter, interviews are conducted with the respondents involved in the ERBPI within private sector. As shown in Fig. 1, the methodology of this study consists of four stages: conducting an intensive literature review, extracting the factors, interviewing the experts regarding the factors of the ERPBI and conducting a thematic analysis to explore the resulted factors of the study.

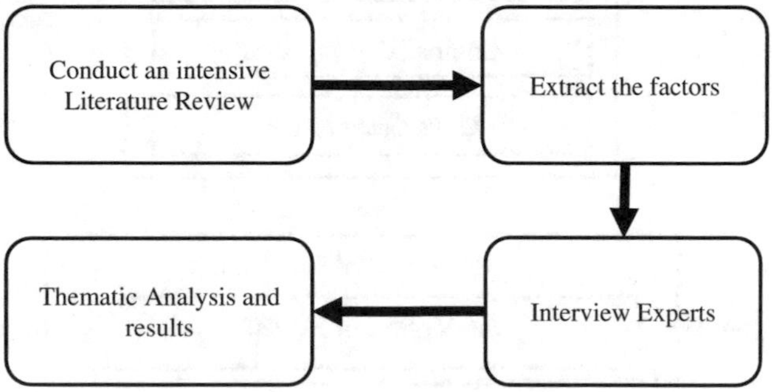

Fig. 1. Methodology adapted from [40]

Semi-structured interviews were conducted with 30 different participants from the private sector. Interviews lasted between 10 and 20 min, for an average of 15 min per interview. Recordings were made of all the interviews with the exception of those with the consultants who requested not to be recorded.

The interviews were carried out using VOIP applications. Each interview began with a self-introduction by the interviewer and an explanation of the purpose of the study. Confidentiality and privacy of the participant were ensured. Focus was on the issues that either facilitated or hindered the influence of the ERPBI factors to enhance the behavioral intentions to use such system as well as the strategies of the private sector to ensure the successful implementation of the ERPBI. This methodology was adopted from [40] as seen in Fig. 1.

The model is presented in Fig. 2. The model attempts to align the relationship of different examined parameters and classified the factors based on the TOE dimensions. The technological dimension is based on the DeLone and McLean's IS Success model factors of system quality, service quality and information quality. The organizational dimension encapsulates change management effective communication and training (derived from literature and experts), while the environmental dimension covers clear vision and planning, competitive pressure and government role (derived from literature and experts). The UTAUT, on the other hand, is adopted as the underpinning model addressing the organizational and environmental dimensions.

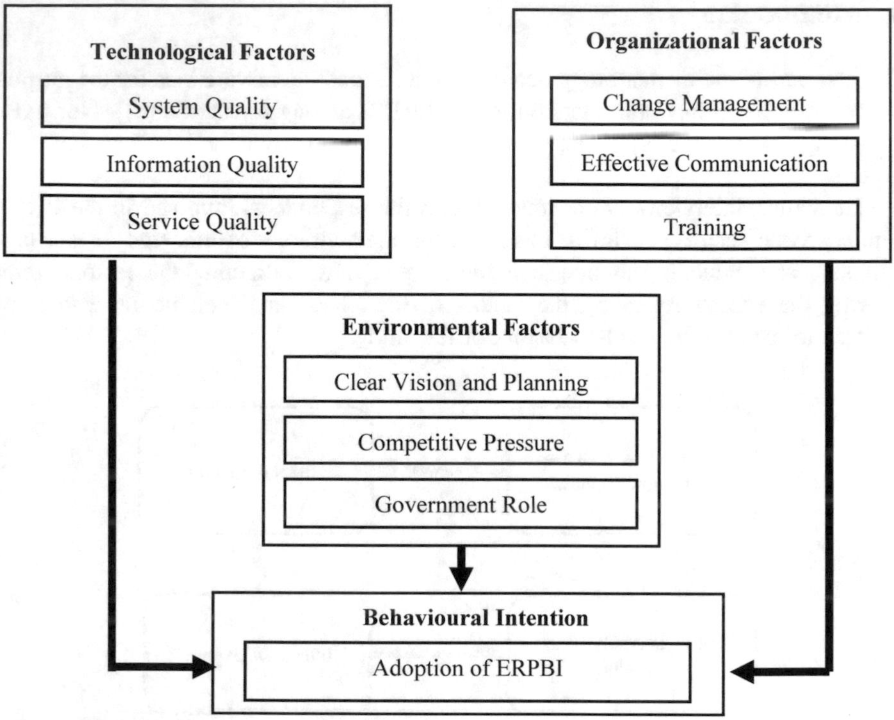

Fig. 2. The proposed model for ERPBI adoption in the private Saudi sector

5 Results and Discussion

The list of factors were forwarded to experts for their feedback on the selection of factors that affected the integrated ERPBI system as a result of which, nine factors were deemed to be significant in behavioral intention towards ERPBI adoption as seen in Table 3, and in turn actual usage. The determination of the factors significance was recommended by [41].

In the present study, the factors that are important for an ERPBI behavioral intention are extracted and recommended by experts. The qualitative data through the interviews confirmed the importance of the extracted factors and experts were in a mutual agreements that the list includes the most important factors for ERPBI adoption. The experts, however, confirm that ERPBI adoption will lead to enhance the overall performance of KSA private sector. Therefore, the model that contains the factors is proposed where factors were categorized into three groups based on the three dimensions of technology, organization and environment as seen in Fig. 2.

Table 3. List of factors recommended by experts

No	Factor	Number of recommenders experts
1	System quality	30
2	Government role	30
3	Competitive pressure	30
4	Change management	28
5	Effective communication	28
6	Clear vision and planning	27
7	Information quality	27
8	Service quality	25
9	Training	25

6 Implication of the Study

The factors that determine ERPBI use, specifically in the developing nations, have to be confirmed and highlighted to ensure effective utilization of IT resources. In this regard, literature on the subject has been few and far between in this context, particularly in Saudi Arabia. It is therefore pertinent to have a deeper insight into the ERPBI principles in developing nations to benefit software suppliers and organizations. This would also assist decision makers in designing a suitable ERPBI set of strategies and software designers and suppliers in designing products/services that could meet the requirements of the end-users.

The understanding of the factors and their relationship with the successful use of ERPBI would assist top management to steer clear of or mitigate organizational losses while using the system for increased cash flow, controlled expenses, management of headcount and enhanced employee performance, streamlined spend-supply chain operations, tracking the financial performance of key projects, enhancing customer relationships and reporting systems that will ultimately result in overall organizational performance. The above elements all contribute towards the possibility of successful use with management overseeing, controlling and supporting all the significant areas.

7 Conclusion

This paper highlights the lack of studies dedicated to ERPBI use in literature and its key role in supporting and enhancing the performance of organizations. The paper laid emphasis on the limitations of prior studies when it comes to using ERPBI and therefore, this study brings forward a model of ERPBI. The importance of ERPBI to organizations requires a proper robust model for its successful implementation. On the basis of the review of literature, a relationship exists between ERPBI implementation and effective guidance in the form of a study model.

This paper reviewed past related works that investigate the factors in the ERPBI. The identified factors were extracted and recommended by experts and classified into three dimensions, namely technological, organizational and environmental based on their interaction level. These factors cannot be applied to ERPBI systems without giving careful consideration to the relevant contextual issues. This paper adds to the body of knowledge by identifying the factors that impact on ERPBI.

Also, this findings could assist senior management by optimizing their scarce resources on those key areas that would improve the implementation process.

References

1. Olszak, C.M., Ziemba, E.: Approach to building and implementing business intelligence systems. Interdiscip. J. Inf. Knowl. Manag. **2**, 135–148 (2007)
2. Olszak, C.M., Ziemba, E.: Critical success factors for implementing business intelligence systems in small and medium enterprises on the example of upper Silesia, Poland. Interdiscip. J. Inf. Knowl. Manag. **7**(12), 129–150 (2012)
3. Ram, J., Wu, M.-L., Tagg, R.: Competitive advantage from ERP projects: examining the role of key implementation drivers. Int. J. Project Manag. **32**(4), 663–675 (2014)
4. Mukred, M., Yusof, Z.M.: The role of electronic records management (ERM) for supporting decision making process in Yemeni higher professional education (HPE): a preliminary review. Jurnal Teknologi. **73**(2), 117–122 (2015)
5. Mukred, M., Yusof, Z.M.: The DeLone–McLean information system success model for electronic records management system adoption in higher professional education institutions of yemen. In: International Conference of Reliable Information and Communication Technology. Springer (2017)
6. Ameen, A.A., Ahmad, K.: Information systems strategies to reduce financial corruption. In: Leadership, Innovation and Entrepreneurship as Driving Forces of the Global Economy, pp. 731–740. Springer (2017)
7. Mukred, M., Yusof, Z.M.: The performance of educational institutions through the electronic records management systems: factors influencing electronic records management system adoption. Int. J. Inf. Technol. Project Manag. (IJITPM) **9**(3), 34–51 (2018)
8. Erp, K.J., et al.: Empowering public service workers to face bystander conflict: enhancing resources through a training intervention. J. Occup. Org. Psychol. **91**(1), 84–109 (2018)
9. Rouhani, S., Lecic, D.M.: Business intelligence impacts on design of enterprise systems. In: Encyclopedia of Information Science and Technology, 4th edn, pp. 2932–2942. IGI Global (2018)
10. Helmy, Y.M., Marie, M.I., Mosaad, S.M.: An integrated ERP with web portal. Adv. Comput. **3**(5), 1 (2012)
11. Chen, H., Chiang, R.H., Storey, V.C.: Business intelligence and analytics: from big data to big impact. MIS Q. **36**(4), 1165–1188 (2012)
12. Yeoh, W., Popovič, A.: Extending the understanding of critical success factors for implementing business intelligence systems. J. Assoc. Inf. Sci. Technol. **67**(1), 134–147 (2016)
13. Saleh, T., Thoumy, M.: The impact of ERP systems on organizational performance: in Lebanese wholesale engineering companies. In: 2018 7th International Conference on Industrial Technology and Management (ICITM). IEEE (2018)
14. Richards, G., et al.: Business intelligence effectiveness and corporate performance management: an empirical analysis. J. Comput. Inf. Syst. **57**(1), 1–9 (2017)
15. Albu, C.-N., et al.: The impact of the interaction between context variables and enterprise resource planning systems on organizational performance: a case study from a transition economy. Inf. Syst. Manag. **32**(3), 252–264 (2015)

16. Ziemba, E.: Conceptual model of information technology support for prosumption. In: Proceedings of the International Conference on Management, Leadership and Governance (2013)
17. Chan, J.O., Abu-Khadra, H., Alramahi, N.: ERP II readiness in jordanian industrial companies. Commun. IIMA **11**(2), 5 (2011)
18. Yusof, Z.M., et al.: The relationship between user engagement and business intelligence system effectiveness. World Appl. Sci. J. **28**(7), 978–984 (2013)
19. Al-Mobaideen, H.O.: The impact of change management on the application enterprise resource planning system (ERP) effectiveness: field study in Jordan Bromine company. J. Manag. Res. **6**(4), 79–98 (2014)
20. Alhirz, H., Sajeev, A.: Do cultural dimensions differentiate ERP acceptance? A study in the context of Saudi Arabia. Inf. Technol. People **28**(1), 163–194 (2015)
21. Badewi, A., Shehab, E., Zeng, J., Mohamad, M.: ERP benefits capability framework: orchestration theory perspective. Bus. Process Manag. J. **24**(1), 266–294 (2018)
22. AlBar, A.M., Hoque, M.R.: Factors affecting cloud ERP adoption in Saudi Arabia: an empirical study. Inf. Dev. 0266666917735677 (2017)
23. Chou, D.C., Bindu Tripuramallu, H., Chou, A.Y.: BI and ERP integration. Inf. Manag. Comput. Secur. **13**(5), 340–349 (2005)
24. Chou, P.: Workplace social support and attitude toward enterprise resource planning system: a perspective of organizational change. Int. J. Inf. Syst. Soc. Change (IJISSC) **9**(1), 58–76 (2018)
25. Velić, M., Padavić, I., Lovrić, Z.: Model of the new sales planning optimization and sales force deployment ERP Business Intelligence module for direct sales of the products and services with temporal characteristics. In: Proceedings of the ITI 2012 34th International Conference on Information Technology Interfaces (ITI). IEEE (2012)
26. Samander, B.A., et al.: ERP acceptance in airline industry of Saudi Arabia with mediating effect of job security. Int. J. **11**(2), 226–240 (2017)
27. Khandewal, V., Miller, J.: Information system study. In: Opportunity Management Program (1992)
28. Khandelwal, V.K., Ferguson, J.R.: Critical success factors (CSFs) and the growth of IT in selected geographic regions. In: Proceedings of the 32nd Annual Hawaii International Conference on Systems Sciences, HICSS-32. IEEE (1999)
29. Kock, N., Jenkins, A., Wellington, R.: A field study of success and failure factors in asynchronous groupware supported process improvement groups. Bus. Process Manag. J. **5** (3), 238–254 (1999)
30. Bullen, C.V., Rockart, J.F.: A primer on critical success factors (1981). https://dspace.mit.edu/bitstream/handle/1721.1/1988/SWP-1220-08368993-CISR069.pdf?sequence=1
31. Barat, J.: Scenario playing for critical success factor analysis. J. Inf. Technol. **7**(1), 12–19 (1992)
32. Tishler, A., et al.: Identifying critical success factors in defense development projects: a multivariate analysis. Technol. Forecast. Soc. Chang. **51**(2), 151–171 (1996)
33. Esteves, J., Pastor, J.: Enterprise resource planning systems research: an annotated bibliography. Commun. Assoc. Inf. Syst. **7**(1), 8 (2001)
34. MacCarthy, B., Atthirawong, W.: Critical factors in international location decisions: a Delphi study. In: The Proceedings of 12th Annual Meeting of the Production and Operations Management, 30 March–2 April 2001
35. Brancheau, J.C., Wetherbe, J.C.: Key issues in information systems management. MIS Q. **11** (1), 23–45 (1987)
36. Lawley, M., et al.: Critical Success Factors for Regional Community Portals: A Preliminary Model. ANZMAC, Massey University, Massey (2001)

37. Parr, A., Shanks, G.: A model of ERP project implementation. J. Inf. Technol. **15**(4), 289–303 (2000)
38. Finney, S., Corbett, M.: ERP implementation: a compilation and analysis of critical success factors. Bus. Process Manag. J. **13**(3), 329–347 (2007)
39. Law, C.C., Ngai, E.W.: ERP systems adoption: an exploratory study of the organizational factors and impacts of ERP success. Inf. Manag. **44**(4), 418–432 (2007)
40. Mukred, M., et al.: Electronic records management system adoption readiness framework for higher professional education institutions in Yemen. Int. J. Adv. Sci. Eng. Inf. Technol. **6**(6), 804–811 (2016)
41. Ahmad, M.M., Cuenca, R.P.: Critical success factors for ERP implementation in SMEs. Robot. Comput. Integr. Manuf. **29**(3), 104–111 (2013)

Internet of Things (IoT): Issues, Techniques and Applications

Towards Semantic Interoperability for IoT: Combining Social Tagging Data and Wikipedia to Generate a Domain-Specific Ontology

Mohammed Alruqimi[1(✉)], Noura Aknin[1], Tawfik Al-Hadhrami[2], and Anne James-Taylor[2]

[1] Information Technology and Modeling Systems Research Unit, Abdelmalek Essaadi University, Tétouan, Morocco
m.alruqimi@uae.ma, aknin@ieee.org

[2] School of Science and Technology, Nottingham Trent University, Nottingham, UK
{tawfik.al-hadhrami,anne.james-taylor}@ntu.ac.uk

Abstract. Handling large-scale heterogeneous data in the IoT and processing it in real time will be a key factor towards realizing IoT services in users' daily lives. In this regard, semantic domain ontologies are increasingly seen as a solution for enabling interoperability across heterogeneous data and sensor-based applications. Several ontologies have been proposed with the aim of addressing interoperability issues and various aspects of IoT device observation. However, for most of these ontologies, they are either in the domain of sensor networks or the much broader domain of the IoT. Furthermore, these ontologies have shown slow improvement, as they have been developed by limited groups of domain experts. This paper proposes a model that exploits the collective intelligence which emerges from social tagging systems to generate up-to-date domain-specific ontologies. The evaluation of the proposed model, using a dataset extracted from BibSonomy, demonstrated that the model could effectively learn a domain terminology, and identify more meaningful semantic information for the domain terminology. Furthermore, the proposed model introduces a simple and effective method for the common problems related to tag ambiguity.

Keywords: Social tagging · Folksonomies · Semantic web ontology IoT

1 Introduction

Semantic domain ontologies are increasingly seen as a key factor in automation of information processing and are being integrated to into the IoT to support interoperability among heterogeneous systems [1–3]. Applying ontology-based approaches to IoT can better enable "things" to work in co-operation and could also enable autonomous interaction between "things" [4–7]. However, the development of ontologies by domain experts is a time-consuming and expensive process. On one hand, the ontologies deployed in the current sensor-based applications are developed by

F. Saeed et al. (Eds.): IRICT 2018, AISC 843, pp. 355–363, 2019.
https://doi.org/10.1007/978-3-319-99007-1_34

restricted groups of domain experts and not by semantic web experts, while social tagging data are contributed by millions of online users. Therefore, social tagging systems represent a continuous source for developing and enriching ontologies. [8, 9]. The ontologies derived from social tagging systems can represent the collective intelligence of online communities rather than the perception of a limited group of experts. Such ontologies would be able to capture changes derived from a more diverse user population. Therefore, they would become semantically richer and thus handier for logical reasoning tasks [10]. However, social tagging systems share the problems inherent to all uncontrolled vocabularies, such as ambiguity, synonymy, and the lack of hierarchy. Thus, knowledge extraction from social tagging data remains a challenge not yet solved. In this paper, a model has been introduced for inducing domain-specific ontology from social tagging data. The experimental results on a snapshot of a dataset from BibSonomy showed that the proposed model could effectively capture domain-specific concepts and enrich these concepts with semantic information extracted from Wikipedia.

2 Social Tagging Systems

Social tagging websites enable users to assign freely chosen tags to categorize their digital content (such as websites, pictures and videos over the Web), forming the so-called folksonomies [11]. Currently, many web-based services foster the concept of tagging. These systems can be differentiated according to the kind of resources they support: for instance, Delicious for sharing bookmarks, Flicker for photos, BibSonomy for publications and bookmarks and YouTube for sharing videos. The basic principle of these services is simply to allow the registered users generating the content to classify it in their own unique way by assigning arbitrary tags to this content. Researchers attribute the success of tagging to the fact that no specific prior knowledge is required to tag, and the immediate benefit obtained from tagging [12, 13]. From a knowledge organization point of view, folksonomies have two main advantages: social tagging systems provide a vast number of user-generated annotations and directly reflect users' vocabularies and interests; they are relatively cheap to develop and harvest, as they emerge from end users' tagging [12–14]. These advantages have turned social tagging systems into an interesting source of data for Semantic Web applications [9, 14, 15].

3 Related Work

An overview of the existing ontologies for the IoT domain has been provided by [3], while many examples of ontologies aiming to achieve semantic interoperability in the IoT domain are presented in [1, 2]. However, much work has also been done to capture semantics [16–18] and to extract ontologies from social tagging systems [19–21]. The early studies (e.g. [22–24]) explored means of leveraging the co-occurrence statistics of tags and the tripartite structure of folksonomies to measure tag relatedness. More recent studies (e.g. [20, 25]), have proposed ways of making tags' semantics explicit by grounding them to corresponding entries in online knowledge bases and socially linked data. In general, most of these approaches focus on grouping related terms rather than developing formal ontologies, much less domain-specific ontologies. Our model produces baseline domain ontologies from tags in folksonomies. The proposed model collects domain-relevant terms from tags, relying on a set of domain keywords extracted automatically from titles of Wikipedia pages. It then identifies the exact meaning of the terms and retrieves semantic information about each term.

4 Building Domain-Specific Ontology

This model takes the name of a specific domain and a prepared folksonomy dataset as inputs and produces a corresponding domain ontology as output. The general structure of the proposed model is shown in Fig. 1. Implementation of this model goes through four main stages as follows.

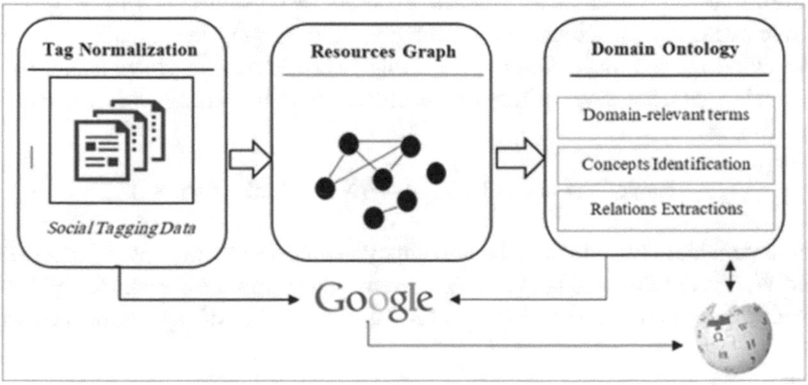

Fig. 1. The proposed model architecture

4.1 Tag Normalization

This activity starts with a pre-processing process which includes deleting special characters, duplicated tags and prepositions. Next, for each term the appropriate Wikipedia article that represents its intended meaning has been identified. By linking tags to corresponding Wikipedia articles, standard names of the terms assigned, capture their alternative names, thanks to Wikipedia redirects, and exclude non-objective tags.

4.2 Building a Resources Graph

The folksonomy resources represented as an undirected graph, *G,* in which *V* represents the set of resources, and *E* represents the set of edges [24]. Two resources (*ri, rj*) will be connected if they have at least one tag in common. In the following section, In depth description how to traverse the graph *G* to collect the relevant domain terms. To implement this phase, *JGraphT library* has been used, which is a free Java class library that provides mathematical graph-theory objects and algorithms (http://jgrapht.org/).

4.3 Extracting Domain-Relevant Terms

In this activity, our model starts by extracting a list of *Domain Keywords* from the titles of Wikipedia articles and redirection pages contained in the main Wikipedia category corresponding to the domain in hand. Based on this *Domain Keywords* list, a set of resources have been selected as starting points (SP) for traversing the graph. For a resource to be marked as starting point, at least two-thirds of the tags assigned to this resource could be mapped in the *Domain Keywords*. Next, the model traverses the graph *G* many times (starting from SPs) looking for resources that are relevant to the domain in hand. Throughout the traversing process by applying a ranking function over each visited vertex. The ranking function rates the relevance of a vertex to the given domain, based on the number and weight of the paths coming to it from the different SPs (Eq. 1). Resources with a ranking value greater than the *h* threshold will be marked as domain-relevant resources and, hence, all their associated tags will be gathered as domain-relevant terms. However, to traverse the graph, the breadth first search (BFS) method is used; once the graph is being traversed, starting from a starting point, the traversing process stops whenever reaching another starting point or reaching a terminal vertex.

$$\text{Rank (Vi)} = \text{Rank (Vi)} + \text{W/NT} * \text{Rank (PVi)} * \text{d} \tag{1}$$

Let us consider *PVi*, which is the previously visited vertex from which current point reached *Vi*; d is the distance between the current vertex and the current *SP*, *W* is weight of the edge that connects *Vi* and *PVi* and *NT* is the number of tags associated with *Vi*.

4.4 Concept Dentification

By concept identification involves making explicit the meaning and semantic information for the domain-relevant terms extracted in the previous activity, as shown in the example depicted in Fig. 2. The advantage of using Wikipedia as a reference to map terms is that Wikipedia is a community-driven knowledge base, much as folksonomies are, so that it rapidly adapts to accommodate new terminology. In addition, the redirect pages in Wikipedia provide synonyms and morphological variations for a concept. For example, when searching the tag '*nyc*' in Wikipedia, the entry for New York City is returned. To perform the linking tasks term to Wikipedia entries in the example shown in Fig. 2. Google was used as an intermediary to retrieve the appropriate corresponding Wikipedia article for each term. First the model passed to Google a term enclosed between the domain name (in this example: "Web Development") as a context and the word ("Wikipedia") to bring Wikipedia pages to the top. Then, it searched for a morphological matching between the term and the titles of the top four retrieved Wikipedia pages. The simplest case occurs when a term can be matched directly to the first Google result. In fact, querying Wikipedia through Google allows us to take advantage of techniques embedded in Google, such as stemming and lemmatization, so there is a high chance of finding the correct corresponding Wikipedia articles. As an example, passing the term 'CSS' to Google resulted in retrieving the Wikipedia article entitled 'Cascading Style Sheets', since CSS represents a redirect page to this article in Wikipedia. In the case of disambiguated terms, (for instance the term "Ajax" that could refer to a programming language or a mythological Greek hero), the Wikipedia article that represents its intended meaning comes first in the Google results, due to using the domain name as context. However, the information available on the returned Wikipedia articles employed to enrich the terms. These include redirect pages as alternative names, and Wikipedia categories containing that page that are listed at the bottom of each article.

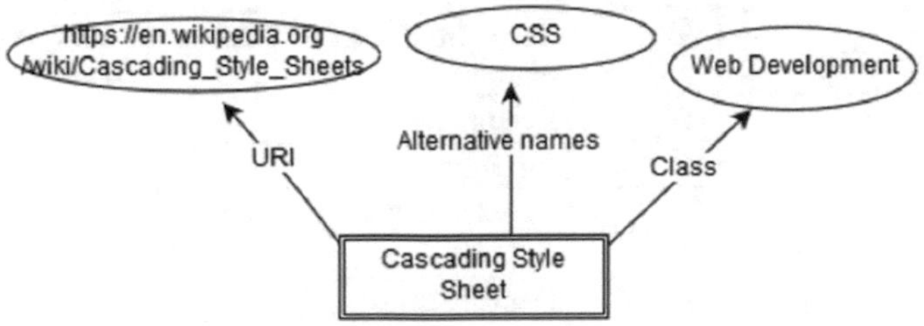

Fig. 2. An example of applying the Concept Identification process for the term "CSS"

In the case of disambiguated terms, (for instance the term "Ajax" that could refer to a programming language or a mythological Greek hero), the Wikipedia article that represents its intended meaning comes first in the Google results, due to using the domain name as context. However, the information available on the returned Wikipedia articles employed to enrich the terms. These include redirect pages as alternative names, and Wikipedia categories containing that page that are listed at the bottom of each article.

4.5 Concept Relations Extraction

Unlike most approaches in the literature, semantic relations have been extracted and captured between concepts from the free-text content of Wikipedia articles rather than using the semantic relations defined in the existing ontologies such as WordNet and DBpedia. Aiming to extract our own relationships of interest that are not provided by these ontologies; In this study, the domain-dependent relationships are considered that can be expressed by the particular verbs of an area of interest.

Pre-processing

In this activity, the sentences containing pairs of domain concepts have been collected. A 'tag string' for each sentence has been generated, as shown in the example in Table 1. In summary, the pre-processing activity starts by elimination of non-alphabetic sentences (such as codes and equations), semi-structured sentences (such as tables, info-boxes, and numeric menus) and the expressions in parentheses. Next, part-of-speech tagging has been applied in order to create a tag string for each sentence. The tag string is composed of the part-of-speech tags defined in OpenNLP, along with our own tags that represent domain concepts and the set of pre-defined verbs and keywords that represent our types of relations, respectively. To accomplish this activity, some of the existing tools reused. First, to access Wikipedia data, Java Wikipedia Library (JWPL) [26] is needed; and then Apache OpenNLP for part-of-speech tagging is employed.

Table 1. POS-tagging definition with examples

Word	POS tags	Word class	Notes
Joomla	**DTs**	*Domain Term*	DTs (Domain concept)
is	**VBZ**	Verb, 3rd person singular present	PDR (pre-defined relation)
a	DT	Determiner	
free	JJ	Adjective	
and	*CC*	Coordinating conjunction	
Open-source	*JJ*	Adjective	
Content management system	***DTs***	Domain term	DTs (Domain concept)

Relations Extraction

To capture the relations of interest between concepts, a set of regular expressions adapted from [27, 28] have been used in order to detect our defined linguistic patterns. The regular expressions to the "tag string" for each sentence has been applied in order to detect linguistic patterns that define our pre-defined relations. As an example, Table 1 shows a sentence with its tag string created by our method. Mainly, the position of the domain concepts in the sentences is important, as our algorithm focuses on detecting directed relations between pairs of concepts in the form (*Concept1*, relation pattern, *Concept2*).

5 Evaluation and Discussion

Due to the lack of evaluation frameworks and electronic resources that can be used as a gold standard, the process of evolution of an ontology is difficult [17]. Moreover, folksonomy tags are uncontrolled vocabularies that contain many slang words and abbreviations, while the electronic resources often use formal and compound terms. Thus, only three domains of computer science have been selected randomly for the experiments: Semantic Web, Computer Networks, and Web Development. The experiments were performed on a dataset captured from BibSonomy [29], composed of 20,000 resources annotated by 85,006 tags (11,865 unique tags). To evaluate the obtained terminology, majority voting of five researchers has been chosen who were asked to make judgments regarding domain relevancy (how strongly a term is relevant to the given domain) for all the obtained terms by associating a label "relevant", "irrelevant", or "uncertain" with each term. Table 2 summarizes the obtained, where the "Distinct Terms" column shows all obtained terms after removing duplicated items, and the "Relevant Terms" column shows the terms marked as domain-relevant terms. The precision has been calculated using the obtained results as follows: Precision = (|relevant|) *100/(|distinct terms|), where "distinct terms" refers to all the unique terms that has been obtained, and "relevant" refers to the terms voted as domain relevant terms. Table 3 shows the number of captured relations for each domain. Although a high recall has not obtained, as simple patterns has been used with limited pre-defined types of relations, nevertheless, it can be seen that our patterns were in the most cases correctly detected, leading to a precision of as much as of 94%.

Table 2. Concepts identification

Domain	Computer networks	Web development	Semantic web
Distinct concepts	43	55	52
Precision	79%	88%	86%

Table 3. Extraction of relations

Domain	Computer networks	Web development	Semantic web
Distinct concepts	28	42	38
Precision	94%	92%	92%

6 Conclusion

Ontology provides a support service for the development of semantic-based heterogeneous resource searches and services in the IoT [1]. However, the current IoT domain ontologies do not meet the rapid growth and evolution of IoT applications. Constructing ontologies automatically from the collective intelligence which emerges from the social web would provide a rich set of concepts for IoT ontologies. This paper has addressed the problem of how to harvest and exploit embedded semantics in social tagging systems for developing semantic ontologies. The evaluation of the model, using a dataset extracted from BibSonomy, demonstrated that the model could effectively learn domain ontology concepts and identify meaningful semantic information for the extracted concepts. Furthermore, the proposed model could help reducing common problems related to tag ambiguity.

References

1. Xu, Y., Zhang, C., Ji, Y.: An upper-ontology-based approach for automatic construction of IoT ontology. Int. J. Distrib. Sens. Netw. **10**, 594782 (2014)
2. Agarwal, R., Fernandez, D.G., Elsaleh, T., Gyrard, A., Lanza, J., Sanchez, L., Georgantas, N., Issarny, V.: Unified IoT ontology to enable interoperability and federation of testbeds. In: 2016 IEEE 3rd World Forum on Internet of Things (WF-IoT), pp. 70–75. IEEE (2016)
3. Bajaj, G., Agarwal, R., Singh, P., Georgantas, N., Issarny, V.: A study of existing ontologies in the IoT-domain. CoRR. abs/1707.0 (2017)
4. Atzori, L., Iera, A., Morabito, G., Nitti, M.: The Social Internet of Things (SIoT): when social networks meet the Internet of Things: concept, architecture and network characterization. Comput. Netw. **56**, 3594–3608 (2012)
5. Barnaghi, P., Wang, W., Henson, C., Taylor, K.: Semantics for the Internet of Things. Int. J. Semant. Web Inf. Syst. **8**, 1–21 (2012)
6. Gyrard, A., Serrano, M., Atemezing, G.A.: Semantic web methodologies, best practices and ontology engineering applied to Internet of Things. In: 2015 IEEE 2nd World Forum on Internet of Things (WF-IoT), pp. 412–417. IEEE (2015)
7. Psomakelis, E., Aisopos, F., Litke, A., Tserpes, K., Kardara, M., Campo, P.M.: Big IoT and social networking data for smart cities: algorithmic improvements on big data analysis in the context of RADICAL City applications. In: Proceedings of the 6th International Conference on Cloud Computing and Services Science, pp. 396–405. SCITEPRESS - Science and Technology Publications (2016)
8. Gruber, T.: Ontology of folksonomy: a mash-up of apples and oranges. Int. J. Semant. Web Inf. Syst. **3**, 1–11 (2005)
9. Shirky, C.: Ontology is overrated: categories, links, and tags. http://www.shirky.com/writings/ontology_overrated.html?goback=.gde_1838701_member_179729766
10. Mikroyannidis, A.: Toward a social semantic web. Computer (Long. Beach. Calif.) **40**, 113–115 (2007)
11. Vander Wal, T.: Folksonomy coinage and definition (2007). http://vanderwal.net/folksonomy.html
12. Mathes, A.: Folksonomies: cooperative classification and communication through shared metadata. http://www.adammathes.com/academic/computer-mediated-communication/folksonomies.html
13. Hotho, A., Jäschke, R., Schmitz, C., Stumme, G.: Information retrieval in folksonomies: search and ranking. In: Proceedings of the 3rd European conference on The Semantic Web: Research and Applications, pp. 411–426. Springer (2006)
14. Szomszor, M., Cattuto, C., Alani, H., O'Hara, K., Baldassarri, A., Loreto, V., Servedio, V.D.P.: Folksonomies, the semantic web, and movie recommendation (2007)
15. Al-Khalifa, H.S., Davis, H.C.: Towards better understanding of folksonomic patterns. In: Proceedings of the 18th Conference on Hypertext and Hypermedia - HT 2007, p. 163. ACM Press, New York (2007)
16. García-Silva, A., Corcho, O., Alani, H., Gómez-Pérez, A.: Review of the state of the art: discovering and associating semantics to tags in folksonomies. Knowl. Eng. Rev. **27**, 57–85 (2012)
17. Alruqimi, M., Aknin, N.: Semantic emergence from social tagging systems. Int. J. Organ. Collect. Intell. **5**, 16–31 (2015)
18. Jabeen, F., Khusro, S., Majid, A., Rauf, A.: Semantics discovery in social tagging systems: a review. Multimed. Tools Appl. **75**, 573–605 (2016)

19. Hamdi, S., Lopes Gancarski, A., Bouzeghoub, A., Ben Yahia, S.: Enriching ontologies from folksonomies for eLearning: DBpedia case. In: 2012 IEEE 12th International Conference on Advanced Learning Technologies, pp. 293–297. IEEE (2012)
20. García-Silva, A., García-Castro, L.J., García, A., Corcho, O.: Social tags and linked data for ontology development: a case study in the financial domain. In: Proceedings of the 4th International Conference on Web Intelligence, Mining and Semantics (WIMS14), pp. 32:1–32:10. ACM, New York (2014)
21. Wang, S., Wang, W., Zhuang, Y., Fei, X.: An ontology evolution method based on folksonomy. J. Appl. Res. Technol. **13**, 177–187 (2015)
22. Begelman, G., Keller, P., Smadja, F.: Automated tag clustering: improving searching and exploration in the tag space. In: WWW2006 (2006)
23. Schmitz, P.: Inducing ontology from Flickr tags. In: Proceedings of the Workshop on Collaborative Tagging at WWW2006, Edinburgh, Scotland (2006)
24. Mika, P.: Ontologies are us: a unified model of social networks and semantics. Presented at the 2005
25. Angeletou, S.: Semantic enrichment of folksonomy tagspaces. In: Sheth, A., Staab, S., Dean, M., Paolucci, M., Maynard, D., Finin, T., Thirunarayan, K. (eds.) The Semantic Web: ISWC 2008, pp. 889–894. Springer, Berlin (2008)
26. Zesch, T., Müller, C., Gurevych, I.: Extracting lexical semantic knowledge from wikipedia and wiktionary. In: Calzolari, N., Choukri, K. (eds.) Proceedings of the Sixth International Conference on Language Resources and Evaluation (LREC 2008). European Language Resources Association (ELRA), Marrakech, Morocco (2008)
27. Arnold, P., Rahm, E.: Extracting semantic concept relations from Wikipedia. In: Proceedings of the 4th International Conference on Web Intelligence, Mining and Semantics (WIMS14) - WIMS 2014, pp. 1–11. ACM Press, New York (2014)
28. Stoutenburg, S., Kalita, J., Hawthorne, S.: Extracting semantic relationships between Wikipedia articles. In: Proceedings of 35th International Conference on Current Trends in Theory and Practice of Computer Science, Spindelruv Mlyn, Czech Republic (2009)
29. Knowledge & Data Engineering Group: University of Kassel: Benchmark Folksonomy Data from BibSonomy, version of September 30, 2008. https://www.kde.cs.uni-kassel.de/bibsonomy/dumps/

IoT-Based Smart Waste Management System in a Smart City

Nibras Abdullah[1(✉)], Ola A. Alwesabi[2], and Rosni Abdullah[1]

[1] National Advanced Ipv6 Center, Universiti Sains Malaysia, 1800 Penang, Malaysia
nibras@usm.my, rosni@nav6.usm.my

[2] Faculty of Computer Science and Engineering, Hodeidah University, Hodeidah, Yemen
ola_osabi@yahoo.com

Abstract. The revolution of the Internet and the Internet of Things (IoT) has led to the development of numerous devices, such as radio-frequency identification tags, sensors, and other intelligent devices. Devices with significant computational capabilities and those that are transformed into intelligent objects are used to monitor and gather information about the environment of a city, thereby leading to smart cities. The most important problem that is currently experienced by smart cities is waste management problem. The following factors directly affect this problem: an increase in urban areas and rapid population growth. Intelligent services can function as the frontline in obtaining information regarding every aspect of human activities. A typical example of a service provided by smart cities is waste management supported by IoT. Waste management involves different responsibilities, such as the collection, disposal, and utilization of waste in relevant facilities. A review of common waste management models is presented in this paper. Then, an enhanced waste management design is proposed. This design considers and deals with population and urban growth by using different truck sizes according to waste type and IoT devices that ease communication among system entities, such as smart bins, waste source areas, waste collection trucks, and waste management centers.

Keywords: IoT · Smart city · Waste management

1 Introduction

The 2012 World Bank report indicates that global municipal solid waste (MSW) is composed of 46% organic waste. The percentage of organic waste composition is 62% in the East Asia-Pacific region and 61% in the Middle East and North Africa. Organic waste is a global problem that should be acknowledged. The following technologies involved in organic waste recycling can solve this problem: composting technology for producing fertilizers and anaerobic digestion technology for generating energy [1].

The United Nations Environment Program highlights the increasing demand for food with higher quantity, quality, and diversity given that the world population is predicted to grow from 7.2 billion in 2010 to over 9 billion in 2050, and the corresponding food demand is expected to increase by 60%. Approximately 30% of the food

F. Saeed et al. (Eds.): IRICT 2018, AISC 843, pp. 364–371, 2019.
https://doi.org/10.1007/978-3-319-99007-1_35

produced worldwide is lost or wasted annually [2]. Hence, the source of organic waste is a serious concern. The results of a survey conducted by the National Solid Waste Management Department (2013) found that food waste, rather than paper waste, accounts for the highest proportion of the institutional waste stream [3].

Organic waste is difficult to recycle. However, it can be recovered and reuse. Food waste is vital to soil fertility and plant growth. The common treatment method is composting. In a campus, for example, animal bedding and food from the cafeteria can be composted and processed by worms, thereby considerably decreasing the amount of waste stream ending up in landfills, providing nutrients to the landscaping operations of the campus [4], improving soil structure, and reducing the requirement for fertilizers [5]. However, composting entails an unhealthy devotion and requires high cost for utilities [6].

One of the most difficult tasks for authorities during public events and celebrations is waste sorting and disposal. In spiritual and social events, such as during pilgrimage and Ramadan, a substantial increase in the amount of wastes, including food and plastic wastes, is recorded within a short period. Most of these wastes are disposed in landfills, which are not equipped with leachate and gas collection systems, thereby resulting in soil and water contamination, along with the emission of huge amounts of greenhouse gases into the atmosphere [7, 8].

This study aims to review Internet of Things (IoT) techniques for waste management systems and to propose an efficient waste management method by using IoT to improve the activities of collection systems in a source area, which is the most important stage of waste management, particularly with regard to waste recycling in urban areas worldwide.

The enhanced capability of applications, cloud services, and databases to communicate with other IoT devices will produce a vast number of new interconnections between the current system and the new system. The resulting information networks will reduce costs and risks and improve waste management processes. IoT devices are expected to reduce the cost of waste collection by using the information exchanged among bins, containers, and trucks. These devices enable automation and accelerate the identification of waste for processing.

The future adoption of Internet technologies is enhanced by the Internet Protocol, which provides various sensors (e.g., wireless sensors) to an IoT system. Different numbers of sensors form a wireless network, which is called a wireless sensor network (WSN). WSNs process and collect waste data from a city to upgrade its infrastructure; such city is subsequently called a smart city [9]. A smart city is defined as follows: "A Smart City is a city well performing in a forward-looking way in the following fundamental components (i.e., Smart Economy, Smart Mobility, Smart Environment, Smart People, Smart Living, and Smart Governance), built on the 'smart' combination of endowments and activities of self-decisive, independent, and aware citizens" [10].

At present, static models are used to convert the waste collection process into a service that supports the online dynamic routing and scheduling of collection trucks. A dynamic waste management system is presented as an online deduction process that is used to decide when trucks should collect waste from smart bins in a source building and to determine the optimum paths that the collection trucks should use. Many achievements and equipment have been applied to waste management, including various management approaches and information collected in this area. With the emergence of computer equipment and innovations, ordinary objects are transformed into smart devices. IoT innovation focuses on gadgets that can receive, collect, and exchange environment data through radio-frequency identification (RFID) tags, the Global Positing System (GPS), near-field communication (NFC), cameras, sensors, and actuators [9].

RFID tags are utilized for labeling containers and distinguishing proofs. Sensors are the basic components of IoT-based waste management, and they account for a considerable number of investigations in this field. Sensors are specialized with a set of physical quantities that include capacity, weight, temperature, stickiness, and functional responses [9, 11–14].

The sensors used in NFC and WSNs are connected to enable them to exchange information within the foundation. They are connected to one another either wirelessly or through a spinal arrangement framework. The actuator is utilized by a framework that is associated with garbage; it can bolt the cover of a garbage bin once the bin is full [13]. To measure garbage volume, cameras are used as uncommon sensors. Although cameras are not considered conventional sensors, they are part of the sensing system, which are investigated independently. The most utilized device in IoT innovations is the GPS, which provides adequate information regarding collection areas. The GPS is fundamental for area following amid energetic directing [13, 14].

2 Related Works

Different waste management models have been developed using certain IoT devices. This section reviews the existing IoT-based waste management studies.

Distinctive models use sensors, such as temperature, chemical, capacity, and weight sensors, which concur with the functions of IoT gadgets.

In [15], the researcher introduced a stage of municipal waste management using the collection of recycled waste data based on IoT innovation. Several waste management activities, such as waste collection, transport, and recycling, are also presented. However, an active optimization technique was proposed by [12] for efficient waste recycling. An active selection that presents descriptions by identifying factors was developed. It compares the characteristics of wastes in a container on a day-to-day basis.

A specific monitoring system was proposed in [12] to monitor garbage bins and trucks using RFID and information and communication technology. In addition to cameras, GPS is utilized in this system [12].

Garbage bins used for different types of waste are placed outside doors. Homogenous and heterogeneous trucks are used to collect waste [12, 15]. In [16], the collection of bins is recorded to enhance the waste collection process. The design and implementation of the waste disposal system is explained to emphasize the importance and variability of the facility. This system supports areas that can improve the effectiveness of soil nutrients and the population of consumers during different periods of the year. It has an intelligent framework and structure for data used in statistical communication procedures.

The authors of [17] presented a solid waste management system to illustrate the driver training process using a WSN. The WSN model, which provides an effective solution for the problem of waste collection in a city, is presented. This system improves performance and supervises waste transportation to a site during the collection process.

A collection monitoring model was proposed in [18]. This model can detect and evaluate waste through sensors installed in bins during the early stage. In [18], which is an extension of [16], a new application was presented to monitor MSW based on a distributed sensor technology and a geographic information system.

An energy utilization model for optimizing solid waste collection in large cities and other areas was presented in [19]. This work is an extension of [18], in which three models were developed for dynamic scheduling and routing optimization.

In [20], a model for analyzing the impact of solid waste source sorting on fuel consumption and collection costs was proposed. In [21], an economic performance comparison was conducted between pneumatic and door-to-door waste collection systems in urban sites. This study presents an analysis on how a hypothetical stationary pneumatic waste collection system can be economically compared with a standard truck-operated door-to-door collection system. In intensively populated urban areas, the pneumatic and door-to-door waste collection systems will experience different shortcomings. The deployment costs of a pneumatic system in existing residential areas will definitely be increased by buildings and fixed city infrastructure. In [22], the authors presented an automated waste collection system based on a ubiquitous sensor network. A new method for collecting MSW in populated areas and business buildings that applies IoT technology was presented. However, capacity sensors are the most required devices in IoT technology, followed by RFIDs, and then weight sensors.

From the preceding review, most of the proposed models did not consider waste type and the effects of each type on the health of the people. Moreover, the researchers believe that waste cannot be reused. Although, not all waste types can be reused, some types can be utilized as crude materials. Private waste or garbage produced by households, such as vegetables peels, natural products, paper, and polythene, can be reused. Vegetable peels can be utilized as fertilizer, whereas polythene can be sold to advertisers. Recycling occurs when we assume liability in sorting our waste. Similarly, sorting becomes simple when initially conducted at home. A garbage collector experiences difficulty in sorting waste, which may result in health problems. Thus, sorting waste from the source is important. Waste type can be easily identified from the source. For example, organic waste can be found in restaurants, hypermarkets, and households, whereas solid waste can typically be found in construction sites. In addition, waste transportation does not use intelligent algorithms for routing and collection [9, 14].

3 Proposed Design

The architecture of the proposed system consists of waste resources to solve and manage various waste problems. Smart bins are used to improve waste collection and provide information for statistical analysis, waste collection trucks are used to improve the transportation of collected waste, and waste management centers are utilized for data analysis and decision-making. Figure 1 shows the proposed architecture.

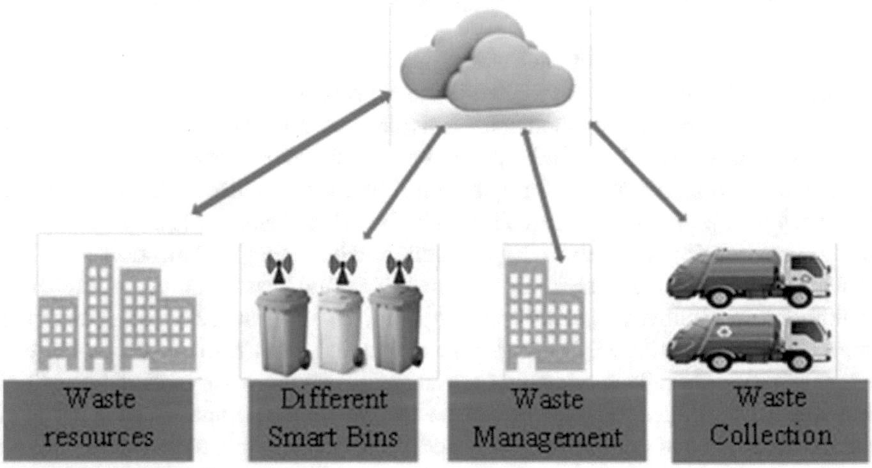

Fig. 1. Architecture of waste management system

Waste Source Building: Each source has a unique ID and priority rate, which depend on the status of waste and the importance of the place, such as hospitals and schools.

Smart Bins: Each bin is connected to the cloud system according to waste type via either GSM or GPRS, which sends the following information to the waste management center: waste type, weight, and volume (only if the bin reaches its maximum weight or if the bin is full). Once a level of 90% is exceeded, the system sends bin information to the transportation to collect waste. Simultaneously, the system provides a weight priority in case the result exceeds the threshold value. Then, the system determines the best truck with an optimal route to collect waste by using an intelligent routing algorithm. Figure 2 shows the smart bin processing information based on the following equations:

$$\begin{gathered} \textit{If } vm \geq tvm \rightarrow \textit{then send the information to perform an action; else, read volume} \\ \textit{or} \\ \textit{if } wg \geq twg \rightarrow \textit{then send the information to perform an action; else, read weight;} \end{gathered}$$

where vm is the volume measure, tvm is the threshold volume measure, wg is the weight measure, and twg is the threshold weight measure.

Waste Collection Trucks (Transportation): Smart cities are densely populated and urbanized; thus, collecting waste using standard trucks, particularly during peak hours, is difficult. The size of collection trucks is determined by the type of waste. Small trucks reduce traffic and can easily move around a city. Trucks are used to collect wastes from waste source buildings and then bring them to a landfill site or to a recycling center. Each truck is connected to GPS/GPRS. This method proposes different truck sizes according to the area and the expected waste generation capacity. Trucks receive commands from the waste management center based on truck position and the available information in the center. Figure 3 shows the procedure of calling trucks for waste collection.

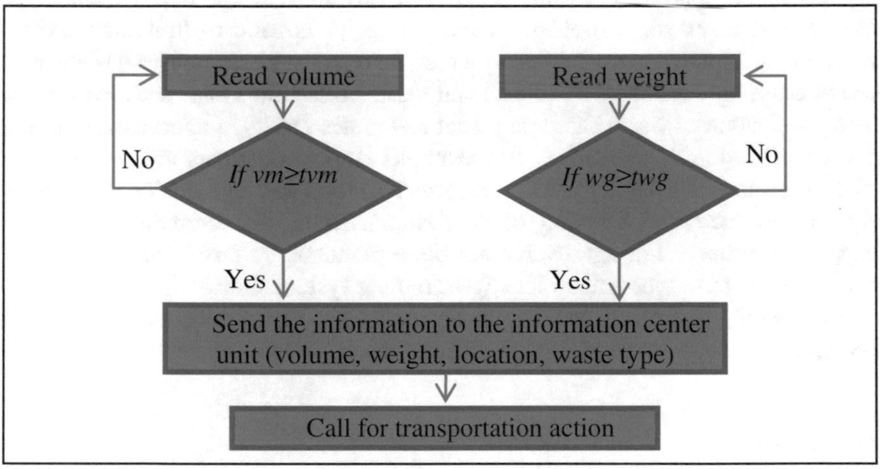

Fig. 2. Smart bin processing information

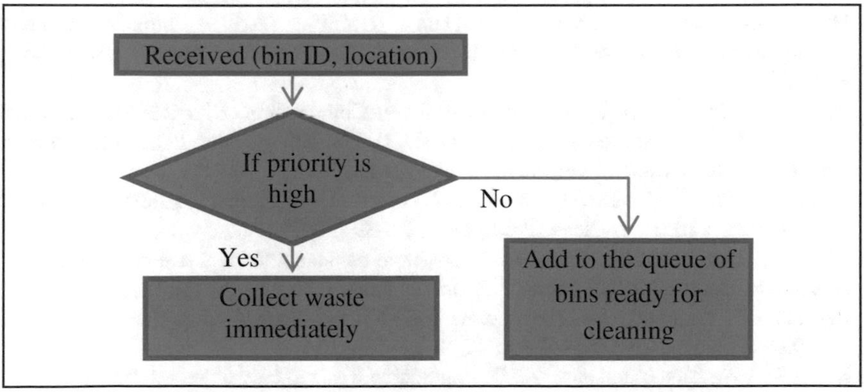

Fig. 3. Procedure for collection decision-making

The smart bin database is hosted on a cloud computing infrastructure that provides high data availability. The devices installed in the bins provide real-time data collection. The waste management center analyzes the collected data from the IoT devices and provides useful information for decision-making within the intelligent waste management system.

4 Conclusions

This study focuses on the most efficient IoT models used for waste management. Collection devices with high time cost and inefficient operation are presented. Collection processes in heavily populated urban areas, such as pneumatic and door-to-door waste collection operations, exhibit different disadvantages. The deployment costs of these systems in existing populated areas are increased due to buildings and fixed infrastructure. Moreover, the limited space in the area for waste transportation and bin location causes problems in truck-operated waste collection systems. However, planning the collection of waste based on predefined routes results in unnecessary costs and the underutilization of equipment. For example, empty containers are frequently collected, whereas full containers spill onto streets. Moreover, IoT devices can immediately lead to increased cleaning costs, health hazards, and complaints from irate citizens. The proposed design overcomes these problems by providing different truck sizes based on waste type and an intelligent routing system. Our future work will focus on investigating effective IoT-based devices for waste collection and energy consumption.

References

1. Boonrod, K., Towprayoon, S., Bonnet, S., Tripetchkul, S.: Enhancing organic waste separation at the source behavior: a case study of the application of motivation mechanisms in communities in Thailand. Resour. Conserv. Recycl. **95**, 77–90 (2015)
2. United Nations Environment Program (UNEP) (2017). http://www.unep.org/resourceefficiency/what-we-do/sustainable-lifestyles/food-and-food-waste. Accessed 12 Sept 2017
3. NSWMD: Final Report: Survey on Solid Waste Composition, Characteristics & Existing Practice of Solid Waste Recycling in Malaysia, p. 171. Kementerian Kesejahteraan Bandar, Perumahan dan Kerajaan Tempatan, Putrajaya, Malaysia (2013)
4. Barnes, P., Jerman, P.: Developing an environmental management system for a multiple-university consortium. J. Clean. Prod. **10**(1), 33–39 (2002)
5. Dahle, M., Neumayer, E.: Overcoming barriers to campus greening: a survey among higher educational institutions in London, UK. Int. J. Sustain. High. Educ. **2**(2), 139–160 (2001)
6. Carvalho, P., Marques, R.C.: Economies of size and density in municipal solid waste recycling in Portugal. Waste Manag. **34**(1), 12–20 (2014)
7. Khan, M.S.M., Kaneesamkandi, Z.: Biodegradable waste to biogas: renewable energy option for the Kingdom of Saudi Arabia. Int. J. Innov. Appl. Stud. **4**(1), 101–113 (2013)
8. Nizami, A.S., Shahzad, K., Rehan, M., Ouda, O.K.M., Khan, M.Z., Ismail, I.M.I., Almeelbi, T., Basahi, J.M., Demirbas, A.: Developing waste biorefinery in Makkah: a way forward to convert urban waste into renewable energy. Appl. Energy **186**, 189–196 (2017)
9. Anagnostopoulos, T., Zaslavsky, A., Kolomvatsos, K., Medvedev, A., Amirian, P., Morley, J., Hadjiefthymiades, S.: Challenges and opportunities of waste management in IoT-enabled smart cities: a survey. IEEE Trans. Sustain. Comput. **2**, 275–289 (2017)
10. Centre of Regional Science: Smart cities. Ranking of European medium-sized cities, Vienna University of Technology (2007). http://www.smart-cities.eu. Accessed 13 Aug 2015

11. Tao, C., Xiang, L.: Municipal solid waste recycle management information platform based on internet of things technology. In: 2010 International Conference on Multimedia Information Networking and Security (MINES), pp. 729–732. IEEE (2010)
12. Hannan, M.A., Arebey, M., Begum, R.A., Basri, H.: Radio frequency identification (RFID) and communication technologies for solid waste bin and truck monitoring system. Waste Manag. **31**(12), 2406–2413 (2011)
13. Al Mamun, M.A., Hannan, M.A., Hussain, A., Basri, H.: Wireless sensor network prototype for solid waste bin monitoring with energy efficient sensing algorithm. In: 2013 IEEE 16th International Conference on Computational Science and Engineering (CSE), pp. 382–387. IEEE (2013)
14. Rajkumar, M.N., Abirami, P., Kumar, V.V.: Smart garbage collection monitoring systems. Int. J. Sci. Res. Comput. Sci. Eng. Inf. Technol. (IJSRCSEIT) **2**(1), 13–17 (2017)
15. Anghinolfi, D., Paolucci, M., Robba, M., Taramasso, A.C.: A dynamic optimization model for solid waste recycling. Waste Manag. **33**(2), 287–296 (2013)
16. Vicentini, F., Giusti, A., Rovetta, A., Fan, X., He, Q., Zhu, M., Liu, B.: Sensorized waste collection container for content estimation and collection optimization. Waste Manag. **29**(5), 1467–1472 (2009)
17. Longhi, S., Marzioni, D., Alidori, E., Di Buo, G., Prist, M., Grisostomi, M., Pirro, M.: Solid waste management architecture using wireless sensor network technology. In: 2012 5th International Conference on New Technologies, Mobility and Security (NTMS), pp. 1–5. IEEE (2012)
18. Rovetta, A., Xiumin, F., Vicentini, F., Zhu, M., Giusti, A., He, Q.: Early detection and evaluation of waste through sensorized containers for a collection monitoring application. Waste Collect. **29**(12), 2939–2949 (2009)
19. Fan, X., Zhu, M., Zhang, X., He, Q., Rovetta, A.: Solid waste collection optimization considering energy utilization for large city area. In: 2010 International Conference on Logistics Systems and Intelligent Management, vol. 3, pp. 1905–1909. IEEE (2010)
20. Di Maria, F., Micale, C.: Impact of source segregation intensity of solid waste on fuel consumption and collection costs. Waste Manag. **33**(11), 2170–2176 (2013)
21. Teerioja, N., Moliis, K., Kuvaja, E., Ollikainen, M., Punkkinen, H., Merta, E.: Pneumatic vs. door-to-door waste collection systems in existing urban areas: a comparison of economic performance. Waste Manag. **32**(10), 1782–1791 (2012)
22. Arbab, W.A., Ijaz, F., Yoon, T.M., Lee, C.: A USN based automatic waste collection system. In: 2012 14th International Conference on Advanced Communication Technology (ICACT), pp. 936–941. IEEE (2012)

Architecture for Latency Reduction in Healthcare Internet-of-Things Using Reinforcement Learning and Fuzzy Based Fog Computing

Saurabh Shukla[1(✉)], Mohd Fadzil Hassan[1], Low Tan Jung[1], and Azlan Awang[2]

[1] Department of Computer and Information Sciences, Universiti Teknologi PETRONAS, 32610 Seri Iskandar, Perak Darul Ridzuan, Malaysia
saurabhshkl.shukla@gmail.com, {mfadzil_hassan,lowtanjung}@utp.edu.my
[2] Department of Electrical and Electronic Engineering, Universiti Teknologi PETRONAS, 32610 Seri Iskandar, Perak Darul Ridzuan, Malaysia
azlanawang@utp.edu.my

Abstract. Internet-of-Things (IoT) generate large data that is processed, analysed and filtered by cloud data centres. IoT is getting tremendously popular: the number of IoT devices worldwide is expected to reach 50.1 billion by 2020 and from this, 30.7% of IoT devices will be made available in Healthcare. Transmission and analysis of this much amount of data will increase the response time of cloud computing. The increase in response time will lead to high service latency to the end-users. The main requirement of IoT is to have low latency to transfer the data in real-time. Cloud cannot fulfill the QoS requirement in a satisfactory manner. Both the volume of data as well as factors related to internet connectivity may lead to high network latency in analyzing and acting upon the data. The propose research work introduces a hybrid approach that combines fuzzy and reinforcement learning to improve service and network latency in healthcare IoT and cloud. This hybrid approach integrates healthcare IoT devices with the cloud and uses fog services with Fuzzy Reinforcement Learning Data Packet Allocation (FRLDPA) algorithm. The propose algorithm performs batch workloads on IoT data to minimize latency and manages the QoS of the latency-critical workloads. It has the potential to automate the reasoning and decision making capability in fog computing nodes.

Keywords: Internet-of-Things · Fog computing · Cloud computing Reinforcement learning · Fuzzy inference system

F. Saeed et al. (Eds.): IRICT 2018, AISC 843, pp. 372–383, 2019.
https://doi.org/10.1007/978-3-319-99007-1_36

1 Introduction

IoT can be defined as a system of interrelated devices comprises of an object having a unique identifier and can share information or transfer the data over a network without the involvement of human beings. The number of IoT devices has exploded and huge numbers of applications are processed in the cloud that makes many service calls by the end users for the request of services over the network. Moving all data from the network edge to the cloud data center for processing adds latency. IoT requires the transfer of data in real-time. Data travels a longer hazardous path from mobile device to the mobile cloud during offloading and thus consumes huge network bandwidth.

When IoT device generates data, it may not necessarily require transferring the data to the cloud. The main requirements for IoT devices are minimum latency, conserve network bandwidth and moving data to the best place for processing. However, there are few challenges in healthcare IoT and cloud. These challenges are high network latency and high service latency. The cloud cannot meet all the requirements of QoS (Quality of Service) in healthcare IoT, so that a new architecture is needed. For a continuous 24/7 remote health monitoring system, rapid decision making and agile responses are essential for several acute diseases and emergencies, where data processing and transmission time should be minimized [1]. In cloud computing where raw data is transferred from sensor nodes to cloud, if network condition is unpredictable, it may cause uncertainty in response latencies [1].

The situation is more critical when streaming-based data processing is needed (e.g., signal processing on ECG or EEG signals) [1]. Healthcare IoT should closely monitor the status of patient health as the health condition changes continuously and it becomes nearly impossible to record the patient health state in real-time [1]. The data collected by the healthcare IoT becomes less authentic as there is constant change in patient physiological state with time [7]. This situation makes the data useless and unreliable for users because it does not reflect the patient's physical condition in real-time.

The issue of delay in healthcare IoT can be from milliseconds to microseconds [11]. Many interactive experiences require response times of under 20 ms [11]. Good real-time online applications require a latency of fewer than 80 ms [11]. Some of the real-time applications cannot be accessed when the latency is higher than 200 ms [11]. In addition, a large amount of data stored in the sensors depletes its energy; wireless body sensor nodes require large and safe storage of real-time data with better communication from the cloud and users. It is challenging in healthcare IoT to have decisions in real-time and to track patient records. Therefore features of healthcare IoT data requires reinforcement learning and fuzzy inference system to track patient past experiences, history and to make decisions in the real-time environment. Here, the gateway should be intelligent enough to stop uploading the data to the cloud.

In January 2014, Cisco launched its own version of fog computing, designed with the idea of bringing cloud computing facilities and network side capabilities to the edge of devices. After a while, Cisco unveiled the company's IOX platform, which has been designed to deliver speed computing. Moving computing resources to the edge of the IoT devices are done by fog computing which generates very little service latency [2]. Fog computing gateway works as a middleware between physical devices like IoT and the cloud. The primary function of fog computing is to transfer the data in real-time by acting as a gateway between IoT and cloud [2]. Fog computing nodes have very close proximity to IoT due to this fog node can process the most time-sensitive data in real-time mode and send the data to cloud for further processing. Fog computing is not aimed to substitute cloud computing but to complement it, which can ease bandwidth burden and reduce transmission latency.

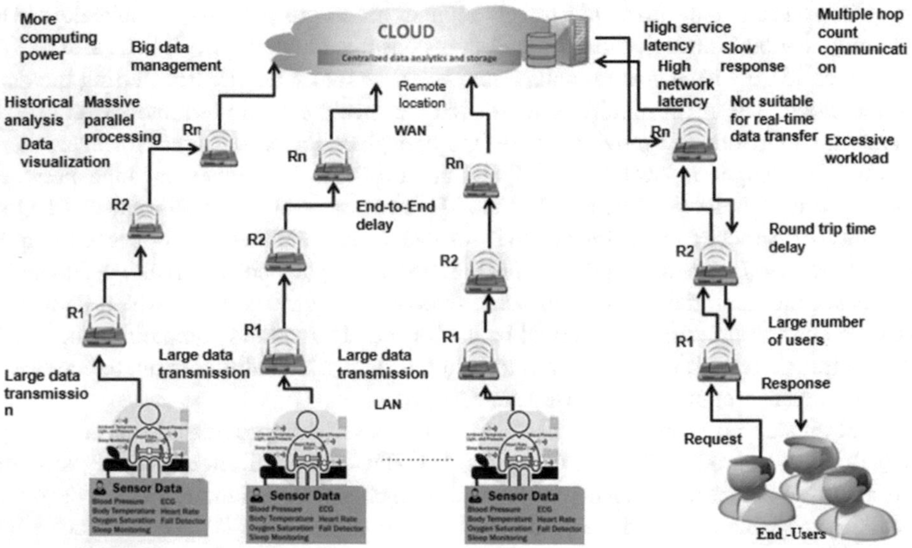

Fig. 1. Conventional data transmission between healthcare IoT and cloud

Here Fig. 1 shows the conventional method of data transmission between healthcare IoT and cloud. All the conventional fog computing deals only with reducing the burden of the cloud and minimizing latency in IoT. In this, an infrastructure is required that can collect patient health data in real-time by minimizing high network and service latency. Enhanced fog computing, by acting as a middleware, will able to remove the limitations of healthcare IoT and improves the services in the network. Fog computing, which performs on network edges, is a front-end distributed computing archetype of centralized cloud computing. Enhanced intelligent based fog is a special purpose computing architecture which leverages the time-sensitive data to end-users by reducing high latency.

Hence we propose a novel solution for the above-mentioned problem using a hybrid approach that uses fog computing architecture that integrates healthcare IoT devices with the cloud and uses fog services with FRLDPA algorithm. The hybrid mechanism will allow the collocation of latency-critical workloads to maximize the data center utilization. Here jobs and resources allocated to latency critical workloads are enough to meet the QoS target. Reinforcement learning helps here to make the best usage of the available resources by assigning the various process to different processors without violating the QoS constraints for latency critical workloads. In practical settings, reinforcement learning can provide a substantial advantage over other regression techniques by eliminating the need to train the model with large amounts of offline labelled-data. Here reinforcement learning will help in getting feedback from the past patient health records in the propose algorithm where a decision-making process will learn the best course of action to maximize its total reward over time for end-users. This will further help in minimizing high service and network latency.

2 Background and Related Work

In [1], a Smart e-Health Gateway is proposed to solve the issue of high network latency in IoT by exploiting the concept of fog computing and forming a Geo-distributed layer of intelligence between sensor nodes and cloud. The main limitation of this work is that the smart fog gateway lacks the decision making capability. However, it can cope with many challenges like mobility, energy efficiency, and scalability and reliability issues. In [4], a Smart Gateway with fog computing is proposed to solve the issue of handling a large amount of data generated by IoT which further leads to high service latency. Here the fog nodes are not able to store the heterogeneous data. In [7], the fog computing as a smart gateway is proposed to provide the techniques and services as embedded data mining, distributed storage, and notification service and to reduce high latency. The main limitation of this work is that it cannot perform data aggregation and filter at the network edge. In [10], a technology called Hermes is proposed to minimize latency for delay-sensitive applications and optimal task assignment for resource-constrained mobile devices using fog computing. The proposed technique used to perform offloading of the computational task along with the formulation of an NP-hard problem to minimize the application latency. They have use novel fully polynomial time approximation scheme (FPTAS). There are few challenges like a decision in real-time mode and extracting previous historical data. In [12], the concept of iFogstar is introduced. iFogstar helps in minimizing service latency, network latency of cloud and reduces network traffic. It helps in reducing the burden of the cloud. It acts in a distributed manner and suitable for heterogeneous devices. iFogstar has fog infrastructure and uses heuristic approach. Solutions proved very good performance as they reduced the latency by more than 86% as compared to a Cloud-based solution and by 60% as compared to a naive Fog solution. It has few challenges like storing heterogeneous data, tracking previous records and to make decisions in real-time mode.

In [13], a reinforcement learning code offloading mechanism is proposed using mobile fog to provide low latency services to mobile service consumers. They have used distributed reinforcement learning algorithm to offload basic blocks in a decentralized fashion to deploy mobile codes on geographically distributed mobile Fogs. The approach consists of basic block migration policy along with multi-agent distributed method. Agents learn from the environment through reinforcement. However, it has its own limitations that it can be used for mobile fog and distributed computing environment only. The proposed method significantly reduces the execution time of mobile services. There are few limitations in the proposed work like virtualization, decision-making capability, tracking previous record, mobility, and end-users issues. In [14], a Hipster, is introduced that combines heuristics and reinforcement learning to manage latency-critical workloads in the cloud. Hipster's goal is to improve resource efficiency in data centers while respecting the QoS of the latency-critical workloads. Hipster, which is a hybrid management solution combining heuristic techniques and reinforcement learning to make resource-efficient allocation decisions, specifically deciding the best mapping of latency-critical and batch workloads on heterogeneous cores. The proposed technique has limitation for decision making in the real-time environment.

All the mention of approaches and techniques used in conventional fog computing environment are unable to fulfill the requirement of healthcare IoT. Since healthcare IoT requires decision making capability, tracking previous patient health records and to transfer the patient health data (PHD) in real-time mode. Existing contributions made to reduce latency in the cloud and healthcare IoT devices defined a minimum role for fog computing to transmit the data in real-time mode to end-users. The discussed systems basically dispense conventional method for collecting data from nodes and sending these data to remote servers using fog computing. More precisely, none of these works have considered covering the deep learning approach with fog computing to address all the issues related to latency minimization and transmission of data in real-time mode. The physiological state of patient health data (PHD) changes with time so there is an urgent need to track the patient past experience to make the decisions in real-time mode by minimizing high network and service latency. This can be done using reinforcement learning and fuzzy inference system. There are also few challenges in above-mentioned works and techniques like software virtualization, mobility, packet loss and transmission of dynamic data in real-time mode. Hence, there is a scope to carry out research into this domain. Due to the lack of existence of possible solutions to the above-mentioned challenges in cloud computing and IoT, this research leads to the possibility of development of a novel approach.

3 Reinforcement Learning with Fuzzy Inference System Hybrid Approach

In this section, the approach of latency reduction is discussed in healthcare IoT using fog nodes. Reinforcement learning with fuzzy inference system is used in fog nodes for latency reduction in healthcare IoTs. The basic way to operate and interact with IoT nodes, fog nodes, and cloud nodes is as follows: IoT nodes can process requests locally, send it to a fog node and then send it to the cloud. Fog nodes can request, forward to other fog nodes in the same domain or forward the request to the cloud. Cloud nodes process request and send back feedback to IoT nodes. Dynamic nature of applications in modern data centers, where reinforcement learning can customize/tune the model at runtime for different applications and servers, making it easy to dynamically learn the behavior of an application and decrease the latency. It has the ability to improve QoS Guarantee. A decision-making process will have to learn the best practice over time to maximize its total reward. At each discrete instant, this process can see its current "state", and it must choose "action" from one set of options. Based on the chosen action and current situation, it enters the next state and receives the reward. To perform inferencing from the collected data, membership functions are designed for fuzzification and defuzzification processes (Fig. 2).

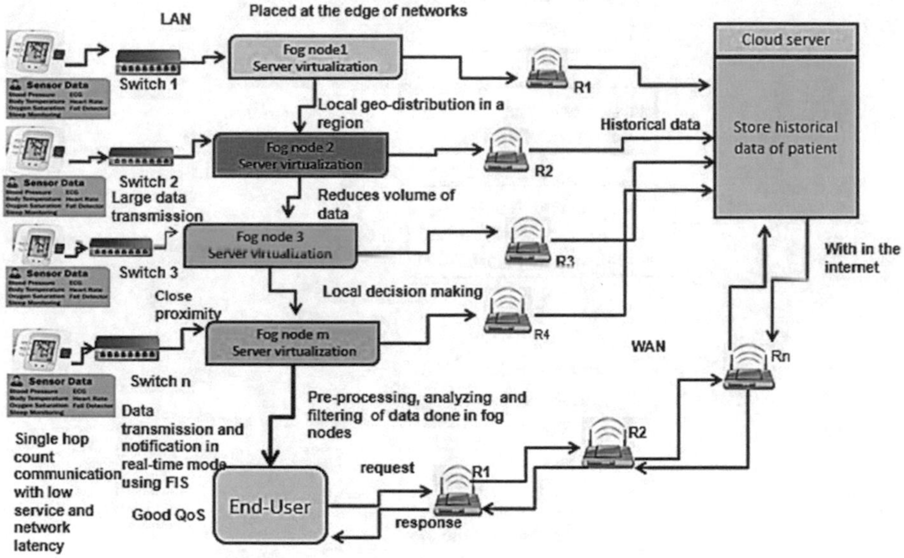

Fig. 2. Enhanced Data Transmission in real-time mode using enhanced fog computing to reduce high network and service latency in healthcare IoT and cloud.

3.1 FRLDPA Algorithm

The propose algorithm performs batch workloads on healthcare IoT data to minimize latency and manages the latency-critical workloads. The FRLDPA algorithm performs the function in two ways i.e. from pseudo code 1 to 12, it allocates the data packet to processors of multiple virtual machines in fog server using virtualization. Then the data are processed simultaneously in different processors. The pscudo code from 13 to 28 shows the patient health data generated by healthcare IoT can be transferred in real-time mode to medical agencies and doctors using reinforcement learning and fuzzy inference system in fog computing environment. The fog gateway first receive the sensor data. Then divide the fog server into virtual machines using virtualization.

Then to check the availability of processors in virtual machines of fog server using reinforcement learning. After this allocation of data packets to various processors of fog server for local distributed data processing at the edge using reinforcement learning. Then send the same patient health data in real-time mode using SPARK in fog node. Patient id is created and patient health data are merged in fog node. Then fuzzy seta are created for the final values of patient. Use of fuzzy inference system and kalman filter to measure and update the different physiological state of patient health data and to send the notification of PHD in real-time mode to end-users from fog node. This algorithm is expected to minimize the high service latency in healthcare IoT and cloud by transferring the healthcare IoT data in real-time mode (Fig. 3).

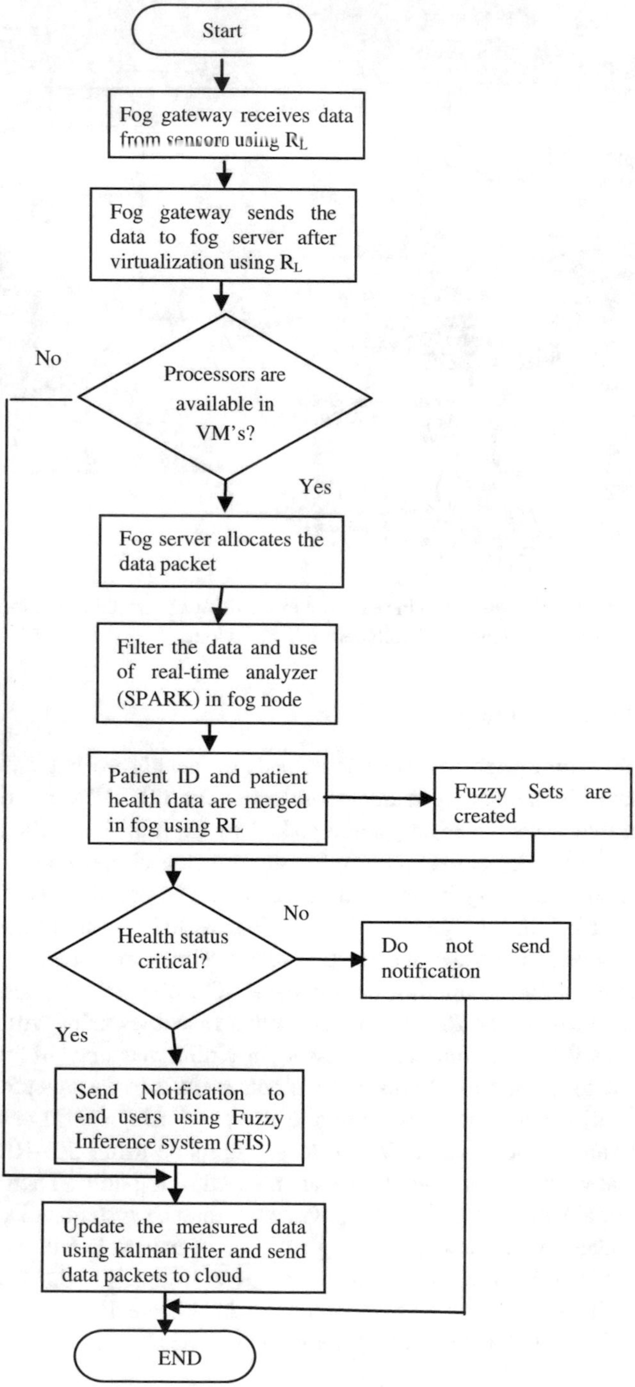

Fig. 3. FRLDPA algorithm process flowchart

FRLDPA Algorithm Symbol Notations
d_p: data packet
C_D: Cloud data center
f_G: Fog gateway for allocation of data packet
f_S: Each fog gateway contains fog server
V_M: Virtual machines at fog server
M_i: Minimum time to release the processor
Input: Streams of data from various body sensors
Output: Patient report and health status
P_{id}: Patient id
R_L: Reinforcement Learning
$\mu 1$: Fuzzy Inference System membership function
$\mu 2$: Fuzzy Inference System membership function

FRLDPA Algorithm Steps
Requirement: Fog layer and Cloud layer consists of gateways and data centers. Fog layer has fog gateway (f_G). Whereas cloud layer consists of cloud data center (C_D).
Step 1: (f_G) has fog server (f_S) that is going to check the availability of the processors for allocating data packets. It also manages virtual machines using R_L feedback controller
Step 2: The body sensor sends its data packet (for allocation) to fog gateway (f_G). Next fog gateway uploads the data packet to its fog server (f_S) using R_L.
Step 3: (f_S) allocates the data packet to the processors of virtual machines after virtualization in following cases:

1) If the processors for data packet allocation available to virtual machines (V_M) of fog server, it allocates the data packet.
2) If (fG) is in the initial state of allocation, then the data packet allocation in virtual machines is stopped for some Mi (minimum time); then after some time, sensor sends its data packet for allocation to a virtual machine (VM) of fog server.
3) If the processors for data packet allocation are available at one fog server, but few fail during allocation, then sensor again sends the data packet for allocation
4) If no processor is available for data packet allocation in virtual machines of fog server, then the data packet is sent to a cloud for allocation.

Step 4: Next is to transfer the patient medical data in real time through Spark (as a real-time analyzer).
Step 5: Consider the historical data of the patients in the system. Real-time data and historical data both will be merged using Pid(Patient id) as the primary key.
Step 6: Tuples are merged in a fuzzy system which is used as input to the fuzzy system.
Step 7: Fuzzy sets are created for the final values.
Step 8: Next to use Fuzzy Inference system for the notification of patient health status to end users.
Here the role of VM's is to consider the request for data packet allocation and transfer it to fog server. Whereas the role of fog server is to handle the data for allocation to available processors by using the virtualization technique.

FRLDPA Algorithm Pseudo Code

1. For each data packet d_p.
2. Each data d_p is sent to nearest location fog gateway.
3. Each f_G will allocate the data packets to f_S using R_L
4. f_S will allocate the data packets to the various processors of virtual machines in following cases.
5. IF all processors for data packet allocation are available to VM of f_G
ELSE
6. THEN f_G uploads the data packet for allocation to virtual machines using R_L
ENDIF
7 IF f_G in the initial state of allocation.
8. THEN stop the data packet allocation for some M_i (minimum time)
9. IF for allocation of data packet processors are available at one f_G but some fail during allocation of data packet
10. THEN again send d_P for allocation using R_L
11. IF no processors are available for allocation in V_M of f_G
12. THEN d_P is sent for allocation to C_D through a network.
ENDIF
13. For Patient ID, Pid Σ (Pid (Healthcare IoT data))
14. Σ (Pid (Healthcare IoT Data)) ->SPARK (RTA) in fog
15. In SPARK,
IF (Pid) registered in the cloud.
Row (new) <=Σ (Pid (Healthcare IoT Data)) + Pid
ELSE
register and create a new Pid.
Row (new1) <=Σ (Pid (Healthcare IoT data))
}
(Fuzzy system) the system is created with inputs and their functions μ1
16. With the function μ1 (Blood Pressure1), μ1 (HeartRate1), μ1 (ECG1), μ1 (Blood Sugar1) get the condition of health as μ1 (normal) or μ1 (low risk) or μ1 (high risk)
17. IF (Health Condition = μ1 (high risk1))
THEN get geo-location
Send a notification from fog based gateway to the cloud storage as the user can access from the cloud and fog itself
ELSE IF (Health Condition = μ1 (low risk)
THEN the prescriptions are sent to the patients
18. Kalman _State
19. Initialization kalman_init μ1 (Blood Pressure1), μ1 (HeartRate1), μ1 (ECG1), μ1 (Blood Sugar1))
20. Kalman_State result
21. Kalman_update μ2 (Blood Pressure2), μ2 (HeartRate2), μ2 (ECG2), μ2 (blood sugar2)

22. Predict and Measure Update
23. state (μ1 (Blood Pressure1)) ->μ2 (Blood Pressure2)
24. state (μ1 (Heartrate1)) -> μ2 (HeartRate2)
25. state (μ1 (ECG1)) ->μ2 (ECG2)
26. state (μ1 (Blood Sugar1) ->μ2 (Blood Sugar2)
27. Insert the NEW Tuple in fog server along with the condition of health which is further sent to cloud storage.
28. The client can access the CD and fG directly

Finally, the collected patient data are classified as a normal, low-risk and high-risk. Whenever a doctor enters a query based on the symptoms of a disease for checking the list of patients, the query is executed in two modes-serial and parallel with an aim to reduce the overall high network and service latency of the proposed scheme (Fig. 4).

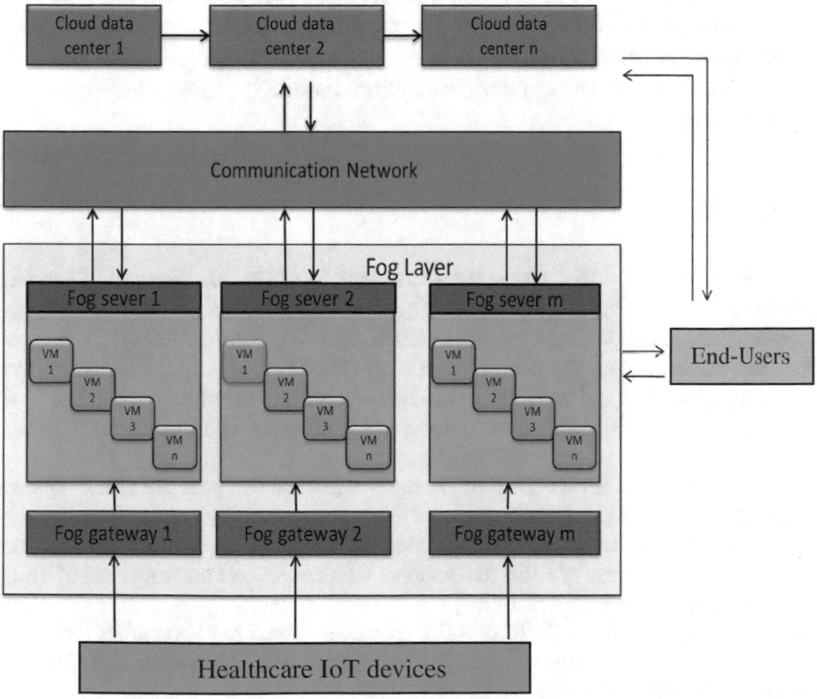

Fig. 4. Fog computing architecture representing fog layer, fog server and fog gateway for real-time data transmission between healthcare IoT devices and cloud data center.

4 Conclusion

To overcome this limitation, an enhanced fog computing architecture has been proposed, where cloud services are extended to the edge of the network to decrease the latency and network congestion. To realize the full potential of fog and IoT paradigms for real-time analytics, several challenges need to be addressed. Here fog computing covers all the latency minimization parameters which makes it more efficient for real-time data transmission in IoT devices and cloud. A fuzzy-based multi-agent based method is proposed using reinforcement learning. Agents learn from the environment through reinforcements. The propose FRLDPA algorithm is expected to allocate the data packets to various processors of virtual machines available in fog gateway by using the technique of virtualization. Using virtualization, fog gateway can tentatively reduce the volume of data to be sent to into data centers of the cloud. A biomedical data analysis of a patient is yet to be done with the early medical warning using FIS to transfer the patient's health data in real-time mode by reducing high latency to medical agencies and doctors. It is expected to show the improved performance of the proposed hybrid method in respect to network and service latency. The main motivation of this paper is to enhance IoT-based systems by providing a smart fog computing architecture. There are some challenging issues in the successful development of Fog including mobility we continue our works on virtualization, mobility, privacy and end-user issues of fog computing.

References

1. Rahmani, A.M., Gia, T.N., Nash, B., Anzanpur, A., Azimi, I., Jiang, M., Liljeberg, P.: Exploiting smart e-Health gateway at the edge of healthcare Internet-of-Things: a fog computing approach. J. Future Gener. Comput. Syst. **78**, 641–658 (2017)
2. Constant, N., Borthakur, D., Abtahi, M., Dubey, H., Mankodiya, K.: Fog-assisted wIoT: a smart fog gateway for end-to-end analytics in wearable internet of things. In: The 23rd IEEE Symposium on High-Performance Computer Architecture HPCA, Austin, Texas, USA, pp. 1–5 (2017)
3. Sethi, P., Sarangi, S.R.: Internet of Things: architectures, protocols, and applications. J. Electr. Comput. Eng. 2 (2017)
4. Aazam, M., Huh, E.N.: Fog computing and smart gateway based communication for cloud of things. In: Proceedings of IEEE International Conference on Future Internet of Things and Cloud, Barcelona, pp. 464–470 (2014)
5. Lee, W., Nam, K., Roh, H.G., Kim, S.H.: A gateway based fog computing architecture for wireless sensors and actuator networks. In: Proceedings of IEEE International Conference on Advanced Communication Technology (ICACT), Pyeongchang, pp. 210–213 (2016)
6. Al-Doghman, F., Chaczko, Z., Ajayan, A.R., Klempous, R.: A review of fog computing technology. In: Proceedings of IEEE International Conference on Systems, Man, and Cybernetics (SMC), Budapest, pp. 001525–001530 (2016)
7. Gia, N., Jiang, M., Rahmani, A.M., Westerlund, T., Liljeberg, P., Tenhunen, H.: Fog computing in healthcare internet of things: a case study on ECG feature extraction. In: Proceedings of IEEE International Conference on Computer and Information Technology, Liverpool, pp. 356–363 (2015)

8. Divi, K., Liu, H.: Modelling of WBAN and cloud integration for secure and reliable healthcare. In: Proceedings of the 8th International Conference on Body Area Networks (BodyNets 2013). ICST (Institute for Computer Sciences, Social-Informatics and Telecommunications Engineering), ICST, Brussels, Belgium, pp. 128–131 (2013)
9. Bibani, O., Mouradian, C., Yangui, S., Glitho, R.H., Gaaloul, W., Hadj-Alouane, N.B., Morrow, M.J., Polakos, P.A.: A demo of IoT healthcare application provisioning in hybrid cloud/fog environment. In: Proceedings of IEEE International Conference on Cloud Computing Technology and Science (CloudCom), Luxembourg City, pp. 472–475 (2016)
10. Kao, Y.-H.: Hermes: latency optimal task assignment for resource-constrained mobile computing. IEEE Trans. Mob. Comput. Trans. Mob. Comput. **16**, 3056–3069 (2017)
11. Datapath.io Homepage. https://datapath.io/resources/blog/aws-network-latency-map. Accessed 4 Sept 2017
12. Naas, M.I., Parvedy, P.R., Boukhobza, J., Lemarchand, L.: iFogStor: an IoT Data placement strategy for fog infrastructure. In: IEEE 1st International Conference on Fog and Edge Computing (ICFEC), pp. 97–104 (2017)
13. Alam, M.G.R., Tun, Y.K., Hong, C.S.: Multi-agent and reinforcement learning based code offloading in mobile fog. In: IEEE ICOIN, pp. 285– 290 (2016)
14. Nishtala, R., Carpenter, P., Petrucci, V., Martorell, X.: Hipster: hybrid task manager for latency-critical cloud workloads. In: IEEE International Symposium on High Performance Computer Architecture, pp. 409–420 (2017)
15. Skorin-Kapov, L., Matijasevic, M.: Analysis of QoS requirements for e-health services and mapping to evolved packet system QoS classes. Int. J. Telemed. Appl. 1–18 (2010)

Security Factors Based Evaluation of Verification Algorithm for an IoT Access System

Abbas M. Al-Ghaili[1(✉)], Hairoladenan Kasim[2], Marini Othman[1,2], and Zainuddin Hassan[2]

[1] Institute of Informatics and Computing in Energy (IICE), Universiti Tenaga Nasional (UNITEN), 43000 Kajang, Selangor, Malaysia
abbasghaili@yahoo.com, abbas@uniten.edu.my

[2] College of Computer Science and Information Technology (CSIT), Universiti Tenaga Nasional (UNITEN), 43000 Kajang, Selangor, Malaysia
{hairol,marini,zainuddin}@uniten.edu.my

Abstract. This paper proposes a verification algorithm (VA) used as an access tool for Internet-of-Things (IoT) systems. From a security aspect's point of view, the VA has been evaluated in terms of security factors. To perform this evaluation, a Multiple-Layer Security Architecture (MLSA) has been used in order to attain the three security objectives which are: confidentiality, integrity, and availability. The VA aims to enable an authorized access to smart IoT applications such as smart city applications. The VA has adopted the Quick Response (QR) tag to encrypt values being used as a smart-key. Performance of the proposed algorithm has considered the security factors in order to strengthen the security side of such an IoT system. To evaluate the VA accuracy, a total number of 70 QR tags have been generated. Results have shown a high percentage of accuracy in terms of security for verified QR tags used as smart keys for IoT systems.

Keywords: IoT · QR tag · Security objectives · Smart access

1 Introduction

Many researches from a broad variety of fields have exploited the technique of Quick Response (QR) in order to design a smart system. Lots of proposed researches have shown that many recently published studies related to the QR tag topic existent. Thus, when comparing the number of researches for the current decade to last three decades, it is rapidly increasing as shown in Fig. 1. It is used as an access card or a door-key for a wide range of applications since a QR-tag is easy-to-scan [1]. Thus, it has been exploited by many Internet-of-Things (IoT) applications [2]. QR tag can be used with IoT based smart applications, access systems [3, 4], authentication systems [5], smart home applications [6], smart cities related applications [7], identification systems [8].

However, these IoT based applications require a strong QR-tag encryption and verification procedures implemented to secure data [9]. There have been several security factors considered by many IoT research studies and applications.

F. Saeed et al. (Eds.): IRICT 2018, AISC 843, pp. 384–395, 2019.
https://doi.org/10.1007/978-3-319-99007-1_37

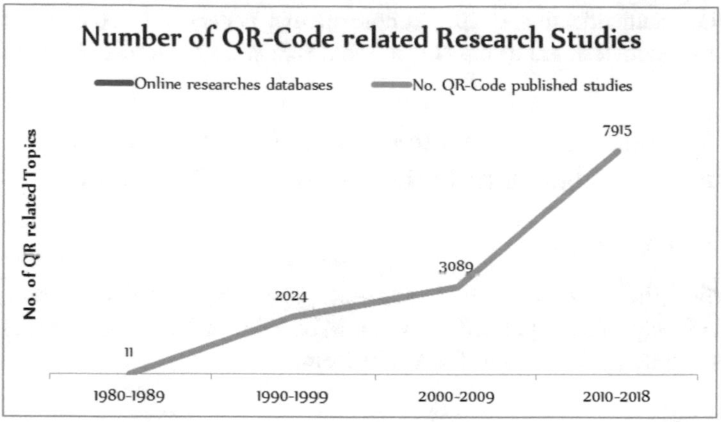

Fig. 1. Number of QR-code related research studies and topics

Even though QR tag based IoT algorithms dealing with smart applications still face challenges to cover security factors. Nevertheless, QR tag features are exploited by many researches to achieve a limited level to which the smart application is expected to reach. For example, the proposed work in [10] has designed a QR tag based authentication method for users achieving good performance in terms of threats prevention. Some other methods [3] used remote user authentication process with smart cards. Once, QR tag contents have been extracted, the verification procedure is an importantly performed step to make sure that contents are identical to original ones. Thus, the QR tag verification is essential in IoT applications, because it affects the system accuracy and privacy.

Many QR tag related researches have used a simple layer of encryption. These researches have used a single layer of security either to use a secure QR code without adding further verification algorithm to validate the time the QR tag has been generated. An authentication process [11] has been proposed in order to verify the document authentication. The existence of a single layer is however suitable with smart applications but those which include sensitive data are vulnerable to threats and attacks because they consider one or two security objectives. This paper considers five security objectives to verify. Therefore, to make the Verification Algorithm (VA) to achieve a high level of privacy, a Multiple-Layer Security Architecture for VA (MLSAVA) is designed to allow a safe access to the IoT application. In this paper, the proposed MLSAVA considers a number of security issues including integrity, authentication, and privacy. Additionally, its verification procedure has several layers to increase the security. The MLSAVA is useful to verify requests that need permissions to access such IoT automatic access systems by using a smart time-based label (TL) to calculate a QR tag age to decide whether the QR tag is authenticated or not and how authenticated it is.

This paper is organized as follows: Sect. 2 gives a simple introduction into the proposed MLSAVA. In Sect. 3, the proposed VA is discussed in detail to attain security objectives. In Sect. 4, the proposed MLSA for making a decision on QR tag security (i.e., authentication is an example in this paper) is presented. Performance analysis and evaluation are discussed in Sect. 5. Section 6 summarizes Conclusion.

2 The Proposed Multiple-Layer Security Architecture Verification Algorithm (MLSAVA) for IoT Systems

2.1 MLSAVA Definition

In this section, the VA has adopted to use multiple layers to increase the security of the encrypted QR tag. Meaning, MLSAVA is a MLSA-based VA which verifies user data to allow an authorized access to the IoT system.

2.2 MLSAVA Block-Diagram

The proposed MLSAVA block-diagram consists of three simple processes, as depicted in Fig. 2 which are as follows: the first one receives information from the QR tag. The second one alters verified results between VA and database that stores original values. The third one sends results to MLSA to verify their identities to make a decision whether to approve the request or disapprove.

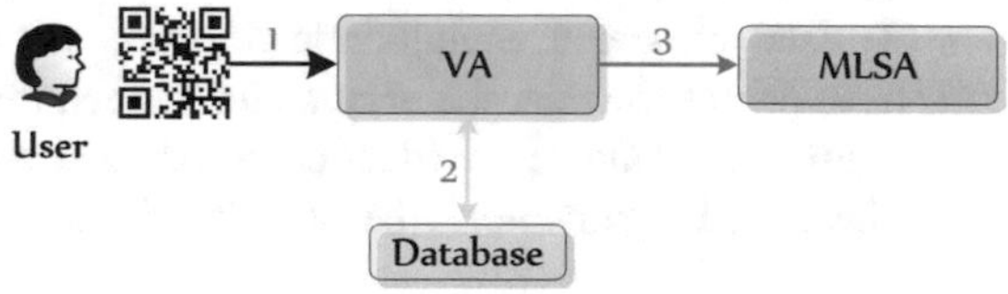

Fig. 2. The proposed MLSAVA block-diagram

2.3 Introduction to MLSAVA

This sub-section introduces three units of which the MLSAVA consists. The proposed architecture depicted in Fig. 2 includes three units. The first and second units are the processing units whereas the third one is the decision making unit. The first unit receives inputs from the QR tag made by a user who is interested to access an IoT system. The user request is verified by applying a number of verification procedures, will be discussed in detail in the following section. The user request is processed between VA and database units. The second unit feeds the VA with updated results. The obtained results will be sent to the third unit in order to validate and verify the user inputs' values. Finally, if the verification result is true, the MLSAVA accepts the request and then allows the system access.

3 The Proposed Verification Algorithm (VA)

3.1 VA Security Procedure

There are three procedures applied in order to implement the verification process which are as follows: QR Tag Verification is dedicated to verify the MLSAVA integrity of QR contents. The second verification process verifies the authentication and confidentiality of QR tag (QR tag is the key of the smart IoT application). The third one is Offline Data Base (ODB) verification. ODB is adopted in order to make sure the MLSAVA keep the QR tag available and responsive as a key during each time an access is needed.

3.2 QR Tag Verification

In this type of verification, three procedures are applied which are biometrics images-, serial ID-, and SQ- based verification procedures. The proposed architecture is illustrated in Fig. 3. Once the QR tag is scanned, its contents (e.g., biometrics pre-stored and pre-encrypted) are extracted and verified. The user is asked to provide a serial ID and also needs to answer a security question. The result of a comparison between original and entered values is sent to the MLSA.

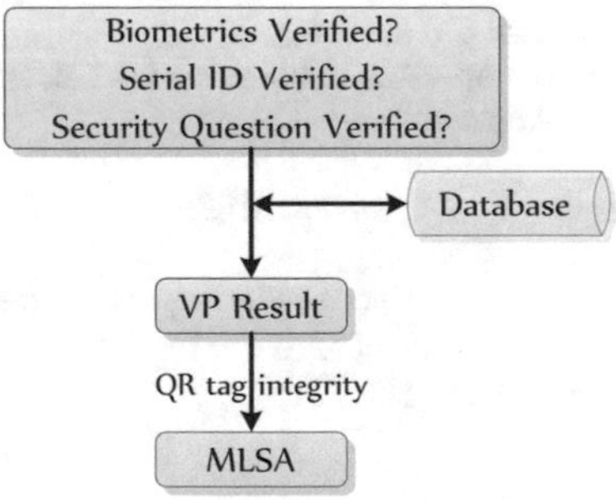

Fig. 3. QR tag verification

Biometrics Based Face Access Verification Procedure. The main VA step is to collect user information, after scanning the QR tag. The purpose is to do a successful comparison between QR tag and original pre-stored information as illustrated in Fig. 4. Once information asked by the user and information stored in the database are identical, the system can be accessed. Otherwise, the access is denied.

In this figure, certain information stored in the database will be read to check whether they are identical to information collected from the user or not. This comparison simply checks values of images of face and fingerprint with QR tag values. If they are correct, the process goes forward to a further verification step; otherwise, the process is halted.

Fig. 4. Biometrics verification procedure

Serial ID Verification Process (ID VP). New users are given distinctive IDs each of which consists of a hash-based encrypted ID. To implement ID VP, the user is going to key-in his/her ID first. Then, a series of decryption-based math operations is applied to extract the hash value. After that, several related hash values are extracted. A comparison is applied to check whether they are identical or not. Finally, approval result is sent to MLSA to do a further decision. This is simplified in Algorithm 1.

```
Start
QR tag for the user #i is scanned
Related Serial ID is extracted using decryption procedure
Its related hash value is calculated → HashQR(i)
Hash value stored in DB is recalled → HashDB(i)
Compare HashQR(i) to HashDB(i)
Return the comparison value (True OR False)
        Approved?
           Then;
                Send to MLSA
           Else
                The access request is Rejected
End
```

Algorithm 1. Serial ID VP

If the comparing result is approved; the values are identical. Thus, the system is integrated. Any mismatch denotes for the user ID is wrong. Then, it might be that the user has used a different QR tag and/or the QR tag might be expired.

Security Question. This process verifies two parts which are as follows: the first one extracts the real values from the DB whereas the second one extracts the SQ message (*M*) from the user's QR tag and applies a similar procedural encryption scheme, with the real encryption scheme applied previously, in order to find the hash values. Then, both hash values are verified and compared. The aim is to guarantee that the SQ is continuously updated and therefore its related values and mathematical operations are changing. Thus, the SQ is expired once it has been used for a short period of time, as shown in Fig. 5 marked by green dotted lines. The SQ is generated using a pseudo-random number generator (*PRNG*) based on user activity history (UAH) values obtained from database. The UAH is a changeable variable with the rule: "each time the user i, user(i), has successfully performed an activity, new values could be created". Hence, the SQ is updated in the database.

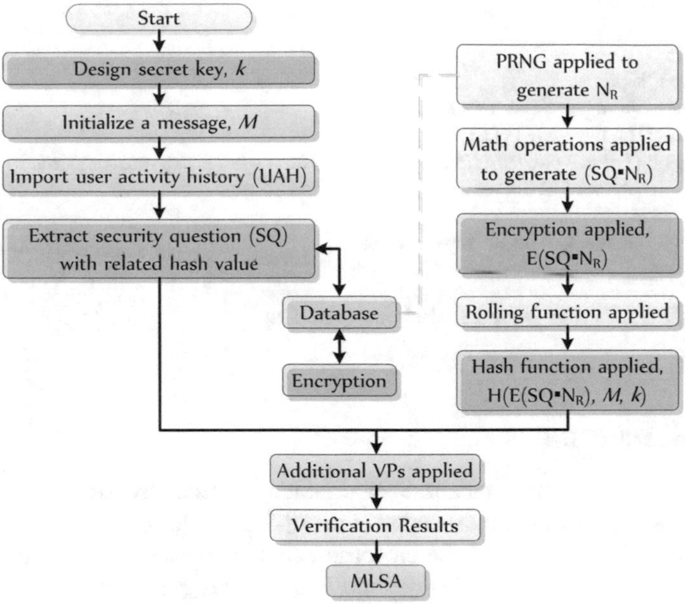

Fig. 5. SQ VP flowchart

In Fig. 5, on the right side, in order for the mathematical procedure based encryption be implemented. There will be a pre-process for the SQ; that is producing a random number (N_R) to be combined with the SQ using a series of mathematical operations is applied on SQ and N_R producing a combined random number: $SQ{\blacksquare}N_R$. The produced number will be one input of encryption algorithm applied subsequently. This is followed by a rolling function that aims to increase the security and privacy of SQ. Finally, a more secure procedure is applied: $H(E(SQ{\blacksquare}N_R), M, k)$.

The proposed pseudo-code of Fig. 5 verifies results taken from both sides of the abovementioned flowchart (left-side and right). If it is approved, the result is passed to MLSA; otherwise, the SQ VP is rejected.

3.3 Periodically Generated QR Tag VP

Usually, each generated QR tag is assigned a time-based label (TL) to denote QR tag's validity starting for the time specified. This is done in encryption scheme. But, in this type of verification, the TL will be verified. QR tag's expiration date is controlled by the label. The verification here focuses on the time the QR tag has been generated and not its contents. This is shown in Algorithm 2. Finally, the TL has considered a periodically generated QR tag policy every 24 h.

Start
Scan QR tag
Match QR tag with its user #i
Extract the time label (TL)
Apply: $(Time.Now\_function() - TL) \leq 24\ hours$? // read current time to compare
    Allow QR tag scan procedure
  Else
    Reject QR tag scan procedure
End

Algorithm 2. Time-based Label for QR Tag

3.4 Offline DB (ODB)

ODB considers the system database modification caused by an authorized access. ODB, periodically and in-offline mode, stores updated values. To reduce access to the database, ODB has determined the database access upon necessary requests. That is, once the process has been completely done, i.e., *process_requests_access ==0*, certain values are stored in such a way to reduce the data size and access times. ODB is summarized in Algorithm 3.

```
1: While (process_requests_access==0 AND connect_status==0) {
2:     Close resources;
3:     Calc_Time();
4:     Assign a resource;
5:     Disconnect();
6:     Update offline database;
7:   }
```

Algorithm 3. ODB

In this algorithm, in line 3 and 5, two important functions are recalled, which are as follows: the first function is a time, *Calc_Time()*, to assign a certain period of time. The second one is *Disconnect()* function to convert the connection status to 'Disable' and also to update database while there is no connectivity.

4 Multiple-Layer Security Architecture (MLSA)

4.1 Introduction into MLSA

The proposed MLSA concerns on how the VA design is secure and attains the security objectives therefore the relation between VA and MLSA is illustrated in Fig. 6.

This figure shows the relation between abovementioned verification procedure algorithm (VA) and MLSA. It is clear that, to achieve a security objective e.g., *Integrity* (marked by a blue line coming from VA), the QR Tag Verification is needed thru a database-based verification. Similarly, by applying an Offline DB (track orange colored

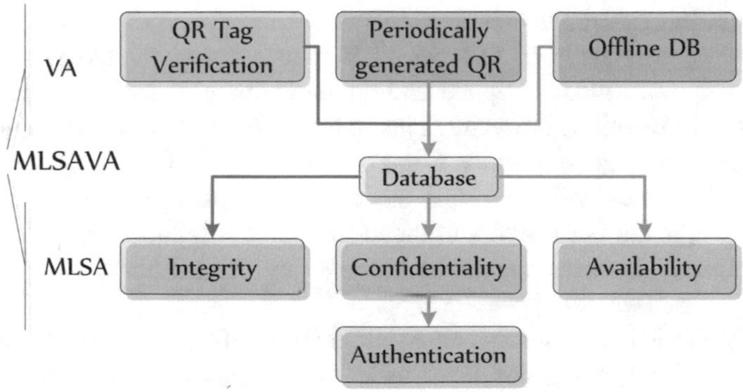

Fig. 6. MLSA versus VA

lines), the security objective i.e., *Availability* is achieved. However, these integrated processes between VA and MLSA are discussed in the following sub-sections.

4.2 The Proposed MLSA for Decision Making

The decision making factor verified is denoted by: *vote*. One of the measures that *vote* needs to verify is the Serial ID predefined earlier. Other required steps to verify their values are however attributes of images of biometrics, SQ, and TL-based expiration date of QR tag. Their hash values are compared to original ones stored in the database. This is shown in Algorithm 4 to guarantee the authority of QR tag issue.

```
//initialization process: vote=0;
  Scan QR tag;
   verify Biometrics_attributes();
   If (verified_value==1)
      vote1=vote1+20;
   verify Serial_ID();
   If (verified_value==1)
      vote1=vote1+20;
   verify SQ(); //TL
   If (verified_value==1)
      vote1=vote1+20;
   verify expiry_date(); //TL in hours
   switch case (TL)
   {
         1≤TL≤6 ? vote=vote1+24;
         7≤TL≤12 ? vote=vote1+18;
         13≤TL≤18 ? vote=vote1+12;
         19≤TL≤24 ? vote=vote1+6;
         24<TL ? vote=--vote1;
   }
```

Algorithm 4: QR Tag contents based Decision Making Procedure

This algorithm assigns values to each procedure approved. Every assigned value increases the validity scale of QR tag. If it is recently generated within 6 h, $1 \leq TL \leq 6$, a value of '24' is assigned and added to '*vote*' therefore '*vote* = 84' if the previous three verifications are valid. This value is the highest value a QR tag gets to ensure a high percentage of validity. Suppose that, $19 \leq TL \leq 24$, then: '*vote* = 66' to point out the lowest value the QR tag can get to pass the authentication test. If '*TL* > 24', the QR tag is expired and therefore '*vote*' is reduced by 1, '*vote* = 59', to ensure its expiry. This case cannot be accepted by the MLSAVA and the access is rejected.

To verify the MLSAVA robustness, all three verifications must be correct none some. Thus, if the QR tag has been used successfully and one or more of other verifications was wrong or mismatched to original values, the user can't access. Meaning, suppose that the two verification processes '*Serial_ID*()' and '*SQ*()' are correct but '*Biometrics_attributes*()' isn't; then the '*vote1* = 40' value and if $1 \leq TL \leq 6$, then '*vote* = 64 < 66' will be rejected. Thus, this procedure is used to test the MLSAVA robustness. Hence, a more robustness level against unauthorized attempts to prevent threats and actions is achieved.

4.3 MLSA Based Security Objectives Verification

In this sub-section, four security objectives which are: integrity, confidentiality, authentication, and availability are addressed. While the VA always verifies the user entries and contents of QR tag to ensure that contents are true as well as a hash based verifications for biometrics and ID, the obtained results could verify the integrity of user information thru database to compare between original values and inputs.

In regard to confidentiality, once user's information has been collected, it is encrypted and bounded by a specific time which is TL. To disallow any information leakage, the VA is designed to make sure that such a decryption attempt needs time more than the TL; i.e., $TL < Time\ of\ a\ decryption\ attempt$ Additionally, UAH, for example, is considered to increase the information system privacy whereas UAH is a personal history and has private values by which it is difficult to track the user.

As for authentication, by updating user information frequently and periodically, the QR tag is generated; every 24 h in order to authenticate both user and access process. All related values, e.g., SQ are frequently updated in advance in order to authenticate the system database. That is, with specific interval of time, the system is given a new QR tag to reduce vulnerability in the system design. During the next 24-hours, a new updated QR tag is generated again using different inputs, i.e., SQs to guarantee more security.

Finally, the availability of the QR tag and responsiveness is addressed. The policy of offline database storage is considered and connected to the VA. Every encrypted value is stored and updated accordingly. The QR tag is generated using updated values re-called from the offline database to guarantee an authorized access to database by other VPs anytime and also to prevent an unauthorized access caused by attacks.

5 Performance Analysis and Evaluation

This section evaluates the MLSAVA performance. First, security factors are subjectively discussed. Then, MLSAVA security objectives are evaluated (authentication is an example). Finally, the proposed MLSAVA accuracy is evaluated.

5.1 Security Factor Based Analysis

Confidentiality. The QR tag is periodically re-generated using a temporary secret key to increase information confidentiality. The secret key is designed using a very long series of bits to ensure the decryption time being lengthily increased. If the key has successfully decrypted the QR, the time needed > TL indicating an expired QR tag.

Integrity and Availability. QR tag contents are verified in order to ensure the integrity. If there is any mismatch between original values (database) and scanned (QR tag), then the MLSAVA has no integrity and therefore the verification process is stop. Thus, a third party has modified the QR tag contents or it might be the original QR tag is expired. Based on that, the user QR tag and other contents are protected by rejecting any attempt to access.
There will be no access by a third party but only one authorized source is allowed to access thru the offline database given a certain period of time.

Authentication and Robustness. This evaluation performs a vote-based process in order to measure the authority of the QR tag. That is, a several steps of processes are verified to make sure that the QR tag is issued by an authentically original source. Additionally, it measures the percentage of QR tag authentication.

5.2 MLSA Based Security Objectives Evaluation

In this evaluation part, Algorithm 4 is applied on a series of QR tags to validate authentication status. Every QR tag is considered for approval to be used as an access key to an IoT system based on the TL value. Thus, results for which the access is approved are provided in Table 1. The '*vote*' varies based on TL age. That is, '*vote*' has an inversely proportional relation with the TL age as shown in Fig. 7.

Table 1. Approved verified procedure statuses.

Bio. Attr.	ID	SQ	*vote1*	TL	*vote* (updated)	Status
+20	+20	+20	60	$1 \leq$ TL $\leq 6 = 24$	84	Approved
				$7 \leq$ TL $\leq 12 = 18$	78	Approved
				$13 \leq$ TL $\leq 18 = 12$	72	Approved
				$19 \leq$ TL $\leq 24 = 6$	66	Approved
+20	+20	+20	60	TL > 24 = -1	59	Disapproved

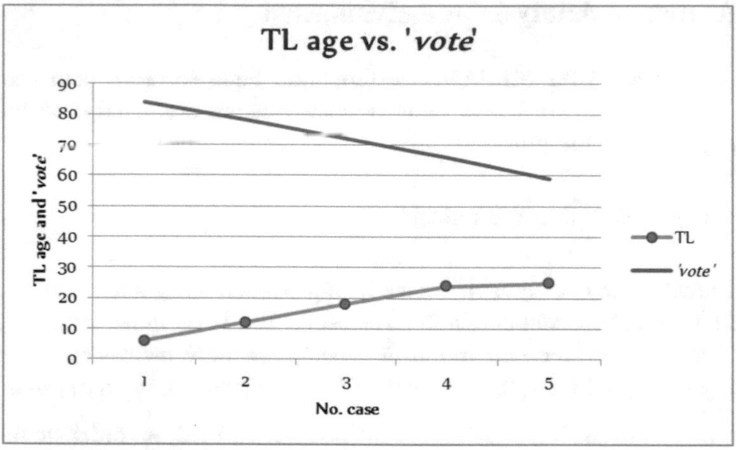

Fig. 7. Relation between TL age and 'vote'

In this table, every QR tag is considered for approval to be used as an access key to an IoT system based on the TL value. There are five cases four of which are considered as 'approved' status and one of which is disapproved.

This figure illustrates the rule on which the decision is made. That is, when the key age is shorter, its chance to be used as a smart key for IoT system is higher (i.e., TL = 6 h; '*vote*' = 84). Therefore, the '*vote*' value stands for the authentication strength.

5.3 MLSAVA Accuracy

In this evaluation, two groups of QR-tags are considered whereas the MLSAVA accuracy is measured based on QR readability and verified values e.g., *Serial_ID* and *SQ*, as shown in Table 2.

Table 2. MLSAVA accuracy vs. Rate Approved Access (RAA).

Group (#QR tags)	*QR readability*	*Biometrics Attributes*		*Serial_ID*		*SQ*		Accuracy %	RAA %
		V	R	V	R	V	R		
G1 = 30	30/30	30	0	28	2	29	1	100%	90–93%
G2 = 40	39/40	38	2	39	1	39	1	97.5%	90–95%

V = verified, R = refused

In this table, the QR readability is used to calculate the accuracy using Eq. (1).

$$QR\,readability = \frac{No.\,QR\,Generated}{No.\,QR\,Scanned} \tag{1}$$

This table shows that when QR tags have more RAA values, they can be used as smart keys for IoT. In other words, for Group 1 (G1), 90–93% of QR tags are approved and therefore they are authenticated to be used as smart keys.

6 Conclusion

A verification algorithm (VA) is proposed to authenticate a QR tag and measure how authenticated it is. The VA assigns a value to measure the expiry time of QR tag; if the time > 24 h it is considered expired. This aims to authenticate the QR tag is a smart key used to access an IoT system. The VA design has adopted a special Multiple-Layer Security Architecture (MLSA) to evaluate its overall performance in terms of security factors. Evaluation section has discussed different factors such as integrity and availability. The MLSAVA accuracy for 70 QR tags has been evaluated in terms of key's validity for IoT systems e.g., smart home applications and security gates.

Acknowledgement. This research is funded by an internal grant (RJO10289176/B/1/2017/12) from Universiti Tenaga Nasional.

References

1. Lin, S.S., et al.: Efficient QR code beautification with high quality visual content. IEEE Trans. Multimed. **17**(9), 1515–1524 (2015)
2. Rane, S., Dubey, A., Parida, T.: Design of IoT based intelligent parking system using image processing algorithms. In: 2017 International Conference on Computing Methodologies and Communication (ICCMC) (2017)
3. Huang, H.-F., Liu, S.-E., Chen, H.-F.: Designing a new mutual authentication scheme based on nonce and smart cards. J. Chin. Inst. Eng. **36**(1), 98–102 (2013)
4. Hung, C.H., et al.: A door lock system with augmented reality technology. In: 2017 IEEE 6th Global Conference on Consumer Electronics (GCCE) (2017)
5. Tkachenko, I., et al.: Two-level QR code for private message sharing and document authentication. IEEE Trans. Inf. Forensics Secur. **11**(3), 571–583 (2016)
6. Kanaris, L., et al.: Sample size determination algorithm for fingerprint-based indoor localization systems. Comput. Netw. **101**, 169–177 (2016)
7. Ota, K., et al.: Smart infrastructure design for smart cities. IT Prof. **19**(5), 42–49 (2017)
8. Kavitha, K.J., Shan, B.P.: Implementation of DWM for medical images using IWT and QR code as a watermark. In: 2017 Conference on Emerging Devices and Smart Systems (ICEDSS) (2017)
9. Kirkham, T., et al.: Risk driven smart home resource management using cloud services. Future Gener. Comput. Syst. **38**, 13–22 (2014)
10. Kim, Y.G., Jun, M.S.: A design of user authentication system using QR code identifying method. In: 2011 6th International Conference on Computer Sciences and Convergence Information Technology (ICCIT) (2011)
11. Warasart, M., Kuacharoen, P.: Paper-based document authentication using digital signature and QR code. In: 4th International Conference on Computer Engineering and Technology. International Proceedings of Computer Science and Information Technology (2012)

Security in Internet of Things: Issues, Challenges and Solutions

Hanan Aldowah[1(✉)], Shafiq Ul Rehman[2], and Irfan Umar[1]

[1] Universiti Sains Malaysia (USM), Penang, Malaysia
hanan_aldwoah@yahoo.com, irfan@usm.my

[2] Singapore University of Technology and Design (SUTD), Tampines, Singapore
shafiq_rehman@sutd.edu.sg

Abstract. In the recent past, Internet of Things (IoT) has been a focus of research. With the great potential of IoT, there comes many types of issues and challenges. Security is one of the main issues for IoT technologies, applications, and platforms. In order to cover this key aspect of IoT, this paper reviews the research progress of IoT, and found that several security issues and challenges need to be considered and briefly outlines them. Efficient and functional security for IoT is required to ensure data anonymity, confidentiality, integrity, authentication, access control, and ability to identify, as well as heterogeneity, scalability, and availability must be taken into the consideration. Considering these facts, by reviewing some of the latest researches in the IoT domain, new IoT solutions from technical, academic, and industry sides are provided and discussed. Based on the findings of this study, desirable IoT solutions need to be designed and deployed, which can guarantee: anonymity, confidentiality, and integrity in heterogeneous environments.

Keywords: Internet of Things · Security threats · Challenges
Solutions

1 Introduction

Internet of Things (IoT) is the emerging technology and it is considered to be the future of Internet [1, 2]. By allowing the devices/things self-configuring capabilities based on standard and interoperable communication protocols to identities, and use intelligent interfaces, over the dynamic global network infrastructure [1, 2]. The concept of IoT can be considered as an extension of the existing interaction between the humans and applications communicating through a new dimension [3]. Due to the advancements in mobile communication, Radio Frequency Identification (RFID) innovation, and Wireless Sensor Networks (WSNs), things and mechanisms in IoT can communicate with each other regardless of time, place or form [4]. The major breakthrough of IoT is in the formation of smart environments: smart homes, smart transport, smart items, smart cities, smart health, smart living, and etc. [5, 6]. Furthermore, in business perspective, IoT has enormous potential for various types of organizations and companies, including IoT applications and service providers, IoT platform providers and

F. Saeed et al. (Eds.): IRICT 2018, AISC 843, pp. 396–405, 2019.
https://doi.org/10.1007/978-3-319-99007-1_38

integrators, telecom operators and software vendors [2, 3]. Moreover, IoT will have a major impact in learning experience; especially in higher education system [7].

With the rapid increase in IoT application use, several security issues have raised sharply. As devices and things are becoming part of Internet infrastructure, therefore these issues need to be considered. When almost everything will be connected on Internet, these issues will become more prominent; with continuous Internet global exposure will literally disclose more security vulnerabilities. Such security flaws will be subsequently exploited by hackers, and later can be misused in uncontrolled environments with billions of IoT devices [2]. In addition, the IoT will also increase the potential attack surfaces for hackers and other cyber criminals.

A study conducted by Hewlett Packard [8] revealed that 70% of the most commonly used IoT devices contain serious vulnerabilities. IoT devices are vulnerable to security threats due to their design by lacking certain security features such as insecure communication medium, insufficient authentication and authorization configurations. As a matter of fact, when IoT become everywhere, everyone whether individuals or companies will be concerned. Additionally, crosslinking of objects presents new potentials to influence and to exchange. This leads to a variety of new potential risks concerning information security and data protection, which should be considered. Further, lack of security will create resistance to adoption of the IoT by companies and individuals.

Security issues and challenges can be addressed by providing proper training to the designers and developers to integrate security solutions into IoT products and thus, encouraging the users to utilize IoT security features that are built into the devices [2]. Our motivation to conduct this study is that most of the previous studies had focused on academic solutions only and had ignored other type of solutions from technical and industrial sides. However, these three sectors should work cooperatively and synchronously in order to reach integrated solutions as well as all the considerations from the three aspects should be taken into an account. Therefore, this paper provides a review on the main issues of IoT in terms of security as well as addresses some considerations that must be taken into account before and during the design stages to fill the gap of the literature regarding this issue by providing some solutions from three aspects including technical, academic, and industrial solutions. The main contribution of this paper is to provide the necessary insights on how certain utilization of such technologies can be facilitated by certain mechanisms and algorithms. This is believed to guide future studies to the use of certain solutions for certain problem based on the suggested algorithms and mechanisms by academic researchers with attention to technical and industrial solutions too.

The structure of the paper is organized as follows: the main issues and challenges related to IoT technology and its considerations are discussed in Sect. 2. The solutions proposed by academic researchers, technician, and industry experts are described in Sect. 3 whereas the discussion of the review findings is provided in Sect. 4. And finally, the conclusion, recommendations, and future work are outlined in Sect. 5.

2 Security Issues, Challenges and Considerations

IoT started to gain new momentum in current years as the consequence of the rapid growth of internet connected devices. However, security remains one of the major issues of the IoT [9] and the foremost concern raised by different stakeholders in Internet of Things which has the potential to slow down its adoption [10]. Therefore, it is considered one of the major issue which needs to be addressed to promote the IoT in real world [11]. Security is a fundamental quality of an IoT system and it is related to specific security features which are oftentimes a basic prerequisite for enabling Trust and Privacy qualities in a system [9]. IoT Security is the area to focus on securing the connected devices, protecting data, and networks in the Internet of things [12]. The computing devices and embedded sensors used in machine-to-machine (M2M) communication, smart home systems and in wearable devices are the main driving forces of IoT [13].

Weak security and poor security behaviors need to be considered from the outset and resilience designed in, in both individual devices and whole systems. Billions of additional connected devices in new locations and applications mean that the IoT world has increased the complexity of systems [14]. As the number of connected IoT devices continually increase, security issues are exponentially multiplied and there are many security concerns need to be considered as an entire system [15]. Moreover, traditional security mechanisms cannot be directly implemented to IoT technologies due to their designed system i.e. limited power as well as these large number of connected devices raise heterogeneity and scalability issues [9]. The security and safety of such systems can be endangered by a wide range of risks, both predictable and unpredictable, and therefore system elasticity should be a strong consideration.

Heterogeneity is one of the most critical issue, alongside with the security mechanisms that should be integrated into the IoT and has a considerable impact over the network security services that have to be implemented in the IoT [11]. Constrained devices will interact with various heterogeneous devices either directly or through gateways [16, 17]. Heterogeneity needs security to overcome the impossibility of implementing effective algorithms and protocols on all the devices in the IoT application fields [9]. In this case, it is essential to implement effective cryptographic algorithms that can provide a high throughput and adapt lightweight security protocols that offer an end-to-end secure communication channel. These protocols require credentials, thus optimal key management systems must be implemented to distribute these credentials and to help in establishing the necessary session keys between peers [11].

Addressing scalability for a large scale IoT deployment is another key issue. A significant challenge is to provide reliable solutions, which are scalable for the billions of things linked to many different local or global networks [9, 18]. Additionally, lots of them are mobile objects and finding the location of and verifying the correct identity of a specific item will be a major problem for the IoT infrastructure [9, 19]. Therefore, the development of applicable techniques that support heterogeneity and scalability, to anonymize users' data are key issues [20]. Moreover, providing flexible subscription schemas and events management while ensuring scalability with respect to things and users is still considered an open issue.

Security threats are problematical issue for the IoT deployment as the minimum capacity of devices being used, as a matter of fact, physical accessibility to sensors, devices, and the openness of the systems, considering the devices/things will communicate wirelessly [21]. Security concerns like DoS/DDoS attacks, man-in-the middle attacks, heterogeneous network issue, application risk of IPv6, WLAN application conflicts also hinders the deployment of IoT security [22–24] as well as the application security issues including information access and user authentication, information, platform management and so on [15, 25–27].

According to the research [15], data security issues can be classified into four types as: confidentiality, integrity, authenticity, and data availability. These security issues can be resolved by employing security measures: Data confidentiality ensures data protection from unauthorized users, while data integrity maintains correctness/accuracy of data. Moreover, authenticity makes sure that only authorized entities can access network resources to restrict any invalid users from the networks, and data availability guarantees that there is no restrains of authorized access to network resources, services and applications [28].

Furthermore, a larger number of IoT applications and services are increasingly vulnerable to attacks or data theft. To secure IoT against such attacks, advanced technology is required in several fields. The security of information and network should be equipped with properties such as identification, confidentiality, integrality and availability [29]. More precisely, authentication, confidentiality, and data integrity are the key problems related to IoT security [30]. Authentication is required for building a connection between devices and the exchange of number of public and private keys through the node to prevent steal data. In addition, confidentiality ensures that the data inside an IoT device is concealed from unauthorized objects, while data integrity prevents any modification to data in the middle by safeguarding that the data which arrived at the receiver node is unchanged and remains as transmitted by the source (sender) [2].

3 Securing New IoT Solutions from Technical, Academic, and Industry Sides (Architectures, Approaches, and Mechanisms)

In this section, we discuss some of the solutions proposed by Academic researchers, Technician, and Industry experts as counter measures to IoT security threats as follows:

3.1 Academic Solutions

Academics researchers have proposed some solutions in the field of network security. These solutions came in the form of architecture, new approaches and models, and mechanisms through which they endeavor to raise the quality of security in IoT environment. Some of these proposed solutions are:

One of the security solution proposed by [31] namely Dynamic Prime Number Based Security Verification (DPBSV). This solution is desirable for big data streams; it uses the concept of sharing a common key which is updated periodically by yielding a synchronized pair of prime numbers for real time security verification on big data stream. The study has conducted theoretical analyses and experimental evaluations to show the efficiency of its approach and to prove that DPBSV technique requires less processing time and can prevent malicious attacks on big data streams.

While most of the security challenges are often addressed by centralized approaches, a recent research work carried out by [32] have proposed an entirely distributed security approach for IoT. For the design and implementation of this security mechanism and its application in IoT environments, authors used an optimized Elliptic Curve Cryptography approach. Based on a lightweight and flexible design, this work presents an optimum solution for resource-constrained devices, providing the benefit of a distributed security approach for IoT in terms of end-to-end security. According to authors, this solution has already been tested and validated by using AVISPA tool and had been implemented on a real scenario over the Jennic/NXP JN5148 chipset based on a 32-bit RISC CPU [32]. The results have proved the feasibility of this work. Therefore, DCapBAC can be considered a security solution for IoT environments.

Sicari, Rizzardi [9] emphasized on design and deployment of appropriate solutions, which are platform independent and can provide resilient security measures. Considering the authentication and access control an approach has been proposed by [33], to establish the session key it uses an Elliptic Curve Cryptography (ECC) which is a lightweight encryption algorithm [34]. This technique specifies access control policies, managed by an attribute authority, which ensures to maintain mutual authentication among the user and the sensor nodes. Hence, can resolve the resource restrained issue at application level in IoT.

Particularly, in order to maximize the IoT benefits, it is mandatory to reduce the risks involved with security concerns. For that purpose, [35] proposed a comprehensive architectural design named as (ARMY) which proposed based on the Architectural Reference Model (ARM) to analyze the main security prerequisites during the design of IoT devices. The proposed architecture has been designed and implemented within different European IoT enterprises; to initiate and promote the development of security based IoT-enabled services.

Recently in 2017 researchers [36] proposed a Secure IoT (SIT) based on 64-bit block cipher. The architecture of the designed algorithm is integration of feistel cipher and a substitution-permutation network. Authors claim that SIT is a lightweight encryption algorithm and it can be deployed in IoT applications.

Moreover, emerging techniques such as software defined networking (SDN) and blockchain techniques are being adopted to provide security solution for IoT in heterogonous environments. For instance, In [37] researchers proposed an OpenFlow based SDN architecture for IoT devices. According to the researches the proposed architecture can perform anomaly detection to figure out the compromised devices in a network. To do so, network gateway executes dynamic traffic analysis. In case any abnormal traffic behavior is detected it will take the mitigation measures accordingly. Similarly, researchers in [38] proposed a multi-layered security architecture based on blockchain techniques to share and store the heterogeneous IoT data related to the smart city environment. The proposed architecture is designed to address the scalability and reliability issues that are very challenging in heterogeneous environments

3.2 Technical Solutions

In order to mitigate ever-expanding security threats to companies, organizations, and governments have to change their perspective towards security. This paradigm shift is the one that addresses security through an essentially broader scope at every level of the interaction. Organizations must emphasize the nature of the challenges, risks, and technological advantages and disadvantages unique to the product or service environments. They must understand the internal skills, existing practices, strategies, governance, and controls related to security, what is lacking, and where the gaps lie. To support this change, Harbor Research [39] has developed a new approach consists of a three-step process to guide and help companies, organizations, and governments in their approach to IoT security. The design of such process is to aid companies in designing and implementing a comprehensive approach to security in IoT solutions, including conducting an impact assessment, considering five primary security functions, and defining lifecycle controls are as follows:

Step 1: Impact of Security Assessment in Heterogeneous Environments
Addressing the impact of security in diverse environments must be the foremost consideration during the solution design process. The proposed solutions should be compatible with various applications and platforms. The foundation of any IoT solution deployment depends on a proper security mechanism. Therefore, before designing and deploying an IoT security solution for heterogeneous environments, proper information of customers' environment, sensitive data, risk assessment, infrastructure etc. related to organizations need to be considered.

Step 2: Application of Primary Security Functions
To ensure a secure IoT deployment across the entire organizational network, the designed IoT security solutions must possess five key functionalities as: data encryption, network security, identification, user access and management, and analytics. By doing so, a secure end-to-end communication between the IoT devices, data centers and cloud architectures can be ensured.

Step 3: Lifecycle Controls
The entire lifecycle of IoT devices in each phase need to be considered as: during the Deployment phase of security solution, IoT devices need to be authenticated by verifying its software via digital signatures, certifications, and other security methods to ensure a secure communication across the network. During Operation phase, IoT devices need to be continuously monitored by the network which is responsible for penetration testing and vulnerability assessment. The network should provide real-time monitoring operation and response during the event of an attack. In Incident and Remediation phase, IoT devices need to be integrated with system-wide incident response policies. And finally, in Retirement and Disposal phase, IoT devices that possess sensitive data, information, certifications etc. must be deleted securely.

In 2014, an online study was conducted by Zebra Technologies on global wide companies, corresponding to various industrial sectors [40]. The study focused on identifying the organizations interested in IoT solutions. The results showed that companies are taking initiatives in deployment of IoT solutions. Moreover, many organizations consult the IoT experts in deployment of IoT solutions and applications. IoT solutions provide new opportunities for companies to transform their strategic, operational, and business activities.

Nevertheless, deploying these IoT solutions are challenging for the companies as IoT solutions relies on various technical elements one of them is deploying end-to-end IoT security solutions. Mostly companies design IoT solutions meant for specific purpose within an organization. Considering IoT as emerging technology, there is a need for interoperability, so that a unified standard is set to enable seamless integration across IoT devices, applications, and services among different vendors.

3.3 Industrial Solutions

Security is critical to IoT and need to be taken care of at every stage [10]. Through the literature review and the recommendations of many workshops and conferences that emphasized on implementing the proper security measures while designing the IoT devices. They came to a consensus that while designing the IoT devices companies should consider three major aspects. First, adopting security by design; second, engaging in data minimization; and third, increasing transparency among the consumers with notice and choice for unexpected services.

Security by Design

Security for IoT devices depends on various elements, such as the amount of sensitive data collection and mitigating costs of security vulnerabilities. Ramirez [41] presented some ideas to address these key issues, as suggested companies should consider follow key points: (1) perform security risk assessment during the design process; (2) test device security measures; (3) consider protection of sensitive data while transmission or storage; and (4) monitor IoT devices and regular software updates. Moreover, to ensure a desirable security measures, organizations must deploy administrative and technical privileges by conducting security training sessions for employees. According to [42], security measures that are considered from initiating a device, establishes a secure computing environment which are tamper resistant. The article affirmed that security for IoT device must be addressed for its entire lifespan process, from design to the operational phases including: (1) Secure booting, (2) Access control, (3) Device authentication, (4) Firewalling and IPS, and (5) Updates and patches.

Data Minimization

Data minimization is the strategy that organizations can use to maintain the data repository within organizations by defining its duration. As security and privacy solution, organizations that gather the personal information should follow this data minimization concept. In other words, organizations should obtain the data required for specific purpose and period only and should safely discard it after. Gathering and maintaining large data repository can induce a risk of data breach.

Notice and Choice

The aim of a privacy notice is to ensure that customers and users are aware of data practicing involves personal information. Moreover, users should be aware about the personal information sharing, gathering, processing, and its retention [43].

4 Discussion

From this study, we found that Internet of Things (IoT) need to be designed in a user-friendly manner yet with consideration of its security measures. We observed that security is one of the main issues and challenges in the deployment of IoT in heterogeneous environments. The concept of IoT is to connect everything to the global Internet and allow the devices and things remotely to communicate with each other, which raises new security problems related to the confidentiality, integrity, and authenticity of data being exchanged between the IoT devices. To restrain the adversaries to obtain sensitive information while allowing authentic users to share and gather this information. We found that further research is needed that can focus on designing security measures in IoT environments. In maintaining the data confidentiality, integrity, and authenticity proper encryption algorithms need to be used, which not only fulfills these security measures yet consumes less data processing time.

In brief, security in the IoT technologies is very important and full of challenges and intuitively usable solutions are needed as well as these solutions should seamlessly integrate into the real world.

5 Conclusion, Recommendations and Further Work

The aim of this study was to provide a review of the most critical aspects of IoT with specific focus on the security issues and challenges involved with IoT devices. Several problems and challenges related to the security of the IoT are still being faced. Research focuses are much needed in this area to address these security issues and challenges in IoT heterogeneous environments so that users can confidently use IoT devices to communicate and share information globally with safety assurance. In addition, this paper recommended some solutions from academic, technical, and industrial aspects. These solutions came in the form of architecture, new approaches and models, and mechanisms through which they aim to increase the quality of security in IoT environment. Furthermore, data security, and data protection must methodically be considered and addressed at the design stage. For this, there are three key aspects that organizations should take into consideration to enhance security in IoT devices: security by design, data minimization, and providing users with notice and choice for unexpected services. There are nevertheless remaining numbers of potential gaps in the overall 'security' framework where further research will be potentially beneficial. As conclusion, there are still many open questions and problems that need further thinking and harmonization. The IoT includes a complex set of technological, social, and policy considerations across various set of stakeholders. The technological developments that enables the use of IoT are real, growing, and here to stay. Efforts by governments, engineering, production, industry, and academic world to provide processes for the effective and safe use of these developments clearly need further research work.

Acknowledgment. Authors would like to thank the Institute of Postgraduate Studies (IPS), Universiti Sains Malaysia (USM) for the financial support through the USM Fellowship.

References

1. Van Kranenburg, R.: A Critique of Ambient Technology and the All-seeing Network of RFID. The Netherlands Institute of Network Culture, Amsterdam (2008)
2. Abomhara, M., Køien, G.M.: Security and privacy in the Internet of Things: current status and open issues. In: International Conference on Privacy and Security in Mobile Systems (PRISMS). IEEE (2014)
3. Sundmaeker, H., et al.: Vision and Challenges for Realising the Internet of Things. Cluster of European Research Projects on the Internet of Things. European Commision, Brussels (2010)
4. Bandyopadhyay, D., Sen, J.: Internet of things: applications and challenges in technology and standardization. Wirel. Pers. Commun. **58**(1), 49–69 (2011)
5. Ul Rehman, A., Manickam, S.: A study of smart home environment and its security threats. Int. J. Reliab. Qual. Saf. Eng. **23**(03), 1640005 (2016)
6. Miorandi, D., et al.: Internet of things: vision, applications and research challenges. Ad Hoc Netw. **10**(7), 1497–1516 (2012)
7. Aldowah, H., et al.: Internet of Things in higher education: a study on future learning. In: Journal of Physics: Conference Series. IOP Publishing (2017)
8. Gen, H.P.-C.S.A. Controllers, R.: Hewlett-Packard Enterprise Development LP. Citeseer (2015)
9. Sicari, S., et al.: Security, privacy and trust in Internet of Things: the road ahead. Comput. Netw. **76**, 146–164 (2015)
10. Jha, A., Sunil, M.: Security considerations for Internet of Things. L&T Technology Services, Vadodara (2014)
11. Roman, R., Zhou, J., Lopez, J.: On the features and challenges of security and privacy in distributed internet of things. Comput. Netw. **57**(10), 2266–2279 (2013)
12. Yue, X., et al.: Cloud-assisted industrial cyber-physical systems: an insight. Microprocess. Microsyst. **39**(8), 1262–1270 (2015)
13. Minoli, D.: Building the Internet Of Things with IPv6 and MIPv6: The Evolving World of M2M Communications. Wiley, Hoboken (2013)
14. Jan, S., et al.: Applications and challenges faced by internet of things-a survey. Int. J. Eng. Trends Appl. **ISSN**, 2393–9516 (2016)
15. Jing, Q., et al.: Security of the internet of things: perspectives and challenges. Wirel. Netw. **20**(8), 2481–2501 (2014)
16. Vasilomanolakis, E., et al.: On the security and privacy of internet of things architectures and systems. In: 2015 International Workshop on Secure Internet of Things (SIoT). IEEE (2015)
17. Botta, A., et al.: Integration of cloud computing and internet of things: a survey. Future Gener. Comput. Syst. **56**, 684–700 (2016)
18. Issarny, V., et al.: Service-oriented middleware for the future internet: state of the art and research directions. J. Internet Serv. Appl. **2**(1), 23–45 (2011)
19. Gubbi, J., et al.: Internet of Things (IoT): a vision, architectural elements, and future directions. Future Gener. Comput. Syst. **29**(7), 1645–1660 (2013)

20. Jara, A.J., Kafle, V.P., Skarmeta, A.F.: Secure and scalable mobility management scheme for the Internet of Things integration in the future internet architecture. Int. J. Ad Hoc Ubiquitous Comput. **13**(3–4), 228–242 (2013)
21. Stankovic, J.A.: Research directions for the internet of things. IEEE Internet Things J. **1**(1), 3–9 (2014)
22. Haitao, L.B.C.H.W., Ying, F.: Security analysis and security model research on IOT. Comput. Digital Eng. **11**, 006 (2012)
23. Tan, Y., Han, J.: Service-oriented middleware model for internet of things. Comput. Sci. **38** (3), 23–45 (2011)
24. Henze, M., et al.: A comprehensive approach to privacy in the cloud-based Internet of Things. Future Gener. Comput. Syst. **56**, 701–718 (2016)
25. Suo, H., et al:. Security and privacy in mobile cloud computing. In: 2013 9th International Wireless Communications and Mobile Computing Conference (IWCMC). IEEE (2013)
26. Wan, J., et al.: From machine-to-machine communications towards cyber-physical systems. Comput. Sci. Inf. Syst. **10**(3), 1105–1128 (2013)
27. Wan, J., et al.: VCMIA: a novel architecture for integrating vehicular cyber-physical systems and mobile cloud computing. Mob. Netw. Appl. **19**(2), 153–160 (2014)
28. Ning, H., Liu, H.: Cyber-physical-social based security architecture for future internet of things. Adv. Internet Things **2**(01), 1 (2012)
29. Kim, J.T.: Requirement of security for IoT application based on gateway system. Communications **9**(10), 201–208 (2015)
30. Kim, J.T.: Analyses of requirement for secure IoT gateway and assessment. International information institute (Tokyo). Information **19**(3), 833 (2016)
31. Puthal, D., et al.: A dynamic prime number based efficient security mechanism for big sensing data streams. J. Comput. Syst. Sci. **83**(1), 22–42 (2017)
32. Hernández-Ramos, J.L., et al.: DCapBAC: embedding authorization logic into smart things through ECC optimizations. Int. J. Comput. Math. **93**(2), 345–366 (2016)
33. Ye, N., et al.: An efficient authentication and access control scheme for perception layer of internet of things. Appl. Math. Inf. Sci. **8**(4), 1617 (2014)
34. Szczechowiak, P., et al.: NanoECC: testing the limits of elliptic curve cryptography in sensor networks. In: Wireless Sensor Networks, pp. 305–320. Springer, Berlin (2008)
35. Hernandez-Ramos, J.L., Bernabe, J.B., Skarmeta, A.: ARMY: architecture for a secure and privacy-aware lifecycle of smart objects in the internet of my things. IEEE Commun. Mag. **54**(9), 28–35 (2016)
36. Usman, M., et al.: Sit: a lightweight encryption algorithm for secure internet of things. arXiv preprint arXiv:1704.08688 (2017)
37. Bull, P., et al.: Flow based security for IoT devices using an SDN gateway. In: IEEE 4th International Conference on Future Internet of Things and Cloud (FiCloud). IEEE (2016)
38. Biswas, K., Muthukkumarasamy, V.: Securing smart cities using blockchain technology. In: IEEE 18th International Conference on High Performance Computing and Communications; IEEE 14th International Conference on Smart City; IEEE 2nd International Conference on Data Science and Systems (HPCC/SmartCity/DSS). IEEE (2016)
39. Harbor White Paper: Security for the internet of things. Harbor Res. **16**, 1–16 (2016)
40. Zebra Internet-Of-Things Solution Deployment Gains Momentum Among Firms Globally, A Forrester Consulting Thought Leadership Paper Commissioned By Zebra Technologies, October 2014
41. Ramirez, E.: Privacy and the IoT: Navigating Policy Issues. US FTC, Washington (2015)
42. Shipley, A.: Security in the Internet of Things. Wind River, September 2014 (2015)
43. Schaub, F., et al.: A design space for effective privacy notices. In: Eleventh Symposium on Usable Privacy and Security (SOUPS 2015) (2015)

Cloud Computing and Internet of Things Integration Systems: A Review

Ayman Amairah[1], Bassam Naji Al-tamimi[2(✉)], Mohammed Anbar[1(✉)], and Khalid Aloufi[2]

[1] National Advanced IPv6 Center of Excellence (NAv6), Universiti Sains Malaysia, 11800 Gelugor, Penang, Malaysia
aalamir@nyit.edu, anbar@usm.my
[2] College of Computer Science and Engineering, Taibah University, Medina, Saudi Arabia
mr.altamimi@gmail.com, koufi@taibahu.edu.sa

Abstract. Cloud Computing (CC) is established as an important and widely used technology in enabling users to access to a shared pool of online resources. While alongside, the Internet of Things (IoT) has emerged as a new technology to enable multiple network-based objects to communicate with each other over wired or wireless networks. In this paper, the reference architectures and definitions of these two technologies are reviewed together with their security concerns, and their proposed solutions. Firstly, the most common references for CC and IoT are reviewed in this paper. Secondly, the most recent research trends for both technologies are presented. Finally, future research directions of the CC and IoT are suggested.

Keywords: Cloud Computing · Internet of Things
Integration of cloud computing and IoT
Cloud computing reference architecture · IoT architecture

1 Introduction

Cloud Computing is defined by the National Institute of Standards and Technology (NIST) as a new paradigm for enabling users to access to a shared computing resources pool (i.e., storage, servers, networks, applications, and services) in a convenient manner. CC has also been expected to be used as a user's utility service, like electricity and water. More recently, a new technology, called the Internet of Things (IoT) has emerged as one of the worldwide technologies. The IoT enables billions of smart objects and things to communicate with each other over the Internet [1] and has been considered as one of the significant technologies that will impact the United States interests towards 2025 [2].

The integration of these two technologies (i.e., CC and IoT) has been intensively studied and implemented in many fields. Nevertheless, it is still an open and promising research area to explore. At present, many applications have already begun relying on the integration of CC and IoT, including smart cities, agricultural applications, video surveillance and healthcare applications [3]. Along with the rapid growth of the CC and

F. Saeed et al. (Eds.): IRICT 2018, AISC 843, pp. 406–414, 2019.
https://doi.org/10.1007/978-3-319-99007-1_39

IoT have come a number of challenges and issues that have to be tackled. One of the most significant challenges facing this innovative integration is security. Distributed Denial of Service attacks are more disruptive in the applications of these technologies and lead to resource consumption.

The rest of this paper is organized as follows: definitions of the cloud computing and Internet of Things are provided in Sect. 2. Section 3 consists of a review and background on the integration of cloud computing and Internet of things. Research trends in cloud computing and the Internet of Things are explored in Sect. 4. Finally, Sect. 5 concludes the paper.

2 Definitions and Architectures

The following subsections discuss the most broadly used definitions for CC and IoT.

2.1 Cloud Computing Definitions

As explained above, CC is defined by NIST as a new paradigm for enabling users to access to a shared computing resources pool [4]. Cloud computing services are categorized into three broad delivery models, which are Software as a Service (SaaS) [5], Platform as a Service (PaaS) [6] and Infrastructure as a Service (IaaS) [7]. The essential purpose of such service models is to deliver services to cloud clients on a pay-as-you-go basis. Thus, CC is defined in [8] as a group of services available that are delivered over the Internet including software, platforms, and infrastructure, each group providing specific services, such as SaaS, PaaS and IaaS.

According to [9], CC is defined as a new way to develop the Information Technology (IT) industry to become more practical, through offering multiple services, applications, and infrastructure. However many issues concerning CC, are still unresolved and need to be investigated in depth; specifically, these are related to security implications. Cloud computing may therefore be considered a destructive technology, as suggested by [10], unless solutions and technologies are developed to mitigate its security impacts.

Many individuals and companies are keen to use CC technology to back up their data and explore new ways to market their services and products. However, cloud security concerns remain the main challenges that make many companies reluctant to move their businesses to the cloud. It has been recently reported by Cloud Security Alliance [11] that many companies and IT sectors are very keen to adopt CC in their businesses, but the security concerns and the absence of a compliance environment for using this new technology is having a limiting impact on the growth of CC.

2.2 Internet of Things Definitions

The IoT includes any physical object, which could be buildings, vehicles or devices with sensors which include embedded software, and electrical circuits that connect with each other over a network to achieve some significant effect [8]. The IoT is defined by [12] as a world where physical objects are integrated with the information network in a smooth manner, and these objects become intelligent objects and be able to participate in a business process. Thus, the IoT is defined by Botta et al. [8] as a new paradigm which is based on smart and self-configuring things that are interconnected over a global network infrastructure.

The word "Things" in IoT is defined as virtual objects which represent a set of items that are connected to the Internet. According to Gubbi et al. [13], IoT is an "interconnection of sensing and actuating devices providing the ability to share information across platforms through a unified framework, developing a common operating picture for enabling innovative applications. This is achieved by seamless ubiquitous sensing, data analytics and information representation with Cloud computing as the unifying framework." Lu et al. [14] propose a looser definition of the IoT as a new model that facilitates the connection for anything and anyone at anytime and anywhere which in turn led to the emergence of new innovative applications and services. Concisely, the idea behind the IoT can be summarised as a way to link the physical and the digital world [15].

A recent formal definition of IoT was provided by the IEEE IoT Initiative in their white paper [16] as "a wired or wireless network of uniquely identifiable connected devices which are able to process data and communicate with each other with or without human involvement". However, despite many research efforts in this area of interest, there is no comprehensive and standard definition that can cover all the IoT facilities, features, and usages.

2.3 Reference Architectures (CC and IoT)

Reference architectures provide a description of a specific technology and presents entire structure and integrations which can be referenced for best practices to obtain solutions. Figure 1 presents the NIST cloud computing reference architecture described in [17] with the five major actors, including their roles and functions. The figure shows

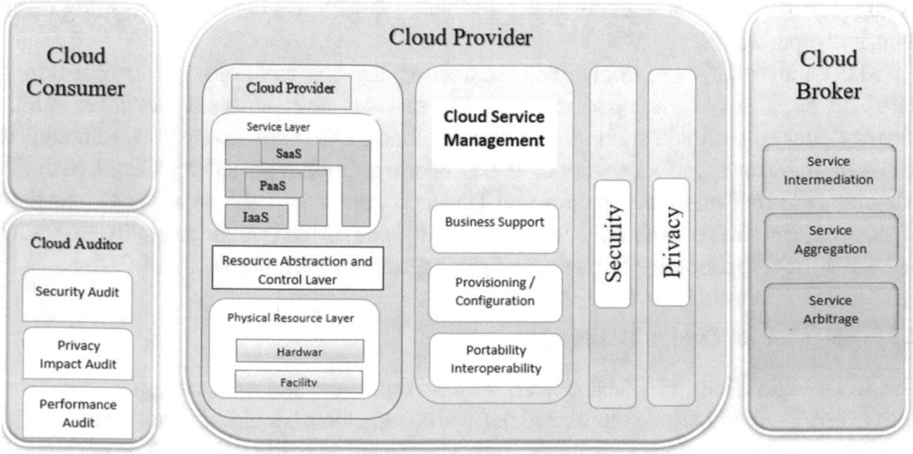

Fig. 1. NIST conceptual reference model for cloud computing.

a generic high-level conceptual reference model which aims to provide a standard and easy way for understanding the significant relationships between the actors, services, and the elements of the cloud environment.

Many researchers have contributed to the development of the IoT architectural reference model. Figure 2 depicts the IoT reference architecture model presented in [18], which contains several sub-models, including the IoT Information Model, the IoT Functional Model and the IoT Communication Model, as well as the IoT Trust, Security, Privacy Model. Further elaboration on the conceptual framework of the IoT is provided in [18].

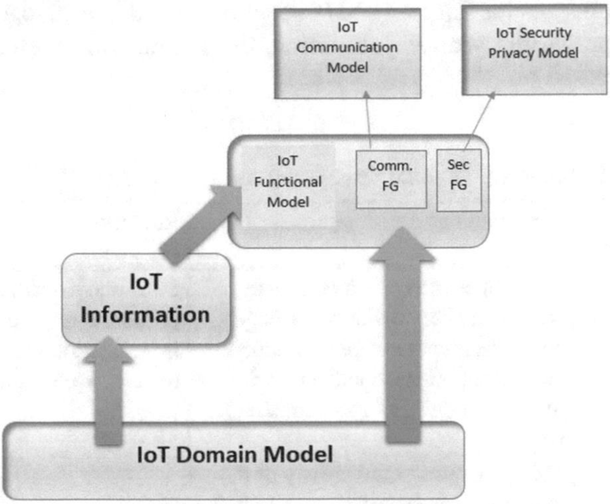

Fig. 2. IoT reference model.

3 CC and IoT Integration Systems

There have been several studies concerning the integration of CC with IoT. For example, [19] suggests using cloud computing features to support connectivity and real-time applications and services in smart cities. It also presents a framework for data obtained from real and virtual devices managed by cloud based services. A survey of the security risks threatening the cloud is presented in [20], which also surveyed other security issues resulting from the models of the service delivery of a cloud computing system, while [21] explores roadblocks and solutions to provide a reliable cloud computing environment. An approach to join the IoT and CC under the name CloudThings Architecture is proposed by [22], who also examine the most advanced integration approaches of CC and IoT, in addition to analyzing the IoT application requirements for smart homes. [23] presents the main difficulties related to Cloud of Things and proposes a smart gateway based communication.

On a larger scale, [13] suggests a cloud centric vision of IoT, to be applied internationally. This integration is referred to as the cloud of things, since anything can become part of the Internet and generate data. However, with the explosive growth in development of IoT applications and devices, the conventional cloud centric is facing serious challenges, such as High Latency, Bandwidth, and Quality of Service [21]. Fog Computing is initiated by Cisco in 2012 [25] as another paradigm that extends CC and its services to the edge of a network. Similar to Cloud, Fog provides data, compute, storage, and application services to end-users. To decrease the latency, a new architectural technology known as the Cloudlets is proposed in [26]. The Cloudlets supports the low latency requirements for processing massive data produced from intensive IoT-based applications.

Table 1 summarizes the findings and concepts reviewed in each paper together with information regarding the year of publication, the authors, and the contributions and remarks they focused on.

Table 1. Summary of some survey articles on integration of CC and IoT

Ref.	Authors and year	Contributions	Remarks
[24]	Ai et al. (2017)	This paper provides a complete survey of the popular three edge computing technologies (mobile edge computing, cloudlets, and fog computing) in a systematic way	Still many challenges and problems require further investigation from the perspective of techniques and solutions
[27]	Stergiou et al. (2018)	This paper presents a survey of the most relevant security issues of the CC and IoT integration system. Also, the benefits of such integration are presented in this study. Moreover, the study shows how the CC technology improves the IoT functionality	

(continued)

Table 1. (*continued*)

Ref.	Authors and year	Contributions	Remarks
[28]	Thota et al. (2018)	The authors of this study have deployed a new centralized secure architecture for IoT based healthcare system based on the integration of both technologies (CC and IoT). They deployed their system in cloud environment associated with IoT medical sensor nodes. Medical sensor nodes transfer the collected data to the cloud for seamless access by healthcare professionals. The study focuses on secure the crucial patients data to be move among IoT-based devices and cloud environment	CC and IoT integration system is implemented in this study
[29]	Díaz et al. (2016)	This survey analyzes existing integration components such as infrastructures and platforms of cloud computing. Also in this survey, a set of integration proposals as well as integrations challenges and research directions are pointed out	Information security in the integration CC and IoT environment is still a major concern for many organizations

4 Research Trends

Both technologies (CC and IoT) are now working together and this integration is expected to grow rapidly over the coming years. CC service providers are increasingly planning to provide services and solutions to cover "things". Another important advantage of creating an integration platform for CC and IoT is to facilitate the connectivity between humans and things. Figure 3 shows the exponential growth of CC and IoT in the last five years, and the number of worldwide research studies published on both technologies, which can be considered as evidence of their popularity.

Reviewing and exploring the state of the art in this domain, indicates that, both CC and IoT appear as one of the most promising technologies. This is due to the fact that they have many advantages which can be summarized by improving the quality of life. The terms (CC and IoT) were search using the popular online databases: IEEE, Springer Link Online Libraries, and ACM Digital Library as shown in Figs. 4 and 5 respectively.

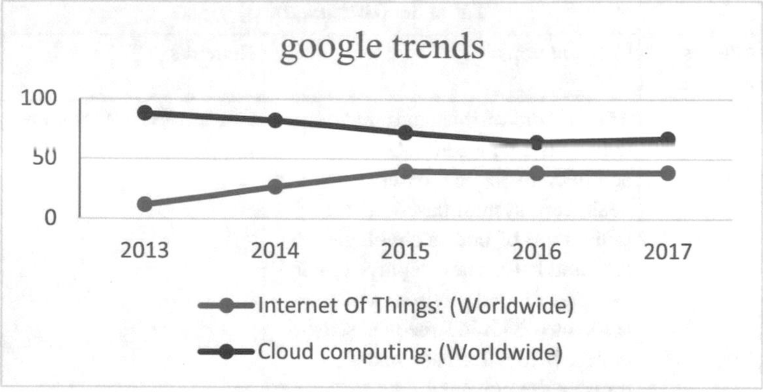

Fig. 3. CC and IoT research trends (Google based).

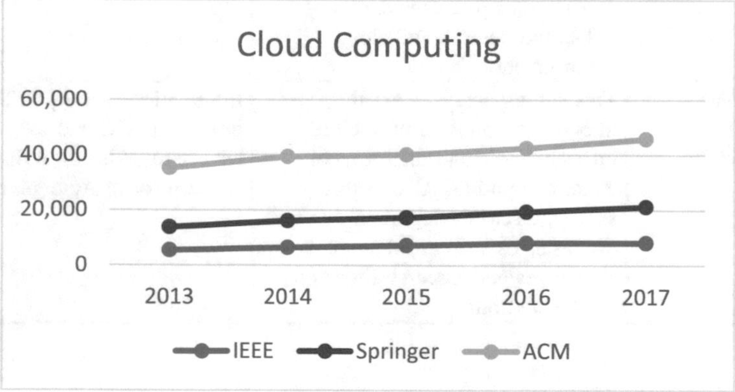

Fig. 4. Recent publication trends for cloud computing.

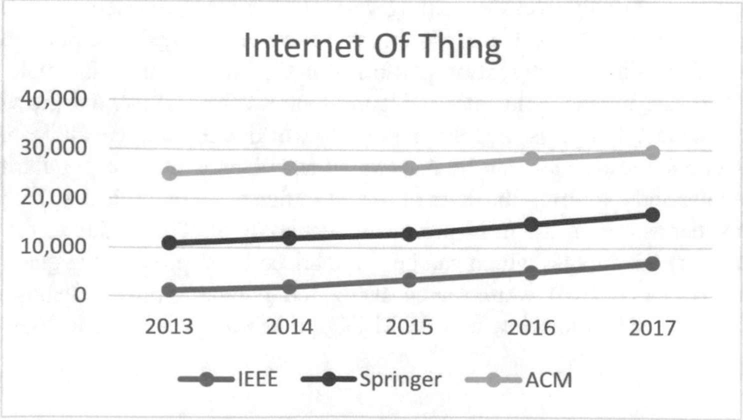

Fig. 5. Recent publication trends for IoT.

5 Conclusion

In this paper we have reviewed cloud computing and IoT technologies separately and also as an integrated system. We have also presented the popular reference architectures for both technologies. Based on this review, it can be seen that there is still a need for a considerable amount of research to study and develop applications, devices, and platforms that rely on cloud computing and IoT. There has also been considerable attention devoted to the publications regarding cloud computing and IoT, as noticed by the popular publishers in computer sciences and engineering such as IEEE, Springer, and ACM, who include many articles on these topics. We are confident that the next few years will witness a spectacular integration of cloud computing and IoT resulting from significant solutions to the problems of applications and services, as well as security challenges. Thus, further research is encouraged to investigate and address the security gaps in the integration system as new research directions in this domain.

References

1. da Cruz, M.A., Rodrigues, J.J.P., Al-Muhtadi, J., Korotaev, V.V., de Albuquerque, V.H.C.: A reference model for internet of things middleware. IEEE Internet Things J. **5**(2), 871–883 (2018)
2. Nic, N.I.C.: Disruptive civil technologies: six technologies with potential impacts on US interests out to 2025. Technical report (2008)
3. Malik, A., Om, H.: Cloud computing and internet of things integration: architecture, applications, issues, and challenges. In: Sustainable Cloud and Energy Services, pp. 1–24. Springer, Cham (2018)
4. Mell, P., Grance, T.: The NIST definition of cloud computing (2011)
5. Seethamraju, R.: Adoption of software as a service (SaaS) enterprise resource planning (ERP) systems in small and medium sized enterprises (SMEs). Inf. Syst. Front. **17**(3), 475–492 (2015)
6. Kolb, S., Wirtz, G.: Towards application portability in platform as a service. In: IEEE 8th international symposium on service oriented system engineering (SOSE), pp. 218–229. IEEE (2014)
7. Manvi, S.S., Shyam, G.K.: Resource management for Infrastructure as a Service (IaaS) in cloud computing: a survey. J. Netw. Comput. Appl. **41**, 424–440 (2014)
8. Botta, A., De Donato, W., Persico, V., Pescapé, A.: Integration of cloud computing and internet of things: a survey. Future Gener. Comput. Syst. **56**, 684–700 (2016)
9. Armbrust, M., et al.: Above the clouds: a berkeley view of cloud computing. vol. 4. Technical Report UCB/EECS-2009–28, EECS Department, University of California, Berkeley (2009)
10. Khajeh-Hosseini, A., Sommerville, I., Sriram, I.: Research challenges for enterprise cloud computing. arXiv preprint arXiv:1001.3257 (2010)
11. Cloud Security Alliance (CSA). http://www.cloudsecurityalliance.org. Accessed 26 Apr 2018
12. Xu, K., Qu, Y., Yang, K.: A tutorial on the internet of things: from a heterogeneous network integration perspective. IEEE Netw. **30**(2), 102–108 (2016)

13. Gubbi, J., Buyya, R., Marusic, S., Palaniswami, M.: Internet of Things (IoT): a vision, architectural elements, and future directions. Future Gener. Comput. Syst. **29**(7), 1645–1660 (2013)
14. Lu, Y., Papagiannidis, S., Alamanos, E.: Internet of Things: a systematic review of the business literature from the user and organisational perspectives. Technological Forecasting and Social Change (2018)
15. Guth, J., Breitenbücher, U., Falkenthal, M., Fremantle, P., Kopp, O., Leymann, F., Reinfurt, L.: A detailed analysis of IoT platform architectures: concepts, similarities, and differences. In: Internet of Everything, pp. 81–101. Springer, Singapore (2018)
16. Minerva, R., Biru, A., Rotondi, D.: Towards a definition of the Internet of Things (IoT). In: IEEE Internet Initiative 1 (2015)
17. Bohn, R.B., Messina, J., Liu, F., Tong, J., Mao, J.: NIST cloud computing reference architecture. In: 2011 IEEE World Congress on Services, pp. 594–596. IEEE (2011)
18. Bauer, M., et al.: IoT reference model. In: Bassi, A., et al. (eds.) Enabling Things to Talk, pp. 113–162. Springer, Berlin, Heidelberg (2013)
19. Suciu G., et al.: Smart cities built on resilient cloud computing and secure Internet of things. In: 2013 19th International Conference on Control Systems and Computer Science, pp. 513–518. IEEE (2013)
20. S.C.B. Intelligence.: Disruptive civil technologies. In: Six technologies with potential impacts on US interests out to 2025 (2008)
21. Hassan, T., Joshi, J.B.D.: Security and privacy challenges in cloud computing environments. IEEE Comput. Reliab. Soc. 24–31 (2010)
22. Zhou, J., et al.: CloudThings: a common architecture for integrating the Internet of things with cloud computing. In: Huazhong University of Science and Technology, Wuhan (2013)
23. Ibáñez, J.A.G., et al.: Integration challenges of intelligent transportation systems with connected vehicle, cloud computing, and Internet of things technologies. In: IEEE Wireless Communications, pp. 122–128 (2015)
24. Ai, Y., Peng, M., Zhang, K.: Edge cloud computing technologies for internet of things: a primer. Digit. Commun. Netw. (2017)
25. Bonomi, F., et al.: Fog computing and its role in the internet of things. In: Proceedings of the First Edition of the MCC Workshop on Mobile Cloud Computing, pp. 13–16. ACM (2012)
26. Ceselli, A., Premoli, M., Secci, S.: Cloudlet network design optimization. In: IFIP Networking Conference (IFIP Networking), pp. 1–9. IEEE (2015)
27. Stergiou, C., Psannis, K.E., Kim, B.G., Gupta, B.: Secure integration of IoT and cloud computing. Future Gener. Comput. Syst. **78**, 964–975 (2018)
28. Thota, C., Sundarasekar, R., Manogaran, G., Varatharajan, R., Priyan, M.K.: Centralized fog computing security platform for IoT and cloud in healthcare system. In: Exploring the Convergence of Big Data and the Internet of Things, pp. 141–154, IGI Global (2018)
29. Díaz, M., Martín, C., Rubio, B.: State-of-the-art, challenges, and open issues in the integration of Internet of things and cloud computing. J. Netw. Comput. Appl. **67**, 99–117 (2016)

Intelligent Communication Systems

Densification of 5G Wireless Access Network for Urban Area at Taiz City, Yemen

Redhwan Q. Shaddad, Amer A. Al-Mekhlafi(✉), Haithm M. Al-Gunid, Osama M. Al-Kamali, Hussain A. Al-Thawami, Hussum A. Mohammed, Mohammed A. Abduljalil, Nasher M. Sallam, Saber F. Al-Ariqi, and Taha M. Assag

Faculty of Engineering and Information Technology, Taiz University, Taiz, Yemen
{rqs2006, amralmkhlafy07, eng.hitham2014, osamaalkamalii, husseinl2128, hussumalhashimy945, m3a2016, nasher.absil991}@gmail.com

Abstract. The 5G wireless networks will cause a huge revolution in wireless communications because of the evolution of smart devices and rising technologies and applications as well as its enormous ability to send and transfer millions of data very quickly and in a short period of time. In this paper, the planning of 5G wireless access network is proposed in urban area of Taiz city, Yemen where the cells within the city divided into small cells and focused on the densest areas in city. This network densification can support various networks and multiple cells within the primary network, which are called Heterogeneous Networks (Het-Nets). The proposed wireless network uses the millimeter wave (mm-Wave) with the 28 GHz band which achieved data rate up to 7 Gbps at each cell. There are 212 micro cells to covered urban area and offer acceptable services to 189624 users in the city to satisfy densification network.

Keywords: 5G wireless communication · Wireless access network Het-Nets · Network densification · mm-Wave

1 Introduction

The current 4G cellular technologies cannot confront the required data rates due to using of bandwidth-hungry devices and the lack of the available spectrum [1, 2]. The recent advancements of wireless networks have seen a change to include picocells, femtocells, and ultra-dense small cells. Furthermore, it is widely agreed that increasing the number of cells with small radii will be the main contributor of the next generation cellular system to increase the network capacity and to provide required data rates [3]. Indeed, reducing cell sizes through the use of heterogeneous networks (Het-Nets), and employing higher bandwidth efficiency at each base station (BS) with additional wide spectrum are considered as solutions to meet the demands of next generation wireless networks [4, 5]. These are classed as network densifications which are provided by spatial densification and spectral aggregation [6]. The network densification in 5G is

F. Saeed et al. (Eds.): IRICT 2018, AISC 843, pp. 417–426, 2019.
https://doi.org/10.1007/978-3-319-99007-1_40

rapid development of wireless communication, with an innovative air interface called New Radio (NR), which is purposed mostly for new spectrum bands such as millimeter wave bands [6]. A prefect interworking between LTE and NR access technologies is required to guarantee working the 5G functionality well [6]. Millimetre-wave (mm-Wave) communication is a favorable technology to support multi-gigabit transmission rate, and offer a flexible and cost-effective candidate for 5G back-hauling [7, 8].

In mm-Wave deployments, network densification is not an option, but a necessity because of the limited coverage radius, as well as the opportunity to provide a vast capacity increase in the back-haul or in the access [9]. The Key Performance Indicators (KPIs) are indicators of the wireless network performance [9]. The KPIs are taken as basis to estimate the radio link parameters such as traffic volume density, experienced end-user throughput, latency, reliability, availability and mobility [9].

The 5G wireless technology is able to provide intelligent planning and optimization based on services and users awareness. In addition, the 5G will improve energy and cost efficiency by over a hundred of times [10]. The densification of existing cellular networks is implemented by a massive addition of small cells and a provision for peer-to-peer (P2P) communication such as device-to-device (D2D) and machine-to-machine (M2M) communication-enabled multi-tier Het-Net 20 s [11–15].

A lot of studies at 5G wireless network are observed and comparisons are made using the data obtained from some measurements from 28 GHz to 80 GHz in the band of mm-Wave that extended from 30 GHz to 300 GHz. Most of the current research and studies are focused on the 28 GHz band, the 38 GHz band, the 60 GHz band, and the E-band (71–76 GHz and 81–86 GHz) [13, 16–18]. Extensive propagation measurements have been performed at 28 GHz, 38 GHz, and 73 GHz in urban microcellular, urban macro-cellular, and/or indoor scenarios.

Densification of 5G wireless network at 28 GHz frequency have been proposed based on small cells and Het-Net 20 s [13]. The proposed wireless network supported data rate up to 1500 Mbps. The densification of this network offered small latency, low cost, high mobility and low energy consumption. When smaller cells were used at the 38 GHz frequency, the wireless network offered greater data rate that up to 1800 Mbps. Ultra-high densification of 5G wireless network have been implemented using the 60 GHz frequency [16]. The 60 GHz frequency suffers an additional atmospheric loss of 10-20 dB/km as compared to other mm-Wave bands. Although 60 GHz suffers from increasing of path loss compared with 28, 38 GHz, it enhances capacity and data rate. These studies have been concluded that consistent coverage can be achieved by having base stations with a cell-radius of 200 m at 28 GHz and 38 GHz [13, 17], and when using higher frequencies such 60 GHz or 72 GHz the cell radius will decrease and can be lower than 100 m [16].

The main challenges of ultra-dense networks include the inter-tier and intra-tier interferences due to dense deployment of multi-tier heterogeneous, the spectrum aggregation, the user scheduling, and hand- off among cells. In this paper, the densification of 5G wireless access network is applied on the dense areas in the urban area of Taiz city, Yemen. The densification of the wireless access networks increases the scalability of the network in the terms of the throughputs at the end users and the coverage of network.

2 5G Wireless Network Architecture

Architectures for 5G networks, namely multi-tier, Cognitive Radio Access Network (CRAN)-based, D2D communication based, mm-Wave, massive multiple-input multiple-output (M-MIMO) and the cloud-RAN architectures, multi-tier architecture consisting of several types of small-cells, relays, and D2D communication for serving users with different Quality of Service (QoS) requirements in spectrum-efficient and energy-efficient manners. Figure 1 illustrates a prefect 5G deployment including of a macro-cell, under-laid with a heterogeneous mix of small cells: microcells, Picocells, Relay Station (RS), Remote Radio Heads (RRH), femtocells, Control Data Separation Architecture (CDSA), Base Band Unit (BBU), Control Base Station (CBS) and D2D. The 5G networks will consist of nodes/cells with heterogeneous characteristics and capacities for examples macro-cells, small cells such as femtocells and picocells, and D2D user equipment's (UEs) [19].

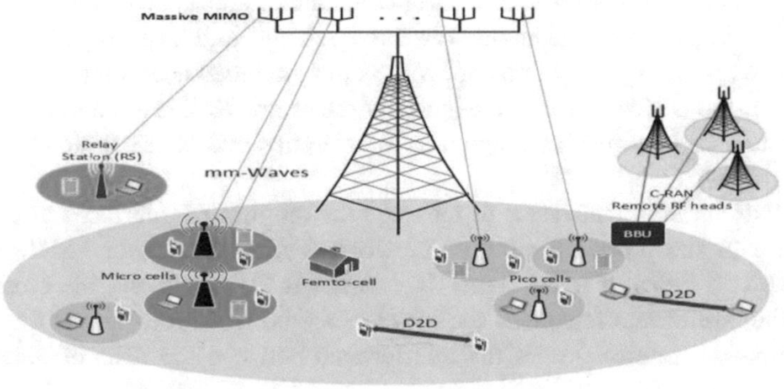

Fig. 1. 5G wireless network architecture.

The back-haul is a wired (fiber optics) or wireless (mm-wave/microwave) connection that connects base stations with other base stations and base stations connects to the core network. The front-haul is a wired (fiber) or wireless (mm-wave/microwave) connection that connects between RRHs and BBU pools in C-RAN scenario for interference management in a multitier infrastructure network [20]. The back-haul connects the RAN to the rest of the network, where the baseband processing takes place at the cell site.

3 Planning of 5G Wireless Access Network

Planning of wireless network is a complex process but it is the most important stage in the network design phase. It requires focus on many data and inputs, taking into consideration the appropriate coverage, capacity, possible interference, network performance efficiency and expected cost.

In general, the objectives of the cell planning process for the wireless access network are to support one of the following outcomes: optimum number of base stations, prefect locations to install base stations, suitable types of base station for each location, appropriate configuration of parameters, frequency reuse pattern, and capacity dimensioning such as number of carriers or carrier components per sector.

Planning by using mm-Wave frequencies in 5G wireless access network offers high performance capacity, throughput and QoS compared with microwave frequencies which applied in the current networks such as Long Term Evaluation (LTE) wireless networks (2 GHz or 3.6 GHz).

One of the main emerging requirements of 5G networks is the support of a large number of heterogeneous devices. The group of smaller elements for Het-Nets is including picocells, femtocells, and distributed antenna elements. The densification of wireless network will serve a significant role in enhancing network performance (High data rate, low latency, more scalability).

The 5G wireless access network planning is simulated to determine the coverage and capacity evaluation, radio link budget calculation, and inter and intra interference, as shown in Fig. 2. After planning for the coverage and capacity of the proposed wireless network, the design of this network is implemented according the environment nature. Next, the performance of designed wireless network is evaluated according to the parameters such as received Signal to Interference and Noise Ratio (SINR) levels, and throughputs.

The SINR is important factor to measure the amount of interfaces to estimate the performance of the proposed wireless network. The heterogeneous SINR takes into account the two types of interference: the co-tier interference and the crosstier interference. The co-tier interference is the interference between the cells of the same type and the cross-tier interference is the interference between the cells of different types [21].

The coverage planning can be summarized in the following steps: (1) determine the maximum allowable path loss for downlink (DL) and uplink (UL), (2) study the parameters involved in the link budget, (3) determine the distance between base station and mobile device, and (4) measure the cell radius to UL and DL.

The capacity planning also can be summarized in following steps: (1) predict traffic demand in time at small cell level, (2) requirement of bandwidth by small cell, and (3) modified version of the actual network deployment and spectrum allocation.

In this paper, network densification simulator OMNET++ is used to model a dense urban wireless network through coverage area of 21.97 km^2. This wireless network supports 189624 subscribers of 541744 persons (population of Taiz city, Yemen). This wireless network includes 212 Remote Radio Heads (RRHs) spanning over a set of 18 smart macro base stations operating on the band of 28 GHz.

From planning of wireless network, the coverage area is divided to 42 regions, and it is distributed along urban Taiz city area as shown in Fig. 3. The distribution of base stations on urban area in Taiz city is shown in Fig. 4. This distribution is non-uniform from site to site, there will be some sites have population density more dense than others. Additional small cells also is design in the dense areas to satisfy densification network. The red points in Fig. 4 presents small cells in the dense areas. In the dense areas, additional 75 small cells are distributed, and it operating on the band of 28 GHz.

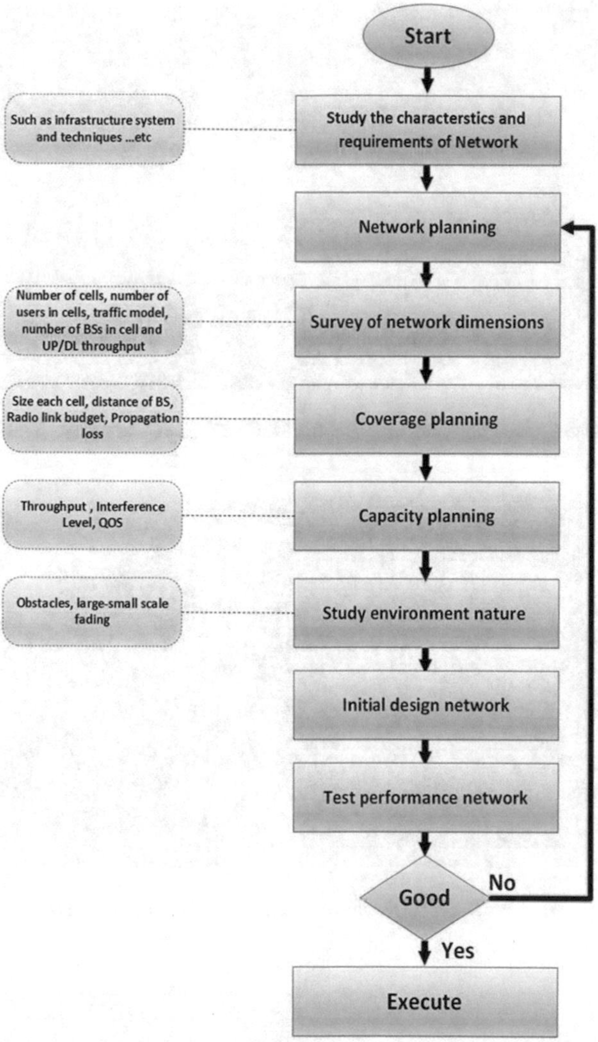

Fig. 2. Flow diagram for planning network.

The densification of wireless network produces Het-Nets which increases the QoS for users. The number of micro cells in each region is shown in Fig. 5. The general parameters of the proposed 5G wireless network is given in Table 1.

4 Results and Discussions

The UL and DL throughput are calculating from each site around urban area of Taiz city. Figure 6 shows the resultant values of throughput in each region. Each region has many sites inside it according to population density. We notice that the throughput

Fig. 3. Geographical map for divided region at Taiz city, Yemen.

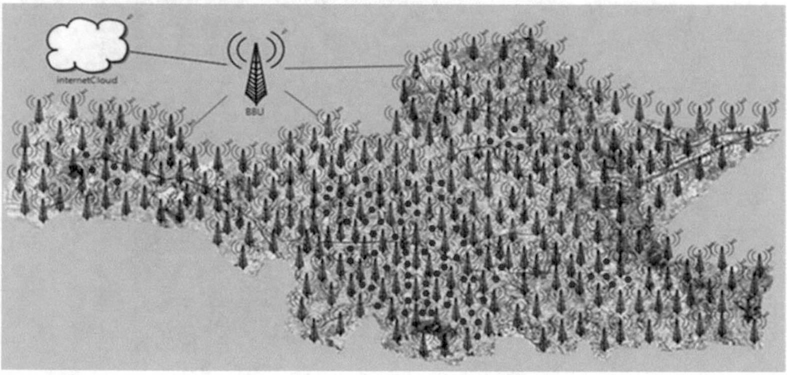

Fig. 4. Distribution of base stations on urban area in Taiz city.

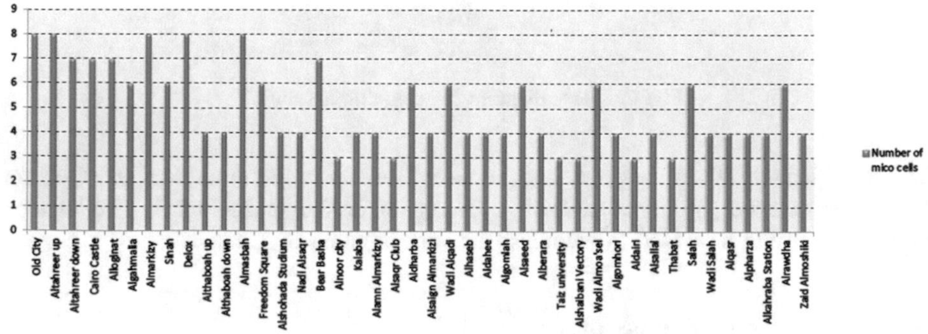

Fig. 5. Number of micro cells in each region.

Table 1. General parameters of the proposed 5G wireless network

Parameter	Specification	Parameter	Specification
Frequency	27.925 GHz	Path loss exponent	1.9
Channel band	520 MHz	Macro BS TX power	46 dBm
Channel model	Dense urban	Micro BS TX power	20 dBm
Cell radius	200 m	Device power	1 mW
Antenna gain	16 dBi	Area	21.97 km^2
Penetration	20 dB	Number of sites	212

affected by the geographical nature of covered area. The regions that have many abstracts and barriers has low throughput than other areas which haven't a lot of barriers. As shown from Fig. 6, regions such as Old city, Bear Basha, Algomlah, Alshohada'a stadium have the maximum downlink throughput that reach for more than 6000 Mbps because these regions are free areas and have little barriers compared with other regions. The success rate exceeds 95% because the using of densification networks which that the users may have many ways to connect to the network, and because using of Het-Nets, which make the probability of blocking service very low.

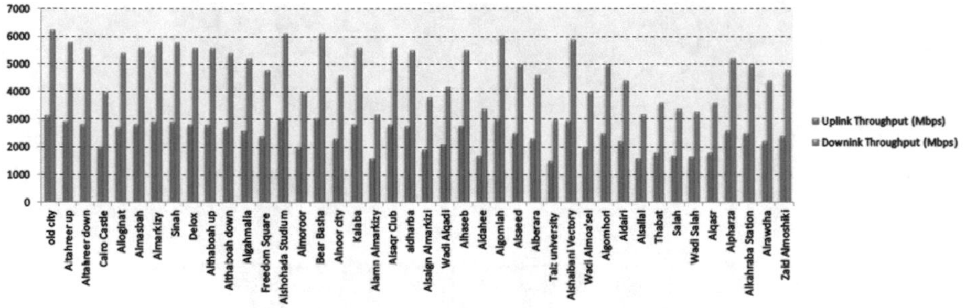

Fig. 6. Uplink and downlink throughput at each region.

Figure 7 illustrates the relationship between the mean of SINR (DL and UL) and the distance from transmitter to receiver. The requirements for the DL must have high level of SINR, since the DL supports higher data rate than the UL. Figure 8 shows the throughput in Gbps versus the SINR. There is adaptive modulation to support higher throughputs at the shorter distances, so the required SINR in this case must be high.

Because of using densification networks, the cells will be smaller and smaller, and hence the path loss will decrease due to smallness of the covered area. Figure 9 shows the path loss versus distance. The acceptable level of path loss is 110 dB, so the typical radius of cell is 200 m.

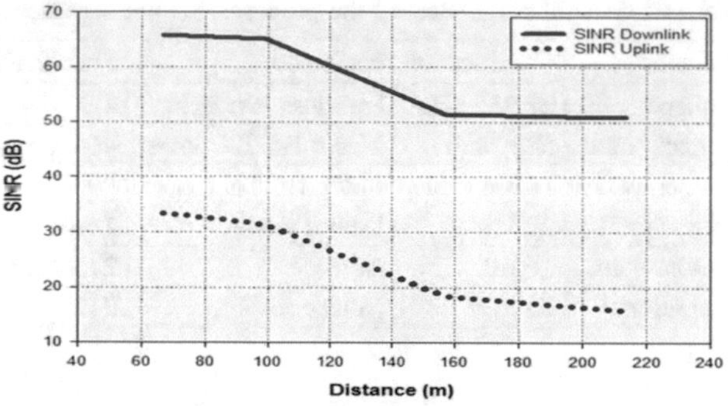

Fig. 7. Mean SINR versus the distance between Tx and Rx.

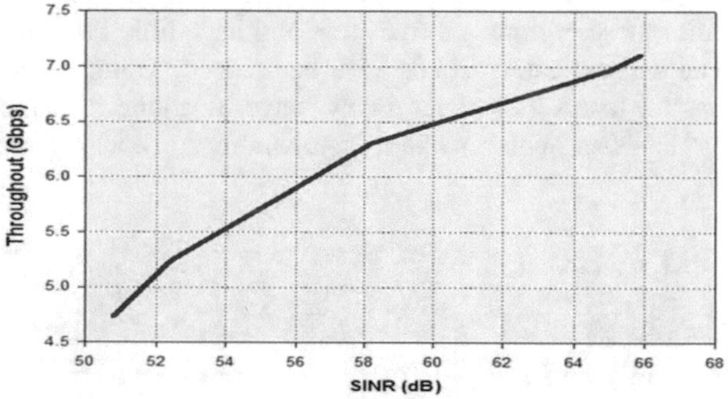

Fig. 8. Network throughput versus the SINR.

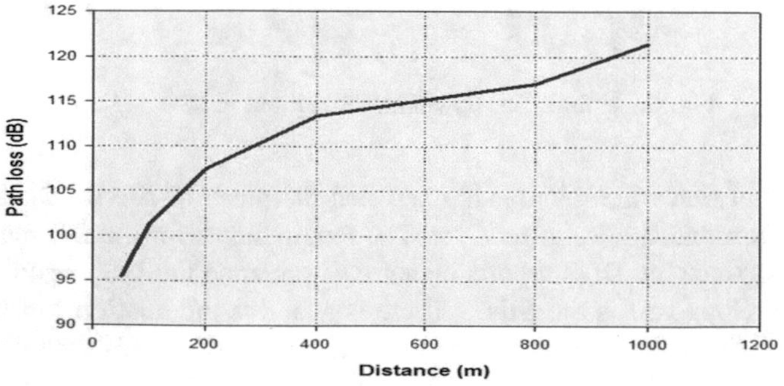

Fig. 9. Path loss versus the distance.

5 Conclusions

The 5G network planning for urban Taiz city has been proposed by using mm-Wave wireless access network technique. In addition, the densification of 5G wireless access network is proposed in the dense areas in Taiz city, Yemen using Het-Nets. The results of planning network indicated that we need 212 micro cells to covered urban area in Taiz city and offer acceptable services to 189624 users in city to satisfy densification network. Clearly, through this work some of the KPIs are achieved standards such as throughput, SINR, cost and energy consumption. The success of the designed network depends on many factors such capacity, coverage, cost and QoS.

References

1. Viswanathan, H., Weldon, M.: The past, present, and future of mobile communication. Bell Labs Tech. J. **19**, 8–21 (2014)
2. Pirinen, P.: A brief overview of 5G research activities. In: International Conference on 5G for Ubiquitous Connectivity (5GU), pp. 17–22 (2014)
3. Andrews, J., et al.: What will 5G be? IEEE J. Sel. Areas Commun. **32**(6), 1065–1082 (2014)
4. Ghosh, A., et al.: Millimeter-wave enhanced local area systems: a high-data-rate approach for future wireless networks. IEEE J. Sel. Areas Commun. **32**(6), 1152–1163 (2014)
5. Bhushan, N., et al.: Network densification: the dominant theme for wire-less evolution into 5G. IEEE Commun. Mag. **52**(2), 82–89 (2014)
6. Ericsson: 5G radio access: Research and vision, white paper (2014)
7. Elkashlan, M., Duong, T., Chen, H.: Millimeter-wave communications for 5G: fundamentals: part I [Guest Editorial]. IEEE Commun. Mag. **52**(9), 52–54 (2014)
8. Elkashlan, M., Duong, T., Chen, H.: Millimeter-wave communications for 5G–Part2: applications. IEEE Commun. Mag. **53**(1), 166–167 (2015)
9. Fallgren, M., Timus, B.: Scenarios, requirements and KPIs for 5G mobile and wireless system Scenarios requirements and KPIs for 5G mobile and wireless system. In: METIS (2013)
10. Hu, R.: Towards spectrum efficient, energy efficient and QoE aware 5G wireless systems. In: IEEE ICC2015 Trends for 5G Wireless Networks (2015)
11. Shaddad, R., Mohammad, A., Al-Gailani, S., Al-hetar, A., Elmagzoub, M.: survey on access technologies for broadband optical and wireless networks. J. Netw. Comput. Appl. **41**, 459–472 (2014)
12. Hossain, E., Rasti, M., Tabassum, H., Abdelnasser, A.: Evolution towards 5G multi-tier cellular wireless networks: an interference management perspective. IEEE Wirel. Commun. **21**(3), 118–127 (2014)
13. Sulyman, A., Nassar, A., Samimi, M., MacCartney Jr., J., Rappaport, T., Alsanie, A.: Radio propagation Path loss models for 5G cellular networks in the 28 GHz and 38 GHz Millimeter-Wave bands. IEEE Commun. Mag. **52**(9), 78–86 (2014)
14. Shaddad, R., Abdulwadood, M., Kuradah, N., Alsharaie, S., Qaid, M., Alramesi, A., Rassam, A.: Planning and optimization of LTE radio access network for urban area at Taiz City, Yemen. In: IRICT 2017, Johor, Malaysia (2017)

15. Zhao, H., Mayzus, R., Sun, S., Samimi, M., Schulz, J., Azar, Y., Wang, K., Wong, G., Gutierrez, F., Rappaport, T.: 28 GHz millimeter wave cellular communication measurements for reflection and penetration loss in and around buildings in New York City. In: IEEE International Conference on Communications (2013)
16. Roh, W., Seol, J., Park, J., Lee, B., Lee, J., Kim, Y.: Millimeter-wave beamforming as an enabling technology for 5G cellular communications: theoretical feasibility and prototype results. IEEE Commun. Mag. **52**, 106–113 (2014)
17. Hyytia, E., Virtamo, J.: Random waypoint mobility model in cellular networks. Springer Wirel. Netw. **13**(2), 177–188 (2006)
18. El-Beaino, W., El-Hajj, A., Dawy, Z.: A proactive approach for LTE radio network planning with green considerations. In: IEEE International Conference on Telecommunications (2012)
19. Zakrzewska, A., Ruepp, S., Berger, M.: Towards converged 5G mobile networks - Challenges and current trends. In: ITU Kaleidoscope Academic Conference, pp. 39–45 (2014)
20. Lee, W., Choi, J., Kim, Y., Lee, J., Kim, S.: Adaptive sector coloring game for geometric network information-based inter-cell interference coordination in wireless cellular networks. IEEE Trans. Netw. **26**(1), 288–301 (2018)
21. Guey, J., Liao, P., Chen, Y., Hsu, A., Hwang, C., Lin, G.: On 5G radio access architecture and technology. IEEE Wirel. Commun. **22**(5), 2–5 (2015)

Intelligent Offset Time Algorithm for Optical Burst Switching Networks

Abdulsalam A. Yayah[1(✉)], Abdul Samad Ismail[1], Tawfik Al-Hadhrami[2(✉)], and Yahaya Coulibaly[3]

[1] Department of Computer Science, Faculty of Computing, Universiti Teknologi Malaysia, Johor, Malaysia
salam.yayah@gmail.com, abdsamad@utm.my

[2] School of Engineering and Computing, University of the West of Scotland, Paisley, UK
tawfik.al-hadhrami@ntu.ac.uk

[3] Institut Universitaire de Formation Professionelle (IUFP), Université de Ségou, Ségou, Mali
cyahaya@gmail.com

Abstract. Optical Burst Switching (OBS) is an optical network that is efficient and reliable. However, there are some critical issues in OBS such as burst loss ratio, which occur due to contending bursts as optical buffers are lacked at the core node. Many techniques have been proposed to tackle the problem such as using offset time. Offset time is managed to allow Burst Header Packet (BHP) to reserve the required resources necessary for a successful transmission process. If the added extra time is long, it reserves the resources for a long time. In contrast, setting low offset time results in burst loss. As a consequence, the techniques that manages offset time need to be developed. This paper proposes an Intelligent Offset Time (IOT) algorithm that adapts offset time based on the condition of the network and the traffic. Three parameters are used as IOT's fuzzy input that are burst size, destination and burst queuing delay. The proposed algorithm is evaluated against the conventional offset time and adaptive offset time algorithms. The results of IOT's simulation show that it has effectively reduced burst loss ratio and end-to-end delay in comparison with the conventional and adaptive algorithms.

Keywords: Offset time · Optical burst switching · Fuzzy logic Burst delay · Burst loss

1 Introduction

Optical Burst Switching (OBS) is a Wavelength Division Multiplexing (WDM) network, which is an attractive and preferable choice over Optical Packet Switching (OPS) and Optical Circuit Switching (OCS), as it is able to handle the dramatic increase of multimedia applications' traffic [1, 2]. Its data transmission speed is high and its bandwidth is huge in comparison with other networks. Each optical fibre can support up to 50 THz [3]. Unlike OCS and OPS, data transmission is done in bursts instead of packets. The bursts are grouped and sent based on their destination. The assembly and

F. Saeed et al. (Eds.): IRICT 2018, AISC 843, pp. 427–439, 2019.
https://doi.org/10.1007/978-3-319-99007-1_41

disassembly of bursts are carried out in the ingress and egress nodes [4, 5]. A BHP is sent prior to sending its corresponding burst with the burst's properties such as its destination and size. The burst is scheduled based on either void-filling or horizon techniques in the core nodes [6]. Setting a suitable offset time is crucial for burst's delivery and reducing burst loss.

This paper is organized as follows: in Sect. 2, current offset time algorithms are reviewed. Section 3 describes the contribution of the paper where the proposed algorithm is elaborated. Simulation environment and parameters are discussed in Sect. 4. Evaluation results are presented and discussed in Sect. 5. The paper is concluded in Sect. 6.

2 Offset Time

Burst transmission in OBS requires initiating and sending a BHP first. The data burst is assigned a period of delay (offset time) before it is sent. Incurring an offset time after sending the BHP is essential in order to allow the nodes to configure the switches along the path [2, 7]. It also allows the BHP to reserve the required resources (wavelength) at the core nodes before forwarding the burst. Reserving the wavelength requires the conversion of the BHP to an electronic signal and then to an optical signal. This process is repeated in each core node in the reserved path. By converting the BHP, the data burst is sent to its destination as an optical signal without an Optical-Electronic-Optical (OEO) conversion along the path [8]. Figure 1 illustrates the algorithm of offset time in OBS network.

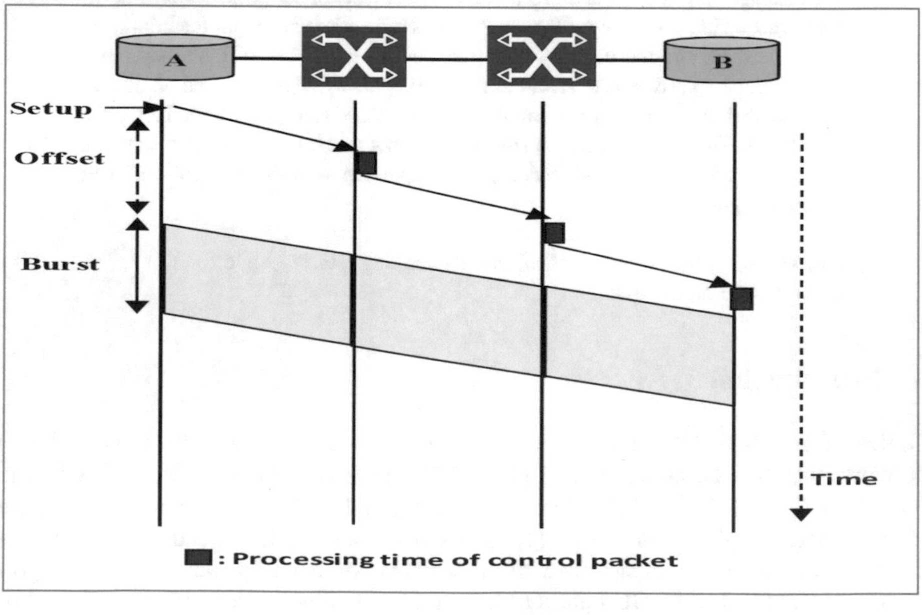

Fig. 1. Offset time algorithm

The configuration and reservation processes in OBS require the assignment of an extra time. As a result, the offset time to be incurred has to be longer than the sum of BHP processing time at the core nodes. Assigning an offset time that is shorter than the required processing time results of dropping the data burst as it arrives before its BHP. Thus, the parameter of burst loss probability is essential to measure the effectiveness of any implemented offset time algorithm.

One of the offset time algorithms is fixed offset time that comes from OBS protocol Just-Enough-Time (JET) [9]. This protocol is based on a fixed offset time that equals the sum of the total processing time of all the intermediate OBS hops that is the switch fabric configuration time of the egress OBS node. For the purpose of estimating the suitable offset time, the number of hops between the edge nodes (ingress and egress) is needed and such information is mostly provided by the edge OBS node. The drawback of fixed offset time is that it is fixed for all bursts either small or big. Small bursts do not require the same offset time as big bursts and can be sent earlier while big bursts might need an additional time to be sent successfully.

Unlike fixed offset time algorithm, adaptive offset time algorithm assigns extra offset time to big bursts in order to get a high isolation degree among bursts of different sizes [10]. Also, an isolation degree of 1 can be achieved if the offset time differentiation equals the burst size. Incurring extra offset time to bigger bursts decreases the blocking probability while the overall good put of the network is increased. However, assigning an extra delay for big bursts results of bigger loss penalty in case of burst loss.

In [11], the BHP is sent with the burst characteristics before the completion of burst assembly process. Offset time delay in this method is not high and the burst is transmitted faster than the conventional method. Nonetheless, sending the BHP before the assembly process completion requires a length estimation of the burst being assembled to be included in the header of the BHP. For this purpose, the authors used the Jacobson/Karels algorithm in order to calculate the retransmit time in Transmission Control Protocol (TCP) [12]. However, in case of inaccurate estimation, the burst loss probability is increased.

A proposed algorithm by Hirata and Kawahara [13] avoids contention of multicast bursts by assigning priority to multicast bursts and assigning additional offset times to unicast bursts. After prioritizing multicast bursts, the contention of unicast bursts is avoided with deflection routing. As a result, burst loss of both multicast and unicast bursts is improved.

In [14], an algorithm is proposed to get a moderate offset time, which is under target Insufficient Offset Time (IOT) drop ratio's restriction in order to maintain a trade-off between them. This balance is achieved by assigning offset time dynamically based on probe burst's drop ratio that is used as observing variable. Once the background traffic in the core nodes is measured, the offset time is dynamically assigned.

The new burst scheduling algorithms proposed in the last decade are focused toward maximizing the utilization of local links unlike the conventional scheduling algorithms. For example, the proposed algorithm in [15] is aimed to utilize the local links without increasing the burst loss rate. The number of voids between the bursts is reduced by connecting the coming bursts with existing bursts. Instead of setting an offset time value, a minimum and a maximum limit of offset time is set, which allows a variable value of offset time. The bursts are also aligned at the beginning or at the end of the selected void. However, there is no indication of the effect of this algorithm on the value of offset time.

An integrated algorithm that aims to reduce loss and utilize resources is also proposed [16]. The proposed algorithm requires full wavelength conversion capabilities of the nodes where routing information is exchanged among all nodes in OBS network. The edge nodes compute best routs based on links' average availability. The average availability of the links is updated and so the average traffic load. The offset time adaptation algorithm considers the state of wavelength availability. After searching for a slot of a specific burst length that is estimated to result of a minimum burst loss ratio. Offset time of the burst is selected based on the start time of the chosen composite slot. The performance of this algorithm was found to be superior to other compared algorithms; although, utilization results were not as significant as burst loss results under low load conditions.

The researchers in [17] studied the effect of the process of offset time calculation on blocking probability and fairness in OBS. The researchers found that blocking probability is highly reduced under variable offset time conditions. Offset time was calculated based on two algorithms. Firstly, it was calculated and managed based on the Load-adaptive Offset Time algorithm (LOT) proposed in [18] that calculates offset time based on the variable BCPs sojourn times. It is used to ensure achieving low dropping probability. On the other hand, Link Scheduling state based Offset Selection (LSOS) proposed in [19] was chosen for its efficiency in achieving fairness in OBS network. LSOS choses offset time based on link states and the bursts different hop lengths. In general, all bursts travelling through one or two hops should have the same blocking probability in order to achieve fairness that supports Quality of Service (QoS) in OBS network.

The authors in [20] studied the effect of offset time on burst loss ratio. Offset time was used as the tuning parameter in order to guarantee zero burst loss ratio. The researchers proposed an adaptive offset time using closed loop feedback control technique. Offset time is adaptively changed based on the received feedback. The model also ensures using the lowest offset time value needed to deliver the BHP before its corresponding burst. Hence, low end-to-end delay and zero burst loss ratio are achieved with the use of automatic and dynamic adjustment of offset time in the actual network condition.

A BHP processing model was proposed in [21] at the Switch Control Units (SCU) of Optical Burst-Switched Ring (OBSR) network. The architecture of OBSR network includes receiving module, forwarder module, scheduler module, and sending module. Three methods, central limit, large deviation, and phase-type distribution theorem-based, were proposed based on the distribution model. By using the three methods, the distribution function of the total BHP processing time is derived to configure offset time according to the derived distribution function as well as the allowed BLR in case of insufficient offset time. The proposed mechanisms, especially large deviation and PH distribution, reduce BLR for insufficient offset time to the allowed limit.

The authors in [22] argued that the mechanism of adding extra offset time to high priority packets does not necessarily improve its QoS, i.e. its blocking probability. The aforementioned mechanism is analysed in respect of TCP over OBS network. Based on the analysis, high class bursts with high propagation delay (10 ms) has higher Round Trip Time (RTT) than high class bursts with lower propagation delay (1 ms).

Consequently, high class bursts with 1 ms propagation delay has a better chance of blocking lower classes bursts than high class bursts with 10 ms propagation delay. Thus, assigning extra offset time to high class bursts based QoS mechanism does not achieve the required results when RTT is short. In short, the mechanism of extra offset time is unable to guarantee QoS of TCP traffic under the condition of short RTT.

Service differentiation was also investigated in [23] in response to the need for traffic differentiation. A feedback-based offset time selection method was proposed for the achievement of proportional QoS in regards of burst loss. Proportional QoS is achieved when the performance spacing of a traffic class is adjusted in relativity to other classes. The proposed method is aimed to achieve end-to-end proportional QoS. Offset time is dynamically adjusted based on the feedback information from egress nodes and probe packets that are sent periodically to collect link state of each node. The offset times are adjusted to satisfy the predefined proportional QoS among traffic classes for different ingress-egress node pairs. This method is supposed to maintain fairness among the different node pairs regardless of their hop lengths.

On the on hand, [24] extended and modified the work in [9] where they introduced the concept of prioritized OBS protocol called pJET that is based on Just-Enough-Time (JET) protocol and the work in [25] where they proposed a randomised offset time algorithm as a form of traffic shaping. The protocol of pJET categorises the burst based on class which resolves interclass contention; however, it does not resolve intraclass contention where two bursts of the same class collide. The condition is called systematic synchronization. On the other hand, randomised offset time algorithm [25] is proposed for reducing blocking probability resulting from collision of bursts of the same class. Nevertheless, it doesn't resolve interclass contention and the value of randomised offset time may be short. [24] introduces a randomized offset time for intraclass contentions. In this method, each class has two offset times; i.e. fixed offset time and randomised offset time. The fixed offset time is used to resolve interclass contention while randomised offset time is used when intraclass contention occurs. The randomised offset time is set by adding, not deducting, a value to the fixed offset time to ensure the delivery of the burst before its header packet. The limitation of this method is that setting higher randomised offset time increases end-to-end delay; although, it improves bursts' blocking probability.

The issue of assigning an appropriate offset time was also addressed in [26] in the proposed Variable offset time for Composite burst assembly with Segmentation (VCS) algorithm. Assigning the suitable offset time for composite burst algorithm was investigated as the traditional composite algorithm does not cater for the number of high priority packets in a given burst. The higher priority packets in a burst, the higher the loss due to the insufficient offset time chosen. The traditional variable offset time adds offset time to high priority packets which increases end-to-end delay. The proposed VCS algorithm filters the packets based on their priority and places high priority packets in the tail of the burst. Offset time is assigned based on the number of high priority packets in each burst. As the number of high priority packets increases, extra offset time is added to ensure its delivery. Once contention occurs in the core nodes, the head of the burst is dropped. If the segmentation process was successful, the bursts are sent normally. However, if the segmentation process was not successful in contention resolution, the contending burst is dropped. In comparison to traditional variable and composite algorithms, burst loss probability is improved. Nevertheless, delay is improved in comparison with variable offset time algorithm.

In order to avoid setting a fixed offset time value that might not be sufficient for the BHP to configure the switches and reserve the resources or adding an extra delay that increases the end-to-end delay, a new algorithm is proposed and tested.

3 Proposed Algorithm

The drawback of setting a fixed offset time in the conventional algorithm is avoided in the proposed method. The proposed method adapts the value of offset time based on the variables of burst size, distance (in this method, distance is the number of hops between the source and the destination) and the time that the burst will take in queuing at the edge node before it is sent to the core nodes. Thus, the Intelligent Offset Time (IOT) process is modelled as a Multiple-Input-Single-Output (MISO) fuzzy logic control process.

The three control variables, which are burst size, distance, and burst queuing delay, are the inputs of the fuzzy logic controller while the output is an offset time value. The fuzzy logic controller adapts the value of offset time as it has the inbuilt intelligence to execute the modelling process. An input of new offset time value is produced every time the modelling process of the fuzzy offset time algorithm is executed.

Figure 2 shows the flowchart of the Intelligent Offset Time algorithm.

The fuzzy rules of the fuzzy logic controller are tested, developed and then stored in the controller. After that, required fuzzy logic operations, the tested fuzzy rules and the three inputs are used to calculate the suitable value of offset time. The new offset time value produced after the modelling process is important for the BHP for configuring the switches and reserving the resources in the core nodes. The fuzzy logic of IOT algorithm computes the new offset time value through using the fuzzy inputs (Burst size, Distance, and Q.Delay) and the fuzzy rules. The fuzzy logic control process consists of three steps that are fuzzification, inferencing and defuzzification processes:

The first step, which is fuzzification process, is used for converting the crisp input and output control variables and values to linguistically accepted variables by executing the suitable membership functions. The converted fuzzy variables are used by the fuzzy logic controller.

The crisp input control variables are burst size (B.Size), distance (Dist.), and burst queuing delay (Q.Delay) at the edge node. The crisp output control variable is new Intelligent Offset Time (IOT). The triangular membership function is used in this study. Table 1 shows the summary of the input and output variables. Figure 3 shows a graphical representation of the input and output linguistic variables and their membership functions.

The second step in IOT algorithm is fuzzy inferencing process. Inferencing process is used to compute and collect the input fuzzy values. After that, the required fuzzy rules are activated. Then, the activated fuzzy rules are accumulated to produce a single crisp output variable that is used in the third step. The fuzzy rules used in this algorithm

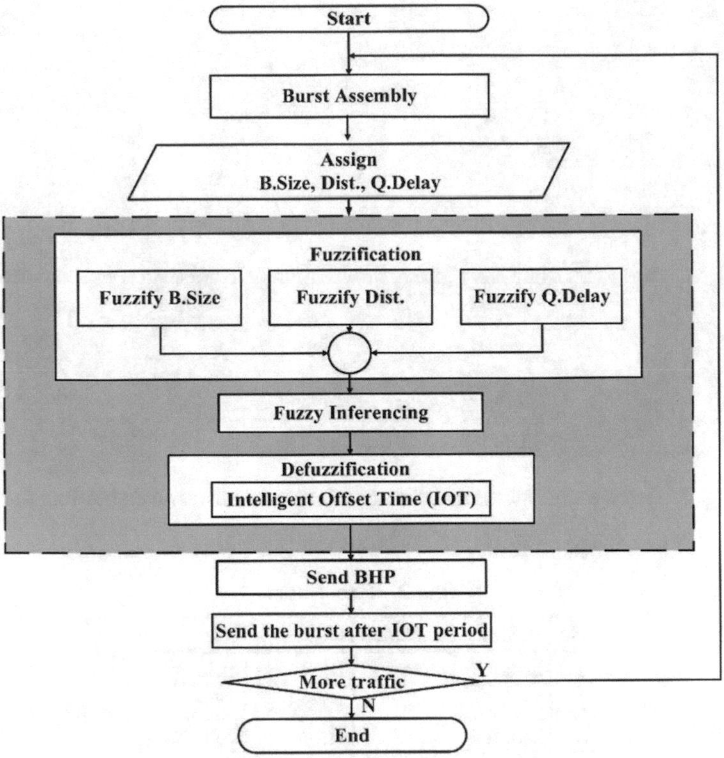

Fig. 2. Flowchart of Intelligent Offset Time algorithm

Table 1. Crisp and fuzzy input and output variables

	Crisp variable	Fuzzy variable	Operation range (Min–Max)
Input	Burst size	B.Size	0–60 (kbytes)
	Number of hops	Dist.	1–8 (hop)
	Burst queuing delay	Q.Delay	0–400 (μs)
Output	Intelligent offset time	IOT	0–100 (μs)

are 27 and are illustrated in Table 2. Each fuzzy rule shown in the table has three inputs and an output. The illustrated fuzzy rules have the following syntax: IF (B.Size AND Dist. AND Q.Delay) THEN (IOT).

Defuzzification is the final stage in the fuzzy logic control process. The output fuzzy value is converted through defuzzification process in an output crisp value, IOT.

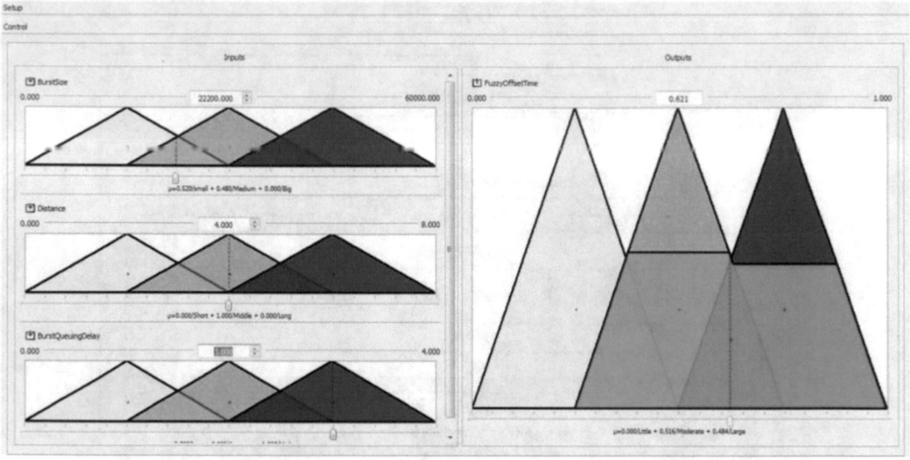

Fig. 3. Input and output linguistic variables and their membership functions.

Table 2. Fuzzy rules.

No	B.Size	Distance	Q.Delay	IOT
1	Small	Short	Low	Little
2	Small	Short	Average	Little
3	Small	Short	High	Moderate
4	Small	Middle	Low	Moderate
5	Small	Middle	Average	Moderate
6	Small	Middle	High	Moderate
7	Small	Long	Low	Large
8	Small	Long	Average	Large
9	Small	Long	High	Large
10	Medium	Short	Low	Little
11	Medium	Short	Average	Little
12	Medium	Short	High	Little
13	Medium	Middle	Low	Moderate
14	Medium	Middle	Average	Moderate
15	Medium	Middle	High	Moderate
16	Medium	Long	Low	Moderate
17	Medium	Long	Average	Moderate
18	Medium	Long	High	Moderate
19	Big	Short	Low	Moderate
20	Big	Short	Average	Little
21	Big	Short	High	Little

(*continued*)

Table 2. (*continued*)

No	B.Size	Distance	Q.Delay	IOT
22	Big	Middle	Low	Moderate
23	Big	Middle	Average	Little
24	Big	Middle	High	Little
25	Big	Long	Low	Large
26	Big	Long	Average	Large
27	Big	Long	High	Large

4 Simulation Setup and Parameters

The proposed algorithm was evaluated through Omnet++ [27]. OBS Modules [28] and fuzzylite [29] were used to create the simulation environment. OBS modules are used to create a similar network environment to OBS. Fuzzylite is used to provide the fuzzy logic control library needed for the simulation. The parameters and values set for configuring the nodes and simulating the OBS network are detailed in Table 3.

Table 3. Simulation parameters and values

No	Parameter	Value
1	Network topology	NSFNET
2	Number of channels per link	4(3 date and 1 Control)
3	Bandwidth per channel (Gbps)	1
4	Packet size (bytes)	1250
5	Packet interval time	Exponential
6	BCP processing time (μ)	10
7	Timer threshold (s)	0.0005
8	Burst threshold minimum (B)	1500
9	Burst threshold maximum (B)	60000
10	Load (min)	0.1
11	Load (Max)	1
12	Conjunction	Minimum
13	Disjunction	Maximum
14	Activation	Minimum
15	Defuzzification	Centre of gravity (CoG)

The NSFNET, which is the network topology used, consists of 14 bidirectional source/destination pairs. All the nodes in the topology were configured to transmit and receive uniformly distributed traffic. Traffic with exponential inter arrival time was used to generate the network offered load.

Offered load is incremented in steps of 0.1 for every point of measurement. Latest available unused channel scheduling with full wavelength conversion is used in the core node. Our evaluation metrics are average burst end-to-end delay, and average packet loss ratio (PLR). The average burst end-to-end delay is the burst delay from the ingress node to the egress node. IOT in comparison with the conventional and adaptive offset time algorithms are used to represent the proposed and the existing offset time algorithms respectively.

5 Result and Discussion

The results of the simulation process are demonstrated in Figs. 4 and 5 in relation to end-to-end delay and burst loss ratio that are the main issues discussed in OBS network.

Figure 4 plots average delay against offered load where the proposed algorithm, IOT, is compared with the conventional and adaptive algorithms. The figure demonstrates the highly improved performance of IOT against the adaptive algorithm. Average burst delay increases dramatically using the adaptive algorithm even under low traffic load. However, the results show that the performance of the conventional algorithm is slightly better than IOT algorithm's performance.

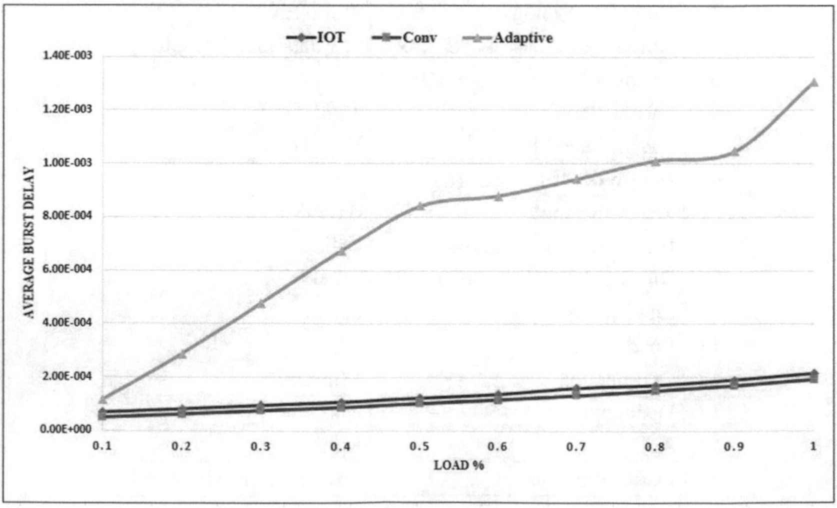

Fig. 4. Average burst delay versus offered load

Since the adaptive algorithm works with the variable of size only, the increase of burst size means adding extra offset time; thus, it experiences a higher average burst delay in comparison with other algorithms. IOT algorithm considers the three variables namely, B.size, dist and Q.Delay; thus, it chooses the appropriate offset time value with taken the three variables into consideration. On the other hand, the value of average delay of IOT algorithm is slightly higher than the conventional algorithm. The main reason is that the conventional algorithm sets a fixed value of offset time. In general, IOT algorithm maintains a stable average delay value regardless of the increased burst size and high traffic load.

Figure 5 plots average burst loss ratio versus offered load. For testing the performance improvement, IOT algorithm is compared with the conventional and adaptive algorithms. The results show that the conventional algorithm suffers the highest burst loss ratio unlike the results discussed under burst delay section where the adaptive algorithm suffered the highest burst delay. Moreover, the results strongly suggest that IOT algorithm has the best burst loss ratio among the compared algorithms.

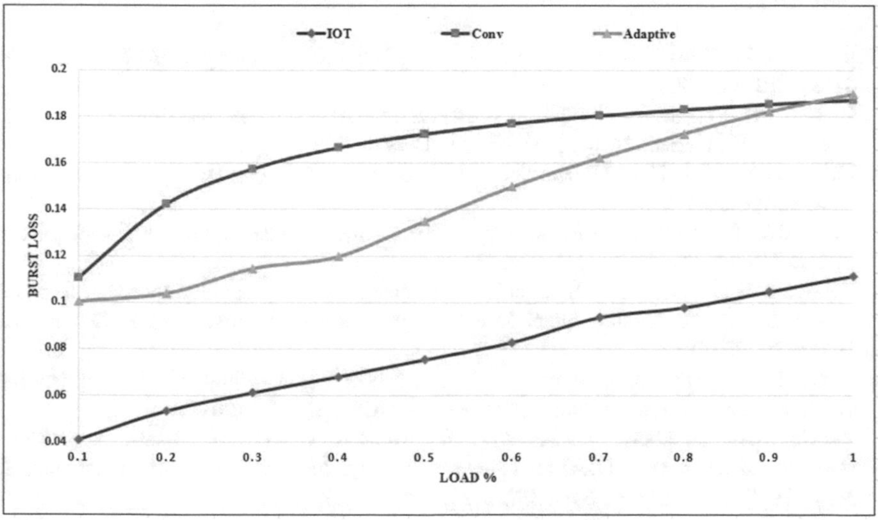

Fig. 5. Burst loss ratio versus offered load.

Based on the results depicted in Fig. 5, IOT algorithm has the best performance in terms of burs loss ratio. The poor performance of the conventional algorithm versus the adaptive and IOT algorithms is due to its fixed offset time with no consideration of other parameters affecting burst delivery. Thus, conventional algorithm suffers a high burst loss ratio. On the other hand, the adaptive algorithm adapts offset time based on the size of the bursts. Thus, burst loss ratio is improved. IOT algorithm; however, considers burst size, distance and burst queuing delay values. As a result, IOT algorithm has a good performance in general and the best burst loss ratio in specific. IOT algorithm outperforms both the adaptive and the conventional algorithms.

6 Conclusion

This paper presented the proposed Intelligent Offset Time algorithm that is modelled using the fuzzy logic control for minimizing end-to-end delay and reducing burst loss ratio. IOT is evaluated and compared to the conventional and adaptive algorithms. The simulation results demonstrate significant improvement of burst loss ratio in comparison with both the conventional and adaptive algorithms. This improvement is achieved without incurring a high increase in end-to-end delay results that indicated approximation between IOT and conventional algorithms. In future works, the proposed method will be applied with different defuzzification and aggregation methods.

References

1. Battestilli, T., Perros, H.: An introduction to optical burst switching. IEEE Commun. Mag. **41**(8), S10–S15 (2003)
2. Qiao, C.M., Yoo, M.S.: Optical burst switching (OBS) - a new paradigm for an optical Internet. J. High Speed Netw. **8**(1), 69–84 (1999)
3. Zheng, J., Mouftah, H.T.: Optical WDM Networks: Concepts and Design Principles. Wiley, New York (2004)
4. Venkatesh, T., Murthy, C.S.R., Murthy, C.S.R.: An Analytical Approach to Optical Burst Switched Networks. Springer, Berlin (2010)
5. Rajab, A., Huang, C.-T., Al-Shargabi, M.: Decision tree rule learning approach to counter burst header packet flooding attack in optical burst switching network. Opt. Switch. Netw. **29**, 15–26 (2018)
6. Adgaonkar, R., Sharma, S.: A review of burst scheduling algorithm in WDM optical burst switching network. Int. J. Comput. Sci. Issues (IJCSI) **8**(6), 75 (2011)
7. Yayah, A.A., Coulibaly, Y., Ismail, A.S., Rouskas, G.: Hybrid offset-time and burst assembly algorithm (H-OTBA) for delay sensitive applications over optical burst switching networks. Int. J. Commun. Syst. **29**(2), 251–261 (2016)
8. Li, J.K., Qiao, C.M., Xu, J.H., Xu, D.H.: Maximizing throughput for optical burst switching networks. IEEE-ACM Trans. Netw. **15**(5), 1163–1176 (2007)
9. Qiao, C., Yoo, M.: Choices, features and issues in optical burst switching. Opt. Netw. Mag. **1**(2), 36–44 (2000)
10. Xiaoyuan, C., Jian, W., Xiaobin, H., Jintong, L.: An adaptive offset time scheme in OBS network. In: 2008, pp. 1–2 (2008)
11. Mikoshi, T., Takenaka, T.: Improvement of burst transmission delay using offset time for burst assembly in optical burt switching. In: 2008 7th Asia-Pacific Symposium on Information and Telecommunication Technologies, vol. 268, pp. 13–18 (2008)
12. Peterson, L.L., Davie, B.S.: Computer Networks: A Systems Approach, 3rd edn. Elsevier, New York (2003)
13. Hirata, K., Kawahara, M.: Contention resolution considering multicast traffic in optically burst-switched WDM networks. Photon Netw. Commun. **23**(2), 157–165 (2012)
14. Lai-xian, N.D.-W.P., Zhi-chao, Y.W.-B.M., Hai, Z.W.-D.W.: Control plane measurement based dynamic offset time algorithm in optical burst switching networks. J. Electron. Inf. Technol. **4**, 004 (2012)

15. Choudhury, S., Nair, V., Biswas, P., Mal, A., Choudhury, B.: A family of flexible offset-time based wavelength schedulers for OBS edge-nodes. In: Photonics (ICP), IEEE 3rd International Conference, pp. 280–284 (2012)
16. Choudhury, S., Nair, V., Howlader, J., Choudhury, B., Mal, A.K.: An integrated routing and offset-time adaptation scheme for OBS network. In: Proceedings of the International Conference on Distributed Computing and Networking (ICDCN). ACM (2015). Article no: 28
17. Rashed, A.N.Z., Mohammed, A.E.N.A., Dardeer, O.M.: Offset time management for fairness improvement and blocking probability reduction in optical burst switched networks. Int. J. Adv. Res. Electron. Commun. Eng. (IJARECE) **2**(11), 846–857 (2013)
18. Martínez, A., Aracil, J., De Vergara, J.L.: Optimizing offset times in optical burst switching networks with variable burst control packets sojourn times. Opt. Switch. Netw. **4**(3), 189–199 (2007)
19. Tan, S.K., Mohan, G., Chua, K.C.: Link scheduling state information based offset management for fairness improvement in WDM optical burst switching networks. Comput. Netw. **45**(6), 819–834 (2004)
20. Aly, W.H.F., Zhani, M.F., Elbiaze, H.: Adaptive offset for OBS networks using feedback control techniques. In: Computers and Communications. ISCC . IEEE Symposium, pp. 195–199 (2009)
21. Dai, W., Wu, G., Qian, W., Li, X., Chen, J.: Offset time configuration in optical burst switching ring network. J. Lightwave Technol. **27**(19), 4269–4279 (2009)
22. Kim, M.-G., Jeong, H., Choi, J., Kim, J.-H., Kang, M.: Performance analysis of the extra offset-time based QoS mechanism in TCP over optical burst switching networks. In: Advanced Communication Technology. ICACT. The 8th International Conference, vol. 2, p. 1251. IEEE (2006)
23. Tan, S.K., Mohan, G., Chua, K.C.: Feedback-based offset time selection for end-to-end proportional QoS provisioning in WDM optical burst switching networks. Comput. Commun. **30**(4), 904–921 (2007)
24. Liu, J., Zeng, Q., Wang, J., Huang, J.: Algorithm of offset time in IP-over-WDM optical network. In: Optical Switching and Optical Interconnection II. pp. 274–279. International Society for Optics and Photonics (2002)
25. Verma, S., Chaskar, H., Ravikanth, R.: Optical burst switching: a viable solution for terabit IP backbone. IEEE Netw. **14**(6), 48–53 (2000)
26. Al-Habieb, Omar H.: Variable offset time for composite burst assembly with segmentation scheme in optical burst switching networks. Int. J. Adv. Res. Comput. Sci. Softw. Eng. **3** (12), 491–496 (2013)
27. Varga, A., Hornig, R.: An overview of the OMNeT++ simulation environment. In: Proceedings of the 1st international conference on Simulation tools and techniques for communications, networks and systems & workshops (SIMUTools), p. 60. ICST (Institute for Computer Sciences, Social-Informatics and Telecommunications Engineering) (2008)
28. Espina, F., Armendariz, J., García, N., Morató, D., Izal, M., Magaña, E.: OBS network model for OMNeT++: a performance evaluation. In: Proceedings of the 3rd International ICST Conference on Simulation Tools and Techniques (SIMUTools), p. 18. ICST (Institute for Computer Sciences, Social-Informatics and Telecommunications Engineering) (2010)
29. Rada-Vilela, J.: Fuzzylite: a fuzzy logic control library written in C++ (2013)

Planning and Optimization of LTE Radio Access Network for Suburban and Rural Area at Taiz City, Yemen

Redhwan Q. Shaddad[1], Nabil Y. Saleh[1(✉)], Aziz A. Mohammed[1], Zahi A. Saeed[1], Afif A. Shaif[1], and Samir A. Al-Gaialni[2]

[1] Faculty of Engineering and Information Technology, Taiz University, Taiz, Yemen
rqs2006@gmail.com, eng.nabil20185@gmail.com, aziz.alhammadi1000@gmail.com, zahiab1990@gmail.com

[2] School of Electrical and Electronic Engineering, Universiti Sains Malaysia, Penang, Malaysia
samer.algailani@usm.my

Abstract. Long Term Evolution (LTE) networks are designed with overlap between cells to support mobility along the network cells and to avoid out of coverage area; especially in suburban and rural area because high mobility. The overlapping results in interferences which reduce network capacity so this paper aims to bring customer demands for high-quality networks. This paper involves a good understanding of radio network planning and optimization of LTE and perform a case study in Taiz governorate with a selected suburban and rural area of 86.846 km^2. Self-organizing networks (SONs) are widely considered to improve the end users' quality of experience. The simulation was performed using ATOLL software. The radio frequency (RF) optimization involves Inter-Cell Interference Coordination (ICIC). The downlink (DL) LTE coverage area is enhanced and increased to 84 km^2 with probability of 97%. In addition, the overlapping zone is reduced to 0.4% and the block error rate (BLER $\geq$ 0.2) is reduced to 2.1% in the DL and (BLER $\geq$ 0.2) to 0.6% in the uplink (UL).

Keywords: LTE · LTE-A · 4G wireless communication
Access network optimization · ICIC

1 Introduction

LTE transmission is implemented using Orthogonal Frequency Division Multiple Access (OFDMA) which is suitable for fading channels. The LTE supports downstream rates at least 100 Mbps, an upstream at least 50 Mbps and radio access network (RAN) round-trip times of less than 10 ms [1–4].

A self-Optimization Network (SON) is dedicated to dynamically adjust its network control parameters according to varying conditions such as: a traffic demand that varies in intensity and location, or changing propagation conditions [5]. There is the possibility to adjust many different parameter types: azimuths, tilts, antenna height, evolved NodeB (eNB) transmit power and handover parameters [6].

F. Saeed et al. (Eds.): IRICT 2018, AISC 843, pp. 440–450, 2019.
https://doi.org/10.1007/978-3-319-99007-1_42

The key performance indicators (KPIs) must be balanced to get a perfect wireless network. The KPIs indicate the level of network performance such as cell loads, data throughput and coverage [7]. This paper proposes planning of LTE network and improving of network coverage and capacity within selected suburban and rural regions around Taiz city. Different parameters are used to maximize the network's performance for cell outage compensation.

One of the challenging problems in the SON is coverage-capacity optimization. In other words, self-optimizing algorithms are designed to achieve optimal trade-offs between coverage and capacity. It can be seen as a form of fairness, a service in which users have a minimum bit rate requirement is considered to ensure good quality of service (QoS). We also try to satisfy the maximum number of users while minimizing the corresponding capacity losses [8, 9].

This paper involves a good understanding of radio network planning and optimization of LTE and perform a case study in Taiz governorate, Yemen with a selected suburban and rural area of 86.846 km^2. The planning and optimization of LTE radio access network is proposed. The results of LTE analysis planning indicated that we need 36 base stations (20 base stations at suburban area and 16 base stations at rural area) and the numbers of total subscribers are equal to 88,460 users.

The remaining of this paper is arranged as: Sect. 2 presents the architecture of LTE network and review the key features of LTE. Section 3 explains the basic algorithms which are used in planning and optimizations of LTE proposed network. Section 4 discusses the results. Finally Sect. 5 concludes the paper.

2 Architecture of LTE Network

Figure 1 shows the Evolved Packet System (EPS) architecture. The EPS system consists of the User End (UE), the E-UTRAN and the Evolved Packet Core (EPC). The EPC provides access to external data network (e.g., Internet, Corporate Networks) and operator services. It also performs functions related to security (authentication, key agreement), subscriber information, charging and inter-access mobility [10].

The UE equipment consist of three main components: the mobile termination (MT) which handles all the communication functions, the Terminal Equipment (TE) which terminates the data streams, and the universal integrated circuit card (UICC) which is a smart card. The mobile termination might be a plug-in LTE card for a laptop. The UICC runs an application known as the universal Subscriber Identity Module (USIM) [11]. In addition, Fig. 1 illustrates that the EPC consists of a control plane node called Mobility Management Entity (MME) and two user plane nodes called Serving GW (S-GW) and Packet Data Network Gateway (PDN-GW) [11].

3 LTE Network Planning and Optimization

Radio network planning and optimization contains number of phases: (1) Site survey-which includes collection of pre-planning information that will be used in the link budget preparation and coverage and capacity planning calculations, (2) Frequency and

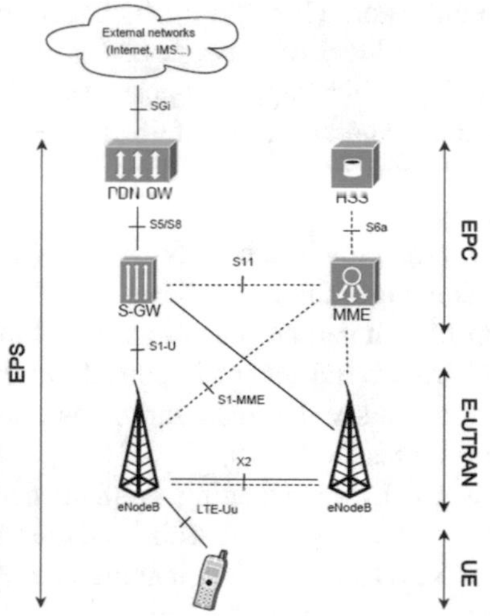

Fig. 1. LTE network architecture.

spectrum planning - in this phase a variety of parameters values will be chosen, and according to these parameters the rest of the calculation is processed, (3) Link budget and coverage planning, (4) Capacity planning, (5) Optimization process by use some parameter such as: Antenna Azimuth, Antenna Height, Antenna Tilt, Handover and Inter-Cell Interference Coordination (ICIC).

Taiz city in Yemen is with population of 573607 and having an average building height of 10 m. The area of Taiz is 86.846 km^2. The Fig. 2 shows that Taiz has been divided into two main sections according to the population distribution over the city. These sections are suburban and rural area. This paper will focus on suburban which is 36.965 km^2 and Rural which is 49.877 km^2 [12].

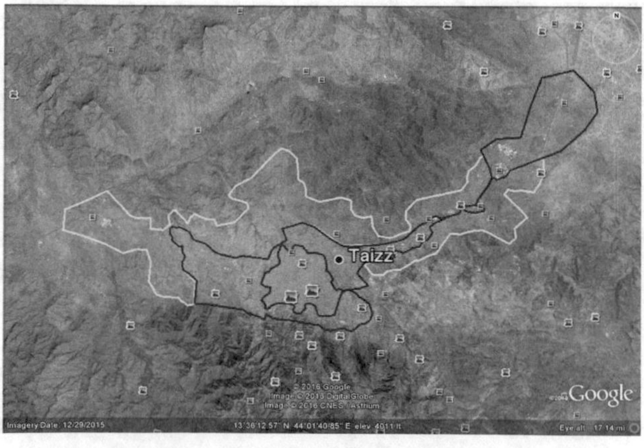

Fig. 2. Area of planning and optimization.

The available frequency band is chosen to be the frequency band for LTE network, the bandwidth, duplex mode, SFR, and cyclic prefix are specified also: Frequency band of 2100 MHz is used at suburban area and 1800 MHz is used at rural area. System bandwidth is 20 MHz. The duplex mode is frequency division duplexing (FDD). Soft frequency reuse is used. Cyclic prefix is chosen to be normal.

Taiz is mountain's city and it is considered an overpopulated city compared to the rest of most of the country. In this paper, coverage analysis along with link budget preparation and capacity analysis have been performed. Calculations have been made specifically for Taiz city. Maximum Allowed Path Loss (MAPL) has different values for suburban and rural area (UL and DL). So the calculation must be done to every condition and scenario apart, and from these results the cell radius can be calculated for each case. At the end, the minimum cell radius from UL and DL cell radiuses is chosen for each scenario. There are two different cell radiuses; each scenario has its own cell radius [13, 14].

Link budget and coverage planning is calculated for each scenario separately, for both cases "UL and DL". The procedure steps are: Step 1: Calculate the MAPL for DL and UL. Step 2: Calculate the DL and UL cell radiuses by the propagation model equation and the MAPL. Step 3: Determine the appropriate cell radius by balancing the DL and UL radiuses. Step 4: Calculate the site coverage area and the required sites number. In order to calculate the MAPL; the EIRP, MRRSS, Extra Gain, and Extra Margin and Loss must be calculated first as follows [13–15]:

$$\mathbf{EIRP} = \text{Max Tx Power} + \text{Total Tx Gain} - \text{Total Tx Loss} \tag{3.1}$$

$$\mathbf{MRRSS} = \text{Rx Sensitivity} - \text{Total Rx Gain} + \text{Total Rx Loss} \tag{3.2}$$

$$\mathbf{Extra\ Gain} = \text{Hard Handoff Gain} + \text{MIMO Gain} + \text{Other Gain} \tag{3.3}$$

$$\mathbf{Extra\ Margin}\ \&\ \mathbf{Loss} = \text{Shadow Fading Margin} + \text{Penetration Loss} + \text{Other Loss} \tag{3.4}$$

Then the MAPL is calculated by this equation:

$$\mathbf{MAPL} = \text{EIRP} - \text{MRRSS} + \text{Extra Gain} - \text{Extra Margin}\ \&\ \text{Loss} \tag{3.5}$$

Using Cost231-Hata model equations, the maximum distance between the terminal and the base station is calculated, which is the cell radius.

$$\mathbf{Total} = \text{L} - \alpha(\text{HSS}) + \text{Cm} \tag{3.6}$$

$$\mathbf{L} = 46.3 + 33.9 \times \log(\text{f}) + 13.82 \times \log(\text{HBS}) + (44.9 - 6.55 \log(\text{HSS})) \times \log(\text{d}) \tag{3.7}$$

$$\boldsymbol{\alpha}(\mathbf{Hss}) = [1.1 \log(\text{f}) - 0.7]\text{Hss} - [1.56 \log(\text{f}) - 0.8] \text{ For suburban or rural areas} \tag{3.8}$$

After determining the cell radius for each scenario, sites number and sites coverage areas are calculated by the equations below:

$$\text{Site coverage area} = 9/(8) \times \sqrt{(3)} \times \text{R}^2 \tag{3.9}$$

$$\textbf{Required sites number} = (\text{Area to be covered})/(\text{Site covered area}) \tag{3.10}$$

Capacity planning procedure is described in the following steps: Step 1: calculate the total average throughput per subscriber (UL + DL). Step 2: calculate the average throughput per subscriber for both UL & DL. Step 3: calculate the peak and average throughput per site for both UL & DL. Step 4: determine the maximum number of subscribers per site by calculating the number of subscribers for both UL & DL and taking the lowest one. Step 5: calculate total sites number required for each scenario.

$$\textbf{Avg.Throughput per subscriber for UL} = \text{Total Avg.Throughput per sub(UL} + \text{DL)} \times \text{ULtraffic} \tag{3.11}$$

$$\textbf{Avg.Throughput per subscriber for DL} = \text{Total Avg.Throughput per sub(UL} + \text{DL)} \times \text{ULtraffic} \tag{3.12}$$

$$\textbf{peak throughput per site} = (\text{data RE/sec} \times \text{bits pet RE} \times \text{MIMO effect} \times \text{coding rate}) \tag{3.13}$$

$$\textbf{Average throughput per site} = \sum(\text{peak throughput per modulation scheme} \times \text{sub number percentage}) \tag{3.14}$$

$$\textbf{Max Sub No.per site} = (\text{Average throughput per site} \div \text{Average throughput per subscriber}) \tag{3.15}$$

This paper provides an overview of different network parameters that the network operator can modify to influence the coverage and capacity. Figure 3 shows the steps of optimization process. After that problem analysis, one or more parameters is selected to get optimum service. The RF optimization involves optimization and adjustment of antenna system hardware and neighbor sites. The Atoll software is used to simulate this work.

4 Results and Discussions

The DL + UL throughput (kbps) is calculated from all sites as shown in Fig. 4. The red color shows Peak RC Cumulated Throughput (DL) (kbps) while the blue color shows Peak RC Cumulated Throughput (UL) (kbps). The DL throughput reached maxima

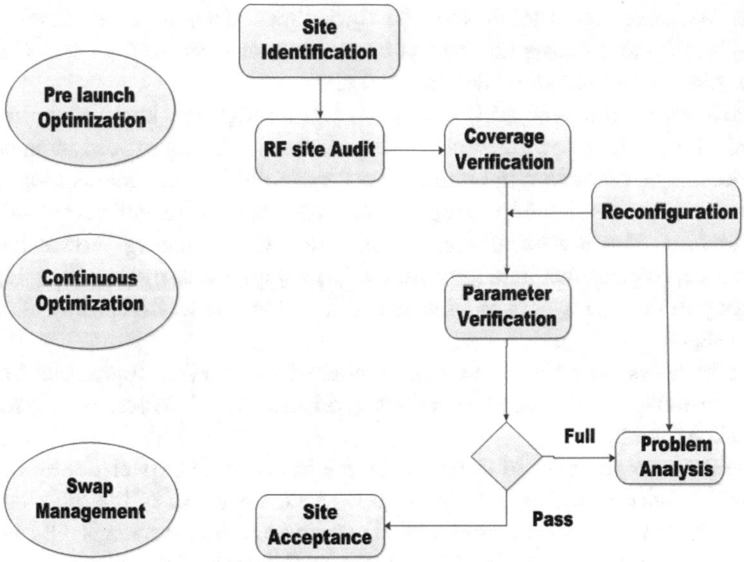

Fig. 3. LTE optimization flow diagram.

(above 60 Mbps) at the sites of Maweh_1 and Almasanaa_1 because the topography free area of barriers which are affect on the signal, unlike other area that landforms and obstructions on the propagation path such as buildings and trees, the mountains, losses, hills build are affect them must be considered. Signals fade at varying rates in different area. Figure 5 shows peak RC cumulated throughput (DL) (kbps) at different traffic load (10% (blue color), 20% (brown color) and 30% (green color) for different

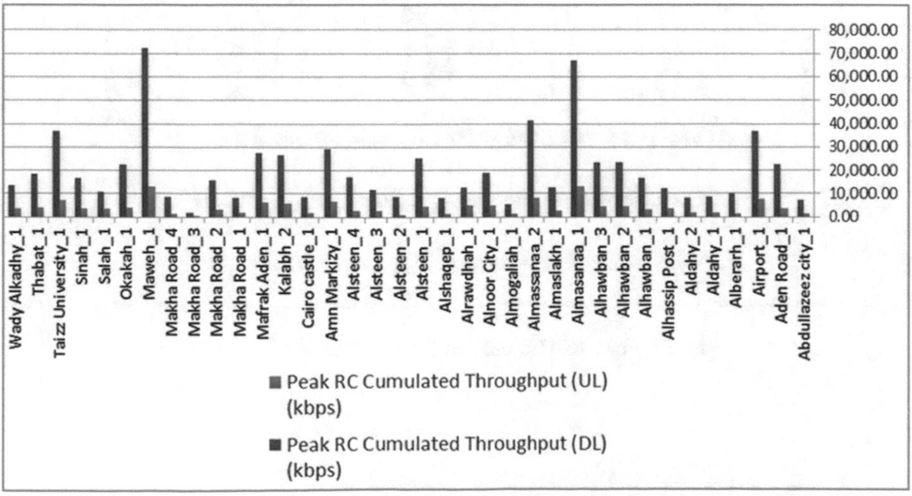

Fig. 4. Peak RLC cumulated DL + UL throughput (kbps).

locations. We note that for all site, the throughput (DL) in more traffic load are increasingly. Figure 6 shows the ratio of connection success from all site. Connection success achieved rate above 97.8% from all sites.

Figure 7 shows the peak RLC cumulated DL throughput (kbps) according to service for all site as the following: VoIP, video conferencing, high speed internet, mobile internet access. We note that the throughput for mobile internet access more than any other service because of the growing demand for users in recent years On Internet-related services. The lowest throughput of Video Conferencing service because of advanced video codecs facilitate improved video compression by providing high video quality at significantly lower capacities, in addition that this service is only allocated for business users.

Figure 8 shows the ratio of the total number of users trying to connect to the total number of users not connected. This ratio is good and enough to connect the most users by the LTE network.

Figure 9 shows the ratio of different service between all expected users. The most widely used service is Mobile Internet Access service because of the growing demand for users in recent years On Internet-related services and this service is allocated for all users on the contrary, the service is High Speed Internet allocated only for business users.

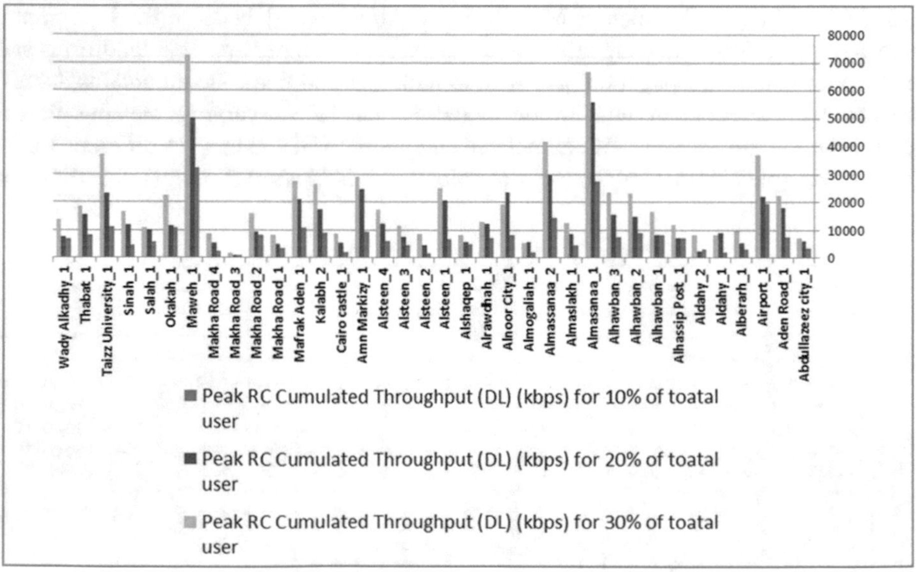

Fig. 5. Peak RC Cumulated Throughput (DL) (kbps).

Figure 10 shows the compression between Peak RLC Cumulated Throughput (DL) and Peak RLC Cumulated Throughput (UL). Figure 10 shows Peak RLC Cumulated Throughput (DL) + (UL) for different service.

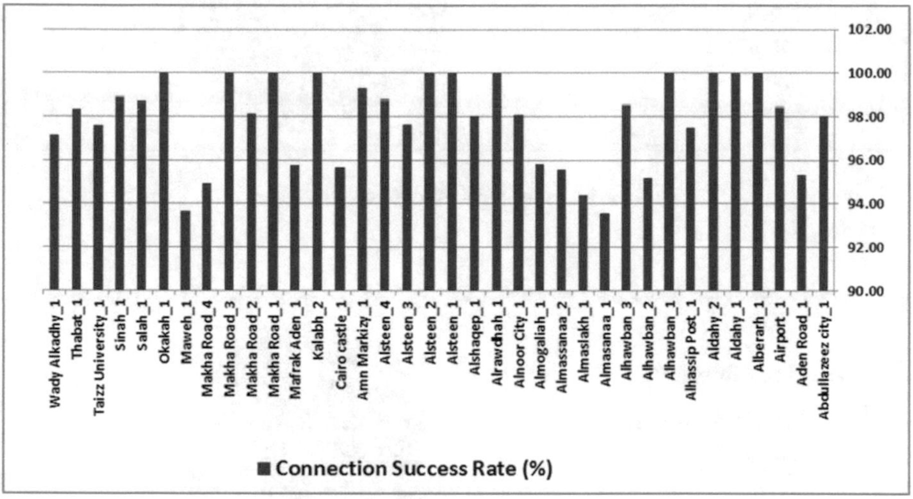

Fig. 6. Connection success rate (%).

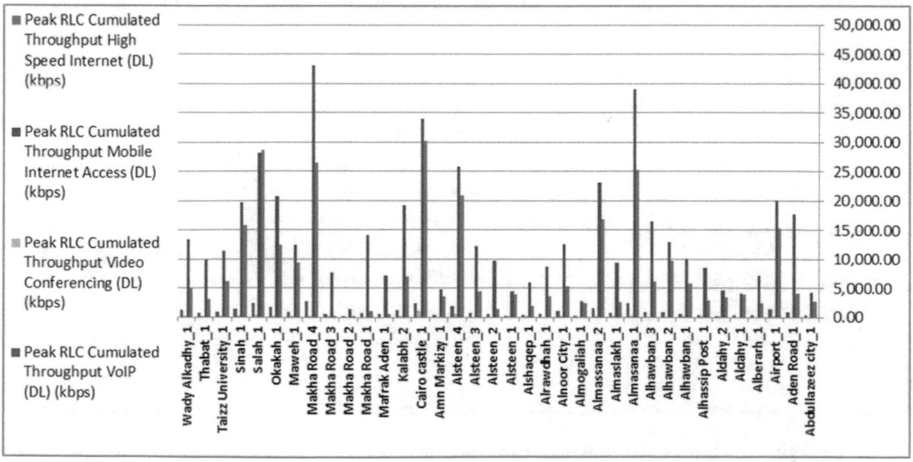

Fig. 7. Peak RLC Cumulated DL Throughput (kbps) for service.

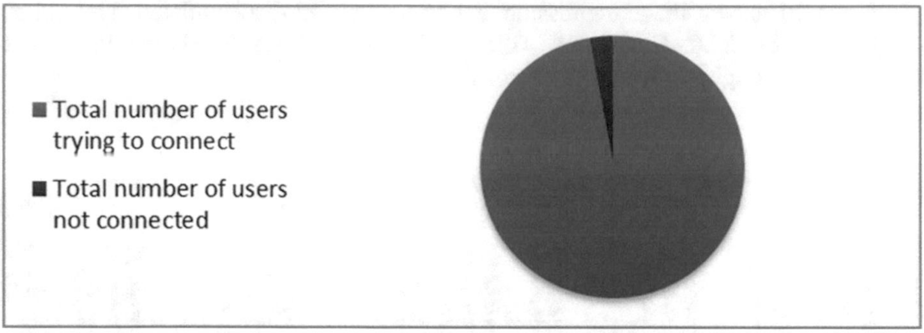

Fig. 8. Statistic of simulation by user.

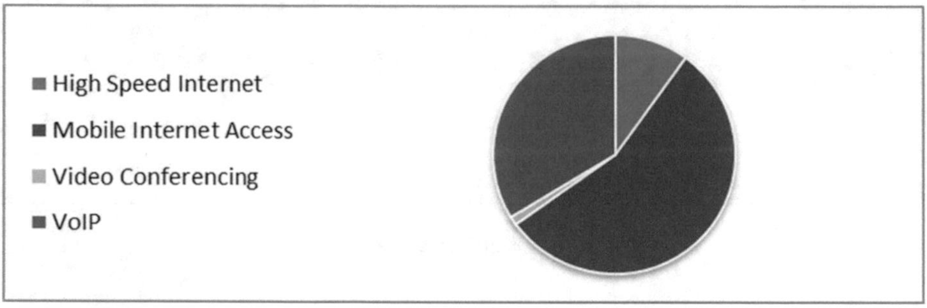

Fig. 9. The ratio between different service.

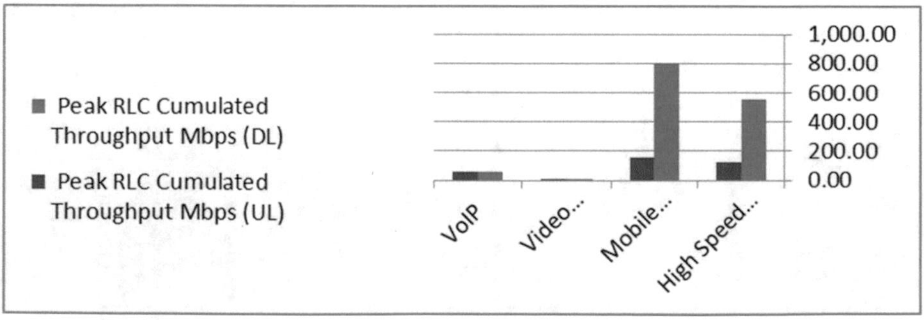

Fig. 10. Peak RLC Cumulated Throughput (DL) and (UL) for different service.

5 Conclusions

This paper has been proposed optimal LTE wireless network at suburban and rural areas in Taiz, Yemen. The results of LTE analysis planning indicated that we need 36 base stations (20 base stations at suburban area and 16 base stations at rural area) and the numbers of total subscribers are equal to 88,460 users. Max Sub number per site (DL) = 1657.724 sub/site and max Sub number per site (UL) = 1097.1 sub/site. After LTE optimization by using some of parameter such as: Antenna Azimuth, Antenna Height, Antenna Tilt, Handover and ICIC. The coverage, signal level, throughput, Connection Success Rate and CINR have been improved. The downlink (DL) LTE coverage area is enhanced and increased to 84 km^2 with probability of 97%. In addition, the overlapping zone is reduced to 0.4% and the block error rate (BLER $\geq$ 0.2) is reduced to 2.1% in the DL and (BLER $\geq$ 0.2) to 0.6% in the uplink (UL).

References

1. Sesia, S., Toufik, I., Baker, M.: LTE-the UMTS Long Term Evolution. Wiley, New York (2015)
2. Liu, H., Jiao, Y., Gao, Y., Sang, L., Yang, D.: Performance evaluation of flexible duplex implement based on radio frame selection in LTE heterogeneous network. In: ICT IEEE, pp. 308–312 (2015)
3. Shaddad, R., Mohammad, A., Al-Gailani, S., Al-hetar, A., Elmagzoub, M.: A survey on access technologies for broadband optical and wireless networks. J. Netw. Comput. Appl. **41**, 459–472 (2014)
4. Takpor, T., Idachaba, F.: Analysis and simulation of LTE downlink and uplink transceiver. In: Proceedings of the World Congress on Engineering 2014 (2014)
5. Shaddad, R., Abdulwadood, M., Kuradah, N., Alsharaie, S., Qaid, M., Alramesi, A., Rassam, A.: Planning and optimization of LTE radio access network for urban area at Taiz City, Yemen. In: IRICT 2017, Johor, Malaysia (2017)
6. Yassin, M., Aboulhassan, M., Lahoud, S., Ibrahim, M., Mezher, D.: Survey of ICIC techniques in LTE networks under various mobile environment parameters. Wirel. Netw. **23** (2), 403–418 (2017)
7. Saxena, A., Sindal, R.: An optimized LTE han over model on quality and margin with key performance indicator. Wirel. Pers. Commun. **98**(2), 2389–2401 (2018)
8. Buenestado, V., Toril, M., Luna-Ramirez, S., Ruiz-Aviles, J.M., Mendo, A.: Self-tuning of remote electrical tilts based on call traces for coverage and capacity optimization in LTE. IEEE Trans. Veh. Technol. **66**(5), 4315–4326 (2016)
9. Ghazzai, H., Yaacoub, E., Alouini, M., Dawy, Z., Abu-Dayya, A.: Optimized LTE cell planning with varying spatial and temporal user densities. IEEE Trans. Veh. Technol. **65**, 1575–1589 (2016)
10. E-UTRA 3GPP TS 36.331 E-UTRA Radio Resource Control (RRC) protocol specification. http://www.3gpp.org/dynareport/36331.htm. Accessed 06 June 2016

11. Ericsson. Long Term Evolution (LTE), an introduction, October 2014
12. Taiz Information Center (2015). http://www.taizgov.com/new/
13. HUAWEI TECHNOLOGIES, Long Term Evolution (LTE) Radio Access Network Planning Guide CO., LTD
14. Abdul, B.S., Dimensioning of LTE network, description of models and tool, coverage and capacity estimation of 3GPP long term evolution radio interface (2009)
15. www.nanocellnetworks.com

A Survey of Millimeter Wave (mm-Wave) Communications for 5G: Channel Measurement Below and Above 6 GHz

Ahmed M. Al-samman(✉), Marwan Hadri Azmi, and Tharek Abd Rahman

Wireless Communication Centre, Universiti Teknologi Malaysia, Johor Bahru, Malaysia
ahmedsecure99@gmail.com

Abstract. As the demand for higher speed data transmissions continues to increase exponentially beyond the speed-limit of the fourth generation (4G) wireless networks due to the rapid spectrum depletion of the microwave frequency bands below 6 GHz, it has become quite evident that the existing wireless communication systems will eventually be constrained from meeting the huge throughput requirements for various emerging applications beyond 4G wireless networks. In order to sustain the future market dominance of wireless communications, the narrowness of wireless bandwidths in the existing systems, which has become a key issue for the upcoming wireless systems, needs to be addressed by looking beyond the traditional microwave spectrum domain through the exploitation of the huge bandwidths available in the millimeter wave bands. In this work, we have conducted the review for a series of measurements at various microwave bands, below 6 GHz and above, to study the behavior of ultra-wideband (UWB) channels, typically in different indoor and outdoor environments. These measurements have been used to gain a useful insight into the path loss and time dispersion parametric behaviors of the 5G channel and to investigate the channel characterization of the UWB signals within spatially restricted locations. Moreover, these measurements have been used to evaluate the newest channel model and channel prediction which have proposed for 5G.

Keywords: 5G networks · Millimeter wave bands · Path loss
Power delay profile · Root mean square delay spread

1 Introduction

Wireless communication has become more prevalent in our today's rapidly advancing world of technology. In the contemporary times, wireless networking is virtually applied in every area of our day-to-day works through the means of the latest advancements in wireless technologies, such as, wireless local area networks (WLAN), wireless personal area networks (WPAN), wireless body area networks (WBAN), and cellular networks, all of which are meant to routinely improve the daily productivity of man in various areas of human endeavors. The ever-increasing demand for better performance of the wireless technology, in terms of data rates, has caused the manufacturers, device makers

F. Saeed et al. (Eds.): IRICT 2018, AISC 843, pp. 451–463, 2019.
https://doi.org/10.1007/978-3-319-99007-1_43

and infrastructure developers to dwell in a state of perpetual search for wider radio spectrum to satisfy the emerging product offerings. In fact, wireless communications have tremendously transformed the way we work and study, making it possible for people to socialize with one another globally without any physical connections. In the near future, wireless communications will be able to connect everything (Internet of Everything), anywhere (pervasive communications), at any time (ubiquitous communications) [1]. Wireless communications at ultra-wideband (UWB) microwave frequencies presents the most recent milestone in short-range wireless communications technology. In wireless communications, devices are considered as UWB if they have the ability to emit signals with more than 500 MHz of absolute bandwidth (AB-B), or over 20% of relative bandwidth (RE-B) during each transmission time interval (TTI) [2].

Recently, significant research attentions have been channeled to the fifth generation (5G) wireless communications. Early discussions and indication of interests in a possible 5G standard had begun since 2013, and today, 5G concept has become one of the most discussed topics that has captured the attention of wireless researchers and engineers around the world [3]. Over the past decades, the high adoption rate of the numerous emerging communications technologies, such as tablets, smart phones and various wearable devices for health and fitness purposes, is a clear indication that the changing trend in the growth and development of wireless technologies will continue to grow at extreme rates over the coming decades. For instance, the universal mobile telecommunications system (UMTS) has predicted that the mobile traffic will rise above 800 Mb/subcarrier daily by 2020 [4], which implies that some 130 exabits (10^{18}) of data traffic per year will be attainable for some operators [5]. Beside the huge volume of data, the population of devices are also expected to grow exponentially [3]. The massive proliferation of emerging applications justifies the assertion that the number of devices could possibly hit tens or if not hundreds of billions of devices by the time the 5G is fully deployed [6]. In order to prepare ahead of time, a new wireless technology is required that can support orders of magnitude increase in capacity for the future wireless communications. Since high capacity means more bandwidth, massive antenna mechanism is considered one of the solutions to meeting the stringent signal-to-noise ratio (SNR) requirements.

Due to the fact that the current spectrum allocation plan does not allow the carriers to offer the much needed wide bandwidth irrespective of the advanced modulation schemes and multiple-inputs multiple-output (MIMO) antenna systems used, the problem of capacity crises in the microwave bands, i.e., below 6 GHz, can be addressed by using two well-known schemes, namely, carrier aggregation and MIMO techniques [7]. With carrier aggregation, a maximum bandwidth up to 100 MHz can be achieved, and to fulfill the maximum capacity requirements of the 4G cellular network, MIMO technique was introduced. In LTE-A systems, up to 8×8 MIMO downlink transmission, and 4×4 uplink transmission are possible [7]. However, considering the magnitude of traffic load that are expected to hit the electromagnetic space of the cellular networks by 2020 and beyond as forecasted by UMTS, the next generation of cellular networks are expected to deliver as much as 1000 times in capacity higher than what is currently obtainable. Likewise, the wireless connectivity will be ubiquitous and ever accessible to the emerging smart devices predicted at about 50 billon devices by 2020 [8, 9]. The shortage of bandwidth in the current below 6 GHz cellular systems has limited the chances of meeting the broadband requirements for future media-rich communication in the microwave systems. Virtually all the existing mobile communication systems are operated in the traditional below 6 GHz microwave bands, hence, in the next few years, no frequency allocation will be possible within this range of spectrum bands as it must have reached the theoretical limit [10]. Thus, attention has been shifted to the millimeter wave band as the main candidate for the next generation wireless networks (5G wireless technology) due to the huge bandwidth (up to 10 times as much as today's cellular networks) that available in the millimeter wave spectrum space [5]. Moreover, ongoing development related to the compatibility of mobile broadband systems with frequency bands above 6 GHz has been reported in academic and industrial research entries [3, 11, 12]. The availability of huge bandwidths in the millimeter wave bands has prompted the random development of millimeter wave technologies in all areas including circuits, antennas and communication protocols. The spectrum of 3–300 GHz are often being referred to as the millimeter-wave (mm-wave) bands with wavelengths ranging from 1 to 100 mm [13].

This paper features the outcomes of the latest measurements that have been carried out on different below the 6 GHz microwaves bands and above the 6 GHz bands (also known as the mm-wave bands) for 5G networks. This work covers a comprehensive study of indoor and outdoor measurements below and above 6 GHz bands conducted in different environments under different scenarios at various frequencies using the latest state-of-the-art channel sounding equipment to study, model, select, and develop optimum operation bands for future network. Our analysis is focused on path loss, power delay profile, root mean square delay spread, and angle of arrival and departures of different bands in different scenarios.

2 Channel Measurement Techniques

The most direct method of studying radio wave propagation is by channel measurements, which will achieve statistical models and verify propagation theory. Different measurement techniques with various experimental setups have been used to study the different aspects of the RF channel. In general, they can be classified as narrowband, wideband and ultra-wideband techniques based on the probing signal bandwidth and the channel delay spread. Narrowband measurement techniques are used to measure the path loss, narrowband fading and Doppler spread. The simplicity of this technique is the main advantage of narrowband measurements. In narrowband techniques, a continuous wave (CW) is transmitted. The received power is measured over space or time. The main drawback of the narrowband measurement technique is that the received signal represents only the envelope of the vector summation of the multipath components (MPCs); no quantitative multipath information is available. Hence, if the bandwidth of the signal is high, wideband measurement techniques should be used to estimate the time dispersion parameters of the channel. For this reason, wideband and ultra-wideband techniques are used in the UWB channel measurements of mm-wave bands for 5G wireless communication networks. The wideband and ultra-wideband channel sounders are used to resolve each MPC and provide the time dispersion parameters. The large bandwidth in the ultra-wideband signal makes the symbol rate relatively high. This implies that the channel is frequency selective, hence, the time delay spread cannot be neglected. In order to extract the time dispersion parameters for the wideband channel, the channel sounding measurement is applied.

The wideband and ultra-wideband channel sounders are provided by either frequency or time domain, where the Fourier transform is used to convert between frequency and time domain. In the frequency domain, the channel transfer function is measured using a vector network analyzer (VNA). The complex frequency response of the channel is measured by the S21 parameter [14, 15]. Using the VNA system, the channel is measured at different frequency tones, along the bandwidth of the system, by using stepped frequency sweeping. Hence, the large bandwidth results in slower measurement of the channel. So, the VNA system cannot be used to measure a time-variant channel; implying that it can be used only for a slowly varying channel. The time domain measurement provides a more direct characterization approach [16]. These measurement techniques use a pulse generator to transmit short pulses of the order of nano-seconds representing the large bandwidth of wideband channels, while a digital sampling oscilloscope (DSO) is used to record the received signal [17]. This measurement technique is fast and suitable for rapid channel variations. However, is it is a challenge to generate short pulses with adequate power to achieve good-quality received signal. A more prominent measurement technique in the time domain approach uses correlation channel sounders. In this method, a sequence of pulses such as a pseudo noise (PN) sequence is sent by the transmitter, while in the receiver part the cross correlation between the transmitted and received signals is used to extract the channel. Since the transmitter and the receiver are separated in the time domain approach, it can be used for long distance, although with a synchronization challenge. The correlation channel sounder approach is based on the assumption that the clock rate of the pulses sequence is the same at the transmitter and receiver for real-time correlation processing [18, 19]. To avoid the digital signal processing required for a correlation sounder, the clock rate of the receiver pulses sequence should be lower than the clock rate of the transmitted pulses. This type of channel sounder is referred to as the sliding correlation channel sounder (SCS) [20]. Although the hardware connection between the transmitter and receiver is not required by the SCS measurement technique, the separation PN generators at the transmitter and receiver require different references from the frequency oscillators. Moreover, the dynamic range of the measurement is restricted by the clock rate of PN sequence differences [21, 22].

3 Study of Channel Measurement Below 6 GHz

The signal propagation study relative to the channel is important for developing robust wireless communication systems. Over the past couple of decades, a considerable amount of effort has been dedicated to studying the propagation of wireless channels at various frequency bands in different indoor and outdoor environments, under line of sight (LOS) and non-LOS (NLOS) scenarios. In narrowband and wideband, many measurement campaigns have investigated propagation in different environments [23–30]. The interest on UWB channel measurements has increased since the first measurement campaign results were published in 1997 [31, 32].

3.1 Indoor Environment

Win et al. [31, 32] conducted and reported on the first measurement campaigns to characterize the UWB channel within the bandwidth of 1 GHz and 1.3 GHz for indoor office and outdoor forest environments, respectively. UWB channel measurement campaigns were conducted in different indoor environments for various frequency range of frequency range below 6 GHz. Table 1 gives an overview of different reported indoor UWB channel measurements in frequency domain and time domain below 6 GHz. This indoor measurements were taken in offices, residential areas, laboratories, parking lots and corridors, for different range of frequencies and Tx–Rx separation distance. In 2004, Chuan et al. [33] has reported all conducted measurement from 2002 until 2004 for indoor office.

It can be shown from Table 1 that the longest distance range of time domain measurement is greater than the longest one in frequency domain due to the limitation of frequency domain techniques. The path loss (PL) model has been studied using different models [22, 34–37, 39, 40]. The time dispersion parameters have been investigated based on RMS delay spread (τ_{rms}) and mean excess delay (τ_{m}) in [22, 34–38, 40]. Few researchers have investigated the interdependencies of time dispersion parameters with path loss and Tx–Rx separation distance [22,37].

3.2 Outdoor Environment

Subsequently, UWB channel measurements have been performed in different outdoor LOS and NLOS scenarios to characterize the frequency bands below 6 GHz in different frequency and time domains. The outdoor measurements were taken in the forest [32, 41, 42], on the roadway [35], in an outdoor office [43], a suburban area [44], a fuel station [14], an outdoor parking lot, and other outdoor environments as listed in Table 2. The path loss model and time dispersion parameters have been studied extensively for most of these environments. However, there are few studies on tracking the variation of the UWB channel under these parameters in small spatial locations [45, 46].

Based on the extensive literature review on UWB measurements addressed in this work, it was shown that the UWB channels are generally time varying. Moreover, different case studies performed by different authors and researchers have shown that the UWB channels are sparse channels.

4 Channel Measurements Above 6 GHz

With the explosive growth of demand for wireless communications, the discrepancy between capacity requirements and shortage of spectrum has become increasingly apparent. The congestion of the current frequency band (below 6 GHz) and the narrowness of wireless bandwidth is a key problem for fifth generation wireless networks. Yet huge bandwidth is available in the millimeter wave band, so mm-wave communications have been proposed as an important part of the 5G mobile network, in order to provide multi-gigabit communication services such as ultra-high definition video (UHDV) and high definition television (HDTV) [50]. As with any wireless communication, the study of

Table 1. Comparison of indoor UWB channel measurement below 6 GHz frequency bands.

Source	Environment and setting	Domain	Frequency range (GHz)	Distance range (m)	Parameters of study
Win et al. [31]	Indoor office (LOS)	Time	BW = 1 GHz	5.5–17.5	Received signal power
Ciccognani et al. [34]	Indoor laboratory (LOS)	Time	3.6–6	1–11	PL, τ_{rms}, τ_m
	Indoor laboratory (NLOS)			≤ 5.3	
Lee [35]	Indoor parking (LOS)	Time	3.1–6.3	2.5–20	PL, τ_{rms}, τ_m and MPCs
Ghassemzadeh et al. [36, 37]	Indoor office (LOS)	Frequency	4.375–5.625	1–15	PL, τ_{rms}
	Indoor office (NLOS)				
Keignart and Daniele [38]	Indoor laboratory (LOS)	Frequency	2–6	1–6	τ_{rms}
	Indoor laboratory (NLOS)			3–10	
Wang et al. [39]	Indoor office (LOS)	Frequency	4.3–7.3	1–12	PL
	Indoor office (NLOS)				
Al-samman et al. [22, 40]	Indoor corridor and office	Time	3.1–5.3	1–67	PL, τ_{rms}, τ_m

signal propagation is important for designing and modeling the mm-wave systems. Recently, most research has focused on the 28 GHz band, the 38 GHz band and the E-band (71–76 GHz and 81–86 GHz) [51]. In the past two decades, measurement campaigns were conducted in 28 GHz and 38 GHz mm-wave bands for local multipoint distribution systems (LMDS) [52, 53]. In addition, wideband NLOS measurements were performed by Violette et al. at the 9.6, 28.8, and 57.6 GHz bands in downtown Denver, where the results showed significant signal attenuation due to obstruction by large buildings [54]. Foliage attenuation measurements at the 35 GHz band were conducted through approximately 15 m of red pine trees, showed a large loss in excess of free space [55]. Propagation through a canopy of orchard trees was measured using continuous wave (CW) signals at 9.6, 28.8, and 57.6 GHz [56]. The 60 GHz band is one of the well-studied mm-wave bands. This band is important to offer multiple gigabits per second data rates for short range indoor communications. Many measurements were conducted in indoor environments due to the earliest intended-use cases (WLAN) [57–59]. However, 60 GHz outdoor communication for unlicensed backhaul applications recently garnered great

Table 2. Comparison of outdoor UWB channel measurement below 6 GHz frequency bands.

Source	Environment and setting	Frequency range (GHz)	Distance range (m)	Domain	Parameters of study
Kim et al. [47]	Outdoor office (LOS)	3–6	7–14	Frequency	τ_{rms} and τ_m
	Outdoor office (NLOS)				
Molisch et al. [48]	Outdoor (LOS) Outdoor (NLOS)	3–6	5–17	Frequency	PL, τ_{rms} and MPCs
Lee [35]	Outdoor roadway (LOS)	3.1–6.3	2.5–20, 30–80	Time	PL, τ_{rms} and MPCs
Santos et al. [14]	Outdoor–fuel filling station (LOS)	3.1–10.6	–	Frequency	MPCs
Win et al. [32]	Outdoor forest	BW = 1 GHz	3–18.3	Time	τ_{rms} and MPCs
Francesco et al. [43]	Outdoor suburban area	3–6	–	Time	PL and τ_{rms}
Anderson et al. [41]	Outdoor forest	0.83–4.2	4–50	Frequency	Ric K-Factor and MPCs
Richardson et al. [49]	Outdoor–inside vehicle (LOS)	3.1–6	3–45.1	Time	PL, τ_{rms} and Capacity
	(NLOS)		3–35.5		
Renzo et al. [42]	Outdoor-forest, sub-urban	3–6	1–25, 10–51	Time	PL, τ_{rms}, τ_m

interest when the 2013 FCC part 15 rule changed, greatly expanding the effective radiated power of WLAN devices in the 60 GHz band from 40 to 82 dBm [15]. The measurement campaigns at 59 GHz were carried out in Oslo city streets for LOS and obstructed environments over 7 different street scenarios [60]. Other outdoor measurements were conducted in a city street environment at 55 GHz through narrow and wide city streets [61]. Recently, 60 GHz wideband propagation measurements have been investigated in cellular peer-to-peer outdoor environments and in-vehicle scenarios [62].

For 71–76 GHz and 81–86 GHz bands, which are a part of the millimeter-wave E-band (60–90 GHz), some measurements were taken for microwave link and point-to-point applications (LOS propagation) [63, 64].

Nowadays, the measurements of mm-wave bands are exploring the candidate frequencies for 5G wireless networks, including the microwave range above 6 GHz. Extensive channel sounding campaigns at 28, 38, and 60 GHz were conducted on the campus of the University of Texas at Austin (UTA) over the summer of 2011 and on the campus of New York University (NYU) in New York City during the summers of 2012 and 2013 [50]. In 2011, the University of Surrey, England, set up a world research hub for 5G mobile technology with the goal of expanding UK telecommunications research and innovation [65]. Ben-Dor et al. [62] conducted measurements at 38 GHz and 60 GHz with a 1.5 GHz first null-to-null bandwidth. The measurements were taken in cellular peer-to-peer outdoor environments and in-vehicle scenarios. The results present the measurement environment setting, the range of measurement distance, frequency range, the channel sounder measurement technique (i.e., frequency domain or time domain), and the parameters that have been studied such as PL, delay spread, angle of arrival (AOA) and angle of departure.

The initial tests by Samsung were performed at 28 GHz and 40 GHz for a penetration losses study of common obstructions such as wood, water, hands, and leaves [66]. For LOS outdoor measurements at 28 GHz, the path loss was measured for distances up to 100 m [66]. In NLOS environments, the researchers at UTA conducted many measurements to study different channel propagation parameters [20, 67]. In May 2013, Samsung Electronics conducted measurements to test next-generation equipment at 28 GHz over distances up to 2 km, using an adaptive array transceiver with multiple antenna elements [68]. This early work did not include extensive measurements, so a channel sounder that measured power delay profiles (PDP) from multiple directions of arrival to create omnidirectional-like wideband channel measurements was used [50]. NYU Wireless research center, one of the most active participants in supporting mm-wave technologies, has done extensive measurements in New York City. For the outdoor environment, many measurement campaigns were undertaken in NYU for different scenarios that studied different aspects and parameters for 5G propagation. The most inclusive reference on the outdoor propagation channels can be found in [50]. The candidate frequency bands investigated by UTA and NYU were 28 GHz band, 38 GHz band, 60 GHz band and 73 GHz band. With a 1 ns multipath resolution, a set of mm-wave propagation measurements was conducted at 10.5, 15, 19, 28, and 38 GHz center frequencies in outdoor environment in tropical region i.e. Malaysia [69].

For indoor environment, some measurements have been carried out by NYU in the 28 GHz and 73 GHz frequency bands in a typical indoor office environment [70–73]. Also, indoor measurements for wireless and backhaul communications have been conducted in the frequency band of 72 GHz [74]. At Finland's Aalto University, mm-wave measurements have been conducted in the 60 GHz and 70 GHz frequency bands [75]. Hur et al. and Rajagopal et al. have conducted measurements in the 28 GHz and 40 GHz bands in the indoor environment [76]. Moreover, different academic researchers have conducted other measurements for different types of indoor environments at different 5G candidate bands [19, 76–78].

5 Conclusion

This paper presents a general overview of UWB channel, millimeter wave bands and the 5G system. It also presents a detailed overview of previous works on UWB channel measurement in mm-wave bands below and above 6 GHz for 5G wireless networks. It covers a comprehensive study of indoor and outdoor measurements below and above 6 GHz bands conducted in different environments under different scenarios at various frequencies using the latest state-of-the-art channel sounding equipment to study, model, select, and develop optimum operation bands for future net-work.

Acknowledgment. We would like to thank the Research Management Centre (RMC) at Universiti Teknologi Malaysia for funding this work under Grant Number Q.J130000.21A2.03E69. Also, the authors would like to acknowledge UTM research Grant (Vot 4J218), Universiti Teknologi Malaysia.

References

1. Gupta, A., Jha, R.K.: A survey of 5G network: architecture and emerging technologies. IEEE Access **3**, 1206–1232 (2015)
2. Molisch, A.F.: Ultra-wide-band propagation channels. Proc. IEEE **97**, 353–371 (2009)
3. Andrews, J.G., Buzzi, S., Choi, W., et al.: What will 5G be? IEEE J. Sel. Areas Commun. **32**, 1065–1082 (2014)
4. UMTS Forum: Mobile traffic forecasts 2010–2020. UMTS Forum Rep. 44 (2011)
5. Akdeniz, M.R., Liu, Y., Samimi, M.K., et al.: Millimeter wave channel modeling and cellular capacity evaluation. IEEE J. Sel. Areas Commun. **32**, 1164–1179 (2014)
6. Liu, D., Wang, L., Chen, Y., et al.: User association in 5G networks: a survey and an outlook. IEEE Commun. Surv. Tutor. **18**, 1018–1044 (2016)
7. Akyildiz, I.F., Gutierrez-Estevez, D.M., Balakrishnan, R., Chavarria-Reyes, E.: LTE-advanced and the evolution to beyond 4G (B4G) systems. Phys. Commun. **10**, 31–60 (2014)
8. Talwar, S., Choudhury, D., Dimou, K., et al.: Enabling technologies and architectures for 5G wireless. In: 2014 IEEE MTT-S International Microwave Symposium, pp. 1–4. IEEE (2014)
9. Morant, M., Prat, J., Llorente, R.: Radio-over-fiber optical polarization-multiplexed networks for 3GPP wireless carrier-aggregated MIMO provision. J. Light Technol. **8724**, 1 (2014)
10. Wang, P., Li, Y., Song, L., Vucetic, B.: Multi-gigabit millimeter wave wireless communications for 5G: from fixed access to cellular networks. IEEE Commun. Mag. **53**, 168–178 (2015)
11. Ofcom: Spectrum above 6 GHz for future mobile communications (2015). http://stakeholders.ofcom.org.uk/binaries/consultations/above-6ghz/summary/spectrumabove6GHzCFI.pdf
12. ITU: The technical feasibility of IMT in the bands above 6 GHz (2015). https://www.itu.int/pub/R-REP-M.2376
13. Pi, Z., Khan, F.: An introduction to millimeter-wave mobile broadband systems. IEEE Commun. Mag. **49**, 101–107 (2011)
14. Santos, T., Karedal, J., Almers, P., et al.: Modeling the ultra-wideband outdoor channel: Measurements and parameter extraction method. IEEE Trans. Wirel. Commun. **9**, 282–290 (2010)
15. Rappaport, T.S., Heath, R.W., Daniels, R.C., Murdock, J.N.: Millimeter wave wireless communications, 1st edn. Prentice Hall, Upper Saddle River (2015)
16. Dezfooliyan, A., Weiner, A.M.: Experimental investigation of UWB impulse response and time reversal technique up to 12 GHz: omnidirectional and directional antennas. IEEE Trans. Antennas Propag. **60**, 3407–3415 (2012)
17. Muqaibel, A., Safaai-Jazi, A., Attiya, A., et al.: Path-loss and time dispersion parameters for indoor UWB propagation. IEEE Trans. Wirel. Commun. **5**, 550–559 (2006)

18. Oudin, H., Wen, Z.: mmWave MIMO channel sounding for 5G: technical challenges and prototype system. In: 1st International Conference on 5G for Ubiquitous Connectivity. ICST, pp. 192–197 (2014)
19. Al-Samman, A.M., Rahman, T.A., Azmi, M.H., et al.: Statistical modelling and characterization of experimental mm-wave indoor channels for future 5G wireless communication networks. PLoS ONE **11**, e0163034 (2016)
20. Rappaport, T.S., Mayzus, R., Azar, Y., et al.: Millimeter wave mobile communications for 5G cellular: it will work! IEEE Access **1**, 335–349 (2013)
21. Hao, Xu, Rappaport, T.S., Boyle, R.J., Schaffner, J.H.: Measurements and models for 38-GHz point-to-multipoint radiowave propagation. IEEE J. Sel. Areas Commun. **18**, 310–321 (2000)
22. Al-Samman, A.M., Rahman, T.A., Hadri, M., et al.: Experimental UWB indoor channel characterization in stationary and mobility scheme. Measurement **111**, 333–339 (2017)
23. Hashemi, H.: Indoor radio propagation channel. Proc. IEEE **81**, 943–968 (1993)
24. Abouraddy, A.F., Elnoubi, S.M.: Statistical modeling of the indoor radio channel at 10 GHz through propagation measurements Part I: narrow-band measurements and modeling. IEEE Trans. Veh. Technol. **49**, 1491–1507 (2000)
25. Athanasiadou, G.E., Nix, A.R.: A novel 3-D indoor ray-tracing propagation model: the path generator and evaluation of narrow-band and wide-band predictions. IEEE Trans. Veh. Technol. **49**, 1152–1168 (2000)
26. Greenstein, L.J., Ghassemzadeh, S.S., Erceg, V., Michelson, D.G.: Ricean K-factors in narrow-band fixed wireless channels: theory, experiments, and statistical models. IEEE Trans. Veh. Technol. **58**, 4000–4012 (2009)
27. Yu, K., Bengtsson, M., Ottersten, B., et al.: Modeling of wide-band MIMO radio channels based on NLoS indoor measurements. IEEE Trans. Veh. Technol. **53**, 655–665 (2004)
28. Joshi, G.G., Dietrich, C.B., Anderson, C.R., et al.: Near-ground channel measurements over line-of-sight and forested paths. IEE Proc. Microw. Antennas Propag. **152**, 589 (2005)
29. Kivinen, J., Zhao, X., Vainikainen, P.: Empirical characterization of wideband indoor radio channel at 5.3 GHz. IEEE Trans. Antennas Propag. **49**, 1192–1203 (2001)
30. Liang, J., Liang, Q.: Outdoor propagation channel modeling in foliage environment. IEEE Trans. Veh. Technol. **59**, 2243–2252 (2010)
31. Win, M.Z., Scholtz, R.A., Barnes, M.A.: Ultra-wide bandwidth signal propagation for indoor wireless communications. In: Proceedings of ICC 1997—International Conference on Communications. IEEE, pp. 56–60 (1997)
32. Win, M.Z., Ramirez-Mireles, F., Scholtz, R.A., Barnes, M.A.: Ultra-wide bandwidth (UWB) signal propagation for outdoor wireless communications. In: 1997 IEEE 47th Vehicular Technology Conference Technology in Motion. IEEE, pp. 251–255 (1997)
33. Chuan, C.L., Chin, F.: UWB Channel Characterization in Indoor Office Environments. In: IEEE P802.15 Wireless Personal Area Networks. IEEE P802, pp. 1–10 (2004)
34. Ciccognani, W., Durantini, A., Cassioli, D.: Time domain propagation measurements of the UWB indoor channel using PN-sequence in the FCC-compliant band 3.6–6 GHz. IEEE Trans. Antennas Propag. **53**, 1542–1549 (2005)
35. Lee, J.-Y.: UWB Channel modeling in roadway and indoor parking environments. IEEE Trans. Veh. Technol. **59**, 3171–3180 (2010)
36. Ghassemzadeh, S.S., Jana, R., Rice, C.W., et al.: Measurement and modeling of an ultra-wide bandwidth indoor channel. IEEE Trans. Commun. **52**, 1786–1796 (2004)
37. Ghassemzadeh, S.S., Greenstein, L.J., Sveinsson, T., et al.: UWB delay profile models for residential and commercial indoor environments. IEEE Trans. Veh. Technol. **54**, 1235–1244 (2005)

38. Keignart, J., Daniele, N.: Subnanosecond UWB channel sounding in frequency and temporal domain. In: 2002 IEEE Conference on Ultra Wideband Systems and Technologies (IEEE Cat. No. 02EX580), pp. 25–30. IEEE (2002)
39. Wang, Y., Lu, W., Zhu, H.: An empirical path-loss model for wireless channels in indoor short-range office environment. Int. J. Antennas Propag. **2012**, 1–7 (2012)
40. Al-Samman, A.M., Rahman, T.A.: Experimental characterization of multipath channels for ultra-wideband systems in indoor environment based on time dispersion parameters. Wirel. Pers. Commun. **95**, 1713–1724 (2016)
41. Anderson, C.R., Volos, H.I., Buehrer, R.M.: Characterization of low-antenna ultrawideband propagation in a forest environment. IEEE Trans. Veh. Technol. **62**, 2878–2895 (2013)
42. Di Renzo, M., Graziosi, F., Minutolo, R., et al.: The ultra-wide bandwidth outdoor channel: from measurement campaign to statistical modelling. Mob. Netw. Appl. **11**, 451–467 (2006)
43. Di Francesco, A., Di Renzo, M., Feliziani, M., et al.: Sounding and modelling of the ultra wide-band channel in outdoor scenarios. In: 2nd International Workshop on Networks with Ultra Wide Band Work, Ultra Wide Band Sensors Networks, Network with UWB 2005, pp. 20–24. IEEE (2005)
44. Molisch, A.F., Balakrishnan, K., Cassioli, D., et al.: A comprehensive model for ultrawideband propagation channels. In: Global Telecommunications Conference 2005, GLOBECOM 2005, vol. 6, p. 3653. IEEE (2005)
45. Al-Samman, A.M., Rahman, T.A., Nunoo, S., et al.: Experimental characterization and analysis for ultra wideband outdoor channel. Wirel. Pers. Commun. **83**, 3103–3118 (2015)
46. Al-Samman, A.M., Azmi, M.H., Rahman, T.A.: Window-based channel impulse response prediction for time-varying ultra-wideband channels. PLoS ONE **11**, e0164944 (2016)
47. Ko, J., Cho, Y.-J., Hur, S., et al.: Millimeter-wave channel measurements and analysis for statistical spatial channel model in in-building and urban environments at 28 GHz. IEEE Trans. Wirel. Commun. **2**, 1 (2017)
48. Molisch, A.F., Cassioli, D., Emami, S., et al.: A comprehensive standardized model for ultrawideband propagation channels. IEEE Trans. Antennas Propag. **54**, 3151–3166 (2006)
49. Richardson, P.C., Xiang, W., Stark, W.: Modeling of ultra-wideband channels within vehicles. IEEE J. Sel. Areas Commun. **24**, 906–912 (2006)
50. Rappaport, T.S., MacCartney, G.R., Samimi, M.K., Sun, S.: Wideband millimeter-wave propagation measurements and channel models for future wireless communication system design. IEEE Trans. Commun. **63**, 3029–3056 (2015)
51. Niu, Y., Li, Y., Jin, D., et al.: A survey of millimeter wave communications (mmWave) for 5G: opportunities and challenges. Wirel. Netw. **21**, 2657–2676 (2015)
52. Elrefaie, A.F., Shakouri, M.: Propagation measurements at 28 GHz for coverage evaluation of local multipoint distribution service. In: Proceedings of 1997 Wireless Communications Conference, pp. 12–17. IEEE (1997)
53. Xu, H., Rappaport, T.S., Boyle, R.J., Schaffner, J.H.: 38 GHz wideband point-to-multipoint radio wave propagation study for a campus environment. In: 1999 IEEE 49th Vehicular Technology Conference (Cat. No. 99CH36363), pp. 1575–1579. IEEE (1999)
54. Violette, E.J., Espeland, R.H., DeBolt, R.O., Schwering, F.K.: Millimeter-wave propagation at street level in an urban environment. IEEE Trans. Geosci. Remote Sens. **26**, 368–380 (1988)
55. Wang, F., Sarabandi, K.: An enhanced millimeter-wave foliage propagation model. IEEE Trans. Antennas Propag. **53**, 2138–2145 (2005)
56. Schwering, F.K., Violette, E.J., Espeland, R.H.: Millimeter-wave propagation in vegetation: experiments and theory. IEEE Trans. Geosci. Remote Sens. **26**, 355–367 (1988)
57. Hao, Xu, Kukshya, V., Rappaport, T.S.: Spatial and temporal characteristics of 60-GHz indoor channels. IEEE J. Sel. Areas Commun. **20**, 620–630 (2002)

58. Anderson, C.R., Rappaport, T.S.: In-building wideband partition loss measurements at 2.5 and 60 GHz. IEEE Trans. Wirel. Commun. **3**, 922–928 (2004)
59. Peter, W.K.M., Keusgen, W., Felbecker, R.: Measurement and ray-tracing simulation of the 60 GHz indoor broadband channel: model accuracy and parameterization. IET Semin. Dig. 432–432 (2007)
60. Lovnes, G., Reis, J.J., Raekken, R.H.: Channel sounding measurements at 59 GHz in city streets. In: IEEE International Symposium on Personal, Indoor and Mobile Radio Communication. Wireless Networks—Catch. Mob. Futur., pp. 496–500. IOS Press (1994)
61. Thomas, H.J., Cole, R.S., Siqueira, G.L.: An Experimental study of the propagation of 55 GHz millimeter waves in an urban mobile radio environment. IEEE Trans. Veh. Technol. **43**, 140–146 (1994)
62. Ben-Dor, E., Rappaport, T.S., Yijun, Q., Lauffenburger, S.J.: Millimeter-wave 60 GHz outdoor and vehicle AOA propagation measurements using a broadband channel sounder. In: 2011 IEEE Global Telecommunications Conference-GLOBECOM 2011, pp. 1–6. IEEE (2011)
63. Kyrö, M., Kolmonen, V.M., Vainikainen, P.: Experimental propagation channel characterization of mm-wave radio links in urban scenarios. IEEE Antennas Wirel. Propag. Lett. **11**, 865–868 (2012)
64. Kyrö, M., Ranvier, S., Kolmonen, V.-M., et al.: Long range wideband channel measurements at 81–86 GHz frequency range. In: 2010 Proceedings of Fourth European Conference on Antennas and Propagation (2010)
65. Chen, Y., De, S., Kernchen, R., Moessner, K.: Device discovery in future service platforms through SIP. In: 2012 IEEE Vehicular Technology Conference (VTC Fall), pp. 1–5. IEEE (2012)
66. Rajagopal, S., Abu-Surra, S., Malmirchegini, M.: Channel feasibility for outdoor non-line-of-sight mmWave mobile communication. In: 2012 IEEE Vehicular Technology Conference (VTC Fall), pp. 1–6. IEEE (2012)
67. Rappaport, T.S., Ben-Dor, E., Murdock, J.N., Qiao, Y.: 38 GHz and 60 GHz angle-dependent propagation for cellular & peer-to-peer wireless communications. In: IEEE International Conference on Communications, pp. 4568–4573. IEEE (2012)
68. Samsung Electronics Millimeter Waves May Be the Future of 5G Phones. http://spectrum.ieee.org/telecom/wireless/millimeter-waves-may-be-the-future-of-5g-phones
69. Al-Samman, A.M., Rahman, T.A., Azmi, M.H., Hindia, M.N.: Large-scale path loss models and time dispersion in an outdoor line-of-sight environment for 5G wireless communications. AEU Int. J. Electron. Commun. **70**, 1515–1521 (2016)
70. Deng, S., Samimi, M.K., Rappaport, T.S.: 28 GHz and 73 GHz millimeter-wave indoor propagation measurements and path loss models. In: 2015 IEEE International Conference on Communications Work, pp. 1244–1250. IEEE (2015)
71. MacCartney, G.R., Deng, S., Rappaport, T.S.: Indoor office plan environment and layout-based mmWave path loss models for 28 GHz and 73 GHz. In: 2016 IEEE 83rd Vehicular Technology Conference (VTC Spring), pp. 1–6. IEEE (2016)
72. Nie, S., MacCartney, G.R., Sun, S., Rappaport, T.S.: 72 GHz millimeter wave indoor measurements for wireless and backhaul communications. In: 2013 IEEE 24th Annual International Symposium on Personal, Indoor and Mobile Radio Communications, pp. 2429–2433. IEEE (2013)
73. Maccartney, G.R., Rappaport, T.S., Sun, S., Deng, S.: Indoor office wideband millimeter-wave propagation measurements and channel models at 28 and 73 GHz for ultra-dense 5G wireless networks. IEEE Access **3**, 2388–2424 (2015)

74. Zhang, N., Yin, X., Lu, S.X., et al: Measurement-based angular characterization for 72 GHz propagation channels in indoor environments. In: 2014 IEEE Globecom Workshops (GC Wkshps), pp. 370–376. IEEE (2014)
75. Haneda, K., Jarvelainen, J., Karttunen, A., et al Indoor short-range radio propagation measurements at 60 and 70 GHz. In: 8th European Conference on Antennas and Propagation (EuCAP 2014), pp. 634–638. IEEE (2014)
76. Hur, S., Cho, Y.-J., Lee, J., et al: Synchronous channel sounder using horn antenna and indoor measurements on 28 GHz. In: 2014 IEEE International Black Sea Conference on Communications and Networking, pp. 83–87. IEEE (2014)
77. Al-Samman AM, Rahman TA, Azmi MH: Indoor corridor wideband radio propagation measurements and channel models for 5G millimeter-wave wireless communications at 19 GHz, 28 GHz and 38 GHz Bands. Wirel. Commun. Mobile Comput. **2018**, 12 (2018)
78. Oyie, N.O., Afullo, T.J.O.: Measurements and analysis of large-scale path loss model at 14 and 22 GHz in indoor corridor. IEEE Access **6**, 17205–17214 (2018)

QoS Based Independent Carrier Scheduling Scheme for Heterogeneous Services in 5G LTE-Advanced Networks

Haitham S. Ben Abdelmula(✉), M. N. Mohd Warip, Ong Bi Lynn, and Naimah Yaakob

Embedded, Network and Advanced Computing Research, School of Computer and Communication Engineering, University Malaysia Perlis, Level 1, Pauh Putra Main Campus, 02600 Arau, Malaysia
hsaa8383@gmail.com,
{nazriwarip,drlynn,naimahyaakob}@unimap.edu.my

Abstract. Efficient user-level QoS provisioning for heterogeneous services is of vital importance in future 5G Long-Term Evolution-Advanced (LTE-A) with Carrier Aggregation (CA) mobile system. The CA is very vital technique which invented by 3GPP for creating the huge virtual carrier bandwidths to support the 5G LTE-A network with unprecedented speed of data transmission and minimal latency. To the best of our knowledge, in order to support and fulfill the user level QoS obligations for diversified services "Real-Time (RT) and Non-Real-Time (NRT)", one of the main challenges in the presence of CA is to offer robust and suitable resource scheduling scheme. In this paper, the authors propose an efficient scheduling scheme, termed as "QoS based Independent Carrier (QoS-IC)" by designing Round-Robin with Service (RR-S) method for ensuring the load balancing based on the user's service in 5G LTE-A system. In addition, a user-level Quality of Service aware packet scheduling for OFDMA resource blocks suitable to multi-services system is proposed by introducing the service utility factor. The design objective is to guarantee the user quality of service (QoS) and equitable allocation of radio resources among users. Extensive simulation outcomes demonstrate that our scheduling algorithm offers significantly better QoS performance in terms of packet drop rate (PDR), variance of packet drop rate, and the average packet delay for RT traffic as well as satisfying the minimum transmission rate demand for NRT traffic.

Keywords: 5G · LTE-Advanced · CA · QoS-IC · OFDMA
Service utility factor

1 Introduction

Nowadays, tremendous number of mobile users accesses various wireless broadband applications "Real-time and Non-Real time" such as web TV, video conference, video streaming, social network applications and so forth [1]. Since diversified wireless services have a variety of quality of service (QoS) obligations like as "tolerable packet

F. Saeed et al. (Eds.): IRICT 2018, AISC 843, pp. 464–476, 2019.
https://doi.org/10.1007/978-3-319-99007-1_44

drop rates and constrained packet delivery time" for real-time users, and "minimum transmission rates" for Non real-time users and demand enough bandwidth for achieving [2]. For ensuring the adequate user-level quality of service (QoS) for mobile users which were crucial concern in the preceding generation of mobile network, the obligation for enormous bandwidth to the mobile users is imperative [3]. Consequently, the third generation partnership project (3GPP) has proposed the aggregating frequency spectrum technique, termed as 'Carrier Aggregation' for the LTE-Advanced wireless system to support the outburst of higher peak downlink transmission rate. Meanwhile, the sole objective here is to offer extreme spectrum, i.e., up to 100 MHz in order to meet high transmission rates of 100 Mbps and 1 Gbps for high and low mobility respectively within the new generation of mobile systems [4]. As a result, the carrier aggregation is exploited as a solution for obtaining the bandwidth expansion, which supports (dis-)contiguous spectrum aggregation. Thus, the mobile user (MU) can be potentially programmed to aggregate numerous Component Carriers (CCs) that are in the same or various bands to exploit frequency spectrum efficiently.

According to the literature survey, several research studies have been conducted on the downlink radio resource scheduling in the presence of CA technique and proposed many schemes in order to achieve the LTE-A network performance. Going by [5–7], the authors recommended load balancing schemes to assign CCs to MUs, while the radio resources of each CC are scheduled amongst MUs based on the packet scheduling strategy in order to achieve the coverage functionality and resource distribution fairness among users. An improved Proportional Fair packet scheduler in multi carrier system was introduced by [8, 9] to address the radio scheduling problem for optimizing multi-user multi-service performance and pursues to prevent starvation for any MU in the system. In [10], the author proposed the joint radio resource reserved PF scheme with power allocation algorithm in order to attain the mobile system capacity. While the study presented in [11] has proposed the joint CC, RB and power allocation (JCRPA) scheme that assigns CCs and Physical Resource Blocks (PRBs) dynamically with fixed power distribution in order to optimize the system utility. Moreover, it exploited the multilevel water-filling power allocation scheme in order to improve the power of a specified CC and RB assignment for achieving the system fairness. Authors in [1] introduced a novel greedy-based algorithm which allocates PRBs of each CC with its associated Modulation and Coding Schemes (MCSs) to users at each transmission time interval (TTI)for maximizing the system-level QoS. The novel cross-CC proportional fair was proposed in [12] which satisfy the fairness among various kinds of users "LTE and LTE-A users" by designing a weight factor that is based on the number of CCs assigned to users for RBs scheduler. As well as, it is achieving the system throughput. Finally, an Efficient RB Allocation Algorithm (ERAA) was proposed in [13] which consider the link adaptation jointly with the CC selection and RB assignment. ERAA calculates the utility function of all combinations of users, CCs, and MCSs at each TTI, resulting significantly enhances for the system throughput and fairness while computational complexity increases a little bit.

Most above studies always strive to optimize the system-level QoS performance i.e., system throughput and the fairness among all users while in the presence of various services "RT and NRT" in the system, the user-level QoS performance has not been researched in depth. Therefore, by considering the user-level QoS demands for various applications in this paper, the novel QoS-IC Scheduling scheme is designed by exploit the features of IC Carrier Scheduling.

The rest of this paper is organized as follows: the system model is presented in Sect. 2. After a brief review of the two conventional CS algorithms, Sect. 3 analyzes the proposed radio resource scheduling scheme in detail. In Sect. 4, the simulation setup and outcomes are explained. Finally, conclusions are made in Sect. 5.

2 System Model

For excellent immunity to frequency selective fading and completely omit the chance of inter-symbol interference of the radio channels, the OFDMA based LTE-A system along with CA feature is considered for downlink transmission, where the base station (eNB) is able to serve K Mobile Users (MUs) on N CCs of frequency bandwidth [14]. Each MU is capable to receive data from a single or numerous CCs. Regarding the signal computational intricacy and the power saving at MU, it is efficient to have MU jointing to CCs as less as possible [15]. According to the time domain as depicted in Fig. 1, 10 consecutive sub-frames with 1 ms long for each form the physical frame structure of each CC. Furthermore, each sub-frame is splitted into equally 2 time slots which are able to convey 6 or 7 OFDM symbols for each slot based on whether the normal or extended

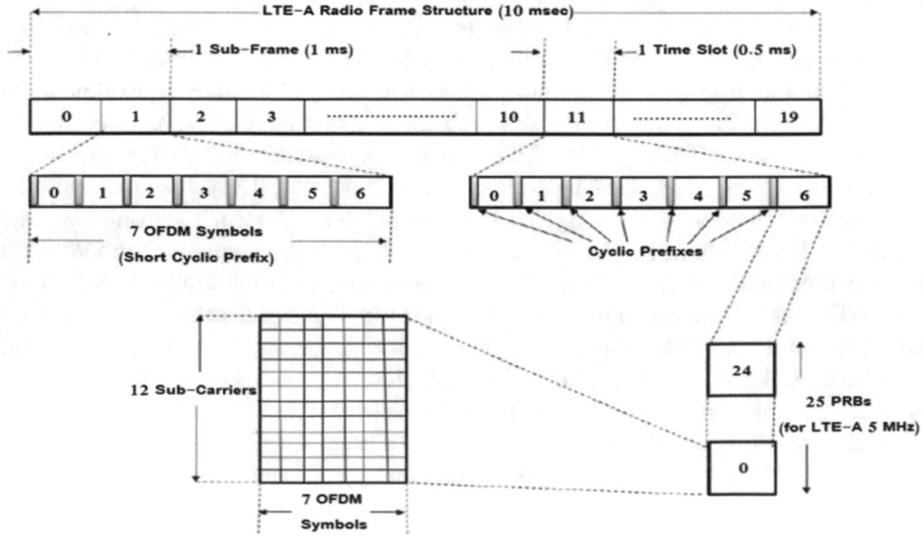

Fig. 1. OFDMA technology in LTE-advanced system

cyclic prefix is employed. Whereas in frequency domain, each time slot consists a numerous of PRBs depending on the utilized bandwidth which specified by LTE-A system. The PRB is the smallest unit of resource distribution allocated by scheduler which spans 12 consecutive subcarriers spacing of 15 kHz for one time slot [16].

All PRBs occupied by connected MUs are managed by the resource scheduler (RS) as demonstrated in Fig. 2. Each RS has merely one Resource Pool (RP) which composed of PRBs that belong to the CCs handled by this RS. Based on the applied CS scheme, the number of CCs which controlled by each RS can be one or numerous. Specially, the PRB of each component carrier can only be allotted to one of the RSs. For each MU, there is a Serving Queue (SQ) in each RS to accumulate data separately in provisional buffers. Once a new MU is allocated to the RS, the buffer is established, and removed when the transmission is completed. At each Transmission Time Interval (TTI), RS schedules the PRBs in its RP by packet scheduling policy such as Best CQI, Proportional Fair (PF), and so on [15].

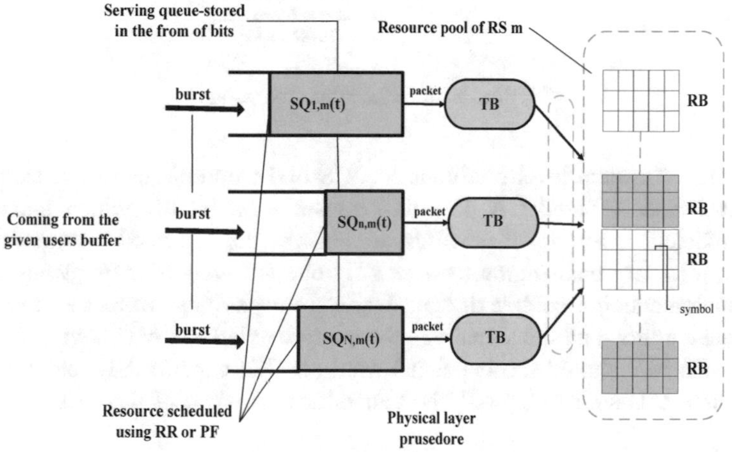

Fig. 2. Framework of the resource scheduler with CA

3 Analysis on the Proposed CS Schemes

In this part of the paper, firstly, Joint Carrier Scheduling (JCS) and Independent Carrier Scheduling (ICS) as the two straightforward CS schemes are briefly reviewed [17], and then the authors emphasis on the proposed scheduling algorithm by designing a multi-service utility factor.

3.1 Joint Carrier Scheduling (JCS)

JCS as the one of conventional CS schemes, it controls the numerous of CCs by placed the PRBs of all CCs into the RP of single RS which schedules radio resources among all mobile users, as depicted in Fig. 3. Due to completely joint processing of all CCs together as a single carrier, JCS only needs a single-level scheduling "packet scheduling", which is managed by RS for maps the data to PRBs of all CCs.

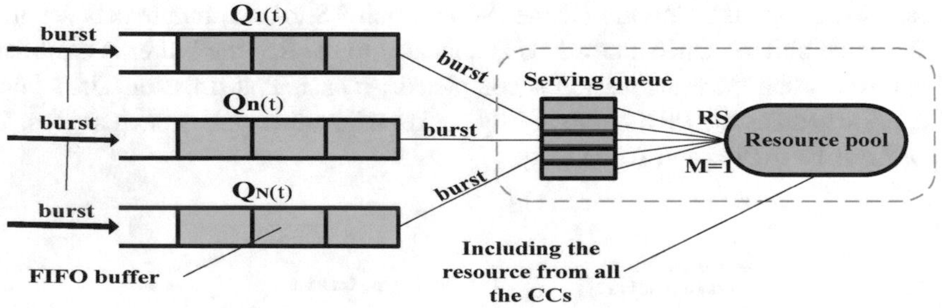

Fig. 3. JC scheduling technique

In terms of system level performance, JCS is the optimal carrier scheduling algorithm for OFDMA based CA downlink transmission model, which maximizes the spectral efficiency. In the other words, it achieves the saturated resource utilization. However, it is very challenging to work only one RS on carrier frequency of all CCs due to implementation complexity and hardware capability. Moreover, the additional power consumption and the signaling processing overhead at MU have been increased because each MU should receive signal from all CCs concurrently and continuously, although one MU's data may only be transmitted on some of the CCs.

3.2 Independent Carrier Scheduling (ICS)

Various from JCS, ICS manages each CC autonomously as demonstrated in Fig. 4, so each CC in the system should have separate RS to store its own PRBs. Consequently, ICS restrictions the MUs to receive from a single CC. Obviously, ICS demands two-level radio resource scheduling, "CC-level scheduler and PRB-level scheduler". The former scheduler is in charge of the user assignment among CCs, while the later is actually the same as single-carrier PRBs scheduling which controlled by each RS. The CC that allocates to a single MU is chosen randomly and cannot be altered any more until the end of session.

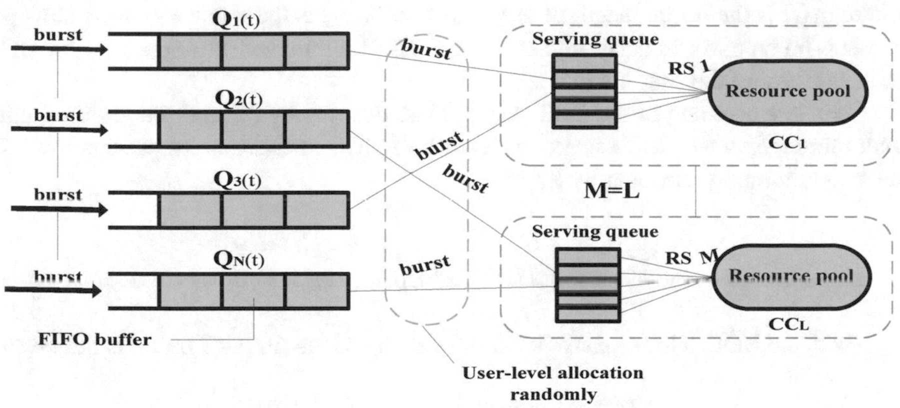

Fig. 4. IC scheduling technique

As demonstrated above, JCS is able to achieve the system performance, but its feasibility is not high due to the additional signal overhead, whereas ICS is not likely to growth the signal processing intricacy and consumption of power. Thus, ICS is more practical algorithm for LTE-A network with CA. Nevertheless, at PRB-level phase, the selectable MU of each RS is just a subset of the entire MUs, resulting in deficiency of the spectral efficiency lower than JCS. Hence, the authors propose optimizing related technology of ICS, to obtain the compromise of performance and complexity.

3.3 Design of Multi-service Utility Factor

In recent years, the modern mobile communication technology has emerged various applications with a variety of QoS obligations. According to delay-sensitive level of distinct applications and maintain low level of complexity, the authors partitions two application classifications, namely, RT-applications and NRT-applications. In view of multi-service QoS demands, the authors format the application utility factor for each MU to assess user-level QoS performance. Assuming these utility factors are independent and can acquire by calculation, which exactly expressed through two steps:

Step 1: Calculate the RT-utility factor that each user maintains on the service. The HOL delay of packets i in the serving queue for real-time user k is defined as:

$$D_{k,i}^{RT}(t) = T_{current} - T_{k,i}^{arrival} \tag{1}$$

In which $T_{current}$ is the current transmission time and $T_{k,i}^{arrival}$ is the arrival time of packets i in the user queue, which very crucial to be considered because RT flows is highly delay-sensitive. Then, the RT-utility factor can be expressed as:

$$D_k^{RT}(t) = \frac{D_{k,i}^{RT}(t) * q_k(t) * n_{TS}}{\sqrt{HD_k^{RT} - D_{k,i}^{RT}(t)}} \tag{2}$$

where $q_k(t)$ is the queue length of user k at time t, n_{TS} is the number of time slots per one sub-frame and the delay highest bound HD_k^{RT} is utilized to normalize the HOL packets delay in the user k's queue.

Step 2: According to the NRT packet size, denoted by M; and the required minimum throughput for NRT service, denoted by R. The arrival time of packets i will be calculated virtually, denoted by $T_{k,i}^{virtual}$.

$$T_{k,i}^{virtual} = \begin{cases} 0, & if\ q_k(t) = 0 \\ T_{k,i}^{arrival} + {}^{M}/_{R}, & if\ q_k(t) \neq 0 \end{cases} \tag{3}$$

Then, the HOL delay of packets i in the serving queue for NRT user k is defined as:

$$D_{k,i}^{NRT}(t) = \max(T_{current} - T_{k,i}^{virtual}, 0) \tag{4}$$

Consequently, the NRT-utility factor can be expressed as:

$$D_k^{NRT}(t) = \frac{D_{k,i}^{NRT}(t) * q_k(t)}{HD_k^{NRT}} \tag{5}$$

3.4 QoS-IC Scheduling Scheme

As mentioned above, the mobile users could access different applications "RT and non-RT" even encounter homogenous channels, the application of conventional ICS scheduling is not a promising for accommodate the RT and Non-RT time QoS requirements. To solve this issue, a novel user-level QoS aware packet scheduling based on ICS scheme is proposed, named as QoS-ICS. Similar to ICS, the two-phases resource scheduling still adopted.

CC-Phase Scheduler: Based on the preliminary designed of ICS scheme, the eNB allots CCs to the MUs in random manner, resulting inefficient network utilization and performance. Thus, the authors design adopting carrier scheduling scheme for load balancing which take into account the network load. In order to keep the proposed approach with low-level computational complexity, the metric to evaluate the load of each CC is the number of RT-MUs and NRT-MUs, which are assigned to the carrier. In this scheme, there are M CCs and each MU able to allocate at one CC. At initial time, eNB runs the CC assignment algorithm, which detailed in algorithm 1for allocating CCs to MUs.

Algorithm 1. CC-level Scheduling

Require: *1 out of M CCs to be assigned to each of K MUs.*
Create zeros array of CCsfor RT-MUs
Create zeros array of CCsfor NRT-MUs
1:***for****mobile users k = 1 ……. K****Do***
2:***if*** *MU(k) is RT-mobile user*
3: *Choose the $CC_m$with the least load from CCs array for RT-MUs;*
4:*Allocate the MU(k) to CC_m and mark it as allocated ;*
5:*Update the load of CCs in RT-array;*
6:***else***
7:*Choose the CC_m with the least load from CCs array for NRT-MUs;*
8:*Allocate the MU(k) to CC_m and mark it as allocated ;*
9:*Update the load of CCs in NRT-array;*
10:***end if***
12:***end for***

PRB-Phase Scheduler: For PRB-level scheduler, the multi-carrier Proportional Fair communication system based on traffic QoS is employ in order to achieve the overall resource allocation than conventional PF. Based on the authors' strategy, it alters the existing ICS scheduling by introducing the application utility factor for each traffic user to priority function as presented in Algorithm 2.

Algorithm 2: PRBs-level Scheduling

Require:*M physical resource blocks to be allocated to (K) MUs*
1:***for*** *resource blocks m = 1 …….. M****Do***
2: *Choose the userjwith the largest priority metric*
3:*Priority metric =*
$$\boldsymbol{argmax_k\ (application\ utility\ factor\ \times\ channel\ rate\ of\ resource\ block\ m\ for\ usr\ j\ \div\ averagethroughputforuser\ j)}$$
4:*Allocate the resource block m to user j;*
5:*Update the average throughput for user j;*
6:***end for***

Based on the priority function of the scheduling algorithm, the resource scheduler assigns the PRBs to the users with the highest priority on all CCs:

$$k_{i,n} = arg_{k=1,2,\ldots,K} \max\{P_{k,i,n}\} \quad (6)$$

Where $k_{i,n}$ is the selected user k at i^{th} CC of the nth PRBs, K is the total number of users and the resource scheduler performs the priority metric for user k on the ith CC and the nth PRBs at each TTI:

$$P_{k,i,n}(t) = \log\left(\frac{R_{k,i,n}(t) * D_k(t)}{(T_{PF} - 1)\overline{R_{k,total}(t)} + \sum_{n=1}^{N} \rho_{k,n} * R_{k,i,n}(t)}\right) \tag{7}$$

$R_{k,i,n}(t)$ is estimated transmittable data rate for user k at the nth PRB group of the ith CC; $D_k(t)$ is the service utility factor for user k; $\sum_{n=1}^{N} \rho_{k,n} * R_{k,i,n}(t)$ is the overall throughput of user k at time slot t; $\rho_{k,n}$ is the assignment indicator variable for user k and subcarrier n that is equal to 1 when the subcarrier n is allocated to the user k, while 0 otherwise; $\overline{R_{k,total}(t)}$ is user k throughput; T_{PF} is the average proportional window length.

The average user throughput, $\overline{R_{k,total}(t)}$ is defined by:

$$\overline{R_{k,total}(t)} = \begin{cases} \overline{R_{k,i}(t)}, & k \in RT \\ \overline{R_{k,total}(t)} - \delta, & k \in NRT \end{cases} \tag{8}$$

Where $\overline{R_{k,i}(t)}$ is the average throughput for user k on the ith CC; δ is the minimum throughput for NRT traffic “240 kbps”.

In which,

$$\overline{R_{k,i}(t+1)} = \begin{cases} \left(1 - \frac{1}{T_P}\right)\overline{R_{k,i}(t)} + \frac{1}{T_P}\overline{R_{k,i}(t)}, & k = k^* \\ \left(1 - \frac{1}{T_P}\right)\overline{R_{k,i}(t)}, & k \neq k^* \end{cases} \tag{9}$$

Repeat the priority function until all N radio resources (PRBs) allocated to users, and then update $\overline{R_{k,total}(t)}$.

4 Simulation Outcomes and Performance Evaluation

4.1 Simulation Model

The OFDMA downlink transmission system based on the ICS scheme is considered and simulated by LTE-A downlink system level Vienna simulator. Three CCs with frequency of each carrier is 2.14 GHz, 2.4 GHz and 2.6 GHz are proposed in this simulation. Each CC has 10 MHz bandwidth, which includes 600 subcarriers grouped into 50 PRBs. All component carriers possess uniformly power transmitting. According to PF scheduling algorithm, the window length average is $T_{PF} = 25$. In order to attain high data transmission rate, the adaptive Modulation and Coding is very important. The number of both RT-users and non RT-users is assumed to be equally. The delay upper bound for RT-traffic and non RT-traffic is equal, which is taken 20 ms. Additional simulation settings are listed in Table 1.

Table 1. LTE-A system level simulation parameters

Parameter	Setting and value
System structure	7 cell, 3 sectors per cell
Site-to-site distance	500 m
MIMO configuration	2 × 2 CLSM
Path-loss model	Hata Model
Shadow fading model	Longnormal fading model
CC bandwidth	10 MHz
Traffic model	Full Buffer
Traffic type	RT and NRT
Number of users per sector	70, 78, 86, 94, 102, 110
MU distribution	Uniform

4.2 Performance Evaluation

In this section, the count values of user-level QoS, "average packet drop rate, average packet delay and variance of average PDR" for RT-users and "minimal transmission rate" for Non RT-users are investigated for classical ICS and the proposed QoS-ICS algorithms.

Average PDR for RT-Users: Figure 5 demonstrates the average PDR for RT-users, which significantly increases when the load of system increases. However, QoS-IC result shows good achieving around 66% compared to classical ICS technique because it adapts the resource scheduler for giving more resources to the users, which have more packets in the queue with higher delay for minimizing the average delay of users. As a result, packets in the queues don't exceed their delay budget leading to inferior average PDR.

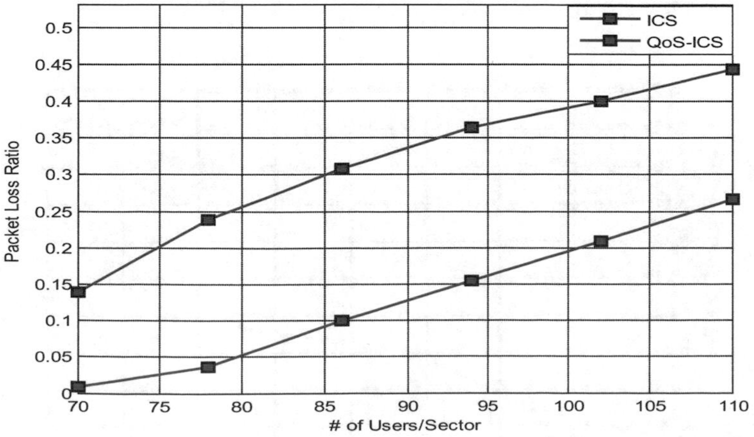

Fig. 5. Packet drop rate for RT service

Average Packet Delay for RT-Users: Figure 6 illustrates the corresponding average packet delay for the designed QoS-IC compared to conventional ICS scheme in different cross-system loads. Due to the proposed QoS-IC jointly utilize channel state information and the queue status information in terms of queue length, it is ascertained that QoS-ICS performance better than classical ICS algorithm. The result shows the QoS-ICS scheme has a good performance in terms of packet delay which gives 25.37% less packet delivery as compared to ICS scheme.

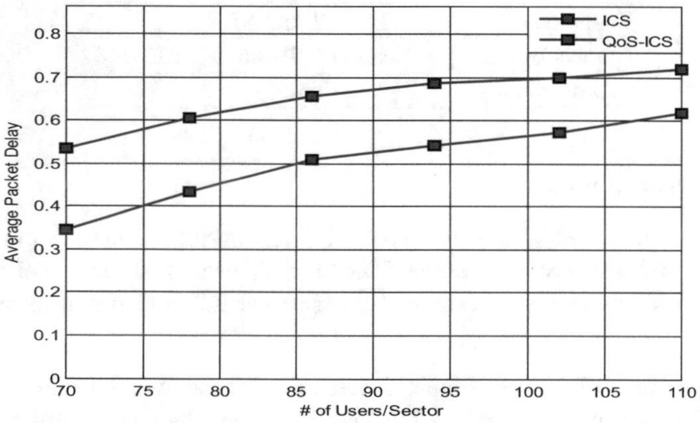

Fig. 6. Average delay of RT service

Variance of Average PDR of RT-Users: Figure 7 illustrates the variance of PDR for all RT-users under different cross-system loads. Due to the proposed QoS-IC which takes into consideration the HOL packet delay and queue status, it demonstrations good fairness for all RT users under various system loads as compared to conventional ICS scheme by stabilizing their queues.

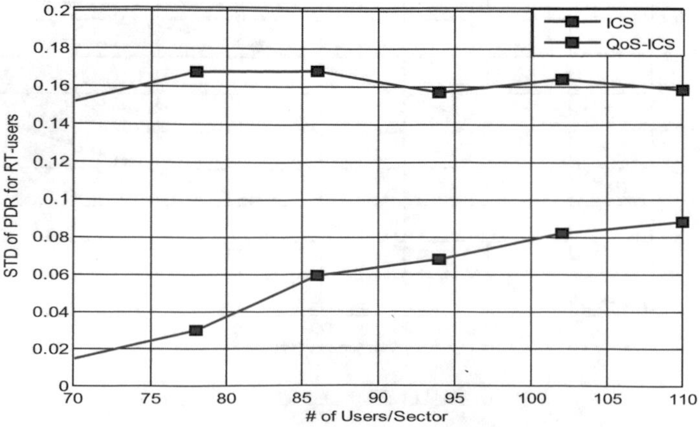

Fig. 7. Fairness for all RT-users

Minimum Throughput (TP) for Non Real-Time Users: Figure 8 shows the minimal data rate among all NRT-users versus different system loads. As the system load maximizes, the NRT-users' minimal throughput reduces for classical and proposed QoS-IC algorithms. Both algorithms guarantee the NRT users' minimal throughput "higher than 240 kbps". However, QoS-IC shows the highest minimal throughput as compared to conventional IC, which is 0.309 Mbps for QoS-IC and 0.244 Mbps for IC at a network load of 110 mobile users.

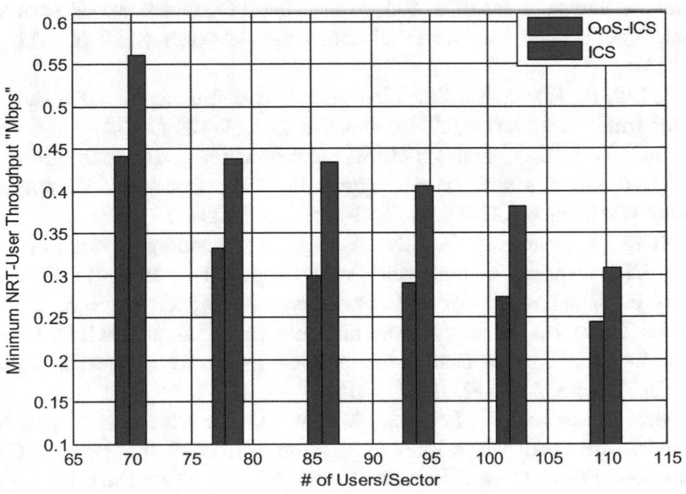

Fig. 8. NRT user's minimal throughput

5 Conclusion

As presented in this finding a new allotment physical resource blocks for the OFDMA technique based on user-level QoS obligations, i.e., QoS-IC is used. QoS-IC technique adaptively adjusts the proportion of radio resource that allocates to heterogeneous services based on the service utility factor for each service. Numerical outcomes show that QoS-IC scheme can optimize a good level of user-level QoS performance in reducing packet delivery time, packet drop rate and variance of average PDR for RT-users, while maintain the minimal throughput for Non-RT users.

References

1. Liao, H.-S., Chen, P.-Y., Chen, W.-T.: An efficient downlink radio resource allocation with carrier aggregation in LTE-advanced networks. IEEE Trans. Mobile Comput. **13**, 2229–2239 (2014)
2. Wang, C., Huang, Y.-C.: Delay-scheduler coupled throughput-fairness resource allocation algorithm in the long-term evolution wireless networks. IET Commun. **8**(17), 3105–3112 (2014)

3. Rostami, S., Arshad, K., Rapajic, P.: Resource allocation algorithms for OFDM based wireless systems. In: IEEE 26th Annual International Symposium on Personal Indoor and Mobile Radio Communications, PIMRC 2015, pp. 1105–1110. IEEE (2015)
4. Lee, H., Vahid, S., Moessner, K.: A survey of radio resource management for spectrum aggregation in LTE-advanced. IEEE Commun. Surv. Tutor. **16**, 745–760 (2014)
5. Wang, Y., Pedersen, K.I., Sorensen, T.B., Mogensen, P.E.: Carrier load balancing and packet scheduling for multi-carrier systems. IEEE Trans. Wirel. Commun. **9**(5), 1780–1789 (2010)
6. Cheng, X., Gupta, G., Mohapatra, P.: Joint carrier aggregation and packet scheduling in LTE-advanced networks. In: IEEE 10th Annual IEEE Communications Society Conference on Sensor, Mesh and Ad Hoc Communications and Networks, SECON 2013, pp. 469–477. IEEE (2013)
7. Zheng, K., Liu, F., Xiang, W., Xin, X.: Dynamic downlink aggregation carrier scheduling scheme for wireless networks. IET Commun. **8**(1), 114–123 (2014)
8. Chung, Y.-L., Jang, L.-J., Tsai, Z.: An efficient downlink packet scheduling algorithm in LTE-advanced systems with carrier aggregation. In: Consumer Communications and Networking Conference (CCNC), pp. 632–636. IEEE (2011)
9. Fu, W., Kong, Q., Zhang, Y., Yan, X.: A resource scheduling algorithm based on carrier weight in LTE-advanced system with carrier aggregation. In: 2013 22nd International Conference on Wireless and Optical Communication (WOCC), pp. 1–5. IEEE (2013)
10. Bai, J., Hao, Z., Du, H., Wang, Y.: A resource reserved PF algorithm for LTE-A with power allocation. In: 2015 International Conference on Estimation, Detection and Information Fusion (ICEDIF), pp. 344–347. IEEE (2015)
11. Wu, F., Mao, Y., Huang, X., Leng, S.: A joint resource allocation scheme for OFDMA-based wireless networks with carrier aggregation. In: 2012 International Conference on Wireless Communications and Networking (WCNC), pp. 1299–1304. IEEE (2012)
12. Al-Shibly, M.A., Habaebi, M.H., Islam, M.R.: Novel packet scheduling algorithm based on cross component carrier in LTE-advanced network with carrier aggregation. In: 2014 International Conference on Computer and Communication Engineering (ICCCE), pp. 213–216. IEEE (2014)
13. Rostami S., Arshad K., Rapajic P.: A joint resource allocation and link adaptation algorithm with carrier aggregation for 5GLTE-Advanced network. In: 2015 22nd International Conference on Telecommunications (ICT), pp. 102–106. IEEE (2015)
14. Ali, S.G.A., Baba, M.D., Mansor, M.A.: RB preserver downlink scheduling algorithm for real-time multimedia services in LTE networks. In: IEEE Symposium on Computer Applications and Industrial Electronics, ISCAIE2015, pp 21–26. IEEE (2015)
15. Abdelmula, H.S.B., Warip, M.M., Lynn, O.B., Yaakob, N.: Technical review of RRM for carrier aggregation in LTE-advanced. J. Theoret. Appl. Inf. Technol. **91**(2), 397–410 (2016)
16. Abdelmula, H.S.B., Warip, M.M., Lynn, O.B., Yaakob, N.: Performance investigation of joint user-proportional fair scheduling algorithm in LTE-advanced system with carrier aggregation. Int. J. Eng. Trends Technol. **50**(4), 198–202 (2017)
17. Ali, A.A., Nordin, R., Ismail, M., Abdullah H.: Impact of feedback channel delay over joint user scheduling scheme and separated random user scheduling scheme in LTE-A system with carrier aggregation. J. Comput. Netw. Commun. (2014)

Literature Survey on Enhancement of MOM for Intelligent Cross-Platform Communications in Service Oriented Architecture

Najhan Muhamad Ibrahim[1(✉)] and Mohd Fadzil Hassan[2(✉)]

[1] Department of Information Technology, Universiti Sultan Azlan Shah, Bukit Chandan, 33000 Kuala Kangsar, Perak, Malaysia
hanfast@gmail.com

[2] Department of Computer and Information Sciences, Universiti Teknologi PETRONAS, 32610 Bandar Seri Iskandar, Perak, Malaysia
mfadzil_hassan@utp.edu.my

Abstract. Presently, numerous of different middleware available to connect different types of application, namely Message Oriented Middleware (MOM), Message Broker, Enterprise Service Bus (ESB) and Content-Centric Middleware. MOM is the most widely used as a tool to enable the communications between distributed applications where the execution or event notifications are delivered between applications through message. MOM provides program-to-program connection by message passing. Most of MOM environments are implemented using queued message store-and-forward capability, which is Message Queuing Middleware (MQM). In addition, an intelligent cross-platform communications can be defined as the computer system that enables self-coordination to distribute computing resources and adaptive to the new environment while hiding fundamental complexity from the users. Generally, intelligent system has the competence of self-management to overcome the rapidly growing complexity in computing application. The application makes decisions on its own by using the rules that have been identified. Intelligent system also provides an adaptive and dynamic environment for the process of communication. The framework of intelligent system, fundamentally consist of intelligent elements. Each element performs a fixed function and interacting with other elements in a very dynamic environment.

Keywords: Service Oriented Architecture (SOA)
Message Oriented Middleware (MOM) · Web services
Intelligent cross-platform communications

1 Introduction

Service-oriented architecture (SOA) can be considered as software architecture to deploy software application, which the main components are services [1]. SOA is also referred as a collection of services which incorporate with each other. The incorporate not only engage a message passing but also attach two or more services to coordinate some activities. In addition, SOA is an architectural style for building software

F. Saeed et al. (Eds.): IRICT 2018, AISC 843, pp. 477–487, 2019.
https://doi.org/10.1007/978-3-319-99007-1_45

applications that provide services to other components via a communications protocol, such as web applications. It encourages distributed among software components or services that supports them to be reusable. A service is an employment of well-defined software application functionality. Clients in different applications or partner can then consume. It also allows for the reuse of existing services or objects where new services can be created from an existing application system. In other words, it allows applications to leverage present object by letting them to reuse existing services with assurances interoperability between heterogeneous applications [2–4].

At this point of time, SOA has provided a new direction for a software application where the architecture is determined at runtime and the framework can be deployed to meet the new software requirements. One of the substantial attributes for executions and communications in SOA application is web service. Web service was implemented with a general goal to construct elements of business logic and services, which can be used and reused by other applications. The services themselves hide the complexity of their business logic from the consumers through simple interfaces allowing the services to be reused in many different applications. The service and the consumer are described as being distributed by an approach that allows complex composite solutions to be developed through leveraging multiple Web Services. There are two types of web services, simple web services and complex web services. Simple web services comprise of only simple data types (string, integer, and other types) which are sent or received as arguments and values that are returned from methods. Complex web services consist of Web Service Deployment Descriptor (WSDD) that is able to send and receive more than simple data types. It also enables authoring and providing of the field in the web service console or through the web service [5–9].

Furthermore, web services can be described as a growing set of standard protocols, which mainly based on SOAP specifications and developed for an extensive industry need. The goal is to support common messaging requirements between systems. Web service also can be referred as a software system designed to assist the communication between application-to-application over distributed network. Web service standards and specifications are used to implement SOA [10, 11].

2 Background and Motivation

Numerous of research works have been found in literature regarding cross-platform frameworks in different areas. Some of these frameworks are domain specific while some are general and not specific for any one particular domain. The existing research works have been conducted to solve different cross-platform problem for a variety of domains, such as resource management, enterprise systems and e-Government. One of the main research works that focus on cross-platform framework [13], which models and discusses the requirement attributes for the collaboration between different application systems. Chituc et al. [13] strongly argued that the development of cross-platform communications framework to support communications among diverse and geographically distributed system are significantly important. They have proposed six main requirement attribute with twenty-seven sub-attributes, which is a general requirement for distributed and collaboration application as the following. Firstly is description and publication. Second is potential business partner and opportunity identification. Third is collaboration that includes messaging, inter-organizational collaboration activities, negotiation, agreement and semantics. Fourth is management that includes policy, CN management and dissemination. Fifth is performance assessment and lastly is non-functional aspect. In the sub-requirements, several significant requirements for cross-platform communications have been highlighted. Nevertheless, the main proposed requirements for cross-platform in this research work are too general to be incorporated in the concrete cross-platform communications framework. This research work is most likely focus on non-functional requirement attributes. Therefore, the comprehensive literature study need to be conducted to evaluate and consolidate others requirement attribute to be included in the proposed cross-platform communication.

Consequently, in this research work, the comprehensive literature review on cross-platform has been conducted. Based on the literature review, which has been examined the existing cross-platform frameworks. There are eight most rely requirements attribute for cross-platform communication in SOA environment, which are Communications type, Availability, Adaptability level, Message type, Software failure recovery, Guaranteed transmission, Scalability and Specification compliance. The requirement attributes were mainly selected from generic specification requirements for seamless cross-platform communication. The rest of requirements were selected from existing research works as presented in requirements comparison in Table 1. The existing research works can be divided into two categories, domain specific and non-domain specific. The frameworks were selected from several areas of cross-platform frameworks, in which some of them are very theoretical but they have consisted of significant highlighted requirement attributes.

3 Research Problem

Over the last decade, artificial intelligent (AI) has shown great potential for solving complications in large scale, distributed and cross-platform applications. The reason for the growing success of AI in this area is the inherent distribution allows a regular decomposition of application into multiple interaction to achieve the global goal by using agent technology [14]. Agent-based technology can significantly enhance the performance in the following three conditions. Firstly, the domain problem of geographically distributed and cross-platform communications. Secondly, the sub-application exists in a dynamic environment. Lastly, the sub-application require more flexibly interactive with each other [15]. There are several types of artificial intelligent, one of them is agent-based.

Agent-based approach is well suited for the domain of cross-platform communications because of its geographically distributed and dynamic changing in nature [16]. The literature study shows that the techniques and methods resulting from the field of agent-based application have been applied into the many aspects of distributed and cross-platform applications, including modeling and simulation for cross-platform communication. In this section, will be discuss some of existing research works on enhancement of agent-technology for cross-platform [17–19]. Such as Agent-based

Table 1. Comparisons table

No	Authors	Objectives	Message type	Intelligent cross-platform communications at runtime	Gaps
1	Goel et al. [26]	The work for loosely integrating cross-platform and distributed systems.	SOAP	Do not support	The proposed communication model requires on time respond to complete the transaction
2	Xu et al. [27]	To resolve issues of asynchronous communications systems	Compensating messages	Do not support	Do not support multiple types of cross-platform communications in runtime and only focus on Compensating messages web services message
3	Pietzuch and Bhola [28]	Provide dependable and accessible MOM by allowing publish/subscribe communications style	DCQ and UCA	Do not support	The proposed research work implemented by using different type of web services message
4	Laumay [29]	Provide guarantee scalability through matrix clocks in Message Oriented Middleware (MOM)	ACL	Do not support	Using different message type for communications, which may have a difficulty to communicate between them
5	Parkin et al. [30]	Provides reliable messaging across organizational limitations while implementing appropriate mechanisms for non-repudiation	SOAP	Do not support	This research work emphasis on clients and servers that use SOAP RPC for interaction and WSDL to descript services
6	Yuan and Jin [31]	Enables the integration of services and resources within and across enterprises or other organizations	AMQP	Do not support	cMOM using AMQP message type for communications

(continued)

Table 1. (*continued*)

No	Authors	Objectives	Message type	Intelligent cross-platform communications at runtime	Gaps
7	Ahn and Chong [32]	To make use of an intelligent connection for information exchange in heterogeneous network environment	SOAP	Do not support	Do not support one-to-many cross-platform communications and only focus on SOAP web services message
8	Tai et al. [33]	Comprehensive MOM for reliable web service messaging	SOAP	Do not support	Do not support multiple types of cross-platform communications and only focus for reliable SOAP web services message

Middleware, Agent platform for reliable asynchronous distributed application and Multi-agent System as a new paradigm for distributed and cross-platform.

The increasing number of collaboration in multiple types of system derives the significant challenges for cross-platform communication to be more adaptive. Variety of existing solution based on self-regulating and automation components have been proposed. An abundant research studies have recently discovered on the significant research in the intelligent systems to accommodate cross-platform interaction and communications, which will be discuss in the next section. Nevertheless, these research studies are typically conceived with the specific issues or isolate solutions in mind and mostly address the need of reducing management costs rather than the need of enabling complex cross-platform communication issues. The existing frameworks are poorly designed to support intelligent interaction between multiple platforms. Traditional framework have been constructed and coordinated based on manual process for single goal to resolve a specific issue. On the other hand, next-generation systems are expected to grow more rapidly with no centrally-control, and no specific message type to be used. By release of central control over system has the potential to release an innovation for the future business process [20].

4 Achieving Cross-Platform via MOM

Middleware is a tool to assist the communication between distributed applications, it is a piece of software lying between the operating system and the application. It can be achieved by hiding a number of complexities involved in building application systems. SOA may consist different components that described by different web services and running on the different deployment style. By using middleware, it will make the communication much more efficient. Most of the process failures can be handled by the middleware without requiring any manual process from user. At present, several types of middleware are available in transparent accessing and making a remote operation as a local application. Application transparent is another important feature provided by middleware in which components can be migrated between applications without any changes required to other components [21]. Middleware is also software that allows sharing data between distributed and cross-platform application for instance the communications between different types of application-based programs. In general, middleware will be located in the middle between application and operating system. All the application partners will be attached to the middleware as a medium for the communications.

In order to suit intelligence systems, some middleware has also been designed to be robust and fault tolerant, keeping the messages exchanged as short as possible. There are many types of middleware available, which provided different connectivity methods. According to the literature study, middleware can be divided into three different main categories [24]. Firstly, Message Oriented Middleware (MOM) is a type of Message-queuing middleware (MQM) that combines a message send/receive and a queuing (message storage) service. Message queues provide an asynchronous communications protocol in which the sender and receiver of the message do not required to interact with the message queue at the same time. The messages placed in the queue and stored until the recipient retrieves them. Many implementations of message queues are implemented within an operating system and between different applications, such as applications allowing the passing of messages among different computer systems and potentially connecting multiple applications systems [25].

Secondly, Remote Procedure Calls (RPC) is a significant practice to construct distributed and client-server based applications. It is based on the extension of conventional and local procedure calling that the called procedure does not need to exist in the same location as the calling procedure. The two processes may be on the same or different application systems which connected by a network. By using RPC, programmers of distributed applications can be hiding the communication of application with the interface. Such as, a process on Client calls a procedure on Server, the calling process on Client is suspended, and execution of the called procedure takes place on Server. Information can be transported from the caller to the caller in the parameters and can come back in the procedure result. No message passing at all is visible to the user. Thus, the transport independence of RPC isolates the application from the physical and logical elements of the data communications mechanism and allows the application to use a variety of transports [23].

5 The Finding

It is necessary to conduct a systematic research and bring forward a practical solution for corresponding cross-platform communications issues that have become the main barrier to interact between multiple types of SOA applications. At present, there are a limited research works on enhancement of MOM that directly address this issue.

Existing solutions can be divided into two categories, synchronize and asynchronous communications. The solutions were selected from several kinds of extensions, in which some of them are the basic extension of MOM, which consist of significant highlighted point of view for cross-platform communications. The attributes used in the evaluation were mainly selected from cross-platform communications requirement.

One of the early study was conducted by Goel et al. [26]. The study examined the work for loosely integrating distributed systems by enhancing Message Oriented Middleware messaging. The authors provide the fundamental monitoring of the technology that widely distributed, often on real-time, such as task execution, offers to buy and sell and weather reports. They proposed internal tasks, such as orders, shipments, deliveries, and manufacturing that can form the pattern to connect applications within and distributed organizations are also consolidating in this work.

Xu et al. [27] proposed a method to overcome issues of asynchronous communications systems. WMOM system model has been presented in this research work where WMOM system associates workflow mechanism and MOM. This research work does not only make applications to be loosely couple from communications system, but also enhances the flexibility and scalability. The authors also proposed model that to reduce application code and application programming complexity, as most of the required functions are provided by the middleware. However, this research work did not support failure recovery that may occur in any transaction.

Pietzuch and Bhola [28] presented blocking control mechanisms for dependable and ascendable MOM by allowing the publish/subscribe communications style. They have classified two main requirements key of blocking control in this research work, which are the combination of two congestion control methods, (1) driven by a publisher-hosting broker, (2) driven by a subscriber-hosting broker. Nevertheless, this research study also implemented the work by using synchronous type of communications.

Another research work conducted by Laumay et al. [29]. The authors proposed the research work to guarantee scalability through matrix clocks in Message Oriented Middleware (MOM). This solution is based on a breakdown of the MOM in platform of causality such as small groups of distributed servers connected by router servers. They have proved that, provided the domain interconnection graph has no cycles. The global causal order on message delivery is guaranteed through purely local order.

Besides that, Parkin et al. [30] presented a system that provides trustworthy messaging across organizational limitations while implementing appropriate mechanisms for non-repudiation. This research work highlighting on clients and servers that use SOAP RPC for interaction and WSDL to descript services. They have tackled the problem by using a new approach of employing MOM technologies to achieve interorganizational communications. Nevertheless, both of research works are using different message type for communications, which may have a difficulty to communicate between them.

In [31], the authors have proposed cMOM that facilitae the integration of services and resources within and between enterprises or other organizations. They have pointed that the semantics is a key to capture the concrete meaning of events in real life. Same as composite event configurations may indicate very different semantic method in different context. On the other hand, Ahn and Chong [32] have proposed message-oriented middleware to make use of an intelligent connection for information exchange in heterogeneous network environment. Both of research works have implemented using asynchronous type of communication, which is the best suit for cross-platform and distributed application. However, cMOM using AMQP and heterogeneous MOM using SOAP message type for communications.

Another study in [33] has attempted to suggest a comprehensive MOM for trustworthy web service messaging. Tai S. Makalsen T.A. and Rouvellou have presented options to implement trustworthy messaging for web services using the existing messaging technology. Here, they discussed two types of middleware communications. Firstly, from endpoint-to-endpoint where once a message is delivered from the process application to the messaging middleware, it is guaranteed to be available for the receiver. Another one is from the application-to-middleware in which the middleware's API is used to send and receive messages. This research work supports trustworthy properties, such as guaranteeing message transmission, message determination and messaging transactions. Furthermore, they showed how trustworthy messaging, when combined with classical transaction processing techniques (such as direct transaction processing and queued transaction processing), can be applied to support basic application-to-application reliability.

Although most the existing framework are able to deal with complex tasks and provide the capability to expire the message after a certain amount of time. However, majority of existing solutions are still lack of consideration on requirement attributes for cross-platform communications. From the comparison study, there are two main significant viewpoint can be concluded. Firstly, none of existing research works are supported intelligent cross-platform communications. Secondly, even though SOAP is widely used to communication, there are also numerous of different message type has been used. Hence, it would be very difficult for these extensions to interact with each other or different application, which were build based on different web services standard and specifications. To overcome these limitations, the proposed research work suggests an agent-based message oriented middleware that extracts the communication requirements by parsing web services message, which is based on web service integration gateway (WSIG) specification. The agent technology used in the framework to provide an intelligent cross-platform communication that includes the translation model that support multiple web service and interruption predication.

6 Conclusion

Cross-platform mechanism is essential for different types of SOA applications that ensure the communications will be well maintained. The need for cross-platform communications within SOA has attracted different research efforts as reported in various literatures. This research interest is specifically focused on intelligent cross-platform communications for multiple types of SOA applications. The literature survey includes several areas, such as SOA, Web services, MOM and Agent-based technology. The aim of this paper is to introduce and review these technologies and to present the current efforts as reported in various literatures regarding cross-platform communications within SOA environments.

Interoperability framework plays an important facilitate seamless role to communication and share of services for SOA-based system. The lack of interoperable awareness, motivate researchers to focus on seamlessness attributes of web services and SOA of different deployment style. We found that some researchers looked at a particular perspective while others have attempted to integrate several specifications such as: enterprise coordination, business process integration, semantic applications, syntactical applications, physical integration, etc. We believe that more research in this area is still needed. Currently, we are working to develop a generic framework for seamlessness interoperability of SOA based on the different interoperability requirements which support seamlessness for multi SOA platform.

References

1. Hamzah, M.H.I., Baharom, F., Mohd, H.: A conceptual model for service-oriented architecture adoption maturity MODEL. In: The 6th International Conference on Computing and Informatics, ICOCI 2017, 25–27 April, Kuala Lumpur (2017)
2. Serrano, N., Hernantes, J., Gallardo, G.: Service-oriented architecture and legacy systems. IEEE Softw. **31**(5), 1–4 (2014)
3. Hustad, E., De Lange, C.: Service-Oriented Architecture Projects in Practice: A Study of a Shared Document Service Implementation. IEEE, Piscataway (2014)
4. Serrano, N., Hernantes, J., Gallardo, G.: Service-oriented architecture and legacy systems. J. Syst. Softw. **31**(5), 141–171 (2014)
5. Louridas, P.: Web services: a comparison of soap and rest services. Can. Center Sci. Educ. **12**(3), 363 (2017)
6. Halili, F., Ramadani, E.: Web services: a comparison of soap and rest services. Modern Appl. Sci. **12**(3), 175 (2018). ISSN 1913-1844, E-ISSN 1913-1852, Published by Canadian Center of Science and Education
7. Manuaba, I.B.P., Rudiastini, E.: API REST web service and backend system of lecturer's assessment information system on Politeknik Negeri Bali (2017)
8. Feng, X., Shen, J., Fan, Y.: REST: an alternative to RPC for web services architecture. In: 1st International Conference on Future Information Networks. IEEE (2009)
9. Wagh, K., Thool, R.: A comparative study of SOAP vs REST web services provisioning techniques. J. Inf. Eng. Appl. SSN 2224-5782 (print) ISSN 2225-0506 (online) 2(5) (2012)

10. Chakaravarthi, S.: Secure model for SOAP message communication in web services. Int. J. Adv. Intell. Paradig. Arch. **8**(4), 437–442 (2016)
11. Kumar, R.: Introduction to middleware: web services, object components, and cloud computing (2014)
12. Ibrahim, N.M., Hassan, M.F., Balfagih, Z.: Agent-based MOM interoperability framework for integration and communication of different applications. Int. J. New Comput. Arch. Appl. (IJNCAA) (2011)
13. Chituc, C.-M., Azevedo, A.R., Toscano, S.: A framework proposal for seamless interoperability in a collaborative networked environment. Comput. Ind. 22 (2009)
14. Blair, G., Schmidt, D., Taconet, C.: Middleware for Internet distribution in the context of cloud computing. Ann. Telecommun. **71**(3–4), 87–92 (2016)
15. Albano, M., Ferreira, L.L., Miguel, P., Alkhawaja, A.R.: Message-oriented middleware for smart grids. Technical report CISTER-TR-140803 (2014)
16. Sachs, K., Kounev, S., Appel, S.: Benchmarking of message-oriented middleware. In: DEBS 2009. ACM, Nashville (2009)
17. Dobson, S., Denazis, S., Fernandez, A., Gaiti, D., Gelenbe, E., Massacci, F., Nixon, P., Saffre, F., Schmidt, N., Smirnow, M., Zambonelli, F.: A survey of automate communications. ACM Trans. Auton. Adapt. Syst. **1**(2), 223–259 (2006)
18. Ibrahim, N.M., Hassan, M.F.: A survey on different interoperability frameworks of SOA systems towards seamless interoperability. In: ITsim 2010. IEEE, Kuala Lumpur (2010)
19. Nahvi, B., Habibi, J.: A model based approach on multi-agent system and genetic algorithm to improve the process management in service oriented architecture. J. Telecommun. Electron. Comput. Eng. **8**(5), 33–40 (2016). ISSN: 2180-1843, e-ISSN: 2289-8131
20. Belmabrouk, K., Bendella, F., Bouzid, M.: Multi-agent based model for web service composition. (IJACSA) Int. J. Adv. Comput. Sci. Appl. **7**(3) (2016)
21. Ibrahim, N.M., Hassan, M.F., Hussin Abdullah, M.: Agent-based service oriented architecture (SOA) for cross-platform. Int. J. Soft Comput. Softw. Eng. (JSCSE) (2013)
22. Adnan, M.H.M., Hassan, M.F., Aziz, I., Paputungan, I.V.: Protocols for agent-based autonomous negotiations: a review. In: Computer and Information Sciences (ICCOINS), 2016 3rd International Conference. IEEE, Kuala Lumpur (2016)
23. Atif Nazir Raja, M., Farooq Ahmad, H.: SOA compliant FIPA agent communication language. IEEE (2008)
24. Lin, A., Maheshwari, P.: Agent-based middleware for web service dynamic integration on peer-to-peer networks. In: AI 2005, LNAI 3809, pp. 405–414. Springer, Berlin (2005)
25. Bellissard, L., De Palma, N., Freyssinet, A., Herrmann, M., Lacourte, S.: An Agent Platform for Reliable Asynchronous Distributed Programming. IEEE, France (1999)
26. Goel, S., Sharda, H., Taniar, D.: Java message-oriented-middleware in a distributed environment. In: IICS 2003, pp. 93–103. Springer, Berlin (2003)
27. Xu, Y.-Z., Liu, D.-X., Huang, F.: Design and implementation of a workflow-based message-oriented middleware. In: APWeb Workshops, pp. 842–845. Springer, Berlin (2006)
28. Pietzuch, P.R., Bhola, S.: Congestion control in a reliable scalable message-oriented middleware. In: International Federation for Information Processing, IFIP USA, pp. 202–221 (2003)
29. Laumay, P., Bruneton, E., Palma, N., Krakowiak, S.; Preserving causality in a scalable message-oriented middleware. In: Middleware 2001, LNCS 2218, pp. 311–328. Springer, Berlin (2001)

30. Parkin, S., Ingham, D., Morgan, G.: A message oriented middleware solution enabling non-repudiation evidence generation for reliable web services. In: ISAS 2007, pp. 9–19. Springer, Berlin (2007)
31. Yuan, P., Jin, H.: A composite-event-based message-oriented middleware. In: GCC 2003, LNCS 3032, pp. 700–707. Springer, Berlin (2004)
32. Ahn, S., Chong, K.: A case study on message-oriented middleware for heterogeneous sensor networks. In: IFIP International Federation for Information Processing, IFIP, pp. 945–955 (2006)
33. Tai, S., Mikalsen, T.A., Rouvellou, I.: Using message-oriented middleware for reliable web services messaging. In: WES 2003, LNCS 3095, pp. 89–104. Springer, Berlin (2004)

A Survey and Future Vision of Double Auctions-Based Autonomous Cloud Service Negotiations

Muhamad Hariz Adnan[1(✉)], Mohd Fadzil Hassan[1], Izzatdin Abdul Aziz[1], and Nuraini Abdul Rashid[2]

[1] Department of Computer and Information Sciences, Universiti Teknologi PETRONAS, Bandar Seri Iskandar, 31750 Tronoh, Perak, Malaysia
m.hariz_g03017@utp.edu.my
[2] Department of Computer Sciences, College of Computer and Information Sciences, Princess Nourah bint Abdulrahman University, Riyadh, Kingdom of Saudi Arabia

Abstract. Double auction-based mechanisms have gained considerable attention as autonomous cloud service negotiation because it has been proven to be economically efficient and possesses the ability to accommodate multiple buyers and sellers. The main objective of this paper was to outline the limitations of the practically tested double auction mechanisms in cloud service negotiations; and subsequently suggest future vision. Ten practically proven double auction-based mechanisms have been surveyed from the research database, whereby most of them were found to employ the double auction protocol and two variations of the double auction protocol, namely continuous double auction and combinatorial double auction. Comparisons on the types, objectives, parameters, and results between the selected mechanisms are presented in this paper. Moreover, it was identified that the surveyed double auction mechanisms were unable to address multi-attributes heterogeneous cloud services because it required a long execution time. Nevertheless, future vision managed to be suggested to solve these issues.

Keywords: Double auction · Continuous double auction
Combinatorial double auction · Autonomous negotiation · Cloud service
Cloud service negotiation · Heterogeneous service · Multi-attributes

1 Introduction

A negotiation protocol refers to a set of rules that defines the interaction boundaries between participants [1]. On the other hand, double auction is described as many-to-many (M-N) negotiation protocol that supports simultaneous multiple agents negotiations, which results in wider size, greater complexity, and greater heterogeneity [1]. The double auction-based protocols and mechanisms have gained considerable attention because it has been proven to be economically efficient in some cloud service negotiation

F. Saeed et al. (Eds.): IRICT 2018, AISC 843, pp. 488–498, 2019.
https://doi.org/10.1007/978-3-319-99007-1_46

even for a large number of customers and providers [2]. The cloud service negotiations have recently received increased attention among scholars for research purposes [3, 4]. The aim is to potentially facilitate customers and providers in obtaining better contracts and avoiding resource overprovisioning, respectively [5–9]. However, it should be understood that it is not easy to realize a real-world implementation of such complex mechanisms that is deemed autonomous. Hence, the limitations of the current mechanisms must be identified and addressed prior to a successful implementation. In relation to this matter, the present study aims to investigate these limitations and suggest future vision. The process of data collection involves the study of several journals and proceedings obtained from the research databases such as IEEE Xplore, Science Direct, Scopus, Springer Link, and Thomson Reuters Web of Science. As a result, a total of ten proven double auction-based mechanisms for cloud service negotiation managed to be filtered and selected. The selection was performed based on the performances, originality, and uniqueness of the mechanisms. On another note, the implementation and parameters were also important to be considered. In Sect. 2, the survey carried out on double auction mechanisms are summarized, while the advantages and limitations of the mechanism are presented in Sect. 3. Next, the discussion is continued with matters regarding future vision. Finally, a conclusion is provided in Sect. 4.

2 Double Auctions for Cloud Service Negotiations

The double auction protocols are capable of handling dynamic pricing and a large number of negotiations through the process of managing the symmetrical bids (requests) and asks (offers) [2, 10–12]. The double auction protocols provide many variations such as continuous double auction (CDA) and combinatorial double auction [13–16]. In the survey of the current research, the protocols were distinguished into three variations, namely double auction, CDA, and combinatorial double auction.

2.1 Double Auction

The double auction-based mechanism was proposed by [17] for dynamic VM trading and scheduling. The aim of the mechanism is to maximize the time-averaged profit (revenue minus cost) of each cloud federation over the long run as well as for the purpose of fulfilling the resource and SLA requirements of each job. A double auction-based mechanism which is known as Double Multi-Attribute Auction (DMAA was proposed by [18] for cloud resource allocation. In particular, the mechanism uses the Support Vector Machine (SVM) for price prediction, while Neural Network (NN) is adopted to calculate the Quality Index.

2.2 Continuous Double Auction

The CDA protocol allows multiple buyers to send bids, while sellers are permitted to submit asks in the auction market [13, 19]. A double auction-based mechanism known as MANDI was proposed by [20] for market exchange infrastructure. Specifically, MANDI is adopted as a combination of the double auction which consists of first bid sealed auction and commodity market. The first bid sealed auction is described as a common auction protocol where all bidders submit simultaneous unrevealed bids. As a result, the commodity market directly trades goods/services. On the other hand, a double auction-based mechanism known as Nash Equilibrium Continuous Double Auction (NECDA) was introduced by [21] for cloud resource allocation and performance optimization. It is important to note that NECDA is benchmarked with the Max-min algorithm, Min-min algorithm, and CDA which manages to produce promising results.

In addition, a double auction-based mechanism was proposed by [22] for cloud service negotiation in future market and spot market which adopts a knowledge-based CDA. Meanwhile, an auction-based mechanism that was introduced by [23] for cloud resource allocation and strategic pricing tend to employ a multi-unit CDA. Apart from that, a double auction-based mechanism was proposed by [24] for cloud service allocation, which is based on a Parallel CDA (PDCA). The resource is allocated by the mechanism based on a parallel method known as PCM and PREZ algorithm. On the other hand, a double auction-based mechanism was proposed by [25] for the purpose of decentralizing resource allocation in cloud markets. In this case, the mechanism adopts the CDA protocol for the homogeneous market which utilizes a new bidding strategy to improve profit known as Belief based Hybrid Strategy (BH-strategy).

2.3 Combinatorial Double Auction

The combinatorial double auction refers to a protocol that is suitable for cloud services negotiation which has been widely used to solve various kinds of cloud services combination [2, 26]. A double auction-based mechanism was proposed by [27] for the allocation of service combinations (bundling) in order to reserve future services from a forward market as well as current services from the spot market. Apart from that, a double auction-based mechanism known as Intelligent Economic Approach for Dynamic Resource Allocation (IEDA) was proposed by [2] for IaaS resource allocation which employs enhanced combinatorial double auction protocol. In this case, the two mechanism agents used are known as Backpropagation Neural Network (BPNN) algorithm and Paddy Field Algorithm (PFA). Next, the survey on double auction protocols is summarized in the following subsection.

2.4 Summarization

The type, objectives, parameters, and results of the double auction-based mechanisms are summarized in Table 1.

Table 1. Summarization of the double auction-based mechanisms.

Work	Type	Objectives	Parameters
[17]	Basic double auction	VM trading and scheduling	Job types VM types CPU
[18]	Basic double auction	Cloud resource allocation	Price range Buy amount Sell amount Resource Delivery time Payment time
[20]	CDA	Market exchange infrastructure. Combine auctions and commodity market	Price Expected bid
[21]	CDA	Cloud resource allocation and performance optimization	Task length Budget Task deadline
[22]	CDA	Cloud service future and spot market Knowledge-based	Computing capacity Price
[23]	CDA	Cloud resource allocation and strategic pricing	Resource type Amount Budget Negotiation deadline Price range Resource maximum capacity
[24]	CDA	Cloud service allocation	Resource type Resource specification Utilization length Workload Deadline Price range Budget
[25]	CDA	Decentralized resource allocation. BH-Strategy	Reserve price
[27]	Combinatorial	Allocation of service combinations in forward and spot market	Price Service type Amount Time slot

(*continued*)

Table 1. (*continued*)

Work	Type	Objectives	Parameters
[2]	Combinatorial	Cloud resource allocation	Price Service type Service time CPU Memory and storage Network bandwidth Size of task Volume of data Reputation Service begin and end time

3 Discussion

Table 2 presents the advantages and limitations of the discussed mechanisms in terms of protocol (technical aspects), efficiency, negotiation outcome, and service heterogeneity.

Table 2. Advantages and limitations of the double auction mechanisms.

Work	Advantages	Limitations
[17]	1. Higher profit 2. Better social welfare 3. Lower average response delay	1. Heterogeneous setting not defined 2. Negotiation parameters are only price and service duration
[18]	1. Multi-attributes negotiation 2. Historical data increase performance for non-price attributes 3. Higher resource utilization rate	1. Does not accommodate heterogeneous cloud services 2. Execution times were long due to utilizing price prediction and winner determination methods
[20]	1. Scalable than CDA 2. Platform heterogeneity 3. Fault tolerance 4. Short-live auction thread 5. Able to discover and choose a resource from three markets	1. Does not accommodate heterogeneous cloud services 2. Negotiation parameter is only price
[21]	1. Higher successful execution rate and fairness deviation with a higher number of users 2. Scalable than CDA	1. Does not accommodate heterogeneous cloud services 2. The negotiation parameter is only price

(*continued*)

Table 2. (*continued*)

Work	Advantages	Limitations
[22]	Stable overall profit	1. Does not accommodate heterogeneous cloud services 2. The negotiation parameter is only computing capacity and price 3. A more flexible pricing policy was suggested
[23]	Multi-unit	1. Does not accommodate heterogeneous cloud services 2. The negotiation parameter is only price 3. Not scalable 4. Unable to address complex user requirements
[24]	1. Higher average trading price in higher system load 2. Able to cope with market supply and demand 3. Improved auctioneer performance	1. Does not accommodate heterogeneous cloud services 2. Negotiation parameter is only price
[25]	1. High buyers' surplus and total surpluses. 2. High average market efficiency 3. Small daily price volatility	1. Does not accommodate heterogeneous cloud services 2. Negotiation parameter is only price 3. High execution time
[27]	1. Properly allocate service 2. Scalable for a large number of users	1. Does not accommodate heterogeneous cloud services 2. Negotiable parameters are only price and service duration 3. Can improve negotiation speed

3.1 Advantages and Limitations

According to the results of the survey, there are several main limitations of the double auction mechanisms which include single-attribute homogeneous cloud services, limited negotiation parameters, and high execution time. In addition, the heterogeneity of cloud services were also considered. Moreover, it should be noted that the cloud service market is growing fast especially the IaaS market which involves different sizes, capabilities, and types of the machines that are added by cloud providers [28]. However, most of the double auction mechanisms adopted in the survey were not designed to address the heterogeneous cloud services.

Meanwhile, according to [29], cloud adoption decisions tend to involve multiple attributes which include conflicting technical, economic, and strategic criteria or issues. Multi-attribute negotiation is defined as a negotiation that involves multiple issues which are required to be negotiated simultaneously [30]. A multi-attribute negotiation is more complex and challenging compared to a single-attribute negotiation due to several factors which include the process of identifying preferential dependency as well as the solution space that is n-dimensional ($n > 1$) instead of a single dimensional line and the existence of "Win-Win" situations [30]. In addition, it is arguably important for the cloud service negotiation mechanism to negotiate multiple attributes from multiple alternatives for the purpose of satisfying bidder's preferences [29, 31]. Nevertheless, these attributes were argued to have no common standard of measurements which can cause the cloud services heterogeneity [29]. Therefore, it is strongly recommended for the mechanism to address multi-attributes heterogeneous cloud services and dynamic negotiation parameters.

On another note, the negotiation speed of the mechanisms must also be considered. In this case, it was argued that the speed of the double auction mechanism negotiation speed can be increased [27]. It is important to note that the mechanism execution times are long due to the utilization of various algorithms [18]. In addition, there were also arguments that the mechanism is not scalable [2, 23].

3.2 Future Vision

As presented in Table 2, it can be identified that a double auction mechanism which addresses the multi-attributes heterogeneous cloud services, dynamic negotiation parameters, and high negotiation speed is necessary. Furthermore, the combination of the CDA and combinatorial double auction protocol are yet to be practically emphasized. The CDA which is the dominant form of marketplaces and trading possesses the ability to provide a dynamic and efficient approach that can be used to decentralize the allocation of scarce resources [19]. Moreover, it allows the cloud customers and cloud providers to submit their bids anytime during the trading period, whereby it would be immediately matched with compatible bids to ensure that a trade can be executed [25]. Meanwhile, the combinatorial double auction allows users to bid on a combination of discrete items, express their complex preferences, and obtain higher values in negotiation [26, 31]. Hence, the combination of both protocols is believed to be able to partially solve the issue of multi-attributes heterogeneous cloud services as well as the dynamic negotiation parameters.

In the future, further studies should be carried out on several parameters. First, an enhancement of the current double auction mechanism by combining the continuous and combinatorial double auction protocols in a single mechanism. Second, the components of the double auction framework can be further enhanced. The components of the traditional double auction framework for cloud service negotiation is illustrated in Fig. 1, while the foresighted enhanced double auction framework is presented in Fig. 2.

Cloud services clustering is able to assist the double auction protocol in classifying the heterogeneous resources. In addition, several automatic clustering techniques, namely evolutionary algorithms and swarm algorithms can be used [32, 33]. Apart from that, enhanced algorithm can be used to improve service discovery and selections. For example, hybrid optimization algorithms [34–36] can be employed to dynamically negotiate multiple attributes and objectives for both the customers and providers. Meanwhile, enhanced utility function can be used to accommodate multi-attributes and dynamic negotiation parameters. Finally, machine learning can be added by studying the historical data in order to improve the agents' outcomes.

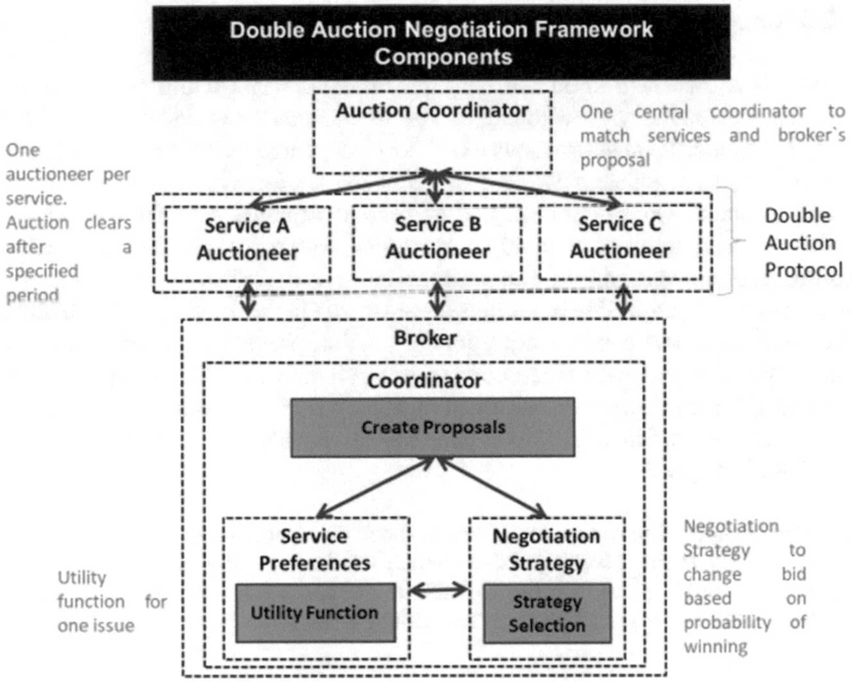

Fig. 1. The traditional double auction negotiation framework's components for cloud service negotiations.

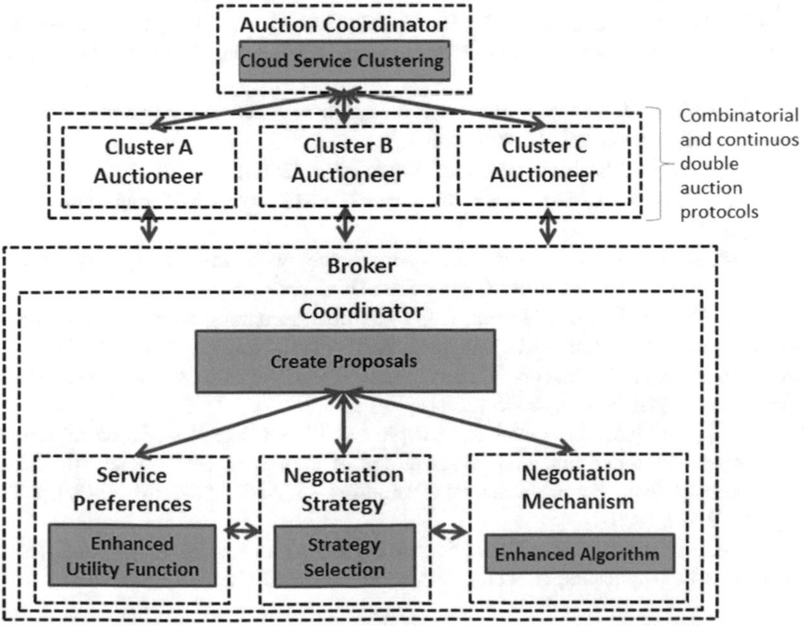

Fig. 2. The foresighted enhanced double auction negotiation framework's component.

4 Conclusion

This paper had managed to outline both the limitations and future vision for double auction-based cloud service negotiation. The first limitation is described by the difficulty to accommodate heterogeneous cloud services, while the second limitation refers to the inability to negotiate a dynamic or wider range of parameters such as the price, services, duration, QoS, discounts, and additional benefits. Arguably, cloud service negotiations tend to involve multiple attributes which include different technical, economic, and strategic criteria. The third limitation is described by the long execution time caused by the utilization of various algorithms. On the other hand, a combination of the continuous and combinatorial double auction protocols was proposed for the future vision. The proposed framework of the current research is equipped with the components that can accommodate multi-attributes heterogeneous cloud services and dynamic parameters. Finally, it was suggested that machine learning can be applied to the proposed framework.

Acknowledgements. The authors would like to thank the Centre for Research in Data Science (CeRDaS) of Universiti Teknologi PETRONAS, Ministry of Higher Education (MOHE) Malaysia (MyBrain15, Grants No: 0153AB-I27 and 0153AB-K47), POSTECH, South Korea (Grant No. 0153AB-M20) and Universiti Teknologi PETRONAS for supporting this study.

References

1. Adnan, M.H., Hassan, M.F., Aziz, I.A.: Protocols for agent-based autonomous negotiations: a review. In: International Conference on Computer & Information Sciences
2. Wang, X., Wang, X., Che, H., Li, K., Huang, M., Gao, C.: An intelligent economic approach for dynamic resource allocation in cloud services. IEEE Trans. Cloud Comput. **3**, 275–289 (2015)
3. Dang, J., Huhns, M.N.: Concurrent multiple-issue negotiation for internet-based services. IEEE Internet Comput. **10**, 42–49 (2006)
4. Luecke, R.: Harvard business essentials: negotiation (2003)
5. Financier Worldwide. http://www.financierworldwide.com/negotiating-cloud-contracts/#.Vzv3J_l9600
6. Hasan, M.H., Jaafar, J., Hassan, M.F.: Monitoring web services' quality of service: a literature review. Artif. Intell. Rev. 1–16 (2014)
7. Lopes, F., Wooldridge, M., Novais, A.Q.: Negotiation among autonomous computational agents: principles, analysis and challenges. Artif. Intell. Rev. **29**, 1–44 (2008)
8. Aknine, S., Pinson, S., Shakun, M.: An extended multi-agent negotiation protocol. Auton. Agent. Multi Agent Syst. **8**, 5–45 (2004)
9. Di Nitto, E., Di Penta, M., Gambi, A., Ripa, G., Villani, M.: Negotiation of service level agreements: an architecture and a search-based approach. In: Krämer, B., Lin, K.-J., Narasimhan, P. (eds.) Service-Oriented Computing – ICSOC 2007, vol. 4749, pp. 295–306. Springer, Berlin (2007)
10. Adnan, M.H., Hassan, M.F.: Dynamic pricing for cloud service negotiation. In: Control Theory and Its Application (CNTIA) 2015. MALTESAS (2016)

11. Ibrahim, N.M., Hassan, M.F., Shamsudin, A.F.: Intelligent-based translation model for cross = platform communications in distributed systems. Al-Qimah Al-Mudhafah J. Manag. Sci. 2 (2017)
12. Kumar, D., Baranwal, G., Raza, Z., Vidyarthi, D.P.: A systematic study of double auction mechanisms in cloud computing. J. Syst. Softw. **125**, 234–255 (2017)
13. Bichler, M.: A roadmap to auction-based negotiation protocols for electronic commerce. In: Proceedings of the 33rd Annual Hawaii International Conference on System Sciences, vol. 11, pp. 10. IEEE (2000)
14. Kraus, S.: Automated negotiation and decision making in multiagent environments. In: Multi-agent Systems and Applications, pp. 150–172. Springer, Berlin (2001)
15. Li, B., Lau, H.C.: A Combinatorial Auction for Transportation Matching Service: Formulation and Adaptive Large Neighborhood Search Heuristic, pp. 127–142. Springer, Berlin (2017)
16. Wang, J., Liu, A., Yan, T., Zeng, Z.: A resource allocation model based on double-sided combinational auctions for transparent computing. Peer-to-Peer Netw. Appl. **11**, 679–696 (2018)
17. Li, H., Wu, C., Li, Z., Lau, F.: Virtual machine trading in a federation of clouds: individual profit and social welfare maximization (2013)
18. Wang, X., Wang, X., Wang, C.-L., Li, K., Huang, M.: Resource allocation in cloud environment: a model based on double multi-attribute auction mechanism. In: 2014 IEEE 6th International Conference on Cloud Computing Technology and Science (CloudCom), pp. 599–604. IEEE
19. Vytelingum, P., Cliff, D., Jennings, N.R.: Strategic bidding in continuous double auctions. Artif. Intell. **172**, 1700–1729 (2008)
20. Garg, S.K., Vecchiola, C., Buyya, R.: Mandi: a market exchange for trading utility and cloud computing services. J. Supercomput. **64**, 1153–1174 (2013)
21. Sun, D., Chang, G., Wang, C., Xiong, Y., Wang, X.: Efficient Nash equilibrium based cloud resource allocation by using a continuous double auction. In: 2010 International Conference on Computer Design and Applications (ICCDA), pp. V1-94–V91-99. IEEE (2011)
22. Shang, S., Jiang, J., Wu, Y., Yang, G., Zheng, W.: A knowledge-based continuous double auction model for cloud market. In: 2010 6th International Conference on Semantics Knowledge and Grid (SKG), pp. 129–134. IEEE (2010)
23. Lan, Y., Tong, W., Liu, Z., Hou, Y.: Multi-unit continuous double auction based resource allocation method. In: 2012 3rd International Conference on Intelligent Control and Information Processing (ICICIP), pp. 773–777. IEEE (2012)
24. Farajian, N., Ebrahim pour-komleh, H.: Parallel continuous double auction for service allocation in cloud computing. Comput. Eng. Appl. J. 133–142 (2015)
25. Shi, X., Xu, K., Liu, J., Wang, Y.: Continuous double auction mechanism and bidding strategies in cloud computing markets. arXiv preprint arXiv:1307.6066 (2013)
26. Zaman, S., Grosu, D.: Combinatorial auction-based dynamic vm provisioning and allocation in clouds. In: 2011 IEEE 3rd International Conference on Cloud Computing Technology and Science (CloudCom), pp. 107–114. IEEE (2011)
27. Fujiwara, I., Aida, K., Ono, I.: Applying double-sided combinational auctions to resource allocation in cloud computing. In: 2010 10th IEEE/IPSJ International Symposium on Applications and the Internet (SAINT), pp. 7–14. IEEE (2010)
28. Rightscale: State of the Cloud Report (2016)

29. Saripalli, P., Pingali, G.: Madmac: Multiple attribute decision methodology for adoption of clouds. In: 2011 IEEE International Conference on Cloud Computing (CLOUD), pp. 316–323. IEEE (2011)
30. Lai, G., Li, C., Sycara, K., Giampapa, J.: Literature review on multi-attribute negotiations. Robotics Institute, Carnegie Mellon University, Pittsburgh, PA. Technical Report CMU-RI-TR-04-66 (2004)
31. Nassiri-Mofakham, F., Nematbakhsh, M.A., Baraani-Dastjerdi, A., Ghasem-Aghaee, N., Kowalczyk, R.: Bidding strategy for agents in multi-attribute combinatorial double auction. Expert Syst. Appl. **42**, 3268–3295 (2015)
32. José-García, A., Gómez-Flores, W.: Automatic clustering using nature-inspired metaheuristics: a survey. Appl. Soft Comput. **41**, 192–213 (2016)
33. Hsieh, F.-S.: A Discrete Particle Swarm Algorithm for Combinatorial Auctions, pp. 201–208. Springer, Berlin (2017)
34. Adnan, M.H., Husain, W., Rashid, N.A.A.: Hybrid approaches using decision tree, Naive Bayes, means and euclidean distances for childhood obesity prediction. Int. J. Softw. Eng. Appl. **6**, 99–106 (2012)
35. Adnan, M.H.B.M., Husain, W., Rashid, N.A.: A hybrid approach using Naive Bayes and Genetic Algorithm for childhood obesity prediction. In: 2012 International Conference on Computer & Information Science (ICCIS), pp. 281–285 (2012)
36. Adnan, M.H., Husain, W., Rashid, N.A.: Implementation of hybrid Naive Bayesian-decision tree for childhood obesity predictions. Network **40**, 84–87 (2018)

RFID Smart Shelf Based on Reference Tag Technique

Wan Nurfatin Faiqa Wan Abdullah, Muhammad Rafie Mohd. Arshad, Kamal Karkonasasi(✉), Seyed Aliakbar Mousavi, and Hasimah Mohamed

School of Computer Sciences, Universiti Sains Malaysia, USM, 11800 Pulau Pinang, Malaysia
wnfaiqa.ucoml2@student.usm.my, {rafie,hasimah}@usm.my, asasi.kamal@gmail.com, pouyaye@gmail.com

Abstract. Radio Frequency Identification (RFID) is an automatic identification technology that can track and identify people, animals, and objects. It uses the radio frequency signal to send data from RFID reader to RFID tags. A common RFID system will involve RFID tag, RFID reader, and antenna as its components. Nowadays, RFID technology has been widely used for indoor positioning especially for the smart shelf in locating and tracking the object on the shelf. But, the existing smart shelf needs more than four antenna and more than two readers to increase the accuracy of locating the location and position of the misplaced item at the certain covered area. This research paper will compare a few algorithms which are Fingerprinting (Block Distance), Trilateration (Least Square Estimate) and Trilateration (Two-Border) to show the different accuracy among them and to increase the accuracy in locating the exact position of missing object by using Real Time Locating System (RTLS) by proposing a new algorithm. Besides, Received Signal Strength Indication (RSSI) is the chosen method of getting data from different points of the shelf to find the location of the missing object on the shelf. This indoor positioning use reference tag technique to get the RSSI data from the shelf. Thus, we can identify the missing object and detect if there is misplaced object occur or track the real-time location of the object.

Keywords: RFID · RTLS · RSSI · Smart shelf · Algorithm · Reference tag Indoor positioning

1 Introduction

Radio Frequency Identification (RFID) is widely used for automatic identification and tracking with associate an object to a univocal identification code such as in animal tracking, asset management, industrial automation, electronic article surveillance and access control applications. It is the most famous technology used to track and determine objects for the indoor environment [1]. By tracking the location of the objects, the system provides retailers with information that can be analyzed and translated into customer preferences. But, in order to track the location, the existing RFID based smart shelf used more readers and antenna to increase the accuracy of

F. Saeed et al. (Eds.): IRICT 2018, AISC 843, pp. 499–509, 2019.
https://doi.org/10.1007/978-3-319-99007-1_47

locating the location and position of the misplaced object which is costly. The antennas need to be embedded at every level of the shelf. The existing algorithm used also still lacks accuracy in locating the object. This application can be extended to include searching for relevant items more effectively, thereby minimizing time spent by staff managing paperwork, conducting searches and checks and reducing human error. Thus, by testing this application using different algorithms and locating using the RSSI method they can help in locating target object with different accuracy and increase the accuracy of positioning by proposing a new algorithm in order to get the most accurate data. The existing indoor positioning of RFID based smart shelf system is being implemented in such expensive way since at least four antennas are needed to support a shelf and a common portable reader has their own limit to support the number of antennae. So, to implement this RFID technology approach to any shelf will be unacceptably expensive. Apart from that, the existing method lacks the accuracy to determine the location of an object from the antenna through tracking devices and also there are many algorithms have been used to get the position of an object in a shelf but it is still in unclear which one of them giving the most accurate ones. Either the existing algorithms are accurate enough to get the exact position or need to propose an improved algorithm so that the accuracy will be increased. So, this paper has studied about exploring on how to reduce the cost of implementing smart shelf. At the same time, it will test the system with different algorithms to find which algorithm give the exact position of the target object in the shelf and increase the accuracy of locating by proposing a new algorithm.

The main motivation of the current research is to reduce the cost of implementing RFID technology based smart shelf. At the same time, to increase the accuracy of locating the exact location and position of the target object which can help the retailers to invest in the small amount of money in implementing this technology system while still maintain or increase the accuracy of locating.

This research study will contribute in reducing the cost of implementing RFID based smart shelf and determining the exact location of a target object from the antenna by using reference tag technique which will be used to get the RSSI value. This RSSI value later will be converted into the distance of target object from antennas. Besides, this study intends to contribute to knowledge by comparing different algorithms to get the most accurate positions of tracking object on the shelf. Next, a new algorithm is proposed to increase the accuracy of locating the position target object. Thus, this will help a lot in getting a precise information technology with a low cost of implementation.

2 Existing RFID Based Smart Shelf

The current implementation of RFID based smart shelf has widely explored to get the most accurate location and position of the target object. There are many studies showing that researchers have tried to use various types of methods and algorithm to achieve their objectives. In getting the best accuracy of object location, methods like Angle of Arrival (AoA), Time Difference of Arrival (TDoA), Time of Arrival (ToA) and Received Signal Strength Indication (RSSI) has been used.

2.1 Sensing Technologies for RFID Estimation Distance

The existing system has used numerous of different techniques to indicate the distance of the object from the antenna. It gives the different accuracy of object position depends on the techniques chosen.

Angle of Arrival (AOA): it is a method used to decide on the RF signal's direction (propagation) accepted from a tag to a reader and hence, the direction of the transmitter (Tag) can be obtained. This is shown in Fig. 1. This is done by measuring the angle between lines that run from the reader to tag and the line from the reader with a predefined direction [2].

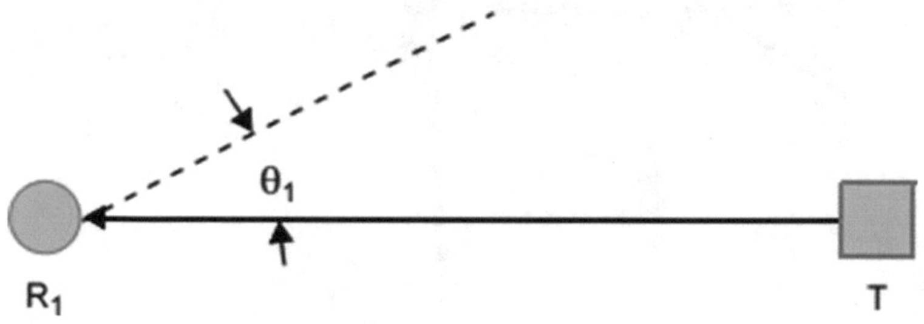

Fig. 1. Angle of Arrival method.

Time Difference of Arrival (TDoA): By measuring the delay in the transmission time recorded between signals obtained from each transmitter to a tag, as in Fig. 2, the Time Difference of Arrival (TDoA) technique can be applied as the intersection of hyperbolas in a 3 dimension system. TDOA requires that the receiver (or readers) register when the signal is received, either simultaneously or with some established delay between signal transmissions.

Time of Arrival (ToA): As in the Fig. 3 in between a transmitter (Tag) and one or more receivers (readers), the Time of Arrival (ToA) is one approach which depends on the propagation delay measurement of the radio signal whereby, the propagation delay is calculated as $t_i - t_0$. As such, it is the time lag of the departure conveyed by a signal from a source to a destination station. It is defined as the amount of time that a signal requires to go from the transmitter to receiver [2].

Received Signal Strength Indication (RSSI): This technique uses several wireless access points (AP) simultaneously to track the location of a device. The signal strength obtained from received signals from at least three accesses points is used to determine the location of object or person being tracked. In order to increase accuracy, more sophisticated RSSI methods using a map called RF fingerprint, which is based in calibrations of the strength of WLAN (Wi-Fi) signals at various points in a predefined area is used. Also, the distance between a tag (object or person) and a reader is determined by converting the value of the signal strength at the reader (receiver) into a distance measurement based on the known signal output power at the tag (transmitter) and on a particular path-loss model in an RSSI system. Figure 4 shows Determining tag position with RSSI. The RSSI technique can be illustrated as follows:

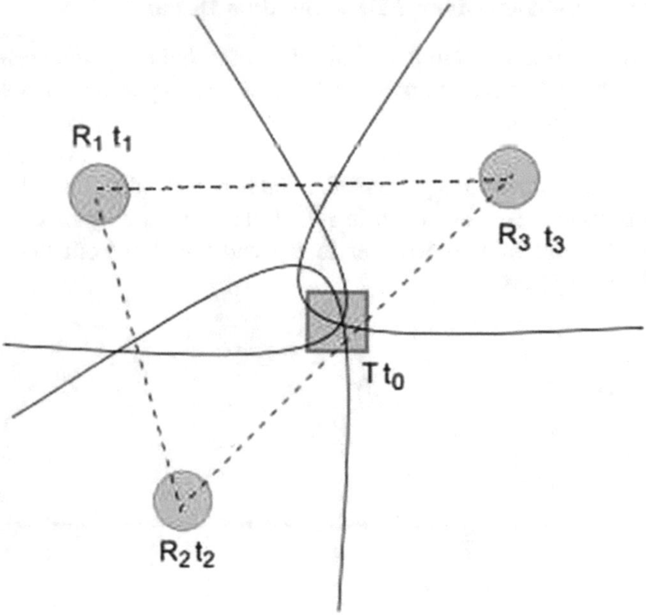

Fig. 2. Determining tag position with TDoA.

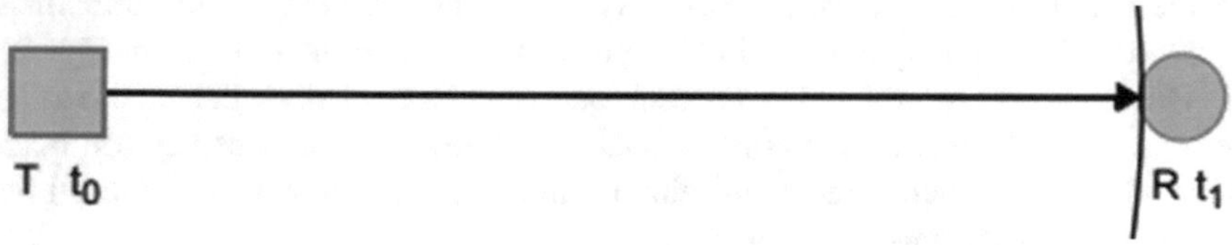

Fig. 3. Time of Arrival.

Virtual Reference Tag: Wang et al. also work on the improvement of locating object position by using two methods [3]. The first method is by using VireBys indoor positioning method based on virtual reference tags and Bayesian estimation is proposed. The idea is to obtain the initial coordinates of tags to be positioned by making use of VIRE algorithm, and then to process these initial coordinates by a Bayesian principle so as to obtain a more accurate coordinate position. Originally covered by 4 reference tags, a small grid is further divided into N × N virtual grid cells with the same size (the V region in Fig. 5 is a virtual grid cell).

2.2 Sensing Technologies for RFID Estimation Position

This section will discuss a few algorithms that usually are used to calculate the position of the object in the smart shelf. There are two types of algorithm involve which are Fingerprinting and Trilateration algorithm. The accuracy of locating the object position will be different depends on the chosen algorithm.

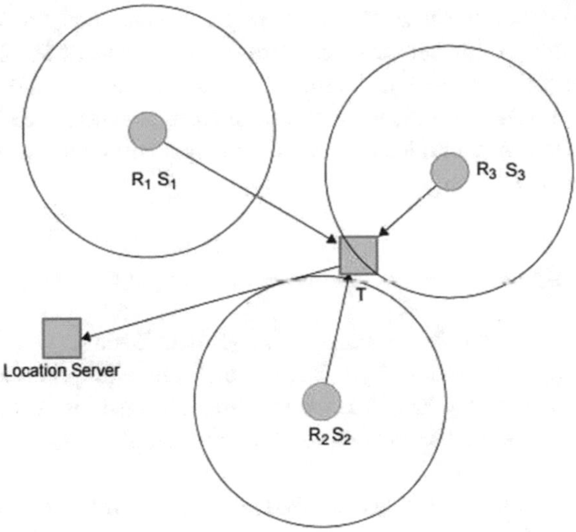

Fig. 4. Determining tag position with RSSI.

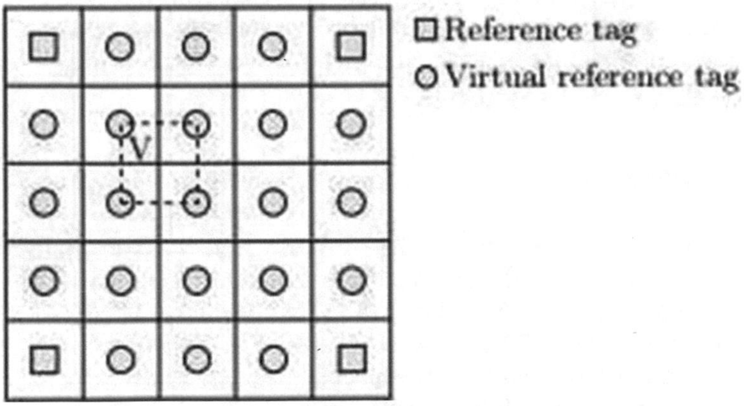

Fig. 5. The layout of VIRE system.

Fingerprinting: This type of algorithm requires the database of RSSI values at each compartment of the shelf location. Similar to how fingerprinting used to detect criminals. It is as follows:

i. a database of criminal fingerprints is required.
ii. i. The fingerprint of the criminal is taken and compared with all fingerprints in the database. ii. If a match is found, then the criminal has been identified.

Trilateration: Involve calculating object position by solving simultaneous quadratic equations. Trilateration (Least Square Estimation): Li et al. discuss least square methods based a low-cost 3D indoor positioning with Bluetooth smart device [4].

Various linear and nonlinear least square methods and their theoretical basis and application performance for indoor positioning have been studied. The nonlinear least square algorithm will be used to estimate the parameters of the Bluetooth signal propagation model. Least square, weighted least square, generalized least square, total least square and weighted total least square algorithms will be utilized to determine the location form coarse to fine.

3 Methodology

The proposed work is based on Fingerprint (Block Distance), Trilateration (Least Square Estimation), Trilateration (Two-Border) and a new proposed algorithm that use the concept of references tag to locate the location and position of the target object in real time. Figure 6 shows the flowchart of overall implementation involves in this research.

In Proposed Algorithm, distance is used to measure the distance between target object and antenna. Then, the distance is used to calculate the RSSI value. The RSSI value will be compared with the RSSI value stored in the database as the reference to find the exact location of the target object. Figure 6 shows the flowchart of Proposed Algorithm. The Algorithm can locate the position of the target object and overcome the problem when the target object is located in between or at the border of the bookshelf. This algorithm increases the accuracy of object position. Hence, it will perform better than the existing algorithm that has higher error possibility.

4 Implementation

Various simulation tools are used by the RFID research works to justify the RFID research-based environment. Many simulation tools are taken into account in this study. Such tools are NS-2, NCTUns, MATLAB Simulink, OMNET++ and many more. Particularly, MATLAB Simulink tool is applied in this research study. Simulink provides a graphical editor, customizable block libraries, and solvers for modeling and simulating dynamic systems. It is integrated with MATLAB, enabling to incorporate MATLAB algorithms into models and export simulation results to MATLAB for further analysis [5]. The only simulation is used to test the data in this research study because to implement this research by hardware is costly. The RFID reader that can support RSSI is very expensive that could not be afforded to buy. That is why simulation has become the best approach for testing and analyzing the data in this study. Figure 7 shows the simulation interface and Fig. 8 shows the position of RSSI value interface.

The boxes of Distance 1 (feet) and Distance 2 (feet) are to enter the distance between the object and both of the antennas. Distance 1 (feet) will present distance from x-axis while Distance 2 (feet) will present distance from the y-axis. At the Model combo box, shows the FRIIS Transmission Equation for no obstacles calculation. The function of 'Estimate RSSI' button is to calculate the RSSI value of the object location by using the distance entered at Distance 1 (feet) and Distance 2 (feet). The calculated

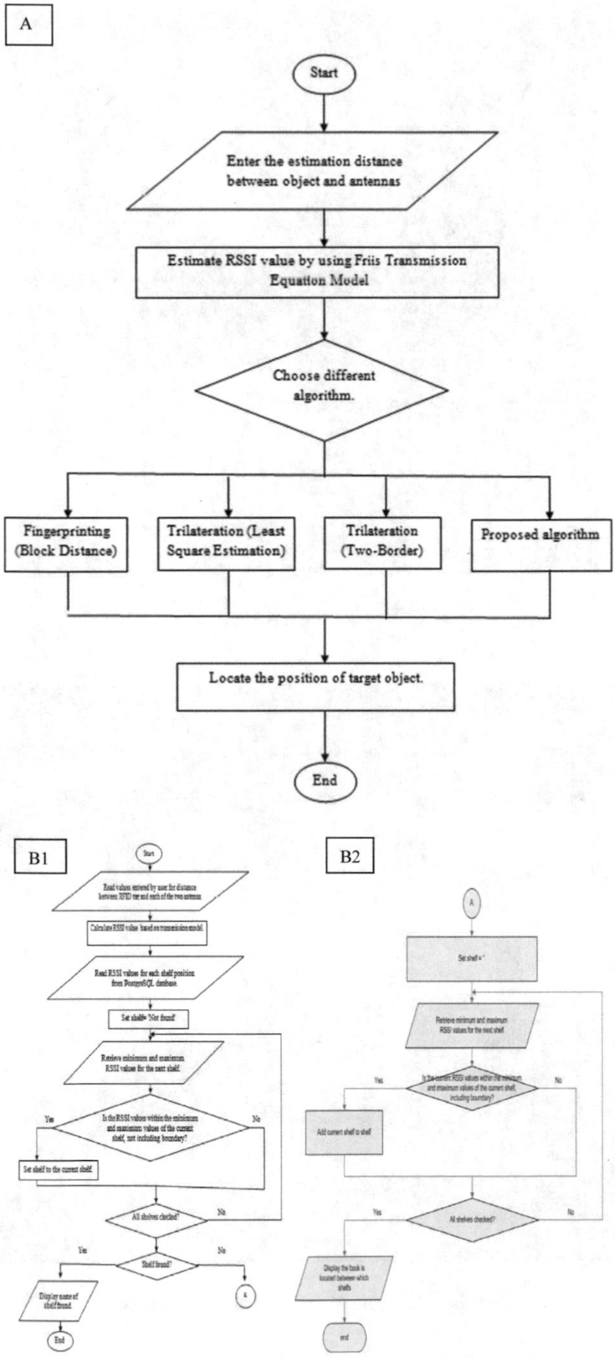

Fig. 6. Flowchart of overall implementation and B1 & B2 Proposed Algorithm.

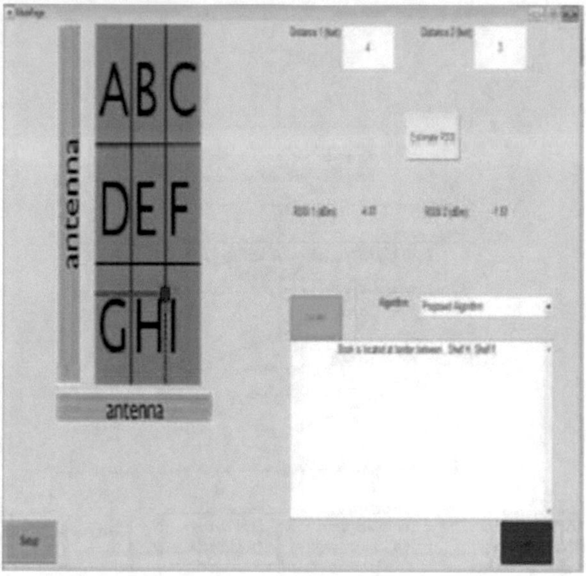

Fig. 7. Simulation interface.

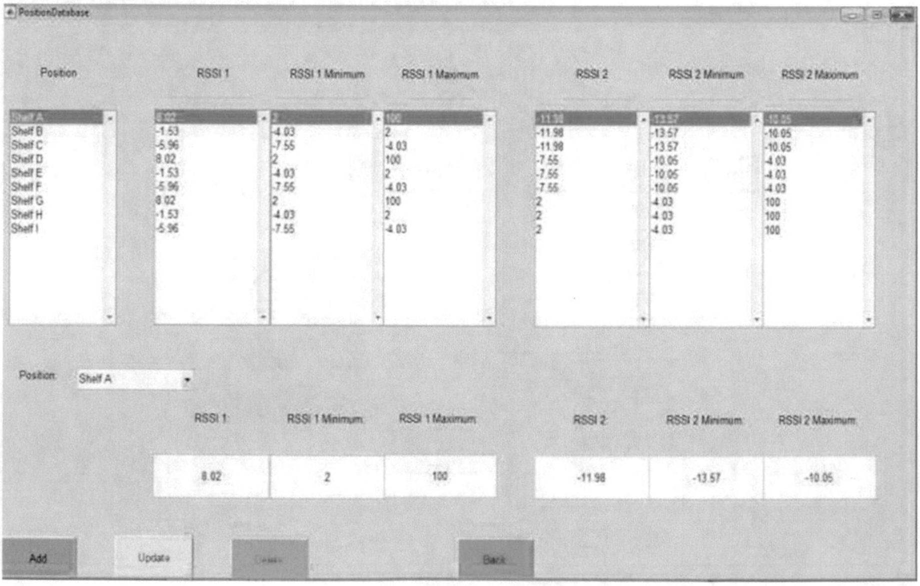

Fig. 8. Position of RSSI value interface.

RSSI value will appear at the RSSI 1 (dBm) and RSSI 2 (dBm). After that, we can choose the Algorithm to be tested in the Algorithm combo box. There are lists of the algorithms which are Fingerprint (Block Distance), Trilateration (Least Square

Estimation), Trilateration (Two-Border) and a new proposed algorithm. The 'locate' button will use the calculated RSSI value to be implemented in the chosen algorithm. The result will appear in the empty box below the 'locate' button. The 'setup' button will show setup interface where all the reference RSSI values were being entered and stored in the database. There it can be added, edited or deleted.

5 Analysis and Evaluation

The smart shelf is represented by coordinate and each of the coordinates is increased by 0.5 at each point as shown in Fig. 9. As shown, 0.5, 1.5, 2.5 and 3.5 are representing both X and Y coordination at the border of the shelf. At the axis, 0.5 coordinate represents the border of shelf A, shelf D, and shelf G. 1.5 coordinate represent border of shelf between shelf A and shelf B, shelf D and shelf E, and shelf G and shelf H. 2.5 coordinate represent border of shelf between shelf B and shelf C, shelf E and shelf F and shelf H and shelf I. Lastly, 3.5 coordinate represent the border of shelf C, shelf F, and shelf I. 1 coordinate represent area in the shelf A, shelf D and shelf G. 2 coordinate represent area in shelf B, shelf E and shelf H. 3 coordinate represent the area in the shelf C, shelf F, and shelf I. Next, at the Y axis, 0.5 coordinate represent border of shelf A, shelf B and shelf C. 1.5 coordinate represent border of shelf between shelf A and D, shelf B and shelf E, and shelf C and shelf F. 2.5 coordinate represent border of shelf between shelf D and shelf G, shelf E and shelf H, and shelf F and shelf I. While 3.5 coordinate represent border of shelf G, shelf H and shelf I. 1 coordinate represent area in the shelf A, shelf B, and shelf C. 2 coordinate represent area on the shelf D, shelf E and shelf F. 3 coordinate represent area on the shelf G, shelf H and shelf I. First, we took the actual coordinate (Actual X and Actual Y) of object position which involves 91 positions. The distance of each of the positions is increased by one (feet). Then, by using the same distance, we test with a different algorithm to locate the position and the result is converted it into a measured coordinate (Measured X and Measured Y). The result is compared with the actual ones to get the percentage of accuracy.

These evaluations tell us the percentage of accuracy in getting the position of target objects by using a different algorithm. Figure 10 provides a comparison of accuracy for every algorithm. The result of accuracy for each for the object position is shown in the Table 1.

6 Discussion

The accuracy is one of the main attributes for the good RFID technology, our proposed algorithm is able to increase the accuracy over the existing implemented algorithm which are Block Distance algorithm, Least Square Estimation algorithm, and Two-Border algorithm. Thus, we have conducted a coordination test of data obtained, after conducting the experiment shown that the proposed algorithm has better accuracy compared to other implemented algorithm. Proposed Algorithm has shown that even the object is located exactly at the middle between two shelves or at the order of them,

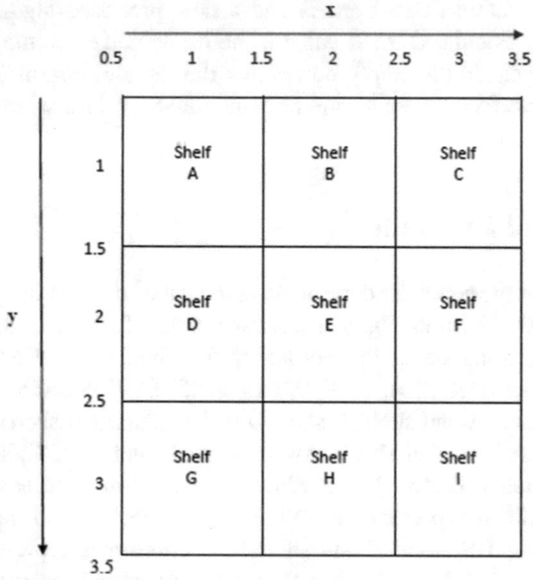

Fig. 9. Coordinate for each compartment of the smart shelf.

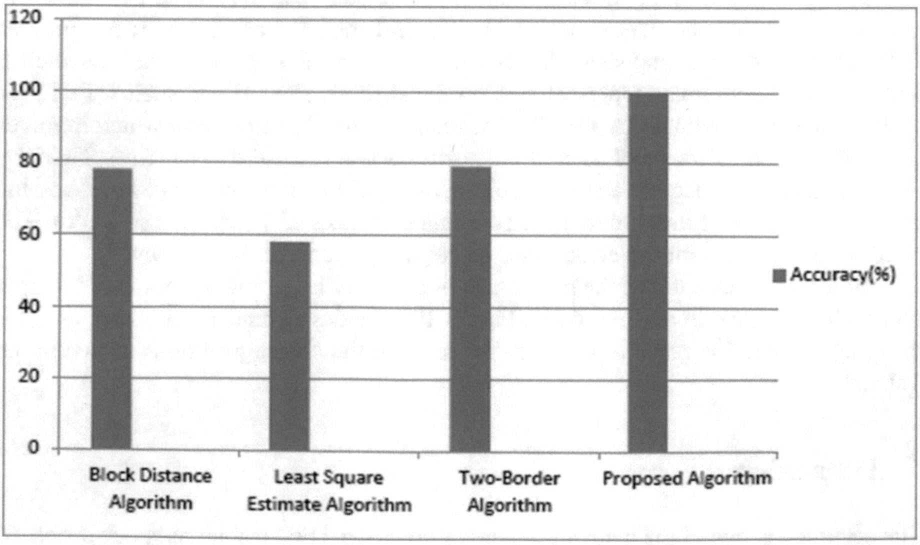

Fig. 10. Comparison of accuracy for every algorithm.

by using this algorithm they can be located. But, among the existing implemented algorithm, Two-Border algorithm shows the highest accuracy among them.

Table 1. The result of accuracy for each for the object position.

Algorithm	Accuracy (%)
Fingerprinting (Block Distance)	78.022
Trilateration (Least Square Estimate)	58.132
Trilateration (Two-Border)	79.121
Proposed method	100.000

7 Conclusion

In this research, we have proved that this indoor RTLS of RFID based smart shelf concept is realistic. The system, rather than uses more than 3 antennas, uses only two antennas. The Proposed algorithm employs the idea of having minimum and maximum RSSI of x-axis and y-axis at each compartment of the shelf to precisely indicate the border of the shelf. To get the more accurate location of the object, reference tag technique is effective to be implemented. And, the Two-Border algorithm also has proved that it is the best-implemented algorithm in this research that has the highest accuracy of locating. As RFID technology is one of the technologies that bring many benefits to the people especially those who are retailers, it is such a waste if this kind of technology is not being implemented or used during this time as there is so much effort have been done by the researchers in order to find a solution that can provide a system which is low cost of implementation but giving the precise information to the users. Besides, the low cost of implementation research may give the opportunity to every family to have their own mini smart shelf in their house in future.

Acknowledgment. The authors would like to thanks the Ministry of Higher Education for financially supporting this research under the Fundamental Research Grant Scheme (FRGS) 2017 with account number 203/PKOMP/6711580.

References

1. Silva, P., Paralta, M., Caldeirinha, R., Rodrigues, J., Serodio, C.: TracemeUindoor real-time location system. In: 35th Annual Conference of IEEE Industrial Electronics, IECON 2009, pp. 2721–2725. IEEE (2009)
2. Vossiek, M., Wiebking, L., Gulden, P., Wieghardt, J., Hoffman, C., Heide, P.: Wireless local positioning. Microw. Mag. IEEE **4**(4), 77–86 (2003)
3. Wang, D.: RFID indoor positioning method based on virtual reference tags and Bayesian estimation. J. Inf. Comput. Sci. **11**(17), 6141–6149 (2014)
4. Li, H.: Low-cost 3D bluetooth indoor positioning with least square. Wirel. Pers. Commun. **78** (2), 1331–1344 (2014)
5. Mathworks.com: Simulink - Simulation and Model-Based Design (2016). http://www.mathworks.com/products/simulink/?requestedDomain=www.mathworks.com. Accessed 2 Mar 2016

Proposed Mechanism Based on Genetic Algorithm to Find the Optimal Multi-hop Path in Wireless Sensor Networks

Mohammed Al-Shalabi[1(✉)], Mohammed Anbar[1(✉)], and Tat-Chee Wan[1,2]

[1] National Advanced IPv6 Centre (NAV6), Universiti Sains Malaysia, 11800 Gelugor, Penang, Malaysia
{moshalabi,anbar}@nav6.usm.my, tcwan@usm.my
[2] Computer Science Department, Universiti Sains Malaysia, 11800 Gelugor, Penang, Malaysia

Abstract. In widespread wireless sensor networks, where the cluster heads (CHs) and the base station (BS) are far from each other, data transmission is considered a critical factor because of its influence on the network efficiency in terms of power consumption and lifetime. Direct transmission problem, when a CH sends its data directly to the BS, causes a rapid drain of energy in CHs that are far from the BS. Several studies have been proposed to address this problem by finding intermediate nodes between the CH and the BS, but they suffer from many problems such as the complexity and considering just one parameter to select the intermediate nodes. This paper proposed a new mechanism to find the optimal multi-hop path for data transmission from a CH to the BS by utilizing the Genetic Algorithm (GA). The proposed mechanism aims to increase the network lifetime by reducing the consumed energy in the data transmission process.

Keywords: LEACH protocol · Genetic Algorithm · Data transmission Multi-hop path

1 Introduction

In WSNs, there are a lot of sensor nodes which they are randomly distributed over a large area, and the role of these nodes is to sense a specific event and send the sensed data to the main device which is called Base Station for further analysis [1, 2]. But due to the large number of nodes, the wide area of sensing, and to the limitation of nodes in terms of power, it is difficult to send data from nodes to the BS directly. Thus, data should be sent via some intermediate nodes, this is called a route from a source node to the destination [3].

To deal with this technique, many routing protocols were proposed in WSNs to find an efficient route between the sensor node and the BS. These routing protocols are divided into three classes based on network structure, which are flat routing, hierarchical-based routing, and location-based routing [4], as shown in Fig. 1.

F. Saeed et al. (Eds.): IRICT 2018, AISC 843, pp. 510–522, 2019.
https://doi.org/10.1007/978-3-319-99007-1_48

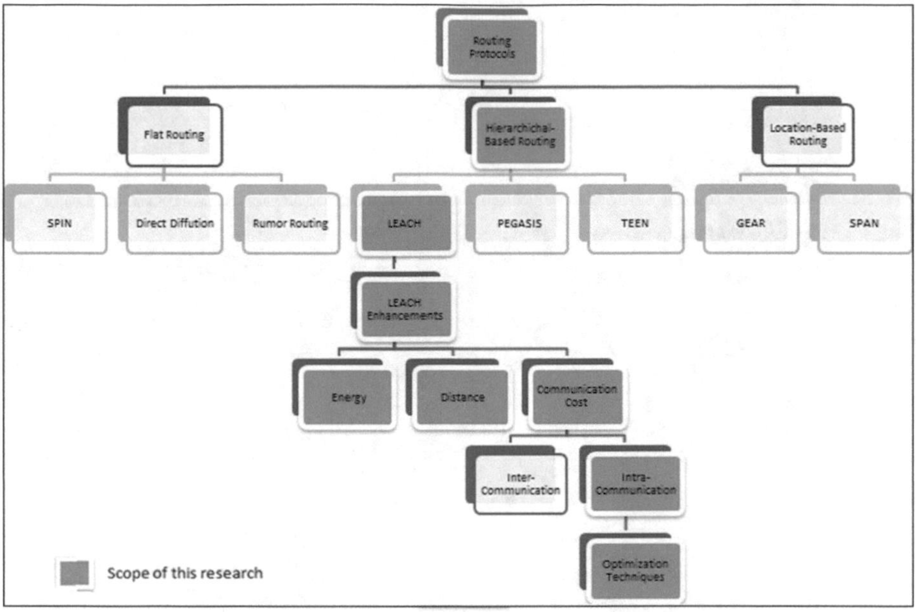

Fig. 1. Routing protocols for WSN

The Hierarchical-based routing protocols will be considered in the proposed mechanism in this paper due to the advantages of this type of routing protocols as will be clarified in the following paragraph.

The purpose of these protocols is to increase the lifetime of networks by dividing the network into a set of clusters, and each cluster consists of a number of nodes [5]. This technique leads to reduce the number of transmissions which be sent to the base station by using data aggregation in a single node inside a cluster called cluster head (CH) instead of transmitting data directly from each node to the base station [6]. Another purpose is that the task of a CH will be distributed among senor nodes inside a specific cluster to minimize the energy consumption. Figure 2 shows the structure of the hierarchal protocols.

Many hierarchical-based routing protocols have been proposed to increase the network lifetime. The Low-Energy Adaptive Clustering Hierarchy (LEACH) [8] is considered as the main protocol in this type of routing protocols and from the first proposed protocols. The following section describes the LEACH protocol in details because the proposed mechanism in this paper inspires it.

Data transmission from a CH to the BS is a very important issue in WSNs due to its significant effect on the lifetime of a whole network. In addition, data transmission plays a significant role in the hierarchical-based routing protocols, especially if the distance between the CH and the BS is far, in this case, the direct transmission of data will cause the energy of the CH drain very fast.

Many techniques were utilized to deal with the data transmission in order to increase the performance of the network in terms of the energy consumption. The GA

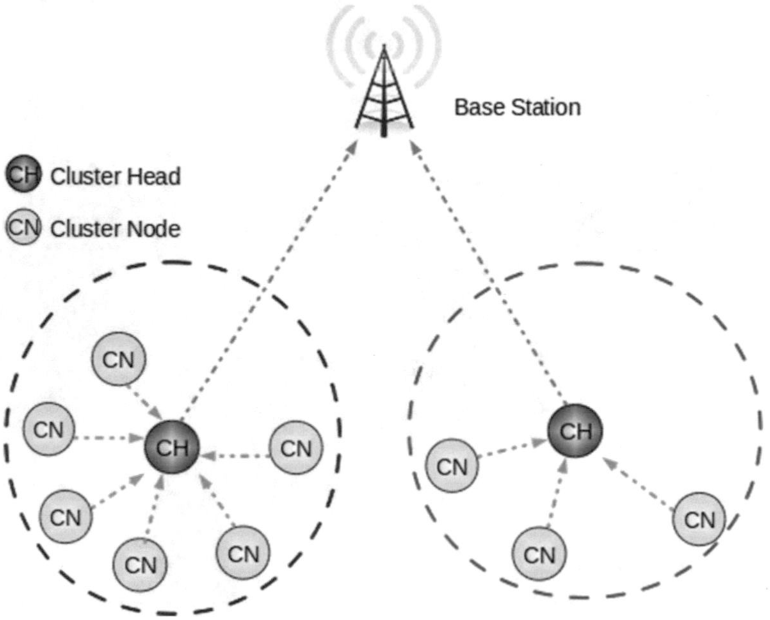

Fig. 2. Hierarchical-based routing [7]

is a technique that was used in many researches to solve the direct transmission of data. The GA also is considered in this paper to propose a new mechanism in order to find a multi-hop path between a CH and the BS for data transmission instead of the direct data transmission.

Furthermore, the LEACH protocol, which is considered the mainly hierarchical-based routing protocol, is considered as the inspiration of the proposed mechanism in this paper due to use the same clustering approach and the same parameters which were proposed and used in the LEACH protocol.

2 Related Work

Many protocols have been proposed to enhance the LEACH protocol in terms of the data transmission. This section highlights the most related works which based on the GA and other evolutionary algorithms to find the multi-hop path and their limitations.

The authors of the WST-LEACH protocol (Weighted Spanning Tree clustering routing algorithm based on LEACH) [21] deal with the two problems in LEACH protocol, which are the selection of CHs and the data transmission to the BS. A modified threshold has been proposed to select the cluster heads based on different parameters such as the distance to the BS, the residual energy of nodes, and the number of neighbours of each node. After selection of the CHs, this protocol constructs a weighted spanning tree to establish a communication path with the BS. This protocol increases the lifetime of compared to the LEACH protocol, but it has control packets overhead. The comparison with this protocol did not set due to the missing of the important factor in simulations which is the location of the BS, because it plays a significant role, especially in the communication protocols.

In GAEMW protocol (Genetic Algorithm for Energy-entropy based Multipath routing in WSNs) [20], authors proposed a GA for multipath routing in WSNs, the idea is to find the minimal node residual energy of each route in selecting multipath routing. To initialize population, a routing path is encoded by a set of integers represent IDs, energy, and other node's information. The first gene represents a source node and the last gene represents the BS, then several paths (chromosomes) will be created from the source to the destination. The fitness function is used to evaluate the quality of each chromosome (path) by finding the least cost path between the source and destination. The fitness function of a chromosome Ci is defined as follows [20]:

$$f(C_i) = \frac{1}{G}\sum_{k=1}^{path}\left((E_{tr} + E_{rec}) \times P_{ki} \times \left|P_{ki}^{s,d}\right| - E_{avg}\right) \tag{1}$$

After that, the two operators in GA (crossover and mutation) are applied to the population and the next generation is created. This protocol provides an efficient method for evaluating the route stability in WSNs.

The literature that has been proposed by [22] describes an improved protocol using GA called ROS-IGA (Routing Optimization Strategy based on Improved GA) to find a suitable route between a sensor node and the BS, the new protocol called ROS-IGA. In this protocol, authors tried to solve the problem of generating invalid individuals; this problem negatively affects the efficiency of GA, for example, Some routes that have been generated by the crossover and mutation operations may be not valid based on the connections in the network due to the existence of some nodes in the generated rout which do not have direct connection to other nodes. In the proposed protocol, the locations of nodes and other information such as energy consumption between adjacent nodes, accumulated distance, and the remaining energy of nodes are considered in the fitness function to find the suitable and practical path. The fitness function is maximization function, thus, the higher fitness value means more paths [22].

The crossover and mutation operations are improved in this protocol based on the neighboring information and the structure of the network. This improvement of crossover and mutation operations provides better routing and practical routes in WSNs.

FUCHAR (Fuzzy-Based Unequal Clustering and ACO-Based Routing Hybrid Protocol) is proposed by [23] to maximize the network lifetime by using a fuzzy logic to select a CH based on five parameters which are the distance to the BS, the residual energy, the distance to neighbors, the node degree and the node centrality. Moreover, the proposed protocol used the Ant Colony Optimization (ACO) evolutionary algorithm for routing data from CHs to the BS. The limitation of this protocol is the choosing of just one relay node between a CH and the BS and based on the distance and the residual energy only. Moreover, this protocol suffers from the control message overhead and the increasing in network traffic due to the many sending messages from neighbors to the candidate nodes, and using a lot of control messages between nodes.

In [24], a new protocol, which depends on GA to improve the energy efficiency, is presented by optimizing the selection of CHs and minimizing energy consumption in the communication process between a CH and the BS, namely, GA based on distance-aware LEACH (GADA-LEACH). In [24], after selecting CHs on the basis of the GA, a relay node between a CH and the BS and several factors, such as the total energy of nodes, CH energy, distance between sensor nodes and CH, and the distance between each CH and the BS were used to compute fitness function, as shown as follows:

$$Fitness = [(0.3 * Fit1) + (0.35 * Fit2) + (0.35 * Fit3)] \tag{2}$$

$$Fit1 = \frac{Total\ Energy\ of\ all\ nodes}{Energy\ of\ CHs} \tag{3}$$

$$Fit2 = \frac{Eucledian\ distance\ from\ CH\ to\ its\ nodes}{Number\ of\ nodes\ in\ cluster} \tag{4}$$

$$Fit3 = \frac{Eucledian\ distance\ from\ all\ CHs\ to\ the\ BS}{Number\ of\ CHs} \tag{5}$$

On the basis of fitness value, the top individuals are selected from the population. Then, the crossover and mutation operations will be used to select the best CHs. Subsequently, a relay node is introduced to minimize the distance from a CH to the BS [24]. This protocol performs better than the LEACH protocol on the basis of network lifetime by improving the CH selection process and introducing an intermediate node in the network. Table 1 summarizes the proposed protocols.

Table 1. Summary of the proposed protocols

Protocol	Modeling parameter	Drawbacks and limitations
WST-LEACH (2010)	Distance, number of members, number of neighbors and residual energy	Control packets overhead
GAEMW (2015)	Residual energy of nodes	Fitness function concentrates only on the residual energy
ROS-IGA (2016)	Locations of nodes, energy consumption between adjacent nodes, accumulated distance, remaining energy of nodes	Concentration on the invalid individuals, it is based on the locations of nodes which got during the deployment
		Does not consider the CH selection
FUCHAR (2017)	CH selection and inter-cluster communication	Control message overhead
		Increasing network traffic
GADA-LEACH	Energy, distance	GA used to select the CHs
		One relay node

Based on the abovementioned discussion, the related works suffer from many drawbacks such as the complexity and extra overhead. Moreover, most of the related works have many limitations such as considering only the path with one intermediate node, and using limited parameters to find the multi-hop path which are not efficient in terms of the optimality of the path.

To conclude, there are still gabs in the related works in the ways of finding the optimal multi-hop path and also to increase the efficiency of selecting the CHs instead of keeping them randomly selected. This paper attempts to propose a new mechanism which can find an optimal multi-hop path in WSNs instead of just using a single relay node such as in the existing works, and also the proposed mechanism selects the CHs in an efficient way instead of the random selection in the existing works. Moreover, the proposed mechanism increases the performance of the GA in terms of the execution time and the quality of the chromosomes by contrast the existing works which used the standard GA. The following section explains the related methods (i.e. the LEACH protocol and the GA) and the proposed mechanism in this paper.

3 Methods

This section presents the LEACH protocol and the GA in details to give an overview of them before starting with the proposed mechanism.

Moreover, this section presents the main goals of the proposed mechanism in details. Moreover, the proposed pre-processes are also explained in this section. Furthermore, the stages and the workflow of the proposed mechanism are also explained.

3.1 Leach Routing Protocol

The LEACH protocol has been proposed by [8] to increase the network lifetime by reducing the consumed energy. The idea of the LEACH protocol is to distribute the load of energy consumption among the nodes in the network by grouping the sensor nodes in groups called clusters, with a special node in each cluster called Cluster Head (CH). The main duties of the CHs are: collecting the data from the sensor nodes in the cluster, fusion the data, and sending that data directly to the Base Station (BS) [9]. Figure 3 shows the structure of the LEACH protocol.

The LEACH protocol has many advantages [10]. It balances power consumption among all nodes in each cluster due to the rotation of the CH selection process. This procedure can increase the network's lifetime. By contrast, several disadvantages exist. For example, LEACH selects the CHs randomly and it does not take into account the residual energy of a node in the selection process of a CH. Another problem is the direct data transmission from a CH to the BS using single-hop communication. This problem is considered a serious issue if the CH and the BS were located in far positions because the CH, in this case, uses more energy than a CH when it is close to the BS, resulting in energy holes in which isolated nodes are unable to transmit their data. This situation considerably affects the performance of the network [11, 12].

To overcome the problem of data transmission in the LEACH, many improvements have been proposed by modifying the LEACH protocol to increase its efficiency by finding a multi-hop path for sending the data based on the Genetic Algorithm (GA). An overview of the GA is presented in the following subsection.

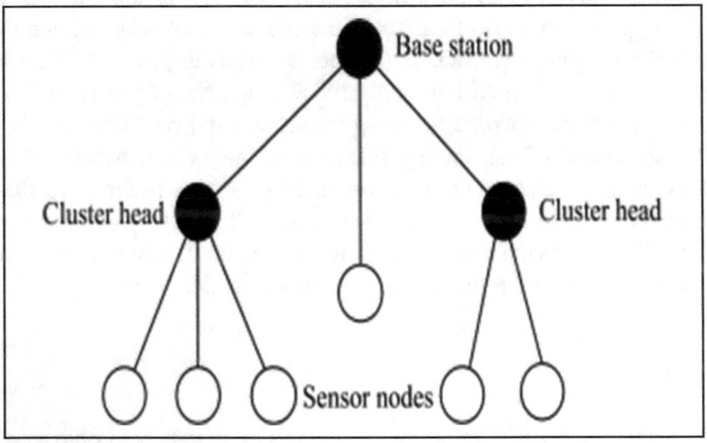

Fig. 3. Structure of LEACH protocol [26]

3.2 Genetic Algorithm

GA is an optimization algorithm that simulates a nature evolution, and it uses the same protocols in the natural selection to find the optimal solutions [13, 14]. GA is used in optimization and search problems to find the global best solution.

The main operations in GA are selection, crossover, and mutation [15]. At first, the initial population is randomly generated with a set of individuals, each of them will be evaluated using the fitness function, and then a new population will be selected from the initial population depending on its fitness value and will contribute in the generation of the population [16]. The fitness function is very important in GA and depends on the problem, it is used to evaluate the quality of individuals and select the best of them to contribute to the next generation [17]. The crossover operator is used to generate new children from the parents by combining them in different ways, and these children will have a combination of the two parents [18].

In mutation, random swap is applied to each child in order to form a new solution. The algorithm stops when a specific condition is reached [19, 20].

3.3 The Design Goal of the Proposed Mechanism

The current existing protocols are limited in finding an optimal multi-hop path for transmission the data from the source CH to the BS in terms of power consumption. This paper introduces a new mechanism to find such multi-hop path with several intermediate hops. The existing protocols suffer from limitations in finding the multi-hop path.

Given that these multi-hop paths consist of just one hop for sending the data or the selection of the intermediate hops based only on one parameter such as the energy or the distance. Moreover, the random selection of the CHs still exists in the existing protocols, which may also cause a fast drain of the energy in the CH if a node with little energy is selected as a CH. This paper proposes a mechanism to increase the network lifetime by reducing the energy consumption in the data transmission process. This reduction in the energy consumption based on finding an optimal multi-hop path that consists of more than one hop. The optimality of the path does not mean the shortest of it, but its ability to reduce the energy consumption based on several parameters. The following section explains the proposed mechanism.

The main goals of the proposed mechanism are as follows:

- Select the most proper nodes to be CHs instead of the random selection.
- Enhance the GA in terms of the execution time and the quality of the chromosomes.
- Find an optimal multi-hop path from a source CH to the BS.

The following subsections describe how the proposed mechanism works.

3.4 The First Pre-processing

This pre-process is important in the proposed mechanism, because if the energy in the CHs is considered as the main target, then the selection of the efficient nodes to be CHs significantly affects the overall lifetime of the network.

This pre-process is the CHs selection process. Many protocols used different parameters individually to select the CHs such as the energy average of the nodes, the distance between the node and the BS and other parameters. In this paper, the residual energy and the distance to the BS are just considered in the CHs selection process.

The modified threshold which is proposed in this paper divides the CHs selection process to three levels. The first level of the modified threshold value is used to start with the nodes that have energy greater than the average energy and their distances to the BS are less than the average distance to be the CHs from the first round to the round number n1. The second level is set after n1 rounds to select nodes with maximum residual energy and minimum distance to the BS to be the CHs from round number n1 to the round number n2. Finally, in the third level, and after the round number n2, the threshold value is set to be as in the LEACH protocol [8]. The values of n1 and n2 are set based on the applications and on the parameters which are set on the networks. For example, the value of n1 is maximized if it is important in the application to increase the stable period of the network, the round where the first node dies, or if the initial energy of nodes is high, i.e. 1 J. The value of n2 is always maximized because in the third level, the number of dead nodes will increase and the selection of CHs will be done randomly as in the LEACH protocol.

3.5 The Second Pre-processing

This pre-process is used in the proposed mechanism before starting with the GA to increase the efficiency of the GA in terms of the execution time and the quality of the chromosomes. This pre-process based on the energy of the CHs in each round, where the CHs with the energy more than the average energy of all CHs will be involved in the GA to find the optimal multi-hop path. The idea behind this pre-process is to minimize the number of the CHs that will be involved in the GA to decrease the execution time of it and minimize the complexity. Moreover, the selected CHs have energy level more than others, and this will increase the optimality of the selected multi-hop path. The following example illustrates the proposed pre-process:

The number of the selected CHs in the current round is 7 with IDs from 1 to 7 and the expected energy for each CH as follows:

ID	CH1	CH2	CH3	CH4	CH5	CH6	CH7
Energy	0.4	0.5	0.3	0.3	0.4	0.5	0.2

The average energy of these CHs is 2.6/7 = 0.37, so the number of CHs with energy greater than the average is 4, which are CH1, CH2, CH5, and CH6. These CHs will be considered in the initial population of the GA to formulate the chromosomes that represent the path from each source CH to the BS. In other words, the multi-hop path that will be used to send the data from each source CH (CH1, CH2, …., CH7) to the BS will be formulated using just the selected 4 CHs in the above example.

After ending the pre-process, the GA will start to select the optimal multi-hop path that consists of the selected CHs or some of them based on the used fitness function and the fitness values of the chromosomes.

All required information for the selected CHs will be stored in the array data structure (G0), and then the BS will formulate the initial individuals from G0 depending on the source CH and the destination (BS).

After initialization of the individuals, one of the GA operators which is the crossover will be applied on each pair of the selected individuals (parents) which will be selected randomly based on the crossover rate (Cr) to produce new offspring in order to vary the programming of the chromosomes [13, 25].

Then the mutation operation will be applied on each one of these offspring. After that, the fitness function will be calculated to select the best chromosomes.

3.6 The Fitness Function

In this paper, several parameters in the fitness function are proposed. The distance from each participated CH to the BS is one parameter in the fitness function. Moreover, the number of intermediate CHs in a path plays an important role in the fitness function, so, the ratio of the suggested CHs in the path to the total number of the participated CHs is another parameter in the fitness function. The last parameter in the fitness function is the number of the participant for each suggested CH. The shortest distance between the source and the destination in the transmission process results in greater energy efficiency, and longest distance consumes more energy, but the distance itself dos not enough to select the optimal multi-hop path.

These parameters are included in the following fitness function which will be used to find the optimal multi-hop path by minimizing the value of it:

$$\boldsymbol{F(c) = d(CH_s, BS) + \left(\frac{N_{par}}{N}\right) + No\ of\ Part} \tag{6}$$

where *F*(*c*) is the fitness value of the *cth* chromosome, *d*(*CHs, BS*) is the average distance between the source CH to the BS through the intermediate CHs, *Npar* is the number of participated CHs in a multi-hop path, *N* is the total number of participated CHs and *NoOfPart* is the total number of participation in the transmission process of all the participated CHs in the multi-hop path.

In the first section of the fitness function, the average distance from the source CH to the BS is proposed to distribute the energy consumption among all CHs. And the second section represents the percentage of the participated CHs in the transmission process from all participated CHs in the GA, because of its important in finding the optimal multi-hop path. When the percentage is low, the optimality of the path will increase considering that the proposed fitness value should be minimized, and when the percentage decreases, the optimality increases. The last section represents the total number of participation in the transmission process of all CHs in the path because if one CH participates many times in the transmission process, it will consume its energy very fast. So, this parameter will increase the value of the fitness function when its value increases and will give inefficient multi-hop path.

3.7 Workflow of the Proposed Mechanism

The proposal of this mechanism based on three stages to find the optimal multi-hop path in the WSNs as illustrated below:

- CH Selection Stage (Stage One): In this stage, the most efficient nodes will be selected as CHs in each round in the proposed mechanism. This selection of the CHs will increase the efficiency of the network in terms of the energy consumption because the nodes with high energy and close to the BS will be selected as CHs, especially in the first n1 rounds as explained in Subsect. 3.4. These efficient CHs when involved in the multi-hop path, the energy consumption will decrease and the lifetime of the whole network will increase.
- Filtering the CHs (Stage Two): Before starting with the GA, the number of the CHs in each round should be reduced to increase the quality of the chromosomes, and to decrease the execution time of the GA. In this stage, the CHs with energy more than the average energy of the all CHs will be selected to formulate the population in the GA.
- Finding the Optimal Multi-hop Path (Stage Three): Finally, the GA will start to find the optimal multi-hop path from the source CH to the BS for all CHs in each round. The selection of the optimal solution (path) based on the parameters in the fitness function.

Figure 4 represents the flowchart of the proposed mechanism. The flowchart represents the steps of the proposed mechanism using the GA.

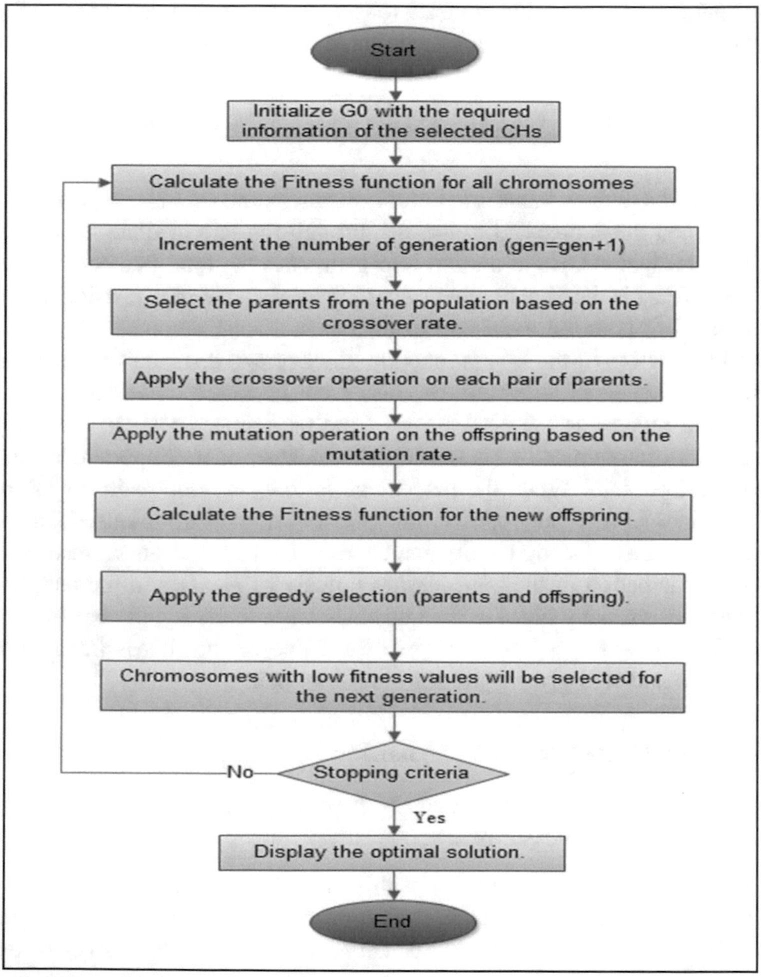

Fig. 4. Flowchart of the proposed mechanism

4 Conclusion and Future Work

This paper proposes a new mechanism to find the optimal multi-hop path for data transmission from the CHs to the BS. The GA is utilized to find the optimal multi-hop path by proposing a fitness function that consists of several efficient parameters to achieve the optimality of the multi-hop path in terms of energy consumption. Moreover, a pre-processing is proposed to increase the efficiency of the GA by selecting the CHs with high energy to involve in it. This paper expects that the proposed mechanism can find a multi-hop path that is more optimal than those of other existing protocols. The future work will consider the implementation of the proposed mechanism using the MATLAB simulation tool, in addition to the evaluation of it compared with the related works.

References

1. Abed, A., Alkhatib, A., Baicher, G.S.: Wireless sensor network architecture. In: International Conference on Computer Networks and Communication Systems (CNCS 2012), vol. 35, pp. 11–15. IACSIT Press, Singapore (2012)
2. Pantazis, N.A., Nikolidakis, S.A., Vergados, D.D.: Energy-efficient routing protocols in wireless sensor networks: a survey. IEEE Commun. Surv. Tutor. **15**(2), 551–591 (2013). https://doi.org/10.1109/SURV.2012.062612.00084
3. Pino-Povedano, S., Arroyo-Valles, R., Cid-Sueiro, J.: Selective forwarding for energy-efficient target tracking in sensor networks. Signal Process. **94**, 557–569 (2014)
4. Wu, Y., Xu, D., Gao, J.: Comparison study to hierarchical routing protocols in wireless sensor networks. Proc. Environ. Sci. **10**, 595–600 (2011)
5. Munir, A.: Cluster based routing protocols: a comparative study. In: IEEE Fifth International Conference on Advanced Computing and Communication Technologies (ACCT) (2015)
6. Lotf, J.J., Hosseinzadeh, M., Alguliev, R.M.: Hierarchical routing in wireless sensor networks: a survey. In: 2010 2nd International Conference on Computer Engineering and Technology (ICCET), vol. 3, pp. V3–650–V3–654, April 2010
7. Kole, S., Vhatkar, K.N., Bag, V.V.: Distance based cluster formation technique for LEACH protocol in wireless sensor network. Int. J. Appl. Innov. Eng. Manag. (IJAIEM) **3**(3), 334–338 (2014)
8. Heinzelman, W., Chandrakasan, A., Balakrishnan, H.: Energy-efficient communication protocol for wireless microsensor networks. In: Proceedings of the 33rd Annual Hawaii International Conference on System Sciences, 2000, vol. 2, p. 10, January 2000
9. Yadav, S., Yadav, S.S.: Review for LEACH protocol in WSN. Int. J. Recent Dev. Eng. Technol. **2**(6), 69–71 (2014)
10. Li, Y., Ding, L., Liu, F.: The improvement of LEACH protocol in WSN. In: Proceedings of 2011 International Conference on Computer Science and Network Technology, ICCSNT 2011, vol. 2, pp. 1345–1348 (2011)
11. Li, X., Wang, J., Bai, L.: LEACH protocol and its improved algorithm in wireless sensor network. In: International Conference on Cyber-Enabled Distributed Computing and Knowledge Discovery (CyberC), pp. 418–422 (2016)
12. Li, Y., Yu, N., Zhang, W., Zhao, W., You, X., Daneshmand, M.: Enhancing the performance of LEACH protocol in wireless sensor networks. In: 2011 IEEE Conference on Computer Communications Workshops (INFOCOM WKSHPS), September 2015, pp. 223–228 (2011)
13. Kumar, M., Husian, M., Upreti, N., Gupta, D.: Genetic algorithm: review and application. Int. J. Inf. Technol. Knowl. Manag. **2**(2), 451–454 (2010)
14. Maulik, U., Bandyopadhyay, S.: Genetic algorithm-based clustering technique. Pattern Recognit. **33**(9), 1455–1465 (2000). https://doi.org/10.1016/S00313203(99)00137-5
15. Zhou, R., Chen, M., Feng, G., Liu, H., He, S.: Genetic clustering route algorithm in WSN. In Proceedings of 2010 6th International Conference on Natural Computation, ICNC 2010, vol. 8, pp. 4023–4026 (2010)
16. Norouzi, A., Zaim, A.H.: Genetic algorithm application in optimization of wireless sensor networks. Sci. World J. **2014**, 286575 (2014). https://doi.org/10.1155/2014/286575
17. Liu, W., Wu, Y.: Routing protocol based on genetic algorithm for energy harvesting-wireless sensor networks. IET Wirel. Sens. Syst. **3**(2), 112–118 (2013). https://doi.org/10.1049/iet-wss.2012.0117

18. Sivanandam, S.N., Deepa, S.N.: Genetic algorithm implementation using matlab. In: Introduction to Genetic Algorithms, p. 453 (2008). https://doi.org/10.1007/978-3-540-73190-0_2
19. Bhatia, T., Kansal, S., Goel, S., Verma, A.K.: A genetic algorithm based distance-aware routing protocol for wireless sensor networks. Comput. Electr. Eng. (2016). https://doi.org/10.1016/j.compeleceng.2016.09.016
20. Song, Y., Gui, C., Lu, X., Chen, H., Sun, B.: A genetic algorithm for energy-efficient based multipath routing in wireless sensor networks. Wirel. Pers. Commun. **85**(4), 2055–2066 (2015)
21. Zhang, H., Chen, P., Gong, S.: Weighted spanning tree clustering routing algorithm based on LEACH. In: 2010 2nd International Conference on Future Computer and Communication, vol. 2 (2011)
22. Yao, G., Dong, Z., Wen, W., Ren, Q.: A routing optimization strategy for wireless sensor networks based on improved genetic algorithm. J. Appl. Sci. Eng. **19**(2), 221–228 (2016)
23. Arjunan, S., Sujatha, P.: Lifetime maximization of wireless sensor network using fuzzy based unequal clustering and ACO based routing hybrid protocol. Appl. Intell. 1–18 (2017)
24. Bhatia, T., Kansal, S., Goel, S., Verma, A.K.: A genetic algorithm based distance-aware routing protocol for wireless sensor networks. Comput. Electr. Eng. (2016). https://doi.org/10.1016/j.compeleceng.2016.09.016
25. Hengpraprohm, S., Chongstitvatana, P.: Selective crossover in genetic programming. In: ISCIT International Symposium on Communications and Information Technologies (2001). http://citeseer.ist.psu.edu/536164.html
26. Kochhar, A., Kaur, P., Singh, P., Sharma, S.: Protocols for wireless sensor networks: a survey. J. Telecommun. Inf. Technol. **1**, 77–87 (2018). https://doi.org/10.26636/jtit.2018.117417

Advances in Information Security

Enhancement of Text Steganography Technique Using Lempel-Ziv-Welch Algorithm and Two-Letter Word Technique

Salwa Shakir Baawi(✉), Mohd Rosmadi Mokhtar, and Rossilawati Sulaiman

Center for Cybersecurity, Faculty of Information Science and Technology, National University of Malaysia, 43600 Bangi, Selangor, Malaysia
salwa_sh2001@yahoo.com, {mrm, rossilawati}@ukm.edu.my

Abstract. Text steganography is regarded as the most challenging type to hide secret data to a text file because it is not enough unnecessary information compared with other carrier files. This study aimed to deal with the capacity (how much data can be hidden in the cover carrier) and security (the inability of disclosing the data by an unauthorized party) issues of text steganography. Generally, the data hiding capacity of text steganography is limited, and imperceptibility is very poor. Therefore, a new scheme was suggested to improve the two-letter word technique by using the Lempel-Ziv-Welch algorithm. This scheme can hide 4 bits in each position of a two-letter word in the cover text by inserting a nonprinting Unicode symbol. Each two-letter word can have four different locations in the text. Some experiments were conducted on the proposed method (enhancement) and our previous method (two-letter word) to compare their performance applying twelve secret message samples in terms of capacity and Jaro-Winkler. On the other hand, the performance of our proposed method was compared with other related studies in terms of capacity. The results show that the proposed method not only had a high embedding capacity but also reduced the growing size ratio between the original cover and stego cover. In addition, the security of the proposed approach was improved through improving imperceptibility and by using stego-key.

Keywords: Compression · Information security · Steganography Capacity

1 Introduction

Communication has massively expanded with the rapid development of new technologies, such as the Internet, mobile phones, and computers. Consequently, accesses, tampering and illegal copying can be done due to information being easily available to anyone. Therefore, information security has gained particular significance. Hiding

F. Saeed et al. (Eds.): IRICT 2018, AISC 843, pp. 525–537, 2019.
https://doi.org/10.1007/978-3-319-99007-1_49

exchanges of information is a critical area of information security that includes cryptography and information hiding methods. The privacy and authenticity of secret data can be ensured through information hiding techniques, such as steganography [1].

Steganography is the most common sub-discipline for hiding information. This term originated from the Greek word Steganographia (στεγανογραφία), which was recorded as early as 1462 to 1516 and literally means "covered writing". Secure communication ensures that unauthorized users cannot obtain secret messages for their private use [2, 3]. Steganography can be accomplished using various file formats, such as text, image, audio or video. Among these formats, text files are considered the most challenging carriers of secret data because they have insufficient redundant information in comparison with other formats [4]. Furthermore, text files which are frequently used over the Internet have different characteristics that enable the user to insert secret data [3].

The general requirements for safely hiding information and measuring the strengths and weaknesses of text steganography techniques are capacity, robustness, and imperceptibility. Capacity refers to the maximum quantity of secret messages that can be planted into the cover text without reaction to the text quality and is generally expressed as bit by bit. Security relates to the ability of an eavesdropper to figure out the hidden information easily. Robustness is concerned about the resist possibility of modifying or destroying the unseen data [5]. Security of steganography method is measured by imperceptibility which figures out the level of perception of the existence of hidden information in the cover file [6]. Imperceptibility of the hidden information refers to the text transparency and quality. Thus, the stego text should be nearly identical to the original text. The imperceptibility of the stego text can be evaluated by a similarity measure that is based on the Jaro-Winkler distance between the cover and stego files.

Although, several successful attempts of research have been done in text steganography. However, most of the existing text steganography methods focus on attaining a high hiding capacity rather than a high imperceptibility of the stego cover. For instance, [7] hide the secret message by an invisible character in Microsoft Word file. In this work, concealing secret message growths the size of stego file than the cover file. While [8] worked on compressing the cover file before hiding process. As a result, their work also suffered from poor imperceptibility. Furthermore, Our previous study [3] proposed new text steganography which is focused on the use of a set of two-letter words, by inserting the nonprintable Unicode characters (ZWNJ, ZWJ) to hide 2 bits of secret data in the cover text. We encountered issues involving the increasing size of the stego text files.

The present study handle capacity and security issues by proposing an approach of text steganography based on compression. Namely, the present study aimed to develop the text steganography method based on a set of the two-letter word with high capacity and security. This work presented a high imperceptibility text steganography method that is based on the Lempel-Ziv-Welch (LZW) algorithm for encoding secret messages. The LZW data compression technique is used for compressing hidden data to improve the quality of stego text through reducing the increasing size ratio between the original and stego documents and also increasing the capacity of the cover file. To improve security, this study proposed an algorithm for generating and using stego key on the basis of size of the secret message. The capacity metric and Jaro-Winkler metric are used to evaluate the quality of the proposed method. The remainder of this paper is organized as follows. Section 2 discusses previous work relevant to the proposed algorithm. Section 3 elaborates the proposed scheme and implemented framework details. Section 4 describes the experimental results. Section 5 summarizes the overall work and recommends future enhancements.

2 Related Work

This section highlights the most commonly utilized text steganography techniques. Por et al. [7] established the format-based text steganography method that uses open space coding. This approach is based on a space character manipulation technique called UniSpaCh, which is recommended for text steganography using Microsoft Word documents containing Unicode space characters. UniSpaCh uses the white spaces present in any document by hiding the payload in the inter-sentence, inter-word, end-of-line, and inter-paragraph spacing with Unicode space characters. The manipulation of white spaces is a suitable technique because it presents an insignificant effect on the overall appearance of the document. Contrary to conventional methods, UniSpaCh is undetectable, has a high embedding efficiency, and is robust against DASH attack. Moreover, the contents of the cover document only minimally influence the embedding efficiency of UniSpaCh. Satir and Isik [8] presented a steganographic scheme that aims to solve capacity and security issues. This approach uses the LZW data compression algorithm to hide information in text documents. The authors selected email as their communication medium and a mail forwarding platform as the stego cover. The embedding process uses two secret keys: a global stego key comprising numerous email addresses and a set of selected email addresses that have been modified. The final second key value in the embedding process is built on the basis of the original message's bits, and the algorithm generates the secret message bits. In the extraction phase, each element of the first key is compared with the second key value to decode the message. The experimental results revealed that the algorithm performed 6.92% better than existing methods. However, the stego file had poor imperceptibility and was thus prone to attack. Moreover, according to [6] the codewords must be recalculated if the content of the cover file would be changed. In 2013, Wang et al. [9] introduced a high-performance reversible data hiding scheme for LZW codes. This scheme embeds the secret data into LZW codes by altering the value of codes either by adding the dictionary size to code value or leaving it unchanged on the basis of the secret data bits. This scheme had a high data hiding capacity and low computation complexity. However, it aroused suspicion as it abruptly changed the LZW code value. Kumar et al. [10] discussed a high-capacity email based text steganography method that uses combinational compression. This method utilizes a forwarding email platform to hide the secret data in email addresses. This approach adopts a combined Burrows-Wheeler transform + moves to forwarding + LZW coding algorithm to increase the hiding capacity to 7.03%. The number of characters of email ID is also used to refer to the secret data bits while enhancing capacity. Random characters are added immediately before the "@" symbol of the email IDs to increase the randomness. Malik et al. [11] introduced a high capacity text steganography scheme based on LZW compression and color coding that directly uses the LZW technique on the secret data and the obtained bitstream is embedded into email addresses and also in the message email, that raises the capacity around 13.43%. This scheme used the color to hide hidden data in email text following some color coding, which is the main weakness.

Table 1 summarizes the related works of text steganography, and their strong and weak points.

Table 1. Related work

Refs.	Methodology	Strength	Weaknesses
[7]	An inter-sentence, inter-word, end-of-line and inter-paragraph spacings	Higher embedding efficiency. Robust on DASH attack	It has increasing the size of stego file
[8]	LZW data compression algorithm	Provided a significant increment in terms of capacity	The stego file had poor imperceptibility and was thus prone to attack
[9]	LZW data compression algorithm	High data hiding capacity	It aroused suspicion as it abruptly changed the LZW code value
[10]	Combined Burrows–Wheeler transform + moves to forwarding + LZW coding algorithm	The hiding capacity is 7.03%	Poor imperceptibility
[11]	LZW compression and color coding	Higher capacity around 13.43%	It used the color to hide hidden data and change the format of stego text

As shown in Table 1, the main weaknesses of these methods are the size of the hidden data depending on the size of the cover file and that any change made is easily detected (poor imperceptibility). Thus, the present study aims to address these weaknesses by proposing enhancing the two-letter word technique with the use of the LZW algorithm. Capacity and security issues are considered in this work. Improving the two-letter word techniques is used to hide the secret information bits within the cover text to avoid modifying the content. The proposed method increases hiding capacity and improves security by using stego key generated on the basis of secret message and enhancing the quality of stego text. In the next section, the proposed scheme is discussed.

3 Proposed Scheme

3.1 Lempel-Ziv-Welch

Terry Welch created the LZW algorithm in 1984. LZW is a general compression algorithm that works on nearly any type of data by referencing the table of strings commonly occurring in the data being compressed and replacing the actual data with references to the table. The table is formed during compression while the data are encoded and decoded. LZW is a well-known method for data compression. The essential steps for this technique are as follows. First, the file is read, and a code is provided to all characters. If the same characters are found in a file, then no new code will be assigned, and an existing code from a dictionary will be used instead. The process is continued until the characters in the file are null [12].

User-Defined Code. The table is proposed for building the dictionary symbols in the LZW algorithm. In this step, the secret message must be coded and converted to equivalent bits. Two methods of data coding are used: UDC, which maps each character to equivalent 6 bits (English alphabet [capital and lowercase], numbers, dots, and spaces) to increase security and reduce storage. Therefore, each character requires 6 bits rather than 8 bits, as shown in Table 2. Initialization is performed by assigning 6-bits codes into the dictionary, thus resulting in 64 initialization codes (0–63) in dictionary. The conversion of character into 6-bit ASCII code allows for three locations of two-letter word to embed two letters of compressed secret message within the cover because each location can conceal 4 bits. Therefore, the secret message can be written in scrambling alphabet before the LZW algorithm is applied. UDC is used in this work to represent the symbol of the secret message instead of using the coding data in ASCII code to convert each character to equivalent 8 bits, thereby enhancing the security and capacity of the system.

Table 2. Proposed Use-Defined Code using 6-bits

#	Char	Code	#	Char	Code	#	Char	Code	#	Char	Code
0	Sp	000000	16	e	010000	32	M	100000	48	.	110000
1	S	000001	17	t	010001	33	w	100001	49	8	110001
2	A	000010	18	f	010010	34	3	100010	50	7	110010
3	M	000011	19	T	010011	35	R	100011	51	F	110011
4	i	000100	20	h	010100	36	y	100100	52	L	110100
5	r	000101	21	l	010101	37	N	100101	53	J	110101
6	a	000110	22	B	010110	38	j	100110	54	q	110110
7	W	000111	23	o	010111	39	Q	100111	55	V	110111
8	n	001000	24	v	011000	40	C	101000	56	K	111000
9	b	001001	25	D	011001	41	G	101001	57	U	111001
10	O	001010	26	u	011010	42	X	101010	58	Y	111010
11	c	001011	27	p	011011	43	K	101011	59	2	111011
12	g	001010	28	z	011100	44	Z	101100	60	1	111100
13	x	001011	29	H	011101	45	0	101101	61	6	111101
14	d	001100	30	P	011110	46	E	101110	62	5	111110
15	s	001101	31	I	011111	47	4	101111	63	9	111111

3.2 Text Steganography Method

For text steganography, our previous study can be categorized under format-based methods. This work utilized a set of two-letter words from the Oxford Dictionary, which is used for hiding a significant amount of data in English writing. Character shapes in English writing are independent of their position in words, unlike Arabic or Persian characters, which have different forms depending on location. This method utilizes the feature of Arabic and Persian languages is reflected in the two-letter word in the English language for information hiding. The proposed method uses nonprinting characters, such as zero-width non-joiner and zero width joiner to hide 2 bits of hidden data in the cover text. Based on the location of these words in the text, we randomly applied four location scenarios to hide the secret data. This method achieved the hiding capacity, still suffers from the lack of the quality of stego cover text. Due to the increasing size of the stego text files owing to the use of Unicode in place of ASCII in the embedding process and the application of UTF-16 encoding to each character.

3.3 Techniques for Embedding and Extracting Text Steganography

This work focuses on enhancing the two-letter word method, which allows 4 bits to be concealed in each location of the two-letter word in the cover file. The hiding of the bits is based on four invisible Unicode symbols after the secret message has been compressed using LZW coding before embedding. This process improves capacity and increases the level of security without affecting the quality of stego text. Figure 1 shows the block diagram of the proposed scheme.

The two processes required in the proposed method, namely, embedding and extracting of messages, are explained as follows.

Embedding Algorithm. Embedding Algorithm with the proposed method to improve a two-letter word method (TLW) by LZW coding (Table 3).

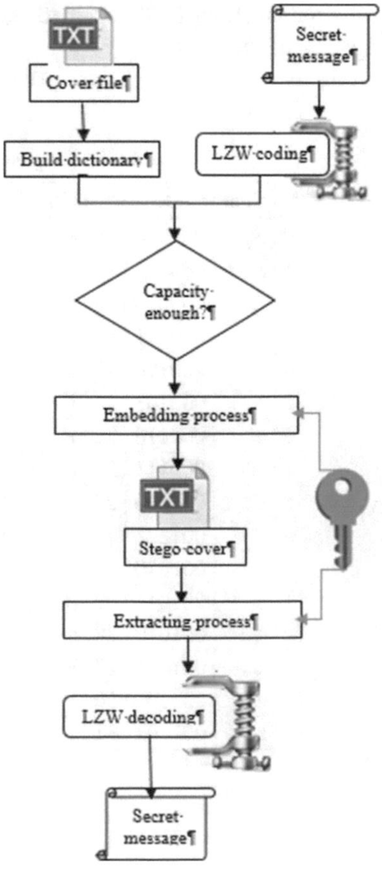

Fig. 1. Block diagram of the proposed scheme

```
Input: Cover text file
(C) and secret message
(M)
Output: Stego text file
(S) and Stego-key(K).
Process:
Step1: Open the cover
text file.
Step2: Build the dictionary for each two-letter word
      found in the cover text and those used in the Ox-
      ford Dictionary.
Step3: Scan the secret message (M).
Step4: Compressed secret message(CM)← Call LZW coding(M).
Step5: K ← Call Generate key(CM, Dic).
Step6: Convert the(CM) into binary bits, and then divide
       these bit streams into blocks of 4 bits each.
Step7: Compute the number of two-letter words in the
       cover text to check the capacity of embedding.
Step8: For each block in a CM has 4 bits, where the first
       2 bits represent the location of the two-letter
       word in the text. Each position has four block op-
       tions (i.e., 00, 01, 10, and 11) that are availa-
       ble for the embedding process. The latter 2 bits
       represent the nonprintable character that is
       inserted before a two-letter word that belongs to
       STLW, as shown in Table 3.
Step9: Return the stego text file (S) and Stego-key(K).
```

To increase the security level of the proposed method, a new algorithm is proposed to generate the stego key, which is used to determine the position to be utilized to embed data. This algorithm works on the basis of secret message bits and the position that is available in the cover file.

Table 3. A sample of hidden bits by using nonprintable characters

First 2 bits	Location	Last 2 bits	Nonprintable char
00	At the beginning	00	U + 200B
01	In the middle	01	U + 200D
10	At the end	10	U + 200E
11	Isolated	11	U + 200C

Generating Stego-Key Algorithm

Input: Dictionary of two-letter words (Dic) and com pressed secret message (CM).
Output: Stego key (K)
Process:
Step1: Read the dictionary of the two-letter word (Dic)
Step2: Input the secret message bits
Step3: Set gen_k ← null, R ← 0
Step4: For each two-letter word in a dictionary, there are four position options (i.e., B, M, E and I) are representation the location of the current two-letter word in cover-text, according to **Table 3**. Then based on the position (B, M, E, and I) will be added (00, 01,10 and 11) at the end of gen_K.
Step5: R ←(R +1)% 4
Step6: For each value to R, there are four value options (i.e., 0, 1, 2 and 3) are representation the two-bits such are an (i.e., 11, 10, 01, and 00) respectively and then added to gen_k.
Step7: If it is not (get the end of the dictionary and the length of gen_k < the length of the secret message bits) repeat the steps 5 to step 9.
Step8: Stego key ←Mixture gen_k with Secret message bits
Step9: Return Steg key

The proposed data extraction process is described in the following section.

Extraction Process. Enhanced Extraction algorithm by LZW coding.

Input: Stego text file (S), Stego-key (K)
Output: secret message (M).
Process:
Step1: Open the stego text file, segment it into lines, and then split each line into words as follows:

$$W = \{w_0, w_1, \ldots, w_{n-1}\}$$

Step2: Select the word (w_i), $0 \le i < n$ and separate it into characters. Then, check each character if it is an invisible Unicode character and if it is one of the following: {(U+200B), (U+200C), (U+200D), or (U+200E)}. Also, test the position of the current character to find the identical secret message bits on the basis of the scenarios in **Table 3**.
Step3: Repeat Step2 for each word in each line
Step4: Extracting stream bits from stego text using key.
Step5: Decode the compressed secret message (CM)
Step6: Return the secret message (M).

4 Experimental Results

In this section, we conducted a set of experiments on our previous method of two-letter word and the proposed method of the enhancement of two-letter word based on LZW algorithm to evaluate the performance both of them regarding hiding capacity and imperceptibility. The hiding capacity is a primary decisive parameter for the performance analysis of a text steganography algorithm. As in [8, 10], we define the hiding capacity or bit rate as the size of the hidden data relative to the size of the stego cover. The hiding capacity can be calculated as formulated in Eq. (1):

$$Capacity\,(C) = \frac{Size\ of\ secret\ message\,(bits)}{Size\ of\ stego\ cover\,(bits)} \tag{1}$$

To verify the imperceptibility of the proposed enhanced method, we used the similarity measure (Eq. (2)) concerning the cover text and stego cover through the Jaro-Winkler distance, as used in [13]. The similarity metrics were used to measure the similarity between two strings, where (0) indicated a dissimilarity and (1) suggested equal matching or imperceptibility of strings.

$$Jaro\ Winkler\,(S, C) = Jaro\_score + (L * p * (1 - Jaro\_score)) \tag{2}$$

$$Jaro\_score = 1/3 * \left(\frac{m}{length(s1)} + \frac{m}{length(s2)} + \frac{(m-t)}{m}\right) \tag{3}$$

Where m is the number of matching characters, s1 is the first string, s2 is the second string, t is the number of transpositions. First of all, we have applied our methods over twelve secret message samples to embed into the cover file shown Fig. 2. Then we put the results in a table which includes the secret messages, size secret message (n), hiding capacity ratio (CR) and Jaro-Winkler (JW) for the method of two-letter word and the proposed approach based on LZW in order to compare them (see Table 4).

Table 4. Results of previous method (TLW) and the proposed method based on LZW code

Secret messages	n	Two-letter word method		Proposed method using LZW	
		CR %	JW	CR %	JW
The import	10	0.678	0.99	0.721	1
The importance and s	20	1.255	0.98	1.411	1
The importance and size of tex	30	1.750	0.98	2.073	0.99
The importance and size of text data hav	40	2.181	0.97	2.633	0.99
The importance and size of text data have increase	50	2.559	0.96	3.209	0.99
The importance and size of text data have increased at an ac	60	2.893	0.95	3.778	0.98

(*continued*)

Table 4. (*continued*)

Secret messages	n	Two-letter word method		Proposed method using LZW	
		CR %	JW	CR %	JW
The importance and size of text data have increased at an accelerating	70	3.191	0.95	4.294	0.98
The importance and size of text data have increased at an accelerating pace beca	80	3.457	0.94	4.80	0.98
The importance and size of text data have increased at an accelerating pace because the re	90	3.698	0.94	5.297	0.98
The importance and size of text data have increased at an accelerating pace because the reliance on	100	3.915	0.93	5.726	0.97
The importance and size of text data have increased at an accelerating pace because the reliance on text based	110	4.114	0.93	6.264	0.97
The importance and size of text data have increased at accelerating pace because the reliance on text based web01234.	117	4.245	0.92	6.507	0.997
Average		2.828	0.953	3.893	0.983

As shown in Table 4 the empirical findings indicate that the proposed method based on Lempel-Ziv-Welch coding can be applied to the secret message to embed in the cover file. The hiding capacity became 3.893% in comparison with two-letter word technique 2.828%. At the same time, the Jaro-Winkler similarity metric grew 0.983 in comparison with two-letter word technique 0.953.

> The loss of tree cover as a result of forests being cleared for other land uses such as farming and logging is called deforestation. Deforestation activities affect carbon fluxes in the soil, vegetation, and atmosphere. However, logging can also lead to carbon emissions if the surrounding trees and vegetation are damaged. Deforestation is defined as the destruction of forested land. It is a major problem all over the world. The causes of deforestation vary from place to place. The most common causes, however, are logging, agricultural expansion, wars, and mining. Deforestation has been the cause of many problems facing the world today such as erosions, loss of bio diversity through extinction of plant and animal species, and increased atmospheric carbon dioxide. In this paper, I will present that we can reduce deforestation by moving from physical letter mail to electronic mail. From the ancient era physical letter mail has come, till now it is going on, but, on the other side due to this everyday trees are being cut i.e., deforestation is taking place by the paper industry, hence increasing CO2 emission and global warming. In place of physical letter mail, we can use electronic mail which will definitely do some reduction in deforestation. There are critical effects of deforestation.

Fig. 2. Cover text-1

In order to compare the performance of our proposed approach based on LZW code with other related studies in the literature in terms of capacity, we assumed the secret message (see Fig. 3) and the cover file (see Fig. 4). Therefore, the proposed method satisfies the hiding capacity requirement. Table 5 compares the capacities of the proposed method and existing techniques. The proposed scheme achieved 12.02% capacity in comparison with that in [8] which produced only 6.92% for the same secret message and cover text given in Figs. 3 and 4. The proposed scheme also performed better than the method in [10] devised by Rajeev et al. for the same cover text and the secret message given in Figs. 3 and 4. By using Eq. (1), the capacity was calculated as 12.02% for this example.

Table 5. A compression of hiding capacity ratio

Refs.	C %	Explanation
[8]	6.92	Evaluated by utilizing the cover text and sample secret message of Figs. 3 and 4 of this article
[10]	7.03	Evaluated by utilizing the cover text and sample secret message of Figs. 3 and 4 of this article
Proposed method	12.02	Evaluated by utilizing the cover text and sample secret message of Figs. 3 and 4 of this article

behind using a cover text is to hide the presence of secret messages the presence of embedded messages in the resulting stego text cannot be easily discovered by anyone except the intended recipient

Fig. 3. Secret message

in the research area of text steganography, algorithms based on font format have advantages of great capacity, good imperceptibility and wide application range. However, little work on steganalysis for such algorithms has been reported in the literature. based on the fact that the statistic features of font format will be changed after using font-format-based steganographic algorithms, we present a novel support vector machine-based steganalysis algorithm to detect whether hidden information exists or not. this algorithm can not only effectively detect the existence of hidden information, but also estimate the hidden information length according to variations of font attribute value. as shown by experimental results, the detection accuracy of our algorithm reaches as high as 99.3% when the hidden information length is at least 16 bits.

Fig. 4. Cover file-2

In general, the proposed method can hide 4 bits of the compressed secret message by using one location of the two-letter word; moreover, it can conceal two times the number of bits of the two-letter word.

Figure 5 below presents the hiding capacity ratio and a comparison between our proposed algorithm and previous methods listed in Table 5.

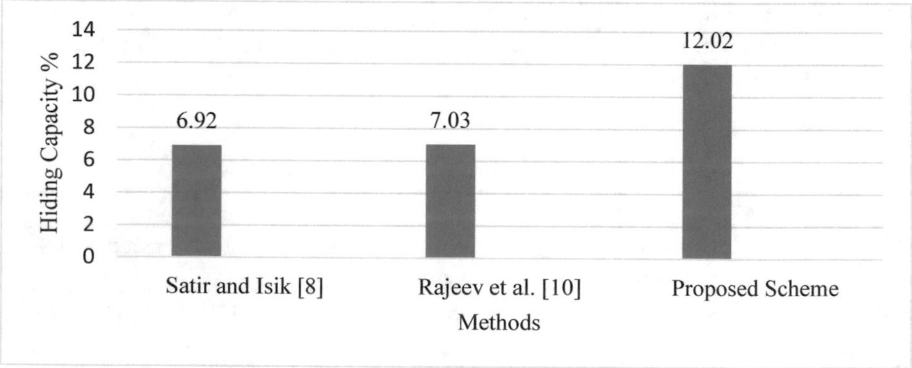

Fig. 5. A comparison between three different algorithms

5 Conclusion and Future Work

In this work, an enhancement of the two-letter word technique has been introduced. The proposed scheme compresses the secret message using the LZW algorithm. This scheme enhances the two-letter word technique such that 4 bits can be embedded at each time. In addition, it achieved a better compression ratio than existing methods did. Experimentally, this work improved data hiding capacity and security by adopting a stego key that is generated by the secret message and locations of a two-letter word that is available in the cover text. Moreover, the proposed technique produces a small size increase ratio between the original and stego documents. This scheme is expected to apply to any language. The limitation of the scheme is the user-defined code designed to English text only. Future work should focus on optimizing the highest capacity and security of the algorithm by combining the proposed scheme with data compression techniques, such as LZW and Huffman coding, to address the increase in the stego file size.

References

1. Obeidat, A.A.: Arabic text steganography using unicode of non-joined to right side letters. J. Comput. Sci. **13**, 184–191 (2017)
2. Bhattacharjee, A.K., Bej, T., Agarwal, S.: Comparison study of lossless data compression algorithms for text data. IOSR J. Comput. Eng. **11**, 15–19 (2013)
3. Baawi, S.S., Mokhtar, M.R., Sulaiman, R.: New text steganography technique based on a set of two-letter words. J. Theor. Appl. Inf. Technol. **95**, 6247–6255 (2017)

4. Vidhya, P.M., Paul, V.: A method for text steganography using malayalam text. Proc. Comput. Sci. **46**, 524–531 (2015)
5. Satir, E., Isik, H.: A Huffman compression based text steganography method. Multimed. Tools Appl. **70**, 2085–2110 (2014)
6. Khalil, I.: Highly imperceptible and reversible text steganography using invisible character based codeword. In: Twenty First Pacific Asia Conference on Information Systems, Langkawi, p. 230 (2017)
7. Por, L.Y., Wong, K., Chee, K.O.: UniSpaCh: a text-based data hiding method using unicode space characters. J. Syst. Softw. **85**, 1075–1082 (2012)
8. Satir, E., Isik, H.: A compression based text steganography method. J. Syst. Softw. **85**, 2385–2394 (2012)
9. Wang, F., Huang, L., Chen, Z., Yang, W., Miao, H.: A novel text steganography by context-based equivalent substitution. In: 2013 IEEE International Conference on Signal Processing, Communication and Computing (ICSPCC), pp. 1–6 (2013)
10. Kumar, R., Chand, S., Singh, S.: An email based high capacity text steganography scheme using combinatorial compression. In: 25th Internal Conference Confluence The Next Generation Information Technology Summit (Confluence), 25–26 September, pp. 336–339. IEEE (2014)
11. Malik, A., Sikka, G., Verma, H.K.: A high capacity text steganography scheme based on LZW compression and color coding. Eng. Sci. Technol. Int. J. **20**, 72–79 (2017)
12. Kapoor, S., Chopra, A.: A review of Lempel Ziv compression techniques. Int. J. Comput. Sci. Technol. **4**, 246–248 (2013)
13. Kingslin, S., Kavitha, N.: Evaluative approach towards text steganographic techniques. Indian J. Sci. Technol. **8**, 79–86 (2015)

Acquiring RFID Tag Asymmetric Key from IOT Cyber Physical Environment

Muhammad Thariq Abdul Razak, Nurul Azma Abdullah(✉), and Nurul Hidayah A. Rahman

Information Security Interest Group (ISIG), Faculty of Computer Sciences & Information Technology, Universiti Tun Hussein Onn Malaysia (UTHM), 86400 Parit Raja, Johor, Malaysia
{hi160058,azma,hidayahar}@uthm.edu.my

Abstract. Radio Frequency Identification (RFID) is the example of current technology that enable the IoT environment to identify and locate objects and record metadata. RFID is the typical, important technology that first creates the term of IoT and most recent technology discussed. Since the close relationship of the evolution between technology and crime, the need of understanding RFID data is inevitable. There is some researcher discussed of IoT forensic but there is no specific work related to the RFID data acquisition standard procedure in IOT environment. Therefore, this research is to propose a methodology for acquisition of RFID tag asymmetric key for IoT forensic purpose. Acquisition is the initial step in IoT forensic to acquire digital evidence from IoT cyberphysical environment. Later, the key acquired will be used to extract useful information from RFID tag memory for further investigation.

Keywords: Radio frequency identification (RFID) · Internet of Thing (IoT) IoT forensic · RFID tag · Asymmetric keys

1 Introduction

In a world of Internet of Thing (IoT), everything is moving to the cloud using the internet as a link in communication, saving, retrieving and processing data. IoT refers to the use of intelligently connected devices to the system to leverage data gathered by embedding sensors in machines and other physical objects [1]. According to [1–3], IoT is the paradigm that considers presence in the environment of a variety object through wired and wireless connection. "Things" refers to the different devices that can communicate each other by exchanging the data through the internet. IoT consist physical smart devices that combined with sensor, software and network connectivity to collect and analysis data to produce the useful information [4]. RFID is one of the current technology that support and prerequisite to the IoT environment. The aim of RFID in the IoT environment is to collect, identify objects based on tag attached to the object. An RFID tag can store many types of data such as name, address and identification number. The data is stored in the RFID tag memory with the specific storage size and protected by two asymmetric keys.

F. Saeed et al. (Eds.): IRICT 2018, AISC 843, pp. 538–547, 2019.
https://doi.org/10.1007/978-3-319-99007-1_50

In this regard, critical security and privacy issue will be raised such as physical attacks, malicious code injection and data tempering. In fact, in 2016 there were Distributed Denial of Service (DDoS) attacks to the DNS provider through a botnet consisting a lot of IoT devices such as RFID and NFC technology [5]. With this issue in IoT technology, it is worth to consider the possible forensics analysis. Inside the RFID tag, there has a specific information that might help during an IoT forensic investigation. Thus, this research will focus on the IoT forensic acquisition of evidence from an RFID tag based on the proposed methodology.

IoT forensic is one of the example of digital forensic branches that execute digital forensics procedures in the IoT environment [6]. IoT forensic acquisition is the initial step procedure to acquire the digital evidence. Since IoT devices connected to the internet for sharing data, then there have several problems faced by the forensic expertise such as lack of standard guidelines [7] and approach. Therefore, this research paper proposes a methodology for acquiring RFID asymmetric key for IoT forensic purposes.

2 Background and Related Work

Most of review studies [8–10] defines that, IoT can be classified into three main layer which is Perception layer, Transportation layer and Application layer. Perception layer is the lowest layer in the IoT environment and consist various types of technologies and sensors to collect the data from the physical world. RFID technology is one of the examples of device that operate in the perception layer. All data collected from the perception layer will be transmitted to the transport layer. Then transport layer will send the data to the specific system in the application layer through the communication network such as 3G/LTE and Wi-Fi.

According to [11] classified Perception layer has the highest tendency for the security breaches and cybercrime. This is because the perception layer consists a large number and many types of devices that connect together for collect and transfer data. Previous literature [12] illustrates the main security threat in this layer such as physical attack, DDOS attack and data transit attack. Therefore, the penetrator can implement those attack to gain access to the IoT application system or steal data for the specific purpose. In terms of RFID security attack in IoT environment, there were several attacks related to RFID tag in order to gain the symmetric key A and key B. In a study by [13] proposes a Keystream Recovery Attack to record the communication between the reader and the tag. Besides that, another example of study conducted in [14] develop a ghost device which is a custom programmable RFID tag. The author stores a malicious firmware that can allowing to take control the communication between the reader and tag. Therefore, the data collected from the communication can be used to recover the secret key of the tag. Based on the facts above, the IoT forensic investigation is important for RFID technology to acquire possible information from RFID tag based on proposed methodology to find any possible evidence related to the specific crime.

Mifare Classic is one of the contactless card used by the RFID technology and owned by the NXP Semiconductor. According to the NXP, more than 200 million

Mifare card in use around the world and covering more than 80% of the contactless smart card market. NXP offers several types of contactless card and each type consists different size of memory and structure. In this research, we will focus on Mifare classic 1k because in term of cost and widely used by the organization. Mifare classic 1k consist 1024 bytes of data and split into 16 sectors and each sector consist 4 blocks [15]. Each sector consists 16 bytes of storage and secured by the two-different symmetric key which is key A and key B. Those keys are used to authenticate the reader before gain access to the specific sector for the specific operation such as read or write function [15].

Previous works in IoT forensics were focusing on reviewing the challenges and research opportunities, conducting empirical studies, and proposing the specific IoT digital forensics frameworks. The embedded IoT functionality that includes on-board sensors and IoT devices' logs has opened challenges in a digital forensics area specially to acquire and analyze the evidence artefacts. Most of review studies [16–20] suggested the need of specific procedures in IoT forensics. It has been pointed out the lack of IoT forensic tools as most of the existing forensics tools, are customized for conventional computers operating systems and file systems and may not meet the IoT devices heterogeneous requirements such as RFID technology.

3 Proposed Acquisition Methodologies

In this section, the proposed methodology to acquire the asymmetric key from the RFID tag are illustrated in Fig. 1. The proposed acquisition methodology consists three phases which is collection of RFID tag phase, Pre-processing phase and Acquisition RFID asymmetric key phase.

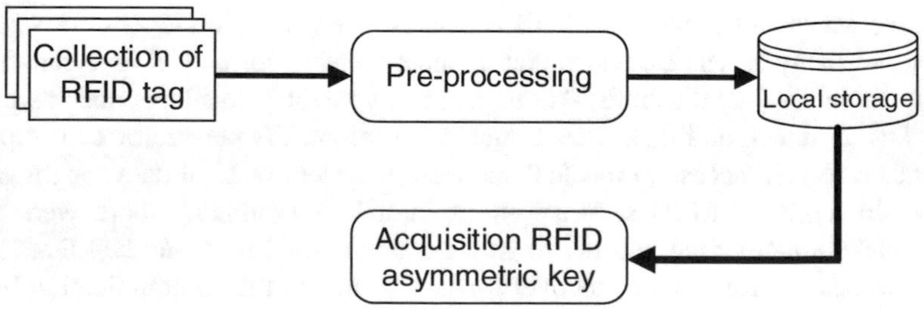

Fig. 1. Propose acquisition methodology

Collection of RFID Tag

In the experiment, a predefined dataset was created which consists of four RFID tags collected and from four different application which is transportation payment application, smart parking application, smart door hotel application and electronic identity application as shown in Table 1.

A transportation payment application such as road tolling and public transport used RFID tag to make a contactless payment. Basically, user needs to reload the RFID tag

Table 1. Dataset of RFID tag

RFID application	Total tag collected
Transportation payment	1 tag
Smart parking	1 tag
Smart door hotel	1 tag
Electronic identity	1 tag

with specific amount of money before it can be used. RFID tag will store the amount of money reloaded by user and RFID tag unique identification number to the RFID tag memory. The second application of RFID is in smart parking application where the tag is used to store the specific data of the car like plate number in the RFID car [21]. The data will be used for verification process before the vehicle is permitted to enter the designated parking. The third application of RFID is in smart door hotel which store specific information like room id for the authentication process [22]. Finally, Electronic identity application is the physical identity card that can be used for online and offline personal identification and authentication for citizen and organization such as campus card. A campus card can store specific information about specific student such as name, identification number and course of study. All these cards are collected to create a dataset for the experiment. Once the dataset is ready, it will undergo pre-processing procedure.

Pre-processing

In this phase, the list of possible keys generated in order to gain access the memory of the RFID tag. As we know, RFID tag consists two passwords or key which is key A and key B and both key stored in the RFID sector of memory for the authentication process. With the valid key, only then the memory of the RFID tag can be accessed to collect the useful evidence for forensic purpose. After the key generation completed, then the list of possible keys will be saved in a text file then stored in local storage. Table 2 shows the example list of possible asymmetric keys of RFID tag.

Table 2. Example of asymmetric key

Example of asymmetric key	
FRQS….WX8 V	NAZ2….N3C8
9YC5….CWXI	7TW5….GEK7
5RCB….0FPV	YDY3….V6NC
9ASY….J82Z	UORN….TQWG
FFFF….FFFF	BV80….30K4

After the generation of possible keys, the next step is to acquire the RFID asymmetric key.

Acquisition RFID Asymmetric Key

In the acquisition phase, forensic expertise will try to find the exact key store in the sector in the RFID tag memory using a specific technique. After the acquisition process done, the key found will be stored to the local storage for the next purpose. the steps to find the exact key for the authentication of RFID tag are shown in Fig. 2.

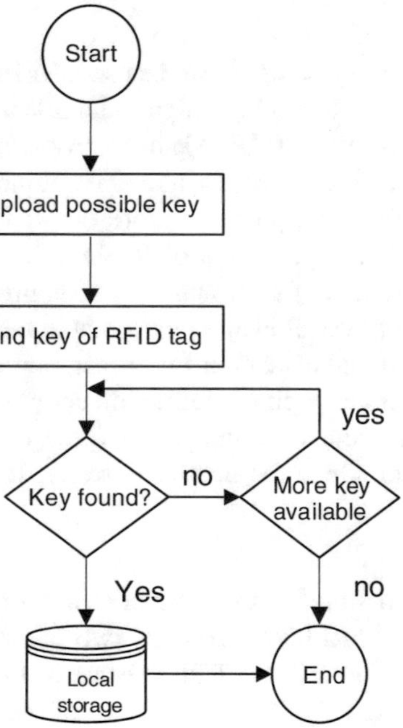

Fig. 2. Workflow to acquire the key of RFID tag

Firstly, forensic expertise needs to upload the list of the possible keys from the local storage to the Mifare Classic Tools (MCT) application. Then, place the RFID tag at the smartphone reader. MCT application will display the tag UID number if the tag is detected successfully. Then, MCT application will try to find a match key with the uploaded text file using a brute force technique. Brute force is the technique to find the exact key or password by trial and error [23]. If the key found, then the key is stored to the local storage for the future action. If the key is not found, then the acquisition of asymmetric key is failed.

3.1 Experimental Setup

Table 3 outlines the tools specification that will be used in the experiment with their function. Predefined dataset will be used in this experiment to acquire the asymmetric key of the RFID tag.

Table 3. Tools specification

Tools	Function
NFC smartphone reader	To read RFID tag
Mifare Classic tool	To find the asymmetric key A and key B for authentication of RFID tag.

In this research, NFC smartphone will be used because according to [24], NFC is based on RFID technology and both operate at 13.56 MHz. Therefore, the HTC Butterfly S smartphone will be use in the experiment. This smartphone will used android operating system and version 4.2.2 (Jelly Bean). Besides that, a Mifare Classic tool (MCT) is an android based application version 2.2.3 will be used to read and acquire the key from the RFID tag sector. Later, the key acquired can be used to gain access to the RFID tag memory to extract useful information for IoT forensic purposes. This application will be installed on the specific smartphone before starting the experiment.

Before carrying out the experiment, the specific smartphone must be in airplane mode in order to prevent the data traffic in and out from the internet and to protect the acquired data from predefined dataset. This follows standard procedures by [25]. After the specific smartphone in airplane mode, forensic investigator will install MCT application. After the installation complete, digital expertise must make sure to enable the NFC adapter in the setting because by default smartphone will disable the NFC function when the airplane mode is running. After the airplane mode is active and the MCT application is installed only then the forensic investigator can start the experiment to acquire the asymmetric key from the collccted dataset based on proposed methodology.

4 Findings

In this section, will be shown the result data that has been acquired from the RFID tag using MCT tools. Table 4 shows the experiment result for acquisition of RFID tag asymmetric key.

Table 4. The result from the experiment for acquisition of RFID tag asymmetric key

RFID application	Tag	UID number	Key A	Key B
Transportation	Tag 1	√		
Smart parking	Tag 2	√	√	√
Smart door hotel	Tag 3	√	√	√
Electronic identity	Tag 4	√	√	√

As we can see from Table 4, unique identification (UID) of all tag is found or can be read by the RFID reader. UID number only consists unique number of the tag. Therefore, UID number can be read by any reader without any encryption key. Besides that, after doing the experiment, only key A and Key B of tag 2, tag 3 and tag 4 can be found but not tag 1. This is because the list of possible keys that been generated do not contain the exact key for tag 1.

5 Discussion

The result from this study suggest that, without the exact asymmetric key the tag memory cannot be accessed to extract the valuable information. Figure 3 shows the summary of acquisition of RFID tag asymmetric key.

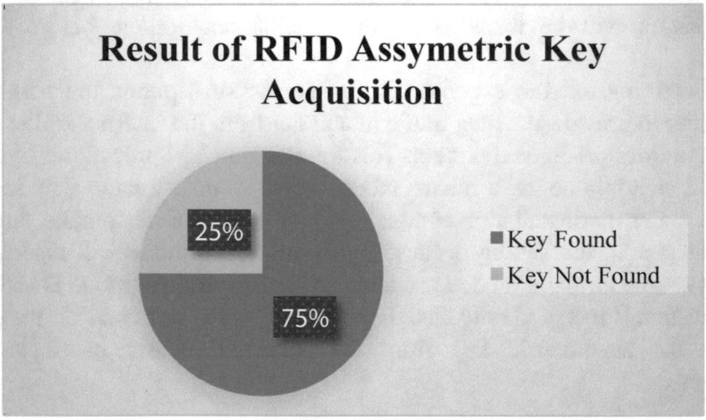

Fig. 3. Result for asymmetric key acquisition

From the pie chart above, we can see that, in order to gain access to the tag memory, forensic expertise must find the exact asymmetric key A and key B. From the result, 75% of the RFID tag dataset key A and key B is found successfully and stored in the local storage. On the other hand, only 25% of tag is failed to acquire the asymmetric key. The reasonable reason why the key is not found because the randomness of key follows the specific standard or policy based on organization requirement. For example, a card issued by the company must have a different secret key differentiated each year to ensure the confidentiality and integrity of data stored. After the asymmetric key is found only then expertise can further the investigation to extract the valuable information from the specific sector of RFID tag. This information could provide additional clues of what when, who and where the crime happens and can be used in court.

Based on Fig. 4, 67% of the RFID tag asymmetric key applies the default key compared to only 33% uses the random key to secure the tag. In normal practice, the manufacturer set the default key A and key B such as FFFF….FFFF to ensure the security of the tag. It depends on the organization that purchased the tag to change the default key to maintain the integrity of data stored in the RFID tag memory. In order to modify the default key, the organization need to gain access to the memory block by using the default key. Only then the user is allowed to change the default key A and key B. If the default key is remained unchanged, the attacker can gain access to the tag memory by using default key and manipulate the data stored for the specific purposes such as tag cloning and data alteration. However, in other hand when user did not change the default key, it will benefit the forensic investigator which ease the tasks to be undertaken.

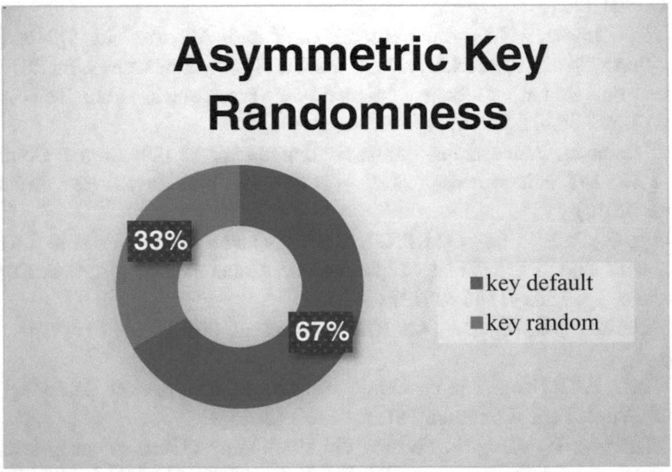

Fig. 4. Result for asymmetric key randomness

6 Conclusion

The advancement in IoT technology has become an attraction to human and organization to transmit and receive data over the internet with the shortest time. RFID is one example of technology that support IoT environment to collect the data from the physical world and transmit the data to the cyber world via the internet connection. As long as RFID is placed in the physical area where human can touch and see it, then RFID has a high probability for the cybercrime attack. The purpose of this research is to develop a methodology to acquire the asymmetric key from the RFID tag for the IoT forensic investigation. The key acquired later can be used to gain access of RFID tag memory to find valuable information for the forensic analysis purpose.

Acknowledgement. The authors express appreciation to the Ministry of Higher Education (MOHE) and Universiti Tun Hussein Onn Malaysia (UTHM). This research is supported by FRGS grant (Vot 1640). The author thanks to the anonymous viewers for the feedback. The author also would like to thank Universiti Tun Hussien Onn Malaysia for supporting this research.

References

1. GSMA Association: Understanding the internet of things (IoT). Gsma Connect. Living, no. July, p. 15 (2014)
2. Gershenfeld, N., Krikorian, R., Cohen, D.: The internet of things. Sci. Am. **291**(4), 76–81 (2004)
3. Xu, J., Yao, J., Wang, L., Ming, Z., Wu, K., Chen, L.: Narrowband internet of things: evolutions, technologies and open issues. IEEE Internet Things J. **PP**(99), 1 (2017)
4. Mitchell, S., Villa, N., Stewart-Weeks, M., Lange, A.: The internet of everything for cities. Connect. People, Process, Data, Things to Improv. 'Livability' of Cities Communities, Cisco, pp. 1–21 (2013)
5. Khan, W.Z., Zangoti, H.M., Aalsalem, M.Y., Zahid, M., Arshad, Q.: Mobile RFID in internet of things: security attacks, privacy risks, and countermeasures. In: 2016 International Conference on Radar, Antenna, Microwave, Electronics, and Telecommunications (ICRAMET), pp. 36–41 (2016)
6. Liu, J.: IoT forensics issues strategies and challenges. In: 12 IDF Annual Conference (2015)
7. Rughani, P.H.: IoT evidence acquisition – issues and challenges. Res. India Publ. **10**(5), 1285–1293 (2017)
8. Jia, X., Feng, Q., Fan, T., Lei, Q.: RFID technology and its applications in Internet of Things (IoT). In: 2012 2nd International Conference on Consumer Electronics, Communications and Networks, pp. 1282–1285 (2012)
9. Rose, K., Eldridge, S., Lyman, C.: The internet of things: an overview. Internet Soc. 53 (2015)
10. Hota, J., Sinha, P.K.: Scope and challenges of internet of things: an emerging technological innovation. Wirel. Pers. Commun. **58**(1), 46–69 (2015)
11. Frustaci, M., Pace, P., Aloi, G., Fortino, G.: Evaluating critical security issues of the IoT world: present and future challenges. IEEE Internet Things J. **4662**(c) (2017)
12. Frustaci, M., Pace, P., Aloi, G.: Securing the IoT world: issues and perspectives. In: 2017 IEEE Conference on Standards Communications and Networking, no. table I, pp. 246–251 (2017)
13. Kamesh, S.P.N., Priya, S.: Security enhancement of authenticated RFID generation. Int. J. Appl. Eng. Res. **9**(22), 5968–5974 (2014)
14. Morbitzer, M.: The MIFARE Hack. no. Ccc
15. Garcia, F.D., Van Rossum, P., Verdult, R., Schreur, R.W.: Wirelessly pickpocketing a Mifare Classic card. In: Proceedings - IEEE Symposium on Security and Privacy, pp. 3–15 (2009)
16. Oriwoh, E., Jazani, D., Epiphaniou, G., Sant, P.: Internet of things forensics: challenges and approaches. In: Proceedings of the 9th IEEE International Conference on Collaborative Computing: Networking, Applications and Worksharing, pp. 608–615 (2013)
17. Watson, S., Dehghantanha, A.: Digital forensics: the missing piece of the internet of things promise. Comput. Fraud Secur. **2016**(6), 5–8 (2016)
18. Hossain, M.M., Fotouhi, M., Hasan, R.: Towards an analysis of security issues, challenges, and open problems in the internet of things. In: 2015 IEEE World Congress on Services, pp. 21–28 (2015)
19. Wang, Y., Uehara, T., Sasaki, R.: Fog computing: issues and challenges in security and forensics. In: 2015 IEEE 39th Annual Computer Software and Applications Conference, vol. 3, pp. 53–59 (2015)

20. Grispos, G., Glisson, W.B., Storer, T.: Recovering residual forensic data from smartphone interactions with cloud storage providers. In: The Cloud Security Ecosystem, pp. 347–382. Elsevier (2015)
21. Joshi, Y., Gharate, P., Ahire, C., Alai, N., Sonavane, S.: Smart parking management system using RFID and OCR. In: 2015 International Conference on Energy Systems and Applications, pp. 729–734 (2015)
22. Farooq, U., ul Hasan, M., Amar, M., Hanif, A., Usman Asad, M.: RFID based security and access control system. Int. J. Eng. Technol. **6**, 309–314 (2014)
23. Patil, A.: A multilevel system to mitigate DDoS, brute force and SQL injection attack for cloud security. In: International Conference on Information, Communication Instruments Control (ICICIC – 2017) (2017)
24. Asaduzzaman, A., Mazumder, S., Salinas, S.: A security-aware near field communication architecture. In: 2017 International Conference Networking, Systems Security (2017)
25. Valjarevic, A., Venter, H., Petrovic, R.: ISO/IEC 27043:2015 - role and application. In: 24th Telecommunications Forum, TELFOR 2016 (2017)

SDN/NFV-Based Moving Target DDoS Defense Mechanism

Chien-Chang Liu, Bo-Sheng Huang, Chia-Wei Tseng, Yao-Tsung Yang, and Li-Der Chou(✉)

Department of Computer Science and Information Engineering, National Central University, Taoyuan, Taiwan
c341cd@yahoo.com.tw, 104522093@cc.ncu.edu.tw, cwtseng@g.ncu.edu.tw, yao.vct@gmail.com, cld@csie.ncu.edu.tw

Abstract. The rapid development of internet technology makes the hacker's attack more mature and diversified. One of the most serious security problems is distributed denial of service (DDoS) attack. In order to cope with information security issues, a new form of defensive thinking, moving target defense (MTD), has been proposed. The emergence of new network architecture software defined network (SDN) and network function virtualization (NFV) - has also changed the future of network security schemes. In this paper, an SDN/NFV-based moving target DDoS defense mechanism using multiple fuzzy systems and a proxy virtual network function (VNF) is proposed to achieve DDoS detection and mitigation. The experimental results show that the proposed mechanism can redirect suspicious traffic and quarantine it, thereby shifting the attack surface.

Keywords: Distributed denial of service · Moving target defense
Software defined network · Network function virtualization · Fuzzy logic

1 Introduction

With the development of internet technology, the hackers' attacks have become diversified and can be switched to different distributed denial of service (DDoS) attacks. How to effectively detect DDoS attacks and mitigate them is an important research topic [1, 2]. Software definition network (SDN) [3] separates the control plane of the network decision from the data plane of transferred packets. The control layer uses SDN controllers to make dynamic adjustments by acquiring the status of the entire network and managing a programmable network. Network function virtualization (NFV) virtualizes network function hardware devices into programs to provide the flexibility and scalability required by managers and to reduce the cost of network physical equipment procurement and network construction maintenance costs [4].

The purpose of a moving target defense (MTD) is to constantly change the system information (e.g. IP address, communication port, routing, or service source) to delay the attacker's detection and probe scheduling [5, 6]. Fuzzy logic is widely used in automatic control, decision analysis and network pattern analysis, e.g., DDoS detection

F. Saeed et al. (Eds.): IRICT 2018, AISC 843, pp. 548–556, 2019.
https://doi.org/10.1007/978-3-319-99007-1_51

[7, 8]. The proxy harvest attack problem involves an attacker collecting all available relay node lists by continuously requesting services; if a certain percentage of the relay node list is collected and the relay node is attacked, the relay node's MTD mechanism will disintegrate. Previous research has been unable solve a proxy harvest attack problem effectively. In this paper, we design and implement a moving target DDoS defense mechanism based on SDN/NFV, and utilize the programmability of SDN and the flexibility and scalability of NFV for DDoS detection and defense. The contribution of this paper is the design and implementation of a moving target DDoS defense mechanism to solve a proxy harvest attack problem in the MTD mechanism. The rest of the paper is organized as follows: Sect. 2, the background and related works are addressed. Section 3 describes the system module and operation flow of each module. Section 4 presents experiment results. The last section concludes.

2 Background and Related Works

A denial of service attack (DoS) is a type of cyber attack. Its purpose is to make the attacked server unable to provide services to normal users. The goal of a distributed denial of service attack (DDoS) is the same. The biggest difference from DoS is that DDoS attacks come from more than one source. DDoS attacks can be divided into two types: bandwidth consumption and resource consumption. Bandwidth consumption blocks the server by continuously sending a large number of packets, such as user datagram protocol (UDP) flooding. Resource consumption uses the characteristics of the network protocol to occupy the target server's computing resources or memory resources, such as transmission control protocol (TCP) synchronized (SYN) flooding. As DDoS attacks continue to evolve, defensive tactics continue to change, such as traffic cleansing: traffic is directed to the DDoS traffic cleaning center [9], which separates malicious traffic from normal traffic.

The purpose of MTD is to reduce the advantage of the attacker by employing a network system architecture in which multiple dimensions are constantly changing. On study [10] has pointed out that MTD is a mechanism that increases the complexity and cost of attacks by creating, analyzing and deploying diversified, time-varying mechanisms. On study [11] has proposed Network Address Space Randomization (NASR) to prevent network worm attacks by using DHCP update frequency to randomize network addresses. On study [12] uses the communication frequency hopping to perform frequency hopping for the TCP/UDP port used by the service.

In the SDN network architecture, the communication channel between the control and data layers is called the SDN control to data-plane interface (CDPI). There are a variety of communication protocols, the most commonly used of which is the OpenFlow protocol [13]. The SDN centralizes the control plane as an SDN controller, so that the controller can grasp the global topology of the entire network and flexibly control the network packets with programmable features. NFV virtualizes network function hardware devices as virtual network functions (VNF). The framework of NFV was proposed by the European Telecommunications Standards Institute (ETSI) [14]. Fuzzy theory was first proposed by American scholar Zadeh in 1965 [15]. It allows computers to efficiently handle ambiguous states in the real world. In general, the operation flow of fuzzy logic can be divided into four parts: fuzzification, fuzzy database, fuzzy inference and defuzzification [16]. In [17], Iyengar et al. use fuzzy logic to detect DDoS traffic and set a fuzzy system at the border of an area network. Packets are discarded when attack traffic is detected. In [18], Trung et al. combines the concept of hard threshold and SDN to provide more flexible control of network packets.

Previous work [19] has proposed an MOTAG mechanism to use the hidden proxy server node group to perform relay transmission. The DOSE mechanism in [20] also makes use of the concept of relay nodes, using the content delivery network (CDN) node group for relay transmission and using puzzle authentication to allow users to obtain node information.

3 Methodology

The system architecture is divided into four parts: SDN data plane, SDN control plane, SDN application plane and reverse proxy VNF pool. The SDN data plane is an OpenFlow switch. The SDN control plane is a module group for controlling the data plane and implementing network manager software applications. The SDN application plane is used to place software applications written by network managers to achieve a programmable network architecture and is the focus of the defense mechanism for building MTD. There are many reverse proxy VNFs in the reverse proxy VNF pool to provide relay transmission between users and service servers as shown in Fig. 1.

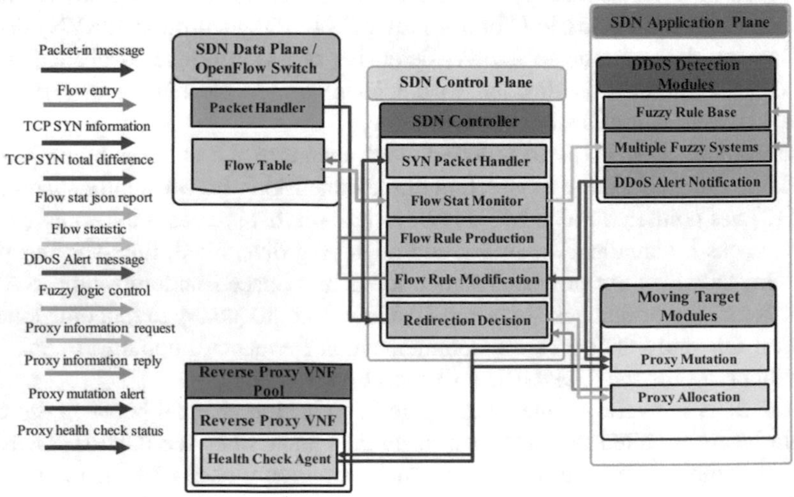

Fig. 1. System architecture.

The primary goal of the defense mechanism proposed in this paper is to combine fuzzy logic detection and moving target defense to mitigate DDoS attacks. It uses a fuzzy logic analysis system to detect whether traffic from an external network is DDoS traffic. In addition, to prevent DDoS attack traffic from directly affecting the target server, NFV technology is used to implement a virtual relay node as the base of the moving target defense method. Reverse proxy VNF is used at a relay transmission between the user and the target server, thereby transferring the DDoS attack surface, isolating suspicious traffic from normal traffic, and redirecting suspicious traffic to a restricted node with SDN technology so that normal users can use the service without affecting the system under attack.

The operational flow chart of the DDoS attack detection mechanism in this paper is shown in Fig. 2.

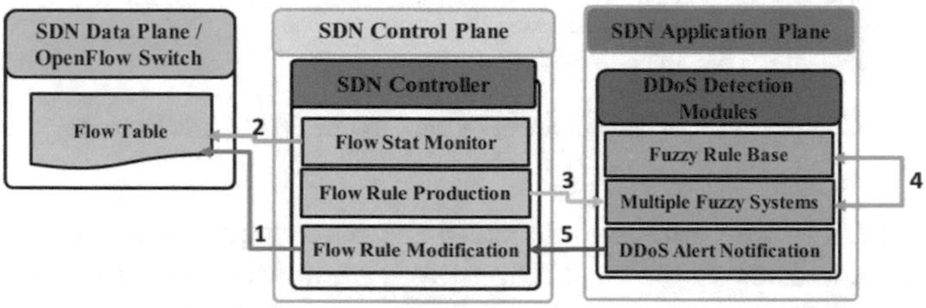

Fig. 2. The operational flow chart of the DDoS attack detection mechanism.

1. When the OpenFlow switch is initialized, the OpenFlow controller will take the initiative to create a default Flow Entry with an empty Match Field.
2. The flow stat monitor module will periodically submit a FlowStat Request and receive the FlowStat Reply returned by the OpenFlow switch.
3. The flow stat monitor module sends the traffic characteristics to multiple fuzzy systems module for fuzzy decision.
4. The fuzzy rules in the fuzzy rule base are used for fuzzy inference and analyze the possibility that the Flow Entry is a DDoS.
5. The DDoS alert notification module will perform corresponding processing according to the probability. If the probability is high, the flow rule modification module is notified to modify Flow Entry to discard the Flow packet.

The operational flow chart of the MTD mechanism is shown in Fig. 3.

1. When the service request packet from the user reaches the OpenFlow switch and Flow Entry has not yet been established, it is first sent to the SDN controller SYN packet handler for SYN flooding detection.
2. After detecting that the user is a normal user, reverse proxy VNF is requested from the proxy allocation module through the redirection decision module.
3. The redirection decision instructs the flow production module to issue Flow Entry and directs the packet to the reverse Proxy VNF.
4. After the DDoS detection module analyzes the possibility of one-item Flow Entry for DDoS in the OpenFlow switch, the DDoS alert notification module will perform corresponding processing according to the probability that it is a DDoS attack.
5. If there is a possibility that Flow Entry is a DDoS, the DDoS detection module will notify the proxy mutation module and the flow modification module in the SDN controller. The flow modification module will set the Match Field to match the Flow

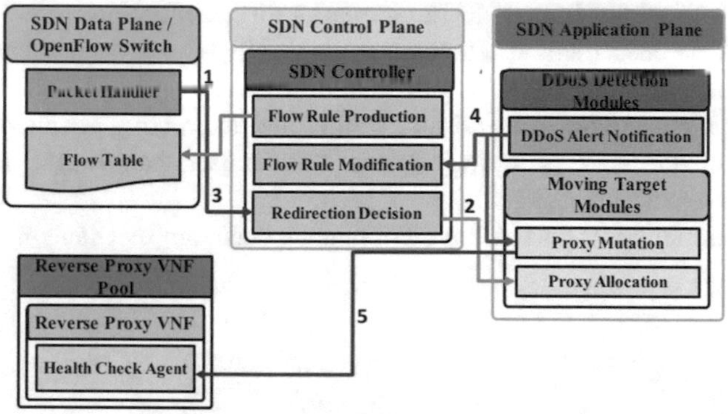

Fig. 3. The operational flow chart of the MTD mechanism.

Entry, and the Action Field will be set to the port where the restricted proxy VNF resides. This isolates suspicious traffic from normal user traffic so that normal users and the proxy they serve can continue to operate unaffected.

The system uses an OpenFlow controller VM and an OpenFlow switch to forward network packets. The back end consists of two Xen servers acting as reverse proxy VNF pools and webservers providing web services. It is responsible for providing reverse proxy VNF and web services. The user who wants to obtain the service will be directed to the reverse proxy VNF and the reverse proxy VNF is responsible for the relay transmission.

4 Experiments and Discussions

In this section, the performance of the proposed DDoS defense mechanism based on SDN/NFV and a moving target defense scheme will be compared with MOTAG [19]. The MOTAG mechanism uses a hidden proxy server node group to perform relay transmission. When a proxy server node is affected by a DDoS attack, the system can redistribute normal users served by the proxy server node to other nodes. The detection rate is defined as the percentage of correctly detection instances (TP + TN)/(TP + TN + FP + FN), where TP, FN, FP and TN represent the number of true positives, false negatives, false positives and true negatives, respectively. The experiments are designed with a Pica8 OpenFlow switch and Ryu controller. In the absence of a SYN flooding detection mechanism, TCP SYN packets are directed to the target service server. In the case of using a SYN flooding detection mechanism, SYN packets are not directly forwarded, but instead the time difference of retransmissions is calculated. If the time difference of retransmissions is too short, SYN packets will be identified as SYN flooding, and therefore these SYN packets will not reach the backend network as shown in Fig. 4.

This experiment uses Linux's built-in kernel module: pktgen (Packet Generator) as the source of traffic for DDoS attacks. In the DDoS detection mechanism proposed in this paper, the Flow Stat Monitor requests and returns once per second, so there will be

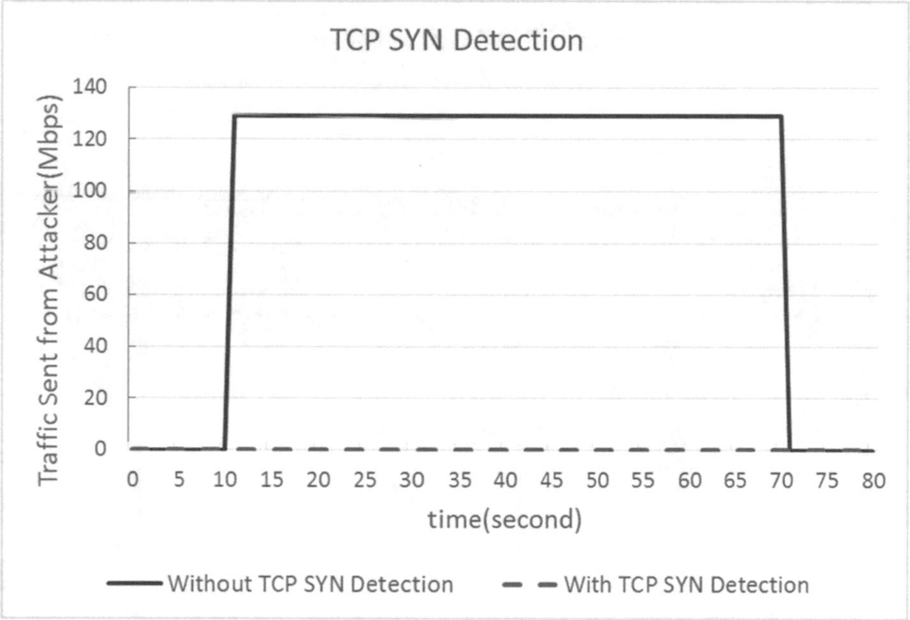

Fig. 4. SYN flooding detection traffic comparison.

a difference of about one second between an attacker launching, and detection of a DDoS. The results show that the detection rate is 99.4% as shown in Fig. 5, which is 12.4% higher than in a previous study [17].

MOTAG [19] is to distribute proxy IP to users through authentication. The number of attempts needed for the attacker to collect all the IP under the random distribution mechanism will vary due to the probability. With an increase in the number of proxies, attackers need to spend more time for proxy IP collection. When the number of relay nodes is five, it takes an average of 11 attempts to collect all the proxy IPs. A comparison chart of IP statistics collected by MOTAG and the mechanism of this paper is shown in Fig. 6. In the MOTAG mechanism, as the number of attempts needed for the attacker to collect all the IP increases, the number of proxy IPs collected increases. However, because the network architecture of this paper is an SDN architecture, it is possible to modify the packet content and redirect it when the packet passes through the OpenFlow switch. Therefore, the redirected operation is on the server side, not the user side, and the user cannot obtain the proxy IP information. The attacker should not be able to obtain proxy IP information through continuous authentication, and proxy harvest attack cannot be successful against in the mechanism proposed in this paper.

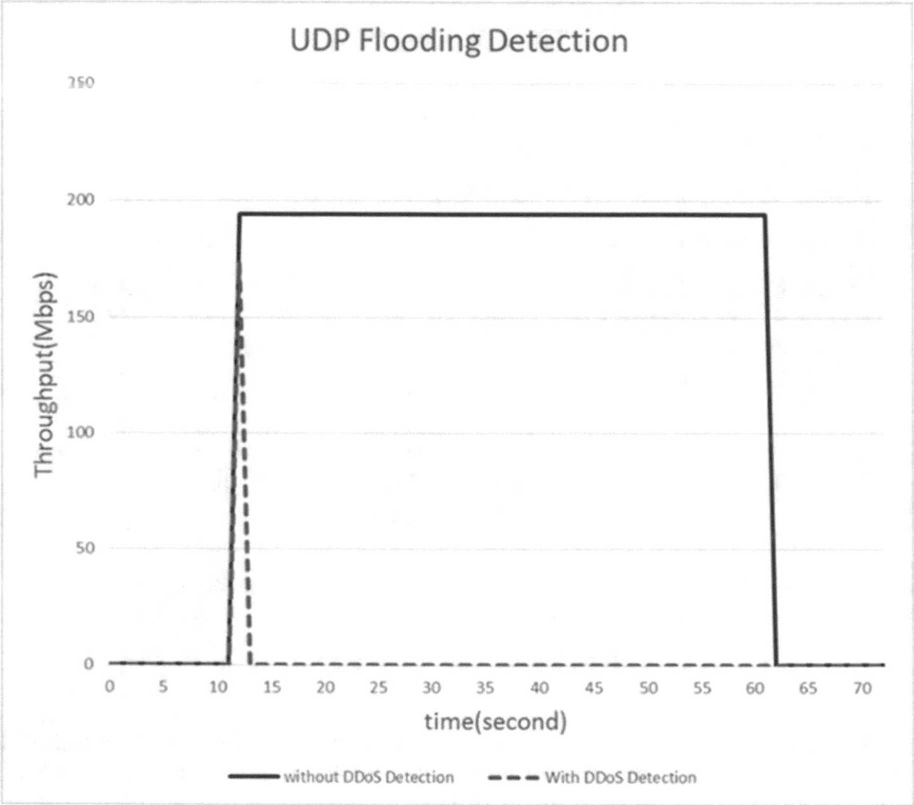

Fig. 5. UDP flooding detection traffic comparison.

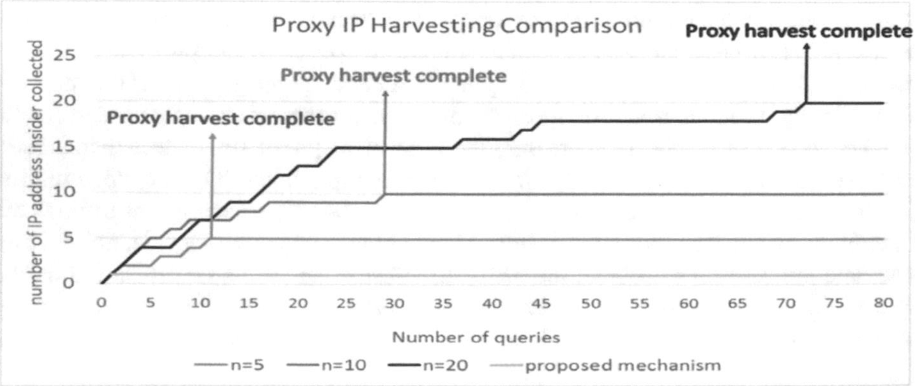

Fig. 6. The number of proxy IPs which collected (proxy number n = 5, 10, 20).

5 Conclusions

In this paper, a DDoS defense mechanism based on SDN/NFV and moving target defense is proposed, which uses the fuzzy decision system and threshold to detect DDoS attack, uses OpenFlow switch's Flow Stat request/return mechanism to effectively obtain the latest network traffic information, and use the fuzzy decision system to detect flooding attack. This is the first paper to propose the SDN/NFV concept implementing the relay node moving target defense mechanism. When an attack is not occurring, the reverse proxy VNF can be used for relay transmission to prevent attack traffic from directly attacking the target server to achieve the effect of transferring the attack surface. When an attack occurs, the mechanism proposed in this paper can use SDN's modified packet technology to isolate suspicious traffic or drop Flow Entry to discard packets. Experimental results show that with the DDoS defense mechanism proposed in this paper, the detection rate of UDP flooding attacks reaches 99.4%.

Acknowledgment. The work described in this paper was supported in part by Ministry of Science and Technology of the Republic of China (Project No. 104-2221-E-008-039-MY3, 105-2221-E-008-071-MY3 and 107-2623-E-008-002-D).

References

1. Ali, S.H.A., Ozawa, S., Ban, T., Nakazato, J., Shimamura, J.: A neural network model for detecting DDoS attacks using darknet traffic features. In: 2016 International Joint Conference on Neural Networks (IJCNN), pp. 2979–2985 (2016)
2. Fuertes, W., Morales, M., Aules, H., Toulkeridis, T.: Software-based computing platform as an experimental topology assembled to detect and mitigate DDoS attacks using virtual environments. In: 2016 International Symposium on Performance Evaluation of Computer and Telecommunication Systems (SPECTS), pp. 1–8 (2016)
3. Mcdysan, D.: Software defined networking opportunities for transport. IEEE Commun. Mag. **51**(3), 28–31 (2013)
4. Wikipedia: Network Function Virtualization (2017). https://en.wikipedia.org/wiki/Network_function_virtualization. Accessed 27 May 2019
5. Hong, J.B., Kim, D.S.: Assessing the effectiveness of moving target defenses using security models. IEEE Trans. Dependable Secure Comput. **13**(2), 163–177 (2016)
6. Bhop, S.K., Dongre, N.M.: Study of dynamic defense technique to overcome drawbacks of moving target defense. In: 2015 International Conference on Information Processing (ICIP), pp. 637–641 (2015)
7. Zhuang, R., Zhang, S., Bardas, A.G., DeLoach, S.A., Ou, X., Singhal, A.: Investigating the application of moving target defenses to network security. In: 2013 6th International Symposium on Resilient Control Systems (ISRCS), pp. 162–169 (2013)
8. Zhuang, R., Bardas, A.G., Deloach, S.A., Ou, X.: A theory of cyber attacks: a step towards analyzing MTD systems. In: CCS 2015 MTD Workshop, pp. 11–20 (2015)
9. Cisco: Cisco DDos Protection Solution-Delivering "Clean Pipes" Capabilities for Service Providers and Their Customers. Cisco Systems White Paper, pp. 4–16 (2016)
10. Guilin, C., Baosheng, W., Tianzuo, W., Yuebin, L., Xiaofeng, W., Xinwu, C.: Research and development of moving target defense technology. J. Comput. Res. Dev. **53**(5), 1 (2016)

11. Antonatos, S., Akritidis, P., Markatos, E.P., Anagnostakis, K.G.: Defending against hitlist worms using network address space randomization. In: 2005 ACM workshop on Rapid malcode (WORM 2005), pp. 30–40 (2005)
12. Lee, H.C.J., Thing, V.L.L.: Port hopping for resilient networks. In: Proceedings of 60th IEEE Vehicular Technology Conference, 2004-Sept (2004)
13. McKeown, N., Anderson, T., Balakrishnan, H., Parulkar, G., Peterson, L., Rexford, J., Shenker, S., Turner, J.: OpenFlow: enabling innovation in campus networks. SIGCOMM Comput. Commun. Rev. **38**(2), 69–74 (2008)
14. Network Functions Virtualisation (NFV); Architectural Framework. ETSI GS NFV 002 v1.2.1 (2014)
15. Zadeh, L.A.: Fuzzy sets. Inf. Control **8**, 338–535 (1996)
16. Seising, R., Trillas, E., Moraga, C., Termini, S.: On Fuzziness: A Homage to Lotfi A. Zadeh. Springer, Heidelberg (2013)
17. Iyengar, N.Ch.S.N., Banerjee, A., Ganapathy, G.: A fuzzy logic based defense mechanism against distributed denial of services attack in cloud environment. Int. J. Commun. Netw. Inf. Secur. **6**(3), 233–245 (2014)
18. Trung, P.V., Huong, T.T., Tuyen, D.V., Duc, D.M., Thanh, N.H., Marshall, A.: A multi-criteria-based DDoS-attack prevention solution using software defined networking. In: 2015 Advanced Technologies for Communications, pp. 308–313 (2015)
19. Jia, Q., Sun, K., Stavrou, A.: MOTAG: moving target defense against internet denial of service attacks. In: 22nd International Conference on 2013 Computer Communications and Networks (ICCCN), Nassau, Bahamas (2013)
20. Wood, P., Gutierrez, C., Bagchi, S.: Denial of Service Elusion (DoSE): keeping clients connected for less. In: IEEE 34th Symposium on 2015 Reliable Distributed Systems (SRDS), Montreal, Canada (2015)

Biometric Based Signature Authentication Scheme for Cloud Healthcare Data Security

Gunasekar Thangarasu[1(✉)], P. D. D. Dominic[2(✉)], Kayalvizhi Subramanian[1(✉)], and Sajitha Smiley[1(✉)]

[1] Linton University College, Mantin, Negeri Sembilan, Malaysia
{dr.t.gunasekar,kayalvizhi,sajitha}@ktg.edu.my
[2] University Technology PETRONAS, Seri Iskandar, Perak, Malaysia
dhanapal_d@utp.edu.my

Abstract. Cloud Computing is a recent and fastest growing area of development in healthcare. Cloud-based healthcare management services are considered as an effective way to manage the data related to healthcare. However, the major problem associated with the healthcare management service is data security. It leads to tax, bank, insurance and medical fraudulence. Hence, retrieval of data onto more secured access is required to improve the security of medical records of healthcare services. In this research, proposed a biometric based signature authentication scheme with neural network for cloud healthcare data security. The neural network is used in this scheme for the accuracy of retrieving the clinical data onto the cloud. The neural network acquires biometric signature through biometric sensor processed with quality checker for effective authentication. This network also has support in terms of statistical learning of the clinical datasets. After various experiments, it is concluded that the proposed method provides faster results with higher sensitivity, specificity with accuracy.

Keywords: Biometric · Healthcare · Neural network · Data security
Hadoop MapReduce

1 Introduction

In recent years, advancement in technological innovation paves a path of the growth of computing infrastructure like cloud computing. Cloud computing offers a significant platform for users with its eminent infrastructure and lend software to an end user based on lease. Nowadays, cloud computing is emerging from healthcare with professional empowerment and effective management of medical data for treatment of illness [1]. Recently, Personal Health Record (PHR) derived a patient-centric model using cloud technology for health information logistic. It provides a tool for the correlation between patients and healthcare providers by maintaining their medical history. PHR allows patients to manage their individual health records of cloud server [2].

Besides, the healthcare service, Electronic Health Record (EHR) is considered as an effective approach to improve the quality of healthcare services in maintaining patient's health records. EHR improves the quality of treatment by providing safe patient data access, improving efficiency and reducing delivery cost in healthcare. Patient's EHR

F. Saeed et al. (Eds.): IRICT 2018, AISC 843, pp. 557–565, 2019.
https://doi.org/10.1007/978-3-319-99007-1_52

medical data are stored in digital format on the server. The primary purpose of the EHR is to provide patient details to the health care providers or clinicians. Alternately, EHR provides communication between clinicians and patients regarding medical care and processing. EHR has the ability with the full potential for delivering high quality services in affordable cost of interoperability fundamental technology [3]. When compared to paper records the EHR also maintains a huge amount of medical records which make use of performance in aspects like efficiency, instant updating and accuracy [4].

The primary focus of this research is to provide data security in the cloud based medical data sharing. Data security refers to any form of safeguard tools to protect healthcare data onto unauthorized access in terms of human, technological and administration. The fundamental information security parts are protection, privacy, honesty and accessibility. Securing medical images can be more challenging compared to other aspects of electronic protected health information as they are often transmitted between providers and clinicians. Data security is about the assurance and cautious utilization of the individual data of patients.

1.1 Problem Statement

Medical data theft is a serious issue of healthcare sector. Health data are the most vulnerable, easy and tempting target for hackers. According to a survey conducted by Ponemon Institute, 2.32 million Americans have been victims of medical data theft [5]. Damages occurring from medical data theft are graver than any other types of thefts like a credit card or bank ATMs. The bank can be reported on the theft and immediate remedial actions may be taken. But in the health sector, data theft fraud detection is rather tricky and slow. Health data have become a very tempting target for frauds due to a number of reasons. It is very easy to access health data compared to any other kind of financial data due to lack of security. Further, this information is used by the adversaries for creating fraud, credit card accounts and file fake tax returns. In addition, the health information can be misused to purchase drugs or medical equipment illegally or even to claim false medical care [6, 7].

1.2 Objectives

The objective of this study is to protect healthcare data from accessing unauthorized parties by proposing a biometric based signature authentication scheme for cloud healthcare data security using neural networks. The biometric signature from people is examined with quality checker for evaluating quality and processed with the Hadoop MapReduce framework of effective authentication. The biometric system helps to access more accurate health information about security as well as accountability.

1.3 Research Motivation

Biometric technique is increasing being used for medical monitoring and mobile health care. The growing demand for biometrics solutions is mainly driven by the need to combat fraud, along with the imperative to improve patient privacy along with healthcare safety. This has motivated the researcher to study about biometric authentication in healthcare management system.

2 Related Study

Advances for cyber-physical security have accomplished rapid development in the course of recent decades. There are mounting evidences that they are vital in biometric-based access control. Nowadays the fingerprints, footprints, pistol bullets with their ballistic traits, yet in addition hair, voice, blood, semen, DNA and the fibber from garments are generally employed in crime detection. Some of this crime detection can be used in biometrics [8, 9]. Due to advancement in healthcare services, several IT applications and infrastructure requires regular security. Utilization of this IT services in the healthcare system is very expensive and this causes the healthcare organization to enforce their IT expenses on patients. Some healthcare organizations have purchased or developed their own IT system for supporting operations. In the case of small-medium sized healthcare organizations perform operations either manually or paper-based since the implementation of IT equipment's are very expensive. This diversification in healthcare organizations leads to difficulty in accessing patients' medical data and information [10]. Due to the complexity of the patient's data retrieval, the healthcare providers require a platform to distribute computing efficiency in data sharing and effective software with minimum cost [11]. In later years, cloud computing technology has been developed with the objective of offering the increased capability of services and minimal cost. In 2006, Amazon Web Services (AWS) has been launched. In 2007, google and IBM started their research for large scale utilization of cloud technology. In addition, Microsoft also announced their cloud system in 2009 termed as Windows Azure. Cloud health data storage and access are a highly complicated service, even though the public is not much aware about it [12, 13]. Lack or incomplete knowledge among public leads to security threats like data theft, manipulation of information and fail to proper diagnosis. In order to preserve patient's medical data securely, medical information system needed to be properly integrated with security architecture.

3 Research Methodology

This researches, proposed a biometric signature authentication scheme for healthcare data management security. The proposed scheme uses Hadoop MapReduce framework for biometric signature processing. The biometric signature from doctor or healthcare providers and patients are obtained via signature capturing software installed in mobile or other electronic devices. Capturing of biometric signature from people is examined with quality checker for evaluating quality of signature and processed with the Hadoop MapReduce framework. Biometric signature is processed under two phases which are enrollment phase and authentication phase. In enrollment phase, the patients' medical data onto their biometric signature will be collected. In authentication phase, Hadoop MapReduce framework processes the acquired biometric signature and stores them in parallel [14]. The MapReduce framework is processed using the neural network of testing and training scenario to increase computation cost and convergence rate. When a user requires access to particular medical data they need to provide a biometric signature with the neural network fed biometric signature is verified if biometric signature matched with the stored data than the person is provided with access otherwise they need to enroll themselves. Figure 1 shows the flow of Biometric signature authentication scheme.

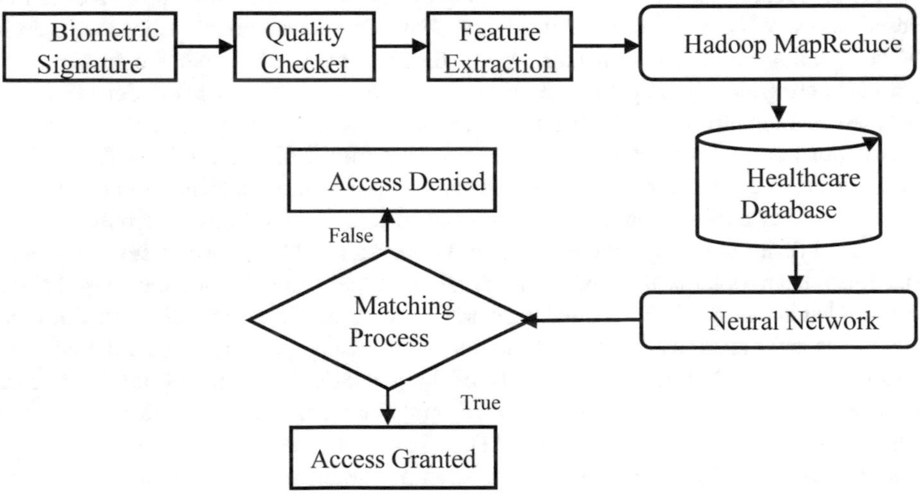

Fig. 1. Flow of biometric signature authentication scheme

3.1 Healthcare Data Processing with Hadoop MapReduce

The healthcare data is process with neural network of signature authentication. Data is trained in neural network based on the amount of data stored in the healthcare database via Hadoop MapReduce framework. The signature prediction for template is necessary to analyze data center history for authentication. Biometric signature pattern recorded for medical data is processed and recorded using the Hadoop MapReduce framework of individual pattern. An extracted signature pattern of data is used to forecast and predict when authentication is required for particular medical data. In this research, the researchers considered 'n' as previously recorded biometric signature sample using MapReduce environment. To predict the biometric signature sample of accessing the medical data linear mathematical model is given in Eq. (1) as follows [15].

$$y^{\Delta}_{t+1} = a_1 \times x_1 + a_2 \times x_2 + \ldots . + a_n \times x_n \tag{1}$$

In this x_i is an actual biometric signature, t is signature load and a_i defines the importance of x_i in predicting signature y^{Δ}_{t+1}.

3.2 Process of Neural Network

The neural network is one of the intricate mathematical problems solving model and it provides tremendous effectual and supportive contrivance in distinct areas, mainly in medical diagnosis, prognosis and therapy. The network models must be properly formulated and trained in using good quality and reliability data [16]. The neural networks are made of preliminary processing components known as neurons. The neurons are intersected though synaptic weights establishing the network. In common, the biological neural networks can accommodate many inputs in parallel and encode the data onto an appropriated manner. It will endure the partial destruction of its structure and still be ready to perform well [17].

3.3 Cloud for Healthcare Data

The Healthcare services with cloud computing platform enable users for significant modeling on data onto prominent sharing and scalability for detailed analysis without additional investment in computational infrastructure [18]. Cloud computing environment with the EMR application model is hosted internally for multi tenant's subscription within the single database application.

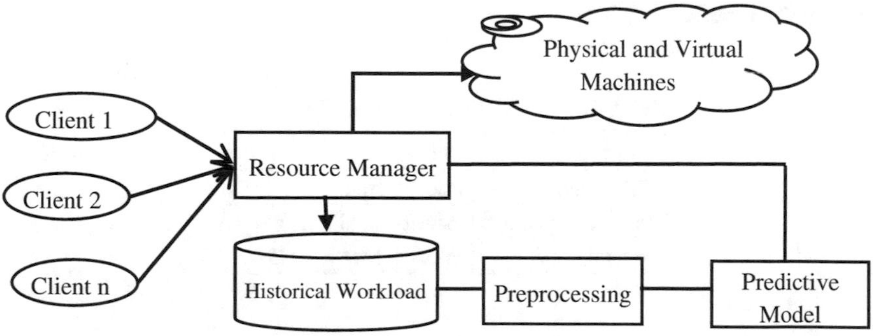

Fig. 2. Flow of medical data processing

The above Fig. 2 illustrates generic healthcare data processing in the cloud. The resource manager receives the data onto cloud to perform pre-processing workload and maintenance.

4 Process of Biometric Signature Authentication

4.1 Healthcare Cloud Biometric Signature in Neural Network

The Biometric signature with Neural Network (BIOsig_NN) schemes is used for biometric signature authentication to improve healthcare data security. The BIOsig_NN stores the medical data in the cloud platform with a biometric signature to ensure the security of healthcare data onto the cloud. The step in the proposed scheme contains the biometric signature processes and quality checking, MapReduce for data storing and neural network for biometric signature verification. In appropriate biometric traits several factors are involved in processing such as application-specific, environment, identification verification, user profile, accuracy, throughput, convergence rate and overall system cost. The biometric signature from doctor or healthcare providers and patients are obtained via signature capturing software installed in mobile or other electronic devices. Capturing of biometric signature from people is examined with quality checker for evaluating quality of signature and processed with the Hadoop MapReduce framework. The MapReduce framework is processed using neural network of testing and training scenario to increase computation cost and convergence rate. When a user requires access to particular medical data they need to provide biometric signature with the neural network fed biometric signature is verified if biometric signature matched with the stored data than the person is provided with access else they need to enroll themself. Based on the expert guidance healthcare biometric data are processed using High 'H', medium 'M', and low 'L'. Through the NN cloud computing convergence and computation rate are reduced. Table 1 shows the data sets used for MapReduce process.

Table 1. Created data sets using MapReduce

User type	Priority	Value
VIP patients	Priority high	H
Privileged patients	Priority medium	M
Regular staff	Priority low	L

4.2 Flow of Biometric Signature in Neural Network

As mentioned earlier the biometric authentication system contains two phases namely enrollment module and authentication module [12]. In Enrollment module, the biometric signature is obtained using a suitable biometric signature reader. The data processed using Hadoop MapReduce framework of extracted features. The characteristic of Biometric features is shown in Table 2. In authentication module uses user claims uniqueness of biometric signature data onto appropriate technique for accuracy improvisation.

Table 2. Characteristic of biometric features

Biometric identifier	Maturity	Accuracy	Uniqueness	Failure-to-enroll rate	Record size	Universality	Durability
Face	M	M	M	L	H 84 - 2,000	H	M
Finger print	H	H	M	L - M	M 250 - 1,000	M	M
Hand	M	L	L	L	L 9	M	M
Iris	M	M	H	L	M 688	M	H
Signature	L	L	M	L	M 500 - 1,000	M	M
Vascular	M	M	H	L	M 512	H	H
Voice	L	L	M	M	H 1,500 - 3000	H	L

5 Performance Evaluation

This research uses smartphones, phablets, tablets and different handheld devices for data acquiring. Biometric signature samples were collected from 300 people that involve both patients and staff. Among those, 200 were patient's biometric signature and 100 are staff biometric signature. For critical information processing to comply ethics are involved in consent participants with clause identity anonymous for various clauses.

For Neural network training purpose input raw data were preprocessed using Hadoop MapReduce framework. Primarily, Hadoop distributed file system are used through Principal Component Analysis (PCA) [17]. MapReduce framework is executed for parallel computation manner. Based on MapReduce framework obtained healthcare data are stated in the Table 3. The medical data contains patient Id, room Id, location, medicine Id and status.

Table 3. Created medical data

Patient ID	Room ID	Location	Medicine ID	Status
457	23	Rack	2342	Room occupied
156	47	Ground floor	4576	Claimed
127	75	Terrace	4848	Rejected

For experimental analysis of proposed approach Hadoop2.x is utilized for testing. In simulation analysis three factors evaluated whether the system is robust against failure of Hadoop cluster execution property robustness of system is evaluated. The comparison of proposed method of traditional methods factors such as True Positive (TP), False Positive (FP), True Negative (TN) and False Negative (FN) are created for analysis. Performance of proposed scheme is evaluated under four parameters such as sensitivity, accuracy, F-measure and precision. Evaluation of proposed scheme is examined with conventional methods like Naïve Bayes, Support Vector Machines (SVM) with Average Inverse Rule Frequency (AIRF), SVM with Genetic Algorithm (GA), and SVM with Fuzzy logics. Performance comparison is performed via both present and absent of feature selection. Among all classifiers accuracy is considered as major factor of top classification process. As shown in Table 4 proposed scheme is comparatively evaluated using SVM-AIRF, SVM-GA, and SVM-Fuzzy.

Table 4. Performance comparison of proposed scheme with existing approaches

Prediction algorithm	Accuracy	F-measure	Precision	Sensitivity
Proposed approach	0.996361	0.881124	0.825497	0.943898
SVM - AIRF	0.995378	0.880592	0.825347	0.943764
SVM - GA	0.996133	0.901165	0.869621	0.935083
SVM - Fuzzy	0.996146	0.90153	0.869147	0.93642

The simulation result analysis exhibits that proposed scheme provides effective performance in terms of sensitivity, F-measure, accuracy and precision which are 0.996361, 0.881124, 0.825497 and 0.943898 respectively. Performance improvement in proposed scheme is obtained via optimization rule and unstructured dataset in larger format. In larger dataset incomplete data are reason for accuracy minimization in proposed scheme incomplete data are handled effectively in testing and training phase. Further it minimizes variance rather than decision tree for decision on classification. Feature selection processes in proposed scheme more effective than conventional method of medical data record process.

6 Conclusion

In this research, proposed a biometric based signature authentication scheme for healthcare data security. The proposed scheme uses the Hadoop MapReduce framework for biometric signature template formation. This research used 300 biometric signature samples acquired from the patients and staff in stratified random sampling method. The collected biometric signature samples were processed using the neural network of improving performance after quality checking. Through various experiments, it has been concluded that the proposed method provides faster results with higher sensitivity and specificity than the existing systems.

References

1. Au, M.H., Yuen, T.H., Liu, J.K., Susilo, W., Huang, X., Xiang, Y., Jiang, Z.L.: A general framework for secure sharing of personal health records in cloud system. J. Comput. Syst. Sci. **90**, 46–62 (2017)
2. Kaur, P.D., Chana, I.: Cloud based intelligent system for delivering health care as a service. Comput. Methods Prog. Biomed. **113**(1), 346–359 (2014)
3. Scott, K., Richards, D., Adhikari, R.: A review and comparative analysis of security risks and safety measures of mobile health apps (2015)
4. Deshmukh, P.: Design of cloud security in the EHR for Indian healthcare services. J. King Saud Univ. Comput. Inf. Sci. **29**(3), 281–287 (2017)
5. Vengadapurvaja, A.M., Nisha, G., Aarthy, R., Sasikaladevi, N.: An efficient homomorphic medical image encryption algorithm for cloud storage security. Procedia Comput. Sci. **115**, 643–650 (2017)
6. Liu, X., Xia, Y., Yang, W., Yang, F.: Secure and efficient querying over personal health records in cloud computing. Neurocomputing **274**, 99–105 (2018)
7. Shini, S.G., Thomas, T., Chithraranjan, K.: Cloud based medical image exchange security challenges. Procedia Eng. **38**, 3454–3461 (2012)
8. Wrobel, K., Doroz, R., Porwik, P., Naruniec, J., Kowalski, M.: Using a probabilistic neural network for lip-based biometric verification. Eng. Appl. Artif. Intell. **64**, 112–127 (2017)
9. Pardamean, B., Rumanda, R.R.: Integrated model of cloud-based E-medical record for health care organizations. In: 10th WSEAS International Conference on e-Activities, pp. 157–162 (2011)
10. Gao, S., Mioc, D., Yi, X., Anton, F., Oldfield, E., Coleman, D.J.: Towards Web-based representation and processing of health information. Int. J. Health Geogr. **8**(1), 3 (2009)
11. Bisbal, J., Berry, D.: An analysis framework for electronic health record systems. Methods Inf. Med. **50**(02), 180–189 (2011)
12. Hu, C.M., Jian, W.S., Chang, P.L., Hsu, C.Y.: A web based prototype system for patient use confirming Taiwan electronic medical-record templates. In: AMIA, Annual Symposium Proceedings. AMIA Symposium, American Medical Informatics Association, pp. 987–997 (2005)
13. Khan, S.H., Akbar, M.A., Shahzad, F., Farooq, M., Khan, Z.: Secure biometric template generation for multi-factor authentication. Pattern Recognit. **48**, 458–472 (2015)
14. Darve, N.R., Theng, D.P.: Comparison of biometric and non-biometric security techniques in mobile cloud computing. In: 2nd International Conference on Electronics and Communication Systems (ICECS), pp. 213–216. IEEE (2015)
15. Babaeizadeh, M., Bakhtiari, M., Mohammed, A.M.: Authentication methods in cloud computing: a survey. Res. J. Appl. Sci. Eng. Technol. **9**, 655–664 (2015)
16. Castelo, A., Kritski, A.L., Werneck, A., Lemos, A.C., Ruffino, N.A.: Diretrizes Brasileiras para Tuberculose. Jornal Brasileiro de Pneumologia **35**(10), 1018–1048 (2009)
17. Kumar, V.P., Velide, L.: A data mining approach for prediction and treatment of diabetes disease. Int. J. Sci. Inven. Today **3**(1), 073–079 (2014)

Novel Risk Assessment Method to Identify Information Security Threats in Cloud Computing Environment

Ganthan Narayana Samy[1(✉)], Sameer Hasan Albakri[1], Nurazean Maarop[1], Pritheega Magalingam[1], Doris Hooi-Ten Wong[1], Bharanidharan Shanmugam[2], and Sundresan Perumal[3]

[1] Advanced Informatics Department, Razak Faculty of Technology and Informatics, Universiti Teknologi Malaysia, Kuala Lumpur, Malaysia
ganthan.kl@utm.my
[2] School of Engineering and Information Technology, Charles Darwin University, Casuarina, Australia
[3] Faculty of Science and Technology, Universiti Sains Islam Malaysia, Nilai, Malaysia

Abstract. Cloud computing model brought many technical and economic benefits, however, there are many security issues. Most of the common traditional information security risk assessment methods such as ISO27005, NIST SP800-30 and AS/NZS 4360 are not fit for the cloud computing environment. Therefore, this study applies medical research approach to assess the information security threats in the cloud computing environment. This study has been conducted as a retrospective cohort study and the collected data has been analyzed by using the survival analysis method. The study has been conducted on the software as a service (SaaS) environment that has more than one thousand and seven hundred cloud customers. The survival analysis method is used to measure the significance of the risk factor level. The information security threats have been categorized into twenty-two categories. This study has proven that the medical research approach can be used to assess the security risk assessment in cloud computing environment to overcome the weaknesses that accompany the usage of the traditional information security risk assessment methods in cloud computing environment.

Keywords: Cloud computing security
Information security risk assessment method
Medical research design and method

1 Introduction

The cloud computing model bargains many economic and functional advantages, for small and medium-sized businesses (SMBs). The economic benefits include but not limited: Low cost, availability of resources, energy savings, and increased focus on business objectives [1–3]. International Data Corporation IDC in 2014 conducted a

F. Saeed et al. (Eds.): IRICT 2018, AISC 843, pp. 566–578, 2019.
https://doi.org/10.1007/978-3-319-99007-1_53

survey on cloud-related topics and published it in April 2015. The survey was conducted on 3,464 organizations across North America, Latin America, Europe, and Asia. In sub-report named 'IDC's European Enterprise Communications Survey' with sample size consisting of 933 interviews, IDC expected that cloud connectivity services market in Western Europe would grow from less than $100 million in 2013 to almost $1 billion by 2019. However, even with this large rate of growth, concerns over security in cloud computing are the main inhibitors of public cloud [4].

Cloud computing adds new challenges the ordinary information security challenges, as its model architecture was designed to outsource the essential services of the IT systems to a third party. Guaranteeing the data confidentiality, integrity, authenticity, auditability, availability and compliance in outsourcing scheme is a difficult task to achieve [5]. Furthermore, cloud computing virtualization environment requires the determination of new risks and the re-evaluation of well-known risks [6]. In addition, introducing multi-tenants or sharing resource services in the virtual environment of cloud computing adds new security challenges [7]. Moreover, cloud computing distinguished characteristics have raised many security risks and make the traditional risk assessment methods unsuitable for cloud computing environment. It is difficult to use most of the common traditional risk assessment methods (such as ISO27005, NIST SP800-30 and AS/NZS 4360) to assess the security risks in cloud computing environment due to its design and structure. These methods are designed for the traditional computer model; thus, it has some assumptions and risk level calculation approaches that are not suitable for the cloud computing model.

Many studies have investigated the similarity between the medical environment and the computing environment. The study has been designed as a retrospective study and the collected data has been analyzed by using the survival analysis method. The second section of this paper review the related work. The research methodology that we used in this study been discussed in Sect. 3. Section 4 presents the suggested categorization by this study for some of the security threats. The obtained results have been presented and discussed in Sect. 5. Finally, Sect. 6 concludes the paper.

2 Related Work

The main features of the cloud computing (i.e. Resource pooling, broad network access, rapid elasticity, on-demand self-service, and measured service) of the cloud computing model [8] have raised several new security risks and call for many past, well defined risks to be re-evaluate and redefine according to the cloud computing model [9]. Several studies have been conducted that define cloud computing risks. In this section, we present some of these studies to address the security risks in cloud computing environment. Munir and Palaniappan (2013), listed some of the potential security threats they found in cloud computing environment. Examples of these threats include changes to business models, abusive use of cloud computing, insecure interfaces and APIs, malicious insiders, shared technology issues and the nature of multi-tenancy nature, data loss and leakage, service hijacking, risk profiling, and identity theft. They also mentioned security attacks such as zombie service injection Man-in-the Middle, Metadata spoofing, Phishing, and Backdoor channel attacks as well as attacks on virtualization, VM Escape, and Rootkit in Hypervisor, [10].

Tanimoto et al. in 2014, covered some cloud computing risks in their research on assessing cloud computing risks. They evaluated twenty-three risks including wrongly used data, data being deleted after cloud service use, regulatory non-compliance by the service provider, and service providers limiting information disclosure [7]. Alruwaili and Gulliver in 2014, listed eight types of the security threats that must be assessed during the security risk assessment process for cloud computing environment. These security threats were hardware failure or errors, software failure or errors, quality of service and policy deviation, compromise of intellectual property (IP), deliberate software or hardware attacks, human error or failure, obsolete technology, and acts of nature [11]. Al-Anzi et al. in 2014, mentioned some of the prominent security threats they found in cloud computing environment. Those threats included, abuse and nefarious use of cloud computing, insecure application programming interfaces, malicious insiders, customer-data manipulation, data loss/leakage, account, service and traffic hijacking, data scavenging and malicious VM creation [12]. Jafarpour and Yousefi in 2016, listed some of the security risk in cloud computing such as difficulty to guarantee data privacy and data integrity, losing of control of data, lack of trust, inadequate of security control, malicious or ignorant tenants, single point of failure due to the sharing services, controls misconfigurations, commingled tenant data, and performance risks [13].

Most of the popular risk assessment standards such as ISO27005, NIST SP800-30 and AS/NZS 4360 are designed with main assumption, which is the organization's assets exist in the organization's data center and the information security risk assessor can grant full access to the information assets by the organization itself [14, 15]. However, the cloud computing model has some distinguishing characteristics that make this assumption unfit for the cloud computing [9]. For example, cloud service provider is not the real owner for the information assets but the cloud customer. There is a sharing for the hardware and software ownership, access and control authorities, and security responsibilities between the cloud service provider and the cloud customers.

3 Research Methodology

Most statistical methods used in medical research have better and more accurate results. For instance, survival analysis is more efficient and provides more accurate results compared to other methods such as neuron network, fuzzy logic, and decision trees [16, 17]. The medical studies can be classified as a primary study which usually conducted be using primary data or as a secondary study which use a secondary data that has been produced by other studies [16, 18]. One of the most popular medical studies approaches is the epidemiological studies, this type of studies focuses on specific population and investigate the patterns and frequencies, and relationship between risk factors. There are four forms of the epidemiological studies: cohort, case control, cross-sectional, and ecological studies [18, 19]. The term 'cohort' in medical studies refers to a part of pre-defined population with common characteristics and the 'cohort study' is a study that depends on the observation as a method for data collection to answer the research questions by selecting appropriate samples [20].

There are two approaches to conduct a cohort study; prospective and retrospective approach. Prospective approach starts the observation (i.e. data collection) at baseline time and follow up to the specific time in the future or until specific condition is satisfied. Retrospective approach starts the observation (i.e. data collection) at baseline time and follow up to the specific time in the past [21]. There are many advantages for this approach such as it is required short time and less expenses because it depends on historical data. Besides, it is efficient to discover new findings based on existing data and it is able to combine the data from different sources [22, 23].

This study has been designed as retrospective study, the historical raw data that collected by our previous study for the information security risks in cloud computing environment [24]. The original study has been conducted on the software as a service (SaaS) environment that has more than one thousand and seven hundred cloud customers. In this study, the information security threats have been categorized into twenty-two categories as explained in the next section. All the collected data has been analyzed by the survival analysis method within R software. The survival analysis method is used to decide the significance of the risks' factors.

4 Threats Categorization

In this study, eighty-one information security threats have been identified during literature review for the previous studies. These information security threats have been grouped into twenty-two categories. These security threats categories include Natural Disasters, Environment, Accidental Accidents, Hardware Problems, Software Problems, Application Design, Human Sabotage, Human Errors, Users Awareness, Unauthorized Access, Unauthorized actions, Security Attack on the Server, Security Attack on the Clients, Application Security Risks, Security Attack on the Administration, Security Attack on the Network, Loss of Communication Services, Loss of Essential Services, Cloud Risks, Organizational Risks, Administration Problems, and Location.

The security threats T08 (Freezing), T09 (Flood), T10 (Climatic phenomenon), T11 (Volcanic phenomenon), T12 (Meteorological phenomenon), and T13 (Seismic phenomenon) have been categorized as 'Natural Disasters'. This category includes all the security threats that beyond the human control and usually have catastrophic effects. This category is equivalent to 'Natural events' category in ISO27005 [25]. The 'Environment' category includes the security threats T02 (Water damage), T03 (Pollution), T06 (Dust), and T07 (Corrosion). This category involves the security threats that related to the environments and its effects can be controlled or limited by human action. This category is equivalent to 'Physical damage' category in ISO27005 [25]. The 'Accidental Accidents' category includes the security threats T01 (Fire), and T04 (Major accident). This category includes the security threats that occurred accidentally such as fire or any physical damage in way that cause partially or completely stop the cloud service provider system. The 'Hardware Problems' category includes the security threats T18 (Equipment failure), and T20 (Equipment malfunction). This category involves the security threats that usually effect on the hardware in the cloud service provider's system without any human interfering such as failure and malfunction.

This category is equivalent to 'Hardware failure or error' category in Alruwaili and Gulliver (2014) study.

The 'Human Sabotage' category involves the security threats T05 (Destruction of equipment or media), T51 (Tampering with hardware), T52 (Tampering with software), T33 (Backups lost, stolen), T35 (Theft of computer equipment), T46 (Theft of media or documents), and T48 (Retrieval of recycled or discarded media). This category includes the damage that my occurred to the hardware in the cloud service provider system because the human interfering. It also involves all the stealing actions that might occur for the cloud service provider assets. The category 'Software Problems' contains the security threats T19 (Software malfunction), T72 (Outdated application software), T28 (Use of counterfeit or copied software), and T47 (Operating System Failure). This category includes the security threats that accompany the software in the cloud service provider system. The 'Application Design' category includes T21 (Saturation of the information system) and T80 (Using Known Vulnerable Components). The 'Application Design' category involves the security threats that accompany the improper software design (i.e. it may work perfect but the security threat come from its design), while the 'Software Problems' category covers the well-known software problems that may accompany any software [26].

The category 'Human Errors' includes the security threats the occurred by human unintentionally such as T56 (Error in use), regardless the level of the user experience. This category is equivalent to 'human error or failure' category in Alruwaili and Gulliver (2014) study. The category 'Users Awareness' involves the security threats T58 (Loss of encryption keys), T59 (Loss authentication keys), and T60 (Lack of user technical expertise). These security threats usually occurred because the lake of the user awareness, it may be happened intentionally or unintentionally. The 'Unauthorized Access' category includes the security threats T26 (Unauthorized use of equipment), T32 (Forging of rights), and T34 (Unauthorized access to premises). This category involves the unauthorized access to the cloud service provider's assets. The 'Unauthorized actions' category contains the security threats T25 (Abuse of rights), T27 (Fraudulent copying of software), T30 (Illegal processing of data), T31 (Denial of actions), and T49 (Disclosure). This category involves any actions that violate the cloud service provider's rules. This category is equivalent to 'Unauthorised actions' category in ISO27005 [25].

The category 'Security Attack on the Server' includes the security threats, T44 (Remote spying), T61 (Distributed denial of service (DDoS)), T62 (Economic denial of service (EDOS)), T63 (Undertaking malicious probes or scans), T64 (Compromise service engine), T67 (Loss or compromise of operational logs), T68 (Loss or compromise of security logs), and T69 (SQL Injection). This category involves the security attacks that might occur against the cloud service provider's system. The 'Security Attack on the Clients' category includes the security attacks on the cloud customers such as T65 (Social engineering attacks) [27]. The 'Application Security Risks' category involves the security threats T50 (Data from untrustworthy sources), T29 (Corruption of data), T75 (Insecure Direct Object References), T78 (Cross-Site Scripting (XSS)), T79 (Cross-Site Request Forgery (CSRF)), and T81 (Un-validated Redirects and Forwards). This category includes the security attacks that targeting the application interface. The category 'Security Attack on the Administration' contains the security attacks that targeting the system administrators, including T55 (Breach of personnel availability), T57 (Administrator's email attack), T70 (Account lockout attack), and T71 (Login brute force attack). The category 'Security Attack on the Network' involves T45 (Eavesdropping), T54 (Interception of compromising interference signals), and T66 (Modifying network traffic). This category include the security attacks that targeting the network and the data traffic within the system network [26]. The 'Loss of Communication Services' category includes the security threats that cause the communications failure such as T17 (Failure of telecommunication equipment). The category 'Loss of Essential Services' involves the security threats that cause losing one of the essential services for the data center such as T14 (Loss of power supply), T15 (Failure of air-conditioning system), and T16 (Failure of water supply system).

The category 'Cloud Risks' includes the security threats that accompany the cloud computing implementation such as T22 (Resource exhaustion), T23 (Isolation failure), T24 (Conflicts between customer hardening procedures and cloud environment), and T74 (Insecure or Ineffective Deletion of Data). The category 'Organizational Risks' involves T36 (Lock-in), T37 (Loss of governance), T38 (Compliance challenges), T39 (Loss of business reputation due to co-tenant activities), T40 (Cloud service termination or failure), T41 (Cloud provider acquisition), T42 (Supply chain failure), and T43 (Risk from changes of jurisdiction). This category includes the security threats that accompany the organization's sustainability, legitimacy and how it manages the cloud infrastructure. This category is equivalent to 'Policy and organizational' in the categories proposed by [27]. The category 'Administration Problems' includes the security threats T73 (Long time for system recovery), T76 (Security Misconfiguration), and T77 (Missing Function Level Access Control). This category involves the security threats that occur because of the administration actions. The last category 'Location' includes the security threats that related to the organization location such as T53 (Position detection).

5 Results and Discussion

This section presents the results that obtained by this study. It also discusses how adapting the medical research design and methods to assess the security risk in cloud computing environment fit for cloud computing security field. There are two main popular data layout for the survival data; basic data layout and counting process [21], in this study, we used counting process (CP). The survival analysis method has been used to analysis the collected data. This section presents the results that obtained from data analysis according to their categorization as discussed in the previous section. Multiple security incidents have been recorded for seventeen security threats, these security threats are T18 Equipment failure, T19 Software malfunction, T22 Resource exhaustion, T28 Use of counterfeit or copied software, T29 Corruption of data, T37 Loss of governance, T49 Disclosure, T56 Error in use, T57 Administrator's email attack, T60 Lack of user technical expertise, T63 Undertaking malicious probes or scans, T64 Compromise service engine, T68 Loss or compromise of security logs, T71 Login brute force attack, T72 Outdated application software, and T73 Long time for system recovery. There are some categories have no recorded security incidents. These categories are Natural Disasters, Environment, Accidental Accidents, Application Design, Human Sabotage, Unauthorized Access, Security Attack on the Clients, Security Attack on the Network, Loss of Communication Services, Loss of Essential Services, and Location.

In the category of hardware problems only T18 Equipment failure has some recorded incidents. As shown in the Table 1 below, it is statistically significant where p-value is less than 0.05 and it has a positive regression coefficient value, which is higher 5.51 higher than zero. Moreover, the hazard ratio value is quite high, 245 which means the group that exposed for the security threats has a potential to get affected, 245% higher than the unexposed group. Usually the security incident of T18 Equipment failure will cause many other security threats.

Table 1. Hardware Problems

ID	Security threats	Regression coefficient	Proportional hazard ratio	Significance
T18	Equipment failure	5.51	246.00	<0.05
T20	Equipment malfunction	NA	NA	NA

In the category of software problems, T19 Software malfunction, T28 Use of counterfeit or copied software, and T72 Outdated application software have recorded incidents. As shown in the Table 2 below, T72 Outdated application software is statistically significant where p-value is less than 0.05 and it has a positive regression coefficient value, which is higher 12.56 higher than zero. Moreover, the hazard ratio value is extremely high, 284900. The T72 Outdated application software leads to many other security threats such as software malfunction and data disclosure. The security threat T19 (Software malfunction) also is statistically significant where p-value is less than 0.05 and it has a positive regression coefficient value, which is higher 3.91 higher than zero. Moreover, the hazard ratio value is 49.73 which means the group that exposed for the security threats has a potential to get affected, 48.73% higher than the unexposed group. The security threat T28 Use of counterfeit or copied software is not statistically significant where p-value is greater than 0.05.

Table 2. Software Problems

ID	Security threats	Regression coefficient	Proportional hazard ratio	Significance
T19	Software malfunction	3.91	49.73	<0.05
T28	Use of counterfeit or copied software	1.31	3.72	>0.05
T47	Operating System Failure	NA	NA	NA
T72	Outdated application software	12.56	284900.00	<0.05

In the category of human errors, the T56 Error in use has some recorded incidents. As shown in the Table 3 below, it is statistically significant where p-value is less than 0.05 and it has a positive regression coefficient value, which is higher 11.61 higher than zero. Moreover, the hazard ratio value is extremely high, 110400 which means the group that exposed for the security threats has a potential to get affected higher than the unexposed group.

Table 3. Human Errors

ID	Security threats	Regression coefficient	Proportional hazard ratio	Significance
T56	Error in use	11.61	110400.00	<0.05

In the category of user awareness only T60 Lack of user technical expertise has some recorded incidents. As shown in the Table 4 below, it is statistically significant where p-value is less than 0.05 and it has a positive regression coefficient value, which is higher 0.65 higher than zero. Moreover, the hazard ratio value is slightly high, 3.15 which means the group that exposed for the security threats has a potential to get affected, 2.15% higher than the unexposed group.

Table 4. Users Awareness

ID	Security threats	Regression coefficient	Proportional hazard ratio	Significance
T58	Loss of encryption keys	NA	NA	NA
T59	Loss authentication keys	NA	NA	NA
T60	Lack of user technical expertise	0.65	1.91	<0.05

In the category of unauthorized actions, T49 Disclosure has some recorded incidents. As shown in the Table 5 below, it is not statistically significant where p-value is equal 0.05, even it has a positive regression coefficient value, which is higher 1.15 higher than zero. Moreover, the hazard ratio value is slightly high, 1.91 which means the group that exposed for the security threats has a potential to get affected, 0.91% higher than the unexposed group.

Table 5. Unauthorized actions

ID	Security threats	Regression coefficient	Proportional hazard ratio	Significance
T25	Abuse of rights	NA	NA	NA
T27	Fraudulent copying of software	NA	NA	NA
T30	Illegal processing of data	NA	NA	NA
T31	Denial of actions	NA	NA	NA
T49	Disclosure	1.15	3.15	0.05

In the category of security attack on the server, only T63 Undertaking malicious probes or scans, T64 Compromise service engine, and T68 Loss or compromise of security logs have recorded incidents. As shown in the Table 6 below, T64 Compromise service engine is statistically significant where p-value is less than 0.05 and it has a positive regression coefficient value, which is higher 6.80 higher than zero. Moreover, the hazard ratio value is extremely high, 899.70. The occurrence of T64 Compromise service engine leads to many security consequences. The security threat T63 Undertaking malicious probes or scans also is statistically significant where p-value is less than 0.05 and it has a positive regression coefficient value, which is higher 2.25 higher than zero. Moreover, the hazard ratio value is 9.51 which means the group that exposed for the security threats has a potential to get affected, 8.51% higher than the unexposed group. The security threat T68 Loss or compromise of security logs is not statistically significant where p-value is greater than 0.05.

Table 6. Security Attack on the Server

ID	Security threats	Regression coefficient	Proportional hazard ratio	Significance
T44	Remote spying	NA	NA	NA
T61	Distributed denial of service (DDoS)	NA	NA	NA
T62	Economic denial of service (EDOS)	NA	NA	NA
T63	Undertaking malicious probes or scans	2.25	9.51	<0.05
T64	Compromise service engine	6.80	899.70	<0.05
T67	Loss or compromise of operational logs	NA	NA	NA
T68	Loss or compromise of security logs	1.34	3.81	>0.05
T69	SQL Injection	NA	NA	NA

In the category of application security risks, T29 Corruption of data has some recorded incidents. As shown in the Table 7 below, it is not statistically significant where p-value is greater than 0.05, even it has a positive regression coefficient value, which is higher 6.69 higher than zero. Moreover, the hazard ratio value is extremely high, 801.10 which means the group that exposed for the security threats has a potential to get affected.

Table 7. Application Security Risks

ID	Security threats	Regression coefficient	Proportional hazard ratio	Significance
T29	Corruption of data	6.69	801.10	>0.05
T50	Data from untrustworthy sources	NA	NA	NA
T75	Insecure Direct Object References	NA	NA	NA
T78	Cross-Site Scripting (XSS)	NA	NA	NA
T79	Cross-Site Request Forgery (CSRF)	NA	NA	NA
T81	Un-validated Redirects and Forwards	NA	NA	NA

In the category of security attack on the administration, T57 Administrator's email attack, and T71 Login brute force attack have recorded incidents. As shown in the Table 8 below, T57 Administrator's email attack is statistically significant where p-value is less than 0.05 and it has a positive regression coefficient value, which is higher 9.86 higher than zero. Moreover, the hazard ratio value is extremely high, 19210. T71 Login brute force attack is statistically significant where p-value is less than 0.05, and it has a positive regression coefficient value, which is higher 1.90 higher than zero. Moreover, the hazard ratio value is slightly high, 6.66 which means the group that exposed for the security threats has a potential to get affected, 5.66% higher than the unexposed group.

Table 8. Security Attack on the Administration

ID	Security threats	Regression coefficient	Proportional hazard ratio	Significance
T55	Breach of personnel availability	NA	NA	NA
T57	Administrator's email attack	9.86	19210.00	<0.05
T70	Account lockout attack	NA	NA	NA
T71	Login brute force attack	1.90	6.66	<0.05

In the category of cloud risks only T22 Resource exhaustion has some recorded incidents. As shown in the Table 9 below, it is statistically significant where p-value is less than 0.05 and it has a positive regression coefficient value, which is higher 5.09 higher than zero. Moreover, the hazard ratio value is quite high, 161.90 which means the group that exposed for the security threats has a potential to get affected, 160.90% higher than the unexposed group.

Table 9. Cloud Risks

ID	Security threats	Regression coefficient	Proportional hazard ratio	Significance
T22	Resource exhaustion	5.09	161.90	<0.05
T23	Isolation failure	NA	NA	NA
T24	Conflicts between customer hardening procedures and cloud environment	NA	NA	NA
T74	Insecure or Ineffective Deletion of Data	NA	NA	NA

In the category of organizational risks only T37 Loss of governance has some recorded incidents. As shown in the Table 10 below, it is statistically significant where p-value is less than 0.05 and it has a positive regression coefficient value, which is higher 1.62 higher than zero. Moreover, the hazard ratio value is slightly high, 5.06 which means the group that exposed for the security threats has a potential to get affected, 4.06% higher than the unexposed group.

Table 10. Organizational Risks

ID	Security threats	Regression coefficient	Proportional hazard ratio	Significance
T36	Lock-in	NA	NA	NA
T37	Loss of governance	1.62	5.06	<0.05
T38	Compliance challenges	NA	NA	NA
T39	Loss of business reputation due to co-tenant activities	NA	NA	NA
T40	Cloud service termination or failure	NA	NA	NA
T41	Cloud provider acquisition	NA	NA	NA
T42	Supply chain failure	NA	NA	NA
T43	Risk from changes of jurisdiction	NA	NA	NA

In the category of administration problems application security risks, T73 Long time for system recovery has some recorded incidents. As shown in the Table 11 below, it is not statistically significant where p-value is greater than 0.05, even it has a positive regression coefficient value, which is higher 4.14 higher than zero. Moreover, the hazard ratio value is high, 62.71 which means the group that exposed for the security threats has a potential to get affected.

Table 11. Administration Problems

ID	Security threats	Regression coefficient	Proportional hazard ratio	Significance
T73	Long time for system recovery	4.14	62.71	>0.05
T76	Security Misconfiguration	NA	NA	NA
T77	Missing Function Level Access Control	NA	NA	NA

6 Conclusion

This study used the medical approaches to assess the security risks in the cloud computing environment. The study has been designed as retrospective study and the collected data has been analyzed by using the survival analysis method. The collected data were analyzed using R statistical analysis software, which identified a list of information security risks. The regression coefficient and proportional hazard ratio has been calculated by using survival analysis method. The regression coefficient and proportional hazard ratio give an estimation for the future potential occurrence for each security threat. This study confirms that the medical research design and method can be adapted into cloud computing environments to overcome the weaknesses of the traditional risk assessment methods.

Acknowledgements. The authors would like to thank Universiti Teknologi Malaysia (UTM) for supporting this work through the Tier 1 GUP Grant Scheme under Grant vote number Q.K130000.2538.14H18.

References

1. Amini, A., et al.: A fuzzy logic based risk assessment approach for evaluating and prioritizing risks in cloud computing environment. In: International Conference of Reliable Information and Communication Technology. Springer (2017)
2. Li, J., Li, Q.: Data security and risk assessment in cloud computing. In: ITM Web of Conferences. EDP Sciences (2018)
3. Ali, K.E., Mazen, S.A., Hassanein, E.E.: Assessment of cloud computing adoption models in e-government environment. Int. J. Comput. Intell. Stud. **7**(1), 67–92 (2018)
4. Bakkers, J.H., Eibisch, J.: Cloud Connectivity Services in Europe in Industry Developments and Models. International Data Corporation IDC (2015)
5. Xuan, Z., et al.: Information security risk management framework for the cloud computing environments. In: 10th IEEE International Conference on Computer and Information Technology (CIT 2010), Bradford (2010)
6. Fito, J.O., Macias, M., Guitart, J.: Toward business-driven risk management for cloud computing. In: 2010 International Conference on Network and Service Management (CNSM), Niagara Falls. IEEE (2010)
7. Tanimoto, S., et al.: A study of risk assessment quantification in cloud computing. In: 2014 International Conference on Network-Based Information Systems, Salerno (2014)

8. Mell, P., Grance, T.: The NIST definition of cloud computing. NIST Spec. Publ. **800**(145), 7 (2011)
9. Zissis, D., Lekkas, D.: Addressing cloud computing security issues. Future Gener. Comput. Syst. **28**(3), 583–592 (2012)
10. Munir, K., Palaniappan, S.: Framework for secure cloud computing. Int. J. Cloud Comput. Serv. Archit. **3**(2), 21–35 (2013)
11. Alruwaili, F.F., Gulliver, T.A.: Safeguarding the cloud an effective risk management framework for cloud computing services. Int. J. Comput. Commun. Netw. (IJCCN) **4**(3), 6–16 (2014)
12. Al-Anzi, F.S., Yadav, S.K., Soni, J.: Cloud computing: security model comprising governance, risk management and compliance. In: International Conference on Data Mining and Intelligent Computing (ICDMIC), New Delhi (2014)
13. Jafarpour, S., Yousefi, A.: Security Risks in Cloud Computing: A Review (2016)
14. Almorsy, M., Grundy, J., Ibrahim, A.S.: Collaboration-based cloud computing security management framework. In: 2011 IEEE International Conference on Cloud Computing (CLOUD), Washington, DC (2011)
15. Zhao, G.: Holistic framework of security management for cloud service providers. In: 2012 10th IEEE International Conference on Industrial Informatics (INDIN), Beijing. IEEE (2012)
16. Samy, G.N.: Analysing information security threats in healthcare information systems using survival analysis method. Faculty of Computer Science and Information Systems Universiti Teknologi Malaysia (2012)
17. Ma, Z., Krings, A.W.: Competing risks analysis of reliability, survivability, and prognostics and health management (PHM). In: 2008 IEEE Aerospace Conference. IEEE (2008)
18. Röhrig, B., et al.: Types of study in medical research: part 3 of a series on evaluation of scientific publications. Deutsches Arzteblatt Int. **106**(15), 262–268 (2009)
19. Allen, L.A., Horney, J.A.: Methods: study designs in disaster epidemiology. In: Disaster Epidemiology, pp. 65–74. Elsevier (2018)
20. Bhopal, R.S.: Concepts of Epidemiology an Integrated Introduction to the Ideas, Theories, Principles and Methods of Epidemiology, vol. 38, 1st edn. Oxford University Press, New York (2002)
21. Kleinbaum, D.G., Klein, M.: Survival Analysis: A Self-Learning Text, 3rd edn. Springer, Cham (2012)
22. Van Stralen, K.J., et al.: Case-control studies—an efficient observational study design. Nephron Clin. Pract. **114**(1), c1–c4 (2009)
23. Cox, D.R.: Analysis of Survival Data. Routledge, Abingdon (2018)
24. Albakri, S.H., et al.: Security risk assessment framework for cloud computing environments. Secur. Commun. Netw. (2014)
25. BS.ISO/IEC27005:2011: Information Technology-Security Techniques-Information Security Risk Management: The British Standards Institution (2011)
26. Owasp, T.: The Ten Most Critical Web Application Security Risks (2013)
27. ENISA: Cloud computing: benefits, risks and recommendations for information security. The European Network and Information Security Agency (ENISA) (2009)

Proposing a New Approach for Securing DHCPv6 Server Against Rogue DHCPv6 Attack in IPv6 Network

Ayman Al-Ani(✉), Mohammed Anbar(✉), Rosni Abdullah(✉), and Ahmed K. Al-Ani(✉)

National Advance IPv6 Centre (NAv6), Universiti Sains Malaysia, 11800 Gelugor, Palau Penang, Malaysia
{ayman,anbar,ahmedkhallel91}@nav6.usm.my, rosni@usm.my

Abstract. Internet Protocol version 6 (IPv6) is becoming increasingly entrenched, especially as the shortage of IPv4 address has recently become obvious. However, IPv6 faces many security issues with its new design, such as a rogue Dynamic Host Configuration Protocol for IPv6 (DHCPv6) server attack. The DHCPv6 responsible for assigning the IPv6 address to clients and provide the client with network configuration parameters. A rogue DHCPv6-server can divert client to rogue servers such as Domain Name System (DNS), Network Time Protocol (NTP) servers. Therefore, security issues are vital. This paper proposes an approach to secure a DHCPv6-server against rogue DHCPv6-servers with a complete analysis of security challenges that are facing DHCPv6. The proposed approach uses Edwards-curve Digital Signature Algorithm (EdDSA) to prevent a rogue DHCPv6-server attack through verifying the source and the integrity of DHCPv6-server message whether it comes from a legitimate or rogue DHCPv6-server. The implementation and evolution of the proposed approach will be included in the authors' future work.

Keywords: IPv6 local link network · DHCPv6 server
Rouge DHCPv6 server attack · DHCPv6

1 Introduction

The number of internet nodes is significantly increasing, making limitations in Internet Protocol version 4 (IPv4) addressing space a critical and serious worldwide problem. IPv6 is the next generation of Internet Protocol intended to replace IPv4 [1, 2]. Researchers have shown that IPv6 is currently in use. Currently, clients accessing Google over IPv6 exceed 21% based on Google statistics [3]. Although IPv6 is intended to improve the network security and quality of service, it still faces security challenges such Daniel of Service (DoS), Man-in-The-Middle (MITM) attacks [4].

DHCPv6 is designed to assign IPv6 addresses, prefixes, and other network configuration parameters to hosts in IPv6 network [5, 6]. Thus, DHCPv6 can manage the assignment of multiple addresses, potentially assigned over a period of time as cited in RFC3315 [7]. The DHCPv6 can be exploited by the attacker to divert the client to rogue

F. Saeed et al. (Eds.): IRICT 2018, AISC 843, pp. 579–587, 2019.
https://doi.org/10.1007/978-3-319-99007-1_54

servers such as Domain Name System (DNS), Network Time Protocol (NTP) servers. Thus, securing DHCPv6 is a vital issue in IPv6 link-local network [8].

This study aims to highlight the main security challenges faced by the DHCPv6 process in IPv6 Network and propose an approach for securing DHCPv6 process by verifying the source and the integrity of the DHCPv6-server message. The rest of this paper organized as follows: Sects. 2 and 3 explain the process of DHCPv6 and security issues of DHCPv6, respectively. Section 4 discusses related work. Sections 5 and 6 explains the proposed approach, expected resulted and the conclusion and options for future work are given in Sect. 7.

2 Overview of DHCPv6

DHCPv6 is used by IPv6 network to assignee IPv6 address and distribute other network configuration information such as DNS and, NTP address [9, 10]. IPv6 network can use either DHCPv6 protocol or Stateless auto-configuration (SLAAC) mechanism to configure client's addresses [11, 12]. The basic scenario for how a new client communicates with DHCPv6-server as follows: when the client joins IPv6 network, the client will receive Router Advertisement (RA) message. Based on RA message, the client will communicating with DHCPv6-servers that located on the same link via multicasting a DHCPv6-Solicit Message to all DHCPv6-servers requests IPv6 address and the network configuration parameters. Hence, DHCPv6-server will send these network configuration parameters by the DHCPv6-Advertise Message as a response to client's message. Then, the client sends a request message to confirm the configuration, and the server will reply to confirm the configuration, Fig. 1 shows the DHCPv6 process [7].

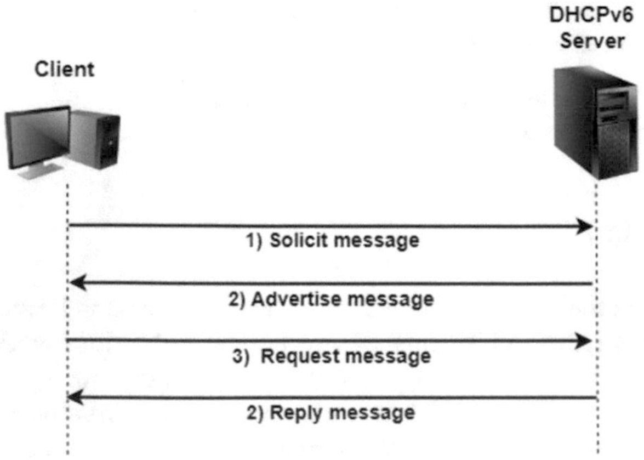

Fig. 1. DHCPv6 processes.

The DHCPv6-server is the one sending configuration information to all the clients located on the same link-local network, Therefore, DHCPv6 is considered one of the main components in the IPv6 network and its protection is needed.

3 Security Challenge of DHCPv6 Server

As mentioned above, upon receiving the DHCPv6-server messages, the client will configure its IP address and other network configuration without any verification about whether this message comes from a legitimate or rogue DHCPv6-server [13]. Therefore, any attacker on the same network can impersonate a legitimate server by sending a fake DHCPv6-server message to the clients when the client sends a Solicit Message asking the server to reply. If there is an attacker on the network, the attacker will respond via an Advertise Message containing incorrect configuration information. As the client does not have any mechanism to verify the source of this DHCPv6-server message, it will accept the message and configure its IP address and other network settings with an incorrect configuration, as shown in Fig. 2. Hence, the client is under attack such as DoS or redirect the user to rogue DHCPv6-servers [14–16]. In addition, the attacker may also provide clients with partially modified information that allows the attacker to route traffic through a specified host in which information could be monitored and collected [8].

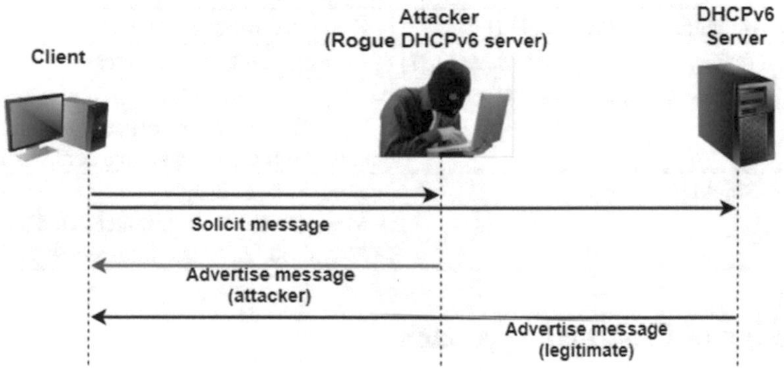

Fig. 2. DHCPv6 security challenge

4 Related Work

Several mechanisms have been proposed to secure DHCPv6-server messages between client and server the main current existing mechanisms are discussed below:

RFC 3315 [17] defines two mechanisms to secure DHCPv6-server messages, namely Delayed Authentication Protocol to authenticate client Solicit Message and Reconfigure Key Authentication Protocol to protect the client against misconfiguration. The two mechanisms required a shard key distributed between client and server. The first mechanism does not provide any key distribution mechanism. Therefore, the shared keys are usually distributed manually, conflicting with the goal of minimizing the configuration data needed at each host. Besides, it is difficult to deploy in a wide-scale network with mobile nodes [18]. The second mechanism is used for protection against misconfiguration of client caused by a reconfigure message send by a rogue DHCPv6-server. The shared key is sent as a plain text to the client in the initial exchange of DHCPv6 message for future use. As the key is transmitted in plaintext, an attacker can hijack the initial messages that contain the key and used the key for authenticating a malicious reconfiguration message [13].

Furthermore, some approaches such as [8, 15] may be used to enforce authentication with an RSA signature algorithm. The RSA has heavy computation and time complexity issues which may induce DoS attack [19, 20]. In addition, these approaches required to distribute certifications to each DHCPv6-client manually. Thus, making these approach difficult to be deployed in a wide-scale network and mobility nodes.

The DHCPv6-Shield approach is considered the best current approach to prevent a rogue DHCPv6-server attack. However, this approach requires a special type of switch and cannot work to prevent rogue DHCPv6-server in IPv6 tunneled traffic. In addition, it may drop legitimate fragmented packets and cause DoS IP fragmentation attack [21]. Table 1 summarizes the limitations in related works.

Table 1. Related works and its limitations

Approach	Limitations
Delayed Authentication Protocol [17]	- Required to distribute the key manually
Reconfigure Key Authentication Protocol [17]	- Transmitted key as plaintext
Secure DHCPv6 mechanisms [8, 15]	- Required pre-configure with certificate - Used RSA signature algorithm which has heavy computation and time complexity issues
DHCPv6-Shield [21]	- Switch-dependent - May cause DoS IP fragmentation attack - May drop legitimate fragmented packets

5 Design of Proposed Approach

The main problem in existing approaches is not providing a sufficient method to distribute the key between the client and DHCPv6-server or the need for external resources to protect the IPv6 network against the rogue DHCPv6-server. The core goal of this study is to propose an approach to secure DHCPv6-server in IPv6 link-local network by proposing a mechanism to distribute the key securely among clients and DHCPv6-server. In addition, it aims to reduce the risk of a replay attack. Below are the main stages of the proposed approach.

5.1 Generating and Distributing Public/Private Key (Stage-One)

The aim of this stage is to generate public and private keys, to distribute key between servers and routers, and to enable the router to distribute public key to DHCPv6-clients. The proposed approach utilizes Digital Signature Algorithm (DSA) [22] for verifying the integrity and source of DHCPv6-server message whether it comes from the legitimate server or rogue server. The DSA used public and private key for signing and verifying the DHCPv6-server message. The private key will be install at the DHCPv6-server and public key will be install at the router. Further, the proposed approach assumed the router is fully secured to distribute the public key to all clients. To achieve the aim of this stage, this stage has been divided into three steps.

Generate Public and Private Keys. In this step, the public and private keys must be generated by the DHCPv6-server that exist on the link-local network. Generating these keys are based on public-key cryptosystems algorithm. The proposed approach utilized Ed25519 algorithm to provide a digital signature algorithm. Ed25519 is designed to provide high performance on a variety of platforms and have a small public key size compared to non- Ed25519 algorithms. This makes Ed25519 more efficient for security purposes, as mentioned in RFC8032 [23].

As mentioned above, Ed25519 will use to generate the public and private keys. After the public and private keys are generated by the DHCPv6-server, the generated public key must be installed at each router that used to configure DHCPv6-clients. Further, In case of the network has more than one DHCPv6-server located on the same link-local network the private key must be installed on each of these servers. Installing the private key on each DHCPv6-server allowing the server to sign on each message. The distributing of public and private keys must be conducted manually to routers and DHCPv6-servers.

The output of this step is the public and private keys are generated and securely installed to each router and server, respectively. After the router got the public key from the server, now it should inform all the client about public key by sending it via RA messages with other parameters. In order to do that, the proposed approach utilized NDP option to transfer the public key to the client [24], this study named the option DHCPv6 Public Key (D-PK). The following step elaborates the DPK option in details.

DHCPv6-Public Key (D-PK) Option. In this step, the new D-PK option is designed for conveying the public key to all the client on the IPv6 network by appending it to RA messages sent out to the clients. Each IPv6 client joining the network has to get network configuration information such as address mode (i.e. Stateless or Stateful) and MTU [25, 26]. These configurations can be sent only by the router through RA message. Therefore, the proposed approach intends to use the RA message for distributing the public key to IPv6 clients, rather than using a third party, by introducing a new option named D-PK attaching on each RA message conveying the public key. The D-PK option design format follows RFC 4816 option format [24], D-PK option consists of four main fields and its size is 40-bytes shown in Fig. 3.

Type (1-byte)	Length (1-byte)	
Reserved (6-bytes)		
Public Key (32-bytes)		

Fig. 3. D-PK option format

Extracting Public Key from D-PK Option. When a client joins, the IPv6 network should firstly send an RS message to the router asking for the network configuration.

The router will reply by RA message convey the network information and the public key held by D-PK option. The client should extract the public key found in the D-PK option and use it later for communication with the DHCPv6-server. Figure 4 shows RA message transmission between client and router.

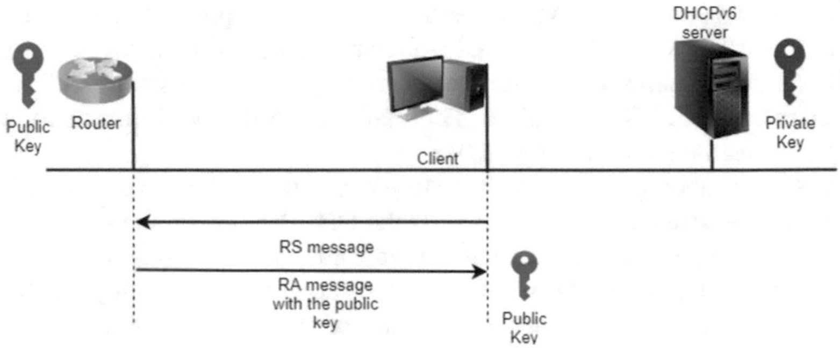

Fig. 4. RA messages transmission between client and router

5.2 Signature DHCPv6 Server Message (Stage-Two)

This study aims to verify the DHCPv6-server messages without change the DHCPv6-server message format. Therefore, the proposed approach follows the RFC 3315 authentication option format by introducing a new option named Signature Authentication (SA), which will attach to each DHCPv6-server message for verification process [7]. The main purpose of using SA option is to convey the DHCPv6 message signature. Figure 5 SA Option Format shows the SA option format.

Option_Auth (2-bytes)		**Option_Len (2-bytes)**
Protocol (1-byte)	**Algorithm (1-byte)**	**RDM (1-byte)**
Replay Detection (8-bytes)		
Signature (64 bytes)		

Fig. 5. SA option format

Each message sent out by the DHCPv6-server message should sign and insert the signature in SA option that will be appended to each message for verifying the source and the integrity of the DHCPv6-server message. As mentions above, the proposed approach utilized Ed25519 to produce the signature. Further, the signature is applied over the whole DHCPv6 message, except the SA option signature fields, which must be set to zero for the signature computation because the signature does not have a signature value yet. The signature consists of two steps, hashing and encrypting. SHA512 is used to hash the message [27], Formula 1 is used to hash a DHCPv6-server message.

$$Hash = SHA512\,(M) \tag{1}$$

where *Hash* is the hash value, *M* is the DHCPv6-server message, and *SHA*512 is the SHA512 Hash Function Algorithm. The next step is to sign the hash value by utilizing Ed25519 algorithm and public key [28]. Formula 2 is used to sign hash value.

$$S = \text{Ed25519}\,(\text{Hash}, \text{PK}) \tag{2}$$

where *S* is the Signature, *Hash* is the hash value of step 1, *PK* is the public key and Ed25519 is the signature algorithm. After the message is signed, the signature value must be appended to the DHCPv6 message by using SA option.

5.3 Verifying DHCPv6 Server Messages (Stage-Three)

The aim of this stage is to verify the source and integrity of the DHCPv6-server message by validating the signature of the SA option. After the DHCPv6-server message is signed in the previous stage, the client must verify the signature to prevent a rogue server attack. This stage achieves the main goal of this study which is verifying DHCPv6-server messages and preventing a rogue DHCPv6-server attack. This stage must be conducted every time when the client receives a message from the DHCPv6-server. This stage has been divided into two main steps. The first step is verifying the reply attack and the second step is to verify the signature.

Verifying Replay Detection Field. A reply attack is when an attacker uses old DHCPv6-server messages to configure the client with an old configuration which may cause a DoS attack or divert user to a rogue DHCPv6-server. In order to prevent replay attacks, the value of the Replay Detection field in the SA option appended to the DHCPv6 message should be verified by the client. If the Replay Detection field has a value below the threshold time, the process continues, or else the DHCPv6-server message is discarded.

Verifying Signature. After the verification of replay detection field, the client should perform verification to detect whether the DHCPv6-server messages come from a legitimate or rogue DHCPv6 server with the following steps: Firstly, the client needs to set SA option signature in DHCPv6-server message to zero. This is followed by hashing the DHCPv6-server message using the SHA512 algorithm. Formula 3 is used to hash a DHCPv6-server message.

$$\text{Hash} = \text{SHA512}\,(\text{M}) \tag{3}$$

where *Hash* is the hash value, *M* is the DHCPv6-server message, and *SHA*512 is the SHA512 HASH function Algorithm. Next, the client needs to verify the signature by using the public key that sent by the RA message at stage 1, and the hash value of step 2. If this verification check is successful, this means the message has been signed by legitimate DHCPv6-server, the message must process as described by RFC 3315 [7]. If the verification check is unsuccessful, the message must be discarded. By doing so, the client verifies that the DHCPv6-server message belongs to a legitimate DHCPv6-server.

6 Expected Result

The proposed approach expects to be an easy method to deploy in a wide-scale network and mobility nodes, due it is not required pre-configuring for the client. Comparing with existing approaches such as [8, 15], which are need to pre-configure the client with keys and certifications. In addition, most of the current approaches used RSA to generate the signature which it causes CPU overhead, whereas, the proposed approach relied on Ed22519 that it consider fast for generating a signature which it can consume less CPU. Moreover, the proposed approach is switch-independent and does not drop a legitimate packet. In contrast, the DHCPv6 shield is switch dependent which it may drop a legitimate packet.

7 Conclusion and Future Work

This paper presents the DHCPv6-server process in terms of its security challenges. The paper shows that most current mechanisms fail to provide an efficient method to secure the DHCPv6-server process on an IPv6 network. Therefore, the paper proposes an approach to secure IPv6 network from rogue DHCPv6-server attack using Ed25519 digital signature algorithm which is considered faster than the RSA digital signature in signature generation used in most previous approaches. Ed25519 can reduce the probability of exploit digital signatures to conduct a Distributed Denial-of-Serve (DDoS) attack, which results in a huge wastage of CPU resources. Finally, the study expects that the proposed approach would be easy to implement and deploy in a wide network and mobility nodes, without using any third-party device, and switch independently. Future work should implement the proposed approach and evaluate it against currently existing techniques.

References

1. Al-Ani, A.K., Anbar, M., Manickam, S., Al-Ani, A., Leau, Y.-B.: Proposed DAD-match security technique based on hash function to secure duplicate address detection in IPv6 link-local network. In: Proceedings of the 2017 International Conference on Information Technology, pp. 175–179. ACM (2017)
2. Al-Ani, A.K., Anbar, M., Manickam, S., Al-Ani, A., Leau, Y.-B.: Proposed DAD-match mechanism for securing duplicate address detection process in IPv6 link-local network based on symmetric-key algorithm. In: International Conference on Computational Science and Technology, pp. 108–118. Springer, Cham (2017)
3. Google IPv6: IPv6—Google (2018)
4. Elejla, O.E., Anbar, M., Belaton, B.: Icmpv6-based dos and ddos attacks and defense mechanisms: review. IETE Tech. Rev. 1–18 (2016)
5. Tirkkonen, L.: Utilising configuration management node data for network infrastructure management (2016)
6. Tripathi, N., Hubballi, N.: Detecting stealth DHCP starvation attack using machine learning approach. J. Comput. Virol. Hack. Tech. 1–12 (2017)

7. Troan, O., Droms, R.: IPv6 prefix options for dynamic host configuration protocol (DHCP) version 6 (2003)
8. Su, Z., Ma, H., Zhang, X., Zhang, B.: Secure DHCPv6 that uses RSA authentication integrated with self-certified address. In: 2011 Third International Workshop on Cyberspace Safety and Security (CSS), pp. 39–44. IEEE (2011)
9. Brzozowski, J., Van de Velde, G.: Unique IPv6 Prefix per Host
10. Horley, E.: IPv6 and DHCP. In: Practical IPv6 for Windows Administrators, pp. 191–207. Springer, Cham (2017)
11. Ruiz, J.M.V., Cardenas, C.S., Tapia, J.L.M.: Implementation and testing of IPv6 transition mechanisms. In: 2017 IEEE 9th Latin-American Conference on Communications (LATINCOM), pp. 1–6. IEEE (2017)
12. Yousheng, G., Lingyun, Y., Lijing, H.: Addressing scheme based on three-dimensional space over 6LoWPAN for internet of things. In: 2017 13th IEEE International Conference on Electronic Measurement & Instruments (ICEMI), pp. 59–64. IEEE (2017)
13. Shen, S., Lee, X., Sun, Z., Jiang, S.: Enhance IPv6 dynamic host configuration with cryptographically generated addresses (2011), undefined. (n.d.). ieeexplore.ieee.org
14. Gont, F., Liu, W., Van de Velde, G.: DHCPv6-Shield: Protecting Against Rogue DHCPv6 Servers (2015)
15. Li, L., Ren, G., Liu, Y., Wu, J.: Secure DHCPv6 mechanism for DHCPv6 security and privacy protection. Tsinghua Sci. Technol. **23**(1), 13–21 (2018). https://doi.org/10.26599/TST.2018.9010020
16. Alangar, V., Swaminathan, A.: IPv6 security: issue of anonymity. Int. J. Eng. Comput. Sci. **2**(8), 2486–2493 (2013)
17. Droms, R., Arbaugh, W.: Authentication for DHCP messages (2001)
18. Su, Z., Ma, H., Zhang, X., Zhang, B.: Secure DHCPv6 that uses RSA authentication integrated with self-certified address (2011), undefined. (n.d.). ieeexplore.ieee.org
19. Annamalai, A., Yegnanarayanan, V.: Secured system against DDoS attack in mobile adhoc network. WSEAS Trans. Commun. **11**(9), 331–341 (2012)
20. van Rijswijk-Deij, R., Jonker, M., Sperotto, A.: On the adoption of the elliptic curve digital signature algorithm (ECDSA) in DNSSEC. In: 2016 12th International Conference on Network and Service Management (CNSM), pp. 258–262. IEEE (2016)
21. Velde, G., Gont, F.: DHCPv6-Shield: Protecting against Rogue DHCPv6 Servers (2015). https://tools.ietf.org/html/rfc7610
22. Dinu, D.D., Togan, M.: DHCP server authentication using digital certificates. In: 2014 10th International Conference on Communications (COMM), pp. 1–6. IEEE (2014)
23. Josefsson, S., Liusvaara, I.: RFC 8032-edwards-curve digital signature algorithm (EdDSA) (2017)
24. Malis, A., Martini, L., Brayley, J., Walsh, T.: Pseudowire emulation edge-to-edge (PWE3) asynchronous transfer mode (ATM) transparent cell transport service (2007)
25. Podermanski, T., Grégr, M., Švéda, M.: Deploying IPv6-practical problems from the campus perspective. In: Terena Networking Conference (2012)
26. Atlasis, A., Rey, E.: IPv6 Router Advertisement Flags, RDNSS and DHCPv6 Conflicting Configurations
27. Kelly, S., Frankel, S.: Using hmac-sha-256, hmac-sha-384, and hmac-sha-512 with ipsec (2007)
28. Romailler, Y., Pelissier, S.: Practical fault attack against the Ed25519 and EdDSA signature schemes. In: 2017 Workshop on Fault Diagnosis and Tolerance in Cryptography (FDTC), pp. 17–24. IEEE (2017)

A Review of Ransomware Families and Detection Methods

Helen Jose Chittooparambil[1], Bharanidharan Shanmugam[1](✉), Sami Azam[1], Krishnan Kannoorpatti[1], Mirjam Jonkman[1], and Ganthan Narayana Samy[2]

[1] College of Engineering, IT and Environment, Charles Darwin University, Casuarina, Australia
Bharanidharan.shanmugam@cdu.edu.au
[2] Advanced Informatics School, Universiti Teknologi Malaysia, Kuala Lumpur, Malaysia

Abstract. Ransomware has become a significant problem and its impact is getting worse. It has now become a lucrative business as it is being offered as a service. Unlike other security issues, the effect of ransomware is irreversible and difficult to stop. This research has analysed existing ransomware classifications and its detection and prevention methods. Due to the difficulty in categorizing the steps none of the existing methods can stop ransomware. Ransomware families are identified and classified from the year 1989 to 2017 and surprisingly there are not much difference in the pattern. This paper concludes with a brief discussion about the findings and future work of this research.

Keywords: Ransomware classifications · Peta · Virus · WannaCry Cybersecurity

1 Background

Ransomware, also known as crypto virus, has received significant attention among cyberspace researchers in the last few years. Intruders use these malwares to steal people's private information with the help of vulnerabilities, including previous malware attacks, and threaten the victim to agree to their demand or they will lose their valuable files, data or private information. The ransom or demand may be e-gold, cryptocurrency or demands to purchase from designated stores. Ransomware is not a new concept, the first ransomware appeared in 1989 named as the PC CYBORG (AIDS) Trojan. It was delivered electronically through a floppy disk and the floppy disk was then used to attack a system [2]. Modern ransomware attacks started in 2005 with "Trojan.Gpcoder".

A malware is considered ransomware only if a minimum of three anti-virus vendors have assigned malware labels and categorised it as ransomware. The ransomware and malware can be differentiated by their attacking behaviour [1] and time taken for the attack. Malware attempts to hide behind applications, but ransomware initiates an attack immediately after installation. Ransomware is created for direct revenue generation, it often uses a countdown clock to alert the victim about the remaining time for paying the

F. Saeed et al. (Eds.): IRICT 2018, AISC 843, pp. 588–597, 2019.
https://doi.org/10.1007/978-3-319-99007-1_55

ransom. The first ransomwares displayed advertisements, but advancements in malware made it possible to block services, disable keyboards and spy on user activities [3].

During a ransomware attack, attackers use system vulnerabilities to intrude into the compromised device. The attacking probability of already attacked devices is always high. In most cases, major ransomware attacks occur on windows machines, but they can also attack android, Linux and IOS devices [4]. The ACSC 2015 Threat Report states that ransomware campaigns against Australian organizations will continue to be prominent. Every sector experienced a cyber security incident, which demonstrates the indiscriminate targeting and the sophistication of this type of threat. The number of ransomware incidents has drastically increased and is now four times higher than recorded in 2013. In 2015 the FBI received around 992 complaints regarding ransomware with the victims losing an estimated $18 billion.

Existing research's [13, 14, 21, 23] indicates that attacks of ransomware are difficult to detect or stop from occurring. The scope of this research is to understand operation of ransomware within Microsoft Windows operating systems by analysing the five-stages of operation of ransomware. In the next section, the evolution of ransomware and how the ransomware operates within a device is discussed. Section three of this paper will discuss the three families of ransomware and the five stages of the attack process. Section four discusses the evolution of ransomware and section four will elucidate the different ransomware detection methods. The last section will conclude with findings obtained from this study and recommendations for future work.

2 Literature Review

2.1 Review Stage

The main motive of introducing ransomware into the cyber world was to obtain financial gain by threatening people. There is no connection between the ransomware and the victims, but previous research shows that people who browse through unwanted websites are vulnerable to ransomware attack. The ransomwares can apply any kind of infection method that the malwares can use [5, 6, 21]. Ransomwares could be broadly classified as follows (Fig. 1) and the scope of this research is highlighted.

Scareware: This is a threat which takes advantage of people's fear. It is also known as rouge security software and was initially not considered a member of ransomware family. However, the increasing level of popularity of this malware has led to the classification of scareware within the family of ransomware. The scareware shows a message which states that the virus attacked the victim's system and it attempts to convince the victim to buy antivirus software which should remove the virus. Usually, these viruses are fake, and the antivirus software is non-functional malware [1, 8].

Lock Screen Ransomware: The lock screen ransomwares lock the victim's system until a ransom is paid for the lock key to retrieve it. It generally locks the desktop of a victim and creates a new window [9] which is used to communicate with the victim. The command and control server then control the victim's system and the new window shows messages about the ransomware threat and possible ways to unlock the system.

Crypto Ransomware: The most dangerous ransomware family is the crypto ransomware family. The ransomware encrypts the victim's files using strong encryption algorithms. In most cases cyber criminals make their own crypto systems. It is very difficult to retrieve the data back if the ransom is not paid within the period. CRYPTOLOCKER [10] is an example for the crypto ransomware which encrypts user files using a private encryption key.

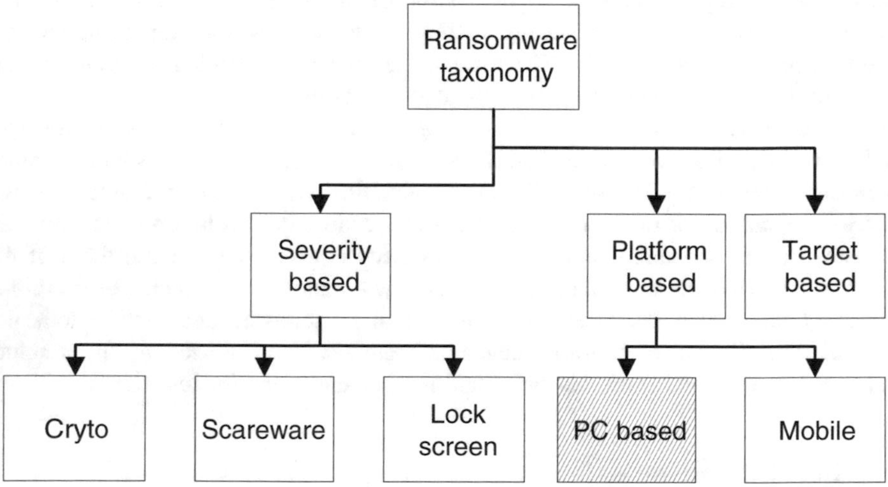

Fig. 1. Ransomware taxonomy [21]

However, the main advantage of this ransomware is that it stores the private key within the victim's system, so that further analysis can be used to access this private encryption key and retrieve the data. Reveton, otherwise also called Trojan: W32/Reveton, is an example of scareware which appeared to take control of the infected machine until the victim accepted the demand [20]. Using the RSA-2048 algorithm, file encryption ransomware encrypts the victim's documents and displays a message demanding money on the screen of the attacked device. In addition, it will also create a new instance of EXPLORER.EXE AND SVCHOST.EXE to make communication with the command and control server of the criminals [19]. Once the installation has occurred, the crypto wall will start to delete the shadow copies of files and install spyware to steal passwords saved on the victim's device as well as bitcoin wallets [22]. No machine is free from ransomware attack; mobile devices are vulnerable too. Fake messages imitating legitimate sources, which offers gifts and vouchers traps people to download ransomware executables. The latest and most effective Wannacry ransomware appeared in 2017 [18] and targeted Microsoft Windows operating systems. A ransomware attack of a system is generally defined five stages [11], refer to Fig. 2.

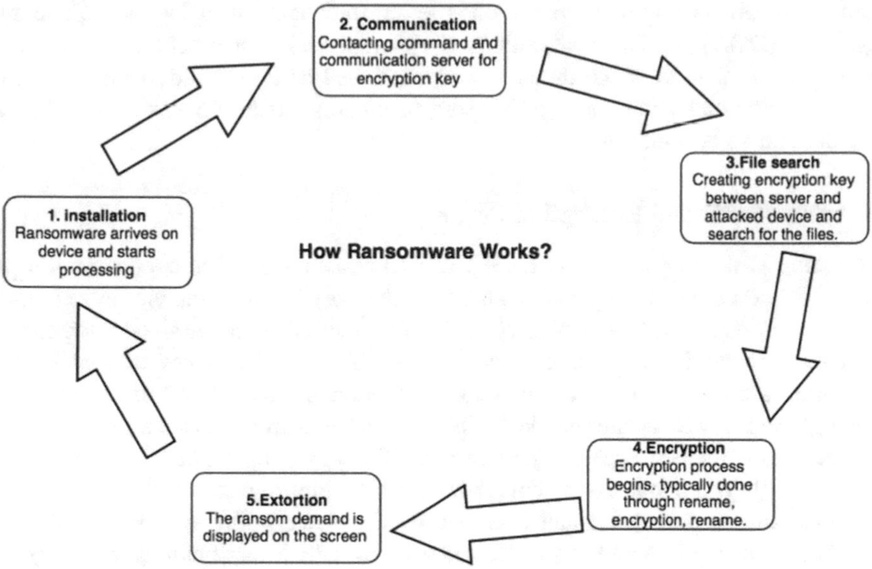

Fig. 2. Five stage process of Ransomware

2.2 Installation

The installation sets up the ransomware in victim's system and creates an environment to work in. Most ransomware hide behind the root path of the AppData or in the local AppData folder. During the installation execution, it randomly creates a file name inside the AppData folder [7]. Prior to that it will delete the original executable files available in the system. After creating the new file, it will then update the registry key files, containing information about tuning parameters, system configurations and user preferences, with the new registers on the device.

The modification of registry key files is done to obtain persistence of the device unless it reboots. Such persistence is obtained by executing an auto run process of registry keys that enables the malware to perform their execution while the device runs in a safe mode. Finally, to delete the shadow copies, the ransomware hijacks most of the executable files of that specific device.

2.3 Communication

For a complete ransomware attack, the ransomware needs to contact a command and control server. One possible way to establish the connection between the victim's device and command and control server is through a TOR network. After the installation of the ransomware file, the device will communicate with the command and control server to obtain the encryption key. Then using this encryption key, the ransomware will encrypt the selected files on the victim's device. To initiate the process, a communication setup has to be executed on the victim device to begin the communication between the command and communication server and the victim's device.

Then a domain is hosted to reach out the ransomwares by a Domain Generating Algorithm (DGA) [7]. These algorithms can generate a list of pseudo random domain names and are able to create thousands of command and control domains on a daily basis. The main advantage to the attackers of creating such domains is that they are very difficult to be shut down.

2.4 File Search and Encryption Process

Encryption is the major process for the crypto ransomware. For the encryption procedure, the server needs to generate an encryption key between the victim's computer and the command and control server. This differs in symmetric and asymmetric encryption, where in the symmetric encryption the same secret key is used for both encryption and decryption. In asymmetric encryption, a private key known only by the owner is used as well as a public key. The encryption method uses the public key and the decryption method uses the private key. The keys are protected using different methods [10]. The public key is embedded into the ransomware or obtained from the command and control server and it is then used to encrypt the symmetric key. After encryption, the key is stored in the infected system [12]. After obtaining the encryption key, the encryption procedure begins. The sender encrypts the plain text using the public key and on the receiver side the text is decrypted using the private key. The files are encrypted using the server generated encryption key and removes the original file. The encryption method is used to secure the communication channel between the ransomware and the command and control server. The dual encryption method is used because of performance and convenience. In some processes the ransomware uses individual keys per file [8], so breaking that key would only give access to that file. After initiation, the ransomware will create an encrypted copy of the original file and attempts to destroy all copies of the original ones. It renames the encrypted copy with the original file name with some extensions added.

2.5 Extortion

In the final stage, a ransom is demanded and displayed on victim's system [6]. There will be a time limit to pay the ransom before the criminals destroy the decryption key. They might use different payment methods for ransom payment. Certain ransomware will allow you to decrypt one file for free to assure you that the key is valid on your device. One type of payment is through Bitcoins, which may be an anonymous payment. More recent ransomware proceeds to delete files to threaten you in paying the ransom quickly. But there is no guarantee that the key they provided will help you to decrypt your files. There are ways in which the time for paying is set in the victim's device. The CMOS can be used as countdown for payment so that even if the user goes offline it will monitor the time limit. The ransom often increases automatically after a given time limit for the payment and updates this with the command and control server.

3 Ransomware Evolution and Detection Methods

The attack of ransomware is not new, and the first ransomware attack occurred in 1989 by ransomware named "PC cyborg", which belongs to the crypto ransomware family. The next major attack occurred five years later in 1994 by the ransomware named the OneHalf virus. In 2004, the "GP coder" marked the beginning of the crypto ransomware attack era. From 2011 to 2013. RansomLock however, was from the so called locky ransomware family. The attack rate of ransomware steadily increased in the following years. In 2014, Bucbi, TorrentLocker, CryptoWall, CryptoDefence, Cryakl, Reactor, CTB-Locker, CryptoGraphiLocker, Cryptowall2, CryFile, KetBTC, BandarChor, Virlock, KEYHolder and Operation Global 3 appeared. In 2015 thirty-five ransomwares appeared. The major ransomware attacks occurred in 2015, 2016 and 2017 as illustrated in Fig. 3 below. Ransomware attacks and the evolution of new ransomware families have increased due to the success of ransomware attacks.

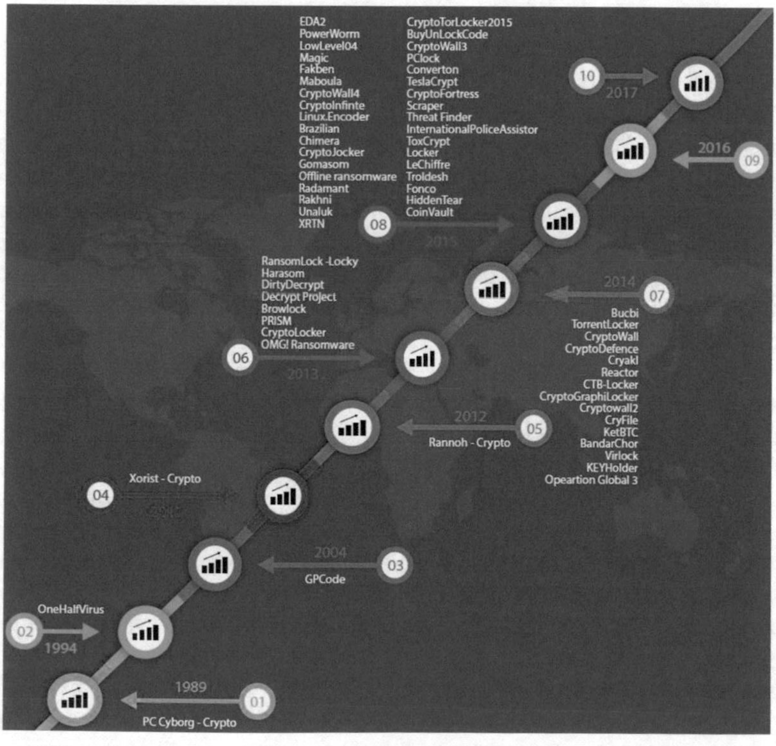

Fig. 3. Ransomware timeline

3.1 Study of Existing Methods

In the UNVEIL method [13], researchers analysed and detected ransomware attacks and modelled the behaviour using a dynamic analysis methods. To monitor how the ransomware interacts, an artificial but realistic execution environment is used which is created by the system automatically [13]. Then the interaction process with the file system is closely monitored to precisely characterize the cryptographic behaviour of the ransomware. The UNVEIL method flags suspicious behaviour as "like ransomware" and the operation of the ransomware is then identified to differentiate different classes of ransomware. UNVEIL uses two steps: the first step is to generate a user environment and the second step is to monitor file system activity [13]. Wecksten et al. [14] analysed four crypto ransomwares in a virtual machine with a Windows 7 operating system. For the investigation, the Cryptowall, Tesla Crypt, CTB Locker and Locky ransomware were used and the infection has been analysed using zeltzers [14]. In addition, the researchers used the software process monitor and regshots for tracking the process activity, file system activity and registry manipulation. This experiment identified that the crypto ransomware attacks are dependent on software named vssadmin.exe in the victim's device [14]. They claim that the attack can be easily prevented by avoiding access to the vssadmin.exe software. The remaining experiment was conducted by renaming the vssadmin.exe with shadow copies being created. Because of this prevention method, the ransomware could not find the restore point which was created before infection and that made it possible to restore all the information [14]. Moore [15] tried to detect the ransomware using a honeypot folder. The honeypot folder is created to monitor the changes occurring in the folder [15]. In addition, the Microsoft File Server Resource Manager feature was used for the file screening service and the Event Sentry was used to manipulate the Windows security logs [15]. This alert can then assist the system administrator to prevent further damage to the system [15].

The main aim was to determine a suitable ransomware detection method and to deploy an additional layer of security to the network to protect network actions rather than shutting down a server for the user when trying to update a file. When the first detection of a change has been encountered, a trigger would occur [15]. This was used to look for the email that caused the change in the monitored folder. If more activity was then encountered, the second hierarchical level was triggered. In the third hierarchical stage, the intensity of activity results in terminating the network service [15]. And finally, the fourth stage occurs when a final threshold of changes has been encountered. At this point, alerting the administrator and blocking the users access was not enough to stop the spread of damages, and therefore the ultimate protection would be to shut down the server [15].

Scaife et al. [16] present an early detection system, CryptoDrop, to alert a user during suspicious file activity. It can halt a process that appears to be tampering with a large amount of the data using a set a of behaviour indicators. Then the system can be parameterized for rapid detection with low false positives by combining a set of indicators common to the ransomware [16]. This research indicated that careful analysis of the behaviour of ransomware can detect it and can mitigate significant loss of data. By doing so, an early warning system has been created that can detect the malicious program through monitoring of the user's data [16]. Each ransomware uses their own algorithm to perform these specific activities. The character of ransomware can be divided into three classes [16]. The first class includes the generic behaviour of ransomware. The renaming behaviour can also be added within this class. In the second class, the ransomware moves the files out of the user's documents directory, then it reads the contents within the file, writes the encrypted contents for each file, and moves the file back to the directory [16]. Renaming can also occur with this second class of ransomware, i.e. when the files are moved back into the directory, the name of the file may be different. The third class of ransomware reads the original file to create a new independent file containing the encrypted contents. It then deletes or overwrites the original file and its shadow copies [16]. For the analysis, a set of data was built representative of measured user document directories. The distribution of file types over the entire file system and document directories was examined. A document corpus of 5099 files with 511 total directories was constructed [16]. Data was then randomly selected from different data sets and placed in different directories. For placing the directory tree, a user's document folder was placed in a cuckoo sandbox running on a Windows 7 machine. The amount of data loss was used as metric to detect the ransomware attack [16]. By detecting the attack early, further attacks were controlled, and the remaining unaffected data was saved. Also, the frequency of file format attacks was calculated. The results showed that ransomware preferred to attack productivity files rather than the media files such as pictures and music [16].

3.2 Evaluation of Existing Methods

Using the UNVEIL method [13], attackers can easily identify the automatically generated user environments and this approach will make it easier for the ransomware to hook with the documents and specific operations in the operating system. Also, most of the ransomware will never encrypt the files in the same locations. The ransomwares will try to shuffle the files and will change the master file system. By making the file unreadable, changing the file names and extensions, it is very difficult for the victim to identify which files are encrypted. The ransomwares can also lock the monitor and will take the rights from the victim. Clearly, it is not possible to analyse and stop the ransomware using file system activity. The novel method described in [14] is only effective when it repairs the damaged structure. The local hard drive would be unusable or permanently damaged at file system level. And the researchers also declared in this article that the presented solution might not be the most suitable one. They highly recommended implementing a proper backup at regular intervals to save all the information in another place.

The honey pot technique [15] can be used to detect a ransomware attack, however this method only focusses on the honeypot folders. In the early detection method [16, 23], the high latency creates an operation so that the file is often locked and cannot be opened by the ransomware. So, the possibility of saving the data files from further attack is also reduced using this technique. Table 1 presents the ransomware detection of the reviewed methodologies over the processing of ransomware. Various approaches have been used to detect ransomware attacks, however there is no method detected all stages. Table 1 represents the ransomware detection of the reviewed methodologies over the processing of ransomware. All these methodologies are focused on ransomware detection based at system file level. Monitoring changes from file levels has significant disadvantages. The ransomware attack process is going through five different stages and the file encryption process starts in the fourth stage. So, further investigation is required to identify the best solution that meets the requirements such as identifying the ransomware attacks from its previous stages.

Table 1. Ransomware detection methods

Process → Existing works	Installation	Communication	File search	Encryption	Extortion
UNVEIL [13]	N	N	Y	N	N
Mattias et al., [14]	N	N	Y	N	N
Honeypot Techniques [15]	N	N	Y	N	N
Cryptolock [16]	N	N	N	Y	N
Behavioral analysis [23]	N	Y	Y	N	N

4 Conclusion and Future Work

Ransomware attacks will continue to increase because of the increased use of the Internet. In this paper, three different families of ransomware and five-stages of operation of the ransomware are identified. Section three portraits the evolution of ransomware by showing the time line from 1989 to 2017. The monitoring of ransomware processes and operating systems activity is an effective method to find solutions for ransomware attacks. But from the evaluation of the existing methods we can conclude that none attempt to detect or stop the ransomware in the initial two stages. Detection at a later stage will increase the impact of ransomware attack, so that a solution to detect ransomware from its initial two stages is desirable to reduce the impact of ransomware. Future work of this research focuses on closely monitoring each stage of live ransomwares which attack Microsoft Windows operating systems in a controlled test environment. By allowing the ransomware to attack a device we can identify the ransomware processes, how it attacks operating system and finally the impact of the ransomware attack.

References

1. Luo, X., Liao, Q.: Awareness education as the key to ransomware prevention. Inf. Syst. Secur. **16**(4), 195–202 (2007)
2. Kharraz, A., Robertson, W., Balzarotti, D., Bilge, L., Kirda, E.: Cutting the Gordian Knot: a look under the hood of ransomware attacks. In: Detection of Intrusions and Malware, and Vulnerability Assessment, DIMVA (2015)
3. Monika, P.Z., Lindskog, D.: Experimental analysis of ransomware on windows and android platforms: evolution and characterization. Procedia Comput. Sci. **94**, 465–472 (2016)

4. Sjouwerman, S.: Ransomware In YouTube Ads, Techtalk.pcpitstop.com, 2014 (2016). http://techtalk.pcpitstop.com/2014/08/26/ransomware-youtube-ads/. Accessed 14 June 2018
5. ACSC 2015 report, 2015 Cyber Security Survey: Major Australian Businesses, Australian Government, Canberra (2015)
6. Mercaldo, F., Nardone, V., Santone, A.: Ransomware inside out. In: 2016 11th International Conference on Availability, Reliability and Security (ARES). SBA Research, Austria (2016)
7. Melendez, M.: Ransomware: An Analysis of a Growing Threat Landscape, Order No. 1605452. Utica College, Ann Arbor (2015)
8. Haas, P.D.: Ransomware Goes Mobile: An Analysis of the Threats Posed by Emerging Methods, Order No. 1586729. Utica College, Ann Arbor (2015)
9. Puodzius, C.: How encryption molded crypto-ransomware, WeLiveSecurity (2017). https://www.welivesecurity.com/2016/09/13/how-encryption-molded-crypto-ransomware/. Accessed 14 June 2018
10. Hampton, N., Baig, Z.A.: Ransomware: emergence of the cyber-extortion menace. In: 13th Australian Information Security Management Conference, Australia, pp. 47–56 (2015)
11. Wyke, J., Ajjan, A.: The Current State of Ransomware, SophosLabs (2015)
12. Brewert, R.: Ransomware attacks: detection, prevention and cure network security, pp. 5–9 (2016)
13. Kharaz, A., Arshad, S., Mulliner, C., Robertson, W., Kirda, E.: 25th USENIX Security Symposium (USENIX Security 16), 1st edn., pp. 757–772. USENIX Association, Berkeley (2016)
14. Wecksten, M., Frick, J., Sjostrom, A., Jarpe, E.: A Novel Method for Recovery from Crypto Ransomware Infections, p. 1354 (2016)
15. Moore, C.: Detecting Ransomware with Honeypot Techniques. In: 2016 CCC, Jordan, p. 77 (2016)
16. Scaife, N., Carter, H., Traynor, P., Butler, K.R.B: CryptoLock (and Drop It): stopping ransomware attacks on user data. In: 36th International Conference on Distributed Computing Systems (ICDCS), p. 303 (2016)
17. Sterling, B.: Ransomware: the basics, WIRED (2017). Accessed 14 June 2018
18. Mathews, L.: WannaCry Ransomware Situation Gets Worse as Copycats and Fake Decryptors Appear, Forbes (2017)
19. Constantin, L.: IDG News Service, "Widespread exploit kit, ransomware program, and password stealer mixed into dangerous malware cocktail Cybercrime group combines Pony, Angler and CryptoWall 4.0 in a single campaign" (2015). https://www.pcworld.com/article/3012112/security/widespread-exploit-kit-password-stealer-and-ransomware-program-mixed-into-dangerous-cocktail.html. Accessed 14 June 2018
20. Bacani, A.: REVETON Ransomware Spreads with Old Tactics, New Infection Method. In: TrendMICRO (2014). Accessed 14 June 2018
21. Al-rimy, B.A.S., Maarof, M.A., Shaid, S.Z.M.: Ransomware threat success factors, taxonomy, and countermeasures: a survey and research directions. Comput. Secur. **74**, 144–166 (2018)
22. Shanmugam, B., Azam, S., Yeo, K.C., Jose, J., Kannoorpatti, K.: A critical review of Bitcoins usage by cybercriminals. In: 2017 International Conference on Computer Communication and Informatics (ICCCI), Coimbatore, pp. 1–7 (2017)
23. Hampton, N., Baig, Z., Zeadally, S.: Ransomware behavioural analysis on windows platforms. J. Inf. Secur. Appl. **40**, 44–51 (2018)

A Survey on SCADA Security and Honeypot in Industrial Control System

Kuan-Chu Lu, I-Hsien Liu, Meng-Wei Sun, and Jung-Shian Li(✉)

Department of Electrical Engineering, Institute of Computer and Communication Engineering, National Cheng Kung University, No. 1, University Road, Tainan City 701, Taiwan (R.O.C.)
{kclu,ihliu,mwsun}@cans.ee.ncku.edu.tw, Jsli@mail.ncku.edu.tw

Abstract. Nowadays, the infrastructure employs SCADA (supervisory control and data acquisition) system for monitoring, such as power plants, reservoir, manufacturing, and logistics industry. With the rapid development of the internet, IOT (internet of things) makes every massive infrastructure more convenient to be control and monitoring. However, the convenience of remote controlling to the infrastructure may cause serious damage which could affect people's livelihood. Consequently, an effective and protective mechanism to protect the SCADA system is necessary. We have investigated the related attacks and defensive methods. This paper surveys varying SCADA system and the honeypot system of SCADA. Then we proposed a defensive architecture with honeypot system to protect the SCADA system.

Keywords: SCADA · Infrastructure security · Industrial control system Honeypot

1 Introduction

With the rising of critical infrastructure and the thriving Internet, most of the energy production and distribution (such as nuclear power plants, power grids, and gas pipelines power grids as well) to utilities (such as water and wastewater processing, water transport, transportation, and manufacturing) rely on Industrial Control Systems, ICS. And the other special type of ICS called Supervisory Control and Data Acquisition, SCADA system (for remote control, monitoring, and data collection as well). Its main orientation is for the wide area of industrial system [1]. The data processing and communication between devices and the SCADA system are developed through the development of network infrastructure. With the help of remote control and monitoring, management of infrastructure and data analysis become more convenient and efficient. However, the convenience of internet also brings multi-types of network security threats and challenge to protect our system. In addition, attacks from hackers are rising these years. In 2014, a report from Industrial Control Systems Cyber Emergency Response Team, ICS-CERT pointed out that a total of 257 attacks had been conducted on key infrastructure, such as energy facilities, water resources facilities, nuclear power plant equipment, transportation facilities, information technology systems, communication equipment and government departments in United States in 2013. Among above, energy facilities, key

F. Saeed et al. (Eds.): IRICT 2018, AISC 843, pp. 598–604, 2019.
https://doi.org/10.1007/978-3-319-99007-1_56

manufacturing equipment, water, transportation, science and technology and power plants are account for a relatively high proportion [2]. In 2016, Trend Micro company and the TippingPoint Zero Day Initiative (ZDI) pointed out that vulnerability founded in SCADA system grew 421% then 2015 [3]. As attacks become more sophisticated, cyber security no longer can depend on supervised, pattern-based detection algorithms to guarantee continuous security monitoring [23]. Traditional firewalls, authentication mechanisms, cryptographic algorithms, and other protocols are not enough secure SCADA systems and the underlying industrial processes from cyber-attacks [22]. To enhance the security of SCADA systems, one approach is to reduce the risk to assets being protected through defense-in-depth model and tiered security control. We can see the importance of defensiveness to critical infrastructure from these reports. Therefore, we surveyed the related attacks and defenses about the SCADA system and we proposed a defensive architecture with honeypot system to protect the SCADA system.

2 Introduction to SCADA

2.1 System Architecture of SCADA

Supervisory Control and Data Acquisition, SCADA, is a data collection, remote monitoring and control system. It is a structure system dominated by information network technology in Industrial Control Systems, ICS. The SCADA is mainly composed of Human Machine Interface (HMI), SCADA control center and database system, a set of Programmable Logic Controller (PLC) or Remote Terminal Unit (RTU) which connect with Intelligent Electronic Device (IED) as shown in Fig. 1. The IEDs collect environment information from the industrial environment and send the data to the database system through the PLC/RTU. Then the managers can monitor the environment and notice while an exception occurred through the HMI web. Managers also can issue control commands to PLC/RTU through the HMI.

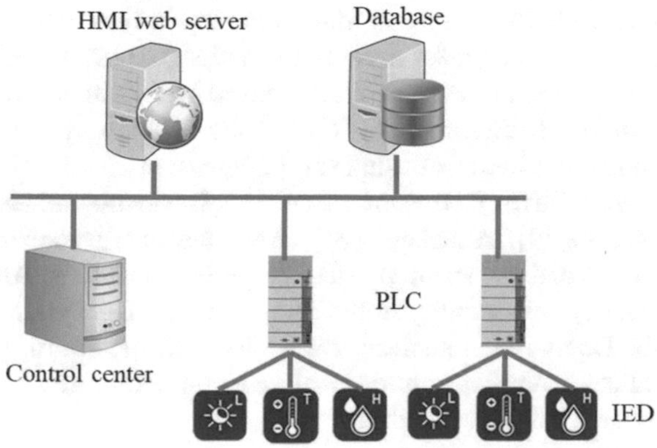

Fig. 1. Basic SCADA system architecture.

2.2 Protocols Used in SCADA

CIP. CIP (Common Industrial Protocol) encompasses a comprehensive suite of messages and services for the collection of manufacturing automation applications - control, safety, synchronization, motion, configuration and information. The greatest significance lies in how to define the object and communication standard in the OSI model, so that different layers can communication with each other [4].

DNP3. DNP3 (Distributed Network Protocol) is a set of communications protocols developed by Harris in 1993 used between components in process automation systems. It's primarily used for communications between a master station and PLCs or IEDs in utilities such as electric and water companies [5].

Modbus. Modbus is a communication architecture specifically support the communication between PLCs and other networks developed by Modicon in 1979. It's a request/reply protocol and there are several versions are using. Such as Modbus Plus, Modbus over TCP/IP, Modbus RTU etc. they are usually used to enable the HMI, SCADA control center, and PLCs been integrated over an Ethernet or other networks [6].

2.3 Threats in SCADA

In recent years, many countries have been faced serious threats to their infrastructure. With the convenience of remote monitoring and control, the network has also brought SCADA system a big weakness that can be attacked. The serious impact of SCADA attacks on country's economy and people's livelihood makes SCADA system a key target of hackers or other countries.

Attacks on SCADA. The use and application of SCADA system increase with the level of industrial automation as it can effectively reduce the cost of operation and promote the global economy. However, with the rapid development of Internet of Things (IoT), a traditional standalone SCADA system transforms into a system that can be remotely operated which becomes a target for hackers to challenge. For example, in June 2010, Stuxnet [7] became the first malware to attack control system and also the first to break SCADA vulnerabilities [8]. Before that, although the SCADA system was vulnerable, it had not yet become the target of a hacker's aggressive attack. The following will introduce more attacks. Havex (2014) influence the software downloads on the SCADA manufacturer's website [9]. Attackers [10] target the energy department primarily through malware tools and use spam to infect target organizations. Attacks from these malware stress security vulnerability in the design of SCADA systems [11]. Slammer virus attacks at the Davis-Besse nuclear power plant left the security staff aloof [12]. There are more attack events occurred these years and some researchers have been investigated the related defensive methods.

Defensive Methods of SCADA Attacks. The environment of SCADA system is a relatively new environment over the traditional network. Thus, some traditional protect mechanisms are not suitable for SCADA system, such as the encryption technology is limited by the computing power of the devices [13]. The National Institute of Standards and Technology (NIST) and the European Network Information Security Agency (ENISA) and other agencies developed a standard on network security specifically for the SCADA system to control the disaster. The Guide to industrial control systems (ICSs) also mentioned how to protect the network security of SCADA systems [14]. Not only the investigation and standard setting, SCADA network security researchers mainly rely on the simulation software and the development of hardware to study about the SCADA attacks. As shown in Table 1. There are some defensive methods used in other researches. In the next section, we will introduce our proposed defensive mechanism with honeypot system which can trap attackers and learn how the attackers issue an attack to the SCADA system.

Table 1. Defensive methods of SCADA.

Topic	Tool	Use
Machine learning	SVM	Detection of telecom network [15]
	SVDD, kernel PCA	Analysis of water treatment plant malware invasion [16]
Intrusion detection system	Modbus, DNP3	Set a firewall in the Industrial Cyber Security Lab [17]
Honeypot	Honeyd	Detection of botnet invasion [18]
	Honey-well CDROM	Protect SCADA system with honeypot technology [19]

3 Proposed Mechanism

In this paper, we focus on the honeypot used in SCADA system which can lure the attackers to attack it. We can log and study how the attackers issue an attack to the SCADA system.

Honeypot. Honeypot is a trap system which pretend as the target of attackers and lure the attackers to attack on it. The honeypot system then logs the attacks and been used to analysis. There have been a lot of researches about the honeypot system for years. With the popularity of network virtualization technology, cost and difficulty of deploying honeypot systems have been significantly reduced.

Honeyd. Honeyd is an ideal framework developed by Niels Provos. It's a free and open source software from GNU. Users do not need to deploy dozens of expensive equipment and can set up in their own computer networks and operate multiple virtual machines. Virtual machines can be configured multiple and simulate a variety of different types of server at the same time, allowing users to simulate a whole network environment [20]. Figure 2 Shows the basic architecture of honeyd system.

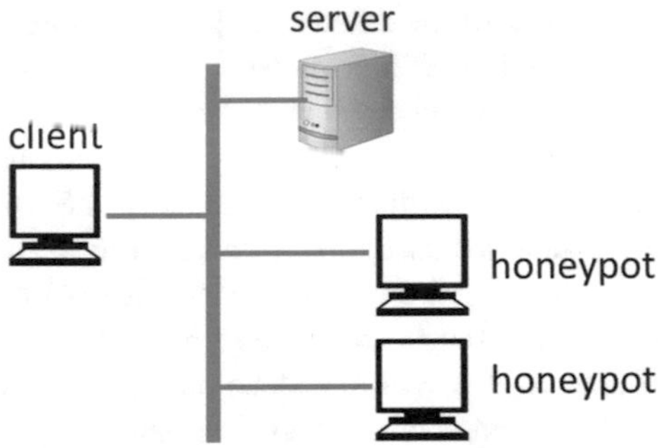

Fig. 2. Basic honeyd system architecture.

Honeypot in Industrial Environment. The traditional honeypots cannot be built in the industrial control environment due to the difference of protocols. With the rising of SCADA system, honeypot systems for SCADA have been development such as conpot, gaspot, etc.

In this paper, we choose to employ the conpot system which is developed firstly by the honeypot project [21] based on the basic SCADA architecture. This system aims to simulate the PLC in the industrial environment and interact with the control center. Figure 3 Shows our system architecture. Almost all of the attacks now come from the HMI web which may connect with other networks. There are several instances that attacks been issued from the HMI web to control the PLCs in the industrial environment. Thus, we deploy several conpot systems in the SCADA system which pretend as PLCs in the industrial control environment. While the attacker breaks into the SCADA system, we can lure the attacker to attack our conpot system to collect and log the information of the attacks. After analysis, we may prevent the attacks or reduce the damage from them.

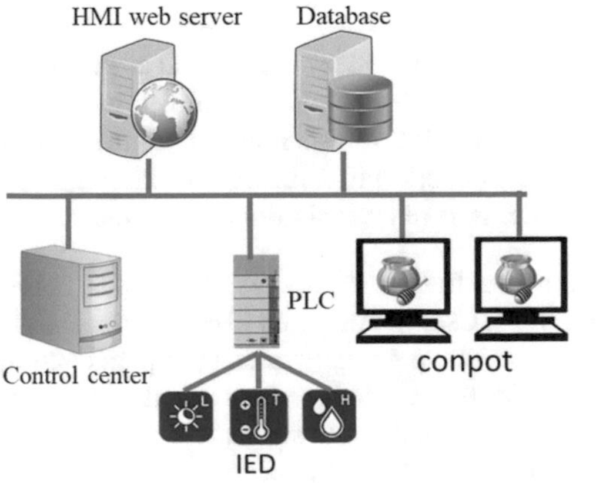

Fig. 3. SCADA system architecture with conpot system.

4 Conclusion

With the rising of industrial automation and the use of SCADA system, the industry has become more efficient. The Internet of Things brings more economic benefits to the country. At the same time, the infrastructure in every countries becomes a key target of attacking by hackers or other countries. Hackers launch attacks through various technique such as embedding malware to update files provided by service provider and sending phishing emails to relevant personnel. Thus, it's important to protect our infrastructure. Most of the approached defensive methods used intrusion detection system, machine learning and honeynet which focused on defensing against attacks from extranets. We have proposed a defensive mechanism which set honeypots in the intranet to pretend to be PLCs. By confusing the attacks, We can trap the attackers and aim to study the behavior of attackers. After collecting the information, it can be used to analysis and trace the attackers. Finding out the location of attackers 'IP, we can make more targeted defenses to enhance the security of our systems.

References

1. Serbanescu, A.V., Obermeier, S., Yu, D.Y.: A flexible architecture for industrial control system honeypots. In: International Joint Conference on e-Business and Telecommunications (ICETE), International Joint Conference, vol. 4, pp. 16–26 (2015)
2. ICS-CERT: ICS-CERT Year in Review (2014). https://ics-cert.us-cert.gov/sites/default/files/Annual_Reports/Year_in_Review_FY2014_Final.pdf. Accessed 18 Jan 2018
3. TrendLabs 2016 Security Roundup: A Record Year for Enterprise Threat. http://www.trendmicro.tw/cloud-content/tw/pdfs/security-intelligence/reports/trendlabs_2016_annual_information_security_review.pdf. Accessed 18 Jan 2018
4. Ronald, L.K.: Securing SCADA Systems. Wiley, New York (2006)
5. Ixia: SCADA Distributed Network Protocol (DNP3) (2015). https://www.ixiacom.com/company/blog/scada-distributed-network-protocol-dnp3. Accessed 18 Jan 2018
6. National Instruments: Learn more about the Modbus protocol (2014). http://www.ni.com/white-paper/52134/zht/. Accessed 18 Jan 2018
7. Langner, R.: Stuxnet: dissecting a cyberwarfare weapon. IEEE Secur. Priv. **9**(3), 49–51 (2011)
8. Disso, J.P., Jones, K., Bailey, S.: A plausible solution to SCADA security honeypot systems. In: Eighth International Conference Broadband and Wireless Computing, Communication and Applications (BWCCA), pp. 443–448 (2013)
9. Constantin, L.: New Havex malware variants target industrial control system and SCADA users (2014). http://www.pcworld.com/article/2367240/new-havex-malware-variants-target-industrial-control-system-and-scada-users.html. Accessed 17 Jan 2018
10. Symantec Security Response: Dragonfly: Cyberespionage attack against energy suppliers (2014). www.symantec.com/content/en/us/enterprise/media/security_response/whitepapers/Dragonfly_Threat_Against_Western_Energy_Suppliers.pdf. Accessed 17 Jan 2018
11. Genge, B., Siaterlis, C.: Physical process resilience-aware network design for SCADA systems. Comput. Electr. Eng. **40**(1), 142–157 (2014)
12. Kesler, B.: The vulnerability of nuclear facilities to cyber attack. Strateg. Insights **10**(1), 15–25 (2011)

13. Igure, V.M., Laughter, S.A., Williams, R.D.: Security issues in SCADA networks. Comput. Secur. **25**(7), 498–506 (2006)
14. Stouffer, K., Pillitteri, V., Lightman, S., Abrams, M., Hahn, A.: Guide to Industrial Control Systems (ICS) Security. NIST Special Publication 800-82, Revision 2 (2015)
15. Jiang, J., Yasakethu, L.: Anomaly detection via one class SVM for protection of SCADA systems. In: International Conference on Cyber-Enabled Distributed Computing and Knowledge Discovery, pp. 82–88 (2013)
16. Nader, P., Honeine, P., Beauseroy, P.: lp-norms in one-class classification for intrusion detection in SCADA systems. IEEE Trans. Ind. Inform. **10**(4), 2308–2317 (2014)
17. Fovino, I.N., Carcano, A., Masera, M., Trombetta, A.: An experimental investigation of malware attacks on SCADA systems. Int. J. Crit. Infrastruct. Prot. **2**(4), 139–145 (2009)
18. Pham, V.H., Dacier, M.: Honeypot trace forensics: the observation viewpoint matters. Future Gener. Comput. Syst. **27**(5), 539–546 (2011)
19. Disso, J.P., Jones, K., Bailey, S.: A plausible solution to SCADA security honeypot systems. In: Eighth International Conference on Broadband and Wireless Computing, Communication and Applications, pp. 443–448 (2013)
20. Provos, N.: A virtual honeypot framework. In: USENIX Security Symposium, vol. 173, pp. 1–14 (2004)
21. Conpot: ICS/SCADA Honeypot. http://conpot.org/. Accessed 24 Jan 2018
22. Cherifi, T., Hamami, L.: A practical implementation of unconditional security for the IEC 60780-5-101 SCADA protocol. Int. J. Crit. Infrastruct. Prot. **20**, 68–84 (2017)
23. Maglaras, L.A., Kim, K.H., Janicke, H., Ferrag, M.A., Rallis, S., Fragkou, P., Maglaras, A., Cruz, T.J.: Cyber security of critical infrastructures. ICT Express **4**(1), 42–45 (2018)
24. Wood, A., He, Y., Maglaras, L.A., Janicke, H.: A security architectural pattern for risk management of industry control systems within critical national infrastructure. Int. J. Crit. Infrastruct. **13**(2–3), 113–132 (2017)

Review on Feature Selection Algorithms for Anomaly-Based Intrusion Detection System

Taief Alaa Alamiedy[1], Mohammed Anbar[1](✉), Ahmed K. Al-Ani[1], Bassam Naji Al-Tamimi[2], and Nameer Faleh[3]

[1] National Advanced IPv6 Centre, Universiti Sains Malaysia, 11800 USM, Penang, Malaysia
taiefalaa@gmail.com,
{anbar, ahmedkhallel91}@nav6.usm.my
[2] College of Computer Science and Engineering, Taibah University, Medina, Saudi Arabia
btamimi@taibahu.edu.sa
[3] School of Electrical and Electronic Engineering, USM, 14300 Nibong Tebal, Penang, Malaysia
namerfaluh@gmail.com

Abstract. As Internet networks expand, the amount of network threats and intrusions increased, the demand for an efficient and reliable defense system is required to detect network security vulnerabilities. Intrusion Detection Systems (IDS) are a vital constituent of security of a network to avert data illegal usage and misappropriation. IDS deal with massive amount of data movement that comprises repetitive and inappropriate features. The detection rate implementation is frequently affected by these inappropriate features which also munch up intrusion detection system resources. A significant portion in the removal of dissimilar and not used features in IDS is done by the feature selection method. Methods included are data mining techniques, machine learning, statistical analysis, support vector machine models and neural networks. In this paper, we provide review of several algorithms used for anomaly-based intrusion detection systems to improve performance of machine learning classifiers. This paper first summarizes the theoretical basis of IDS, and then discusses the feature selection techniques and their types.

Keywords: Intrusion detection system · Feature selection · Genetic algorithm
Bees algorithm · Support vector machine · Principle components analysis
Artificial bee colony

1 Introduction

In recent years, the Internet has become an essential part of people's daily activities, such as paying the bills online, sending and receiving money, booking online tickets and other services. In addition to that, government organizations store valuable data over the network and always aim to keep their information protected, available and reliable [1].

F. Saeed et al. (Eds.): IRICT 2018, AISC 843, pp. 605–619, 2019.
https://doi.org/10.1007/978-3-319-99007-1_57

Nowadays, Internet networks have been increased and extended, which has led to an increase in the number of people using the Internet. The demand for the use services will surge, and the more sensitive information is switched across the network [2]. Thus, in providing safe connections as well as conserving information that will be shared over the Internet, there need to be accurate, reliable and configurable safety systems.

In spite of, there are numerous existing protection techniques such as user authentication, information encryption, access controls and firewalls are utilized as the primary line of defense for computer security, but none of them are able to protect network completely. Several researchers are working on developing a new security system to detect attacks, Intrusion Detection System (IDS) is one of the significant ways to obtain higher security in computer networks and used to thwart attacks report by Anderson in 1980 [3].

The main objective of this paper is to provide a theoretical basis of IDS. In addition, this paper provides a comprehensive review on various algorithms that utilized as feature selection technique to enhance the performance of IDS. Finally, this paper suggests the possible future research directions.

The remaining part of this paper is prearranged in the subsequent order. Section 2 introduces the intrusion detection system. Section 3 focuses on feature selection techniques and their methods. Section 4 describes existing works using bio-inspired as well as non-bio-inspired algorithms castoff in feature selection. Section 5 presents discussion. Section 6 explains some of the future research directions. Finally, Sect. 7 is the conclusion.

2 Intrusion Detection System

An Intrusion Detection System (IDS) is the procedure of observing the actions occurred in a computer or networks and inspecting them for sign of probable intrusions [4, 5]. IDS can be classified into dual chief categories, either based on the location of installation in the network or by detection method, as shown in Fig. 1.

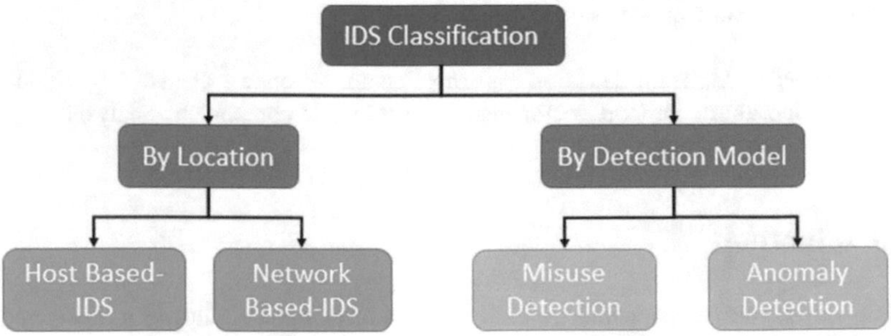

Fig. 1. Intrusion detection system classification [4]

In terms of the location model, IDS can be ordered into two classifications: Host-based IDS and Network-based IDS, as shown in Fig. 2.

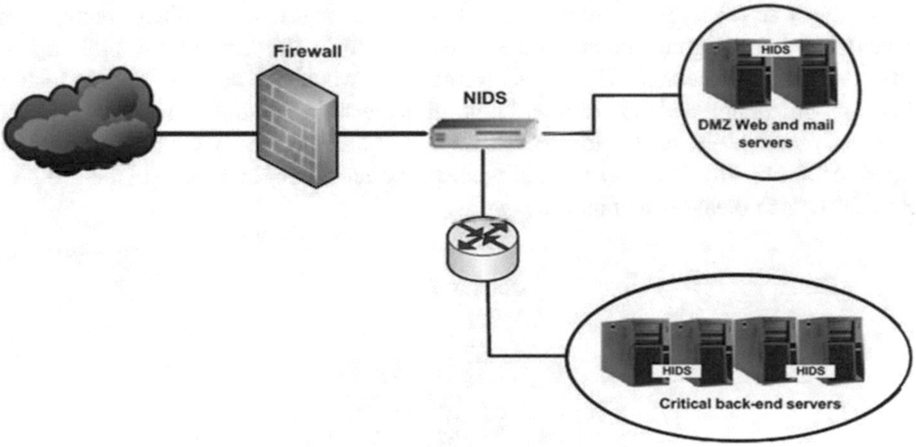

Fig. 2. IDS based on location [6]

- Host-Based IDS: This runs straightforwardly on the client PC also reviewing the information put away on it, for example, log documents, running procedures, and sign in clients. If there are some modification sin important files of the user or operating system, then an alarm will be sent to the administrator to take action [6].
- Network-Based IDS: This will monitor and examine the packets moving across a network to spot activities that are not proper like the denial of services [7, 11].

An IDS can also be separated into double types built on detection method into the following: misuse and irregularity detection.

- Misuse Detection works by comparing the client activity with a stored signature knowledge base of known attacks. It checks an incoming connection against a stored knowledge base if there is any matching then it stops the connection and blocked it. This type has a high accuracy rate to detect known attacks [9].
- Anomaly Detection recognizes interruptions via looking for irregular practices in network traffic that may specify attack, the abnormal behavior can be defined either as the infringement of acknowledged edges for recurrence of events in connection or as client's infringement of the genuine profile produced for normal behavior. This approach can be characterized as a statistical, data-mining, learning based method [8].

3 Feature Selection Technique

The intrusion detection system works in various stages which include the collection of data, preprocessing, classification of the data features and its careful selection. The feature selection stage being a challenging task, in line with the IDS need in handling the gigantic mass of information over the network.

Features selection is a procedure of choosing a subset of significant features by evacuating the insignificant and excess features from the dataset for building an effective learning method. The dataset dimension would be reduced during feature selection processing which in turn will minimize the entire volume of training data that need to process and thus mitigate the time of computation to yield more precise classification [9, 10]. Feature selection process contains four basic steps in the ordinary element choice methods as Fig. 3 shows.

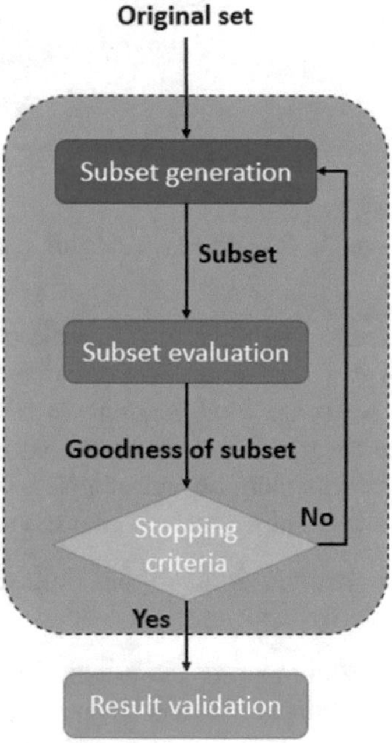

Fig. 3. Feature selection procedure [6]

The first step is the generation procedure to ensure a leaf or part of the features originating from the working data sets. Evaluation function is utilized to evaluate the subset using a learning procedure or by giving cognizance to the underlying data characteristics.

The stopping criteria is to decide when to terminate and the validation step is to check whether the subset is valid [12]. The feature choice algorithms can be characterized into three kinds as follows: wrapper, hybrid method and filter.

3.1 Wrapper Method

The strategy of the wrapper is used as learning procedure in assessing the subcategory of chosen features. Thus, this resulting algorithm used in learning is utilized as an alternate path for the inquiry and search. The target work which is a condition with specific requirements is required in an optimized form. This is used to estimate the subcategory of entire feature based on their prescient precision. This is termed and known as the rate of detection. This resultantly provides two methods of feedback namely: the evaluation and the search. From the search perspective, careful parameter settings are used that are then assessed utilizing the components of evaluation. It utilizes an algorithm of learning calculation to assess the non-valuable highlights and in this manner delivers better component subsets. The resulting features derived are thus, chosen in light of the precision of the classifier. The wrapper technique is broadly utilized in many fields due to its acceptability, despite the fact that it is slower than other different strategies. Figure 4. Depicts the procedure of wrapper technique [13].

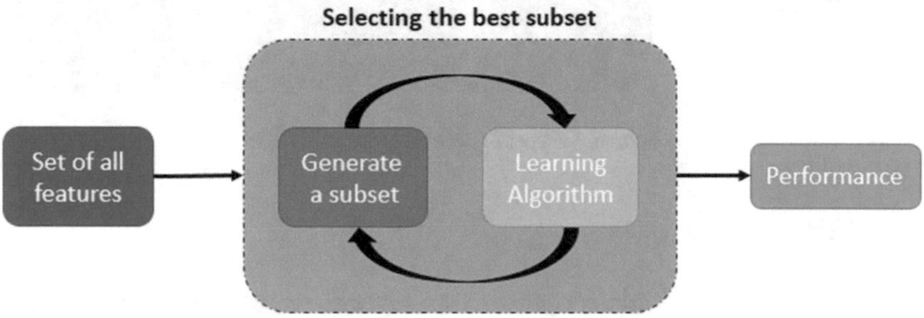

Fig. 4. Wrapper method process [13]

3.2 Filter Method

This method filters insignificant features that have a little option in the analysis of data. The selection of features is independent of any machine learning algorithm. The features that are carefully selected are resultantly assessed in accordance to the characteristics of the data. The filter selection is based on the degree of features (meaning the data are selected according to the resulting numbers of features). However, in some cases all feature totality is the consideration. Therefore, to appropriate the threshold, it is vital for a subset selection to use. The method resultantly has high computation efficiency. Figure 5 Shows the process of filter method [14].

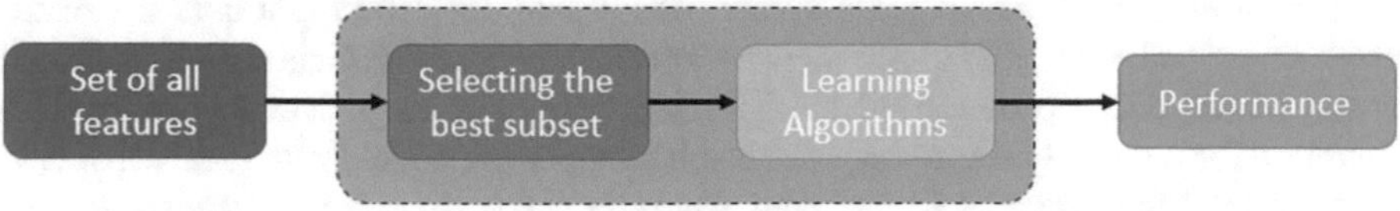

Fig. 5. Filter method process [14]

3.3 Hybrid Method

The hybrid method just as the name implies is defined as the blending of the wrapper and filter approaches to obtain the greatest performance with a particular learning algorithm. It implicitly uses the feature selection algorithm; it searches for an ideal set of structures that are built into the classifier construction [15]. Samples of a hybrid technique are the decision tree with naive Bayes classifier together. Figure 6 shows the hybrid method process.

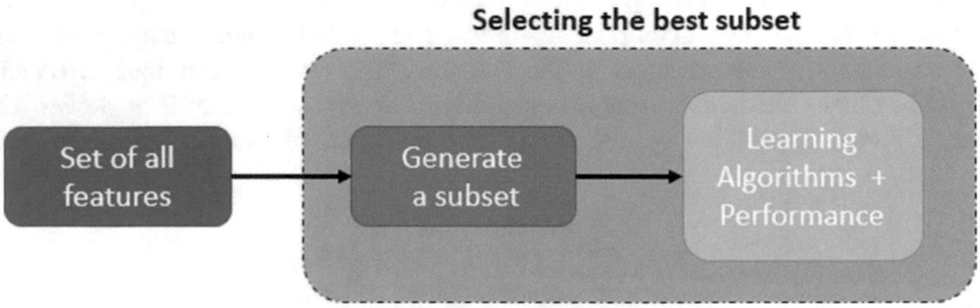

Fig. 6. Hybrid method process [15]

4 Existing Works

This section explains the various optimization algorithms used as feature selection to solve various problems, for example in intrusion detection system, the feature selection technique used to moderate a few issues like precision, huge datasets, lopsided circulation of information and troublesomely to distinguish limits among typical and unusual practices. This survey is divided into two parts, the first one explains feature selection built mainly on an algorithm that is bio-inspired, and the second part is concerned with non-bio-inspired algorithms used as feature selection.

4.1 Feature Selection Based on Bio-Inspired Algorithms

Bio-inspired as it relates and carefully defined by its name, are algorithms that are driven by the characteristics of the nature directives. It is used to formulate and proffer solutions to problems by the use of methods found in nature. These algorithms are further grouped in three classes, namely: evolutionary, swarm-based, and ecology-based [14].

Ahmad et al. [16] anticipated a system of intrusion detection that uses the feature subset selection based on MLP. In their approach they combine Genetic Algorithm (G. A) and Principal Component Analysis (PCA), they applied the genetic algorithm (G.A) to explore for the main feature space for a subcategory of features with ideal sensitivity, and they used the Principal Component Analysis (PCA) to scheme features space to main feature space and select features consistent to the maximum eigenvalues, they utilized multilayer perception classifier in their approach, and they used a technique for ideal features subcategory choice to overwhelm performance problems by means of PCA, G.A and Multilayer Perception (MLP). The KDD-Cup dataset was used and the proposed approach achieved an optimized feature of 12 from 41 rows, the result showed that the accuracy reach to 99% and ideal features improved precision, diminished preparing and computational overheads streamlined the engineering of intrusion analysis engine and maximized the detection rates.

In the work of Kuang et al. [17]. An innovative Support Vector Machine (SVM) prototype merging Kernel Principal Component Analysis (KPCA) with Genetic Algorithm (GA) was an intrusion discovery system proposal. Relative to N-KPCA-GA-SVM model, an adoption of KPCA became achievable, upon which foremost characteristics playing a key role in intrusion detection data was extricated. To guesstimate if an action of a particular manner is an attack, an employment of multi-layer SVM classifier became handy. Gaussian kernel based function became the core for a new radial basis kernel function (N-RBF), which was created with the view to be shortening the time consumed in training and improving the performance of the model of SVM classification. In selecting appropriate parameters, GA was deployed in the SVM classifier. This procedure prevents under-fitting or over-fitting the SVM model emerging as a result of indecorous determination of parameters. The investigational outcomes flaunted that the accuracies gathered from the KPCA-SVM model are grander and more sizable than that of the SVM classifiers characterized by randomly selected parameters. The results also flaunted that outcomes of SVM classifiers upon which the extraction of their features was done using PCA, KPA proved more robust in terms of performance generalization than those lacking extraction of their embedded feature. Furthermore, the investigational outcomes also revealed that KPCA performs better on sets of intrusion detection data than the PCA. This outcome was geared due to the fact that KPCA has the ability of higher information exploration on initial inputs as aligned with PCA. KPCA covertly puts into justification higher intelligence on initial inputs through the employment of kernel technique of generalizing PCA to nonlinear.

In KPCA, extraction of increased size of primary features can be achieved, eventually yielding outcomes whose performances are of better generalization.

In Aslahi et al. [18] research a hybrid technique of SVM and GA was proposed for intrusion detection system. The proposed mix algorithm was utilized in decreasing the quantity of highlights from 41 to 10. The features were ordered into three priorities utilizing GA algorithms as the most significant and of highest significance and the least essential is put in the lower significance category level. The dispersal is done such that 4 features are set in the highest significance, 4 included in the next, and 2 included in the third significance. The outcomes show that the anticipated hybrid algorithms can achieve a genuine positive estimation of 0.973, while the false-positive value was (0.017).

Alomari et al. [19] anticipated a wrapper based component choice approach utilizing the Bees Algorithm (BA) as a exploration technique for subcategory generation, and also utilizing SVM as the classifier. The analyses utilized four arbitrary subsets gathered from KDD-Container 99. Every subset contains around 4000 records. The performance of the anticipated method is assessed by means of the standard IDS estimations. The result shows that the detection accuracy achieved (99%) and the feature set reduced to (8) features, while the false alarm rate was (0.004).

Alternatively, ant colony (ACO) and SVM choice feature weighting of network interruption recognition strategy proposed by et al. [20]. They combined ant colony algorithm to choose the components with a component weighting SVM. In the first place, they utilized SVM grouping precision and highlight subset measurement to develop a complete fitness weighting index. subsequently, they utilized the ant colony algorithm for an optimization that is global and numerous exploration capabilities to accomplish for the optimal solution feature search feature. Afterwards, chose the key element of network data and figured data access different highlights weights and calculated information was used varying features weights and heavy weights to bolster vector machine classifier considering the attributes of network attacks right. Finally, this refined the last plan of the nearby search strategies to make the component determination comes about without repetitive features while enhancing the convergence resistance, and confirmed the data set by KDD-CUP99 adequacy of the algorithm. The results exhibited that the proposed approach can successfully reduce the dimension of features and have enhanced network intrusion detection accuracy to (95.75%)

Rani et al. [21] proposed a detection system that is hybrid intrusive. The system is likewise centered on C5.0 Decision Tree, which also uses a One-Class SVM. C5.0 is used to train the misuse discovery model in the hybrid intrusion detection system. The mismanagement detection model can distinguish recognized attacks with a low false alarm rate. One-class SVM was applied to the anomaly detection (trained using normal training traffic). During the training procedure, decision boundaries are chosen normal data contained in the origin dataset. The outliers are detected as using decision function and the model classify outliers as attack connection. The Experimental results were performed on NSL-KDD Dataset, we were able to show that overall performance of the planned method was enhanced the in terms of discovery rate and low false alarms rate in evaluation to the current available methods.

Ghanem et al. [22] anticipated a novel method based on multi-objective Artificial Bee Colony (ABC) for feature selection, particularly for intrusion-detection systems. This approach is classified into two stages: generating the feature subsets of the Pareto front of non-dominated solutions in the first stage and using the hybrid ABC and particle swarm optimization (PSO) with a Feed-Forward Neural Network (FFNN) as a classifier to evaluate feature subsets in the second stage. Thus, the anticipated method consisted of twofold steps: the first one, using a new feature selection technique called multi-objective ABC feature selection to diminish the number of features of network traffic data and the second one, used a new classification technique called hybrid ABCPSO optimized FFNN to classify the output data from the previous stage, determine an intruder packet, and detect known and unknown intruders. The proposed approach did not only provide a new approach for feature selection but also proposed a new fitness function for feature selection to diminish the number of features and achieve the minimum rate of classification errors and false alarms.

Acharya et al. [23] proposed an Intelligent Water Drops (IWD) algorithm that is based on the feature selection technique. The method is characterized by IWD algorithm. Inspired by nature, the method being an optimization algorithm, is applicable in the selection of feature subset while a vector machine plays the role of a classifier in the evaluation process of selected features. Support Vector Machine (SVM) was the classifier used. Amongst parameters used as evaluation were size of feature subset, false alarm rate and the rate of the classifiers detection. The IWD, being a Meta heuristic optimization algorithm, yielded and optimized procedure of selecting feature for SVM. From 41 to 9, the model substantially reduced the features. Parameters found to have been better improved as presented in the new model with a proposed method (IWD + SVM) are precision, false alarm rate accuracy and rate of detection. This outcome measured improved than other prevailing models. A precision rate of 99.4075% and an accuracy score of 99.0915% were recorded on the new model. While a low false rate of 1.405% and a precision rate of 99.108% were also recorded. The period used by this prototype to do the train was remarkable minimized to 1.15 min.

A likeness of bio-inspired feature selection procedures employed relative to IDS used for IDS is presented in Table 1.

Table 1. A comparison of bio-inspired feature selection algorithms

Ref.	Technique	Dataset used	Feature length	Classifier	Performance evaluation
[16]	G.A + PCA	KDD CUP99	12	Multilayer Perception	Accuracy: 99%
[17]	Hybrid kernel PCA + G.A	KDD CUP99	12	Multilayer SVM	Detection rate: 99.22
[18]	Hybrid method of G.A and SVM	KDD CUP99	10	Support vector machine (SVM)	True positive rate: 0.97 True negative rate: 0.01
[19]	Bees algorithm	KDD CUP99	8	Support vector machine (SVM)	Detection rate: 98.38 False alarm rate: 0.004
[20]	Ant colony optimization + feature weighting SVM	KDD CUP99	–	Support vector machine (SVM)	Detection rate: 95.75
[21]	Cuttle fish algorithm	*NSL-KDD*	–	C 5.0 + one class SVM	Accuracy: 98.20 False alarm rate: 1.405
[22]	Artificial bee colony Optimization	KDD CUP99	–	Feed – forward neural network	–
[23]	Intelligent water Drop (IWD)	KDD CUP99	9	Support vector machine (SVM)	Detection rate: 99.40 False alarm rate: 1.405

4.2 Feature Selection Based on Non-inspired Algorithms

This section reviews different non-bio-inspired algorithms used as feature selection technique. As shown in Table 2, different combinations of algorithms can provide various advantages which improve the performance of IDS.

An rational Conditional Random Field (CRF) founded feature selection for effective intrusion discovery was anticipated by Ganapathy et al. [24]. The new CRF based

Table 2. A comparison of bio-inspired feature selection algorithms

Ref.	Technique	Dataset used	Feature length	Classifier	Performance Evaluation
[24]	Conditional random field based	KDD CUP99	41	Layered Approach (LA) based algorithm	Accuracy: (Dos: 97.62, Probe: 98.83, R2l: 32.43, U2R: 17.14)
[25]	Gradually feature removal method	KDD CUP99	19	Support vector machine (SVM)	Accuracy: 98.62%
[26]	Filter based	KDD CUP99	25	Support vector machine (SVM)	Accuracy: 99.94%
		NSL-KDD	28		
		KYOTO 2006	15		
[27]	Enhanced correlation based model	KDD CUP99	12	–	Accuracy (Probe: 99.98, Dos: 99.3 R2L: 98.10, U2R: 86.91

feature selection algorithm was used to enhance the quantity of features. Furthermore, a current Layered Approach (LA) based algorithm was utilized to perform grouping with these condensed features. This intrusion detection system gave high precision and accomplished more noteworthy productivity in attack detection contrasted with the current methodologies. The real preferences of this proposed framework is in terms of lessening detection time, expanded accuracy in classification and also in reduced false alarm rates. The KDD-CUP99 dataset benchmark was used in this research and the experimental results showed that the accuracy for different types of attack was (probe = 99.98, DOS = 97.62, R2L = 32.43, U2R = 86.91).

Li et al. [25] anticipated an effective intrusion discovery system founded on SVM and progressively feature elimination technique. In their approach, a pipeline of IDS was determined through a sequences of machine learning approaches. They started by developing a built-in dataset by gathering excess information into a minimized set and choosing a suitable small training data with the technique for ACO. They then reduced the feature measurement from 41 to 19 to preserve the main feature of the network visit, get the classifier with SVM, then commence an exhaustive forecast for the total KDD-CUP dataset. The correctness of this proposal IDS reached 98.6249%, and MCC value achieved (0.861161). The corresponding result showed that this IDS is consistent which performs well in terms of precision and efficacy. Importance was placed on research into the feature reduction method. Giving an accurate wrapper-based feature reduction technique, the GFR technique was anticipated. This technique offered realistic assets specific feature selection then showed advantages in the investigational results.

An intrusion detection system that utilizes a filter-based feature selection algorithm was proposed by Ambusaidi et al. [26]. They anticipated a common information-based algorithm that systematically chooses the ideal element for grouping. This common information-based feature selection procedure can deal with linearly and nonlinearly reliant on data features. Its efficiency is assessed in the instances of network discovery. An IDS, termed Least Square Support Vector Machine based IDS (LSSVM-IDS) was assembled using the features chosen by the proposed feature selection procedure. The performance of LSSVM-IDS was assessed utilizing three intrusion detection evaluation datasets, to be specific KDD Container 99, NSL-KDD and Kyoto 2006 dataset. The assessment obtained demonstrated that the feature selection procedure contributed more serious features for LSSVM-IDS to accomplish improved precision and lesser computational cost contrasted and the best in class techniques.

Madbouly et al. [27] proposed an upgraded model to choose an arrangement of the most applicable features to increase attack detection precision and enhance general system performance. In their approach, the KDD99 dataset was utilized, and the chosen important features containing just 12-beyond the 41-full features set. This now decreases the magnitude of the KDD CUP99 workbench dataset by over 70%. They gauged the performance of the proposed model and confirmed its viability and attainability by comparing it with nine-unique models and with a model that utilized the 41-features dataset. The experimental outcomes demonstrated that, their improved prototype could productively accomplish increased detection rate, performance rate, low false alarm rate, and quick and consistent detection process.

5 Discussion

Intrusion Detection System (IDS) deals with large amount of network traffic. IDS work by analyzing network traffic pattern to detect any suspicious behavior. However, the traffic information contains a lot of features, and these features degrade the performance of IDS detection. Feature selection technique assists to diminish the number of features in the dataset by choosing the most relevant and effective features by eliminating irrelevant the redundant and irrelevant features.

The existing algorithms for feature selection technique suffer from many dilemmas. In this study, we divided the existing algorithms into two categories, the first part deals with bio-inspired algorithms and the second one with non-bio-inspired algorithms. For bio-inspired algorithms, as reported in this approach [16], they utilized a combination of two feature selection methods which are wrapper and filter and involves MLP for classification. They were able to reduce the number of features to (12) and achieve a 99% detection accuracy. However, the combination of these two feature methods may increase the computation complexity and need additional time for training. In the work [17] enhanced the combination of feature selection methods by implementing Kernel Principal Component Analysis (KPCA) with Genetic Algorithm (GA). In addition, they used multilayer SVM for classification. The result shows that the implementation of KPCA was valid. Besides that, they also used basis kernel function (N-RBF), which was designed with the view to shortening the time consumed in training and improving the performance of the model of SVM classification. The result of this study was compared to previews work enhanced by achieving (99.20%) for detection accuracy.

The work by in [18], it suggested studying the outcome of a combination of feature selection algorithm with classifier in selecting the significant feature set, their result was better in compression to the result in the works [16, 17], they were able to reduce the feature set to (10) features. In the work in [19], the authors utilized a wrapper feature selection with SVM classifier. In their suggested approach, the feature sets decreased to (8) features and the false alarm rate was equal to (0.004). In association with the previous works in [16–18], they provide the best result in term of the number of selected features. In approach [20], they combined ant colony algorithm to select the components with a component weighting SVM, then used ant colony algorithm to select the optimum feature, in their result, the classification accuracy achieved (95.75). In [21], the authors presented a Cuttle fish algorithm for feature selection with the combination of C.50 and one class SVM for classification, in their approach they used NSL-KDD to evaluate their proposed approach, the experiment result exhibits a high result for classification accuracy. Ghanem and Jantan [22], suggested a novel method based on multi-objective Artificial Bee Colony (ABC) for feature selection, they divided their approach into two stage: generating the feature subsets of the Pareto front of non-dominated solutions in the first stage and using the hybrid ABC and Particle Swarm Optimization (PSO) with a Feed-Forward Neural Network (FFNN) as a classifier to evaluate feature subsets in the second stage. In addition, they proposed a new fitness function for feature selection to diminish the number of features and achieve the minimum rate of classification errors and false alarms. [23], proposed an Intelligent Water Drops (IWD) algorithm for feature selection and used SVM for classification. In their approach, they reduced the number of features to (9) feature set. Meanwhile, the experiment result showed a good result for classification accuracy.

For non-bio-inspired algorithms, various researchers employed different kinds of methods for features selection technique. In the work conducted by [24], they used rational Conditional Random Field for feature selection with the Layered Approach (LA) based algorithm for classification, their result achieved high accuracy for probe attack. In the approach proposed by [25], they utilized Gradually feature removal method with SVM for classification, their approach was able to reduce the number of features to (19) feature and archived (98.62) for classification accuracy. The study of [26], has traditionally relied on three benchmark data set for evaluating their approach, they used filter-based method for feature selection with SVM for classification, the number of selected was (25, 28, 15) for KDD99, NSL-KDD and KYOTO 2006 benchmark datasets respectively. The classification accuracy achieved (99.94). Also, in research [27], they utilized Enhanced correlation-based model for feature selection. In their approach, they were able to reduce the number of features to (12) feature and archived (99.98) for detection accuracy.

Overall, this study provides a strong evidence for the efficacy of using bio-inspired algorithms as a feature selection technique. In terms of feature selection process, bio-inspired algorithms working behavior based on meta-heuristic approach, this method worked on more than one solution at the time which may reduce the processing time. Another benefit of these algorithms is their ability to work on more than one solution at the same time and thus reduce the time required for training. The other techniques are working by select several features set depending on many specifications on the features records, for example, the amount data recorded, then it may ignore the remainder of the feature set that contains less information. However, these features may have an impact on the performance of the IDS in attack detection.

6 Future Research Directions

Based on existing research, the following areas of research are suggested in the field of feature selection for IDS:

(i) Multi-objective function to evaluate the selected feature. Currently, the evaluation of the feature set that represents the initial population is based on a single objective function which is the accuracy detection. Having a multi-objective function such as accuracy and detection rate will contribute to selecting the most significant features to evaluate each feature set.

(ii) For Non-bio-inspired algorithms which include traditional feature selection algorithms such as PCA and IGR, proposing an ensemble feature selection mechanism where two or more feature selection. For example, IGR and PCA can be used together to ensure that the selected features are the most significant features; if a set of features is selected only by IGR or PCA then these features will not be considered as significant features. Therefore, only the features that selected by using both IGR and PCA will be considered to ensure that the selected features are significant.

(iii) For bio-inspired algorithms, enhance the population initialization. In this case, the feature vector includes two parts: (i) the fixed part which includes only the significant features that are used by other researchers (ii) dynamic part which will be generated by the bio-inspired algorithm in each iteration. As result, the computation time of bio-inspired algorithm will be reduced.

7 Conclusion

Intrusion Detection System (IDS) deals with a large quantity of data translated over the network. Since the IDS needs to inspect and analyze all the packets of these data to detect any suspicious activity. However, the data packets contain many features, including redundant and irrelevant features. As a consequence, these features will be effected on the performance of the IDS and consume a lot of system resources. In that case, to resolve these issues, feature selection is a process of selecting significant features by eliminating both the redundant and irrelevant features. This procedure will decrease processing period and increase IDS performance.

This overview presents various feature selection methods and approaches for an IDS. Every algorithm was used with same data and same metrics used for measurements. A few algorithms delivered better outcomes for specific measurements. There is no approach which supersedes in each measure or which delivered a general best outcome. As this investigation proposes, there is a justifiable reason explanation behind the pattern lately toward hybrid strategies, as the best outcomes appear to be gotten after bio-enlivened procedures and data mining procedures are merged, neural networks, rough set theory, and fuzzy logic. The hybridization of bio-inspired procedures with other methods enhance execution and gives operational problem-solving.

References

1. Al-Ani, A.K., Anbar, M., Manickam, S., Al-Ani, A., Leau, Y.-B.: Proposed DAD-match mechanism for securing duplicate address detection process in IPv6 link-local network based on symmetric-key algorithm. In: International Conference on Computational Science and Technology, pp. 108–118 (2017)
2. Al-Ani, A.K., Anbar, M., Manickam, S., Al-Ani, A., Leau, Y.-B.: Proposed DAD-match security technique based on hash function to secure duplicate address detection in IPv6 link-local network. In: Proceedings of the 2017 International Conference on Information Technology, pp. 175–179 (2017)
3. Kim, D.S., Nguyen, H.-N., Ohn, S.-Y., Park, J.S.: Fusions of GA and SVM for anomaly detection in intrusion detection system. In: International Symposium on Neural Networks, pp. 415–420 (2005)
4. Dastanpour, A., Ibrahim, S., Mashinchi, R.: Using genetic algorithm to supporting artificial neural network for intrusion detection system. In: The International Conference on Computer Security and Digital Investigation (ComSec 2014), pp. 1–13 (2014)
5. Anbar, M., Abdullah, R., Al-Tamimi, B.N., Hussain, A.: A machine learning approach to detect router advertisement flooding attacks in next-generation IPv6 networks. Cognit. Comput. **10**, 1–14 (2017)
6. Vithalpura, J.S., Diwanji, H.M.: Analysis of fitness function in designing genetic algorithm based intrusion detection system. Int. J. Sci. Res. Dev. **3**, 86–92 (2015)
7. Kemmerer, R.A., Vigna, G.: Intrusion detection: a brief history and overview. Computer (Long. Beach. Calif.) **35**, 27–30 (2002)
8. Liao, H.-J., Lin, C.-H.R., Lin, Y.-C., Tung, K.-Y.: Intrusion detection system: a comprehensive review. J. Netw. Comput. Appl. **36**, 16–24 (2013)
9. Shahreza, M.L., Moazzami, D., Moshiri, B., Delavar, M.R.: Anomaly detection using a self-organizing map and particle swarm optimization. Sci. Iran. **18**, 1460–1468 (2011)
10. Anbar, M., Abdullah, R., Saad, R., Hasbullah, I.H.: Review of preventive security mechanisms for neighbour discovery protocol. Adv. Sci. Lett. **23**, 11306–11310 (2017)
11. Garcia-Teodoro, P., Diaz-Verdejo, J., Maciá-Fernández, G., Vázquez, E.: Anomaly-based network intrusion detection: techniques, systems and challenges. Comput. Secur. **28**, 18–28 (2009)
12. Kumar, K., Kumar, G.: Analysis of feature selection techniques: a data mining approach. International Journal of Computer Applications (0975 – 8887), 4th International Conference on Engineering & Technology (ICAET 2016) pp. 17–21
13. Kumari, B., Swarnkar, T.: Filter versus wrapper feature subset selection in large dimensionality micro array: a review **2**(3), 1048–1053 (2011)
14. Binitha, S., Sathya, S.S., et al.: A survey of bio inspired optimization algorithms. Int. J. Soft Comput. Eng. **2**, 137–151 (2012)
15. Bolón-Canedo, V., Sánchez-Maroño, N., Alonso-Betanzos, A.: A review of feature selection methods on synthetic data. Knowl. Inf. Syst. **34**, 483–519 (2013)

16. Ahmad, I., Abdullah, A., Alghamdi, A., Alnfajan, K., Hussain, M.: Intrusion detection using feature subset selection based on MLP. Sci. Res. Essays **6**, 6804–6810 (2011)
17. Kuang, F., Xu, W., Zhang, S.: A novel hybrid KPCA and SVM with GA model for intrusion detection. Appl. Soft Comput. **18**, 178–184 (2014)
18. Aslahi-Shahri, B.M., Rahmani, R., Chizari, M., Maralani, A., Eslami, M., Golkar, M.J., Ebrahimi, A.: A hybrid method consisting of GA and SVM for intrusion detection system. Neural Comput. Appl. **27**, 1669–1676 (2016)
19. Alomari, O., Othman, Z.A.: Bees algorithm for feature selection in network anomaly detection. J. Appl. Sci. Res. **8**, 1748–1756 (2012)
20. Xingzhu, W.: ACO and SVM selection feature weighting of network intrusion detection method. Analysis **9**, 129–270 (2015)
21. Rani, M.S., Xavier, S.B.: A hybrid intrusion detection system based on C5.0 decision tree and one-class SVM. Int. J. Curr. Eng. Technol. **5**, 2001–2007 (2015)
22. Ghanem, W.A.H.M., Jantan, A.: Novel multi-objective artificial bee colony optimization for wrapper based feature selection in intrusion detection. Int. J. Adv. Soft Comput. Appl. **8**, 70–81 (2016)
23. Acharya, N., Singh, S.: An IWD-based feature selection method for intrusion detection system. Soft Comput. **22**, 1–10 (2017)
24. Ganapathy, S., Vijayakumar, P., Yogesh, P., Kannan, A.: An intelligent CRF based feature selection for effective intrusion detection. Int. Arab J. Inf. Technol. **13**, 44–50 (2016)
25. Li, Y., Xia, J., Zhang, S., Yan, J., Ai, X., Dai, K.: An efficient intrusion detection system based on support vector machines and gradually feature removal method. Expert Syst. Appl. **39**, 424–430 (2012)
26. Ambusaidi, M.A., He, X., Nanda, P., Tan, Z.: Building an intrusion detection system using a filter-based feature selection algorithm. IEEE Trans. Comput. **65**, 2986–2998 (2016)
27. Madbouly, A.I., Barakat, T.M.: Enhanced relevant feature selection model for intrusion detection systems. Int. J. Intell. Eng. Inform. **4**, 21–45 (2016)

Forensic Use Case Analysis of User Input in Windows Application

Funminiyi Olajide(✉), Tawfik Al-hadrami(✉), and Anne James-Taylor(✉)

School of Science and Technology, Nottingham Trent University, Nottingham, UK
{Funminiyi.olajide,Tawfik.al-hadhrami, Anne.James-Taylor}@ntu.ac.uk

Abstract. Most of digital investigation research on Windows application covers stages of investigation that covers key aspect of forensic image collection, preservation, dumped data and extraction, searching for evidence and possible reconstructions of user input evidence and analysis in a use case environment of business organisations. The user activities on some of the business applications can reveal user involvement on the perception of cyber-human factor. Tangible user information were extracted and reconstructed to determine the forensic artifact in Windows business application. On investigations, the forensic validation process of key stages of digital investigation revealed relevant information on various experiments carried out. The research idea focus on the use cases of MS PowerPoint and MS Excel of Windows business applications. The research determined and formulated the extraction process of user evidence from a sample memory forensics investigation. The quantitative assessment of relevant information was presented to uncovers how user input information are stored and as recovered from the application memory. In this paper, a design methodology to capture, extract and process relevant user information was described on the two most commonly used applications on Windows systems.

Keywords: Application · User · Business · Information · Digital · Use case Forensics · Input · Investigation

1 Introduction

The research work begin by discussing related work, both in digital forensics, digital evidence, seized digital devices, dump memory and the extracted dump memory allocated to user input. It also discusses dump memory are converted into strings of useful evidence, the likelihood ratio of academic reviews. Event of user input on applications are increasingly becoming vital in digital forensic investigations process. Law enforcement agencies were always strives to acquire as much data as possible on digital devices to support any ongoing investigation at hands. A paper of [1] stated that forensic evidence has evidential motives towards security of business or an organisation. This was further discussed on the complex of security schemes in business environment. This enables law enforcement agencies, to invest in deeper knowledge, training and in acquiring state-of-the-art most costly equipment, to perform the advanced forensic

F. Saeed et al. (Eds.): IRICT 2018, AISC 843, pp. 620–629, 2019.
https://doi.org/10.1007/978-3-319-99007-1_58

image acquisition, data extraction and analysis. The motive of this is the ability to present sound and validated useful evidences for use in the court of law.

Government unit [2] and many units of the law enforcement agencies are now investing in the forensic e-discovery/e-disclosure of both classified and unclassified user input activities or fraudulent actions on business applications [3]. As a result, a process of engaging in sound forensic data acquisitions, images preservation, security solution access to systems vulnerabilities, and application security mechanisms become imperative to digital evidence. Digital forensic investigations focus more on recovery of relevant data from digital devices [4]. However, digital forensics tools has become an indispensable for information assurance, solving crime and tracing fraud, whereby, relevant evidences may reside on some Windows Application.

A method of [5] laid emphasis on the importance of forensic live response and event reconstruction methods. The extension of this work relies on the research of application level evidence from physical memory [6]. This approach identified the important aspects of memory analysis and proposed an approach for application level evidence from volatile memory. This output of the research augment the tools and method presented in [7] which is among the few hardware-based memory acquisition that changes memory contents, as little as possible by using a PCI extension card to dump the memory content to an external device. A paper of Garfinkel (2010), a related research of digital forensics, and challenges for the next decade, has resulted into many research extension of user input on business application, including social media, on the knowledge of how digital forensic research should be directed [8]. Digital forensic research investigation is a valid method for processing cybercrime, and for use in continuation of numerous cyber-attacks in many years to come. For example, in business organisations, the research prophecy focuses on image recovery, forensic acquisition, and investigation. This has become increasingly harder as data increases, but a paper of [9], presented on the fact that digital forensics readiness are still at infancy. For example, big data analytical, with the use of information systems are increasing, with ever security solution for many commercial tools. Hence, forensic investigators and law enforcement agencies, faces more of an increasing challenges when it comes to the ability and quick access to forensic data acquisition. The research extension on this can also be verified on the amount of user input on other Web browsers, such as Google [10]. Therefore, user input information on business applications can depends on the activities carried out per time of user access. This includes the event histories of user activities that were routinely log-filed on the application memory of the computer systems [6]. However, it is important to highlight more research on business applications. This is on a process of if user input are contained in other devices, such as mobile phones. It can be said that for a particular user input on any of the commonly business applications, such as MS Word, MS Excel, MS PowerPoints, these are certainly contained logs of user activities, that can typically consists of a list of events and processes for use as evidence.

Each event consists of a date-time-stamp and with some metadata associated with each of these events. For example, with popular Microsoft Office of Windows business applications [11], there are varieties of user events recording, that can be related to user actions, as user are logged on to the applications. This is in a typical business environment of the local network communications of these applications. Therefore, in a temporal, functional and relational analysis of user activities on business applications, user actions may include logs of user input. These logs can uncover what a user is typing at a particular time, what they have been doing and what they are using the applications for. The log files of user activity are also often accessible via Enterprise Systems. The user event relates to functional activities of when user opens and closes applications on Windows systems. The research intention focuses on the need for a forensic investigator access to all these user events. This is to ascertain what a user is doing or typing on these applications and also, determine what the user is using the application for.

The rest of the paper is organised as follows: Sect. 2, focus on the "Related work and contributions" discusses related work and how our contribution are important in relation to other research work in digital forensic investigations. Section 3, discusses the methodology and approach adopted in the forensic image acquisition, memory dumping, extractions, string conversion, and metrics formulated. Section 4, focus simply on research result and evidence analysis and presentation of relevant evidences. Finally, Sect. 5, is the conclusion part of this research work where we discuss the implications of our findings and offer our conclusions with a brief mention of the future work while Sect. 6, is the list of references.

2 Related Work

User input on Windows application and enterprise systems are on the perspective of whether the systems resides either on-site or on cloud-based solution platform. Users interest on the use of Windows Systems and Mobile devices are common nowadays, with the growing use of social media. It can be said that both end users and use of enterprise systems are demanding with keylogging activities, to determine useful information and more secure devices, to help protect sensitive data. However, in a typical business environment, there is need to secure business data on transit of unauthorized users either on mobile devices or Windows applications. This has also resulted in the context of many business perspectives and adoption of BYOD (Bring Your Own Device) policy [12]. These policies are operating without checks in many businesses, and thus, events of user input on using different devices, copying data in and out of the network or enterprises network, must be recorded and monitored. This is to trace-route and can find out what a user is doing, what the user has been doing and what the user is using the application for. A use case of reviews of document information [13] of Facebook [14] and Cambridge Analytical [15] was a typical example of unchallenged user access on Social Media Applications. This includes Windows and Enterprise Systems, which can results into a possible cyber-attack on businesses. This information can be used as political vectors at hostile states for cyber criminals and cyber espionage. It is therefore evident to implement digital forensic readiness in business.

The important of digital forensic research for relevant evidence, can help in promoting digital forensics investigation process and digital forensic readiness. These are useful information security guidelines in business organisations. The focus of this paper is to examine the benefits of integrating digital forensic evidence in readiness and as useful evidence of user input on applications, following ACPO [16] forensic legislation and laws. Understanding events of user data in series of temporal, functional and relational activities including data correlation and events reconstructions processes may be consider as useful tools and as information framework to forensic investigators. The extracted evidence information can also be beneficial to business organisations in any related cases and for use as evidence in the court of law.

Business willingness to implementing digital forensic readiness and information security policies is vital. A range of software-based tools has recently been developed for memory acquisition and memory analysis [17]. A method of [18] reviewed a tool that is capable of revealing hidden and terminated processes and threads. The popularly known commercial tools was considered, such like FTK [19] and EnCase [20], as tools that can capture the physical memory of computer systems and perform all detail analysis required as evidence on crime cases. In addition to these tools, the Volatility Framework [21] are examples of other tools that can perform memory analysis. Volatility Framework is extensive and capable of performing the analysis on a variety of memory image formats, such as DD format, crash dump and Hibernate Dumps. Volatility is able to list OS kernel modules, drivers, open network socket, loaded DLL modules, heaps, stacks and open files. Therefore, this research paper identified the two commonly used business applications such like MS Excel and MS PowerPoint. User input on these applications were recovered and extracted from the memory allocated to each application. As a result of the forensic evidence information, there are further reviews on capable tools for memory analysis. However, recovering user data from Windows devices can be achieved by extracting relevant user information as stored and recovered from volatile memory (RAM). The two sources of data from MS Excel and MS PowerPoint differs both on how data is stored on the application memory. For example, Nathan Scrivens et al. (2017) summarized many of the current options for forensic acquisition of storage on Android mobile devices [22]. Data in long-term storage is often stored and well-structured in file systems. This has to be read by different operating systems and through other tools. This is because data structures in RAM is often less well documented, and the formats are more volatile, as it needs only survive to the next restart of the device. RAM is repopulated each time the device is restarted. Thus, backup applications are rarely accessible to unauthenticated users and are often of limited use for forensics. Based on various reviews and understanding, the research investigated on how much information can be recovered from the memory content of MS Excel and MS PowerPoint applications.

3 Methodology

In this session, we discussed the process of our data collection in corresponding to various user input on business applications. We represent the forensic investigation that happened in a typical business organisation. For example, a use case element of a reported cyber-attack in business environment was considered. A crime case, as invited the involvement of forensic investigators in the business site. As reviewed in the above Session 2, we investigated the seized digital evidence and we hence, engaged in various designs and methodologies to actualize the process of extracting relevant evidence for use as evidence in the law court. During our investigation, the likelihood of user activities in the contexts of determining whether events of user input are applicable, to the user-name access of the applications and as attempted on the seized digital evidence. The motive of our investigation is the ability to record the related user event histories stored on the application memory of digital devices, for example, the Windows Application. This investigation is traceable to the possibility of extracting the relevant evidence and reconstructing evidences, as fragments of useful information. It was found that during the cyber incidence of the reported case, the investigators secured the incidence-site in response. Some digital evidences that were seized can helps in the case at hands. The data gathering process was established and data preservation of tangible evidence was determined. The memory captured and memory dumping using Forensic Volatility was achieved. Relevant evidences were extracted from the memory allocated to the Windows applications. In this paper, reported cases of user input incidences; in particular, of user access to some commonly business applications such as MS Excel and MS PowerPoint were investigated. Incidence response from seized digital evidence or devices indicated various user input and reported user activities that were recovered from the memory allocated to each applications.

Therefore, the theoretical foundation of the user input on Windows applications created avenue for likelihood of temporal information of user access on these two applications. The research was hence, motivated on what the user is typing on the application, what the user has been doing and what the user is using the applications for. This user-based event was done in the context of digital forensics investigation. We then introduce relevant ideas from the user input to determine the systems process associated to these two applications, MS Excel and MS PowerPoint. We formulated and design metrics for the use of statistical information and for evidence gathering based on user input recovered from the memory allocated to these applications. The metric formulated, provided forensically sound framework solution that can widely use to analyze user input in digital evidence. The Table 1: described the method adopted for our investigations. As we designed the strategies for our investigation, we considered components of user input and activities in the memory allocated to the Windows Application. As various users, input information recovered were extracted, the research work proceeded to point out some related user input activities on these applications. The evidence information is presented in sequential to the event of data extracted from the processed applications. In particular, we focus on the use of user input segregation of process data, and mingle various user input recovered, as basis for our user functions on these applications.

As described in Table 1, we use these techniques to evaluate the likelihood events of user input on these two commonly used Windows applications such like MS Excel and MS PowerPoint. This is to determine useful data in relation to the event streams recovered from the Windows systems. The extracted information was reviewed to determine if the information recovered are from the same source of user, per time of events. The information was later reconstructed to actualize the fact of evidence. We apply this methodology to a data set of event histories as recovered from these applications. This is focusing on user activities that are related to the user input on the applications. As indicated below in Session 4, the events of user input activities were recorded and as presented in quantitative graphical approach.

Table 1. Approach of methodology designed

Applications	User action at various interval	User various input on business applications
Excel 2010	Open and Closed Application at Various Interval in a business-like environment	Write paragraph of text, with commas, semi-colon, full stop, user minus sign, question mark, special characters, currency sign, character 0–9, brackets. Long and short sentences with alphanumeric character. Save document or close document and do not save
PowerPoint 2010	Open and Closed Application at Various Interval in a business-like environment	Write a slide, slides of texts with commas, semi-colon, brackets, full stop, insert images, web links, animation, pictures. User input may contain or type alphanumeric, character 0–9, brackets close or open. Long sentences or short sentences. Save document or close document and do not save

4 Results and Findings

In this session, the research work reported the results of user input on Windows applications. As indicated on the user functions, user data recovered from the application process are marked point based on systems processes. The information extracted was from the source of user input per time of events. This was reconstructed to actualize the fact of evidence streams in various experiments carried out. The analysis attempted on the methodology approach of data set of event histories, and as recovered from these applications. We focus ideas on repeated number of times character of user input based on various activity and as related to the amount of relevant data recovered from the application memory. We determined the user process on significant of user access on Excel application. The research ideas not on a discriminative power of event-based datasets of the application, but user access at interval, based on the methodology

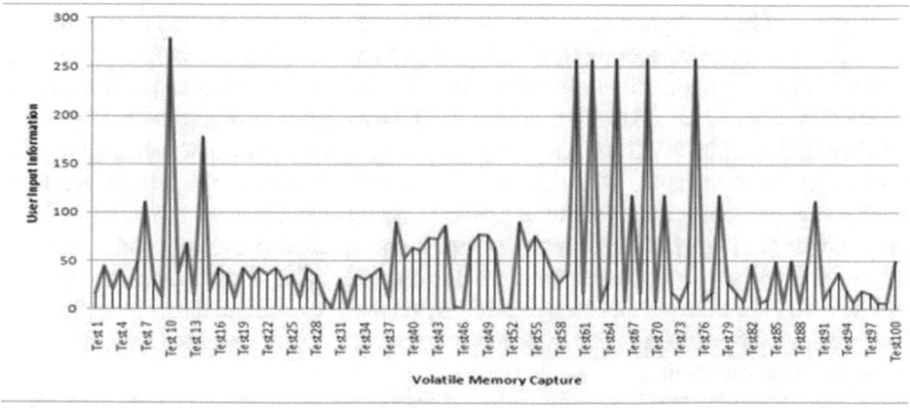

Fig. 1. Number of times a character of user input is repeated

approach adopted in this paper. The scenario of user input access on business applications, for example, in Excel Application, reviewed that it contains a greater mixture of Latin characters and numerical data, when compared to using other Windows applications. Various experiments carried out on this application proved to be very difficult because user input information was hard to recover and when recovered, it was simply to identify the numerical data because of the existence of other numerical data in the memory image of the application. Therefore, on Excel Application, there are variations in the data element. Figures 1 and 2, identified lowest and highest peaks of the repeated number of characters of user input.

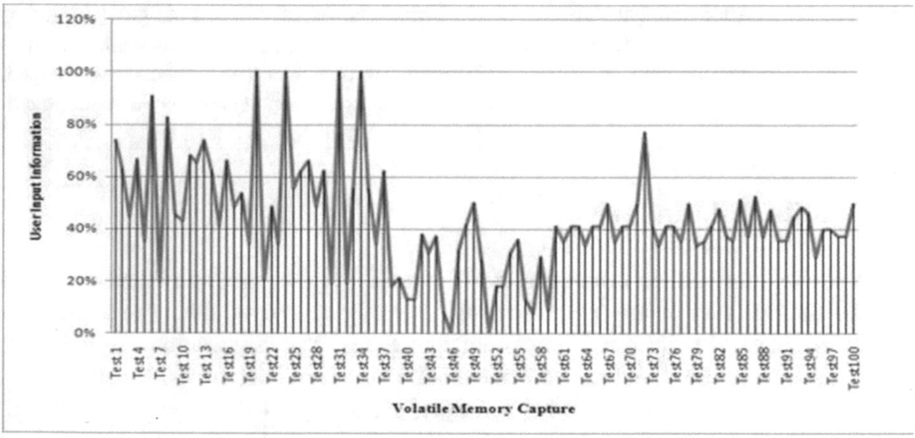

Fig. 2. Percentage of user input found

Figure 1, shown clearly that test10, test59, test62, test65 and test74 reported the highest peak of the repeated number of character of user input. When considering the correlation coefficient between the metrics reported here, there is strong positive association between the length of user input and the number of times character of user input repeated on this application.

Figure 2 of Excel Application, considered the repeated user input on the application based on percentage of user input found, and as extracted from the memory allocated per time of user input on this application. The volatile memory captured per time of user input in this application illustrated in Fig. 2. The user input information indicated that test10, test59, test62, test65 and test74 reported the highest peak of the repeated number of character of user input. When considering the correlation coefficient between the metrics reported here, there is strong positive association between the length of user input and the number of times character of user input repeated on the application memory and as extracted amount of user input found. Figures 1 and 2 of Excel application illustrated the amount of user input recovered from the memory allocated to this application. The evidence information is traceable to the variation of user input in Latin characters that resident in the memory.

In Fig. 3, the PowerPoint application investigated to identify useful information related to user input information in the memory. For example, test34 recorded the highest number of times a character of user input repeated whereas; the lowest amount of data recovered shown in test80. However, the peak of the amount of user input in test34 was higher than the peak time of data recovered in test80 and lower amount in test test82 but lightly shown that the amount of user character repeated per time of user input in this experiment was a bit lower in test85. This means there are no similar characters of user input at various tests carried out in this experiment.

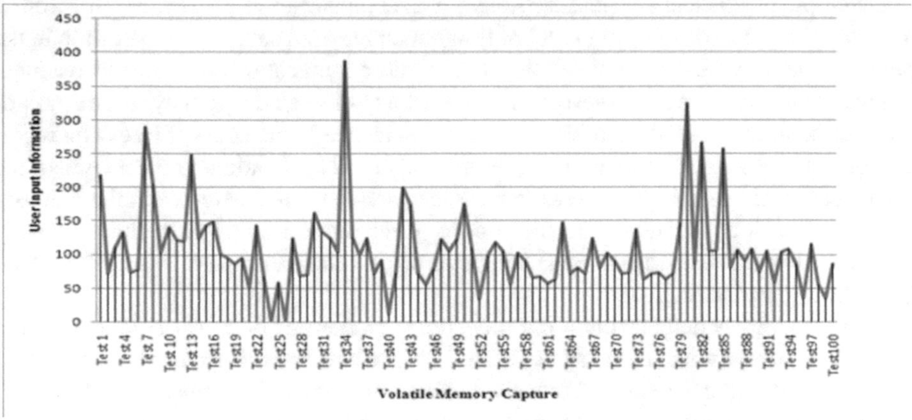

Fig. 3. Number of times a character of user input is repeated

Figure 4 illustrated the percentage amount of information recovered from the memory content of PowerPoint application. This is the repeated amount of user input per time of user input in the application memory. For example, at most of the time of user input based on interval of user access and as experimented in Fig. 4, the percentage of user input was at the high of probably between 90% and 100% of relevant user input that recovered from the memory allocated to this application. This shows clearly that

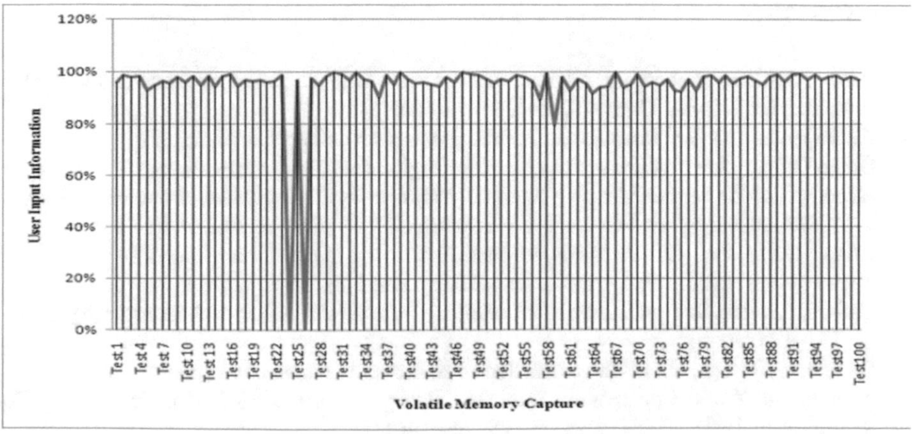

Fig. 4. Percentage of user input found

between 90% and 100% are determined for all tests where there was user input. Again, there was no correlation found between the overall lengths of user input presented here.

5 Conclusions and Summary

In the final section of the paper, we discuss both the promise and challenges involved in developing statistical analysis methods for event histories in the context of forensic investigations. The developed model of how much relevant data stored over time in the physical memory of the memory allocated to these applications are vital to business organisations and forensic investigators. To a law enforcement agency, the results of the user input on these applications are useful evidence in the court of law. The methodology approach focused on two most commonly used applications and data illustrated in this paper laid emphasis on the amount of data evidence recovered from the memory content of each application. The methodology we proposed in this paper is broadly applicable to event data in general and uncovers how much relevant evidence recovered from the application memory.

This model has demonstrated the percentage amount of dispersal evidence stored over time in the physical memory of Outlook, Word and PowerPoint applications. This approach has become part of forensic analysis in digital investigation. The result explained the quantitative assessment of dispersal evidence. We have carried out experiments based on user activities on some of the most commonly used applications.

5.1 Future Work

In the future, the research investigated further on other business applications and social media, through forensic experiments of how much evidence can be recovered when images were captured, while user is still interacting with the system but when the application is closed.

References

1. Brown, C.S.D.: Investigating and prosecuting cyber crime: forensic dependencies and barriers to justice. Int. J. Cyber Criminol. **9**(1), 55–119 (2015)
2. Parliamentary: Digital forensics and crime. Postnote Number 520, March 2016
3. Home Office: Home Office—eDiscovery in digital forensic investigations (2014). https://assets.publishing.service.gov.uk/government/uploads/system/uploads/attachment_data/file/394779/ediscovery-digital-forensic-investigations-3214.pdf. Accessed 14 Feb 2018
4. Brian, D.C., Joe, G.: A hardware-based memory acquisition procedure for digital investigations. Int. J. Dig. Forensics Incid. Response **1**(2), 50–61 (2005)
5. Olajide, F., Nick, S.: Application level evidence from volatile memory. J. Comput. Syst. Eng. **10**, 171–175 (2009)
6. Olajide, F., Savage, N.: On the extraction of forensically relevant information from physical memory. In: World Congress on Internet Security (WORLDCIS-2011), Technically Co-Sponsored by IEEE UK/RI Computer Chapter (2011)

7. Garcia, G.L.: Forensic physical memory analysis: an overview of tools and techniques. In: TKK T-110.5290 Seminar on Network Security (2007)
8. Simson, G.: DFRWS—digital forensics research: the next 10 years. In: DFRWS: From the Proceedings of the Digital Forensic Research Conference, Portland, OR (2010)
9. Mouhtaropoulos, A., Li, C.T., Grobler, M.: Digital forensic readiness: are we there yet? J. Int. Commercial Law Technol. (JICLT) **9**(3) (2014)
10. Google: Google Documents (2018). https://docs.google.com/document/u/0/. Accessed 09 Apr 2018
11. Microsoft: Office Products (2018). https://products.office.com/en-gb/products. Accessed 28 Mar 2018
12. Kilpatrick Townsend, S.L.: BYOD (Bring Your Own Device) Policies and Best Practices. LEXOLOGY, 17 November 2017. https://www.lexology.com/library/detail.aspx?g=46644cf7-b89d-494d-92b4-28463ee94983. Accessed 18 Feb 2018
13. Lapowsky, I.: Cambridge Analytical Could Have Also Accessed Private Facebook Messages. Wired.Com, 10 April 2018. https://www.wired.com/story/cambridge-analytica-private-facebook-messages/. Accessed 13 Apr 2018
14. Facebook: Our Business is to Improve Yours. web.com (2018). https://www.facebook.com/Web.com/. Accessed 09 Apr 2018
15. Cambridge Analytica: Data Drvices All We Do. Data Driven (2018). https://cambridgeanalytica.org/. Accessed 10 Apr 2018
16. ACPO: ACPO Guildelines. 7safe (2007). https://www.7safe.com/about-7Safe/downloads/acpo-guidelines. Accessed 11 Apr 2018
17. Robert, J.M., Cihan, V., Leonardo, C., Lei, C.: In-depth analysis of computer memory acquisition software for forensic purposes. J. Forensic Sci. **61**(S1), 111–116 (2016)
18. ManTech: ManTech International Corporation (2010). http://www.mantech.com/msma/MDD.asp. Accessed 12 Feb 2018
19. FTK: Forensic Toolkit (FTK) Digital Investigations (2018). https://accessdata.com/products-services/forensic-toolkit-ftk. Accessed 19 Feb 2018
20. EnCase: Advanced Forensics Software (2018). http://www.forensicexplorer.com/?gclid=Cj0KCQjwqsHWBRDsARIsALPWMEM2iuU3MtNnvxJOy_rsDSZ76jgRxRrDF37MU8FsVDbMnwE9L8env2saAicoEALw_wcB. Accessed 11 Jan 2018
21. Volatility: Releases—Volatility 2.6 (Windows 10/Server 2016) (2016). http://www.volatilityfoundation.org/releases. Accessed 22 Jan 2018
22. Nathan Scrivens, X.L.: Android digital forensics: data, extraction and analysis. In: ACM TUR-C 2017 Proceedings of the ACM Turing 50th Celebration Conference—China, Shanghai, China (2017)

Digital Investigation and Forensic User Analysis

Funminiyi Olajide(✉), Tawfik Al-Hadrami(✉), and Anne James-Taylor(✉)

School of Science and Technology, Nottingham Trent University, Nottingham, UK
{funminiyi.olajide, tawfik.al-hadhrami, anne.james-taylor}@ntu.ac.uk

Abstract. Today's digital forensic investigation centre towards data collection, preservation, extraction, evidence searching, reconstructions of relevant user input analysis, and evidence presentation of user involvement in business applications. In this research, stages of digital investigation are applied on various experiments carried out on user input information. A typical business environment of security incidence was investigated on use case element of business applications. In this paper, design methodology discussed how images were captured and processed for investigation. The extraction procedure was formulated at interval and dump memory of the extracted information determines artifacts of user evidence per time of user activities on sample memory application. Validated extracted information was analysed and as reconstructed, presented in quantitative assessment of user input, on sample use cases of business applications of MS Outlook and Adobe. The research investigation uncovers how user input information are stored and recovered from the application memory and for use as evidence.

Keywords: Forensic · Evidence · Physical · Memory · Business
User input · Computer · Business

1 Introduction

In the digital world, the use of digital evidence is growing quickly with evidence integrity and validation is the key to any digital crime. Digital evidence investigation and analysis is more revealing on various user input on digital devices. There are different forensic techniques such like open source and commercial tools available in the market. A research work of [1], focused on the use of some popularly known forensic commercial tools EnCase [2] and FTK [3], and presented associated problems of approaches, with proposed solutions of validated integrity procedures for use as evidence in the court of law. However, the evolution of digital technologies involved the need for digital forensic readiness in business organization [4], and proposed solutions for solving issues and challenges posed by cyber criminals. A paper of [5], presented the file type identification using computational intelligence for digital forensics. It was discussed that file type is a very demanding task for forensic examiner

F. Saeed et al. (Eds.): IRICT 2018, AISC 843, pp. 630–640, 2019.
https://doi.org/10.1007/978-3-319-99007-1_59

and proposed a new methodology to identify altered file types, focusing on the popular types of (jpg, png and gif). In the course of digital forensics examination, investigator discussed variety of tools and strategies, to explore the evidence locations and presented relevant artifacts for use in crime cases. While there are research efforts on the problem of anti-forensic by [6], a proposed method of signature-based for triage purposes are useful to assists forensic investigators. It is evident that traceable likelihood of user input information can be recovered from the memory allocated to an application. A review paper of [7], searching through digital evidences to find the clues and tangible evidence. This research showcases of events of user activities, for example, if a device was used or misused and as contained with the evidence required for the purpose of the investigation. In the research work of [8], it was listed evidences, such as documents, pictures, media files that could be used as the immediate target of forensic investigation. In Windows operating systems, physical memory contains parts of address spaces for processes (both user [8] and kernel processes), including cached file system blocks and free pages, that are not currently allocated to any process memory imaging. Analysis in this area is of high ranks of research interest in digital forensics. However, there has been much attention to the acquisition of physical memory in the past few years [9], and carried out investigations on target computer systems or seized devices such as computers, laptops, tablets and smart phones. A literature review of existing research of Windows Registry was investigated as a source of forensic evidence by [10], on digital investigation of Internet usage. The research highlighted issues to Registry inspection in forensic practices, whereas, research extraction by pattern matching techniques of user information by [11], revaluated evidence from application memory in Adobe2007, based on simple user activities of highlighted texts, searching for texts, save action and no action that was made the user when images were captured for this investigation. Investigators can simultaneously, allows Internet connectivity, to provide a tangible ideas of evidence for investigative clues on usage of the seized digital evidence. For example, in the use case evidence of Google document [12], the Internet history reviewed on ordered lists of artefacts that contain date, time and Universal Resource Locator (URL) address of the accessed resources. At the same time, all other user input on business applications can uncovers what a user is typing or doing on the application, what the user has been doing and what the user is using the application for in both temporal and relational analysis of the investigation. Hence, forensic investigators can get holds of the image captured on applications and used the reconstructed information as evidence information. In this paper, more user input information was detailed in Table 1. The two sample applications was investigated based on repeated amount of user input recovered from the memory allocated to these two business application of MS Outlook and Adobe. The user input information investigated detailed more user actions than the research investigated in [10], and research approach of [11]. However, the acquisition techniques have an acceptable degree of automation in [13], and thus, analyzing the extracted memory dump with new approaches is vital. The contribution of the paper presents evidence information of user input on applications based on detailed user activities, while the application is still running. The investigation follows procedures in Table 1, of what a user is doing on the application, what they have been doing and what the user is using the application for. The approach of developing an automated program to extract relevant user evidences, revealed the amount of user input stored on percentage of user access, and per time of user activities, found on the allocated memory of these two applications.

Table 1. Methodology approach designed for this experiment

Applications	User action at various interval	User various input on business applications
MS Outlook 2010	Open and Closed Application at Various Interval in a business-like environment	Write paragraph of text, with commas, semi-colon, full stop, minus sign, question mark, currency sign, character 0–9, and brackets. Send and receive email. Compose email in long and short sentences with alphanumeric character. Long sentences or short sentences. Close and Save document. Save as Draft, Add Web links, Include Signature or Not, Forward Email or Reply Email or Reply All, Group Email, Check Deleted Items, Sent Items, Save to Favourites, Include Archive or Conversation History, Junk Email, Calendar and/or people contact frequently, Mark as Read or Unread
Adobe	Open and Closed Application at Various Interval in a business-like environment	Get an Adobe ID, Sign in, or Use Sign in with Enterprise ID, Edit, Add Signature, Fill a form on Adobe, Write a word or texts with commas, semi-colon, brackets, and full stop. User input may contain alphanumeric, character 0–9, brackets close or open. Long sentences or short sentences, Close and Save document

The process of retrieving digital evidence of document information was presented by [14], and it was understood that most of the digital activities committed by a user can leave definite traces, allowing investigators, to obtain essential evidence in solving criminal cases, obviously, preventing crimes. This user input information recovered from the memory allocated to business applications can enable forensic investigator to understand the user input resident in physical memory of the applications. For example, in the document reviewed in [14], it is evident that in the memory application, history artefacts proved as user evidence to the investigator on this basis of the types of applications accessed. This is whether on specific times of day that the users were genuinely and actively working or illegitimately working on closed or opened applications. In this paper, the research investigated the application memory of MS Outlook and Adobe applications in business. Events of user activities on sample business cases revealed documents contents of information. Therefore, evidence information provided all user interaction on these applications including tracing user access to other applications. The extracted information contained the events of user history in fragments of what user was typing in short and long sentences. In this work, it is evident that more user actions on physical memory of applications can be used to identify more relevant data about user activities. The research identified the extent of information that can be recovered with one important contribution of presenting data in quantitative assessment. This is following, the extraction procedures of forensically relevant information. The rest of the paper is organised as follows: Sect. 2 discussed the related work and Sect. 3 presented on the methodology approach of our investigation. Section 4 is the results of research findings and in Sect. 5, reported on conclusion made based on findings and proposed on future work in Sect. 6 and with the updated list of references used in this research work.

2 Related Work

With the issues of recent digital forensic tools, there is inability to fully address the complexities of recovering data from the physical memory. This led into more research investigation in the area, for example, a paper of [15], explores ways to capture evidence other than those using current digital forensic methods, through investigation on hard disk of virtual machine. The research work of [16], took a different approach of forensic investigation in big companies of Email communication and data analysis, but restricted ideas only to the amount of user information stored in typical inboxes and as recovered reading through all the mails accessed. An investigation procedures and data analysis of Internet history artefacts, is however, time consuming with both irrelevant and relevant information residing on the process allocated to the application memory. The research work of [17] is in support of using different approaches, for the extraction of relevant data such from mobile memory devices. The artefacts often be quite extensive which poses challenges to investigators. The paper of [18] highlighted the R v Schofield case from the United Kingdom, where the prosecution was forced to dismiss charges for possession of unlawful pictures because Trojan Horse software was located on the defendant's computer. However, in the paper reviewed, it was observed that the analysis had not established user responsibility, for the creation of the unlawful pictures that causes the defendant to be proved innocent. It was found impossible to pinpoint that the defendant indeed is the actual person on this case. This has resulted to the use of different usernames, because the evidence was unable to show the Internet history artefacts of the user, an actual person that was interacting with the device. Therefore, such user interaction proved that the user mental component of an action done on digital devices cannot be proved on the person action using the keyboard. According to paper of [19], stated that the forensic investigator may find it difficult to show the intent of the user action on systems, without placing the artefacts into contextual ordering. As related, this research paper presented digital forensic investigation of user input on business application. The identifications of these applications resulted on use case evidence of recent incidence response by forensic investigator. As this paper reports on an ongoing research project, our evaluation and possible conclusions is to review the encouraging results from the experiments carried out and to highlight user input relationship on the two-sample applications of MS Outlook and Adobe applications.

Forensic tool of timeline analysis, Zeitline [20], was introduced by Buchholz and Falk in 2005. The tool purposed on how to reconstruct artefacts and thus enabled investigator to create complex user events, using pattern searching and filtering to populate analysis on the timelines of user activities. These approaches enable different applications and different operating systems to leave behind footprints of user activity. This paper builds on this approach and with the associated review paper of [21], the footprint of applications on a system was based upon the typical artefacts created in normal usage. The analysis of the paper indicated that the features are used to train a neural network which could be used during forensic examination. In the process to attempt a reconstructed timeline of events on applications, in 2009, the Cyber Forensic TimeLab (CFTL) tool was developed by [22], which highlighted the discovery that can parse a hard drive for known predefined artefacts, to produce a histogram timeline. This research does not automatically analysed the artefacts, but requires the analyst to make a visual correlation of different timelines, overlaid to display clusters.

Another tool for forensic investigations was log2timeline [23], reported to creates a super-timeline by placing all the information into a monolithic list which can then be processed. Carbone and Bean endorsed the approach in a new development of timeline creation, but too many irrelevant files were included as tested in [24]. Hargreaves and Patterson developed a tool to reconstruct high-level events from low-level activity using temporal proximity pattern matching in [25], based on the cause and effect of nature of events reconstructed. Of recent, a paper of [26], defined action instances on a state transition model where an action produces a trace. It was argued that if traces can be identified then actions can be implied, because of the causal nature of certain state transitions on computer systems. Therefore, in this paper, related user input information recovered was investigated and artefacts of user evidence were extracted from the memory allocated to the application and reconstructed based user actions on these applications. The extracted user input integrated with the log files of activities. Hence, searches for an activity of interest and uses the relevant evidence to reconstruct the time memory aspect of user input based on repeated character number and presented in percentage of user input to generate useful information.

3 Methodology

A typical flow of the evidence-based procedures was formulated as shown in Table 1. A sample business application was discussed for the purpose of this research based on the methodology approach adopted in Table 1. Within this paper, related work to the analysis of user input in MS Outlook and Adobe reported the history artefacts of user profiling as digital evidence. This research described the methodology approach of user input based on classification and aggregations of user evidence recovered from the memory allocated to the sample applications. The use of Volatility tool was engaged to list the processID and for process scanning including the memory dumping. The extraction procedures followed the standards of digital forensic investigation and relevant user input were reconstructed as converted into strings of user activities on these applications.

The research experiment describes stages of user input per time of activities at interval, resulted into different problems of finding the relevant evidence in sessions of user access on these applications. The user input activities per time was based on the content information recovered from the allocated memory of the applications. Obviously, the state of image captured was on the understanding of a typical business environment and as user input recovered from the allocated memory during the process of digital forensic investigation. It was evident that before the image capturing, Windows machine of the user was shut down and rebooted to ensure that the system memory cache was as clean as possible. The fragments of user information recovered contained both relevant and irrelevant data. However, user actions on each application differ from one period to another. Several testing of user input was engaged, with the aids of algorithm developed for extraction processes.

As shown in Table 1, the methodology approach was applied based on user input and the process of investigation was determined on the amount of repeated number of characters extracted from the memory allocated to two applications. As described, the event of user input were reconstructed and related event histories of user input that was recovered from these applications were fully investigated. The research investigation focused on user activity to recover relevant evidence and presented the evidence in quantitative assessment in next Sect. 4 of this paper.

4 Results and Findings

Session 4, presented the events of user input activities that was recorded and as presented in quantitative assessment as shown in figures below. In this session, the analysis approach was set on the event histories, and as relevant to evidence recovered from the applications. The amount of repeated number of character of user input was recovered from the application memory. The results of user input indicated the user functions based on user data recorded per time of user actions on the application. In various experiments carried out, the amount of information extracted was reconstructed to actualize relevant evidence of user actions. The research determined the user process on significant of user access on MS Outlook application and Adobe. The relevant information reported event-based of user input at percentage interval and with repeated amount of characters per user per time. Various experiments were carried out based on typical scenario of live memory of application running on computer systems, when user input are opened and tested also on when the applications has been closed. Figure 1 illustrated the repeated character of user input recovered from the application and Fig. 2 presented on the percentage of user input.

Figure 1 reported on the number of times a character of user input is repeated in the memory for example, in test42 and test54 recorded in numbers of highest amount of information stored and as extracted from the allocated memory. Almost all amount of user input were recorded at the highest percentage. For example Test 1, was steady at 100% but in other tests, the percentage dropped below 100% and then followed by a peak of experiment shown in Test7 and this is also followed by test16. Many other tests ran was reported at high percentage, whereas, there was a sharp drop at test 34, test50. Also, illustrated in test88, a slight drop of amount of user input recovered.

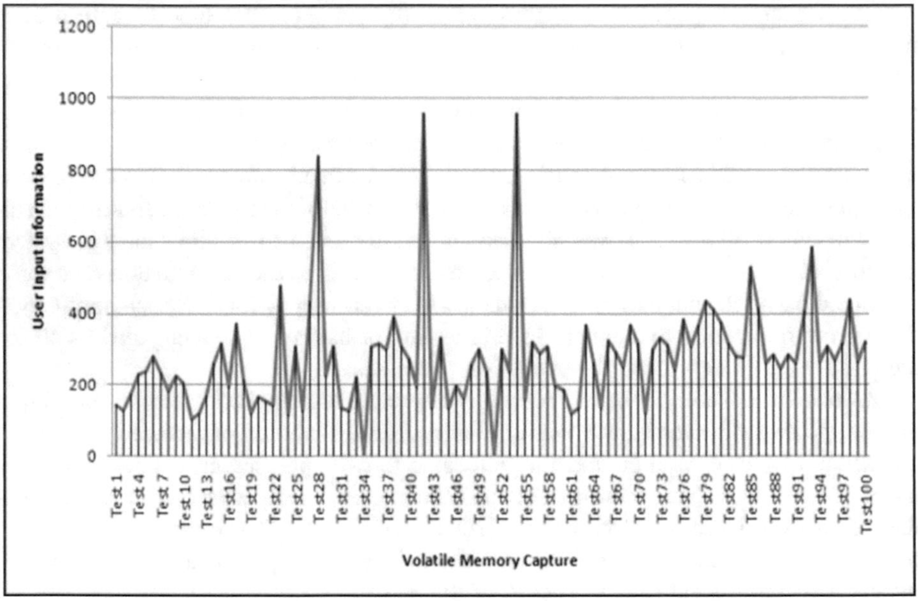

Fig. 1. MS outlook application-number of times a character of user input is repeated

Figure 2, describes high percentage of user input recovered in all tests except few others mentioned above for example in test34 and tesat50. As investigated, in all other tests ran in this experiment, the relevant user input information recorded is determined in-between 99.9% and 100%.

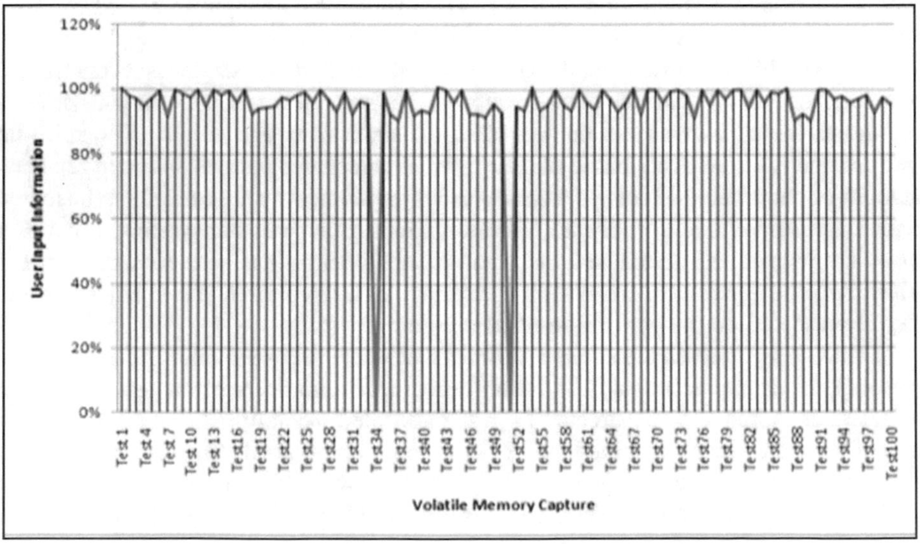

Fig. 2. Outlook application-percentage of user input found

While investigated other application, the case result of investigation was different on Adobe Application. It was visible that user input verified both the number of times the characters of user input was repeated and as recovered from the memory allocated to the business application. Although, the percentage of user input found presented relevant artifacts of user information but almost all activities recorded were not in stable format as shown in Fig. 3. For example, there are least of user input recovered in test35, test46, test67, whereas, there are other slightly event of user input recorded in test16, test74, test83 and so on. Highest amount of user input recovered were visible in three major testing areas of this experiments which resulted above 800, in high amount of user input information, such like in test39, test45 and test53. This means that number of times user repeated character of information was high either by pressing on back-space, or deleted the character of user input information. However, all these character evidence of user input were recorded and as extracted from memory allocated to the application. As shown in Fig. 3, the highest number of repeated character may be useful to forensic investigator and can help to determine further action of user input on this application.

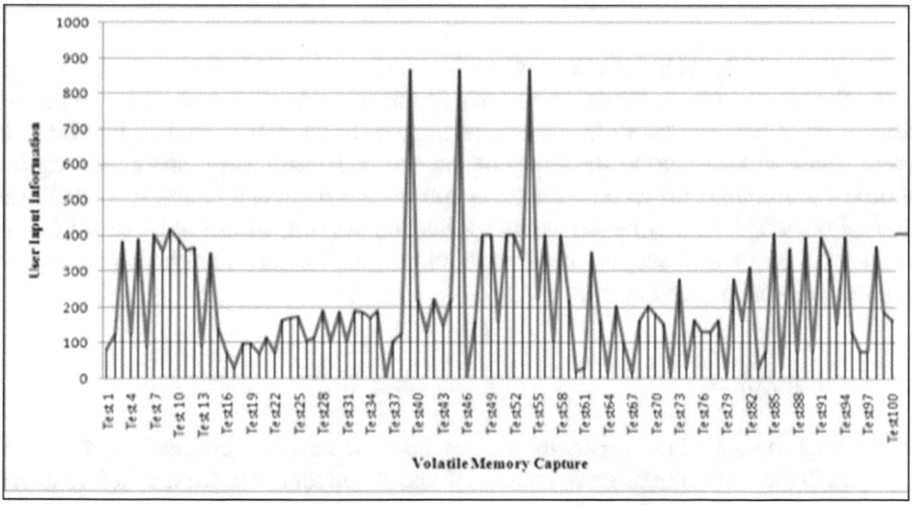

Fig. 3. Adobe application-number of times a character of user input is repeated

Below is the Fig. 4 of the experiment carried in series of user action on Adobe application. As illustrated, Table 1, the volatile memory captured per time of user input in this application recorded at instances of highest peak in some other tests carried out while some at low peak. For example test47, test56 reported the highest peak while the low peak as shown in test41 and test42.

The amount of information found stored on the memory allocated to the application reported on the values that are distinctively larger in one and higher than other in other cases or testing. However, the percentage of user input found in test10 is relatively low

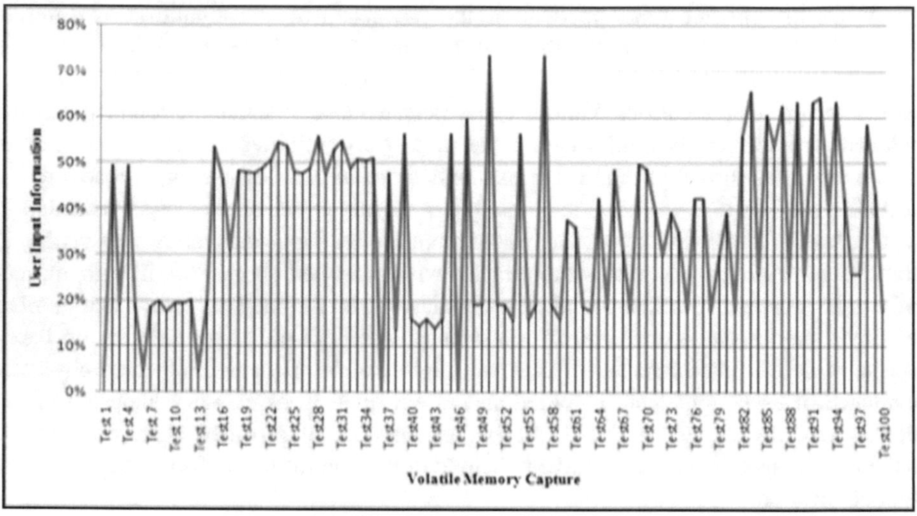

Fig. 4. Adobe-application-percentage of user input found

at just above 10%, with a similar finding for test53 and test59, where the percentage was approximately 20% above. This is due to the differences in user interaction as set time of input between the different tests of the experiments. It is evident that forensic investigator will be able to make use of the relevant information and following the series of procedures, the reconstructed amount of user input can be reconstructed over time of the length of user action on the application per time of use input stored on the memory. Various user actions can be determined and this information can be used to support evidence in the court of law.

5 Conclusions

In the final section of the paper, we discuss both the analysis and methods for event histories in the context of forensic investigations. The model of how much relevant data stored over time in the physical memory of the memory allocated to an application is vital to business organisations and forensic investigator community. The methodology approach focused on two most commonly used applications of MS Outlook and Adobe application. Figures in this papers illustrated emphasis on the repeated amount of data evidence and percentage amount of information recovered from the memory content of each application. The methodology proposed in this paper is broadly applicable to event data in general and can uncover relevant evidence recovered from the application memory. As discussed above in Figs, the model demonstrated the repeated amount of characters of user input and the percentage of evidence stored over time in the physical memory of Outlook and Adobe applications. This research approach has become part of forensic analysis in digital investigation. The result of the research explained the quantitative assessment of evidence and as carried out based on user activities on these applications.

6 Future Work

In the future, research intention will consider other business applications and the use of social media in a typical business environment. The research approach may be similar in order of how much evidence stored on the application memory can be recovered and reconstructed to be used as evidence in scenario of when user is interacting with the systems, when the application is closed.

References

1. Shah, M.S.M.B., Saleem, S., Zulqarnain, R.: Protecting digital evidence integrity and preserving chain of custody. Dig. Forensics Secur. Law (JDFSL) **12**(2) (2017)
2. EnCase: Enfuse 2018. Guidance (2018). https://www.guidancesoftware.com/enfuse-conference/attend. Accessed 22 Apr 2018
3. FTK: Forensic Toolkit (FTK) version 6.4. ACCESSDATA, 01 February 2018. https://accessdata.com/product-download/forensic-toolkit-ftk-version-6.4. Accessed 20 May 2018
4. Karie, N.M., Karume, S.M.: Digital forensic readiness in organisations: issues and challenges. Dig. Forensics Secur. Law (JDFSL) **12**(4) (2017)
5. Karampidis, K., Papadourakis, G.: File type identification—computational intelligence for digital forensics. Dig. Forensics Secur. Law (JDFSL) **12**(2) (2017)
6. Park, K.J., Park, J.-M., Kim, E., Cheon, C.G., James, J.I.: Anti-forensic trace detection in digital forensic triage investigations. Dig. Forensics Secur. Law (JDFSL) **12**(1) (2017)
7. Infosec: Computer forensics investigation—a case study, 6 March 2014. http://resources.infosecinstitute.com/computer-forensics-investigation-case-study/#gref. Accessed 28 Jan 2018
8. Russinovich, M.E.: Microsoft Windows Internal Covering Windows Server 2008 and Windows Vista, 5th edn. Microsoft Press, Washington (2009)
9. Schuster, A.: Searching for processes and threads in microsoft windows memory dumps. In: Digital Forensic Research Workshop (DFRWS) (2006)
10. Mee, V., Sutherland, I.: The Windows Registry as a forensic artefact: illustrating evidence collection for Internet usage. Dig. Investig. **3**(3), 166–173 (2006)
11. Olajide, F., Savage, N.: Extraction of user information by pattern matching techniques in windows physical memory. In: Digital Enterprise and Information Systems: International Conference, DEIS, London (2011)
12. Google: Google Documents (2018). https://docs.google.com/document/u/0/. Accessed 09 Apr 2018
13. Carvey, H.: Windows Forensic Analysis Incident Response and Cybercrime Investigation Secrets, 1st edn. Syngress Publishing, Rockland (2007)
14. Gubanov, Y.: Retrieving digital evidence methods, techniques, and issues, 30 May 2012. https://www.forensicmag.com/article/2012/05/retrieving-digital-evidence-methods-techniques-and-issues. Accessed 20 Mar 2018
15. Tobin, P., Le-Khac, N.-A., Kechadi, T.: Forensic analysis of virtual hard drives. Dig. Forensics Secur. Law (JDFSL) **12**(1) (2017)
16. Stadlinger, J., Dewald, A.: A forensic email analysis tool using dynamic visualization. Dig. Forensics Secur. Law (JDFSL) **12**(1) (2017)
17. Through, J.R., Cantrell, G.D.: Varying instructional approaches to physical extraction of mobile device memory. Dig. Forensics Secur. Law (JDFSL) **12**(4) (2017)

18. Susan, B.C.J.H., Brenner, W.: The Trojan horse defense in cybercrime cases. Santa Clara High Technol. Law J. **21**(1) (2004)
19. Grooty, D.W., Diane Gan, G.L.C.L.: Facilitating forensic examinations of multi-user computer environments through session-to-session analysis of Internet history. In: Digital Investigation—DFRWS 2016 Europe—Proceedings of the Third Annual DFRWS Europe, vol. 16, pp. S124–S133 (2016)
20. Falk, F.B.A.C.: Design and implementation of Zeitline: a forensic timeline editor. In: DFRWS, New Orleans, LA (2005)
21. Khan, M.N., Wakeman, I.: Machine learning for post-event timeline reconstruction. In: First Conference on Advances in Computer Security and Forensics, Liverpool, UK (2006)
22. Olsson, J., Boldt, M.: Computer forensic timeline visualization tool. Dig. Investig. **6**(1), 78–87 (2009)
23. Gudjonsson, K.: SANS reading room—mastering the super timeline with log2timeline (2010). Accessed 01 Apr 2018
24. Carbone, C.B.R.: Generating computer forensic super-timelines under Linux. In: SANS Reading Room, pp. 1–136 (2011)
25. Hargreaves, J.P.C.: An automated timeline reconstruction approach for digital forensic investigations. Dig. Investig. **9**, 69–79 (2012)
26. James, P.G.J.: Automated inference of past action instances in digital investigations. Inf. Secur. Cryptogr. **14**(3) (2014)
27. Beebe, N.L., Clark, J.G.: Digital forensic test string searching: improving information retrieval effectiveness by thematically clustering search results. Dig. Investig. **4**, 49–54 (2007)

Factors Influencing Information Security Awareness of Phishing Attacks from Bank Customers' Perspective: A Preliminary Investigation

Ayman Asfoor[1(✉)], Fiza Abdul Rahim[2], and Salman Yussof[2]

[1] Jubail Industrial College,
Jubail Industrial City 10099, Kingdom of Saudi Arabia
ayman_asfoor@yahoo.com
[2] Universiti Tenaga Nasional,
Jalan Ikram-Uniten, 43000 Kajang, Selangor, Malaysia
Fiza@uniten.edu.my

Abstract. Phishing is the act of sending e-mails pretending to be from genuine financial organizations and requesting private info such as username and password. Information security awareness of phishing attacks becomes the first line of data protection defence. Human factors are recognized as the main factors in addressing the issue of information security in online banking. Thus, identifying the issues and factors that influence information security awareness of phishing attacks are important. A preliminary investigation involving nine (9) professionals and academics using online banking has been conducted to obtain better understanding of the critical factors that may influence the information security awareness of phishing attacks. In-depth interview method is selected in this study. Thematic coding was conducted to characterise the themes and assess the factors found to be most influential. Results from the in-depth interviews with customers experienced in online banking showed that there are six (6) main themes may influence the degree to which information security may be learned. Security concerns, security attentiveness, user competency, computer knowledge, gender as well as the number of years of PC usage are the themes addressed by the key informants. These factors influence information security awareness of phishing attacks from the perspective of bank customers. In this investigation, we expect that the rate of phishing depends on six (6) principal factors likely to be prey to malicious phishers. This can be considered as the correlation between the background of the bank customers and their understanding in deploying online banking safety.

Keywords: Information security · Phishing attacks · Online banking
Security awareness

F. Saeed et al. (Eds.): IRICT 2018, AISC 843, pp. 641–654, 2019.
https://doi.org/10.1007/978-3-319-99007-1_60

1 Introduction

Online banking offers a change from the traditional way of banking to a remote method by Internet connection and provides convenience to its customers through performing activities at any time of the day and from anywhere in the world, which offers many advantages and benefits for bank and customers [1]. However, the use of online banking are growing rapidly and pose a substantial risk to the online bank customers. The attacker always seeks the customer's private information. Phishing attack is one of the techniques used by an attacker to get private information from online banking customers. The phishing attack can be performed by the attacker through a fabricated web site similar to the bank's website used by the target customers. A message (through e-mail or text message) would be sent to the customers to log in into the web site to supply their confidential information. When the customers have completed the submission of their confidential information, they are forwarded to the bank's genuine website from the fake web site in order not to give room for detection [2].

Security of online banking transactions is one of the most significant challenges to the banking sector. Billions of financial data transactions are conducted every day, and bank cybercrimes take place every day by skilled criminal hackers can manipulate a financial institution's online information system, spread malicious bank Trojan viruses that allow remote access to a computer, corrupt data, and impede the quality of an information system's performance. These problems occurred because bank customers are unaware of these risks; they have limited cyber security skills and lack awareness. As a result, they are susceptible to social engineering and phishing attacks that makes them the weakest links in the cybersecurity chain [3].

Based on report by [4], the Anti-Phishing Working Group (APWG) observed that 2016 ended as the worst year in history for phishing. According to the report, the total number of phishing attacks in 2016 was 1,220,523, which is a 65% increase over 2015. Without doubt, banks are the most valuable targets for online fraudsters for the simple reason of their proximity to money. This can be confirmed from both 2016 and 2017 APWG reports, which state that out of the 1,499 phishing attempts in 2016, 886 were directed at banks and bank related accounts. Furthermore, the percentage increase of the attempts over the same period in 2017 was higher than that of all other sites. Having justified that banks are in the red zone, it is then important to understand the techniques used by black hat hackers to gain entry into private online spaces. Usually, the attacks are disguised as phone calls or unidentified email.

Security awareness has always been addressed from the employee's perspective and few studies have been initiated to analyse customer awareness [5]. Based on a study conducted by [6], it has been mentioned that many studies on online banking are not concentrating on human behaviours' security, but instead focus on how to secure online banking systems. For that reason, many banking clients are still afraid of using online banking.

Information security awareness has been the centre of focus for years, past and current efforts to improve information-security practices and behaviours for citizens, consumers and employees have not had the desired impact, resulting in limited or no progress [7]. Primarily, information awareness aims at creating positive change in the staff's behaviour in an organization. Humans remain the weakest link in protecting data in the information security. Therefore, institutions need to improve their customers' security awareness regarding information systems. Based on study [8] it mentions that the institutions should identify how to develop essential competencies and new techniques and methods that may influence or improve human attitudes, behaviour and knowledge towards information security.

It also shows that online banking customers must behave in ways that reduce the online banking risk.

To understand customers' vulnerability to phishing scams online, a previous study [9], identified important factors influencing awareness of information security in phishing attack. These factors are security concerns, security awareness, computer literacy, user competence, demographic variables, motivation, individual difference and personality traits. The previous study conducted a systematic literature review, using publications not earlier than 2010 and thesis publication containing qualitative and quantitative designs.

This paper aims to explore and understand the factors that may influence information security awareness of phishing attacks among online bank customers through an in-depth interviews. This paper is organised into six (6) sections. The first section is the introduction, followed by the details about information security awareness among online banking customers in a phishing attack. The third section describes the research methodology; the fourth section presents in detail the outcomes from in-depth interviews. The fifth section presents the finding and last section is the conclusion.

2 Information Security Awareness of Phishing Attacks Among Online Banking Customers

The years of 2016 through to 2018 have seen a dramatic rise in the number of issues reported in phishing attacks. A report by the FBI showed that since January 2015, the FBI has seen a 270% increase in identified victims and exposed loss occasioning losses of more than $2.3 billion [10].

This spike is also substantiated by APWG's quarterly Phishing Activity Trends Report. The report for Q3 2017 shows that the number of unique phishing reports submitted to APWG during that quarter was 296,208, nearly 23,000 more than the previous quarter. APWG research further reports 1.2 million known phishing attacks in 2016 which represents a marked increase of 65% from 2015 [11].

Drawing from the report by (APWG), banks are prime targets, and are compelled to act proactively [6]. From this report, we deduce three hotspots that are the gateways for fraudsters. The first one are entirely on the side of the bank, while two are with the bank's customers (account holders). The first is whereby computers belonging to the banks are left logged in for a long time. [12] illustrates that in such a case, phishers have a long exposure time to tweak passwords and firewalls. The second is human error, whereby banks send their customers private credentials in an authenticated format. Hackers then have a field day since their spy algorithms effortlessly read login details from client inboxes. On the customer side, the reports identify rather obvious reason, lack of information, lack of reinforcement and negligence with sensitive information.

Based on the study [13], online bank fraud can be executed through different threats including malware, keylogger, identity theft, social engineering, phishing, man-in-the-middle attack and shoulder surfing. Another study by [14] midway through 2017, phishing attacks are very much on the rise, namely because they are too easy to launch and far too lucrative for the attackers.

One of the most prevalent forms of phishing attacks on online banking is referred to as a man-in-the-middle attack which relies enormously on the customer's lack of awareness about phishing techniques intended to breach their security [15]. Lack of customer awareness has been shown to be the primary cause of the success of phishing attacks and consequently the most effective way to protect the end user would simply be to raise their awareness. This was shown through a survey utilizing spear phishing technique as an attack scenario [16]. The level of risk imposed by this lack of awareness is even higher on mobile devices where traditional security indicators are often lacking and on which users have an imprudently false perception of safety.

There are a number of ways through which the attackers get to infiltrate the various online banking platforms. In many instances, the phishing attacks are aimed at gaining access to computer information through a message by using the name of a trustworthy agency. For example, the attacks could be disguised as genuine contacts by phone calls or several emails sent from a help desk that appears to belong to a technician from an Internet service provider, or the bank. According to [17], phishing attacks have become commonplace, and there are various ways by which the operations are performed. The most common way is where the attacker creates a system that fakes that of an existing company, and uses the name to steal personal information of login credentials of the customers of the online banking platforms [18]. The second one is known as spear phishing which involves the attacker using modified emails with the name of the victim, their position, company or work phone number, and any other special information that they may be able to obtain online [19]. The third one is malware-based phishing, whereby an attacker sends an email with an attachment to the victim that utilizes the security vulnerabilities of the user's machine. When the attachment is clicked, it spreads the malicious software that has been embedded within the attached file [20].

Phishing attacks are becoming increasingly difficult to prevent with technology alone. There is, therefore, greater need for empirical analysis of the factors of phishing victimization [21]. There is little attention on the human factors and ergonomics as regards phishing and cybercrime and it shows that there exists a gap in studies analyzing phishing from a systems thinking lens which is inextricably linked to the factors of phishing victimization [22].

The direct consequence of this gap in the available literature is an erroneous understanding of phishing attacks and subsequently deficient protective measures. The burgeoning prevalence of cybercrime and phishing attacks, in particular, create the need for more awareness of the factors likely to increase or decrease the likelihood of victimization [23]. Phishing detection solutions used by financial institutions thereby require a cognitive approach. More attention on the factors of victimization in terms of further research would help in formulation of better solutions as regards educating and protecting online banking customers as well as delaminating the limits of responsibility as rationed between the bank and the customer [24].

In the study conducted by [25], the factor of customer behavior is directly related to online banking security breaches and supported by [26] that security threats will certainly affect Internet customers' behavior in increasing protection of their privacy and minimizing their exposure to security threats. That study mentions that security breach may not necessarily come from that bank system rather it comes from the customer's behavior, but also as a result of inappropriate behavior of customers on online banking platforms, these findings are also supported by [27].

Information security plays an important role in ensuring that all information is protected. Safety information is also referred to as computer security and is defined as the protection granted to automated information systems to achieve the objective of maintaining the confidentiality, integrity and availability of resource information systems [28].

Based on the study [29] the important factor influencing the security of banking is the level of customer knowledge in information security that means more knowledge in information security more safe in using online banking.

Very few of the research papers and studies developed an original conceptual model within which to analyze online phishing victimization. One such paper came up with the Suspicion, Cognition, and Automaticity Model of Phishing Susceptibility [30] Most of the available literature, on the other hand, has instead resulted in using pre-existing models, in particular, the Heuristic-Systematic model and the Routine Activity Theory. The paper [31] is aimed at analyzing the human and psychological factors based on the Heuristic systematic model. Another paper by the same research duo similarly picks out the Heuristic-Systematic model as the ideal theoretical framework within which to analyze psychological factors of phishing victimization [32]. The Heuristic-Systematic Model is also recommended as the most appropriate foundation on which to study psychological factors influencing phishing victimization [33]. Other studies in this relatively scant body of literature have used the routine activity theory for instance to study the risk of victimization [34]. This theory was also used to analyze the effect of digital literacy in reducing phishing victimization [35].

3 Research Methodology

Preliminary investigation was carried out for this research to seek out the influential factors of information security awareness of phishing attacks in online bank customers. The meaning of security awareness mostly refers to the word in the perspective of the individual, and it can be assumed from several explanations. The nature of the description, thus, fits well with the qualitative research, which is intended to assist investigators in comprehending individuals and the social and ethnic settings within which they live [36]. As an outcome, in-depth interviewing method is used to examine this issue. In-depth interviewing is a qualitative investigation technique that includes carrying out thorough individual interviews with a slight sum of respondents to discover their viewpoints on a particular idea, package, or state.

Qualitative research, on the contrary, targets on participant's experiences as well as on the indicated significance they attribute to themselves, to other persons, and to their surroundings [37]. According to [38], qualitative exploration techniques are concerned with gathering data which does not contain numbers. They usually center on a small number of individuals, to be capable of providing significant data, and create large amounts of data regarding these people. The drive of the qualitative features of this investigation is, therefore, to get understandings into respondent's insights and understanding concerning online banking dangers and the safety applications of online banking networks.

The preliminary investigation was carried out in Jordan with a diverse setting of important informants. In this procedure, nine (9) important informers were involved, and detailed interviews were carried out with them. Thematic coding was done to describe the subjects and assess the aspect that found to be significant. The key subjects were picked according to their involvement using any related systems on online banking. Their facts that are involved in this primary investigation are represented in Table 1. Each of the key informants is labelled as K1, K2, K3, K4, K5, K6, K7, K8 and K9 accordingly.

Table 1. Key informants' detail on preliminary investigation

Designation	Specialisation	Organisation	Years of using online banking
Academician (K1)	Online banking	University	10
IT Specialist (K2)	Retail customer	Government sector	5
Academician (K3)	Online banking	University	7
Security Specialist (K4)	Security policy developer	Bank	12
Academician (K5)	Online banking	University	9
Pharmacist (K6)	Social work	Hospital	4
Medical Doctor (K7)	Hospital information system user	Hospital	12
Accountant (K8)	Financial Manager	Government sector	8
Engineer (K9)	Construction	Private sector	7

The purpose is to gather data regarding bank customers' approaches, opinions and conducts so as to recognize the thinking of the group as a whole and their knowledge about phishing attacks on Internet banking systems [39]. Findings from the in-depth interviews with customers experienced in online banking indicated that six main themes might impact the extent to which information security may be refined. Security concerns, security awareness, user competence, computer literacy, gender, and the number of years of PC usage are the main themes. The in-depth interviews were carried out based on research questions derived from preceding research on information security awareness of phishing attacks [40].

This research uses three significant processes following [41] in analysing the in-depth interview. The processes are data reduction, data show, and follow up by conclusion. These processes are performed to guarantee that the themes which emerged were validated properly. Atlas.ti is the software used in qualitative data analysis.

The in-depth interviews were conducted based on research questions obtained from former study on information security awareness in phishing attacks as listed in Table 2. The interview constructs items were selected as information security awareness in phishing attacks that were still ambiguous in terms of its own definition [42]. Table 2 illustrated the interview construct and relevancy that shows justification on why the interview questions was constructed. From the preliminary investigation, the argumentation on the interview outcomes is split according to the research question developed in the next section. Table 2 depicted the interview construct and relevancy that indicates justification on why the interview question was constructed.

Table 2. Interview constructs items

Interview construct	Relevancy
1-What are your opinions on online banking systems?	To find advantage and disadvantage of using online banking system compared to traditional [43]
2-As a customer of online banking, are you taking suitable steps to secure your online connections?	To identify awareness regarding their computers, other devices, plus the security applications that had been set up [44]
3-What levels of knowledge do you have about phishing?	To identify if they have heard about online banking attacks as well as other associated dangers from their banks [45]
4-Do you believe that banks are taking suitable steps to void phishing attacks?	To find out if they had any information about the steps their banks are taking to prevent the attacks [46]

4 Discussion on Interview Outcomes

This research entailed a survey of in (9) individuals from different groups, according to gender, career, age, area of living, and salary. Female respondents who were in the 30–39 age group, who came from the inner area and had an average yearly income of around $60,001–$100,000 were the victims of online banking attacks. Male respondents of the same age group were not much affected by online banking phishing attacks. Both male and female respondents who were in the age group of 20–29 had also been victims of these attacks at some point.

Research question 1: What are your opinions on online banking systems?

In relation to the research, informants recognized the essentials of the functions of the online banking schemes. Their opinions included: “My bank should use multi security guard, like login and password and SMS mobile confirmation code and digital signature…” (K6). “My bank should never permit more than one PC access to the same online banking account at the same time…” (K4). “My bank should close down my account spontaneously after I have been logged on for 45 min…" (K2). "The supplier of online banking services should be more responsive to security requirements, and make the online transaction have a layered protection against security threats…" (K8). It is obligated to provide safe banking online environments based on the advanced security procedures (K9). They also said that, although lots of them were employing online banking, they truly desired not to employ it since they were cognizant of the amount of risks as well as fraud on Internet financial transactions. Related research also established that improved customer trustworthiness was reflected to the positive lasting financial performance.

Research Question 2: As a customer of online banking, are you taking adequate steps to secure your online transactions?

Respondent's behaviors plus their activates concerning wired banking were assessed to observe their awareness regarding their PCs, other devices, plus the safety applications that had been set up. Regarding the safety system, some respondents identified the differences between the types of safety that were installed on their PCs. A large number of them recognized that the PC would not have been protected if there was the presence of an anti-malware within the computer system [43] they presented their opinions as follows: "It should be the responsibility of banks to protect customers with their proper security without asking them to take the responsibility …" (K1). "yes, customers should be aware of the risk, and therefore, I take my own steps to ensure that my online transactions are safe…" (K3). I contact my bank immediately if I get any email regarding my online banking…" (K4). Customers have to make sure that they set up both antivirus and antimalware in your system.…" (K8). It is so important and at all time remember to utilize the virtual keyboard (if available by your bank) present in your banking website login page.…" (K9).

Research Question 3: What levels of knowledge do you have about phishing?

Five of the nine subjects that were interviewed recognized or were conversant with online banking attacks as well as other associated dangers from their banks. Their first opinion showed that their banks sent cautionary mails to them whereas other message networks were not so frequently recognized. They conveyed this message as follows: "Scammers enter into customer's account and takes/transfers money to the scammer's account…" (K1). "There are scam mail that may look like valid business requesting individual information such as username & password…" (K3). "I have heard from somebody's experience where their credit card number or money has been moved/used/taken…" (K5). "Checking whether the Internet connection with the bank's website is secure, for example by checking for https and a closed padlock" (K2)."If the message asks for personal information" (K7) Nevertheless, the results showed that the more refined the kinds of scams were, the less awareness of them the informants consumed. A large number of respondents recognized or had overheard regarding online banking extortions and the ways on which private info can be taken from the Internet [47].

Research question 4: Do you trust that banks are applying suitable actions to avoid phishing assaults?

Starting with the verification schemes given, it can be perceived that techniques can be classified into 2 major sets of a sole verification structure and a dual aspect validation system [48]. Most of their opinions were positive but there were also some negative opinions. Most informants had been delivered with a particular verification network which included a username (login) as well as PIN. This verification system was presented by banks more than two-aspect authorization ways, like login with PIN and SMS telephone confirmation, that was greater than the demonstration procedures, biometric substantiation organizations, network pass, or login as well as PIN with an undisclosed query [49]. But not all the banks presented these two aspects of authentication to their customers. Further complication for customers was added layers by the numerous styles in the more refined devices [50].This is what the respondents had to say: "Banks should increase or improve level of online banking security…" (K5). "My bank encourages their customers to create strong and long passwords of like eight password lengths…" (K3). "Banks should have a call back policy to contact their clients immediately over the phone…" (K2). "banks should always send a warning messages to its customers via a mobile phone and email …"(K6).

5 Findings

The interview outcomes were analysed and visualized in Fig. 1. The themes derived from the interviews are identified as factors that influence Information Security awareness on phishing attacks from bank Customers. Six (6) themes have been identified that may impact the information security awareness of phishing attacks from bank customers. These are security concerns, security attentiveness, user competency, computer knowledge, gender as well as the number of years of PC usage.

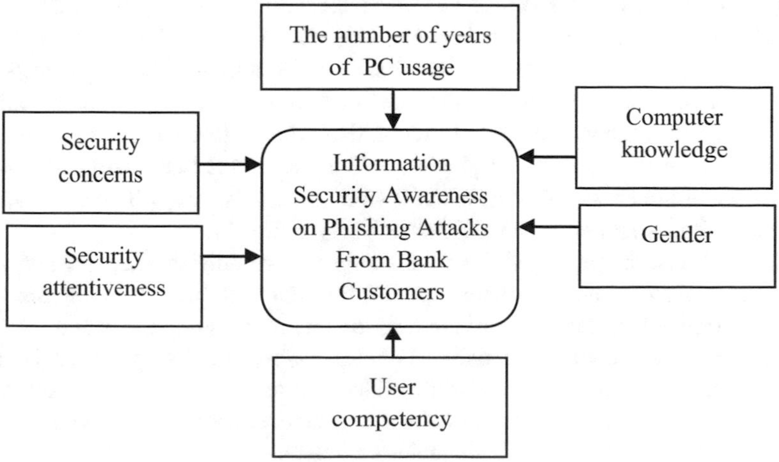

Fig. 1. Emerging Themes from Interview

These themes will be used to design a conceptual model that shows factors influencing information security awareness of phishing attacks from bank customers. From in-depth interviews conducted, this research summarises that security concerns, security attentiveness, user competency, computer knowledge, gender as well as the number of years of computer usage are positively related with information security awareness of phishing attacks from bank customers.

The study and interview have shown some informative findings, and will assist to better comprehend the matters surrounding customers security awareness as well as behavior. Findings from the in-depth interview with experienced online bank customers indicated that six (6) main themes might influence the degree to which information security may be sophisticated. They include security concerns, security attentiveness, user competency, computer knowledge, gender as well as the number of years of computer usage. The most impact subject in this research is safety concerns. This theme found that customers are worried about online banking safety. Approximately 64% of the respondents said they were truly concerned about the security of Internet funding, but are still willing to use the service. Phishing website is a huge effect on the financial and online commerce, detecting and preventing this attack is an important step towards protecting against website phishing attacks. There are several approaches to detect these attacks [51]. One respondent said he has lessened the rate with which he employed online banking and he justified it by the security concerns and lack of adequate technological knowledge as being two main obstacles to the adoption of online banking

The second theme identified is security awareness. Not every customer keenly cares about online banking security [52]. More than 50% of respondents showed that they certainly did not ask for facts regarding online banking safety. This means that a large number of customers never go to the website, read flyers or contact the call center. This shows that customers are expected to be the inactive receivers of security information about awareness actions, which seems to challenge the conviction that if the person is worried about security, he/she may be enthusiastically pursuing appropriate information [53]. The third theme which is user competence shows that customers are not actually accustomed to Internet investment safety. Even though 66.7% of respondents appeared to know more or less regarding online banking security, it was revealed that almost half of the respondents did not recognize what active passwords or numerical credentials are or the meaning of phishing. This advocates that customers are apt to imagine they are more aware of this data than they actually are, which can make them ease their alertness and open themselves to likely risk [54].

The last theme that emerged from this study is computer literacy, gender, and the number years of PC usage. Customers are not familiar with basic technologies and the key risks of online banking. None of the respondents in the study knew enough to reply all the basic queries concerning online banking safety [46]. Approximately 20% of respondents did not recognize the website address of their online bank and almost 45% did not recognize the number of the bank's call centre. Respondents with an IT education were acquainted with the basic technologies required for online banking and the key dangers, but even they were uncertain regarding the measures they should follow when facing serious issues [56]. Nearly half of the respondents with monetary or other qualifications did not distinguish the technologies, the core dangers, or the processes. This advocates that customers are not entirely educated regarding online banking security [57].

6 Conclusion

This ongoing research has explained initial findings based on a preliminary investigation carried out through in-depth interview. Rising numbers of customers have started to use online banking amenities due to their excessive accessibility and speediness. Nevertheless, these big benefits have not gone undetected by those desiring to scam. The similar advantages that customers enjoy can also be shared with those wishing to compromise safety. Guaranteeing the safety of Internet investment is among the significant responsibilities and obligations for banks. Aside from experience, the human aspect is another crucial component in the prevention of scam. Customer's awareness of online banking security is one effective technique that should be considered. Nonetheless, the investigation shows that, in spite of extensive scams, the common level of customer awareness in respect to online banking security is still low. There is a pressing requirement for enhancement here. Moreover, even the high consciousness alone does not guarantee likelihood that customers modify their information security behaviour.

The effects of this study are dual. Scholastically, not only did it take the lead in comprehending the current issue of security awareness and behavior, but it has also primarily explored another potential connection between this cognizance and behavior, as well as empirically stressing a likelihood of the negative relation. This helps to counterpart the up-to-date research activities. Basically, investigation sheds light on the online banking approach that the banks can implement to enrich their customers' awareness and inspire rational behavior. Some banks might not support the customers to employ their online banking service safely, but as an alternative according to the findings, they should deliver them with the sufficient and required awareness and skills. This study has delivered descriptions and valuable perceptions into customers' self-awareness of data security hazards about personality qualities as well as communicative objectives towards common information safety risks.

Moreover, it has contributed to exploring the critical area and factors impacting in information security awareness of phishing attacks, by combining the connected fields of information technology and societal attitude into a sole theoretical context. It has also presented a thorough discussion regarding the key findings of the research work. However, the investigation does have some limitations which is that the number of respondents is not adequately large, which may influence the overview of the results.

Our research came up with various constructs to gather insights from online banking customers, and these will form the survey questionnaires. By gathering their insights, the level of knowledge they have, the efforts they have put, and the perception of the efforts of banks regarding phishing attacks of online banking platforms, it would be easy to identify the current state of the situation, evaluate against the past, and develop a forecast of the future, considering the trend maintains a regular course. Ideally, the best tool for unearthing the current challenge faced world over is by getting to the grassroots and getting first-hand information from the users of such services that are directly affected.

References

1. Alwan, H.: Al-Zu'bi, A.: Determinants of internet banking adoption among customers of commercial banks: an empirical study in the Jordanian banking. Int. J. Bus. Manag. **11**, 95–97 (2016)
2. Fadi, A.: The need for effective information security awareness. J. Adv. Inf. Technol. **3**, 176–183 (2012)
3. Shewangu, D.: Financial consumer protection: internet banking fraud awareness by the banking sector. Banks Bank Syst. **4**, 128–130 (2016)
4. APWG: Phishing Activity Trends Report (2016). Accessed 8 May 2018
5. Zhu, R.: An initial study of customer internet banking security awareness and behaviour in China. In: Pacific Asia Conference on Information Systems (PACIS) (2015). http://aisel.aisnet.org/pacis2015/87
6. Ameme, B.K., Yeboah-Boateng, E.O.: Internet banking security concerns: an exploratory study of customer behaviors based on health belief model. Int. J. Emerg. Sci. Eng. (IJESE) **4** (3) (2016). https://www.researchgate.net/publication/298794389_Internet_Banking_Security_Concerns_An_Exploratory_Study_of_Customer_Behaviors_based_on_Health_Belief_Model
7. Bada, M., Sasse, A.: Cyber security awareness campaigns: why do they fail to change behaviour? In: International Conference on Cyber Security for Sustainable Society 2015 (2015). https://www.sbs.ox.ac.uk/cybersecurity-capacity/content/cybersecurity-awareness-campaigns-why-do-they-fail-change-behaviour
8. Veseli, I.: Measuring the effectiveness of information security awareness program. Masters. Gjøvik University College (2017)
9. Asfoor, A., Rahim, F.: The potential factors influencing information security awareness on phishing attacks from various industries: a systematic literature re-view (SLR). Unpublished manuscript. Manuscript submitted for publication (2017)
10. FBI Warns of Dramatic Increase in Business E-Mail Scams (2018). Federal Bureau of Investigation. https://www.fbi.gov/contact-us/field-offices/phoenix/news/press-releases/fbi-warns-of-dramatic-increase-in-business-e-mail-scams. Accessed 17 Apr 2018
11. Phishing Activity Trends Report 4th Quarter 2016. Phishing Activity Trends Report (2017)
12. Symantic: Internet Security Threat report (2017). https://www.symantec.com/content/dam/symantec/docs/reports/istr-22-2017-en.pdf
13. Olalere, M., Waziri, V.: Assessment of information security awareness among online banking customers in Nigeria. Int. J. Adv. Res. Comput. Sci. Softw. Eng. 13–15 (2014)
14. Charles, B.: Phishing: the speed of attack. Retrieved from Bank info security (2017)
15. Alsayed, A., Bilgrami, A.: E-banking security: internet hacking, phishing attacks, analysis and prevention of fraudulent activities. Int. J. Emerg. Technol. Adv. Eng. **7**(1), 110 (2017). https://www.researchgate.net/publication/315399380_E-Banking_Security_Internet_Hacking_Phishing_Attacks_Analysis_and_Prevention_of_Fraudulent_Activities
16. Bakhshi, T.: Social engineering: revisiting end-user awareness and susceptibility to classic attack vectors. In: 2017 13th International Conference On Emerging Technologies (ICET) (2017). http://dx.doi.org/10.1109/icet.2017.8281653
17. Dakpa, T., Augustine, P.: Study of phishing attacks and preventions. Int. J. Comput. Appl. **163**(2) (2017)
18. Moreno-Fernández, M.M., Blanco, F., Garaizar, P., Matute, H.: Fishing for phishers. Improving internet users' sensitivity to visual deception cues to prevent electronic fraud. Comput. Hum. Behav. **1**(69), 421–436 (2017)

19. Ilyas, M., Sarah, S.: Qualitative exploration of trust towards online banking for phishing attack victims/Siti Sarah Md Ilyas. Ph.D diss., Universiti Teknologi MARA (2017)
20. BissonFollow D. 6 Common Phishing Attacks and How to Protect Against Them (2016). https://www.tripwire.com/state-of-security/security-awareness/6-common-phishing-attacks-and-how-to-protect-against-them/
21. Canfield, C., Fischhoff, B., Davis, A.: Quantifying phishing susceptibility for detection and behavior decisions. Hum. Factors J. Hum. Factors Ergon. Soc. **58**(8), 1158–1172 (2016). https://doi.org/10.1177/0018720816665025
22. Lacey, D., Glancy, P., Salmon, P.: Taking the bait: a systems analysis of phishing attacks. Sciencedirect **1111** (2015)
23. Van de Weijer, S., Leukfeldt, E.: Big five personality traits of cybercrime victims. Cyberpsychol. Behav. Soc. Netw. **20**(7), 407–412 (2017). https://doi.org/10.1089/cyber.2017.0028
24. Jansen, J., Leukfeldt, R.: Phishing and malware attacks on online banking customers in the Netherlands: a qualitative analysis of factors leading to victimization. Int. J. Cyber Criminol. **10**(1), 89 (2016)
25. Ameme, B., Boateng, E.: Internet banking security concerns: an exploratory study of customer behaviors based on health belief model. Int. J. Emerg. Sci. Eng. **4**, 1–7 (2016)
26. Chin, A., Wafa, S.: The effect of internet trust and social influence towards willingness o purchase online in Labuan, Malaysia. Int. Bus. Res. **2**(2), 72 (2009)
27. Jassal, R., Sehgal, R.: Online banking security flaws: a study. Int. J. Adv. Res. Comput. Sci. Softw. Eng. **3**(8), 1016–1021 (2013)
28. Zafar, H.: Human resource information systems: information security concerns. Hum. Resour. Manag. Rev. 105–113 (2013). https://www.sciencedirect.com/science/article/pii/S1053482212000538
29. Zanoon, N., Gharaibeh, N.: The impact of customer knowledge on the security of e-banking. Int. J. Comput. Sci. Secur. (IJCSS) **7**, 81–92 (2013)
30. Vishwanath, A.: Examining the distinct antecedents of e-mail habits and its influence on the outcomes of a phishing attack. J. Comput. Mediat. Commun. **20**(5), 570–584 (2015). https://doi.org/10.1111/jcc4.12126
31. Luo, X., Zhang, W., Burd, S., Seazzu, A.: Investigating phishing victimization with the heuristic-systematic model: a theoretical framework and an exploration. Comput. Secur. **38**, 28–38 (2013). https://doi.org/10.1016/j.cose.2012.12.003
32. Zhang, W., Luo, X., Burd, S., Seazzu, A.: How could I fall for that? Exploring phishing victimization with the heuristic-systematic model. In: 45th Hawaii International Conference on System Sciences (2012)
33. Xu, Z., Zhang, W.: Victimized by phishing: a heuristic-systematic perspective. J. Internet Bank. Commerce **17**(3), 28–38 (2013)
34. Leukfeldt, E.: Phishing for suitable targets in the Netherlands: routine activity theory and phishing victimization. Cyberpsychol. Behav. Soc. Netw. **17**(8), 551–555 (2014). https://doi.org/10.1089/cyber.2014.0008
35. Graham, R., Triplett, R.: Capable guardians in the digital environment: the role of digital literacy in reducing phishing victimization. Deviant Behav. **38**(12), 1371–1382 (2016). https://doi.org/10.1080/01639625.2016.1254980
36. Vrancianu, M., Popa, L.A.: Considerations regarding the security and protection of e-banking services consumers' interests. Amfiteatru Econ. J. Acad. Econ. Stud. **12**(28), 388–403 (2010)
37. Psychology Press Ltd.: Research methods: data analysis (2004). http://onlineclassroom.tv/files/posts/research_methods_chapter/document00/psych%20methods.pdf

38. Conway, D., Taib, R., Harris, M., Yu, K., Berkovsky, S., Chen, F.: A qualitative investigation of bank employee experiences of information security and phishing. In: Thirteenth Symposium on Usable Privacy and Security (SOUPS 2017), pp. 115–129. USENIX Association (2017)
39. Workman, M.: Wisecrackers: a theory-grounded investigation of phishing and pretext social engineering threats to information security. J. Assoc. Inf. Sci. Technol. **59**(4), 662–674 (2008)
40. Mishra, A., Ayesha, A., Elijah, J.: An assessment of the level of an assessment of the level of online banking users in Nigeria. Int. J. Comput. Sci. Mob. Comput. **6**(5), 373–387 (2017)
41. Miles, M.B., Huberman, M., Sladana, J.: Qualitative Data Analysis; A Methods Sourcebook, vol. 28, no. 4. SAGE Publications, Thousand Oaks (2014)
42. Phishing, Prevention, and Examples: INFOSEC INSTITUTE (2017). https://resources.infosecinstitute.com/category/enterprise/phishing/#gref
43. Polasik, M., Piotr Wisniewski, T.: Empirical analysis of internet banking adoption in Poland. Int. J. Bank Market. **27**(1), 32–52 (2009)
44. Greenfield, R.: The internet password paradox. Retrieved from the Atlantic Wire (2011). http://www.theatlanticwire.com/technology/2011/08/irony-inter-netpasswords/41078/
45. Anti-Phishing Working Group: Phishing Activity Trends Report, 2nd Quarter edn. (2010). http://www.apwg.com/reports/apwg_re-port_q2_2010.pdf
46. Haque, A., Hj Ismail, A.Z., Daraz, A.H.: Issues of e-banking transaction: an empirical investigation on Malaysian Customers Perception. J. Appl. Sci. **9**, 1870–1879 (2009)
47. Cao, Y., Han, W., Le, Y.: Anti-phishing based on automated individual whitelist. In: Proceedings of the 4th ACM Workshop on Digital Identity Management, Alexandria, Virginia, USA (2008)
48. Anti-Phishing Working Group: Phishing activity trends report for the month of December (2007). http://www.antiphishing.org/re-ports/apwg_report_dec_2007.pdf
49. Banks Increase Security Measures: Retrieved from the Australian Banking and Finance website (2010). http://www.australianbankingfinance.com/technology/banksin-crease-security-measures/
50. Lichtenstein, S., Williamson, K.: Understanding consumer adoption of internet banking: an interpretive study in the Australian banking context. J. Electron. Commerce Res. **7**(2), 50 (2006)
51. Alkhozae, M.G., Batarfi, O.A.: Phishing websites detection based on phishing characteristics in the webpage source code. Int. J. Inf. Commun. Technol. Res. **1**(6) (2011)
52. Jansen, J., Leukfeldt, R.: Phishing and malware attacks on online banking customers in the Netherlands: a qualitative analysis of factors leading to victimization. Int. J. Cyber Criminol. **10**(1), 79 (2016)
53. Help Net Security: Enhanced phishing methods on the rise. Retrieved from the Help Net Security website (2012). http://www.net-security.org/secworld.php?id=11317
54. Chen, J.Q., Schmidt, M.B., Phan, D.D., Arnett, K.P.: E-commerce security threats: awareness, trust and practice. Int. J. Inf. Syst. Change Manag. **3**(1), 16–32 (2008)
55. Dimitriadis, S., Kyrezis, N.: Linking trust to use intention for technology-enabled bank channels: the role of trusting intentions. Psychol. Market. **27**(8), 799–820 (2010)
56. Afroz, S., Greenstadt, R.: Phishzoo: detecting phishing web-sites by looking at them. In: 2011 Fifth IEEE International Conference on Semantic Computing (ICSC), pp. 368–375. IEEE (2011)
57. Kessem, L.S.: What makes phishing so successful?. Retrieved from the InformationWeek website (2012). http://www.informationweek.in/Security/12-05-08/What_makes_phishing_so_successful.aspx

Computational Vision: Detection and Simulation Techniques

Exploiting DOM Mutation for the Detection of Ad-injecting Browser Extension

Azreen Zaini(✉) and Anazida Zainal(✉)

Information Assurance and Security Research Group (IASRG), Universti Teknologi Malaysia, 81310 Skudai, Johor, Malaysia
ayinbzaini@gmail.com, anazida@utm.my

Abstract. Browser extensions provide additional browserfunctionalities which include blocking advertisements, managing passwords, translating pages, and providing dictionaries. Although the extensions provide functionality to user environment, the extensions have been abused to inject advertisements into web pages without user concern. However, identifying ad-injecting extension is fundamentally difficult since differentiating the content modification is related to ad-injecting extension. Thus, a robust technique that has the ability to identify and highlight the Document Object Model (DOM) element that related to ad-injecting extension in DOM tree is needed. To resolve this dilemma, this study proposed to implement DOM Mutation Observer as the core technique to track the changes in DOM tree. The results show that the selected approach managed to identify which DOM element that can be used as an indicator to track the source of modification on DOM tree that is related to ad injection characteristic.

Keywords: Ad injection · Browser extension
Document Object Model (DOM)

1 Introduction

Web browser is one of the most popular applications. This is because it is easy for users of limited technological expertise to use browsers. Also, developing and deploying web applications are far easier than desktop or mobile applications. All major browsers currently support browser extension system whereby third parties can add functionality or modify the behavior of the browser or web-pages by modifying DOM [1]. DOM is a programming API for HTML and XML documents which define defines the logical structure of documents and the way a document is accessed and manipulated. With the Document Object Model, programmers can create and build documents, navigate their structure, and add, modify, or delete elements and content. In addition, extensions are often granted the privilege to run within sensitive browser tabs that enable banking, social media, and medical web-pages, as well as email and communication applications.

Browser extensions contain content script where a JavaScript file can be used to inject any script into the user web environment such as to modify DOM [2]. Given the huge benefits offered by the extensions, it has been abused where web pages are injected with advertisements to distract revenue from content publishers and risk users to malware [3–5].

F. Saeed et al. (Eds.): IRICT 2018, AISC 843, pp. 657–669, 2019.
https://doi.org/10.1007/978-3-319-99007-1_61

Recently, it was reported that several firms acquired chrome extensions with excellent user acceptance just to inject malware and ads[1]. In other words, some firms target chrome extension that is complete with a large trusting user base so that the inserted 'ads' look benign and can be trusted by the users. Besides that, some ad networks offer capabilities to generate income to the extension developer by allowing ads to be published from their extension[2]. This new business trend caused user safety and privacy to be more obsolete when browsing the Internet.

Numerous approaches have been proposed to identify malicious characteristics which consist of ad injections browser extensions. A pioneering work in this field is Hulk [1] which is an independent system that focuses on distinguishing malicious behavior from benign extension characteristic based on a dynamic analysis which monitors JavaScript execution and corresponding network activity. Using Hulk, Kapravelos et al. identified 130 malicious extensions that perform various malicious behaviors such as affiliate fraud, credential theft, and ad replacement, and also social network abuse. In a similar study, Jagpal et al. proposed WebEval [6] which is an internal automated system that is responsible for reviewing chrome extension for Chrome Web Store (CWS) using both static and dynamic analyses in their study. WebEval's static analysis module extracts different parts of extension such as extension's HTML, Javascript, and manifest permission. As for dynamic analysis, WebEval picks up every chrome DOM call, API call, and network request for advanced analysis. However, there is no guarantee that studying the malicious behavior on browser extension for a short period of time can lead to an understanding of ad injection behavior. Thus, this study proposed to develop a client-side detection methodology that focuses on DOM structure to detect ad injection. Xing et al. proposed Expector [3] which compares legitimate DOM structure with modified DOM structure (extension loaded) to identify the differences between two DOM structures. By using Longest Common Subsequence (LCS), the algorithm output nodes that are only present in the webpage that is loaded with the browser extension for further investigation. In addition to client-side detection, Arshad et al. present OriginTracer [7], an approach to track the DOM content modification by annotating the web page DOM elements that have the ability to add or modify DOM tree. OriginTracer helps users to make a decision whether to allow the changes on DOM that are reported by the system. While ad injection possessed security risk to the user environment, the main issue is to verify whether the source of the modification is based on the browser extension. Some users might legitimately allow the content modification without knowing the source of the modification. Thus, a robust technique that has the ability to identify and highlight the DOM elements that are related to ad-injecting extension in DOM tree is crucially needed.

To resolve this dilemma, this research is focused on detecting modification in individual DOM elements using Mutation Observer as the main core technique. Cypher

[1] Firms buy popular Chrome extensions to inject malware. http://www.zdnet.com/article/firms-buy-popular-chrome-extensions-to-inject-malware-ads/.

[2] Get ready, Chrome users – You're about to start seeing ads inside of extensions.https://thenextweb.com/google/2012/07/03/get-ready-chrome-users-youre-about-to-start-seeing-ads-inside-of-extensions/.

et al. [8] mention that DOM modification is emphasized because the only possibility for ad injection to occur is by mutating the DOM tree. Therefore, by focusing on the mutated DOM elements, the content modifications origins can be identified and the content from the publisher and third parties as for this case, extensions can be distinguished. Our technique is also tested with a method that has the capability to modify DOM element in order to evaluate the effectiveness of the proposed research.

2 Related Work

In the following, we introduce background information on ad injecting extension by explaining the definition and how the extension using specific method to modifying DOM for advertisement injection.

2.1 Ad-injecting Extension

Ad injection is a method where ads are secretly added to web pages without the authorization or acknowledgement of site owners let alone being paid by them. Ad-injecting extension has the capability to replace existing ads that have appeared on the web page [3]. To perform via a browser extension, the developer extension relies on a content script that has the capability to modify DOM. According to Jagpal et al. [6], 3496 chrome extensions were identified to contain ad injection behavior between 2012 and 2012. Due to the ability of the ad injection to monetize illegal revenue for developer extension, this action has become a common unrequested third-party content injection [9].

In order to perform ad injection via chrome extension, the developer extension crafted a simple browser extension that is loaded with ad injection library. The developer extension will provide simple applications such as calculator, weather forecasting, and ability to customize browser theme to help make the extension look benign and trusted.

The extension must employ methods which are capable to mutate DOM tree such as Node replaceChild, insertBefore, appendChild or removeChild methods to inject advertisements [8]. This method can insert, replace, or remove any nodes inside the DOM tree. All items in the DOM are defined as nodes. There are many types of nodes such as element, text, and comment nodes. When an HTML element is an item in the DOM, it is referred to as an element node. Any lone text outside of an element is a text node while an HTML comment is a comment node. In addition to these three node types, the document itself is a document node which is the root of all other nodes. Figure 1 demonstrates the ad injection process.

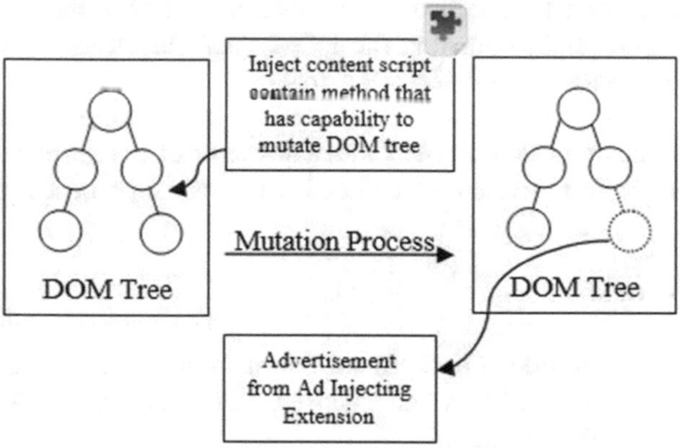

Fig. 1. Ad injection process

Many extensions target websites that contain a large number of users. Thus, ad-injecting extensions target shopping websites such as Amazon and eBay. Besides that, these websites contain advertisements which indirectly helps the ad-injecting advertisement look benign to the user. Figure 2 shows an example of advertisement injection.

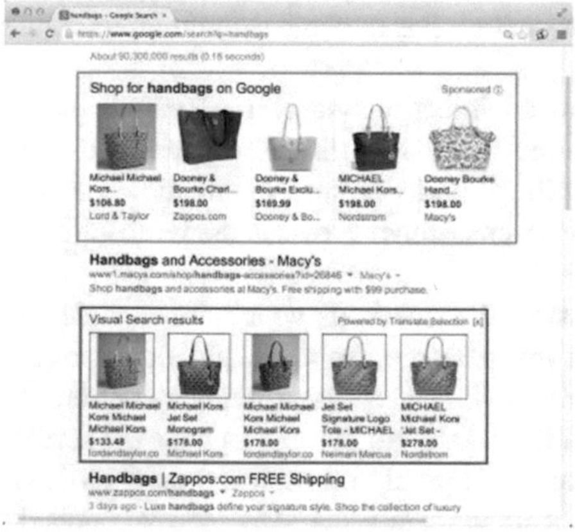

Fig. 2. Advertisement injection from "Translate Selection" Chrome extension (red border) [3].

3 Design and Methodology

This section discusses the design and methodology used in this research in detecting ad-injecting extension. Basically, there are 3 different phases involve in this section which is Phase 1 (Triggering Website and Events), Phase 2 (Collecting DOM Mutation using Mutation Observer) and Phase 3 (Filtering DOM Mutation using Pyhton Diflib Library). Figure 3 show the pseudocode that summarize all the phase involve in this research.

```
PROGRAM TrackingDOMMutation
    Retrieve Dataset;
    Open Browser;
    Run Selenium WebDriver;
        Install Extension;
        Visit Targeted Website;
            If Targeted Website is Shopping Website;
                Visit any advertisement Iframe;
                Collect DOM Mutation using Dom Mutation Observer;
            Else if Targeted Website is Search Result Page;
                Stimulate the searching process on search result pages;
                Collect DOM Mutation using Dom Mutation Observer;
        Repeat the process until all 10 targeted websites are finish
    Open Python IDLE;
            Open ('Orignal_Website') as text1;
            Open('Modified_Website') as text2:
                Compare text 1 and text 2 to find differences between two list;
                Save result as Compare_Result ;
            Repeat the experiment until all 10 targeted websites are finish;
                For each Compare_Result on different website conduct similarity operation;
            Save the similar DOM Mutation that contain DOM Value;
    Open Chrome DevTools
            Insert DOM Value that repeated more than two times;
                If DOM Value highlight specific element on webpage;
                    Identifying whether the element contain link or image;
                    Mark the extension contains ad injector characteristic;
                Else
                    Mark the extension as clean;
END
```

Fig. 3. Tracking DOM mutation pseudocode

3.1 Triggering Website and Events

A crawler was built based on Selenium WebDriver[3] to automatically visit the list of 10 targeted websites (9 shopping websites and 1 search result page) and perform two techniques to trigger the chrome extension events which is visiting advertisement iframe on the targeted website and stimulate the searching process on search result pages. However, it is observed that some extensions only inject the content script if the extension's icon is clicked, therefore, selenium API was instructed to perform the action.

Visiting Advertisement Iframe on Targeting Website

Kapravelos et al. [1] found an extension script which contained a large conditional block that looked for iframe elements of particular sizes such as 728 × 90 pixels and replaced them with new banners of the same size. Therefore, by visiting ads iframe on these 9 targeted websites, there is high potential for our technique to trigger the ad injection.

[3] Selenium Web Driver. https://www.seleniumhq.org/projects/webdriver/.

Stimulate the Searching Process on Search Result Pages

According to Xing et al. [3], ad injection also occurred in search result pages. Therefore, the searching process was stimulated on search result pages to improve the triggering event. In this research, the Google homepage was used as the testing area and the word "Shopping" was used when conducting the searching process.

3.2 Collecting DOM Mutation Using Mutation Observer

The DOM mutation observer[4]was implemented as the main technique to track the changes in DOM elements. Mutation Observer provides a function that is called asynchronously when certain parts of the DOM change such as adding a child to a node, changing an attribute on a node, or changing the text on a node. As these changes happen, the Mutation Observer records them as MutationRecords and then calls a user to provide callback at a later time with all the MutationRecords that are pending As the DOM mutation observer is focused on DOM structure, the probability to miss any changes is low.

To achieve the above operation a lightweight chrome extension called DOM observer was implemented which is loaded before other extensions are installed. The DOM observer extension contains the content script that injects mutation observer API into DOM tree. When the script is injected is controlled using array run_at = document_start inside the manifest file. This action ensures that the files are injected after any files from CSS, but before any other DOM is constructed or any other script is run. As soon as the script is successfully injected, the mutation observer API creates an observer. For instance, by passing it as a function that would be called every time a mutation has occurred. The observer is responsible for listening to any changes in DOM structure by receiving two arguments where the targeted DOM node needs observer and config object. In this research, the observer was set to observe the document node which contains the body and head of DOM structure. For config object, in order to listen for all changes, the true for all config was set as shown Fig. 4.

```
var observerConfig = {
attributes: true,
childList: true,
characterData: true,
subtree: true,
attributeOldValue: true,
characterDataOldValue:true
}
var targetNode =    document;
observer.observe(targetNode, observerConfig);
```

Fig. 4. Config object

[4] Mutation Observer. https://developer.mozilla.org/enUS/docs/Web/API/MutationObserver.

```
   case "attributes":
   console.log("The attribute " + mutation.attributeName
   + " of the node " + target.nodeName
   +"changed to '" + target.getAttribute(mutation.attributeName)    +
"'" );
   break;

   case "childList": Array.prototype.forEach.call
   (mutation.addedNodes, function(node)
      {
   console.log("Node " + node.nodeName + " added to " + target.nodeName
+ node.textContent);
   });
   Array.prototype.forEach.call(mutation.removedNodes, function(node) {
   console.log("Node " + node.nodeName + " removed from " + tar-
get.nodeName + node.textContent);
   });
   break;

   case "characterData":
   console.log("Content of " + target.nodeName + " changed    to '" +
target.data + "'" );
   break;
```

Fig. 5. Mutation categories

The mutation type was differentiated into three different categories which are attribute, childst, and characterData as shown Fig. 5. Attribute mutation involves observing any changes of information which determines the properties of a field or tag in a database or a string of characters in a display. Childst mutation involves detecting any new or deleted node that occurs in the DOM structure. CharacterData involves detecting any changes in the content of the targeted node.

In order to identify which mutation is related to ad-injecting extension, two different DOM mutations were collected which are before and after the extension that needs to be analyzed was installed. All the mutations are recorded and saved in console.log file for further investigation

3.3 Filtering DOM Mutation Using Pyhton Diflib Library

Pyhton Diflib Library algorithm[5] was used to compute the differences between two DOM mutation list as shown below. The algorithm output DOM elements that are present only on the webpage is loaded with the extension. Prefix '+' indicates that the line is unique between two lists.

[5] Python Difflib. https://docs.python.org/2/library/difflib.html.

```
import difflib
with open('Original_Website') as text1: with
open('Modified_Website') as text2:
d = difflib.Differ()
diff=list(d.compare(text1.readlines(),text2.readlines()))
for line in diff:
            for prefix1 in (' '):
                if line.startswith(prefix1):
                        break
            for prefix2 in ('-'):
                if line.startswith(prefix2):
            break
            for prefix3 in ('+'):
                if line.startswith(prefix3):
open('Compare_Result.txt', 'a') as diff_file:
                  print ('Processing')
                  diff_file.write(line)
```

To improve the labelling process, each Compare_Result which was collected was computed to compared the similarities in order to identify DOM elements that are present in different websites using the similar algorithm as shown above. For this process, we use prefix (' ') to indicate the line common to both sequences.

To verify whether the collected DOM mutation is related to ad injection practice, the DOM element that repeated more than 2 times was injected on Chrome's DevTools. Chrome's DevTools provides a function to highlight specific element on webpage. If the element is found and highlighted, the entire content that is associated with the DOM element was checked to identify graphical representations such as image of ads and URL that point to DOM element. If the image of ads and URL are found, the extension containing ad injection practice was marked.

4 Evaluation

The aim of this research is to evaluate the effectiveness of the proposed method. A total of 3 different categories were used, namely, 200 chrome extensions that were collected from Chrome Web Store, 292 chrome extensions that were involved in ad injection practice from previous work [3], and different method that has the capability to modify DOM structure using payload that consists dummy link and image advertisement.

4.1 200 Chrome Extension from Chrome Web Store

The purpose of this dataset is to understand and identify the characteristic of benign modification that is used in chrome extension. The dataset was categorized as benign due to popularity, rating, and the number of user in the chrome web store. In this dataset, the DOM operation that is used in the content script was observed.

This research differentiated the DOM mutation into 2 categories which are the new DOM node and changed attribute as shown in Table 1.

Table 1. Types of DOM mutation

Operation	Type	Popularity (%)
New DOM node	<script>	94.5
	<iframe>	62.5
	<div>	86.5
	<input>	30
	<style>	89.5
Changed Attribute	<style>	90.5
	<class>	71.5
	<id>	50.5

Table 1 clearly shows that most extensions inject script into the DOM structure. Script node is widely used to execute and load JavaScript to perform any function. For changed attribute, the changes in style tag were identified to be the highest in terms of popularity. Style tag contains information of CSS file that had been executed such as position, opacity, and size. Our approach is able to detect the DOM element that changed during the mutation process. This DOM element is used as an indicator to track the source of modification on DOM tree for the validation process. Figure 6 shows an example of DOM mutation that is successfully retrieved by our approach.

```
body.js:11 The attribute id of the node DIV changed to 'container-4280e47re7-f6f1-4ac5-9313-4bd16102696d-panel
```

Fig. 6. Example DOM mutation

Based on the result above, we relate the finding with others method that use to track changes in DOM tree which is by transverse the DOM tree that using by previous work [3]. Our method works more effectively by successfully identifying which type DOM mutation involve and retrieve the specific ID that related to an element. This is important key because by identifying specific ID, the content modifications can be verified whether the mutation contain any ad injecting characteristic. Besides that, our method also retrieves more details such as ability to identifying type of operation involve in DOM mutation compare by previous work [10]. This can be used to provide more information to understanding any changes in DOM tree. Based on our observation, it is determined that benign modification did not affect user browsing activities such as popping up an HTML document without user concern. Besides that, we also noticed that when the user closed the extension, the content script is completely removed from the DOM structure.

4.2 292 Ad Injection Extension

The purpose of this dataset is to understand and identify the characteristic of ad-injecting modification that is used in chrome extension. The dataset is collected from a research that was conducted by Xing et al. [3]. Their research identified 292 chrome extensions out of the 18030 extensions downloaded from the chrome web store. When the dataset was tested using our approach, an error was received which indicated that the extension code is corrupted. Besides that, errors such as the connection refused from console log were received which indicated that the link to download the advertisement has already expired. Based on this evidence, the collected chrome extension was rechecked with chrome web store whether the extension still exists or not. Surprisingly, only 4 out of the 292 chrome extensions still exist in the chrome web store. This populate that these 288 extensions may have disabled the ad injection functionality after being flagged as adware and malicious. The chrome web store removed the chrome extension due to security policy. Then, a manual review was conducted on each extension code to identify the method that is used in ad injector practice. Below is an example of extension code that was identified which has potential as an ad injector (Fig. 7).

```
        }
        var superfishScript = document.createElement("script");
        superfishScript.setAttribute("src",
"//www.superfish.com/ws/sf_main.jsp?dlsource=brandthunder&userId
=" + items.BTMahnetricsBID + "&CTID=" + items.tid);  docu-
ment.head.appendChild(superfishScript);
    });
```

Fig. 7. Potential ad injector script

Based on the code above, it can be concluded that chrome extension that contains ad injector using script element Script element technique will create a new script node which is document.createElement (''script''). Then, by using script.src method, the URL or directory of ad injector is loaded. Document.body.append Child(script) is used to append the script to DOM tree for the execution process. Based on the finding above, we compare the script with Top 10 DOM operation performed during dynamic execution from previous work [6], we noticed that 2 operation which is Document.createElement and Node.appendChild listed in the list thus prove that both method is widely used in extension to inject script.

4.3 Method That Has Capability to Mutate DOM

Although the dataset in Sect. 4.2 shows the extension code that contains ad injector, its only highlights the technique was used to fetch and execute script. In this research, to evaluate the effectiveness our research, this study proposed to investigate the method that has the capability to modify DOM tree as shown in Table 2. By studying this area, how this method modifies DOM to insert any advertisement banner can be further understood.

Table 2. DOM mutation method

Method	Justification
Node replaceChild insertBefore appendChild removeChild	This method has the ability to mutate DOM tree
Element.setAttribute Element.removeAttribute	This method has the capability to change and remove the attribute of targeted node
Document writeln	This method has the capability to replace the entire HTML document on the webpage

To study the related method, a new chrome extension that is loaded with the method above was created. The content script was injected into DOM tree and the mutation process was monitored. The content script that was injected contained method, dummy link, and image advertisement. The results show that our proposed approach is able to detect each method that has the capability to mutates the DOM tree. For example, when our approach was tested using Document writeln, DOM element that mutate the DOM structure was detected as shown in Fig. 8. Then, when the DOM element was inserted on Chrome's DevTools, the tools found and highlighted the content that was associated with the DOM element. The evidence confirms that our approach is successful in detecting methods that have the capability to mutate DOM.

Node CENTER added to BODY **Testing For Ads**
body.js:18 Node A added to CENTER **Testing For Ads**
body.js:18 Node IMG added to A
body.js:18 Node BR added to A
body.js:18 Node SMALL added to A **Testing For Ads**
body.js:18 Node #text added to SMALL **Testing For Ads**

Fig. 8. Example result using document writeln method

5 Conclusion

This paper implemented DOM Mutation Observer API as the core technique to detect the mutation in DOM tree that is derived from chrome extension. It is commonly known that the ad injection is only possible if an extension can mutate the DOM. Thus, by focusing on the client-side detection, DOM element that can be used as an indicator was successfully identified to track the source of modification on DOM tree that is related to ad injection characteristic. The approach used in this study was evaluated through three different datasets to show the capability to detect any DOM modification. The results show that our approach is sufficiently robust to detect any method that has the capability to mutate DOM. As part of future work, we plan on extending our approach to detect other threats which are involved in DOM mutation such as user tracking in extension.

Acknowledgements. This work is supported by the Ministry of Higher Education (MOHE) and Research Management Centre (RMC) at the Universiti Teknologi Malaysia (UTM) under UTM GUP Grant 16H56.

References

1. Kapravelos, A., Grier, C., Chachra, N., Kruegel, C., Vigna, G., Paxson, V.: Hulk: eliciting malicious behavior in browser extensions. In: USENIX Security Symposium, pp. 641–654 (2014)
2. Barth, A., Felt, A.P., Saxena, P., Boodman, A.: Protecting browsers from extension vulnerabilities. In: NDSS (2010)
3. Xing, X., Meng, W., Lee, B., Weinsberg, U., Sheth, A., Perdisci, R., Lee, W.: Understanding malvertising through ad-injecting browser extensions. In: Proceedings of the 24th International Conference on World Wide Web, pp. 1286–1295. International World Wide Web Conferences Steering Committee (2015)
4. Thomas, K., Bursztein, E., Grier, C., Ho, G., Jagpal, N., Kapravelos, A., McCoy, D., Nappa, A., Paxson, V., Pearce, P., Provos, N.: Ad injection at scale: assessing deceptive advertisement modifications. In: 2015 IEEE Symposium on Security and Privacy (SP), pp. 151–167. IEEE (2015)
5. Arshad, S., Kharraz, A., Robertson, W.: Include me out: in-browser detection of malicious third-party content inclusions. In: International Conference on Financial Cryptography and Data Security, pp. 441–459. Springer, Berlin (2016)
6. Jagpal, N., Dingle, E., Gravel, J. P., Mavrommatis, P., Provos, N., Rajab, M. A., Thomas, K.: Trends and lessons from three years fighting malicious extensions. In: USENIX Security Symposium, pp. 579–593 (2015)
7. Arshad, S., Kharraz, A., Robertson, W.: Identifying extension-based ad injection via fine-grained web content provenance. In: International Symposium on Research in Attacks, Intrusions, and Defenses, pp. 415–436. Springer, Cham (2016)

8. Cypher, M., Pietzuch, P., McBrien, P.: Intercepting suspicious chrome extension actions (2017)
9. Marvin, G.: Google study exposes "tangled web" of companies profiting from ad injection (2015). http://marketingland.com/ad-injector-study-google-127738
10. Starov, O., Nikiforakis, N.: XHOUND: quantifying the fingerprintability of browser extensions. In: 2017 IEEE Symposium on Security and Privacy (SP), pp. 941–956. IEEE (2017)

ESCAPE: Interactive Fire Simulation and Training for Children Using Virtual Reality

Wan Fatimah Wan Ahmad[1,2](✉), Aliza Sarlan[1,2], and Nurfadzira Ahmad Rauf[1]

[1] Department of Computer and Information Sciences, Universiti Teknologi PETRONAS, 32610 Seri Iskandar, Perak, Malaysia
{fatimhd, aliza_sarlan}@utp.edu.my
[2] Software Quality and Quality Engineering Research Cluster, Universiti Teknologi PETRONAS, 32610 Seri Iskandar, Perak, Malaysia

Abstract. Fire accident is a destructive disaster that has caused many property damages, injuries and even fatal. Lack of fire safety knowledge is one of the factors that lead to the horrendous experience of fire especially among young children and senior citizens. Conventional training usually does not assess the ability of the children to do practical skills on escaping out of fire but rather of fire drill on remembering escape route to fire point. Therefore, this project focuses on developing fire training and simulation for children by using virtual reality. The objective of the paper is to present a fire drill in the form of simulation and interactive training by using Virtual Reality. The simulation has been developed using Unity 3D and HTC Vive. Rapid application development was chosen for this development. A usability testing has been conducted with 10 participants. An interview has also been conducted with one of the school teachers. The results show that this project is beneficial and gives addition to the approaches of fire safety practice and allow children to learn better by improving their knowledge on fire safety.

Keywords: Interactive fire simulation · Virtual reality · Fire drill for children Fire safety · Usability

1 Introduction

Fire drill is a fire emergency procedure that intend to acknowledge the occupant on a building to escape fire breakout with no involvement of actual fire and element of stressor. Fire evacuation or mostly known as fire drill is an important education that people need to be considered to give full attention. Act such as Fire Service Act 1988 is intended to provide protection to people and property from risks and other purposes. It has been noticed that lack of awareness and understanding of fire safety lead especially children and senior adults exposed to harmful fire hazard [1]. Programs such as fire drills with fire department usually will be executed at the school at least once a year. Fire Certificate is a certificate that specially for appointed buildings accordance with the Fire Services Act & Regulations Act 341 [2]. Fire safety practices in buildings

F. Saeed et al. (Eds.): IRICT 2018, AISC 843, pp. 670–679, 2019.
https://doi.org/10.1007/978-3-319-99007-1_62

comprise fire extinguisher, fire hose reel fire, smoke detector, exit door hardware and smoke control. For school, management duties, housekeeping, fire drill training, alarm system, escape route, place of assembly, staircases, portable fire extinguishers are the element should be considered when it comes to fire precautions at school [3].

Nonetheless, dealing with children is much more different from the adults. Children are more complicated as they are not able to make correct decision without assistive of an adult on event. Conventional training usually does not assess the ability of the children to do practical skills on escaping out of fire but rather of fire drill on remembering escape route to fire point. Hands on training is a significant tool for children to learn and for cognitive development. A practical training using real world environment can be developed using Virtual Reality. By using simulated school environment, it mimics the real environment of the school and real fire situation.

The problem triggering to build virtual reality for fire simulation and training due to lack of real experience of fire situation in schools. Children are not adjusted to any fire situation during fire drill conducted by school management and school management cannot replicate real fire situation which could be harmful for student's health and which also require extreme caution on using real fire [4]. The absent of real fire on buildings is not acceptable just for the sake of giving the real experience of fire situation. Therefore, it is difficult to mimic the fire that could occur, and it is also the main barrier of making the frill drill more effective especially for younger generation.

Feng and Zhao's [5] claimed that common approach of fire safety practices which is fire drill are more likely difficult and the level of danger of real smoke and fire is hard to control. Meanwhile, Nguyen [6] clarifies that the drawbacks of conventional fire drill is hard to duplicate, safety issues and due to its cost. The current practice of fire drill is unreasonably formal for children to understand the importance of practicing fire drills. Some children tend to feel bored and looking puzzled which means only certain information will be processed. In today's digital world, children are surrounded with technologies hence it would be interactive and enrich the learning style if the element of technology is implement to gadgets such as virtual reality. Nguyen's [6] research clarifies the ability of Virtual Reality to make students more active in doing cognitive thinking activities is a good a way for student to perceive information compared learning theoretically from a book or lectures.

Hence, the objectives of this paper is to present are three folds. First objective is to identify the suitable theories and method for developing fire training and simulation for children; second objective is to develop a prototype of fire training simulation that incorporates few scenarios and stages; and finally, the third objective is to evaluate the application by conducting user acceptance test.

2 Literature Review

2.1 Fire Safety

Fire has been known as an important element in daily life of human since centuries ago. Rapid urban community in this era is exposed to fire hazard with several factors such as development of innovation, industrial and another activity. It is known to be as a destructive hazard to human kind which can lead to fatal and property loss if it is not properly utilize and manage in term of its safeness. Due to all the activities and rapid growth of world, extra precaution needs to be executed in making sure of the safety of the people when it comes to the present of hazard. Based on the statistics of cause of fire, it was reported that out of 9610, 6890 were categorized under structured fire and the rest of it are categorized under vehicle fire [7]. Structured fire is fire that happened at premises such as residential, hostel, higher education institution, hospital etc. while vehicle fire is unexpected burning involving vehicle which lead to fire-related property damaged [7]. For structured fire cases, 93.92% was caused by accidental which hold the highest percentage among the causes reported for structured fire. Second place with 4.95% is from incendiary and the rest are from natural and undetermined which hold 0.89% and 0.25% from the overall percentage of the structured fire causes. This can be shown in Table 1 where the breakdown of cause and total of fire incident according to the cause.

Table 1. Number of structured fire breakout in Malaysia for 2015

Cause/Year	2015
Natural	61 (0.89%)
Accidental	6471 (93.92%)
Incendiary	341(4.5%)
Undetermined	17 (0.25%)
TOTAL	6890

It is reported by Fire & Rescue Department Malaysia, the number of death victims is slightly increased on year 2013 by 24 victims while it dropped to 3 victims on year 2014 which gives the total of death 139. In 2015, there is a growth of number in death victim by addition of 14 victims which accumulated the total of death which was 158 victims as shown in Fig. 1. The increase number of death can be caused by several factors which is lack of public awareness about fire safety and no serious awareness about the dangerous of fire hazard. Each human behaves differently from others while interpreting situation and risk occurred to them. Human performs an action in a situation based on his/her decision-making process instead of unconscious decision [8]. There is no guarantee that people would act appropriately during emergency cases because of many variables that can affect the judgement and respond. Thus, only small percentage of people encountered in emergency event would be able to take proper and convincing action [9].

Inadequate level of knowledge of safety practices exposed human to serious risk involving fire. In another study, it was found that young and older person with less knowledge about fire safety tend to be vulnerable which could cause them to serious injuries or death [10] due to their limited capability and processing information caused them to respond to hazard inaccurately. It is reported that a high number of deaths occurred to older people due to cognitive and mobility issues. Meanwhile, children are prone to expose to hazard due to lack of understanding and need supervision of adults. Thus, children become more curious on things that are restricted such as matches

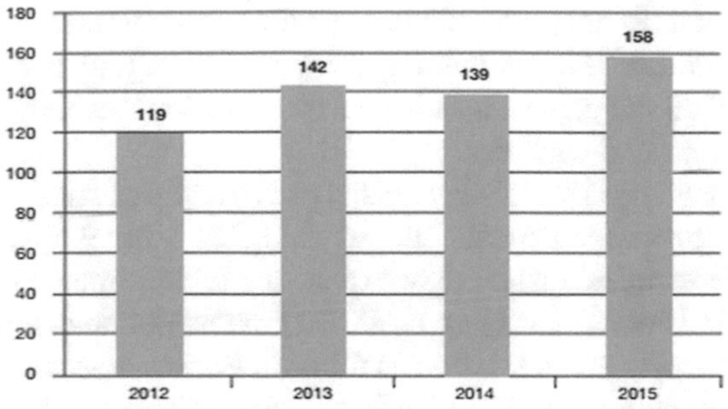

Fig. 1. Statistic of victim of fire in 2012–2015 [7]

candles, fireworks etc. Lack of preparedness can be defined as lack of knowledge on the subject matter which lead to poor decision making. Hence, these situations are few of the reason of acquiring proper fire safety knowledge.

2.2 Virtual Reality Technology

Virtual Reality (VR) is a computer technology that is widely used for various purposes by making an artificial environment. VR is defined as a realistic and immersive simulation of three-dimensional environment, created using software and hardware, and experienced or controlled by movement of the body [14, 15]. Onyesolu and Eze [17] pointed out that VR is being utilized in many ways and based on human activity such as interaction and human behavior. These resulted significant evolvements in culture, economic and worldview as well with heritage, mass media, health care and others. Its capacity to provide real environment through three dimensional allowing user to understand better when they interact with information and environment displayed by VR [18]. VR has become an alternative for human to deal with hazardous situation and environment which offer advantages to such hazardous environment, equipment or situation over conventional form of training [17]. VR is mostly used in medical field for pre-surgeries and rehabilitation, military, firefighters, air craft and pedestrian safety for children [18]. For example, Mandal [16] highlighted that out technologies has aided medical to evolve and improve with the intention of practicing medicine in accurate and effective way. VR in medical field comprises many categories; Medical and Dental Surgical Training, Interactive 3D Diagnostic Imaging, Rehabilitation and Sports Medicine. The rising of technology like VR enable doctors to conduct surgeries effectively in the way of training and rehearsing surgical procedure by using realistic simulation apart from using conventional training method which is becoming inefficient and unsuitable. Moreover, Virtual Reality can assist less experience doctor to reduce error which might resulted fewer complications [19].

For the past two decades, the arrival of digital games was a hit for entertainment. The element of Virtual Reality where it is computer generated simulations fits in into the world of games. Graphic rendering, digital surrounding sound and interaction with interface is the main element that lead to the gaming experience [20]. Rapid movement of technology has resulted game developers to explore the usage of Virtual Reality by incorporating Virtual Reality into digital gaming. The immersion level of VR is intense, but the emotion level is missing compared to digital games. By combining Virtual Reality and digital games, it adds up the immersion feeling with combination of emotions such as fear and curiosity.

Many studies have been widely conducted on Virtual Reality as an educational tool for learning since over two decades. It was said that Virtual Reality can prove the increase in user performance when they experience three dimensional to comprehend abstract problem. However, Sanchez et al. [21] point out that there are few issues regarding Virtual Reality and education as follows: learner de-socialization; evaluation of Virtual Reality experiences; disturbance of task centered learning and HMD usage can cause possible harm through long vision exposure of it. A study conducted by Knox [22] assumed that visual discomfort and fatigue is caused by the eyes trying to make the images clear and single, but no participant reported any discomfort after the research had been carried out. Adapting VR into this era provides more chances to reach out more students and their interests in studying. Some study conducted previously support that VR can help with academic performance and affective quality based on VR environment.

3 Methodology

This project uses the Rapid Application Development (RAD) approach which consists of few phases – analysis, design and implementation. The project starts by the analysis phase to understand and gather information on fire drill education program done at school level, any existing system available, technology used other than Virtual Reality and few benefit of implementing through this technology. As the project focuses on primary school, information gathering is carried out by having an interview session with two school teachers to understand their current fire safety educational programs and activities. Thus, a flowchart of the fire safety training at school has been developed during this phase. Figure 2 shows the flowchart of the training.

During the design phase, storyboard is developed to show the situation and scene that could be in the training and simulation. The design includes the important requirement such as present of sign of fire hazard, simple workflow and story line yet efficient design. A simple flow of several activities was decided to be included in the design as the main objectives is to train children how to escape from a building. In the implementation phase, we focus on constructing and integrate the design with Virtual Reality technology including the 3D environment, the storyline and code the programing. The construction of 3D environment is carried out using Unity 3D software.

During the design phase, storyboard is developed to show the situation and scene that could be in the training and simulation. A simple flow of several activities was decided to be included in the design as the main objectives is to train children how to escape from a building. In the implementation phase, we focus on constructing and integrate the design with Virtual Reality technology including the 3D environment, the storyline and code the programing. The construction of 3D environment is carried out using Unity 3D software.

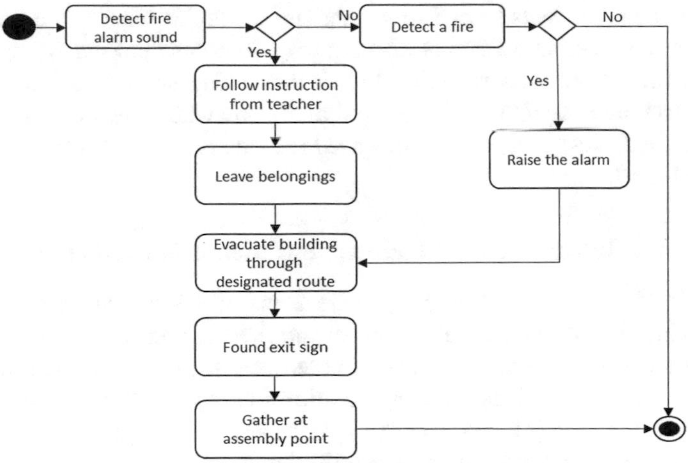

Fig. 2. Flowchart of the training

Finally, cutover phase of RAD, the whole process was installed to create live production environment on fire evacuation training and simulation. As a result, it enables the program to fully run a testing which help to save time to produce a good quality of product. Through user acceptance test, any improvement can be made throughout the testing phase until it meets the satisfactory outcome before the product is being released.

4 Results and Discussion

4.1 Analysis of Interview

An interview is conducted with a safety supervisor at a primary school located at Perak. There are 10 questions in the interview. The objective of the first section of the questions was about fire drill that is conducted at school. According to the teacher, the school is involved with fire drill twice a year and each session aims different objective.

The first session of fire drill, fire fighter will be involved to demonstrate correct way of practicing fire safety such as demonstration on how to use fire extinguisher, water demonstration and firefighter truck exhibition. Invitation of fire fighter to school is one of the method that attract students to learn more about fire safety. Meanwhile, the second session of fire training is by doing building evacuation in term of fire alarm alertness, right route to assembly point and speeches about fire safety during the gathering at the assembly point. Apart from having various activities involving fire safety practices, the school never had any fire mock-up as a demonstration to show to students.

As for student's behavior, interviewee expressed that students from 10 years to 12 years old easier to handle compared to students from the range 7 to 9 years old. It is because students from the age 10 to 12 years old, they are more mature to understand the importance compared to 7 to 9 years old students which still need close supervision from the teachers. Students from 7 to 9 years old recognize what is fire drill based on the common sense taught by the teachers but they do not understand the importance of it. Most of the students focus during fire drill is being conducted but some of the students especially in age 7 were in shocked and panic when they hear the fire alarm is being rang. Thus, these group of students need the supervision of teachers when the fire drill is conducted.

4.2 ESCAPE: Interactive Fire Training and Simulation Prototype

The following figures (Figs. 3, 4, 5, 6, 7) are some of the user interfaces snapshot of ESCAPE. Figure 3 shows the main menu of the fire training with two options in this page which are Play & Menu. Only two buttons were shown in this main menu screen to avoid any other distractions. Play button allows user to start the fire simulation and training and Exit button is to halt and exit to normal screen. Figure 4 shows the school environment created for this training since the target users are students. The building consists of many types of classroom and hallways for ground floor and second floor.

Fig. 3. Main menu screen

Fig. 4. School environment

Once user click the Play button, users will be in the predefined location where they are located at the classroom as shown in Fig. 5 User will be alerted by a fire alarm and need to evacuate within the predefined time. While trying to find an exit, user will encounter few fires and need to avoid it or use another way to evacuate the buildings. Fires will be generated at few places on the building. Figure 6 shows example of fire simulation at the hallway. If the user unable to evacuate within the predefined time, Game Over screen will pop out and give user two options whether the user wants to start over again or to exit the training as shown in Fig. 7.

Fig. 5. Predefined location as first scenario

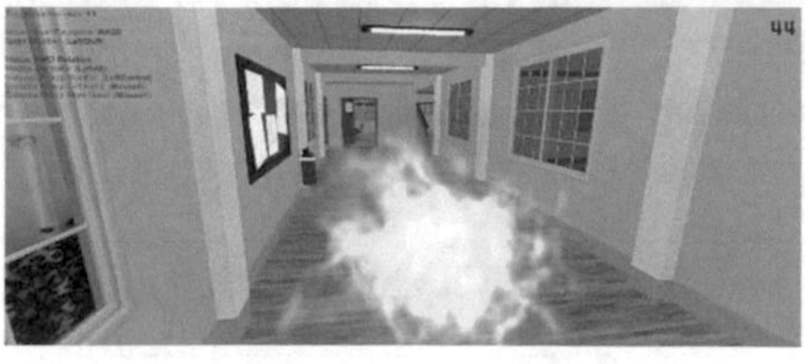

Fig. 6. Fire simulation at hallway

Fig. 7. Game over screen

4.3 Usability Testing

Usability testing was conducted with a total number of ten respondents which consist of eight students and two teachers at the school. The questionnaire consists of six questions as shown in Table 2.

Table 2. User Testing questions

No	Questions
1	The user interface is clear and pleasant to see
2	The colour scheme is attractive for kids
3	The flow of the game is smooth
4	ESCAPE Fire with Virtual Reality is suitable to learn fire safety for kids
5	It was easy to use the Virtual Reality tool with the game
6	It was easy to understand the direction of the game

Based on the following Fig. 8, it can be seen that most of the user agreed that the user interface is clear and pleasant to see. It also indicates that the main menu screen and game over screen are direct to the point and the design is pleasant to see. For the second question, most of the user voted for neutral and 3 of them agreed that the colour scheme is attractive especially for kids. Some user somehow agreed on the flow of the game and some are not as the smoothness of the game is based on the HTC Vive setup and due to some technical error on the controller which disable the movement of the first-person controller. Next question, most of the user agreed that ESCAPE Fire with Virtual Reality is suitable to learn fire safety for kids as it really let them to experience the real situation at school and enhance the decisions making skill. For the fifth question, only 3 users disagree that Virtual Reality tool is easy to use and 7 of them agreed that it was easy to use the Virtual Reality Tool. Some of the user might not familiar with such of tool and does not really understand on how to use it. Lastly, the survey of the last question indicates that most of the user understand the direction of game. Users are able to understand the objectives that need to be achieve by playing this game.

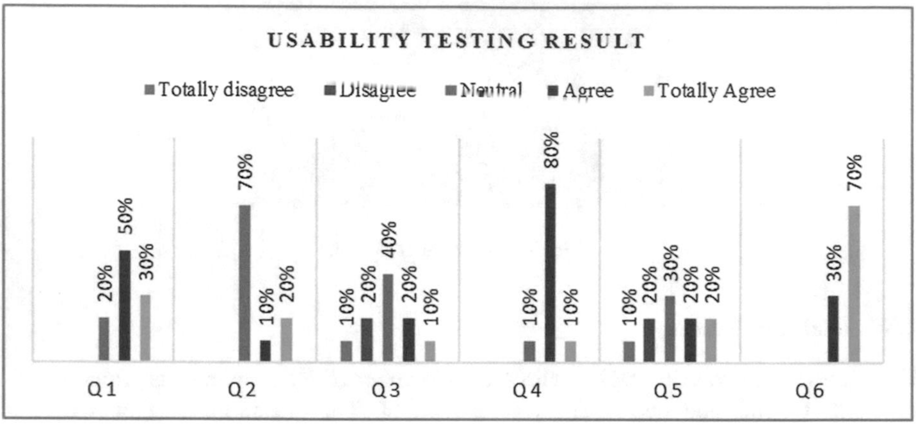

Fig. 8. User testing result

5 Conclusion and Recommendation

In conclusion, instead of implying fire safety practices to only adults, children also are important to be included when it comes to safety. There are a lot of approaches such as campaign talk, and exhibition implemented to create public awareness about fire safety practices. The advancement of technology creates many opportunities to prevent any potential hazard that can cause injuries and fatalities. Hence, Virtual Reality is one of the technology that can help to expose children more about fire safety practices. By using Virtual Reality, user can experience themselves being in the fire situation where allows them to think spontaneously on how to evacuate the building instead of relying on fire. Fire training using virtual reality really helps student to imagine what it is like to be in the real situation. Most of the teachers agreed and fully support of the development of the simulation and training game. We believed that ESCAPE helps children to be more spontaneous and enhance their decision making skills in the event of fire.

References

1. Huseyin, I., Satyen, L.: Fire safety training: its importance in enhancing fire safety knowledge. Aust. J. Emerg. Manag. **21**, 48–53 (2006)
2. Fire & Rescue Department Malaysia: Retrieved from 2015 Annual Report (2015). http://www.bomba.gov.my/resources/index/user_1/UploadFile/Penerbitan/Laporan%20Tahunan%20Bomba%202015.pdf
3. Taib, M., Nadzim, N.: Appraisal of fire safety management systems at educational buildings. In: Sustainable Built Environment Symposium (2014)
4. Fournier, H., Kondratova, I., Lapointe, J.-F., Emond, B.: Crossing the barrier: a scalable simulator for course of fire training. In: Interservice/Industry Training, Simulation, and Education Conference (I/ITSEC), pp. 10–16. National Training and Simulation Association, Orlando (2012)

5. Feng, X., Zhao, J.: The effectiveness of virtual reality for studying human behaviour in fire. In: Shumaker R., Lackey, S. (eds.) Virtual, Augmented and Mixed Reality (2015)
6. Nguyen, A.: Virtual reality with virtual humans simulation for an emergency training in a public sector (2015)
7. Vehicle Fire (2016). https://en.wikipedia.org/wiki/Vehicle_firc
8. Kuligowski, E.: The proccss of human behavior in fires. National Institute of Standards and Technology Technical Note 1632 (2009)
9. Robison, S.J.: When disaster strickes: human behaviour in emergency situation. J. Inst. Civ. Prot. Emerg. Manag. (2012)
10. Fire Risk in 2014: Topical Fire Report Series, pp. 1–14 (2016)
11. Raising fire safety awareness in Malaysia. Retrieved from The Sun Daily (2016). http://www.thesundaily.my/news/1728009
12. Muhammading, M.: About 6,000 premises destroyed by fire annually in Malaysia. Retrieved from New Straits Times (2016). http://www.nst.com.my/news/2016/
13. Kelab 3K: Retrieved from Jabatan Bomba dan Penyelamat Malaysia. http://www.bomba.gov.my/index.php/pages/view/246. Accessed 22 July 2016
14. Kinateder, M., Ronchi, E., Nilsson, D., Kobes, M., Müller, M., Pauli: Virtual reality for fire evacuation research. In: Federated Conference on Computer Science and Information Systems, pp. 319–327 (2014)
15. Virtual Reality. Retrieved from Wikipedia (2016). https://en.wikipedia.org/wiki/Virtual_reality
16. Mandal, S.: Brief introduction of virtual reality & its challenges. Int. J. Sci. Eng. Res. 304–309 (2013)
17. Onyesolu, O.M., Eze, F.U.: Understanding virtual reality technology: advances and applications. Adv. Comput. Sci. Eng. 54–70 (2011)
18. Hsu, E.B., et al.: State of virtual reality based disaster preparedness and response training. PLOS Curr. Disasters (2013)
19. Greenleaf, W. J.: Medical applications of virtual reality (n.d.)
20. Bouvier, P., Sorbier, F.D., Chaudeyrac, P., Biri, V.: Cross benefits between virtual reality and games. In: International Conference on Computer Games, Multimedia and Allied Technology (2008)
21. Sanchez, J., Lumbereras, M., Silva, J.P.: Virtual reality and learning: trend and issues (2001)
22. Knox, P.C.: The effect of a projected virtual reality training environment on vision symptoms in undergraduates (2015)

DWT-CT Image Watermarking Algorithm with Shuffling and Circular Rotation of Bits

Majdi Al-qdah(✉) and Malik Al-qdah

University of Hafir Al Batin, Hafar Al Batin, Kingdom of Saudi Arabia
majdi.qdah@gmail.com, mqdah@uohb.edu.sa

Abstract. This paper presents a novel approach of hiding data in images. The newly designed and implemented method uses two transforms: digital wavelet transform (DWT) and contourlet transform (CT). A two-level DWT is first applied on one standard image and six medical images before applying the CT transform on each HL2 subband. Then, for as an added layer of security, shuffling and circular rotation of bits are applied to the watermark image before inserting it into the coefficients of the directional contourlet sub band based on a randomly generated keys. The extraction of the watermark is achieved by the reversal procedure of the embedding procedure. The method was evaluated using the Peak Signal to Noise Ratio (PSNR) and the Normalized Correlation Coefficient (NC). The results indicate high robustness and imperceptibility of the combined DWT-CT approach as the watermark image was visibly extracted even after applying three serious types of attacks to the watermarked images.

Keywords: Wavelet transform · Contourlet transform · Security

1 Introduction

The spread of multimedia data through the Internet has increased the need to safeguard sensitive information. Various image watermarking methods have been proposed [1]. In general, data hiding has become an important tool to authenticate medical and non medical images and protect the copyrights of those images. Image watermarking techniques can be classified into two main techniques: spatial domain [2, 3] and transform domain [4, 5]. Spatial techniques embed watermarks in the least significant bits of the image pixels but they are not robust against attacks while transform domain methods embed the watermark in the coefficients of some transformed image and they are more robust against attacks.

There are various spatial domain methods such as the Least Significant Bit substitution. Lin et al. [6] introduced a spatial watermarking algorithm where the watermark is fused with noise bits before XORing it with the feature value of the image by 1/T rate forward error correction (FEC). The T is the times of data redundancy. The extraction of the watermark bits are done by taking majority voting.

On the other hand, there are many transform domain watermarking techniques: discrete wavelet transform (DWT), discrete cosine transform (DCT), contourlet transform (CT), and singular value decomposition (SVD). Horng et al. [7] proposed a robust adaptive watermarking method based on DCT, SVD and Genetic Algorithm

F. Saeed et al. (Eds.): IRICT 2018, AISC 843, pp. 680–689, 2019.
https://doi.org/10.1007/978-3-319-99007-1_63

(GA). Rosiyadi et al. [8] proposed a hybrid watermarking method using both the Discrete Cosine Transform and Singular Value Decomposition. Singh et al. [9] suggested another hybrid watermarking technique using a combination of DWT, DCT, and SVD transforms. The host image is first decomposed by DWT and the Low frequency band (LL) and watermark image are transformed using DCT and SVD. The S vector of watermark image is embedded in the S component of the host image and the watermarked image is generated by inverse SVD on modified S vector and original U, V vectors followed by inverse DCT and inverse DWT.

The discrete wavelet transform (DWT) divides images into four multi-resolution components: approximation (LL), vertical (LH), horizontal (HL), and diagonal (HH). Also, multiple sub bands decompositions can be applied. Usually, researchers embed sensitive information such as medical diagnostic data into the high level sub-bands since the detail levels carry most of the energy of the image [10]. Wavelet transform methods achieve high robustness since they localize information both in space and frequency. In short, DWT are attractive because they have multi-resolution representation, multi-scale analysis, adaptability and linear complexity [11]. The Contourlet Transform can be considered as an improvement to the wavelet transform and is suitable to decompose images that have many contours and curves. The Contourlet transform (CT) captures the edges and contours of images much better than the wavelet transform where one can embed watermarks into those edges since humans are visually less sensitive to edges and more sensitive to the smooth areas of images [12]. Also, the Contourlet Transform is a technology that uses similar base structure of Contour segment to approximate images. The support of base is a rectangle structure which can change length-width ratio with change of scale, and has directionality and anisotropy. Sy C. Nguyenet. al. presented an image watermarking technique by decomposing the host image into subbands. Then, the mid frequency subbands are chosen to embed watermark with suitable embedment factors [13].

Using a Contourlet Transform, an image is directionally decomposed using a Laplacian Pyramid (LP) and Directional Filter Bank (DFB). The LP decomposes the image into low pass and band pass image components. Then each band pass image is further decomposed by the DFB.

Numerous image based processing techniques have been proposed in literature [14–16] and a many of these techniques have used a hybrid of two or more transforms in order to compensate for the shortcomings of using a single transform; i.e. in image compression [17], image denoising [18], or image coding [19], and watermarking [14, 20]. In this paper, a new hybrid technique of DWT and CT is designed and implemented with the added security of rotation and shuffling of watermark bits.

2 Proposed Scheme

The requirements of any watermarking algorithm is imperceptibility, robustness, and security. Imperceptibility means that the original and the watermarked images are indistinguishable. Robustness dictates that it would be hard to remove or alter the watermark without significant degradation of the watermarked image. This paper presents a newly implemented hybrid approach of two transforms (DWT & CT) applied on selected images while including some extra encryption security features of shuffling and circular rotation of the watermark bits. The combination of DWT and CT compensate for each algorithm shortcomings. The strength of encryption depends on the keys' sixteen alphanumerical character length in order to break the rotation and shuffling procedures of the watermark image. The embedding and extraction algorithms are listed below:

Embedding algorithm

1: Apply DWT to decompose the image into four components: LL1, HL1, LH1, HH1.
2: Apply DWT to get four smaller sub-bands then select the HL2 sub-band
3: Apply 2-level CT to the HL2 sub-bands; then select directional sub-band
4: Apply shuffling of bits to the watermark based on a randomly generated key1
5: Apply circular rotation of bits based on randomly generated key 2
6: Change the shuffled- rotated watermark matrix into a vector of zeros and ones.
7: Generate *pn_sequence* to be used in embedding watermark bit 0.
8: Modify the coefficients of the chosen sub-band by embedding *pn_sequence* with some gain factor k according to: Coeff. = Coeff. + k*pn-sequence
9: Perform ICT using the selected sub-bands
10: perform IDWT to get a watermarked image.

Extraction algorithm

1: Apply DWT to decompose the watermarked image into four sub-bands: LL1, HL1, LH1, and HH1.
2: Apply DWT again to get four smaller sub-bands then select the HL2 sub-band
3: Apply 2-level CT to HL2 sub-band then select directional sub-band
4: Change the watermark matrix image into a vector of zeros and ones.
5: Generate a pn_sequence using with the same seed as in the embedding procedures.
6: Calculate the correlation of the selected sub-band with the pn_sequence
7: Calculate correlation mean and compare it with each correlation value.
8: The extracted watermark is equals to 0 if the correlation values are greater than the mean, otherwise it is equal to 1.
9: Reconstruct the watermark using the extracted bits,
10: Apply circular rotation of watermark bits based on key 2
11: Apply shuffling of watermark bits based on key1 to get original watermark

Figure 1 shows the block diagram of embedding a copyright watermark image into a cover image; while Fig. 2 shows the extraction approach of the watermark image.

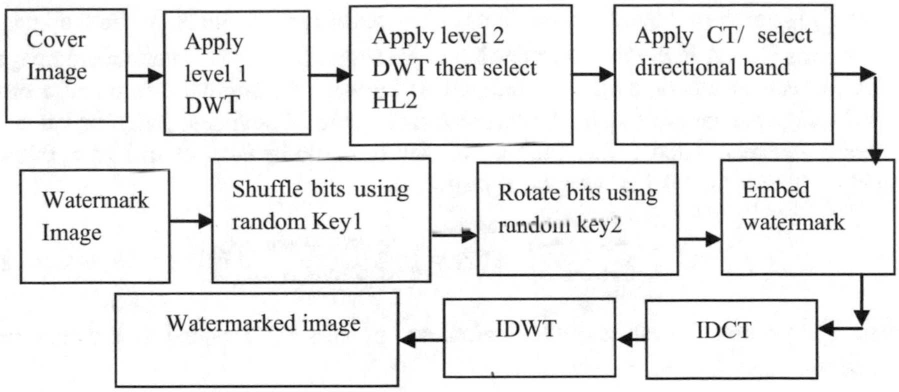

Fig. 1. Embedding process

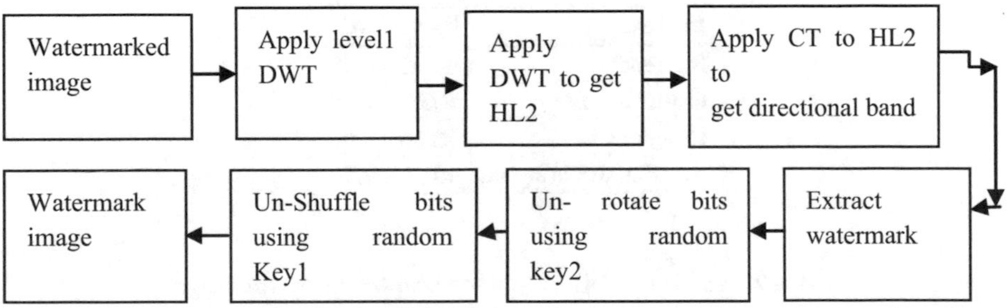

Fig. 2. Extraction process

3 Experimental Results

The newly proposed algorithm was implemented and tested in Matlab simulation environment using one standard (house) image of size 512 × 512 pixels, five medical images of size 512 × 512 pixels, and copyright watermark of size 64 × 64 pixels. Then, three of the most serious attacks that affect the extraction of a watermark are selected for analysis: Gaussian filter, Cropping and Rotation & Cropping attacks.

The imperceptibility was tested using the Peak Signal to Noise Ratio (PSNR) while robustness was measured using the Normalized Correlation Coefficient (NC).The PSNR and Mean Square Error (MSE) values are calculated for an original image and a processed image according to the equations below:

$$\mathrm{PSNR} = 10\log\left(\frac{255^2}{\mathrm{MSE}}\right) \tag{1}$$

$$\mathrm{MSE} = \sum\nolimits_{i=1}^{N}\sum\nolimits_{j=1}^{M}\frac{\left(I_{ij}^{d} - I_{ij}^{n}\right)^2}{NM} \tag{2}$$

where I^d_{ij}, I^n_{ij} are the values of the original and processed pixels and N, M are the image matrix sizes [15]. A high PSNR means high imperceptibility of the watermarked image.

Also, usually the similarity and differences between an original image and a processed image is measured by the Normalized Correlation Coefficient (NC) Its value is generally between 0 and 1. Ideally, The NC value should be as close to 1 as possible; usually a higher than 0.7 is acceptable [20].

$$NC = \sum_{i=1}^{X} \sum_{j=1}^{Y} (I^d_{ij} X\ I^n_{ij}) / \sum_{i=1}^{X} \sum_{j=1}^{Y} \left(I^d_{ij}\right)^2 \tag{3}$$

where I^d_{ij}, I^n_{ij} denote the values of the original and processed pixels and X, Y denote the size of the image matrix.

Table 1. PSNR and NC for the standard house image

Image	PSNR	NC
No attack	87.32	0.997
Gaussian filter	50.45	0.70
Cropping	46.86	0.63
Rotation & Cropping	40.31	0.44

Table 2. PSNR and NC for the medical Mammogram image

Image	PSNR	NC
No attack	90.16	0.996
Gaussian filter	51.82	0.75
Cropping	47.88	0.65
Rotation & Cropping	42.17	0.46

Table 3. PSNR and NC for the medical Head image

Image	PSNR	NC
No attack	89.32	0.997
Gaussian filter	49.23	0.62
Cropping	48.87	0.63
Rotation & Cropping	40.22	0.43

Table 4. PSNR and NC for the medical Eye image

Image	PSNR	NC
No attack	84.33	0.995
Gaussian filter	50.49	0.66
Cropping	45.66	0.67
Rotation & Cropping	39.92	0.41

Table 5. PSNR and NC for the medical Hand image

Image	PSNR	NC
No attack	91.27	0.996
Gaussian filter	52.32	0.77
Cropping	49.73	0.66
Rotation & Cropping	40.19	0.47

Table 6. PSNR and NC for the medical Fingerprint image

Image	PSNR	NC
No attack	90.33	0.998
Gaussian filter	53.45	0.77
Cropping	49.47	0.65
Rotation & Cropping	41.78	0.49

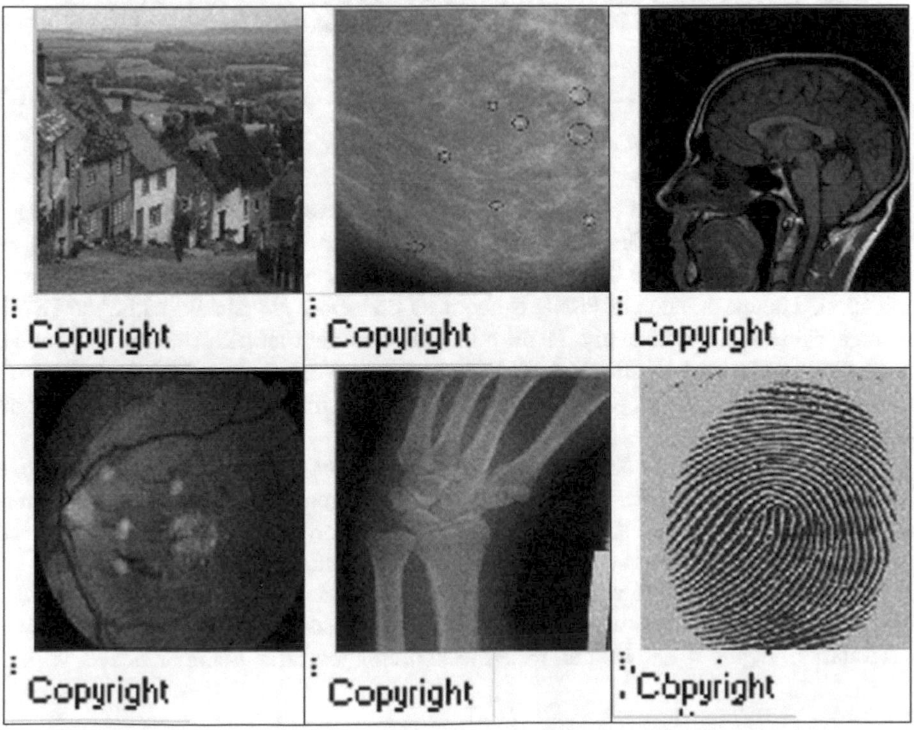

Fig. 3. Watermarked images and the recovered watermark before any attack

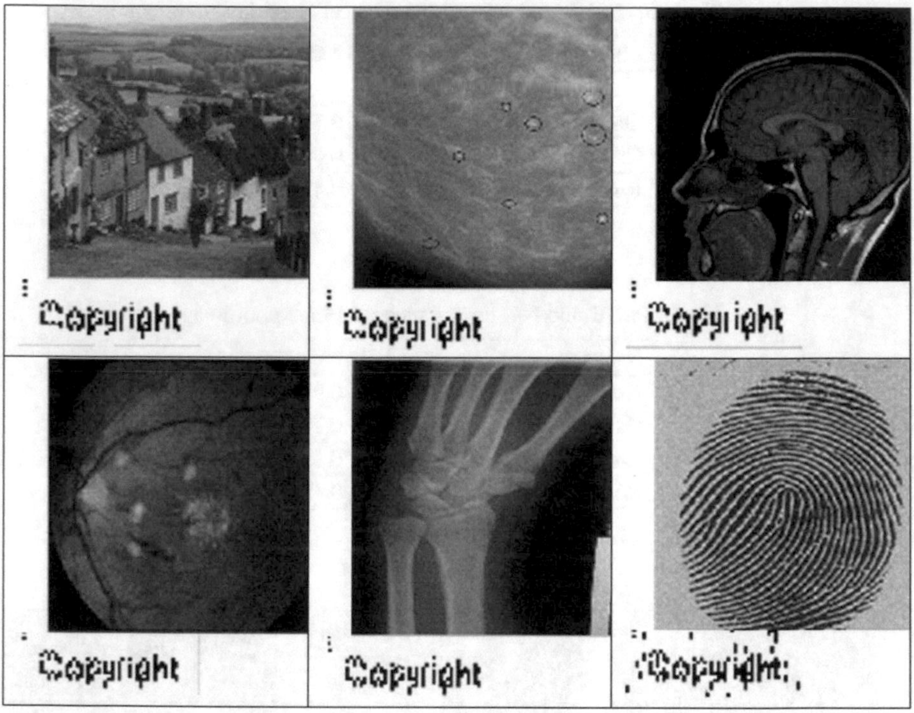

Fig. 4. Watermarked images and the recovered watermark after the Gaussian Filter attack

Tables 1, 2, 3, 4, 5 and 6 show the PSNR and NC values for watermarked images before and after attacks. The PSNR and the NC values are similar among all images before and after attacks; but the numerical values decrease after performing some attack operation. The mean value of PSNR is about 87 db before any attack; while the PSNR average value is approximately 51 db after the Gaussian attack. The PSNR average value drops to around 48 after the cropping attack and drops sharply to around 41 db after the rotation & cropping attack. The extracted watermark image is distorted after an attack operation.

On the other hand, the NC value before attacking the watermarked image is above 0.995 value which indicates a high correlation between the original and the watermarked images. But, the NC value drops sharply to around 0.45 after attacking the watermarked images with a rotation & cropping operation.

Figure 3 shows the six watermarked images and the extracted watermarks before any attack. The extracted watermarks are almost exactly the same as the original watermarks. Figure 4 shows the watermarked images after being attacked with a

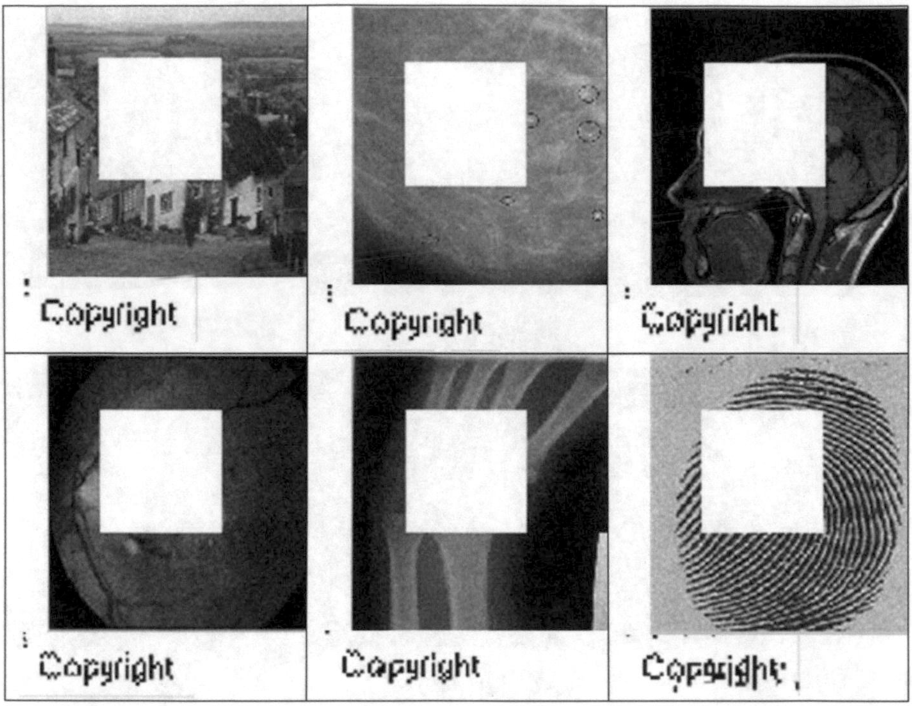

Fig. 5. Watermarked images and the recovered watermark after the cropping attack

Gaussian Filter attack. The Gaussian filter is a non-uniform low pass with Gaussian distribution that smoothens the image and removes both noise and details. In these experiments, the extracted watermarks are distinguishable but distorted after applying the filter. Figure 5 shows that the extracted watermarks after applying a rectangular cropping window in the middle of the images; most of the extracted images are visible and clearly distinguishable except for the fingerprint image which has too many details and contours. Figure 6 shows the extracted watermarks after 45° rotation and cropping of the edges in the watermarked images. Again, the extracted watermark is most noticeably distorted after applying the rotation & cropping attack but it is still distinguishable.

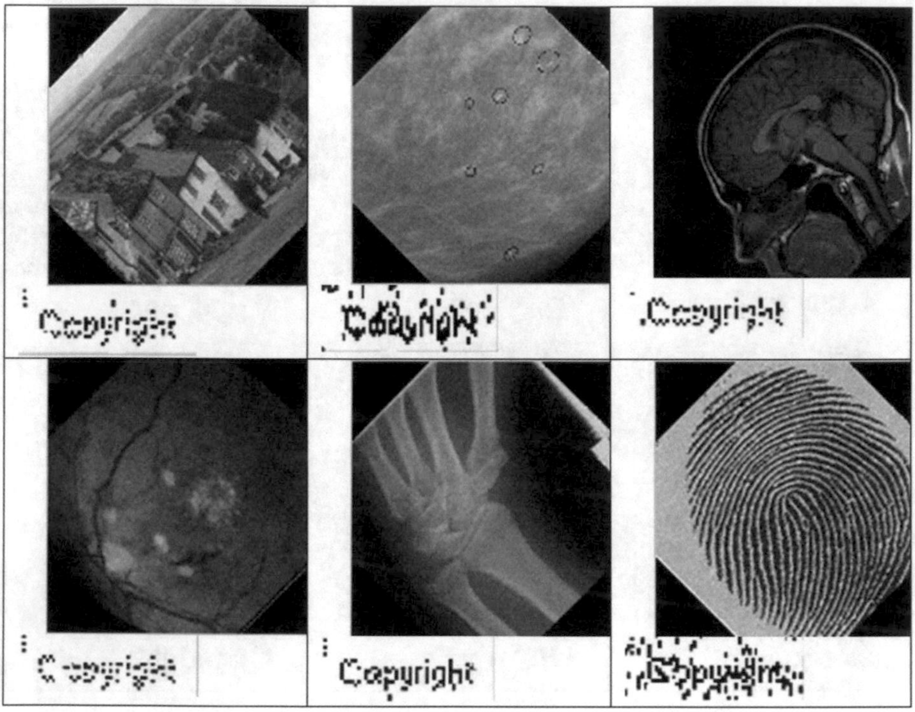

Fig. 6. Watermarked images and the recovered watermark after the rotation & cropping attack

4 Conclusions

A combined DWT-CT watermarking algorithm has been designed and implemented. It is robust and imperceptible with an added encryption security of shuffling and rotation of bits. The encrypted watermark gives an added security layer to watermarking based on the strength of the encryption keys. The performance of the system was evaluated using PSNR and NC and it demonstrated high robustness to attacks. The DWT-CT protects the patients' personal data inside a medical image and provides a good technique for protecting patients' data inside medical images.

References

1. Fahmy, G., Fahmy, M.F., Mohammed, U.: Nonblind and Quasiblind natural preserve watermarking. EURASIP J. Adv. Sign. Process. **2010** (2010). Article ID 452548, 13 pages
2. Ahmed, S.: Intelligent watermark recovery using spatial domain extension. In: International Conference on Intelligent Information Hiding and Multimedia Signal Processing, IIHMSP (2008)
3. Lai, C.C., Tsai, C.C.: Digital image watermarking using discrete wavelet transform and singular value decomposition. IEEE Trans. Instrum. Meas. **59**(11), 3060–3063 (2010)

4. Soliman, M.M., Hassanien, A.E., Ghali, N.I., Onsi, H.M.: An adaptive watermarking approach for medical imaging using swarm intelligence. Int. J. Smart Home **6**, 37–50 (2012)
5. Zain, J., Clarke, M.: Security in telemedicine: issue in watermarking medical images. In: International Conference: Science of Electronic, Technologies of Information and Telecommunications (2011)
6. Lin, W.-H., Horng, S.-J., Kao, T.-W., Chen, R.-J., Chen, Y.-H., Lee, C.-L., Terano, T.: Image copyright protection with forward error correction. Expert Syst. Appl. **36**(9), 11888–11894 (2009)
7. Rosiyadi, D., Horng, S.-J., Fan, P., Wang, X.: Copyright protection for e-government document images. IEEE Multimed. **19**(3), 62–73 (2012)
8. Shi-Jinn, H., Rosiyadi, D., Fan, P., Wang, X., Khan, M.K.: An adaptive watermarking scheme for e-government document images. Multimed. Tools Appl. **72**(3), 3085 (2014)
9. Singh, A.K., Dave, M., Mohan, A.: Hybrid technique for robust and imperceptible image watermarking in DWT-DCT-SVD domain. Natl. Acad. Sci. Lett. **37**(4), 351–358 (2014)
10. Giakoumaki, A., Pavlopoulos, S., Koutsouris, D.: Secure and efficient health data management through multiple watermarking on medical images. Med. Biol. Eng. Comput. **44**, 619–631 (2006)
11. Lin, W.-H., Wang, Y.-R., Horng, S.-J., Kao, T.-W., Pan, Y.: A blind watermarking method using maximum wavelet coefficient quantization. Expert Syst. Appl. **36**(9), 11509–11516 (2009)
12. El rube', I., Abou el Nasr, M., Naim, M., Farouk, M.: Contourlet versus wavelet transform for robust digital image watermarking technique. World Acad. Sci. Eng. Technol. **60**, 288–292 (2009)
13. Nguyen, S.C., Ha, K.H., Nguyen, H.M.: A new image watermarking scheme using contourlet transforms. In: 3rd International Conference on Information Technology, Computer, and Electrical Engineering (ICITACEE), 19–20 October 2016
14. Ponomarenko, N., Lukin, V., Zriakhov, M., Egiazarian, K., Astola, J.: Estimation of accessible quality in noise image compression. In: Proceedings of European Signal Processing Conference (EUSIPCO 2006), Florence, Italy, pp. 1–4 (2006)
15. Chang, S.G., Yu, B., Vetterli, M.: Adaptive wavelet thresholding for image denoising and compression. IEEE Trans. Image Process. **9**(9), 1532–1546 (2000)
16. Shaick, B.Z.: A hybrid transform method for image denoising. In: 10th European Signal Processing Conference (2000)
17. Wang, Z., Bovik, A.C.: Modern Image Quality Assessment. Morgan and Claypool Publishing Company, New York (2006)
18. Wang, Z., Bovik, A.C., Sheikh, H.R., Simoncelli, E.P.: Image quality assessment: from error visibility to structural similarity. IEEE Trans. Image Process. **13**(4), 600–612 (2004)
19. Wang, Z., Bovik, A.C.: Mean squared error: love it or leave it? A new look at signal fidelity measures. IEEE Sign. Process. Mag. **26**(1), 98–117 (2009)
20. Ponomarenko, N., Battisti, F., Egiazarian, K., Astola, J., Lukin, V.: Metrics performance comparison for color image database. In: CD-ROM Proceedings of the 4th International Workshop on Video Processing and Quality Metrics, Scottsdale, Ariz., USA, pp. 1–6 (2009)

Two Stages Haar-Cascad Face Detection with Reduced False Positive

Abdulaziz Ali Saleh Alashbi[1(✉)], Mohd Shahrizal Bin Sunar[1], and Qais Ali AL-Nuzaili[2]

[1] Media and Game Innovation Centre of Excellence (MaGICX), Institute of Human Centre Engineering, Universiti Teknologi Malaysia, 81310 Skudai, Johor, Malaysia
{asaabdulaziz2,shahrizal}@utm.my

[2] School of Computing, Faculty of Engineering, Universiti Teknologi Malaysia, 81310 Skudai, Johor, Malaysia
alnuzaili.qais@gmail.com

Abstract. Face detection is one of the top hottest research topics in Computer vision. Human face remains the robust un-cloned biometric identity recognition which is widely used to provide person identity and has many applications for example security access, face recognition and surveillance. A common issue in face detection is that face detection rate is maximized with low threshold but this in contrast increase the false positive rate. In this paper we present two stage framework haar cascade detection algorithm where in the first stage the detected faces are cropped and re-detected by the second stage. The result is a noticeable improvement with false alarm reduction when compared to the pure algorithm alone.

Keywords: Face detection · Human face classification and localization
Computer vision and machine learning

1 Introduction

Face detection has been for more than two decades and still an important and active area in computer vision research [1]. It is the essential part and the first step to all face applications related. In face verification and face recognition for instance, prior to verify and recognize faces face detection is the first step which is classifying and localizing faces in a given image [2]. It is the building block for more sophisticated systems developed for consumer product like for instance digital cameras, social networks smart phone apps and etc. [3, 4].

Hundreds of research papers have been published which had proposed several approaches for detections methods. The advancement prior to the year of 2001 had been surveyed and grouped by [1, 5] into four categories as summarized in Fig. 1.

F. Saeed et al. (Eds.): IRICT 2018, AISC 843, pp. 690–695, 2019.
https://doi.org/10.1007/978-3-319-99007-1_64

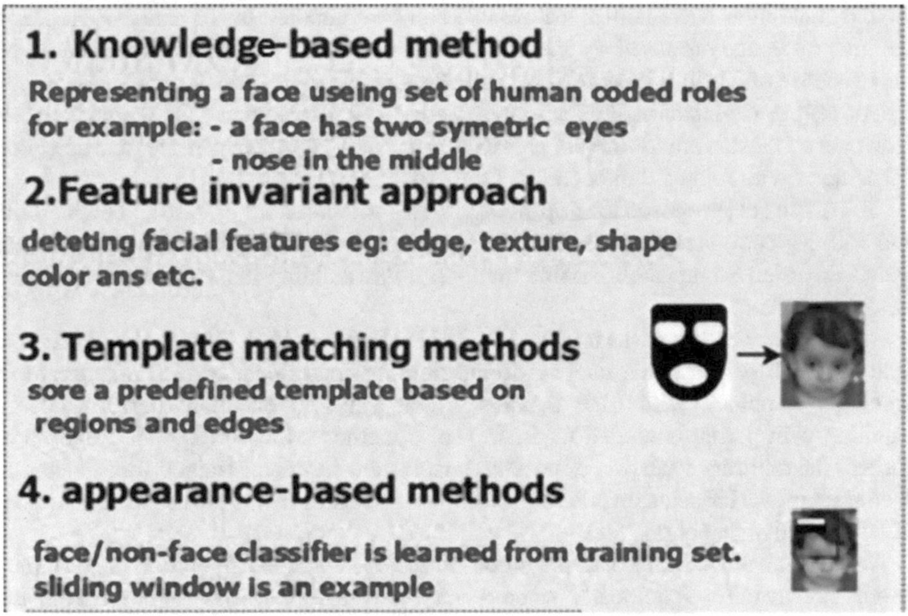
1. Knowledge-based method
Representing a face useing set of human coded roles
for example: - a face has two symetric eyes
- nose in the middle
2.Feature invariant approach
deteting facial features eg: edge, texture, shape
color ans etc.
3. Template matching methods
sore a predefined template based on
regions and edges
4. appearance-based methods
face/non-face classifier is learned from training set.
sliding window is an example

Fig. 1. A four approaches of face detection as grouped by [1, 2].

All researches done within that period were unable to achieve real-time results and couldn't be implemented in real-world applications until the seminal work of Viola and Jones' face detector [6]. It is the most well-known face detection algorithm, that has great impact and made face detection feasible in the real world usage such as digital cameras and photo software organization due to the brilliant ideas which used successful combination of machine learning techniques with feature invariant techniques [1, 3].

In Haar-like features and cascade Adaboost classifier it involves the searching within a given image for faces, using multiple scales explored at each pixel using Haar-like rectangle features with boosting classification. Despite the great success of this algorithms the number of false detection is high. Decreasing the rate of false detection is a tradeoff between low threshold which leads to more computation and/or high threshold but with higher positive rate. In this paper we implemented two stages of haar cascad algorithms to reduce the false positive rate.

2 Literature Review

The seminal work of Viola and Jones et al. [6] consisted of four main ingredients that was behind the great achievement to its work, haar-like features along with adaboost machine learning algorithm, integral image and cascading of classifiers.

Three types of features were defined in the detection sub-window proposed by [7] to extend the feature set and use flexible rectangular regions that could be able to deal with non-frontal faces.

Diagonal filter was another feature that can be computed by 16 array reference to the integral image proposed by Viola and Jones [8]. Joint Haar-like features based on the existence of multiple haar features and was capable to capture human faces characteristics [9]. An efficient method was presented by [10] to calculate 45° rotated features. However all these methods are not able to perform and get more accurate face detection particularly with rotated and occluded faces [9].

A work in [11] proposed look up table with real Adaboost which can deal with profile and 360-degree rotated faces. In [12] they proposed a histogram based weak classifier called Bayesian Stump with flexible multi-split thresholding that improved the detectors' performance.

Local binary patterns was used by [13, 14] for face detection to overcome Haar-like features weak performance under extreme lighting conditions. In [15] Histogram of gradients combined with LBP features which achieved excellent performance in handling with partial occluded faces. In [16] a method called NPD was proposed to address the arbitrary variation of pose and occlusions in unconstrained face detection. Two stage model basseted on NPD and DPM was presented by [17] to reduce the number of false positive detection.

Despite the availability of those proposed methods for face detection, the accuracy is still not matured specifically when a very low false alarm is required from the detector [16].

3 Method

It is clear that the increase of false detection rate has sever effect towards the accuracy of the detection. The high rate of FP in haar like feature algorithm of Viola and jones is due to the low performance of the algorithm in multi-view faces and faces in unconstrained environment where faces are in large variation of pose, illumination variation and in low image resolution. Reducing the FP rate is one way of enhancing its performance. We proposed a face detection method for illuminating the high rate of false alarms consists of two stages based on haar cascade as below.

(a) First stage consists of:
   1. Scan multi faces image.
   2. Localize face/faces if found.
   3. Label each detected face.
   4. Crop the labeled face
   5. Save every cropped face as a single image.
   6. Count the total of detected faces (initial true positive), in this initial detection the false positive rate is high.

(b) Second stage:
   In this stage the goal is to reduce the high false positive rate through the following steps:
   1. Read every single face image (which is saved from the first stage)
   2. Re-detect the face.
   3. Discard the false positive

4. Relabel the confirmed face (final true positive).
5. Count the number of confirmed faces
6. Compare the final true positive in this stage with the initial true positive in the first stage.

When we did the comparison, the high false positive rate is significantly dropped down. Figure 2 below shows the flow of the two stages.

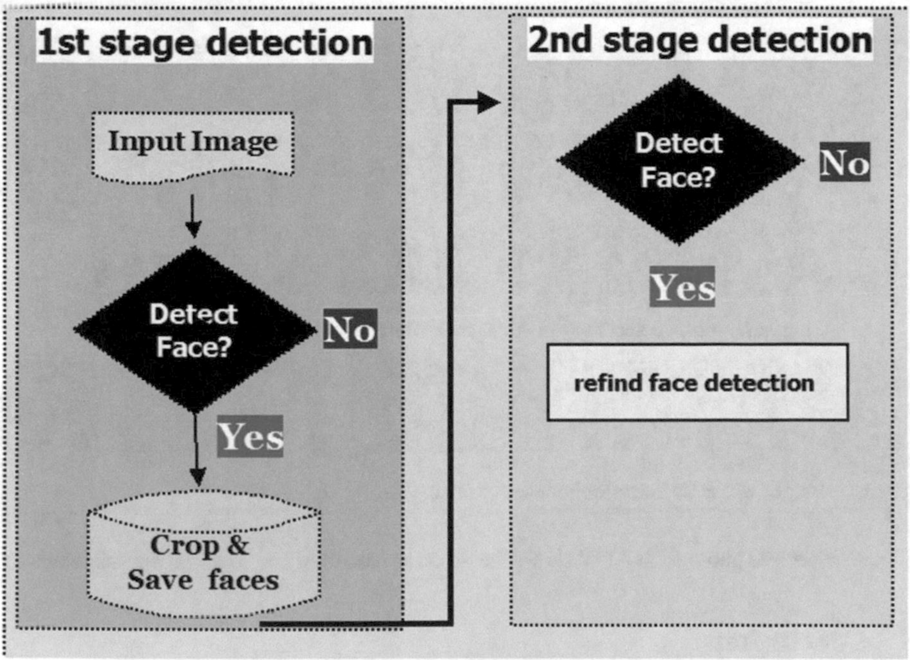

Fig. 2. The two stages of face detection design

4 Experimental Results and Discussions

Annotated Faces in the Wild is a face detection benchmark dataset from Flickr images. It is a publicly available dataset, it contains 205 of images with 468 labeled faces, it is a challenging dataset due to its large variation in occlusion, illumination and resolution [18].

We ran the first stage haar-cascad detector on the AFW dataset. The following results were obtained: 93 faces were not detected (FN) among 468 faces and 793 image regions were detected as a face FP among 205 images and 375 faces detected as TP.

In the second stage, the number of false image region (FP) has been significantly decreased to 313. From Table 1 it is noted that in the second stage the recall and precision are better than the first stage results. From our observation the reduction of FP in second stage is because of the cropped image with single face has clear features and the search region is smaller (Fig. 3).

Table 1. The results of the experiments performed on AFW dataset.

Dataset	AFW [18]				
Number of images	205				
Number of faces	468				
Method	FN	FP	TP	Precision	Recall
Single stage haar FD	93	793	375	0.321	0.801
Two stage haar FD	43	313	332	0.514	0.885

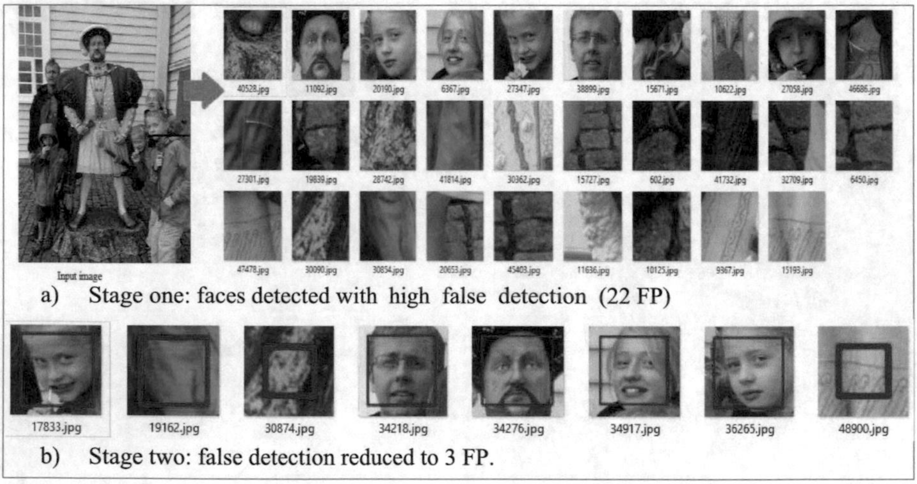
Input image
a) Stage one: faces detected with high false detection (22 FP)
b) Stage two: false detection reduced to 3 FP.

Fig. 3. Example photo form AFW shows the detection result of stage one (a) and stage two (b)

5 Conclusion

A method of two stages of detection using OpenCV haar cascade face detection was proposed. The result showed that the proposed method significantly reduced the false positive rate which enhance the overall detectors performance. In the further work different detectors will be explored and combined to increase the performance of face detection.

Acknowledgements. This research was supported by Q.J130000.2409.03G74 (Flagship Grant) and UTM-IRDA MaGICX (Media and Games Innovation Centre of Excellence) Universiti Teknologi Malaysia 81310 Skudai Johor MALAYSIA.

References

1. Zhang, C., Zhang, Z.: A survey of recent advances in face detection. Technical report, Microsoft Research (2010)
2. Yang, M.-H., Kriegman, D.J., Ahuja, N.: Detecting faces in images: a survey. IEEE Trans. Pattern Anal. Mach. Intell. **24**(1), 34–58 (2002)

3. Zafeiriou, S., Zhang, C., Zhang, Z.: A survey on face detection in the wild: past, present and future. Comput. Vis. Image Underst. **138**, 1–24 (2015)
4. Tikoo, S., Malik, N.: Detection of face using viola jones and recognition using back propagation neural network. arXiv preprint arXiv:1701.08257 (2017)
5. Hjelmås, E., Low, B.K.: Face detection: a survey. Comput. Vis. Image Underst. **83**(3), 236–274 (2001)
6. Viola, P., Jones, M.: Rapid object detection using a boosted cascade of simple features. In: Proceedings of the 2001 IEEE Computer Society Conference on Computer Vision and Pattern Recognition, CVPR 2001. IEEE (2001)
7. Li, S.Z., et al.: Statistical learning of multi-view face detection. In: European Conference on Computer Vision. Springer (2002)
8. Jones, M., Viola, P.: Fast multi-view face detection. MERL—Mitsubishi Electric Research Laboratories (2003)
9. Mita, T., Kaneko, T., Hori, O.: Joint haar-like features for face detection. In: Tenth IEEE International Conference on Computer Vision, ICCV 2005. IEEE (2005)
10. Lienhart, R., Maydt, J.: An extended set of haar-like features for rapid object detection. In: Proceedings of 2002 International Conference on Image Processing. IEEE (2002)
11. Wu, B., et al.: Fast rotation invariant multi-view face detection based on real Adaboost. In: Proceedings of Sixth IEEE International Conference on Automatic Face and Gesture Recognition, 2004. IEEE (2004)
12. Xiao, R., et al.: Dynamic cascades for face detection. In: IEEE 11th International Conference on Computer Vision, ICCV 2007. IEEE (2007)
13. Zhang, L., et al.: Face detection based on multi-block LBP representation. In: International Conference on Biometrics. Springer (2007)
14. Yan, S., et al.: Locally assembled binary (LAB) feature with feature-centric cascade for fast and accurate face detection. In: IEEE Conference on Computer Vision and Pattern Recognition, CVPR 2008. IEEE (2008)
15. Wang, X., Han, T.X., Yan, S.: An HOG-LBP human detector with partial occlusion handling. In: 2009 IEEE 12th International Conference on Computer Vision. IEEE (2009)
16. Liao, S., Jain, A.K., Li, S.Z.: A fast and accurate unconstrained face detector. IEEE Trans. Pattern Anal. Mach. Intell. **38**(2), 211–223 (2016)
17. Marčetić, D., Hrkać, T., Ribarić, S.: Two-stage cascade model for unconstrained face detection. In: 2016 First International Workshop on Sensing, Processing and Learning for Intelligent Machines (SPLINE). IEEE (2016)
18. Zhu, X., Ramanan, D.: Face detection, pose estimation, and landmark localization in the wild. In: 2012 IEEE Conference on Computer Vision and Pattern Recognition (CVPR). IEEE (2012)

An Enhanced Context-Aware Face Recognition Alert System for People with Hearing Impairment

Maythem K. Abbas[1(✉)], Bie Tong[2], and Raed Abdulla[3]

[1] Faculty of Science and Information Technology, Universiti Teknologi PETRONAS, Seri Iskander, Malaysia
eng_maythem_84@yahoo.com

[2] Neighbor Network Technology Co. Ltd., Shanghai, Zhejiang Sheng, China
bietong.cool@gmail.com

[3] Faculty of Computing, Engineering and Technology, Asia Pacific University, Kuala Lumpur, Malaysia
dr.raed@apu.edu.my

Abstract. The life of deaf people or People with Hearing Impairments (PHIs) is difficult as they have to face challenges in identifying sounds like a visitor pressing the doorbell. The traditional doorbell uses an audio signal to notify people. However, it does not notify PHIs because PHIs are insensitive to sound. Due to the lack of help from support services, PHIs may not have normal daily lives. Therefore, it is necessary to propose an efficient doorbell solution in helping PHIs to be notified, and the intangible benefit for PHIs is that it will reduce their dependency on their families in their daily lives. This paper proposes a hybrid (Image Processing and Context-Aware Computing) alert system that would notify PHIs by using visual communication signals and tactile cues instead of audio signals. Compared with the existing solutions, the enhanced face recognition feature included in the proposed system design, made it smart enough to know when the visitor's image has to be snapped or not and how to reduce the transmission time of the image. In addition, what is context aware, and how the context aware feature will benefit the system design is also discussed. On the other hand, the intangible benefit of this research would be helping PHIs by reducing their dependency on others and building their self-confidence.

Keywords: Context aware · Image Processing · Face recognition
Alert system · People with Hearing Impairments

1 Introduction

Although it is extremely hard to apprehend exactly what it's like to be deaf, majority of people are conscious that hearing impairment is accompanied with a unique set of daily life problems which forces them to experience the world from a different angle than the rest of the population, and not always how you'd expect.

F. Saeed et al. (Eds.): IRICT 2018, AISC 843, pp. 696–708, 2019.
https://doi.org/10.1007/978-3-319-99007-1_65

Even though difficulties attendant with a total hearing loss might seem obvious, there are countless daily obstacles People with Hearing Impairment (PHI) have to deal with, such as not getting announcements at public places and events, and/or having to rely on touch to get the attention of others, or communicating in the dark, and many others. Regularly, these situations can be the toughest parts of being a PHI.

Nowadays, according to the World Health Organization facts sheets in 2017 [1], more than 360 Million people (over 5%) from around the world living a life with hearing loss. In this project, deaf people will be referred to as People with Hearing Impairments (PHI).

With insufficient services, PHIs live a life full of inconvenience. For example, the traditional doorbell is not able to alert PHIs because they cannot be notified by audio signals. The existing Internet-of-Things based alert system can help them to be notified when a visitor presses the doorbell [2]. However, the photo duplication and long transmission time of the image in the existing system will cause a poor user experience.

The proposed solution aims to notify the PHI with visual signals and tactile cues; in addition, the system will recognise the visitor and avoid taking photos of the same person repeatedly, reduce the transmission time of the image by adding a context aware feature, and maximise the use of those devices which the PHI already has.

2 Related Work

Many researchers had created different solution; each of them got upsides and downsides. For instance, one of the existing alert systems is an IoT-based Wireless Alert System that proposed in [2]. It is installed on the front door and connected with the doorbell. The whole system is based on the Raspberry pi, and the system also contains a camera, wireless GSM, and Bluetooth. It has the ability to keep an uninterrupted connection with those people's wearable devices via Bluetooth, and could also send an SMS to those people's mobile phones through the wireless GSM.

Similar face recognition approach; Aarhus Face recognition algorithm [3], was created by Kévin Letupe in 2013. It uses the Law of Cosines to recognise a person however its accuracy was low according to the little number of landmarks. Till the paper's publication date, no other works had been done using Internet of Things (IoT) nor any related intelligent technologies.

3 Methodology

The proposed system is able to minimise the total cost for the user and the visitor by adding the context aware feature. It avoids making the visitor learn how to snap the picture and how to send the message. The only interaction needed by the visitor is to press the doorbell (the doorbell button can be an image displayed on a screen). The experience for the visitor is almost as simple as pressing the traditional doorbell. The developed Context-aware model [4] will minimise the studying cost for the user also. For example, the Level of the Context Awareness will define how smart the system should be, whether the system should be executed fully automatically. Context

Acquisition will define how to get the context data, should it be calculated or sensed. Context Modeling will specify how to store those context data, then the Context Reasoning will determine what the system should do after the analysis of the context data.

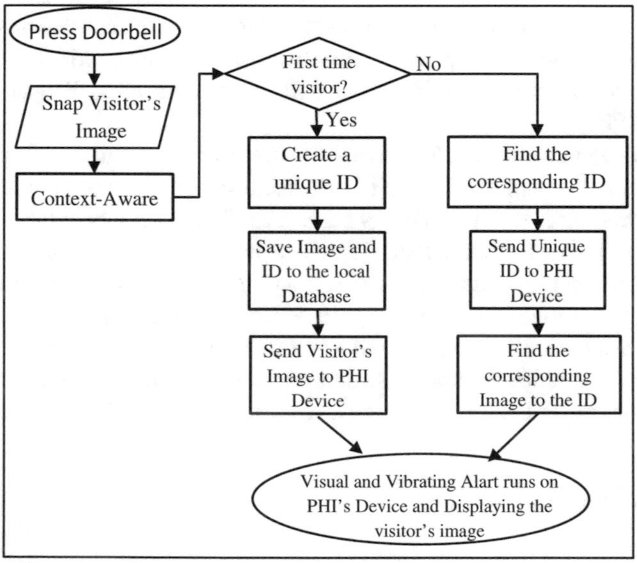

Fig. 1. The flow chart of the proposed system

3.1 The Proposed System Design

The existing system [2] takes a visitor's image every time when the visitor presses the doorbell, even if the visitor's picture had already been stored in the database of the system.

The proposed system has the ability to distinguish between first time visitors and second time visitors. The system will compare the visitor's face with an existing database of the people who had previously. If the visitor is not a first-time visitor (the database has a copy of their image), accordingly, the system will recognise the visitor and not take a photo of the visitor again to save as a new visitor. Otherwise, the system will take the photo of the visitor and save it in the database for future use.

The flow chart of the proposed system has been shown in Fig. 1. The proposed system takes the picture of the first-time visitors and stores the data in the database. Then, when the visitor comes to visit again, the system will detect the face of the visitor, extract the features of their faces, and finally, identify whether the visitor had visited before or not. If he/she has, the system will not snap the image of the visitor again. It will just notify the PHI of who has come over for a visit, and then retrieve the picture of the visitor in the database of the PHI's phone.

3.2 Face Recognition

Face recognition is a technology based on a person's facial feature information. According to [5], the face recognition statement can be formulated as follows:

"Given still or video images of a scene, identify or verify one or more persons in the scene using a stored database of faces". The face recognition system contains three procedures; they are face detection, face extraction, and face recognition as shown in Fig. 2.

In the proposed system, face detection and face extraction were done by the deployment of "Google Mobile Vision API" and "Google Vision Face API", respectively [6, 7]. The (x, y) position of the landmark where (0, 0) is the upper-left corner of the image will be returned once the face is detected. The face recognition stage deployed "Aarhus" approach after enhancing it. The original Aarhus approach uses three landmarks on the face; the position of the two eyes and the position of the mouth [3].

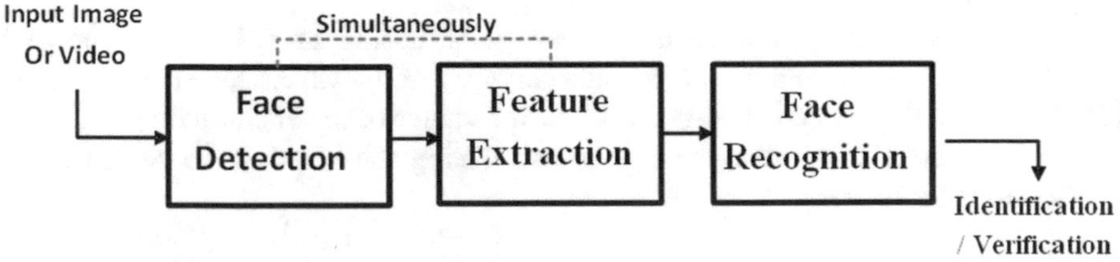

Fig. 2. Three procedures of face recognition [5]

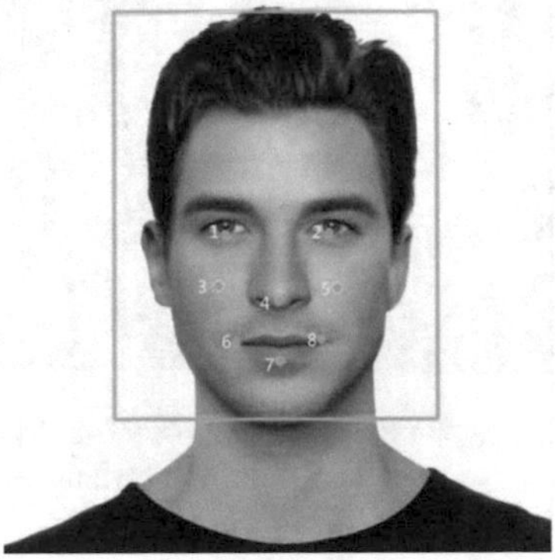

Fig. 3. The 8 facial landmarks of the proposed system

The researchers of this project have decided to add five more facial landmarks aiming the increment of face recognition accuracy. The total facial landmarks became eight, those are shown in Fig. 3 and each is described in Table 1. The enhanced face recognition procedure consists of the following three steps:

(1) Triangulations: the eight landmarks will be linked to form seven triangles, as shown in Fig. 4. Based on those 8 landmarks, there are 7 triangles can be formed, shown in Fig. 4.
(2) Every sides of the triangle: Once the position of two points are given, the distance between two points in a coordinate system can be calculated using Pythagoras equation, shown in (1).

$$AB = \sqrt{(X_A - X_B)^2 + (Y_A - Y_B)^2} \tag{1}$$

Based on (1), all the sides of each triangle shown in Fig. 4 can be calculated. Once the three sides of a triangle are given, the cosine value of any angle can be calculated by the Law of Cosine, shown in (2):

$$COS(C) = \frac{a^2 + b^2 - c^2}{2ab} \tag{2}$$

Therefore, the cosine value from cosine 1 to cosine 14 (14 angles, shown in Fig. 5) can be calculated and produce an array of 14 values for every person.
(3) Face matching: By comparing the 14 cosine values of the visitor with those stored in the system's database, the system can distinguish whether the person is first time visitor or not.

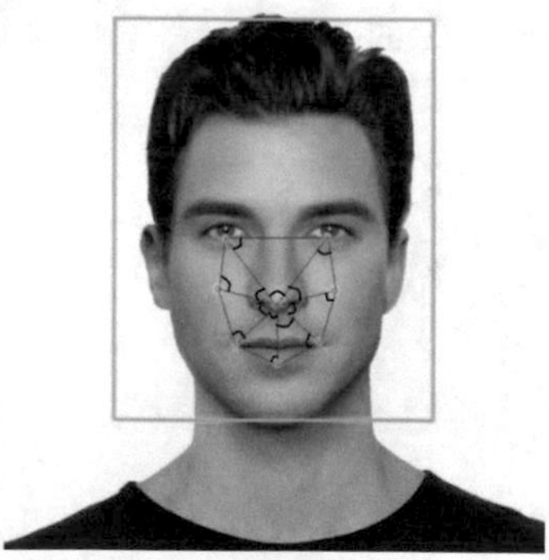

Fig. 4. The formed triangles by the 8 landmarks

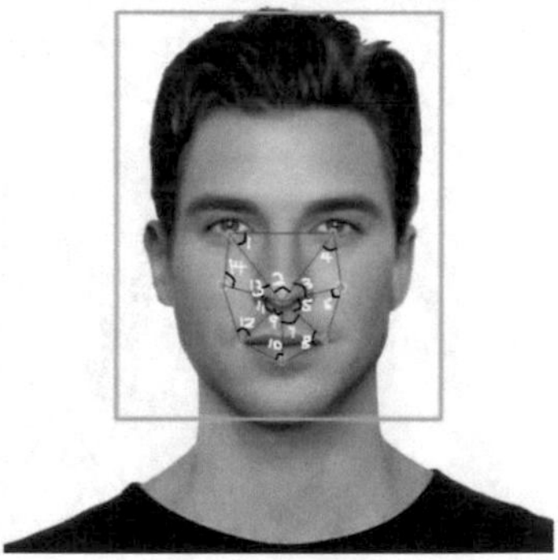

Fig. 5. The 14 angles of the seven formed triangles

3.3 Context-Aware Feature in the Proposed System

"A system is context-aware if it uses context to provide relevant information and/or services to the user, where relevancy depends on the user's task", said in [8].

The level of context awareness can be personalization, passive context awareness or active context awareness [9]. The main purpose of setting up the level of context awareness is to determine how many times the user must interact with the system. In this project, it means what kind of interaction the user should do before the system alert PHI that visitor is front of the door.

Initially, the passive context awareness was used to provide more functions. For instance, the system can give the option to the visitor if they want to put their name attached with their image, so when the second time the visitor comes over, their face picture will be displayed automatically on the screen with their name on and easy to be classified by the system. Whatever the functions that the system provides to the visitors, it involves lots of interaction with the system from the visitor's side and the PHI side. However, for untrained visitors, it takes a long time to get used with the doorbell system if the system offers too many options or functions to the visitor. Therefore, active context awareness was chosen, which will allow the system to continuously monitor the visitor's face and snap the visitor's face image, analyzing autonomously and send the result to PHI's device.

(1) Context Acquisition: Many people, mistakenly consider the visitor's image as context data. In fact, it is not, as it cannot be used for Context Reasoning directly. In other words, the system needs to extract/derive the landmarks from the visitor's image first then it can do reasoning to recognize the visitor. The values of each angle in different triangles are considered as context data.
(2) Context Reasoning: The context data should go through Context Reasoning process after being derived.

Table 1. Description of the proposed 8 facial landmarks

No	Name of the landmark	Description of the landmark
1	LEFT_EYE	The left eye cavity center
2	RIGHT_EYE	The right eye cavity center
3	LEFT_CHEEK	The midpoint between the left mouth corner and the outer corner of the left eye
4	NOSE_BASE	The midpoint between the subject's nostrils where the nose meets the face
5	RIGHT_CHEEK	The midpoint between the right mouth corner and the outer corner of the right eye
6	LEFT_MOUTH	The left mouth corner
7	BOTTOM_MOUTH	The bottom lip's center
8	RIGHT_MOUTH	The right mouth corner

The angles values will go through context reasoning. The process of context reasoning can be roughly explained as the following sentences:

IF the values of the angles from the new visitor are as same as a stored record in SQLite database, THEN the corresponding unique ID of the visitor will be sent to PHI. ELSE, the visitor's image will be snapped and sent to PHI with a unique ID [4].

4 Experimental Work

In real scenario, the doorbell has to be pressed by the visitor, then the image of the visitor will be analyzed by the system and relevant information will be sent PHI. Finally, once their phone received the data from doorbell, PHI will be notified by vibration of the phone and the flash blinking on the phone.

4.1 Experimental Setup

The experiment had used a list of environmental tools for developing the proposed system:

1. Android Studio 2.2.3.
2. Google Nexus-6 works as the doorbell and interacts with the visitor, and Google Nexus-5 works as the device which notifies PHI.
3. Android 7.0 Nougat [10].
4. Wireless network connection (Wi-Fi).
5. SQLite database.

The proposed system is running on Android platform, therefore, there are two permissions needed for the system, those are (Fig. 6).

```
<uses-permission android:name="android.permission.CAMERA" />
<uses-permission android:name="android.permission.INTERNET" />
```

Fig. 6. Android permission request code

The first permission is the Camera Access permission, which permits the system to use the vision hardware of the testing device. The second permission is the Internet Access permission, which allows the system to transfer data over the Internet.

```
mCameraSource = new CameraSource.Builder(context, detector)
        .setRequestedPreviewSize(240, 380)
        .setFacing(CameraSource.CAMERA_FACING_FRONT)
        .setRequestedFps(30.0f)
        .build();
```

Fig. 7. Android image resolution setting code

The doorbell's camera is set to capture small picture of the visitor's (size 240 * 380 Pixels), as shown in Fig. 7, to ease and fasten the transfer of the image across the network.

4.2 Experimental Case Study

The enhanced face recognition algorithm is based on the position of landmarks. The accuracy rate is subjected to several aspects, such as visitor smile or face direction when capturing the picture.

One of the testing devices was set up as the doorbell on the front door, while the second device was placed remotely to act the as the deaf people alert system. Five people were invited to act as the visitors. Every visitor would be asked to have no facial expressions and to face the camera while testing. Initially, the system should be able to snap the visitor's image, store the context data (cosine value of each angle) to the database, assign a unique ID to this visitor, and then send the image with the unique ID to the other test device carried by the deaf person. The same experiment was repeated by the same 5 visitors, the system supposed to recognize the visitors and then forward their corresponding unique IDs only to the remote testing device. The test cases of the system are shown in Table 2.

5 Simulation Results

The first batch of tests showed that the system was not able to recognise the visitor at the second time. No matter how many times the same person pressed the doorbell, the system was still not able to recognise the visitor. The same person was asked to press the doorbell six times and all the cosine values of the same person were recorded. Six sets of data were recorded as shown in Table 3.

At first, it was believed that only if all the values of the angles were exactly the same, it could be regarded as the same person. For example, the cosine value of angle1 when the visitor visited for the first time had to equal the value of angle1 when the visitor visited for the second time. All the cosine values of the angels from the same person have been shown in Table 3, however, when we look at the results, none of each single piece of data is exactly the same as any other one.

Table 2. Experimental cases studies

TC#	Test cases	Expected result
No. 1	Visitor press the doorbell at the first time	(1) The system should snap the visitor's image, the unique ID of this visitor and image will be sent to PHI side
		(2) The phone will vibrate, the flash will blinking, and the image will be displayed at PHI side
No. 2	Visitor press the doorbell at the second time	(1) The unique ID of this visitor will be sent to PHI side with
		(2) The phone will vibrate, the flash will blinking, and the image will be displayed at PHI side

If all of the decimal digits were kept and compared for the different angles, the precision range would have been too high and the system would not have been able to recognise the visitors if they had come before, which means that the system may not have been able to recognise anyone. On the contrary, if only the first decimals were used to compare the different cosine value of the angles, the system might have regarded different visitors as the same person. Because all of the values were from the same person, when analysing the results shown in Table 3, it was noticed that for the same angles, the first two decimal places were basically the same. The first 4 angles were picked up and a list of its first two decimal places has been presented in Table 4.

According to Table 4, the first two decimal places which were from the same angle were basically the same, except for the third record in the value of angle 2. In addition, by looking at the first two decimal places from different angles, for example, the first two decimal digits of angle1 were '94', then they were '82' of angle2, '88' of angle3, and so on. So, two decimal places for each angle were considered; therefore, in the face recognition algorithm, two different angle1's were regarded as the same if their first two decimal places were the same.

There were five people who took this test, and each of the 5 people pressed the doorbell, which was displayed on the screen by Google Nexus-6; and, the testing device Google nexus-5 worked as in the PHI's side to notify the PHI that a visitor is in front of their house. According to the results shown in Table 5, the transmission time of the image: The resolution of each image was 240 $\times$ 380 pixels. The range of the image size was from 33.607 kb to 43.646 kb. The duration from the face being detected to the face's picture being displayed on the PHI's phone was recorded.

After analysing the results, it was found that there were two main reasons why the image transferring was very fast. One was because the internet speed was fast enough; secondly, the size of the image was very small, this will also benefit the system in image transferring, and image encoding and decoding.

1. According to the tests results shown in Table 5, there were 5 people who took this test. The system could recognise 4 out of the 5 visitors. Currently, the accuracy rate of our developed system is 80%. However, the accuracy rate might be changed with more and more people taking the test.
2. Visual communication signals and tactile cues: During the test, Google Nexus 5 vibrated with the flash blinking, and the image was displayed on the screen; the visual communication signals and tactile cues were successfully delivered to the PHI
3. Visual communication signals and tactile cues: During the test, Google Nexus 5 vibrated with the flash blinking, and the image was displayed on the screen; the visual communication signals and tactile cues were successfully delivered to the PHI.

Table 3. The Cosine values for the same person's facial landmarks Angles for six experiments

The Cosine values for the same person's facial landmarks			
Angle#	1st try	2nd try	3rd try
Angle 1	**0.94**42447420026289	**0.94**91096537605782	**0.94**42317240766819
Angle 2	0.8263487159628947	0.8232580914414143	0.8290028921316951
Angle 3	0.8821984445899151	0.8825733881088309	0.8839629982473993
Angle 4	0.9011228725221136	0.9005966325411988	0.8987984271108583
Angle S	0.9075006753805460	0.9085804666947459	0.9078642839393195
Angle 6	0.8777745848556628	0.8788758051460337	0.8770446799616107
Angle 7	0.9132441662144773	0.9149218568928524	0.9126468869055389
Angle 8	0.8531665420995307	0.8517322421600855	0.8521680424304833
Angle 9	0.8932201595019398	0.8928796375343150	0.8909721679793629
Angle 10	0.8370548023791096	0.8334792568844180	0.8369154752254730
Angle 11	0.8481424894721209	0.8447894408051412	0.8451866381967075
Angle 12	0.8564610194088633	0.8580022832943226	0.8592089277971863
Angle 13	0.8359053445265926	0.8343930815656758	0.8330821906715410
Angle 14	0.8768602310891603	0.8754244741198609	0.8800313362535435
The Cosine values for the same person's facial landmarks			
Angle#	4th try	5th try	6th try
Angle 1	**0.94**00086022568210	**0.94**29221946217272	**0.94**53709225047882
Angle 2	0.8317540930172648	0.8269008656427103	0.8288888550465133
Angle 3	0.8818620491748750	0.8810219923648400	0.8837130512438154
Angle 4	0.8973529298443575	0.9007048021126273	0.8974798511257296
Angle S	0.9046009986559939	0.9058178294966966	0.9069322688232350
Angle 6	0.8787148351633371	0.8784730908725835	0.8777198433632770
Angle 7	0.9096114445104722	0.9118947952367761	0.9120880962993670
Angle 8	0.8542660627698051	0.8546995598624582	0.8530014713943436
Angle 9	0.8890968750024542	0.8923457918867733	0.8902430280160694
Angle 10	0.8405643273989615	0.8379956292554791	0.8366337005454679
Angle 11	0.8476965064511812	0.8488268821531257	0.8445343354900411
Angle 12	0.8620566786262907	0.8582122127706232	0.8606417280902617
Angle 13	0.8391276512827417	0.8383394564400052	0.8348953589845500
Angle 14	0.8775718717004356	0.8754266390158760	0.8783399637487866

Table 4. The first two decimal places of each angle

Doorbell Press	Angle 1's Cosine	Angle 3's Cosine	Angle 3's Cosine
1st Press	**0.94**42447420026289	**0.82**63487159628947	**0.88**21984445899151
2nd Press	**0.94**91096537605782	**0.82**32580914414143	**0.88**25733881088309
3rd Press	**0.94**42317240766819	**0.82**90029921316951	**0.88**39629982473993
4th Press	**0.94**00086022568210	**0.83**17540930172648	**0.88**18620491748750
5th Press	**0.94**29221946217272	**0.82**69008656427103	**0.88**10219923648400
6th Press	**0.94**53709225047882	**0.82**88888550265133	**0.88**37130512438154

Table 5. Experimental results finding

Test Person	Test 1 cases	Test 2 cases	Image size (kb)	Transmission time/image	Face recognition
P1	1. PHI's Device vibrated with flash blinking 2. The Image was displayed on screen	1. PHI's device vibrated, flash blinking, and visitor's picture shown in screen 2. The log of the doorbell android device shows that the visitor's image was matched in its local database (SQLite).	36.322	Immediate	Recognised
P2			33.607	Immediate	Recognised
P3			39.605	Immediate	Recognised
P4			43.646	Immediate	Recognised
P5			39.605/36.61	Immediate	Failed

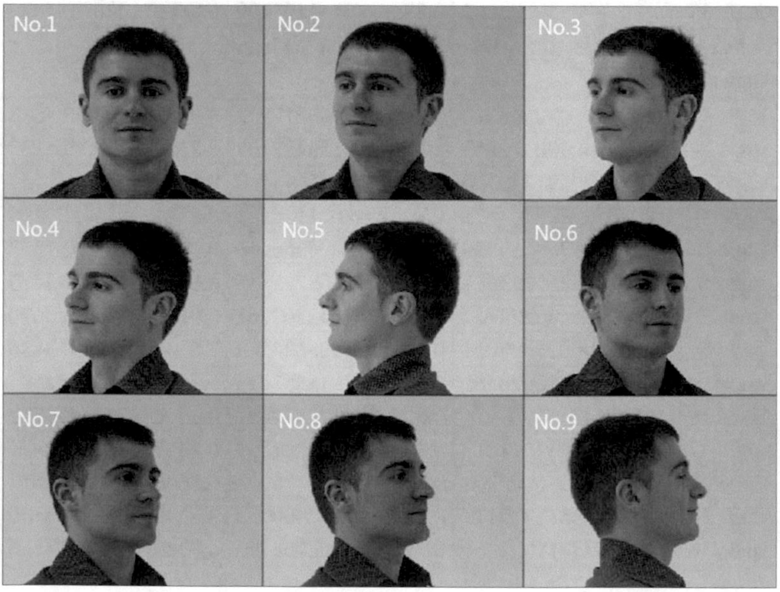

Fig. 8. Angle possibilities for face recognition

6 Conclusion and Findings

The developed system is user-friendly and able to notify PHIs with visual signals and tactile cues, and can also reduce the transmission time of the image by adding a context aware feature into the alert system.

After analysing the cosine values of the same person, it was found that the first 2 decimal places of each cosine value of different angles were the most important for face matching (compared with 'Aarhus face recognition'). Based on this conclusion, the system is able to recognise the visitors by capturing the visitors' images, extracting landmarks, and analysing the context data.

The limitations of the developed system: The system is able to notify PHIs regardless of how the visitor behaves. However, the face recognition accuracy rate greatly depends on how the visitor behaves. The results shown in Table 4 only existed when the visitor was standing statically in front of the doorbell without any facial expressions. In a real scenario, the visitor may have different facial expressions depending on their mood, or sometimes they may not be facing the camera directly (shown in Fig. 8). Those factors may lead the system to fail in face recognition.

7 Future Recommendation

The developed system has its own limitations which can be overcomed by the recommendations of some enhancements to the project in this section. The first limitation is the Self-learning of the system itself. After discussing and analysing the limitations of the system in the conclusion and findings, it will improve the performance of the developed system in recognising the same person regardless of how they face the camera and even with different facial expressions if the system allows the PHI or the visitor to classify the visitor's image by them. For example, 'visitor A' had visited twice already, but the system did not recognise the 'visitor' at the second visit. When the PHI sees the picture and realises those two visitors are actually the same person, then the PHI can classify those two pictures into one record of the same person, or after the system has snapped the visitor's image, it will allow the visitor to put his/her name on the photo; just like when a person has two mobile phone numbers. Therefore, if a visitor's facial data can be matched with any of the records, the system will consider this visitor as 'visitor A'. The person in Fig. 8 has 9 sets of context data, and even if this visitor is not facing the camera, the system would still be able to recognise this visitor.

References

1. Who.int: WHO|World report on disability (2017). http://www.who.int/disabilities/world_report/2011/report/en/. Accessed 4 Jan 2018
2. Kumari, P., Goel, P., Reddy, S.R.N.: PiCam: IoT based wireless alert system for deaf and hard of hearing. In: 2015 International Conference on Advanced Computing and Communications (ADCOM), pp. 39–44. IEEE (2015)
3. Letupe, K.: Android Application for Face Recognition. Aarhus University, Aarhus (2013)
4. Perera, C., Zaslavsky, A., Christen, P., et al.: Context aware computing for the Internet of Things: a survey. IEEE Commun. Surv. Tutor. **16**(1), 414–454 (2014)
5. Zhao, W., Chellappa, R., Phillips, P., Rosenfeld, A.: Face recognition. ACM Comput. Surv. **35**(4), 399–458 (2003)
6. Mobile Vision|Google Developers (2014). https://developers.google.com/vision/. Accessed 2 Feb 2017

7. Landmark|Google APIs for Android|Google Developers (2017). https://developers.google.com/android/reference/com/google/android/gms/vision/face/Landmark. Accessed 2 Feb 2017
8. Dey, A.K., Abowd, G.D.: Towards a Better Understanding of Context and Context-Awareness. Georgia Institute of Technology, Atlanta (1999)
9. Barkhuus, L., Dey, A.: Is context-aware computing taking control away from the user? Three levels of interactivity examined. In: Fifth Annual Conference on Ubiquitous Computing (UBICOMP 2003) (2003)
10. Nougat. Android (2016). https://www.android.com/versions/nougat-7-0/. Accessed 1 Feb 2017

Advances in Health Informatics

Self-reflective Visualizations of Patient-Centered Health Information Systems

Archanaa Visvalingam(✉), Jaspaljeet Singh Dhillon, Saraswathy Shamini Gunasekaran, and Alan Cheah Kah Hoe

Universiti Tenaga Nasional, Putrajaya, Malaysia
achavisva0196@gmail.com,
{jaspaljeet,sshamini,alancheah}@uniten.edu.my

Abstract. Patient-centered systems have great potential to empower patients to take proactive actions in preventing or managing serious diseases and health problems and they can help to reduce healthcare costs. Commonly, these applications are equipped with visuals or graphical representations that enable users to comprehend their health progress. In this paper, we review and analyze the different types of health information visuals that are presented in popular self-monitoring applications and devices, i.e. web-based health support applications, mobile health support applications, stand-alone health monitoring devices, and wearable health monitoring devices. The aim is to gain better insights into designing self-reflective and comprehensible visualizations of health information in patient-centered systems. Results indicate that most of these applications present trends of monitored health data via visuals. Web-based applications feature more comprehensive and complex visuals than applications that are found in other platforms. Stand-alone and wearable devices are found to be more user-friendly and present data in the most simplified manner. There is a need to develop a guideline to aid developers to design effective visuals in patient-centered systems.

Keywords: Patient-centered systems · Health monitoring
Graphical representation of health information · Self-reflective visuals

1 Introduction

Advances in information technology (IT) have introduced new design approaches that support healthcare delivery and patient education. Such advances enable a fundamental redesign of healthcare processes based on the use and integration of electronic communication at all levels. Healthcare IT has the potential to empower patients and support a transition from a role in which the patient is the passive recipient of care services to an active role in which the patient is informed, has choices, and is involved in the decision-making process. These health data are often referred to as big data the reason being its volume, velocity and variety. In healthcare, big data analytics allows significant replies to queries on the rise in the costing in the healthcare industry, the geographical variance in health care, and fraudulent claim detection [13].

F. Saeed et al. (Eds.): IRICT 2018, AISC 843, pp. 711–725, 2019.
https://doi.org/10.1007/978-3-319-99007-1_66

Patient-centered systems are defined as tools that enable a partnership among practitioners, patients, and their families (when appropriate) to ensure that procedures and decisions respect patient's needs and preferences. In order for the patients to have control over their health, it is vital for them to be engaged in patient-centered systems, as their main goal is to serve patients with patient-centered care. These applications and systems cater for users with different health needs. Most of these applications are self-care tools developed to be used independently by the patients themselves.

Data obtained via the aforementioned applications are ideally presented visually to have a higher impact on action rather than having words that may be difficult to interpret [3]. Hence, it is common to find trends of health data presented via visuals or graphical representations that help to educate users on their health progress. There are a few types of graphs that are commonly used to represent health data e.g. bar graphs, pie charts, histograms and line graphs. Health data are presented in the form of graphs to enable healthcare consumers to better understand their health trends with less confusion and complication. These graphs are the evidence of Human-Computer Interaction (HCI) that focuses on the human, social, organizational and technical features of communication between human and machines. Therefore, HCI can be considered to have a large cognitive factor where processing of information by humans is closely related with computer systems. Adding to that, HCI in healthcare increases the usability of a healthcare product or application. This is to achieve specified goals with effectiveness, efficiency and satisfaction when healthcare consumers use the product or application. In return, such information would educate them on their health as well as motivate them in being proactive towards their healthcare (e.g. follow the diet and fitness plans) [1].

In order to better understand one's own health data, he must have health literacy, which is key in comprehending any information presented by health professionals and healthcare providers. The lack of health literacy can result in limitations with patients in interpreting health information and keep track of any medical instructions given and to communicate with their doctors [7]. For example, a person who often falls sick and visits the hospital, is more likely to know more about his medical history correctly. He would be knowledgeable enough to possibly treat himself the next time he falls sick. This shows that the more often a person uses the medical services that are available to them the higher their health literacy. Despite having such advancement in technology, many people do not realize the risks in misinterpretation of graphical displays of risk and associated terminology [5]. The problems in misinterpreting a set of data can lead to major confusion within a patient himself, between the patient and his family members or both the parties with the physicians. Moreover, a certain type of graph can also cause confusion and a longer time is taken to understand it. Graphs like circular and area-based graphs are difficult to understand quickly and accurately [6]. The reasons being the area-based charts are not as efficient in the means of comparing quantities and visualizing as tree maps and pie charts.

Self-reflective visuals in health information systems are main components in enabling healthcare consumers understand their health data without referring to any medical related books or browsing through the Internet. Comprehensible visual on the other hand means the visual is simple and easily understandable. Self-reflective and comprehensible visualization is the idea of having visuals that are self-explanatory and easily understandable. In this paper, we review and analyze different types of health information visuals presented in popular self-monitoring applications and devices. We aim to gain better insights into designing self-reflective and comprehensible visualizations of health information in patient-centered systems. Well-published guidelines that emphasized the importance of Human Computer Interaction (HCI) and usability in healthcare were employed as the basis in conducting the review [11].

Section 2 reviews the visuals in patient-centered health information systems. Section 3 presents the analysis and discussion on how these applications have represented health data graphically and Sect. 4 concludes the paper.

2 Visuals in Patient-Centered Health Information Systems

In this section, we review selected patient-centered health information systems (with a focus on the visuals they provide to users) that are categorized into (1) web-based health support applications, which enable patients to manage their conditions through online healthcare systems; (2) mobile health support applications that help patients to track their health progress conveniently with their smartphones; (3) stand-alone health monitoring devices, which help patients to monitor a specific health condition; and (4) wearable health monitoring devices that help users to monitor and record real-time vital signs by wearing a device with sensors as an accessory [10].

2.1 Web-Based Health Support Applications

Web-based health support applications existed before they evolved into mobile applications for healthcare users. There are few health support websites that have been created with the aim of delivering disease-related information to healthcare users. These applications are designed in a way where patients who are experiencing, had experienced and survived a certain type of diseases, symptoms and treatments to share their thoughts to other patients who have the same health conditions as them. The conditions and symptoms are often presented with understandable charts or graphs that are based on the health data gathered from the patients. Some websites provide a forum space that allow patients to connect and share with other patients.

PatientsLikeMe[1] is an online platform with more than 600,000 registered members that provide online peer support community with treatments, symptoms and health outcomes that are reported by patients themselves.

PatientsLikeMe portrays comprehensive graphs to reflect the patient health progress in the system. For an instance, Fig. 1 presents a graph showing the details of treatments the patient has undergone. The meanings of the horizontal bars according to the colors are given on the left pane of the layout. The horizontal bars are aligned to the right and the dosage amount taken is stated below every horizontal bar. The date bar on top of the horizontal bars indicates the years of the same month. The graph could confuse first time

[1] Heywood, J.: Live better, together! | PatientsLikeMe, https://www.patientslikeme.com/, last accessed 2017/01/17.

users as most people are used to reading sentences or numbers from left to right as how most people write as well. Hence, it may take some time for them to capture everything that is happening around the graph and then later getting the meaning of the graph itself. On the right pane of the screen presents details on the drug prescriptions, procedures, supplements and equipment used by the patient. The details on the right pane co-relate to the graphs presented. The dosages stated below the graphs are the dosages taken according to the medications and equipment shown on the right pane. Patients may not

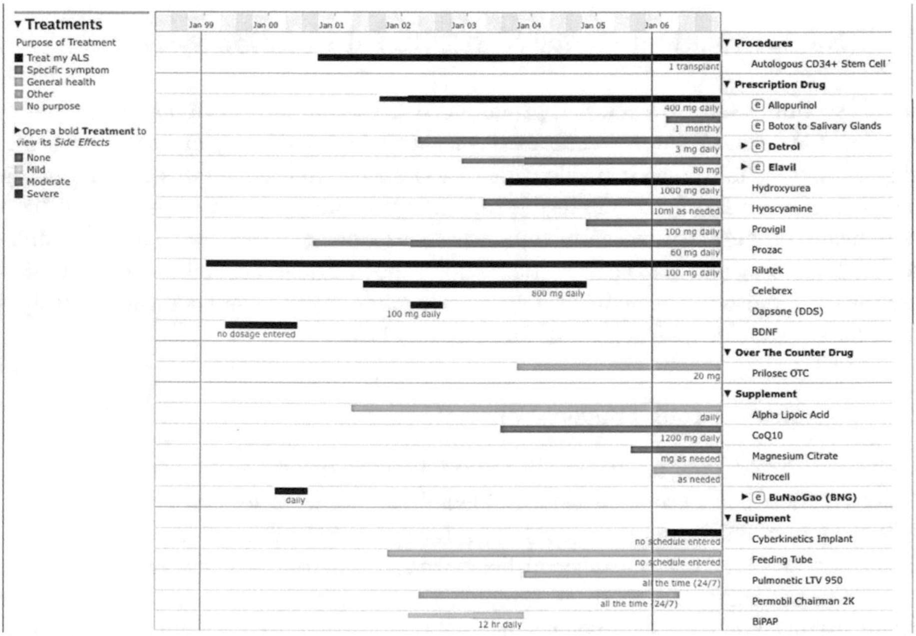

Fig. 1. Screenshot of a section of a patient's report of *PatientsLikeMe* showing the treatments and purposes of the treatments received.

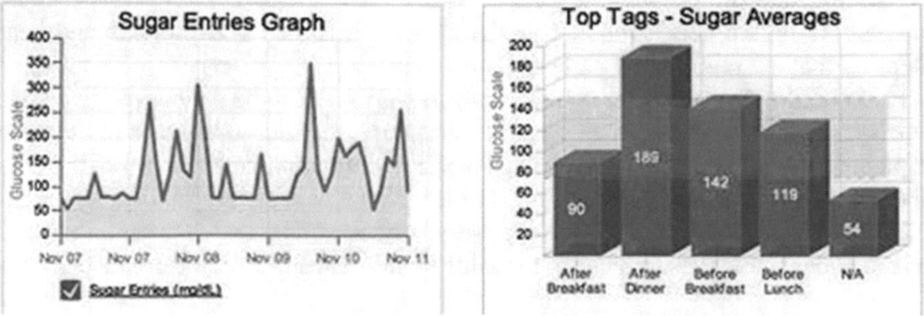

Fig. 2. Screenshot of the graphs displayed on a user's dashboard of *Sugarstats* showing the sugar entries and averages for a few days and time.

be able to relate the graph with the data given on the right pane of the layout at first glance.

Sugarstats[2] is a website that functions as a blood-sugar tracker and a diabetes management system. Users are offered a simple interface to keep track, monitor, and have access to their glucose levels and diabetic statistics to identify the critical changes in the trends to better cope with their health.

Added to that, the site allows users to input data not only for blood glucose and medications but also for glycosylated hemoglobin (A1C), food intakes and exercise. There are two types of graphs (see Fig. 2) that are displayed for the users through their accounts. The graphs show sugar entries and top tags with sugar averages. Both the graphs are confusing because there are no further details explanations given about the graphs. The sugar entries graph shows the glucose level of the user for a few days in the form of a line graph. The x-axis shows the date and the y-axis shows the glucose level, but it does not show the range or level that they are supposed to be at considering their age and severity of their diabetes. On the other hand, the top tags with sugar averages are plotted in a vertical bar graph. The x- axis shows the timing of sugar entries and the y-axis shows the glucose level. In this graph, there is a green shaded area that is just left shaded in the graph without any explanation. This can mislead users not fully knowing the purpose of the graph.

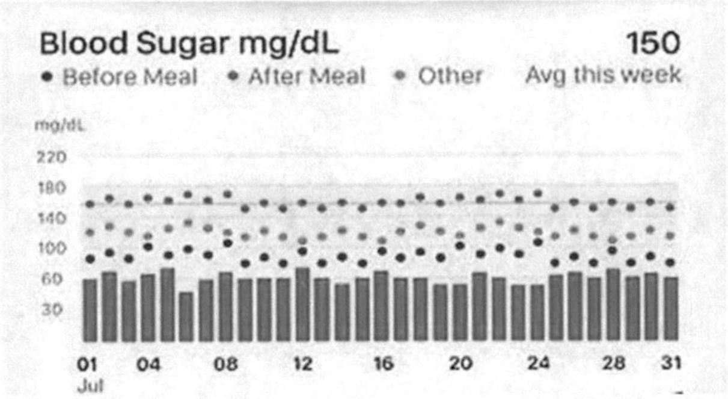

Fig. 3. Screenshot of a section of a user's Insights tab of *Glucose Buddy* showing the blood sugar reading for the entire month.

2.2 Mobile Health Support Applications

Whereas the sites described in the previous section are website based where the users have to have a personal computer or a laptop to check on their health status, the ones described in this section focuses on health support provided via mobile applications for patients. Smartphones with mobile applications are becoming a necessity for everyone, be it among children, adults and the elders. In this fast-moving world, it is a huge

[2] SugarStats.com – Simple, Online Blood Sugar Tracking for Diabetes Management, https://sugarstats.com/, last accessed 2017/01/17.

advantage to have mobile applications that provide health support for healthcare users. There are few mobile health support applications that have been widely used for diabetes management and weight management. These are among the common medical conditions that everyone can relate to easily. The features in the applications are similar to the ones that are available in desktop versions.

Glucose Buddy[3] is a diabetes tracker application. This app has been the number 1 among diabetes management applications for over 9 years. The users are provided with an easy interface layout to keep track of their average blood glucose level, the number of insulin intakes, food calories count, and the number of calories burned from active activities done throughout a week.

The graph (see Fig. 3) that explains the blood sugar level of the user for the entire month is a crucial and important detail for the user. The dates for the whole month (in the interval of four days) are displayed at the x-axis and the glucose level on the y-axis. The colors of the dots that are being used were explained clearly at the top before the graph. This gives users a heads-up on the meaning of the dots that are plotted in the graph. The bar graph shows the blood sugar average for the whole month. At the top right corner of the graph, the blood sugar average for that particular week is displayed as a single number, big and clear.

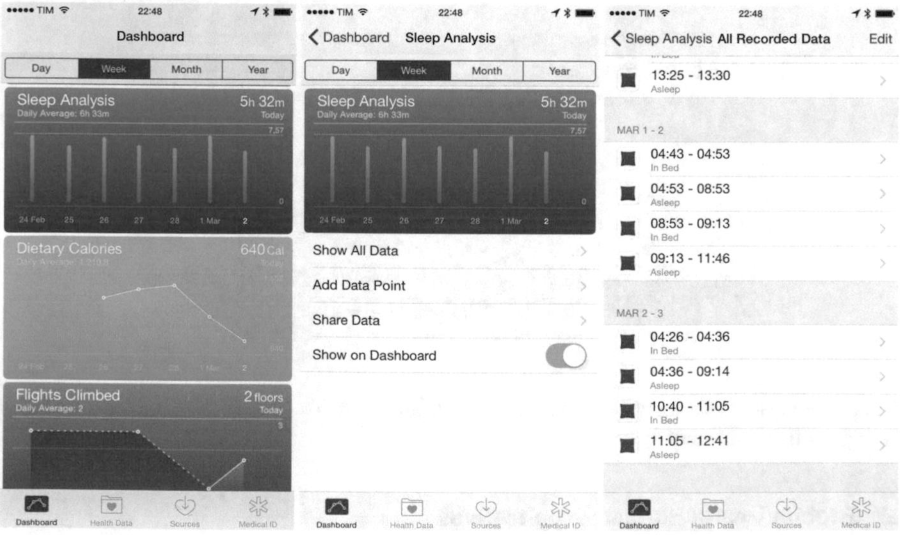

Fig. 4. Screenshot of a user's Dashboard tab of *Health App* showing a few different graphs and data that are plotted for the span of one week.

Health App[4] is a popular application among smartphone users. This app is usually installed in smartphones before users have even purchased it. This is a built-in application by Apple Inc. that can be found in any updated versions of the iPhone series. The

[3] Glucose Buddy, https://www.glucosebuddy.com/, last accessed 2017/01/17.
[4] iOS - Health - Apple (MY), https://www.apple.com/my/ios/health/, last accessed 2018/04/11.

Health app makes it easy for iPhone users to know about their health and start reaching their goals.

This app merges health data from iPhone, the Apple Watch, and any third-party apps that the user has been using, so that they can view their progress in one convenient place. Added to that, the app recommends users with other popular health apps to move their health forward. The *Health app* displays a simple layout of graphs (see Fig. 4) for the users. At the Dashboard tab, the top pane shows options for the users as Day, Week,

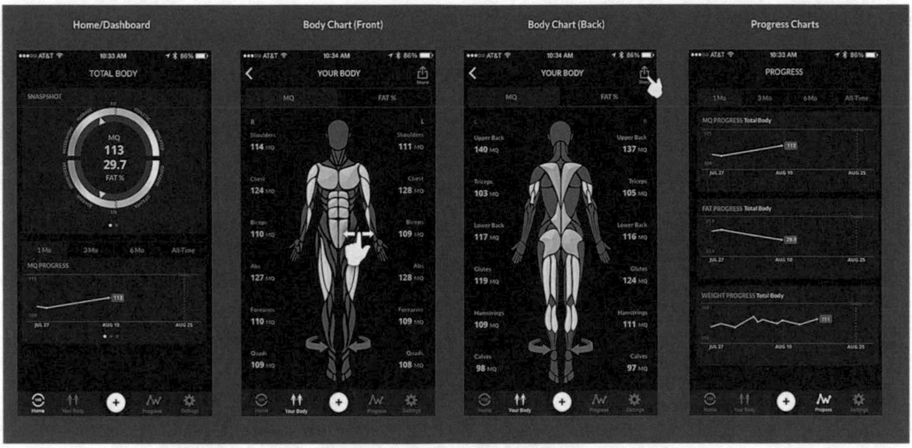

Fig. 5. Screenshot of a few tabs of a user's profile of *Skulpt* showing the body fat percentage and muscle quality of the user.

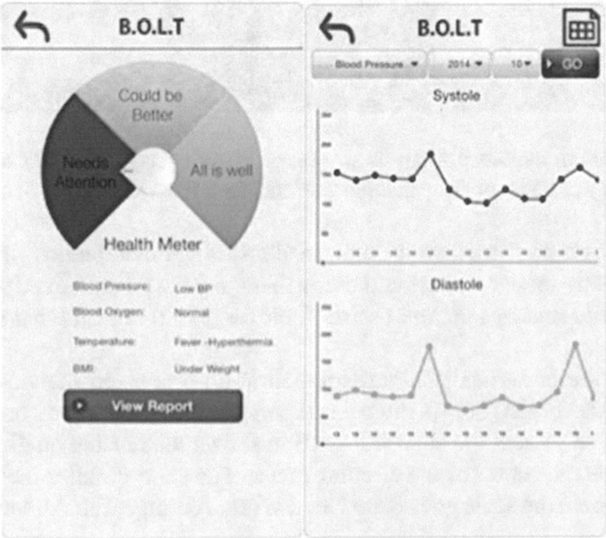

Fig. 6. Screenshot of two sections of a user's profile of *AMI B.O.L.T* showing the health meter reading and the blood pressure report.

Month and Year. In the Week option, three different graphs that displays different details. The first would be the Sleep Analysis graph that has vertical bars that shows the amount of sleep that the user had according to the dates that is on the x-axis. The daily average of hours of sleep and the number of hours of sleep for today is displayed larger at the top corners of the graph. The Dietary Calories graph and the Flight Climbed graph are plotted using lines. No doubt the lines drawn are clear but the values for each plotted point are not given on the graph (Fig. 4).

2.3 Stand Alone Health Monitoring Devices

Health monitoring devices aid users to keep track of health-related factors and visualize them within the device itself. The devices enable patients to track their progress towards their health goal. Most devices nowadays require being paired with mobile applications to save and keep track of the data collected from the devices.

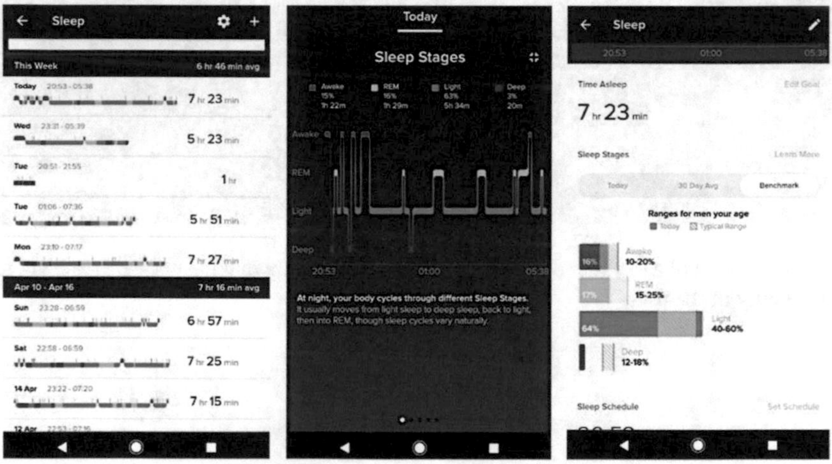

Fig. 7. Screenshot of the Sleep Analysis section of a user's profile of *FitBit* showing the sleep stages and number of hours of sleep that the user had for a few weeks.

SKULPT[5] is a performance training system. This device helps fitness-addicts to measure their body fat percentage and helps them to get a personalized workout plan to burn fat and build muscles. *Skulpt* assists their users to focus on what matters to train smarter.

In the app, the body analysis is displayed in an organized manner. The home or dashboard tab (see Fig. 5) shows the overall perspective of the user's body analysis. At the top half of the screen, the snapshot section shows the muscle quality and the body fat percentage as the center of the circular meter. The semi-circular meter reflects how fit the user is, where the scale goes from Need Work, Average, Fit, Athletic till Skulpted.

5 Skulpt® - Measure Body Fat Percentage and Muscle Quality, https://www.skulpt.me/, last accessed 2018/04/11.

The scale goes from dark orange till white showing the current stage of the user's body condition. The bottom half of the screen shows the muscle quality line graph that can be seen from the past one month, three month, six months and all time. The x-axis of the graph indicates three dates; today's date, the date a month ago and the mid-date for the month. The y-axis shows an average range of muscle quality. In the Progress tab, the Fat Progress and Weight Progress graphs are line graphs as well. The body chart (front and back) can be seen at the Your Body tab where the user can see their muscle quality and body fat percentage at every part of their body.

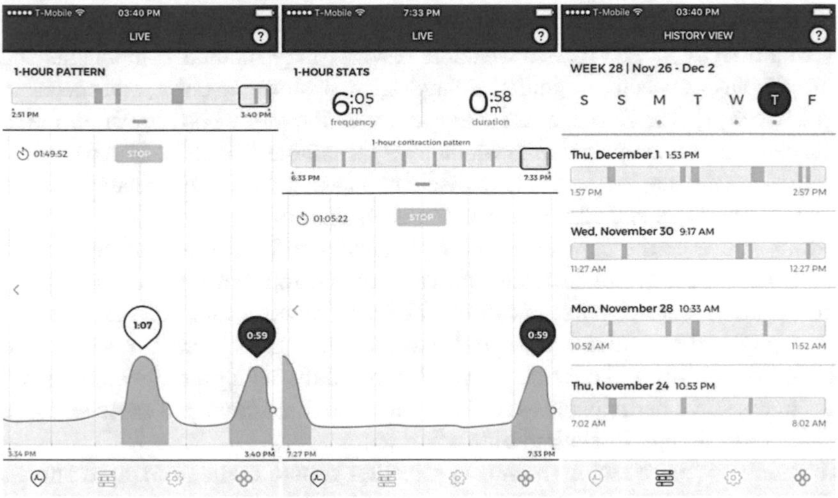

Fig. 8. Screenshot of the Live of contraction of a user of *Bloomlife* showing the contraction patterns and the statistics.

AMI B.O.L.T[6] also known as the *B.O.L.T* is a one-touch wireless health tracker. This device is portable and can be used anywhere to measure vitals of users. Whether the user is a fitness buff or they just want to keep better tabs on their health, *B.O.L.T* is the answer to all your queries related to your health. *B.O.L.T* is a compact and wireless home health monitor that measures blood pressure, blood oxygen levels, temperature and your body vitals without a prick.

The device has to be connected to the *AMI B.O.L.T* app in order to pull put the reports and analysis of a user. There are two types of graphs (see Fig. 6) that are available in the app to help users know about their health progress. The graph on the left is a health meter that indicates the condition of the user when being measured with the device. It shows whether the user's health needs attention, could be better or all is well. Below the health meter, the blood pressure level, the blood oxygen level, body temperature and condition and the BMI of the user is listed. The graphs on the right show the user's blood

[6] 3 In 1 Health Monitoring Kit | Blood Pressure | Blood O2 Level | Body Temperature, http://www.amibolt.com/campaigns/bolt/primary/product, last accessed 2018/04/11.

pressure report. Users are able to go from one date to another using the options available at the top of the screen. The systole graph shows the contraction rate of the heart whereas the diastole graph shows the refilling rate of the blood in the heart. Both the graphs use the line graph method to clearly indicate the high and the low when the blood is pumped from the heart to the entire body.

2.4 Wearable Health Monitoring Devices

Wearable health monitoring devices have some similarities with stand-alone health monitoring devices. They function to keep track on the health factors of the users, as the users would wear it. The classic wearable devices may include components such as sensors, displays, speech recognition technologies, positioning chips, and special monitoring devices [4]. The close contact of the device on the user's body transmits the sensor signals to the mobile application to which every wearable device is often paired. When a user uses the device, they are able to detect changes in their body, related to a disease or illness as it occurs and alert them or their physician [8].

Fitbit[7] is one of the popular wireless physical activity trackers in the market and there has been a study conducted on its validity and reliability standards [2]. This is because primary care physicians can use *Fitbit* to monitor and keep an eye on their patient's physical activities accurately between clinic visits. During a study that was conducted on getting to know the perception of adults using mobile apps for weight management, some participants were favorable about the automated exercise data entry via *Fitbit*, even though it required little effort [9].

The tracker is paired with a mobile application known as the *Fitbit* app. The tracker and the app connect via Bluetooth and send all the data from the tracker to the apps within a few minutes. Besides that, the device's sleep tracking feature is among the most popular features of the device. The sleep tracker gets the sleeping depth of the user and analyses it (see Fig. 7). The number of hours of sleep is calculated accurately from the time the user falls asleep till the time they are fully awake. This is shown by days in a week and the average number of hours of sleep for that particular week. The depth of the sleep is indicated in sleep stages, which are Awake, REM, Light and Deep. The fluctuations of the depth while sleeping are shown in a line graph with the x-axis as the time in hours and the y-axis as the sleep stages. The sleep stages are differentiated with different colors and are labelled before the graph with the number of hours of sleep during each sleep stage. Added to that, the benchmark or the ideal sleep stages are shown with comparison of the current reading in a horizontal graph.

Bloomlife Smart Pregnancy Tracker[8]. A pregnancy tracker is an application that assists women in their pregnancy every day. The women have full access to personalize the content as per their need and receive the latest parenting news and health information

[7] Fitbit Official Site for Activity Trackers and More, https://www.fitbit.com/my/home, last accessed 2017/01/17.

[8] Contraction Monitor | How It Works | Bloomlife, https://bloomlife.com/how-it-works/, last accessed 2017/01/17.

[12]. *Bloomlife Smart Pregnancy Tracker* is a device that tracks contractions of the user during their pregnancy and connects with their body.

In the app, *Bloomlife* shows the Contraction Patterns, Live view of the contractions (see Fig. 8), Contraction Stats on how far apart the contractions are and Trends View to identify the natural rhythm of their contractions. The Live feed shows the contraction pattern in a line graph. The peaks of the graphs are marked as the duration the contractions were high. The statistics for an hour shows the frequency of the contractions and the average duration of it. The stacked horizontal graphs show the start and the finish time of tracking the contractions in a day and the nth week of the pregnancy.

3 Discussion

The four categories of health-related applications and devices presented above differ in terms of portability, interactivity, and connectivity of the applications with healthcare consumers. Some of the devices and applications are popular among consumers while the rest are newly introduced technology that have great potential and are picking up their pace in the market.

Table 1 summarizes the purposes, interactivity, connectivity, target users, type of graphs, and the average cost of the applications and devices. Most of the applications are free of charge except for the device-based applications. All the applications and devices have a common feature that forms informative graphs and reports from the data provided by or captured of its users. Each category follows a different objective and has distinctive abilities and constraints that governs the presentation of the visuals.

Visuals in web-based health support applications consist of many components. The applications would normally function with the input of the users into the system frequently. Users have to be self-disciplined to key in their vital signs or health data into the system using a computer. It is common to find many web-based systems made accessible to users via smartphones apps. However, most of the features are compromised due to limited screen size. The actual system is designed to be displayed on a desktop computer. The large screen enables a rich presentation of visuals that would allow the users to obtain a holistic view about their health progress. However, a bigger screen size may contribute to the complexity of the visuals presented in the application. Rich visuals may increase the cognitive overload for users to digest the information once. The visuals should be designed carefully and kept simple for the users to understand about their own health progress and conditions. The alignment of other items associated to the graph also plays an important role as it may affect the apprehensibility of the visual as a whole.

The visuals found in mobile health support applications have a few common features such as dashboards and graphs that are widely presented. Dashboard is a graphical summary of various components of the application that are represented with icons that serve different functionalities available in the application. In addition, the conciseness of graphs gives a bigger impact on mobile users. The reason being is the limited screen space that sets a constriction for the graphs to be presented in a simple manner. Added to that, it is more convenient and practical for the users to key in their health data frequently into the application as they carry a smartphone with them all the time which could increase better adaptation of the application. Most smartphones are equipped with sensing technology such as accelerometers that extend the ability of mobile apps to capture users' data effortlessly. In addition, mobile apps are also leveraged as tools to display graphs based on data captured via stand-alone and wearable devices.

Table 1. Comparison of health-related applications and devices.

Criteria	Categories			
	Web-based health support applications	Mobile health support applications	Stand-alone health monitoring devices	Wearable health monitoring devices
Objective	Provides rich features that enables informed decision making, tracks vital signs via human input and provides social support	Presented in the form of mobile apps that enables convenient monitoring of vital signs, educates users on their health progress and motives users via social engagement	Offers to check their health condition wherever they are without needing to go to clinics and hospitals via simple devices	Enables users to keep track on their health conditions and alert them on any changes at any time via wearable devices
Type of Input	Human	Human/Natural	Natural	Natural
Type of Graphs	All types of graphs. Mostly, big and complex since the screens are wide in width and height	Simple line and bar graphs that can fit the smartphone screens	Line, bar and meter graphs that can be read easily from the device itself	Simple line and bar graphs that are simple and understandable
Target Users	Users preferably aged 10 and above who are computer literate that have common medical condition, such as diabetes, cancer, depression and anxiety	Users preferably aged 10 and above who have common medical condition such as diabetes that would prefer keeping track on their health using a smartphone	Users of all ages who prefer not to fill any information into the systems but want to heck on their health condition at any time	Users of all ages and in any medical condition who wants to keep track on their health condition closely at all time
Connectivity	Via Websites on desktops	Via Apps and sometimes through websites	Via Devices with sensors and app that gives further details and reports	Via Sensor trackers and apps that give further details and reports
Cost of applications	Mostly, the websites are free for users but sometimes freemium, where basic services are provided for free while advanced featured must be paid for	Mostly, the apps are free for users but sometimes freemium, where basic services are provided for free while advanced featured must be paid for	The devices are ranged from $99 to $400	The devices are ranged from $120 to $500

Visuals in stand-alone health monitoring devices that are developed to track specific vitals such as blood pressure, blood glucose, and body temperature are presented differently from the other applications. Generally, the devices offer dashboards as they have to provide simple information as icons in a confined space within the device. The visuals are presented colorfully to attract users in engaging with the device. Adding to that, colorful visuals can clearly show the differences in severity of the health level of the users. Certain reports on the user's health progress can be retrieved via mobile application connected to the device. The graphs are presented clearly with straightforward information which increases the user's understanding on their health progress. These devices do not require input from the users, instead it collects data from the users via natural sensors that are on the devices.

With regards to visuals presented in wearable devices, these visuals are brief and present real-time readings of the user's vitals. It is displayed within the wearable devices by using relatable icons that can be easily identified and understood by the users. Usually the colors are kept minimal among a few options to avoid confusion. Moreover, a user's data are collected using the natural sensors that are attached on the devices. These data are accurate and have so many perspectives to it, making it easier to classify the user's information to be displayed in a simple manner. The graphs and change in health progress can be seen via the mobile applications connected to the device. The data presented in the graphs are well organized and kept clear as possible.

The visuals offered in all 4 categories can be related with a few dimensions to show their similarities and differences. In general, the visuals presented overall have reasonable fidelity or level of detail and realism in meeting the goal of the application or devices. The focus is to provide their users with as much information about their health in the best way as possible with hassling them. Although all the visuals are created for a specific purpose, there are high chances of over-crowding a given space with too much of information that can lead to confusion or inability to grasp the content presented. Visuals in web-based health support application are the most complex compared to the visuals in the other categories of applications and devices. As mentioned above, this is because of the display screen size of the application that provides opportunity to application developers to optimize the space available. Besides that, visual configuration or the arrangement of parts or elements in an application is essential. It is apparent that each different category of applications follows a different style of interaction and presentation of the information that forms a careful flow within the application. This determines the overall appearance and usability of the application. Visuals presented in mobile-based health support applications, stand-alone and wearable health monitoring devices are more structured and user-friendly as compared to the web-based health applications. The large screen size of the web-based solutions enables applications developers to explore different styles of presenting the information. Hence, these applications should be carefully designed in meeting the objective of the application.

4 Conclusion and Future Work

It is crucial that visuals presented in patient centered applications be comprehensible and enable patients to self-reflect their health. We have reviewed different categories of patient-centred applications, with a focus on the graphical representations presented in these applications. It is noted the visualisations presented via web-based applications are more complex as compared with mobile applications, stand-alone devices, and wearable devices. The large screen size of web-based solutions enables developers to offer rich content with various types of visuals. Different categories of applications follow a distinctive format of health information representation. Some of the applications present visuals more effectively than the others. Our review of existing health applications and devices indicate the necessity to develop a general guideline to aid developers to design effective self-reflective visuals that would meet the expectations of users towards novel patient-centered systems.

Acknowledgement. This study is funded by Universiti Tenaga Nasional under the UNITEN Internal Research Grant (UNIIG 2017).

References

1. Bantug, E.T., Coles, T., Smith, K.C., Snyder, C.F., Rouette, J., Brundage, M.D.: Graphical displays of patient-reported outcomes (PRO) for use in clinical practice: what makes a pro picture worth a thousand words? Patient Educ. Couns. **99**, 483–490 (2016)
2. Diaz, K.M., Krupka, D.J., Chang, M.J., Peacock, J., Ma, Y., Goldsmith, J., Schwartz, J.E., Davidson, K.W.: Fitbit®: an accurate and reliable device for wireless physical activity tracking. Int. J. Cardiol. **185**, 138–140 (2015)
3. California Health Care Foundation: Worth a Thousand Words: How to Display Health Data (2014)
4. Jeong, S.C., Kim, S.-H., Park, J.Y., Choi, B.: Domain-specific innovativeness and new product adoption: a case of wearable devices. Telemat. Inform. **34**, 399–412 (2017)
5. Johnson, C.M., Shaw, R.J.: A usability problem: conveying health risks to consumers on the Internet. In: AMIA Annual Symposium Proceedings 2012, pp. 427–435 (2012)
6. Laubheimer, P.: Dashboards: Making Charts and Graphs Easier to Understand. https://www.nngroup.com/articles/dashboards-preattentive/. Accessed 28 May 2018
7. Lee, S.-Y.D., Arozullah, A.M., Cho, Y.I.: Health literacy, social support, and health: a research agenda. Soc. Sci. Med. **58**, 1309–1321 (2004)
8. Lee, S.Y., Lee, K.: Factors that influence an individual's intention to adopt a wearable healthcare device: the case of a wearable fitness tracker. Technol. Forecast. Soc. Change (2018)
9. Lieffers, J.R.L., Arocha, J.F., Grindrod, K., Hanning, R.M.: Experiences and perceptions of adults accessing publicly available nutrition behavior-change mobile apps for weight management. J. Acad. Nutr. Diet. **118**, 229–239.e3 (2018)
10. Prabha, M.S., Sarojini, B.: IOT in healthcare: a study. Sarojini Int. J. Innov. Adv. Comput. Sci. IJIACS **6**, 2347–8616 (2017)

11. Sandblad, B.: Human-computer interaction and usability in health care
12. Silva, B.M.C., Rodrigues, J.J.P.C., De La, I., Díez, T., López-Coronado, M., Saleem, K.: Mobile-health: a review of current state in 2015. J. Biomed. Inform. **56**, 265–272 (2015)
13. Sukumar, S.R., Natarajan, R., Ferrell, R.K.: International journal of health care quality assurance quality of big data in health care. Int. J. Health Care Qual. Assur. **28**, 621–634 (2015)

Potential Measures to Enhance Information Security Compliance in the Healthcare Internet of Things

Premylla Jeremiah[1(✉)], Ganthan Narayana Samy[1], Bharanidharan Shanmugam[2], Kannan Ponkoodalingam[3], and Sundresan Perumal[4]

[1] Advanced Informatics Department, Razak Faculty of Technology and Informatics, Universiti Teknologi Malaysia, Kuala Lumpur, Malaysia
pam_jeremiah@yahoo.com, ganthan.kl@utm.my

[2] School of Engineering and Information Technology, Charles Darwin University, Casuarina, Australia
Bharanidharan.Shanmugam@cdu.edu.au

[3] Faculty of Information Technology and Science, INTI International University, Nilai, Malaysia
pon.kannan@newinti.edu.my

[4] Faculty of Science and Technology, Universiti Sains Islam Malaysia, Nilai, Malaysia
sundresan.p@usim.edu.my

Abstract. Healthcare organisations are particularly vulnerable to information security threats and breaches due to the highly confidential nature of their patients' medical information. Now, with the emergence of the Internet of Things (IoT) in healthcare that can vary from diagnostic devices to medical wearables, the industry has indeed become more vulnerable to malicious exploitation. One of the reasons that malicious attacks continue to occur at an alarming rate is due to the poor compliance of information security policies. This study investigates the issues that are associated with the causes for poor compliance within the private healthcare organisations in Malaysia. Data was collected through interviews from various healthcare respondents and findings have revealed that often, poor security compliance is mainly caused by behaviour issues and the severe lack of security awareness which requires immediate attention and mitigation. Potential measures to cultivate information security awareness and to safeguard the IoT-based medical devices are proposed to achieve compliance.

Keywords: Awareness · Behaviour · Healthcare · Internet of Things
Information security compliance

F. Saeed et al. (Eds.): IRICT 2018, AISC 843, pp. 726–735, 2019.
https://doi.org/10.1007/978-3-319-99007-1_67

1 Introduction

We live in a digitally connected world, in an age where the speed of technology is advancing rapidly and in which our environment and communities are seamlessly interwoven through the implementation and usage of the latest technologies and facilities. One such emerging cutting-edge technology that has penetrated its way globally is the Internet of Things (IoT) technology which are smart-enabled embedded computing devices [1] and thus, has been hailed as the revolutionary intelligent technology that will offer much potential for business ventures, innovate the way industries operate and even causing our lifestyles to be changed positively. Inevitably, one of the most critical mission industries that has prominently implemented and reaped benefits from IoT is the healthcare industry. It may still be in its infancy stages in many healthcare organisations worldwide [2], but it is indeed a promising trend and is expected to redefine medical technology and healthcare, profoundly impacting the industry in the near future.

However, with such acceleration in the IoT-based healthcare technology and applications, the ensuing risks and vulnerabilities specifically in the area of information security also emerges and could prove to be a socio-technological and organisational challenge [3, 4] if not mitigated during its initial stages. The lax of security in IoT technology for the growing number of healthcare and medical devices could involve cyber risks [5] that may cause both virtual as well as physical harm [4] i.e. increased risk of bodily harm from a patient's breached clinical records or hacked medical devices whose information is now exposed to the criminals who can use it for unscrupulous reasons. In order to mitigate or avoid such vulnerabilities or risks from happening all together, several information security measures such as education, training, awareness campaigns, and technological safeguards [6, 7] must be introduced, enforced and adhered to, in ensuring that the privacy, confidentiality and integrity of the patients' healthcare information remains secured and not jeopardized.

Thus the aim of this paper is to investigate the causes that leads to IoT information security non-compliance within the healthcare industry and the information security awareness measures that can be recommended and enforced to help raise and achieve the levels of security compliance in order to mitigate or reduce the IoT-based vulnerabilities. This paper is organized into seven sections. The next section describes the literature review related to this research. Section 3 discusses the research problem while Sect. 4 explains the research methodology that is applied in this research. Section 5 presents the findings of the studies, Sect. 6 presents the discussion, and followed by a conclusion in Sect. 7.

2 Internet of Things (IoT)

Originally defined during the mid-1990s, the term 'Internet of Things' was used to describe the networked radio-frequency identification (RFID) infrastructures and sensor technologies [8]. Today, the usage of the IoT is wide and diversified, extending towards more and more domains and fields which are totally transformed by the capabilities, efficiency and effectiveness that IoT technology brings [9]. Some of the more prominent domains that have been dynamically implementing IoT-enabled applications would include transportation, automotive, healthcare, manufacturing, agriculture, smart cities and homes, and etc. Powered by the latest enabling technologies such as smart sensors and Near Field Communication (NFC) as well as the interplay with cloud computing, big data analytics and fog computing, the IoT is anticipated to aggressively fuel the transformation of many emerging intelligent devices, applications and technological innovations to match user requirements, compatibility, as well as market demands in terms of effectiveness and efficiency, for any moment and any place [10] in the imminent future.

3 IoT in Healthcare

Among the various domains that have deployed IoT technology and applications, medical and healthcare has become one of the most widespread and well-represented within the last few years [3] in which many highly significant changes have been brought about. These timely interventions in healthcare have not only brought a sharp surge in improved patient clinical care, management of chronic diseases, hospital administration, and medical services such as payment applications to a geographically larger and even remote demographics but also at much reduced costs [11]. More groundbreaking IoT healthcare innovations are looming over the horizon and the integration of communication and network technologies could prove to be a game changer in the healthcare domain as they harness the myriad of possibilities in pervasive and cutting-edge technologies in which complex, intelligent medical devices, sensors, implants or wearables [12] could be developed with the capabilities to store health records and which could be used to save a patient's life during emergency situations [13, 14]. Thus, by observing the huge potential and opportunities for the growth and expansion that could cause astounding paradigm shifts in the healthcare domain, the IoT can be further exploited in order to address the emerging needs of healthcare providers as well as elevating the standard and quality of patients receiving improved clinical care in the future.

However, the risks of information security incidents and vulnerabilities have also seen an alarmingly steady rise [15] whereby healthcare information in particular, is being increasingly targeted by criminals [15]. There have been many published studies on IoT security and privacy issues and challenges [15, 16] and these problems remain at large especially for IoT-based medical devices [16–18] following the endless string of disclosures on major information breaches [19, 20] that have occurred in many healthcare organisations globally. A significant number of researches have shown [17, 18, 20] that these vulnerabilities have opened avenues to criminals who devise new methods to attack, steal, destroy or manipulate the medical devices and its configurations including the network structure misconfigurations, software design flaws, third-party vulnerabilities, absence of tamper proofing, coding defects, authentication, access control, secure middleware, policy enforcement and etc. [17, 18, 20]. Severe ramifications could occur for the patients if such malicious attacks exposes their health information, presenting a pronounced risk and harmful effects to the patients' privacy and safety [17]. Personal clinical information is critical in the healthcare sector and the confidentiality, integrity and availability of healthcare information continues to be a grave concern to all stakeholders within the ecosystem. Therefore, it is imperative to ensure that healthcare information is always safeguarded against both malicious and non-malicious attacks and its security and privacy must be given utmost priority at all times.

4 Research Methods

A preliminary investigation study was carried out for the purpose of information gathering and to gauge the feedback and opinions from selected participants regarding the IoT information security non-compliance issues in healthcare organisations.

4.1 Field Settings

The preliminary investigation involved interviewing respondents who are experts and are currently or previously based in healthcare organisations or facilities and have exposure in healthcare information security issues, healthcare systems and operations, and healthcare information security management. The selection criteria for selecting the participants was each of them must have related working experience in their respective field of expertise for a minimum number of two years.

4.2 Qualitative Research: Semi-structured Interviews

The qualitative research design was considered a suitable approach for the preliminary investigation study because it covers a gap in the healthcare environment as they provide researchers with a comprehensive outlook on work procedures, behaviours, actions, operations, perceptions, and culture in a way that quantitative research alone is unable to achieve [21] The rationale behind the selection of the semi-structured interview method was twofold: Firstly, to gauge the feedback and opinion of the respondents regarding the non-compliance issues that were faced by the healthcare organisations and secondly, to investigate on the potential mitigation measures and how to deliver them. The semi-structured interview method was applied as it provided flexibility with the interview questions, thus allowing an unhindered, detailed line of questioning based on the interviewees' responses. Furthermore, the semi-structured interview was also deemed suitable for the research topic because it allowed the researcher to tailor the questions into the study's context and perspective [21]. Before the interviews were conducted, an interview guide was provided to all the respondents to give them an overall idea about the preliminary investigation study as well as the interview procedure.

4.3 Participants' Background

In total, sixteen individuals who are experts in healthcare information security from a spectrum of organisations were selected for the preliminary investigation study. Two participants were from well-established medical IT solutions companies, and fourteen were from various prominent private hospitals in Malaysia. The participants' job designations among others, were several clinical academicians, healthcare information systems managers, medical doctors, health inspectors, nursing staff, hospital IT administrators and etc. while their work experiences ranged from two to forty years. All identities remained anonymous in order to maintain the confidentiality of the participants' information during data analysis and in reporting the findings for the preliminary investigation study.

4.4 Data Collection

Three research questions were formulated as follows

i. What are the issues faced by your healthcare organisation pertaining to IoT information security compliance?
ii. What are the possible reasons behind the occurrence of these issues?
iii. Which potential measures can be enforced in order to cultivate and increase the compliance levels?

The first question is oriented to identify which issues could lead to IoT information security compliance being compromised in private healthcare organisations. Realizing that these issues could possibly be stemmed from multiple causes, a second question was formulated as we were also interested to find the reasons behind information security non-compliance. Finally, a third question was framed as we also wanted to identify potential measures that can be enforced to raise the levels of compliance in healthcare organisations from independent points of view.

5 Initial Findings

The preliminary study sought to ascertain if the healthcare organisations in Malaysia faced challenges and issues in achieving information security compliance in managing IoT-based medical devices and networks. Based on the findings, the study also suggests several potential means of increasing the compliance levels among the employees of healthcare organisations within the Malaysian private healthcare sector.

5.1 Derivation of Information Security Compliance Issues

According to the findings shown in Table 1, most of the issues and challenges that were identified and revealed as the main causes of non-compliance which have contributed to the lack of proper enforcement of the information security policies, standards and governance in the healthcare organisations are inexorably linked to the Human or People dimension i.e. the healthcare employees rather than technology, as supported by findings from earlier researches [22, 23]. Secondly, through the responses, it has been generally discovered that many of the healthcare employees based in different healthcare organisations with varying job categories and positions tend to share similar workplace attitude and behavioural characteristics or problems as displayed in Table 1, which leads to non-compliance of information security policies.

Table 1. Identification of information security non-compliance issues in Malaysian private healthcare organisations

Poor work attitude	Lack of management commitment	Malicious intentions
Irresponsible behaviour	Overloaded with job tasks	Habits
Lack of IS awareness	Lack of education and training	Unintentional errors or threats
Information security fatigue	Absence of organisational measures	Compromise
Negligence	Unclear policy guidelines	Non-stringent IS policies
Ignorance	Outdated enforcement of policies	Emotions
Ego/ Superiority/ Pride/ Annoyance	Non-malicious causes	Absence of IS work culture
Lack of motivation	Social Norms	Lack of ethical conduct
Lack of disciplinary actions	Lack of employee monitoring	Lack of safeguards and defences
Lack of involvement	Fear towards technology (phobia)	Carelessness
Poor integrity	Resistance towards change	Complexity of policies

6 Discussion

Information security compliance is not a new phenomenon nor is it an emerging trend but it has only been in the last several years that it has started to gain significant acknowledgement across the major industry players globally. The healthcare domain especially would be exposed to a plethora of vulnerabilities and breaches if compliance of information security is overlooked, increasing the security and privacy risks and threats to healthcare organisations. Therefore, it is imperative that information security compliance must be well-managed through implementing a practical, precautionary approach to privacy and security, which should include technological and more importantly, organisational controls in order to create a valuable strategy that healthcare organisations can apply to manage the threats and breaches.

6.1 Organisational Measures

Among the many administrative measures that were proposed in order to motivate and engage behaviour for achieving compliance, every single respondent collectively agreed that the most crucial one would be a comprehensive and continuous Security Education and Training (SETA) which requires that all healthcare employees be given sufficient education and training [7, 22] especially on IoT information security which ought to continue periodically [6, 22] throughout the employment tenure at their respective organisation. In corresponding to their answers, we would like to further suggest that the organisation should identify the different category of users and learners so that the SETA package could be customized accordingly to better suit the needs and requirements of each group.

Besides providing SETA, we would like to suggest that the employees who suffer from information security fatigue and phobia to undergo motivational talks and counselling sessions with the aim of overcoming their resistance towards compliance of information security. Issues dealing with human attitude and behaviour are never easy to handle and these motivational talks and counselling sessions ought to be conducted discreetly and appropriately as not to offend them further as it could cause unnecessary embarrassment or more consternation among themselves due to the sensitivity of the issue. The counsellors must be very wise, cautious and patient in dealing with this category of end-users so that they would be able to overcome their intense dislike, exasperation or fear with computer systems and devices.

Additionally, having a responsible management which is committed to the information security compliance cause is also another crucial factor in motivating and persuading the healthcare employees [24]. Without proper commitment, it will be very difficult to engage or persuade the workforce to adopt an information security workplace culture as they would tend to think that these information security issues are not the crux of the matter, based on their observation of the non-committal attitude of the management. Provision of incentives and stringent enforcement of appropriate sanctions were also proposed by several of the respondents as a means of shaping, engaging and motivating the healthcare employees' attitude and behaviour to practice compliance [6]. The rewards-based measures would increase the motivation and compliance levels [6, 25] while the sanctions-based measures would instill a sense of caution and could determine the success rate of compliance [6, 25].

Other measures that were proposed and discussed during the interview sessions were providing effective technology support from the IT team, efficient communication channels and allowing more room for employee empowerment.

6.2 Technological Measures

The findings from the preliminary investigation also revealed that the healthcare infrastructure requirements for IoT needs to be strengthened or re-evaluated to ascertain its capability in mitigating the latest malware and other types of cyber-attacks that could exploit or cause damage to the IoT-based medical information.

Therefore, many of the respondents proposed that an advanced malware protection software be installed to ensure the organisation is able to fend itself from internal or external malicious attacks on the IoT medical devices [26]. Another safeguard mentioned by the respondents would be the embedding of a hospital Virtual Private Networks (VPNs) within a private Access Point Network (APN) to add another layer of security for the IoT medical devices [5]. The majority of the respondents also mentioned that healthcare organisations should consider implementing a cutting-edge IoT device encryption technology as an option to safeguard sensitive information such as Personally Identifiable Information (PII) [5, 27] which is often used by healthcare organisations and their working associates such as insurance companies, legal firms and etc. Despite being a relatively simple measure, having robust encryption standards would still create difficulties in decrypting the stored information should the device gets hacked [27]. The micro-segmentation of information through sandboxing is also another measure that was stated by the interview respondents. Sandboxing is considered as the most secured state for a networked IoT medical device because it is completely isolated in a "sandbox" which means the location where it is locked down and secured by the network it's attached to [26].

Furthermore, the respondents also suggested that a two-factor authentication controls on IoT medical devices and security servers such as locking the device with biometrics as well as using a passcode should be enforced to mitigate inadequate security controls. Additionally, the healthcare organisations could also apply remote/automatic lock and wipe capabilities after an excessive number of incorrect login attempts on a particular IoT medical device has been made [27]. A baselining strategy could also be implemented to capture the analytics of access control mechanisms information flow in order to detect unauthorized intrusions [28]. According to the respondents, this measure can help eliminate the Denial of Service (DoS) attacks on the information servers of the IoT platform.

Other potential technological measures that were identified and mentioned by the respondents also included keeping security patches up to date, replace aging legacy networks as they can create significant security vulnerabilities, have appropriate logging and monitoring controls, conduct more frequent vulnerability assessments and penetration testing, perform frequent security risk analysis, testing product data security features, enlisting third-party audits of security procedures [5, 26–28] and etc. Having many layers of technical security measures or safeguards can create enough barriers to help reduce information security threats and vulnerabilities and makes hacking into the system complicated and time-consuming. IoT provides the medical industry with immense benefits in many ways, especially in delivering more efficient healthcare service to the patients as well as reducing expenses. Thus, the healthcare providers and their affiliates should ensure that information security policies and procedures must be enforced strictly in order to protect the healthcare information as well as to achieve compliance.

7 Conclusion

Based on the preliminary investigation that was conducted, this paper presents the initial findings on the causes of information security non-compliance among the employees of the private healthcare organisations in Malaysia. Potential organisational and technological measures as safeguards were further identified by the respondents and proposed to be enforced in order to help raise compliance levels and reduce information security threats and vulnerabilities. Further research could include an empirical study with bigger demographics, as well as, to study the best organisational measures that can effectively lead to better awareness and commitment among the healthcare employees.

References

1. Roman, D.H., Conlee, K.D., Sachs, G.: The Digital Revolution Comes to US Healthcare; Technology, Incentives Align to Shake Up the Status Quo, Internet of Things, vol. 5 (2015)
2. Mora, H., Gil, D., Munoz Terol, R., Azorin, J., Szymanski, J.: An IoT-based computational framework for healthcare monitoring in mobile environments. Sensors (Basel, Switzerland) **17**(10), 2302 (2017). https://doi.org/10.3390/s17102302
3. Dimitrov, D.V.: Medical Internet of Things and Big Data in Healthcare. Healthc. Inform. Res. **22**(3), 156–163 (2016). https://doi.org/10.4258/hir.2016.22.3.156
4. Garg, A.: The Internet of Things: impacts on healthcare security and privacy, Litmos Healthcare Division, Berkeley Research Group (2016). http://www.litmos.com/wp-content/uploads/2016/06/webinar-IoT.pdf
5. Al-Siddiq, W.: IoT innovator, Internet of Things news, medical technology and IoT: mitigating security risks (2018). http://iotinnovator.com/medical-technology-and-the-internet-of-things-addressing-and-mitigating-security-risks/
6. Puhakainen, P., Siponen, M.: Improving employees' compliance through information systems security training: an action research study. MIS Q. (2010). https://doi.org/10.2307/25750704
7. Pham, H.C., Pham, D.D., Brennan, L., Richardson, J.: Information security and people: a conundrum for compliance. Aust. J. Inf. Syst. **21** (2017)
8. Atzori, L., Iera, A., Morabito, G.: The Internet of Things: a survey. Comput. Netw. **54**, 2787–2805 (2010). https://doi.org/10.1016/j.comnet.2010.05.010
9. Gil, D., Ferrandez, A., Mora-Mora, H., Peral, J.: Internet of Things: a review of surveys based on context aware intelligent services. Sensors (Basel, Switzerland) **16**(7), 1069 (2016). http://doi.org/10.3390/s16071069
10. Manyika, J., Chui, M., Bisson, P., Woetzel, J., Dobbs, R., Bughin, J., Aharon, D.: The Internet of Things: mapping the value beyond the hype. McKinsey Global Institute, June 2015
11. Wortmann, F., Fluchter, K.: Internet of Things—technology and value added. Bus. Inf. Syst. Syst. Eng. **57**(3), 221–224 (2015)
12. Kulkarni, A., Sathe, S.: Healthcare applications of the Internet of Things: a review. Int. J. Comput. Sci. Inf. Technol. **5**(5), 6229–6232 (2014)
13. Aliverti, A.: Wearable technology: role in respiratory health and disease. Breathe **13**(2), e27–e36 (2017). https://doi.org/10.1183/20734735.008417
14. Jeong, J.-S., Han, O., You, Y.-Y.: A design characteristics of smart healthcare system as the IoT application. Indian J. Sci. Technol. **9**(37) (2016). https://doi.org/10.17485/ijst/2016/v9i37/102547
15. Mamlin, B.W., Tierney, W.M.: The promise of information and communication technology in healthcare: extracting value from the chaos. Am. J. Med. Sci. **351**(1) (2016)
16. Qi, J., Yang, P., Xu, L., Min, G.: Advanced Internet of Things for personalised healthcare system: a survey. Pervasive Mob. Comput. **41** (2017). https://doi.org/10.1016/j.pmcj.2017.06.018
17. Baker, S.B., Xiang, W., Atkinson, I.: Internet of Things for smart healthcare: technologies, challenges, and opportunities. IEEE Access **5**, 26521–26544 (2017). https://doi.org/10.1109/ACCESS.2017.2775180
18. Harbers, M., van Berkel, J., Bargh, M.S., van den Braak, S., Pool, R., Choenni, S.: A conceptual framework for addressing IoT threats: challenges in meeting challenges. In: Proceedings of the 51st Hawaii International Conference on System Sciences (2018)

19. Talkin Cloud: IoT past and present: the history of IoT, and where it's headed today (2016). http://talkincloud.com/cloud-computing/iot-past-andpresent-history-iot-and-where-its-headed-today?page=2
20. Sicari, S., Rizzardi, A., Grieco, L.A., Coen-Porisini, A.: Security, privacy and trust in IoT: the road ahead. Comput. Netw. 146–164 (2015)
21. Creswell, J.W.: Research Design: Qualitative, Quantitative, and Mixed Methods Approachcs, 4th edn. SAGE Publications, Inc., Thousand Oaks (2013)
22. Safa, N.S., Von Solms, R., Furnell, S.: Information security policy compliance model in organisations. J. Comput. Secur. **56**(C), 70–82. Elsevier Advanced Technology Publications Oxford, UK
23. Fernandez-Aleman, J.L., Sanchez-Henarejos, A., Toval, A., Sanchez-Garcia, A.B., Hernandez-Hernandez, I., Fernandez-Luquec, L.: Analysis of health professional security behaviors in a real clinical setting: an empirical study. Int. J. Med. Inform. **84**, 454–467 (2015)
24. Soomro, Z.A., Shah, M.H., Ahmed, J.: Information security management needs more holistic approach: a literature review. Int. J. Inf. Manag. **36**, 215–225 (2016)
25. Bulgurcu, B., Cavusoglu, H., Benbasat, I.: Information security policy compliance: an empirical study of rationality-based beliefs and information security awareness. MIS Q. **34** (3), 523–548 (2010)
26. CISCO Whitepaper. Cybersecurity in the age of medical devices, connected & protected: an industry point of view (2017). https://www.cisco.com/c/dam/m/digital/healthcare/cybersecurity-in-the-age-of-medical-devices.pdf
27. Wright, KOnramp: Control the risks of IoT and BYOD in healthcare—part II (2017). https://www.onr.com/blog/control-the-risks-of-iot-and-byod-in-healthcare-part-ii/
28. Blowers, M., Iribarne, J., Colbert, E., Kott, A.: The future Internet of Things and security of its control systems (2016). https://arxiv.org/ftp/arxiv/papers/1610/1610.01953.pdf

A Proposed Conceptual Framework for Mobile Health Technology Adoption Among Employees at Workplaces in Malaysia

Hasan Sari[1(✉)], Marini Othman[2], and Abbas M. Al-Ghaili[2]

[1] College of Computer Science and Information Technology, Universiti Tenaga Nasional, 43000 Kajang, Selangor, Malaysia
hassan_sari@yahoo.com

[2] Institute of Informatics and Computing in Energy, Universiti Tenaga Nasional, 43000 Kajang, Selangor, Malaysia
{marini,abbas}@uniten.edu.my

Abstract. With the rising incidence and cost of lifestyle diseases among working population, many organizations are becoming increasingly concerned with health promotion and protection. Although many organizations established workplace health promotion (WHP), the uptake of these interventions is low. The emergence of mobile health technologies (MHTs), such as smartphone health applications and connected wearable devices, provide an excellent opportunity for increasing WHP effectiveness. Despite the potential, the benefits of this initiative could not be fully utilized without understanding the employees' perception towards adoption and use of MHTs. While mHealth acceptance has been studied in different contexts, the adoption among employees at workplace received limited attention. Therefore, this study attempts to bridge the gap in the literature by identifying the factors that influence employees at the workplace to adopt and use MHTs. The study addresses the adoption factors from technology acceptance and health behavior perspectives. The proposed conceptual framework incorporates technology acceptance determinants from UTAUT2: performance expectancy, effort expectancy, social influence, facilitating condition, hedonic motivation and health-related determinants (perceived vulnerability, perceived severity). Furthermore, perceived privacy risk, and personal innovativeness found to influence the behavioral intention. Personal innovativeness (PI) and demographic factors (age & gender) found to moderate the relationship between main constructs, behavioral intention, and the actual use. The finding of the study help both policy makers the application developers to understand the critical factors that might influence employees to use mobile technologies for improving their health and enhancing productivity and performance at work.

Keywords: Mobile health technology · Workplace health promotion
Technology acceptance model · UTAUT2 · Health behavior change

F. Saeed et al. (Eds.): IRICT 2018, AISC 843, pp. 736–748, 2019.
https://doi.org/10.1007/978-3-319-99007-1_68

1 Introduction

The prevalence of non-communicable diseases (NCDs) such as diabetes, obesity, cancer and cardiovascular diseases is becoming a serious challenge to the society and healthcare providers. These diseases (known as lifestyle diseases) are the leading cause of premature deaths among adults in Malaysia [1]. Employers at workplaces facing growing healthcare costs caused by the prevalence of these diseases among their workers. World Health Organization (WHO) declared workplace an ideal place for health protection and health promotion [2]. As a result, many employers have established workplace health promotion (WHP) programs to promote health-related knowledge and behaviors among their workers. Several studies confirmed that continuous and frequent usage of worksite health interventions could influence health-related behaviors in a variety of domains such as promoting physical activity, eating healthy and controlling weight [3].

Although WHP interventions have a significant impact on employees' health and well-being, the uptake reported were modest [4]. This calls for exploring new approaches and methods. Recently, mobile health technologies (MHTs) such as mobile health apps and connected wearable devices provide an excellent opportunity for health promotion in the organizational settings.

MHTs have the potential to provide instant feedback [5], self-monitoring, and personalized content [2]. Thus, the employers could keep track of employee's health behavior, develop healthier habit and improve overall productivity [2]. Despite the potential, the benefits of MHTs could not be fully achieved without employees' recognition and adoption to using them.

Testing acceptability and usability of the end user is a critical prerequisite for implementing health interventions programs. A relevant study on using smartphone apps in workplace health promotion, stated, some issues hinder mobile health app usage for example, lacking technological skills, security and privacy concerns, as well as the quality and credibility of these digital interventions [6]. The acceptance of mobile health technology has been studied amongst different contexts such as patients with chronic diseases [7], younger adults [8], and healthcare professionals [9]. However, insufficient consideration for the adoption of mHealth technology among employees at the workplace. Although previous works on health technology acceptance have successfully investigated the adoption factors from the technology perspective, few of them investigated the health behavior's adoption factors [10, 11].

Therefore, this work aims to bridge the gap in the literature by examining MHTs adoption determinants among the working population from both technology and health behavior perspectives. To achieve that, the study will shed light on technology acceptance and health behavior theories and models, then propose an integrated framework that combines all the possible constructs and relationships between them.

2 Literature Review

2.1 Mobile Health Technologies (MHTs)

Mobile health is a growing area in supporting health-related behavior change [12]. Mobile health (mHealth) defined as "the emerging mobile communications and network technologies for healthcare" [13]. There are thousands of mobile health apps available at major mobile app stores provide low-cost, evidence-based, and high-quality content for managing diseases and health-related behaviors [14, 15].The top categories of mobile health apps are wellness management (such as the apps in lifestyle modification and fitness) and chronic diseases management (such as diabetes and cardiovascular diseases apps) [15]. Wearable devices such as Fitbit, Polar, Apple Watch, and Samsung Gear Fit provide users with data automation and instant feedback [16]. These devices, connected with mobile apps increase the motivation for changing health behavior and improve health management [17].

In this study, we define mobile health technologies (MHTs) as the mobile health applications and the connected wearable devices used by an individual for health promotion and protection. MHTs have the potential to revitalize health promotion and occupational safety in an organizational context. Some companies started using this initiative to overcome the limitation in worksite health promotion [17, 18]. However, the benefits of these technologies cannot be fully utilized without assessing the employee's acceptance and adoption of using them.

2.2 Technology Acceptance Models and Theories

The term "technology acceptance" used to examine different factors that determine the user's intention to adopt a technology and the actual user behavior [19]. Different theories and models are used to explain the acceptance of the technology. The most frequently used model is the Technology Acceptance Model (TAM) [20]. TAM states that perceived usefulness (PU) and perceived ease of use (PEOU) are the determinants of the individual's intention to adopt new technology. Perceived usefulness (PU) refers to "the degree to which a person believes that using a particular system would enhance his or her job performance" [20]. Perceived ease of use (PEOU) refers to "the degree to which a person believes that using a particular system would be free of effort" [20]. An extension model to TAM, named TAM2, has been proposed, which incorporates subjective norm as another significant factor of the user's intention to use the technology [21].

Another widely used theory to explain technology acceptance is the Unified Theory of Acceptance and Use of Technology (UTAUT) [19]. UTAUT merges constructs from several dominant IS theories and models, which are: the theory of reasoned action (TRA), the technology acceptance model (TAM), the motivational model, TAM2, the model of PC utilization, the innovation diffusion theory, and social cognitive theory (SCT). UTAUT suggests that three constructs: performance expectancy, effort expectancy, and social influence act as the direct determinants of users' intention to use the technology. Further studies added the facilitating conditions as an additional predictor of intention. Performance expectancy, which is equivalent to PU in TAM, and refers to "the degree to which an individual believes that using the system will help him or her to attain gains in job performance" [19]. Effort expectancy, which is equivalent to PEOU in TAM, refers to "the degree of ease associated with the use of the system" [19]. Social influence defined as "the degree to which an individual perceives that important others believe he or she should use the new system" [19].

The added construct, facilitating conditions (FC), refers to "the degree to which an individual believes that an organizational and technical infrastructure exists to support the use of the system" [19]. Another extension of UTAUT, known as UTAUT2, was added by [22]. The authors added three more constructs that influence the user's intention to use the technology: hedonic motivation, price value, and habit. The authors also proposed that the effect of the primary determinants on the behavioral intention be moderated by four individual differences, which are age, gender, experience, and habit. Hedonic motivation (conceptualized as perceived enjoyment) refers to the "fun or pleasure derived from using technology" [22].

2.3 Health Theories and Models

The acceptance of health information technology differs from technology acceptance in other domains, as there are further health-related factors that might influence users when they decide to adopt health-related behaviors [11, 23]. Several theories and models used to describe health behavior such as the theory of reasoned action (TRA), protection motivation theory (PMT), health belief model (HBM), and subjective expected utility theory (SEU). However, the most widely used theory and model in health behavior research are PMT and HBM. The health belief model (HBM) proposes that an individual's decision on whether or not to take a health-related action is based on his/her perception about their perceived threat of not taking action and the net benefits of taking action [24].

Perceived threat includes perceived susceptibility and perceived severity, while net benefits are calculated based on perceived benefits and perceived barriers. Perceived susceptibility is defined as an individual's perception of his/her vulnerability to health threats, while perceived severity refers to an individual's assessment of the degree to which a health threat is severe or dangerous [25].

Protection motivation theory (PMT) proposes that the individual's decision to take a specific action is based on their perception of threat appraisals (which include perceived vulnerability and perceived severity) and coping appraisals (which includes response efficacy, response costs, and self-efficacy) [24]. The constructs perceived vulnerability, perceived severity, response efficacy, and response costs are equivalent to perceived susceptibility, perceived severity, perceived benefits, and perceived barriers in HBM, respectively. However, PMT includes a new construct: self-efficacy. Hence, PMT is considered better than HBV for understanding health behavior [26].

3 Methodology

To achieve a theoretical overview, an in-depth investigation carried out to identify relevant articles for mobile health technology adoption in workplace settings. A combination of keywords and phrases were used to collect related articles such as "mobile health", "m-health", "smartphone health apps", and other keywords like "workplace health promotion", "employees", "workers", "acceptance", "adoption", "usage", "success", and "barriers" in all possible combinations. The search was conducted in different online databases: ScienceDirect, Web of Science, IEEEXplore, Emerald, PubMed, and Google Scholar. Also, some reports and statistical studies were included. The selection of the article chosen based on the relevance of factors related to the study. Non-relevant and non-significant articles were excluded. Based on the theoretical overview and analysis, this study proposed a conceptual model of mobile health technology adoption among employees at a workplace the literature review for this study includes both conceptual (non-empirical) and empirical study. Several previous mobile health adoption models were analyzed to select relevant factors as well as their relationship.

4 Related Works

The adoption of mobile health technology in the workplace context received little attention in the literature. [27] Proposed a framework for mobile health application adoption at the workplace. The authors claimed that perceived severity and perceived susceptibility are two critical health-related determinants. On the other hand, the perceived usefulness, perceived ease of use, perceived enjoyment, and facilitating condition are the technology-related determinants of the behavioral intention. However, the proposed framework did not investigate the moderating factors between the main constructs and behavioral intention. Another recent study on construction workers' acceptance of wearable technologies [28] reported: perceived usefulness (PU), social influence (SI), and perceived privacy risk (PR) are associated with workers' intention to adopt wearable devices in their work. The study examined the moderating role of wearable technology experience between the main predictors and behavioral intention.

Several studies investigated mobile service adoption among the public, those studies based on TAM model. PU and PEOU found to be the most important determinants for behavioral intention [29, 30].

Other studies examined mHealth adoption based on UTAUT [31, 32]. Those studies stated that the main determinants of mHealth adoption are performance expectancy, effort expectancy, social influence, and facilitating conditions. However, a few of those studies addressed the health-related factors and proposed that threat appraisals is a predictor for mHealth adoption among public users.

Some other studies examined the factors based on UTAUT2 [33, 34] and proposed that: performance expectancy, effort expectancy, social influence, hedonic motivation, and habit are a predictor for mHealth apps adoption.

One other notable factor related to technology adoption is the privacy concern. Perceived privacy influences workers' intention to use wearable devices in occupational health safety [28]. Privacy concern hurts user's intention to adopt mobile technology [35]. Perceived privacy risk (PR) has been proposed as a critical determinant for mHealth adoption [36, 37]. [38] Claimed that Personal innovativeness has a moderating effect on the four UTAUT determinants factors (performance expectancy, effort expectancy, social influence, and facilitating conditions). Employees' experience on using wearable devices could moderate the association between technology determinants and the behavioral intention to adopt wearable devices in the organizational context [28].

Table 1. List of the most frequent related constructs

Factor name	Underpinning Theory/Model	Author (s)	Total
Performance Expectancy (PE) equivalent to perceived usefulness (PU)	UTAUT	[32–34, 38, 42, 43]	15
	TAM	[8, 10, 27–31, 42, 44]	
Effort Expectancy (EE). equivalent to Perceived Ease of Use (PEOU)	UTAUT	[11, 32, 33, 38, 43]	13
	TAM	[10, 27, 29, 30, 34, 42, 44]	
Social Influence (SI) equivalent to Subjective Norm (SN)	UTAUT	[11, 28, 31–33, 38, 43, 45]	10
	TAM 2	[30, 42]	
Hedonic Motivation (HM) or Perceived Enjoyment.	UTAUT 2	[27, 33]	2
Perceived Privacy Risk (PR)		[28, 32, 36, 37]	4
Facilitating Conditions (FC) equivalent to self-efficacy (SE) and Response Cost (RC)	UTAUT	[11, 27, 38]	4
	PMT	[44]	
Perceived Vulnerability (PV) equivalent to perceived susceptibility	PMT	[11, 42]	3
	HBM	[27]	
Perceived Severity (PS)	PMT, HBM	[11, 27, 42]	3
Personal innovativeness (PI)		[28, 31, 32, 38]	4
Age, Gender	UTAUT 2	[23, 39]	2

Few studies addressed the moderating role of demographic differences (e.g., gender and age). [23] Claimed that young adults have a positive attitude towards using new technologies, whereas older adults often show a slower technology evolvement. Different impacts on technology adoption among different age groups have also been reported [39]. A study conducted by [23] stated that age differences have a noticeable moderating impact on threat appraisal and coping appraisal (PMT constructs). Some research reported men are more interested in technology than women [40], whereas women are more concerned about health related conditions and tend to seek medical advice and preventive treatment compared to men [41]. Table 1 lists the most frequent factors investigated and related theories in the previous studies.

5 Conceptual Framework

This study proposes the conceptual framework based on the literature review and the recent related works. The proposed framework incorporates technology acceptance factors, health behavior factors, and other related factors. Table 2 illustrates the constructs and the definition for each of them. Figure 1 illustrates the proposed conceptual framework.

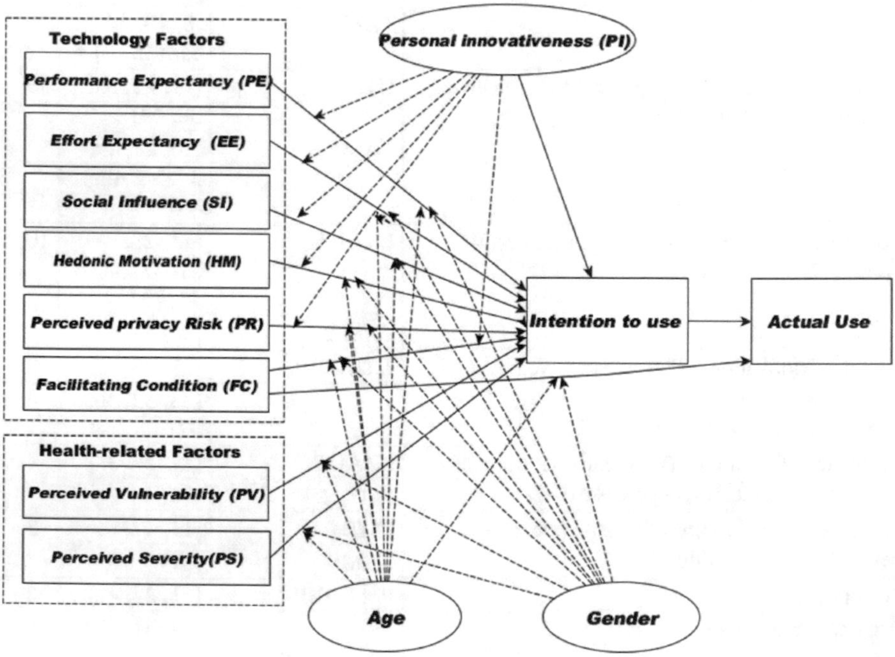

Fig. 1. The proposed conceptual framework

Table 2. The main constructs of the proposed framework

Construct	Definition	Theories/Models
Behavioral Intention	The degree to which a person has formulated conscious plans to perform or not perform some specified future behavior	UTAUT
Performance Expectancy (PE)	The degree to which an individual believes that using the system will help him or her to attain gains in job performance	UTAUT
Effort Expectancy (EE)	The degree of ease associated with the use of the system	UTAUT
Social Influence (SI)	The degree to which an individual perceives that important others believe he or she should use the new system	TPB & UTAUT
Facilitating Condition (FC)	The degree to which an individual believes that an organizational and technical infrastructure exists to support the use of the system	UTAUT
Hedonic Motivation (HM)	The fun or pleasure derived from using technology	UTAUT 2
Perceived Vulnerability (PV)	The judgment that one will feel that his or her health is being threatened	PMT
Perceived Severity (PS)	The measurement of one is perceived health risks if he or she will not adopt mobile health services	PMT
Perceived privacy Risk (PR)	The extent to which an individual believes that using mobile health service is secure and has no privacy threats	
Personal Innovativeness (PI)	The willingness of an individual to try out any new IT plays an important role in determining the outcomes of user acceptance of the technology	IDT

6 Discussion and Findings

This study adopts the most critical user acceptance and behavior analysis models: the Unified Theory of Acceptance and Use of Technology (UTAUT2). UTAUT is widely used in different technology adoption areas including eHealth and mHealth. It is broadly accepted that the UTAUT model could explain as much as 70% of the variance in intention [46]. The study also incorporates the health behavior theories to address health-related factors affecting MHTs adoption. Health behavior is defined in the literature as "any activity undertaken by an individual, regardless of actual or perceived health status, for the purpose of promoting, protecting or maintaining health." [47].

As a result, the study on mobile health technology adoption should consider the health-related behavior constructs. HBM and PMT are the most used health behavior theories, but PMT is more widely used in mobile health adoption [11, 42]. Thus, this study incorporated the concept of "threat appraisal" which consists of perceived vulnerability (PV) and perceived severity (PS). The other concept in PMT is "coping appraisal", which consists of response efficacy, self-efficacy, and response cost.

Self-efficacy and response cost in PMT can be regarded as facilitating conditions in UTAUT and response efficacy can be seen as performance expectancy [11].

We also propose that, personal innovativeness moderate the relationship between the technology constructs and behavioral intention and have direct influence on behavioral intention. Moreover, the demographic differences such as age and gender moderate the relationship between the main constructs from UTAUT2 and PMT. Table 1 lists all the constructs in previous studies. We grouped each equivalent construct under one category. For instance, performance expectancy (PE) in UTAUT is equivalent to perceived usefulness in TAM and TAM2, effort expectancy is equivalent to perceived ease of use (PEOU) in TAM and TAM2, and social influence in UTAUT is equivalent to subjective norm in TAM2. In health technology constructs, perceived vulnerability (PV) in PMT is equivalent to perceived susceptibility (PS) in HBM. Table 2 lists all the main constructs, underpinning theories/models, and the definition for each of them.

7 Implications

For the successful implementation of MHT in the context of the workplace, policy-makers and mobile health developers must be aware of the factors that influence employees' intention and adoption to using this technology. The present study has theoretical and practical implications.

First, about the theoretical implications, the current study provides an integrated model covering both technological and health factors which have a direct influence on employees' perception towards MHT.

Second, the current study has practical implications for both managers and mobile health interventions developers in the context of the workplace.

Second, about the practical implication, the present study provides deep understanding of the importance of mobile health technologies and the possibility to integrate them in the workplace health promotion interventions. Integrating MHT in worksite health promotion will open a new opportunity for designing of low- cost interventions for improving employees' health outcome and enhancing their productivity and performance at work.

For the application developers, the study provides insights into understanding the features and behavior change techniques that should be taking in consideration while designing and building context-aware mobile health interventions.

8 Conclusion and Further Remarks

Employee's health and well-being are importance for both employees and organizations. Many organizations established workplace health interventions to improve their workers' health.

This in-progress research has successfully identified the most significant factors of MHT adoption among workers and proposes a conceptual model based on technology and health behavior theories. Based on the review and related studies, we identify a list of technological health factors which fund to influence employees adoption and use of mobile health technologies. The finding of this work will help both applications developers and policy makers in the workplace to implement a cost-effective and scalable approach to better monitor the workers' health and to address the lower in productivity and performance which driven from workers with unhealthy condition and status.

Future studies are required to examine and validate the proposed framework statistically, as well as to understand the relationship between the different factors.

Acknowledgment. The Funding for this work was provided by the Innovation & Research Management Center (IRMC) at Universiti Tenaga Nasional, project title "Mobile Health Technology Adoption Model at Workplace: the Case of TNB, Malaysia", and project number J510050743.

References

1. National Strategic Plan for Non-Communicable Disease NSP_NCD 2016–2025
2. Balk-Møller, N.C., Larsen, T.M., Holm, L.: Experiences from a web-and app-based workplace health promotion intervention among employees in the social and health care sector based on use-data and qualitative interviews. J. Med. Internet Res. **19**(10), 350–3501 (2017)
3. Bardus, M., Blake, H., Lloyd, S., Suzanne Suggs, L.: Reasons for participating and not participating in a e-health workplace physical activity intervention. Int. J. Work Heal. Manag. **7**(4), 229–246 (2014)
4. Mattila, E., Orsama, A.L., Ahtinen, A., Hopsu, L., Leino, T., Korhonen, I.: Personal health technologies in employee health promotion: usage activity, usefulness, and health-related outcomes in a 1-year randomized controlled trial. J. Med. Internet Res., **15**(7) (2013)
5. Zhao, J., Freeman, B., Li, M.: Can mobile phone apps influence people's health behavior change? An evidence review. J. Med. Internet Res. **18**(11), 1–12 (2016)
6. Dunkl, A., Jimenez, P.: Using smartphone-based applications (apps) in workplace health promotion: the opinion of German and Austrian leaders. Health Inform. J. **23**(1), 44–55 (2017)
7. Siddiqui, M., Ul Islam, M.Y., Mufti, B.A.I., Khan, N., Farooq, M.S., Muhammad, M.G., Osama, M., Kherani, D., Kazi, A.N., Kazi, A.M.: Assessing acceptability of hypertensive/diabetic patients towards mobile health based behavioral interventions in Pakistan: a pilot study. Int. J. Med. Inform. **84**(11), 950–955 (2015)
8. Lim, S., Xue, L., Yen, C.C., Chang, L., Chan, H.C., Tai, B.C., Duh, H.B.L., Choolani, M.: A study on Singaporean women's acceptance of using mobile phones to seek health information. Int. J. Med. Inform. **80**(12), e189–e202 (2011)
9. Koehler, N.: Healthcare professionals' use of mobile phones and the internet in clinical practice. J. Mob. Technol. Med. **2**(1), 3–13 (2013)

10. Zhu, Z., Liu, Y., Che, X., Chen, X.: Moderating factors influencing adoption of a mobile chronic disease management system in China. Inform. Health Soc. Care **00**(00), 1–20 (2017)
11. Sun, Y., Wang, N., Guo, X., Peng, Z.: Understanding the acceptance of mobile health services: a comparison and integration of alternative model. J. Electron. Commer. Res. **14** (2), 183–201 (2013)
12. Kumar, S., Nilsen, W.J., Abernethy, A., Atienza, A., Patrick, K., Pavel, M., Riley, W.T., Shar, A., Spring, B., Spruijt-Metz, D., Hedeker, D., Honavar, V., Kravitz, R., Craig Lefebvre, R., Mohr, D.C., Murphy, S.A., Quinn, C., Shusterman, V., Swendeman, D.: Mobile health technology evaluation: the mHealth evidence workshop. Am. J. Prev. Med. **45** (2), 228–236 (2013)
13. Istepanian, R.S.H., Jovanov, E., Zhang, Y.T.: Introduction to the special section on m-health: beyond seamless mobility and global wireless health-care connectivity. IEEE Trans. Inf. Technol. Biomed. **8**(4), 405–414 (2004)
14. Fogg, B.: A behavior model for persuasive design. In: Proceedings of the 4th International Conference on Persuasive Technology—Persuasive 2009, p. 1 (2009)
15. Kao, C.-K., Liebovitz, D.M.: Consumer mobile health apps: current state, barriers, and future directions. Pm&R **9**(5), S106–S115 (2017)
16. Weichelt: Health in your hand: assessment of clinicians' readiness to adopt mHealth into rural patient, October (2016)
17. Yu, J., Abraham, J.M., Dowd, B., Higuera, L.F., Nyman, J.A.: Impact of a workplace physical activity tracking program on biometric health outcomes. Prev. Med. (Baltim.) **105**, 135–141 (2017)
18. Simons, D., De Bourdeaudhuij, I., Clarys, P., De Cocker, K., Vandelanotte, C., Deforche, B., al Jmir Mhealth, Xsl, U., RenderX, B.F.: A Smartphone App to Promote an Active Lifestyle in Lower-Educated Working Young Adults: Development, Usability, Acceptability, and Feasibility Study, vol. 6, no. 2 (2018)
19. Venkatesh, V., Morris, M., Davis, G., Davis, F.: User acceptance of information technology: toward a unified view. MIS Q. **27**(3), 425–478 (2003)
20. Davis, F.D.: Perceived usefulness, perceived ease of use, and user acceptance of information technology. Source MIS Q. **13**(3), 319–340 (1989)
21. Venkatesh, V., Davis, F.D.: A theoretical extension of the technology acceptance model: four longitudinal field studies. Manag. Sci. **46**(2), 186–204 (2000)
22. Venkatesh, V., Thong, J., Xu, X.: Consumer acceptance and user of information technology: extending the unified theory of acceptance and use of technology. MIS Q. **36**(1), 157–178 (2012)
23. Guo, X., Han, X., Zhang, X., Dang, Y., Chen, C.: Investigating m-health acceptance from a protection motivation theory perspective: gender and age differences. Telemed. JE Health **21** (8), 661–669 (2015)
24. Becker, M.H.: The health belief model and personal health behavior. Health Educ. Monogr. **2**, 324–473 (1974)
25. Becker, M.H., Maiman, L.A.: Strategies for enhancing patient compliance. J. Community Health **6**(2), 113–135 (1980)
26. Prentice-Dunn, S., Rogers, R.W.: Protection motivation theory and preventive health: Beyond the health belief model. Health Educ. Res. **1**(3), 153–161 (1986)
27. Melzner, J., Heinze, J., Fritsch, T.: Mobile health applications in workplace health promotion: an integrated conceptual adoption framework. Procedia Technol. **16**, 1374–1382 (2014)

28. Choi, B., Hwang, S., Lee, S.H.: What drives construction workers' acceptance of wearable technologies in the workplace?: Indoor localization and wearable health devices for occupational safety and health. Autom. Constr. **84**(2016), 31–41 (2017)
29. Deng, Z., Zhang, J., Zhang, L.L.: Applying technology acceptance model to explore the determinants of mobile health service : from the perspective of public user. In: Wuhan International Conference on E-business, pp. 406–411 (2012)
30. Cho, J., Quinlan, M.M., Park, D., Noh, G.Y.: Determinants of adoption of smartphone health apps among college students. Am. J. Health Behav. **38**(6), 860–870 (2014)
31. Cilliers, L., Viljoen, K.L.-A., Chinyamurindi, W.T.: A study on students' acceptance of mobile phone use to seek health information in South Africa. HIM J. (2017). https://doi.org/10.1177/1833358317706185
32. Hoque, R., Sorwar, G.: Understanding factors influencing the adoption of mHealth by the elderly: an extension of the UTAUT model. Int. J. Med. Inform. **101**, 75–84 (2017)
33. Woldeyohannes, H.O., Ngwenyama, O.K.: Factors influencing acceptance and continued use of mHealth apps. **10294**, 239–256 (2017)
34. Krishnan, S.B., Nasional, U.T., Dhillon, J.S., Nasional, U.T., Lutteroth, C., Zealand, N.: Factors Influencing Consumer Intention to Adopt Consumer Health Informatics Applications, pp. 653–658 (2015)
35. Li, H., Wu, J., Gao, Y., Shi, Y.: Examining individuals' adoption of healthcare wearable devices: an empirical study from privacy calculus perspective. Int. J. Med. Inform. **88**, 8–17 (2016)
36. Gao, Y., Li, H., Luo, Y.: An empirical study of wearable technology acceptance in healthcare. Ind. Manag. Data Syst. **115**(9), 1704–1723 (2015)
37. Khatun, F., Hanifi, S.M.A., Iqbal, M., Rasheed, S., Rahman, M.S., Ahmed, T., Hoque, S., Sharmin, T., Khan, N.U.Z., Mahmood, S.S., Peters, D.H., Bhuiya, A.: Prospects of mHealth services in Bangladesh: recent evidence from Chakaria. PLoS ONE **9**(11), e111413 (2014)
38. Okumus, B., Ali, F., Bilgihan, A., Ozturk, A.B.: Psychological factors influencing customers' acceptance of smartphone diet apps when ordering food at restaurants. Int. J. Hosp. Manag. **72**, 67–77 (2018)
39. Lv, X., Guo, X., Xu, Y., Yuan, J., Yu, X.: Explaining the mobile health services acceptance from different age groups: a protection motivation theory perspective. Int. J. Adv. Comput. Technol. **4**(3), 1–9 (2012)
40. Adas, M.: Machines as the measure of men: science, technology and the ideologies of western dominance. IEEE Technol. Soc. Mag. **12**(3), 4–6 (1993)
41. Fitzgerald, J.T., Anderson, R.M., Davis, W.K.: Gender differences in diabetes attitudes and adherence. Diabetes Educ. **21**(6), 523–529 (1995)
42. Zhao, Y., Ni, Q., Zhou, R.: What factors influence the mobile health service adoption? A meta-analysis and the moderating role of age. Int. J. Inf. Manag., August 2017
43. Quaosar, G.M.A.A., Hoque, M.R., Bao, Y.: Investigating factors affecting elderly's intention to use m-health services: an empirical study. Telemed. e-Health **24**(4), 111 (2017)
44. Zhang, X., Han, X., Dang, Y., Meng, F., Guo, X., Lin, J.: User acceptance of mobile health services from users' perspectives: the role of self-efficacy and response-efficacy in technology acceptance. Inform. Health Soc. Care **42**(2), 194–206 (2017)
45. Becker, D.: Acceptance of mobile mental health treatment applications. Procedia Comput. Sci. **58**, 220–227 (2016)

46. Nuq, P.A., Aubert, B.: Towards a better understanding of the intention to use eHealth services by medical professionals: the case of developing countries. Int. J. Healthc. Manag. **6** (4), 217–236 (2013)
47. Nutbeam, D.: Health promotion glossary. Health Promot. Int. **1**(1), 113–127 (1986)
48. Tsu Wei, T., Marthandan, G., Yee-Loong Chong, A., Ooi, K., Arumugam, S.: What drives Malaysian m-commerce adoption? An empirical analysis. Ind. Manag. Data Syst. **109**(3), 370–388 (2009)
49. Agarwal, R., Prasad, J.: A conceptual and operational definition of personal innovativeness in the domain of information technology. Inf. Syst. Res. **9**(2), 204–215 (1998)

Problematic of Massively Multiplayer Online Game Addiction in Malaysia

Muhammad Muhaimin[1](✉), Norshakirah Aziz[2], and Mazeyanti Ariffin[3]

[1] Centre for Research in Data Science (CERDAS), Computer and Information Sciences Department, Universiti Teknologi PETRONAS, 31750 Tronoh, Perak Darul Ridzuan, Malaysia
muhammad_16000052@utp.edu.my

[2] High Performance Cloud Computing Centre (HPC3), Computer and Information Sciences Department, Universiti Teknologi PETRONAS, 31750 Tronoh, Perak Darul Ridzuan, Malaysia
Norshakirah.aziz@utp.edu.my

[3] Computer and Information Sciences Department, Universiti Teknologi PETRONAS, 31750 Tronoh, Perak Darul Ridzuan, Malaysia
Mazeyanti@utp.edu.my

Abstract. This study aimed to find causes of online game addiction among adolescents in Malaysia specifically focus on massively multiplayer online games (MMOG) in Malaysia. The relationship between causes and the addiction components will be investigated. This study used focus group discussion (FGD) as research methodology. Results showed that 3 main causes have been revealed which include; poor time management in which adolescents failed to manage their time wisely, social life in which they make more friends in virtual life compared to in real life and psychological and behavior which affect adolescent's emotion and physical. Addiction components that had been discussed are linked to 3 main factors identified from previous study. The identified components that found influence gaming addiction among adolescents are; salience, mood modification, harm, loss of control and physical health are related to main causes of adolescents addicted to MMOG in Malaysia. Evidences that have been revealed by both medical experts and game experts had explored the highly engaging relationship between adolescent and MMOG addictive behavior which can be seen by observation through physical health.

Keywords: Online gaming · Adolescents · Addiction · MMOG FGD

F. Saeed et al. (Eds.): IRICT 2018, AISC 843, pp. 749–760, 2019.
https://doi.org/10.1007/978-3-319-99007-1_69

1 Introduction

MMOG is a popular type of online game which is capable to support vast number of players playing at one time. Through this era of globalization, adolescents tend to play video games by using online connection and communicate at the same time. This is called "relationship network" which can be considered as a new platform in making friends and not only for entertainment. Adolescents who are addicted to MMOG will reveal several signs of behavior. The evolution of computer games has shown the tendency of people paying for computer games has increased year by year. For instance, some people are addicted to watch cartoon on television, play games on smartphones, watch online Korean dramas or movies, play video games using PlayStation, Nintendo Wii, and Xbox. In addition, the highest digital addiction was found to be in computer games especially in MMOG. MMOG has become most people's choice of computer game in this era. The innovative technology of computer games has allowed people to communicate via online internet with other people and the three-dimension resolution has attracted more people to play MMOG. As a result, people are addicted to online MMOG games rather than other form of digital games. However, the addiction to computer games, especially MMOG was found to be difficult in finding recovery model due to it is related to human's behavior and psychological aspects. Besides, (Fioravanti et al. 2012) it has been proved that there are no recognized clinical entity/diagnosis assessment tools to detect the addiction of adolescent in massively multiplayer online game (Tzavela et al. 2017). The updated DSM-5 result has shown that the screening instrument lacking theoretical basis, existing assessment tools and procedure failed to consider the complex underlying process that linked to high online engagement. Besides, none of the practices that have been recommended tailored with adolescent (Snodgrass et al. 2017; Van Rooij et al. 2011).

The capability to access the Internet has made a vast number of adolescent in Malaysia addicted to MMOG, which sees them spend more time playing and connected with people online without limit. As a result, this will affect their real-world environment relationship, the way of thinking or behavior towards other things and results in poor time management. In addition, majority of young adults who addicted to the Internet may have been driven by some socio-economic factor such as to communicate on important matters, getting sex-oriented materials, and making money (Adiele and Olatokun 2014). This has been proved with other negative impact that drive players to excessive gaming which refers to relationship environment in game platform that provoke a sense of awe and wander. The second is online sexual preoccupation which can be seen by online viewing, downloading adult content materials for trading purposes and the third is for emailing and texting to express their feelings for those who are addicted has suffered from loneliness, lack of confidence and social retreat (Kapahi et al. 2013).

1.1 Motivation of the Study

As reported by Malaysian Communications and Multimedia Commission (MCMC), the current statistic shows that Malaysian Internet users represent approximately 24.1 million (77.6%) of the total population of 31.7 million (MCMC 2016). About 76.1% of the Internet users are youth/adolescent with 43.7% or 13.8 million of them playing online computer games (MCMC 2016). These adolescents are reportedly spending more than 17.9 h a week playing online computer games. Due to the increasing percentage of adolescents playing online computer games, there is a need to investigate further on the impacts of MMOG addiction on Malaysian adolescents. The aim is to identify the factor and addiction components among adolescents in Malaysia to figure out the necessary controls and treatments for their addiction.

2 Literature Review

It was discovered further that approximately 2.4 million Malaysians are involved as e-sports enthusiasts and has contributed to approximately USD 589.4 million revenues to the global game market (Pannekeet 2016). This has made Malaysia being listed as one of the fastest growing regions in e-sports. There are 191 million e-sports enthusiasts around the world. E-sports is known as a tournament for interactive video games that shows gaming skill and professionalization (Pannekeet 2016). Gaming addicts may be finding escape from the problems they cannot handle through computer games. There are many individuals who cannot manage the stress and pressure they face, so they go driving, shopping or do extreme sports.

2.1 Internet Game Disorder

MMOGs are video games that are played online with a highly graphical 2D or 3D form. It allows players to interact with their characters not only with the gaming software but with other players' character as well. Characters in games are self-created digital characters to represent the players. There are millions of people playing online-games without any serious effects on their lives but somehow a minority of players seem to have problems with a healthy amount of gaming hence American Psychiatric Association (APA) introduced the "Internet Gaming Disorder" in latest edition of the Diagnostic and Statistical Manual of Mental Disorders (DSM5).

Internet Game Disorder (IGD) which is most related to quality of life (QOL) and cognitive dysfunction has increasingly growing and becoming a social problem (Lim et al. 2016). The most exciting parts in online gaming are power of the characters, developing their game characters from a beginner to an expert level and reward achieved after winning or advancing to the next level has led to negative impact on game social. A recent study by (Leménager et al. 2016) reports that addiction of online computer games has affected players' brain activation leading to the changes in psychology and behavior. Obsession towards internet gaming can slowly lead to the right brain giving negative social feedback that leads to reduced self-satisfaction.

2.2 Digital Addiction Factors and Components

The term DA is referring to the behavioral phenomenon related to obsessive and problematic usage of digital media (Alrobai et al. 2016) MMOG is listed as DA. The evolution of games has shown an increase of players investing their time playing games for long hours. This addiction is negatively affecting human behavior in terms of psychology, time management and social culture (Alrobai et al. 2016).

Digital addiction can be classified as recent technology addiction of playing, watching or listening by using electronic or machine to play for entertainment purpose. A longitudinal study of MMOG addiction by (Latif et al. 2014) reports that six important predictors have been found which involves 50% of gaming costs expenditures, 23% involves average weekday gaming time, 13% offline internet gaming community meeting attendance, average weekend and holiday gaming time with 7%, marital status 4% and self-perception of addiction to internet game use with 3%. Based on this result, there are three factors that have been identified that led to addiction of massively multiplayer online game (MMOG): social life, psychology and behavior and time management (Fig. 1).

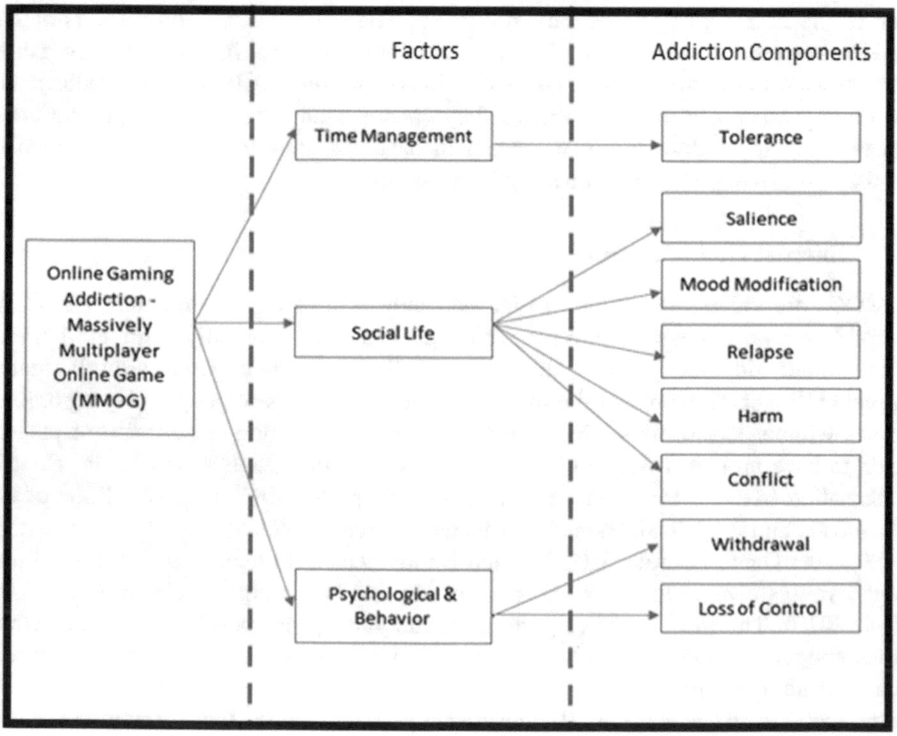

Fig. 1. Initial MMOG Addiction model- association between factors and addiction components

Based on extensive literature review, one initial addiction model was designed as illustrated in Fig. 2 above. It was also found that online game players' attitudes and actual usage are affected by the flow experience, enjoyment, and interaction in the games. Enjoyment and social influence are becoming convincing evidence to measure actual game playing. (Alzahrani et al. 2016).

Undeniably, the capability to access the Internet has made the adolescents in Malaysia addicted to MMOG, causing them to spend more time playing and connecting with people in the virtual world without limit. As a result, it affects their real-world relationships (social life), way of thinking and behavior towards others (psychological and behavior), leading to negative consequences such as poor time management. The table below shows eight addiction components consistently mentioned from previous research. This preliminary stage referred the literature from year 2006 to year 2018. Total 196 papers have been reviewed and found related to this research. There are eight addiction components identified from the extensive literature review. The addiction components relation with the three factors identified presented in Table 1 below.

Table 1. Addiction components relation with factors

	Addiction Components	Definition	Remarks
1.	Relapse	Refer to tendency of the player to make a repeated game (Caplan et al. 2009; Khan and Muqtadir 2014; Lee et al. 2014; Lemmens et al. 2017)	Physical and behaviour (Factor 3)
2.	Mood Modification	Changes in players mood (happy, calm or moody) (Dieris-Hirche et al. 2017; Khan and Muqtadir 2014; Lee et al. 2014; Lemmens et al. 2017; Van Rooij et al. 2011)	Physical and behaviour (Factor 3)
3.	Withdrawal	Behaviour of player if suddenly they stop playing MMOG, including frustration, and moody (Dieris-Hirche et al. 2017; Khan and Muqtadir 2014; Lee et al. 2014; Lemmens et al. 2017; Van Rooij et al. 2011)	Physical and behaviour (Factor 3)
4.	Tolerance	More time has been allocated to play MMOG compared to sleep time (Dieris-Hirche et al. 2017; Khan and Muqtadir 2014; Lee et al. 2014; Lemmens et al. 2017; Van Rooij et al. 2011)	Time management (Factor 1)
5.	Salience	Related to emotions and behaviour of the player which means playing MMOG is the most important thing in their life (Dieris-Hirche et al. 2017; Khan and Muqtadir 2014; Lee et al. 2014; Lemmens et al. 2017; Van Rooij et al. 2011)	Psychological and Behaviour (Factor 3)

(*continued*)

Table 1. (*continued*)

	Addiction Components	Definition	Remarks
6.	Loss of control	Player of MMOG unable to control themselves in terms of time, fail to schedule their daily activities (Dieris-Hirche et al. 2017; Khan and Muqtadir 2014; Lee et al. 2014; Lemmens et al. 2017; Van Rooij et al. 2011)	Time Management (Factor 1)
7.	Harm	Refer to the culture or negative impact to other people, less communication and create more conflict between family, friends and working environment (Caplan et al. 2009; Khan and Muqtadir 2014)	Social life (Factor 2)
8.	Conflict	Extent of excessive behaviour arises both between the addicted person, fight with online friends while playing MMOG (Dieris-Hirche et al. 2017; Khan and Muqtadir 2014; Lee et al. 2014; Lemmens et al. 2017; Van Rooij et al. 2011)	Social life (Factor 2)

3 Methodology

3.1 Focus Group Discussion-FGD

FGD was considered to be the most suitable method for this research. The FGD sessions involved 2 groups which consisted of medical doctors group and professional gamers group.

3.2 Selection of FGD

Medical Expert's Profile

In this research, the first group for FGD consists of medical experts that have dealt with Malaysian adolescents who are addicted to online game. In this group, the medical expert's minimum qualification are at least more than 10 years in government Hospital, as Consultant Child and Adolescent Psychiatrist. The Medical experts group leader chosen based on recommendation of a director of a Malaysian Government Hospital, has an excellent record and consistently being awarded by Malaysian Government for her achievement, hence leading to the decision of nominating her as our leader in FGD team. This study was conducted through six FGD sessions with the medical experts to discuss in deep on factors and addiction components of MMOG among adolescents, besides to confirm and verify what are the addiction components existed in Malaysia. The medical expert's participants also have main interest in:

1. Internet and game addiction,
2. Infant Mental Health,
3. Parent-child relationship and attachment issues,
4. Early childhood emotional development, and
5. Children and Trauma.

One of the participants in medical experts group is the president of Malaysian Society of Internet Addiction Prevention (MYSIA). She shared her fifteen years of experience dealing with adolescents who are addicted to online game. This group were asked on definition of addiction, differences between game addiction and other addiction i.e. drugs, alcohol and gambling.

Game Expert's Profile
The second group was a group that consists of experts in computer game area, and all the game experts contributed to verify and confirm on MMOG addiction components that were found in the preliminary study phase. This group minimum qualification was at least more than ten years' experience of playing MMOG. The game experts shared their experience of playing MMOG and experience of joining MMOG competition. All the game experts in this group were members of the only team who represented Malaysia for DOTA 2. Game experts in this group was known as "Fnatic" Malaysia team. They had three tremendously successful major campaigns including a fourth-place finish at The International DOTA 2 in 2017. In this research, they were responsible to explain how the addiction components impact Malaysian adolescents. All responses from the game experts were important to see how MMOG impact adolescents in Malaysia. The detail research flow is shown in Fig. 2.

Figure 2 shows research activities that are involved in this study. In the beginning of this study, extensive literate review is performed to identify factors of addiction and its components. Then, all data collection will be confirmed and verify by medical experts which are psychiatrics and game experts which are professional game experts. The results from FGD sessions done with both experts will identify factors of addiction and addiction components that are found in Malaysia. Trans-scripted FGD result will be analyzed using Atlas.TI 8. From identification factors of addiction and addiction components in Malaysia, a network diagram will be created. The results will be used to design MMOG addiction recovery model.

The preliminary research of this study allows researchers to understand the challenges, causes, and impacts of MMOG on Malaysia adolescents. The three main factors identified by previous researchers (social life, psychological and behaviour, and time management) are also referred. This study details out the components influencing the identified factors and helps to design MMOG addiction recovery model to benefit the identified MMOG addicts.

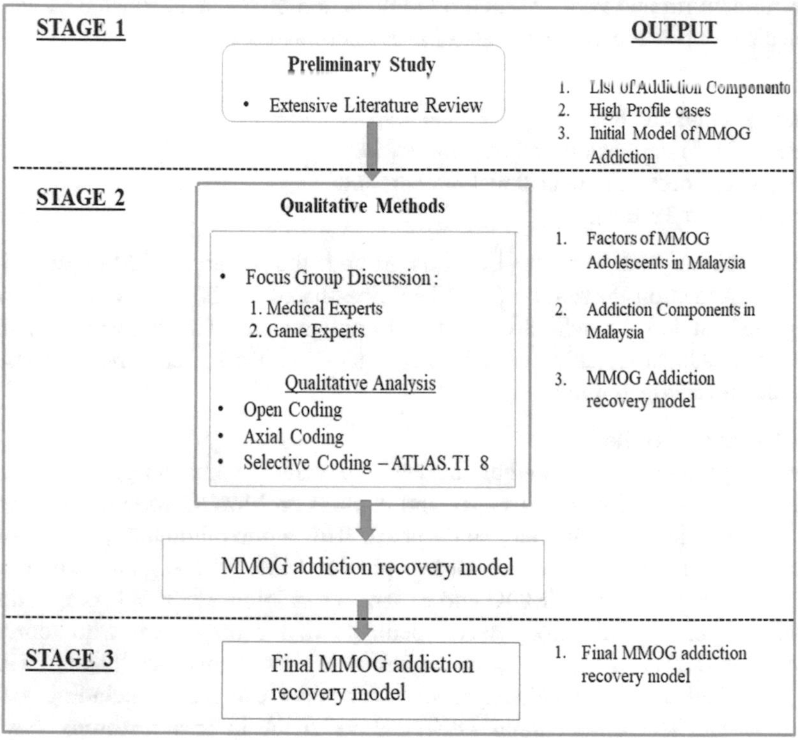

Fig. 2. Research flow

4 Results and Discussion

4.1 Focus Group Discussion Result - FGD 1 and FGD 2

According to the medical experts, they have difficulties to identify main reasons and causes of addiction to online games from players. MMOG players who are addicted will meet psychiatrics because they have difficulties to sleep, attention loss and experiencing sight problems. After in-depth interview with some of the players, they have admitted that the main reason behind all these descriptions was because they were addicted to playing online games. The medical experts have confirmed and verified that all the components strongly exist in Malaysia through relationship diagram but the most obvious are 4 factors listed in Table 2.

Table 2. Critical Addiction Components in Malaysia

	ADDICTION COMPONENT	DEFINITION	LEVEL
1.	Salience	Addiction of MMOG has become one of the most important part in an adolescent's life and it has controlled the attention and craving behavior to play MMOG non-stop	Critical
2.	Mood Modification	The effects that adolescents feel while playing MMOG; happy and satisfied if they achieve or advance to the next level or win a match. Feel sad if they lose the MMOG game	Critical
3.	Harm	Referring to the culture of the negative impact with surrounding people, less communication and create more conflicts between family, friends and working environment	Critical
4.	Loss of control	Adolescents loss of control in time management, more time spent playing MMOG rather than doing outdoor activities which can lead to poor academic achievement, failure to control their behavior while playing MMOG until negatively impacting other people, family and friends	Critical

The team leader of medical experts confirms that new addiction components should be added to the list; which is the Physical component. Physical component is very important, and it is related to:

1. Obesity – adolescent who gain weight consistently and continue eating while playing games
2. Back and neck pain – adolescents who addicted to games was observed having difficulties to stand longer and prefer to sit at one place and continue playing games
3. Orthopedic/joint muscle – because of prolonged usage of keyboard
4. Eye sights – eye sight problems
5. Hearing problems – play games using headphone loudly
6. Less physical activity – adolescents decided to spend more activities in a room instead of going for outdoor activities

New cases also have been revealed by medical experts that parents are reporting their kids have skipped school for about 1 ½ years. The medical expert also added that some of them even went to the extent of throwing things towards their parents when they tried to stop their kids, and this is considered a critical stage of MMOG playing addiction.

According to Malaysia Dota team, from FGD 2 (who have played Dota for 11 years), they started to play Dota since the age 13 years old. In average they spent more than five hours a day to play Dota every night from 11pm–3am. They admitted the obsession of playing online gaming especially when Dota offered more connections with online friends as compared to offline friends. According to them, the powerful attraction of online gaming has made them lost concentration and attention to do other things. Most of the days they will spend time with each other to play and practice. Gamers admitted that they felt happy and excited to play online gaming and at the same time felt disappointed and stressful fighting against another team. They agreed with all addiction components of online gaming but there was an argument made on "Harm" component that it should not be listed in addiction components.

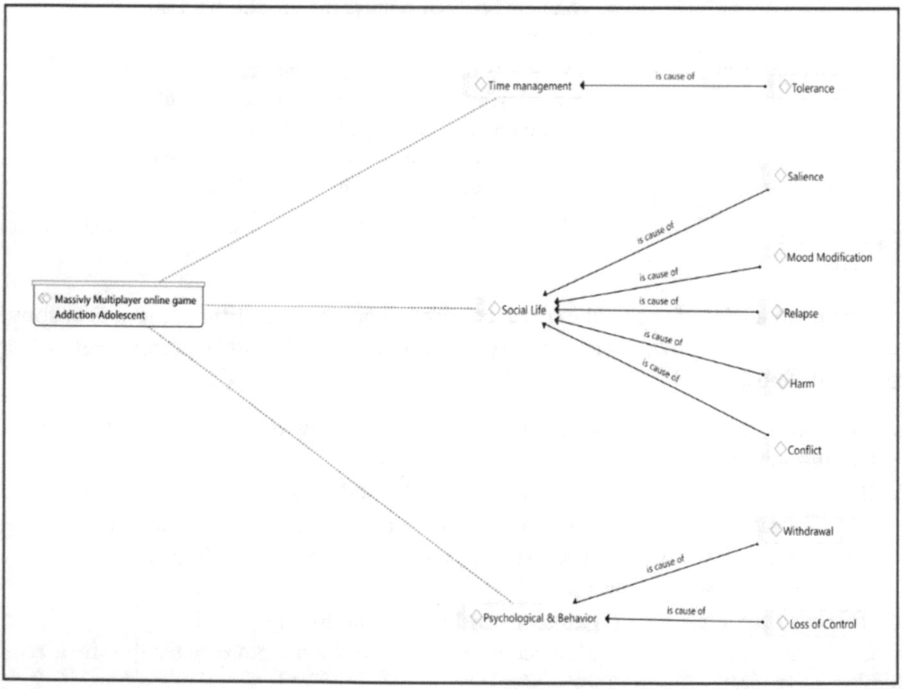

Fig. 3. Relationship diagram

Figure 3 shows the relationship diagram from data analysis using ATLAS.TI 8 qualitative analysis software. The analysis yielding in similar results as the initial model design at preliminary study using extensive literature review. The results showed that the factors of addiction are associated with the identified addiction components in Malaysia. Under qualitative method, all factors and addiction components are verified and confirmed. This includes (salience, loss of control, relapse, tolerance, withdrawal, harm, mood modification, and physical) associated with addiction factor (time management, social life and psychological and behaviour).

5 Conclusion

In summary, this finding has identified addiction of MMOG factors which are time management, psychological and behavior and social life and eight addiction components have been identified and verified. This was achieved through focus group discussions with medical experts and online game experts. Qualitative data on game addiction is very useful for clinical conceptualization of gaming issues. Furthermore, from the expert's viewpoint, this result reflects good potential of the study in helping to detect addiction among adolescents. Implementing technology in diagnostic assessment tools could be considered for future work to come up with final model of MMOG addiction in Malaysia. It is hoped that his study will pave a way for a solution for MMOG addiction among adolescents in Malaysia. All identified addiction factors and components can be used as measurable parameters that can be use by medical experts as principle guidelines of MMOG addiction to detect early stage of addiction among adolescents. As mentioned by medical experts, this research has identified four critical addiction components that can be further analyzed and used as measurement parameters to design and develop any IT automation, tools or apps to detect the addicts. This could help to detect and prevent adolescents who are addicted with MMOG to reach critical level.

References

Adiele, I., Olatokun, W.: Prevalence and determinants of internet addiction among adolescents. Comput. Hum. Behav. **31**(1), 100–110 (2014). https://doi.org/10.1016/j.chb.2013.10.028

Alrobai, A., Mcalaney, J., Dogan, H., Phalp, K., Ali, R.: Exploring the requirements and design of persuasive intervention technology to combat digital addiction (2016)

Alzahrani, A.I., Mahmud, I., Ramayah, T., Alfarraj, O., Alalwan, N.: Extending the theory of planned behavior (TPB) to explain online game playing among Malaysian undergraduate students. Telemat. Inform. 1–13 (2016). https://doi.org/10.1016/j.tele.2016.07.001

Caplan, S., Williams, D., Yee, N.: Problematic Internet use and psychosocial well-being among MMO players. Comput. Hum. Behav. **25**(6), 1312–1319 (2009). https://doi.org/10.1016/j.chb.2009.06.006

Dieris-Hirche, J., Bottel, L., Bielefeld, M., Steinbchel, T., Kehyayan, A., Dieris, B., te Wildt, B.: Media use and internet addiction in adult depression: a case-control study. Comput. Hum. Behav. **68**, 96–103 (2017). https://doi.org/10.1016/j.chb.2016.11.016

Fioravanti, G., Dèttore, D., Casale, S.: Adolescent internet addiction: testing the association between self-esteem, the perception of internet attributes, and preference for online social interactions. Cyberpsychol. Behav. Soc. Netw. **15**(6), 318–323 (2012). https://doi.org/10.1089/cyber.2011.0358

Kapahi, A., Ling, C.S., Ramadass, S., Abdullah, N.: Internet addiction in Malaysia causes and effects, 72–76, June 2013

Khan, A., Muqtadir, R.: Problematic online gaming in Pakistan **3**(6), 2522–2525 (2014)

Latif, R.A., Aziz, N.A., Ab-, M.T.: Impact of online games among undergraduate students (028), 523–532 (2014)

Lee, Z.W.Y., Cheung, C.M.K., Chan, T.K.H.: Explaining the development of the excessive use of massively multiplayer online games: a positive-negative reinforcement perspective (2014). https://doi.org/10.1109/HICSS.2014.89

Leménager, T., Dieter, J., Hill, H., Hoffmann, S., Reinhard, I., Beutel, M., et al.: Exploring the neural basis of avatar identification in pathological internet gamers and of self-reflection in pathological social network users. J. Behav. Addic. **5**(3), 485–499 (2016). https://doi.org/10.1556/2006.5.2016.048

Lemmens, J.S., Valkenburg, P.M., Peter, J.: Development and validation of a game addiction scale for adolescents development and validation of a game, vol. 3269, December 2017. https://doi.org/10.1080/15213260802669458

Lim, J.-A., Lee, J.-Y., Jung, H.Y., Sohn, B.K., Choi, S.-W., Kim, Y.J., et al.: Changes of quality of life and cognitive function in individuals with Internet gaming disorder. Medicine **95**(50), e5695 (2016). https://doi.org/10.1097/MD.0000000000005695

Snodgrass, J.G., Dengah, H.J.F., Lacy, M.G., Bagwell, A., Van Oostenburg, M., Lende, D.: Online gaming involvement and its positive and negative consequences: a cognitive anthropological cultural consensus approach to psychiatric measurement and assessment. Comput. Hum. Behav. **66**, 291–302 (2017). https://doi.org/10.1016/j.chb.2016.09.025

Tzavela, E.C., Karakitsou, C., Halapi, E., Tsitsika, A.K. Adolescent digital profiles: a process-based typology of highly engaged internet users. Comput. Hum. Behav. **69**, 246–255 (2017). https://doi.org/10.1016/j.chb.2016.11.032

Van Rooij, A.J., Schoenmakers, T.M., Vermulst, A.A., Van Den Eijnden, R.J.J.M., Van De Mheen, D., Van Rooij, A.J., Van De Mheen, D.: Online video game addiction: Identification of addicted adolescent gamers. Addiction **106**(1), 205–212 (2011). https://doi.org/10.1111/j.1360-0443.2010.03104.x

Exploring Barriers that Affect Telerehabilitation Readiness: A Case Study of Rehabilitation Centre in Malaysia

Tiara Izrinda Jafni[1], Mahadi Bahari[1](✉), Waidah Ismail[2,3], and Muhammad Hafiz Hanafi[4]

[1] Information Service Systems and Innovation Research Group, Faculty of Computing, Universiti Teknologi Malaysia, Johor Bahru, Malaysia
mahadi@utm.my

[2] Faculty of Science and Technology, Universiti Sains Islam Malaysia, Nilai, Malaysia

[3] Green Science and Technology Research Centre, Islamic Science Institute, Universiti Sains Islam Malaysia, Nilai, Malaysia

[4] Rehabilitation Medicine Unit, School of Medical Sciences, Universiti Sains Malaysia, Kota Bharu, Malaysia

Abstract. TeleRehab enables the rehabilitation services to be delivered in distance by providing information exchange between patient with disabilities and the clinical professionals. The readiness step in any adoption of healthcare services should always be one of the requirements for a successful implementation of an innovation. However, little scholarly has been undertaken to study its influence on TeleRehab and the various barrier factors that influence its adoption. This research explores the barrier factors that influence the readiness of healthcare institution to adopt TeleRehab. This paper presents a semi-structured interview involving 23 clinical professionals of a case study on the issues of TeleRehab readiness in one rehabilitation centre in Malaysia. By applying thematic analysis, the study uncovers seven barriers that affect the TeleRehab readiness. This includes barriers of no urgency to change, less awareness, less involvement in planning, not enough exposure on e-Healthcare knowledge, resistance to change, low usage of hardware and software, and less connectivity. The study contributes to both TeleRehab management and technology readiness research in hospitals.

Keywords: TeleRehab · TeleRehab readiness · Clinical professionals
Thematic analysis

1 Introduction

The successful introduction of telerehabilitation (TeleRehab) services has the potential to solve crucial problems faced by healthcare institution, including geographical barrier and quality of care [1, 2]. TeleRehab seems to be the best healthcare service that can be provided to allow rural patients to get the best treatment [3]. There are few studies conducted on TeleRehab adoption in healthcare domain. For instance, a study conducted to the patients with chronic obstructive pulmonary disease (COPD) which aims to

F. Saeed et al. (Eds.): IRICT 2018, AISC 843, pp. 761–771, 2019.
https://doi.org/10.1007/978-3-319-99007-1_70

explore the factors affecting satisfaction and potential for services improvements in long-term telerehabilitation implementation [4]. In addition, another study was also conducted to the patients with multiple sclerosis (PwPS) to measure the level of acceptance and patients' ability to carry out exercises using telerehabilitation system without supervision [5]. However, in order to adopt TeleRehab successfully, the research on readiness aspect is necessary since readiness shall be one of the requirements for a successful implementation of an innovation [6].

Nonetheless, most studies in TeleRehab were focused on its applications (e.g., telemonitoring, teleconsultation, teletherapy) [7–9], benefits (e.g., cost-effectiveness, overcomes barriers) [10, 11], and methods in delivering TeleRehab services [11, 12]. The study on TeleRehab readiness that uncovers the factors dominating the clinical professionals (e.g., physicians, therapists, nurses) in adopting TeleRehab, however, is still limited. As "readiness is an integral and preliminary step in the successful adoption of an innovation" [13], thus, a thorough study on TeleRehab readiness is important to prevent failure.

The purpose of this study is, therefore, to explore the factors that influence the readiness of healthcare institution to adopt TeleRehab. In doing so, one case study in rehabilitation centre in Malaysia is developed through the utilization of case study research approach. The paper is structured as follows: the following section presents the brief overview of background related studies; the third section will describe the methodology employed by this study; the fourth section reports the findings and discussions. The final section concludes the study and provides some possible future research direction.

2 Literature Review

In delivering good quality health service over distance, a strong engagement between advance telecommunication technologies and healthcare is needed [13]. TeleRehab is the example of modern medical rehabilitation delivery that minimizes the loss and maximizes the patient's abilities with the use of strong internet connection [14]. Generally, TeleRehab is defined as the practice of effective communication and information technologies solution to deliver clinical rehabilitation services from a distance [15]. It offers the advantages over the old-style way of face-to-face rehabilitation sessions to more organized, specialized, and efficient service delivery. Although the service usually involves a long period of time, it becomes popular due to economic potential in reducing cost [16]. It is the best choice for patient doing rehabilitation at home [17] by offering cost-effectiveness solutions [11, 18] with the use of internet connection as the platform to enhance rehabilitation delivery.

Although TeleRehab offers such benefits which can be utilized by stakeholders, healthcare institutions, and healthcare system itself, it is important to study the readiness aspect to make sure all the benefits can be useful. Furthermore, Abouzahra [19] posits that failure to identify stakeholders who involved in healthcare IT project is one of the reasons for its implementation failure. To adopt TeleRehab, the healthcare institution's stakeholders such as clinical professionals need to be mentally and physically prepared. This includes their behavior to resist or support an innovation which is illustrated as readiness towards the innovation's failure or success [20]. This means that clinical professionals need to be ready to open for change and toleration with regards to TeleRehab.

According to Schein [21], the clinical professionals who work directly with patients should consist of physicians, therapists, and nurses. However, existing studies on TeleRehab only concern with the perspective of the patient [22], physician [23], and service provider [24]. For instance, Tan [23] discussed the role of the physician in conducting the assessment of video data transfer. The study shows that physician's readiness has managed to minimize the burden of delivering TeleRehab services. In another study, Kairy [22] explored the perception of patients regarding TeleRehab services. The result of this study shows that a better understanding of patients in TeleRehab service allows the program to be continued and implemented wisely. Next, Cottrell [24] in her study provided the critical steps in determining the readiness of service providers for the implementation of TeleRehab. In this regard, it can be said that existing studies on TeleRehab readiness only concern with physicians, while neglecting therapists' and nurses' perspectives. It is also interesting to highlight that the scope of all these studies did not cover the factors influencing those stakeholders in adopting TeleRehab.

Thus, it is necessary to combine these three groups of profession – physicians, therapists, and nurses, in one study towards TeleRehab readiness since the high level of their engagement and responses to change is the key for the success of system implementation. This leads the researchers to identify all possible factors that need to be considered as important for clinical professionals in adopting TeleRehab.

3 Research Methodology

In this study, the researchers conducted a single case study at one rehabilitation centre in Malaysia. A single case study was chosen because it is a good way to describe case for an easier understanding [25, 26]. In addition, since this study focused on a group of clinical professionals, [27] single case was suitable as it is less time consuming and less expensive [28]. In addition, the chosen rehabilitation centre works in ensuring patients to get the suitable treatment for better recovery. The centre is divided into three units which are prosthetic and orthotic, physiotherapy, and occupational therapy. Few technological innovations related to rehabilitation matters such as Shock Wave (i.e., low energy placed on the skin surface to restore blood flow), ROM (i.e., game base for hand movement) and self-innovated rehabilitation product (i.e., exercise tools to assist in the recovery process of all limbs) can be seen and applied in the centre. All data from the prescribed treatments were recorded in the Folder (i.e., manual). Besides that, there is a system that allowed the patients to book their appointment (i.e., computerized appointment system) for a treatment session.

One-to-one interview sessions with clinical professionals including physician (i.e., 1 interview), therapists (i.e., 10 interviews), and nurses (i.e., 12 interviews) were conducted throughout this case study. The details of the respondents are shown in Table 1.

Table 1. Characteristics of the respondents.

Demographics	Number
Gender	
Male	4
Female	19
Age	
<=30	10
<=40	10
>41	3
Range (year)	22–52
Education	
Secondary	4
Tertiary (Diploma)	8
Tertiary (Degree)	10
Tertiary (Master)	1
Working experience	
<=10 years	15
<=20 years	5
<=30 years	3

Once the data has been collected, next, thematic analysis approach is applied [29]. As claimed by [30], thematic analysis is useful method for exploring the perspective of different research participants. In this case, the participants are clinical professionals which consists of physicians, research participants, thus can produce similarities and differences, and create unanticipated insights. All the interviews' data were then systematized and structured according to categories. In doing so, the interview transcripts were read in search of incidents and facts. The incidents were then compared with each other to discover or redevelop the code.

4 Findings

There are seven issues emerged from the data analysis.

4.1 No Urgency for Change

The result from the case study shows that most of the interviewees are satisfied with the current system. As one of the therapists expressed,

> *"In my opinion, the need to change is important as we should deliver the best services to patients. If there is a new innovation suitable to be implemented, that offers many advantages to patients and also to us (therapists), it could be great for us to use it. But, we are comfortable with what we have now (current system)"*

In addition, another therapist said,

> "*So far, this department leads other departments by providing numbering system. This system allows patients to book their appointment with the therapists. So, the patients do not have to wait long to get the treatment*"

They tend to feel comfortable with the current system causing the clinical professionals not have desire to change. It means that they are still not ready to adopt TeleRehab or there are some limitations such as individual attitudes and acceptance that restrain them from accepting changes. The need to change is an important aspect in convincing the clinical professionals to advance with the service delivery.

4.2 Less Awareness

Awareness is the state where those affected by the IT project at the healthcare institution is aware of the existence of new technology that can be used to advance the current system [31]. The result from the case study shows that only a few of clinical professionals are actually aware of TeleRehab and other healthcare technologies as an advancement for the rehabilitation ecosystem such as telehealth and telemedicine. Majority of them are still not aware of any of the healthcare technologies. As the physician mentioned,

> "*TeleRehab is a good medium for better service delivery. However, all the therapists and nurses here prefer to use manual system because it has been practiced since this unit was established*"

Besides, one of the therapists also mentioned,

> "*I do not aware of advantages and functions of TeleRehab. This is my first time get know about it*"

The less awareness towards TeleRehab causes the lack of exposure to its advantages. The need to always keep updated with current technology with an advancement is important because it can deliver positive impact to the health field and allow the clinical professionals to provide the best services.

4.3 Less Involvement in Planning

In the case study, only a few clinical professionals involved in generating ideas of new innovation and committed in the planning. This means that no one acted as the key champion. In fact, majority of them prefer to use current technology or use small innovation brought from their top management. However, they are free to bring out any ideas for improvement if there would be a meeting or discussion conducted by staff and the top management. As one of the nurses stated,

> "*Normally, only the top management and few specialists involved in the planning for adopting any new technologies. The rest of nurses and therapists only use it. However, we are free to give any ideas for improvement*"

A proper of budget allocation is also an important element in the planning. Although the involvement of the clinical professional in the innovation is very encouraging, they still not be able to use that achievement for the organization and patients. A physician also mentioned this,

"The low budgetary allocation is one of the reasons for the lack adaptation of the new innovation. Hence we were not involved greatly in the planning"

4.4 Not Enough Exposure on E-Healthcare Knowledge

The case study shows that only a few numbers of clinical professionals are exposed to the knowledge of TeleRehab. This knowledge was learned during their studies before they were appointed as clinical professionals in the rehabilitation area. However, this only involved the younger interviewees while the older one such as therapists and the nurses who have been working for more than 15 years claimed that they are not exposed to this knowledge. This may be due to the different curriculum that they followed while doing their courses. As one of the therapists stated,

"During my study, I never learn about any healthcare services such as telehealth, telemedicine, and telerehabilitation. Due to that, I am not exposed to the knowledge"

Another therapist also said,

"I only know the basic knowledge about healthcare services and do not really exposed to the new innovation such as TeleRehab"

However, this situation is not applicable to both physicians as they are already knowledgeable about TeleRehab technology.

4.5 Resistance to Change

The case study reported that most of the clinical professionals agreed that they have the feeling to resist the changes. This is due to budget constraint and financial prioritisation, self-innovated rehabilitation product need to be creatively financially sustainable and less dependent to the budget given by the management. As a consequence, not all of them have the opportunity to use the innovation and apply to their rehab patients.

One of the therapists mentioned that patient's acceptance towards the technology will also influence their attitude; either to resist it or not. As she stated,

"At our rehabilitation centre, before any new innovation will be implemented, it needs to be tested and used wisely to the patients. The patients' acceptance leads to our attitude acceptance"

Another therapist also mentioned,

"The patients treated here live in rural area with less exposure to the internet connection. This caused them to resist changes"

This means that if the tendency of the patient's acceptance is low, definitely such innovation will not be implemented.

4.6 Less Use of Hardware and Software

Hardware and computers are necessary components for any technology adoption in an organization [32]. Certain criteria such as level of computer usage and the availability of compatible computer in accordance to the need of software need to be measured prior to the adoption of new innovation. However, the result from the case study shows low availability of computers in the centre. As mentioned by one therapist,

"The quantity of computer here is limited. We need to share it in order to do daily report"

The usage of hardware and software is also at medium level. This is supported by one of the therapists,

"We only use the computers and internet when necessary. In daily routine, we update the patient's record in the folder and prepare the overall report by the end of the day"

This is because the rehabilitation services in the case study are still delivered manually without using any computers. The low availability of computers and low usage level indirectly show low readiness of clinical professionals to adopt TeleRehab. Both indicators give the rough perspectives about clinical professional's readiness towards TeleRehab. If their level of usage for hardware and software is high, it is easier for them to adopt TeleRehab and vice versa.

4.7 Low Connectivity

The high speed and good quality of internet connection are necessary for TeleRehab adoption since the data need to be transferred from the patient's house to the rehabilitation centre. Strong internet connection is needed from both parties to ensure all the data could be transferred and analyzed [33]. In the case study, all clinical professionals stated that internet connection in the rehabilitation is in very good quality with high speed for access. However, they are worried about the one in the patient's house. As one of the therapists stated,

"There is no problem with internet and phone connection at our rehabilitation unit as the line is above than average level every day. However, I was thinking that it would be a problem with the patient's side as most of them live in rural areas. So, the internet connection may not be available at all or in poor condition"

A nurse also stated,

"If the connectivity issue can be solved, it allows the patients to accept TeleRehab"

5 Discussion and Conclusions

This study helps to identify the main issues concerning TeleRehab readiness from qualitative perspective. The study is novel in that it provides description of clinical professional views on TeleRehab with a focused on readiness aspect. An exploration of why clinical professionals is using rehabilitation delivery services and the ways in which they use it, what they see and feel as the most persistent barriers, are useful in informing TeleRehab readiness. Our study suggests that although few clinical professionals are beginning to understand the nature of TeleRehab, the data analysis from the case study discusses seven (7) issues that hinder for the TeleRehab readiness. This includes no urgency for change, less awareness, less involvement in planning, not enough exposure on e-Healthcare knowledge, resistance to change, less use of hardware and software, and low connectivity. These barriers can be clustered into four (4) themes as shown in Fig. 1.

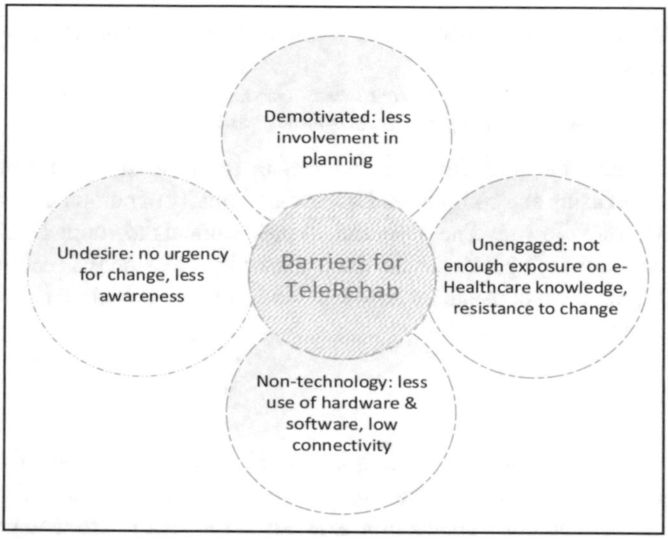

Fig. 1. Classification of TeleRehab barriers into themes

The undesired barrier refers to the clinical professionals' expressions of no urgency and unawareness that leads to the undesired to adopt TeleRehab. In this regard, the undesired barriers include factors of no urgency for change and less awareness about TeleRehab. Respondents expressed that they are not ready to adopt for TeleRehab services and still comfortable with the current system provided by the healthcare institution. They also not concern or less awareness towards new technology needs in the rehabilitation delivery services. These situations cause them not have desire to change. This is contradicting with Yusif [34] study who argued when the desire of readiness at individual level is high, individual will tend to accept for technology innovation. Hence, this theme is the most important theme because it is the root of not to change for adopting TeleRehab. Next, the demotivated barrier refers to the clinical professionals' expression of not showing interest to use technology innovation in the future. The only barrier that falls under motivation theme is less involvement in planning. Respondents stressed that no one acted as the key champion of generating ideas for TeleRehab. They are not totally invited in making a proper plan towards providing a better rehabilitation delivery services, and thus demotivated them to use TeleRehab in the future.

Next, the unengaged barrier refers to the clinical professionals' expressions of not actively engaged with the idea of innovation. This includes factors of not enough exposure on e-Healthcare knowledge and resistance to change. Respondents expressed that only small numbers of them are exposed to e-Healthcare knowledge. This makes them less aware towards the advantages of TeleRehab and just prefer to use current technology instead of becoming an innovator or key champion. In addition, respondents tend to resist the changes due to felt that TeleRehab technology will not suitable for patient. Issue such as less exposure to the internet connection for patients who are living in rural areas affected for the TeleRehab readiness. Finally, the non-technology barrier is the context of the clinical professionals' expressions of not meeting the necessary technological requirements in adopting TeleRehab. This includes factors of less use of hardware and software and less connectivity. Respondents stressed that the usage of hardware and software, and the availability of computer in the case study are at low level. The low availability of computers and low usage level indirectly show low readiness of clinical professionals to adopt TeleRehab. In addition, the quality of internet between the healthcare institution and patient's house is another issue. Although the case study is coverage with good internet connection, but not the case for patients especially those who are in rural areas. This is supported by Ross [35] who stated that complexity (i.e., connectivity issue, unplanned downtime, software and hardware difficulty, data handling), and adaptability (i.e., acceptance and use, technical adjustment) could become the main barrier to adoption and implementation of innovation.

The classification above leads to the conclusion that these barriers avert the case study towards adopting TeleRehab. There is still a lot of attention and hard work to be done by the top management to ensure their clinical professionals are more ready and willing to adopt TeleRehab. Not only the top management, the clinical professionals themselves need to seek initiatives to be more physically and mentally prepared for TeleRehab.

The strength of this study was using interviews that allowed attitudes and perceptions of clinical professionals towards TeleRehab to be explored in greater depth. This study used a relatively small number of respondents, but included a diverse group of clinical professionals which encompasses of physician, therapist and nurse positions, and by age, gender, education tertiary and working experience. Since the study was conducted at one healthcare institution in Malaysia, so findings may not reflect all healthcare institutions in the country.

Future studies shall extend these barriers that impact TeleRehab adoption to higher than just individual level; this issue shall be discussed further in organizational level or national level. Thus, future research shall be considered in recently published TeleRehab studies such as the barriers issues from this study for the purpose of knowledge sharing. This will help increase awareness in the field among researchers and clinical professionals especially with the continuous advancement of new technology used for TeleRehab implementation.

Acknowledgment. The author wants to appreciate the Editor and anonymous referees for their constructive comments and criticism. This work was supported by the International Grant USIM/INT-NEWTON/FST/IHRAM/053000/41616 under Newton-Ungku Omar Fund.

References

1. Cason, J., Otr, L.: Telerehabilitation: an adjunct service delivery model for early intervention services. Int. J. Telerehabil. **3**(1), 19–30 (2011)

2. Jafni, T.I., Bahari, M., Ismail, W., Radman, A.: Understanding the implementation of telerehabilitation at pre-implementation stage: a systematic literature review. Procedia Comput. Sci. **124**, 452–460 (2017)
3. Frederix, I., Hansen, D., Coninx, K., Vandervoort, P., Van Craenenbroeck, E.M., Vrints, C., Dendale, P.: Telerehab III: a multi-center randomized, controlled trial investigating the long-term effectiveness of a comprehensive cardiac telerehabilitation program - rationale and study design. BMC Cardiovasc. Disorder **15**(1), 1–8 (2015)
4. Hoaas, H., Andreassen, H.K., Lien, L.A., Hjalmarsen, A., Zanaboni, P.: Adherence and factors affecting satisfaction in long-term telerehabilitation for patients with chronic obstructive pulmonary disease: a mixed methods study. BMC Med. Inform. Decis. Mak. **16**(26), 1–4 (2016)
5. Jeong, I.C., Finkelstein, J.: Introducing telerehabilitation in patients with multiple sclerosis with significant mobility disability: pilot feasibility study. In: International Conference on Healthcare Informatics, USA, pp. 69–75. IEEE (2015)
6. Jennett, P., Yeo, M., Pauls, M., Graham, J.: Organizational readiness for telemedicine: implications for success and failure. J. Telemed. Telecare **9**(Suppl. 2), 27–30 (2003)
7. Parmanto, B., Saptono, A., Saptono, A.: Telerehabilitation: state-of-the-art from an informatics perspective. Int. J. Telerehabil. **1**(1), 73–84 (2009)
8. Yan, Z., Guo, X., Vogel, D.R.: Understanding dynamic collaboration in teleconsultation. Inf. Technol. Dev. **22**(1), 152–167 (2013)
9. Tse, Y.J., McCarty, C.A., Vander Stoep, A., Myers, K.M.: Teletherapy delivery of caregiver behavior training for children with attention-deficit hyperactivity disorder. Telemed. e-Health **21**(6), 451–458 (2015)
10. Cason, J., Otr, L.: Telehealth: a rapidly developing service delivery model for occupational therapy. Int. J. Telerehabil. **6**(1), 29–37 (2014)
11. Shamsuddin, S., Yussof, H., Mohamed, S., Hanapiah, F.A., Ainudin, H.A.: Telerehabilitation service with a robot for autism intervention. In: International Symposium on Robotics and Intelligent Sensors, Malaysia, pp. 349–354. IEEE (2015)
12. Altilio, R., Liparulo, L., Panella, M., Proietti, A., Paoloni, M.: Multimedia and gaming technologies for telerehabilitation of motor disabilities [leading edge]. IEEE Technol. Soc. Mag. **34**(4), 23–30 (2015)
13. Jennett, P., Jackson, A., Ho, K., Healy, T., Kazanjian, A., Woollard, R., Haydt, S., Bates, J.: The essence of telehealth readiness in rural communities: an organizational perspective. Telemed. e-Health **11**(2), 137–145 (2005)
14. Brennan, D.M., Barker, L.M.: Human factors in the development and implementation of telerehabilitation systems. J. Telemed. Telecare **14**, 55–58 (2008)
15. Kairy, D., Lehoux, P., Vincent, C., Visintin, M.: A systematic review of clinical outcomes, clinical process, healthcare utilization and costs associated with telerehabilitation. Disabil. Rehabil. **31**(6), 427–447 (2009)
16. Pramuka, M., Van Roosmalen, L.: Telerehabilitation technologies: accessibility and usability. Int. J. Telerehabil. **1**(1), 25–36 (2015)
17. Zahid, Z., Atique, S., Saghir, M.H., Ali, I., Shahid, A., Malik, R.A.: A commentary on telerehabilitation services in pakistan: current trends and future possibilities. Int. J. Telerehabil. **9**(1), 71–76 (2017)
18. McCue, M., Fairman, A., Pramuka, M.: Enhancing quality of life through telerehabilitation. Phys. Med. Rehabil. Clin. N. Am. **21**(1), 195–205 (2010)
19. Abouzahra, M.: Causes of failure in healthcare IT projects. In: 3rd International Conference on Advanced Management Science, pp. 46–50. IACSIT Press, Singapore (2011)

20. Ojo, S.O., Olugbara, O.O., Ditsa, G., Adigun, M., Xulu, S.S.: Formal model for e-healthcare readiness assessment in developing country context. In: International Conference on Innovations in Information Technology, Canada, pp. 41–45. IEEE (2007)
21. Schein, E.H.: The Clinical Perspective in Fieldwork, vol. 5. Sage Publications, New York (1987)
22. Kairy, D., Tousignant, M., Leclerc, N., Côté, A.M., Levasseur, M.: The patient's perspective of in-home telerehabilitation physiotherapy services following total knee arthroplasty. Int. J. Environ. Res. Public Health **10**(9), 3998–4011 (2013)
23. Tan, K.K., Narayanan, A.S., Koh, C.H., Caves, K., Hoenig, H.: Extraction of spatial information for low-bandwidth telerehabilitation applications. J. Rehabil. Dev. **51**(5), 825–840 (2014)
24. Cottrell, M.A., Hill, A.J., O'Leary, S.P., Raymer, M.E., Russell, T.G.: Service provider perceptions of telerehabilitation as an additional service delivery option within an Australian neurosurgical and orthopaedic physiotherapy screening clinic: a qualitative study. Musculoskelet. Sci. Pract. **32**, 7–16 (2017)
25. Cousin, G.: Case study research. J. Geogr. High. Educ. **29**(3), 421–427 (2005)
26. Gustafsson, J.: Single case studies vs. multiple case studies: a comparative study. Academy of Business, Engineering and Science, Halmstad University Sweden, pp. 1–15 (2017)
27. Yin, R.K.: Case study research: design and methods. Educ. Res. Eval. **24**(3), 221–222 (2011)
28. Baxter, P., Jack, S.: The qualitative report qualitative case study methodology: study design and implementation for novice researchers. Qual. Rep. **13**(4), 544–559 (2008)
29. Braun, V., Clarke, V.: Using thematic analysis in psychology. Qual. Res. Psychol. **3**, 77–101 (2006)
30. Nowell, L.S., Norris, J.M., White, D.E., Moules, N.J.: Thematic analysis: striving to meet the trustworthiness criteria. Int. J. Qual. Methods **16**(1), 1–13 (2017)
31. Biruk, S., Yilma, T., Andualem, M., Tilahun, B.: Health Professionals' readiness to implement electronic medical record system at three hospitals in Ethiopia: a cross sectional study. BMC Med. Inform. Decis. Mak. **14**(1), 1–8 (2014)
32. Overhage, J.M., Evans, L., Marchibroda, J.: Communities' readiness for health information exchange: the national landscape in 2004. J. Am. Med. Inform. Assoc. **12**(2), 107–112 (2005)
33. Nchise, A., Boateng, R., Mbarika, V., Saiba, E., Johnson, O.: The challenge of taking baby steps-preliminary insights into telemedicine adoption in Rwanda. Health Policy Technol. **1**(4), 207–213 (2012)
34. Yusif, S., Hafeez-Baig, A., Soar, J.: E-health readiness assessment factors and measuring tools: a systematic review. Int. J. Med. Inform. **107**, 56–64 (2017)
35. Ross, J., Stevenson, F., Lau, R., Murray, E.: Factors that influence the implementation of e-health: a systematic review of systematic reviews (an update). Implement. Sci. **11**(1), 1–12 (2016)

A Study on Mobile Applications Developed for Children with Autism

Naziatul Shima Abdul Aziz, Wan Fatimah Wan Ahmad(✉), and Ahmad Sobri Hashim

Computer and Information Science Department, Universiti Teknologi PETRONAS, 32610 Seri Iskandar, Perak, Malaysia
nsaa71@yahoo.com, {fatimhd,sobri.hashim}@utp.edu.my

Abstract. The emerging of mobile technology leads to the extensive used of mobile application for learning purposes of the children with autism. Autism is a neurological disorder that affects the children's behavior and their ability to communicate and interact socially. A lot of studies have been conducted on using mobile application to assist the children with autism to increase their social and communication skills. Mobile applications are now widely used, not only for entertainment and social networking, but also for education. The used of mobile applications in education has extend from dictionaries to special purpose education. This paper reviews six mobile applications developed to assist the children with autism for various purposes. This review will provide a summary of initial studies and preliminary findings for future development of enhanced application.

Keywords: Mobile application · Mobile learning · Autism
Educational application

1 Introduction

Autism is a lifelong development difficulty [1] that affects their intellectual ability [2], communication, social interaction and patterns of restricted or repetitive behavior [3]. These impairment results difficulties in the teaching and learning process of the children with autism. Autism is the fastest growing neurobiological conditions in the world that typically appears by three years of biological age [4]. The rate of autism occurrence rose up to 1 in every 110 children in the year 2009 and the number doubled in the year 2013 [5], thus it is important to give special attention to assist these children in teaching and learning process.

The advent of mobile technology has increased the chances for the children with autism to learn in an effective way. The children with autism need an intermediator either human or a mechanical robot to socially interact with other people [4]. A lot of courseware, mobile application, and other technological medium have been developed to help the children with autism and to improve their impairments. As stated by a world-renowned autism expert named Lovaas, "If a child cannot learn in the way we teach, we must teach in a way the child can learn" [6].

This paper objectively to review six mobile application that have been developed for the children with autism for various purposes. This review will provide a summary

F. Saeed et al. (Eds.): IRICT 2018, AISC 843, pp. 772–780, 2019.
https://doi.org/10.1007/978-3-319-99007-1_71

of initial studies and preliminary findings for future development of enhanced mobile application.

2 Methodology

For the purpose of this research, six scholarly articles were reviewed. Each article was discussing a mobile application developed for the children with autism. These application were developed for various purposes. These application were reviewed and tabled. Among details identified were the types of learning usage, types of mobile device, targeted age of the children using the application, and the development approach used.

3 Mobile Applications

The used of mobile applications in education has become more popular since the applications can be accessed anytime and anywhere. Through mobile applications, the learning process occurs without being tied to a situation and environment.

This section reviewed six mobile applications developed by previous researchers. The comparison of the mobile application have been conducted according to types of learning usage, types of mobile device, age of the children with autism and the development approach used.

3.1 BIUTIS

BIUTIS is a mobile application developed for vocabulary learning of the children with autism [7]. The application is developed using hardware-based mobile Android 2.1. The application consists of learning materials for the children to manipulate menu items by using an acquaintance study. This application is a client server application on the client side. The contents are stored in the server and can be downloaded at anytime from anywhere as additional learning material. An authentication process is required for parents to add or delete data that is mostly used by children, categories, images and sounds of the vocabulary. No authentication process is required for the children to start learning, and the children are free to choose learning based on desired category.

The application was evaluated and the results showed the majority or 60% of the respondents strongly agreed and another 40% agreed that the application can be recommended as a medium of learning for the children with autism. However, this application cannot be used worldwide since the language used is in Indonesian Language. Figure 1 displays the user interface of BIUTIS.

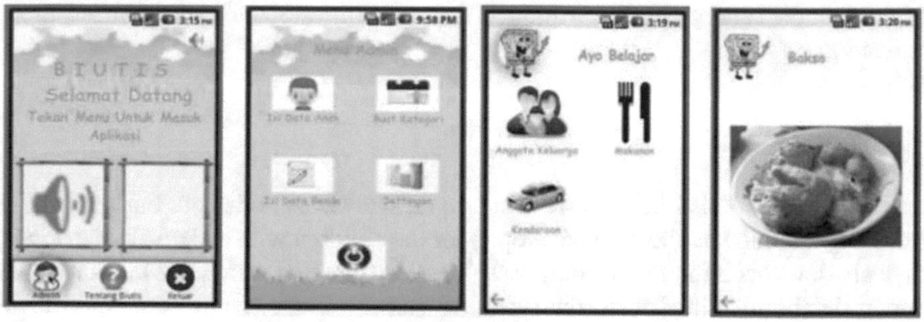

Fig. 1. User interface display for BUITIS

3.2 Autism

Autism [4] is an android mobile application developed for the children with autism. The application is aimed to help the children with autism with basic communication and social interaction skills. The application is designed based on the characteristics of echolalia which is common characteristics among the children with autism. The application consists of four modules. First module is to register with the database, second module is for speech and text conversion, third module is to retrieve and iterate the details, and relevant image and the fourth module tracking the person's location. The parents can enter the details of the children with autism into the application and the details will be kept in the database. When a person ask a question to the children, the person's voice is captured and converted to text. Google speech engine recognizes the question and retrieves the answer from the database. The corresponding sound is produced and the children can then replies to the question asked to them. However, the application was developed only for one language which is English. Figure 2 shows some of the application user interface; the main screen, the registration screen.

Fig. 2. User interface display for Autism

3.3 Educational App

Educational App [8] is another application developed to improve communication skills between the children with autism with others. The children can use the application by interacting with the picture of an object. They need to follow the process of learning by listening to the audio sound related to the picture of the object shown. The application can be accessed by both parents and the children but with different functions. Parents can add activities that suit the needs of their children by key in the any text for activities in the “Text To Say’s” textbox. The text must be related to the picture selected from the photo gallery. The paired text and picture are kept in the application’s database. The children can select the picture that illustrated with text to express what they are trying to speak out.

The application includes four kinds of educational activities; exploration, association, puzzle and sorting. Children and parents will have different access to the application. Figure 3 shows the user interface display for the application.

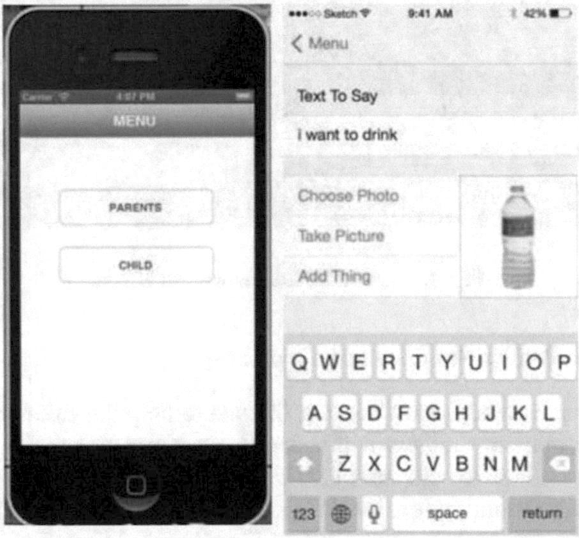

Fig. 3. User interface display for Educational App

3.4 iCAN

iCAN is a tablet-based application developed to improve communication skills of the children with autism [9]. iCAN is a pedagogical system with the intention to eliminate the difficult and complex process of creating and handling collections of paper-based picture cards, and also for better progressing their cognitive, language, and communication learning. iCAN replaced the traditional paper-based picture cards.

iCAN is developed for parents and teachers to teach cognitive, language and communication skills. Studies show that iCAN stimulate the children auditory, tactile, and visual senses, and thus it is more appealing to the children. Since the children can use their hands to interact with the tablet, iCAN is described as attractive and this makes the children learning attitude be more proactive. Figure 4 illustrates the user interface of iCAN.

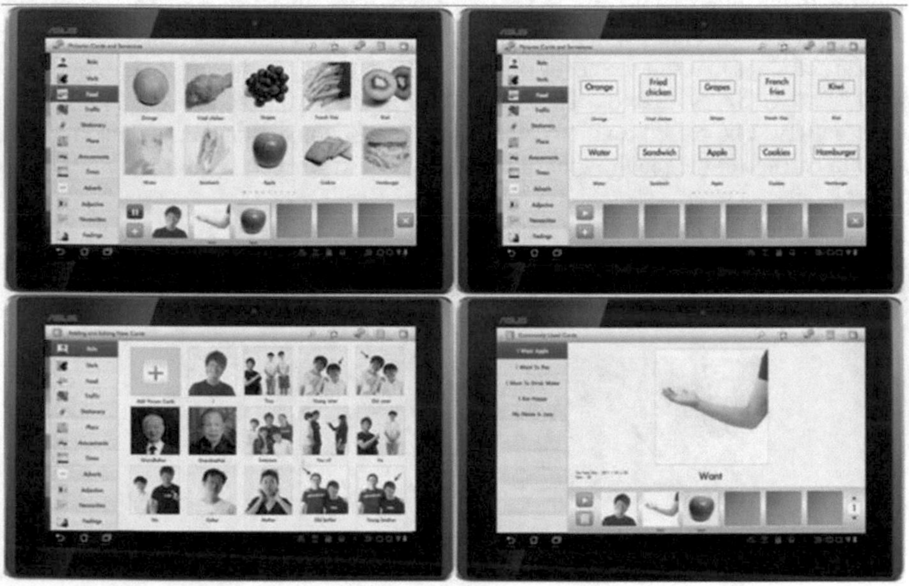

Fig. 4. User interface display for iCAN

3.5 Learn with Rufus: Numbers and Counting

Learn with Rufus: Numbers and Counting intends to help the children with autism to prepare for kindergarten. Children will learn the sequence of numbers, both forwards and backwards, and how to compare quantities either more or less, of various objects. The game is highly customizable to meet the needs of children with varying skills, ability levels, and learning styles. Learn with Rufus: Numbers and Counting was designed by Dr Holly Gastgeb. Dr Holly is a licensed clinical and developmental psychologist with over ten years of experience working with typically developing children and children with autism spectrum disorder. The application is available in both English and Spanish. It consists of three levels of difficulty; easy (numbers go from 1 through 10), medium (numbers go from 5 through 15) and hard (numbers go from 10 through 20).

Learn with Rufus: Numbers and Counting is an application to develop mathematical skills [10]. It can be used under various platform; Android, iPad, iPhone and also iTouch. The application include features such as positive verbal reinforcement once the child answers correctly and if the child answers incorrectly, the correct answer will be restated. Figure 5 presents the user interface of this application.

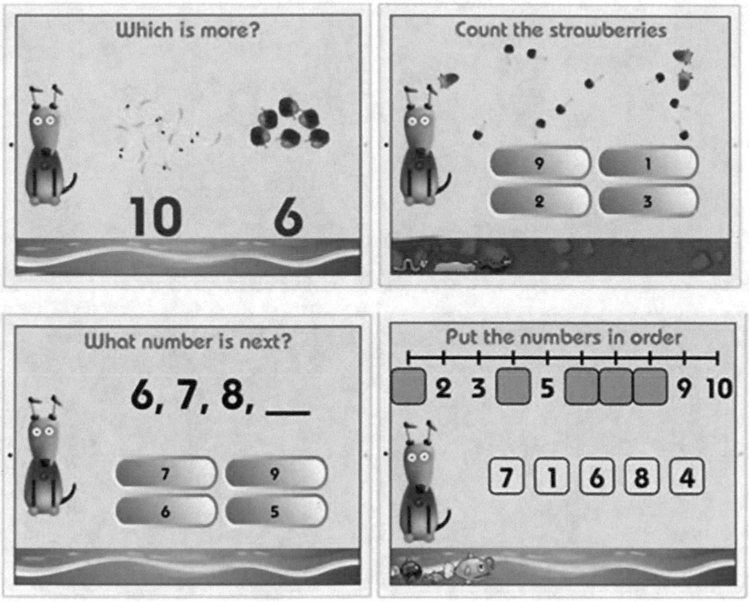

Fig. 5. User interface display for learn with Rufus: numbers and counting

3.6 Mental Imagery Therapy for Autism (MITA)

MITA is a distinctive, early-intervention application developed for children with autism. MITA comprises of cheerful, interactive puzzles. MITA is designed to help the children learn on how to mentally combine several features of an object, and the results supported to substantial enhancements in general learning. Achievement with MITA might strongly result in significant improvements in a child's overall development, particularly in the terms of language, attention and visual skills. MITA is developed by Dr. A. Vyshedskiy, a neuroscientist from Boston University; R. Dunn, an early-child-development specialist from Harvard; J. Elgart from MIT and a group of award-winning artists and developers working alongside experienced therapists.

MITA is developed based on ABA technique of visual-visual and auditory-visual conditional discrimination, language therapy technique of following directions with increasing complexity, and Pivotal Response Treatment that targets development of response to multiple cues [11]. Figure 6 shows the user interface of this application.

Fig. 6. User interface display for MITA

4 Finding and Discussion

The use of mobile application is a promising strategy for direct intervention to the children with autism. This paper aims to discuss on the comparative study conducted on mobile application that have been developed by researchers from previous studies. The mobile application have been compared on four criteria including types of mobile device, age of the children with autism, types of learning usage and also the development approach.

The ability to count and compare quantities is a basic skill each child should have before starting school. Many school systems consider this to be one of the skills essential for school readiness and success. It has been generally shown that children who have early successes in school gain the confidence to achieve even greater successes later in life. However, there are limited numerical application available for the children with autism. Thus there is a need to develop a mobile application for these children to learn mathematical skills.

All six discussed application were developed using Android as a platform. This is because most mobile device are using Android. Therefore, for the purpose of this project, the prototype of numerical mobile application will also be developed using Android as a platform. The prototype will include ABA technique of visual-visual and auditory-visual conditional discrimination, positive verbal reinforcement, and also play therapy.

All applications discussed were developed using one language. To develop a mobile application for the children with autism in Malaysia, it is more relevant if the application offers more than one language. Two most spoken language in schools in Malaysia are English Language and Bahasa Malaysia (Table 1).

Table 1. Comparison of mobile learning application.

Name	Types of learning usage	Types of mobile device	Age of the children with autism	Development approach
BUITIS	Vocabulary Learning	Android	Not available	ABA (applied behavior analysis) Speech therapy Play Therapy Visual therapy Biomedical therapy
AUTISM	Communication Skills Social Interaction Skills	Android	Not available	Not available
EDUCATIONAL APP	Communication Skills	Not specified	Not available	Voice recognition Visual support
iCAN	Cognitive skills Language communication Skills	Tablet	Not available	Not available
Learn with Rufus: Numbers and Counting	Mathematical Skills	Android iPad iPhone iTouch	Preschool (2–5 years old) Children (6–12 years old)	Positive verbal reinforcement
MITA	Language	Android iPad iPhone iTouch	Preschool (2–5 years old) Children (6–12 years old) Adolescents (13–17 years old)	ABA technique of visual-visual and auditory-visual conditional discrimination Language therapy technique of following directions with increasing complexity Pivotal Response Treatment that targets development of response to multiple cues

References

1. Aziz, N.S.A., Ahmad, W.F.W., Zulkifli, J.: User experience on numerical application between children with down syndrome and autism. In: Proceedings of CHI UX Indonesia 2015 (CHIuXiD 2015), pp. 26–31 (2015)
2. Matson, J.L., Goldin, R.L.: Diagnosing young children with autism. Int. J. Dev. Neurosci. **39**, 1–5 (2014)
3. Bernardini, S., Porayska-Pomsta, K., Smith, T.J.: ECHOES: an intelligent serious game for fostering social communication in children with autism. Inf. Sci. (Ny) **264**, 41–60 (2014)
4. Mohanaprakash, T.A.: Assisting echolalia (repetitive speech patterns) in children with autism using android mobile app. IJAICT **1**(12), 928–933 (2015)
5. Fatimah, W., Ahmad, W., Hashim, A.S., Nadia, A.: Development of mobile application for autistic children using augmentative and alternative communication technique. In: 6th International Conference on Computing and Informatics, ICOCT, no. 13, pp. 262–267 (2017)
6. Eder, M.S., Diaz, J.M.L., Madela, J.R.S., Mag-usara, M.U., Sabellano, D.D.M.: Fill me app: an interactive mobile game application for children with autism. iJIM **10**(3), 59–63 (2016)
7. Husni, E., Budianingsih, M.: Mobile applications BIUTIS: let's study vocabulary learning as a media for children with autism. Procedia Technol. **11**, 1147–1155 (2013)
8. Aziz, M.Z.A., Abdullah, S.A.C., Adnan, S.F.S., Mazalan, L.: Educational app for children with autism spectrum disorders. Procedia Comput. Sci. **42**(c), 70–77 (2014)
9. Chien, M., Jheng, C., Lin, N., Tang, H., Taele, P., Tseng, W., Chen, M.Y.: iCAN: a tablet-based pedagogical system for improving communication skills of children with autism. J. Hum. Comput. Stud. **73**, 79–90 (2015)
10. Kagohara, D.M., van der Meer, L., Ramdoss, S., O'Reilly, M.F., Lancioni, G.E., Davis, T.N., Rispoli, M., Lang, R., Marschik, P.B., Sutherland, D., Green, V.A., Sigafoos, J.: Using iPods(®) and iPads(®) in teaching programs for individuals with developmental disabilities: a systematic review. Res. Dev. Disabil. **34**(1), 147–156 (2013)
11. Dunn, R., Vyshedskiy, A.: Mental imagery therapy for autism (MITA)—an early intervention computerized brain training program for children with ASD. Autism Open Access **5**(3) (2015)

Heuristic Evaluation of the Smartphone Applications in Supporting Elderly

Hasanin Mohammed Salman, Wan Fatimah Wan Ahmad(✉), and Suziah Sulaiman

Computer and Information Sciences Department, Universiti Teknologi PETRONAS, 32610 Seri Iskandar, Perak, Malaysia
{hasanin_g03421,fatimhd,suziah}@utp.edu.my

Abstract. Elderly people often experience deterioration in cognitive and physical traits, which impacts their ability to use smartphone applications. Because of this age-related deterioration, designers should pay special attention when it comes to designing user interfaces (UIs) for elderly users. Nevertheless, the typical modern UI apparently is not optimised for this group of users. Our primary objective is to answer the research question: what are the usability problems that could be potentially experienced by elderly as they interact with the user interfaces of smartphones? A further objective is to recommend improvements in the design of UIs that can particularly benefit elderly users. Five individual experts applied a heuristic evaluation method within a controlled environment to determine the anticipated usability problems in the two case study applications i.e., alarm and camera. Analysing the identified usability problems revealed that they were attributed to three issues that the UI failed to adhere to; promote recognition over recall, improve user's efficiency of use, and achieve an intuitive and consistent interface. Experts have suggested solutions that would help tackle these issues and provide insights regarding potential improvements for designing elder-friendly UIs.

Keywords: Smartphone · User interface · Usability · Heuristic evaluation Elderly

1 Introduction

Smartphones comprise a generic class of mobile communications devices that feature independent operating systems. These allow users to install different third-party applications [1]. Both work-related and leisure applications, that assist and entertain users during daily life activities, are combined into a handy, within user's grasp smartphone that could be taken everywhere.

Despite the numerous strengths of mobile apps, several difficulties remain for the typical older user [2, 3]. The typical younger user is usually able to quickly overcome various learning challenges (as in the instance of a non-instinctive multiple-finger gesture), via successful trial and repetition effected by their mental models to gel with newer advances. The difficulties can be greater for elderly adults [4]. Ageing triggers a decrease in the intellectual level, which includes loss of memory, the mental speed of

F. Saeed et al. (Eds.): IRICT 2018, AISC 843, pp. 781–790, 2019.
https://doi.org/10.1007/978-3-319-99007-1_72

processing information, and the learning capability [5]. The majority of the elderly face a worsening of their vision and hearing perception [6]. Furthermore, because of aging, the nerves and muscles are unable to function with full efficiency past the age of consent [7, 8]. Due to these potentials for deterioration in physical and cognitive characteristics, user interface (UI) design should address the difficulties encountered by the typical older user [9]. Nevertheless, most current UIs are seemingly not optimised for seniors [10, 11]. The neglect of "elderly-friendly" UIs may result in reluctant smartphone usage among this particular group [10], whereas a properly devised UI that meets their needs will address this issue effectively [10, 12]. This observation was the motivating basis for our evaluation of the UI usability of smartphones in support of older users.

According to Lewis [13], the two major conceptions of usability are commonly referred to as; (1) Summative, and (2) Formative. Formative usability focusses on the detection of usability problems and the design of interventions to reduce or eliminate problems impact [13]; which is also the focus of this study. In this research, usability evaluation relies on the heuristic evaluation method. The primary objective is to answer the research question: what are the usability problems that could be potentially experienced by elderly as they interact with the user interfaces of smartphones? A further objective is to propose improvements in the designs of smartphone UIs that can benefit older users in particular.

2 Background

In broad terms, usability evaluation techniques can be classified into two groups [14, 15]; (1) user-based (testing methods), and (2) expert-based (inspection methods).

User-based methods search for and detect usability problems via the observation of representative individuals, who utilise the interface while providing comments [15, 16]. Conversely, expert-based methods entail expert examination of UIs in order to predict problems that the typical user might encounter during interactions with UI [17]. Findings in expert-based evaluations can comprise a formal report that highlights a recognised problem or recommends changes [18]. These aspects follow the formative concept of usability implemented in this research.

Of the different expert-based assessment methods, heuristic evaluation is among the most broadly applied methods [19], with many papers utilising it to assess the usability of different information systems [20–23]. Heuristic evaluation method offers many benefits versus other assessment techniques. Heuristic evaluation is relatively affordable, fast [15, 17], and readily implemented [15]. The time consumed and complexity in examination are lessened due to the fewer numbers of expert evaluators (some 3–5) needed to identify the majority of usability problems [24, 25]. Also, the entire process can be conducted in days [24, 25]. Not every identified problem is treated the same, some can result from defective elements in aesthetic designs, whereas others may influence the performance of key system functions [26]. One of the more efficient techniques used to classify the severity of a usability problem is Nielson's multi-level scale, which comprises 5 levels: 0 = no usability problems, 1 = cosmetic problems, 2 = minor usability problems, 3 = major usability problems, and 4 = usability catastrophes [26].

Applying the heuristic evaluation technique, "the expert reviewers critique an interface to determine conformance with a short list of design heuristics" [18]. When an expert encountered a usability problem, then one or more of the heuristics listed were violated [18]. In general, heuristics can be classified into two categories; (1) Generic heuristics - which help assess the majority of UIs [27], e.g., Jakob Nielsen's Ten generic principles of interaction design [28], and (2) Specific heuristics – which are required for particular domains, making sure that specific usability issues are detected [27], e.g., 12 usability heuristics for smartphone and mobile applications (SMASH) [19].

Among the current papers that apply heuristic evaluation of mobile apps that are meant to be utilised by elderly are the reports of Al-Razgan et al. [29], Watkins et al. [20], and Silva et al. [30]. The research of Al-Razgan et al. [29] defines a strategy that adapts design guidelines and recommendations for various smartphones into a set of 13 heuristics that covers the usability of touch-based mobile launchers designed for older users. Three apps and three launchers for Android smartphones were evaluated by four undergraduate student seniors who comprised the usability evaluator group. They carried out assessments on all six packages to detect and prioritise usability problems based on Nielson's scale [26]. In the list, catastrophic ratings (16%) number second overall in severity, while major ratings (12%) number third. The outcome indicates alarm on usability issues for older users found in the evaluated software. In the reports of Watkins et al. [20], and Silva et al. [30], expert evaluators assessed apps that comprised the healthcare class of software. The heuristics used in both evaluations were synthesised and modified from current heuristics and guidelines. The assessments show that the evaluated software is not in compliance with standard requirements set for effective usage by seniors. Notwithstanding the benefits of these researches, two critical gaps were still identifiable. First, applied usability heuristics require validation to sustain their effectiveness. Second, the research focused on UI evaluations for certain categories of applications and neglected UI evaluation of apps that are linked to the performance of daily tasks. Regardless of effort in improving the UI design of a smartphone, its ownership among the elderly users has seen a modest growth in recent years [31], this indicates that there is a need to come up with designs that cater to the needs of the elderly populace.

3 Method

As defined by Rogers et al. [17], heuristic evaluation has three stages: briefing session, evaluation period, and debriefing session. The heuristic evaluation procedure in this study is comprised of these three stages.

3.1 Briefing Session

During this stage, each expert was briefed by the observer (normally called the "experimenter" [32]) regarding the:

1. Case study applications: The study was conducted on alarm and camera applications installed on a Samsung Galaxy J7 device running Android v6.0.1 OS (Marshmallow). Smartphones are replacing more of individual's gadgets such as alarm

clocks and digital cameras [33] and using an alarm clock and taking pictures (using a camera) becoming important even for elderly people [34].
2. Heuristics: A set of 12 SMArtphone's uSability Heuristics (SMASH) [19], was applied to evaluate the case study applications. SMASH was experimentally validated, and the results supported its utility and effectiveness [19]. Applying specific heuristics, SMASH in this article, can potentially detect exact usability issues related to the application's domain [19, 27].
3. Deteriorations Checklist: The age-related limitations which may affect the ability of elderly adults to use a smartphone include; vision, hearing, cognitive and motor function limitations [10]. These limitations were summarised in a checklist to help the experts envision the target users (elderly) of smartphones during the evaluation period.

3.2 Evaluation Period

Five experts separately inspected UI of the case study applications to determine conformance with SMASH and reported the detected potential usability problems. Two experts are academic staff who possess 11 years' experience in the HCI area. While the other three experts are: one graphic designer, game developer (including mobile games), and web and graphics designer. The participation of five experts can be productive to detect about 75% of usability problems [17, 18, 32]. Experimenter read through all the problems identified by the five experts, eliminate duplicates, and compiled into a single master sheet.

3.3 Debriefing Session

This phase started when experts provided with the master sheet, rated the severity of the problems based on applying the Nielsen scale [26], then the mean was taken. At the end of this phase, experts proposed devised design solutions for the identified usability problems.

4 Results and Discussion

4.1 Assessment of the Smartphone UI

The compliance of alarm and camera applications to SMASH was inspected while the experts performed the assessment. Six usability problems, P1 to P6, with the applications were identified.

Under alarm uncommon single letter abbreviation is used to present the seven weekdays (P1), i.e., M presents Monday, T presents Tuesday, etc. (refer to Fig. 1(a)). Such notation for presenting days' abbreviations would not be well known by the elderly who would be more familiar with presenting weekdays with a full word. This design violates "Match between system and the real world" heuristic that indicates the language (text or icons) should be related to the real world and/or recognizable concepts [19]. Furthermore, due to the elders limited or failing memory [10], elderly could have difficulties to remember the abbreviation's meaning. This will increase elderly's memory load, and violate SMASH 6 that suggest to "minimize the user's memory load". Thus, P1 caused violations to two heuristics simultaneously.

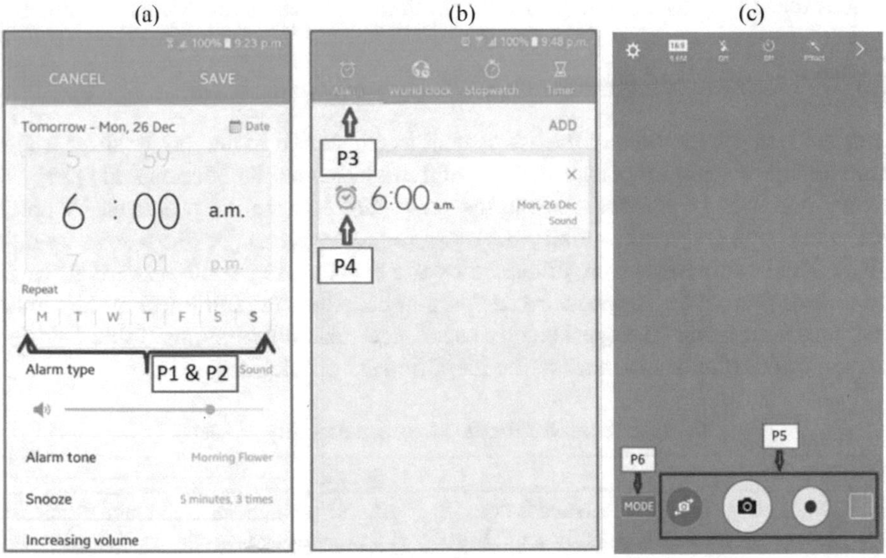

Fig. 1. The detected usability problems.

Elderly user should tap on all days to activate repeat the alarm for every day; seven taps are required to select the seven days (P2). This pattern violates "Efficiency of use and performance" heuristic which suggests that the smartphone should be able to minimize the necessary number of steps to perform a task [19]. Refer to Fig. 1(a).

"Minimize user's memory load" heuristic suggests the sensitive information should be placed in a visible spot [19]. Violations of this suggestion have been detected since the alarm function is hidden as a tab under the clock application (P3), refer to Fig. 1(b). Due to the elders failing vision [10], presenting UI elements in hidden/unobserved spots does not promote recognition for elderly users and increase their memory load since they would need to recall how to find UI elements presented in hidden spots.

Under the alarm application, tapping on the alarm icon will deactivate the alarm instantly without warning the elderly user about the consequence of this action (P4), refer to Fig. 1(b). Given the elderly user's motor functions limitations, an accidental interaction with the UI can perform this critical action immediately. This usability problem violated "error prevention" heuristic that requires the system to "warn users about critical actions" [19].

According to "consistency and standards" heuristic, the smartphone should follow established conventions, allowing users to do tasks in a familiar, standard and consistent manner [19]. The camera application has five buttons to execute the main camera functions; take pictures, record video, switch camera, mode, and go to the gallery. Four of these buttons (all except the "MODE" button) have icons configured that are not consistent in size and shape (P5), refer to Fig. 1(c). "MODE" button labelled with text only, unlike the adjacent four buttons (e.g., switch camera) those labelled with the icon (P6). Like the previous problem (P5), this problem (P6) violates "consistency and standards" heuristics (refer to Fig. 1(c)). The violation of consistency and standards heuristic reduces the elderly user's ability to detect buttons and realize their functionalities.

4.2 Debriefing Session Results

After analysing the results of the assessment as illustrated in the master sheet, experts estimated the severity of each usability problem, based on the Nielsen scale [26], with the average (mean) recorded. Among the six identified usability problems, P2 and P4 were catastrophic (severity rating ≥ 3.5), and the rest were major ($3.5 >$ severity rating ≥ 2.5). The six problems were discussed by the experts. As a result, experts suggested solutions that would help produce a design that considers the fulfilment of the violated heuristics definition and respect elderly users' age-related limitations. Table 1 presents the suggested solutions for each of the identified six problems.

Table 1. Solutions for the identified usability problems.

ID	Problem	Solution
P1	Day's single letter abbreviation is not well known by the elderly. Elderly could have difficulties to remember the abbreviation's meaning	Avoid ambiguous uncommon terms, i.e., present weekdays with a full word or three letter abbreviations
P2	Elderly user should tap on all days to activate repeat the alarm for every day	Minimize the required number of steps to do the task. Provide a selection that directly activates repeat alarm every day
P3	Presenting alarm as a tab under the clock application does not promote recognition for elderly users and increase their memory load, i.e., elderly would need to recall how to find the hidden alarm function	Present the frequently needed UI elements in a visible spot, i.e., the alarm should be presented as an independent application with a representative icon
P4	Given the elderly's motor functions limitations, an accidental pressing on the alarm icon will instantly deactivate the alarm without warning the elderly about the consequence of their action	Provide a descriptive confirmation message for each critical action indicating the action's consequence
P5	Take pictures, record video, switch to front camera, and go to gallery icons configurations are not consistent in size and shape, which reduces the elderly user's ability to detect buttons	Apply a standard (well-known) icon design configuration, which is consistent in size and shape, for all UI elements playing related functions
P6	"MODE" button labelled with text only, unlike the adjacent four buttons (e.g., switch camera) those labelled with the icon. This reduces the elderly user's ability to realize buttons functionalities	Strive for consistency in labelling UI elements, i.e., label "MODE" and its adjacent buttons with icon and text

4.3 Discussion

Through applying heuristic evaluation technique, experts identified six potential usability problems in the alarm and camera applications that could be encountered by elderly. Five heuristics, out of 12, were violated during the study. This limited number of the identified usability problems and unfulfilled heuristics can be attributed to the scope of the study that considered the inspection of the UI of alarm and camera applications that are essentially designed to perform certain specific functions: take pictures/ record video and set alarm. Regardless of this limited number of the discovered usability problems, study findings reveal that the case study applications in some instances failed to provide UI design that can; (i) Promote recognition over recall, (ii) Improve user's efficiency of use, and (iii) Achieve an intuitive and consistent interface.

- Promote Recognition Over Recall.
  Showing user objects he can recognize improves usability over needing to recall items from scratch because the extra context helps a user get back information from memory [35]. The significance of promoting recognition over recall becomes more influential when it comes for designing UIs for elderly because of their deteriorated memory. Problem 3 (P3) which is related to hiding the frequently needed alarm application as a tab under the clock application violates this principle. Hence, experts suggested to present such important UI elements in a visible spot, so the elderly would recognize where and how to find these elements.
- Improve User's Efficiency of Use.
  Even though this study focused on evaluating the UI design of alarm and camera in supporting elderly users. Certain usability problems could have a similar impact on users, irrespective of their age. For example, problem 2 (P2) that requires tapping on all days to activate repeat an alarm for everyday will affect user efficiency because of the time and number of steps required to perform the task. All users would be exposed to this problem, and user abilities and experience, would not help to immunize this problem's impact. While other problems could have a similar impact on both elderly and novice users. Like instant alarm deactivation after tapping on alarm icon without giving a warning message (P4). By letting error to happen, the user will lose time-solving it, which can lead to less efficiency. The effect of such problems on a young novice user will be diminished with a gradual improvement in smartphone usage experience propped by the young user's mental ability to practice their usage experience to deal with such conditions [4].
- Achieve an intuitive and consistent interface.
  Consistency in interface design is concerned with making UI elements uniform [19]. Consistency plays a significant role in helping users become familiar with the software product thus they can achieve their goals more easily [18]. Problem 5 (P5) and problem 6 (P6) show an example of a violation of consistency where the related icons under one application (camera) are not consistent in size, shape, and label. Hence, achievement of an intuitive and consistent interface is not fulfilled.

5 Conclusion and Future Work

In this study, the usability of the smartphone applications in supporting elderly users was evaluated through applying a heuristic evaluation technique. Experts identified six usability problems that elderly would probably face while interacting with the UI. Analysing the identified problems revealed that they were attributed to three issues UI failed to adhere to; (i) Promote recognition over recall, (ii) Improve user's efficiency of use, and (iii) Achieve an intuitive and consistent interface. Solutions suggested by experts consider providing alternative UI design that promoting recognition, improving efficiency, and achieving consistency. The limitation of this study was that it lacked the implementation of the solutions recommended by experts. As a future work these solutions will be implemented as a high-fidelity prototype, and it will be a subject for a user-based test to assess its usability.

Acknowledgement. This work was supported by the Fundamental Research Grant Scheme (FRGS) from the Ministry of Higher Education of Malaysia under Grant FRGS/1/2017/ ICT04/UTP/02/2. The work of H. M. Salman was supported by Universiti Teknologi Petronas.

References

1. Linghao, Z., Ying, L.: On methods of designing smartphone interface. In: IEEE International Conference on Software Engineering and Service Sciences (ICSESS), pp. 584–587. IEEE (2010)
2. Loureiro, B., Rodrigues, R.: Design guidelines and design recommendations of multi-touch interfaces for elders. In: Proceedings of the ACHI, pp. 41–47 (2014)
3. Strengers, J.: Smartphone interface design requirements for seniors. In: Information Studies. University of Amsterdam, Amsterdam (2012)
4. Harada, S., Sato, D., Takagi, H., Asakawa, C.: Characteristics of elderly user behavior on mobile multi-touch devices. In: Kotzé, P., Marsden, G., Lindgaard, G., Wesson, J., Winckler, M. (eds.) Human-Computer Interaction–INTERACT 2013, pp. 323–341. Springer, Berlin (2013)
5. American Psychological Association: Memory and aging. http://www.apa.org/pi/aging/memory-and-aging.pdf. Accessed 23 Apr 2013
6. Fozard, J.L.: Vision and hearing in aging. In: Handbook of the Psychology of Aging, vol. 3, pp. 143–156 (1990)
7. Chaparro, A., Rogers, M., Fernandez, J., Bohan, M., Sang Dae, C., Stumpfhauser, L.: Range of motion of the wrist: implications for designing computer input devices for the elderly. Disabil. Rehabil. **22**, 633–637 (2000)
8. Walker, N., Philbin, D.A., Fisk, A.D.: Age-related differences in movement control: adjusting submovement structure to optimize performance. J. Gerontol. B Psychol. Sci. Soc. Sci. **52**, P40–P53 (1997)
9. Clarkson, P.J., Coleman, R., Keates, S., Lebbon, C.: Inclusive Design: Design for the Whole Population. Springer, Berlin (2013)
10. Balata, J., Mikovec, Z., Slavicek, T.: KoalaPhone: touchscreen mobile phone UI for active seniors. J. Multimodal User Interfaces **9**, 263–273 (2015)
11. Kalimullah, K., Sushmitha, D.: Influence of design elements in mobile applications on user experience of elderly people. Procedia Comput. Sci. **113**, 352–359 (2017)

12. Piper, A.M., Garcia, R.C., Brewer, R.N.: Understanding the challenges and opportunities of smart mobile devices among the oldest old. Int. J. Mobile Hum. Comput. Interact. (IJMHCI) **8**, 83–98 (2016)
13. Lewis, J.R.: Usability: lessons learned… and yet to be learned. Int. J. Hum. Comput. Interact. **30**, 663–684 (2014)
14. Dillon, A.: The Evaluation of Software Usability. Encyclopedia of Human Factors and Ergonomics. Taylor and Francis, London (2001)
15. Yáñez Gómez, R., Cascado Caballero, D., Sevillano, J.-L.: Heuristic evaluation on mobile interfaces: a new checklist. Sci. World J. **2014** (2014). Article ID 434326, 19 pages
16. Inostroza, R., Rusu, C., Roncagliolo, S., Rusu, V.: Usability heuristics for touchscreen-based mobile devices: update. In: Proceedings of the Chilean Conference on Human-Computer Interaction, pp. 24–29. ACM (2013)
17. Rogers, Y., Sharp, H., Preece, J.: Interaction Design: Beyond Human-Computer Interaction. Wiley, New York (2015)
18. Shneiderman, B., Plaisant, C., Cohen, M.S., Jacobs, S., Elmqvist, N., Diakopoulos, N.: Designing the User Interface: Strategies for Effective Human-Computer Interaction. Pearson, New York (2016)
19. Inostroza, R., Rusu, C., Roncagliolo, S., Rusu, V., Collazos, C.A.: Developing SMASH: a set of SMArtphone's uSability Heuristics. Comput. Stand. Interfaces **43**, 40–52 (2016)
20. Watkins, I., Kules, B., Yuan, X., Xie, B.: Heuristic evaluation of healthy eating apps for older adults. J. Consum. Health Internet **18**, 105–127 (2014)
21. Almarashdeh, I., Alsmadi, M.: Heuristic evaluation of mobile government portal services: an experts' review. In: 11th International Conference for Internet Technology and Secured Transactions (ICITST), pp. 427–431. IEEE (2016)
22. Santana, P.C., Anido, L.E.: Heuristic evaluation of an interactive television system to facilitate elders home care. IEEE Latin Am. Trans. **14**, 3455–3460 (2016)
23. Lilholt, P.H., Jensen, M.H., Hejlesen, O.K.: Heuristic evaluation of a telehealth system from the Danish TeleCare North Trial. Int. J. Med. Inf. **84**, 319–326 (2015)
24. Salman, Y.B., Cheng, H.-I., Kim, J.Y., Patterson, P.E.: Medical information system with iconic user interfaces. JDCTA **4**, 137–148 (2010)
25. Allen, M., Currie, L.M., Bakken, S., Patel, V.L., Cimino, J.J.: Heuristic evaluation of paper-based Web pages: a simplified inspection usability methodology. J. Biomed. Inf. **39**, 412–423 (2006)
26. Nielsen, J.: Usability inspection methods. In: Conference Companion on Human Factors in Computing Systems, pp. 413–414. ACM (1994)
27. Hermawati, S., Lawson, G.: Establishing usability heuristics for heuristics evaluation in a specific domain: is there a consensus? Appl. Ergonom. **56**, 34–51 (2016)
28. Nielsen, J.: 10 Usability Heuristics for User Interface Design. Nielsen Norman Group, California, CA, USA (1995)
29. Al-Razgan, M.S., Al-Khalifa, H.S., Al-Shahrani, M.D.: Heuristics for evaluating the usability of mobile launchers for elderly people. In: International Conference of Design, User Experience, and Usability, pp. 415–424. Springer (2014)
30. Silva, P.A., Holden, K., Nii, A.: Smartphones, smart seniors, but not-so-smart apps: a heuristic evaluation of fitness apps. In: International Conference on Augmented Cognition, pp. 347–358. Springer (2014)
31. Bai, Y.-W., Chan, C.-C., Chia-Hao, Y.: Design and implementation of a simple user interface of a smartphone for the elderly. In: 2014 IEEE 3rd Global Conference on Consumer Electronics (GCCE), pp. 753–754. IEEE (2014)

32. Nielsen, J.: How to Conduct a Heuristic Evaluation. Nielsen Norman Group, Fremont, CA, USA (1995)
33. Nagarkoti, B.: Factors influencing consumer behavior of smartphone users. Helsinki, Finland: ARCADA (2014)
34. Ropponen, J.-O.: Usability of mobile devices and applications for elderly users. Vantaa, Finland: Laurea University of Applied Sciences (2016)
35. Budiu, R.: Memory Recognition and Recall in User Interfaces. Nielsen Norman Group, California, USA (2014)

Information Systems: Modeling the Adoption of Innovations

The Application of Computational Thinking and TRIZ Methodology in Patent Innovation Analytics

Zulhasni Abdul Rahim(✉), Shazlinda Md Yusof, Nooh Abu Bakar, and Wan Md Syukri Wan Mohamad

Malaysia-Japan International Institute of Technology, Universiti Teknologi Malaysia Kuala Lumpur, Kuala Lumpur, Malaysia
zulhasni@utm.my

Abstract. Patent is the most high-level document that involved technical inventive solution and legal value for intellectual property protection. Every patent documented will explained about their system in solving a specific problem. This document has critical information that can be used for various purposes such as patent innovation, technology forecasting and developing patent strategy. This paper is focusing to explore the methodology in using the innovation to enhance patent innovation and competitiveness. The proposed methodology uses computational thinking model and TRIZ methodology. It consists of innovative processes such as anatomization, schematization and tabulation of a patent. This research shows new approach to create innovation and develop analytics study of case study patent for continuous improvement and strategy development.

Keywords: Computational thinking · TRIZ · Patent · Innovation · Analytics

1 Introduction

One of the most significant elements in patent development is the unchanging requirements of a patent. Those common requirements are novelty, inventive steps and industrial application [1]. Recently, the European Patent Convention (EPC) revised the requirement that focus on inventive steps, which become a great challenge related to central legal provision [2]. The recommended solution made by The Boards of Appeal of the European Patent Office (EPO) is to introduce new approach called the "problem-and-solution approach" [3]. This open up opportunity for strategic methodology called Theory of Inventive Problem Solving or commonly known as TRIZ in Russian acronyms, to enhance the new approach. TRIZ is founded by Genrich Altshuller, who was been assigned as patent officer and discovered that problem-and-solution extracted from patent database that repeated across industries [4]. The fundamental of TRIZ methodology expands in the domain of patent analysis with many strategy and techniques such as patent strategy, patent circumvention, patent deconstruction, etc. [5].

The challenges we are facing currently that there is still limited information about the structured methods for patent innovation activities. One of the critical barriers is the

F. Saeed et al. (Eds.): IRICT 2018, AISC 843, pp. 793–802, 2019.
https://doi.org/10.1007/978-3-319-99007-1_73

multidisciplinary that involved in patent research, such as engineering and law [4]. This create a complex relationship in terms of data processing and carry out analysis. The common tools and approach in managing complex data processing and analysis is by using support system such as computer and software. The data processing needs to use computational model to improve the structure of the patented big data and help in the development of patent innovation systematically [5]. Using four elements of computational thinking (CT) concept such as, decomposition, abstraction, pattern recognition and algorithm, the patent can be transform into a form that support the process of patent innovation activities, as shows in Fig. 1 [6]. However, it is a critical challenge to use CT concept independently and focus on the innovation impact. Introducing the integration model between TRIZ and CT could provide an opportunity in improving the performance of patent innovation program. The objectives of this paper are to explore the concept of integrating computational thinking and TRIZ methodology through a proposed model for patent innovation and develop a case study to validate the model.

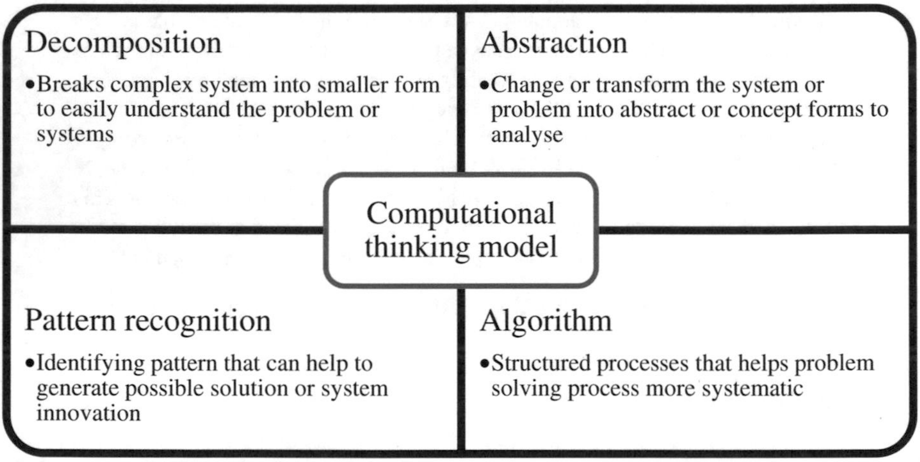

Fig. 1. Concept of computational thinking

1.1 Patent Circumvention Methodology

One of the patent analysis process that used TRIZ as its fundamental mechanism is called patent circumvention. This mechanism able to provide support in decomposition and abstraction of a patent using TRIZ tools called function analysis [7]. The process starts by taking the independent claims of a patent and converted them into a diagram called function model [8]. The function model consists of boxes that can be refer to each component present in the patented system. The model also includes interconnected arrows between boxes which represent the functions that carried by the components [9]. Figure 2 shows an example on how function model is decomposing and converting the independent claim to abstract form.

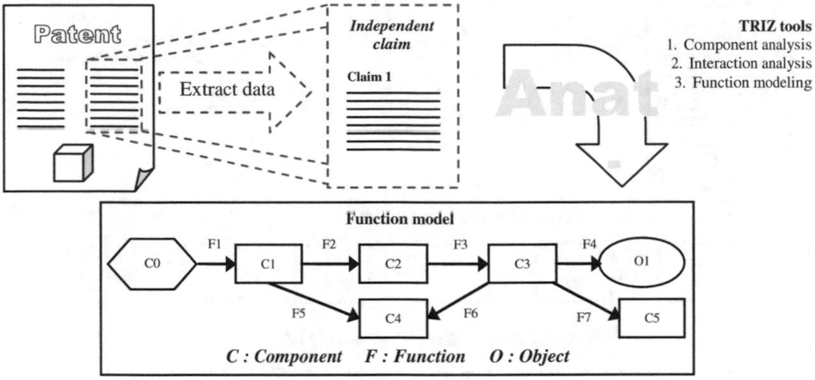

Fig. 2. Anatomization strategy using decomposition and abstraction techniques

The independent claims are going through decomposition process by changing and rearranging the text form into components and functionality forms. The function analysis able to show all relevant components at system level that existed in the independent claim. The technical word used in the patent in explaining functionality of the system might be too complex to comprehend, but the function analysis helps to generalize the word to a basic verb in developing the function model [10]. The abstraction focuses on converting the independent claims, which is in the text form and change it into the functionality form with boxes and arrows. The function model provides overview of system components that mentioned and protected by the independent claim of the original patent. By reducing the number of components while maintaining or changing the system functionality, the original patent can be innovating or circumventing to a new improved patent [11].

1.2 Patent Deconstruction Methodology

The next approach of TRIZ patent analytics is patent deconstruction, which not involving function analysis as tool for patent analysis [12]. This approach analysed the patent's independent claim to a pattern flow and identify the sequence as algorithm of patented system operation according to the patent's inventive steps. Figure 3 shows an example on patent is schematize using pattern flow and algorithm.

Information about inventive steps is very crucial to understand how patented system operates and what level of technology used for the system [13, 14]. This information was not visible in the patent circumvention approach. However, the patent deconstruction approach is putting less priority on the components and functions, which is critical in developing a strong patent protection [15]. By changing the pattern flow or the algorithm elements of inventive steps, a new innovative patent can be developed around the original patent [16].

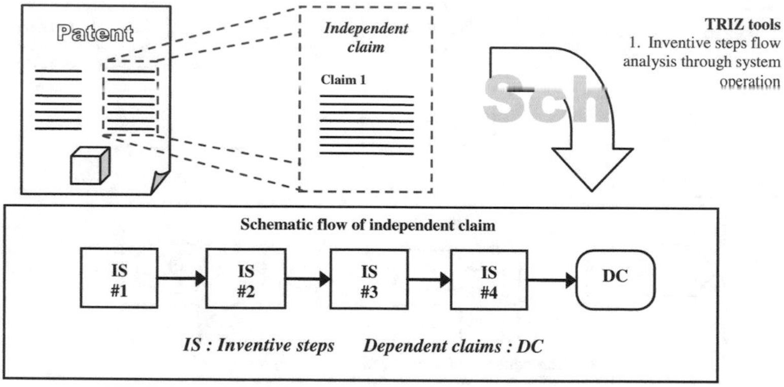

Fig. 3. Schematization strategy using pattern flow and algorithm techniques

2 Integration Framework of Computational Thinking Model and TRIZ Methodology for Patent Analytics

The study proposed a model that includes advantages of both strong patent analyses with TRIZ methodology. The advantages from patent circumvention are the visualization of components and its function using function analysis [17, 18]. This technique is significant in anatomizing the independent claim to a function model. Meanwhile, the advantages from patent deconstruction are the visualization of inventive steps mentioned in the independent claim, which equally important elements for schematize a patent [19]. Therefore, hybridization of both advantages creates a model called 'Anatomize-Schematic-Tabulate' (AST model). Figure 4 shows the proposed model for new approach in patent innovation using integrated CT concept and TRIZ methodology.

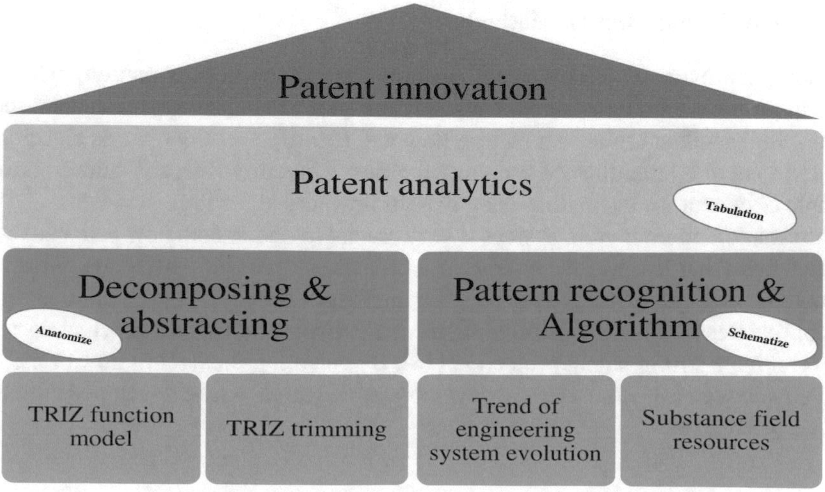

Fig. 4. Proposed AST model for patent innovation

The application of proposed TRIZ tools brings the innovation elements to the targeted patent. The TRIZ function model breaks the system into component level and convert each components based on functionality. This give advantages to the patent anatomization strategy with CT's decomposition and abstracting concept. The TRIZ's trimming tools is paired in the strategy to change the targeted patent at component level using the same functionality concept to achieved the patent innovation goals. Meanwhile TRIZ tools such as Trend of engineering system evolution is used to propose changes based from conceptual technology trends established under TRIZ methodology and Substance field resources is used to explore vertical and horizontal resources that existed in the current patented system [20]. The proposed TRIZ tools able to complement the patent innovation method that aligned with CT's patent recognition and algorithm concept.

The final activities in the AST model is the tabulation of schematize and anatomize patent innovation. The tabulation involved reviewing all elements after the patent has been schematize and anatomize in visual presentation. The tabulation for schematic inventive steps provides opportunity to change the flow or make it shorter for improve patent description [21]. The simpler and shorter your independent claim, the stronger the scope of patent protection is created. Within the each schematize inventive steps, the identified components and functions provide opportunity to reduce the components or substituting another technology [22]. The functions also can be change with the guidance of TRIZ's trimming rules [23]. For each component, there is also opportunity to change the engineering parameters, design, inventive steps, technology, function, performances, etc. to improve the original patent, as shown in Fig. 5.

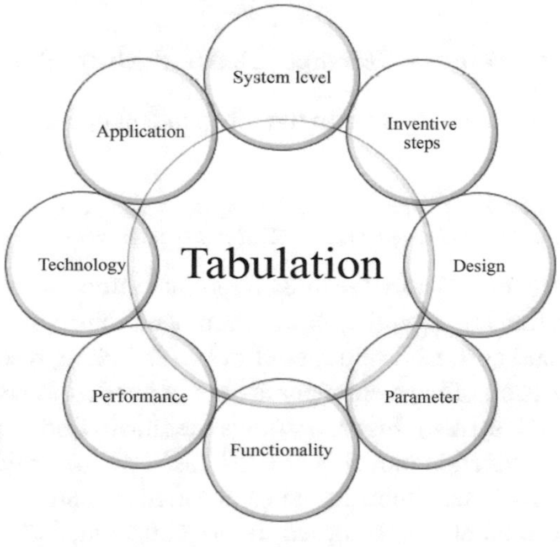

Fig. 5. Tabulation elements for patent analytics

3 Case Study: Application of AST Model on Patented Product (US20050156085A1)

A simple patented product called 'Magnetic paint brush holder' with patent number US20050156085A1 is selected for the case study of the application of AST model, as show in Fig. 6 [24]. According to the background information, the invention is an accessory for painting work on solving problem related to providing the solution to store a paint-laden brush after used. The inventor has cited six patents that similar product application and highlighted his patent originality compared to others. Understanding the background of the selected patent is important because it provides the context of the problem that the inventor is focusing on.

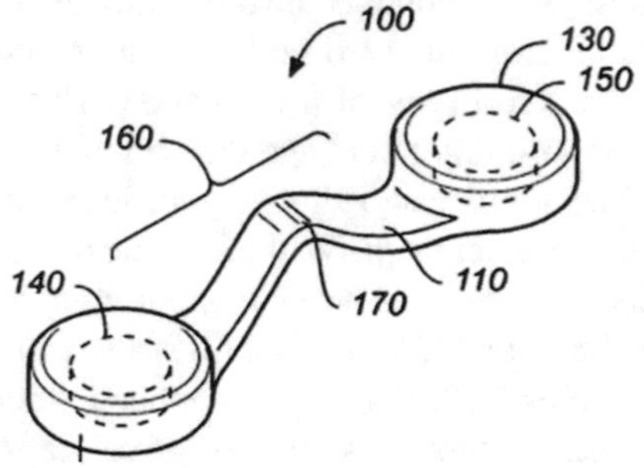

Fig. 6. Illustration of patented system: US20050156085A1

3.1 Anatomization Strategy – Decomposition and Abstraction

In AST model, the first step is to confirm the independent claim of selected patent. The claim mentioned;

> "*1. A paint brush holder, comprising: a strap having a first end, a second end, and a medial portion; at least one permanent magnet disposed at one of said first or second ends.*"

The independent claim provides the most important information of patented system. Using the anatomization strategy, the information need to use decomposition concept at component level and convert it into model of boxes that represent the components. Furthermore, the function of each components are converted to abstract forms, which represented by as arrow symbol. Figure 7 shows the illustration of patented system for visualization. A TRIZ function model is created based on the critical components and functions that have strong legal implication of patented system.

Using the independent claim, the function model decomposes the patented system and convert it into abstract form. The abstract forms in shapes and linked with arrows represents components such as the magnetic paint brush holder and includes all related components that mentioned in the dependent claims, as shown in Fig. 7. The grey zone represents the component and function stated in the independent claim.

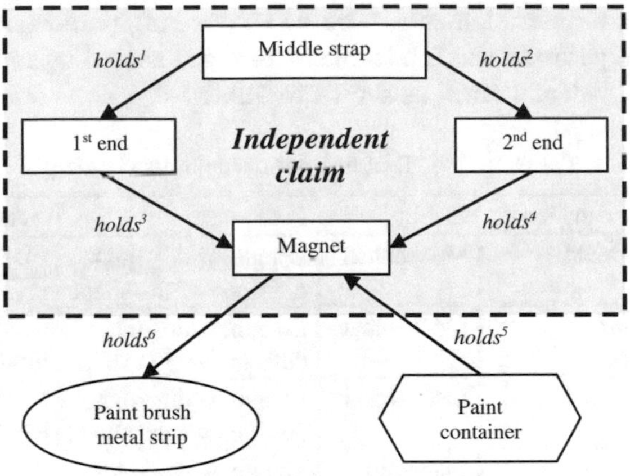

Fig. 7. Function model of patented system

3.2 Schematization Strategy – Pattern Recognition and Algorithm

The second strategy of AST model used is schematization. Each function was numbered according to the sequence of inventive steps mentioned in the independent claim. This helps to schematize the function model for pattern recognition and developing algorithm. The patent file also showed the illustration of the application of magnetic paint brush holder that provide physical image and specific application of the patented system, as show in Fig. 8.

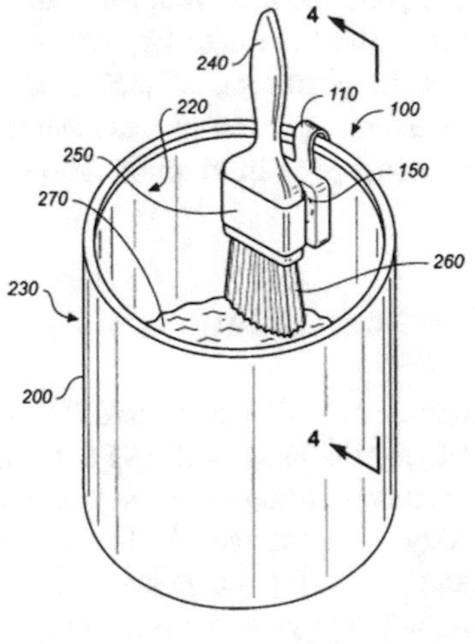

Fig. 8. Application of patented system

The tabulation analysis reviewed all identified components in the patented system in a TRIZ tool called 'Tabulation - Substances Field Resources' (T-SFR) table. The T-SFR table contains the list of components at system level and each component is review according to their elements, such as engineering parameters, functionality, fields, design and performance. The table is arranged according to the sequences of inventive steps of patent claims, as shown in Table 1.

Table 1. T-SFR of magnetic paint brush holder

IS	Component	Function	Field	Parameters	Action
1	Middle strap	Hold	Mechanical	Length, width, thick, deform, weight, shape	- Delete component or inventive steps - Review all TRIZ trends - Change field or parameter
2	1st end	Hold	Mechanical	Length, width, thick, deform, weight, shape	
3	2nd end	Hold	Mechanical	Length, width, thick, deform, weight, shape	
4	Magnet	Hold	Magnetic	Length, width, thick, deform, weight, force, area, shape	
5	Paint container	Hold	Magnetic	Length, width, thick, deform, weight, force, area, shape	
6	Paint brush metal strip	Hold	Magnetic	Length, width, thick, deform, weight, force, area, shape	

In the T-SFR table, Trend of engineering system evolution is imposed on each parameters and forms for each components. This exploration of trend pattern recognition opens wide opportunity to innovate the function, form and shape of identified components in the table. For example, the field category will explore the changes between, mechanical, acoustic, thermal, chemical, electric, magnetic, etc. using TRIZ MATChEM trends. This pattern recognition process using TRIZ trends help to speed up the process in developing concept of innovation with systematic and guided methodology.

4 Conclusions and Recommendations

This paper presented the integration of CT concept and TRIZ methodology in developing patent innovation processes using the proposed AST model. The purpose of this study introduces new perspective in patent innovation and analytics in patent study using CT concept and TRIZ methodology. The proposed AST model has explored the opportunity to innovate the existing patent with CT concept such as decomposition using function model, abstraction using function analysis, pattern recognition using T-SFR table and algorithm from inventive steps. The case study on selected patent showed how AST model is used in anatomizing, schematizing and tabulating the information of patented system. The present study provides additional evidence with respect to patent analytics using CT concept and TRIZ methodology through the proposed AST model. Further work needs to be done to establish whether the proposed model able to enhance faster and better approach in patent coding such as for big data [22]. This might uncover the potential enhancement of patent analytics and add significant value to the body of knowledge in patent study.

Acknowledgement. This paper is based on research grant (Ref. No.: PY/2017/01434), which have been awarded by Universiti Teknologi Malaysia (UTM).

References

1. Verhoeven, D., Bakker, J., Veugelers, R.: Measuring technological novelty with patent-based indicators. Res. Policy **45**(3), 707–723 (2016)
2. Cope, M.: The inventive step requirement in United Kingdom Patent Law and Practice. The Patent Office (2006)
3. Seville, C.: EU intellectual property law and policy. Elgar European Law Series, October 2016. http://dx.doi.org/10.4337/9781781003480
4. Liu, K., Siying, H.: An analysis of patent comprehensive of competitors on electronic map & street view. 3(10) (2016)
5. Jhuang, A.C., Sun, J.J., Trappey, A.J.. Trappey, C.V., Govindarajan, U.H.: Computer supported technology function matrix construction for patent data analytics. In: 2017 IEEE 21st International Conference on IEEE Computer Supported Cooperative Work in Design (CSCWD), pp. 457–462 (2017)
6. Trappey, A.J., Trappey, C.V., Govindarajan, U.H., Chuang, A.C., Sun, J.J.: A review of essential standards and patent landscapes for the Internet of Things: a key enabler for industry 4.0. Adv. Eng. Inf. **33**, 208–229 (2017)
7. Chechurin, L.: TRIZ in science. Reviewing indexed publications. Procedia CIRP **39**, 156–165 (2016)
8. Zulhasni, A.R., Nooh, A.B.: Innovative cases of TRIZ application in the automotive industry. Appl. Mech. Mater. **735**, 326–330 (2015)
9. Zulhasni, A.R., Issac, L.S., Nooh, A.B.: TRIZ methodology for applied chemical engineering: a case study of new product development. Chem. Eng. Res. Des. **103**, 11–24 (2015)
10. Zulhasni, A.R., Nooh, A.B.: Complexity planning for product design using TRIZ. Adv. Mater. Res. **903**, 396–401 (2014)
11. Miao, L., Ming, X., Zheng, M., Xu, Z., He, L.: A framework of product innovative design process based on TRIZ and patent circumvention. J. Eng. Des. **24**(12), 830–848 (2013)
12. Zanten, V., Veldhuijzen, J., Wits, W.W.: Patent circumvention strategy using TRIZ-based design-around approaches. Procedia Eng. **131**, 798–806 (2015)
13. Zulhasni, A.R., Nooh, A.B.: Technology development and assessment to market using TRIZ. Int. J. Bus. Anal. (IJBAN) **3**(4), 83–97 (2016)
14. Yanhong, L., Tan, R., Ma, J.: Patent analysis with text mining for TRIZ. In: IEEE Conference on Management of Innovation and Technology, ICMIT 2008, pp. 1147–1151 (2008)
15. Li, M., Ming, X., He, L., Zheng, M., Xu, Z.: A TRIZ-based trimming method for patent design around. Comput. Aided Des. **62**, 20–30 (2015)
16. Ilevbare, I.M., Probert, D., Phaal, R.: A review of TRIZ, and its benefits and challenges in practice. Technovation **33**(2), 30–37 (2013)
17. Loh, H.T., Cong, H., Lixiang, S.: Automatic classification of patent documents for TRIZ users. World Patent Inf. **28**(1), 6–13 (2006)
18. Zulhasni, A.R., Nooh, A.B.: Elevate design-to-cost innovation using TRIZ. In: Research and Practice on the Theory of Inventive Problem Solving (TRIZ), pp. 15–33 (2016)

19. Verhaegen, P.A., Joris, D., Joris, V., Dewulf, S., Joost, R.D.: Relating properties and functions from patents to TRIZ trends. CIRP J. Manuf. Sci. Technol. **1**(3), 126–130 (2009)
20. Cascini, G., Russo, D.: Computer-aided analysis of patents and search for TRIZ contradictions. Int. J. Prod. Dev. **4**(1–2), 52–67 (2006)
21. Zulhasni, A.R., Nooh, A.B.: Technology forecasting using theory of inventive problem solving: a case study of automotive headlamp system. PERINTIS eJournal **6**(1) (2016)
22. Gao, C.Q., Wei, C., Mi, S.M., Zhao, F.: Research and implementation on patent knowledge modeling based on TRIZ. Appl. Mech. Mater. **775**, 479–483 (2015)
23. Hyunseok, P., Ree, J.J., Kim, K.: Identification of promising patents for technology transfers using TRIZ evolution trends. Expert Syst. Appl. **40**(2), 736–743 (2013)
24. Radovan, T.: Inventor: Magnetic paint brush holder. United States patent application, US 11/021,826 (2004)

Determinant Factors of Customer Satisfaction for E-Hailing Service: A Preliminary Study

Nur Athirah Nabila Mohd Idros(✉), Hazura Mohamed, and Ruzzakiah Jenal

Faculty of Information Science and Technology, National University of Malaysia (UKM), Bangi, Selangor, Malaysia
athirahnabila.idros@gmail.com, {hazura.mohamed, ruzzakiahjenal}@ukm.edu.my

Abstract. E-hailing is an online transportation service that has emerged as the passengers' favourite in transportation industry that involves interaction between customers and service provider. However, as customers in this era have more information and choice to choose any facilities, the service provider needs to satisfy and provide better service to their large number of customers especially for e-hailing service. Maintaining customer satisfaction is crucial for e-hailing's service provider to sustain competitive advantages. Company-customer interaction is one of the definitions for value co-creation. Consequently, value co-creation approach has been used to increase customer satisfaction as the customers take part in creating the experiences and contribute value-in-context that will improve company's performance. Therefore, this study aims to determine technological factors that influence customer satisfaction for e-hailing service by embedding value co-creation elements. Using design science research in information system methodology, the factors of customer satisfaction for e-hailing service are identified and classified based on outline of the study phases in sequences. In this paper, the components of value co-creation will be embedded by comparing the similarities between value co-creation model and existing customer satisfaction models from past researchers.

Keywords: Customer satisfaction model · E-hailing · Value co-creation

1 Introduction

The rising demand for e-hailing or online transportation service has become a phenomenon of mobility all over the world as Uber introduced in 2009 while Lyft arrived in 2012 [1]. E-hailing is an online service that allow the customers to request for a ride by using computer or mobile-phone, then, along with the information given, the driver will be guided to the customers' location to get them and go to the customer's destination and the fare will be charged based on distance travel [2].

Khalaf Ahmad and Ali Al-Zu'bi [3] stated that a system should be well designed and make an ease of use for customer to access as it is very important in influencing user satisfaction. Currently, e-hailing allowing the customers to rate their experience on a scale of 1 to 5 stars to keep track of customer satisfaction only towards the driver [4].

F. Saeed et al. (Eds.): IRICT 2018, AISC 843, pp. 803–811, 2019.
https://doi.org/10.1007/978-3-319-99007-1_74

It is clear that e-hailing service provider needs a customers' feedback in terms of technology provided. Hence, the purpose of this study will present the technological factors that have an impact on customer satisfaction towards e-hailing system based on value co-creation elements.

Value co-creation has been introduced in early 2000s as the most important strength in industrialist economies that carried a general concept between companies and customers to generate the value through interaction [5]. This paper also aimed to adapt value co-creation theories from the user's perspective. Precisely, this study sought to achieve the following objectives:

- To explore the theoretical framework of customer satisfaction, e-hailing and value co-creation.
- To determine technological factors of customer satisfaction for e-hailing service by embedding value co-creation.

The conceptual framework for this paper is shown in Fig. 1. In general, the framework consists of three steps of process which are input, process and output. Input covers the preliminary study and literature review that convey the same purposes, which were to determine the knowledge gaps in this study. After that, for process and output, the factors based on value co-creation elements will be identified to boost the level of customer satisfaction for e-hailing service.

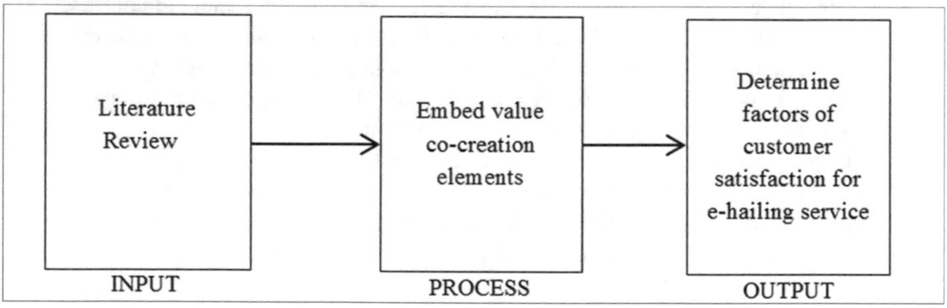

Fig. 1. Research conceptual framework

2 Related Works

This section will explain previous studies related to e-hailing service and discover factors that affecting customer satisfaction. Moreover, value co-creation elements will be highlighted in this study.

2.1 Customer Satisfaction Model

Satisfying the customer is the key point for an enterprise to increase company's profit that will lead to success and help to maintain the growth of country's economy [6]. Previous researchers have evaluated the factors that influencing customer satisfaction and the factors are now being used by the companies as a method of measurement to determine the level of customer satisfaction.

Parasuraman et al. [7] developed a service quality framework named as SERVQUAL model with a purpose to measure the scale of quality in the diverse services in the service sector. SERVQUAL model consists of five dimensions which are reliability, assurance, tangibles, empathy, and responsiveness.

Ojasalo [8] studied a theoretical foundations of e-service and proposed a quality model for e-service. This model has eight factors such as ease of use, website design, personalization, information, responsiveness, communication, security and reliability.

Meanwhile, Silalahi et al. [9] designed a service quality model for online transportation services in Indonesia. This online transportation service model included twenty factors which are website design, reliability, responsiveness, trust, personalization, perceived risk, perceived cognitive, privacy, compensation, contact, billing, punctuality, valence, content usefulness, content adequacy, ease of use, accessibility, interactivity, perceived website innovativeness, system availability. Basically, this model was designed based on previous researchers such as Salameh and Hassan [10], Huang et al. [11], and Stiakakis and Georgiadis [12].

These three customer satisfaction models are used as references in identifying factors of customer satisfaction for e-hailing service because these models covered areas of service quality, e-service and online transportation service. Service quality is measured to compare customer's before-service expectation with their actual-service experience [13]. Selvakumar [14] eloquently stated that a good quality of service leads to increase the customer satisfaction and obtain customer loyalty.

2.2 E-Hailing Service

E-hailing with the other names also known as ride-booking and ridesharing [15] become well-known in transportation industry throughout the world and now getting popular in Malaysia. Malaysian Institute of Language and Literature [16] defined e-hailing as an online process of ordering and booking a vehicle by using a computer or mobile device. Uber has been introduced in Malaysia on January 2014 while Grab introduced on May 2014 [17].

Parliament of Malaysia [18] approved two bills, the amendments to Malaysia's Land Public Transport Act and the Commercial Vehicles Licensing Board (CVLB) which under this law, Malaysia is allowing e-hailing services such as Uber and Grab to operate within the law. E-hailing has become urban transportation because the service expanding in many cities [1].

The fundamental function of e-hailing started with visited the website of e-hailing operators or installed the e-hailing application by downloading from online application store. Next, customer is required to register an account before using this service. After that, customer has to activate GPS location and order the vehicle so that the driver will move to customer's location and send the customer to the destination. This application allowed customer to view the information about the vehicle and driver included the ratings from other passengers before arrival. Then, process of booking the car can be done by selecting the car with one click and through the system, customer will be informed about the estimated price of ride, traffic information and approximated time of arrival. During the journey, this system will let the customer to look after the map to ensure the driver used the optimal travel route. The payment can be made either by cash or link the credit card with e-hailing application [2].

2.3 Value Co-creation

A value is produced when customer decided to purchase the product then started to compare the advantages and disadvantages of purchasing it, in order for making a better decision [19]. Value in this paper was focusing on value co-creation approach. Value co-creation is defined as an interaction between companies', customers' and other related actors' that engaged together through activities and united by using available resources to achieve the goals that set by value-in-context [20].

Customers in this era have more information and options to choose any products or services everywhere which made customers play an active role and very influential in value co-creation process, thus forcing the company to abandon the traditional firm-centric view to a new customer-centric view for success [21]. Hoyer et al. [22] stated that application of value co-creation help to increase customer satisfaction because when the customer participated to contribute for the product it will resulting value-in-context which remained as a resource that will integrate service activities.

Brodie et al. [23] suggested that to improve companies' performance, the customers must have an important role by informing other potential customers about product, service or brand including new service while creating the experiences together and shared the value. Companies' support for customers have been proved to have an impact on degree of customer co-creation as company-customer interactions for service provided will reflect on customers' overall satisfaction [24].

Value co-creation also been used by transportation provider to encourage participation from customers. According to Nunes et al. [25] public transport providers will offer rewards or incentives such as travel discount cards on monthly basis for their customers to encourage participation because it would be trustworthy to give award for the customers who actively co-create the value with the provider.

Ferris et al. [26] studied that getting customers joined in co-creating into public transport will return great potential since customers will retrieve the travel information from the station or via electronic devices and help to disclosure information about service delay, hence will benefit the public transport providers to manage customers' expectations for great travel experience and earned compliments.

DART Model

Prahalad and Ramaswamy [27] introduced a value co-creation model named DART model that contains four building blocks: dialogue, access, risk assessment and transparency.

- Dialogue – Dialogue is a process of communicating, deep involvement, and readiness to take an action on both sides. Dialogue is more than listening to customers as it is involving learning and interaction between two equal problem solvers that will produce a loyal community.

- Access – Access starts with information and tools. The company will provide a platform for customers to access the data and process in the best way. Access gives customers a variety of experiences based on customers' individualize needs.
- Risk assessment – Risk here refers to the possibility of potential danger to customers. Since customers have huge responsibility in co-creation, company needs to inform the customers fully about risk that associated with service provided and customers must willing to deal and manage the risk.
- Transparency – Transparency means the trust that has been evolved between customers and company. Customers tend to firm belief in service when the company provided them a detailed information and made data visible.

3 Method

In identifying the key factors of customer satisfaction for e-hailing service, this research paper used design science research in information systems methodology that has been designed by Kuechler and Vaishnavi [28]. Referring to Fig. 1 as a basic concepts, this study specifically elaborated the activities in study phases. This design science research methodology has been adapted and three phases were included which are awareness of problem, suggestion, and conclusion to make it compatible as this paper focused on preliminary study of discovering elements that influence customer satisfaction for e-hailing service.

First, awareness of problem phase, started with conducting a literature review to have deeper understanding for current issues on customer satisfaction models, e-hailing and value co-creation elements. After that, in suggestion phase, the factors of customer satisfaction for e-hailing service will be proposed by embedding value co-creation. Specifically, this paper will compare the similarities between DART model and factors from customer satisfaction models in related works and then, matched all of those factors with DART model. Finally, for conclusion phase, this study will be summarized by mentioning the purpose of this study and for future works. Table 1 indicates the outline of the phases and findings for this research.

Table 1. Study phases and findings

Study phases	Findings
Phase 1: Reviewed the literature on customer satisfaction models, e-hailing service and value co-creation theory	• Literature review
Phase 2: Determined the factors of customer satisfaction for e-hailing service by embedding value co-creation	• Embedded DART model • Matched the similarities between suggested factors of customer satisfaction models with DART model • Identified the final factors of customer satisfaction for e-hailing service
Phase 3: Concluded the study	• Summarized the study by highlighting the purpose of this study and future works

4 Result

The final result included the identified factors of customer satisfaction for e-hailing service by comparing the similarities between factors of customer satisfaction [7–9] from related works to match with DART model [27] as shown in Table 2.

Table 2. Comparing the similarities between factors from existing customer satisfaction models to match with dart model

DART model [27]	Factors from existing customer satisfaction models
Dialogue	• Empathy [7] • Responsiveness [7–9] • Communication [8] • Personalization [8, 9] • Contact [9] • Interactivity [9] • Valence [9] • Perceived website innovativeness [9]
Access	• Tangibles [7] • Information [8] • Ease of use [8, 9] • Website design [8, 9] • Accessibility [9] • Content adequacy [9] • System availability [9]
Risk assessment	• Perceived risk [9] • Perceived cognitive [9] • Compensation [9]
Transparency	• Assurance [7] • Reliability [7–9] • Security [8] • Privacy [9] • Content usefulness [8] • Trust [9] • Punctuality [9] • Billing [9]

Empathy [7] is the degree to which the employees and company pay attention to understand customer's need. Personalization [8, 9] gives individualized attention to answer customer's question and comment. Then, valence [9, 12] is involving customer's final impression towards service and to produce good experience with service provider. Empathy, personalization and valence are related to individual attention and caring impression that company give to the customers. Responsiveness [7–9] is a method to help and deliver prompt service to customers through the site. Communication [8] is keeping customers properly informed and communicating with them in language they can understand by using various communication method while contact

[9, 11] is the degree to which the company offers live contact through telephone assistance and online representatives. Interactivity [9, 10] is an interactive relationship that has been built between company and customers. Communication [8], contact [9, 11] and interactivity [9, 10] convey the same function as a communication method to build a close relationship between provider and customers by offering live contact and real-time support. Perceived website innovativeness [9, 10] involved customers' action by responding through feedback to the service provider and evaluating the website quality. All of these factors matched with dialogue because this block implies attention and interaction process that occurred between company and customers by using various method of communication [27].

Tangible [7] is an equipment and platform provided for customers. Ease of use [8, 9] implies the site is easy for customers to conduct external search as well as internal navigation while website design [8, 9] comprise aspects of navigation, functionality, and accessibility of information for customers' experience. Next, information [8] is the content displayed being appropriate for the customers while content adequacy [9, 10] has the level of information completeness regarding to materials and services offered. System availability [9, 11] is accessible and the site will able to provide technical function easily. Accessibility [9] is offering the customers to reach the website quickly. Therefore, tangibles, ease of use, website design, accessibility, content adequacy and system availability equivalent to access as this block allows users to retrieve the information completely while the site acts as a tool to process the service provided [27].

Frequently, perceived risk [9, 10] links with system failure and customers' perception about loss when purchasing the products. Perceived cognitive [9, 10] is an acceptance that requires customers to understand service flow and predict the implications aftermath. Compensation [9, 11] is a level which the site recompense the customers for any problems they encounter. Risk assessment block is compatible with perceived risk, perceived cognitive and compensation because the company needs to notify the customers about the probability harm and act of customer willingness to handle the risk [27].

Assurance [7] is the company's style to represent the knowledge and has a character to inspire customer's trust and confidence. Besides, trust [9, 10] is the customers' firm belief in the reliability of online transaction. Reliability [7–9] refers to company's potential to conduct the promised service accurately and consistently. Security [8] keeps the customers freedom from danger or doubt while privacy [9, 11] is customers' desire to check if the site is safe and trust that their personal information is being protected. Furthermore, content usefulness [9, 10] refers to the accuracy and reliability of information provided. Punctuality [9] involves performing the service at the designated time. Billing [9, 11] is the process of sending invoices with accurate billing and provides suitable payment procedures. Transparency is matched with all of these factors because transparency block is compulsory to develop the customers' trust as nowadays company started to provide detailed information related to prices, costs and profit margins to create new level of clearness for the customers [27].

Solakis et al. [29] used dialogue, access, risk assessment and transparency to study customers' perception by using quantitative method and DART model has been proved to enhance customer satisfaction and help customers to obtain positive experience. As a result to embed value co-creation elements, this research paper aimed to well-matched the similarities between DART model and factors of customer satisfaction from related works.

Next, this study will validate the matched-comparison of factors as shown in Table 2 with DART model by referring to experts' reviews in value co creation field. In the future, this study will develop a measurement system for survey approach as various measurement methods have been designed to test DART model quantitatively based on qualitative approach [30].

5 Conclusion

In conclusion, this paper suggests an adoption of value co-creation elements to identify the factors of customer satisfaction for e-hailing services by comparing and well-matched the similarities between factors of customer satisfaction from related works and DART model. A review of the related works showed that there is a need for more research on customer satisfaction and value co-creation in transportation industry. It is essential to highlight the role of customers in value co-creation as they participate actively in the service provided that will lead to increase customer satisfaction because e-hailing services are quickly expanding and significantly influenced the transportation industry. Considering that this study is not fully completed and still on-going study, thus, as a way forward, the similarities between factors from existing customer satisfaction models with DART model will be validated and well-matched based on experts' reviews and feedbacks. The future research is recommended to perform more empirical studies quantitatively in value co-creation field.

Acknowledgements. This study was funded by the National University of Malaysia through the Research Grant Scheme (grant number GGP-2017-021).

References

1. Flores, O., Rayle, L.: How cities use regulation for innovation: the case of Uber, Lyft and Sidecar in San Francisco. Transp. Res. Procedia **25**, 3760–3772 (2017)
2. Stalmašeková, N., Genzorová, T., Čorejová, T., Gašperová, L.: The impact of using the digital environment in transport. Procedia Eng. **192**, 231–236 (2017)
3. Khalaf Ahmad, A.M., Ali Al-Zu'bi, H.: E-banking functionality and outcomes of customer satisfaction: an empirical investigation. Int. J. Mark. Stud. **3**(1), 50–65 (2011)
4. Grab: How grab's star rating works (2017). https://www.grab.com/sg/blog/driver/thank-lucky-star-rating-grabs-star-rating-works/. Accessed 12 Mar 2018
5. Galvagno, M., Dalli, D.: Theory of value co-creation: a systematic literature review. Manag. Serv. Qual. Int. J. **24**(6), 643–683 (2014)
6. Golder, P.N., Mitra, D., Moorman, C.: What is quality? An integrative framework of processes and states. J. Mark. **76**(4), 1–23 (2012)
7. Parasuraman, A., Zeithaml, V.A., Berry, L.L.: SERVQUAL: a multiple-item scale for measuring consumer perceptions of service quality. J. Retail. **64**(1), 12–40 (1988)
8. Ojasalo, J.: E-service quality: a conceptual model. Int. J. Arts Sci. **3**(7), 127–143 (2010)

9. Silalahi, S.L.B., Handayani, P.W., Munajat, Q.: Service quality analysis for online transportation services: case study of GO-JEK. Procedia Comput. Sci. **124**, 487–495 (2017)
10. Salameh, A.A., Bin Hassan, S.: Measuring service quality in M-commerce context : a conceptual model. Int. J. Sci. Res. Publ. **5**(3), 1–9 (2015)
11. Huang, E.Y., Lin, S.W., Fan, Y.C.: M-S-QUAL: mobile service quality measurement. Electron. Commer. Res. Appl. **14**(2), 126–142 (2015)
12. Stiakakis, E., Georgiadis, C.K.: A model to identify the dimensions of mobile service quality. In: 2011 10th International Conference on Mobile Business, pp. 195–204 (2011)
13. Naik, C.N.K., Gantasala, S.B., Prabhakar, G.V.: Service quality (Servqual) and its effect on customer satisfaction in retailing. Eur. J. Soc. Sci. **16**(2), 231–243 (2010)
14. Selvakumar, J.J.: Impact of service quality on customer satisfaction in public sector and private sector banks. J. Manag. Ethics Spiritual. **8**(1), 1–12 (2015)
15. Shaheen, S., Chan, N., Rayle, L.: Ridesourcing's impact and role in urban transportation. Access Magazine, pp. 1–9 (2017). https://www.accessmagazine.org/spring-2017/ridesourcings-impact-and-role-in-urban-transportation/
16. Malaysian Institute of Language and Literature: E-hailing definition. Malay Literary Reference Center, (2018). http://prpm.dbp.gov.my/Cari1?keyword=e-hailing&d=69812&#LIHATSINI. Accessed 12 Mar 2018
17. Ahmad, A.: Regulating Uber and Grab Drivers. New Straits Times (2017)
18. Parliament of Malaysia, Land Public Transport (Amendment) 2017, pp. 1–22 (2017)
19. Siltaloppi, J., Vargo, S.L.: Reconciling resource integration and value propositions—the dynamics of value co-creation. In: Proceedings of the Annual Hawaii International Conference on System Sciences, pp. 1278–1284 (2014)
20. Bettencourt, L.A., Lusch, R.F., Vargo, S.L.: A service lens on value creation: marketing's role in achieving strategic advantage. Calif. Manag. Rev. **57**(1), 44–66 (2014)
21. Sashi, C.M.: Customer engagement, buyer-seller relationships, and social media. Manag. Decis. **50**(2), 253–272 (2012)
22. Hoyer, W.D., Chandy, R., Dorotic, M., Krafft, M., Singh, S.S.: Consumer cocreation in new product development. J. Serv. Res. **13**(3), 283–296 (2010)
23. Brodie, R.J., Hollebeek, L.D., Jurić, B., Ilić, A.: Customer engagement: conceptual domain, fundamental propositions, and implications for research. J. Serv. Res. **14**(3), 252–271 (2011)
24. Grissemann, U.S., Stokburger-Sauer, N.E.: Customer co-creation of travel services: the role of company support and customer satisfaction with the co-creation performance. Tour. Manag. **33**(6), 1483–1492 (2012)
25. Nunes, A.A., Galvão, T., e Cunha, J.F.: Urban public transport service co-creation: leveraging passenger's knowledge to enhance travel experience. Procedia—Soc. Behav. Sci. **111**, 577–585 (2014)
26. Ferris, B., Watkins, K., Borning, A.: OneBusAway: results from providing real-time arrival information for public transit. In: Proceedings of the 28th International Conference on Human Factors in Computing Systems—CHI 2010, pp. 1–10 (2010)
27. Prahalad, C.K., Ramaswamy, V.: Co-creating unique value with customers. Strateg. Leadersh. **32**(3), 4–9 (2004)
28. Kuechler, W., Vaishnavi, V.: A framework for theory development in design science research: multiple perspectives. J. Assoc. Inf. Syst. **13**(6), 395–423 (2012)
29. Solakis, K., Peña-Vinces, J.C., Lopéz-Bonilla, J.M.: DART model from a customer's perspective: an exploratory study in the hospitality industry of Greece. Probl. Perspect. Manag. **15**(2), 536–548 (2017)
30. Mazur, J., Zaborek, P.: Validating dart model with data from the survey of Polish manufacturing and service companies. In: Global Marketing Conference at Singapore, pp. 2038–2055, September 2014

Searching for Suitable Face of Quality in Crowdsourcing - A Personality Perspective

Shakir Karim[1,2(✉)], Umair Uddin Shaikh[1], and Zaheeruddin Asif[1]

[1] Institute of Business Administration, Karachi, Karachi, Pakistan
{sbuksh,ushaikh,zasif}@iba.edu.pk
[2] Sir Syed University of Engineering and Technology, Karachi, Karachi, Pakistan

Abstract. In crowdsourcing it is a challenge to get quality work from crowd-workers. To get better quality of work two approaches i.e., run-time and design-time are used. Literature suggests that design-time approach is cost-effective as compared to run-time approach. In design-time approach, characteristic of crowd-worker is one of the components used. Previous research advocates that personality can be used to reflect the characteristic of a crowd-worker. However, there is little research available using personality to study which task a crowd-worker would like to work on. This study aims to explore which personality trait of an individual is suitable to perform tasks in crowdsourcing domain using survey questionnaire based on Big Five Inventory. Big Five Inventory (BFI). BFI is a measurement tool for five dimensions of personality. Findings reveal the relationship between task-related interest of participants and their respective personalities. It is also observed that workers with antagonist and neurotic personalities may not be suitable for the crowdsourced task.

Keywords: Cowdsourcing · Personality · Quality of work · Big Five Inventory

1 Introduction

The term Crowdsourcing was coined by Howe for the first time in the article "the rise of crowdsourcing" [1]. Crowdsourcing is a process of connecting with large groups of people via the Internet who contribute their knowledge, expertise, time or resources. The crowd is a large undefined group of individuals [2] i.e., the crowd is a group of workers temporarily connected to perform a task. Since its inception, crowdsourcing has evolved to become an important business model. Some popular platforms to execute crowdsourcing are American Mechanical Turk, oDesk, CrowdFlower, Wikipedia.

Researchers have studied different features of crowdsourcing such as cost effectiveness [3], expertise development, skills and labor [4, 5], techniques to address performance quality [6]. However, research has shown concerns regarding quality of work in crowdsourcing. Though crowdsourcing offers both time-efficient as well as cost-efficient solution, the quality of crowdsourcing is always under question [6, 7]. More than 30% of the work submitted on the Amazon Mechanical Turk (AMT) was reported to be of low quality [8]. Quality, in particular, is reported to be the key concern

F. Saeed et al. (Eds.): IRICT 2018, AISC 843, pp. 812–819, 2019.
https://doi.org/10.1007/978-3-319-99007-1_75

of personnel from every platform, with AMT being "plagued by low-quality results [9]". A recent survey reports that only half of the crowdsourcing websites offer some form of quality control of work undertaken [10]. Nevo and colleague [11] revealed that one of the key challenges is quality of submissions in crowdsourcing. The management of the quality involves asking questions such as what efforts should be rewarded [12] and how to generate a capable crowd that can provide quality outcome [13]. This is also challenging since organizations (requesters or solution-seekers) do not have the same level of control over crowd's behavior as they would have over their own employees [2]. A possible reason may be the temporary connectedness of crowd [13]. To get better quality of work two approaches namely, run-time approach and design-time approach, are used. Redundancy is one of a technique used in run-time approach. Redundancy technique requires allocating same task to number of workers, from which the best is selected. Literature suggests that design-time approach is cost-effective as compared to run-time approach. In design-time approach, characteristic of crowd-worker is one of the components used. Previous research advocates that personality can be used to reflect the characteristic of a crowd-worker. There is limited research regarding the characteristic aspect of the crowd-worker.

To fill this gap in the literature it is of interest to study the interest of crowd to acquire the task that match their personality. Knowing the interest of workers will help in recruiting and managing the crowd in a better way. Saxton and colleagues [13] suggested that utilization of management strategies, a group of capable workers can be generated which may produce quality work. Personality is assumed, almost universally, to have a marked effect on the performance by management officials [14] and revealed to have a relationship with career choice [15]. In crowdsourcing domain, the career choice can be considered as the choice for a task a crowd-worker feels interest to work on.

The current study is an attempt to explore the relationship between the personality of a crowd-worker and the task which s/he is interest to work on, in crowdsourced setting and aims to answer the following research question:

- *Which personality trait of crowd-worker is suitable for crowdsourcing?*

2 Literature Review

The concept of crowdsourcing can be understood by observing the theme of Wikipedia. Wikipedia is a multilingual, web-based, free-content encyclopedia which is supported by the "crowd" mostly contributing voluntarily. Any individual, who has access to the Internet, can easily contribute to this online encyclopedia by writing an article from scratch or updating already shared content available online. The major advantage of this approach besides generation of a huge online encyclopedia is its continuous updating feature which is not available in the traditional printed versions of encyclopedia. The overall objective is to acquire the services of individuals from all over the world – the crowd – to do a task.

In crowdsourcing process, there are four main components [16]. First is the company that requests for a particular work, termed as the requestor. Second component is the task, for example, logo designing, translations, audio transcriptions, map annotation. The requestor can provide tasks with or without the help of crowdsourcing platforms such as AMT, oDesk, CrowdFlower, and Threadless. These platforms are the third component. Finally, the people come as the fourth component in crowdsourcing i.e., the crowd.

Literature reveals several types of crowdsourcing. Recently, Prpic and colleague [17] have described two types of crowdsourcing at the broad level namely content-based crowdsourcing and contribution-based crowdsourcing, which are further divided into two sub-categories each. Content-based crowdsourcing is divided on the basis of subjective and objective content whereas contribution-based crowdsourcing is divided as aggregated contribution and filtered contribution. Another categorization, similar to the suggestions of Prpic and colleague [17], is provided by Geiger and colleague [18], according to which crowdsourcing can be viewed as one among four types which are Crowd Creation, Crowd Processing, Crowd Solving and Crowd Rrating. This is illustrated in Table 1.

Table 1. Types of crowdsourcing

Category	Sub-category	
Prpic [17]	Prpic [17]	Geiger [18]
Content-based	Subjective	Crowd rating
	Objective	Crowd creation
Contribution-based	Aggregated	Crowd processing
	Filtered	Crowd solving

Almost universally it is assumed that personality has a profound effect on the worker's performance by management officials [14] and worker's career choice is related to personality [15]. While hiring for a job, personality of a candidate matters. In crowdsourcing domain, the crowd-worker seeks a job for many reasons. Some individuals among the crowd prefer intrinsic rewards while the other are inclined towards the financial (extrinsic) rewards [3–5]. It is very important to investigate which personality traits are favorable in performing what kind of task(s).

In crowdsourcing literature, personality is reported as the real face of the crowd-worker [20]. It is suggested that task matching can be achieved by combining behavioral aspects with the workers' personality, which further help in developing methods to attract them. The research, however, is limited to specific task. Kazai [21] reported that personality affect the performance of a crowd-worker [21], along with effort and cognitive skills [22, 23]. Using crowdsourcing, large groups of people can solve a problem in a time efficient manner by sharing ideas. A task, if distributed to the crowd, may be completed in cheaper and time effective manner. Personality analysis is helpful in team formation of crowd-workers to get better performance where collaboration is required [24].

However, still it is not known which personality is suitable for crowdsourcing. This study is step forward to answer this question.

3 Research Design

3.1 Research Method and Approach

A quantitative approach is followed in the current study. Quantitative research is a formal, objective, and systematic process to describe and examine the causal relationships among variables [25]. Survey may be used to conduct a quantitative study for descriptive and explanatory research.

On the crowdsourcing platforms, there are several tasks available for the crowd-workers. Translation, Transcription, Photo-tagging, and Tagline suggestions are examples of such tasks. In the current study, to investigate the personality suitability of crowd-worker for crowdsourcing, authors are interested to identify the tasks in which most of the crowd-workers feel interest. A few tasks were identified that were mostly the part of crowdsourcing and human resource management.

A questionnaire was adopted as data collection instrument. A questionnaire may be a printed or online self-report form. The purpose to design a questionnaire is to obtain the information regarding the subjects through their responses.

3.2 Pilot Study

Research regarding the demographic of crowd-workers reveals that around 70% are the undergraduate students [27]. In the current study, considering [27], sample mostly consists of undergraduate students are requested to fill up the survey. To do the pilot study, the official university online groups are used to approach the students and alumni for this study. Around 160 participants filled the forms. From 160 filled forms, 136 *(with an acceptable response rate of* >80%*)* were filled correctly. The remaining 24 were incomplete and therefore rejected.

3.3 Instrument

Survey for personality assessment is based on the Big Five Inventory (BFI) [26]. The questions of the questionnaire were modified and framed in simple easy English. Closed-ended questions were used due to their easy to administer and analyze nature. Further, closed-ended questions are efficient as compared to their counterpart i.e., open-ended questions because a respondent can complete more close-ended questions in a given time period [27].

The first section of the questionnaire is related to demographics. Second section is related to personality assessment, based on BFI, having 10 questions. There are five questions focusing five personality traits termed as OCEAN where O stands for Open to Experience, C stands Conscientiousness, E stands for Extrovert, A stands for Agreeableness and N stands for Neuroticism. For each five traits there are five reverse questions. For example, for Extrovert it is asked "I see myself as Reserved, Quiet" as reverse question in addition to normal question "I see myself as Extroverted, Enthusiastic". In the third section students were asked to opt for the tasks among the grouped tasks with respect to their interest.

Further, to conform to ethical practices, the rights of respondents to confidentiality, anonymity and self-expression have been observed. Researchers have made sure to avoid any kind of manipulation in data collected through the survey to maintain honesty.

4 Findings

The first section of the survey asked the descriptive statistics of respondents. The gender breakdown, age distribution, and educational background of participants are shown in Figs. 1, 2, and 3 respectively.

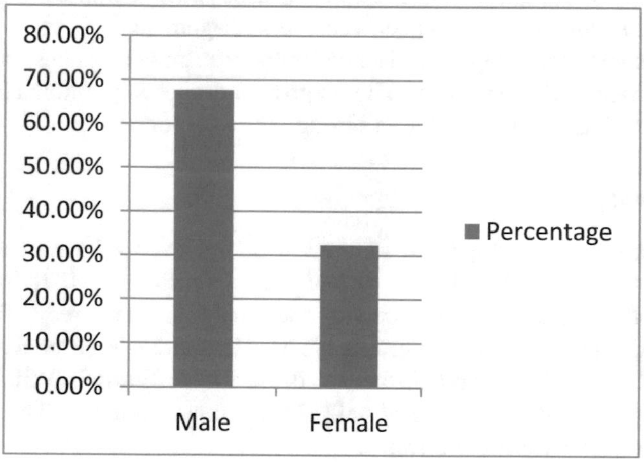

Fig. 1. Gender breakdown

The students as well as the alumni of the local universities were approached online for the survey. It is observed that undergraduate students dominate the results amongst which 66% of the participants were male. Among the participants 68% were youngsters with the age below 25 years and 29% were below 35 years of age. This shows majority of the participants interested in crowdsourcing were young graduates or/and undergraduates verifying the findings of [27].

Further, the task matching with respect to personality is depicted in Table 2, according to which a greater proportion of the sample were open to experience.

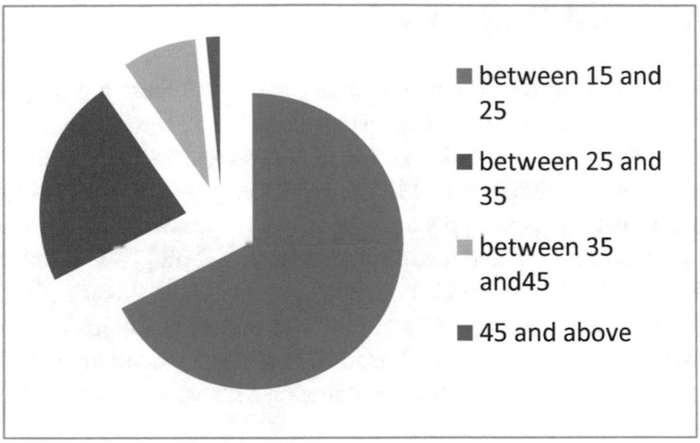

Fig. 2. Age distribution (in years)

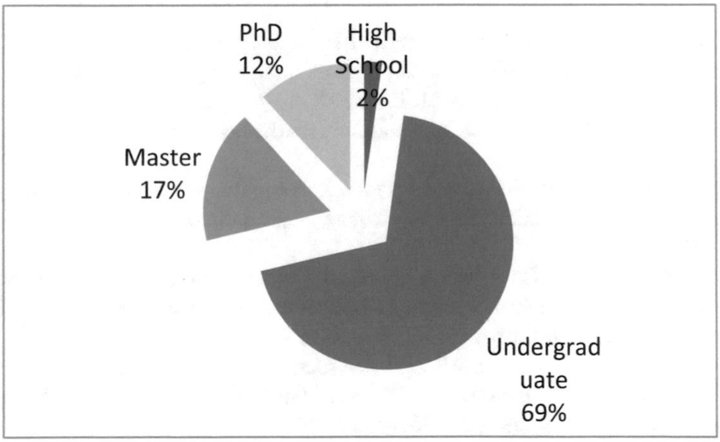

Fig. 3. Educational background

Table 2. Task matching with personality.

Personality	Tasks	Match (%)
Extravert & conscientiousness	Translation	12
	Audio transcription	13
Open to experience	Quick ideas suggestion	15
	Creative campaign	20
	Tagline suggestion	40

5 Conclusion and Future Work

The preliminary results of the conducted survey show that there are five tasks in which majority of the respondents shown interest as depicted in the Table 2.

The research contribution consists of the observed relationship between task related interest shown by the participants and their respective personalities. Table 2 represents the task matching with respect to personality of the respondents. It is observed that the participants with antagonist and neurotic personalities may not be suitable for the crowdsourced task. Further research is necessary to support this observation.

Like other research, this work is also not free from limitation. In this study, only university students and alumni were selected for the survey as participants. Future work will use a crowdsourcing platform such as Amazon Mechanical Turk (AMT) for further investigation.

References

1. Howe, J.: The rise of crowdsourcing. Wired Mag. **14**, 1–4 (2006)
2. Thuan, N.H., Antunes, P., Johnstone, D.: Factors influencing the decision to crowdsource: a systematic literature review. Inf. Syst. Front. **18**, 47–68 (2016)
3. Whitla, P.: Crowdsourcing and its application in marketing activities. Contemp. Manag. Res. **5**, 15–28 (2009)
4. Brabham, D.C.: Moving the crowd at iStockphoto: the composition of the crowd and motivations for participation in a crowdsourcing application. In: First Monday, vol. 13 (2008)
5. Vukovic, M., Bartolini, C.: Towards a research agenda for enterprise crowdsourcing. In International Symposium on Leveraging Applications of Formal Methods, Verification and Validation, pp. 425–434 (2010)
6. Kittur, A., Nickerson, J.V., Bernstein, M., Gerber, E., Shaw, A., Zimmerma, J., Lease, M., Horton, J.: The future of crowd work. In: Proceedings of the 2013 Conference on Computer Supported Cooperative Work, pp. 1301–1318 (2013)
7. Allahbakhsh, M., Benatallah, B., Ignjatovic, A., Motahari-Nezhad, H., Bertino, E., Dustdar, S.: Quality control in crowdsourcing systems. IEEE Internet Comput. **17**, 76–81 (2013)
8. Kittur, A., Chi, E.H., Suh, B.: Crowdsourcing user studies with mechanical turk. In: Proceedings of the SIGCHI Conference on Human Factors in Computing Systems, ACM, pp. 453–456 (2008)
9. Vakharia, D., Lease, M.: Beyond mechanical turk: an analysis of paid crowd work platforms. In: Proceedings of the iConference (2015)
10. Mourelatos, E., Frarakis, N., Tzagarakis, M.: A study on the evolution of crowdsourcing websites. Eur. J. Soc. Sci. Educ. Res. **11**(1), 29–40 (2017)
11. Kirsch, L.J.: Deploying common systems globally: the dynamics of control. Inf. Syst. Res. **15**, 374–395 (2004)
12. Straub, T., Gimpel, H., Teschner, F., Weinhardt, C.: Feedback and performance in crowd work: a real effort experiment. In: ECIS Proceedings (2014)
13. Saxton, G.D., Oh, O., Kishore, R.: Rules of crowdsourcing: models, issues, and systems of control. Inf. Syst. Manag. **30**, 2–20 (2013)
14. Blumberg, M., Pringle, C.D.: The missing opportunity in organizational research: some implications for a theory of work performance. Acad. Manag. Rev. **7**(4), 560–569 (1982)

15. Kemboi, R.J.K., Kindiki, N., Misigo, B.: Relationship between personality types and career choices of undergraduate students: a case of Moi University, Kenya. J. Educ. Pract. **7**(3), 102–112 (2016)
16. Hosseini, M., Phalp, K., Taylor, J., Ali, R.: The four pillars of crowdsourcing: a reference model. In: 2014 IEEE Eighth International Conference on Research Challenges in Information Science (RCIS), pp. 1–12 (2014)
17. Prpić, J., Shukla, P.P., Kietzmann, J.H., McCarthy, I.P.: How to work a crowd: developing crowd capital through crowdsourcing. Bus. Horiz. **58**, 77–85 (2015)
18. Geiger, D., Seedorf, S., Schulze, T., Nickerson, R.C., Schader, M.: Managing the crowd: towards a taxonomy of crowdsourcing processes. In: AMCIS Proceedings, Paper 430 (2011)
19. Sun, Y., Wang, N., Yin, C., Che, T.: Investigating the non-linear relationships in the expectancy theory: the case of crowdsourcing marketplace. In: AMCIS Proceedings, Paper 6 (2012) http://aisel.aisnet.org/amcis2012/proceedings/VirtualCommunities/6
20. Kazai, G., Kamps, J., Milic-Frayling, N.: Worker types and personality traits in crowdsourcing relevance labels. In: Proceedings of the 20th ACM International Conference on Information and Knowledge Management. pp. 1941–1944 (2011)
21. Kazai, G., Kamps, J., Milic-Frayling, N.: The face of quality in crowdsourcing relevance labels: demographics, personality and labeling accuracy. In: Proceedings of the 21st ACM International Conference on Information and Knowledge Management, pp. 2583–2586 (2012)
22. Karim, S., Shaikh, U.U., Asif, Z.: Motivating crowdworkers by managing expectations. In: 2017 International Conference on Research and Innovation in Information Systems (ICRIIS), Langkawi, Malaysia. IEEE (2017). https://doi.org/10.1109/icriis.2017.8002539
23. Mourelatos, E., Tzagarakis, M.: Worker's cognitive abilities and personality traits as predictors of effective task performance in crowdsourcing tasks. In: Proceedings of 5th ISCA/DEGA Workshop on Perceptual Quality of Systems (PQS 2016), pp. 112–116 (2016)
24. Lykourentzou, I., Antoniou, A., Naudet, Y., Dow, S.P.: Personality matters: balancing for personality types leads to better outcomes for crowd teams. In: Proceedings of the 19th ACM Conference on Computer-Supported Cooperative Work and Social Computing, pp. 260–273 (2016)
25. Creswell, J.W.: Research Design: Qualitative & Quantitative Approaches. Sage Publications Inc., Thousand Oaks (1994)
26. John, O.P., Naumann, L.P., Soto, C.J.: Paradigm shift to the integrative big-five trait taxonomy. In: Handbook of Personality, Chap. 4, pp. 114–212. Guilford Press, New York (2008)
27. Ipeirotis, P.: Mechanical Turk: The Demographics (2008). http://behind-the-enemy-ines.blogspot.com/2008/03/mechanical-turk-demographics.html.2008

Proposing a Comprehensive IS Continuance Model and Its Factors

Mohd Zuhan Mohd Zain(✉), Ab Razak Che Hussin(✉), and Nittee Wanichavorapong(✉)

Universiti Teknologi Malaysia, Johor Bahru, Malaysia
mohdzuhan@gmail.com, abrazak@utm.my, nittee@hotmail.com

Abstract. Continuous use of information systems (IS) has become crucial for an organization's survival as it provides efficiency and effectiveness of managing business transactions. Lacking continuance usage of IS poses an obstacle in the advancement of IS in an organization. Previous studies have examined continuance intention using the Expectation Confirmation Model (ECM) as it provides a basis of investigating IS continuance. However, the expansion in IS role in today's business requires a further integration with other factors such as social support, experience, technology fit and self-efficacy. Therefore, the aim of this study is to develop a comprehensive IS continuance model through the extension of ECM by integrating new factors from other related theories. Upon developing the model, we extensively review the literature in order to understand the theories used, then extracted the relevant factors to be used in the model. The outcome of this study would provide the richness of knowledge in IS continuance domain and provides an opportunity for businesses to develop an effective plan of IS continuance in the organizations.

Keywords: Information systems · Information systems continuance Information systems theory

1 Introduction

Nowadays, IS plays a critical role in an organization's success, where globalization, digital economics, and digital organizations take place. IS can facilitate organization through several courses of process of information including gathering, processing and disseminating information in an organization. Therefore, the ineffectiveness of IS creates a risky situation to organizational success and survival (Bhattacherjee 2001). The positive impacts through the existence of IS in organization is shown through the organizational effectiveness and individuals job performance. This will motivate organizations to increase their investments in the IS field. IS in an organization should be considered as a vital resource, as those without an IS implementation are considered as a competitive disadvantage (Chen et al. 2015).

IS continuance has not yet received a comparable attention as IS acceptance (Chen et al. 2015). It can be said that, the IS continue research still lacking in-depth understanding of its drivers (Mouakket 2015). Thus, this will provide a significant

F. Saeed et al. (Eds.): IRICT 2018, AISC 843, pp. 820–830, 2019.
https://doi.org/10.1007/978-3-319-99007-1_76

input to investigate the need for a comprehensiveness of IS continuance. Previous studies have mainly examined variables that motivate individuals to accept a new system, and how to use it (Limayem and Cheung 2011; Venkatesh et al. 2011). Most of these studies investigated IS continuance were mostly based on Expectation-Confirmation Model (ECM) (Chen et al. 2015; Mouakket 2015; Terzis et al. 2013). They employed mainly the three (3) factors to explain behavioral intention, namely confirmation, perceived usefulness, and satisfaction. However, behavioral intention towards adopting IS would also be affected by other factors, such as the ability of the individual to use the IS, organizational tasks, and its suitability. This is because with the evolution of a new business model focusing on consumers and an open-access system, other factors related to consumer characteristics such as self-efficacy, prior experience, and utilization may need to be included for the investigations of its impact on IS continuance. Therefore, this study explores the possibility of integrates other models including Task-Technology Fit (TTF), Social Cognitive Theory (SCT), Social Support Theory (SST), and Unified Theory of Acceptance and Use of Technology (UTAUT) with ECM to develop a comprehensive IS continuance model. Integrating these factors will provide an extended view of IS continuance research.

This paper is organized as follows. First is conducting a literature review on IS continuance. Second is a derivation of IS continuance factors and model development. The last section is about the conclusion and opportunity for further research.

2 Literature Review

Various definitions of Information Systems (IS) are provided by different researchers through their own understanding based on their study domains. Generally, IS is described as a complete system that is designed for storing, producing, processing and distributing information for organizations or institutions. IIS have become the important backbone that supports in most organizational activities such as planning and decision making. In this study, IS is defined as a group of correlated process including a combination of software, hardware, infrastructure and trained personnel to help organizational activities to improve efficiency and effectiveness in performing organizational functions and achieving its objectives (adapted from Murcko 2012).

Continuance is a type of post-adoption behavior which it can also relate to IS research domains. Limayem and Cheung (2011) suggested that IS continuance is a similar term of post-adoption phase. However, there is an argument with the term of post-adoption phase within IS research such as adoption, continuance, diffusion, and compliance. Bhattacherjee (2001) moved beyond post-adoption studies by exploring the long-term consequences of continued use from the first-time use of IS. It is found that, IS continuance research concentrated more on exploring the role of satisfaction and trust in user continuance behaviors compared to acceptance research. Also, IS continuance is concerned the relationship between actual usage of an IS system with the formation of trust, confidence, and satisfaction. IS continuance or post-adoption is the decision to continuously use a certain product or service which is a kind of "acceptance" behavior at a later stage (Bhattacherjee 2001). In the perspective of an individual, IS continuance is a center of survival of several types of businesses such as banking, online retailers, and online travel agencies.

According to Table 1, several studies have been conducted on IS continuance focusing on different study domains. All these studies use quantitative research approach as their methodology. Based on the existing literature, there is a lack of a theoretical model of IS continuance as a comprehensive model. In addition, the search for factors that may explain the continuous use of IS remains an important effort for the success of firms.

Table 1. Previous studies of IS continuance.

No	Author	Title	Theory used
1.	Venkatesh et al. (2011)	Extending the two-stage information systems continuance model: incorporating UTAUT predictors and the role of context	UTAUT, ECT
2.	Tseng (2015)	Exploring the intention to continue using web-based self-service	TAM, IS Success Model
3.	Mouakket (2015)	Factors influencing continuance intention to use social network sites: The Facebook case	ECM
4.	Chen et al. (2015)	Why do teachers continue to use teaching blogs? the roles of perceived voluntariness and habit	ECM
5.	Lu and Yang (2014)	Toward an understanding of the behavioral intention to use a social networking site: An extension of task-technology fit to social-technology fit	SCT, TTF
6.	Terzis et al. (2013)	Continuance acceptance of computer-based assessment through the integration of user's expectations and perceptions	ECM
7.	Ramayah et al. (2012)	International Forum of Educational Technology & Society An Assessment of E-training Effectiveness in Multinational Companies in Malaysia	TAM, ECM, IS Success Model
8.	Lin (2012)	Perceived fit and satisfaction on web learning performance: IS continuance intention and task-technology fit perspectives	IS continuance theory, TTF
9.	Chang and Zhu (2012)	The role of perceived social capital and flow experience in building users' continuance intention to social networking sites in China	ECM
10.	Shin et al. (2011)	Smartphones as smart pedagogical tools: Implications for smartphones as u-learning devices	UTAUT, ECM
11.	Kang et al. (2009)	Exploring continued online service usage behavior: The roles of self-image congruity and regret	ECM

(*continued*)

Table 1. (*continued*)

No	Author	Title	Theory used
12.	Akter et al. (2013)	Continuance of mHealth services at the bottom of the pyramid: the roles of service quality and trust	ECM
13.	Chen et al. (2010)	Confirmation of Expectations and Satisfaction with the Internet Shopping: The Role of Internet Self-efficacy	SCT, ECT
14.	Liang et al. (2011)	What Drives Social Commerce: The Role of Social Support and Relationship Quality	SST
15.	Shanmugam et al. (2015)	A theoretical extension and empirical investigation for continuance use in social networking sites	SST
16.	Hajli (2015)	A study on the continuance participation in on-line communities with social commerce perspective	SST

3 The Purposed Model

3.1 Theoretical Foundation

Various theories were employed by researchers to investigate IS continuance in different contexts as shown in Table 1. This study uses five (5) theories namely; ECM, TTF, SST, UTAUT, and SCT as a theoretical foundation for extending the IS continuance model as they provide the core factors and cited by many previous studies. The explanations of four selected models are presented in the following paragraphs.

Expectation Confirmation Model (ECM). Expectation Confirmation theory—IS continuance predominantly described to predict and understand the continuance intention of users in the consumer behavior research which includes post-purchase behaviors and satisfaction (Chang and Zhu 2012; Limayem and Cheung 2011; Venkatesh et al. 2011). In another word, it aims as to understand the post-adoption behavior by utilizing satisfaction. Firstly, individuals initialize their expectation towards the product or service. Next, individuals accept and utilize the product or service, then individuals weight the product or service performance with the initial expectation which bring to confirmation. It is important to note that IS continuance is not simply an extension of adoption behavior. Bhattacherjee (2001) has empirically shown that the antecedents or initial of IS continuance usage are different considerably from those of initial adoption behavior. He further indicated that the relationship between satisfaction and the earlier use of products or services is an important element in the determination of individuals' intention to continuously use the products or services. In consumer behavior literature, ECM theory is widely applied to study users' satisfaction and post-purchase behavior. ECM remains as one of the important models in continuance intention studies which specifically usually commonly applied in IS, shopping and marketing contexts (Bhattacherjee 2001).

Task-Technology Fit (TTF) Model. The task-technology fit (TTF) model is a model extensively used as a theoretical model to examine the ways of information technology can lead to performances and examine the match between the task and technology characteristics (Wu and Chen 2017). For an IS to positively influence the use of technology, the technology should fit the task where it can support to have a performance impact. The fit of the technology to task is the degree which the technology features correspond to the task requirements (Lu and Yang 2014). The impact of TTF is not only covered to predict current utilization and performance intensity but TTF also helps in predicting future utilization and performance (Aljukhadar et al. 2014). TTF has been widely used and applied in several range of IS studies such as to examine the behavioral intention in using social networking sites (Lu and Yang 2014), online context (Aljukhadar et al. 2014), and continuance intention (Wu and Chen 2017).

Social Cognitive Theory (SCT). This theory mainly concentrated on the concept of self-efficacy, which is practically one of the most beneficial concept formulated in modern psychology. SCT postulates that, the combination of internal self-influence factors and external social system able to motivate and regulate individuals' behavior ((Bandura 2012). As theorized from self-influence factors, it is found that self-efficacy (SE) is the main component that refers to individuals' perception of their capabilities to organize, manage and execute series of action that required to achieve desired performance. Self-efficacy has a strong influence on individuals output in an organization. Once individuals exhibit higher self-efficacy in an organization, it perceived that the individuals are more likely to execute the job more effective and easily achieve the satisfaction level compared to those with lower self-efficacy.

Unified Theory of Acceptance and Use of Technology (UTAUT). UTAUT has included four main factors that are performance expectancy, effort expectancy, social influence, and facilitating conditions to directly and indirectly investigate and predict intention to use and use behavior. Besides, the use of 4 moderators including gender, age, experience, and voluntariness of use to study the impact on intention to use and use behavior (Venkatesh et al. 2003). The theory considerably inherited some quality from the theory of planned behavior (Ajzen 1991), especially the model's structure. The idea is that technologies can help users to complete and achieved their tasks more quickly (performance expectance), people who are important to them use the system and expect them too (social influence), systems are easy to use and learn as such it should be unambiguous and simple, (effort expectancy), and there are knowledge, resources, and system supports that can be found to assist users (facilitating conditions).

Social Support Theory (SST). Social support is a huge concept with multidimensional aspects including emotional, instrumental, appraisal, and informational. Precisely, emotional support has some connection with listening and expressing sympathy or trust (Shumaker and Brownell 1984). Colloca and Colloca (2016) identified social support as an important mechanism in maintaining the quality of life. This theory has been used in IS research domain such as investigating role of social support in social commerce (Liang et al. 2011) and continuance use of networking sites (Shanmugam et al. 2015).

3.2 Factor Derivation

After an extensive review, the researchers identify the relevant factors that always used in IS continuance study. All these factors appear to be scatted and not in one comprehensive model as shown in Table 2. Therefore, this study compiled the selected factors in one model and will be empirically validated in the next stage of research. The definition for each of the factor as modified in this study is presented in Table 3.

Table 2. Factors of IS continuance

Theory/model	Factors	Selected factors
Unified Theory of Acceptance and Use of Technology (UTAUT)	Performance expectancy, Effort expectancy, Social influence, Facilitating condition, Behavioral intention, Use behavior, Gender, Age, Experience	Prior Experience Self–efficacy Utilization Perceived task technology fit Confirmation Perceived usefulness Perceived support Is continue intention User satisfaction
Expectation confirmation theory (ECT)	Expectation, Perceived performance, Disconfirmation of beliefs, Satisfaction	
Technology Acceptance Model (TAM)	Perceived usefulness, Perceived ease of use, Behavioral intention to use, Actual system use	
IS Success Model (ISM)	Information quality, System quality, Service quality, Intention to use, User satisfaction, Net benefit	
Information Systems Continuance Theory (ISCT)	Perceived usefulness, Perceived ease of use, Confirmation, Satisfaction, Attitude, IS continue intention	
Social cognitive theory (SCT)	Observing, Outcome expectancies, Self–efficacy, Identification	
Task Technology fit (TTF)	Task characteristics, Technology characteristics, Task Technology fit, Performance impact, Utilization	
Expectation confirmation model (ECM)	Perceived usefulness, Confirmation, Satisfaction, IS continue intention	
Social Support Theory (SST)	Perceive Support, Receive Support, Social Network	

It is clear that the previous literature employed the ECM (Bhattacherjee 2001) which is an integration and modification of Technology Acceptance Model (TAM) and Expectation-Confirmation Theory (ECT). The ECM hypothesizes that an individuals' degree of satisfaction and perceived usefulness of an innovation determine its continuous use. Thus, an individuals' degree of satisfaction with an innovation is determined by the level of the individuals' confirmed expectations and perceived usefulness. The ECM will be extended with other factors such as prior experience, task-technology fit on individuals continuous use of an IS. Individuals mainly utilize technologies to aid in their task performance. Individual features such as computer experience, motivation could influence how effectively an individual use a technology (Goodhue and Thompson 1995). These factors have been shown to impact individual behavior in several studies. Thus, this study is motivated to consider the integration of these factors with ECM in order to improve understanding of the continuance use of IS in an organization. The explanations of the selected factors are as follows:

Prior Experience (PE)
Experience contributes to better a user's confidence in employing a system. People with such a long experience using a system will lead to greater ability to deal with it (Venkatesh et al. 2003). Individual gains experience by learning and acquiring knowledge as such the increase in understanding with a product or service will lead to the sustainable use. Therefore, this study perceives the usefulness of prior experience that drives individual to adhere to the long-term use.

Utilization (UT)
From the literature, the increase in utilization can result in positive performance impacts (Goodhue and Thompson 1995). Utilization is a complex result based on other factors apart from technology fit such as social norms, habit, and situational factors. As seen the functionality of job search website (JSW) utilization, it had a statistical influence on unemployment duration (Huang and Chuang 2016). The integration of utilization can be beneficial to increase the understanding of CI on this part.

Perceived Support (PS)
It is necessary to receive the adequate level of social support for people in a society (Shumaker and Brownell 1984). Besides, there exists that the need to investigate social support that forces users' intention to continue using an online-service thus this will ensure the successful implementation (Hajli 2015). As a result, this study found it legitimate to include perceived support to improve the prediction and understanding of CI from the social dimension.

Perceived Task Technology Fit (PTTF)
If a task that would be performed is not compatible with the technology, it is said to be a poor fit. Goodhue and Thompson 1995 There are many digital-services that users encountered, and they fit the tasks that users want to perform so they will continue to use these services. Lin and Wang (2012) employed TTF to investigate confirmation and usefulness in continuing use of e-learning. Further, they reported TTF had positive impacts on both confirmation and usefulness. Therefore, TTF should be used to investigate CI.

Self-Efficacy (SE)
Self-efficacy involves an individual's convictions in his or her capacities to execute significant courses of action to fulfill situational demands (Bandura 1986). Self-efficacy has been included in many studies of CI. Hsu and Chiu (2004) used web specific self-efficacy which was analyzed having statistical impacts on both intention and E-service usage. Bhattacherjee (2001) integrated IT self-efficacy to investigate CI and found IT self-efficacy could make a positive correlation to CI. As such, self-efficacy will be one of the factors that helps research to enhance the comprehension of CI.

Confirmation (Con)
Confirmation refers to a cognitive mechanism that is the result of the operation between expectation and perceived performance (Kim 2014; Hsu et al. 2006). Confirmation has been discovered to influence satisfaction levels (Kim 2014) and, in turn, influences IS continuance Intention. Moreover, confirmation was found to have a significant impact on PU in the post-use stage (Bhattacherjee 2001). Confirmation remains some extent of important in predicting and understanding CI as such the inclusion of confirmation is imperative.

IS Continuance Intention (ISCI)
The benefits of successfully understanding continuance behavior will improve the success of an IS implementation. Goods and services cannot only rely on pre-consumption (adoption); to be specific, stakeholders must find the best possible channels maintain continuance usage (Chiu et al. 2007). Abandoning the goods and services is usually the result of dissatisfaction which prompts scholars to thoroughly understand the characteristics of ISCI to avoid failure. Thus, it is mandatory to include ISCI in this study as the dependent variable (DV).

Perceived Usefulness (PU)
Usefulness, a cognitive factor, signifies the extent of being possible to use as a person senses that using goods or services can increase the performance of tasks (Davis et al. 1989; Davis 1993). In IS continuance, usefulness is the result of cognitive beliefs and has greatly and steadily influenced attitude in first-time use as well as the post-adoption processes (Kim 2014). Al-Maghrabi et al. (2011) concluded that usefulness has a quantity of influence on ISCI. These are the reasons that perceived usefulness should be included in the proposed model.

User Satisfaction (SAT)
Measuring satisfaction is considered as one of the most important measures of IS success and satisfaction has strong connection with the feelings about like or please with a system. Satisfaction also associates with distinct cognitive and emotional states, and that these different states influence future (Stauss and Neuhaus 1997). The goal of the ECM is to understand the post-adoption behavior by using satisfaction. ECM exists in the process that satisfaction is the key determinant of the model. In general, prior models paid huge attention on satisfaction, thus, this makes satisfaction a huge space in this study (Fig. 1).

Table 3. Factors and their definitions

Factors	Definition	Source
Prior Experience	It is defined as an individual's psychological views based on previous experience with an IS	Venkatesh et al. (2003)-UTAUT
Self-efficacy	It refers to person's beliefs in their abilities to manage series of action that required to achieve the desire type of performance	Bandura (2001)-SCT

(*continued*)

Table 3. (*continued*)

Factors	Definition	Source
Utilization	It is defined as the behavior of employing an e-waqf system in completing a task	Goodhue and Thompson (1995)-TTF
Perceived Task-Technology Fit	It is defined as the degree of e-waqf fits or assists users in completing their work	Goodhue and Thompson (1995)-TTF
Confirmation	It is defined as the degree to which users' expectation of the performance of the e-waqf system is acknowledged in the course of actual use	Bhattacherjee (2001)-ECM
Perceived Usefulness	It refers to users' perception that using a particular system able to enhance the job performance	Bhattacherjee (2001)-ECM
Perceived Support	It is defined as how much help is accessible when users face challenges with an e-waqf system	Shumaker and Brownell (1984)-SST
User Satisfaction	It is defined as a positive emotional state results from the utilization of an e-waqf system	Bhattacherjee (2001)-ECM

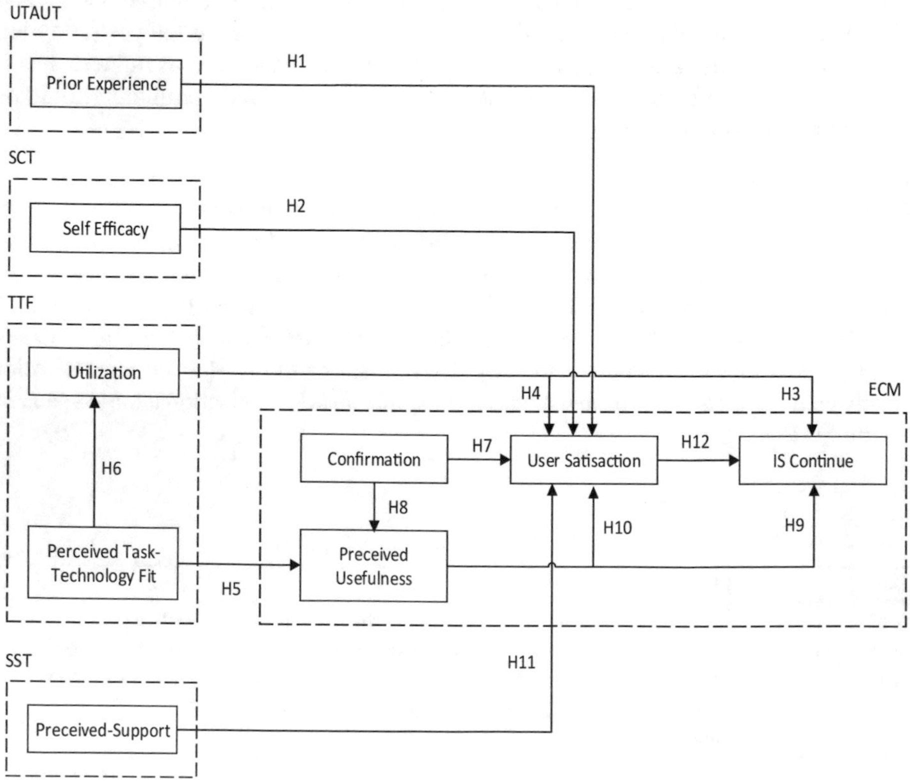

Fig. 1. The proposed IS continuance model

4 Conclusion and Future Work

Most of the studies considered in this research were conducted using different approaches and also obtained varying results. The ultimate goal of the present study is to develop comprehensive IS continuance model. With useful information derived from the literature review, the study used several well-known IS theories related to IS continuance and factors. This will increase the knowledge in IS continuance field. The proposed model will used as a guideline for the future research direction which hypotheses, the survey instrument, and reliability and validity test in the e-waqf context.

References

Aljukhadar, M., Senecal, S., Nantel, J.: Is more always better? Investigating the task-technology fit theory in an online user context. Inf. Manag. **51**(4), 391–397 (2014)

Al-Maghrabi, T., Dennis, C., Vaux Halliday, S.: Antecedents of continuance intentions towards e-shopping: the case of Saudi Arabia. J. Enterp. Inf. Manag. **24**(1), 85–111 (2011)

Bandura, A.: Social cognitive theory: an agentic perspective. Annu. Rev. Psychol. **52**(1), 1–26 (2001)

Bandura, A.: On the functional properties of perceived self-efficacy revisited. J. Manag. **38**(1), 9–44 (2012)

Bhattacherjee, A.: Understanding information systems continuance: an expectation-confirmation model. MIS Q. **25**(3), 351–370 (2001)

Chang, Y.P., Zhu, D.H.: The role of perceived social capital and flow experience in building users' continuance intention to social networking sites in China. Comput. Hum. Behav. **28**(3), 995–1001 (2012)

Chen, Y.Y., Huang, H.L., Hsu, Y.C., Tseng, H.C.: Confirmation of expectations and satisfaction with the internet shopping: the role of internet self-efficacy. Comput. Inf. Sci. **3**(3), 14–122 (2010)

Chen, C.P., Lai, H.M., Ho, C.Y.: Why do teachers continue to use teaching blogs? The roles of perceived voluntariness and habit. Comput. Educ. **82**(1), 236–249 (2015). https://doi.org/10.1016/j.compedu.2014.11.017

Chiu, C.M., Chiu, C.S., Chang, H.C.: Examining the integrated influence of fairness and quality on learners' satisfaction and web-based learning continuance intention. Inf. Syst. J. **17**(3), 271–287 (2007)

Colloca, G., Colloca, P.: The effects of social support on health-related quality of life of patients (2016)

Davis, F.D.: User acceptance of information technology: system characteristics, user perceptions and behavioral impacts. Int. J. Man Mach. Stud. **38**(3), 475–487 (1993)

Davis, F.D., Bagozzi, R.P., Warshaw, P.R.: User acceptance of computer technology: a comparison of two theoretical models. Manage. Sci. **35**(8), 982–1003 (1989)

Goodhue, D., Thompson, R.L.: Task-technology fit and individual performance. MIS Q. **19**(2), 213–236 (1995)

Hajli, N.: Social commerce constructs and consumer's intention to buy. Int. J. Inf. Manage. **35**(2), 183–191 (2015)

Hsu, M.H., Chiu, C.M.: Internet self-efficacy and electronic service acceptance. Decis. Support Syst. **38**(3), 369–381 (2004)

Huang, K.Y., Chuang, Y.R.: A task–technology fit view of job search website impact on performance effects: an empirical analysis from Taiwan. Cogent Bus. Manag. **3**(1), 1253943 (2016)

Kim, J.: Analysis of health consumers' behavior using self-tracker for activity, sleep, and diet. Telemed. E Health **20**(6), 552–558 (2014)

Liang, T.-P., Ho, Y.-T., Li, Y.-W., Turban, E.: What drives social commerce: the role of social support and relationship quality. Int. J. Electron. Commer. **16**(2), 69–90 (2011)

Limayem, M., Cheung, C.M.K.: Predicting the continued use of Internet-based learning technologies: the role of habit. Behav. Inf. Technol. **30**(1), 91–99 (2011)

Lin, W.S., Wang, C.H.: Antecedences to continued intentions of adopting e-learning system in blended learning instruction: a contingency framework based on models of information system success and task-technology fit. Comput. Educ. **58**(1), 88–99 (2012)

Lin, W.S.: Perceived fit and satisfaction on web learning performance: IS continuance intention and task-technology fit perspectives. Int. J. Hum. Comput. Stud. **70**(7), 498–507 (2012)

Lu, H.-P., Yang, Y.-W.: Toward an understanding of the behavioral intention to use a social networking site: an extension of task-technology fit to social-technology fit. Comput. Hum. Behav. **34**, 323–332 (2014)

Mouakket, S.: Factors influencing continuance intention to use social network sites: the Facebook case. Comput. Hum. Behav. **53**, 102–110 (2015)

Ramayah, T., Ahmad, N.H., Hong, T.S.: An assessment of e-training effectiveness in multinational companies in Malaysia Thurasamy. J. Educ. Technol. Soc. **15**(2), 125–137 (2012)

Shumaker, S.A., Brownell, A.: Toward a theory of social support: closing conceptual gaps. J. Soc. Issues **40**(4), 11–36 (1984)

Stauss, B., Neuhaus, P.: The qualitative satisfaction model. Int. J. Serv. Ind. Manag. **8**(3), 236–249 (1997)

Terzis, V., Moridis, C.N., Economides, A.A.: Continuance acceptance of computer based assessment through the integration of user's expectations and perceptions. Comput. Educ. **62**, 50–61 (2013)

Tseng, S.M.: Exploring the intention to continue using web-based self-service. J. Retail. Consum. Serv. **24**(C), 85–93 (2015)

Venkatesh, V., Morris, M.G., Davis, G.B., Davis, F.D.: User acceptance of information technology: toward a unified view. MIS Q., 425–478 (2003)

Venkatesh, V., Thong, J.Y.L., Chan, F.K.Y., Hu, P.J.H., Brown, S.A.: Extending the two-stage information systems continuance model: incorporating UTAUT predictors and the role of context. Inf. Syst. J. **21**(6), 527–555 (2011)

Wu, B., Chen, X.: Continuance intention to use MOOCs: integrating the technology acceptance model (TAM) and task technology fit (TTF) model. Comput. Hum. Behav. **67**, 221–232 (2017)

A Proposed Tourism Information Accuracy Assessment (TIAA) Framework

Sivakumar Pertheban[1(✉)], Ganthan Narayana Samy[1], Bharanidharan Shanmugam[2], and Sundresan Perumal[3]

[1] Advanced Informatics School, Universiti Teknologi, Malaysia, Kuala Lumpur, Malaysia
sivakumar84972709@gmail.com

[2] School of Engineering and Information Technology, Charles Darwin University, Casuarina, Australia
Bharanidharan.Shanmugam@cdu.edu.au

[3] Faculty of Science and Technology, Universiti Sains Islam Malaysia, Nilai, Malaysia
sundresan.p@usim.edu.my

Abstract. In the tourism business, there is an urgent need to provide better information accuracy levels. The information accuracy gaps and issues are affecting the tourism stakeholders from making informed decisions. The proposed Tourism Information Accuracy Assessment (TIAA) framework strongly underscores the need for accurate tourism information. The proposed framework is adapted from the ISO/IEC 9126 standard for tourism information resource, environment, assessment and quality process identification and the existing methodologies of information quality assessment in determining the framework information accuracy dimensions, accuracy characteristics and indicator identification. In this article, we will explain the proposed framework in terms of the information accuracy factors, sub-factors, dimensions, accuracy characteristics and accuracy assessments. The framework will be evaluated by tourism experts. The tourism experts' reviews and findings will be discussed in the analysis and results presentation sections.

Keywords: Tourism information · Information accuracy
Information assessments · Information quality

1 Introduction

The tourism industry is well known for its mass and complex information [1]. The industry, in line with the digital information age, plays a vital role in most of the country's economic development [2]. Increases in information needs in tourism have given rise to an increased need for information accuracy by the tourism stakeholders in their decision making processes [1, 3]. Generally, information accuracy can be defined as the capability of the existing processes or procedures or systems to provide precise information to the users for their own needs and requirements [4]. Information accuracy also can be defined as the precise level of delivered information compared to the actual source [5]. Information accuracy which is supported by quality dimensions or attributes

F. Saeed et al. (Eds.): IRICT 2018, AISC 843, pp. 831–839, 2019.
https://doi.org/10.1007/978-3-319-99007-1_77

is able to trigger the proper management of the information [5]. In tourism, the information sources and information providers play an important role in providing accurate information to tourism stakeholders for their decision making processes [3]. In the tourism information environment, there is always room for questioning the information accuracy of the available tourism information [1]. There is a distinct possibility that the tourism organizations could mistakenly provide misleading, confusing, outdated and statements that lack credibility because of information source accuracy issues [1, 3, 6, 7]. Furthermore, the inaccurate information through the information sources can cause confusion in the tourism stakeholders' decision making process [8]. The information gaps in tourism are often suspicious. Therefore, correct information sources are greatly needed. The tourism information accessibility and its sources are provided by tourism agencies, tourism commissions, regulatory bodies, tourism information centers and other sources. The quality and accuracy of the available information can be questioned as well [8]. Information accuracy is an important aspect that must be encouraged, managed, observed and maintained by the tourism information providers in order to enable the tourism stakeholders to make precise and informed decisions. The tourism information accuracy problems are triggered by mass, complex and inaccurate tourism information that are disseminated in a tourism environment without quality information accuracy assessment methods. Thus, the tourism information that has been disseminated has been low in quality, inaccurate, inconsistent and incomplete and this has affected the tourism information stakeholders' decision making abilities [1, 9]. Furthermore, according to (Kourouthanassis et al. [1], Li et al. [3]), information credibility is one of the decisive factors in the decision making process. Taking into consideration the gaps in tourism information accuracy, a vital information accuracy assessment framework is therefore essential for the tourism business to improve its information accuracy.

2 Tourism Information Accuracy Assessment (TIAA) Framework

The TIAA framework methodology and framework assessment details will be explained in the following sections.

2.1 Framework Development Methodology

The framework development considers the relevant and existing common generic information assessments practices. The relevancy of the existing common generic information assessment practices has been referred as guides in developing the TIAA framework. The ISO/IEC 9126 standard has been referred during the tourism information resource, environment, assessment and quality processes [10–14]. The quality model framework approach such as the quality process and the quality attributes in ISO/IEC 9126 standard has been referred in the development of the framework as well. The ISO/IEC 15504 standard has been referred during the assessment process in assigning the accuracy indicators, assessment indicators and accuracy levels with this standard "Capability Levels of ISO/IEC 15504" [15]. As for the rating information accuracy, we have followed the KAPPA statistical techniques in measuring the accuracy assessment results [16]. Thus, with the referred information, the TIAA framework consists of information accuracy factors, sub-factors, dimensions, accuracy characteristics, accuracy assessments and assessment results interpretations as the general conceptual architecture (see Fig. 1).

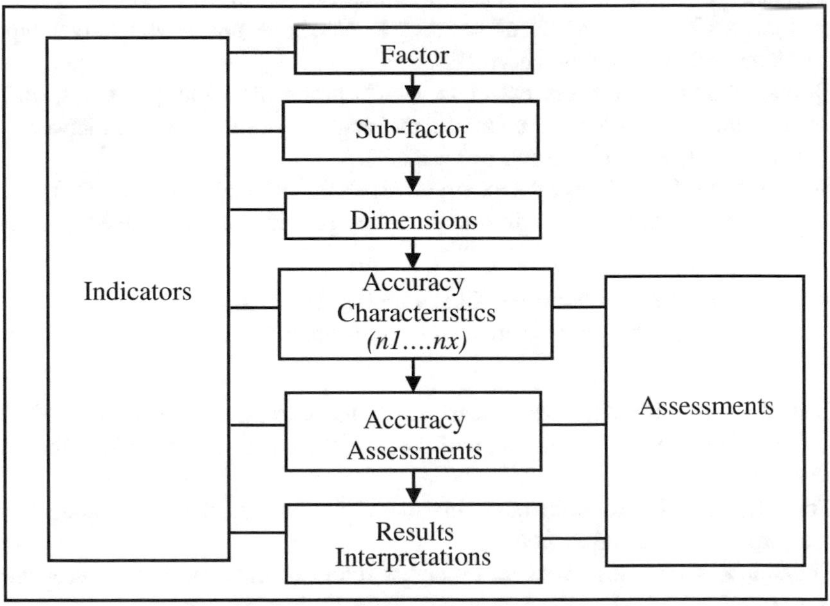

Fig. 1. TIAA framework architecture

Figure 2 illustrates the TIAA framework and the details of the framework have been explained in the following sections.

2.1.1 Factor

Factor is referring to the tourism information "Accuracy" in the development of the framework.

2.1.2 Sub Factors and Dimensions

The TIAA framework sub factors consist of tourism institutional and information environments. The tourism institutional environment refers to the tourism institutional information organizational and business dimensions. The tourism information environment contains data quality dimensions.

2.1.3 Tourism Institutional Environment

The tourism institutional environment is defined in terms of the information organizational & business dimensions. Appended are the characteristics of organizational & business dimensions:

- **Fairness and objectivity:** the initiation of tourism information is undertaken in an objective, professional and transparent manner.
- **Professional independence:** the extent to which the information providers producing information are independent or dependent from other tourism policies, regulatory or administrative departments and bodies and potential conflicts of interest.
- **Adequacy of resources:** the extent to which the resources available to the information providers are sufficient to meet its needs in terms of the development, capturing and coverage of information.
- **Quality commitment:** the extent to which procedures and processes, staff and facilities are in place to ensure that the information produced is commensurate with their internal quality objectives or measures.
- **Statistical confidentiality:** the extent to which the privacy of information providers and the confidentiality of the information they provide are guaranteed (if relevant).

2.1.4 Information Environment Data Quality Dimension

The following are the characteristics of information environment data quality dimensions:

- **Completeness:** Breadth, depth and scope: the extent to which information has sufficient breadth, depth and scope of the task at hand in accordance to the tourism perspective.
- **Consistency:** integrity constraints involve differences in the information received compared to the actual source.
- **Timeliness, Currency, Traceability:** this refers to the time lag between the reference period and when the information actually becomes available.
- **Credibility, Correctness, Reputation, Reliability:** Information source and content: representing whether a source and content can provide the right information, is free from errors, is trusted or is highly regarded in terms of the source and content.
- **Volatility—Time length:** time length for which information remains valid and relevant.

2.1.5 Accuracy Assessment

The accuracy assessments consist of accuracy and assessment indicators.

Accuracy Indicators

The information content and source indicators have been identified as the accuracy indicators. Appended below are the details of the accuracy indicators:

- Information source and content refer to the tourism information.
- Information source assessment and results validation.

Table 1. Assessments indicators

Assessment indicators	Information assessments focal
Information coverage	Source and content
Information capture and collection	Collection and gathering
Information depth	Mismatch and consistency
Information edit and imputation	Logical and consistent
Information processing and estimation	Processing of raw information
Information document and currency	Information state
Information standard	Information elements
Information standardization	Completeness and originality
Information historical comparability	Historical changes
Information adaptability	Traceability
Information value	Information usage and gap analysis

Assessments Indicators

Table 1 provides the list of these indicators measures:

2.1.6 Assessments Stages and Process

The information accuracy assessments have four distinct stages and the assessments of these stages are explained below:

- **Preliminary**—The "Preliminary" stage involves the gathering of tourism information which includes tourism information capture, collection and coverage assessments.
- **Administration**—The "Administration" stage involves tourism information processing, editing and accessing the input.
- **Measurement**—The "Measurement" stage evaluates the accuracy of the information.
- **Verification**—The "Verification" stage verifies the accuracy of the information standards, standardization, accessibility and adaptability.

2.1.7 Information Accuracy Ratings

The information is rated as met (Yes = 1), not met (No = 0) or partially met (P = 0.5). With the scores from these assessments, a percentage (%) of the scores for each of information accuracy assessments will be computed and will be compared with the Kappa Statistic-agreement (see Table 2) in order to determine the information accuracy level (see Table 3).

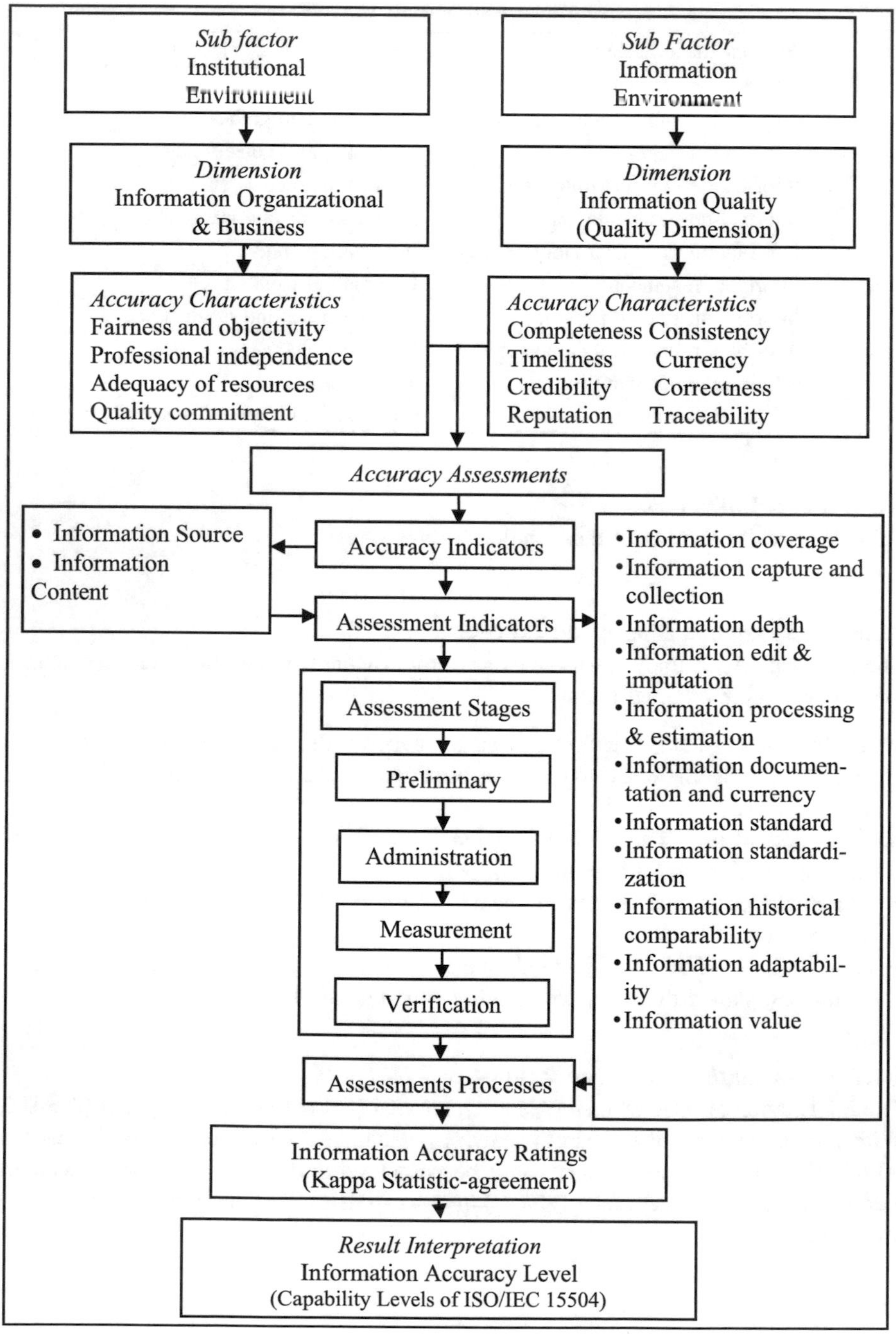

Fig. 2. Tourism information accuracy assessment (TIAA) framework

Table 2. Suggested rating information accuracy

Verification information accuracy rate (%)	Kappa statistic	Kappa agreement	Information accuracy
80% to 100%	0.80 to 1.00	Very good	Level 5
60% to 80%	0.60 to 0.80	Good	Level 4
40% to 60%	0.40 to 0.60	Moderate	Level 3
20% to 40%	0.20 to 0.40	Fair	Level 2
Less than 20%	Less than 0.20	Poor	Level 1

Table 3. Information accuracy levels and descriptions

Information accuracy	Description information accuracy level
Level 5	Meets the information accuracy measures
Level 4	Meets the information accuracy consistency
Level 3	The accuracy measures are not well performed and managed
Level 2	None existence of evidences of information accuracy measures
Level 1	General failure to accomplish the tourism information accuracy

2.2 Analysis and Result

This section presents the results of the TIAA framework evaluation. As an extension of the tourism information, the TIAA framework evaluation is aimed at evaluating the usability of the TIAA framework in a tourism information environment in terms of framework suitability, efficiency, satisfaction, adaptability and safety as proposed by Larusdottir [17], Shackel [18], Riihiaho [19] in the usability evaluation in software development practice. The selection framework usability attributes are based on factors that can influence and determine the information accuracy levels in tourism information and institution environments. The usability evaluation has been conducted using the expert review evaluation methodology proposed by Ayyub [20]. The framework's expert review forms were furnished to the identified tourism experts in performing the evaluation. The framework's suitability, efficiency, satisfaction, adaptability and safety were evaluated by the tourism experts using a Likert-type scale, with point scores of 1 (very Low), 2 (low), 3 (average), 4 (high) and 5 (very high). Table 4 provides the results of the expert review.

Table 4. Framework expert score framework usability

Usability evaluation	R1	R2	R3	R4	R5	SD	Mean	Median
Suitability	3	2	3	3	3	0.5	2.8	3.00
Efficiency	3	3	3	3	3	0.39	2.95	3.00
Satisfaction	3	3	3	3	4	0.47	3.11	3.00
Adaptability	3	3	3	3	3	0.49	3.00	3.00
Safety	3	3	3	3	4	0.49	2.97	3.00

3 Conclusion

In this article, we have explained the TIAA framework in supporting the tourism information providers in ensuring tourism information accuracy levels. The proposed TIAA framework is based on prior information and research that have been carried out to sustain and maintain tourism information accuracy. In the proposed TIAA framework, we have covered a broad range of areas and have developed information accuracy assessments so that the tourism information provided is reliable, accurate and relevant.

References

1. Kourouthanassis, P.E., et al.: Explaining travellers online information satisfaction: a complexity theory approach on information needs, barriers, sources and personal characteristics. Inf. Manag. **54**, 814–824 (2017)
2. Schloegel, C.: Sustainable tourism. J. Sustain. For. **25**(3–4), 247–264 (2007)
3. Li, Y., et al.: The concept of smart tourism in the context of tourism information services. Tour. Manag. **58**, 293–300 (2017)
4. Romero-Rodríguez, L.M., de-Casas-Moreno, P., Torres-Toukoumidis, Á.: Dimensions and indicators of the information quality in digital media (Dimensiones e indicadores de la calidad informativa en los medios digitales). Comunicar **24**(49), 91–100 (2016)
5. Batini, C., Scannapieco, M.: Data Quality Dimensions. Data and Information Quality, pp. 21–51. Springer, Berlin (2016)
6. Liang, S., et al.: Consumer motivation in providing high-quality information: building toward a novel design for travel guide websites. Asia Pac. J. Tour. Res. **22**, 1–15 (2017)
7. Coromina, L., Camprubí, R.: Analysis of tourism information sources using a Mokken Scale perspective. Tour. Manag. **56**, 75–84 (2016)
8. Lam, C., McKercher, B.: The tourism data gap: the utility of official tourism information for the hospitality and tourism industry. Tour. Manag. Perspect. **6**, 82–94 (2013)
9. Ben-Elia, E., et al.: The impact of travel information's accuracy on route-choice. Transp. Res. Part C Emerg. Technol. **26**, 146–159 (2013)
10. Jung, H.-W., Kim, S.-G., Chung, C.-S.: Measuring software product quality: a survey of ISO/IEC 9126. IEEE Softw. **21**(5), 88–92 (2004)
11. Al-Kilidar, H., Cox, K., Kitchenham, B.: The use and usefulness of the ISO/IEC 9126 quality standard. In: 2005 International Symposium on Empirical Software Engineering (2005)

12. Zeiss, B., et al.: Applying the ISO 9126 quality model to test specifications. Softw. Eng. **15** (6), 231–242 (2007)
13. Kanellopoulos, Y., et al.: Code quality evaluation methodology using the ISO/IEC 9126 standard. arXiv preprint arXiv:1007.5117 (2010)
14. Padayachee, I., Kotze, P., van Der Merwe, A.: ISO 9126 external systems quality characteristics, sub-characteristics and domain specific criteria for evaluating e-Learning systems. The Southern African Computer Lecturers' Association, University of Pretoria, South Africa (2010)
15. Mesquida, A.-L., Mas, A., Barafort, B.: The project management SPICE (PMSPICE) process reference model: towards a process assessment model. In: European Conference on Software Process Improvement. Springer (2015)
16. Viera, A.J., Garrett, J.M.: Understanding interobserver agreement: the kappa statistic. Fam. Med. **37**(5), 360–363 (2005)
17. Larusdottir, M.K.: Usability evaluation in software development practice. In: IFIP Conference on Human-Computer Interaction. Springer (2011)
18. Shackel, B.: Usability-context, framework, definition, design and evaluation. Interact. Comput. **21**(5–6), 339–346 (2009)
19. Riihiaho, S.: Experiences with usability evaluation methods. Licentiate thesis, Helsinki University of Technology, Laboratory of Information Processing Science (2000)
20. Ayyub, B.M.: A Practical Guide on Conducting Expert-Opinion Elicitation of Probabilities and Consequences for Corps Facilities. Institute for Water Resources, Alexandria (2001)

A Framework for Electronic Records Management System Adoption in the Higher Professional Education: Individual, Technological and Environmental Factors

Muaadh Mukred(✉), Zawiyah M. Yusof, Umi Asma' Mokhtar, and Fariza Fauzi

Faculty of Information Science and Technology, Universiti Kebangsaan Malaysia, 43600 Bangi, Selangor, Malaysia
mmkrd@yahoo.com, zawiy@ukm.edu.my

Abstract. Majority of organizations, with the inclusion of higher education institutions, strive to establish strategic, efficient and timely information management. In the context of higher education institutions, valuable information is required to run the business efficiently and effectively. Although the role of information in running educational institutions has become critical, higher professional education (HPE) institutions face a challenge as they seek to adopt technologies to help in their information management. Electronic Records Management System (ERMS) helps the organizations to manage the information that are needed to plan and make well informed decisions, thus improving their competencies. In relation to this, ERMS is a relatively new addition to organizations and as such, its contributions have not been explored and because of this, organizations are still hesitant to adopt ERMS initiatives. Therefore, this study proposed a framework containing factors that influence ERMS adoption in HPEs. There were ten primary factors extracted through the reviewed relevant literature of 75 articles, with UTAUT and TOE that were used as the basis for the framework development. This framework is expected to assist HPEs institutions to determine and understand various aspects (individual, technological and environment) during ERMS adoption.

Keywords: Electronic records management system
Higher professional education · Framework · Adoption

1 Introduction

Organizations have been exerting effort to implement information management that is efficient and timely. This holds true in higher education institutions. The importance of information has affected several organizations, particularly in the developed nations like the U.K. and the U.S., where information management is appropriated a budget to make sure that the country's mandatory laws are complied with and the survival of the organization is supported. In relation to this, there has been a notable shift within several developing countries to knowledge-based economy from an agriculture-based

F. Saeed et al. (Eds.): IRICT 2018, AISC 843, pp. 840–849, 2019.
https://doi.org/10.1007/978-3-319-99007-1_78

one. Such a shift indicates that information is increasingly accepted as a resource, with its direct effect on the development of the nation and its optimization of production and uses of resources—an overall contribution to the nation's development [1–5].

To be of value information must be recorded. This information contains evidence of the organization's activities or transactions. This type of recorded information is used to support business functions and as such, it is important in the evaluation of the performance of the organization. Lack of reliable records leaves agencies and governments unable to manage their resources, revenues, while civil service cannot deliver services like healthcare and education. This highlights the importance of records in effective governance [6, 7].

In the case of HPE institutions, e-records are mostly used as a strategic resource to contribute to the institutions' competency, enhance the evaluation of performance, and ultimately, achieve effective performance. Several institutions around the globe in the form of HPE indicate the importance of Electronic Records Management System (ERMS) implementation and use [8–10]. Moreover, there is a continuous exponential growth of information generated by electronic media and this is coupled with the technological innovations constantly changing the way individuals, organizations and institutions and governments are run. This can be exemplified by the technology advances that have managed to transform the storage of records and archives. In addition, information and communication technologies (ICT) have resulted in an alternative way e-records are produced through office automated and digitalized mechanisms [9, 11–14].

In prior literature, researchers have been noted to ignore the use of ERMS in educational field [15, 16], which means that schools have largely been left out when it comes to studies examining the relationship between computers, communication skills, teaching experience and the effective technologies that are utilizable to overcome that challenges that arise out of the integration of the above. Also, research in the same area would enable schools to identify the way technology can be used to improve the students' personal development and their educational outcomes via strategic policies. Added to this, research in the field would also lead to developing technologies geared towards educational institutions needs and allow its members to experience technology changes. Relevant technologies and support provided to the user ensure the effective use of technology [17, 18].

In the case of HPEs institutions, the generation of records takes place in periodic terms and because records can make or break organizations, they need to be effectively and efficiently managed and optimally leveraged [19]. The considerable records volume produced in HPEs are ripe for management guidance to assist in the educational performance and quality provision. In this regard, the effective e-records management requires a framework for guiding the process in its entirety.

The minimal adoption of such system could threaten its successful implementation [12, 20] and therefore, the factors influencing ERMS adoption have to be addressed, particularly as researches dedicated to the ERMS adoption in HPE are still few and far between. Once the issue is addressed, a framework can be created to assist successful ERMS adoption in HPEs.

This paper identifies the influence of the individual, technological and environmental factors on ERMS adoption. The structure of the paper is as follows: The need of

HPE institutions to adopt ERMS is presented in Sect. 2. The benefits of ERMS is included in Sect. 3 followed by related work on ERMS adoption factors in Sect. 4. The research methodology of this research is described in Sect. 3, and the framework development presented in Sect. 4. Lastly the conclusion is outlined in Sect. 5.

2 Related Works on ERMS

Employees working in different organizations have been evidenced to steer clear of using technology when they can [21] and this has led researchers in the IS field to determine factors that contribute to the laggard adoption and use of IS within organizations [22–24]. The question arises as to what the primary factors are for successful ERMS adoption. Therefore, in this study, a research framework is developed and proposed to investigate factors that affect the adoption of ERMS, in light of three dimensions (individual, technological and environmental).

Studies on ERMS, specifically in the case of HPEs, are still lacking compared to the innumerable ones in healthcare [25–29]. While the healthcare records are managed with the assistance of electronic health records (EHR) and are specific to the sector, the concepts and methods and management may be superimposed to the management of other fields' records.

The adoption of IT refers to the level to which potential adopters believe that adopting IT would lessen their efforts in task completion [30]. Added to this, adoption is referred to as the innovation use that appears to be the top optimum course of action [31] and it is described as the initial decision and intention towards using the innovation in question [32]. In this regard, ERMS adoption has a major role in delivering educational services by ensuring effective services implementation and quality delivery. The process is initiated by the awareness of the potential adopters of the technology and it ends by the actual adoption and usage of the technology [33].

In the field of education, intention towards the adoption of ERMS is viewed all over the globe as a method that could mitigate the gap between the demand and supply of education. In fact, majority of nations have become increasingly aware of the importance of investing in innovation to produce improvements to the educational system [34]. Also, adoption of ERMS can contribute to providing education through a set of processes that could enhance HPE effectiveness, performance and goals achievement. In prior studies, several barriers have been mentioned to hinder ERMS adoption, specifically in the context of developing nations [2, 5, 9, 12]. Additionally, according to studies in the educational sector, ERMS adoption is still in its infancy [10, 12, 33, 35–37], with a considerable proportion of studies dedicated to the healthcare sector. Such studies found the barriers to be under the categories of technological and organizational barriers [38, 39].

According to [40], a good way to shed light on the current ERMS adoption state among public institutions is to determine the content and context of record dimensions and to illustrate the move from legacy shared-drive systems to ERMS. Therefore, ERMS has been the focus of empirical studies, with different aims and outcomes, and different examined factors influencing the adoption of the system. The next sections provide discussions on the importance of the factors (individual, technological and environmental) influencing ERMS adoption.

To date, ERMS adoption and implementation has been negligible in developing nations but this does not negate the fact that such system increases the efficiency and effectiveness of governments, facilitates transparency and accountability, and improves efficient public service provision. Challenges of ERMS implementation are numerous as noted by countries all over the globe and do not depend on the country's nature.

3 Methodology

This paper primarily aims to examine and determine the factors affecting ERMS adoption in the educational sector, particularly in HPEs. The study conducted an explorative literature review and used the outcome as the basis to analyze both opportunities and challenges in HPEs' adoption of ERMS.

More specifically, the paper made use of content analysis to determine the factors that influence the adoption of ERMS. The relevant papers were chosen through a search of research databases, open search and authentic websites, using the keywords, ERMS adoption, e-records system factors, factors influencing technology adoption, and factors for ERMS adoption. Added to this, only English articles were chosen, after which the papers' contents were analysed and categorized to acquire the related input. More than 75 related articles had been reviewed. The factors were accordingly extracted and confined. The top cited factors were selected to bring forth the study framework. Table 1 shows the list of top cited factors.

Table 1. List of top cited factors

No	Factor	No. of frequencies
1	Perceived ease of use	55
2	Perceived usefulness	51
3	ICT infrastructure	50
4	Knowledge and skills	45
5	Policies	42
6	Law and legislation	37
7	Self-efficacy	36
8	Competitive pressure	29
9	Adaptability	28
10	Satisfaction	26
11	Attitude	25

The methodological steps followed by the paper are depicted in Fig. 1.

According to Ngulube and Ngulube [41], models are the main road for researchers to conceptual frameworks, while theories lead to theoretical frameworks. Researchers in Social science start out with models and then progress to concepts that represent an identified research problem within a subject matter, and collect data to understand and

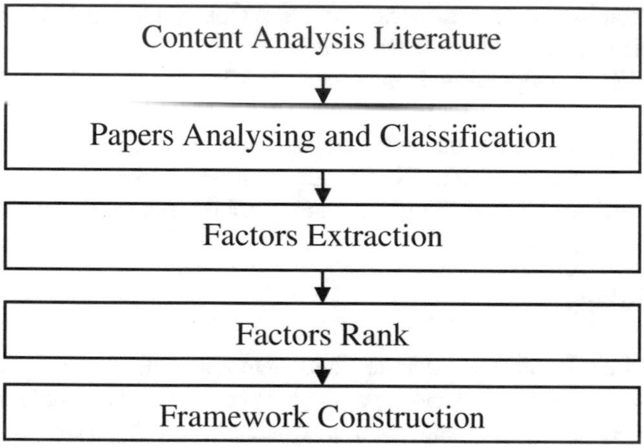

Fig. 1. Methodology of the paper

establish linkages between these concepts. Concepts become theoretical structures, as they are the building blocks of theory. Concepts are measurable, and measurement is the essence of operationalization. Unlike theories, which have the power to explain and predict, models merely describe the phenomenon. Theories are tested through propositions or hypotheses using a methodology that fits with the model or theory.

Prior studies have adopted various theories to explain the influence of technology use and adoption, which include Technology Acceptance Model (TAM), Diffusion of Innovation (DOI), Theory of Reasoned Action (TRA), Unified Theory of Acceptance and Use of Technology (UTAUT) and Theory of Planned Behaviour (TPB). In this study the UTAUT, as an integrated model, and Technology-Organization-Environment (TOE) were employed for the assessment of behavioural intention towards the adoption of ERMS.

In the context of education, prior studies primarily used UTAUT or TAM to base their theoretical models on in terms of technology adoption [42–46]. Nevertheless, the above mentioned models are not extensive in their prediction and therefore attempts should be made to integrate other theories [47, 48]. TAM is capable of explaining 40% of the variance in technology acceptance/adoption in the workplace and UTAUT has 70% capability of doing the same [49], therefore, this study uses UTAUT as the base for the conceptual framework.

4 Framework Development

In order to enhance education, new technologies adoption should be adopted in educational institutions. Successful new technologies adoption requires examining the factors that affect the adoption and thus, this paper identified the factors important for ERMS adoption in the educational sector. Accordingly, the paper examined the level of relationship among the relevant adoption factors, which are individual, technological and environmental factors.

Therefore, the identification of the content and context of records dimensions provided an approach to shedding light on the ERMS adoption state in the current public sector educational institutions, and in providing an insight into the hindrance that arises in the shift from legacy shared-drive systems to ERMS.

Literature classified factors that influence the adoption of ERMS into several categories, including technological factors, organizational factors, environmental factors and individual factors, but this paper focused on the individual, technological and environmental factors only. The ERMS adoption in the educational sector calls for a comprehensive robust framework that encompasses all the three categories in terms of ERMS adoption.

The reviewed literature shows that individual dimension, technological dimension as well as environmental dimension could determine ERMS adoption. The empirical studies in literature showed several sub-important factors that could impact the intention towards EMRS adoption relating to mentioned dimensions. Specific to this study, the individual dimension refers to a set of four factors, namely attitude, knowledge and skills, satisfaction and self-efficacy, while the technological dimension refers to a set of four factors, namely perceived usefulness, perceived ease of use, adaptability and ICT Infrastructure. Meanwhile, the environmental factors refer to a set of three factors including law and legislation, policies and competitive pressure.

As mentioned, the proposed framework was developed using UTAUT and TOE (see Fig. 2).

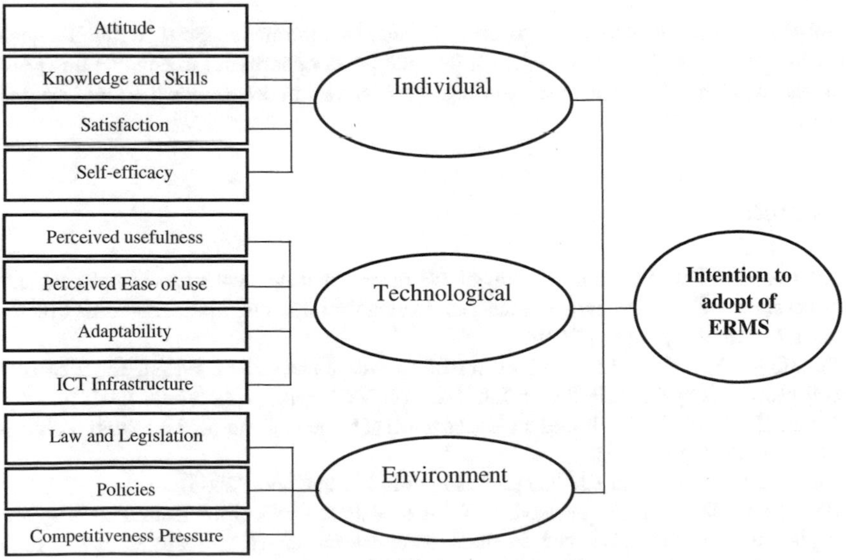

Fig. 2. Proposed conceptual framework for electronic records management system adoption

5 Conclusion

The relevance of e-records has driven researchers to examine the factors influencing ERMS adoption in order to improve management and decision making in organizations in light of efficiency and effectiveness. Several researches have been utilized for such examination in the hopes of promoting ERMS adoption and in this paper, the author provided an understanding of the factors that influence ERMS adoption from the HPEs perspective.

This study's findings are expected to promote awareness among public records management of the system's importance in managing educational records and to lay emphasis on the requirement for the sufficient appropriation of individual, technological and environmental factors needed for successful ERMS adoption.

On the basis of this paper's results, HPE institutions should be aware of the factors that can drive the development of strategies and formulate guidelines for employees to accept ERMS and use the system consistently in their daily tasks. It is hoped that the findings can provide a deeper insight into organizations of other domains as to the factors that have to be focused on for the successful system adoption. The paper urges more studies to examine the ERMS role in decision making, and the formation of a robust framework for the process of its adoption. In this paper, a framework is brought forward as a guide to system implementation and adoption. The reviewed prior literature indicated the relationship between successful ERMS adoption and the proper framework to guide such adoption.

Acknowledgment. We would like to thank Faculty of Information Science and Technology, Universiti Kebangsaan Malaysia by giving the authors an opportunity to conduct this research. This research is funded by Universiti Kebangsaan Malaysia under Research Grant Code: GUP-2017-046.

References

1. Ab Aziz, A., et al.: Electronic document and records management system implementation in Malaysia: a preliminary study of issues embracing the initiative. In: International Conference on Information. Springer (2018)
2. Eusoff, R., Yusof, Z.M.: Development of records management system for matriculation colleges in Malaysia. Asia Pac. J. Inf. Technol. Multimed. **11**(1), 23–28 (2011)
3. Mohd, Z., Chell, R.W.: Issues in Records Management. Penerbit Universiti Kebangsaan Malaysia, Selangor (2005)
4. Yusof, Z.M.: Issues and Challenges in Records Management (2005)
5. Baron, J.R., Thurston, A., Mcleod, J.: What lessons can be learned from the US archivist's digital mandate for 2019 and is there potential for applying them in lower resource countries? Rec. Manag. J. **26**(2), 206–217 (2016)
6. Dastan, Đ., Çiçek, M., Naralan, A.: The effects of information technology supported education on strategic decision making: an empirical study. Procedia Soc. Behav. Sci. **24**, 1134–1142 (2011)
7. Yoon, D., et al.: Adoption of electronic health records in Korean tertiary teaching and general hospitals. Int. J. Med. Inform. **81**(3), 196–203 (2012)

8. Mukred, M., Yusof, Z.M.: Factors influencing the adoption of electronic records management (ERM) for decision making process at higher professional education (HPE)'s institutions (2015)
9. Adade, A., Quashigah, A.Y., Eshun, P.: Academic records management in Ghanaian basic schools: a study of basic schools in the Ashiedu Keteke sub-metro in the greater Accra region. Br. J. Educ. **6**(4), 33–49 (2018)
10. Adu, K.K.: Framework for digital preservation of electronic government in Ghana (2015)
11. Eze Asogwa, B.: The readiness of universities in managing electronic records: a study of three federal universities in Nigeria. Electron. Libr. **31**(6), 792–807 (2013)
12. Mosweu, O., Bwalya, K.J., Mutshewa, A.: A probe into the factors for adoption and usage of electronic document and records management systems in the Botswana context. Inf. Dev. **33** (1), 97–110 (2017)
13. Nwaomah, A.E.: Records information management practices: a study on a faith based university. Int. J. Innov. Educ. Res. **5**(11), 87–102 (2017)
14. Nyathi, T., Dewah, P.: Towards implementing a records management policy at the National University of Science and Technology. ESARBICA J. **36**, 12 (2017)
15. Gröger, S., Decker, J., Schumann, M.: Do universities get the hang of working efficiently? A survey of the influencing factors on the adoption of electronic document and workflow management in German-speaking countries (2014)
16. Tzeng, W.-S., et al.: Do ePortfolios contribute to learners' reflective thinking activities? A preliminary study of nursing staff users. J. Med. Syst. **39**(9), 1–10 (2015)
17. Merigó Lindahl, J.M., López-Jurado, M.P., Gracia Ramos, M.C.: A decision making method for educational management based on distance measures. Revista de Métodos Cuantitativos para la Economía y la Empresa 8 (2009)
18. Mukred, M., Yusof, Z.M.: The performance of educational institutions through the electronic records management systems: factors influencing electronic records management system adoption. Int. J. Inf. Technol. Proj. Manag. (IJITPM). **9**(3), 34–51 (2018)
19. Hase, S., Galt, J.: Records management myopia: a case study. Rec. Manag. J. **21**(1), 36–45 (2011)
20. Moseti, I.: Digital preservation and institutional repositories: case study of universities in Kenya. J. S. Afr. Soc. Arch. **49**, 137–154 (2016)
21. Schaper, L.K., Pervan, G.P.: ICT and OTs: a model of information and communication technology acceptance and utilisation by occupational therapists. Int. J. Med. Inform. **76**, S212–S221 (2007)
22. Ameen, A.A., Ahmad, K.: Information systems strategies to reduce financial corruption. In: Benlamri, R., Sparer, M. (eds.) Leadership, Innovation and Entrepreneurship as Driving Forces of the Global Economy, pp. 731–740. Springer, Berlin (2017)
23. Ameen, A.B., Ahmad, K.: A conceptual framework of financial information systems to reduce corruption. J. Theor. Appl. Inf. Technol. **54**(1), 59–72 (2013)
24. Ismail, M.B., Yusof, Z.M.: The relationship between knowledge sharing, employee performance and service delivery in public sector organizations: a theoretical framework. Public Sect. ICT Manag. Rev. **3**(1), 37–45 (2009)
25. Mukred, M., Yusof, Z.M.: The DeLone–McLean information system success model for electronic records management system adoption in higher professional education institutions of Yemen. In: International Conference of Reliable Information and Communication Technology. Springer (2017)
26. Mukred, M., et al.: Electronic records management system adoption readiness framework for higher professional education institutions in Yemen. Int. J. Adv. Sci. Eng. Inf. Technol. **6**(6), 804–811 (2016)

27. Mukred, M., Yusof, Z.M.: The role of electronic records management (ERM) for supporting decision making process in Yemeni higher professional education (HPE): a preliminary review. J. Teknol. **73**(2), 117–122 (2015)
28. Mukred, M., Yusof, Z.M.: Factors influencing the adoption of electronic records management (ERM) for decision making process at higher professional education (HPE)'s institutions. In: 1st ICRIL-International Conference on Innovation in Science and Technology (IICIST 2015), Kuala Lumpur, pp. 399–403 (2015)
29. Mukred, M., Yusof, Z.M.: Electronic records management and its importance for decision making process in Yemeni higher professional education (HPE): a preliminary review. In: 1st International Conference of Recent Trends in Information and Communication Technologies (IRICT 2014), Johor Bahru, pp. 105–114 (2014)
30. Phillips, L.A., Calantone, R., Lee, M.-T.: International technology adoption: behavior structure, demand certainty and culture. J. Bus. Ind. Mark. **9**(2), 16–28 (1994)
31. Rogers, E.M.: Diffusion of Innovations. Simon and Schuster, New York (2010)
32. Proctor, E., et al.: Outcomes for implementation research: conceptual distinctions, measurement challenges, and research agenda. Adm. Policy Ment. Health Ment. Health Serv. Res. **38**(2), 65–76 (2011)
33. Phiri, M.J.: Managing University Records and Documents in the World of Governance, Audit and Risk: Case Studies from South Africa and Malawi. University of Glasgow, Glasgow (2016)
34. Cheon, J., et al.: An investigation of mobile learning readiness in higher education based on the theory of planned behavior. Comput. Educ. **59**(3), 1054–1064 (2012)
35. Atulomah, B.C.: Perceived records management practice and decision making among university administrators in Nigeria (2011)
36. Asogwa, B.E.: The challenge of managing electronic records in developing countries: implications for records managers in sub Saharan Africa. Rec. Manag. J. **22**(3), 198–211 (2012)
37. Tarhini, A., Hone, K., Liu, X.: Factors affecting students' acceptance of e-learning environments in developing countries: a structural equation modeling approach. Int. J. Inf. Educ. Technol **3**(1), 54 (2013)
38. Mosse, E.L.: Understanding the Introduction of Computer-Based Health Information Systems in Developing Countries: Counter Networks, Communication Practices, and Social Identity–A Case Study from Mozambique. University of Oslo, Oslo (2004)
39. Hassall, C., Lewis, D.I.: Institutional and technological barriers to the use of open educational resources (OERs) in physiology and medical education. Adv. Physiol. Educ. **41** (1), 77–81 (2016)
40. Lewellen, M.J.: The impact of the perceived value of records on the use of electronic recordkeeping systems (2015)
41. Ngulube, P., Ngulube, B.: Mixed methods research in the South African Journal of Economic and Management Sciences: an investigation of trends in the literature. S. Afr. J. Econ. Manag. Sci. **18**(1), 1–13 (2015)
42. Gruzd, A., Staves, K., Wilk, A.: Connected scholars: examining the role of social media in research practices of faculty using the UTAUT model. Comput. Hum. Behav. **28**(6), 2340–2350 (2012)
43. Buchanan, T., Sainter, P., Saunders, G.: Factors affecting faculty use of learning technologies: Implications for models of technology adoption. J. Comput. Higher Educ. **25**(1), 1–11 (2013)
44. Findik, D., Özkan, S.: A model for instructors' adoption of learning management systems: empirical validation in higher education context. TOJET Turk. Online J. Educ. Technol. **12** (2), 13–25 (2013)

45. Grover, S.: Predicting the adoption of video podcast in online health education: using a modified version of the technology acceptance model (health education technology adoption model HEDTAM). In: E-Learn: World Conference on E-Learning in Corporate, Government, Healthcare, and Higher Education. Association for the Advancement of Computing in Education (AACE) (2015)
46. Tosuntaş, Ş.B., Karadağ, E., Orhan, S.: The factors affecting acceptance and use of interactive whiteboard within the scope of FATIH project: a structural equation model based on the unified theory of acceptance and use of technology. Comput. Educ. **81**, 169–178 (2015)
47. Venkatesh, V., Zhang, X.: Unified theory of acceptance and use of technology: US vs. China. J. Glob. Inf. Technol. Manag. **13**(1), 5–27 (2010)
48. Venkatesh, V., Sykes, T.A., Zhang, X.: 'Just what the doctor ordered': a revised UTAUT for EMR system adoption and use by doctors. In: 44th Hawaii International Conference on System Sciences (HICSS), 2011. IEEE (2011)
49. Ibrahim, R., Khalil, K., Jaafar, A.: Towards educational games acceptance model (EGAM): a revised unified theory of acceptance and use of technology (UTAUT). Int. J. Res. Rev. Comput. Sci. **2**(3), 839–846 (2011)

Intention to Use a Cloud-Based Point of Sale Software Among Retailers in Malaysia: The Mediating Effect of Attitude

Kamal Karkonasasi(✉), Lingam Arusanthran, G. C. Sodhy, Seyed Aliakbar Mousavi, and Putra Sumari

School of Computer Science, Universiti Sains Malaysia, USM, 11800 Pulau Pinang, Malaysia
asasi.kamal@gmail.com

Abstract. The on-premise point of sale (POS) system has been using by retailers for years. However, this system causes some user pains and problems, i.e. high capital investment, setup and maintenance hassle, lack of flexibility and reliability, and data security and privacy concern. Cloud computing as an emerging technology is used for developing a cloud-based POS software named RetailNow. A cloud service provider is in charge of delivering and maintenances of the software to retailers based on their environments. The provider is also responsible for security and privacy of data on its servers. Due to the pay-per-use subscription model of the cloud, retailers also save in their finance. The retailers only require a basic computer and an internet connection in order to connect to RetailNow. RetailNow is targeted for all scales and environments of retail outlets. The purpose of the study is to understand the influential factors that impact on intention to use RetailNow among retailers in Malaysia. The proposed model is the combination of the TOE and the TAM to obtain applicability and predictive power. In order to test the proposed model, a quantitative study using a self-administrated survey will be directed. A research method which is based on the multi-analytical approach of Partial Least Squares Structural Equation Modeling (PLS-SEM) and Artificial Neural Networks (ANNs) is applied which enables extra verification of the results provided by the PLS-SEM for direct relationships in the model. We only apply PLS-SEM for testing the mediation effect.

Keywords: The Point of Sale (POS) system · Cloud-based POS SaaS · RetailNow · TAM · TOE · PLS-SEM · ANN

1 Introduction

The use of computer application for retail operations has been around as early as 1973 with IBM's 3650 store system as a mainframe controlling IBM's 3653 POS registers [1]. Now, On-premise POS system is being used commonly by many retail businesses with various sizes and environments to manage their operations. The fundamental functions of the system are to record sales and manage inventory. The usage of the system in the retail environment improves supply change management, customer

F. Saeed et al. (Eds.): IRICT 2018, AISC 843, pp. 850–860, 2019.
https://doi.org/10.1007/978-3-319-99007-1_79

experience, and inventory control [2, 3]. Many obstacles in retail forecasting, such as fluctuating demand and unreliable assumptions can be overcome through analysis of POS data [4]. However, there are some concerns with the system.

The first concern is its cost for retailers. Since the data processing must be done on the retailers' sites, they require buying computers with high processing power and storage which can be costly for them. It is also possible that they buy a computer which does not fulfill their needs. In that case, it is almost impossible for them to return it to the vendor. They also need to buy a license for the system's software in order to be installed on each computer. The vendors also charge for installation, setup, and upgrades to the system. Moreover, disaster recovery efforts for the system may require highly skilled IT professional which can be costly [5]. Heart and Pliskin [6] also stated that implementation of the system requires network infrastructure for client terminal and server connection, which need additional licensing for operating systems, backup utilities, security applications, and database alongside the POS system's license itself. The authors further pointed out that besides the above-mentioned fixed costs, there are also recurring costs for regular maintenance including backup, security management, and software upgrades which are also exhausting as each terminal and server needs to be upgraded separately.

The second concern is that the system's installation and maintenance can be complex for the retailers. The system implementation uses RS-485 cabling for network access which is complex, costly, and difficult to upgrade. Retail outlets are not IT based organizations. Therefore, retail managers need to divert attention into planning and execution of the system implementation. Setting up an application server including the installation of software on all client machines requires lots of effort and time which delays the process of setting up a new outlet [6]. Moreover, disaster recovery efforts for the system may require highly skilled IT professional [5].

The third concern is that the system is unreliable and inflexible. The software requires downtime during initial setup, patching, and upgrading which affect retail operation. Young-Gun and Lim [7] stated that upgrades might require the whole software to be replaced and cause operations downtime. Young-Gun and Lim [7] stated that the POS software is bound to extensibility and scalability limitation.

The last concern is its data security and privacy. Weil [8] stated that in the case of theft in a retail outlet, devices which carry POS data including customer information could be damaged or lost. Besides, POS software mostly runs on an old operating system which less frequently is patched and therefore leaves the device vulnerable to data or information breaches [9]. The report by Close-Up Media showed that the 50 percent of top 10 global retailers have already been breached [9].

In order to overcome the above-mentioned concerns, a cloud-based POS software named RetailNow is proposed. The National Institute of Standards and Technology (NIST) referred to cloud computing as a model that enables ubiquitous, convenient and on-demand network access to computing resources that are promptly supplied and released with less effort [10]. Cloud service models are categorized based on the computing requirements of the customers and represent three distinct layers of architecture: Infrastructure as a Service (IaaS), Platform as a Service (PaaS) and Software as a Service (SaaS). In IaaS as the basic level of cloud services, the primary devices of computing power, network equipment, and storage are fully cloud-based and ready-to-use on demand over a network (e.g. internet) (e.g., Amazon Elastic Compute Cloud (EC2), Rackspace, Amazon simple storage service (S3), and GoGrid) [11–13]. In PaaS as the second level of cloud computing, the service provides an integrated platform for developing and deploying applications (e.g. Salesforce, Google AppEngine, and Microsoft Azure). This model offers all components of software development (design, testing, version control, maintenance, and hosting) over the internet [11]. SaaS emerges as a new model to deliver software applications (e.g., Joyent and SalesForce CRM) centrally hosted inside the cloud via a network (e.g. the internet) using a thin client (e.g. a web browser or mobile application) [11, 14, 15]. SaaS providers set up software applications on cloud servers for customers to order on-demand based on their necessities and pay for the services based on their actual usage [15, 16]. This "on-demand" service delivery model is similar to utility service mode that a customer just subscribes an application without the need to purchase, install and maintain the software, like getting power from the grid instead of one's own generator. Moreover, SaaS improves the quality of software services through regular and automatic application upgrade and data backup [15, 17] with minimal customers' interaction.

SaaS permits organizations to outsource business applications (e.g. POS software). Cloud computing also allows organizations to outsource their IT infrastructures (e.g. storage, backup, and computing) in form of Infrastructure as a Service (IaaS) and IT platforms (e.g. database and business intelligence) in form of Platform as a Service (PaaS) [15, 18]. Among the three, SaaS is taken into account as the most promising because it provides business customers different real benefits, such as reduced IT costs and improved IT performance [15, 19, 20]. SaaS has reformed the concept of software as a product to that of a service instead [12].

In term of functionality, RetailNow has all the features of the traditional POS system which includes sales, inventory, supplier, and customer management. Some traditional system has separate client application for a different component of POS whereby components required at cashier terminal and components required by store supervisor come in a separate installation. However, RetailNow will be a single system with all the functional components. Availability of these components differs based on the type of user, i.e. cashier or supervisor. The software will be delivered as SaaS from a cloud service provider and retailers subscribe to the software. Retailers get access to the software on the internet. Moreover, retailer's data will be stored in the cloud service provider's servers.

The level of adoption of SaaS among retailers in Malaysia is still low [21–25]. A few studies explored the factors which impact on the SaaS adoption decisions at the organizational level [23]. Therefore, the purpose of the study is to find out the influential factors that impact on intention to use RetailNow among retailers in Malaysia. Moreover, we examine the mediating role of attitude. This study contributes to the theoretical body of knowledge about Malaysian retailers' intention to use SaaS application, i.e. RetailNow and provides practical implementations for cloud service providers and retailers.

2 The Proposed Research Model

This study considers two of the technology adoption models, i.e. TAM [26] and TOE [27] which have been widely adopted for studies in organizational context [33]. Both models have limitations. The limitation of extended models of TAM is that external variables are not clearly defined yet [13]. On the other hand, the TOE is not considering the decision maker's attitude about and intention of an innovation usage in an organization. Therefore, in order to enrich the proposed model in explaining organizational behavior in adoption studies, individual factors should be considered alongside the TOE's factors [28, 29]. In an organization, it is crucial to consider the decision maker's attitude about and intention of an innovation usage, as his/her decision is influential for adoption and organizational behavior [30]. We theorize that the TOE's factors impact on attitude and intention of decision maker toward using RetailNow directly and indirectly respectively. Meanwhile, the attitude mediates the relationships between the independent variables and intention.

The proposed model is the mixture of the TOE and the TAM to acquire applicability and predictive power [13] to study significant factors that impact on intention to use RetailNow. The proposed model is shown in Fig. 1.

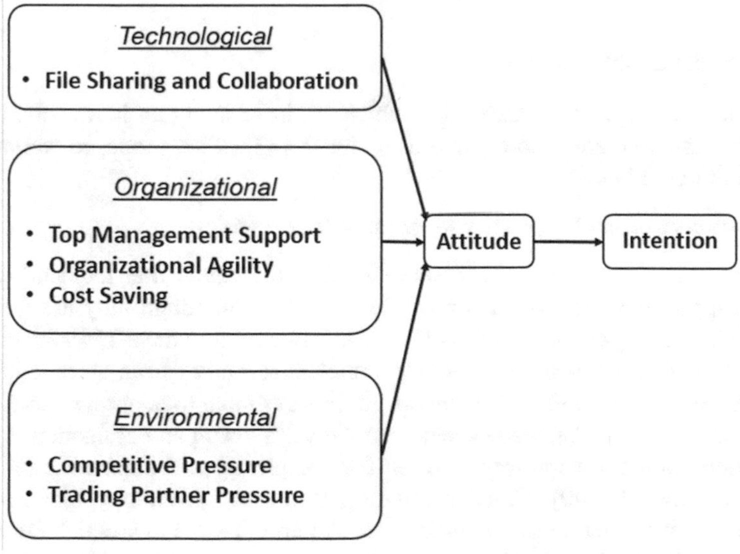

Fig. 1. The proposed research model.

3 Hypotheses Development

3.1 Technological Context

Technology context emphasizes how features of the technologies of innovations affect the adoption process [27]. One variable has been proposed:

(a) The Perceived Easiness of File Sharing and Collaboration and Attitude (IVs → MV)

The easiness of file sharing and collaboration is defined as the extent that retailers believe that by using the RetailNow, they are capable to share their work with their staff and partners easier [44]. A cloud-based software makes file sharing easy for retailers. The software uploads their files to cloud and, then shares between their staff and partners on the internet. Cloud service provider will take care of data by keeping back up. Moreover, only authorized users have access to the data. Therefore, retailers are not needed to back up data in several storages and to worry about security issues and integrity of them. Consequently, they can save their time and money [31]. Therefore, the following hypothesis is proposed:

Hypothesis H1. Easiness of File Sharing and Collaboration will have a positive impact on attitude toward using the RetailNow.

3.2 Organizational Context

The internal factors for an organization which are influencing an innovation adoption and implementation are studied in this context [32]. The proposed variables are described in detail below:

(a) Top Management Support and Attitude (IVs → MV)

Top management support (TMS) is defined as the extent that retailers' top management support the implementation of the RetailNow although they are required to take financial and organizational risks [11, 33]. It is crucial to have TMS for building a positive organizational climate to provide sufficient resources from an organization to adopt new technologies [34, 35]. Retailers are less expected to adopt new technologies without TMS [36–38]. Top management can deliver a vision and obligation to create a positive environment for innovation when the complexity and sophistication of technologies increase [39, 40]. Cloud computing as a new technology may require an enterprise to have some changes during its adoption. TMS is essential for applying changes to facilitate the adoption of the cloud among enterprises [41, 42]. The following hypothesis is proposed to study the importance of TMS:

Hypothesis H2. Top Management Support will have a positive impact on attitude toward using the RetailNow.

(b) The Perceived Organizational Agility and Attitude (IVs → MV)

Organizational agility is defined as the extent to the capability of firms to effortlessly and rapidly change their strategy in terms of customer responsiveness, business partnerships, and operations [43] by using the RetailNow. Cloud computing as a cutting-edge technology allows its consumers to get access to elastic IT resources without allocating capital expenses. On the cloud, everything is delivered as a service, which is like traditional public utilities such as electricity [44]. Due to pay per use subscription model of the cloud, retailers can manage their budget efficiently.

Moreover, due to autonomic computing feature of the cloud, they can acquire their required IT resources without or less human interaction with cloud service provider [44]. Then, this feature allows them to fulfill unpredicted market demands. Consequently, they can obtain organizational agility using the cloud [51], which is an essential factor to remain in their business and acts as an influential factor to improve their performance [43, 45]. We hypothesize that:

Hypothesis H3. The Perceived Organizational Agility will have a positive impact on attitude toward using the RetailNow.

(c) Cost Saving and Attitude (IVs → MV)

Cost saving is defined as the extent that retailers believe they can save their upfront and operating costs due to using the RetailNow [11, 46]. Retailers expend a big percentage of their budget on IT infrastructure while less than 10 percent of their servers are really utilized [54]. Cloud computing helps them to make savings on their costs which is related to hardware and software and IT infrastructures and administration expenses [38, 39]. The following hypothesis is proposed to understand the importance of this factor:

Hypothesis H4. Cost saving will have a positive impact on attitude toward using the RetailNow.

3.3 Environmental Context

The environmental context is the field in which a firm runs its business and interacts with its industry, rivals, and government [48]. In this study, several variables have been offered for this context:

(a) The Perceived Competitive Pressure and Attitude (IVs → MV)

In the current study, perceived competitive pressure (CP) is defined as the extent to the observed pressure from business competitors that force a retailer to adopt novel technology to sustain competitiveness [48]. Retailers follow their competitors in accepting an innovation because they do not want to lag behind them in gaining competitive advantage through the usage of the innovation [49]. Besides, they follow other retailers' acceptance of innovation to appear legitimate. The reason is that to be in line with institutional norms in their social interaction process [49]. Low et al. stated that CP obligates firms to adopt cloud computing ($B = 0.842$, $p < 0.05$) [50]. Therefore, the following hypothesis is proposed:

Hypothesis H5. The perceived competitive pressure will have a positive impact on attitude toward using the RetailNow.

(b) The Perceived Trading Partner Pressure and Attitude (IVs → MV)

Trading partner pressure (TPP) is defined as the extent to the observed pressure from business partners that influences a retailer to adopt novel technology to sustain cooperative relationships [48]. An enterprise is encouraged to apply cloud computing by noting that business partner gets to benefit from its adoption [47, 50]. However, retailers may not be knowledgeable about cloud computing services. Therefore, they may follow other organizations to minimize uncertainties about cloud [49]. A business partner can also force a retailer to adopt cloud computing to stay in connection with each other [50].

Low et al. [50] stated that TPP has a significant outcome on cloud computing adoption ($B = 1.834$, $p < 0.01$). Therefore, the following hypothesis is proposed:

Hypothesis H6. The perceived trading partner pressure will have a positive impact on attitude toward using the RetailNow.

3.4 Attitude and Intention (MV → DV)

Attitude toward using a system is defined as the individual's positive or negative feelings about working with a system [51]. Moreover, intention to use a system is defined as the degree of the determination of the individual's aim to use a system [51]. Ajzen and Fishbein [52] stated that attitude impacts behavioral intention. Therefore, the following hypothesis is proposed:

Hypothesis H7. Attitude will have a positive impact on intention.

3.5 The Mediating Role of the Attitude on the Relationship Between IVs and DV (IV → MV → DV)

There is a sign of mediation when a third variable interposes between two other related variables [53]. Davis et al. [51] considered Attitude as a mediator between the predictor variables and intention. Therefore, the following hypothesis is proposed to test the mediation:

Hypothesis H8. Attitude mediates the relationship between File Sharing and Collaboration and intention.

Hypothesis H9. Attitude mediates the relationship between Top Management Support and intention.

Hypothesis H10. Attitude mediates the relationship between Organizational Agility and intention.

Hypothesis H11. Attitude mediates the relationship between Cost Saving and intention.

Hypothesis H12. Attitude mediates the relationship between Competitive Pressure and intention.

Hypothesis H13. Attitude mediates the relationship between Trading Partner Pressure and intention.

4 Research Methodology

In order to test the proposed model, a quantitative study using a self-administrated survey will be conducted among retailers in Malaysia. A research method which is based on the multi-analytical approach of Partial Least Squares Structural Equation Modeling (PLS-SEM) and Artificial Neural Networks (ANNs) is applied. In this regard, after assessment of the measurement models and also analyzing the significance of independent variables on their perspective dependent variable within the PLS-SEM, ANNs is used in the research. Therefore, it enables extra verification of the results provided by the PLS-SEM for direct relationships in the model. For testing the mediation effect, we only apply PLS-SEM.

5 Conclusion

A cloud-based software named RetailNow is proposed in order to overcome the current issues that retailers encounter with during the usage of on-premise POS system. A combination of TAM and TOE is proposed. We expect that the proposed model is applicable and acquires predictive power in order to find out what factors impact on intention to use the software.

References

1. Blokdijk, G.: Point of sale—simple steps to win, insights and opportunities for maxing out success (2015)
2. Margaret Weber, M., Prasad Kantamneni, S.: POS and EDI in retailing: an examination of underlying benefits and barriers. Supply Chain Manag. **7**(5), 311–317 (2002)
3. Tolbert, F.: Why point of sales data matters for demand management. J. Bus. Forecast. **27**(4), 33–35 (2009)
4. Borgos, M.: More power with point of sales data. J. Bus. Forecast. **27**(4), 19–21 (2009)
5. Aberdeen Group: Retail point-of-sale (POS) systems AXIS (2008). www.ncr.com/documents/AberdeenPOSAxisReport_123008.pdf
6. Heart, T.: Who is out there? Exploring the effects of trust and perceived risk on SaaS adoption intentions. Database Adv. Inf. Syst. **41**(3), 49–68 (2010)
7. Young-Gyun, K., Lim, J.: A POS system based on the remote client-server model in the small business environment. Manag. Res. Rev. **34**(12), 1334–1350 (2011)
8. Weil, N.: Security at the point of sale. CIO **22**(3) (2008)
9. Close-Up Media Inc: iSheriff White Paper: 'Point of Sale = Point of Entry: The Achilles Heel in Retail, Hospitality Cyber Security'. Professional Services Close-Up. Close-Up Media, Inc., Jacksonville (2015)
10. Mell, P., Grance, T.: The NIST definition of cloud computing (draft). NIST special publication, 800(145), 7, (2011)

11. Oliveira, T., Thomas, M., Espadanal, M.: Assessing the determinants of cloud computing adoption: an analysis of the manufacturing and services sectors. Inf. Manag. **51**(5), 497–510 (2014)
12. Aharony, N.: An exploratory study on factors affecting the adoption of cloud computing by information professionals. Electron. Libr. **33**(2), 308–323 (2015)
13. Gangwar, H., Date, H., Ramaswamy, R.: Understanding determinants of cloud computing adoption using an integrated TAM-TOE model. J. Enterp. Inf. Manag. **28**(1), 107–130 (2015)
14. Chou, D., Chou, A.: Analysis of a new information systems outsourcing practice: software-as-a-service business model. Int. J. Inf. Syst. Change Manage. **2**(4), 392–405 (2007)
15. Yang, Z., Sun, J., Zhang, Y., Wang, Y.: Understanding SaaS adoption from the perspective of organizational users: a tripod readiness model. Comput. Hum. Behav. **45**, 254–264 (2015)
16. Armbrust, M., Fox, A., Griffith, R., Joseph, A.D., Katz, R., Konwinski, A., et al.: A view of cloud computing. Commun. ACM **53**(4), 50–58 (2010)
17. Xin, M., Levina, N.: Software-as-a service model: elaborating client-side adoption factors. In: Boland, R., Limayem, M., Pentland, B. (eds.) Proceedings of the 29th International Conference on Information Systems, Paris, France, 14–17 December 2008
18. Vaquero, L.M., Rodero-merino, L., Caceres, J., Lindner, M.: A break in the clouds: towards a cloud definition ACM SIGCOMM. Comput. Commun. Rev. **39**(1), 50–55 (2009)
19. Catteddu, D.: Cloud computing: benefits, risks and recommendations for information security. In: Web Application Security, p. 17 (2010)
20. Wu, W.-W.: Developing an explorative model for SaaS adoption. Expert Syst. Appl. **38**(12), 15057–15064 (2011)
21. VMware: The 3rd annual VMware cloud index (2012). http://info.vmware.com/content/APAC
22. VMware.: VMware cloud index 2013: study shows it driving business transformation in Asia (2013). https://www.vmware.com/company/news/releases/2013/vmw-cloud-index-2013-study-shows-it-driving-business-transformation-in-asia.html. Accessed 14 July 2017
23. EMC: Transforming IT study (2013). https://www.emc.com/campaign/global/forum2013/survey.htm. Accessed 15 July 2017
24. RightScale: State of the cloud report (2016). http://assets.rightscale.com/uploads/pdfs/RightScale-2016-State-of-the-Cloud-Report.pdf
25. DSA: Micro survey—appetite for cloud—ASEAN trends in cloud sentiment (2016). http://datastorageasean.com/sites/default/files/Micro%20Survey%20%20ASEAN%20Apetite%20for%20cloud%20-%20Prepared%20by%20Data%20%26%20Storage%20ASEAN%20-%20Supported%20by%20Barracuda%2011-6-12.pdf
26. Davis, F.D.: Perceived usefulness, perceived ease of use, and user acceptance of information technology. MIS Q. **13**(3), 319–340 (1989)
27. Tornatzky, L.G., Fleischer, M.: The Processes of Technological Innovation. Lexington Books, Lexington (1990)
28. Baker, J.: The technology–organization–environment framework. In: Information Systems Theory, pp. 231–245. Springer, New York (2012)
29. Premkumar, G.: A meta-analysis of research on information technology implementation in small business. J. Organ. Comput. Electron. Commer. **13**(2), 91–121 (2003)
30. Lal, P., Bharadwaj, S.S.: Understanding the impact of cloud-based services adoption on organizational flexibility: an exploratory study. J. Enterp. Inf. Manag. **29**(4), 566–588 (2016)

31. Teneyuca, D.: Internet cloud security: the illusion of inclusion. Inf. Secur. Tech. Rep. **16**(3), 102–107 (2011)
32. Pan, M.J., Jang, W.Y.: Determinants of the adoption of enterprise resource planning within the technology-organization-environment framework: Taiwan's communications industry. J. Comput. Inf. Syst. **48**(3), 94–102 (2008)
33. Lian, J.W., Yen, D.C., Wang, Y.T.: An exploratory study to understand the critical factors affecting the decision to adopt cloud computing in Taiwan hospital. Int. J. Inf. Manag. **34**(1), 28–36 (2014)
34. Lin, H.F., Lee, G.G.: Impact of organizational learning and knowledge management factors on e-business adoption. Manag. Decis. **43**, 171–188 (2005)
35. Wang, Y.M., Wang, Y.S., Yang, Y.F.: Understanding the determinants of RFID adoption in the manufacturing industry. Technol. Forecast. Soc. Chang. **77**(5), 803–815 (2010)
36. Lertwongsatien, C., Wongpinunwatana, N.: E-commerce adoption in Thailand: an empirical study of small and medium enterprises (SMEs). J. Glob. Inf. Technol. Manag. **6**(3), 67–83 (2003)
37. Ramdani, B., Kawalek, P.: SME adoption of enterprise systems in the northwest of England: an environmental, technological and organizational perspective. In: McMaster, T. (ed.) IFIP WG 86 Organizational Dynamics of Technology-Based Innovation: Diversifying the Research Agenda, pp. 409–430. Springer, Boston (2007)
38. Ramdani, B., Kawalek, P.: SMEs, IS innovations adoption: a review, assessment of previous research, vol. 39, pp. 47–70. Academia Revista Latinoamericana de Administracioen (2008)
39. Lee, S., Kim, K.J.: Factors affecting the implementation success of internet based information systems. Comput. Hum. Behav. **23**(4), 1853–1880 (2007)
40. Pyke, J.: Now is the time to take the cloud seriously. White paper (2009). www.cordys.com/cordyscmssites/objects/bb1a0bd7f47b1c91ddf36ba7db88241d/timetotakethecloudseroiuslyonline1.pdf
41. Alkhater, N., Walters, R., Wills, G.: An empirical study of factors influencing cloud adoption among private sector organisations. Telematics Inform. **35**, 38–54 (2017)
42. Rahi, S.B., Bisui, S., Misra, S.C.: Identifying the moderating effect of trust on the adoption of cloud-based services. Int. J. Commun. Syst. **30**(11), e3253 (2017)
43. Tallon, P.P., Pinsonneault, A.: Competing perspectives on the link between strategic information technology alignment and organizational agility: insights from a mediation model. MIS Q. **35**(2), 463–486 (2011)
44. Buyya, R., Broberg, J., Gościński, A.: Cloud Computing: Principles and Paradigms. Wiley, Hoboken (2011)
45. Ravichandran, T.: Exploring the relationships between IT competence, innovation capacity and organizational agility. J. Strateg. Inform. Syst. **27**(1), 22–42 (2017)
46. Gupta, P., Seetharaman, A., Raj, J.R.: The usage and adoption of cloud computing by small and medium businesses. Int. J. Inf. Manag. **33**(5), 861–874 (2013)
47. Lin, A., Chen, N.C.: Cloud computing as an innovation: percepetion, attitude, and adoption. Int. J. Inf. Manag. **32**(6), 533–540 (2012)
48. Hsu, P.F., Ray, S., Yu-Yu, L.H.: Examining cloud computing adoption intention, pricing mechanism, and deployment model. Int. J. Inf. Manag. **34**(4), 474–488 (2014)
49. Yu, Y., Li, M., Li, X., Zhao, J.L., Zhao, D.: Effects of entrepreneurship and IT fashion on SMEs' transformation toward cloud service through mediation of trust. Inf. Manag. **55**, 245–257 (2017)

50. Low, C., Chen, Y., Wu, M.: Understanding the determinants of cloud computing adoption. Ind. Manag. Data Syst. **111**(7), 1006–1023 (2011)
51. Davis, F.D., Bagozzi, R.P., Warshaw, P.R.: User acceptance of computer technology: a comparison of two theoretical models. Manag. Sci. **35**(8), 982–1003 (1989)
52. Ajzen, F., Ajzen, I., Fishbein, M.: Understanding Attitudes and Predicting Social Behaviour. Prentice Hall, Englewood Cliffs (1980)
53. Hair, J.F., Hult, G.T.M., Ringle, C.M., Sarstedt, M.: A Primer on Partial Least Squares Structural Equation Modeling (PLS-SEM), 2nd edn. Sage, Thousand Oaks (2017)
54. Marston, S., Li, Z., Bandyopadhyay, S., Zhang, J., Ghalsasi, A.: Cloud computing—the business perspective. Decis. Support Syst. **51**(1), 176–189 (2011)

Conceptualizing a Model for the Effect of Information Culture on Electronic Commerce Adoption

Fadi A. T. Herzallah[1], Mohammed A. Al-Sharafi[2(✉)], Qasim Alajmi[2,3], Muriati Mukhtar[4], Ruzaini Abdullah Arshah[2], and Dirar Eleyan[1]

[1] Faculty of Business and Economic, Palestine Technical University – Kadoorie, P. O. Box 7, Tulkarem, Palestine
f.herzallah@ptuk.edu.ps

[2] Faculty of Computer Systems and Software Engineering, Universiti Malaysia Pahang, Lebuhraya Tun Razak, 26300 Gambang, Pahang, Malaysia
ma_shrafi@yahoo.com

[3] Faculty of Computer Science and MIS, Oman College of Management and Technology, Muscat, Oman

[4] Research Center for Software Technology and Management, Faculty of Information Science and Technology, Universiti Kebangsaan Malaysia, 43600 Bangi, Selangor, Malaysia

Abstract. There are many vital roles and attributes played by the internal organizational environment in IT adoption. Technological innovations are adopted as a result of the reaction of many organizations to change and for influencing the business environment. For this study, a conceptual model was developed to focus on the effect of information culture as internal organizational factors on Electronic Commerce (EC) adoption on small and medium enterprises (SMEs). The proposed model buttressed by four of information culture as internal organizational factors. Specifically, the model constructed from, the information integrity, formality, control and proactiveness variables, evaluated their effects and determined how they can optimally be combined to permit EC adoption by SMEs. To achieve the objectives of the research, a pilot study was conducted with 35 managers of ICT SMEs from Palestine, to assess the proposed model. The proposed model will be assisting SMEs managers better understand and increase predictive capacity on EC adoption. This model will be applied to increase the rate at which EC is adopted among SMEs.

Keywords: E-commerce · IT adoption · Information culture
Small and medium enterprises

1 Introduction

The growing interest of researchers to the adoption of EC buttress the importance of such fast-growing technological innovations in being adopted by many sectors, including Small and Medium-Sized Enterprises (SMEs). Many SMEs have therefore

F. Saeed et al. (Eds.): IRICT 2018, AISC 843, pp. 861–870, 2019.
https://doi.org/10.1007/978-3-319-99007-1_80

embraced the use of technologies as a means to promote their internal and external activities and services [1, 2]. With the need demanding for the application of the importance of technological adoption in the context of various workplace. SMEs being driven by the numerous benefits advocated by the technology, mainly the use of EC in various companies or organizations. New customers find the various advantages offered by EC attractive, because it lowers the costs in carrying out business, reaching suppliers, increasing or enhancing the products and services as well as producing new ways or channels on how the products can be distributed [2, 3]. SMEs and large firms can take advantage of the numerous benefits offered by EC [4, 5]. Being inspired by the role of technologies in SMEs. These roles are essential factors that determine the economic growth of any country. Empirical studies are carried out by both researchers and practitioners to focus on the acceptance and use of EC by SMEs [6–9].

Today's SMEs businesses operate in a rapidly changing and unpredictable environment. Significant role is played by the internal organizational environment in organizations' decisions and their capabilities to adopt advanced technologies. The internal organizational environment consists of several variables that affect the operations and this goes a long way to determine what can possibly be achieved by business organizations [10]. SMEs have control and manipulation of their internal environment to some extent. Every decision to be taken should be anticipated in light of the internal organizational possibilities. In developing countries, the effects of these internal factors are more pronounced because basic amenities for IT implementation are grossly lacking in these countries [9]. According to studies in information and communication technology (ICT) adoption, it has been found that organizational environment is the major determining factor of implementation [11]. These studies suggested a wider evaluation of firm's internal environment when adoption of technology is considered. According to TOE framework [12], the internal environments of firms consists of technological and organizational factors. The implication of these variables in e-business implementation and use are revealed in the authors' findings [9]. As shown by Silvius [13], majority of organizations that use IT more, successfully have a tendency to create a high sense of commitment to IT and to encourage IT values and develop a robust internal information culture. Moreover, Oliver [14] stated that the approaches towards IT and resulting infrastructure as well as trust in information and systems used in managing it reflects the attitudes of the information culture. Davenport and Prusak [15] refers to information culture as behavioral and attitudinal patterns that express the orientation of an organization towards information.

It has been clearly agreed on that the adoption of EC positively affects the quality, efficiency and effectiveness of the services delivered by SMEs [6]. However, looking at this with respect to developing countries, like Palestine, the development of a holistic approach to building a more mature information culture in order to intensify the adoption of technological innovations remains a thought-provoking task [16, 17]. Thus, the specific conditions and capabilities for evaluation, interpretation, and utilization of information culture should be well addressed and deliberated on for successful adoption to occur [16].

Although the results obtained from most of the above cited studies buttress the role of EC in offering diverse potential advantages, its adoption among SMEs in the Palestinian context is still limited [11] and the effect of information culture on EC adoption among ICT organizations has not been precisely addressed by none of these studies. Moreover, none of the above studies have critically looked at IT adoption in all the listed ICT organizations in Palestine. Although the adoption of EC has various advantages for SMEs, there still remain several challenging issues, one of which is the information culture that burden the recognition and adoption of EC among managers of SMEs. In this regard, previous researchers examining the role of EC in SMEs stressed the necessity for investigating managers' information culture, in order to be able to provide clearer understanding on how this will affects their adoption and the use of such EC services [18–20]. Consequently, this current study aims at exploring the impact of information culture factors on the acceptance and use of EC services among managers of SMEs in the context of Palestine through proposing a framework for understanding those relations.

This paper is prepared as follows: the conceptual model development and hypotheses will be discussed in the next section following by methodology and findings and the future direction with conclusions at the end of this study.

2 Conceptual Model Development and Hypotheses

Information culture is defined as a culture in which intellectual resources are converted in line with physical resources [21]. It is debated by Travica [22] that information culture is the part of organizational culture that concentrates on information and technology. Also, Choo et al. [21] defined information culture as a term that signifies the elements of an organization's culture that impact the way in which an organization use and manage information. According to Alrousan and Jones [6], organizational culture is one of the key factors that have influence on the acceptance of EC by SMEs in developing countries, and this factor could be scrutinized further. It is claimed by Travica [22] that information culture is the part of organizational culture that concentrates on information and technology. In addition, information culture is defined by Choo et al. [21] as a term that means the essentials of an organization's culture that affects the way in which an organization manages and make use of information. Based on the literature used for this study, various variables applicable to the information culture and EC adoption are chosen to suggest a conceptual framework and hypotheses development. The proposed research model comprises five constructs: information integrity, information formality, information proactiveness, information control. These constructs function as independent variables and EC adoption as a dependent variable as shown in Fig. 1.

2.1 Information Integrity

Information integration is termed as the information trustworthiness and dependability [23]. It is basically the reliability, accuracy and consistency of the content, system and processes of information. It is also a requirement for the initiative of information management. Within this jurisdiction, if the basic information is lacking integrity, the success of the business activities relying on such information will also be affected [24]. Furthermore, degraded or negative values could lead to manipulation of information,

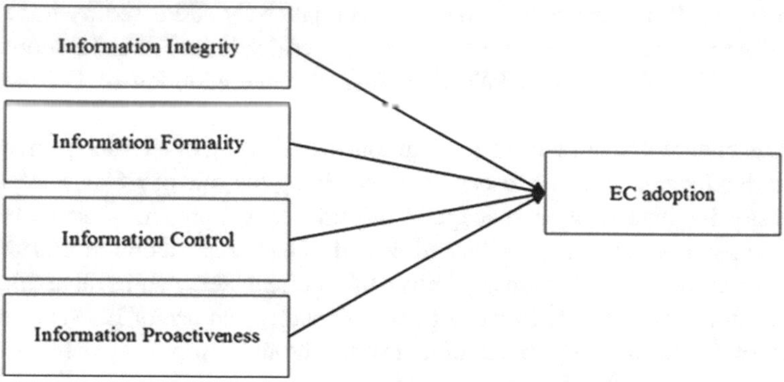

Fig. 1. Conceptual model.

making it confusing or unusable, and this consecutively, could unfavorably influence the process of decision making and lead to information manipulation for individual benefits [25]. Contrastingly, if the values are positive, transparent information will arise and this will replicate the actual happening, encouraging trust in the flow of information and decision making considering that the idea behind the decision is morally and practically suitable [24]. Lastly, a study data was gathered from 650 individuals from various organizations, with majority of them having information integrity. This confirms that it is a required trait to be possessed by managers and decision-makers [21]. Therefore, information integrity will positively have effect on EC adoption and it is hypothesized that:

H1: Information Integrity will positively influence EC adoption in Palestinian SMEs.

2.2 Information Formality

Information formality is the intentional reliance on institutionalized information as contrasting to informal form of resources [23, 24]. The stress on formal communication could lead to wide and detailed documentation of processes and methods, and this will lead to intensified information quality used and strong information formality [25]. There are numerous variables that may impact information formality as specified in past studies namely geographical dispersion, size, and the degree of virtual interactions. It was also discovered that information systems (formal and informal) are mandatory to conduct group activities but information formality is still stressed, and the information quality and consistency are anticipated to increase operations and processes [24]. Finally, the results of perceptions of information are interrelated with the information values perceptions, particularly ones that are associated to information formality [26]. According to Sinitsyna [27], information formality has a positive effect on information quality. According to Ke [28], it is essential for organizations to sustenance formal information for decision making through the usage of formal influences and communications with internal and external people to the company to confirm formal information consistency, or to enhance the available formal information. Therefore, this study anticipates informational formality to positively influence the owners and managers of SMEs in Palestine in their EC acceptance. Consequently, the researcher recommends that:

H2: Information formality will positively influence EC adoption in Palestinian SMEs.

2.3 Information Control

Information control is the information applied in ways that discourage or aids specific behaviors [24]. Presently, there has been a remarkable growth in the determinations from management to generate metrics (financial and non-financial) like balanced scorecard and economic value added, and to carry-out the interpretation of such criteria and measures at each company level. These are established to encourage employees' awareness of the relationships between their job performance and that of the company's [28]. Additionally, the managers' use of information, allows the close observing and guiding of operations, activities and decision making. Such abilities are important for deliberate planning and business performance argumentation to turn out to be optimal. The control mechanisms are mainly associated to the performance of the individual, which is in turn connected to the performance of the organization [25]. Certain rules were established by Simons [29] regarding negative information control that confines opportunity seeking behavior and offers reward to positive behavior and firm attainments. He further explained that positive information control would allow the enthusiasm of employees to be superior through their use of information resources, with the definitive aim of fostering a belief system. This could result into accommodating the value of learning in the organization. The top-down approach to management has always been the traditional choice but information control may also be enacted in a bottom-up approach according to Marchand, Kettinger [24] who indicated that in the latter, goal-setting is conducted with employees as opposed to managers. Furness [25] reported the influence of information integrity and information formality on information control and as such, the following hypothesis is proposed to be tested in the present study:

H3: Information control will positively influence EC adoption in Palestinian SMEs.

2.4 Information Proactiveness

Information proactiveness is an approach of thought on how to gain and use recently gained information in a fast and timely manner in business environment in response to the dynamic environment and by doing this, promoting products and services innovation [24]. Looking for signs of change in the environment and expecting precise conditions are part of being proactive and it often involves information exchange. Furthermore, the willingness and capability of understanding and responding to the knowledge acquired is also covered under proactiveness. Information proactiveness could be subject to the effect of information integrity [25] and consequently, information proactiveness in this study is anticipated to have a positive impact on the adoption of EC as clear from the following proposed hypothesis:

H4: Information proactiveness will positively influence EC adoption in Palestinian SMEs.

3 Methodology

In this pilot study, quantitative research method was employed. For the study, 35 Palestinian managers of SMEs under information and communication sector (ICT) were used as research participants. The survey item used for the survey instrument was taken from earlier studies on information system acceptance and use as indicated in Table 1.

Table 1. Sources of survey instrument items

Constructs	Items	References
Information integrity	5	[21, 27, 30]
Information formality	5	[21, 27, 30, 31]
Information control	5	[21, 27]
Information pro-activeness	5	[21, 27, 30, 31]
EC adoption	6	[32–34]

It consists of a total of (26) number of items adjusted by the researcher as the purpose of the current study demands. They were also interpreted into Arabic language so that they could easily be understood and answered by non-English speaking respondents. In other to be able to answer each of the items correctly, there is need for the respondent to select one of the five-point Likert-scale ranging from (1) strongly disagree to (5) strongly agree.

To eliminate ambiguous questions from the survey instrument of this research and offer valuable feedback and comments, three academic experts from information system field authenticated the structure of survey instrument before carrying out the pilot study. As indicated above the questions used for the instrument were obtained from information system studies, where the validity and reliability of the questions had also been established. To guarantee that the survey instrument is clear and to decide on the reliability of the instrument, there is need to validate the survey instrument. To measure the internal consistency "reliability" Cronbach's alpha is the most regularly used where we have Likert scale [35].

4 Findings

To conduct analysis of reliability of instrument from the pilot study, Statistical Package for the Social Sciences (SPSS) version 25 was used. The concept of reliability is all about determining to what degree the values of measurements are reliable and free from errors [36]. The measurement of construct reliability or internal consistency of a scale between the quotations of survey instrument is carried out in this research to establish the degree to which the items of survey instrument are identical. Survey instrument are divided in this study into five different variables, with each variable containing a set of items. Cronbach's alpha coefficient range falls between the values of 0 and 1.0 [37]. The scale of Cronbach's α used to measure the reliability of items should be more than 0.7 [36, 38] in this study. Table 2 shows the reliability, variance, standard deviation, and mean for measurement items. As previously stated, the Five-point Likert scale beginning from (1) which dignifies strongly disagrees to (5) which signifies strongly agrees are both used to measure variable in each of the model, and its items measurements are focused on its mean, the measurement division can be categorized into the following range. From interval of 1 to 2.33 is considered low, the range of 2.33 to 3.66 is medium, while above 3.66 up to 5 is considered high. The variation from the mean is measured by standard deviation. Table 2 indicates the summary of the reliability analysis for measuring items for all model variables. The results in Table 2 shows that all the variables attained a good reliability.

Table 2. Reliability analysis for model construct

Construct	Items	Mean	Std. deviation	Cronbach's alpha if item deleted	Cronbach's alpha
Information integrity	INI1	3.54	.950	.873	0.902
	INI2	3.40	.914	.880	
	INI3	3.46	.886	.892	
	INI4	3.34	.938	.879	
	INI5	3.43	1.008	.875	
Information formality	INC1	3.49	1.011	.891	0.913
	INC2	3.34	.968	.895	
	INC3	3.37	1.003	.902	
	INC4	3.49	1.067	.888	
	INC5	3.29	1.226	.894	
Information control	INF1	3.40	.976	.908	0.922
	INF2	3.46	1.172	.908	
	INF3	3.34	.906	.901	
	INF4	3.46	.980	.912	
	INF5	3.23	1.003	.892	
Information proactiveness	INP1	3.54	1.039	.921	0.928
	INP2	3.40	.847	.925	
	INP3	3.46	1.067	.910	
	INP4	3.37	.973	.906	
	INP5	3.46	1.120	.894	
E-commerce adoption	ECA1	3.83	.891	.874	0.900
	ECA2	3.63	.877	.884	
	ECA3	3.71	.893	.894	
	ECA4	3.83	.954	.878	
	ECA5	3.74	.980	.877	
	ECA6	3.60	.812	.888	

5 Conclusions and Future Work

The outcomes of this study have important implications and significant value for SMEs in order to develop better strategies for the usage and acceptance of EC. Having an understanding of what factors that affect SMEs acceptance of EC and the relationship between these factors will go a long way to help in evaluating and predicting the success of adoption of EC before applying them. A new conceptual model for EC acceptance centered on information culture factors has been proven in this study. The developed research model is strengthened by four factors of information culture which are Information Control; Information Formality; Information Integrity; and Information Proactiveness. In this study four hypotheses were developed to show the relationship between information culture factors and e-commerce adoption among SMEs. The construction of the proposed research model and hypotheses development is centered on the empirical validation of the concepts gained from various research studies in information system field primarily on e-commerce adoption.

This study outcome will have significant meanings and use for SMEs as signified by their managers to make effort in improving EC services, thus changing them into a system that is user-friendly. ECA should also be established and fostered between the SMEs, precisely between EC services and their consumers or users. For this to be achieved, managers of SMEs should strengthen their determination in the area of increasing public consciousness on the usefulness of EC services among people and organizations.

Acknowledgments. The authors would like to express their sincere gratitude to Universiti Malaysia Pahang (UMP) for supporting this work through Research Grant Scheme RDU180310.

References

1. Al-Sharafi, M.A., Arshah, R.A., Abu-Shanab, E.A.: Factors affecting the continuous use of cloud computing services from expert's perspective. In: TENCON 2017 IEEE Region 10 Conference (2017)
2. Herzallah, F., Mukhtar, M.: Organization information ecology and e-commerce adoption: effect on organizational SMEs performance. J. Comput. Sci. **11**(3), 540 (2015)
3. Hamdan, A.R., et al.: The success factors and barriers of information technology implementation in small and medium enterprises: an empirical study in Malaysia. Int. J. Bus. Inf. Syst. **21**(4), 477–494 (2016)
4. Pham, L., Pham, L.N., Nguyen, D.T.: Determinants of e-commerce adoption in Vietnamese small and medium sized enterprises. Int. J. Entrep. **15**, 45–72 (2011)
5. Al-Sharafi, M.A., Arshah, R.A., Abu-Shanab, E.A.: Factors influencing the continuous use of cloud computing services in organization level. In: Proceedings of the International Conference on Advances in Image Processing—ICAIP 2017, Bangkok, Thailand, pp. 189–194. ACM (2017)
6. Alrousan, M.K., Jones, E.: A conceptual model of factors affecting e-commerce adoption by SME owner/managers in Jordan. Int. J. Bus. Inf. Syst. **21**(3), 269–308 (2016)
7. Al-Alawi, A.I., Al-Ali, F.M.: Factors affecting e-commerce adoption in SMEs in the GCC: an empirical study of Kuwait. Res. J. Inf. Technol. **7**(1), 1–21 (2015)

8. Rahayu, R., Day, J.: Determinant factors of e-commerce adoption by SMEs in developing country: evidence from Indonesia. Procedia Soc. Behav. Sci. **195**, 142–150 (2015)
9. Herzallah, F., Mukhtar, M.: The impact of perceived usefulness, ease of use and trust on managers' acceptance of e-commerce services in small and medium-sized enterprises (SMEs) in Palestine. Int. J. Adv. Sci. Eng. Inf. Technol. **6**(6), 922–929 (2016)
10. Herzallah, F., Mukhtar, M.: The impact of internal organization factors on the adoption of e-commerce and its effect on organizational performance among Palestinian small and medium enterprise (2015)
11. Herzallah, F., Mukhtar, M.: The impact of internal organization factors on the adoption of e-commerce and its effect on organizational performance among Palestinian small and medium enterprise. In: International Conference on e-Commerce, Sarawak, Malysia, p. 104 (2015)
12. Tornatzky, L., Fleischer, M.: The Process of Technology Innovation. Lexington Books, Lexington. (Trott, P.: The role of market research in the development of discontinuous new products. Eur. J. Innov. Manag, **4**, 117–125 (1990))
13. Silvius, A.: Business and IT Alignment in Context, p. 407. Utrecht University, Utrecht (2013)
14. Oliver, G.: Information culture: exploration of differing values and attitudes to information in organisations. J. Doc. **64**(3), 363–385 (2008)
15. Davenport, T.H., Prusak, L.: Information Ecology: Mastering the Information and Knowledge Environment. Oxford University Press, Oxford (1997)
16. Osubor, V., Chiemeke, S.: The impacts of information culture on e-learning innovation adoption in learning institutions in Nigeria. Afr. J. Comput. ICT **8**(1), 17–26 (2015)
17. Mukred, A., Singh, D., Safie, N.: A review on the impact of information culture on the adoption of health information system in developing countries. J. Comput. Sci. **9**(1), 128–138 (2013)
18. Fung, R., Lee, M.: EC-trust (trust in electronic commerce): exploring the antecedent factors. In: Americas Conference on Information Systems (AMCIS) (1999)
19. Iddris, F.: Adoption of e-commerce solutions in small and medium-sized enterprises in Ghana. Eur. J. Bus. Manag. **4**(10), 48–57 (2012)
20. Sila, I.: Factors affecting the adoption of B2B e-commerce technologies. Electron. Commer. Res. **13**(2), 199–236 (2013)
21. Choo, C.W., et al.: Information culture and information use: an exploratory study of three organizations. J. Am. Soc. Inform. Sci. Technol. **59**(5), 792–804 (2008)
22. Travica, B.: Influence of information culture on adoption of a self-service system. J. Inf. Inf. Technol. Organ. **3**(1), 1–15 (2008)
23. AlAjmi, Q., et al.: A conceptual model of e-learning based on cloud computing adoption in higher education institutions. In: 2017 International Conference on Electrical and Computing Technologies and Applications (ICECTA) (2017)
24. Marchand, D.A., Kettinger, W.J., Rollins, J.D.: Information Orientation: The Link to Business Performance. Oxford University Press, New York (2001)
25. Furness, C.D.: Group Information Behavioural Norms and the Effective Use of a Collaborative Information System: A Case Study. University of Toronto (2010). https://tspace.library.utoronto.ca/bitstream/1807/26357/1/Furness_Colin_D_201011_PhD_thesis.pdf
26. Choo, C.W., et al.: Working with information: information management and culture in a professional services organization. J. Inf. Sci. **32**(6), 491–510 (2006)
27. Sinitsyna, A.: Impact of information culture and information behaviour on information quality. Business School Victoria University of Wellington, School of Information Management (2014)

28. Ke, Y.: Applying Marchand's information orientation theory to Sigma Kudos—an information product company. School of Computer Science, Physics and Mathematics Linnaeus University (2011)
29. Simons, R.: Levers of Control: How Managers Use Innovative Control Systems to Drive Strategic Renewal. Harvard Business Press, Boston (2013)
30. Abrahamson, D.E., Goodman-Delahunty, J.: The impact of organizational information culture on information use outcomes in policing: an exploratory study. Inf. Res. **18**(4), 1–16 (2013)
31. Hwang, Y.: Measuring information behaviour performance inside a company: a case study. Inf. Res. **16**(2), 16-12 (2011)
32. Bhattacherjee, A.: Understanding information systems continuance: an expectation-confirmation model. MIS Q. **25**, 351–370 (2001)
33. Venkatesh, V., et al.: User acceptance of information technology: toward a unified view. MIS Q. **27**(3), 425–478 (2003)
34. Teo, H.H., Wei, K.K., Benbasat, I.: Predicting intention to adopt interorganizational linkages: an institutional perspective. MIS Q. **27**, 19–49 (2003)
35. Sekaran, U.: Research methods for business: a skill building approach. J. Educ. Bus. **68**(5), 316–317 (2003)
36. Nie, N.H., Bent, D.H., Hull, C.H.: SPSS: Statistical Package for the Social Sciences. McGraw-Hill, New York (1970)
37. Leech, N.L., Barrett, K.C., Morgan, G.A.: SPSS for Intermediate Statistics: Use and Interpretation. Psychology Press, Abingdon-on-Thames (2005)
38. Lance, C.E., Butts, M.M., Michels, L.C.: The sources of four commonly reported cutoff criteria: what did they really say? Organ. Res. Methods **9**(2), 202–220 (2006)

Information Relevance Factors of Argument Quality for E-Commerce Consumer Review

Nur Syadhila Bt Che Lah(✉), Ab Razak Bin Che Hussin, and Halina Bt Mohamed Dahlan

Information System Department, Faculty of Computing, University of Technology Malaysia, Skudai, Johor, Malaysia
Sya_ila87@yahoo.com.my, {abrazak,halina}@utm.my

Abstract. With the rapid growth of E-Commerce recently, a review platform has been among the important tool in facilitating online decision making. Therefore, research scholars have put much effort in investigating various factors that may contribute to the enhancement of this tool. This study is interested in exploring the subjective characteristics of online reviews with the aim at improving the relevance filter of review content. This can be done through investigating the factors from information retrieval perspective which consider the relevance criteria for information searching in accepting text recommendation from consumer reviews. This study takes the opportunity since the problem that still rises in filtering the useful information features based on the judgment that, not every review exerts the same degree of applicability towards consumer needs. This work contributes to the literature by employing communication theory from ELM perspective to study the role of argument quality in understanding the behavior of information acceptance in E-Commerce consumer review platform. A research model is further proposed to visualize the research investigation.

Keywords: Online decision · Subjective characteristics · Argument quality Relevance judgment · Information acceptance · ELM theory · E-Commerce

1 Introduction

The progressive growth of E-Commerce recently has facilitate the development of consumer review platform which represent the trend of opinion sharing related to product and service consumption. A recent industry survey from Channel Advisor [7] reported that, online reviews may possibly generate the great influences which 90% of online buyers read consumer reviews and 83% of them trust online reviews in making their purchase decision. However, with the abundance of reviews presented online, consumers may not likely to read each of them, instead they may depend on certain information cues to determine the helpfulness reviews and worth for their further reading [11]. Practically, individual used the message output from information system (IS) in making a decision [5]. After that, they will evaluate the significance of the produced output, and determine the credibility of information based on their perceived value [8]. Therefore, the features such as information relevancy, completeness, accuracy,

F. Saeed et al. (Eds.): IRICT 2018, AISC 843, pp. 871–881, 2019.
https://doi.org/10.1007/978-3-319-99007-1_81

timeliness, and consistency are always perceptual as objective measure of message credibility [9]. However, recently there exist literature inconsistencies related to the strength of argument quality conceptualization. Moreover, [10] argued that there are some qualities dimensions of review argument cannot be objectively measured and might be vary within usage context. In addition, most of previous measures of argument quality may not related to or not practical in evaluating the value of subjective features of content information [13]. Thus, scholar further proposed relevance as one of the most salient subjective qualities which the level of relevance is generally difference depending on the decision it is being applied to [24]. Following the concept on information search and retrieval by [37], information relevance perspective can be used as a criterion to provide a judgment as to whether content received is related to one's information needs. For that reason, the main objective of this study is to explore the subjective measure of information credibility by identifying the relevance criteria of review features and hence affect the quality of review argument and contribute to online consumption decision. This study possibly will help electronic marketers in facilitating their review platform as well as surface the way for upcoming interesting research on this topic. The paper is outlined as below. First, we make a review on literature regarding dual process theory and perspective of argument quality. Then we proposed a theoretical model on argument quality and discussed their related factors that might contribute to adoption of online reviews in E-Commerce context.

2 Literature Review

During information processing, the degree of information transfer may vary among online consumer. Different responses may be retrieved for the same content by different recipients depending on the recipient's perception, experiences and sources [12]. This led the scholars in investigating the process of information adoption in understanding the extent of informational influence among people. Besides that, this study assumed that, dual process theories can be used to afford a comprehensive discussion related to individual's information processing, forming its validity assessment and generating their decision outcomes [27]. Previous scholars actively investigated about information seeking behavior among online user and strongly highlight the importance of argument quality and source credibility [6]. Argument quality can be referred to the strength of convincing arguments employ in an informational message [14]. This can be assumed as the value of results produced by the system as perceived by the user [17]. Moreover, online consumer are disposed to make a product judgment based on criteria of their purchase decision when they perceived the given information have meets their specific needs and requirements. Besides that, scholar pointed that argument quality can be defined as the strength or plausibility of persuasive argumentation [12].This definition is parallel with [30] that defined it as perceptions related to the strength and credibility of arguments rather that unreal and weak of information. Following this definition, [6] investigate as to whether argument strength can persuade a person to trust in something or to carry out the change in their online behavior.

Meanwhile source credibility can be defined as reliability perception towards the message source which has no effect on the message itself [12]. It also can be defined as perception towards information source which assume to be realistic, knowledgeable and trustworthy by information's recipients [30]. Information from well credible sources is perceived to be helpful and dependable which facilitate knowledge transfer [26]. Besides that, information that perceived to be high credible may alter a recipient's judgment towards message's communicator [22]. Meanwhile, the total number of reviews found in a review platform may conceal significant marketing effects [29]. This is consistent with findings by [11] which stated that, perceived popularity of online products positively affect consumer decision making. Furthermore, message persuasiveness could be dependable on number of positive attributes possess by the communicators [12]. These previous findings further validate that perceived quantity of reviews is relevant as one of peripheral cue for consumer decision making. In addition, it is effortless for online business to create a numerical indicator to represent overall number of reviews in the platform. Based on this finding, we assume that if consumers perceive there is high enough numbers of reviews for a wish product, then they will more potential to purchase it.

3 Study Design

Elaboration likelihood model (ELM) posits that a message can influence the attitude and behavior of people in two ways; centrally and peripherally. ELM also suggests that, the degree of explanation from both routes is depending on recipient's ability and motivation [14]. The central route represents the nature of message arguments while the peripheral route refers to the issues or themes that are not directly related to the subject matter of the message [30]. In this study, we denote argument quality as a central factor that represent user's perceptions related to content information as the result from systematic processing of consumer reviews. Argument quality is further categorized into two dimensions: perceived informative relevance and perceived motivational relevance which will be the main research investigation [29]. Perceived quantity of review and perceived source credibility are used as peripheral factors to signify two types of non-content related perceptions. With that, we assumed argument quality, perceived quantity of reviews and perceived source credibility will collectively predict the behavioral intention of online consumer. Figure 1 illustrates the proposed research model for this study.

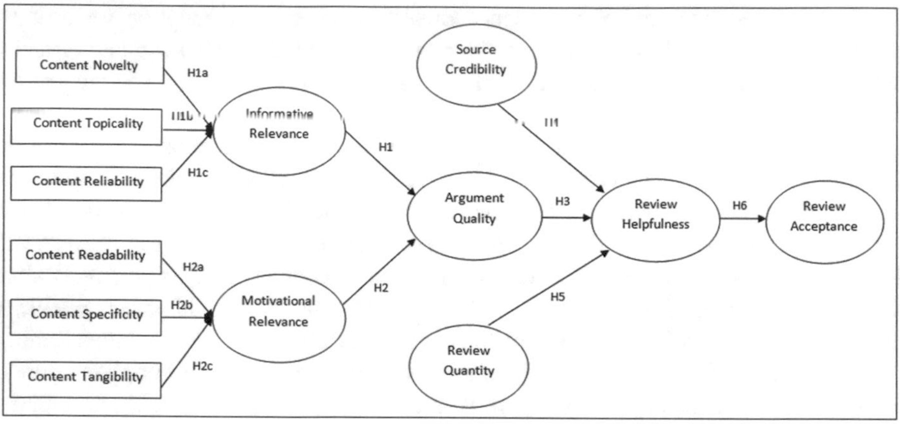

Fig. 1. Proposed research model

3.1 Discussion on Central Cues: Argument Quality and Information Relevancy

In general, most of Internet users will straight away scan the pages to find the related content rather than reading the web pages in detail [28]. This might be related to the time constraint as they desire instant information with little effort. Therefore, it is important to provide only the most relevant information on the web. This can be supported by research finding which suggest that information relevancy can be an important element for online decision making [19]. For that reason, this study assumed that the more relevant the content of information is, the higher perceived information quality of the retrieved message. However, previous scholar pointed that there is an arguable perceptions related to argument quality studies [29]. They further proposed a pacify conclusion by representing two dimensions of argument quality; perceived informativeness and perceived persuasiveness. Perceived informativeness is defined as the whole perceptions of consumer regarding the quality of information related to the characteristics of online review [33]. While perceived persuasiveness can be defined as consumer's perception regarding the strength of persuasive message embedded in online reviews [16].

In summary, scholars expect that the better quality of consumer reviews should be both informative and persuasive. This employs that, online consumer reviews should not just be presented in "standard" form, but to some extent, they must include a certain level of objective product descriptions (informativeness) and supporting with persuasive language including subjective or emotional expressions (persuasiveness) [15]. In extension with previous concern of argument quality, this study assumed that, there are some improvement can be achieved by leveraging the benefits of relevance criteria during information searching process. Thus, we seek to we seek to investigate argument quality from information relevancy perspective to represent broader view of subjective characteristics on content information. Based on communication theory from [18, 20, 21], information relevance criteria can be categorized with two dimensions; informative and motivational relevance perspective. Informative relevance is referring to how information could affect the current state of knowledge in a user [3, 13 and 34]. This definition is regarded as the cognitive correspondence between information and the user's existing knowledge. In a context of information searching for knowledge enhancement, a relevant information needs to be related to a user topic of interest (content topicality), must consists such a novel element (content novelty)and should perceived as reliable (content reliability) [2, 18].

H1: Informative relevance positively affects argument quality of online reviews.

Meanwhile, motivational relevance can be defined as an emotional reaction of people towards information as to whether it satisfies their intention, goals and motivations [34]. This reaction includes the feeling of sadness or enjoyment in consuming any activities provided in given information. In addition, where a user is seeking information for a leisure reading, people might be persuaded from relevance judgment through easy understanding of information (content readability), perceived tangibility of content information (content tangibility) and appropriate scope of information provided (content specificity) [18].

H2: Motivational relevance positively affects argument quality of online reviews.

Perceived content novelty is defined as the degree of dissimilarity between the present perception and past experience [31]. This concept applies in opposition of any familiarity of user existing knowledge. Consumer in the upper stage of product familiarity might posses more inclusive product knowledge or experiences. A unique experience with particular product knowledge may represent unforeseen product attributes that could activate a level of surprising feeling among consumer. The level of surprise is based on the variation of new and previous experiences. The powerful surprise may support an individual to pay their attention on the divergence, and hence generating their exploratory behavior [25]. As a return, the social process of obtaining and sharing consumer's experiences is activated.

H1a: Content novelty has positive affects towards informative relevance of online reviews.

Perceived content topicality is defined as topical relatedness or 'aboutness' of information towards user's queries. In the contexts where a user is looking for information about any essential interest or simple while away time, a relevant content should be described towards the topic of user intention [18]. Any content of information might be perceived as relevance towards user topic if it is substitute with any keywords or terminology related to user topic of interest. Furthermore, an information may possibly perceived as on topic if it is supporting with specific content that could explain the nature of search matter and may possibly expand the topic that a user is interested in.

H1b: Content topicality has positive affects towards informative relevance of online reviews.

Perceived content reliability can be termed as the whole credibility of retrieve information and can be defined as the degree of received content that perceived to be true, accurate or believable [36]. Content reliability also can be achieved if the user perceived that the retrieved information is consistent with real facts of existing information.

Information reliability can be defined with two dimensions; content and source credibility [13]. This study assumes that, consumer would feel very reluctant to read any review content if they perceived that information provided in the review is totally against facts. This feeling may lead to knowledge objection in reading online reviews when the users perceived that those opinions might be fake and not trustable. Reliability, therefore, is an important indicator to determine the informative state of information relevance towards user's needs.

H1c: Content reliability has positive affects towards informative relevance of online reviews.

Based on motivational relevance perspective, perceived content readability is referred to the amount of text to be read and understood with the use of appropriate language and semantic elements that can leads the individual in accepting the given information [37]. The used of practical language and correct product jargon might possibly reduces message transparency and leads to significantly less evaluation of any retrieved message [18]. With the certain level of given information, an information content might be perceived as difficult to be understood which causes a user to feel less happy and fatigue [3]. In a more relaxed context which user might first searching for information with less product knowledge, online user might prefer to be provided with easier rather than complicated information content. To some extent, previous scholar [4] stated that readers are more engaged in easy to read information rather than difficult ones. Vivid and affective psychological states can take place with no trouble when accessing blogs with multimedia types of articles [23]. In opposition, content with less-understandability may generate knowledge insecurity about existing product features, which may unfavorably influence the information search [1].

H2a: Content readability has positive affects towards motivational relevance of online reviews.

While perceived content tangibility represent the relevance judgment related to the presence of content references related to users' needs [2]. In this context, information might perceived as relevance if it is stand in with sufficient amount of any subject utility that may perhaps related to user specific needs. This includes the presence of reliable information with discussion related to the point of reality about subject situation and the inclusion of recent statistics of subject information. Content tangibility might trigger first intention of user in reading online reviews which further convince them towards information acceptance or online behavior change.

H2b: Content tangibility has positive affects towards motivational relevance of online reviews.

Moreover, perceived appropriate scope of review or content specificity is defined as a user judgment regarding information completeness, depth or the level of information specificity [3]. In this study, we assume that for leisure reading, user is more absorbed with the review that provide sufficient amount of product descriptions rather than concrete topic discussion related to their search product information. Consumer will pay more attention on reviews that could narrow down the specific needs of their search and will be happier if their first search could be satisfied.

H2c: Content specificity has positive affects towards motivational relevance of online reviews.
H3: Argument quality positively affects online review helpfulness.

3.2 Discussion on Peripheral Cues: Source Credibility and Review Quantity

In contrast with central route, ELM posits another peripheral route which implies that people do not always elaborate on argument's message only [14]. Peripheral cues are contextually oriented based on information environment and embrace other merit considerations than argument themselves. Considering on previous findings from literatures, this study have identified two peripheral cues which assumed to have important characteristics in evaluating the credibility of online reviews; source credibility and review quantity. Prior scholars highlighted that individual that received peripheral cues can be influenced by information source's attractiveness, likeability and credibility [6]. Within online environment, it is depends on users themselves to determine the level of expertise and trustworthiness of the contributors in order to either adopting or rejecting any presented information. Higher perception of the content usefulness may possibly determine by high degree of information's source expertise and trustworthiness. According to [35], statement written by "experts" is more trustable and "experts know best" are the most acknowledged peripheral cues by online consumer. It entail that the decision regulation from a credible source is expected to be accessible in the structure of consumer's knowledge. In online situation, consumers often exposed to being inconvenient to physically touch desired products and might feeling unsatisfactory to make a high-quality decision by plainly depending on product information provided by online retailers. Thus, seeking online reviews is assumed to be the best alternatives in reducing information uncertainty. Reviews from someone who are perceived as credible are anticipated to provide such a reliable information that help consumer to learn more about products and enhance decision making. Previous researches shown that source credibility exert a significant impact on the change of online behavior [3].

H4: Source credibility positively affects online review helpfulness.

Another important peripheral cues proposed by this study is perceived quantity of reviews. Previous scholars pointed that, perceived quantity of reviews are represented by consumer's perceptions related to the amount of consumer's opinions and the attractiveness of corresponding products within online environments [11]. Scholars also highlighted that the number of arguments or review's length can be an important element and such a common decision rules for users in making their online judgments [1, 35]. Thus, perceive quantity of reviews is assumed to be an important cue that is obtainable in the structure of consumer knowledge. This cue may infer the extent of product popularity from perception that other users have consistently purchased that product before. By consulting this number of reviews, online consumers are easy to get to their peripheral cue of review quantity and hence identify the reputation of certain product on the websites. This sign may possibly help online consumer in triggering their purchase confidence or improve online decision quality by inspecting or simulate other user's purchase behavior [18].

H5: Review quantity positively affects online review helpfulness.

Previous literatures have used dual process theories in explaining how people can be influenced through accepting opinions, knowledge and information from others. Scholars [27] have employed ELM and proposed a research model regarding information acceptance to explain how people can be manipulated in adopting information posted in online contexts. Information acceptance behavior is one of the foremost online activities for users in virtual communities. Scholar [6] suggested that, online users will perform the review scanning posted by others before they make a purchase decision. Likewise, they will plead for a virtual community platform to ask for help by asking questions when they have queries [1]. Hence, the specific cognitive beliefs of information adoption behavior in virtual communities are seen essential in investigating their impact within online communication.

Perceived review helpfulness can be defined as individual's perception related to performance enhancement by implementing the used of new technology. Scholar [32] pointed that information helpfulness as a fundamental constructs of user acceptance, with significant correlations to both current and future self-reported system usage. Within an online platform, varies opinions and shared experiences about products may be uttered. These opinions will be carried by users as the useful information in helping them to compose a better purchase decision. For that reason, if someone thinks that these opinions within online communities are helpful, they will have greater intention in adopting the opinions.

H6: Review helpfulness positively affects online review acceptance.

4 Conclusion

This study is sought to explore the motivation behind the acceptance of product argument from online review platform. The appearance of various patterns among relevance criteria could be occupied for future expansion of implicit feedback mechanisms and the more vibrant display of interfaces for information search results [2]. Inspiring from this discussion, a theoretical research model is proposed based on concept of online opinions adoption behavior. Resulting in relationship between information acceptance, information helpfulness and information relevancy. The related constructs are proposed based on previous literature findings with additional relevance context to test the significant impacts towards online consumer decision making. The proposed research model will be tested with development of survey instruments and will be distributed among experienced online buyer with good reviews exposure. This study is expected to contribute towards existing research studied related to online consumer reviews. Furthermore, this study is sought to explore the significant impacts of perceived information relevancy in enhancing argument quality which assume to be important predictors to invoke perception of review helpfulness and hence act as the significant factors in determining the acceptance of message content within online environments. With that, this study is hopes to give a suggestive understanding towards developing a more convenient consumer review platform and thus making E-Commerce as always the first choice for product buying.

Acknowledgements. This work is supported by the Ministry of Higher Education (MOHE) and Research Management Centre (RMC) at the Universiti Teknologi Malaysia (UTM) under the Research University Grant Category (VOT Q.J130000.2528.11H79).

References

1. Todorov, A., Chaiken, S., Henderson, M.D.: The heuristic-systematic model of social information processing. In: Dillard, J.P., Pfau, M. (eds.) The Persuasion Handbook: Developments in Theory and Practice, pp. 195–212. Sage Publications, Thousand Oaks (2002)
2. Balatsoukas, P., Ruthven, I.: An eye-tracking approach to the analysis of relevance judgments on the Web: the case of Google search engine. J. Assoc. Inf. Sci. Technol. **63**(9), 1728–1746 (2012)
3. Borlund, P.: The concept of relevance in IR. J. Assoc. Inf. Sci. Technol. **54**(10), 913–925 (2003)
4. Britton, B.K., Piha, A., Davis, J., Wehausen, E.: Reading and cognitive capacity usage: adjunct question effects. Mem. Cognit. **6**(3), 266–273 (1978)
5. Mai, J.E.: Looking for Information: A Survey of Research on Information Seeking, Needs, and Behavior. Emerald Group Publishing, Bingley (2016)
6. Cheung, C.M., Lee, M.K., Rabjohn, N.: The impact of electronic word-of-mouth: the adoption of online opinions in online customer communities. Internet Res. **18**(3), 229–247 (2008)
7. ChannelAdvisor: Consumer Survey: Global Consumer Shopping Habits. ChannelAdvisor (2011)
8. Fang, Y.H.: Beyond the credibility of electronic word of mouth: exploring eWOM adoption on social networking sites from affective and curiosity perspectives. Int. J. Electron. Commerce **18**(3), 67–102 (2014)
9. Liang, Y., DeAngelis, B.N., Clare, D.D., Dorros, S.M., Levine, T.R.: Message characteristics in online product reviews and consumer ratings of helpfulness. South. Commun. J. **79**(5), 468–483 (2014)
10. Watts, S., Shankaranarayanan, G., Even, A.: Data quality assessment in context: a cognitive perspective. Decis. Support Syst. **48**(1), 202–211 (2009)
11. Park, D.-H., Lee, J., Han, I.: The effect of on-line consumer reviews on consumer purchasing intention: the moderating role of involvement. Int. J. Electron. Commerce **11**(4), 125–148 (2007)
12. Cappella, J.N., Kim, H.S., Albarracín, D.: Selection and transmission processes for information in the emerging media environment: psychological motives and message characteristics. Media Psychol. **18**(3), 396–424 (2015)
13. Chen, Y.C., Shang, R.A., Li, M.J.: The effects of perceived relevance of travel blogs' content on the behavioral intention to visit a tourist destination. Comput. Hum. Behav. **30**, 787–799 (2014)
14. Cheung, C.M.Y., Sia, C.L., Kuan, K.K.: Is this review believable? A study of factors affecting the credibility of online consumer reviews from an ELM perspective. J. Assoc. Inf. Syst. **13**(8), 618 (2012)

15. Cheung, C.M., Thadani, D.R.: The impact of electronic word-of-mouth communication: a literature analysis and integrative model. Decis. Support Syst. **54**(1), 461–470 (2012)
16. Wu, L., Shen, H., Fan, A., Mattila, A.S.: The impact of language style on consumers' reactions to online reviews. Tour. Manag. **59**, 590–596 (2017)
17. Xiao, B., Benbasat, I.: E-commerce product recommendation agents: use, characteristics, and impact. MIS Q. **31**(1), 137–209 (2007)
18. Xu, Y.: Relevance judgment in epistemic and hedonic information searches. J. Assoc. Inf. Sci. Technol. **58**(2), 179–189 (2007)
19. Dunk, A.S.: Product life cycle cost analysis: the impact of customer profiling, competitive advantage, and quality of IS information. Manag. Account. Res. **15**(4), 401–414 (2004)
20. Grice, H.P.: Logic and conversation. In: Cole, P., Morgan, J. (eds.) Syntax and Semantics, vol. 3, pp. 41–58. Academic Press, New York (1975)
21. Vansteenkiste, M., Aelterman, N., De Muynck, G.J., Haerens, L., Patall, E., Reeve, J.: Fostering personal meaning and self-relevance: a self-determination theory perspective on internalization. J. Exp. Educ. **86**(1), 30–49 (2018)
22. O'Reilly, K., MacMillan, A., Mumuni, A.G., Lancendorfer, K.M.: Extending our understanding of eWOM impact: the role of source credibility and message relevance. J. Internet Commerce **15**(2), 77–96 (2016)
23. Tausczik, Y.R., Pennebaker, J.W.: The psychological meaning of words: LIWC and computerized text analysis methods. J. Lang. Soci. Psychol. **29**(1), 24–54 (2010)
24. Jepsen, A.L.: Factors affecting consumer use of the Internet for information search. J. Interact. Mark. **21**(3), 21–34 (2007)
25. Joëlle, V., Dirk, S.: The role of surprise in satisfaction judgment. J. Consum. Satisf. Dissatisf. Complain. Behav. **14**, 27–46 (2001)
26. Chang, C.M., Hsu, M.H., Hsu, C.S., Cheng, H.L.: Examining the role of perceived value in virtual community's continuance: its antecedents and the influence of experience. Behav. Inf. Technol. **33**(5), 502–521 (2014)
27. Tam, K.Y., Ho, S.Y.: Web personalization as a persuasion strategy: an elaboration likelihood model perspective. Inf. Syst. Res. **16**(3), 271–291 (2005)
28. Madu, C.N., Madu, A.A.: Dimensions of e-quality. Int. J. Quality Reliab. Manag. **19**(3), 246–258 (2002)
29. Zhang, K.Z., Zhao, S.J., Cheung, C.M., Lee, M.K.: Examining the influence of online reviews on consumers' decision-making: a heuristic-systematic model. Decis. Support Syst. **67**, 78–89 (2014)
30. Petty, R.E., Cacioppo, J.T., Kasmer, J.A.: The role of affect in the elaboration likelihood model of persuasion. Commun. Soc. Cognit. Affect 117–147 (2015)
31. Pearson, P.H.: Relationships between global and specified measures of novelty seeking. J. Consult. Clin. Psychol. **34**(2), 199 (1970)
32. Tarhini, A., Arachchilage, N.A.G., Abbasi, M.S.: A critical review of theories and models of technology adoption and acceptance in information system research. Int. J. Technol. Diffus. (IJTD) **6**(4), 58–77 (2015)
33. Jafari, S.M., Taghavi, H., Daryakenari, M.Y., Shahbazi, S.: Impact of trust and perceived content of advertisement on intention to accept mobile advertisement. In: 2016 Third International Conference on Digital Information Processing, Data Mining, and Wireless Communications (DIPDMWC), pp. 154–159. IEEE (2016)
34. Saracevic, T.: The notion of relevance in information science: everybody knows what relevance is. But, what is it really? Synth. Lect. Inf. Concepts Retr. Serv. **8**(3), i-109 (2016)

35. Hussain, S., Ahmed, W., Jafar, R.M.S., Rabnawaz, A., Jianzhou, Y.: eWOM source credibility, perceived risk and food product customer's information adoption. Comput. Hum. Behav. **66**, 96–102 (2017)
36. Zhang, W., Watts, S.: Capitalizing on content: information adoption in two online communities. J. Assoc. Inf. Syst. **9**(2), 73–94 (2008)
37. Xu, Y., Chen, Z.: Relevance judgment: what do information consumers consider beyond topicality? J. Am. Soc. Inf. Sci. Technol. **57**(7), 961–973 (2006)

The Determinants of Customer Knowledge Sharing Behavior: A Review Study

Hasan Sari[1](✉), Marini Othman[2], and Abbas M. Al-Ghaili[2]

[1] College of Computer Science and Information Technology, Universiti Tenaga Nasional, 43000 Kajang, Selangor, Malaysia
hassan_sari@yahoo.com

[2] Institute of Informatics and Computing in Energy, Universiti Tenaga Nasional, 43000 Kajang, Selangor, Malaysia
{marini,abbas}@uniten.edu.my

Abstract. In the current age of dynamic business, knowledge became a vital source of competitive advantage. Traditional literature on knowledge management (KM) focused mostly on knowledge sharing within the organization. In this regard, while the general enablers and inhibitors of knowledge sharing behavior have been studied in prior research, the factors that encourage or hinder customers to share their knowledge have not received enough attention in the field of knowledge management. This study thus reviewed previous literature in the context of customer knowledge management to examine the most cited factors that influence customer knowledge sharing (CKS). The relevant articles were analyzed and screened to identify the factors of customer knowledge sharing behavior. Based on the findings, CKS behavior determinants could be classified into three main categories: personal, social, and technological. Among the personal determinants, the most cited factors are the reciprocal benefit, enjoyment of helping others, and knowledge sharing self-efficacy. On the other hand, sense of belonging shared goal/language and interpersonal trust is the most cited social determinants. Meanwhile, perceived usefulness and perceived ease of use are the most cited technological determinants of CKS behavior. The previous research focused more on personal and psychological factors, and as such, more research is needed to examine other motivation factors that might influence CKS behavior, particularly the technological factors.

Keywords: Knowledge sharing · Customer knowledge
Customer knowledge sharing

1 Introduction

In the current economy, knowledge is being highly considered as a base of sustainable competitive advantage and growth [1]. One of the most significant issues in knowledge management inside a firm is how to build an efficient mechanism for sharing and exchanging knowledge [2]. Customer knowledge is a valuable knowledge asset, which

F. Saeed et al. (Eds.): IRICT 2018, AISC 843, pp. 882–891, 2019.
https://doi.org/10.1007/978-3-319-99007-1_82

allocated beyond the company's boundaries. This type of 'external knowledge' can help the organization to allocate and manage customer's ideas, and insights then utilize the generated knowledge for business growth and success [3]. Recently, many organizations recognize the importance of sharing customer knowledge [2].

Generally, customers are reluctant to share their knowledge with brands and retailers if they do not receive enough motivation and care. If customers have a positive attitude towards the brands and the company, they would have a positive impact on knowledge sharing process [1, 4]. Despite the benefits of customer knowledge, customer contributions in the process of knowledge creation considered low and lacking of active involvement [2].

The majority of studies in the field of KM focused on general enablers and inhibitors of knowledge sharing behavior inside a firm. However, the drivers that influence customers to share their knowledge with the firms and with other customers have not received enough attention [4, 5]. Therefore, this study aimed to review the previous research in customer knowledge context to identify the most cited determinants of customer knowledge sharing with the firms and with other customers.

2 Literature Review

2.1 Customer Knowledge (CK)

Customer knowledge is the knowledge derived from customers through an interactive and mutually beneficial process between the enterprise and its customers [6]. Due to the nature of customer knowledge, which spreads through multiple different channels and sources, the management and integration of this type of knowledge perceived as a complicated knowledge process [6].

[6, 7] classified CK into three categories: (1) - Knowledge 'for' customer: which includes the knowledge directed from the firm to its customers. This includes all the data, information that could be analyzed and converted into knowledge like knowledge about products, services, markets, and suppliers and other related items. (2) - Knowledge 'about' the customer, which includes knowledge about customer transactions and preferences. (3) - Knowledge 'from' customers, which includes knowledge, information, feedback, and ideas about the usage of products/services.

2.2 Customer Knowledge Management (CKM)

Customer Knowledge Management (CKM) defined in some literature as a strategic process, wherein companies transform their customers from being passive recipients of products and services to active knowledge partners [6]. In the current competitive dynamic business environment, many firms have shifted their business strategy from a product-centric to a customer-centric strategy [8], where the managing of customer assets, especially knowledge from customers, becomes a significant way to obtain a sustainable competitive advantage [9].

2.3 Customer Knowledge Sharing (CKS)

The term knowledge sharing (KS) defined in the literature as "activities involved in disseminating or transferring knowledge among individuals, groups or organizations" [10, 11], where people "exchange their explicit and tacit knowledge and generate new knowledge" [11]. Customer Knowledge Sharing (CKS) defined in some studies as "voluntary acts of contributions by providing information or sharing experiences" [12]. [4] defined CKS as "the extent to which customers exchange their information, ideas, and expertise with their suppliers."

How to encourage individuals to share their knowledge and insights is a critical issue in the field of knowledge management [13]. In this regards, Many companies strive to understand and utilize the knowledge of the customers for the company growth and success. For instance, the Sony Corporation has built a "test store," where customers can test the product and share their ideas with the community. Other major companies like Microsoft and Google have built their own "virtual innovation communication"(VIC) to allow their customers to share their knowledge and ideas and to involve them in product and service innovation [2].

[14, 15] stated that there is a lack of research on how firms establish reward and incentive systems that could support customer knowledge sharing. The reasons behind consumers' decision to share knowledge with their suppliers have been discussed in the literature of CKM [4, 12, 13, 16–18]. Customer knowledge sharing behavior could be identified as a response from customers to a variety of motivational such as social, esteem, cognitive, and altruistic needs [12].

3 Methodology

3.1 Search Strategy

This study searched for relevant articles in the fields of customer knowledge and in order to determine the factors that influence CKMS, a combination of keywords and phrases were applied to identify related articles. The combination of the following keywords 1 - "customer", "consumer", "user", 2 - "knowledge contribution" "knowledge participation", "determinants", such as, "drivers", "factors", and "motivation" have been used to conduct in-depth research through the online databases namely, ScienceDirect, IEEXplore, Emerland, Springer Link, and Google Scholar.

3.2 Inclusion and Exclusion Criteria

The articles were selected based on the defined inclusion criteria: (a) The article is written in English; (b) The research employed qualitative, quantitative, or mixed methods; (c) The study aimed to investigate factors that influence customers to share their knowledge. The article selection process excluded duplicated articles, non-English articles, and irrelevant articles out of the scope of this study.

4 Analysis

The analysis of the collected data contains two phases: phase one involved the analysis of the most cited theories and models used in CKS selected studies, while phase two involved the analysis of the most cited determinants of CKS behavior in the selected studies.

4.1 Theories and Models in Previous Studies

The selected articles were analyzed to recognize all theories and models used to examine CKS behavior. Most of the reviewed articles examined the determinants of CKS based on the personal, social, and technology concepts. Table 1 illustrates all the theories and models in prior studies. Based on the analysis, the most used personal theories to examine knowledge sharing behavior is the personal motivation theories, Expectation Confirmation theory, Motivation-Opportunity-Ability theory (MOA), Theory of Reasoned Action (TRA), and Theory of Planned Behavior (TPB). Meanwhile, the most used social theories are the social context-related theories, such as Social Capital Theory, Social Cognitive Theory, and Social Exchange Theory. On the other hand, the Technology Acceptance Model (TAM) and Media Richness theory were used in some studies to analyze the technology determinants of customer knowledge sharing, particularly in online platforms.

Table 1. Previous studies on CKS

References	Topic	Research method	Underpinning theory
[19]	The drivers of CKS in web-based discussion boards	Theoretical FW	Social exchange theory Personal motivational theory Technological model (TAM)
[12]	Motivations of CKS in online communities	Theoretical	Motivation theory, social theory
[13]	Factors fostering CKS in healthcare communities	Theoretical model	TRA TPB
[17]	Intra-organizational factors that enhance CKS in the context of international key accounts management	Conceptual model	Motivational theories technological theory
[18]	CKS behavior in social shopping communities	Theoretical model, survey	Social capital theory, customer knowledge contribution behavior
[4]	The antecedents and consequences of CKS in B2B telecoms.	Theoretical and empirical framework, survey	MOA theory
[16]	The drivers of CKS in online travel communities.	Conceptual model, survey	Consumer psychology theories TAM model

(*continued*)

Table 1. (*continued*)

References	Topic	Research method	Underpinning theory
[20]	The motivation of CKS in virtual food communities	Conceptual model, survey	Self-presentation theory Economic theory
[2]	The factors of CKS in virtual innovation community	Theoretical model	Social cognitive theory and social exchange theory
[21]	The factors influencing CK value	Conceptual model, survey	Nonaka model
[22]	The factors affecting CKS in professional virtual communities	Conceptual model, survey	Social exchange theory Social cognitive theory
[23]	The drivers of CKS in online-consumer opinions	Conceptual model, survey	Social cognitive theory Social identity theory Social exchange theory
[24]	The drivers of CKS in virtual professional community	Conceptual model, survey	Expectation confirmation theory, social exchange theory, personal motivational theories, Social cognitive theory.
[23]	The drivers of CKS in online community	Theoretical model	Maslow's theory, personal motivation theory
[25]	The drivers of CKS in travel online social network	Theoretical model	TRA, TPB, TAM
[26]	The drivers of CKS in virtual communities	Theoretical model	Social influence theory, TRA, TPB
[27]	The drivers of CKS in company-hosted online community		Social capital theory

4.2 CKS Determinants in Previous Studies

The selected articles were analyzed and screened to identify all the mentioned CKS determinants. Based on the analysis, the determinants can be classified into three main themes: personal, social, and technology. The factors mentioned in the selected articles were listed based on the occurrence and number of citation.

During the analysis of the determinants of CKS, we noticed that different articles used different concepts for the same factors. This can be exemplified by the finding that in personal determinants, the term 'reputation' has been used in some other articles using different terms, like "image enhancement," with the same meaning. In social determinants, the term 'sense of community' has been synonymous used as 'sense of belongingness' or 'social identity' in other studies. Therefore, we grouped these different forms under the same category. Table 2 illustrates all the determinants of CKS in the selected articles.

Table 2. The determinants of CKS in previous stud

Theme	Determinant	Citation	References
Personal	Enjoyment of helping (intrinsic benefits)	6	[12, 16, 19, 23, 24, 28]
	Knowledge self-efficacy (intrinsic motivation)	5	[4, 19, 22–24]
	Expected reciprocal benefits (extrinsic benefits)	11	[2, 4, 12, 18, 19, 22–25, 27, 28]
	Reputation/image enhancement (extrinsic benefits/motivation)	5	[2, 12, 18, 23, 28]
Social	Sense of belonging/sense of community/belongingness/social identity	7	[12, 19, 23, 24, 26, 28, 29]
	Social trust (interpersonal trust)	5	[12, 13, 22, 28]
	Subjective norm	3	[13, 25, 26]
	Group norm/shared goal/shared language/shared culture	8	[13, 19, 22, 26, 27, 29]
	Social network, social tie (interaction), community tie	3	[13, 18, 27]
Technological	Perceived usefulness	2	[16, 19]
	Perceived ease of use	3	[16, 19, 25]
	Privacy	1	[19]

Personal Determinants

The first category of CKS determinants is related to the personal characteristics. Majority of the selected articles studied customer knowledge sharing behavior in online communities (web-discussion-board, social community, professional virtual communities, travel community, social shopping communities, and health communities). The study [19], examined the customer knowledge sharing in the web-based discussion board. This was the first study to examine customer knowledge sharing based on the concept of "socio-technical" systems, which comprises interactions among user characteristics (customers), groups (consumer groups), and systems (web-based discussion boards).

The study provided new insights into the field of customer knowledge sharing, by exploring social contexts (social identity) or the sense of community as a significant factor that affects CKS in web discussion boards. Studies conducted by [19, 30] claimed that customer might participate in knowledge sharing with the aim of achieving both personal and community benefits. The authors argued that, when a customer decides to share knowledge with firms and other customers, he/she tends to maximize benefits and minimize costs. Another useful perspective suggests that motivation involves extrinsic and intrinsic factors [31] such as payments, reputation, sense of self- knowledge efficacy, learning, and volunteering to help others [18].

Most of the selected articles emphasized the role of intrinsic and extrinsic motivation as significant drivers inducing customer's knowledge sharing behavior. Based on the analysis of the selected studies, the most cited personal determinants of CKS are reciprocal benefits, enjoyment of helping others, knowledge self-efficacy, reputation.

Social Determinants

The second category of CKS determinants is related to the communities or customer groups, where customers share and exchange knowledge. Aside from achieving personal benefits, the customers tend to share their knowledge for social or community factors, such as, the obligation to the community (sense of community/belongingness) [12, 19], or to confirm social identity [23, 26, 28].

Some studies reported that the interpersonal trust among the community is a significant determinant of customer knowledge sharing [12, 13, 22]. Several other studies found that, when the users share some characteristics like a goal, language, and culture, they tend to exchange the knowledge with each other [13, 19, 29].

Based on the analysis of the selected articles, the most significant social factors of CKS behavior are group norm, sense of belonging, social trust, and social tie.

Technological Determinants

The third category related to the mechanism of knowledge sharing is the technological determinants, particularly, where the sharing mechanism involved online sharing environments such as web-based discussion, online community, or other forms of online sharing platform. In general, perceived ease of use, perceived usefulness, are the most technological factors inducing CKS behavior.

We noticed that little research had been dedicated to the impact on CKS behavior [16, 19]. Therefore, more research is needed to evaluate the impact of technical factors on CKS behavior.

5 Findings

The main finding of this study is related to the obtained results on the most cited determinants of customer knowledge sharing behaviors in previous literature with the company and other customers.

Based on our analysis, the factors that influence customers to share their knowledge can be classified into three main categories: personal, social, and technological factors. The most cited personal factors are a reciprocal benefit, enjoyment of helping others, knowledge sharing self-efficacy and reputation, and the most cited social factors are a sense of belonging, shared goal language, and interpersonal trust. Finally, perceived usefulness and perceived ease of use are the most cited technology determinants.

6 Limitations

This review has some limitations. First, the literature search was limited to searching the concept "customer knowledge sharing," which might have been used citing different terms such as, "customer participation," "member contribution" and other concepts, and thus, it is possible that we missed out some useful articles. We assume this limitation affect the findings.

Second, the resulting articles on customer knowledge sharing behavior in online platforms are limited, because the focus is laid on customer knowledge as a general concept. Further studies, however, are needed to understand the factors that influence customers to share their knowledge in the online platform, particularly, in online communities.

Third, pooling of all the CKS determinants in a variety of contexts is impossible because the determinants vary from each perspective and each context. Therefore, it is recommended to examine CKS based on the context and the type of knowledge sharing platform.

7 Conclusion

One of the most significant challenges in knowledge management is how to encourage people to share their knowledge. This study examines the existing research on customer knowledge sharing behavior to identify the top antecedents that encourage customers to share their knowledge with organizations and other customers or members. Few studies have examined the determinants of customer knowledge sharing behavior in the context of knowledge management. In the current dynamic business environment, the customers play an important role in company's growth and success. Therefore, the factors of CKS should be taken into consideration by the company when building their customer knowledge platform to ensure high participation of customers in knowledge creation process and to satisfy their needs and wants from the products and services.

In the current age of dynamic and competitive markets, the Internet and web technology have a significant impact on the way businesses and clients communicate. Consequently, firms should put more effort on how to encourage their customers to share and exchange their ideas, insights, and experiences. Therefore, a call for more studies is made by the present study to identify other motivational factors that might influence customer knowledge sharing, especially from a technology perspective.

References

1. Al-shammari, M.: Customer Knowledge Management: People, Processes, and Technology. IGI Global, New York (2009)
2. Zhang, D., Zhang, F., Lin, M., Du, H.S.: Knowledge sharing among innovative customers in a virtual innovation community: the roles of psychological capital, material reward and reciprocal relationship. Online Inf. Rev. (2017)
3. Menguc, B., Auh, S., Yannopoulos, P.: Customer and supplier involvement in design: the moderating role of incremental and radical innovation capability. J. Prod. Innov. Manag. **31** (2), 313–328 (2014)
4. Wang, Y., Wu, J., Yang, Z.: Customer participation and project performance: the mediating role of knowledge sharing in the Chinese Telecommunication Service Industry. J. Bus. Bus. Mark. **20**(4), 227–244 (2013)
5. Zhao, J., Wang, T., Fan, X.: Patient value co-creation in online health communities. J. Serv. Manag. **26**(1), 72–96 (2015)

6. García-Murillo, M., Annabi, H.: Customer knowledge management. J. Oper. Res. Soc. **53** (8), 875–884 (2002)
7. Davenport, T., Marchand, D.: Is KM just good information management. Financ. Times Mastering Ser. Mastering Inf. Manag. 1–3 (1999)
8. Campbell, A.J.: Creating customer knowledge competence: managing customer relationship management programs strategically. Ind. Mark. Manag. **32**(5), 375–383 (2003)
9. Tseng, S.-M.: The effect of knowledge management capability and customer knowledge gaps on corporate performance. J. Enterp. Inf. Manag. **29**(1), 51–71 (2016)
10. Lee, J.N.: The impact of knowledge sharing, organizational capability and partnership quality on IS outsourcing success. Inf. Manag. **38**(5), 323–335 (2001)
11. Tong, C., Tak, W.I.W., Wong, A.: The impact of knowledge sharing on the relationship between organizational culture and job satisfaction: the perception of information communication and technology (ICT) practitioners in Hong Kong. Int. J. Hum. Resour. Stud. **3**(1), 9 (2013)
12. Shek, S.P.W.: Understanding the Motivations of Consumer Knowledge Sharing in Online Community Sharing in Online Community (2008)
13. Von Krogh, G., Kim, S., Erden, Z.: Fostering the knowledge-sharing behavior of customers in interorganizational healthcare communities. In: Proceedings of 2008 IFIP International Conference on Network and Parallel Computing NPC 2008, pp. 426–433 (2008)
14. Korhonen-Sande, S., Sande, J.B.: Improving customer knowledge transfer in industrial firms: how does previous work experience influence the effect of reward systems? J. Bus. Ind. Mark. **31**(2), 232–246 (2016)
15. Salojärvi, H., Saarenketo, S.: The effect of teams on customer knowledge processing, esprit de corps and account performance in international key account management. Eur. J. Mark. **47**(5), 987–1005 (2013)
16. Yuan, D., Lin, Z., Zhuo, R.: What drives consumer knowledge sharing in online travel communities? Personal attributes or e-service factors? Comput. Hum. Behav. **63**, 68–74 (2016)
17. Salojarvi, H., Sainio, L.-M., Saarenketo, S., Tarkiainen, A.: What factors enhance intra-organizational customer knowledge sharing in international key account management? In: Industrial Marketing and Purchasing Conference 26th IMP-Conference, Budapest, Hungary (2010)
18. Liu, I.L.B., Cheung, C.M.K., Lee, M.K.O.: Customer knowledge contribution behavior in social shopping communities. In: 2013 46th Hawaii International Conference on System Sciences, pp. 3604–3613 (2013)
19. Lee, M.K.O., Cheung, C.M.K., Lim, K.H., Ling Sia, C.: Understanding customer knowledge sharing in web-based discussion boards. Internet Res. **16**(3), 289–303 (2006)
20. Jacobsen, L.F., Tudoran, A.A., Lähteenmäki, L.: Consumers' motivation to interact in virtual food communities—the importance of self-presentation and learning. Food Qual. Prefer. **62**, 8–16 (2017)
21. Wang, H., Yang, J., Cheng, A.: Research on the factors influencing the customer knowledge value. In: 2010 International Conference Management and Service Science, MASS 2010, no. 70672116, pp. 0–3 (2010)
22. Lin, M.J.J., Hung, S.W., Chen, C.J.: Fostering the determinants of knowledge sharing in professional virtual communities. Comput. Hum. Behav. **25**(4), 929–939 (2009)
23. Cheung, C.M.K., Lee, M.K.O.: What drives consumers to spread electronic word of mouth in online consumer-opinion platforms. Decis. Support Syst. **53**(1), 218–225 (2012)

24. Cheung, C.M.K., Lee, M.K.O.: What drives members to continue sharing knowledge in a virtual professional community? The role of knowledge self-efficacy and satisfaction. Knowl. Sci. Eng. Manag. **4798**, 472–484 (2007)
25. Bilgihan, A., Barreda, A., Okumus, F., Nusair, K.: Consumer perception of knowledge-sharing in travel-related Online Social Networks. Tour. Manag. **52**, 287–296 (2016)
26. Liao, T.H.: Developing an antecedent model of knowledge sharing intention in virtual communities. Univers. Access Inf. Soc. **16**(1), 215–224 (2017)
27. Yang, X., Li, G.: Factors influencing the popularity of customer-generated content in a company-hosted online co-creation community: a social capital perspective. Comput. Hum. Behav. **64**, 760–768 (2016)
28. Lu, H., Hsiao, K.: Understanding intention to continuously share information on weblogs. Internet Res. **17**(4), 345–361 (2007)
29. Chen, C., Du, R., Li, J., Fan, W.: The Impacts of Knowledge Sharing-Based Value Co-creation on User Continuance in Online Communities (2016)
30. Yan, Z., Wang, T., Chen, Y., Zhang, H.: Knowledge sharing in online health communities: a social exchange theory perspective. Inf. Manag. **53**(5), 643–653 (2016)
31. Kankanhalli, A., Tan, B., Wei, K.: Contributing knowledge to electronic knowledge repositories: an empirical investigation. MIS Q. **29**, 113–143 (2005)

Value Innovation in the Malaysian Telecommunications Service Industry: Case Study

Mohammed A. Hajar[1(✉)], Daing Nasir Ibrahim[1], and Mohammed A. Al-Sharafi[2]

[1] Faculty of Industrial Management, Universiti Malaysia Pahang, Lebuhraya Tun Razak, 26300 Kuantan, Pahang, Malaysia
eng.mohammed.hajar@gmail.com

[2] Faculty of Computer Systems and Software Engineering, Universiti Malaysia Pahang, Lebuhraya Tun Razak, 26300 Kuantan, Pahang, Malaysia

Abstract. Wireless telecommunications industry has been widely known for its high levels of innovativeness and competitively dynamic business environment. Globalization, privatization, liberalization and consolidation are the terms mostly associated with this sector. The Malaysian telecommunications industry is of no exception for it has exhibited high levels of competitiveness and dynamic change leading to rapid evolution of wireless technologies and high growth of subscribers and penetration rate. Thus, this paper aims to shed some light on the logic of value innovation in the telecommunication service industry. The present study adopted a qualitative research approach to investigate the value innovation activities on Malaysian telecommunications service sector. Specifically, it concentrates on the motivation of telecommunications companies to shift their strategies from tariff competition to value competition in order to improve customer value and promote customer loyalty which lead to increased business performance and profitable growth. An analytical case study was used to examine services and loyalty packages offered by telecommunications companies. In this regard, the paper discusses the shortfalls and actions that discouraged customer satisfaction and loyalty. The paper also provides recommendations for service providers on how to attain long-term profitable growth.

Keywords: Innovation · Value innovation · Telecommunications Malaysian telecommunications industry

1 Introduction

As an important source of economic growth, innovation has received great attention from both entrepreneurs and scholars. In today's global competitive business environment, innovation has become essential for organizational success and survival [1–5] where companies need to quickly provide the required amount of services to the customers and products that can satisfy customers' changing needs effectively [4]. Indeed, the importance of innovation goes beyond the survival and growth of a firm,

F. Saeed et al. (Eds.): IRICT 2018, AISC 843, pp. 892–901, 2019.
https://doi.org/10.1007/978-3-319-99007-1_83

for it is also an important driver for the economic growth and improvement of a nation and a region [5]. According to Joseph [6, 7], innovation leads to the development and growth of the economy, and eventually, prosperity and wealth. Hence, innovation is indeed becoming a top priority with some significant sectoral and regional developments [8].

Valuc innovation is a key part of a firm's strategic thinking and innovation management efforts. Firms with clearer strategic objectives are expected to be more successful at generating, recognizing, and acting upon new business opportunities that represent significant changes in market, technology and operation [9]. Value innovation adopts business differentiation through the enabling of a quantum leap in buyer value to break out the competition and create uncongested market space [10, 11].

Telecommunication sector depends heavily on innovation as an important source of sustainable growth. New products and services with high degrees of innovativeness leverage a company's performance and profit. For instance, the high technological development and the arrival of 4G and CDMA with high data transmission speed and bandwidth have helped in generating new revenue stream such as Value Added Services (VAS).

In addition, telecommunication service providers have produced a wide range of new services to avoid market growth saturation. These service providers are shifting their strategies to improve customer value by offering new products, not to attract new customers, but to retain existing customers through the promotion of customer loyalty [12]. Breaking through the competition and making a leap in value through innovation are some of the current challenges that companies have to face in order to create uncontested market, make competition irrelevant, and create demand and opportunities for highly profitable growth [13].

This paper aims to shed some the light on the logic of value innovation in telecommunications service industry. Specifically, the paper encourages telecommunications companies to shift their strategies from tariff competition to value competition in order to improve customer value, promote customer loyalty, and hence, increase business performance and profitable growth. Within this context, the present study adopted a qualitative research approach to investigate the value innovation activities on Malaysian telecommunications service sector. In general, it examined the behaviors of companies in enabling customer value. An analytical case study approach was used to examine services and loyalty packages offered by telecommunications companies. Thus, data were collected and obtained through the observation of organizational activities as well as secondary data which were available on the websites of the companies and from their published reports.

2 Related Studies

Innovation has contributed to the growth and development in the telecommunications industry since 1960s when a cluster of radical innovations began to change the traditional technical and institutional set up of the telephone network [14]. Innovation began in the telecommunications industry through research and development (R&D) as a problem solving and technological development. Research in the area of telecommunications technology has improved the telecommunications system radically from automatic to electronic telephone switches, and later to involve, private branch exchanges, mobile telephone systems, television, radio links, telefax, computer networks and data transmission, the potential of satellite technology, cable television and alarm systems [15].

Later, administration was involved in telecommunications R&D projects to minimize cost, improve quality and enhance profits. For instance, the rapid growth of the telecommunications market has resulted in competitive environment which forces telecommunications companies to improve their competitive advantage through innovative practices not only by deploying new technologies but also by providing better products and services packages, using creative marketing strategies and finding innovative ways of financing their projects [16].

According to Ranju and Versha [4], the importance of innovation in the telecommunications industry mainly stems from the recent growth of demand and competitiveness in this sector, and from the rapid advances in the technologies available for telecommunications organizations. Hence, innovation is becoming more pertinent due to three major trends, namely, concentrated international competition, disjointed and challenging markets, and assorted and swiftly changing technologies [3].

Currently, innovation in the telecommunications services is seen to be the key to survival and growth [1–5]. According to Bentley [17], innovation in the telecommunications industry is not a choice, but an imperative. The environmental changes of market liberalization, globalization and privatization make it more challenging and complex for the telecommunications industry in facing fierce competition arising from such situational changes which force telecommunications organizations across the globe to realize that without innovative practices, they may not be able to survive [4]. Thus, telecommunications organizations could cooperate with other organizations in global alliance in order to share knowledge and experience, and hence, enhance their innovation capabilities.

The introduction of 3G (and now 4G and 5G) mobile technology with the efficient features of high data transmission rate and bandwidth has helped in transforming the telecommunications industry from simple voice and messaging services to the rich media and complex services [1]. This has given the telecommunications organizations the opportunities to develop new streams of revenue such as value-added service (VAS) and business-to-business (B2B) services.

3 Case Study: The Malaysian Mobile Telecommunications Service Industry

3.1 Overview

Since early 1990s, the Malaysian mobile telecommunications market has experienced a tremendous growth as the market tended more to privatization and liberalization [18–20]. High rivalry, ease of access, low-priced usage initiation especially for prepaid services, and public awareness as well as rapid acceptance of mobile phones were the main attributes that contributed in the market high growth and development [19]. Besides that, the Malaysian economic policies and regulations regarding the telecommunications industry have significantly assisted the industry transformation and technological development in coping with global market dynamic change. In this regard, apart from Telekom Malaysia (TM), the following six private companies were granted the license to operate: Time Engineering Group (Time Wireless), Mobikom, Celcom, Maxis, Mutiara Swisscom (now rebranded as DiGi), and Sapura Digital [18, 20]. Despite the potential benefits of liberalization added to the market evolution, it had led to intensive tariff competition whereby more challenges in terms of profitability ensued. According to Chuah, Marimuthu [18], the competition among seven operators serving 20 million people at most made telecommunications companies struggle, particularly the new players who stood on small customer base which only had hundreds of thousands customers for more than a decade [18]. The 1997–1998 Asian financial crises had added salt to injury which left even incumbent telecommunications companies, such as Celcom, in debt [18]. That pushed service providers to strive through series of consolidations. For instance, Sapura Digital merged with Time Wireless to create Time Cel which was purchased by Maxis in 2002, while Mobikom merged with DiGi to become DiGi.com and TM Cellular with Celcom to create TM Celcom [18]. Later in 2008, TM Celcom was demerged as part of TM revamp plan to exit from wireless business and let Celcom to continue offering wireless services as a member of the Axiata Group [18]. In the same year, a new service provider entered the market, launching 3G mobile services under the name of U Mobile.

During the last two decades, the Malaysian mobile market witnessed high subscription growth where penetration rate surpassed 100% for most years. According to MCMC (the regulatory body of Malaysian Communications and Multimedia Commission) industry performance report [21, 22], mobile subscribers increased from 23.35 million in 2007 to 42.34 million in 2007 (see Fig. 1). That high increase on number of subscribers returns to the great contribution of mobile broadband subscription rate which posted double digit growth of 23.6% in 2017 only [22]. In other words, the motive for this growth was the robust adoption of smartphones, along with migration of pure voice subscription to the affordable postpaid or prepaid data plans, as well as the unprecedented advancement in wireless broadband technologies, and the usage of mobile Internet and digital content, which thereby have resulted in opening up a sizeable revenue opportunity for wireless service providers [18, 22]. However, having a penetration rate surpassing 100% implies that the mobile telecommunications market in Malaysia has reached its saturation point [18].

3.2 Market Analysis

The telecommunications industry in Malaysia has exhibited high competitiveness and dynamic change environment, which have led to rapid evolution of wireless technologies and high growth of subscribers and penetration rate especially for the more sophisticated wireless services, such as broadband and digital content services. Thus, as the telecommunications industry has been transformed from simple voice and messaging services to rich media and complex services, the traditional business models of wireless service providers have become increasingly obsolete, pushing the Malaysian service providers to shift their business strategies from voice-based subscription to

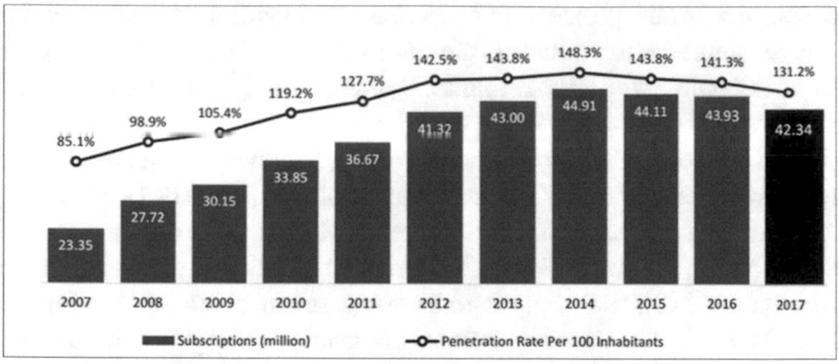

Fig. 1. Mobile cellular subscriptions and penetration Rate from 2007 to 2017. *Source*: MCMC [22]

innovative bundle plans (including voice, messages, data, etc.). As shown in Fig. 1, the market has attained maturity since 2009, having a penetration rate more than 100%, which means that the market has reached its saturation point and customers tend to use more than one SIM card. Accordingly, the gradual moderation of subscriptions for the years from 2015 to 2017 was due to the promoting of competitively priced packages, especially for postpaid, which have made it less advantageous for consumers to use more than one SIM [22]. That can be confirmed from the slight increase in postpaid subscribers from 8.14 million in 2014 to 10.23 million on 2017.

However, the ratio of 16:5 indicated that the number of prepaid subscribers was still higher compared to postpaid subscribers as there were 32.11 million prepaid subscribers compared to 10.23 million postpaid subscribers in 2017 [22]. This suggests that service providers are suffering from a high churn rate and low customer loyalty as subscribers often have no qualms switching between networks, based primarily on pricing and availability [18]. In addition, the introduction of WNP (Wireless Number Portability) service on 15 October 2008, which allows wireless customers to retain their existing phone numbers when switching from one service provider to another, broke old traditions and created a new wireless telecommunications environment in which subscribers have better freedom of choice and control to change service providers. Despite the WNP objectives of mitigating the inconvenience caused by switching networks and reducing switching costs for customers, it has increased the competitive pressure on the service providers. Therefore, service providers have had to intensify their marketing activities as well as improve the overall value being offered to customers by lowering prices, improving the quality of service, and providing new and innovative products and services in a bid to retain and expand their subscriber base. Indeed, it has made service providers struggle to gain profit especially with the high operation and marketing costs. For instance, the strategy of offering smartphone subsidies to customers has led to high operational cost [18].

Another challenge for mobile service providers that has posed a tremendous threat on the traditional voice and messaging business is the introduction and availability of OTT (Over-The-Top) applications such as Viber, WhatsApp, WeChat, Imo, Skype, and Messenger [18, 21–24]. According to OVUM, a London-based research firm, the global telecommunications industry might lose a total of $386 billion between 2012 and 2018 due to the growing impact of OTT applications [23]. Domestically, the Malaysian telecommunications market has been directly affected by the high number of non-traditional competitors among the OTT players. According to an analysis by MCMC [21, 22], the increasing popularity of the OTT services led to lower usage of voice and short message system (SMS) whereby a decline of service providers revenue of 4.5% (RM982.35 million) was recorded in 2016 compared to the previous year.

Consequently, service providers reacted to the OTT threat by calling for re-evaluation and re-invention of their core competence therefore providing advanced data networks with bandwidth efficient technologies, reliable video streaming and wide-spread broadband availability in order to benefit from the OTT applications [25]. In addition, service providers have gone beyond content business and customer transaction expansion strategies to diversify their business models such as by providing OTT services and home shopping to boost revenue [22]. For example, Digi.Com Berhad has a collaboration partnership with WhatsApp service provider from which Digi customers can enjoy unlimited access to the WhatsApp service for affordable fixed fee [24].

As mentioned earlier, the mobile market in Malaysia has been particularly dominated by four main players: Maxis, Celcom, Digi and U Mobile. Based on industry performance reports [21, 22], Maxis had the maximum market share in 2008 (40%), followed by Celcom (32%) and Digi (25%) as well as a very small percentage (3%) for U Mobile, the newly-launched company at that time. Until 2011, the four incumbent companies somehow managed to keep their market shares with a minimal loss of 4% from Maxis share to the other three competitors. During this period, U mobile was still developing its infrastructure and core competencies. After 2011, the Malaysian market had witnessed the emergence of non-traditional players of mobile virtual network operators (MVN), such as Altel, redONE, Tune Talk, XOX, Webe, Smart Pinoy, Merchantrade Asia, FRiENDi, Telin, Tron, XiddiG, ANSAR, SpeakOut, UCSI, MyAngkasa, Buzz Me, and ECI, that offered very lowly-priced competitive packages to attract the existing subscribers of incumbent mobile network operators which led to a cut-through aggressive tariff competition. That exacerbated the challenges for the incumbents such as Maxis and Celcom, to sustain their market shares which were observed to have been declining over the past 10 years to 26% and 22%, respectively, by 2017 [22].

Meanwhile, Digi market share has been almost constant throughout the past 10 years as U Mobile and MVN operators market shared increased gradually from 1% to reach 14% (6.12 million) and 11% (4.79 million), respectively, by the end of 2017 [22]. In addition, this trend has been clearly reflected on the revenue market share [21, 22]. Despite its declining market share from 45% in 2007 to 35% in 2017, Maxis has remained the leader in the revenue market share and maintained strong postpaid subscriptions and higher ARPU (average revenue per user) with its strategy of improving and enhancing customer experience to attract high-end subscribers [22]. Similarly, Celcom revenue market share slightly declined from 30% in 2007 to 27% in 2017 [22]. By contrast, Digi managed to maintain its revenue market share average at 26% during the same period while U Mobile and MVN operators increased their revenue market shares from 11% in 2016 to 14% in 2017 [22]. According to MCMC [22], the high growth in U Mobile and MVN operators market shares was due to the new innovative product offerings, substantial promotional and marketing activities which attracted subscribers to switch from the three major incumbents.

3.3 Shortfalls and Actions Discourage Customer Satisfaction and Loyalty

As the Malaysian telecommunications market became more mature, it was a necessity, not a choice, for service providers to shift their strategies from voice competition base to bundle plan base. This motivated service providers to come up with long-term loyalty packages to enhance customer satisfaction and loyalty in order to maintain or improve their market shares and profitability. However, service providers were still having some shortfalls and actions that disappointed customer satisfaction and make them switch to other operators instead of being loyal.

For instance, the inefficient management of loyalty offers and promotions, of which service providers had been making telesales and telemarketing through third-party companies (such as Star Solutions Sdn Bhd) to attract new or more recent customers, disappointed long-standing loyal customers. The misuse of loyalty packages which offered better deals for new or more recent customers while long-standing customers were still using the normal high tariff for many years without effort to appreciate their loyalty had caused an upset over the unfair treatment, and hence, led to the shift to such new loyalty packages offered by other service providers.

Another example is the auto renewal of add-on or VAS (Value-Added Services) subscriptions such as music, videos, ringtones, and news which customers find themselves to be bonded to the auto renewal fee deduction that undermines customer satisfaction and loyalty. The difficulty of unsubscribing from such services, associated with unsatisfactory customer care from service providers as an efficient source of revenue, would sometimes strongly drive customers to terminate their subscriptions or swap to different operators, especially for prepaid subscribers.

Another action which has been observed that undermines customer trust is the misled rewards and promotional offers. Most of these offers contain confusing terms and conditions which make the offers impractical and give adverse effect. For instance, the SMS reward that gives customers RM 3 credit when they top up or reload their subscription with RM 30 before a specific date makes the customers upset to find its availability period of 1–3 days and only after they finish utilizing the RM 30 credit. In other words, the offer asks customer to use up the purchased RM 30 credit in less than 3 days to enjoy additional RM 3 credit reward which makes it less valuable and a ridiculous reward offer to customers. Likewise, the internet data packages are equally ridiculous when customers are offered with validity periods only during weekends or after midnight.

In addition, the default activation of auto answering voicemail is annoying to many customers, particularly to those who receive international calls frequently. Despite being a major source of revenue, this service leads to customer dissatisfaction; hence, it further causes long-term negative effects on the company's performance and profitability, especially when it is difficult to deactivate such service and bundling it with call alert messages and other such similar services.

3.4 Recommendations

For a highly aggressive competitive market like Malaysia, mere innovation may not be sufficient to maintain a profitable and sustainable growth. The differentiation through providing more innovative packages and affordable bundle plans is just a marketing pioneer and short-term profitable strategy, which can easily be copied and provided by other service providers. It is noticeable that the offered bundle plans for many service providers are quite similar. Therefore, it is strongly recommended that service providers should pursue the strategic logic of value innovation that does not focus on beating the competitors but rather on making the competition irrelevant by creating a leap in value for customers and themselves. According to Chan Kim and Mauborgne [13], value innovation is more than an innovation; it is a strategy that embraces the entire system of a company's activities and orients the whole system towards achieving a leap in value for both buyers and the company itself [13]. Furthermore, value innovation occurs only when companies would be able to align innovation with utility, price, and cost position [13].

In this regard, service providers are required to enhance customer satisfaction and loyalty by providing a superior customer value that meets customer needs and expectations to ensure sustainable growth and long-term survival. Previous studies have pointed out that customer satisfaction contributes to a firm's profitability and customer retention [26–29]. According to Zhao et al. [29], "*As the market becomes more and more mature, mobile services become more homogeneous and the competition for acquiring new customers and retaining existing customers becomes more intense. In this environment, customer satisfaction is a critical factor for mobile service providers to maintain or improve their market share and profitability*". Therefore, service providers have to facilitate more of their existing low-paying prepaid subscribers to migrate to postpaid subscription in order to enhance customer loyalty and mitigate the phenomenon of customer churn.

Furthermore, service providers need to differentiate their offers by tapping the uncongested market segment, creating demand and opening new market space. For example, MVN operators managed to acquire 4.79 million subscribers in a few years through the adoption of new and creative ways to differentiate themselves. Telin Malaysia has led to the creation of KarTuAS, a 2-in-1 SIM card service that can hold both Malaysian and Indonesian mobile numbers while 8 Telco Sdn Bhd under the "Buzz Me" brand name has introduced the hybrid travel SIM using multi-International Mobile Subscriber Identities (multi-IMSI) technology which targets personal and business travelers [22].

It is also recommended that service providers should enhance their relationships and interactions with potential customers through Customer Relationship Management (CRM) software. Thus, service providers should treat their customers as partners and not as followers. This will give customers more options to customize and choose their respective bundle plans. This business model strategy has shown great success for Dell, where customers have the option to customize their own personal computers. Currently, Yes 4G, a brand under YTL Communications Sdn Bhd, has been promoting customer satisfaction and loyalty by giving wider options for customers to optimize their respective bundle plans.

4 Conclusions

The Malaysian telecommunications industry has indeed exhibited high competitiveness and dynamic change in the business environment. These have led to rapid evolution of wireless technologies and high growth of subscribers as well as penetration rate. However, the market also witnessed a cut-through competition in tariff rate and the number of offered services in the form of bundle plans. The situation was exacerbated by the emergence of new untraditional players, namely, the MVN operators, and also by the introduction of OTT applications as well as the availability of WNP service.

The present study has shed some light on the logic of value innovation in the telecommunications service industry. It has sought to motivate telecommunications companies to shift their strategies from tariff competition to value competition as for to improve customer value, promote customer loyalty and hence increase business performance and profitable growth. The current study adopted a qualitative research approach to investigate the value innovation activities in the Malaysian telecommunications service sector. Primary data were collected by making observation of organizational activities while secondary data were obtained from the websites of companies and available published reports. An analytical case study approach was employed to examine services and loyalty packages offered by the telecommunications companies. To sum up, the paper has discussed the shortfalls and actions that discouraged customer satisfaction and loyalty, and provided recommendations for service providers in order to attain sustainable profitable growth and long-term survival of their businesses.

References

1. Birudavolu, S., Nag, B.: A Study of Open Innovation in Telecommunication Services: A Review of Literature & Trends (2011)
2. Sachdeva, M., Agarwal, R.: An innovation experience: what does innovation mean to practising organisations? ANZAM (2011)
3. Letangule, S.L., Letting, D., Nicholas, K.: Effect of innovation strategies on performance of firms in the telecommunication sector in Kenya. Int. J. Manag. Bus. Stud. **2**(3), 75–78 (2012)
4. Ranju, K., Versha, M.: Building technological capability through innovations in telecom sector. Pac. Bus. Rev. Int. **5**(5), 37–49 (2012)
5. Abdi, A.M., Ali, A.Y.S.: Innovation and business performance in telecommunication industry in Sub-Saharan African context: case of Somalia. Asian J. Manag. Sci. Educ. **2**(4), 53–67 (2013)
6. Joseph, S.: The Theory of Economic Development. Cambridge University Press, Cambridge (1934)
7. Schumpeter, J.A.: Business Cycles: A Theoretical, Historical, and Statistical Analysis of the Capitalist Process. McGraw-Hill, New York (1939)
8. Wagner, K., et al.: The Most Innovative Companies 2014: Breaking Through is Hard to do. Boston Consulting Group, Boston (2014)
9. Smith, R.K.: Innovation in telecommunication services. Theses (School of Communication)/ Simon Fraser University (1994)

10. Kim, W.C., Mauborgne, R.: Value Innovation: The Strategic Logic of High Growth. Harvard Business School Publication, Brighton (1997)
11. Kim, W.C., Mauborgne, R.: Strategy, value innovation, and the knowledge economy. Sloan Manag. Rev. **40**(3), 41 (1999)
12. Ahn, H., et al.: A novel customer scoring model to encourage the use of mobile value added services. Expert Syst. Appl. **38**(9), 11693–11700 (2011)
13. Kim, W.C., Mauborgne, R.: Blue Ocean Strategy: How to Create Uncontested Market Space and Make the Competition Irrelevant. Harvard Business School Press, Boston (2005)
14. Davies, A.: Innovation in large technical systems: the case of telecommunications. Ind. Corp. Change **5**(4), 1143–1180 (1996)
15. Hauknes, J., Smith, K.: Corporate Governance and Innovation in Mobile Telecommunications: How Did the Nordic Area Become a World Leader? (2003)
16. Chaturvedi, R.M.: Innovation in telecom sector. BMA E J. Bombay Manag. Assoc. **14**(2), 1–7 (2003)
17. Bentley, C.: Telecom at a crossroad, the role of innovation. In: Innovation 360 Industry Seminar, Dubai, UAE (2014)
18. Chuah, H.W., Marimuthu, M., Ramayah, T.: Wireless telecommunication industry in Malaysia: trends, challenges, and opportunities. In: Marimuthu, M., Hassan, S. (eds.) Consumption in Malaysia: Meeting of New Changes. Universiti Sains Malaysia Publisher, Penang (2015)
19. Said, M.F., Adham, K.A.: Online mobile content innovations and industry structure: implications for firms' strategies. Int. J. Econ. Manag. **4**(1), 101–119 (2010)
20. Yoong, H., et al.: Malaysia's telecommunications sector: an efficiency and productivity analysis. In: The 2011 Australasian Meetings of the Econometric Society (ESAM). The University of Adelaide (2011)
21. MCMC, Digital Connectivity: Industry Performance Report 2016. ISSN 1823–3724
22. MCMC, Connectivity to Facilitate Digital Transformation: Industry Performance Report 2017. ISSN 1823–3724
23. Hossain, M.M.: A Literature Review of OTT-related Policy and Regulatory Issues
24. Sujata, J., et al.: Impact of over the top (OTT) services on telecom service providers. Indian J. Sci. Technol. **8**(S4), 145–160 (2015)
25. Berhad, M.: Annual Report 2012 (2012)
26. Lin, H.-H., Wang, Y.-S.: An examination of the determinants of customer loyalty in mobile commerce contexts. Inf. Manag. **43**(3), 271–282 (2006)
27. Kuo, Y.-F., Wu, C.-M., Deng, W.-J.: The relationships among service quality, perceived value, customer satisfaction, and post-purchase intention in mobile value-added services. Comput. Hum. Behav. **25**(4), 887–896 (2009)
28. Lee, H.S.: Factors influencing customer loyalty of mobile phone service: empirical evidence from Koreans. J. Intern. Bank. Commer. **15**(2), 1–14 (2010)
29. Zhao, L., et al.: Assessing the effects of service quality and justice on customer satisfaction and the continuance intention of mobile value-added services: an empirical test of a multidimensional model. Decis. Support Syst. **52**(3), 645–656 (2012)

The Impact of Trust on the Adoption of Cloud Computing Services by University Students

Abdulwahab Ali Almazroi[1(✉)], Haifeng Shen[2], and Fathey Mohammed[3]

[1] Faculty of Computing and Information Technology, University of Jeddah, Khulais, Saudi Arabia
aalmazroi@uj.edu.sa

[2] College of Science and Engineering, Flinders University, Adelaide, Australia

[3] Faculty of Engineering and Information Technology, Taiz University, Taiz, Yemen

Abstract. Cloud computing is a state-of-the-art technology that delivers computing resources to organizations and educational institutions over the Internet. Cloud computing services such as Google Apps, Microsoft education cloud, Amazon Web services, and IBM cloud academy help universities by facilitating teaching, learning, research, and other development activities. Nevertheless, one of the factors influencing cloud computing adoption is trust. Therefore, this paper aims at assessing the impact of trust on adopting cloud services by undergraduate students in Saudi Arabian universities. The study utilized a modified version of Technology Acceptance Model (TAM) by integrating trust to collect the empirical data. The data were gathered using a questionnaire from 527 undergraduate students in two Saudi Arabian universities. The responses were analyzed using SPSS and AMOS software. The findings showed that perceived usefulness, perceived ease of use, and trust are the significant predictors of intention to adopt cloud services. The findings suggested that perceived usefulness, perceived ease of use, and trust should be considered by decision makers in universities and cloud service providers in order to promote adoption of cloud services among students.

Keywords: Cloud computing · Technology acceptance model
Trust · Perceived usefulness · Perceived ease of use

1 Introduction

Cloud computing is an innovative technology that deliver scalable and flexible services to customers [1, 2]. The services can be IT infrastructure, platform, or software solutions [3]. Users can access cloud computing resources via Internet from anywhere at any time using any device [4]. There is growing interest in cloud computing due to its wide range of application in the area of computer science and Information Technology [5, 6]. Cloud computing offers many benefits to individuals and educational institutions [4]. It can provide efficient services and infrastructure to education institutions without the need to obtain the required IT infrastructure [7, 8]. As shown in Fig. 1, an IDC survey reports the benefits of cloud computing [9].

F. Saeed et al. (Eds.): IRICT 2018, AISC 843, pp. 902–911, 2019.
https://doi.org/10.1007/978-3-319-99007-1_84

The benefits of cloud computing services for students are huge. The benefits include easy and quick access to resources, improvement of student performance, acceleration of scientific research findings, IT related cost savings, aid of independent learning, and collaboration [10–13]. Nevertheless, adoption of cloud services by universities of developing nations is very low, and there is lack of studies conducted to identify the factors that affect the adoption by students. Hence, there is need to explore and investigate these factors in the context of a developing country. Therefore, this study investigates the factors that affect the adoption of cloud services by university students in one of the developing nations, Saudi Arabia. The study adopted Technology Acceptance Model (TAM). TAM is in fact one of the popular IS acceptance theories conceptualized in [14]. The model as adapted from Theory of Reasoned Action (TRA) defined two factors (perceived usefulness and perceived ease of use) as the predictors of intention to adopt a new technology [15].

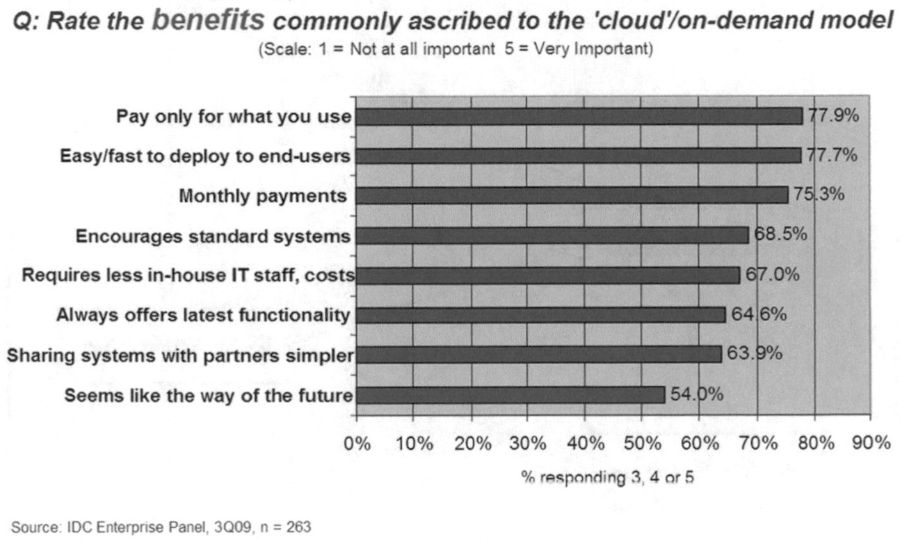

Fig. 1. The benefits of cloud computing services.

This paper is organized as follows: Sect. 2 presents the research methodology which includes the research model, hypotheses, and data collection process. The data analysis and results are presented in Sect. 3. Section 4 discusses the findings of the study in relation to prior studies. Finally, conclusion including future research direction is presented in Sect. 5.

2 Methodology

2.1 Research Model and Hypotheses

The theoretical basis of this study was Technology Acceptance Model (TAM). TAM identified perceived usefulness and perceived ease of use as the predictors of individual's behaviour intention toward adoption of a technology. Thus, behavioural intention is the dependent variable. Perceived usefulness refers to "the degree to which a person believes that using a particular system would enhance his or her job performance" [16], while perceived ease of use is "the degree to which a person believes that using a particular system would be free of effort" [16].

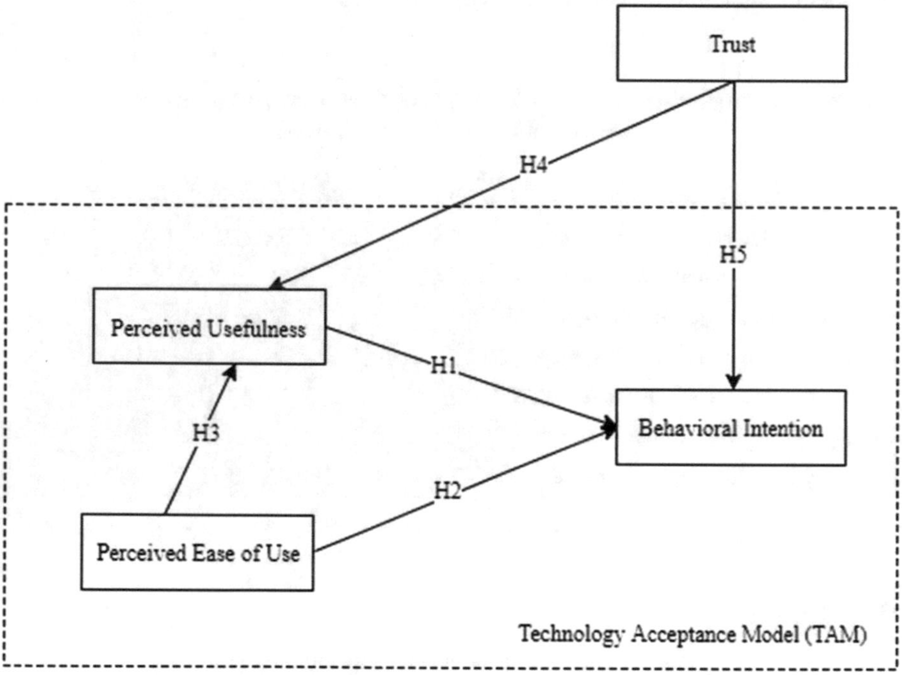

Fig. 2. The research model

In this study, trust was added to the model because it is one of the commonly reported challenges of cloud computing adoption [17], and it has been associated with technology adoption in various studies [18–21]. Trust is defined as "the belief that the other party will behave in a socially responsible manner, and, by so doing, will fulfil the trusting party's expectations without taking advantage of its vulnerabilities" [19]. Therefore, the research model has four factors namely Perceived Usefulness (PU),

Perceived Ease of Use (PEOU), Trust (TR), and Behavioural Intention (BI) as shown in Fig. 2. Based on the literature, we proposed 5 hypotheses:

H_1: *PU positively influences BI.*
H_2: *PEOU positively influences BI.*
H_3: *PEOU positively influences PU.*
H_4: *TR positively influences PU.*
H_5: *TR positively influences BI.*

2.2 Data Collection

A quantitative research approach was employed in this study by using a questionnaire to collect the research data. The questionnaire's items were adapted from related literature as follows.

- PU and PEOU were measured using six items adapted from Davis [16], Davis, Bagozzi [22], and Venkatesh and Bala [23].
- BI was measured using three items adapted from Davis [16], and Venkatesh and Bala [23].
- TR was measured using five items adapted from Pavlou [19], and Pikkarainen, Pikkarainen [24].

All the factors and items were modified based on the feedback from nine experts in the research field. The back translation technique [25] was used to translate the items in the questionnaire from English to Arabic (Saudi official language), and then the two versions were reviewed by an official translator for the authenticity of the translation. The questionnaire was rated using five-point Likert scale (from 1 = strongly disagree to 5 = strongly agree). A total of 527 questionnaires were administered to undergraduate students in two Saudi Arabian universities. 451 responses were returned and used for the analysis after excluding invalid answers.

3 Data Analysis and Results

SPSS (version 22) and AMOS (version 19) were used for the analysis of the empirical data. The demographic profile of the respondents and preliminary reliability of the instrument were analysed using descriptive statistics in SPSS software, while the main reliability and validity of the instrument were assessed using Structural Equation Modelling (SEM) in AMOS software. The latter software was also used to examine the relationships between the model's variables and to test the hypotheses. SEM was conducted in two stages as suggested in Hair, Black [26], including using the measurement model to assess the reliability and validity in the first stage, and using the structural model to test the hypotheses in the second stage. The demographic characteristics of the respondents are presented in Table 1.

Table 1. Demographic profile of the respondents

Variable		Frequency	Percent
Gender	Male	242	53.7
	Female	209	46.3
Major	Arts	237	52.5
	Sciences	214	47.5
Year of study	First year	78	17.3
	Second year	184	40.8
	Third year	126	27.9
	Fourth year	34	7.5
	Other	29	6.4
Years of using the Internet	<1 year	24	5.3
	1–3 years	200	44.3
	4–6 years	137	30.4
	7–9 years	73	16.2
	>9 years	17	3.8

3.1 Reliability and Validity

Reliability is the degree of consistency of an instrument. The preliminary reliability was assessed by Cronbach's alpha and composite reliability, which should be greater than 0.70 [26]. The reliability result for the constructs is presented in Table 2, where the Cronbach's alpha and composite reliability values for all the constructs are above 0.70, confirming the reliability of the instrument.

Table 2. Reliability and validity results

Constructs	Cronbach's Alpha	CR	AVE
PU	0.89	0.89	0.57
PEOU	0.87	0.87	0.54
BI	0.85	0.85	0.66
TR	0.88	0.88	0.60

On the other hand, validity is defined by Hair et al. [26] as the "extent to which a measure or set of measures correctly represents the concept of study". Two types of validity should be reported: content validity and construct validity [27–30]. Content validity is "the extent to which a measuring instrument provides adequate coverage of the topic under study" [30]. A panel of experts is involved to validate the content of the instrument. The experts assess how the questionnaire items are relevant and meet the standards [30]. Construct validity "refers to the extent to which your measurement questions actually measure the presence of those constructs you intended them to measure" [27]. Convergent and discriminant validity are used to assess the construct validity. Convergent validity is obtained if Average Variance Extracted (AVE) for the construct is above 0.50 [26]. Similarly, discriminant validity is attained if the square root of AVE for each construct is higher than the correlations between the construct and the other constructs [31].

Table 3. Discriminant validity results

Constructs	PEOU	PU	BI	TR
PEOU	**0.732**			
PU	0.568	**0.753**		
BI	0.642	0.652	**0.811**	
TR	0.287	0.370	0.305	**0.773**

Diagonal cells denote the square root of AVE

The convergent validity results (see Table 2) show that the AVE values for all the constructs are higher than 0.50. This supports the convergent validity of the model constructs. Likewise, as can be seen in Table 3, the square root of AVE for each construct is higher than its correlations with other constructs which indicates that all constructs have discriminant validity. Therefore, these results have proved the reliability and validity of the constructs.

3.2 Testing of Hypotheses

The hypotheses testing is conducted to determine whether the independent variables have significant contributions to the explanation of the dependent variables [26]. Therefore, the hypothesized paths were investigated in order to find those that are significant. The hypotheses were tested using AMOS in this study, and the hypotheses testing result presented in Fig. 3 shows that four out of the five hypotheses were supported.

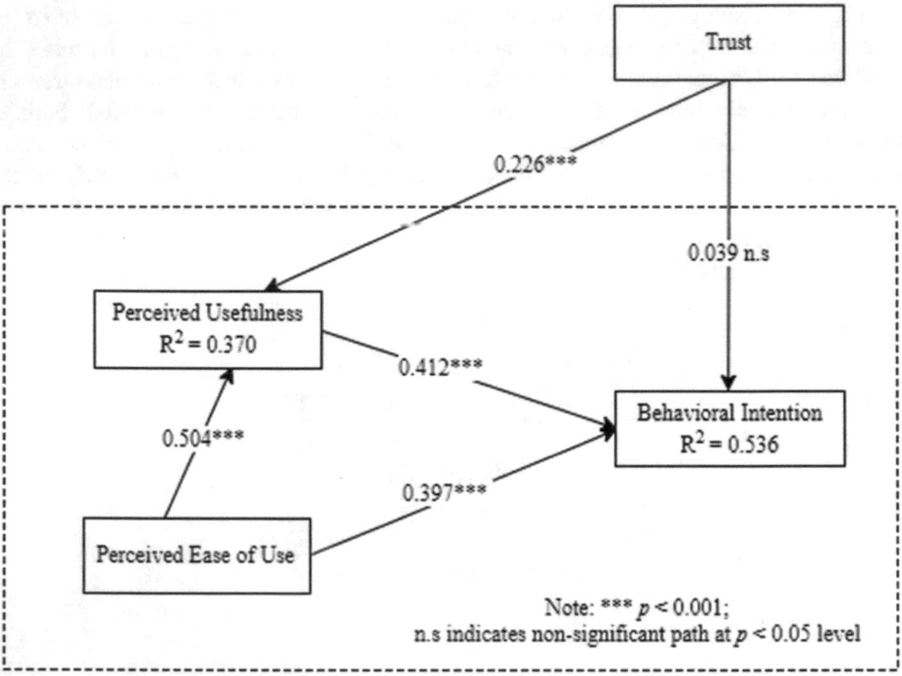

Fig. 3. The research model with hypotheses results

4 Discussion

In this study, factors that affect Saudi Arabian undergraduate students' adoption of cloud computing services were assessed. The factors were examined using TAM. The analysed factors are perceived usefulness, perceived ease of use, trust, and behavioural intention. The results shown in Fig. 3 suggested that the hypotheses H_1, H_2, H_3, and H_4 were significant. In contrast, H_5 was not significant.

The positive significant effect of perceived usefulness on behavioural intention highlights the importance of perceived usefulness on technology adoption. Among the predictors of behavioural intention, perceived usefulness was the strongest as reported in most of the TAM related studies [3, 32]. This means the students would adopt cloud services primarily to improve their learning activities. Perceived ease of use had a positive effect on both perceived usefulness and behavioural intention. Perceived ease of use also shows significant determinant of behavioural intention. These findings are consistent with previous studies [23, 33–35]. Therefore, the ease of using cloud services will lead to realization of its benefits which also increase students' intention to adopt it. The last significant path in this study is the link between trust and perceived usefulness which is in line with findings from various studies [36–38]. This relationship emphasizes the significance of trust in the formation of usefulness perception. This suggests that the students' perception of usefulness of cloud services will increase if they perceive it as trustworthy. Finally, we found an insignificant relationship between trust and behavioural intention. Studies such as in [19, 36, 39] found that trust has significant effect on behavioural intention, however the findings of this study was consistent with finding from Zafiropoulos et al. [40] which indicates that trust had a non-significant effect on behavioural intention. Trust does not have direct effect on behavioural intention; however it has an indirect effect through perceived usefulness. Perhaps the reason for the indirect effect is because the students consider it as part of perceived usefulness.

5 Conclusion

Adoption of cloud services by Saudi Arabian undergraduate students is affected by perceived usefulness, perceived ease of use, and trust as established in this study. Four out of five hypotheses were supported. Implying that the defined factors significantly contribute to the adoption of the cloud services. The findings support the inclusion of trust into the original TAM model to examine adoption of cloud services. Therefore, cloud service providers are recommended to provide more trustable and secure services to promote students in Saudi Arabia universities to adopt their services. In addition, universities should provide information on the possible threats and security and privacy controls and solutions in order to increase students' confidence and trust in cloud services. In the future, the researchers will consider other factors to assess the adoption of cloud services in the same environment.

References

1. Sun, D., et al.: Surveying and analyzing security, privacy and trust issues in cloud computing environments. Proc. Eng. **15**, 2852–2856 (2011)
2. Prabowo, A.H., Janssen, M., Barjis, J.: A conceptual model for assessing the benefits of software as a service from different perspectives. In: Business Information Systems. Springer (2012)
3. Li, Y., Chang, K.-C.: A study on user acceptance of cloud computing: a multi-theoretical perspective. In: Eighteenth Americas Conference on Information Systems. Association for Information Systems AIS Electronic Library (AISeL), Washington (2012)
4. Al Tayeb, A., et al.: The impact of cloud computing technologies in e-learning. Int. J. Emerg. Technol. Learn. (iJET) **8**(2013), 37–43 (2013)
5. Kim, W., et al.: Adoption issues for cloud computing. In: Proceedings of the 7th International Conference on Advances in Mobile Computing and Multimedia. ACM (2009)
6. Gonzalez, N., et al.: A quantitative analysis of current security concerns and solutions for cloud computing. In: 4th IEEE International Conference on Cloud Computing Technology and Science Proceedings. IEEE (2011)
7. Milić, A., Simić, K., Milutinović, M.: Cloud computing environment for E-learning services for students with disabilities. In: Mahmood, Z. (ed.) Continued Rise of the Cloud, pp. 363–381. Springer, London (2014)
8. González-Martínez, J.A., et al.: Cloud computing and education: a state-of-the-art survey. Comput. Educ. **80**, 132–151 (2015)

9. Gens, F.: New IDC IT Cloud Services Survey: Top Benefits and Challenges (2009)
10. Alshuwaier, F.A., Alshwaier, A.A., Areshey, A.M.: Effective use of cloud computing services in education. J. Next Gener. Inf. Technol. **3**(4), 38 (2012)
11. Rosati, P., et al.: Quantifying the financial value of cloud investments: a systematic literature review. In: 2017 IEEE International Conference on Cloud Computing Technology and Science (CloudCom). IEEE (2017)
12. Neelakantan, P.: A Study on E-Learning and Cloud Computing (2018)
13. Yadegaridehkordi, E., et al.: Predicting the adoption of cloud-based technology using fuzzy analytic hierarchy process and structural equation modelling approaches. Appl. Soft Comput. **66**, 77–89 (2018)
14. Sabi, H.M., Uzoka, F.-M.E., Mlay, S.V.: Staff perception towards cloud computing adoption at universities in a developing country. Educ. Inf. Technol. **23**, 1–24 (2018)
15. Davis, F.D.: A Technology Acceptance Model for Empirically Testing New End-User Information Systems: Theory and results. Massachusetts Institute of Technology, Cambridge (1986)
16. Davis, F.D.: Perceived usefulness, perceived ease of use, and user acceptance of information technology. MIS Q. **13**(3), 319–340 (1989)
17. Alharbi, S.T.: Trust and acceptance of cloud computing: a revised UTAUT model. In: 2014 International Conference on Computational Science and Computational Intelligence, Las Vegas, Nevada, USA. IEEE (2014)
18. Van der Schyff, K., Krauss, K.E.: Higher education cloud computing in South Africa: towards understanding trust and adoption issues. S. Afr. Comput. J. **55**, 40–55 (2014)
19. Pavlou, P.A.: Consumer acceptance of electronic commerce: integrating trust and risk with the technology acceptance model. Int. J. Electron. Commer. **7**(3), 101–134 (2003)
20. Pavlou, P.: Integrating trust in electronic commerce with the technology acceptance model: model development and validation. In: Seventh Americas Conference on Information Systems, Boston, USA (2001)
21. Mary, M., Pauline, R.: The impact of trust on the technology acceptance model in business to consumer E-commerce. In: Mehdi, K.-P. (ed.) Innovations Through Information Technology, pp. 921–924. Idea Group Publishing, London (2004)
22. Davis, F.D., Bagozzi, R.P., Warshaw, P.R.: User acceptance of computer technology: a comparison of two theoretical models. Manag. Sci. **35**(8), 982–1003 (1989)
23. Venkatesh, V., Bala, H.: Technology acceptance model 3 and a research agenda on interventions. Decis. Sci. **39**(2), 273–315 (2008)
24. Pikkarainen, T., et al.: Consumer acceptance of online banking: an extension of the technology acceptance model. Intern. Res. **14**(3), 224–235 (2004)
25. Chapman, D.W., Carter, J.F.: Translation procedures for the cross cultural use of measurement instruments. Educ. Eval. Policy Anal. **1**(3), 71–76 (1979)
26. Hair, J.F., et al.: Multivariate Data Analysis. Pearson, London (2010)
27. Saunders, M.N., Lewis, P., Thornhill, A.: Research Methods for Business Students, 5th edn. Prentice Hall, London (2009)
28. Sekaran, U.: Research Methods for Business: A Skill Building Approach, 4th edn. Wiley, New York (2003)
29. Creswell, J.W.: Research Design: Qualitative, Quantitative, and Mixed Methods Approaches, 3rd edn. Sage Publications, Thousand Oaks (2009)
30. Kothari, C.R.: Research Methodology: Methods and Techniques. New Age International, New Delhi (2004)

31. Fornell, C., Larcker, D.F.: Evaluating structural equation models with unobservable variables and measurement error. J. Mark. Res. **18**(1), 39–50 (1981)
32. Behrend, T.S., et al.: Cloud computing adoption and usage in community colleges. Behav. Inf. Technol. **30**(2), 231–240 (2011)
33. Faqih, K.M., Jaradat, M.-I.R.M.: Assessing the moderating effect of gender differences and individualism-collectivism at individual-level on the adoption of mobile commerce technology: TAM3 perspective. J. Retail. Consum. Serv. **22**, 37–52 (2015)
34. Al-Gahtani, S.S.: Empirical investigation of E-learning acceptance and assimilation: a structural equation model. Appl. Comput. Inf. **12**(1), 27–50 (2014)
35. Venkatesh, V., Davis, F.D.: A theoretical extension of the technology acceptance model: four longitudinal field studies. Manag. Sci. **46**(2), 186–204 (2000)
36. Gefen, D., Karahanna, E., Straub, D.W.: Trust and TAM in online shopping: an integrated model. MIS Q. **27**(1), 51–90 (2003)
37. Chircu, A.M., Davis, G.B., Kauffman, R.J.: Trust, expertise, and E-commerce intermediary adoption. In: AMCIS 2000 Proceedings, pp. 710–716 (2000)
38. Belanche, D., Casaló, L.V., Flavián, C.: Integrating trust and personal values into the technology acceptance model: the case of E-government services adoption. Cuad. Econ. Dir. Empresa **15**(4), 192–204 (2012)
39. Carter, L., Weerakkody, V.: E-government adoption: a cultural comparison. Inf. Syst. Front. **10**(4), 473–482 (2008)
40. Zafiropoulos, K., Karavasilis, I., Vrana, V.: Assessing the adoption of E-government services by teachers in Greece. Future Intern. **4**(2), 528–544 (2012)

Adoption of Cloud Computing Services by Developing Country Students: An Empirical Study

Abdulwahab Ali Almazroi[1(✉)] and Haifeng Shen[2]

[1] Faculty of Computing and Information Technology, University of Jeddah, Khulais, Saudi Arabia
aalmazroi@uj.edu.sa

[2] College of Science and Engineering, Flinders University, Adelaide, Australia

Abstract. This study aims to examine the factors that influence the adoption of cloud services by Saudi university students. Based on a modified Technology Acceptance Model 3 (TAM3), a questionnaire was administered to 527 undergraduate students from two universities who were selected using stratified cluster sampling technique. The data collected from the respondents were analysed using Structural Equation Modelling (SEM) technique. Regardless of students' Internet experience, the findings of this study show that the main determinants of behavioral intention of students to adopt cloud computing services are perceived usefulness and perceived ease of use. Additionally, trust and job relevance predict perceived usefulness; subjective norm predicts image; and also perceived ease of use is determined by perceptions of external control, perceived enjoyment and playfulness. These findings will guide decision makers in Saudi Arabian academic institutions on how to facilitate successful deployment of cloud computing services.

Keywords: Cloud computing · Technology Acceptance Model
Software as a Service · Perceived ease of use · Perceived usefulness
Trust

1 Introduction

Cloud computing technology provides computing resources in a scalable and flexible way to customers, which can be accessed anywhere and anytime using any device [1]. Thanks to its broad applicability in the fields of information systems and technologies and its salient benefits of cost saving with improved security, efficiency and portability. There is rapid growing interest and investment in cloud computing in various organizations. In the case of higher education in developed countries, cloud computing, especially through Software as a Service (SaaS), has been widely adopted to make teaching, learning, and research easier while at the same time reducing the cost of acquiring and maintaining computing hardware and software [2].

A great number of studies presented the benefits of cloud computing which include low costs, increased mobility, enhanced security, strong compliance, and facilitated collaboration [3]. Despite the reported benefits of cloud computing to university

F. Saeed et al. (Eds.): IRICT 2018, AISC 843, pp. 912–927, 2019.
https://doi.org/10.1007/978-3-319-99007-1_85

students, there is lack of studies conducted in Saudi Arabia that examined the factors influencing undergraduate students' intention to adopt cloud services for their learning tasks. Thus, this study aims to examine the factors influencing cloud computing adoption from students' perspective. The results of this empirical study can be used as a guideline for the decision makers in academic institutions in general and in Saudi Arabian universities in particular to ensure successful adoption of cloud services among students. The findings will also help the providers of cloud services better understand the factors that influence students' adoption so that they can develop cloud services that would be effectively utilized by students.

While a variety of cloud computing technologies and applications exist, Google Docs was particularly selected for this research because it is a popular SaaS widely used by students across the globe. Technology Acceptance Model 3 (TAM3) was adapted as a theoretical framework of the study to identify and understand the factors that affect the adoption of cloud computing services by university students. TAM3 is an all-inclusive model that focusses on individual behaviors regarding the adoption and use of information technologies [4]. The model was formed as a result of integrating model of the determinants of perceived usefulness and model of the determinants of perceived ease of use.

2 Literature Review

2.1 Technology Acceptance Model 3 (TAM3)

TAM3 is one of the extended versions of the TAM and a comprehensive model that concentrates on the determinants of Perceived Usefulness (PU) and Perceived Ease of Use (PEOU). PU is defined as "the extent to which a person believes that using an information technology will enhance his or her job performance", while PEOU measures "the degree to which a person believes that using an information technology will be free of effort" [4]. PU and PEOU are assumed to be influenced by different factors defined by researchers depending on their studies' context. This enables researchers to further study and validate these factors in different contexts and fields. TAM3 has its roots in TAM, a popular theory that predicts Behavioural Intention (BI) to adopt and use new information technologies by users. TAM was adapted from the Theory of Reasoned Action (TRA) by Davis [5].

An extension of TAM called TAM2 was proposed by Venkatesh and Davis [6] who identified the determinants of PU and moderators of PU and intention. The determinants include Subjective Norm (SN), Image (IMG), Job Relevance (REL), Output Quality (OUT), Result Demonstrability (RES) and Perceived Ease of Use (PEOU), while the moderators are experience and voluntariness. Later, the determinants of PEOU were proposed in order to have an in-depth understanding of the variable, which are grouped into anchors and adjustments [7]. The anchors comprises of Computer Self-Efficacy (SE), Perceptions of External Control (PEC), Computer Anxiety (ANX), and Computer Playfulness (PLAY), while adjustments are Perceived Enjoyment (ENJ) and Objective Usability (OU). Venkatesh and Bala [4] further developed a comprehensive version of TAM, which is TAM3, by combining the models that present determinants of PU (TAM2) and determinants of PEOU. Therefore, TAM3 presents a complete model that can explain behavioural intention to adopt and use of a technology in detail because it integrates determinants of the main antecedents of technology adoption behaviour and use into a single model.

In TAM3, the determinants of PU are categorised as either social influence or technology characteristics. Social influence represents "various social processes and mechanisms that guide individuals to formulate perceptions of various aspects of an IT" [4], including SN and IMG. SN refers to "the degree to which an individual perceives that most people who are important to them think they should or should not use the technology" [8], while IMG is "the degree to which an individual perceives that use of a technology will enhance their status in their social system" [9]. Technology characteristics denote important features that contribute to the development of individuals' perceptions towards the usefulness or ease of using a technology. They include factors of REL, OUT, RES, and PEOU. REL refers to "the degree to which an individual believes that the target technology is applicable to their job" [6]. OUT refers to "the degree to which an individual believes that the technology performs their job tasks well" [4]. RES refers to "the degree to which an individual believes that the results of using a technology are tangible, observable, and communicable" [4]. TAM3 postulates that PEOU, SN, IMG, RES and REL have direct relationship with PU, whereas the relationship of REL with PU is moderated by OUT. PU has two moderating factors of voluntariness and experience.

Similarly, the determinants of PEOU are grouped under technology characteristics related adjustments, individual differences, and facilitating conditions. Individual difference variables are "personality and/or demographics that can influence individuals' perceptions of perceived usefulness and perceived ease of use" [4], which include three anchors of SE, ANX and PLAY. SE refers to "the degree to which an individual believes that they have the ability to perform a specific task/job using the technology" [4]. ANX refers to "the degree of an individual's apprehension, or even fear, when they are faced with the possibility of using the technology" [7]. PLAY refers to "the degree of cognitive spontaneity in the interactions of the technology" [10]. It is worth noting that the judgments of perceived ease of use can initially be motivated by these anchors, but individuals will adjust their judgments as they "gain direct hands-on experience with the new system" [4]. The system characteristics related adjustments are Perceived Enjoyment (ENJ) and Objective Usability (OU). ENJ refers to "the extent to which the activity of using a specific technology is perceived to be enjoyable in its own right, aside from any performance consequences resulting from the use", while OU refers to "a comparison of systems based on the actual level (rather than perceptions) of effort required to completing specific tasks" [7]. Finally, the facilitating conditions are represented by Perceptions of External Control (PEC), which refers to "the degree to which an individual believes that organisational and technical resources exist to support the use of the technology" [4].

Venkatesh [4] further hypothesised three new relationships that were not quantitatively assessed in the studies leading to TAM3. The relationships are: (a) the moderating effect of experience on the relationship between PEOU and PU such that the influence will be stronger with increased experience; (b) the moderating effect of experience on the relationship between ANX and PEOU such that the influence becomes weaker with increased experience; and (c) the moderating influence of experience on the relationship between PEOU and BI such that when the experience increases the influence of PEOU on BI will become weaker.

Venkatesh and Bala [4] conducted longitudinal studies on four different organizations using validated items from previous studies to assess and validate TAM3. The findings were as expected as reported in Al-Gahtani [11]. As reported in prior studies, TAM3 is considered to be a comprehensive, validated and suitable model to investigate technology adoption in different contexts [11, 12]. Thus, it is a suitable theoretical framework for examining the factors affecting Saudi university students' perception toward adopting cloud computing applications. Furthermore, TAM3 considers many important factors that are important to the understanding of factors affecting cloud services adoption by university students such as individual differences, social influence, technology characteristics, and facilitating conditions.

2.2 Related Work

Several studies identify and investigate factors that affect cloud computing services adoption. Alsaeed and Saleh [13] identified around 40 exploratory studies on the adoption of cloud computing published between 2009 and 2014. Some of them focused on adoption by organizations [14, 15], while others examined the adoption by individual users [16, 17]. From the perspective of the end-users, the effect of personal characteristics on individual's behavioral intention towards adopting and using cloud computing services was investigated by Coursaris, van Osch [17]. A model was proposed which incorporated technological, contextual, demographic, and lifestyle factors. Based on 402 valid responses, the study found that individuals' intention to use the innovation was significantly influenced by innovation characteristics including compatibility, relative advantage, triability and observability. Some of the innovation attributes were influenced by the contextual factors such as social influence, and past experience and knowledge.

However, studies that investigate factors affecting adoption of cloud computing services by university students are generally lacking [18]. The limited studies include a study by Behrend, Wiebe [2] who investigated cloud computing services adoption by colleges' students in the South-eastern of USA. The study found that the perceived usefulness of the technology had an impact on background features such as students' capability to travel to the campus. Also, the perceived ease of using cloud computing services was influenced by experience and instructor support. The results also showed that the perceived ease of use had more influence than the perceived usefulness on cloud computing adoption. It can be observed from the previous studies that few studies focused on the end user perspective on using cloud services in higher education context. More specific, studies that examine factors affecting cloud services adoption by university students in Saudi Arabia are lacking. Hence, this study aims at filling this gap by identifying and examining factors that influence Saudi Arabian higher education students to adopt cloud services.

3 Research Model

The research model extended TAM3 by adding a new construct which is Trust (TR) as a direct determinant of PU and BI. TR is expressed as "the belief that the other party will behave in a socially responsible manner, and, by so doing, will fulfil the trusting party's expectations without taking advantage of its vulnerabilities" [19]. The integration of trust is due to its influence on technology adoption processes as reported by Alharbi, [20]. On the other hand, the research model excludes usage and objective usability constructs from TAM3 and changed experience construct to Internet experience. The research model which comprised of 15 constructs and 25 hypotheses are presented in Fig. 1.

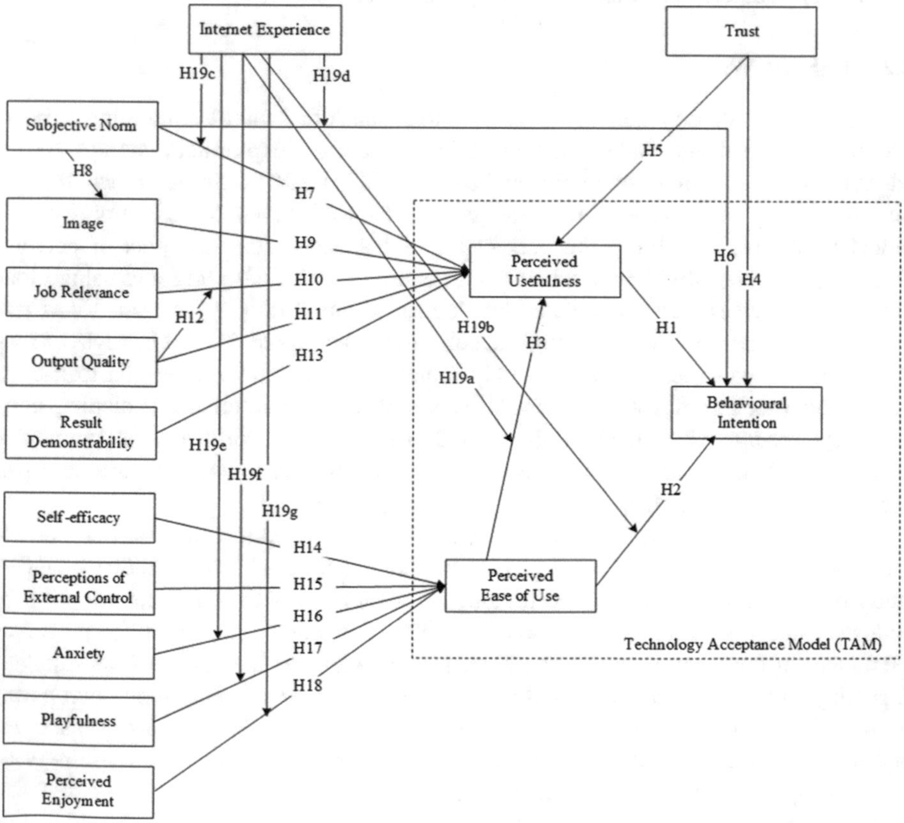

Fig. 1. The research model

4 Research Methodology

4.1 Research Instrument

This study used the five-point Likert scale questionnaire to collect data from the respondents. The questionnaire was adapted from previous studies of TAM [21]. All the variables and the associated items were verified by nine experts from the research field. The questionnaire as accordingly modified based on the experts comments. The questionnaire was originally developed in English and then translated into Arabic (Saudi official language) using back translation technique. Both English and Arabic versions were checked by official translators for the clarity of the questions and the authenticity of the translation.

4.2 Population and Sample

The target population is the undergraduate students in two Saudi universities. As the population was very large and surveying the entire population would require huge amount of budget and time [22], stratified cluster sampling technique was applied to select the sample.

4.3 Pilot Study

Two pilot studies were carried out to get suggestions and comments from the students about the clarity of the instrument, to estimate the time required to do the designed tasks and fill the questionnaire, and to assess the reliability of the items in the questionnaire [22]. The first pilot study was conducted in a computer lab with 6 students. After completing the questionnaire, the students were asked to provide feedback on the following issues: the time needed to complete the given tasks and also the time needed to complete the questionnaire, the clarity of the items and instructions, and the questionnaire layout. The questionnaire was modified accordingly following the feedback received from the pilot study. The second pilot study was conducted with 31 students in the same environment and using the same cloud service as in the first pilot study. A total of 29 valid responses were used after excluding two incomplete ones. The reliability of the instrument was measured using Cronbach's Alpha whose values for all the factors were above the recommended threshold of 0.70 [23].

4.4 Main Study

The questionnaire was distributed to 527 students and a total of 451 responses were usable after excluding the invalid responses. Before conducting data analysis, data screening was performed using SPSS (version 22) to make sure the data is appropriate for analysis. The normality of the data was examined as the first step in the screening process. This process is important because the maximum likelihood estimation technique in Structural Equation Modelling (SEM) requires the data to be normally distributed [23]. Hence, in this study normality was examined using skewness and kurtosis whose values were found to be within acceptable ranges of ± 3 and ± 10 respectively

[24]. Outliers were then screened, and the results of both univariate and multivariate outlier screening show no evidence of outlier detected. Multicollinearity test was further performed using Tolerance and Variance Inflation Factor (VIF). A tolerance value below 0.10 or a VIF value above 10 suggests the possibility of multicollinearity [25]. The Tolerance and VIF results are shown in Table 1. The Tolerance and VIF values for all the constructs were above 0.10 and below 10 respectively. This is an indication of the suitability of data for conducting multiple regression.

Table 1. Multicollinearity test results

Construct	Tolerance	VIF
PU	0.525	1.904
PEOU	0.573	1.746
BI	0.551	1.815
SE	0.812	1.232
PLAY	0.741	1.350
PEC	0.709	1.410
ANX	0.922	1.084
ENJ	0.591	1.693
SN	0.681	1.468
IMG	0.673	1.487
REL	0.613	1.633
OUT	0.743	1.346
RES	0.812	1.232
TR	0.746	1.341

5 Analysis and Results

5.1 Exploratory Factor Analysis (EFA)

EFA was conducted to identify the structure of the items/factors in the proposed model. Prior to conducting EFA, the reliability of the factors was examined using Cronbach's Alpha. Cronbach's Alpha values for all the factors were above the recommended threshold of 0.70 [23]. The Cronbach's Alpha values for the factors of SE and PEC were improved after deleting one item from each factor as presented in Table 2.

Kaiser–Meyer–Olkin (KMO) measure of sampling adequacy and Bartlett's test of sphericity were used to assess whether the data was appropriate for factor analysis. The overall KMO value was 0.891 and Bartlett's test was strongly significant ($p = 0.000$), confirming the sample size was adequate for factor analysis and there were satisfactory relationships between the variables. The principal components analysis with Varimax rotation was then used to extract factors, where 14 factors were extracted based on the eigenvalue greater than 1, and all the factor loadings were above 0.50.

Table 2. Results of reliability and validity

Constructs	Cronbach's Alpha	Construct reliability	AVE
OUT	0.75	0.76	0.51
PU	0.89	0.80	0.57
PEOU	0.87	0.77	0.53
BI	0.85	0.85	0.66
ANX	0.86	0.83	0.63
RES	0.88	0.84	0.64
SN	0.83	0.83	0.62
REL	0.84	0.84	0.64
IMG	0.78	0.79	0.55
ENJ	0.88	0.88	0.72
PEC	0.81	0.79	0.55
PLAY	0.80	0.77	0.52
TR	0.88	0.81	0.59
SE	0.80	0.77	0.53

5.2 Confirmatory Factor Analysis (CFA)

The Goodness of Fit (GOF) for the measurement model was first established during the CFA. As shown in Table 3, all the GOF indices were initially within the recommended ranges [23, 26, 27], except GFI, which was less than the recommended threshold. Therefore, the measurement model was redefined and improved in accordance with 3 recommended methods [23, 24, 28]. These methods include factor loading with threshold of 0.50 for each item, Squared Multiple Correlation (SMC) value with minimum value of 0.30 for each item, and modification indices (MI) which suggest deletion of items with high covariance (above 4.0) between measurement errors. The factor loadings for all the items were found to be above 0.50. Also, the SMC values for all the items were above 0.30. Finally, during the MI test, 13 items were found to have high covariance with other items and after deleting them one after another, all the GOF indices improved and fell within the acceptable ranges. The goodness of fit indices for the final improved measurement model is presented in Table 3.

Table 3. Goodness of fit indices for the measurement and structural models

Fit measures	Recommended threshold	Initial measurement model	Final measurement model	Structural model
X^2/df	<3.00	1.63	1.40	1.44
GFI	>0.90	0.85	0.91	0.90
AGFI	>0.80	0.83	0.89	0.88
CFI	>0.90	0.93	0.97	0.96
TLI	>0.90	0.93	0.96	0.95
RMSEA	<0.08	0.04	0.03	0.03

The constructs validity was evaluated using convergent and discriminant validity during CFA as recommended by Hair, Black [23]. Construct reliability (>0.70), and Average Variance Extracted (AVE) (>0.50) were used in determining the convergent validity [23], which was confirmed with the results in Table 2. In addition, the evidence of convergent validity is the factor loadings to be equal to or greater than the recommended value of 0.50 [23]. This was achieved since the factor loadings were all above 0.50. The discriminant validity requires the square roots of AVE be greater than the correlations between the factors [29]. As shown in Table 4, the square roots of AVE are all greater than the inter-factor correlations, confirming that the discriminant validity was achieved.

5.3 Testing of Hypotheses

Prior to examining the paths between the factors, fitness of the structural model was examined and showed good fit as presented in Table 3. Figure 2 shows the results of the proposed research model consisting of 15 constructs. Table 5 reveals that nine out of the seventeen hypotheses, namely H1, H2, H3, H5, H8, H10, H15, H17, and H18, are statistically significant. The significant relationships are depicted using thick lines while the insignificant paths are presented with thin lines as shown in Fig. 2.

5.4 Testing the Moderation Effect

Factors between dependent and independent factors are called moderator factors [23, 24]. The presence of moderator factors may influence the direction/strength of the relationships between dependent and independent factors. The moderation/interaction effect is to evaluate how a moderator factor influences a relationship. In this research, the moderating effect of output quality and Internet experience is assessed. To examine the effect of output quality on the relationship between job relevance and perceived usefulness, the moderator factor (output quality) and the independent factor (job relevance) are mean-centered prior to creating the interaction term to avoid the problem of multicollinearity [4]. It's found that the output quality does not moderate the relationship between job relevance and perceived usefulness (REL * OUT → PU = (β) 0.05, $p = 0.29$), hence rejecting H12, and implying that students do consider the cloud service relevant to their learning and collaboration tasks regardless of the quality of its output, which is contrary to the previous findings [4, 11].

On the other hand, multi-group analysis is applied to assess the moderating effect of Internet experience as recommended by Byrne [28]. Table 6 presents the change in chi-square following the multi-group analysis on the moderating effect of Internet experience on identified paths.

The result shows that the change in chi-square is not significant ($\Delta\chi^2$ (7) = 12.80, $p = 0.00$), hence rejecting hypotheses H19a… H19 g. Accordingly, this study found that Internet experience has no moderating effect on the defined paths even though, various studies suggested that experience of using a product moderates the relationships between beliefs and behavioral intention [30]. This fiinding implies that students in this study had adequate Internet experience.

Table 4. Discriminant validity for the factors

Construct	ANX	PU	PEOU	SE	TR	PLAY	SN	IMG	REL	OUT	ENJ	BI	PEC	RES
ANX	**0.79**													
PU	−0.01	**0.76**												
PEOU	−0.11	0.61	**0.73**											
SE	0.07	0.33	0.17	**0.73**										
TR	0.02	0.43	0.36	0.27	**0.77**									
PLAY	−0.03	0.42	0.42	0.24	0.36	**0.72**								
SN	0.19	0.35	0.27	0.29	0.36	0.34	**0.79**							
IMG	0.18	0.32	0.27	0.29	0.40	0.36	0.54	**0.74**						
REL	0.06	0.49	0.36	0.28	0.34	0.34	0.40	0.46	**0.80**					
OUT	0.05	0.39	0.27	0.25	0.39	0.23	0.44	0.37	0.45	**0.72**				
ENJ	−0.03	0.51	0.48	0.15	0.38	0.50	0.41	0.38	0.53	0.49	**0.85**			
BI	−0.02	0.66	0.65	0.27	0.37	0.43	0.28	0.29	0.44	0.34	0.42	**0.81**		
PEC	−0.09	0.39	0.39	0.23	0.34	0.36	0.26	0.28	0.47	0.31	0.39	0.34	**0.74**	
RES	−0.12	0.26	0.28	0.14	0.16	0.20	0.16	0.24	0.35	0.29	0.29	0.24	0.34	**0.80**

Note: the diagonal (bold) values are the square root of AVE; while the remaining values are the correlations between the factors.

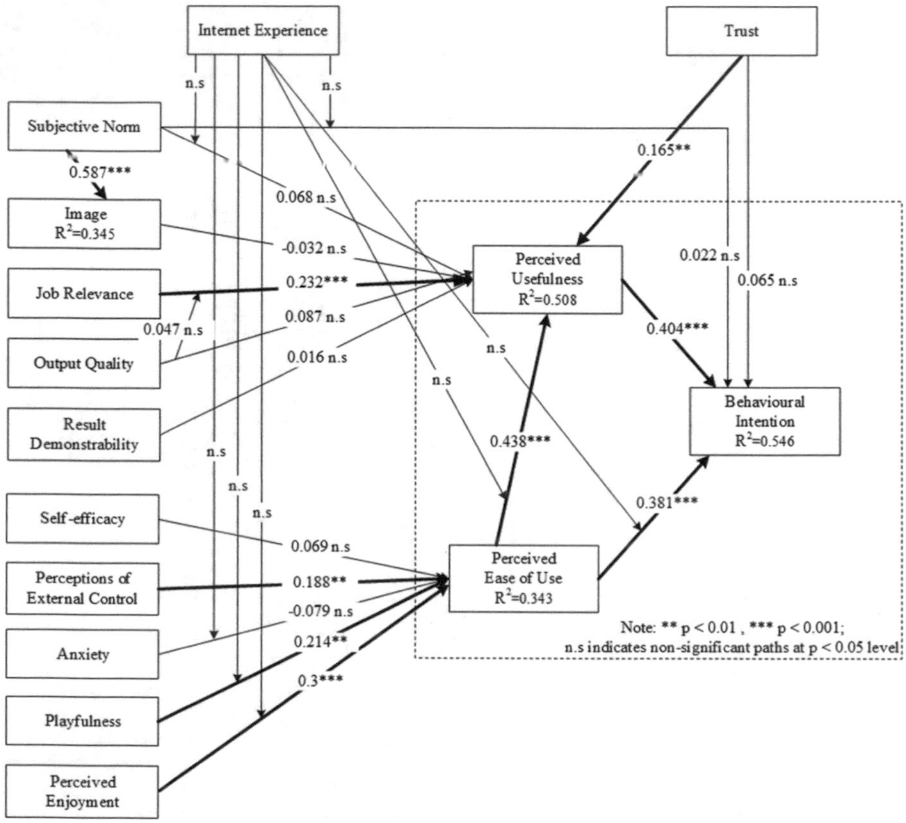

Fig. 2. The structural model

6 Discussion

The findings related to the main TAM constructs revealed that: (a) H1: PU has the strongest positive effect on BI ($\beta = 0.40$; $p < 0.001$), (b) H2: PEOU also has a strong positive effect on BI ($\beta = 0.38$; $p < 0.001$), and (c) H3: PEOU has the strongest positive influence on PU ($\beta = 0.44$; $p < 0.001$). Therefore, hypotheses H1, H2, and H3 are all accepted, echoing the findings from a study on e-Learning in Saudi Arabia [11] and the previous studies on TAM and its extensions [6, 7]. This also provides an evidence that it is suitable to use the original TAM to study the adoption of cloud computing services as it can investigate the influence of PU and PEOU on determining the behavioural intention of Saudi undergraduate students to adopt cloud computing services. Consistent with TAM3, PU and PEOU are the main determinants of BI, which is the main dependent variable which measures the students' intention to adopt cloud services in this study. The findings imply that Saudi university students will adopt cloud services when they perceive that: (a) they require decreasing effort to use cloud

Table 5. Hypotheses testing results

Path	Standardized regression weights	p -value	Hypothesis testing result
H1: PU → BI	0.404	***	Supported
H2: PEOU → BI	0.381	***	Supported
H3: PEOU → PU	0.438	***	Supported
H4: TR → BI	0.065	0.213	Not supported
H5: TR → PU	0.165	**	Supported
H6: SN → BI	0.022	0.657	Not supported
H7: SN → PU	0.068	0.356	Not supported
H8: SN → IMG	0.587	***	Supported
H9: IMG → PU	−0.032	0.607.	Not supported
H10: REL → PU	0.232	***	Supported
H11: OUT → PU	0.087	0.172	Not supported
H13: RES → PU	0.016	0.752	Not supported
H14: SE → PEOU	0.069	0.219	Not supported
H15: PEC → PEOU	0.188	**	Supported
H16: ANX → PEOU	−0.079	0.120	Not supported
H17: PLYA → PEOU	0.214	**	Supported
H18: ENJ → PEOU	0.3	***	Supported

**$p < 0.01$; ***$p < 0.001$

Table 6. Change in chi-square

Model	χ^2	df	$\Delta\chi^2$	Δdf	p -value
Unconstrained model (baseline)	2032.87	1514			
Fully constrained model	2045.67	1521	12.80	7	0.08

services and they become increasingly proficient in using them, and (b) cloud services will help them improve their productivity and effectiveness in their learning activities.

Regarding the newly added construct (trust), it has a significant influence on PU ($\beta = 0.17$; $p < 0.01$), therefore, H5 is accepted; however it does not have a significant effect on BI ($\beta = 0.065$; $p = 0.213$), rejecting H4. This result is contrary to [16] which showed that the trust had a significant influence on behavioural intention to adopt cloud services but an insignificant influence on perceived usefulness. This is probably because students consider trust as part of the benefits of cloud services adoption for leaning and collaboration tasks.

In related to the social influence factors (SN and IMG), the results showed that SN has direct positive effect on IMG and has the strongest path ($\beta = 0.59$; $p < 0.001$) obtained in this study, leading to the acceptance of H8 and the confirmation of similar hypotheses in prior studies [11, 12]. However, SN does not present a significant effect on BI ($\beta = 0.02$; $p = 0.66$) and PU ($\beta = 0.07$; $p = 0.36$), rejecting both H6 and H7. Similarly, IMG has no significant effect on PU ($\beta = -0.03$; $p = 0.61$), rejecting H9 as well. This finding suggest that the influence of peers and teachers in students' social groups only affects the perception of enhancing the students' status, but has no effect on their beliefs and behaviors towards adopting cloud computing services. The insignificant effect of IMG on PU has confirmed the findings from previous studies [31], while the results related to SN have also confirmed other investigations [32].

In addition, the findings related to the cognitive factors REL, OUT, and RES show that only REL has a strong positive influence on PU (β = 0.23; $p < 0.001$), whereas OUT (β = 0.09; $p = 0.17$) and RES (β = 0.02; $p = 0.75$) have no significant effect on PU. Therefore, H10 is accepted, but H11 and H13 are both rejected, suggesting that although students perceived the relevance of the cloud service to their learning activities, they did not believe the results produced by the cloud service were tangible and the cloud service can be used to enhance the performance of learning and collaboration tasks and lead to high quality output. There are empirical evidences in previous studies that support the significant influence of job relevance [11, 31], the insignificant effect of result demonstrability [12], and the insignificant effect of output quality on PU [4].

Regarding the four anchors (SE, ANX, PEC, and PLYA) that are expected to predict PEOU in TAM3, it was found that PEC (β = 0.19; $p < 0.01$) and PLYA (β = 0.2194; $p < 0.01$) both significantly predict PEOU, confirming H15 and H17 are accepted. This means that the availability of resources such as computing devices and Internet connectivity as well as students' intrinsic motivation to use the cloud services such as feeling playful and spontaneous play an important role in the formation of perceived ease of using the cloud services. These findings are consistent with those from previous studies that reported the significant influence of perceptions of external control on perceived ease of use [7], and that of playfulness on perceived ease of use [4].

However, SE (β = 0.07; p = 0.22) and ANX (β = −0.08; p = 0.12) did not significantly influence PEOU, rejecting H14 and H16 and implying that the students don't have confidence to use the cloud services without external assistance, but they are not negative toward using it either. Some previous studies also reported the insignificant effect of self-efficacy [33], and anxiety on perceived ease of use [2, 12]. Further, it is clear that ENJ has a strong positive effect on PEOU (β = 0.30; $p < 0.001$), thus H18 is accepted and this result is supported by Al-Gahtani [11]. This finding implies that the students' perception towards the ease of using cloud services will increase when they perceive the enjoyment of using them. Finally, it can be observed from Fig. 2 that the model explained 51%, 34%, and 55% of the variance in PU, PEOU, and BI respectively.

7 Conclusion, Limitations and Future Work

Cloud computing services are becoming popular means of providing computing resources and applications to users on demand. Although cloud computing offers numerous advantages besides cost savings to users in academic settings, studies that investigate factors affecting university students' intention to adopt cloud services in developing countries are very rare, particularly in Saudi Arabia. Hence, this study proposed a research model by extending TAM3 and empirically examined it to predict factors that affect Saudi Arabian university students' intention to adopt cloud computing services. The data in this study was collected using a questionnaire from undergraduate students. The responses were analyzed using SEM to validate the proposed model and test the research hypotheses. About half of the hypotheses related to the direct effects between the factors were accepted. However, the hypotheses related to the moderating effect of output quality and Internet experience were all rejected.

For future work, it is recommended to examine the reasons behind the insignificant factors found in this study, which are: image, subjective norm, result demonstrability, self-efficacy, output quality, and anxiety, as well as the moderator factors of Internet experience and output quality. In addition, as the cloud service can also be accessed using mobile devices, the effect of mobile devices' specific characteristics on the adoption of cloud services can be investigated. Further, since the data in this research was collected cross-sectionally, the behavioral intention and use of cloud services at different points in time (longitudinal) can be examined in order to have a detailed understanding of the phenomena and investigate if there is any change in the students' behavior after some time. Furthermore, due to time and resource constraints the study was conducted in two Saudi universities. Hence, more studies can be conducted to cover other Saudi universities and even other Arabian Gulf countries to validate the proposed model and generalize the findings. Finally, the perceptions of teachers, decision makers, IT and other relevant personnel in Saudi universities should be studied as they play significant roles in the successful deployment of cloud computing services in universities.

References

1. Prabowo, A.H., Janssen, M., Barjis, J.: A conceptual model for assessing the benefits of software as a service from different perspectives. In: 15th International Conference on Business Information Systems, pp. 108–119. Springer, Vilnius (2012)
2. Behrend, T.S., et al.: Cloud computing adoption and usage in community colleges. Behav. Inf. Technol. **30**(2), 231–240 (2011)
3. Park, S.C., Ryoo, S.Y.: An empirical investigation of end-users' switching toward cloud computing: a two factor theory perspective. Comput. Hum. Behav. **29**(1), 160–170 (2013)
4. Venkatesh, V., Bala, H.: Technology acceptance model 3 and a research agenda on interventions. Dec. Sci. **39**(2), 273–315 (2008)
5. Davis, F.D.: A Technology Acceptance Model for Empirically Testing New End-User Information Systems: Theory and Results. Massachusetts Institute of Technology, Cambridge (1986)
6. Venkatesh, V., Davis, F.D.: A theoretical extension of the technology acceptance model: four longitudinal field studies. Manag. Sci. **46**(2), 186–204 (2000)
7. Venkatesh, V.: Determinants of perceived ease of use: integrating control, intrinsic motivation, and emotion into the technology acceptance model. Inf. Syst. Res. **11**(4), 342–365 (2000)
8. Fishbein, M., Ajzen, I.: Belief, Attitude, Intention and Behavior: An Introduction to Theory and Research. Addison-Wesley Publishing Company, Philippines (1975)
9. Moore, G.C., Benbasat, I.: Development of an instrument to measure the perceptions of adopting an information technology innovation. Inf. Syst. Res. **2**(3), 192–222 (1991)

10. Webster, J., Martocchio, J.J.: Microcomputer playfulness: development of a measure with workplace implications. MIS Q. **16**(2), 201–226 (1992)
11. Al-Gahtani, S.S.: Empirical investigation of E-learning acceptance and assimilation: a structural equation model. Appl. Comput. Inf. **12**(1), 27–50 (2014)
12. Faqih, K.M., Jaradat, M.-I.R.M.: Assessing the moderating effect of gender differences and individualism-collectivism at individual-level on the adoption of mobile commerce technology: TAM3 perspective. J. Retail. Consum. Serv. **22**, 37–52 (2015)
13. Alsaeed, N., Saleh, M.: Towards cloud computing services for higher educational institutions: concepts & literature review. In: 2015 International Conference on Cloud Computing (ICCC), Riyadh, pp. 1–7. IEEE (2015)
14. Gupta, P., Seetharaman, A., Raj, J.R.: The usage and adoption of cloud computing by small and medium businesses. Int. J. Inf. Manag. **33**(5), 861–874 (2013)
15. Opala, O.J., Rahman, S.M.: An exploratory analysis of the influence of information security on the adoption of cloud computing. In: 8th International Conference on System of Systems Engineering. IEEE: Wailea-Makena, Hawaii, pp. 165–170 (2013)
16. Alotaibi, M.B.: Exploring users' attitudes and intentions toward the adoption of cloud computing in Saudi Arabia: an empirical investigation. J. Comput. Sci. **10**(11), 2315–2329 (2014)
17. Coursaris, C.K., van Osch, W., Sung, J.: A "cloud lifestyle": the diffusion of cloud computing applications and the effect of demographic and lifestyle clusters. In: 46th Hawaii International Conference on System Sciences. IEEE: Hawaii, pp. 2803–2812 (2013)
18. Hashim, H.S., Hassan, Z.B., Hashim, A.S.: Factors influence the adoption of cloud computing: a comprehensive review. Int. J. Educ. Res. **3**(7), 295–306 (2015)
19. Pavlou, P.A.: Consumer acceptance of electronic commerce: integrating trust and risk with the technology acceptance model. Int. J. Electron. Commer. **7**(3), 101–134 (2003)
20. Alharbi, S.T.: Trust and acceptance of cloud computing: a revised UTAUT model. In: 2014 International Conference on Computational Science and Computational Intelligence, pp. 131–134. IEEE, Las Vegas (2014)
21. Almazroi, A.A.: An Empirical Study of Factors that Influence the Adoption of Cloud Computing Applications by Students in Saudi Arabian Universities. School of Computer Science, Engineering and Mathematics. Thesis, Flinders University (2017)
22. Saunders, M.N., Lewis, P., Thornhill, A.: Research Methods for Business Students, 5th edn. Prentice Hall, London (2009)
23. Hair, J.F., et al.: Multivariate Data Analysis: A Global Perspective, 7 (Global Edition) edn. Pearson, Prentice Hall, Upper Saddle River (2010)
24. Kline, R.B.: Principles and Practice of Structural Equation Modeling, 3rd edn. Guilford Publications, London (2011)
25. Pallant, J.: SPSS Survival Manual: A Step by Step Guide to Data Analysis Using SPSS, 4th edn. Allen & Unwin, Crows Nest (2011)
26. Hoyle, R.H.: Structural Equation Modeling: Concepts, Issues, and Applications. Sage Publications, London (1995)
27. Suh, B., Han, I.: Effect of trust on customer acceptance of Internet banking. Electron. Commer. Res. Appl. **1**(3), 247–263 (2002)
28. Byrne, B.M.: Structural Equation Modeling with AMOS: Basic Concepts, Applications, and Programming, 2nd edn. Routledge, New York (2010)
29. Fornell, C., Larcker, D.F.: Evaluating structural equation models with unobservable variables and measurement error. J. Mark. Res. **18**, 39–50 (1981)

30. Lin, H.-F.: The effect of absorptive capacity perceptions on the context-aware ubiquitous learning acceptance. Campus Wide Inf. Syst. **30**(4), 249–265 (2013)
31. Chang, S.J., Im, E.-O.: A path analysis of internet health information seeking behaviors among older adults. Geriatr. Nurs. **35**(2), 137–141 (2014)
32. Mathieson, K.: Predicting user intentions: comparing the technology acceptance model with the theory of planned behavior. Inf. Syst. Res. **2**(3), 173–191 (1991)
33. Jung, S., Hwang, J., Ju, D.Y.: Determinants of continued participation in web-based co-creation platforms. In: 1st International Conference on HCI in Business, Crete, Greece (2014)

Social Media Services and Applications

The Interaction of Trust and Social Influence Factors in the Social Commerce Environment

Hasan Beyari(✉) and Ahmad Abareshi

School of Business IT and Logistics, RMIT University, Melbourne, Australia
{hasan.beyari,ahmad.abareshi}@rmit.edu.au

Abstract. Social commerce is a rapidly expanding sub-category of e-commerce that uses social media to support user input and to assist in the sale of services and products. This study explores the interrelationship between trust and social influence in the context of social commerce. It also adds a theoretical framework that help businesses to have a better understanding of the factors that affect consumers when they use their social commerce sites. Using a survey method, data were gathered on the relationships between various key factors associated with these constructs, as well as on the role they play in social commerce. A total of 314 students from Saudi Arabia studying in Australian universities and at Saudi Arabia's University of Business and Technology were surveyed. The study made use of partial least squares (PLS) to test the measurement and structural model, and to study the interconnection between variables. The results of the survey and analysis show that, Information Quality, Reputation, and Word-of-Mouth have significant positive relationships with Trust and Social Influence. Additionally, variables such as Communication, Transaction Safety and Long-term Orientation were found to have a negative impact on both Trust and Social influence. This research recognizes that social commerce is unique, and different factors may affect customer Trust and Social Influence compared to face-to-face transactions. This research is expected to help companies devise effective strategies to grow their businesses. The results obtained from this study will add depth to existing literature relating to social commerce. The results also provide an opportunity for researchers to delve more deeply into the factors that affect Trust and Social Influence in social commerce.

Keywords: Social commerce · Trust · Social influence · Saudi Arabia Australia · PLS

1 Introduction

Dramatic changes have taken place in Information and Communication Technology (ICT) throughout the course of the twentieth and early twenty-first centuries. Over the past few decades, key innovations have included the development of the Internet and dramatic increases in processing speed. Most recent IT advancements are linked to the benefits from Internet growth, while Web 2.0 has given people the chance to interact on the Internet more frictionlessly than before [1].

The origins of Social Commerce (SC) can be traced to the advent of Web 2.0 technology and the first social networks [2]. This study contributes to the existing

F. Saeed et al. (Eds.): IRICT 2018, AISC 843, pp. 931–944, 2019.
https://doi.org/10.1007/978-3-319-99007-1_86

literature on SC by investigating the factors that influence Trust and Social Influence. It aims to help both scholars and business owners develop an understanding of the ways SC businesses can enjoy greater competitive advantages and better performance.

This paper begins with an introductory section, Sect. 1. Section 2 reviews the relevant literature, and Sect. 3 describes the theoretical foundation that was used, along with the hypotheses that were developed for this study. Section 4 describes the research methodology, Sect. 5 presents the data analysis and findings, and Sect. 6 explains the discussion and the major findings. Finally, Sect. 7 offers a conclusion and suggests directions for future research.

2 Literature Review

Social commerce (SC) is a category of electronic commerce (e-commerce) and refers to the use of social media, online media that allows social exchange and customer input, to improve the online buying and selling experience [3]. In addition, social commerce is a new trend in e-commerce, which integrates commercial and social activities through the development of social technologies in e-commerce venues [4].

Risk and uncertainty are often associated with financial transactions online, as sellers and buyers are prevented from having any direct or face-to-face communication. The 'Social Exchange theory' suggests that individuals are keen on financial transactions if trust and confidence are present. According to [5], trust is fundamental to minimizing uncertainty and risk in online business transactions. As indicated by [6, 7], building trust in social commerce businesses is crucial, and dependent on Social Networking Sites (SNS) where users can share a variety of information. In addition, [8] defines the term 'social influence' as changes that arise in an individual's behavior, actions, feelings, thoughts and attitudes as a result of interactions with another individual or group. [9] found a strong relationship between social influence and online interaction when it comes to developing relationships between consumers regarding reviews and shared experiences. Although existing research has incorporated the effect of social influence in social commerce, relevant studies have been restricted to data sources addressing social interaction derived from e-commerce transactions, which represent only a subset of information [10]. Significantly, [11] states that social commerce constructs; for example, reviews, recommendations, forums and ratings, generate 'social word-of-mouth', which is the recent evolution of electronic word-of-mouth and uses social media to provide more opportunities for consumers to have online interaction and communication.

Communication is vital for trust and social influence to be achieved. According to [12], communication reflects relevant and timely information being provided to customers and is crucial in any marketing relationship. Hence, communication ensures that customers obtain the necessary information to guide their purchasing decisions. This entails providing sufficient information about a particular product or service, and details about associated issues such as delivery schedules, payment options and shipping conditions. Privacy is also important to online customers, and includes the safeguarding of their personal information; for example, address, phone number and credit card number. Therefore, consumers may have the propensity to share their personal information if online vendors act responsibly to protect and respect their privacy [13].

For the purpose of this study, 'information quality' refers to the quality of the information available to consumers on social commerce sites, and includes the relevance, accuracy, and usefulness of the information offered. According to [6], the quality of information in the context of social commerce is crucial because it determines the level of trust that the consumers have in social commerce. Information quality is the quality of the content on an e-commerce website that can highly influence a consumers' attitude and their interactions [14].

The term 'reputation' can have various meanings; however, in general, it refers to an individual's perception of people, an organization, or business. In this study, reputation refers to the reputation of social commerce enterprises. [15] notes that reputation has a plethora of definitions in different disciplines. The reputation of a social commerce enterprise is a significant aspect affecting the trust of social commerce users [6]. Moreover, consumers' perceptions of the reputation of a social commerce website can be crucial in determining the level of consumers' trust in the website [6]. Notably, social commerce consumers share information about the reputation of the website with other consumers. Hence, reputation can serve as a powerful antecedent for trust to be developed [15].

Online consumers in today's e-commerce environment possess the knowledge that purchasing goods or services online involves a higher level of risk than traditional commerce transactions due to the anonymity of the online environment. Additionally, the level of fraud in online transactions is higher than telephone or in-person transactions of the same nature [18]. Therefore, transaction safety, which is the security level offered by social commerce sites to the consumers during and after they make a purchase, is very important. According to [16, 17], transaction safety plays a vital role in building trust by offering consumers a high level of security during their interactions online.

Word-of-Mouth (WOM) refers to the exchange of information and experiences among consumers to assist each other in purchase decisions [19]. If consumers are happy with their purchase experience, their WOM exchanges about the company are likely to be positive. Consequently, organizations emphasize the significance of consumers' feedback, as consumers' dissatisfaction may produce negative WOM [20]. Various studies suggest that positive social word-of-mouth spread via social media enhances the level of trust for new products and services among users of social media. This online interaction provides users with the value of trust through the sharing of knowledge and experience of a new product or service [11].

Long-term orientation is one of the six dimensions of national culture, and is determined by the level to which an individual is oriented to gain rewards by persistence [21]. Long-term orientation is related to an individual's attitude toward the future. They can either choose to live in the present or to be future-oriented. This dimension determines a cultures' preferred communication style; for example, collectivistic cultures with short-term orientation usually communicate more freely and more often than collectivistic cultures with long-term orientation, who prefer to be more precise and succinct.

3 Theoretical Background and Hypothesis Development

In this study, trust theory is used to investigate the relationship between consumers' trust in SC, and claims that a higher level of consumer trust leads to an increased intent to purchase products and services [22], especially online. Trust needs to be considered a priority for SC sites when it comes to satisfaction of the consumer, as trust plays a key role in online interactions. Applications on Web 2.0, such as consumer ratings and reviews, may be a good way to promote trust in new customers. Evidently, interactions among the connected users on SNS increase trust [23]. Social influence theory studies an individual's social interactions with others from a behavioral and technological perspective [24]. Social influence theory merges the social information processing theory with research on the adoption and diffusion of communication technologies. Social relationships are a key element that differentiates SC from other forms of online commercial activities.

A strong relationship exists between social influence and trust [25]. A consumer will have a higher level of trust in SC when he or she receives information from family or friends, or from the people with whom he or she interacts on SNS. This means that if a consumer perceives that the information they have received through the groups or individuals on social networking sites with whom they interact, they are likely to gain more trust in social commerce. Therefore, the following hypothesis is proposed:

H1. Trust has a positive impact on Social Influence in SC.

Communication is one of the key factors in enhancing trust. It is the process of creating and sharing information between consumers, through either formal or informal methods [26]. SC sites provide various ways for consumers to share information, such as posting reviews, recommendations and ratings. The proper way of communication improves the relationship and the level of trust between consumers and companies [27]. This means that consumers who interact more on social platforms with regards to their shopping experiences are likely to have greater trust in both online shopping as well as in these social commerce companies. In SC, communication is essential for consumers to be able to trust organizations as they exchange experiences and information. Thus, the following hypothesis is proposed:

H2. Communication has a positive impact on the level of Trust in SC.

[16] argues that in online shopping environments, consumers usually depend on the information provided on the website about a product or service due to having limited information available from other sources. [28] claims that correct and accurate information offered by companies on their websites has a direct impact on online consumer trust. Therefore, the following hypothesis is proposed:

H3. The Information Quality on SC websites has a positive impact on Trust.

When a company has a good reputation or image, its consumers usually have a high level of trust in its operations [29]. Hence, reputation is an important factor for any company if it aims to gain consumer trust. In addition, consumers are likely to share a company's reputation with others, and this plays a primary role in increasing trust [30].

SC site users are likely to consider the company's reputation when evaluating his or her level of trust before making any purchases of services or products. If the reputation of a social commerce company is good, the consumer is likely to have a high level of trust in it, which may lead to a purchase decision. Therefore, the following hypothesis is proposed:

H4. Acompany's Reputation has a positive impact on Trust in SC.

The level of security that any social commerce site provides, pertaining to both the actual purchase transaction as well as to the information related to it, is known as 'transaction safety'. It is clear that if a social commerce website is successful in providing its consumers with a high level of security, both during the purchase transaction and in terms of the information related to it, consumer trust is improved. [31] claims that e-commerce site users are commonly concerned about their online transaction safety. In the context of e-commerce, transaction safety is represented as a key factor that increases consumer levels of trust [32]. Consequently, high-level transaction safety provided to consumers will increase their levels of trust in SC sites and online shopping environments. Therefore, the following hypothesis is proposed:

H5. Transaction Safety has a positive impact on Trust in SC.

Word-of-Mouth (WOM) is defined as consumers' exchanging experiences and information to help one another to make purchasing decisions [19]. Consumers are likely to hear what others have said about a product or a service rather than watching or listening to an advertisement [33]. Therefore, according to [34], virtual communities have and will continue to have an impact on the decision-making process of online consumers. [35] indicates that consumers usually prefer to seek WOM before they purchase a product due to their trust in the people sharing the information. [22] claims that WOM is a key factor that influences a consumer's trust in SNS. According to [36], more and more business enterprises are exploring the potential of social commerce websites to convey their "promotional information to consumers" in order to enhance their brand awareness through the use of e-WOM. Therefore, the following two hypotheses are proposed:

H6. WOM has a positive impact on Trust in SC
H7. WOM has a positive impact on Social Influence in SC.

Consumers of various cultures have different social characteristics that affect how they are influenced by a social commerce site. [37]. Since culture alters consumer behavior and preferences, it can have a major influence on both consumption and the buying decision [38]. Hence, cultural differences define individuals' aptitude regarding short- or long-term orientation. According to [37], long-term orientation is the most important dimension related to trust and the recommendations of others in social commerce. Therefore, the following hypothesis is proposed:

H8. Long-term Orientation has a positive impact on the level of Social Influence in SC.

4 Research Methodology

This study tested various hypotheses and endeavoured to answer the research questions through the development of a questionnaire as a research instrument. The researchers developed the research instrument and the items on it based on previous studies. A seven-point Likert scale (1–7) was applied to measure all items, using 1 = strongly disagree through to 7 = strongly agree to reflect the respondents' level of agreement. All sources of the constructs and related items used for the measurement of each of these constructs are presented in Table 1. The study samples include Saudi Arabian students who are studying either at the University of Business and Technology (UBT) in Saudi Arabia or at other universities in Australia, who have experience with SC sites. Two separate groups were included to see if there is any difference in the use of SC sites between residents of the two countries. For the Australian sample, data were collected from Saudi social groups via Twitter and Facebook, with the help of a survey link distributed on the Qualtrics platform. In total, 164 Australian students started the survey and 72 completed it. For the Saudi sample, data were obtained by distributing paper copies of the survey to students at UBT. The respondents were selected by choosing classes at random and asking each class member to complete the survey. Of the 500 students who were included, completed surveys were obtained from 300 respondents. A total of 314 valid responses were collected in the two locations. The respondents were 64.8% male and 35.2% female. More than half, 50.3%, were between the ages of 26 and 35, while 24.3% fell between the ages of 36 and 45. Respondents' education level was generally high, with 32.8% being graduate students and 61.8% being postgraduate students. This study made use of the Smart PLS software program to analyse the measurement and structural models, particularly for the "Partial Least Squares Structural Equation Modelling" (PLS-SEM) techniques [39].

Table 1. Summary of constructs' Factor loadings, Reliability and the Average Variance Extracted values

Construct	Range of factor loading within constructs	Composite reliability $\alpha > 0.7$	AVE ≥ 0.5	Items' source
Trust	0.93–0.96 (4 Items)	0.92	0.62	[16, 42–44]
Social Influence	0.85–0.91 (6 Items)	0.95	0.74	[45–47]
Communication	0.75–0.81 (4 Items)	0.87	0.63	[48, 49]
Information Quality	0.85–0.92 (8 Items)	0.97	0.80	[16, 48, 50]
Reputation	0.89–0. 96 (5 Items)	0.96	0.86	[16, 51, 52]
Transaction Safety	0.63–0.82 (7 Items)	0.92	0.62	[32, 53, 54]
Word-of-Mouth	0.92–0.88 (7 Items)	0.97	0.82	[44, 55, 56]
Long-Term Orientation	0.62–0.87 (4 Items)	0.85	0.59	[57, 58]

5 Data Analysis Results and Findings

To assess the indicator reliability via the PLS algorithm calculation, the researchers used the common path weighting scheme to set the inner weighting option. As discussed by [40], this scheme is preferable to the centroid weighting and factorial weighting schemes. The high iterations' number is 50. The factor loadings of the measurement items were ascertained with the help of this procedure. To accomplish an item reliability, the item loadings should be at least 0.5, as indicated by [41]. Table 1. shows the summary of construct factor loadings, reliability and Average Variance Extracted (AVE) values.

Another approach used to evaluate the reliability of internal consistency was Composite reliability. As shown in Table 1 above, all constructs displayed composite reliability of more than 0.7, a result which is considered satisfactory [41]. Sufficient convergent validity is achieved when the AVE value of a construct is at least 0.5 [59]. As shown in Table 1, the AVE values for all constructs are between 0.59 and 0.86. They thus fulfill the 0.5 threshold and demonstrate convergent validity.

Discriminant validity helps measure the degree to which an individual construct is related to its own measurement items, in comparison with those intended to measure other constructs. Fornell-Larcker standards and the investigation of the cross loadings are the prevalent approaches used to determine discriminant validity. These measurements are shown in Table 2, alongside the square root of the AVE values.

As shown by [60], the Heterotrait-Monotrait Correlations Ratio (HTMT) can also be used to evaluate discriminant validity. These values are indicated in Table 3 for the real measurement model framework. If the HTMT value is below 0.90, this indicates discriminant validity between two reflective constructs. In the original framework, the HTMT ratio for all the relationships is less than 0.90. However, the HTMT ratio for the relationship between 'Word-of-Mouth' and 'Trust,' is 0.888. For the relationship between 'Trust' and 'Information Quality,' the HTMT value is 0.895. This result is very close to the threshold level of 0.90. Since these instances barely meet the desired level, discriminant validity cannot be assumed.

After analysing the relevance and importance of the path coefficients, the researchers determined their explanatory power. This was done by determining the coefficient, R^2. The R^2 values for Trust and Social Influence are 0.845, 0.682. These are considered strong and substantial respectively [61]. Predictive relevance is used as a measurement of the predictive value of a model. The results obtained are indicated in Table 4. They show that Word-of-Mouth, Information Quality and Reputation, in that order, are supported as predictors of trust. Moreover, Word-of-Mouth is supported as being the strongest predictor for both Trust and Social Influence.

The results can offer meaningful insights to businesses attempting to use information quality, reputation, word-of-mouth, trust and social influence for their competitive advantage. This study contributes to the existing social commerce literature in two major ways. Firstly, this study enhances our understanding of the factors impacting customer experience by adding a tested theoretical framework. Secondly, the results of this study provide social commerce websites with a theoretical model to improve their services and satisfy their customers.

Table 2. The Fornell-Larcker Criterion

Constructs	Communication	Information Quality	Long-Term Orientation	Reputation	Social Influence	Transaction Safety	Trust	Word-of-Mouth
Communication	0.794							
Information Quality	0.134	0.898						
Long-Term Orientation	0.340	0.019	0.770					
Reputation	0.040	0.744	−0.018	0.928				
Social Influence	0.161	0.767	0.035	0.634	0.883			
Transaction Safety	0.079	0.570	0.032	0.576	0.586	0.791		
Trust	0.101	0.862	0.023	0.738	0.750	0.603	0.945	
Word-of-Mouth	0.161	0.772	0.080	0.645	0.821	0.613	0.857	0.905

Table 3. The Heterotrait-Monotrait Correlations Ratio

Constructs	Communication	Information Quality	Long-Term Orientation	Reputation	Social Influence	Transaction Safety	Trust	Word-of-Mouth
Communication								
Information Quality	0.143							
Long-Term Orientation	0.518	0.030						
Reputation	0.051	0.773	0.056					
Social Influence	0.181	0.798	0.038	0.660				
Transaction Safety	0.079	0.449	0.087	0.494	0.439			
Trust	0.103	0.895	0.027	0.768	0.780	0.436		
Word-of-Mouth	0.176	0.798	0.080	0.670	0.858	0.438	0.888	

Table 4. The Results Obtained from the Analysis of the Structural Equation Model

Hypothesis	Path	P-value	Decision
H1	Trust → Social Influence	0.007	Supported
H2	Communication → Trust	0.187	Rejected
H3	Information Quality → Trust	0.000	Supported
H4	Reputation → Trust	0.030	Supported
H5	Transaction Safety → Trust	0.471	Rejected
H6	Word-of-Mouth → Trust	0.000	Supported
H7	Word-of-Mouth → Social Influence	0.000	Supported
H8	Long-Term Orientation → Social Influence	0.566	Rejected

6 Discussion of the Major Findings

This study was aimed at providing a better understanding of the factors that affect Trust and Social Influence, and the links between these factors, in the context of social commerce. The results indicated support for the proposed model of this study. Specifically, it was found that Word-of-Mouth was the dimension which impacted both Trust and Social Influence most strongly.

Trust was found to be the dimension which impacted Social Influence most strongly; hence, it is clear that increasing the trust of consumers in social commerce websites will automatically increase the satisfaction of the consumer. As Trust was most strongly impacted by Word-of-Mouth it is imperative that social commerce enterprises pay special attention to consumer feedback and reviews, as Word-of-Mouth will determine the level of Trust that consumers have in their products and services.

This study and its findings are a valuable addition to the body of knowledge in social commerce which deals with both trust and social influence. It offers evidence to prove that factors such as information quality, reputation and word-of-mouth impact Trust and Social Influence where social commerce is concerned. Social commerce enterprises need to monitor the nature of Word-of-Mouth their businesses are generating online in order to keep track of how the levels of consumer trust and social influence are evolving. This, in the end, will determine the success of their enterprise.

The results obtained fully supported the theoretical model with the exception of one relationship; the relationship between Transaction Safety and Trust. Customers may be unaware of the risks of unsafe transactions or, conversely, they may believe that all SC transactions carry a risk of fraud or of information being compromised. Such explanations could prevent customers from being strongly influenced by knowledge of a specific merchant's safety features.

Ultimately, it was found that Word-of-Mouth was the dimension which most strongly influenced customer experience in the social commerce context.

The findings obtained must be interpreted within the limitations of the study. One such limitation was that the survey included only students; therefore, the findings from this research may not be generalizable to all consumers.

7 Conclusion and Recommendations for Further Studies

This study is intended to improve the business community's understanding of the factors affecting Trust and Social Influence in the social commerce context. It shows the interrelationships between key factors such as Word-of-Mouth, Information Quality and Reputation. In addition, this study provides evidence for aspects of social commerce which have previously seen limited research. The phenomenon of social commerce has been studied from the standpoint of consumer intention and behaviour, but there has been little literature available as far as Trust and Social Influence in social commerce is concerned. This study has worked to bridge this gap by defining the concepts of Trust and Social Influence in social commerce settings.

Suggestions for future studies include: (1) enhancing the study model by adding new antecedents and variables of consumer experience; (2) exploring the hypothesized relationships via new techniques of investigation; and (3) exploring the relationships among the variables in this study.

References

1. Hajli, M.: A research framework for social commerce adoption. Inf. Manag. Comput. Secur. **21**, 144–154 (2013)
2. Yang, C.C., Tang, X., Dai, Q., Yang, H., Jiang, L.: Identifying implicit and explicit relationships through user activities in social media. Int. J. Electron. Commer. **18**, 73–96 (2013)
3. Mardsen, P.: Social commerce: monetizing social media, Germany. Syzygy Deutschland Gmbh, Hamburg (2010)
4. Zhang, J., Liu, H., Sayogo, D.S., Picazo-Vela, S., Luna-Reyes, L.: Strengthening institutional-based trust for sustainable consumption: lessons for smart disclosure. Gov. Inf. Q. **33**, 552–561 (2016)
5. Esmaeili, L., Mutallebi, M., Mardani, S., Golpayegani, S.A.H.: Studying the affecting factors on trust in social commerce. arXiv preprint arXiv:1508.04048 (2015)
6. Kim, S., Noh, M.-J.: Determinants influencing consumers' trust and trust performance of social commerce and moderating effect of experience. Inf. Technol. J. **11**, 1369 (2012)
7. Hajli, N., Sims, J., Zadeh, A.H., Richard, M.-O.: A social commerce investigation of the role of trust in a social networking site on purchase intentions. J. Bus. Res. **71**, 133–141 (2017)
8. Friedkin, N.E.: A Structural Theory of Social Influence. Cambridge University Press, Cambridge (2006)
9. Wu, H.-L., Wang, J.-W.: An empirical study of flow experiences in social network sites. In: 15th Pacific Asia Conference on Information Systems: Quality Research in Pacific, pp. 1–11. PACIS (2011)
10. Ghafourian, M., et al.: Universal navigation through social networking. In: Ozok, A., Zaphiris, P. (eds.) Conference 2009, LNCS, vol. 5621, pp. 13–22. Springer, Heidelberg (2009)
11. Hajli, M., Khani, F.: Establishing trust in social commerce through social word of mouth. In: 7th International Conference on e-Commerce in Developing Countries: With Focus on e-Security (ECDC), pp. 1–22. IEEE (2013)

12. Kim, D.J., Ferrin, D.L., Rao, H.R.: A study of the effect of consumer trust on consumer expectations and satisfaction: the Korean experience. In: Proceedings of the 5th International Conference on Electronic commerce, pp. 310–315. ACM Press, New York (2003)
13. Lubbe, S.: The Economic and Social Impacts of E-commerce. IGI Global, Hershey (2002)
14. Huang, Z., Benyoucef, M.: From e-commerce to social commerce: a close look at design features. Electron. Commer. Res. Appl. **12**, 246–259 (2013)
15. Einwiller, S.: When reputation engenders trust: an empirical investigation in business-to-consumer electronic commerce. Electron. Mark. **13**(3), 196–209 (2003)
16. Kim, D.J., Ferrin, D.L., Rao, H.R.: A trust-based consumer decision-making model in electronic commerce: the role of trust, perceived risk, and their antecedents. Decis. Support Syst. **44**(2), 544–564 (2008)
17. Sharma, S., Menard, P., Mutchler, L.A.: Who to trust? Applying trust to social commerce. J. Comput. Inf. Syst. **16**, 1–11 (2017)
18. Kanyaru, P.M., Kyalo, J.K.: Factors affecting the online transactions in the developing countries: a case of e-commerce businesses in Nairobi County, Kenya. J. Educ. Policy Entrep. Res. **2**, 1–7 (2015)
19. Kim, K., Prabhakar, B.: Initial trust, perceived risk, and the adoption of internet banking. In: Proceedings of the 21st International Conference on Information systems, pp. 537–543. Association for Information Systems (2000)
20. Mikalef, P., Giannakos, M., Pateli, A.: Shopping and word-of-mouth intentions on social media. J. Theor. Appl. Electron. Commer. Res. **8**, 17–34 (2013)
21. Hoftede, G., Hofstede, G.J., Minkov, M.: Cultures and Organizations: Software of the Mind: Intercultural Cooperation and Its Importance for Survival. McGraw-Hill, New York (2010)
22. Kuan, H.-H., Bock, G.-W.: Trust transference in brick and click retailers: an investigation of the before-online-visit phase. Inf. Manag. **44**, 175–187 (2007)
23. Swamynathan, G., Wilson, C., Boe, B., Almeroth, K., Zhao, B.Y.: Do social networks improve e-commerce?: a study on social marketplaces. In: Proceedings of the 1st workshop on Online Social Networks, pp. 1–6. ACM Press, New York (2008)
24. Kelman, H.C.: Further thoughts on the processes of compliance, identification, and internalization. Perspect. Soc. Power 125–171 (1974)
25. Tsai, W., Ghoshal, S.: Social capital and value creation: the role of intrafirm networks. Acad. Manag. J. **41**, 464–476 (1998)
26. Moon, Y., Lee, I.: A study on the performance of online community reputation, social presence, interactivity, playfulness: mediating role of trust and flow. Bus. Stud. **9**, 75–99 (2008)
27. Moorman, C., Zaltman, G., Deshpande, R.: Relationships between providers and users of market research: the dynamics of trust within and between organizations. J. Mark. Res. (JMR) **29**, 314–328 (1992)
28. Fung, R., Lee, M.: EC-trust (trust in electronic commerce): exploring the antecedent factors. AMCIS 1999 Proceedings, pp. 517–519
29. Park, J., Gunn, F., Han, S.-L.: Multidimensional trust building in e-retailing: cross-cultural differences in trust formation and implications for perceived risk. J. Retail. Consum. Serv. **19**, 304–312 (2012)
30. Teo, T.S., Liu, J.: Consumer trust in e-commerce in the United States, Singapore and China. Omega **35**, 22–38 (2007)
31. Koufaris, M., Hampton-Sosa, W.: The development of initial trust in an online company by new customers. Inf. Manag. **41**, 377–397 (2004)
32. Yoon, S.J.: The antecedents and consequences of trust in online-purchase decisions. J. Interact. Mark. **16**, 47–63 (2002)

33. Park, J., Chaiy, S., Lee, S.: The moderating role of relationship quality in the effect of service satisfaction on repurchase intentions. Korea Mark. Rev. **13**, 119–139 (1998)
34. Evans, P., Beardmore, A., Page, K., Osborne, J., O'Brien, P., Willingale, R., Starling, R., Burrows, D., Godet, O., Vetere, L.: Methods and results of an automatic analysis of a complete sample of Swift-XRT observations of GRBs. Mon. Not. R. Astron. Soc. **397**, 1177–1201 (2009)
35. Brown, J.J., Reingen, P.H.: Social ties and word-of-mouth referral behavior. J. Consum. Res. **14**, 350–362 (1987)
36. De Vries, L., Gensler, S., Leeflang, P.S.: Popularity of brand posts on brand fan pages: an investigation of the effects of social media marketing. J. Interact. Mark. **26**, 83–91 (2012)
37. Ng, C.S.-P.: Examining the cultural difference in the intention to purchase in social commerce. In: PACIS Proceedings, Paper 163 (2012)
38. Abălăesei, M.: The influence of culture on the role of social media in decision making, pp. 600–612 (2015)
39. Ringle, C.M., Sarstedt, M., Straub, D.: A critical look at the use of PLS-SEM in MIS quarterly. MIS Q. (MISQ) **36**, 3–14 (2012)
40. Vinzi, V.E., Trinchera, L., Amato, S.: PLS path modeling: from foundations to recent developments and open issues for model assessment and improvement. In: Handbook of Partial Least Squares, pp. 47–82. Springer, Berlin (2010)
41. Hair, J.F., Ringle, C.M., Sarstedt, M.: PLS-SEM: indeed a silver bullet. J. Mark. Theory Pract. **19**, 139–152 (2011)
42. Gefen, D.: E-commerce: the role of familiarity and trust. Omega **28**, 725–737 (2000)
43. Kassim, N., Abdullah, N.A.: The effect of perceived service quality dimensions on customer satisfaction, trust, and loyalty in e-commerce settings: a cross cultural analysis. Asia Pac. J. Mark. Logist. **22**, 351–371 (2010)
44. Hajli, M.: Social commerce: the role of trust. In: Eighteenth Americas Conference on Information Systems, Seattle, Washington, 9–12 August 2012, pp. 1–11 (2012)
45. Xu-Priour, D.-L., Truong, Y., Klink, R.R.: The effects of collectivism and polychronic time orientation on online social interaction and shopping behavior: a comparative study between China and France. Technol. Forecast. Soc. Chang. **88**, 265–275 (2014)
46. Hsu, C.-L., Lin, J.C.-C.: Acceptance of blog usage: the roles of technology acceptance, social influence and knowledge sharing motivation. Inf. Manag. **45**, 65–74 (2008)
47. Liang, T.-P., Ho, Y.-T., Li, Y.-W., Turban, E.: What drives social commerce: the role of social support and relationship quality. Int. J. Electron. Commer. **16**, 69–90 (2011)
48. Kim, S., Park, H.: Effects of various characteristics of social commerce (s-commerce) on consumers' trust and trust performance. Int. J. Inf. Manag. **33**, 318–332 (2013)
49. Kumar, N., Benbasat, I.: Para-social presence and communication capabilities of a web site: a theoretical perspective. E-serv. J. **1**, 5–24 (2002)
50. Barnes, S.J., Vidgen, R.T.: An integrative approach to the assessment of e-commerce quality. J. Electron. Commer. Res. **3**, 114–127 (2002)
51. Jarvenpaa, S.L., Tractinsky, N., Saarinen, L.: Consumer trust in an internet store: a cross-cultural validation. J. Comput. Mediat. Commun. **5**, 0–0 (1999)
52. Doney, P.M., Cannon, J.P.: Trust in buyer-seller relationships. J. Mark. **61**, 35–51 (1997)
53. Shergill, G.S., Chen, Z.: Web-based shopping: consumers' attitudes towards online shopping in New Zealand. J. Electron. Commer. Res. **6**, 79–94 (2005)
54. Kim, W.G., Kim, D.J.: Factors affecting online hotel reservation intention between online and non-online customers. Int. J. Hosp. Manag. **23**, 381–395 (2004)
55. Srinivasan, S.S., Anderson, R., Ponnavolu, K.: Customer loyalty in e-commerce: an exploration of its antecedents and consequences. J. Retail. **78**, 41–50 (2002)

56. Harrison-Walker, L.J.: The measurement of word-of-mouth communication and an investigation of service quality and customer commitment as potential antecedents. J. Serv. Res. **4**, 60–75 (2001)
57. Wang, C.L., Siu, N.Y., Barnes, B.R.: The significance of trust and renqing in the long-term orientation of Chinese business-to-business relationships. Ind. Mark. Manag. **37**, 819–824 (2008)
58. Ganesan, S.: Determinants of long-term orientation in buyer-seller relationships. J. Mark. **58**, 1–19 (1994)
59. Fornell, C., Larcker, D.F.: Evaluating structural equation models with unobservable variables and measurement error. J. Mark. Res. **18**, 39–50 (1981)
60. Henseler, J., Ringle, C.M., Sarstedt, M.: A new criterion for assessing discriminant validity in variance-based structural equation modeling. J. Acad. Mark. Sci. **43**, 115–135 (2015)
61. Chin, W.W.: How to write up and report PLS analyses. In: Handbook of Partial Least Squares, pp. 655–690. Springer, Berlin (2010)

Determining Underlying Factors that Influence Online Social Network Usage Among Public Sector Employees in the UAE

Ali Ameen(✉), Hamad Almari, and Osama Isaac

Lincoln University College, Petaling Jaya, Malaysia
ali.ameen@aol.com

Abstract. Recently, there has been a rising concern over technology dependency, especially with the revolution of online social networking, which becomes the biggest phenomenon. It harnesses a web application to build social relations with other people who share similar personal or career interests, activities, backgrounds or real-life connections. There is a lot of research, which have been done in the field of Online Social Network Usage, but studies indicate the lack of research has been conducted for determining underlying factors that influence online social network usage among public sector employees in the UAE. This study employs structural equation modeling via AMOS to analyze 401 valid questionnaires for the assessment of the proposed model that is based on Unified Theory of Acceptance and Use of Technology to identify the factors that affect the use of online social networks (OSN). The current research focused in one of Abu Dhabi's public organizations, the Tourism Development and Investment Company. The main independent constructs in the model are related to performance, effort, social influence, and facilitating conditions about the use of OSN. The dependent constructs cover actual utilization of OSN. This paper describes the relations among the various constructs. Our work improved our insights on the online social networking model. Results indicated all four independent variables significantly predicted the actual usage of OSN with various percentages. The proposed model explained 50% of the variance in actual usage.

Keywords: Unified theory of acceptance and use of technology
UTAUT · Online social networks · UAE

1 Introduction

Change in information technology (IT), one of the most salient and active factors influencing employees and organizations today, grows substantially and itself becomes a driver of further change [1]. One of the technology applications of interest involves online social networks (OSN) (also known as social networking services, social networking sites, social media, social media sites, social media platforms), which, in turn, constitutes a phenomenon that remains under debate [2]. OSN is among the fast-growing internet platforms which is growing fast. Some studies link the internet and OSN to organization performance [3, 4], and others emphasize how internet platforms

F. Saeed et al. (Eds.): IRICT 2018, AISC 843, pp. 945–954, 2019.
https://doi.org/10.1007/978-3-319-99007-1_87

improve knowledge acquisition, task efficiency, communication quality, and decision quality in organizations [5]. The growth of OSN users worldwide in leading platforms such as Facebook, Twitter, WhatsApp, and LinkedIn need initiatives and investigations to clarify this issue and understand how organizations benefit from such platforms. Extant sources [6, 7] indicate that OSN in the Arab world are perceived as having numerous positive aspects that enhance the quality of life of individuals, business profitability, and governmental interaction with the public.

Furthermore, active social network penetration occurs in the top countries in the world (see Fig. 1). Comparison of the number of active accounts on the top social networks in each country to the population shows that the United Arab Emirates (UAE) ranked first with a social network usage penetration of 99% [8], indicating a huge opportunity that must be exploited in both public and private sectors to improve professional practice, personal development, and quality of working life [9].

The context of this study is the Tourism Development and Investment Company (TDIC), a public organization in Abu Dhabi that faces the challenges of promotion and attracting visitors from all over the world. Investigating the factors that influence OSN use in the tourism industry would be of tremendous benefit to the said UAE government organization. Hence, this research aims to investigate the influence of performance expectancy, effort expectancy, social influence, and facilitating conditions on the actual usage of OSN in the TDIC in Abu Dhabi.

2 Literature Review

2.1 Performance Expectancy (PE)

Performance expectancy is defined as the degree to which an individual believes using the system will help him or her to attain gains in job performance [10]. Many studies have proven that the higher the performance expectancy, the higher the actual system usage [11]. Therefore, the following hypothesis is proposed:

H1: Performance expectancy has a positive effect on online social network usage.

2.2 Effort Expectancy (EE)

Venkatesh [10] has defined effort expectancy or perceived ease of use as the degree of ease associated with the use of the system. Numerous studies explored the influence of effort expectancy on actual system usage. According to Martins, Oliveira, and Popovič [12], a positive relationship seems to exists between effort expectancy and system usage in the context of internet banking. Many investigations in different contexts and technological applications have also emphasized such relationship [13]. Therefore, the following hypothesis is proposed:

H2: Effort expectancy has a positive effect on online social network usage.

2.3 Social Influence (SI)

Another important factor that affects actual system usage is social influence, which is described broadly as the degree to which an individual perceives that important others believe he or she should use the new system [10]. The social influence factor is substantial in the context of technology success [12]. According to Venkatesh, Thong, and Xu [14], in the context of information systems usage, a positive relationship exists between social influence and usage behavior. Therefore, the following hypothesis is proposed:

H3: Social influence has a positive effect on online social network usage.

2.4 Facilitating Conditions (FC)

Facilitating conditions are defined in this study as the degree to which an individual believes organizational and technical infrastructure exists to support the use of the system [10]. In the IS domain, facilitating conditions are essential for influencing individuals to use the system [15, 16]. To guarantee that the technology is utilized fully, appropriate introduction and back-up are required before an organization can expect its employees to adopt any innovations [17]. A previous study showed that facilitating conditions had a positive influence on usage behavior within the context of learning management software in Malaysia [18]. Therefore, the following hypothesis is proposed:

H4: Facilitating conditions have a positive effect on online social network usage.

2.5 Actual Usage (USE)

Actual usage is defined as the degree and manner in which users utilize the capabilities of an information system. Examples of actual usage relates to the amount, frequency, nature, appropriateness, extent, and purpose of use [19]. Kim [20] suggested actual usage reflects the usage frequency of the technology and usage times.

3 Research Methodology

The relationships between the constructs hypothesized in the conceptual framework have been adapted from relevant literature. Figure 1 shows the proposed model for this study. Developing the instrument for this study involved a 17-item questionnaire and a multi-item Likert scale was applied based on the information systems literature [21]. Constructs were measured using a Likert scale with 5 being “Strongly Agree” and 1 being “Strongly Disagree.” Data were collected by delivering self-administered questionnaires personally to employees within the Public Sector in the UAE from July to November 2017. From the total of 750 distributed questionnaires, 443 were returned, among which 401 were considered suitable for the analysis.

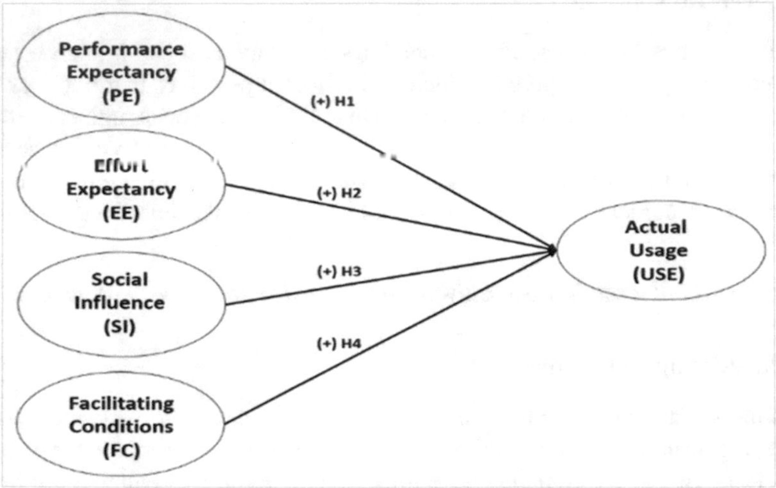

Fig. 1. Proposed conceptual framework

4 Data Analysis and Results

For this study, the main reasons for choosing structural equation modeling (SEM) as an analytical technique is that it offers simultaneous analysis that leads to more accurate estimates [22–24].

4.1 Measurement Model Assessment and Confirmatory Factor Analysis (CFA)

As shown in Table 1, all *goodness-of-fit indices* surpassed acceptance levels as recommended by previous research, thereby indicating that the measurement model displayed a good fit with the data collected.

As regards *construct reliability,* the results show all individual Cronbach's alpha coefficients are bigger than the suggested level of 0.7 [35]. Furthermore, in assessing construct reliability, all values of the composite reliability (CR) were greater than the advised value of 0.7 [25, 36]. This result confirms that construct reliability has been achieved (Table 2). To determine *indicator reliability,* factor loadings were observed [37]. The loadings for all items surpassed the suggested value of 0.5, and thus, the loadings for all items are fulfilled excluding item PE4, which was removed because of low loading. Moreover, to examine *convergent validity* average variance extracted (AVE) was used, and all AVE values were greater than the suggested value of 0.50 [38]. Therefore, sufficient convergent validity was demonstrated successfully (Table 2).

Table 1. Goodness-of-fit indices for the measurement model

Fit Index	Cited	Admissibility	Result	Fit (yes/no)
X^2			110.560	
DF			94	
P value		>.05	.000	No
X^2/DF	[25]	1.00–5.00	**1.176**	**Yes**
RMSEA	[26]	<.08	**.021**	**Yes**
SRMR	[27]	<.08	.031	Yes
GFI	[28]	>.90	.967	Yes
AGFI	[28]	>.80	.952	Yes
NFI	[29]	>.80	.980	Yes
PNFI	[29]	>.05	.768	Yes
IFI	[30]	>.90	.997	Yes
TLI	[31]	>.90	.996	Yes
CFI	[32]	>.90	**.997**	**Yes**
PGFI	[33]	>.50	.668	Yes

Note: X^2 = Chi Square, DF = Degree of freedom, CFI = Comparative-fit-index, RMSEA = Root mean square error of approximation, SRMR = Standardized root mean square residual, GFI = Goodness-of-fit, NFI = Normed fit index, AGFI = Adjusted goodness of fit index, IFI = Increment fit index, TLI = Tucker–Lewis coefficient index, PNFI = Parsimony normed fit index.

- The indexes in bold are recommended because they are frequently reported in the literature [34]

Table 2. Measurement assessment

Constructs	Item	Loading (>0.5)	M	SD	α (>0.7)	CR (>0.7)	AVE (>0.5)
Performance expectancy (PE)	PE1 PE2 PE3 PE4	0.88 0.88 0.89 Deleted	3.405	1.025	0.914	0.914	0.781
Effort expectancy (EE)	EE1 EE2 EE3	0.89 0.87 0.89	3.395	1.037	0.914	0.914	0.781

(*continued*)

Table 2. (*continued*)

Constructs	Item	Loading (>0.5)	M	SD	α (>0.7)	CR (>0.7)	AVE (>0.5)
Social influence (SI)	SI1	0.88	3.259	0.996	0.903	0.904	0.703
	SI2	0.73					
	SI3	0.85					
	SI4	0.89					
Facilitating conditions (FC)	FC1	0.90	3.333	1.091	0.927	0.927	0.808
	FC2	0.90					
	FC3	0.90					
Actual Usage (USE)	USE1	0.83	3.201	0.969	0.886	0.887	0.723
	USE2	0.86					
	USE3	0.86					

Note: M = Mean, SD = Standard deviation, AVE = Average variance extracted, CR = Composite reliability, α = Cronbach's alpha.

4.2 Structural Model Assessment

The hypotheses of this study were tested using SEM via AMOS (Fig. 2). The structural model assessment shown in Table 3 provides indication of the hypotheses tests, with four out of the four hypotheses of this study being supported. Performance expectancy ($\beta = .20$, $p < 0.05$), effort expectancy ($\beta = .17$, $p < 0.05$), social influence ($\beta = .19$, $p < 0.05$) and facilitating conditions ($\beta = .22$, $p < 0.05$) all have a positive impact on actual usage of OSN. Therefore, H1, H2, H3, and H4 are supported. Note that the standardized path coefficient indicates the strengths of the relationship between independent and dependent variables, and thus, the direct effects of facilitating conditions on actual usage of OSN is stronger than those of the other independent variables. The R^2 from the structural model showed all R^2 values are sufficiently high such that the model can fulfill a satisfactory level of explanatory power [39].

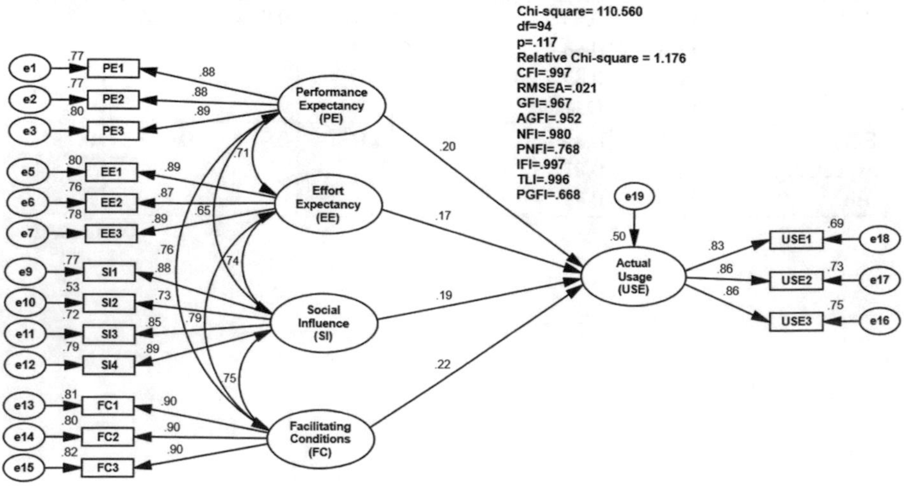

Fig. 2. Structural model results

Table 3. Structural path analysis results

Hypothesis	Dependent variables		Independent variables	Estimate B (path coefficient)	S.E	C.R (t-value)	p-value	Decision
H1	USE	←	PE	.20	.059	2.749	0.006	Supported
H2	USE	←	EE	.17	.072	2.063	0.039	Supported
H3	USE	←	SI	.19	.062	2.571	0.010	Supported
H4	USE	←	FC	.22	.075	2.386	0.017	Supported

S.E = Standard error, C.R = Critical ratio
Key: PE: performance expectancy, EE: effort expectancy, SI: social influence, FC: facilitating conditions, USE: actual usage

5 Discussion

Through the proposed model, this study improves understanding of the role played by the characteristics of technology systems in terms of performance and effort along with the impact of society and facilitating conditions in predicting the actual use of OSN among employees of the TDIC in Abu Dhabi and highlights relevant implications. The discussions are detailed further in the following.

The study found that performance expectancy has a positive effect on the actual usage of OSN among respondents, and this finding is supported by previous studies [40, 41]. This outcome is explained by the fact that the more the OSN are useful for improvements in work, daily life, user productivity, and user performance, the more likely the user is to utilize OSN regularly and employ them as communication platforms, aside from making users more willing to promote OSN to his/her colleagues. Likewise, effort expectancy positively affects actual usage of OSN among respondents, and this outcome is supported by previous studies [12, 42]. This result is explained by the fact that the more OSN are easy to use, flexible to interact with, and perceived as clear and understandable to engage with, the more likely the user is to regularly utilize OSN and employ them as communication platforms, aside from making users more willing to promote OSN to his/her colleagues.

Additionally, social influence was found to have a positive effect on the actual usage of OSN among respondents, and this outcome is supported by previous studies [43, 44]. This result is explained by the fact that the more supervisors promote the use of OSN along with colleagues surrounding the employee, in addition to his/her family and close friends, the more likely the user is to regularly utilize OSN and employ them as communication platforms, aside from making users more willing to promote OSN to his/her colleagues.

Finally, facilitating conditions was found to have a positive effect on the actual usage of OSN among respondents, and this finding is supported by previous studies [45]. This result is explained by the fact that the more the organization is well equipped, dedicates resources to provide necessary hardware and software to use OSN, and provides training and guidelines for its employees to ensure they have the necessary knowledge and skills to use the technology, the more likely the user is to regularly utilize OSN and employ them as communication platforms, aside from making users more willing to promote OSN to his/her colleagues.

6 Conclusion

While the government institutions of the United Arab Emirates are ahead of its regional counterparts in terms of performance, the country is striving to enhance its public organizations' productivity [46]. The findings of this study could be considered as one of the initiatives to serve that aim. The main objective of this research is to determine factors that affect the use of OSN within the TDIC of Abu Dhabi. Despite various constraints to the study, the outcomes have been encouraging, as this work managed to shed some light on a new perspective. This research proposed a model which includes performance expectancy, effort expectancy, social influence, and facilitating conditions as independent variables and actual usage of OSN as the dependent variable. Findings revealed the four independent variables significantly explain 50% of actual usage of OSN. The implications of this study have been deliberated and some directions for future research have been suggested.

References

1. Kassim, N.M., Ramayah, T., Kurnia, S.: Antecedents and outcomes of human resource information system (HRIS) use. Int. J. Product. Perform. Manag. **61**(6), 603–623 (2012)
2. Cocosila, M., Igonor, A.: How important is the 'social' in social networking? A perceived value empirical investigation. Inf. Technol. People **28**(2), 366–382 (2015)
3. Wang, J., Hou F.: Research on the relationship between the internet usages and the organizational performance in the Taiwanese e-commerce business organizations. Inf. Sci. 17–25 (2003)
4. Chen, C.: Study on application of e-commerce and organizational performance in Taiwanese professional sports event promotion organizations. J. Hum. Resour. Adult Learn. **4**(1), 66–73 (2008)
5. Isaac, O., Abdullah, Z., Ramayah, T., Mutahar, A.M., Alrajawy, I.: Perceived usefulness, perceived ease of use, perceived compatibility, and net benefits: an empirical study of internet usage among employees in Yemen. In: The 7th International Conference Postgraduate Education (ICPE7), pp. 899–919 (2016)
6. Arab Social Media Report: Benefits of social media usage, Arab social media influencers summit (2015)
7. Abdulbaqi Ameen, A., Ahmad, K.: The role of finance information systems in anti financial corruptions: a theoretical review. In: 2011 International Conference on Research and Innovation in Information Systems, ICRIIS 2011, pp. 1–6 (2011)

8. Statista: Active social network penetration (2017)
9. Salmeen Al-Obthani, F., Abdulbaqi Ameen, A.: Towards customized smart government quality model. Int. J. Softw. Eng. Appl. **9**(2), 41–49 (2018)
10. Venkatesh, V., Morris, M.G., Davis, G.B., Davis, F.D.: User acceptance of information technology: toward a unified view. MIS Q. **27**(3), 425–478 (2003)
11. Zhou, T., Lu, Y., Wang, B.: Integrating TTF and UTAUT to explain mobile banking user adoption. Comput. Human Behav. **26**(4), 760–767 (2010)
12. Martins, C., Oliveira, T., Popovič, A.: Understanding the Internet banking adoption: a unified theory of acceptance and use of technology and perceived risk application. Int. J. Inf. Manag. **34**(1), 1–13 (2014)
13. Venkatesh, V., Thong, J.Y.L., Chan, F.K.Y., Hu, P.J.-H., Brown, S.A.: Extending the two-stage information systems continuance model: incorporating UTAUT predictors and the role of context. Inf. Syst. J. **21**(6), 527–555 (2011)
14. Venkatesh, V., Thong, J.Y.L., Xu, X.: Consumer acceptance and use of information technology: extending the unified theory of acceptance and use of technology. MIS Q. **36**(1), 157–178 (2012)
15. Guo, Y.: Moderating effects of gender in the acceptance of mobile SNS based on UTAUT model. Int. J. Smart Home **9**(1), 203–216 (2015)
16. Mukred, M., Yusof, Z.M., Mokhtar, U.A., Abdul Manap, N.: Electronic records management system adoption readiness framework for higher professional education institutions in Yemen. Int. J. Adv. Sci. Eng. Inf. Technol. **6**(6), 804 (2016)
17. Al-Maamari, Q.A., Muhammed Kassim, R.-N., Raju, V., Al-Tahitah, A., Ameen, A.A., Abdulrab, M.: Factors affecting individual readiness for change: a conceptual framework. Int. J. Manag. Hum. Sci. **2**(1), 13–18 (2018)
18. Raman, A., Don, Y.: Preservice teachers' acceptance of learning management software: an application of the UTAUT2 model. Int. Educ. Stud. **6**(7), 157–164 (2013)
19. DeLone, W.H., McLean, E.R.: Information Systems Success Measurement. now Publishers Inc., Boston (2016)
20. Kim, H.-W., Chan, H.C., Gupta, S.: Value-based adoption of mobile internet: an empirical investigation. Decis. Support Syst. **43**(1), 111–126 (2007)
21. Lee, B.C., Yoon, J.O., Lee, I.: Learners' acceptance of e-learning in South Korea: theories and results. Comput. Educ. **53**(4), 1320–1329 (2009)
22. Isaac, O., Abdullah, Z., Ramayah, T., Mutahar, A.M.: Internet usage and net benefit among employees within government institutions in Yemen: an extension of DeLone and McLean information systems success model (DMISM) with task-technology fit. Int. J. Soft Comput. **12**(3), 178–198 (2017)
23. Isaac, O., Abdullah, Z., Ramayah, T., Mutahar, A.M.: Internet usage within government institutions in Yemen: an extended technology acceptance model (TAM) with internet self-efficacy and performance impact. Sci. Int. **29**(4), 737–747 (2017)
24. Isaac, O., Masoud, Y., Samad, S., Abdullah, Z.: The mediating effect of strategic implementation between strategy formulation and organizational performance within government institutions in Yemen. Res. J. Appl. Sci. **11**(10), 1002–1013 (2016)
25. Kline, R.B.: Principles and Practice of Structural Equation Modeling, 3rd edn. The Guilford Press, New York (2010)
26. Steiger, J.H.: Structural model evaluation and modification: an interval estimation approach. Multivar. Behav. Res. **25**(2), 173–180 (1990)
27. Hu, L., Bentler, P.M.: Cutoff criteria for fit indexes in covariance structure analysis: conventional criteria versus new alternatives. Struct. Equ. Model. **6**, 1–55 (1999)

28. Jöreskog, K., Sörbom, D.: LISREL 8: Structural Equation Modeling with the SIMPLIS Command Language. Scientific Software International Inc., Chicago (1998)
29. Bentler, P.M., Bonnet, D.G.: Significance tests and goodness of fit in the analysis of covariance structures. Psychol. Bull. **88**(3), 588–606 (1980)
30. Bollen, K.A.: Overall fit in covariance structure models: two types of sample size effects. Psychol. Bull. **107**(2), 256–259 (1990)
31. Tucker, L.R., Lewis, C.: A reliability coefficient for maximum likelihood factor analysis. Psychometrika **38**(1), 1–10 (1973)
32. Byrne, B.M.: Structural Equation Modeling with AMOS: Basic Concepts, Applications, and Programming, 2nd edn. Routledge, Abingdon (2010)
33. James, L.R., Muliak, S.A., Brett, J.M.: Causal Analysis: Models, Assumptions and Data. Sage, Beverly Hills (1982)
34. Awang, Z.: Structural Equation Modeling Using AMOS. University Teknologi MARA Publication Center, Shah Alam (2014)
35. Kannana, V.R., Tan, K.C.: Just in time, total quality management, and supply chain management: understanding their linkages and impact on business performance. Omega Int. J. Manag. Sci. **33**(2), 153–162 (2005)
36. Gefen, D., Straub, D., Boudreau, M.-C.: Structural equation modeling and regression: guidelines for research practice. Commun. Assoc. Inf. Syst. **4**(1), 1–79 (2000)
37. Hair, J.F.J., Hult, G.T.M., Ringle, C., Sarstedt, M.: A Primer on Partial Least Squares Structural Equation Modeling (PLS-SEM), vol. 46(1–2), 2nd edn, p. 328. SAGE, London (2014)
38. Hair, J.F., Black, W.C., Babin, B.J., Anderson, R.E.: Multivariate Data Analysis. Pearson, Upper Saddle River (2010)
39. Urbach, N., Ahlemann, F.: Structural equation modelling in information systems research using partial least squares. J. Inf. Technol. Theory Appl. **11**(2), 5–40 (2010)
40. Chang, C.-C.: Library mobile applications in university libraries. Libr. Hi Tech **31**(3), 478–492 (2013)
41. Toh, C.H.: Assessing adoption of wikis in a Singapore secondary school: using the UTAUT model. In: Proceedings of 2013 IEEE 63rd Annual Conference of International Council of Educational Media, ICEM 2013, pp. 1–9 (2013)
42. Nysveen, H., Pedersen, P.E.: Consumer adoption of RFID-enabled services. Applying an extended UTAUT model. Inf. Syst. Front. (2014)
43. Gonzalez, G.C., Sharma, P.N., Galletta, D.: Factors influencing the planned adoption of continuous monitoring technology. J. Inf. Syst. **26**(2), 53–69 (2012)
44. Lian, J.-W.: Critical factors for cloud based e-invoice service adoption in Taiwan: an empirical study. Int. J. Inf. Manag. **35**(1), 98–109 (2015)
45. Aldholay, A.H., Isaac, O., Abdullah, Z., Alrajawy, I., Nusari, M.: The role of compatibility as a moderating variable in the information system success model: the context of online learning usage. Int. J. Manag. Hum. Sci. **2**(1), 9–15 (2018)
46. Global Innovation Index: Government institutions effectiveness: Yemen versus Arab countries: Rank among 143 countries, Cornell University, INSEAD, and the World Intellectual Property Organization (WIPO) (2016)

A Theoretical Framework to Build Trust and Prevent Fake News in Social Media Using Blockchain

Tee Wee Jing and Raja Kumar Murugesan(✉)

School of Computing and IT, Taylor's University, Subang Jaya, Malaysia
{WeeJing.Tee,RajaKumar.Murugesan}@Taylors.edu.my

Abstract. This study aims to provide an insight of implementing blockchain technology on social media to build public trust on credible news spread via major social media platforms in order to determine the truthfulness of source, and prevent spread of fake news. This research uses blockchain technology with advanced AI in social media platforms to verify news content for its credibility. This study provides high impact preventing negative impacts on individuals, society, and the world that is becoming rampant today.

Keywords: Blockchain · Social media · Trust · Fake news

1 Introduction

In today's age of technological innovation, we embrace change every day by integrating and synergizing with disruptive and exponential digital technologies. In the coming age of Technological Singularity, the cyber arms race of powerful algorithms has become a significant battlefront among countries. For the survival of human race a transparent, trustable and reliable technology is very important. The advanced and common use of social media has enabled easy access and rapid dissemination of information. This on the other hand also enables fake news to be spread extensively. This phenomena is becoming very serious, as it would shake the foundation of human trust resulting in extremely negative impacts on individuals, society, and the world. In recent investigation [1], Cambridge Analytica was found to illegally use personal data of about 87 million Facebook users for a major political campaign during the Donald Trump US presidential election. Therefore, fake news detection and prevention on social media has recently become an emerging interdisciplinary research that needs great attention and solution. Blockchain is an emerging technology that is capable to disrupt the publishing industry and rebuild trust with the media. Blockchain has to do with trust, or the lack of trust, which makes it great especially for the social media industry [2].

In affective computing and sentiment analysis, automatic fake news detection is a challenging problem in deception detection and establish the truth, that is essential in protecting democracy and civil liberties [3, 4]. In this research [5], multiple prediction tasks have been performed on fact-checked statements of varying levels of truth (graded deception) as well as a deeper linguistic comparison of differing types of fake

F. Saeed et al. (Eds.): IRICT 2018, AISC 843, pp. 955–962, 2019.
https://doi.org/10.1007/978-3-319-99007-1_88

news e.g. propaganda, satire and hoaxes. It is shown that fact-checking is indeed a challenging task. However, various lexical features can contribute to our understanding of the differences between more reliable and less reliable digital news sources.

Social media is a double-edged sword, where "the pen is mightier than the sword" [6]. The advance and wide use of social media by generation X, Y and Z has enabled extensive spread of fake news, which will cause extreme negative impacts to our society and the world today. As a result, fake news detection on social media has recently become an emerging and important research area. Fake news detection and prevention on state-of-the-art social media presents unique challenges that make existing algorithms used in current news media not effective. As a result, extensive research need to be conducted in order to find new and effective algorithms to detect and prevent fake news. We humans pride ourselves on being rational and logical thinkers with an inherent sense of morality and consciousness that guides our thoughts and actions towards the greater good. These virtues hold true across all levels of society, but collectively on a global scale, we as human beings often make self-destructive decisions. Among the many challenges related to studying credibility on social networks [7] are as follows:

(a) The complexity of social networks and the web creates difficulty in identifying resources for use in studying and assessing credibility and identifying fake news.
(b) Online Social Networks (OSNs) evolve dynamically over time and become very large in size, with various structures that make it difficult to obtain the relevant information to assess the credibility of users.
(c) The credibility of users are influenced continuously by various factors, for example changes in the topography, user preferences, and context.
(d) Malicious activities can evade and bypass existing spam filters through various means.
(e) The process of evaluating solutions is also a problem in terms of resources, because most researchers are limited in terms of the extent to which they can test their algorithms and solutions.

2 Literature Review

With the advent of Web 3.0 and evolution of social media, people are savvier to express their emotions and share their opinions on web and social media regarding day-to-day activities and global issues. Affective Computing is computing that relates to, arises from, or deliberately influences emotion or other affective phenomena [8]. The emerging fields of affective computing and sentiment analysis leverage on human–computer interaction, information retrieval, and multimodal signal processing for distilling people's sentiments from the exponential growth of big data from online social data. Sentiment analysis aka opinion mining, is used extensively in brand management and product promotion. In recent years, sentiment analysis has shifted from analyzing online product reviews to social media texts from Twitter and Facebook [9]. Many topics beyond product reviews like stock markets, elections, disasters, medicine, software development and cyberbullying extend the utilization of sentiment analysis [10].

Most recent application in affective computing and sentiment analysis is in the context of political fact-checking and fake news detection [11]. Recent analysis shows that there is a branch labeled as the "Truth", which contains research about detecting fake and spam opinions. Media and social media fact-checking remains to be an open research question in order to determine the truthfulness of source. Gartner top strategic predictions for 2018 and beyond is that IT leaders need to quickly develop Artificial Intelligence (AI) algorithm to solve counterfeit reality and fake news [12]. By 2022, it is predicted that the majority of individuals in mature economies will consume more false information than true information [13].

Blockchain is an emerging technology and has disrupted multiple industries including finance and real estate. As the technology becomes more matured in these industries, it will begin to disrupt other industries as well. Publishers can use blockchain technology to verify news content, streamline advertising, and rapidly adjust to trends. The main features of blockchain are trust and transparency, and decentralization, which is uniquely suited to help with social media ecosystems. As fake news spreads across the Internet and people question their trust with the media, blockchain media applications can be used to help ensure the information that is shared and published is credible. This could help organizations come up with new business models for the way publishing can work.

AI has grown exponentially in recent decade and has created many disruptive and innovative opportunity in various fields. Cryptocurrencies and blockchain, are topics of big conversation today. In this study, we provide an insight to converge blockchain and next-generation artificial intelligence technologies to rebuild trust and prevent fake news on social media. The seven reasons blockchain-based approach is beneficial to social networking are as follows:

2.1 Users Data are Protected

The exponential growth of social networks requires heavy investment in infrastructure. In order to sustain growth, platform owners need to use users' data to analyze and predict new market opportunity, esp. through various advertisement sales. Users are not the product, and should not be traded for profits. A decentralized approach to social networking with blockchain technology can ensure better privacy and protection for users. Blockchain enabled social networking platforms have the potential to achieve more outcomes and profits in the right terms and conditions with smart contracts and disruptive applications, and is also highly effective and efficient in e-commerce, m-commerce, u-commerce and crowdfunding transactions.

2.2 User Control Over Contents

Blockchain can also be implemented in social networking platform to give the power and control of the contents back to the users themselves by establishing a decentralized approach to network connectivity. Current social networking uses a centralized approach. Without the central servers, monitoring and exploitation over user-generated contents will not be controlled and owed by companies. With blockchain, users now have the full control over their own contents, and they can decide to share and not to share their information. With decentralized system, companies cannot use users' contents for analytic, prediction, and advertising.

2.3 Improved Privacy and Security

Current social networking platforms have the risks of compromising user's privacy and security. There are serious risks, like eavesdropping on users through smartphone microphones and visual access via smartphone video recorder. Government secret agents and agencies in some countries have used these platforms to monitor and detect criminals and terrorists. To solve this problem, the decentralized system of the blockchain ensures better privacy and security through a distributed consensus mechanism.

2.4 Freedom of Speech

Current social networking platforms are under surveillance by companies and government agencies in the interest of national and international security. Although the monitoring of behavior, activities, or other changing information of users on social networking for the purpose of influencing, managing, directing, or protecting people have some merits, this approach in the same time prevents freedom of speech, and in some occasions, it suppresses basic human rights. In this regards, a blockchain-based approach to social networking offers the benefits of secure authentication whilst still ensuring anonymity and freedom of speech.

2.5 The Silver Bullet for Crowdfunding

The problem with traditional crowdfunding companies is that they are centralized organization, charging high fees and also influencing the projects. Blockchain-based crowdfunding has great potential to facilitate and enhance crowdfunding websites. This include Kickstarter, Indiegogo and other companies as it decentralizes the funding model. In fact, this is a more pure form of crowdfunding because it removes any intermediaries standing between the backers and the startup. Recently, blockchain crowdfunding has grown in popularity for funding innovative startup ideas in comparison with traditional model of seed funding and venture capital.

2.6 A Better Payment Method

Other than communication between peers, another potential business model of social networking is peer-to-peer commerce. Some current social networking platforms also implement some form of payment mechanism through their messaging services. With blockchain-based approach to messaging and social network, users can easily exchange coins or tokens through the same social network because cryptocurrencies are blockchain-based. On top of this, smart contracts and zero-knowledge proofs [14] in blockchain technology makes social networks function as a trusted network, where users can make actual business deals without third party through cryptographically-signed smart contracts with secured and fast peer-to-peer transactions.

2.7 The Verifiable Truth on the Internet

Since the recent election in United States, Facebook has come under intense scrutiny from the technology world, the government, and users alike, due to multiple accusations that Facebook did not actively try to curb the fake news pertaining to the presidential candidates that was spread over the social platform. Facebook CEO, Mark Zuckerberg had recently apologized for the Cambridge Analytica scandal, which was a major breach of trust among Facebook users [15]. While Facebook has pledged to accelerate AI research in order to prevent similar scenarios from becoming regular practices, blockchain capabilities stand out as superior methods for controlling the distribution of unverified information, particularly that has significant real-world consequences when read by users.

3 Research Theoretical Framework and Methodology

In this research, blockchain with advanced AI is used to build trust and prevent fake news in social media. A blockchain is a digitized, decentralized, publicly shared immutable ledger of all transactions that minimizes any security risk [16]. Referring to below theoretical framework of blockchain (see Fig. 1), transactions are contained in blocks, which are linked together through a series of hash pointers. Tampering of a block will cause the hash pointer to be invalid. In blockchain, there is no centralized system, the multiple copies of all valid transactions are held by the users in the community. Users are rewarded for their effort through unique algorithm that can result in payment. The consensus on what types of blocks and transactions is valid can be automatically reached when the majority of the blockchain users accept newly proposed blocks. In this way, a blockchain can provide a trustable, reliable, transparent, verifiable, distributed and secure fundamental platform for all transactions [17].

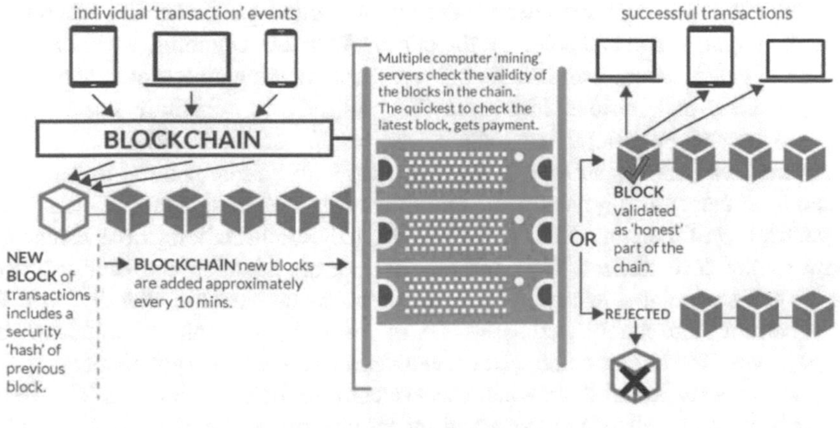

Fig. 1. Theoretical framework of blockchain

The proposed research design is to implement blockchain technology with advanced AI in social media platforms to verify news content, streamline advertising, and prevent fake news. This research aims to develop a new hybrid algorithms that utilizes the principles and methods of blockchain to create transparent and publicly auditable content networks and ranking algorithms that use digital tokens as ranking signals for social media to detect and prevent fake news. A novel theoretical framework will be designed to implement blockchain technology in social platform using simulation. The effectiveness and performance of the hybrid algorithms will be validated in the blockchain-enabled social network in preventing and detecting fake news with numerical simulation results with valid datasets, and results compared with current social network that is not implementing blockchain technology. For further understanding of this new ecosystem of information on blockchain-enabled social network, qualitative survey on focus groups will also be carried out in order to understand users' sentiments on the convergence of blockchain and social media to rebuild trust and prevent fake news.

4 Research Impact and Vision

In this research, when we pair blockchain with advanced AI to fight fake news, there are tremendous opportunities that we can create transparent and publicly auditable content networks. As a result, information or news shared by web or social media users are credible and trustworthy. Startups running news portal on blockchain will ensure a decentralized system which relies on the wisdom of the crowd and return the powers to the people in order to provide people with high quality and verifiable news. It will generate more trust in the quality news ecosystem. This is as opposed to entrenching the popular idea that the government is controlling and totalitarian even further. With the emergence of a decentralized Web 3.0 powered by blockchain, individuals will finally be fully in control of their assets and identity online.

Web 3.0 will overcome the shortcomings of Web 2.0, i.e. the siloes, walled gardened, centralized power, where users' identity are culled and collected, scrutinized, monetized, bought, stolen and sold. On the current Web 2.0 technology, whether users are streaming videos, sharing articles with friends, commenting on a forum, banking, or buying groceries, and various fragments that collectively constitute users' online identity are owned by centralized institutions. In this scenario, all users' favorite platforms and services are subject to legal processes that require them to report the information on demand to governments and other empowered agencies.

In contrast, on Ethereum, NEO and Cardano, the current major global blockchain platforms of the decentralized Web 3.0, any user or organization can establish user-centric or self-sovereign identity. Users can combine user-centric identity with the ability to control blockchain-based assets, i.e. tokens of value, money, capital, and the full scope of what the Internet offers. Users can also selectively encrypt elements of that identity, or choose to reveal them when users deem it in their best interest. This return the power back to the citizens or users and incentivize users to pursue the actions and behavior patterns that deliver the best outcomes to themselves and to the communities in which they are embedded.

These blockchain networks enable us to become more than a citizen of the country where we were born. We can establish our own global virtual country, which we can organize by topics of interest or shared principles. Through code, governance can happen by group consensus. We can outfit our global country with an alternate voting system to reflect the values our community shares and agrees upon, safeguarded from interference by corruption, or dominance by a vocal minority. As for the leadership for the next era in blockchain technology, good governance of such distributed global innovation is not the role of government alone, nor can we leave it to the private companies. In fact, everyone one of us, as a global citizen, should collaborate and provide leadership via principles and consensus.

5 Conclusion

As a conclusion, this study has provided preliminary insights in the convergence of blockchain and social media to build trust and prevent fake news. There are many challenges in verifying news content and prevent fake news on social media. Extensive research is needed in this area to reap the potential benefits. This research area has tremendous real-world political, economy and social impacts. With blockchain technology that are evolving and upgrading continuously, there are many untapped opportunities for disruptive innovation in current and future ecosystem of social media in order to rebuild a more trustable, reliable, transparent, and secure fundamental fabrics for our modern society.

References

1. Valuenex, Inc.: Facebook—A Breach In the Wall—A Technological Approach to the Facebook Cambridge Analytica Scandal. http://www.valuenex.net/te-facebook-cambridge-analytica. Accessed 23 May 2018
2. Andreas, M.: Antonopoulos: Mastering Bitcoin—Programming the Open Blockchain, 2nd edn. O'Reilly Media, Inc, Boston (2017)
3. Kai, S., Amy, S., Suhang, W., Jiliang, T., Huan, L.: Fake News Detection on Social Media—A Data Mining Perspective. arXiv.org, Cornell University Library (2017)
4. William, Y.W.: Liar, Liar Pants on Fire: A New Benchmark Dataset for Fake News Detection. arXiv.org, Cornell University Library (2017)
5. Hannah, R., Eunsol, C., Jin, Y.J., Svitlana, V., Yejin, C.: Truth of varying shades: analyzing language in fake news and political fact-checking, conference on empirical methods. In: Natural Language Processing, Copenhagen, Denmark, 7–11 September 2017
6. Poria, S., Cambria, E., Bajpai, R., Hussain, A.: A Review of Affective Computing: From Unimodal Analysis to Multimodal Fusion, Information Fusion. Elsevier, Amsterdam (2017)
7. Alrubaian, M., Al-Qurishi, M., Hassan, M.M., Alamri, A.: A credibility analysis system for assessing information on twitter. IEEE Trans. Depend. Secure Comput. (2016)
8. Picard, R.W.: Affective Computing. MIT Press, Cambridge (2017)
9. Ravi, K., Ravi, V.: A survey on opinion mining and sentiment analysis: tasks, approaches and applications. Knowl. Based Syst. **89**, 14 (2015)

10. Eichstaedt, J.C., Schwartz, H.A., Kern, M.L., Park, G., Labarthe, D.R., Merchant, R.M., Jha, S., Agrawal, M., Dziurzynski, L.A., Sap, M., et al.: Psychological language on twitter predicts county-level heart disease mortality. Psychol. Sci. **2**, 159–169 (2015)
11. Mäntylä, V., Graziotin, D., Kuutila, M.: The Evolution of Sentiment Analysis—A Review of Research Topics, Venues, and Top Cited Papers. arXiv.org, Cornell University Library (2016)
12. Gartner Report: Gartner Top Strategic Predictions for 2018 and Beyond. https://www.gartner.com/smarterwithgartner/gartner-top-strategic-predictions-for-2018-and-beyond/. Accessed 05 Feb 2018
13. MIT Technology Review: Can AI win the war against fake news? https://www.technologyreview.com/s/609717/can-ai-win-the-war-against-fake-news/. Accessed 05 Feb 2018
14. Kosba, A., Miller, A., Shi, E., Wen, Z., Papamanthou, C.: Hawk: the blockchain model of cryptography and privacy-preserving smart contracts. In: 2016 IEEE Symposium on Security and Privacy. IEEE Computer Society (2016)
15. The Verge: Mark Zuckerberg apologizes for the Cambridge Analytica scandal. https://www.theverge.com/2018/3/21/17150158/mark-zuckerberg-cnn-interview-cambridge-analytica. Accessed 27 Mar 2018
16. Open Blockchain: Researching the potential of blockchains. http://blockchain.open.ac.uk/. Accessed 24 Mar 2018
17. Chen, Y., Li, Q., Wang, H.: Towards Trusted Social Networks with Blockchain Technology. arXiv.org, Cornell University Library (2018)

Social Network Sites (SNS) Utilization in Learning Landscape – Systematic Literature Review

Nani Amalina Zulkanain(✉), Suraya Miskon, Norris Syed Abdullah, Nazmona Mat Ali, and Mahadi Bahari

Faculty of Computing, Universiti Teknologi Malaysia, 81310 Johor Bahru, Johor, Malaysia
amalinanan94@gmail.com, {suraya,norris,nazmona,mahadi}@utm.my

Abstract. Social network sites (SNS) have been widely used around the world to facilitate communication and engagement in the education landscape. The availability of various SNS types and their different functions have provided instructors and students with options in adding value to their learning process. However, there still lack of guidance on how to choose and strategize SNS utilization in learning. This paper explores SNS utilization by instructors and students in supporting the learning environment. The resources of the data are taken from four digital databases and analyzed using thematic analysis. The findings show the current types of SNS most used for learning purposes are Facebook, Twitter, YouTube and WhatsApp. The main uses of these SNS are identified as being for communication, collaboration, sharing information, enhancing learning and social connection.

Keywords: Social network sites · Education · Thematic analysis

1 Introduction

Social network sites (SNS) also known as social media, have been widely used for communication among people regardless of distance or time constraints. SNS defined as a platform for sharing information that can be expressed through text, video, images, documents and music [1]. SNS can create the sense of engagement [2] in educational settings where they provide a sense of presence even with distance constraints. A study by See et al. [2], found that instructors use SNS as tools of communication, collaboration and encouraging students to make friend with each other. In addition, many educational institutes adopt SNS platforms to extend the limits and benefits of learning. SNS as an educational platform has also been proven to contributes to improvement of academic performance [3–5] and to have a positive impact on students' academic experience. The functions include an interface that records the overall activities carried out by the user and provides ability to upload and download any media, share, copy and forward text or media to others as well as the ability to generate comfortable informal conversation. All of these functions provide vital advantages to the instructors and students for the purpose of learning [6].

SNS is generally well-accepted in education shown by both instructors and students as a platform for learning [7], thus leading to encouragement of further development in

F. Saeed et al. (Eds.): IRICT 2018, AISC 843, pp. 963–972, 2019.
https://doi.org/10.1007/978-3-319-99007-1_89

utilizing new approach for learning. The ease of access of SNS using internet connections motivates [5] and enhances effective communication [8] which can facilitate distance learning using different course and materials design and teaching methods [9]. However, the use of SNS may lead to a situation which is uncomfortable for some learners because not all students can adapt SNS to a new learning platform. In addition, there may be a lack of appropriate training provided [10], as different types of SNS provides different functions of learning, thus this research aims to explore the different types of SNS utilization in learning contexts.

1.1 Research Aims

This research aims to identify the types of SNS and their utilization for learning purposes. The outcome of this study will assist instructors and students to choose the most suitable SNS for their learning purpose because as they might have different perceptions or preference depends on the SNS functions. Its contribution will be concerning the development of educational platforms and to both instructors and students, in providing valuable learning sessions that are suitable for their learning purpose.

2 Methods

2.1 Data Collection

This research started with a review of literature contributed by prior studies regarding the use of SNS platforms. To find the types of SNS used for learning purposes, the first step in this research process was to collect papers from selected databases, which were Scopus, Web of Science, Emerald and Ebscohost. All the publications published in these databases were filtered based on the period year 2006 (average number of SNS released to market) until year 2018. All the publications, including journals and conference proceedings were filtered based on "social media" and "social network" and categorized based on educational perspectives only. All publications were carefully identified, read and examined to ensure all are in the scope of education perspectives.

2.2 Data Analysis

All the selected texts were sorted based on the requirements needed for using NVIVO software, by applying the thematic analysis technique. This technique focuses on identifying themes to uncover meanings [11]. This part of the research aimed to uncover the justification given in other research regarding the use of a specific SNS types according to their specific function, thus systematic literature review (SLR) was essential to review all the studies easily and analyzed the data in a manageable way with the use of a thematic analysis technique. All the attributes were coded with first-level coding based on five categories (types of social network, reasons for use of social network, patterns based on year, gaps and issues) to easily understand the patterns of topics. The first-level coding showed the overall statements that related to the categories and contained a brief explanation regarding the topic of the paper. The statements were then reviewed to develop the themes to next level of detail, referred to as second-level coding [12, 13]. The categories represented the points of the results to be further discussed in the next section. After filtration through the first and second-levels of coding, 33 publications in all were found to be suitable accepted for this research. Figure 1 below shows the steps of thematic analysis applied in this research.

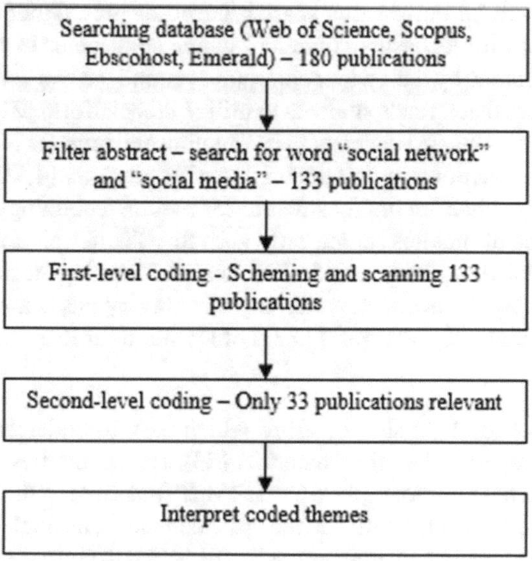

Fig. 1. Steps of thematic analysis

3 Findings and Discussions

3.1 Types of Social Network Sites (SNS)

Many types of SNS were found to have been studied and validated as being used as learning platforms. This study aims to identify the current SNS that have been highlighted in many publications as learning platforms.

Facebook. The use of Facebook as a learning platform has received attention from many researchers and used widely to test and validate the purpose [2, 14], impacts [15] and benefits for students and instructors. There are many perspectives in choosing Facebook in terms of usage, many studies showed that Facebook was the highest ranked choice SNS for instance, in a study regarding the roles of SNS, which involved 489 students, 84.7% reported using Facebook for learning purposes [16]. Other study by Bialy involving with 160 students, 59.8% of the students stated that they chose Facebook to connect with other students to learn [17]. In a study comparing Facebook and Twitter, the highest ranked choice by students with 93.8% was Facebook [18]. This shows the high popularity of Facebook, which can be adapted as platform to communicate and for learning as well. In a study related to the uses of SNS in higher education, which included three institutions and involved 138 students, 93.48% of the students rated Facebook as their main preference for intellectual social interaction [19].

Twitter. This platform created with a limited number of characters and words that the user can write and update [18]. It has a simple interface with the special function of 'retweet', which provides ability for users to copy any status from other tweets and share it in their Twitter wall for information sharing purposes [20]. Other than 'retweet', this platform also famous for its hashtag function, which contributes to easier searching for trending topics discussed by the Twitter community round the world [20], thus providing an interactive feature that attracts students to utilize this platform [21]. Twitter has been included as a research topics to understand its function and impacts for learning purposes [17] because of its functionality to support learning sessions [4, 22]. In addition, the functionality and interface are highly suitable for mobile technology, enabling instructors to adopt the use of mobiles for learning with this Twitter platform [18]. However, in most studies related to the types of SNS used for learning, it has been found that Twitter cannot rival with Facebook in ranking. A study by Fasae and Adegbilero-Iwari [19] found that Twitter only receives 47.83% of attention for learning purpose compare to Facebook.

YouTube. This platform consists of video which easy to understand and learn from and can provide support in learning materials [23]. Recent studies concerning the use of YouTube for learning purposes shown that YouTube is one of the best teaching tools [24] allows social relationships based on the uploaded video, through finding and sharing information [25]. As well as its interactive function, it also provides brief information based on various types of video that contain the graphical information or real life examples from individuals themselves [21, 26]. In a study regarding the impact of YouTube particularly as learning platform, 78% of students reported using YouTube as their main source of learning [27]. However, a study showed that YouTube cannot outrank Facebook, only gaining 43.6% of users [16].

WhatsApp. This SNS basically focuses on instant messaging applications which are very familiar nowadays [28]. WhatsApp has already been part of a recent study regarding its use of it as a learning platform [5]. The most important findings are that this application can improve academic performance [8, 10] as well as assisting students for assessment preparation [29]. A recent study regarding the impact of using WhatsApp found that 57% of students chose WhatsApp for learning purposes [28]. In another recent study regarding the use of Twitter and WhatsApp, 89.9% of the students reported using both applications frequently in their daily communication [30]. In addition, the study also showed that all participants said that they were comfortable in using WhatsApp for communication and sending images or media. In a study regarding the use of instant messages for dental education, WhatsApp was ranked higher compared to Facebook instant message as a communication strategy among students [31], because of its effectiveness and its feature of the ability to check whether the recipients have received the message.

3.2 Uses of SNS for Learning Purpose

Facebook, Twitter, YouTube and WhatsApp are the most currently used platforms and require further study regarding their uses as learning platform. The reasons for using SNS as learning platforms can be categorized into communication, collaboration, sharing information, enhancing learning and social connection.

Communication. Each learning sessions necessarily starts with communication. A study regarding Facebook, YouTube and Twitter showed that these SNS provide the ability to connect with members through communication by receiving and sending messages [25] while another study found that Facebook and Twitter are considered to provide ease of interaction and to be user friendly [17]. This findings is supported by Dafoulas and Shokri [32], who found that the features of Facebook that support social interaction can enrich group discussion sessions. In addition, it has been pointed out that the first function of these platforms when they were created was for communication purposes, before they turned into social networks applying the concept of sharing information to the public [33]. YouTube is focused on sharing information with video, however, it is observed that lack of mention of the communication purpose of YouTube in existing studies. This observation supported by Saw et al. [23], who found that most students use YouTube as a channel for support materials in learning sessions. However, the WhatsApp application focuses on instant message which are specifically directed to communication functions [10], and thus, provide easy access in the form of a communication structure [30].

Collaboration. Communication among communities creates social interactions that enhance collaboration among students via Facebook and Twitter [15]. This is because most SNS involve a community with large numbers. Such collaboration among students may provide advantages to them for sharing knowledge purposes. A recent study by Denker et al. [34], shows that Twitter can provide collaborative learning through comments and questions [35], which specifically known as 'threads', and provide direct feedback based on replies to the respective comments [34]. In addition, Facebook and WhatsApp, particularly provide the facility for creation of groups. A recent study by Khatoon et al. [31] demonstrated the suitability for group work of Facebook and WhatsApp applications that provide functions for sharing files and information directly to all members in the group. It showed the benefits of collaboration, which makes work easier and efficient. WhatsApp has been noted as an application that provides an effective platform for socialization [36] while providing a mobile application of instant messaging that makes it easy to conduct discussions virtually, regardless of constraints of money, time and distance. However, there have been negative results regarding collaboration purposes in YouTube.

Sharing Information. Students tend to gather information through instructors or class members. As mentioned by Dafoulas and Shokri [32], students learn from each other in Facebook by sharing knowledge through tools provided such as video, photos and links to a website. Sharing information enhances teamwork and thus, increases work performance in a group, for instance, a study by Saw et al. [23], indicated that 60% of students chose Facebook as their preferred medium for sharing information. WhatsApp in particular, can manage discussion sessions easily through brief conversation in instant message. Interestingly, it also easy for sharing other resources such as images, video, audio, website links, documents, contacts and location to other users which subsequently provides ease of communication and fosters collaboration [28]. YouTube provides wide access to knowledge through video and can easily use by student to seek information [9] and indicate their interests, as a way for them to gain knowledge [26]. For Twitter, which focuses on a daily and updated status, the functions of 'retweet' makes the shared information easy to spread among all virtual SNS friends or followers [20]. The user can choose either to disseminate all the information in private or publicly, based on the option of privacy provided.

Enhancing Learning. In general, most use of SNS for learning purposes has been found to enhance academic performance [7] because it improves students' satisfaction and their gain of learning experiences. The use of Facebook has shown the enhancement of grades among students through the social interaction that brings about a gain in cognitive presence, which is ability to create their own understanding [32]. Further analysis regarding the use of Facebook has indicated that students tend to easily adapt to the educational setting via SNS by obtaining social support and the social presence of the instructors and class members [37]. Other studies indicated that YouTube provides benefits and enhances potential in classroom learning, which eventually assists students in their studies and improves understanding [23, 24]. Studies regarding Twitter shows fewer results regarding the enhancement of learning using this platform. However, a few studies mention about the ability of Twitter to develop ideas through sharing thoughts [35, 38]. This suggests that Twitter can be a platform to generate more ideas and thus, enhance critical thinking among students. Most studies that relate to Twitter concern engagement and social connection which is explained in the next point. WhatsApp has been used in practice to support students' performance [10], where students gain ideas and knowledge through discussion. A study by Asiri et al. [30], found that most students use WhatsApp more than 5 h per day, and concluded that it was beneficial for students to gain access to each other frequently to communicate and discuss issues related with academic purposes. WhatsApp's features provide the ability to gain learning experience, which eventually supports the development of students' performance in the group created [36].

Social Connection. This point relates to the engagement of students with instructors and class members in SNS. Facebook in particular promotes the ability to find 'friends' and allow students to find their own friends that have same interests, to share ideas and knowledge [2]. A study by Mansour [25] revealed that Facebook, Twitter and YouTube help to connect participants on a student–teacher relationship for learning purposes, thus providing a sense of engagement. In addition, each member get to know the other members better and is able to collaborate more easily [39]. Engagement through social activities can leads to improvement of social skills and social presence in the SNS learning platform. The Twitter platform reveals social support through interaction between students [32]. Moreover, it can become a suitable platform for a conservative community to express idea and make friends in connection with others without facing them physically [30]. WhatsApp also provides a platform for effective socialization [36] through group or private messages. Group members have the ability to put forward ideas and have easy to access members through the contacts list in the group. A study by Smit [29] showed this platform can assist students with preparing for assessments, through sharing documents and links in conversation groups. This suggest that WhatsApp provides an interactive platform which combines both easy instant conversation and the capacity for sharing materials.

4 Conclusions and Future Works

This study has established that SNS have been recently used as learning platforms to support learning purpose. Table 1 illustrates the different uses of SNS for learning purposes.

Table 1. The uses of SNS for learning purposes

	Facebook	Twitter	YouTube	WhatsApp
Communication	/	/		/
Collaboration	/	/		/
Sharing information	/	/	/	/
Enhancing learning	/	/	/	/
Social connection	/	/	/	/

Overall, each SNS has its own characteristics that contribute to different aspects of functions. However, instructors or students can choose the most suitable SNS platform that encompasses the learning objectives. Research concerning the main valuable uses of types of SNS for learning purpose is still vague, since this study has focused on a review of the literature, without specific experiments or observations. Moreover, every learning session will need to refer the guidelines or instruction from the instructor, thus, future work should focus on the role of instructor and his or her capacity to perform teaching through different types of SNS.

Acknowledgements. The authors would like to thank the Ministry of Higher Education (MOHE) and the Universiti Teknologi Malaysia (UTM) for the Research University Grant Scheme (GUP) (Vote Number: I6H76) that had supported this research.

References

1. Brooker, P., Barnett, J., Cribbin, T.: Doing social media analytics. Big Data Soc. **3** (2016). https://doi.org/10.1177/2053951716658060
2. See, J., Lim, Y., Agostinho, S., Harper, B., Chicharo, J.: The engagement of social media technologies by undergraduate informatics students for academic purpose in Malaysia. J. Inf. Commun. Ethics Soc. **12**, 177–194 (2014). https://doi.org/10.1108/JICES-03-2014-0016

3. Tezer, M., Taşpolat, A., Sami, K., Fatih, S.: The impact of using social media on academic achievement and attitudes of prospective. Int. J. Cogn. Res. Sci. Eng. Educ. **5**, 75–81 (2017). https://doi.org/10.5937/ijcrsee1702075T
4. Al-Rahmi, W.M., Zeki, A.M.: A model of using social media for collaborative learning to enhance learners' performance on learning. J. King Saud Univ. Comput. Inf. Sci. **29**, 526–535 (2017). https://doi.org/10.1016/j.jksuci.2016.09.002
5. Dar, Q.A., Ahmad, F., Ramzan, M., Khan, S.H., Ramzan, K., Ahmed, W., Kamal, Z.: Use of social media tool "Whatsapp" in medical education. Ann. King Edward Med. Univ. **23**, 39–42 (2017). https://doi.org/10.21649/akemu.v23i1.1497
6. Maleko Munguatosha, G., Birevu Muyinda, P., Thaddeus Lubega, J.: A social networked learning adoption model for higher education institutions in developing countries. Horizon **19**, 307–320 (2011). https://doi.org/10.1108/10748121111179439
7. Cao, Y., Hong, P.: Antecedents and consequences of social media utilization in college teaching: a proposed model with mixed-methods investigation. Horizon **19**, 297–306 (2011). https://doi.org/10.1108/10748121111179420
8. Wong, C.-H., Tan, G.W.-H., Loke, S.-P., Ooi, K.-B.: Adoption of mobile social networking sites for learning? Online Inf. Rev. **39**, 762–778 (2015). https://doi.org/10.1108/OIR-05-2015-0152
9. Moghavvemi, S., Sulaiman, A., Jaafar, N.I., Kasem, N.: Social media as a complementary learning tool for teaching and learning: the case of youtube. Int. J. Manag. Educ. **16**, 37–42 (2018). https://doi.org/10.1016/j.ijme.2017.12.001
10. Altaany, F.: Usage whatsapp application for e-learning and its impact on academic performance in Irbid national university in Jordan. Int. J. Appl. Eng. Res. **10**, 39875–39879 (2015)
11. Fereday, J., Muir-Cochrane, E.: Demonstrating rigor using thematic analysis: a hybrid approach of inductive and deductive coding and theme development. Int. J. Qual. Methods **5**, 80–92 (2006). https://doi.org/10.1177/160940690600500107
12. Bandara, W., Furtmueller, E., Beekhuyzen, J., Gorbacheva, E., Miskon, S.: Achieving rigour in literature reviews: insights from qualitative data analysis and tool-support. Commun. Assoc. Inf. Syst. **34**, 154–204 (2015)
13. Bandara, W., Miskon, S., Fielt, E.: Association for information systems AIS electronic library (AISeL) a systematic, tool-supported method for conducting literature reviews in information systems. Eur. Council. Inf. Syst. **14**, 1–14 (2011)
14. Hamade, S.N.: Perception and use of social networking sites among university students. Libr. Rev. **62**, 388–397 (2013). https://doi.org/10.1108/LR-12-2012-0131
15. Burbules, N.C.: How we use and are used by social media in education. Educ. Theory **66**, 551–565 (2016). https://doi.org/10.1111/edth.12188
16. Singh, K.P., Gill, M.S.: Role and users' approach to social networking sites (SNSs): a study of universities of North India. Electron. Libr. **33**, 19–34 (2015). https://doi.org/10.1108/EL-12-2012-0165
17. Bialy, S.El, Ayoub, A.R.: The trends of use of social media by medical students. Educ. Med. J. **9**, 59–68 (2017). https://doi.org/10.21315/eimj2017.9.1.6
18. Buzurovic, I., Lj, D.D., Misic, V., Simeunovic, G.: Stability of the robotic system with time delay in open kinematic chain configuration. Acta Polytech. Hung. **11**, 45–64 (2014). https://doi.org/10.4018/IJICTE.2016010106
19. Fasae, J.K., Adegbilero-Iwari, I.: Use of social media by science students in public universities in Southwest Nigeria. Electron. Libr. **34**, 213–222 (2016). https://doi.org/10.1108/EL-11-2014-0205

20. Merrill, N.: Higher Education Administration with Social Media Social media for social research: applications for higher education communications. Cutt. Edge Technol. High Educ. **2**, 25–48 (2015). https://doi.org/10.1108/S2044-9968(2011)0000002005
21. Rueda, L., Benitez, J., Braojos, J.: From traditional education technologies to student satisfaction in management education: a theory of the role of social media applications. Inf. Manag. (2017). https://doi.org/10.1016/j.im.2017.06.002
22. Waycott, J., Thompson, C., Sheard, J., Clerehan, R.: A virtual panopticon in the community of practice: students' experiences of being visible on social media. Internet High. Educ. **35**, 12–20 (2017). https://doi.org/10.1016/j.iheduc.2017.07.001
23. Saw, G., Abbott, W., Donaghey, J., Mcdonald, C.: Social media for international students—it's not all about Facebook. Libr. Manag. J. Serv. Manag. Issue Qual. Mark. Res. Int. J. **34**, 156–174 (2012)
24. Neier, S., Zayer, L.T.: Students' perceptions and experiences of social media in higher education. J. Mark. Educ. **37**, 133–143 (2015). https://doi.org/10.1177/0273475315583748
25. Mansour, E.A.H.: The use of social networking sites (SNSs) by the faculty members of the School of Library and Information Science, PAAET, Kuwait. Electron. Libr. **33**, 524–546 (2015). https://doi.org/10.1108/EL-06-2013-0110
26. Pacheco, E., Lips, M., Yoong, P.: Transition 2.0: digital technologies, higher education, and vision impairment. Internet High. Educ. **37**, 1–10 (2017). https://doi.org/10.1016/j.iheduc.2017.11.001
27. Barry, D.S., Marzouk, F., Chulak-Oglu, K., Bennett, D., Tierney, P., O'Keeffe, G.W.: Anatomy education for the YouTube generation. Anat. Sci. Educ. **9**, 90–96 (2016). https://doi.org/10.1002/ase.1550
28. Jaafar, A.: The impact of using social media and internet on academic performance: case study Bahrain Universities. EAI Endors. Trans. Scal. Inf. Syst. **4**, 1–12 (2017). https://doi.org/10.4108/eai.28-6-2017.152748
29. Smit, I.: WhatsApp with learning preferences? In: Proceedings—Frontiers in Education Conference, FIE (2015)
30. Asiri, A.K., Almetrek, M.A., Alsamghan, A.S., Mustafa, O., Alshehri, S.F.: Impact of Twitter and WhatsApp on sleep quality among medical students in King Khalid University, Saudi Arabia. Sleep Hypn. **20**, 247–252 (2018). https://doi.org/10.5350/Sleep.Hypn.2018.20.0158
31. Khatoon, B., Hill, K., Walmsley, A.D.: Instant messaging in dental education. J. Dent. Educ. **79**, 1471–1478 (2015). https://doi.org/10.5958/2393-8005.2016.00013.9
32. Dafoulas, G., Shokri, A.: Investigating the educational value of social learning networks: a quantitative analysis. Interact. Technol. Smart Educ. **13**, 305–322 (2016). https://doi.org/10.1108/ITSE-09-2016-0034
33. Liu, D., Kirschner, P.A., Karpinski, A.C.: A meta-analysis of the relationship of academic performance and social network site use among adolescents and young adults. Comput. Hum. Behav. **77**, 148–157 (2017). https://doi.org/10.1016/j.chb.2017.08.039
34. Denker, K.J., Manning, J., Heuett, K.B., Summers, M.E.: Twitter in the classroom: modeling online communication attitudes and student motivations to connect. Comput. Hum. Behav. **79**, 1–8 (2018). https://doi.org/10.1016/j.chb.2017.09.037
35. Chawinga, W.D.: Taking social media to a university classroom: teaching and learning using Twitter and blogs. Int. J. Educ. Technol. High. Educ. **14**, 3 (2017). https://doi.org/10.1186/s41239-017-0041-6
36. Robinson, L., Behi, O., Corcoran, A., Cowley, V., Cullinane, J., Martin, I., Tomkinson, D.: Evaluation of whatsapp for promoting social presence in a first year undergraduate radiography problem-based learning group. J. Med. Imaging Radiat. Sci. **46**, 280–286 (2015). https://doi.org/10.1016/j.jmir.2015.06.007

37. Greenhow, C.: Online social networks and learning viewpoint online social networks and learning. Horizon **19**, 4–12 (2013). https://doi.org/10.1108/10748121111107663
38. Gikas, J., Grant, M.M.: Mobile computing devices in higher education: student perspectives on learning with cellphones, smartphones and social media. Internet High. Educ. **19**, 18–26 (2013)
39. Towner, T.L., Lego Muñoz, C.: Facebook and education: a classroom connection? In: Educating Educators with Social Media, pp. 33–57 (2011)

An Empirical Study of How Social Influence Impacts Customer Satisfaction with Social Commerce Sites

Hasan Beyari(✉) and Ahmad Abareshi

School of Business IT and Logistics, RMIT University, Melbourne, Australia
{hasan.beyari,ahmad.abareshi}@rmit.edu.au

Abstract. The term "social commerce" describes a subgroup of e-commerce that uses online media and social media to support social interaction and user input, thus assisting in the sale and purchase of services and products. This study explores the relationships affecting social influence and customer satisfaction in the context of social commerce. It adds a research framework that gives social commerce businesses a better understanding of how the consumer mind works, which may result in competitive advantages. Using a survey method, the researcher measured the relationships between factors associated with social influence and evaluated the role it plays in social commerce. A total of 314 students from Saudi Arabia, some studying in Australia and some enrolled in Saudi Arabia's University of Business and Technology, were surveyed. To test the measurement and structural model, and to analyze the interconnection between variables, the study made use of partial least squares (PLS). The results show that the constructs word-of-mouth (WOM) and Long-term Orientation have a significant relationship with social influence. This research recognizes that social commerce differs in important ways from brick-and-mortar commerce. There are multiple factors that influence customer satisfaction differently in social commerce compared to face-to-face transactions. It is anticipated that the results of this research will help merchants in devising effective strategies to retain customers and grow their social commerce businesses. The results obtained from this study make a significant addition to existing literature relating to social commerce. They also serve as the basis from which other researchers can delve more deeply into the different factors that interact with social influence in this context.

Keywords: Social commerce · Customer satisfaction · Social influence
Saudi Arabia · Australia · PLS

1 Introduction

Rapid changes have taken place in Information and Communication Technology (ICT) during the twentieth and early twenty-first centuries, and this trend seems destined to continue. Over the past few decades, key innovations have included the development of the Internet and increases in processing speed. These provide the basis for new ways to communicate, and have changed the way people buy and sell [1]. The majority of advancements in IT can be linked to the growth of the internet. Web 2.0 has

F. Saeed et al. (Eds.): IRICT 2018, AISC 843, pp. 973–984, 2019.
https://doi.org/10.1007/978-3-319-99007-1_90

given more people than ever the chance to interact on the Internet [2]. With the emergence of this newly developed social media environment, people across the globe routinely interact and exchange information. This technology has been integral to the growth and success of many businesses. It has also granted customers new power. They can both inform each other about their experiences, and ensure they are equipped with enough knowledge to make good purchasing decisions. Customers are also able to affect the success of the businesses' publicity and reputation more directly than ever before, as word-of-mouth information from a trusted source has a strong influence on other consumers. Businesses, therefore, must consistently take customer satisfaction into account, since their growth is based on positive consumer experience.

When the concept of Social Commerce (SC) was developed, it could be traced to the emergence of Web 2.0 and social networks [3]. SC was initially introduced on the Yahoo website in 2005, and came to represent the combination of social networking activities and shopping activities [4]. As little research has been done to investigate the impact of social influence on social commerce, this study helps to investigate how social influence impacts customer satisfaction in a SC environment. It also serves as a contribution to the current literature on SC and to the field's understanding of these factors, as well as highlighting areas where businesses can improve performance to enjoy greater competitive advantages.

This research paper begins with an introduction in Sect. 1, followed by a review of the relevant literature in Sect. 2. A discussion of the theoretical foundation and hypothesis development for this study appears in Sect. 3, and Sect. 4 gives a description of the research methodology adopted. Data are analyzed and findings are presented in Sect. 5. Finally, in Sect. 6, conclusions are presented along with study limitations and possibilities for future research.

2 Literature Review

Social commerce is sub-category of e-commerce that makes use of online media and social media to allow customers to exchange information about services and products [5]. In addition, customer satisfaction is an essential factor to consider when evaluating the success of using social commerce websites. According to [6], consumer satisfaction is defined as a consumer's feelings of pleasure or disappointment resulting from a comparison between the perceived performance and his or her expectations. A dissatisfied customer can very easily switch to a competing social commerce company in order to satisfy needs. Hence, companies have to keep tabs on what the customers expect from their products and services, as well as other factors such as the interactive social features of their websites, customer convenience, flexible delivery schedules and personalized relationships. According to [7], social commerce fuses social networking technology and e-commerce in a way that promotes social trust and interaction. As [8] argues, SC allows e-commerce companies to make use of web-based social communities. Social influence shapes the interaction among many of the consumers within these communities.

2.1 Consumer Satisfaction in Social Commerce

Consumer satisfaction can be defined as a post-evaluation a customer makes of a purchase decision [9]. According to [6], consumer satisfaction is defined as the feelings of either satisfaction or disappointment a customer experiences as a result of comparing the perceived performance of a product or service to the expectations they held before buying it. Consumers desire products that fulfill their needs and wants, and buy products they expect to do so to a reasonable extent [10]. If the product falls short of these expectations, the customer will be disappointed, and if perceived performance exceeds expectations, they will be satisfied. If the customer perceives the performance as being equal to their expectation, they will be indifferent or neutral [11].

A consumer's buying decisions are influenced by many factors relating to their culture, social interactions, personal circumstances and psychological characteristics [11]. There are three aspects of experience that significantly affect consumer satisfaction: consumer need, consumer value and consumer cost. Customers buy products to fill specific needs [10]. Researchers have found that when choosing products, consumers are driven by a desire to fulfil their needs and are influenced by specific expectations about the ways in which a given product can accomplish this [10]. Customer value refers to an individual's evaluation of what is essential in his or her life [12]. Although all customers may have some things in common, customer value is, thus, individual [11]. Customer cost refers to what the customer will pay for a service or product that fulfills a need and provides the things they value [13]. Customers want to buy things that are worth the cost, and the extent to which they consider the cost justified affects satisfaction [11]. While these aspects of satisfaction are true for all forms of commerce, some features of social commerce, including the ways customers obtain information about a product, set it apart from traditional commerce.

2.2 Social Influence

The term "social influence" refers to the ways in which a person's beliefs, attitudes, thoughts and actions change as a result of social interactions [14]. As a person interacts with others, they may accept the others' views and behaviours, coming to think and act similarly [15]. Social influence can take place in the social commerce context, as customers of SC sites interact online to share reviews and recommendations [16]. Online interaction can promote a high level of social influence, creating relationships between consumers who read each other's reviews and learn about their experiences [17]. One of the ways social influence is measured is by quantifying the social impact of the sender on the receiver [18]. This influence comes in two forms: normative social influence and informational social influence [19]. Normative social influence creates pressure on people to think and behave in a manner similar to their peers, as if they do not they will be considered "old-fashioned", regardless of personal preferences [20]. In the context of commerce, this would translate into pressure to buy the same items as the people who influence them. In contrast, informational social influence is influencing behaviour by shaping what people know and believe. Customers learn about the things they intend to purchase, particularly by reading about the experiences and recommendations of others [20]. In order to fully understand what factors affect SC satisfaction, it is important to determine the role of social influence in this context. Online interaction and social influence combine to create relationships among consumers who write reviews and share experiences [17]. The decision-making process in social commerce is strongly impacted by social influence, with the greater the degree of influence (for example, of a person recommending a specific product to another who is considering buying it), the greater the likelihood of a positive reaction (purchase of the product).

2.3 Word of Mouth

For the most part, customers take others' assessments into account when deciding whether to make a purchase on a social commerce site. As indicated by [21], word-of-mouth (WOM) is essentially the communication of information and experiences among customers to help each other make purchase decisions. If a customer is satisfied with the experience he or she had, that customer will share this experience through positive word-of-mouth, saying good things about the organisation. Word-of-mouth, as defined by [21], is the exchange of information between consumers to help each other make purchase decisions. Firms place a high level of importance on consumer feedback, as negative comments by these consumers may generate negative word-of-mouth [22]. Studies have shown that attributes of social media such as social presence and interactivity affect the level of consumer satisfaction related to social commerce [23]. There is credible justification for deeming that trust represents a major challenge for organisations that desire to utilise social e-commerce to attain their objectives. In order to attain their goals, organisations using social commerce need to focus on the expectations as well as trust of their consumers, which is a guarantee for future purchase behaviour. When conducted through social media, social WOM increases social media users' levels of trust in new products.

According to [2], when there is either no knowledge or limited knowledge about a new product in the market, users of social commerce prefer to search for more information than what is provided by the company website—usually through social media, other technological developments, or through offline environment sources such as friends or family. Successful businesses owe their success to the hard work they put into developing their trust-building strategies in online platforms [24]. As per [25], online companies try out different strategies in order to enhance their relationship with consumers and to develop consumer trust in the company's networks. According to [2], they use social media to build such relationships. Moreover, customers usually seek out others' accounts of social commerce websites, including an evaluation of their experiences with them. If a customer is happy with an experience with a given company, he or she is likely to engage in positive word-of-mouth about the company.

2.4 Long-Term Orientation

As [26] emphasises, long-term orientation relates to an individual's attitudes about the future. They can choose either a future-oriented or a more short-term point of view. Individuals with long-term orientation expect their relationships, such as with a company they buy from, to be long-lasting. They are willing to overlook or forgive short-term inconveniences because they believe the relationship will ultimately benefit them [27]. An example of this would be a brand-loyal customer who believes the brand will make restitution for defective products or late orders. By contrast, a customer with a short-term orientation is more focused on the benefits of a single transaction. Long-term versus short-term orientation may be influenced by cultural factors, among others [26]. Researchers have suggested that long-term relationships may benefit customers in numerous ways. Existing work on relational benefits has consumers stay in relationships because they trust the company, expect to save money and time, and receive social benefits [28].

Some researchers have argued that the indicators of relationship effectiveness are trust, communication and commitment [29]. An influential study by [30] explained that customers receive benefits from long-term relationships above and beyond core service performance. They identified three types of relational benefits: confidence, social benefits and special treatment benefits. In other words, customers felt more certain they would continue to receive satisfactory products or services, experienced increased social prestige from being a loyal customer, and received free perks as a reward for loyalty. According to [30], this research has been the basis for many subsequent studies, which have validated their findings in various different service sectors [28].

3 Theoretical Background and Hypothesis Development

The theory of social influence analyses individual social interactions from a behavioral and technological point of view [31]. The social influence theory makes use of insights from other fields, including social information processing theory, and research on the adoption and diffusion of communication technologies. Social influence has an impact on how individuals use social media and what products they buy [32]. Social relationships are a key element differentiating SC from conventional online stores. Consumers check with their sources to gather information before they buy from social network sites, to increase their chances of being happy with a purchase. Consumers gain information from a variety of social sources, such as recommendations, referrals, ratings and reviews.

It is clear that social influence is a significant factor affecting consumers when they use e-commerce sites [33]. If social influence guides purchase decision-making, we can expect it to also affect the level of satisfaction that results from a customer's buying behavior on social commerce sites. In the context of SC, most existing research has focused on purchase intentions [34]. Comparatively few studies have addressed consumers' post-purchase experience evaluations [35]. More research on the subject is needed. Thus the following hypothesis is proposed:

H1. Social influence has a positive impact on consumer satisfaction in SC.

According to [36], business enterprises are increasingly exploring the potential of social commerce websites to transmit their promotional information to consumers. Through such e-WOM, they enhance their brand awareness. Consumers tend to trust others' experiences and build trust in a product or service after hearing positive feedback about it. Social relationships crucially differentiate SC from other forms of online commerce. Traditionally, research that has examined the social element has focused on social influence [37]; that is, how a customer's purchase decision is affected by other people's attitudes and opinions. Consumers want to be sure of satisfaction before they buy from an unfamiliar SC site and, to do this, they rely on others' opinions. Consumers also like to feel they are making a choice that others would consider wise, and gain social support from different factors such as recommendations, referrals, ratings and reviews. Social influence thus plays an important role in consumer satisfaction. Therefore, the following hypothesis is proposed:

H2. WOM has a positive impact on social influence in SC.

[38] define long-term orientation as the degree to which a person is oriented to persist in a relationship from which they expect some sort of benefit or reward. In the context of a buyer–seller relationship, a buyer with long-term orientation is one who expects to benefit from continuing patronage of a seller and is thus willing to overlook short- term setbacks. According to [39], the concept of long-term orientation is founded on the assumption that the relationship between the parties involved is stable and will last long enough for these parties to realize the long-term benefits of their association. Long-term orientation in any buyer-seller relationship is a function of two main factors: mutual dependence and the extent to which they trust one another. Dependence and trust are, in turn, related to environmental uncertainty, transaction-specific investments, reputation, and satisfaction [40]. If a customer knows little about a company from whom they are making a purchase, they may not expect to continue buying from that company. By contrast, if the customer has been influenced by peers to regard the company in a favorable light, we might expect them to be more interested in an ongoing business relationship. Thus, the following hypothesis is proposed:

H3. Long-term orientation has a positive impact on social influence in SC.

Based on the above discussion, the researcher developed the conceptual model presented in Fig. 1 below. This model was used as a theoretical framework to analyze the components that influence customer satisfaction in social commerce.

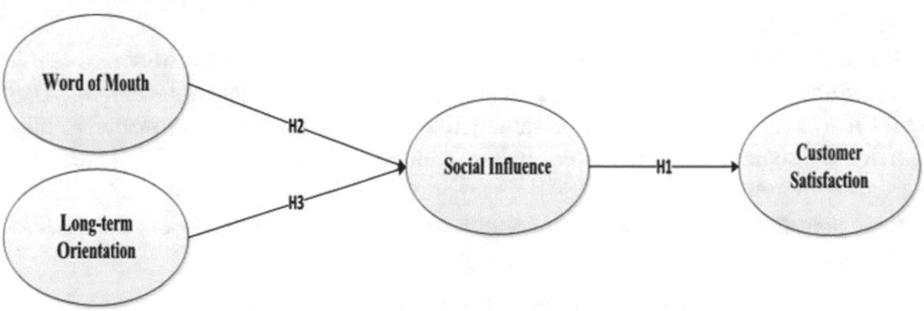

Fig. 1. The research model and hypothesis

4 Research Methodology

The study tested the hypotheses and attempted to answer the research questions through the development of a questionnaire as a research instrument. The items of each construct were adopted from previous literature. All of the items were measured on a seven-point Likert scale where 1 = strongly disagree and 7 = strongly agree.

The study samples focused on students from Saudi Arabia, some currently studying in Australia and others in Saudi Arabia, who have experience with online shopping and SC sites. The study population included students in two different countries to see if there were any differences between the two in their use of SC sites. Data from students studying in Saudi Arabia was obtained by distributing paper copies of the survey to students at UBT University. Participants were selected by choosing classes at random and asking each class member to complete the survey. Of the 500 students who were included, completed surveys were obtained from 300, for a response rate of 60%. In Australia, a survey link was disseminated to Saudi social groups on Twitter and Facebook, and data were collected with the help of Qualtrics software. In the case of the Australian students, it was not feasible to compute the online survey response rate, since the researchers were not sure how many total users saw the link. Overall, a total of 314 valid responses were obtained. The respondents were 64.8% men and 35.2% women. Their ages mostly fell between 20 and 50, with 50.3% of them between 26 and 35 and another 24.3% between 36 and 45. Participants tended to be highly educated; 32.8% were graduate students and 61.8% were postgraduate students having already attained at least one graduate degree. In its analysis of the measurement and structural models, this study makes use of the Smart PLS application program to perform the "Partial Least Squares Structural Equation Modelling" techniques (PLS-SEM) [41].

5 Data Analysis and Findings

The researcher sought to assess the indicator reliability by using the PLS algorithm calculation. To this end, the common path weighting scheme was used to set the inner weighting option. As indicated by [42], this scheme is preferable to the centroid weighting and factorial weighting schemes. The number of high iterations is 50. Through this procedure, the researcher ascertained the measurement items' factor loadings. In order to accomplish the desired item reliability of approximately 0.5, the item loadings should be 0.7 or more, as indicated by [43]. This cut-off was met for all of the relevant items. Table 1 shows the summary of each constructs' factor loadings, along with Reliability and the Average Variance Extracted values.

Table 1. Summary of constructs' factor loadings, reliability and the average variance extracted values

Construct	Range of factor loading within constructs	Composite reliability α > 0.7	AVE ≥ 0.5	Items' sources
Consumer satisfaction	0.96–0.87 (7 Items)	0.98	0.90	[44–46]
Social influence	0.85–0.91 (6 Items)	0.95	0.78	[44, 47, 48]
Word-of-mouth	0.92–0.87 (8 Items)	0.97	0.81	[49–51]
Long-Term-orientation	0.62–0.87 (4 items)	0.85	0.59	[39, 40]

Composite reliability is another approach used to evaluate the internal consistency of constructs. As shown in Table 2 below, all constructs displayed composite reliability of 0.85 or greater, well above the desired minimum of 0.7 [52]. AVE is another measure of item validity; when the AVE value of a construct is at least 0.5, sufficient convergent validity is achieved [53]. As shown in Table 1, the AVE values for all constructs fall within the range of 0.59–0.90. Thus, convergent validity is established. Discriminant validity refers to the degree to which an individual construct is related to its own measurement items, as opposed to those associated with other constructs. The practices most prevalently used to establish discriminant validity are Fornell–Larcker standards and investigation of the cross-loadings. The results are shown in Table 2, alongside the AVE values' square root.

Table 2. The Fornell–Larcker criterion

Constructs	Customer satisfaction	Long-term orientation	Social influence	Word-of -mouth
Customer satisfaction	0.951			
Long-term orientation	−0.018	0.770		
Social influence	0.563	0.035	0.883	
Word-of-mouth	0.566	0.076	0.821	0.900

As discussed by [54], the Heterotrait-Monotrait Correlations Ratio (HTMT) is another form of statistical analysis used to evaluate discriminant validity.

The results are presented on Table 3 for the real measurement model framework.

If the HTMT value is below 0.90, such a result establishes discriminant validity between two constructs. In the case of the framework for the present study, the HTMT ratio for most relationships is significantly less than 0.90. For the relationship between word-of-mouth and social influence, the HTMT value is 0.855. This is very close to the threshold level of 0.90. Since this instance barely meets the threshold level, it does not strongly establish discriminant validity.

Table 3. The Heterotrait-Monotrait correlations ratio

Constructs	Customer satisfaction	Long-term orientation	Social influence	Word-of - mouth
Customer satisfaction				
Long-term orientation	0.055			
Social influence	0.578	0.038		
Word-of-mouth	0.582	0.076	0.855	

After analysing the relevance and importance of the path coefficients, the researcher sought to determine the explanatory power. This was done by determining the coefficient, R2. The R2 value for social influence is 0.675, which is considered strong, and for customer satisfaction is 0.317, which is considered moderate [55]. Predictive relevance, regarded as the measurement of functionality of a model for predictive purposes, was also measured. The results obtained are indicated in Table 4. The strongest predictor of social influence is word-of-mouth; however, all constructs included affect customer satisfaction considerably.

Table 4. The results obtained from the analysis of the structural equation model

Hypothesis	Path	P-value	Decision
H1	Long-term orientation → social influence	0.502	Rejected
H2	Social influence → customer satisfaction	0.00	Supported
H3	Word-of-mouth → social influence	0.00	Supported

The results obtained from the present study can provide meaningful insights to businesses seeking competitive advantage. This study contributes to the existing literature on social commerce in a number of ways. Firstly, by using a tested theoretical framework, namely the social influence theory, it can deepen our understanding of the factors impacted by social influence. Secondly, the results of this research provide social commerce website owners a tool that can assist them in improving their services and satisfying their customers. Ultimately, this study provides knowledge about the antecedents that impact customers' use of social commerce sites.

6 Conclusion

This study aims at improving the field's understanding of how social influence impacts customer satisfaction in the social commerce context. It looks at the interrelationships between multiple factors including Word-of-Mouth and long-term orientation. The results obtained fully supported the theoretical model with the exception of one relationship, the relationship between long-term orientation and social influence. This may reflect customers having difficulties with these websites in terms of after-purchase experience. Customers may demand extra promotions and special offers on products and services to feel a high level of satisfaction with social commerce sites they use.

By contrast, it was discovered that word-of-mouth was a unique dimension and had the strongest influence on social influence, particularly in the social commerce context. It is thus evident that the more a customer is influenced by positive evaluations from others, the greater will be their level of customer satisfaction. It is, therefore, worthwhile for social commerce enterprises to give special consideration to customer feedback and reviews.

The findings obtained must be interpreted within the context of this study's limitations. One such limitation was that the study sample included only students. The findings may thus not be generalizable to all consumers. For future study, areas to be examined in more detail include: (1) enhancing the study model by adding new antecedents and variables to measure customer satisfaction; (2) further exploring hypothesized relationships with new techniques of investigation; and (3) further exploring the relationships that exist between the variables in this study; for example, including different study populations.

References

1. Canavan, O., Henchion, M., O'Reilly, S.: The use of the internet as a marketing channel for Irish speciality food. Int. J. Retail Distrib. Manag. **35**, 178–195 (2007)
2. Hajli, M.: A research framework for social commerce adoption. Inf. Manag. Comput. Secur. **21**, 144–154 (2013)
3. Yang, C.C., Tang, X., Dai, Q., Yang, H., Jiang, L.: Identifying implicit and explicit relationships through user activities in social media. Int. J. Electron. Commer. **18**, 73–96 (2013)
4. Wang, C., Zhang, P.: The evolution of social commerce: the people, management, technology, and information dimensions. Commun. Assoc. Inf. Syst. **31**, 105 (2012)
5. Mardsen, P.: Social Commerce: Monetizing Social Media. Syzygy Deutschland Gmbh, Hamburg (2010)
6. Kotler, P.: Marketing Management, The Millennium edn. Prentice Hall, Upper Saddle River (2000)
7. Zhou, L., Zhang, P., Zimmermann, H.D.: Social commerce research: an integrated view. Electron. Commer. Res. Appl. **12**, 61–68 (2013)
8. Kim, Y.A., Srivastava, J.: Impact of social influence in e-commerce decision making. In: Proceedings of the Ninth International Conference on Electronic Commerce, pp. 293–302. ACM (2007)
9. Churchill Jr., G.A., Surprenant, C.: An investigation into the determinants of customer satisfaction. J. Market. Res. **19**, 491–504 (1982)
10. Parker, C., Mathews, B.P.: Customer satisfaction: contrasting academic and consumers' interpretations. Market. Intell. Plan. **19**, 38–44 (2001)
11. Chi Lin, C.: A critical appraisal of customer satisfaction and e-commerce. Manag. Audit. J. **18**, 202–212 (2003)
12. Kenny, T.: From vision to reality through values. Manag. Dev. Rev. **7**, 17–20 (1994)
13. Best, R.J.: Market-Based Management. Pearson, London (1997)
14. Friedkin, N.E.: A Structural Theory of Social Influence. Cambridge University Press, Cambridge (2006)

15. Kelman, H.C.: Compliance, identification, and internalization: three processes of attitude change. J. Confl. Resolut. **2**, 51–60 (1958)
16. Ng, C.S.P.: Examining the cultural difference in the intention to purchase in social commerce. In: PACIS, p. 163 (2012)
17. Wu, H.L., Wang, J.W.: An empirical study of flow experiences in social network sites. In: PACIS Proceedings (2011)
18. Latane, B.: The psychology of social impact. Am. Psychol. **36**, 343 (1981)
19. Bearden, W.O., Calcich, S.E., Netemeyer, R., Teel, J.E.: An exploratory investigation of consumer innovativeness and interpersonal influences. Adv. Consum. Res. **13**, 77–82 (1986)
20. Kim, Y., Srivastava, J.: Impact of social influence in e-commerce decision making. In: Proceedings of the Ninth International Conference on Electronic Commerce, pp. 293–302. ACM Press, New York (2007)
21. Kim, K., Prabhakar, B.: Initial trust, perceived risk, and the adoption of internet banking. In: Proceedings of the Twenty First International Conference on Information Systems, pp. 537–543. Association for Information Systems (2000)
22. Mikalef, P., Giannakos, M., Pateli, A.: Shopping and word-of-mouth intentions on social media. J. Theor. Appl. Electron. Commer. Res. **8**, 17–34 (2013)
23. Chen, L., Wang, R.: Trust development and transfer from electronic commerce to social commerce: an empirical investigation. Am. J. Ind. Bus. Manag. **6**, 568 (2016)
24. Hajli, M., Khani, F.: Establishing trust in social commerce through social word of mouth. In: 7th International Conference on e-Commerce in Developing Countries: With Focus on e-Security (ECDC), pp. 1–22. IEEE (2013)
25. Füller, J., Bartl, M., Ernst, H., Mühlbacher, H.: Community based innovation: how to integrate members of virtual communities into new product development. Electron. Commer. Res. **6**, 57–73 (2006)
26. Abălăesei, M.: The influence of culture on the role of social media in. Decis. Mak. 600–612 (2015). http://www.upm.ro/ldmd/LDMD-03/Ssc/Ssc%2003%2052.pdf
27. Shin, Y., Thai, V.V., Grewal, D., Kim, Y.: Do Corporate sustainable management activities improve customer satisfaction, word of mouth intention and repurchase intention? Empirical evidence from the shipping industry. Int. J. Logist. Manag. **28**, 555–570 (2017)
28. Kushwaha, T.: An exploratory study of consumer's perception about relational benefits in retailing. Procedia Soc. Behav. Sci. **133**, 438–446 (2014)
29. Graca, S.S., Barry, J.M., Doney, P.M.: Performance outcomes of behavioral attributes in buyer-supplier relationships. J. Bus. Ind. Mark. **30**, 805–816 (2015)
30. Gwinner, K.P., Gremler, D.D., Bitner, M.J.: Relational benefits in services industries: the customer's perspective. J. Acad. Mark. Sci. **26**, 101–114 (1998)
31. Kelman, H.C.: Further thoughts on the processes of compliance, identification, and internalization. Perspect. Soc. Power **15**, 125–171 (1974)
32. Kim, Y., Sohn, D., Choi, S.M.: Cultural difference in motivations for using social network sites: a comparative study of American and Korean college students. Comput. Hum. Behav. **27**, 365–372 (2011)
33. Bhattacherjee, A.: Acceptance of e-commerce services: the case of electronic brokerages. IEEE Trans. Syst. Man Cybern. A Syst. Hum. **30**, 411–420 (2000)
34. Hajli, N., Sims, J., Zadeh, A.H., Richard, M.O.: A social commerce investigation of the role of trust in a social networking site on purchase intentions. J. Bus. Res. **71**, 133–141 (2017)
35. Jones, M.A., Taylor, V.A.: Marketer requests for positive post-purchase satisfaction evaluations: consumer depth interview findings. J. Retail. Consum. Serv. **41**, 218–226 (2018)
36. De Vries, L., Gensler, S., Leeflang, P.S.: Popularity of brand posts on brand fan pages: an investigation of the effects of social media marketing. J. Interact. Mark. **26**, 83–91 (2012)

37. Venkatesh, V.M.: Why don't men ever stop to ask for directions? Gender, social influence, and their role in technology acceptance and usage behavior. MIS Q. **24**, 115–139 (2000)
38. Hoftede, G., Hofstede, G.J., Minkov, M.: Cultures and Organizations: Software of the Mind: Intercultural Cooperation and Its Importance for Survival. McGraw-Hill, New York (2010)
39. Wang, C.L., Siu, N.Y., Barnes, B.R.: The significance of trust and renqing in the long- term orientation of chinese business-to-business relationships. Ind. Mark. Manag. **37**, 819–824 (2008)
40. Ganesan, S.: Determinants of long-term orientation in buyer-seller relationships. J. Mark. **58** (2), 1–19 (1994)
41. Ringle, C.M., Sarstedt, M., Straub, D.: A critical look at the use of PLS-SEM in MIS quarterly. MIS Q. **36**, 1 (2012)
42. Vinzi, V.E., Trinchera, L., Amato, S.: PLS path modeling: from foundations to recent developments and open issues for model assessment and improvement. In: Esposito Vinzi, V., Chin, W., Henseler, J., Wang, H. (eds.) Handbook of Partial Least Squares. Springer Handbooks of Computational Statistics. Springer, Berlin, Heidelberg (2010)
43. Hair, J.F., Ringle, C.M., Sarstedt, M.: Editorial-partial least squares structural equation modeling: rigorous applications, better results and higher acceptance. Long Range Plan. **46**, 1–12 (2013)
44. Liang, T.P., Ho, Y.T., Li, Y.W., Turban, E.: What drives social commerce: the role of social support and relationship quality. Int. J. Electron. Commer. **16**, 69–90 (2011)
45. Eid, M.I.: Determinants of e-commerce customer satisfaction, trust, and loyalty in Saudi Arabia. J. Electron. Commer. Res. **12**, 78 (2011)
46. Doll, W.J., Torkzadeh, G.: The measurement of end-user computing satisfaction. MIS Q. **19**, 259–274 (1988)
47. Xu-Priour, D.-L., Truong, Y., Klink, R.R.: The effects of collectivism and polychronic time orientation on online social interaction and shopping behavior: a comparative study between China and France. Technol. Forecast. Soc. Change **88**, 265–275 (2014)
48. Hsu, C.L., Lin, J.C.C.: Acceptance of blog usage: the roles of technology acceptance, social influence and knowledge sharing motivation. Inf. Manag. **45**, 65–74 (2008)
49. Hajli, M.: Social commerce: the role of trust. In: Eighteenth Americas Conference on Information Systems, Seattle, Washington, 9–12 August, pp. 1–11 (2012)
50. Srinivasan, S.S., Anderson, R., Ponnavolu, K.: Customer loyalty in e-commerce: an exploration of its antecedents and consequences. J. Retail. **78**, 41–50 (2002)
51. Harrison-Walker, L.J.: The measurement of word-of-mouth communication and an investigation of service quality and customer commitment as potential antecedents. J. Serv. Res. **4**, 60–75 (2001)
52. Hair, J.F., Ringle, C.M., Sarstedt, M.: PLS-SEM: indeed a silver bullet. J. Mark. Theory Pract. **19**, 139–152 (2011)
53. Fornell, C., Larcker, D.F.: Evaluating structural equation models with unobservable variables and measurement error. J. Mark. Res. **18**, 39–50 (1981)
54. Henseler, J., Ringle, C.M., Sarstedt, M.: A new criterion for assessing discriminant validity in variance-based structural equation modeling. J. Acad. Mark. Sci. **43**, 115–135 (2015)
55. Chin W.W.: How to write up and report PLS analyses. In: Esposito Vinzi, V., Chin, W., Henseler, J., Wang, H. (eds.) Handbook of Partial Least Squares. Springer Handbooks of Computational Statistics. Springer, Berlin, Heidelberg (2010)

Electronic Word of Mouth Engagement Model in Social Commerce

Yusuf Sahabi Ali[1,2](✉) and Ab Razak Che Hussin[1]

[1] Universiti Teknologi Malaysia, 81310 Johor Bahru, Malaysia
sahabiali@yahoo.com, abrazak@utm.my
[2] Ahmadu Bello University, Zaria, Nigeria

Abstract. Customer engagement in electronic word of mouth (eWOM) has attained a sizable focus from both practitioners and the academia. With the growth of the e-commerce industry and the proliferation of social media as a marketing tool, a new business paradigm has been created termed social commerce (s-commerce). S-commerce is a combination of commercial and social activities, in which, individuals may spread word of mouth (WOM) about their shopping experiences, knowledge, as well as providing information about product and services to their friends. This kind of social interaction among individuals has increased the potential of eWOM communication. Several studies have investigated the influence of eWOM from the perspective of either consumer behavior or information characteristics. However, to effectively understand eWOM engagement and its impact, **researchers** must consider both information characteristic and consumer behavior in an integrated model. Given such a backdrop, the present study intends to look into the impact of eWOM engagement on consumers' purchase intention in s-commerce considering both information characteristics, consumer behavior, technical and social factors, which will complement the current effort of the research community in this field. The present study proposes a theoretical model for eWOM engagement by means of elaboration likelihood model, theory of reasoned action and social support theory. We believe that the proposed model will provide a significant step and an effective foundation for future eWOM research in s-commerce.

Keywords: Customer engagement · Social commerce · WOM · EWOM
Social media

1 Introduction

With the appearance of the web, the impact of word of mouth (WOM) has become even more prominent [1]. Moreover, the evolution of Web 2.0, has changed the web into a social domain, especially with the introduction of social media, in which individuals can connect and create content online [2]. With the aid of these social domains, individuals now become sources of valuable information to their peers. This shared information is alluded to as Electronic Word of Mouth (eWOM) and is defined as any positive or negative statement made by potential, actual, or former customers about a product or

F. Saeed et al. (Eds.): IRICT 2018, AISC 843, pp. 985–994, 2019.
https://doi.org/10.1007/978-3-319-99007-1_91

company, which is made available to a multitude of people and institutions via the internet [1]. eWOM conversations among consumers in s-commerce often revolves around brands [3]; and hence, they are naturally expected to have a strong impact in influencing consumers' behaviour [4]. Recent empirical studies have shown that social collaboration in the form of WOM is a prerequisite element for a successful s-commerce [5]

Industrial statistics have given proof supporting the noteworthy effect of eWOM communication. A study conducted uncovered that 61% of consumers refer to online reviews and blogs before acquiring a new product or service, and 80% of consumers before settling on their buying choice, are willing to purchase online after consulting online customers' reviews [6]. Due to these enormous potentials of eWOM, researchers are now showing great interest in eWOM; in which considerable amount of research has been carried out [1]. With the ever growing market competition in commerce, eWOM communication is now effective in moulding consumers' attitudes and behaviours [7]. Nowadays, online business firms connect with their customers in social networking sites (SNSs) so as to get important feedbacks on their activities [8]. This engagement marks the beginning of a change in ways of doing business which is called s-commerce [9, 10]. S-commerce is defined as an evolution of e-commerce, in which, social factors are the drivers of this business model and consumers are engaged to produce content [11].

Previous studies in relation to the influence of eWOM information were reported to have focused on either the characteristics of eWOM information [12, 13] or consumers' behaviour towards eWOM information [14, 15]. However, according to a latest study, investigating the impact of eWOM depends on both the information characteristics and consumers behavior [16]. The present study concurs with [16] suggestion which is deemed necessary in improving our understanding of eWOM engagement. Nevertheless, we also argue that [16] suggestion may be necessary, but it is not a sufficient requirement in investigating the influence of eWOM engagement in s-commerce. Additionally, investigating the effects of eWOM engagement remains a crucial endeavor [17]. The objective of the research is to investigate the suggestions by [16, 17]. Therefore, the present study contributes to the research community in several fronts, namely, improving our knowledge of users' engagement with eWOM by developing a comprehensive model that further explains this phenomenon in s-commerce. Apart from that, managers may become familiar with the determinants of eWOM engagement and develop a better marketing strategy with eWOM as a marketing tool.

2 Theoretical Foundation

2.1 Social Commerce and EWOM

With the development observed in Web 2.0 and online networking, e-commerce firms have begun to coordinate new innovations on their sites with the aim of providing their customers with more social and engaging shopping experience [18, 19]. The proliferation of social media has given individuals to leeway to participate in virtual discussions in order to share their experiences [20]. Such shared experiences and opinions are termed eWOM. One of the rationales behind s-commerce is that it is capable of benefitting commercial transactions of merchants by means of establishing cordial relationship with customers, improving the nature of that relationship among consumers and businesses [2]. S-commerce emphasize the importance of communities which relies on an information sharing network capable of helping consumers to fulfill their needs [21].

2.2 Elaboration Likelihood Model (ELM)

At present, consumers nowadays are flooded with massive amount of information with regards to products, services and brands. Constrained by time and other factors, individuals may not be able to process every persuasive messages from data sources like firms, colleagues, or forums [22]. Thus, certain aspect of the information will be given a deep thought. The model notably offers a valuable framework for acknowledging the viability of an advocacy [23]. The elaboration likelihood model (ELM) presents two potential routes to persuasion which may impact development and changes in consumer behavior (i.e., the central route (CR) and peripheral route (PR)). Specifically, CR to persuasion is portrayed as the way towards participating in keen, and sound processing of persuasive advocacy [23]. On the other hand, the PR to persuasion is portrayed by low subjective exertion and an emphasis on non-content, and extraneous cues when processing an advocacy [23].

Some recent studies reported in the literature have widely employed ELM to investigate eWOM communication [7, 24–26]. For example, [24] critically discussed the drivers of eWOM when recipients intend to accept and use eWOM.

2.3 Theory of Reasoned Action (TRA)

It has to be noted that attitude is one of the critical factors accounting for human behaviour in various circumstances. It can therefore be defined as a learned disposition to react in a good or horrible way concerning a given object or behaviour [27]. In this regard, Theory of Reasoned Action (TRA) propagates that a person's behaviour is affected largely by their intentions to carry out that behaviour, and that such behavioural intention is considered as an element function of his/her attitude toward the behaviour. TRA has frequently been used by previous researchers to look into the relationship identified between eWOM and purchase intention [1, 28, 15]. [27] contended that as people's attitude towards behaviour is determined by their salient beliefs about the behaviour, their attitude could therefore be changed by means of influencing their primary beliefs. Such view discussed in the literature help justify the relevance of this theory for this study.

2.4 Social Support Theory (SST)

The growing popularity of social media has transformed it into an important platform for encouraging social associations. Social support has been observed to be a noteworthy social value which the Internet users are capable of acquiring from various online groups [29–31]. Therefore, social support is defined as social resources which are readily provided by individuals belonging to social group [32]. Social support is a multidimensional construct under which, the components could differ from one context to another [29]. Considering that the interactions on a virtual settings are often based on message exchange, online support in s-commerce is usually intangible in nature, leaving informational support and emotional support [29, 33]. Interestingly, the fundamental motivation behind why consumers participate in virtual group is for social support and valuable information [34].

3 Research Model and Hypothesis Development

In this study we consider the suggestion by [16] in developing the research model where information quality, information credibility, website quality, social support and attitude towards eWOM determine eWOM engagement which eventually influences consumer purchase intention. Hence, the present study proposes a novel model which integrates ELM, TRA, and SST. The proposed model is presented in Fig. 1.

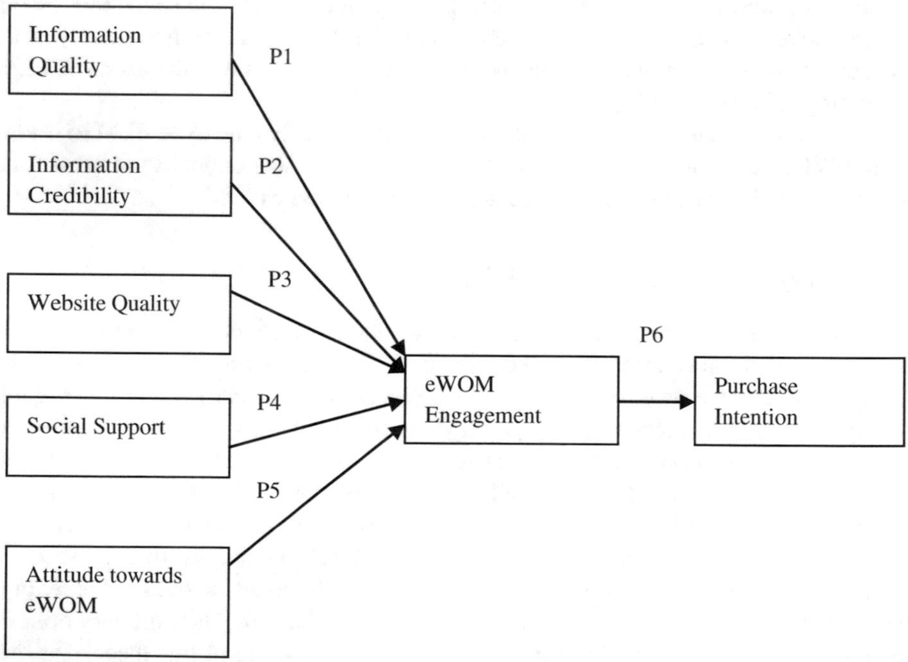

Fig. 1. The proposed eWOM engagement model

3.1 Information Quality

Information quality (IQ) is a generally utilized and established antecedent of the central route as it is referred to as the persuasive strength of a message [1]. In this study, IQ is defined as the persuasive strength of eWOM message. Consumers always attempt to process any given information to ascertain the veracity or otherwise of any given message. IQ is believed to be a vital construct by previous studies [1, 35, 36] and in the context of eWOM [1, 12, 37]. More specifically, previous researchers have reported that IQ may have a strong effect on purchase intention [38, 39]. Accordingly, we propose that:

P1: Information quality is positively related to eWOM engagement.

3.2 Information Credibility

Information credibility (IC) has been a major antecedent of peripheral routes due to the fact that it is the information recipients perception on the credibility of a given advocacy [1]. In this study, we adopt the definition of [1]. According to [1], consumers have the tendency to regard information as valuable when it is adjudged to be credible. Several researchers have demonstrated the importance of IC. For instance, a study found that IC is the main determinant in consumers' decision making process [40]. Moreover, studies also revealed the impact of IC on purchase intention [28]. Consequently, the present study proposes that:

P2: Information credibility is positively related to eWOM engagement.

3.3 Website Quality

Website quality is referred to as the customers perception of a website's performance in information retrieval and delivery [41]. In this study we adopt the definition provided by [41]. Website quality is measured using system and service quality. Good quality of a social networking websites make individuals to adjudge the site as effective in promoting interactions, which leads to continuous use, especially as a platform to exchange information [42]. The quality of a website may affect the users' views of the sites because transactions were carried out through the portal [43]. For this reason, the present study posits that consumers' eWOM engagement is influenced by websites quality. Hence, the following proposition is postulated:

P3: Perceived website quality is positively related to eWOM engagement.

3.4 Social Support

With regards to online communities, SNSs are the platform for users to acquire information through interaction with friends, which may result in positive emotional responses [45, 46]. Cobb considered social support as information leading the subject to believe that he is cared for and loved, esteemed, and a member of a network of mutual obligation [47]. This study adopts [47] definition of social support. It should be noted that the notion of social support has received considerable attention in s-commerce [2, 11, 45]. In an online context, individuals are more likely to receive informational and emotional supports [33, 48, 49]. Thus, this hypothesis is posited:

P4: The perception of social support is positively related to eWOM engagement.

3.5 Attitude Towards EWOM

The attitudes towards eWOM communication is defined as the over-all effect of 'goodness' or 'badness' of eWOM communication and is not specific to a certain episode of satisfaction [50]. This study adopts the definition provided by Fu et al. (2015). Attitude towards information is believed to be an important driver of eWOM information in s-commerce which strongly impacts purchase intentions. It is noteworthy that the attitude of consumers in relation to eWOM has been examined in several studies [28, 15, 50, 51].Consumers with positive attitudes towards eWOM in s-commerce will easily get engaged in eWOM communication. Therefore, it is proposed that:

P5: Attitude towards eWOM is positively related to eWOM engagement.

3.6 eWOM Engagement

Engagement refers to the emotional reactions to situations and to other stimuli from which the consumers form some bond or relationship with the stimulus [52]. Customer engagement (CE) can be defined as the behavioural manifestation of a customer towards firms, for instance, affecting the firms in ways other than the purchase alone [53, 54]. To ascertain the role of eWOM on consumers, we need to as a first step consider user engagement with eWOM communication. This may provide a deeper understanding of what individual users require for eWOM communication. In this study, eWOM engagement is defined as the willingness to request and share eWOM information with other customers. Consumers happen to develop a higher purchase intention when they engage in eWOM communication. Several related studies have looked into the influences of eWOM on consumer purchase intention [55, 56]. For example, it is reported that a significant relationship between eWOM and consumer purchase intention was observed [57]. Therefore, we predict that:

P6: eWOM engagement is positively related to consumers' purchase intention.

4 Discussion and Future Research Direction

The ultimate aim of this study is to investigate the influence of eWOM engagement in s-commerce. It is evident in the previous literature that, despite the growth of s-commerce, more efforts are needed not just to understand the phenomenon, but also to overcome the challenges of this area as it is still at its infancy. Upon reviewing the extant literature, it was found that most of the studies employ either information characteristics or consumers' behaviour in studying eWOM communication. However, it is noteworthy a combination of information characteristics and consumers' behaviour is of utmost importance to effectively understand eWOM communication. Apart from that, technological features or the medium through which eWOM communication is carried out, as well as the social interaction aspect of s-commerce have to be taken into consideration. Along these lines, the present study proposes a model of eWOM engagement on consumers' purchase intention in s-commerce by means of incorporating information characteristics, consumers' behaviour, technological features and social support. This is achieved by integrating the ELM, TRA and SST.

4.1 Conclusions

To conclude, this study investigates the influence of eWOM engagement in s-commerce from the perspective of information characteristics, consumer behavior, social support and technological factors. In doing so, several factors, namely information quality, information credibility, website quality, and social support which may provide significant effects on eWOM engagement in s-commerce were identified and presented in a comprehensive model. The study argued that the influence of eWOM engagement in s-commerce does not depend on either information characteristics or consumer behavior. It is our believe that the present study will encourage further research on eWOM engagement as well as its consequence in s-commerce by shedding light on the variables and propositions that require further investigation. We therefore encourage researchers to explore how other factors such as tie strength, homophily, gender, and consumer skepticism affects eWOM engagement in s-commerce.

Acknowledgement. This work is supported by the Ministry of Higher Education (MOHE) and Research Management Centre (RMC) at the Universiti Teknologi Malaysia (UTM) under the Research University Grant Category (VOT Q.J130000.2528.16H49).

References

1. Cheung, C.M.K., Thadani, D.R.: The impact of electronic word-of-mouth communication: a literature analysis and integrative model. Decis. Support Syst. **54**(1), 461–470 (2012)
2. Hajli, M.N.: The role of social support on relationship quality and social commerce. Technol. Forecast. Soc. Change **87**, 17–27 (2014)
3. Wolny, J., Mueller, C.: Analysis of fashion consumers' motives to engage in electronic word of mouth communication through social media platforms. J. Mark. Manag. **29**(5–6), 562–583 (2013)
4. Wang, X., Yu, C., Wei, Y.: Social media peer communication and impacts on purchase intentions: a consumer socialization framework. J. Interact. Mark. **26**(4), 198–208 (2012)
5. Wang, Y., Yu, C.: Social interaction-based consumer decision-making model in social commerce: the role of word of mouth and observational learning. Int. J. Inf. Manag. **37**, 179 (2015)
6. eMarketer: Online review sway shoppers (2008). http://www.ccsenet.org/journal/index.php/ijms/article/viewFile/18487/13453
7. Shih, H., Lai, K., Cheng, T.C.E.: Informational and relational influences on electronic word of mouth: an empirical study of an online consumer discussion forum. Int. J. Electron. Commer. **17**(4), 137–166 (2013)
8. Hajli, N., Lin, X., Featherman, M.S.: Social word of mouth: How trust develops in the market. Int. J. Mark. Res. **56**(3), 387–404 (2014)

9. Saundage, D., Lee, C.Y.: Social commerce activities—a taxonomy. In: ACIS 2011: Identifying the Information Systems Discipline: Proceedings of the 22nd Australasian Conference on Information Systems (2011)
10. Shanmugam, M., Sun, S., Amidi, A., Khani, F., Khani, F.: The applications of social commerce constructs. Int. J. Inf. Manag. **36**(3), 425–432 (2016)
11. Hajli, N., Sims, J.: Social commerce: the transfer of power from sellers to buyers. Technol. Forecast. Soc. Change **94**, 350–358 (2015)
12. Cheung, C.M.K., Lee, M.K.O., Rabjohn, N.: The impact of electronic word-of-mouth: the adoption of online opinions in online customer communities. Internet Res. **18**(3), 229–247 (2008)
13. Shu, M.L., Scott, N.: Influence of social media on Chinese students' choice of an overseas study destination: an information adoption model perspective. J. Travel Tour. Mark. **31**(2), 286–302 (2014)
14. Plotkina, D., Munzel, A.: Delight the experts, but never dissatisfy your customers! A multi-category study on the effects of online review source on intention to buy a new product. J. Retail. Consum. Serv. **29**, 1–11 (2016)
15. Reichelt, J., Sievert, J., Jacob, F.: How credibility affects eWOM reading: the influences of expertise, trustworthiness, and similarity on utilitarian and social functions. J. Mark. Commun. **20**(1–2), 65–81 (2014)
16. Knoll, J.: Advertising in social media: a review of empirical evidence. Int. J. Advert. **35**(2), 266–300 (2016)
17. Rossmann, A., Ranjan, K.R., Sugathan, P.: Drivers of user engagement in eWoM communication. J. Serv. Mark. **30**(5), 541–553 (2016)
18. Friedrich, T: Analyzing the factors that influence consumers' adoption of social commerce —a literature review. In: Twenty-first Americas Conference on Information Systems, pp. 1–16 (2015)
19. Curty, R.G., Zhang, P.: Social commerce: looking back and forward. Proc. Am. Soc. Inf. Sci. Technol. **48**(1), 1–10 (2011)
20. Cheung, C.M.K., Xiao, B.S., Liu, I.L.B.: Do actions speak louder than voices? The signaling role of social information cues in influencing consumer purchase decisions. Decis. Support Syst. **65**, 50–58 (2014)
21. Wang, T., Yeh, R.K.-J., Chen, C., Tsydypov, Z.: What drives electronic word-of-mouth on social networking sites? Perspectives of social capital and self-determination. Telemat. Inform. **33**(4), 1034–1047 (2016)
22. Miller, N., Maruyama, G., Beaber, R.J., Valone, K.: Speed of speech and persuasion. J. Personal. Soc. Psychol. **34**(4), 615–624 (1976)
23. Petty, R., Cacioppo, J.T.: The effects of involvement on response to argument quality and quantity: central and peripheral routes to persuasion. J. Personal. Soc. Psychol. **46**(1), 69–81 (1984)
24. Kok, S.T., Wei, W.K., Goh, W.A., Chong, Y.L.: Examining the antecedents of persuasive eWOM messages in social media. Online Inf. Rev. **38**(6), 746–768 (2014)
25. Shamhuyenhanzva, R.M., van Tonder, E., Roberts-Lombard, M., Hemsworth, D.: Factors influencing generation Y consumers' perceptions of eWOM credibility: a study of the fast-food industry. Int. Rev. Retail. Distrib. Consum. Res. **26**(4), 435–455 (2016)
26. Beneke, J., de Sousa, S., Mbuyu, M., Wickham, B.: The effect of negative online customer reviews on brand equity and purchase intention of consumer electronics in South Africa. Int. Rev. Retail. Distrib. Consum. Res. **26**(2), 171–201 (2016)
27. Ajzen, I., Fishbein, M.: Attitude-behavior relations: a theoretical analysis and review of empirical research. Psychol. Bull. **84**(5), 888–918 (1975)

28. Prendergast, G., Ko, D., Yuen, S.Y.V.: Online word of mouth and consumer purchase intentions. Int. J. Advert. **29**(5), 2 (2010)
29. Huang, K.-Y., Nambisan, P., Uzuner, Ö.: Informational support or emotional support: preliminary study of an automated approach to analyze online support community contents. In: ICIS 2010 Proceedings, pp. 1–11 (2010)
30. Obst, P., Stafurik, J.: Online we are all able bodied: online psychological sense of community and social support found through membership of disability-specific websites promotes well-being for people living with a physical disability. J. Community Appl. Soc. Psychol. **20**(6), 525–531 (2010)
31. Shaw, L.H., Gant, L.M.: In defense of the internet—the relationship between internet communication and depression, loneliness, self-esteem, and perceived social support. Cyberpsychol. Behav. **5**(2), 157–171 (2002)
32. Gottlieb, B.H., Bergen, A.E.: Social support concepts and measures. J. Psychosom. Res. **69**(5), 511–520 (2010)
33. Coulson, N.S.: Receiving social support online: an analysis of a computer-mediated support group for individuals living with irritable bowel syndrome. Cyberpsychology Behav. **8**(6), 580–584 (2005)
34. Hajli, N.: Social commerce constructs and consumer's intention to buy. Int. J. Inf. Manag. **35**(2), 183–191 (2015)
35. DeLone, W.H., McLean, E.R.: Information systems success: the quest for the dependent variable. Inf. Syst. Res. **3**(1), 60–95 (1992)
36. Sussman, S.W., Siegal, W.S.: Informational influence in organizations: an integrated approach to knowledge adoption. Inf. Syst. Res. **14**(1), 47–65 (2003)
37. Aghakhani, N., Karimi, J.: Acceptance of implicit and explicit eWOM: a factor based study of social networking sites. In: AMCIS 2013 Proceedings, pp. 1–6 (2013)
38. Lee, E.J., Shin, S.Y.: When do consumers buy online product reviews? Effects of review quality, product type, and reviewer's photo. Comput. Hum. Behav. **31**(1), 356–366 (2014)
39. Park, D.-H., Lee, J., Han, I.: The effect of on-line consumer reviews on consumer purchasing intention: the moderating role of involvement. Int. J. Electron. Commer. **11**(4), 125–148 (2007)
40. Awad, N.F., Ragowsky, A.: Establishing trust in electronic commerce through online word of mouth: an examination across genders. J. Manag. Inf. Syst. **24**(4), 101–121 (2008)
41. Yang, Z., Cai, S., Zhou, Z., Zhou, N.: Development and validation of an instrument to measure user perceived service quality of information presenting web portals. Inf. Manag. **42**(4), 575–589 (2005)
42. Teo, T.S.H., Srivastava, S.C., Jiang, L.: Trust and electronic government success: an empirical study. J. Manag. Inf. Syst. **25**(3), 99–132 (2009)
43. Ahn, T., Ryu, S., Han, I.: The impact of Web quality and playfulness on user acceptance of online retailing. Inf. Manag. **44**(3), 263–275 (2007)
44. Sabherwal, R., Chowa, C.: Information success: individual determinants organizational system. Manag. Sci. **52**(12), 1849–1864 (2006)
45. Bai, Y., Yao, Z., Dou, Y.: Effect of social commerce factors on user purchase behavior: an empirical investigation from renren.com. Int. J. Inf. Manag. **35**(5), 538–550 (2015)
46. Lee, J., Pi, S.-M., Kwok, R.C., Huynh, M.Q.: The contribution of commitment value in internet commerce: an empirical investigation. J. Assoc. Inf. Syst. **4**(1), 1–38 (2003)
47. Cobb, S.: Social support as a moderator of life stress. Psychosom. Med. **38**(5), 300–314 (1976)

48. Ommen, O., Janssen, C., Neugebauer, E., Bouillon, B., Rehm, K., Rangger, C., Erli, H.J., Pfaff, H.: Trust, social support and patient type-associations between patients perceived trust, supportive communication and patients preferences in regard to paternalism, clarification and participation of severely injured patients. Patient Educ. Couns. **73**(2), 196–204 (2008)
49. Welbourne, L., Blanchard, A.L., Boughton, M.D.: Supportive communication, sense of virtual community and health outcomes in online infertility groups. In: Proceedings of the fourth International Conference Communities Technology, p. 31 (2009)
50. Fu, J.R., Ju, P.H., Hsu, C.W.: Understanding why consumers engage in electronic word-of-mouth communication: perspectives from theory of planned behavior and justice theory. Electron. Commer. Res. Appl. **14**(6), 616–630 (2015)
51. Ayeh, J.K.: Travellers' acceptance of consumer-generated media: An integrated model of technology acceptance and source credibility theories. Comput. Hum. Behav. **48**, 173–180 (2015)
52. Kapoor, A., Kulshrestha, C.: Branding and Sustainable Competitive Advantage: Building Virtual Presence. IGI Global (2011)
53. van Doorn, J., Lemon, K.N., Mittal, V., Nass, S., Pick, D., Pirner, P., Verhoef, P.C.: Customer engagement behavior—theoretical foundations and research directions. J. Serv. Res. **13**(3), 253–266 (2010)
54. Bijmolt, T.H.A., Leeflang, P.S.H., Block, F., Eisenbeiss, M., Hardie, B.G.S., Lemmens, A., Saffert, P.: Analytics for customer engagement. J. Serv. Res. **13**(3), 341–356 (2010)
55. Baber, A., Thurasamy, R., Malik, M.I., Sadiq, B., Islam, S., Sajjad, M.: Online word-of-mouth antecedents, attitude and intention-to-purchase electronic products in Pakistan. Telemat. Inform. **33**(2), 388–400 (2016)
56. Chen, W., Chen, W.C., Chen, W.K.: Understanding the effects of eWOM on cosmetic consumer behavioral intention. Int. J. Electron. Commer. Stud. **5**(1), 97–102 (2014)
57. Sharifpour, Y., Sukati, I., Noor, M., Bin, A.: The influence of electronic word-of-mouth on consumers' purchase intentions in iranian telecommunication industry. Am. J. Bus. **5**(3), 1–6 (2016)

Social Media for Informal Learning Usage in Malaysia: Barriers and Benefits

Mohmed Y. Mohmed Al-Sabaawi[1,2](✉) and Halina Mohamed Dahlan[1]

[1] Faculty of Computing, Department of Information System, Universiti Teknologi Malaysia (UTM), Johor Bahru, Malaysia
mohutm@gmail.com, halina@utm.my

[2] Department of Management Information Systems, College of Administration and Economics, University of Mosul, Mosul, Iraq

Abstract. Social media (SM) tools have turned into a vital aspect of our day to day activities. As of late, this form of media has garnered a large set of users. There is perceived usefulness on the effect of SM on certain aspect of scholarly activities. However, it is still not clear why some researchers fail to adopt SM for informal learning (IL). The aim of this study is to investigate the barriers and benefits of using SM for IL among Malaysian academic researchers. A total of 170 responses were received through paper-based and online-based questionnaire. To demonstrate the results in this study, a descriptive interpretation of the responses is conducted. The general discovery of this study indicates that lack of encouragement from colleagues, lack of quality information, threat to research materials and data are the barriers affecting SM usage among academic researcher in Malaysia. Additionally, the core benefits of using SM among academic researchers in Malaysia is to communicate with colleagues, share knowledge with other researchers and to also collaborate in the research field.

Keywords: Academic researcher · Social media · Informal learning
Formal learning · Barriers · Benefits

1 Introduction

Recently, the evolution of SM such as Facebook and YouTube has led some academic researchers to consider its use in certain scholarly activities. Facebook as social networking sites (SNSs) provides a stage for academic researchers to collaborate and interact. YouTube is also SNSs that provides a stage for scholars and other users to share multimedia works, update their knowledge. The overwhelming popularity of these SNSs has resulted in their adoption into the academic environment [1–5]. Facebook is founded in 2004 with one million users and now burst of more than 2 billion users [6]. Also, YouTube was founded in 2005 with 8 million users every day and now it burst of more than 1,32 billion subscribers in 2018 [7]. Nevertheless, the low rate of SM adoption for IL calls for investigation [8, 9].

This study empirically examines the barriers and benefits of using SM for IL in Universiti Teknologi Malaysia (UTM) which is one of the foremost research

F. Saeed et al. (Eds.): IRICT 2018, AISC 843, pp. 995–1001, 2019.
https://doi.org/10.1007/978-3-319-99007-1_92

universities in Malaysia. [10] Affirms that research universities are highly valuable in developing nations for them to effectively compete in the knowledge economy. As indicated by [11], innovations that make it easy to transfer knowledge and maximize collaboration among researchers play a major role in research growth and productivity. Researchers have shown that, university productivity in research ultimately result in favorable ranking among other universities [12, 13]. Undoubtedly, high level productivity among researchers always boils down to efficient collaboration and interaction. Hence, the ability of scholars is highly influenced by creating a collaborative atmosphere [14]. Thus, the usage of SM as a communication, interaction and collaboration tool would effectively improve research productivity which will in turn result in favorable ranking among other universities. The aim of this study is to identify the barriers and benefits of using SM for IL among academic research in Malaysia. The main objectives of this study are:

I. To find out the barriers affecting the use of SM for IL by academic researchers.
II. To find out the benefits of using SM for IL by academic researchers.

The paper is structured as follows: a brief review of the literature is presented in Sect. 2, followed by the research methodology in Sect. 3. The research findings are presented in Sect. 4. Section 5 concludes the paper and recommends some future research.

2 Literature Review

Very few studies have investigated the use of SM for IL [15]. This section present previous studies on SM for IL indicating their limitations. SNSs such as Facebook, has transform the Internet into a social environment which eventually encourage learners' to learn informally and disseminate information effectively [16]. The adoption of SNSs in the educational environment is not something new. An increasing number of studies in recent years have investigated the pedagogical potential of SNSs and its effectiveness as a learning tool. With results showing a majority of participants frequently uses Facebook and YouTube for communication and collaboration [17].

[18] Conducted a survey on 711 academics to investigate how they integrate SM into their work activities. A large number of the respondents claim to use SM for their research work. The study conducted by [19] investigates why researchers use SM. 55% of the respondents indicate that they mostly use Facebook. Furthermore, [20] investigate the use of Facebook among different age groups, they found that females and younger population are frequent users. This finding is corroborated by [21] who indicate gender and age differences in the use of SM.

Recent literature by [22] provides insights on how the structure of SM varies across disciplines. The research showed that discipline plays an important role in SM usage. [8] Noticed that a large number of SM features are adopted based on disciplines. Certain literatures indicate these differences according to disciplines [4, 23–25]. As indicated by [26], there exists a large difference in SM membership rate based on disciplines.

Given the extent of literature reviewed, it can be inferred that research has mainly focused on the influence of SM rather than its barriers and benefits to the academic researchers. Hence, the ultimate goal of this study is to identify the barriers and benefits of SM among researchers in Malaysia.

3 Methodology

The present study employed a survey technique to collect data. The questionnaires comprises of twelve questions and all the questions were adapted from prior literatures [24, 26–28].

The convenience sample was employed to investigate the barriers and benefits of using SM for IL by academic researchers in Malaysia. The sample includes the academic researchers consist of Postgraduate Students, Research Fellow and Academic Staff of UTM. A total of 170 responses were collected using paper-based and online-based questionnaire. The collection process began in February, 2018 and lasted one month. The collected data was analyzed using Statistical Package for Social Science for Windows (SPSS for Windows Version 25.0). The level of significance at probability level of 5% was used.

4 Research Findings

In this study, majority of the respondents were male (63.5%) and also 36.5% were female, indicating a good mix among the genders in an academic environment (see Table 1). A larger part of the respondents were aged between 25 to 30 years (34.7). With regards to position, 106 (62.4%) of the respondents were Postgraduate students (PhD and Master). Table 1 shows that a large part of the respondents were found not to use SM (62%, n = 106). Whereas, a lesser figure of the respondent (38%, n = 64) have experience in using SM for IL.

4.1 Barriers to Using Social Media for Informal Learning by Academic Researchers

The barriers that hinder researchers from using SM for IL were identified from the survey (see Table 2). The major barrier as indicated by the researchers was, lack of encouragement from colleagues on the need to use SM for IL (66.04%, n = 70), followed by lack of quality of information (59.43%, n = 63), threat to research materials and data (57.55%, n = 61), lack of necessity to use SM for IL (57.55%, n = 61) and finally time concerns (52.83%, n = 56).

4.2 Benefits to Using Social Media for Informal Learning by Academic Researchers

The evolution of SM has opened a new vista for researchers to collaborate and interact with colleagues and ultimately share resources. The benefits of SM for IL as indicated

Table 1. Characteristics of respondents

	Users		Non-users		Total	
	n	%	n	%	N	%
Gender						
Male	45	26.4%	63	37.1%	108	63.5%
Female	19	11.2%	43	25.3%	62	36.5%
Age						
Less than 25–30 years	19	11.2%	40	23.5%	59	34.7
31–35	14	8.2%	16	9.4%	30	17.6
36–40	13	8.8%	18	9.4%	31	18.2
41–45	11	6.5%	11	6.5%	22	12.9
More than 45	7	4.1%	21	12.4%	28	16.5
Position						
Academic Staff (Lecturer, Senior Lecturer, Associate Professor, Professor)	19	11.2%	33	19.4%	52	30.6%
Research Fellow	5	3%	7	4%	12	7%
Postgraduate Student	40	23.5%	66	38.9%	106	62.4%
Experience as a researcher						
less than 1 year	6	3.5%	14	8.2%	20	11.8
1–3 years	18	10.6%	35	20.6%	53	31.2
3–5 years	15	8.8%	25	14.7	40	23.5
5–10 years	13	7.6%	14	8.2	27	15.9
More than 10 years	12	7.1%	18	10.6	30	17.6

Table 2. Barriers associated with use of SM for IL by academic researchers

Barriers	Respondents	Percentage (%)
There is lack of encouragement from colleagues on the need to use SM for IL	70	66.04
I don't use SM for IL because most information obtained are lack of quality	63	59.43
I don't use SM for IL because it is a threat to my research materials and data	61	57.55
I don't feel any necessity to use SM for IL	61	57.55
I don't have time to use SM for IL	56	52.83

by researchers are presented in Table 3. The findings indicate that researchers need to keep up to date is the most common benefit (81.25%), followed by communicating my research with research partners (81.25%), facilitating interaction with my research partners (81.25%), communicating about my research with researchers globally (79.68%), exchanging knowledge more quickly with other researcher (78.12%), share knowledge with other researcher (75.00%), communicate with the renowned experts in my research field (67.18%), facilitate collaboration with my researcher partners (64.06%), get feedback about my research from other researchers (60.93%),discuss my research method with other researchers(57.81%), discuss my research finding with other researchers(56.25%),facilitate collaboration with my research respondents to collect the required data(53.12%),discuss with other research in conducting literature review(51.56%),and find collaborators for my research projects (50.00%).

Table 3. Benefits of using SM for IL by academic researchers

Benefits	Respondents	Percentage (%)
I use SM to communicate with others researcher to keep up to date with the new information related to my research field	52	81.25
I use SM to communicate about my research with my research partners	52	81.25
I use SM to facilitate interaction with my research partners	52	81.25
I use SM to communicate about my research with researchers globally	51	79.68
I use SM to exchange knowledge more quickly with other researcher	50	78.12
I use SM to share knowledge with other researcher	48	75.00
I use SM to communicate with the renowned experts in my research field	43	67.18
I use SM to facilitate collaboration with my researcher partner	41	64.06
I use SM to get feedback about my research from other researchers	39	60.93
I use SM to discuss my research method with other researchers	37	57.81
I use SM to discuss my research finding with other researchers.	36	56.25
I use SM to facilitate collaboration with my research respondents to collect the required data	34	53.12
I use SM to discuss with other research in conducting literature review	33	51.56
I use SM to find collaborators for my research projects	32	50.00

Note: multiple answers are permitted (No. of user = 64)

5 Discussion

The overall findings in this research showed that a majority of the respondents do not use SM for IL. Respondents who have experienced the use of SM indicate that the most common benefit derived are: keeping up to date, communicating about research activities with other researchers, facilitation of interaction with research partners, communicate about research with researchers globally, exchange and share knowledge more quickly with other researchers. Whereas, the major barrier to the use of SM for IL by academic researchers are: lack of encouragement from colleagues, most information obtained lack quality, threat to research materials and data, lack of any necessity to use SM for IL and time concerns. The findings in this paper will help policy makers in tackling this barriers and maximizing the identified benefits to the research community.

6 Conclusion and Limitations

The research set out to investigate the barriers and benefits of using social media for informal learning by academic researchers in Malaysia. The research identifies the major barriers that affect the adoption of social media for informal learning by academic researchers. It also present the benefits associated with using social media for informal. In conclusion policy and decision makers should help in overcoming the identified barriers in this study, and maximize the benefit inherent in social media use for informal learning by academic researchers. As this will go a long way in reducing the pressure on the limited resources in academic environments.

Though the findings of this paper are far reaching they are not devoid of limitations. Firstly, the survey was conducted in a single university (UTM), generalizing its findings to the whole population must be done with caution. The research used quantitative method; other research could consider using alternative methods as this might provide varying results.

References

1. Manasijević, D., Živković, D., Arsić, S., Milošević, I.: Exploring students' purposes of usage and educational usage of Facebook. Comput. Hum. Behav. **60**, 441–450 (2016)
2. Jaffar, A.A.: YouTube: an emerging tool in anatomy education. Anat. Sci. Educ. **5**(3), 158–164 (2012)
3. Krauskopf, K., Zahn, C., Hesse, F.W.: Leveraging the affordances of Youtube: the role of pedagogical knowledge and mental models of technology functions for lesson planning with technology. Comput. Educ. **58**(4), 1194–1206 (2012)
4. Moran, M., Seaman, J., Tinti-Kane, H.: Teaching, Learning, and Sharing: How Today's Higher Education Faculty Use Social Media. Babson Survey Research Group (2011)
5. Kirschner, P.A., Karpinski, A.C.: Facebook® and academic performance. Comput. Hum. Behav. **26**(6), 1237–1245 (2010)
6. Facebook Homepage. http://www.newsroom.fb.com/company-info. Accessed 22 Jan 2018
7. YouTube Homepage. http://www.statisticbrain.com/youtube-statistics. Accessed 22 Jan 2018
8. Bullinger, A., Renken, U., Moeslein, K.: Understanding online collaboration technology adoption by researchers—a model and empirical study (2011)
9. Church, M., Salam, A.F.: Facebook, the spice of life? In: Proceedings of ICIS, p. 212 (2010)
10. Altbach, P.G.: Peripheries and centers: research universities in developing countries. Asia Pac. Educ. Rev. **10**(1), 15–27 (2009)
11. He, Z.-L., Geng, X.-S., Campbell-Hunt, C.: Research collaboration and research output: a longitudinal study of 65 biomedical scientists in a New Zealand University. Res. Policy **38** (2), 306–317 (2009)

12. Da Silva, N., Davis, A.R.: Absorptive capacity at the individual level: linking creativity to innovation in academia. Rev. High. Educ. **34**(3), 355–379 (2011)
13. Liu, N.C., Cheng, Y.: The academic ranking of world universities. High. Educ. Eur. **30**(2), 127–136 (2005)
14. Abramo, G., D'Angelo, C.A., Murgia, G.: Gender differences in research collaboration. J. Inform. **7**(4), 811–822 (2013)
15. Manca, S., Ranieri, M.: Networked scholarship and motivations for social media use in scholarly communication. Int. Rev. Res. Open Distrib. Learn. **18**(2), 124 (2017)
16. Rashid, R.A., Rahman, M.F.A.: Social networking sites for online mentoring and creativity enhancement. Int. J. Technol. Enhanc. Learn. **6**(1), 34–45 (2014)
17. Nentwich, M., König, R.: Academia goes Facebook? The potential of social network sites in the scholarly realm. In: Opening Science, pp. 107–124. Springer, Berlin (2014)
18. Lupton, D.: Feeling Better Connected: Academics' Use of Social Media. News & Media Research Centre, University of Canberra, Canberra (2015)
19. Nature Publishing Group: NPG 2014 Social Networks Survey. Figshare (2014)
20. Thelwall, M., Kousha, K.: Academia.edu: social network or academic network? J. Assoc. Inf. Sci. Technol. **65**(4), 721–731 (2014)
21. Poellhuber, B., Anderson, T., Racette, N., Upton, L.: Distance students' readiness for and interest in collaboration and social media. Interact. Technol. Smart Educ. **10**(1), 63–78 (2013)
22. Jordan, K.: Academics and their online networks: exploring the role of academic social networking sites. First Monday **19**(11) (2014)
23. Maron, N.L., Smith, K.K.: Current Models of Digital Scholarly Communication: Results of an Investigation Conducted by Ithaka for the Association of Research Libraries. Association of Research Libraries, Washington (2008)
24. Rowlands, I., Nicholas, D., Russell, B., Canty, N., Watkinson, A.: Social media use in the research workflow. Learn. Publish. **24**(3), 183–195 (2011)
25. Al-Aufi, A., Fulton, C.: Impact of social networking tools on scholarly communication: a cross-institutional study. Electron. Libr. **33**(2), 224–241 (2015)
26. Jamali, H.R., Russell, B., Nicholas, D., Watkinson, A.: Do online communities support research collaboration? Aslib J. Inf. Manag. **66**(6), 603–622 (2014)
27. Madhusudhan, M.: Use of social networking sites by research scholars of the University of Delhi: a study. Int. Inf. Libr. Rev. **44**(2), 100–113 (2012)
28. Schöndienst, V., Krasnova, H., Günther, O., Riehle, D., Schwabe, G.: Micro-blogging adoption in the enterprise: an empirical analysis. 931–940 (2011)

Towards the Development of a Citizens' Trust Model in Using Social Media for e-Government Services: The Context of Pakistan

Sohrab Khan(✉), Nor Zairah Ab. Rahim, and Nurazean Maarop

Advanced Informatics School, Universiti Teknologi Malaysia, 541000 Kuala Lumpur, Malaysia
skhan2@live.utm.my, {nzairah,nurazean.kl}@utm.my

Abstract. The recent evolvements in social media attracted governments around the world to utilize its benefits for providing e-government services and rebuild their relationship with citizens' through participation and engagement. However, lack of citizens' trust is still a serious concern that discourages them to participate in these services. Majority of previous studies on relationship with people and government were conducted on e-government websites. Very few studies have attempted to investigate citizens' trust factors in context of using social media for e-government services particularly in a developing country like Pakistan. Therefore, this study is designed to bridge this gap by identifying such factors and propose a model from multiple perspectives of citizens trust. The proposed model will assist government organizations in understanding factors that may influence citizens trust to participate in social media for e-government services.

Keywords: Antecedents of trust · Government social media · e-Government Trust factors

1 Introduction

In recent years the development of e-government services has migrated from static websites to social media [1]. Social media is defined by Harris and Rea [2] as "the perceived second generation of web development and design which facilitates secure online communication, information sharing, interoperability and collaboration on world wide web". Unlike static government websites based on one-way interaction, social media provides more interactive features for citizens' to understand e-government and serves as a two-way dialogue channel between government and public [1]. Social media enables users to generate their content and share up-to-date information in real time environment. Government organizations need to wait for citizens to visit their e-government websites. On the contrary using social media sites can allow government organizations to reach closer to their citizens. Government organizations around the world have started using social media to rebuild their relationship with citizens' and increase their level of engagement and participation. Social media is easily accessible and very cheap to afford now days. This is particularly important in developing countries that do not have a dedicated online portal for public consultation [3, 4]. According to the latest

F. Saeed et al. (Eds.): IRICT 2018, AISC 843, pp. 1002–1012, 2019.
https://doi.org/10.1007/978-3-319-99007-1_93

survey of United Nations, 152 out of 193 UN countries (four out of five) offer social networking features on their national portals [4]. Government organizations use social media for various functions but not limited to communicating current events, disaster and disease alerts, weather reporting, disseminating awareness about various government jobs, education, tourism, law enforcement, political campaigns etc. [3, 5]. However, despite of having the popularity of social media among governments, its acceptance among citizens' is still a serious issue [6]. The efforts of government organizations and potential value of social media cannot be displayed without addressing citizens trust to use these services [1, 3].

Trust is a major concern that develops citizens willingness to participate with government and use its electronic services. Previous studies have confirmed the significant role of trust in citizens' adoption of e-government service [6, 7]. According to Al-Khouri [8] developing trust is the most important parameter to promote citizens' participation in e-government services. However, there is limited prior focus on how to generate citizens' trust to increase their level of participation in these services [7]. Previous studies have not paid much attention in context of using social media as a technology for e-government services, especially in a developing country like Pakistan. Though a few studies have highlighted the significance of trust as an essential determinant for citizen-government engagement through social media [3, 9]. However, there is lack of thorough investigation to identify the factors that can generate citizens' trust in context of using social media for e-government services. This paper fills this gap and examines citizens' trust factors from multiple perspectives. A model will be proposed to identify those factors that may impact citizens' trust to use social media for e-government services. The identification of these factors will assist government organizations in understanding citizens' trust aspects to use these services. This will help to strengthen collaboration and engagement of government organizations with citizen's in a more transparent and cost-effective way.

2 Methodology

An extensive literature review was conducted through several databases including Scopus, IEEE, ACM, Science Direct, Springer link, Taylors and Francis and Google Scholar. Several keywords were chosen to identify relevant set of articles to meet the objectives of this study. Such as "antecedents of trust", "citizens trust", "e-government factors", "government social media" etc. The keywords were also linked together using logical AND. In addition, a detailed analysis of materials from journals, conference proceedings and book chapters were carried out to gather results for this study.

3 Factors Influencing Citizens' Trust

This section discusses different concepts of trust and related studies for establishing the background and context of this paper.

3.1 e-Government in Pakistan

The very first web portal of the Government of Pakistan was launched in 2005 by the Electronic Government Directorate (EGD) of the Ministry of Science and Technology. E-government Directorate (EGD) of Pakistan was set up as a unit within the ministry of IT to focus specifically on e-government. The goals of government are to increase transparency and accountability in decision making and to enhance delivery of public services to the citizens efficiently and cost effectively [10]. Like any other developing country Pakistan is also facing a lot of challenges to implement e-government projects in the country [10]. According to the latest survey by United Nation in 2016 Pakistan was ranked at 159th position with very low E-government development index (EGDI) and E-participation index, which was even below than the average in Asian countries. This indicates that citizens' usage of government portals for e-government services is positioned at a very low level in the country. According to Susanto and Aljoza [11] citizens' in developing countries who are used to dealing with poor public services need to trust on government online services before they decide to use the services. Haider, Shuwen [12] stressed that e-government services are likely to be adopted in Pakistan with having social media presence. The usage of social media has increased rapidly in Pakistan since the introduction of high speed internet and mobile broadband (3G and 4G) in 2014 [13]. A significant majority of Pakistani citizens' who use the internet are using social media websites such as Facebook, Twitter, etc. In recent years government organizations in Pakistan have also realized the importance of maintaining a social media presence to reach to their people. In Pakistan, Federal and Provincial governments, regulatory authorities, military public relations and political parties use social media to deliver messages and get feedback [13]. The official social media presence in all government ministries is increasing and most of them have social media links on their official portals. However, these efforts of government organizations to utilize the benefits of social media strongly depends on citizens' trust to participate in these services. Citizens' need to trust on these services before they decide to use them [11]. Therefore, this draws our attention to study and understand the citizens' trust in using social media for e-government services in context of Pakistan.

3.2 Antecedents of Trust in e-Government Services

Trust is a very rich concept that covers a wide range of relationships with a variety of objects. Trust has been defined as the willingness of one party to depends on unfamiliar party where the trustor does not have credible meaningful information or bonds with trusted party [14]. There are two targets trust in online services. The entity providing services and the mechanism through which service is provided. Therefore, it's important that characteristics of both vendor and the supporting technological platform should be considered before using any service [15]. The literature shows that majority of previous studies have addressed technical aspect of technology and government factors as important antecedents of trust in e-government services [7]. Some previous studies have also reflected perceived risk as an important factor influencing trust, along with technical and government factors [7, 16]. Though citizen aspect was also discussed by some researchers but has not been addressed thoroughly with other trust factors in context of e-government services [7]. There are very few studies that have also used multidimensional approach to highlight factors influencing trust. For example, Beldad, Geest [17] categorized trust in online environment into three dimensions: which are Internet user based, Organization based and Web based determinants of trust in online Environment. Similarly, Alzahrani, Al-Karaghouli [7] added four factors in their framework to identify antecedents of citizens trust in e-government services, which are technology, government agencies, perceived risk and citizens' characteristics. However, those studies were limited on e-government websites using general technological characteristics of internet. Social media involves a variety of information from various sources and is likely to be more sensitive and risky than e-government websites. Although being a relatively new approach of e-government practice, there is some prior research on social media in government [18]. Trust has been highlighted as a significant contributor on citizens intention to use social media for communicating with government organizations. For example ALotaibi, Ramachandran [9] mentioned public trust as an important dimension of government-public relationship in using government social media services. Similarly, Park, Kang [3] also emphasized that building citizen's trust is very critical in communication of government organizations through social media. Table 1 shows antecedents of trust discussed in previous studies from multiple perspectives including technology, risk, government and citizens factors.

Table 1. Multiple antecedents of trust

Trust factors	References
Technology	[11, 19, 20]
Technology + risk	[21, 22]
Technology + government	[6, 9, 23, 24]
Technology + government + risk	[16]
Technology + government + citizens	[15, 17, 25]
Technology + government + citizens + risk	[7, 26]

The literature reflects that that focus of majority of previous studies was only on technological and government factors with a very little attention on other aspects of citizen's trust. In addition, the factors were selected randomly without any proper categorization in most of the previous studies. Though few studies used multiple perspectives of trust, but their research was limited on e- government websites. There are very few studies in context of citizens trust to use social media for government services. Therefore, this can be argued that there is a missing gap of research to examine antecedents of citizen's trust from multiple perspectives in context of using social media for e-government services.

3.3 Important Causal Relationships and Constructs

The most important constructs and their relationship has been identified from literature and illustrated in Table 2. This table shows the most studied significant causal relationship, the corresponding studies or sources and total number of occurrence of causal relationship. The causal relationship is used to develop the research model of this study. Table 2 reflects multiple constructs and their relationships that have been found significant and validated in previous studies. Based upon the number of significantly tested occurrences of the causal relationships, a model is proposed that identifies antecedents of trust in government social media services and shows the influence of citizen's trust towards their intention to participate in social media for e-government services.

Table 2. Causal relationship.

Causal relationship	Validated studies	Total occurrence
DTT ⟶ Trust	[15, 25, 27, 28]	04
ABL ⟶ Trust	[14, 29–31]	04
BEN ⟶ Trust	[14, 29–31]	04
ITG ⟶ Trust	[14, 29–31]	04
PR ⟶ Trust	[16, 22, 26, 27, 32, 33]	06
SR ⟶ Trust	[16, 22, 26, 32, 33]	05
PEOU ⟶ Trust	[25, 32, 34–36]	05
SA ⟶ Trust	[14, 35, 37]	03
Trust ⟶ INT	[6, 7, 11, 15, 25, 30, 35]	07
IQ ⟶ Trust	[6, 16, 20, 26, 32]	05
PEOU ⟶ INT	[10, 24, 26, 28, 38, 39]	06
Trust ⟶ PU	[21, 34, 35, 39, 40]	05
PEOU ⟶ PU	[21, 28, 34, 35, 39]	05
PU ⟶ INT	[21, 26, 28, 34, 35, 39, 40]	07
UA ⟶ Trust	[41, 42]	02

Disposition to Trust (DTT), Ability (ABL), Benevolence (BEN), Integrity (ITG), Intention to use (INT), Security risk (SR), Privacy risk (PR), Perceived Ease of use (PEOU), Perceived Usefulness (PU), Structural assurances (SA), Uncertainty Avoidance (UA) and Information quality (IQ).

4 Proposed Model

An integrated model has been developed to provide a holistic view of influential factors on citizen's trust leading towards their intention to participate in government social media services as illustrated in Fig. 1. The theoretical foundation of this model is based on two streams of literature: i.e. TAM (Technology acceptance model) and the relevant literature on multiple trust aspects. The connections between trust and TAM has been validated in majority of previous literature showing a significant relationship of PEOU (perceived ease of use) and PU (perceived usefulness) with trust in online environment [34, 35]. This paper contributes in identifying antecedents of trust in context of using social media for e-government services where there is hardly any thorough model available in previous research. Furthermore, this paper also adds a factor of culture as antecedent of trust which must not be ignored to ensure the success of an e-government initiative [9]. The model illustrated in Fig. 1 provides a comprehensive understanding of citizen's trust factors from multiple trust perspectives including individual characteristics, government factors, risk factors, culture and social media characteristics. These perspectives are discussed as under:

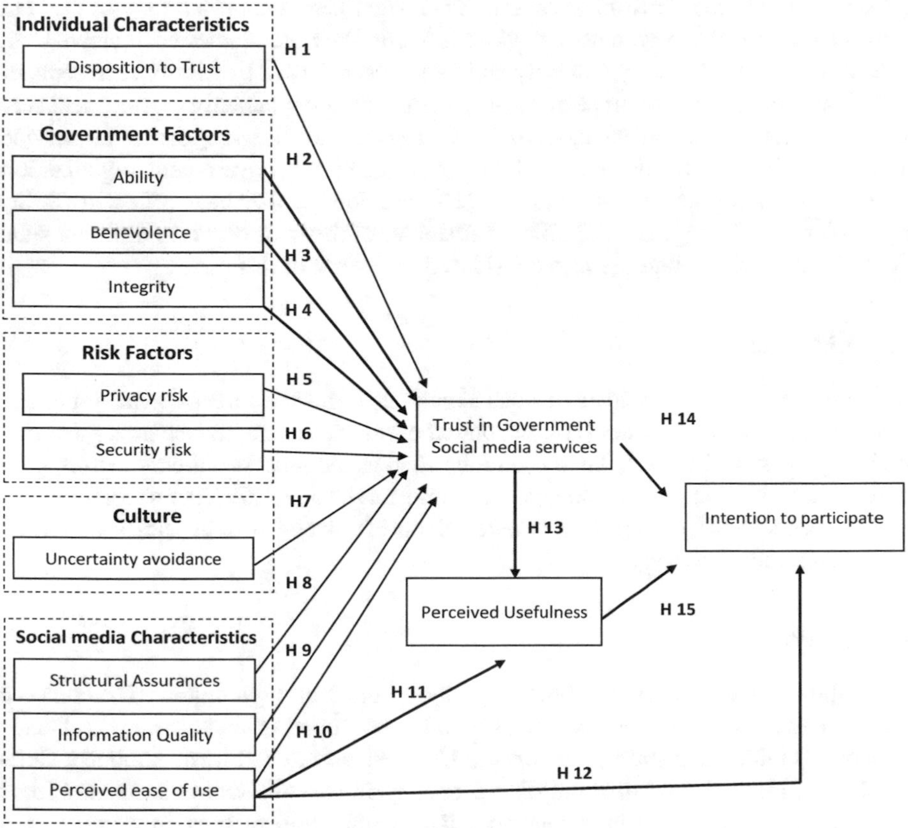

Fig. 1. Research model

4.1 Individual Characteristics

Citizens' characteristics influence their perception to trust on other parties. Previous studies have highlighted disposition to trust as a significant factor that can influence citizen's behavior to trust in e-government services [15, 25, 27]. Disposition to trust refers to a tendency to be willing to depend on others [14]. Some people display a greater disposition to trust on anything, anybody or a technology. They are more likely to trust in online government services despite of their limited information on these services. However, people with low disposition of trust generally require more information about something before deciding to trust on them [7, 17]. Using social media for government services is still in initial stages of development. Therefore, disposition to trust could be an important determinant of trust in using these services.

4.2 Government Factors

Citizen-Government relationship plays an important role in developing trust towards government services. Trust in government is a significant factor that may impact citizens' behavior to use government services [15, 26]. Previous studies have viewed trust in three dimensions: Ability, integrity and benevolence [14, 31]. Ability represents the organizational capability/competency of government to implement e-government services, Benevolence is about having the belief that government is taking care of the services in the best interest of citizens' and Integrity represents government organizations honesty and their promises with citizens [14]. Previous studies have reflected Ability, Benevolence and Integrity as significant attributes of trustworthiness in different situations [14, 29, 30] and thus are also considered in context of this paper.

4.3 Risk Factors

Risk is an important determinant that can negatively influence citizens' behavior to trust in online government services [10]. In context of e-government, the technological risks such as privacy and security have been identified as the most significant factors influencing citizens' trust towards adoption of these services [7, 26]. Therefore, this paper also considers security and privacy as antecedents of citizens' trust in using social media for e-government services.

4.4 Culture

It is believed that citizens' behavior can be explained through culture. Difference in culture may influence the way how trust can be developed [42]. According to Doney, Cannon [41] there is a strong relationship between culture and trust. Similarly Gefen and Heart [43] examined trust building processes in two different cultures and found that culture has a significant effect on citizens' intention to trust in context of e-commerce. Uncertainty avoidance as one of Hofstede [44] five cultural dimensions has been examined in a number of IS studies and found to have a significant influence on citizens' trust behavior [42]. Uncertainty avoidance refers to the level of risk tolerated by individuals when they are in uncertain situations [45]. According to Hofstede, the level of uncertainty avoidance in Pakistan shows high preference for avoiding uncertainty. He argued that people from such high uncertainty avoidance culture have lower uncertainty tolerance and higher structural needs (i.e., rules and regulations) than people from low uncertainty avoidance culture. Given that using social media for e-government services is a new emerging technology so there might be a degree of uncertainty among citizens to use these services. Therefore, this paper considers examining the relationship of uncertainty avoidance towards citizens' trust in using government social media services.

4.5 Social Media Characteristics

In literature, several factors have been identified as important parameters to investigate citizens' level of confidence to use a technology to interact with government organizations. Structural assurances is an institutional based trust that develops confidence of users to believe that there are policies and mechanism in place to keep their information safe from unauthorized access [14]. Doney, Cannon [41] argued that structural assurances play a significant role in building citizens' trust. This makes them to believe that organizations have no intention be act defectively. Similarly, Information quality is an essential determinant to generate trust and has been validated by researchers in using e-government services [16]. The credible exchanges of information between citizens' and government organizations on social media can generate their trust to use these services and thus has been considered in this paper. Majority of previous studies have used Technology Acceptance Model (TAM) and have supported a significant relationship of perceived ease of use and usefulness with trust in e-government services. Therefore, this paper also considers the significance of these technological factors in using social media as a technology in e-government services.

5 Discussion

This paper proposes an integrated model of citizens' trust in context of using social media for e-government services. This model contributes to provide a multidimensional perspective of citizens' trust including individual characteristics, government factors, risk factors, culture and social media characteristics. There is hardly any previous thorough model available in this context. In addition, this model also includes a very important dimension of culture to determine its relationship with citizens' trust towards government social media services.

There are several research and practical implications of this study. This paper fills the research gap due to lack of studies in context of using social media for e-government services. Further the proposed model is intended to be empirically examined in Pakistan. Being a developing country where the citizens' participation with government is at a very low level, government organizations are required to introduce the concept of using social media for government services and encourage citizens to use these services. The results can assist them to gain citizens trust and increase their participation in government social media services. Though the scope of this study is limited in Pakistan, but we believe with confidence that the results can be further enhanced and generalized in other contexts as well.

6 Conclusion

While majority of previous studies in relationship of trust between people and government were conducted on e-government websites, this paper presents multiple antecedents of citizens' trust in context of using social media for e-government services. The constructs were extracted based on their causal relationships that were empirically validated in multiple contexts in previous studies. This paper is an attempt to fill the literature gap due to lack of studies in context of using social media for government services especially in a developing country like Pakistan. The proposed model can be used by both researchers and practitioners to analyze and suggest recommendation to enhance citizens' trust in using government social media services.

References

1. Hao, X., Zheng, D., Zeng, Q.: How to strenghthen social media interactivity of e-government. Online Inf. Rev. **40**(1), 79–96 (2016)
2. Harris, A.L., Rea, A.: Web 2.0 and virtual world technologies: a growing impact on IS education. J. Inf. Syst. Educ. **20**(2), 137–144 (2009)
3. Park, M.J., Kang, D., Rho, J.J., Lee, D.H.: Policy role of social media in developing public trust: twitter communication with government leaders. Publ. Manag. Rev. **18**(9), 1265–1288 (2016)
4. UNPAN: United Nations E-government Survey. The Department of Economic and Social Affairs, New York (2016)
5. Park, M.J., Choi, H., Kim, S.K., Rho, J.J.: Trust in government's social media service and citizen's patronage behavior. Telemat. Inform. **32**(4), 629–641 (2015)
6. Alharbi, A., Kang, K., Hawryszkiewycz, I.: The influence of trust and subjective norms on citizens intentions to engage in E-participation on E-government websites. In: Australian Conference on Information Systems (2015)
7. Alzahrani, L., Al-Karaghouli, W., Weerakkody, V.: Analysing the critical factors influencing trust in e-government adoption from citizens' perspective: a systematic review and a conceptual framework. Int. Bus. Rev. **26**(1), 164–175 (2017)
8. Al-Khouri, A.M.: E-government in Arab countries: a 6-staged roadmap to develop the public sector. J. Manag. Strategy **4**(1), 80 (2013)
9. Alotaibi, R.M., Ramachandran, M., Kor, A.-L., Hosseinian-Far, A.: Factors affecting citizens' use of social media to communicate with the government: a proposed model. Electron. J. e-Gov. **14**(1), 60 (2016)
10. Rehman, M., Esichaikul, V., Kamal, M.: Factors influencing e-government adoption in Pakistan. Transforming Gov. People Process Policy **6**(3), 258–282 (2012)
11. Susanto, T.D., Aljoza, M.: Individual acceptance of e-government services in a developing country: dimensions of perceived usefulness and perceived ease of use and the importance of trust and social influence. Proc. Comput. Sci. **72**, 622–629 (2015)
12. Haider, Z., Shuwen, C., Lalani, F., Mangi, A.A.: Adoption of e-government in Pakistan: supply perspective. Int. J. Adv. Comput. Sci. Appl. **6**, 55–63 (2015)
13. Yusuf, A.: State of social media in Pakistan. https://propakistani.pk/2016/01/26/state-of-social-media-in-pakistan-in-2016/. Accessed 15 July 2016
14. McKnight, D.H., Choudhury, V., Kacmar, C.: Developing and validating trust measures for e-commerce: an integrative typology. Inf. Syst. Res. **13**(3), 334–359 (2002)
15. Bélanger, F., Carter, L.: Trust and risk in e-government adoption. J. Strateg. Inf. Syst. **17**(2), 165–176 (2008)
16. Ranaweera, H.M.B.P.: Perspective of trust towards e-government initiatives in Sri Lanka. SpringerPlus **5**(1), 22 (2016)
17. Beldad, A., Geest, T., Jong, M., Steehouder, M.: A cue or two and I'll trust you: determinants of trust in government organizations in terms of their processing and usage of citizens personal information disclosed online. Gov. Inf. Q. **29**, 41 (2012)

18. Warren, A.M., Sulaiman, A., Jaafar, N.I.: Social media effects on fostering online civic engagement and building citizen trust and trust in institutions. Gov. Inf. Q. **31**(2), 291–301 (2014)
19. Rodrigues, G., Sarabdeen, J., Balasubramanian, S.: Factors that influence consumer adoption of e-government services in the UAE: a UTAUT model perspective. J. Internet Commer. **15**(1), 18–39 (2016)
20. Weerakkody, V., et al.: Are U.K. citizens satisfied with e-government services? Identifying and testing antecedents of satisfaction. Inf. Syst. Manag. **33**(4), 331–343 (2016)
21. Lorenzo-Romero, C., Chiappa, G., Alarcon-del-Amo, M.: The users adoption and usage of social network sites: an empirical investigation in the context of Italy. In: International Marketing Trends Conference, Italy (2013)
22. Mohajerani, S., Shahrekordi, S.Z., Azarlo, M.: The impact of privacy and security concerns, trust in technology and information quality on trust in e-government and intention to use e-government. In: 9th International Conference on e-Commerce in Developing Countries: With Focus on e-Business (ECDC). IEEE (2015)
23. Wang, T., Lu, Y.: Determinants of trust in e-government. In: International Conference on Computational Intelligence and Software Engineering (CiSE) (2010)
24. Zhao, F., Khan, M.S.: An empirical study of e-government service adoption: culture and behavioral intention. Int. J. Publ. Adm. **36**(10), 710–722 (2013)
25. Alsaghier, H., Hussain, R.: Conceptualization of trust in the e-government context: a qualitative analysis. In: Active Citizen Participation in E-Government: A Global Perspective, pp. 528–557. IGI Global (2012)
26. Abu-Shanab, E.: Antecedents of trust in e-government services: an empirical test in Jordan. Transform. Gov. People Process Policy **8**(4), 480–499 (2014)
27. Colesca, S.E.: Understanding trust in e-government. Eng. Econ. **63**(4), 7 (2009)
28. Carter, L., Weerakkody, V., Phillips, B., Dwivedi, Y.K.: Citizen adoption of e-government services: exploring citizen perceptions of online services in the United States and United Kingdom. Inf. Syst. Manag. **33**(2), 124–140 (2016)
29. Colquitt, J.A., Scott, B.A., LePine, J.A.: Trust, trustworthiness, and trust propensity: a meta-analytic test of their unique relationships with risk taking and job performance. J. Appl. Psychol. **92**(4), 909 (2007)
30. Fuller, M.A., Serva, M.A., Baroudi, J.: Clarifying the integration of trust and TAM in e-commerce environments: implications for systems design and management. IEEE Trans. Eng. Manag. **57**(3), 380–393 (2010)
31. Mayer, R.C., Davis, J.H.: The effect of the performance appraisal system on trust for management: a field quasi-experiment. J. Appl. Psychol. **84**(1), 123 (1999)
32. Ayyash, M.M., Ahmad, K., Singh, D.: Investigating the effect of information systems factors on trust in e-government initiative adoption in Palestinian public sector. Res. J. Appl. Sci. Eng. Technol. **5**(15), 3865–3875 (2013)
33. Alam, S.S., Ahmad, M., Khatibi, A.A., Ahsan, M.N.: Factors affecting trust in publishing personal information in online social network: an empirical study of Malaysia's Klang Valley users. Geogr. Malays. J. Soc. Space **12**(2), 132–143 (2016)
34. Belanche, D., Casaló, L.V., Flavián, C.: Integrating trust and personal values into the technology acceptance model: the case of e-government services adoption. Cuadernos de Economía y Dirección de la Empresa **15**(4), 192–204 (2012)
35. Gefen, D., Karahanna, E., Straub, D.W.: Trust and TAM in online shopping: an integrated model. MIS Q. **27**(1), 51–90 (2003)

36. Kivijari, H., Leppanen, A., Hallikainen, P.: Technology trust: from antecedents to perceived performance effects. In: 46th International Conference on System Sciences. IEEE Computer Society, Hawaii (2013)
37. Pennington, R., Wilcox, H.D., Grover, V.: The role of system trust in business-to-consumer transactions. J. Manag. Inf. Syst. **20**(3), 197–226 (2003)
38. Venkatesh, V., Davis, F.D.: A theoretical extension of the technology acceptance model: four longitudinal field studies. Manag. Sci. **46**(2), 186–204 (2000)
39. Tung, F.-C., Chang, S.-C., Chou, C.-M.: An extension of trust and TAM model with IDT in the adoption of the electronic logistics information system in HIS in the medical industry. Int. J. Med. Inform. **77**(5), 324–335 (2008)
40. Yoo, D.-H., Ko, D.-S., Yeo, I.-S.: Effect of user's trust in usefulness, attitude and intention for mobile sports content services. J. Phys. Educ. Sport **17**, 92 (2017)
41. Doney, P.M., Cannon, J.P., Mullen, M.R.: Understanding the influence of national culture on the development of trust. Acad. Manag. Rev. **23**(3), 601–620 (1998)
42. Xin, H., Techatassanasoontorn, A.A., Tan, F.B.: Antecedents of consumer trust in mobile payment adoption. J. Comput. Inf. Syst. **55**(4), 1–10 (2015)
43. Gefen, D., Heart, T.H.: On the need to include national culture as a central issue in e-commerce trust beliefs. J. Glob. Inf. Manag. (JGIM) **14**(4), 1–30 (2006)
44. Hofstede, G.: Cultures and Organizations: Software of the Mind. McGraw Hill, New York (1997)
45. Srite, M., Karahanna, E.: The role of espoused national cultural values in technology acceptance. MIS Q. **30**, 679–704 (2006)

The Influence of Social Presence and Trust on Customers' Loyalty to Social Commerce Websites

Hilal Alhulail(✉), Martin Dick, and Ahmad Abareshi

RMIT University, GPO Box 2476, Melbourne, VIC 3001, Australia
{hilal.alhulail,martin.dick,
ahmad.abareshi}@rmit.edu.au

Abstract. Social Commerce (s-commerce) is a relatively new form of electronic commerce built around the notion of customer-oriented social website content, and made possible by the development of Web 2.0 technology. Businesses in this field, in order to develop effective marketing strategies, must understand the dynamics of how client loyalty is established and maintained. Scholars have explored the issue of customer loyalty from many angles, including social marketing, online environments and eCommerce. Some researchers have focused their studies on the behavioural dimension of customer loyalty, others on its attitudinal dimension, and still others on the composite dimension of customer loyalty. Though the importance of these individual influences has been broadly accepted, the societal aspect of s-commerce is still poorly understood, as is the influence on s-commerce of trust and social presence. The objective of this study is thus to investigate the impact of social presence (SP) and trust on s-commerce customer loyalty. Data were collected from s-commerce website users in Australia using a web-based questionnaire. Based on the results of the survey, social presence and trust both have significant influence on customer loyalty to s-commerce websites. The findings offer insight to business decision-makers who hope to improve their implementation of s-commerce as well as the design of their sites.

Keywords: Social commerce · Social presence · Customer loyalty
Trust

1 Introduction

With the development of eCommerce applications, innovative forms of online commerce have arisen. This new business model has led to the development of many new businesses, and attracted buyers in ever-increasing numbers [1]. Previously, an important difference between offline and online commerce was the lack of social elements in the eCommerce environment. Whereas buying items face to face involved a social interaction and the chance to ask questions of a real person, online shopping did not. More recently, this situation has been changed by integrating Web 2.0 capabilities into the eCommerce website applications. This evolution is usually referred to as the birth of social commerce (s-commerce).

F. Saeed et al. (Eds.): IRICT 2018, AISC 843, pp. 1013–1024, 2019.
https://doi.org/10.1007/978-3-319-99007-1_94

New shopping website features based on social broadcasting and Web 2.0 tools help improve consumer participation. These s-commerce tools allow buyers to gather socially rich information, resulting in a more dependable and sociable shopping environment [2]. Today more brands than ever sell their products via s-commerce websites. In 2015, Dell generated $6.5 million revenue via Twitter sales. While s-commerce has many advantages [3], there are also may cases of ineffective implementation of s-commerce, from both the clients' and vendors' perspectives [4]. If the services of an s-commerce contractor do not offer any value to a company, in the long run, they will decide that s-commerce is not the best way for them to do business.

Hence, it is crucial for companies to understand the dynamics of how client loyalty is established and maintained on the Internet [5]. Zeithaml et al. [6] argue that loyal clients forge bonds with a corporation and behave differently from non-loyal clients. Client loyalty influences behavioural outcomes and, eventually, the ability of a company to generate profit.

After the emergence of eCommerce in the mid-1990s, customer loyalty began to interest the academic community [7]. Since then, scholars have sought to explore the issue of customer loyalty in many areas, including social marketing [7], online environments [8] and eCommerce [9].

There are two prominent approaches to s-commerce. The first approach is for a social networking site to be enhanced with commercial features in order to help people buy while they are socializing. Examples include Facebook and Twitter. The second approach is for traditional e-commerce websites to be enhanced with social features (recommendations, comments, rankings) to help people socialize while they buy. Examples include Amazon and eBay. This paper focuses on the second approach, as most s-commerce transactions around the world are of this type.

Some researchers have focused their studies on the behavioural dimension of customer loyalty [10], others on its attitudinal dimension [11], and still others on the composite dimension of customer loyalty [12]. While the importance of these factors is accepted, the societal aspect of s-commerce is still poorly understood, as is the effect on s-commerce of trust and social presence [13]. Therefore, this study aims to investigate the impact of social presence (SP) and trust on customer loyalty to s-commerce websites.

2 Literature Review

With the rise of social networks such as Facebook and Twitter, people are spending more time than ever before on social media. The phenomenon of s-commerce first appeared in 2005 [14, 15] and has grown exponentially since then. S-commerce is not merely a buzzword for the grouping of electronic commerce and Social Networking Service (SNS) and Web 2.0 [16], as Tim O'Reilly and Dale Dougherty make clear in their description of Web 2.0 [17]. Web 2.0 describes today's user-centric Internet, which more and more people are using. These changes have led business organizations to investigate how to generate Internet traffic [18]. Social Networking Service (SNS) allows businesses to control traffic successfully [19]. It provides traditional blogs that are primarily based on the dimension of website content to track users and fine-tune their content [20]. This idea, of increasing revenue by thinking of the website reader as a consumer whose tastes needed to be catered, to led to the emergence of s-commerce. Zhang et al. [2] investigated the effects of technological features of eCommerce on customers' virtual experiences, and subsequently, on their intention to participate in social commerce.

Zhang et al. [2] and other studies have focused on how s-commerce networking has grown over the decades, and on how these practices contribute to other significant matters studied in the information science field [16]. The literature shows several benefits of customer loyalty in s-commerce. First, so-called brand advocacy arises when a purchaser becomes a company brand advocate. Brand advocates are diehard supporters—evangelists who are so loyal to a company brand that they will go out of their way to promote it, without having to be rewarded. Second, a loyal customer may become price-impervious and is not discourage by a high price. Even if a company raises its prices, the purchaser will still buy because they have faith in its value. Third, direct referrals occur when a person receives a recommendation from a current customer. Fourth, loyal customers tend to be those who have not had any problems with company service. Since they have not had any problems, they probably have not used customer support, an expensive service to provide. Fifth and finally, having a loyal base of clients makes it easier to predict future sales and revenue. The fact that a company has consistent clients who make repeat purchases allows the company to estimate revenue easily [21].

Many previous studies have found strong relationships among social presence, trust and customer loyalty in the s-commerce context [14, 15, 22]. A summary of previous literature on SP, trust and customer loyalty is presented in Table 1.

3 Theoretical Framework and Hypothesis

SP has a close relationship to information richness theory [30], which argues that media differ in their ability to convey information and accomplish tasks due to varying degrees of content ambiguity [31]. SP theory argues that a user compares the SP of a medium to the amount of SP required by a task, thereby assessing how a communicator deals with partners as being psychologically present [31]. In the present study, SP was included in the research model as one of the factors that impact customer loyalty (Fig. 1). Two dimensions of SP were considered: the SP of the web, and the perception of it by others [22]. Multiple previous studies have examined the impact of trust on s-commerce [32, 33]; this study included trust as a variable that influences customer loyalty to s-commerce sites.

Existing literature defines social presence (SP) as the scope to which a medium permits users to experience others as being psychosomatically present. Typically, SP is an inherent ability of a medium to convey a sense of human interaction, sensitivity and sociability. The comparison between user experience in the context of online shopping versus offline shopping might be described as sociability. Online shopping is impersonal and lacks human warmth; being automated and anonymous, it is devoid of personal or face-to-face interactions. Developing customer loyalty in the context of

Table 1. A summary of SP, trust and customer loyalty in previous literature

Paper	Dimensionality of SP	Key contributions
[23]	One dimension (SP of the website)	Makes a case for treating a website as a social actor though using theory of social response
[24]	One dimension (SP of the website)	Embraces a multi-dimensional trust construct and introduces a trust antecedent in the model
[25]	One dimension (SP of the website)	Investigates the impact of recommendations and consumer reviews on the perceived usefulness and social presence of the web
[26]	One dimension (SP of the website)	Examines the impact of manipulating online SP through imaginary interactions, specifically focusing on the impact of picture and text content
[27]	One dimension (SP of the website)	Empirically examines the influence of trust in Social Virtual Worlds (SVW) users on continuous use and purchase behaviour. It also investigates the effect of SP on trust and customer loyalty in SVWs
[28]	One dimension (SP of the website)	Investigates the relationship between past online purchases and purchasing intentions, representing the social context through the notions of SP and trust
[29]	One dimension (SP of the website)	Examines social factors such as social comparison, SP and enjoyment for the specificity of s-commerce applications
[22]	Three dimensions (SP of the Web, perception of others)	Investigates the mediator role of multidimensional trust between multidimensional SP as 1st construct and purchase intention as dependent variable

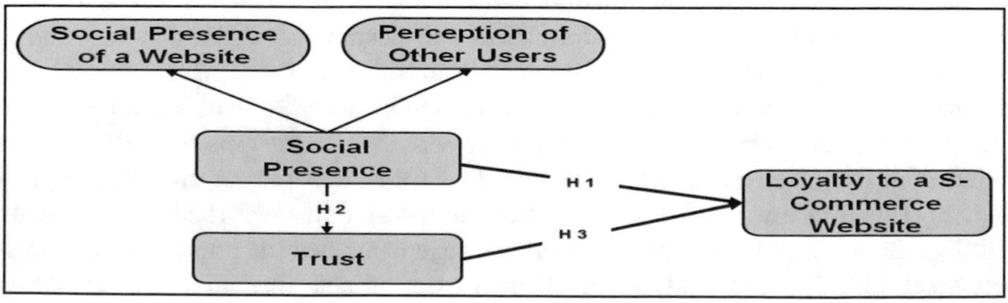

Fig. 1. Theoretical model

online environments is thus a challenge. Cyr et al. [34] have proposed a model for online loyalty, which they use to examine how various aspects of s-commerce affect user loyalty and user evaluations of practicality, enjoyment and trust [34]. Gefen and Straub [35] have investigated the impact of social presence on consumer loyalty in online services. The authors argued that social presence does indeed impact customer loyalty, and that belief of the customer shows a stronger impact on buying intentions as compare to the TAM beliefs. Based on this evidence, Hypothesis 1 below is proposed.

Hypothesis 1. Level of SP positively influences customer loyalty to an s-commerce website.

Building customer trust has always been viewed as a challenge for s-commerce businesses. Hajli [36] analysed customer trust in the context of various s-commerce constructs that play a role in an individual's purchase decision. Hajli's paper suggests a s-commerce model known as Social Commerce Adoption Model (SCAM) [33]. In another study, perceived high SP on an apparel website was also found to positively impact customer trust [34]. Multiple studies on online experiences have similarly suggested that a positive relationship exists between the perception of SP and user trust and intentions [23]. Based on the literature on SP discussed above, it can be assumed that SP positively influences customer trust. Therefore, Hypothesis 2 below is proposed.

Hypothesis 2. Level of SP positively influences customer trust in an s-commerce website.

Developing s-commerce user trust is a key topic impacting the development of social commerce, and is much discussed in current academic studies. Anderson and Srinivasan [32] examined the impact of consumer trust in s-commerce services and the comparative importance of customer trust in judgment, using the widely studied s-commerce framework known as Technology Acceptance Model (TAM). Their study indicates that social presence affects customer trust, and that trust, in turn, has a strong impact on buying intentions and loyalty. In order to manage s-commerce and electronic services, business owners must manage the level of trust that is stimulated in the purchaser on the s-commerce website. Based on the literature on trust discussed above, Hypothesis 3 below is proposed.

Hypothesis 3. Customers' level of trust positively influences customer loyalty to an s-commerce website.

4 Research Methodology

The current study used positivism as its research philosophy. The core feature of positivism is its emphasis on collecting objective data, which are then used to validate hypotheses and gain an understanding of the area of study [37]. Given that this research involved collecting objective data, the quantitative approach was deemed appropriate for it. Partial Least Squares (PLS) structural equation modelling (PLS-SEM) technique was used applying SmartPLS 3 software to test the proposed relationships in the research model. Domain and construct are specified through extensive review of the literature; the dominant domain in this study is s-commerce. After defining the constructs, the researcher further qualified each through identifying items related to it and developing a pool of items by means of this exploration.

The Customer Loyalty (CL) construct initially included eight items. The items used to operationalize customer loyalty are based on research by Liang et al. [38], Kim and Park [39], Wang et al. [40], and Shin [41]. The Trust (TR) construct was initially operationalized with six items. Two items are based on research by Hassanein and Head [26], another two on research by Brown and Jayakody [42], and the remaining items on research by Kim et al. [8]. Social Presence (SP) was operationalized with 11 initial items that measured two dimensions: Social Presence of the Website (SPW) and Social Presence of Other Users (SPOU). The five items used to operationalize SPW are based on research by Cyr et al. [34] and Kumar and Benbasat [25]. The six items used to operationalize the SPOU are based on Lu and Fan [22] and on Caspi and Blau [43]. The study ultimately excluded some items that did not support the recommended value. Table 2 below shows each construct, and the items related to it.

Table 2. Construct and items under s-commerce domain

Construct	Items
Customer loyalty	(1) Continue using the s-commerce website, (2) purchase from the s-commerce website in the near future, (3) say positive things about this website to other people, (4) recommend this website to someone who seeks advice, (5) share purchases with relatives, friends, and others to encourage them to use this website, (6) consider this website to be his first choice for future online shopping for the chosen type of goods/services, (7) provide others with information on this website, and (8) recommend this website to others
Trust	(1) Trustworthy (2) honest (3) reliable (4) this website has his/her best interests in mind, (5) this website is reliable, and (6) this website has his/her information safety in mind
Social presence: social presence of the website	(1) Human contact in this website, (2) personal contact in this website, (3) sociability in this website, (4) human warmth in this website, and (5) human sensitivity in this website
Social presence: social presence of other users	(1) Feel interested in the product, (2) provide information about the seller, (3) provide information about the product, (4) have browsed this website, (5) are disappointed about products or services, and (6) are satisfied with products or services

The study obtained results through a web-based survey. The researcher recruit participants through a marketing research company database. The sample specification was as follows: any Australian adult who regularly used s-commerce websites to purchase products or services. If they below 18 age, participants were asked to click on a specific button which will terminate the survey before starting the survey. The participants were asked about 15 s-commerce websites commonly used in Australia, including Kogan, eBay, Amazon, Target, Booking.com and Gumtree.

5 Findings

A structural equation model entails the comparison of two related models, as a measurement model and a structural model [44]. In the analytical process, first the measurement model is tested, and then the structural relationship among constructs is analyzed.

5.1 Assessment of Measurement Model

In the measurement model, the relationship between the latent constructs and their items are assessed [45]. Before checking the structural relationship among the constructs, the researcher must check the reliability and validity of the items of all constructs in the measurement model [44].

After completing this step, the researcher evaluated internal reliability examining Cronbach's alpha and composite reliability, where a level of 0.70 is an indicator of acceptable internal consistency. Satisfactory convergent validity is indicated when constructs have an average variance extracted (AVE) of at least 0.50, and item loadings are well above 0.50 [46]. Table 3 shows the results for constructs, composite reliability, Cronbach's alpha, and AVE. The Cronbach's alpha value ranged from 0.711 to 0.856, and composite reliability value ranged from 0.839 to 0.893. This is higher than the recommended level of 0.70, and is thus an indicator of strong internal reliability. The AVE, ranging from 0.549 to 0.634, and item loadings, ranging from 0.668 to 0.841, are all greater than the recommended levels; thus the needed conditions for convergent validity were met.

Table 3. Measurement model

Constructs	Cronbach's alpha	Composite reliability	Average variance extracted (AVE)
Loyalty to a s-commerce website	0.856	0.893	0.583
SPO	0.794	0.859	0.549
SPW	0.711	0.839	0.634
Social presence	0.843	0.879	0.577
Trust	0.840	0.887	0.611

Similarly, the square roots of AVE, shown in Table 4, were greater than their corresponding correlation, representing satisfactory discriminant validity.

Table 4. Discriminant validity

	Loyalty to a s-commerce website	SPO	SPW	Social presence	Trust
Loyalty to a s-commerce website	**0.764**				
SPO	0.493	**0.741**			
SPW	0.385	0.628	**0.796**		
Social presence	0.497	0.940	0.856	**0.691**	
Trust	0.756	0.411	0.409	0.454	**0.782**

5.2 Assessment of Structural Model

In the second step, a structural model was built with the purpose of identifying the relationship among the constructs. In this study, bootstrap t-statistics were computed (on the basis of 5000 resamples) [47], with cases set at 797. Table 5 shows the statistical relationship between endogenous and exogenous constructs. It was found that social presence ($t = 7.602$, $\beta = 0.194$, $p < 0.05$) and Trust ($t = 31.519$, $\beta = 0.669$, $p < 0.05$) both have a strong positive impact on loyalty to s-commerce websites. Thus H1 and H3 were supported. Similarly, the relationship between social presence and Trust ($t = 16.111$, $\beta = 0.455$, $p < 0.05$) was significant, and supported H2.

Table 5. Structural model

Path	β	*t*-statistics	P values
Social presence → loyalty to a s-commerce website	0.194	7.602	0.000
Social presence → trust	0.455	16.111	0.000
Trust → loyalty to a s-commerce website	0.669	31.519	0.000

6 Discussion and Conclusion

This study found that a positive association exists between SP and customer loyalty to s-commerce websites. Among the most important factors for online customer loyalty is SP (both of the website and of others users). Some researchers have claimed that websites that are rich in information have more SP than those that do not [48]. Additionally, Gefen and Straub [35] found that SP of a website had an impact on intention to purchase, attitude and loyalty, while Mäntymäki [21] found that SP of a website influenced the constituent factors of customer loyalty. Lu and Fan [22] point out that people can be influenced by the reported experience of other people they know and trust. Godes et al. [49] suggest that social interaction with other users can affect the beliefs, attitudes, behaviours and loyalty of consumers, giving further support for the interrelationship of these factors.

The implication of the findings is important for s-commerce websites to retain the loyalty of the customer. S-commerce websites customer prefer the social presence attributes in its website where they can interact with each other and share their messages. This study found that as like with e-commerce websites customer loyalty, the s-commerce website can also retain the customer loyalty through the social presence. Thus, customer loyalty in a s-commerce website is dependent on the presence of social interaction attributes of the website. Customer are more likely to prefer communicating through the s-commerce and feel free to share suggestions, information to other customers and friends. The s-commerce website should provide the accessibility to the customer featuring the social presence attribute. The communications among the customers should be retained and displayed in the website so that new customers can get access and benefit from the website and eventually become the loyal customer.

The present study also found a positive relationship between the SP and customer trust in and loyalty to an s-commerce website. As with other relationships supported by this study, this result in line with those found by existing research. In a study of trust in s-commerce websites, Gefen and Straub [35] found that trust can be generated through the perception of high SP. Hajli [50] and Mäntymäki and Salo [27] also found that SP has a positive impact on trust.

Using the s-commerce website, customers can interact with other customers and establish communication, which in turn, influence them to believe the website as a trustworthy medium. Through the medium, customers may share information, suggest others to rely on the website, and believe that the website might keep information secret. The s-commerce website provides the platform of social networking among the users. S-commerce websites are facing a severe threat from its competitive websites in gaining and retaining trust of the users. Therefore, customer trust is becoming a crucial challenge for ensuring success of the s-commerce website. The e-commerce websites consists of social and emotional appeal, but absence of human warmth. Several types of online items provide social interaction among the users to share images and descriptions of the items, which influence the users to express their attitudes toward the purchase. The degree of human warmness and friendliness differs in varied online items and services. Customers are more likely to influence and to be influenced by other trusted friends' experience and suggest to have trust on a website for purchase [22]. Moreover, customer preference, beliefs, attitudes, and behaviors are influenced and guided by the social interaction with the other users [49].

The practitioners and managers of the s-commerce website may provide enough space for interaction and communication of messages to achieve consumer trust. As like with the conventional purchase from the retail shop, customer are more likely to take advice, shopping experience from the others. If s-commerce website offers trusted communication platform for the users, the website will be more likely to be accepted and trusted by the user. The user generally prefers to share their experience and suggests the future shoppers to trust on the website. Thus, social presence can also attribute to design customer's belief, attitudes and behavior toward accepting the s-commerce website as trusted one.

The result of the study found that Customer trust had a strong influence on loyalty to s-commerce website. Retaining customers for a sustained period, customer trust becomes important for online shopping sites. Customers are more likely to purchase again from the same s-commerce website, if they have a high level of trust in the service capability of the website. Overall, the existing body of research, including the present study, indicates that customer trust is important in the field of s-commerce [33].

The finding of the study provides implications of the managers of s-commerce website to maintain the customer loyalty. For example, the finding suggests that trust is one of the predictors of customer loyalty in Australian context, which is vital for sustained business. Thus, the managers who run the s-commerce sites should emphasize on retaining the customer trust that influences customer loyalty to the site. Many customers are reluctant to purchase through s-commerce website, because the website seems less trustworthy to them. Thus, managers of s-commerce website should endeavor to utilize all avenues of social networking site, so that the customers can interact and share the shopping experience with other customers before taking purchase decision and eventually show trust in the website for becoming a loyal customer in the future. The managers of s-commerce website can introduce a loyalty rewards program based on the frequent user program, usage levels of the site, and recommendation provided by the users. Such caring program might increase customer trust to a s-commerce website.

Much earlier e-commerce and s-commerce research based on Social Presence Theory has accepted a one-dimensional conceptualization of s-commerce and e-commerce [13]. This study focused on the impact of the SP of websites and the SP of other users on customer loyalty to s-commerce sites. It examined multiple dimensions of the s-commerce experience, thus avoiding the limitations of such one-dimensional conceptions. The findings of this study can offer insight to s-commerce practitioners in improving the implementation of social commerce and designing s-commerce sites.

References

1. Hong, I.B.: Understanding the consumer's online merchant selection process: the roles of product involvement, perceived risk, and trust expectation. Int. J. Inf. Manag. **35**, 322–336 (2015)
2. Zhang, H., Lu, Y., Gupta, S., Zhao, L.: What motivates customers to participate in social commerce? The impact of technological environments and virtual customer experiences. Inf. Manag. **51**, 1017–1030 (2014)
3. Distaso, M.W., McCorkindale, T., Agugliaro, A.: America's most admired companies social media industry divide. J. Promot. Manag. **21**, 163–189 (2015)
4. Pucihar, A.: Model dejavnikov organiziranosti organizacije za uspešno vključevanje na elektronske tržnice (2003)
5. Gommans, M., Krishnan, K.S., Scheffold, K.B.: From brand loyalty to e-loyalty: a conceptual framework. J. Econ. Soc. Res. **3**, 43–58 (2001)
6. Zeithaml, V.A., Berry, L.L., Parasuraman, A.: The behavioral consequences of service quality. J. Mark. **60**, 31–46 (1996)
7. Toufaily, E., Ricard, L., Perrien, J.: Customer loyalty to a commercial website: descriptive meta-analysis of the empirical literature and proposal of an integrative model. J. Bus. Res. **66**, 1436–1447 (2013)
8. Kim, M.-J., Chung, N., Lee, C.-K.: The effect of perceived trust on electronic commerce: shopping online for tourism products and services in South Korea. Tour. Manag. **32**, 256–265 (2011)

9. Jang, H., Olfman, L., Ko, I., Koh, J., Kim, K.: The influence of on-line brand community characteristics on community commitment and brand loyalty. Int. J. Electron. Commer. **12**, 57–80 (2008)
10. Eid, M., Al-Anazi, F.U.: Factors influencing Saudi consumers loyalty toward B2C E-commerce. In: AMCIS 2008 Proceedings, p. 405 (2008)
11. Kwon, W.-S., Lennon, S.J.: What induces online loyalty? Online versus offline brand images. J. Bus. Res. **62**, 557–564 (2009)
12. Hong, I.B., Cho, H.: The impact of consumer trust on attitudinal loyalty and purchase intentions in B2C e-marketplaces: intermediary trust vs. seller trust. Int. J. Inf. Manag. **31**, 469–479 (2011)
13. Hassanein, K., Head, M., Ju, C.: A cross-cultural comparison of the impact of social presence on website trust, usefulness and enjoyment. Int. J. Electron. Bus. **7**, 625–641 (2009)
14. Huang, Z., Benyoucef, M.: From e-commerce to social commerce: a close look at design features. Electron. Commer. Res. Appl. **12**, 246–259 (2013)
15. Lu, B., Fan, W., Zhou, M.: Social presence, trust, and social commerce purchase intention: an empirical research. Comput. Hum. Behav. **56**, 225–237 (2016)
16. Marsden, P.: Social commerce (English): monetizing social media: Grin Verlag (2010)
17. O'Reilly, P., Finnegan, P.: Intermediaries in inter-organisational networks: building a theory of electronic marketplace performance. Eur. J. Inf. Syst. **19**, 462–480 (2010)
18. Lewis, K., Gonzalez, M., Kaufman, J.: Social selection and peer influence in an online social network. Proc. Natl. Acad. Sci. **109**, 68–72 (2012)
19. Strauss, J.: E-marketing. Routledge, London (2016)
20. Shin, J.-K., Park, M., Ju, Y.: The effect of the online social network structure characteristics on network involvement and consumer purchasing intention: focus on Korean social promotion sites. In: The 11th International DSI and the 16th PDSI Joint Meeting (2011)
21. Mäntymäki, M.: Customer loyalty in social virtual worlds (2009)
22. Lu, B., Fan, W.: Social presence, trust, and social commerce purchase intention: an empirical research (2014)
23. Kumar, N., Benbasat, I.: Para-social presence and communication capabilities of a web site: a theoretical perspective. E-service J. **1**, 5–24 (2002)
24. Gefen, D., Straub, D.W.: Consumer trust in B2C e-commerce and the importance of social presence: experiments in e-products and e-services. Omega **32**, 407–424 (2004)
25. Kumar, N., Benbasat, I.: Research note: the influence of recommendations and consumer reviews on evaluations of websites. Inf. Syst. Res. **17**, 425–439 (2006)
26. Hassanein, K., Head, M.: Manipulating perceived social presence through the web interface and its impact on attitude towards online shopping. Int. J. Hum. Comput. Stud. **65**, 689–708 (2007)
27. Mäntymäki, M., Salo, J.: Trust, social presence and customer loyalty in social virtual worlds (2010)
28. Weisberg, J., Te'eni, D., Arman, L.: Past purchase and intention to purchase in e-commerce: the mediation of social presence and trust. Internet Res. **21**, 82–96 (2011)
29. Shen, J.: Social comparison, social presence, and enjoyment in the acceptance of social shopping websites. J. Electron. Commer. Res. **13**, 198–212 (2012)
30. Daft, R.L., Lengel, R.H.: Information richness. A new approach to managerial behavior and organization design. DTIC document (1983)
31. Zhong, Y.: Social commerce: a new electronic commerce. In: Eleventh Wuhan International Conference on e-Business, Paper 49 (2012)
32. Anderson, R.E., Srinivasan, S.S.: E-satisfaction and e-loyalty: a contingency framework. Psychol. Mark. **20**, 123–138 (2003)

33. Hajli, M.: An integrated model for e-commerce adoption at the customer level with the impact of social commerce. Int. J. Inf. Sci. Manag. (IJISM) **2**, 77–97 (2012)
34. Cyr, D., Hassanein, K., Head, M., Ivanov, A.: The role of social presence in establishing loyalty in e-service environments. Interact. Comput. **19**, 43–56 (2007)
35. Gefen, D., Straub, D.W.: Managing user trust in B2C e-services. E-service J. **2**, 7–24 (2003)
36. Hajli, M.: Social commerce adoption model. In: Proceedings of the UK Academy of Information Systems Conference (2012)
37. Walsham, G.: The emergence of interpretivism in IS research. Inf. Syst. Res. **6**, 376–394 (1995)
38. Liang, T.-P., Ho, Y.-T., Li, Y.-W., Turban, E.: What drives social commerce: the role of social support and relationship quality. Int. J. Electron. Commer. **16**, 69–90 (2011)
39. Kim, S., Park, H.: Effects of various characteristics of social commerce (s-commerce) on consumers' trust and trust performance. Int. J. Inf. Manag. **33**, 318–332 (2013)
40. Wang, Y.-S., Wu, S.-C., Lin, H.-H., Wang, Y.-Y.: The relationship of service failure severity, service recovery justice and perceived switching costs with customer loyalty in the context of e-tailing. Int. J. Inf. Manag. **31**, 350–359 (2011)
41. Shin, D.-H.: User experience in social commerce: in friends we trust. Behav. Inf. Technol. **32**, 52–67 (2013)
42. Brown, I., Jayakody, R.: B2C e-commerce success: a test and validation of a revised conceptual model. Electron. J. Inf. Syst. Eval. **11**, 167–184 (2008)
43. Caspi, A., Blau, I.: Social presence in online discussion groups: testing three conceptions and their relations to perceived learning. Soc. Psychol. Educ. **11**, 323–346 (2008)
44. Hair, J.F., Ringle, C.M., Sarstedt, M.: PLS-SEM: indeed a silver bullet. J. Mark. Theory Pract. **19**, 139–152 (2011)
45. Henseler, J., Ringle, C.M., Sarstedt, M.: Using partial least squares path modeling in advertising research: basic concepts and recent issues. In: Handbook of Research on International Advertising, vol. 252 (2012)
46. Hair, J.F., Ringle, C.M., Sarstedt, M.: Editorial-partial least squares structural equation modeling: rigorous applications, better results and higher acceptance. Long Range Plan. **46**, 1–12 (2013)
47. Henseler, J., Ringle, C.M., Sinkovics, R.R.: The use of partial least squares path modeling in international marketing. Adv. Int. Mark. **20**, 277–319 (2009)
48. Simon, S.J.: The impact of culture and gender on web sites: an empirical study. ACM SIGMIS Database **32**, 18–37 (2000)
49. Godes, D., Mayzlin, D., Chen, Y., Das, S., Dellarocas, C., Pfeiffer, B., et al.: The firm's management of social interactions. Mark. Lett. **16**, 415–428 (2005)
50. Hajli, M.: Social commerce: the role of trust. In: AMCIS 2012 Proceedings, 29 July 2012, Paper 9 (2012)

The Influence of Word of Mouth on Customer Loyalty to Social Commerce Websites: Trust as a Mediator

Hilal Alhulail(✉), Martin Dick, and Ahmad Abareshi

RMIT University, GPO Box 2476, Melbourne, VIC 3001, Australia
{hilal.alhulail,martin.dick,
ahmad.abareshi}@rmit.edu.au

Abstract. The term Social Commerce (s-commerce) is referred as a new trend where the sellers and buyers are connected for interacting on social media. Customer loyalty is an important tool for developing future marketing business strategies in s-commerce field. Moreover, the effect of Word-of-Mouth (WOM) on customer loyalty is also very important to the stability and sustainability of an s-commerce platform. While customer loyalty and WOM have been examined widely in various research contexts, a considerable gap is noticed in the literature on the study of customer loyalty. Therefore, the objective of this study was to identify the influence of WOM on customers' loyalty to s-commerce websites. This research gathers survey data and applies Partial Least Squares (PLS-SEM) structural equation modelling to analyze the data. The study found that WOM and trust have strong positive impact on loyalty to s-commerce website. The findings provide insights for s-commerce industries in developing strategies for improved implementation of s-commerce as well as the design of s-commerce sites.

Keywords: Social commerce · Word-of-Mouth · Customer loyalty
Trust

1 Introduction

Over the past decades, the way of transaction on ecommerce have changed due to the advanced use of social media. Today, the use of social media is fulfilling the gap in social interaction that was existing between the ecommerce customers and businesses. Social media as a platform lets the customers to make and share their own opinions and feedbacks conveniently and openly with other people [1–3]. Besides, social media help consumers to collaborate for decision making while purchasing on ecommerce sites. As a result, making transactions using social media has increased significantly. Today, such social media based ecommerce activities are called as s-commerce [2, 4]. In s-commerce, the customers are free to conveniently share their purchasing preferences and experiences as well as making comments and recommendations on any product. Study shows that about 83% of the ecommerce customers tend to share their opinions with others, whereas about and about 67% of such consumers prefer to make their

F. Saeed et al. (Eds.): IRICT 2018, AISC 843, pp. 1025–1033, 2019.
https://doi.org/10.1007/978-3-319-99007-1_95

purchasing decision based on other consumers' reviews and feedbacks. This kind of interaction and communication are called WOM [5].

Today, WOM is a frequently used phrase in the literature of s-commerce. In s-commerce context, it is denoted as sharing of reviews and feedbacks about any online purchase and experience. Although, WOM significantly influences the consumers' online shopping behavior and decision, sufficient study has not found about its influence on customers' loyalty in s-commerce context. However, as a matter of the stability and sustainability of s-commerce websites, it is essential to understand more about the influence of WOM on customer loyalty. Because the increase of customer loyalty also increases repeated ecommerce purchase and visits by customers. As the level of competition in s-commerce market is growing fast, gaining competitive advantage is not possible through offering quality products and service only but also by increasing customer loyalty as well.

Over a long time, trust has been considered as an important element in building customer relationship. Reichheld and Schefter [6] observe that "to gain the loyalty of customers, you must first gain their trust". Trust on s-commerce turns into customer loyalty. While the influence of trust on customer loyalty has been examined widely, very little study has been conducted in this regard in the context of s-commerce. Therefore, the objective of this study is to identify the influence of WOM and trust on customers' loyalty to s-commerce.

2 Literature Review

The term s-commerce is used to refer a newer trend where the buyers and sellers are connected for interaction and communication on social media. It is a type of ecommerce that connects the sellers and buyers socially and let them share reviews and feedbacks on the products and services purchased. Thus, it adds the users' contributions ecommerce. Some big examples of online social media networking in s-commerce are: Amazon's Wish Lists, Facebook, Instagram, Google Play, Etsy etc. On the other hand, WOM is referred as the exchange of information and experiences among customers that assist them to make purchasing decisions [7].

According to Park et al. [8] customers have greater chance to buy a product or service by being influenced by WOM rather than watching an advertisement. A study also shows that customers generally prefer to hear the WOM when going to make any purchasing decision [9]. According to Kuan and Bock [10], WOM influences the process of trust building more critically in online shopping environment than offline environment for customers. Similarly, Lee and Kwon [11] have pointed out that customers' purchasing decisions involve high trust when made based on the purchasing experiences of others. For this reason, based on the above findings on WOM in different contexts, it can be logically presumed that customers of s-commerce site are also expected to trust the WOM (e.g. recommendations and feedbacks) of others on social media.

Previous studies had evaluated the influence of WOM on various factors. For instance, according to the study by Hajli et al. [12], WOM significantly affect consumers' level of trust. Another study also found that WOM can considerably influence the community commitment [5, 13]. Karjaluoto et al. [14] showed interest to test how a customer's intention for online repurchase of a product or service affected by other customers' WOM. On the other hand, trust is also found as a significant factor in the context of s-commerce as a study [15] recommended to use Trust theory to study customer's intention for buy on s-commerce sites. Similarly, the trust factor can determine the customer acceptance of any s-commerce in comparison to other s-commerce sites. Therefore, it is important to study the influence of trust factor on s-commerce platforms where the customers share buying experience with others [16]. As many previous studies used Trust theory to understand the social behaviour in the context of social science, its use ought to be appropriate for testing the influence of trust in shopping on s-commerce platforms.

3 Theoretical Framework and Hypothesis

The Fig. 1 presents the theoretical framework of this study. Hypothesis 1 is proposed to identify if WOM positively impacts the customers' trust. At the same time hypothesis 2 is proposed to identify the mediating influence of trust on the relationship of WOM and loyalty to s-commerce websites.

Fig. 1. Theoretical framework

Brown and Reingen [9] studied extensively on the relationship between customers and s-commerce websites. At the macro level, the relationship implies that the information is transportable. It permits to transfer of information from one group of recommended users to the users of alternative sub group over s-commerce platforms [9]. On the other hand, Kuan and Bock [10] found that the customers' trust on online website grows or diminishes based on their previous experience of visiting that website. Study shows that WOM can influence an offline company client who had never visited a company website to visit and trust that company's website [10]. Positive WOM lets the s-commerce website managers to promote their products with significant positive outcome, for example, building long-term relationship with prospective customers. Studies show that positive WOM shared by consumer is an important determinant for building trust and customers' frequency of repurchasing [14, 16]. Therefore, based on the above-mentioned discussion on WOM, it can be assumed that WOM positively impacts customer's trust level and building strong customer relationship. Therefore, our hypothesis 1 is as below:

Hypothesis 1. High levels of positive WOM influence customer trust in an s-commerce website.

Customer trust is deemed as an important factor for the stakeholders in s-commerce and it is challenging to increase customer trust in the context of s-commerce communications. According to S-commerce Adoption Model (SCAM) Hajli [17], customer trust influences customers in evaluating a s-commerce for an intention to purchase. Therefore, building customer trust in the context of s-commerce itself is a significant factor for growing s-commerce business and as well as an important topic in current academic studies. Using the Technology Acceptance Model (TAM), Anderson and Srinivasan [18] studied the influence of consumer trust on customer loyalty to s-commerce where is the study shows that WOM affects customer trust which in turn affects the consumer buying intentions and loyalty. Besides, it is required for the s-commerce managers to enhance their consumer trust which ultimately would influence the customers to purchase on their s-commerce websites. Therefore, based on the above-mentioned discussion we propose our hypothesis 2 as below:

Hypothesis 2. Customers' level of trust positively influences customer loyalty to an s-commerce website.

4 Research Methodology

The study follows the quantitative approach of research. In this process, at first the constructs were defined and then items were identified through additional literature analysis. Then all the identified items have been grouped as pools of items under each construct. In total, eight items were identified for Customer Loyalty (CL) construct from the studies of Liang et al. [19], Wang et al. [20] and Shin [21] respectively. On the other hand, in total, six items were identified for Trust (TR) construct. Among those, two items were identified from the studies of Hassanein and Head [22]. Another two items were identified from the studies of Brown and Jayakody [23], whereas, the rest of the items were identified from the study of Kim et al. [24].

The WOM construct was comprised of a total of 12 initial items where four of these items were identified from the study of Kim and Park [25]. These four items would help to understand whether the participants had used s-commerce after hearing from others and how frequently. The rest of the items were identified from the study of Edward [26]. These were selected to understand whether how other customers' recommendations influence the participants for shopping online. However, the study has excluded some items fall outside of the recommended value. Table 1 below shows each of those constructs and their related items.

The quantitative survey of online s-commerce customers' opinions regarding the measurement items was based on a probability sample of qualified Australian customers of s-commerce websites. A stratified random sampling was used with all Australian states that constitute the strata of the population of Australia. The population of the study consists of male and female customers of multiple s-commerce websites (e.g., Kogan, eBay, Amazon, Target, Big W, Harvey Norman, Dick Smith, Etsy, OO, Booktopia, Shopping.com Network, Deals Direct, and Gumtree) who live in Australia. Hair et al. [27] stated that the minimum sample size should be 500 if there are more than seven latent constructs in a study (the framework in this study has 11 latent

Table 1. Construct and items under S-commerce domain

Construct	Items
Customer loyalty	(1) Continue using the s-commerce website, (2) purchase from the s-commerce website in the near future, (3) say positive things about this website to other people, (4) recommend this website to someone who seeks advice, (5) share purchases with relatives, friends, and others to encourage them to use this website, (6) consider this website to be his first choice for future online shopping for the chosen type of goods/services, (7) provide others with information on this website, and (8) recommend this website to others
Trust	(1) Trustworthy (2) honest (3) reliable (4) this website has his/her best interests in mind, (5) this website is reliable, and (6) this website has his/her information safety in mind
WOM	(1) Useful, (2) easy to use, (3) reliable, and (4) not worth the effort, (5) are useful to him/her, (6) will affect his/her choice when he/she shops online, (7) will provide him/her with different advisory opinions, (8) will change his/her purchasing motivation, (9) will increase his/her interest in searching for a product, (10) will change his/her purchasing intention, (11) will let him/her make purchase decision, and (12) will change the items he/she intends to purchase

constructs). However, Kline (2011) suggests that sample size should be determined by the rule of thumb, which is 10:1 or N:q, where N is the number of cases and q is the number of parameters. For this study, there are 12 constructs (including dependent constructs) and 58 items. The sample size of the ratio 10:58 would be 580 by multiplying 10 by 58. Based on the above two opinions, and in order to have sufficient sample size, the investigator decided on a sample size of 1000 s-commerce website users in Australia. The researcher used a professional company to distribute the survey link online and then the researcher done the rest. The responses were 997 response. After data cleaning process, 797 survey were ready for analyses.

The survey of the study was web based. Anyone from the Australian population, over 18 years of age, who uses s-commerce platform for purchasing products or services were selected in random manner. After obtaining the ethical approval, the survey was conducted using the Qualtrics web-based system. Partial Least Squares (PLS-SEM) structural equation modelling is the main analytical method used in this study for data analysis.

5 Findings

The survey of the study entails two related models such as measurement model and structural model [27]. In the analytical phrase, the measurement model is tested before analyzing the structural relationship among constructs.

5.1 Assessment of Measurement Model

In the measurement model, the relationship between the innate constructs and their items are tested. Ideally, the reliability and validity of the items of all constructs are examined in measurement model prior to the examination of the structural relationship among all the constructs. The internal reliability is evaluated by Cronbach's alpha and composite reliability, where the level of 0.70 is an indicator for acceptable internal consistency. Convergent validity is satisfactory when constructs have an average variance extracted (AVE) of at least 0.50 and items loading are well above 0.50.

Table 2 shows the constructs, composite reliability, Cronbach's alpha, and AVE. The Cronbach's alpha value ranged from 0.82 to 0.86 and composite reliability value ranged from 0.88 to 0.89, which is greater than the recommended level of 0.70 and thus is an indicator of strong internal reliability. The Average Variance Extracted of the constructs are from the range of 0.58 to 0.65 and it is crystal clear that the convergent validity is satisfactory.

Table 2. Measurement model

Constructs	Cronbach's alpha	Composite reliability	Average variance extracted (AVE)
Loyalty to a S-commerce website	0.86	0.89	0.58
Trust	0.84	0.89	0.61
WOM	0.82	0.88	0.65

The following table shows cross loading, item loading in loyalty to a S-commerce website ranging from 0.70 to 0.84, items loading in the Trust ranging from 0.74 to 0.82 and cross loading in the most significant constructs in this study Word of- Mouth ranging from 0.74 to 0.85 are greater than the recommended levels. Thus, requirements for convergent validity are met (Table 3).

On the other hand, Table 4 represents the Discriminant Validity which was examined by the square root of the AVE and cross-loading matrix. Ideally, the value of square root of the AVE of a construct needs be greater than its correlation with other constructs for satisfactory discriminant validity. Here the calculated square root of AVE was greater than the corresponding correlation, confirms satisfactory discriminant validity of the data.

5.2 Assessment of Structural Model

In the second step, the structural model is built to identify the relationship among the constructs. In this study, bootstrap t-statistics is computed. Table 5 shows the statistical relationship between endogenous constructs and exogenous constructs. It is found that Trust ($t = 51.89$, $\beta = 0.76$, $p < 0.05$) and WOM ($t = 21.56$, $\beta = 0.56$, $p < 0.05$) have strong positive impact on loyalty to s-commerce website. Thus, H1 and H2 are supported.

Table 3. Cross loading

	Loyalty to a S-commerce website	Trust	WOM
CUL1	0.73		
CUL2	0.75		
CUL3	0.84		
CUL6	0.76		
CUL7	0.70		
CUL8	0.81		
TRU1		0.74	
TRU2		0.81	
TRU3		0.82	
TRU5		0.78	
TRU6		0.76	
WOM1			0.74
WOM2			0.82
WOM3			0.85
WOM4			0.79

Table 4. Discriminant validity

	Loyalty to a S-commerce website	Trust	WOM
Loyalty to a S-commerce website	0.76		
Trust	0.76	0.78	
WOM	0.65	0.56	0.80

Table 5. Structural model

Path	*B*	*t*-Statistics	P values
Trust → loyalty to a S-commerce website	0.76	51.89	0.00
WOM → trust	0.56	21.56	0.00

6 Discussion and Conclusion

This research paper shows that consumers are prone to know the real information on what products or services they want to buy. In this regard, the WOM plays role to build trust on consumer's mind that in the nearest future turns into customer long term loyalty. The result implies that while making purchasing decision, the available information on social media is influenced by WOM. Social media also impact the problem identification and post-purchase evaluation process of the customers. The result showed that consumers deem WOM on s-commerce websites as a useful source of information for making purchasing decision and now they increasingly search Facebook for information based on WOM.

The result also shows that trust is an influential factor in s-commerce. It strongly influences consumers' intentions to buy the goods and services. Trust on products or services formed through collecting the information from various sources build the customer loyalty to s-commerce website. WOM here works as the main influence to build the challenging trust on customer loyalty to s-commerce. The customers grow more faith and loyalty when they interact with social media. The findings provide insights for s-commerce industries in developing strategies for improved implementation of s-commerce as well as the design of s-commerce sites.

This paper aims to study the impact of WOM on customer loyalty to s-commerce websites in the existence of trust as a mediator. PLS-SEM method was used for data analysis stage. Results shown that WOM has a significant impact on customers' loyalty to s-commerce websites. In addition, trust has a significant influence on loyalty to social-commerce websites.

References

1. Kempe, D., Kleinberg, J., Tardos, É.: Maximizing the spread of influence through a social network. In: Proceedings of the 9th ACM SIGKDD International Conference on Knowledge Discovery and Data Mining, pp. 137–146 (2003)
2. Baethge, C., Klier, J., Klier, M.: Social commerce—state-of-the-art and future research directions. Electron. Mark. **26**, 1–22 (2016)
3. Turban, E., Strauss, J., Lai, L.: Implementing social commerce systems. In: Social Commerce, pp. 265–289. Springer, Heidelberg (2016)
4. Ly, P., Cho, W.-S., Kwon, S.-D.: Influencing factors of purchase intention on social commerce in cambodia: the moderating roles of experience. JITAM **24**, 129–141 (2017)
5. Ahmad, S.N., Laroche, M.: Analyzing electronic word of mouth: a social commerce construct. Int. J. Inf. Manag. **37**, 202–213 (2017)
6. Reichheld, F.F., Schefter, P.: E-loyalty: your secret weapon on the web. Harv. Bus. Rev. **78**, 105–113 (2000)
7. Kim, Y., Chang, Y., Wong, S.F., Park, M.C.: Customer attribution of service failure and its impact in social commerce environment. Int. J. Electron. Cust. Relatsh. Manag. **8**, 136–158 (2014)
8. Park, J., Chaiy, S., Lee, S.: The moderating role of relationship quality in the effect of service satisfaction on repurchase intentions. Korea Mark. Rev. **13**, 119–139 (1998)
9. Brown, J.J., Reingen, P.H.: Social ties and word-of-mouth referral behavior. J. Consum. Res. **14**, 350–362 (1987)
10. Kuan, H.-H., Bock, G.-W.: Trust transference in brick and click retailers: an investigation of the before-online-visit phase. Inf. Manag. **44**, 175–187 (2007)
11. Lee, Y., Kwon, O.: Intimacy, familiarity and continuance intention: An extended expectation–confirmation model in web-based services. Electron. Commer. Res. Appl. **10**, 342–357 (2011)
12. Hajli, M., Hajli, M., Khani, F.: Establishing trust in social commerce through social word of mouth. In: 2013 7th International Conference on e-Commerce in Developing Countries: With Focus on e-Security (ECDC), pp. 1–22 (2013)
13. Villarejo-Ramos, Á.F., Sánchez-Franco, M.J., García-Vacas, E.M., Navarro-García, A.: Modelling the influence of eWOM on loyalty behaviour in social network sites. In: Strategies in E-Business, pp. 11–28. Springer, Heidelberg (2014)

14. Karjaluoto, H., Munnukka, J., Tikkanen, A.: Are Facebook brand community members really loyal to the brand? In: Bled eConference, p. 28 (2014)
15. Liang, T.-P., Turban, E.: Introduction to the special issue social commerce: a research framework for social commerce. Int. J. Electron. Commer. **16**, 5–14 (2011)
16. Chen, S.-C., Yen, D.C., Hwang, M.I.: Factors influencing the continuance intention to the usage of Web 2.0: an empirical study. Comput. Hum. Behav. **28**, 933–941 (2012)
17. Hajli, M.: Social commerce adoption model. In: Proceedings of the UK Academy of Information Systems Conference (2012c)
18. Anderson, R.E., Srinivasan, S.S.: E-satisfaction and e-loyalty: a contingency framework. Psychol. Mark. **20**, 123–138 (2003)
19. Liang, T.-P., Ho, Y.-T., Li, Y.-W., Turban, E.: What drives social commerce: the role of social support and relationship quality. Int. J. Electron. Commer. **16**, 69–90 (2011)
20. Wang, Y.-S., Wu, S.-C., Lin, H.-H., Wang, Y.-Y.: The relationship of service failure severity, service recovery justice and perceived switching costs with customer loyalty in the context of e-tailing. Int. J. Inf. Manag. **31**, 350–359 (2011)
21. Shin, D.-H.: User experience in social commerce: in friends we trust. Behav. Inf. Technol. **32**, 52–67 (2013)
22. Hassanein, K., Head, M.: Manipulating perceived social presence through the web interface and its impact on attitude towards online shopping. Int. J. Hum. Comput. Stud. **65**, 689–708 (2007)
23. Brown, I., Jayakody, R.: B2C e-commerce success: a test and validation of a revised conceptual model. Electron. J. Inf. Syst. Eval. **11**, 167–184 (2008)
24. Kim, M.-J., Chung, N., Lee, C.-K.: The effect of perceived trust on electronic commerce: shopping online for tourism products and services in South Korea. Tour. Manag. **32**, 256–265 (2011)
25. Kim, S., Park, H.: Effects of various characteristics of social commerce (s-commerce) on consumers' trust and trust performance. Int. J. Inf. Manag. **33**, 318–332 (2013)
26. Edward, C.S.K.: Beyond price: how does trust encourage online group's buying intention? Internet Res. **22**, 569–590 (2012)
27. Hair, J.F., Ringle, C.M., Sarstedt, M.: PLS-SEM: indeed a silver bullet. J. Mark. Theory Pract. **19**, 139–152 (2011)

Advances in Information Technology for Education

A Systematic Literature Review of Augmented Reality Applications in Libraries

Rasimah Che Mohd Yusoff(✉), Azhar Osman, Sya Azmeela Shariff, Noor Hafizah Hassan, Nilam Nur Amir Sjarif, Roslina Ibrahim, Norziha Megat Zainuddin, and Nurazean Maarop

Advanced Informatics School, UTM, Kuala Lumpur, Malaysia
rasimah.kl@utm.my

Abstract. The Augmented Reality (AR) has existed for over five decades, but the growth and progress in the past few years has been exponentially increased and are getting more popular in recent years. Despite the importance and rapid growth of AR applications in a variety of fields, AR applications is not well known in a library setting. This paper aims to provide a systematic literature review (SLR) method to collect and review studies following a predefined procedure on AR application in a library setting. The review studies includes filtering relevant information of AR application from five databases to answer research questions. A total of 23 primary studies published between 2009 to 2017 were used in the analysis. Results from SLR shows that the most common categories of AR applications used in the libraries setting is reading materials and navigational.

Keywords: Augmented reality · Systematic literature review · Library

1 Introduction

Technology has become a crucial part of our lives in the digital age today. The way people think and apply knowledge has changed due to the advancement of the technology. Augmented Reality (AR) has emerged as a technology that has the ability to overlay images, text, video and audio components onto existing images or space. As of today, AR can be applied on many platforms such as computers, tablets, and smartphones. AR can be defined as a situation where a three-dimensional (3D) virtual object is superimposed on top of 3D real environment, thus creating a synthetic environment [1]. AR was a bit different from the virtual reality as its user does not interact with a virtual object. Instead, the user will experience virtual objects appear in the real world [2–4]. Investigating prior research in a field is important, as this reveals the current state of the field and offers guidance to researchers who are seeking suitable topics to explore. Currently, there are many multimedia and online resources provided in the library, but reviews of research on AR technology are less common.

The objective of this paper is to identify the potential AR applications that are available in the libraries. More specifically, the main research question (RQ) addresses: *What type of AR applications are currently being used in the library?*

F. Saeed et al. (Eds.): IRICT 2018, AISC 843, pp. 1037–1046, 2019.
https://doi.org/10.1007/978-3-319-99007-1_96

2 Methodology

In order to answer the research questions, a Systematic Literature Review (SLR) approach was used. SLR aims to search, appraise, synthetize and analyze all the studies relevant for a specific field of research. The methodology utilized is described by Kitchenham in "systematic approaches to a successful literature review". The seven steps utilized to carry out this SLR are: Planning; Define scope; Searching; Screening; Data extraction and synthesis; Analyzing; Writing [6].

2.1 Planning

In this phase, available online scientific databases were used to search for the literature. Five (5) relevant literature databases have been selected: ACM digital library, IEEE Xplore, Science Direct, Google Scholar and Emerald.

2.2 Defining the Scope

Defining the scope actualizes in properly formulate answerable research questions using PICOC (Population, Intervention, Comparison, Outcomes and Context) framework [6]. For this study, the *Population* consists of library users. The *Intervention* considered is the utilization of the AR technology. The *Outcomes* are the application at the library and the *Context* includes the library setting. Inclusion and exclusion criteria have been developed for the selection process.

2.3 Searching

The keywords used to search and find the relevant contents in a paper's title and content is "Augmented reality library" OR "mixed reality library" since the word 'augmented reality' and 'mixed reality is interchangeable.

2.4 Screening

Documents are screened to narrow down the documents found in the search phase to a final number of documents which are relevant for answering the research questions. Articles published from 2009–2017 are taken into consideration for the inclusion in the search criteria. The exclusion criteria were: studies that are not related to the augmented reality as in technological perspective; studies that are not related to library (for example "library" as in "software library", not physical "library"); not in English; repeated articles (by title or content); not available online.

2.5 Data Extraction and Synthesis

The process for selections involved skimming the title and the abstract; skimming the introduction and conclusions; skimming full text; exclude duplicate; and quality assessment (QA). In this review, we developed four QA criteria in order to assess the quality of each study as presented below:

QA1: Is the topic addressed in the paper related to augmented reality in library?
QA2: Is the research methodology described in the paper?
QA3: Is the data collection method described in the paper?
QA4: Are the data analysis steps clearly described in the paper?

Each publication was assessed according to the ratio scale: Yes = 1 point, No = 0 point, and Partially = 0.5 point. The total quality score for each selected studies was measured between 0 (very poor) and 4 (very good). Finally, 23 selected papers were retrieved and chosen which was related to AR applications in the library setting.

2.6 Data Analysis

After reviewing the final selected papers, categorization was made to identify the type of potential AR applications that can be used in the library. Figure 1 shows the SLR process to select the articles.

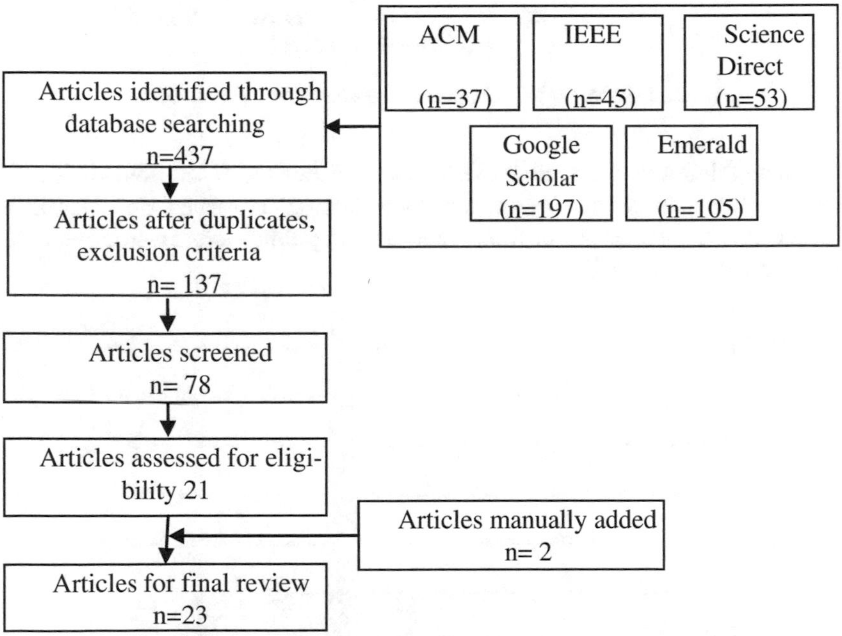

Fig. 1. SLR process to select the primary articles

3 Results

The search process provided a total of 23 primary studies. The following sections summarizes the results of SLR according to the research question designed based on primary studies identified.

Articles related with AR applications for library setting are mostly published in Science Direct and Google Scholar (as shows in Fig. 2). Majority related articles retrieved from these databases because these databases published articles that are related with computer and education.

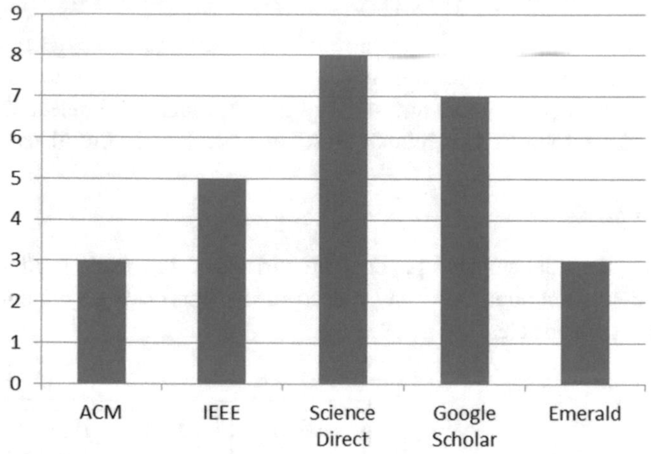

Fig. 2. Databases for SLR process to select the primary articles

Based on this study, the number of articles published increased from 2012 to 2014. The drastic increase may have been caused by technology advances. However, in 2015, 2016, and 2017, there is a consistent number of publications related with AR in libraries (as shows in Fig. 3).

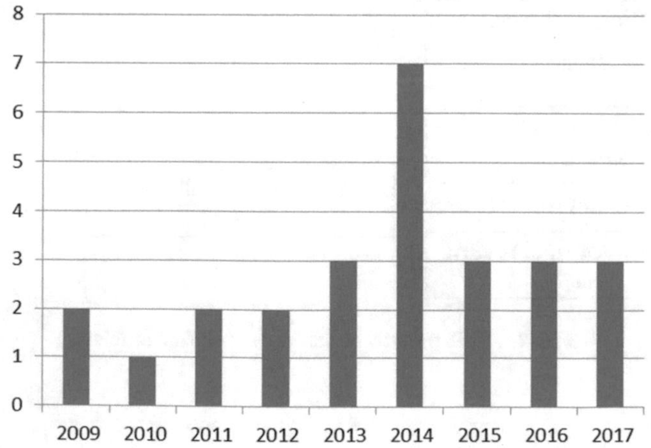

Fig. 3. Publications on AR in library

RQ. What type of AR applications are currently being used in the library?

Results indicated that the most discussed of potential AR applications in the library are in four (4) following categories: reading materials, navigational, tour and promotion, and gaming. AR Applications in reading materials become the main researchable topics followed by navigational, tour and promotion and gaming. Table 1 shows the results from the SLR studies, which are the identified categories of potential AR applications to be used in the library.

Table 1. The categories of AR applications in library setting

No.	Source	Applications	Category	No.	Source	Applications	Category
1	[7]	Special collections	1	13	[19]	Navigation	2
2	[8]	Interactive AR book	1	14	[20]	Navigation	2
3	[9]	Spatial abilities	1	15	[21]	Library tour	3
4	[10]	Special collections	1	16	[22]	Indoor navigation	2
5	[11]	Contextualizing digital contents	1	17	[23]	Exhibition library's art gallery	3
6	[12]	Book stacks browsing, navigation	2	18	[24]	Reader's advisory and reference services	3
7	[13]	AR physic book	1	19	[25]	Book	1
8	[14]	Shelf-reading system	2	20	[26]	Book	1
9	[15]	Game for elementary students	4	21	[27]	Navigating library	2
10	[16]	Scavenger hunt	4	22	[28]	Book tracking	2
11	[17]	Game	4	23	[29]	Shelf searching system	3
12	[18]	Game	3				

Categories: 1 = Reading Materials; 2 = Navigational; 3 = Tour and Promotion; 4 = Game

Based on this SLR findings, majority articles develop the AR applications for reading Materials (n = 8) and Navigation (n = 7). Development AR for tour applications (n = 4) and game (n = 3).

3.1 AR Reading Materials

Books are the most basic, valued asset in the library. However, in the era of technology and 21st century, a physical book has limited capabilities to cope with the younger generations needs and expectations. The lack of interactions, feedback and limited capabilities of demonstrating abstract and spatial concepts are among the major drawbacks of learning by physical books these days [13]. In this case, of course, AR can help by complementing physical book to enhance learning while adding some fun elements. The Augmented Reality Rare Book and Manuscript (AR Rare-BM) for special library collection applications are one of the usage of AR books in the library. The need for such application is raised by researchers and students who often have to face with tedious procedures when they need to access a book or manuscript from the special collection in the library. With AR Rare-BM, the problem can be solved as the students or researcher just need to use a head-mounted display (HMD). HMD in AR is used to track a marker to trigger the virtual book and they can access the contents just like the original artefacts [7].

For low ability readers, remembering content from a text-based book could be a difficult task. They can barely remember the content. However, when they are tested with an AR book, the amount of information remembered are increased. Results in this study indicated that using AR books may help to develop memory recall skills rather than reading on a plain textbook. AR book add values to traditional learning method (e.g. reading books) into something that is interesting, fun, and helpful [8, 10].

The use of AR books can help students to understand complex contents by making sense of the context that is hard to verbally explain in an effective way. Its interactivity capabilities and non-static contents are the attributes that a traditional book lack of. Imagine explaining spatial concepts using 2D images in a textbook versus interactive spatial concept learning through 3D models that enable the student interactively explore the whole concept. This is the advantage of AR books in creating spatial abilities while strengthening the cognitive abilities of the students [9, 13, 27].

A mobile app from the Bavarian State Library has a feature called 3D e-book. The feature lets the user see the related digital contents of medieval manuscripts. When a user scans a page from the manuscript, related objects pertaining to the page will appear and the user can interact with either to zoom it, rotate it or flip a page [11]. AR books would bring the digital nativity to them. It has the features that a traditional book cannot afford to have such as 3D animations, 3D models, audios, videos and website reference that appear instantaneously at the point of enquiry. Those extra elements will improve student's interest in learning and building the independence of self-learning skills. A study shows that the majority of the students who are given a chance to learn using AR book application on top of their textbook agree that it helps them to understand better on the content and increasing their motivation to learn [26].

3.2 Navigational

Application in this category is used for operational tasks in the library, either for library workers or users. This includes the navigation system, usually as a guide for users to get around the library. ARLib is among the earliest AR application that has been designed to cater the need of operational tasks in the library. It uses ATK marker, a laptop that is tied to the user, an input mechanism through a virtual keyboard and an OHMD for display. ARLib is capable of guiding its user to search a book at the shelves from its interface. Once a book is chosen from the search results, the user can see the shelves that holding the book is highlighted, thus, the user can take the book on the shelf. Another application of ARLib is that it can help a user to return a book to the shelf. The user just needs to scan the marker on the book and an indicator at the shelf will appear to show the correct location to return the book. Another concept of AR use for library shows that it is possible for users to get additional information when doing physical book stack browsing. For example, a user will use AR application to scan the book and the additional information on that book will appear. This includes the information on related e-resources about the book, history of that book for example check out counts and similar book recommendations [12, 22, 25].

AR Library Administration can overcome the shortcomings of the old static administration systems by directly providing the physical location of the book in the library. The application can be used to check whether the physical books on the shelf are sorted correctly. A technique called multiple target acquisition is used to identify the target. Then, a comparison will be made to the database to check the correct sorting sequence. If the sorting is not correct, a symbol representing correct and wrong will be displayed at the book marker. An evaluation of the efficiency is made. The results for an experienced user is that there is no efficiency change. However, for inexperienced user, it has shown an increased efficiency of 40% [14]. The application can do tasks such as sorting and searching books [20]. It also has the capabilities to provide information such as missing books or books that are on loan. An evaluation has been made on the efficiency and the results show that some tasks cannot be done faster than human. However, positive results can be seen on improved on the correct results of tasks as well as perceived usefulness and ease of use.

A context-aware library management system using AR had been developed [19]. Their purpose of creating the system is to avoid the books being misplaced by users when browsing the books for selection. They came up with the idea of using AR to have the user check the book contents using a mobile phone. When the user scan at the markers that is put on the book, a page with the table of contents (TOC) of the book appears. The user can now browse the TOC to find out the contents of the book without having to take out the books from its original position Researchers have implemented an AR navigational system to their self-learning app called the "NO Donkey E-Learning" (NODE) [23]. NODE is an application that could help a user for self-regulated learning in the library. First, the user will set a topic for the learning. There are two options from there, either a user use reading path (RP) or learning path (LP). The user who chose RP can see recommended books based on the topic sequence. Then a map will be generated and the user will be able to use the AR navigational arrow to get the books in sequence. If the user chooses LP, they will be able to set their own node or path to get the materials. Once the have set the path, they will be able to use the navigational system just like those who chose RP. Their LP can be shared to the public and another user can just follow the path or flow to learn the same topics.

3.3 Tour and Promotion

In this category, the AR application will be used as a tool for a tour around the library and as a promotional tool to be used in exhibition or special event in the library. The "Ludwig II" app by the Bavarian State Library enables a user to view directions, information and superimposed old picture as well as 360° view of old buildings [11, 18]. The app used geotagging features such as GPS, camera, and compass to represent the near accurate model of the virtual object based on the user's location [12]. Besides that, the Bavarian State Library also used an application called BSB explorer, a gesture-based interaction AR application. The application gave a user the capabilities to interact directly with the augmented virtual object using hand gestures. This application is intended to be used to turn pages, rotate and enlarge the depicted books completely contact free when there are exhibitions or special events at the library [11]. By this, the user can stay at certain distance and enjoy the high quality of interaction hygienic. The augmentation of sculpture in 3D virtual forms and using geotagging to create an AR map that can be used with Google Earth and Wikitude [21].

The development of AR application for this category can be developed using off the shelf (OTS) such as Aurasma and Layar, easy to learn mobile based AR application builder [24]. Videos or animation can be added when a user scans the related promotional posters. This AR applications helps the libraries to accommodate "digital natives" generation as a student-centric approach. This could attract more students to the library and would provide enhance the learning experience for them.

3.4 Gaming

The applications in this category are used as a tool for learning about libraries using the game approach. Games would serve as an excellent approach for the pedagogical model in teaching [17]. The first example in this category is the game-based AR library instruction system (GARLIS). GARLIS is a game for elementary school students, intended for them to learn about library instructions. While providing the element of fun and entertaining, the game also has the feature to test the learner's performance. According to the results, there are no different between the outcome of learners who received instruction from an instructor and those who learn from GARLIS. Therefore, GARLIS has a potential to replace an instructor in teaching about the library instructions [15].

An idea of gaming using the iOS application called Stiktu in the academic law library has been develop [16]. Stiktu is one of those OTS apps that can be used to create an AR effect using mobile phones in an easy way. Their approach is to create a scavenger hunt using Stiktu as a method for publicity. Markers will be put as a target for clues to solve a puzzle at the locations where the library wanted their user to know about. By having such a game, users will go to the locations in the library that they have never known before this. This may give them an idea on where to go to get resources regarding their studies in the future. Game element integrated to AR book able to make learning fun and attractive [25].

4 Conclusions

The results of this SLR aims to answer the main research question: *What type of AR applications are used in the library?* Review on the final 23 articles has been provided. Based on our study, articles related with AR in library setting majority published in Science Direct and Google Scholar. SLR results shows that the highest number of articles related to AR application in library published in year 2014. Results indicated that the most discussed of potential AR applications in the library are in the following categories: reading materials, navigational, tour and promotion, and gaming. In conclusion, by reviewing the current studies on AR applications in library setting, we consider this effort could be valuable for both academic and practitioners.

The finding of this review can be used as foundation for researchers to identify new research equations, and get overview of current research to position their work.

Acknowledgements. This research is supported by GUP research Grant (No. Q.K130000. 2538.15H48) from Universiti Teknologi Malaysia.

References

1. Azuma, R.T.: A survey of augmented reality. Presence Teleoperators Virtual Environ. **6**(4), 355–385 (1997)
2. Abolfazli, S., Sanaei, Z., Gani, A., Xia, F., Yang, L.T.: Rich mobile applications: genesis, taxonomy, and open issues. J. Netw. Comput. Appl. **40**, 345–362 (2014)
3. Wu, H.K., Lee, S.W.Y., Chang, H.Y., Liang, J.C.: Current status, opportunities and challenges of augmented reality in education. Comput. Educ. **62**, 41–49 (2013)
4. Dunleavy, M., Dede, C., Mitchell, R.: Affordances and limitations of immersive participatory augmented reality simulations for teaching and learning. J. Sci. Educ. Technol. **18**(1), 7–22 (2009)
5. Bacca, J., Baldiris, S., Fabregat, R., Graf, S.: Augmented reality trends in education: a systematic review of research and applications. J. Educ. Technol. Soc. **17**(4), 133–149 (2014)
6. Kitchenham, B.: Procedures for performing systematic reviews, vol. 33, pp. 1–26. Keele University, Keele, UK (2004)
7. Parhizkar, B., Zaman, H.B.: Development of an augmented reality rare book and manuscript for special library collection (AR rare-BM). In: Visual Informatics: Bridging Research and Practice, pp. 344–355. Springer, Berlin (2009)
8. Dias, A.: Technology Enhanced Learning and Augmented Reality: An Application on Multimedia Interactive Books (2009)
9. Martín-Gutiérrez, J., Saorín, J.L., Contero, M., Alcañiz, M., Pérez-López, D.C., Ortega, M.: Design and validation of an augmented book for spatial abilities development in engineering students. Comput. Graph. **34**(1), 77–91 (2010)
10. SCARLET: Mapping out a User Journey. Augmented Reality in Education. https://teamscarlet.wordpress.com/2011/09/12/mapping-out-a-user-journey (2011)
11. Ceynowa, K.: Mobile applications, augmented reality, gesture-based computing and more–innovative information services for the internet of the future: the case of the bavarian state library. In World Library and Information Congress: 77th IFLA General Conference and Assembly (2011). http://Conference.ilia.org/past-wlic/2011
12. Hahn, J.: Mobile augmented reality applications for library services. New Libr. World **113** (9/10), 429–438 (2012)
13. Dünser, A., Walker, L., Horner, H., Bentall, D.: Creating interactive physics education books with augmented reality. In: Proceedings of the 24th Australian Computer-Human Interaction Conference, pp. 107–114. ACM (2012)
14. Brinkman, B., Brinkman, S.: AR in the library: a pilot study of multi-target acquisition usability. In: IEEE International Symposium on Mixed and Augmented Reality (ISMAR), pp. 241–242. IEEE (2013)
15. Wang, Y.S., Chen, C.M., Hong, C.M., Tsai, Y.N.: Interactive augmented reality game for enhancing library instruction in elementary schools. In: 2013 IEEE 37th Annual Computer Software and Applications Conference Workshops (COMPSACW), pp. 391–396. IEEE (2013)

16. Barnes, E., Brammer, R.M.: Bringing augmented reality to the academic law library. AALL Spectr. **17**, 13 (2013)
17. Ireton, D., Pitts, J., Ward, B.D.: Library discovery through augmented reality: a game plan for academics. Int. J. Technol. Knowl. Soc. **9**(4), 119–128 (2014)
18. Denton, W.: Libraries and Archives Augmenting the World. ALA TechSource (2014)
19. Malhotra, N., Singh, A., DivyaKrishna, J., Saini, K., Gupta, N.: Context-aware library management system using augmented reality. Int. J. Electron. Electr. Eng. **7**, 923–929 (2014)
20. Shatte, A., Holdsworth, J., Lee, I.: Mobile augmented reality based context-aware library management system. Expert Syst. Appl. **41**(5), 2174–2185 (2014)
21. Fernandez, P.: Through the looking glass: envisioning new library technologies augmented reality in the (real) library world. Libr. Hi Tech News **31**(3) (2017)
22. Huang, T.C., Shu, Y., Yeh, T.C., Zeng, P.Y.: Get lost in the library? An innovative application of augmented reality and indoor positioning technologies. Electron. Libr. **34**(1), 99–115 (2014)
23. Massis, B.: Using virtual and augmented reality in the library. New Libr. World **116**(11/12), 796–799 (2015)
24. Meredith, T.R.: Using augmented reality tools to enhance children's library services. Technol. Knowl. Learn. **20**(1), 71–77 (2015)
25. Lai, A.S., Wong, C.Y., Lo, O.C.: Applying augmented reality technology to book publication business. In 2015 IEEE 12th International Conference on e-Business Engineering (ICEBE), pp. 281–286 (2015)
26. Altinpulluk, H., Kesim, M.: The classification of augmented reality books: a literature review. In Proceedings of 10th annual International Technology, Education and Development Conference (INTED), pp. 4110–4118 (2016)
27. Liu, D.Y.: Combined with augmented reality navigation applications in the library. In International Conference on Advanced Materials for Science and Engineering (ICAMSE), pp. 441–443 (2016)
28. Mahadik, A., Katta, Y., Naik, R., Naikwade, N., Shaikh, N.F.: A review of augmented reality and its application in context aware library system. In: International Conference on ICT in Business Industry and Government (ICTBIG), pp. 1–6. IEEE (2016)
29. Cervera-Uribe, A.A. : The augmented library: an approach for improving users awareness in a campus library. In: 2017 IEEE International Symposium on Mixed and Augmented Reality (ISMAR-Adjunct), pp. 15–19. IEEE (2017)

The Effects of Demographic Characteristics of Lecturers on Individual and Course Challenges of E-Learning Implementation in a Public University in Yemen

Hanan Aldowah(✉), Irfan Umar, and Samar Ghazal

Universiti Sains Malaysia, Gelugor, Penang, Malaysia
hanan_aldwoah@yahoo.com, Irfan@usm.my,
samar_ghzl@yahoo.com

Abstract. Many higher education institutions including universities are adopting the use of e-learning to support the traditional learning approach. However, there are several challenges facing universities in developing countries including Yemen as they seek to implement e-learning. Some of these challenges are associated with the demographic variables of the participants involved in the implementation of e-learning initiatives. Therefore, this paper examines the influence of the demographic factors (age, gender, teaching experience, e-learning experience) on individual and course challenges of e-learning. Four hypotheses were generated from the research questions of the study. A quantitative method was employed to collect the data and 107 participants (lecturers) completed the survey-based questionnaire. *T*-test and ANOVA analyses were employed as statistical tools. Based on the findings, the effects of demographic characteristics on the individual and course challenges of implementing e-learning at the university were identified. Some significant differences were found in terms of the years of lecturers' teaching experience. Conversely, no significant differences were found in terms of age, gender, and e-learning technology related experience. Furthermore, it was found that the experience of lecturers in the use of e-learning technologies was low. University teachers need to be provided with the necessary training to bring them in line with latest learning technologies. Decision makers at the university should be aware of the challenges that lecturers may face during the implementation process and provide the necessary support.

Keywords: E-learning · Demographic characteristics · Challenges
Yemen

1 Introduction

Today's advanced technologies in the field of Information technology and Internet, telecommunication and wireless networking are transferring the learning process from a traditional classroom learning environment to an online environment [1]. E-learning is a newly emerging tool of information technology that has been integrated into many higher education institutions including universities to meet educational needs as well as

F. Saeed et al. (Eds.): IRICT 2018, AISC 843, pp. 1047–1056, 2019.
https://doi.org/10.1007/978-3-319-99007-1_97

to enhance knowledge sharing [2], shifting from the traditional methods of education to an electronic environment. E-learning is defined as a new approach for delivering electronically within an interactive learning environment to anyone, at anytime and anywhere via internet and other technologies with attention paid to the principles of instructional design [3].

In the last decade, e-learning usage has increased in most educational institutions over the world [4]. This is because the integration of e-learning systems has become urgent and the main concern of higher education institutions. Therefore, some of the Yemeni universities are spending funds for the acquisition of these systems, but are unable to achieve purposeful goals [5]. Although higher education in Yemen is committed to implementing and integrating technology into learning and teaching processes, learning in a purely electronic environment is still a difficult step at the moment and needs more effort and ongoing support from the international and local community [6]. The process seems to be influenced by a number of barriers and challenges [6–10].

Most of the universities in developing countries suffer from common problems such as lack of awareness, resources, and infrastructure [11–13]. In some countries, the culture and mindset have become a major problem facing the application of e-learning platforms. Their universities and governments are struggling for fully functional e-learning systems, but all efforts seem to be worthless [14]. The inability of the developing countries including Yemen to achieve the full use of e-learning has led many researchers to encourage further studies to try to understand the reasons behind the unsuccessful implementation of e-learning in their countries [6, 13, 15]. Some public universities are trying to adopt fully functional e-learning systems, but they are unable to achieve all the benefits and goals [16, 17]. Based on literature, these challenges are related to the individual differences and the cultural background of the lecturers as well as their awareness and attitudes towards e-learning [5, 18].

The term "demographic characteristics" refers to the personal characteristics and the circumstantial factors that represent the individual differences which could be attributed to conditions like personal skills, confidence, motivation, previous experiences in teaching, e-training, and e-technology, computer skills, age, gender, and etc., while the term "individual challenges" refers to the problems that are associated with these personal characteristics and differences that arise from these individual circumstances [19, 20]. [21] indicated that the individual characteristics of a lecturer are supposed to be an important issue in the implementation process of e-learning and therefore should be taken into account [22, 23]. Likewise, the term "course challenges" refers to the issues that are related to the course provided such as course design, course content, and so on.

Much research is devoted to studying the effects of demographic differences and e-learning implementation [19, 21, 23, 24]. Examples of demographic differences researched in the e-learning literature include age [21–26], gender [22, 24, 26–29], years of experience in teaching [21, 27], computer skills [23, 30–32], and e-learning technology related experiences [21, 30, 31]. Furthermore, many prior studies have indicated that studying demographic aspects or differences of individual is an important area of research in e-learning environment. In addition researchers in e-learning field are continuously encouraged to recognize the demographic characteristics of individuals involved in the e-learning projects that may affect the successful implementation of the e-learning in educational institutions [20, 26]. Thus, there are constant calls to examine the influence of individual differences of lecturers on the different challenges of implementing e-learning in universities [16, 21, 31]. In addition, most of the previous studies have focused on the influence of students' characteristics in terms of their challenges to the use of e-learning. However, few studies have been focused on the effects of the university teachers' characteristics on the various challenges of e-learning.

Therefore, in order to fill this gap, and extend the body of existing literature especially in the context of Yemen by using Hodiedah University as a case study, this study was carried out to examine the effects of the demographic characteristics of lecturers on individual and course challenges of e-learning implementation based on the selected variables. The demographic characteristics of the lecturers that have been chosen for the investigation in this study includes age, gender, years of experience in teaching, computer skills, e-learning technology related experience. The structure of the paper is as follows: the research objective and the research question are presented in Sect. 2. The research methodology of this research is described in Sect. 3 whereas the findings of the study are presented in Sect. 4. Then, discussions of the results and implications are provided in Sect. 4, and lastly the conclusion is outlined in Sect. 5.

2 Research Objective and Research Question

The paper aims to study the effects of the demographic characteristics of lecturers on the individual and course challenges of e-learning implementation in the university. Therefore, the research question for this study is:

RQ. Do the demographic characteristics of lecturers have any significant effects on the individual and course challenges of e-learning implementation at the university in terms of age, gender, years of experience in teaching, computer skills, and e-learning experience?

3 Research Hypotheses

Based on the research question of the study, the following hypotheses were generated:

Hypothesis 1: There is no significant difference between junior and senior lecturers in terms of individual and course challenges of e-learning implementation.

Hypothesis 2: There is no significant difference between male and female lecturers in terms of individual and course challenges of e-learning implementation.

Hypothesis 3: There is no significant difference between lecturers with more experience in teaching and those with less teaching experience in terms of individual and course challenges of e-learning implementation.

Hypothesis 4: There is no significant difference between lecturers who have vast e-learning experience and those who have less e-learning experience in terms of individual and course challenges of e-learning implementation.

4 Methodology

The objective of this study is to examine if there are any significant differences between the individual and course challenges of implementing e-learning and the demographic characteristics of university teachers. This study involved a quantitative approach in which an inferential statistics is used. This study is conducted at Hodeidah University (public university) in Yemen. This university is chosen for the study because it is the only public university in Yemen that has started the actual implementation of e-learning with the support of the government and the World Bank, but this implementation has faced many challenges from various aspects. Therefore, the focus of the study was on examining the demographic characteristics of the lecturers to understand the challenges they face and hinder the successful implementation of e-learning. The population of the study comprises the lecturers at the university in which 107 out of 497 junior and senior lecturers from all the faculties of the university participated in this study. The sample included 67 male and 40 female teachers. Moreover, data was collected using a questionnaire which consisted of closed-ended questions. A pilot study was first conducted to improve the questionnaire structure and its content and to evaluate the feasibility and time. The questionnaire's reliability and validity were ensured by the researcher and the experts in the field of research. To determine the internal reliability of the instrument, the Cronbach alpha coefficient was calculated. Alpha values for the two challenges exceeded the minimum threshold of 0.70 [33]. For individual challenges construct, the coefficient value is 0.772, and for course challenges, it is 0.786.

The questionnaire was sent to the e-learning center in the university via the Internet and the staff hand-delivered the questionnaire to the participants. They were asked to rate their agreement of each item of the survey on a 4-point Likert Scale, where 1 = Strongly disagree, 2 = Disagree, 3 = Agree, 4 = strongly agree. Approximately 120 questionnaires were distributed to the lecturersin the different faculties of the university. However, only 107 questionnaires have been returned for analysis. The data obtained from the questionnaire has been analyzed using Statistical Package for Social Sciences (SPSS) version 22.0. Then, the inferential statistics were used to examine if there any significant differences in terms of individual and course challenges between demographic variables groups. For this purpose, four demographic variables have been selected including age, gender, years of experience in teaching, and e-learning experience. The individual challenges involved technological confidence, motivation and commitment, qualification and competence, and time. Meanwhile, course challenges involved curriculum, pedagogical model, subject content, teaching and learning activities, localization, flexibility, and availability of educational resources. T-test and ANOVA analysis were used as statistical tools to study the effect of demographic characteristics on the individual and course challenges of the e-learning implementation. The statistical analysis selection was based on the number of groups that are compared and the distributions of the dependent variables. Accordingly, t-tests were conducted to test gender, computer skills, and e-learning experience while ANOVA is used for age and teaching experience. In addition, when the differences are significant, a $\mathbf{p < 0.05}$ will be used.

5 Data Analysis and Findings

To attain the study goal answer the main question of the research, several null hypotheses were examined using t-test and ANOVA.

5.1 Demographic Background of the Lecturers

The descriptive statistics involving frequencies and percentage for the demographic characteristics of the lecturers are shown in Table 1. As shown, 60% of participants were male and 40% were female. The majority of participants ranged in age from 25 to 46 years. In addition, the majority of participants had between 1 and 15 years teaching experience. All but seven participants have basic computer skills. Furthermore, 63 participants did not have any e-learning technology related experience while 44 participants had experience of using e-learning technologies.

Table 1. Demographic background of the participants

Demographic variables	Frequencies	Percentage
Age		
25–35	42	39.3
36–46	40	37.4
47–57	23	21.5
More than 57	2	2
Gender		
Male	67	59.4
Female	40	38.6
Years of teaching experience		
1–5	30	28.0
6–10	35	32.7
11–15	26	24.3
16–20	12	11.9
21–25	2	1.9
26–30	2	1.9
E-learning technology related experience		
Yes	44	41.1
No	63	58.9

5.2 The Effects of Demographics Factors on Individual and Course Challenges of E-learning

Age (Hypothesis 1)

To examine the effect of participants' age on individual and course challenges of e-learning, a one-way ANOVA was conducted and the results are presented in Table 2.

Table 2. Age and e-learning challenges

Challenges	Age group	Sum of squares	df	Mean square	F	Sig. *
Individual challenges	Between groups	0.436	3	0.145	1.822	0.148
	Within groups	8.221	103	0.080		
	Total	8.658	106			
Course challenges	Between groups	736	3	0.245	2.049	0.112
	Within groups	12.338	103	0.120		
	Total	13.074	106			

The results in Table 2 indicate that there are no significant differences in terms of individual and course among the lecturers from the different age groups. Hence, the first hypothesis was also supported.

Gender (Hypothesis 2)
To test the second hypothesis which is related to participants' gender, t-test was employed to examine the differences between male and female groups in terms of individual and course challenges, and the results are presented in Table 3.

Table 3. Gender and e-learning challenges.

Challenges	Gender	n	Mean	SD	Dif	t	Sig*
Individual challenges	Male	67	2.13	0.270	105	0.992	0.323
	Female	40	2.08	0.311			
Course challenges	Male	67	3.16	0.376	105	1.286	0.201
	Female	40	3.07	0.303			

Table 3 shows the result of the significant differences between the individual and course challenges and gender. The results reveal no significant differences in terms of individual and course challenges with regards to gender. In other words, male and female university teachers held similar views about these two challenges. Therefore, the second hypothesis failed to be rejected.

Years of Experience in Teaching (Hypothesis 3)
A one-way ANOVA was used to examine the significant differences between the different teaching experience groups. The results are given in Table 4.

Table 4. Years of experience in teaching and e-learning challenges

Challenges	Age group	Sum of squares	df	Mean square	F	Sig. *
Individual challenges	Between groups	0.436	3	0.145	1.822	0.148
	Within groups	8.221	103	0.080		
	Total	8.658	106			
Course challenges	Between groups	1.470	5	0.294	2.558	**0.032**
	Within groups	11.604	101	0.115		

*Sig $p < 0.5$, 95% confidence interval of the difference.

The findings in Table 4 show significant differences between lecturers with high experience in teaching and those with less teaching experience in terms of course challenges. However, no significant difference in term of individual challenges was observed between the two teaching experience groups. Hence, the third hypothesis was partially supported.

E-learning Technology Related Experience: Hypothesis 4

In order to examine the significant differences between the two groups of e-learning technology experience, a t-test was used for this purpose. The results are presented in Table 5.

Table 5. E-learning technology related experience and e-learning challenges

Challenges	E-learning technology experience	n	Mean	SD	dif	t	Sig*
Individual challenges	Yes	43	2.11	0.238	105	−0.171	0.864
	No	64	2.12	0.316			
Course challenges	Yes	43	3.12	0.341	105	−0.134	0.894
	No	64	3.13	0.361			

As shown in Table 5, the differences between e-learning technology experience groups were found to be not significant in terms of individual and course challenges (all the *p* values are greater than 0.05). This shows that the two challenges are not affected by the experiences of teachers in e-learning. Thus, the null hypothesis was supported.

6 Discussion

The demographic factors involved in this study were age, gender, years of experience in teaching, and e-learning technology related experience. The study attempted to find out the effects of these factors that may be useful in recognizing the various challenges that are expected to affect the implementation of e-learning systems in the university context of Yemen. The outcome of this study proved that there are no significant differences among university teachers' views (male and female) concerning individual and course challenges that impact the implementation and adoption of e-learning. This is attributed to the fact that both groups of lecturers (male and female) acquired the same training and were provided with the same interface design and content. Furthermore, in the context of Yemen, every university teacher regardless of gender is required to take an introductory computer course to improve their computer proficiency and skills. Therefore, computer illiteracy is no longer an issue for those teachers. Because the mentality of using the internet and computers as a required tool has matured, the university teachers' views, skills or efficacy should not be considered an issue in the e-learning context. This result is supported by [20, 21, 34]. However, this result is in contrast with prior studies, which found a significant difference between male and female university teachers' views about individual and course challenges of e-learning [24, 35].

Moreover, the finding of the study indicated that university teachers' age does not significantly influence individual and course challenges of e-learning. Since most of the participants were aged between 25 and 46 years old, they are the most probable groups to use the technology and internet for learning and for communicating with friends and students [36]. Thus, their ability to use the internet will help them to use and adopt the e-learning system effectively, which will affect later their perception of the importance of using e-learning in their teaching. However, these results are not in line with other previous studies [8, 9] that mostly proved that age has a significant impact on e-learning challenges.

The findings revealed that there are significant differences between the lecturers' views from different teaching experience groups regarding course challenges of e-learning implementation. This result implies that the teachers' experience in teaching has an effect on their attitude toward the implementation of e-learning. This finding is in consistent with prior studies which found that length of experience in teaching plays a significant and stronger role on the implementation of e-learning [21, 27]. On the other hand, the findings showed that the university teachers' length of experience in teaching has no significant difference with regards to the individual challenges of e-learning implementation. This finding implies that the length of experience in teaching of lecturers did not have any effect on individual challenges of e-learning implementation at the university. One possible explanation for this finding is that the majority of the participants have less teaching experience, and therefore, they might not face any individual challenges in the implementation of e-learning. These results support the conclusion made by US National Center for Education Statistics (2000) which stated that lecturers with less experience in teaching may face fewer technology challenges in their teaching process. It can be concluded that the fewer years of teaching experience lecturers have, the more probable it is that they apply and use technology in their learning and teaching processes.

With regard to e-learning technology related experience, no significant differences have been observed between individual and course challenges of e-learning implementation and lecturers experience on the use of e-learning technology. To put it another way, lecturers who have experience in e-learning technologies and those with no prior experience have similar opinions regarding the individual and course challenges of e-learning. Thus, whether the lecturers have an experience in e-learning systems or not, they still face other challenges from various aspects in implementing e-learning. This finding is inconsistent with a prior study conducted by [21] who confirmed that there are no statistically significant differences between the challenges of e-learning and the e-learning technology experience levels of the lecturers. Hence, based on the findings of the study and the support of the previous researches, demographic factors were found to play key roles with the individual and course challenges of implementing e-learning environment especially in Yemen [6].

7 Conclusion

This study investigated the effects of demographic characteristics of lecturers on the individual and course challenges of e-learning implementation and documented the critical points that administrators and decision-makers must take into account when implementing e-learning in any university. In general, some significant differences have been found between demographic characteristics and the individual and course challenges of e-learning implementation in terms of length of teaching experience while there were no significant differences in terms of gender, age, and e-learning technology experience. This study confirmed some findings of previous studies, and found several significant new findings. It can be concluded from this finding that the technical skills and expertise needed during the implementation phase of e-learning systems are different from standard ICT such as LMS and course design. In addition, lecturers need intensive training related to e-learning technologies, which improve their basic skills and encourage them to effectively use technology in their teaching.

References

1. Aldowah, H., et al.: The impacts of demographic variables on technological and contextual challenges of e-learning implementation (2017)
2. Aristovnik, A., et al.: Evaluating the impact of socio-economic and demographic factors on selected aspects of e-learning in public administration education (2016)
3. Ajadi, T.O., Salawu, I.O., Adeoye, F.A.: E-learning and distance education in Nigeria. TOJET Turk. Online J. Educ. Technol. **7**(4) (2008)
4. Clark, R.C., Mayer, R.E.: E-Learning and the Science of Instruction: Proven Guidelines for Consumers and Designers of Multimedia Learning. Wiley, London (2016)
5. Thabet, T.S.A., Kalyankar, N.: The effect of e-learning approach on students' achievement in fraction math course level 5 at Yemens Public Primary School. Glob. J. Comput. Sci. Technol. **14**(2), 206–213 (2014)
6. Aldowah, H., Ghazal, S., Muniandy, B.: Issues and challenges of using e-learning in a Yemeni Public University. Indian J. Sci. Technol. **8**(32), 9 (2015)
7. Shormani, M.Q., AlSohbani, Y.A.: Yemen, in e-learning in the middle east and North Africa (MENA) region, pp. 451–482. Springer (2018)
8. Al-Mamary, Y.H., et al.: Adoption of management information systems in context of Yemeni organizations: a structural equation modeling approach. J. Dig. Inf. Manag. **13**(6), 429–444 (2015)
9. Alsurori, M., Salim, J.: Information and communication technology for decision-making in the higher education in Yemen: a review. In: 2009 International Conference on Electrical Engineering and Informatics, ICEEI 2009. IEEE (2009)
10. Muthanna, A., Karaman, A.C.: Higher education challenges in Yemen: discourses on English teacher education. Int. J. Educ. Dev. **37**, 40–47 (2014)
11. Andersson, A.: Seven major challenges for e-learning in developing countries: case study eBIT, Sri Lanka. Int. J. Educ. Dev. ICT **4**(3), 45–62 (2008)
12. Khan, M., et al.: Barriers to the introduction of ICT into education in developing countries: the example of Bangladesh. Online Submiss. **5**(2), 61–80 (2012)

13. Yoloye, E.O.: New technologies for teaching and learning: challenges for higher learning institutions in developing countries. Information Communication Technology (ICT) Integration to Educational Curricula: A New Direction for Africa, p. 250 (2015)
14. Andersson, A., Gronlund, A.: A conceptual framework for e-learning in developing countries: a critical review of research challenges. Electron. J. Inf. Syst. Dev. Ctries. **38**(8), 1–16 (2009). http://www.ejisdc.org/ojs2/index.php/ejisdc/article/viewFile/564/291
15. Olutola, A.T., Olatoye, O.O.: Challenges of e-learning technologies in Nigerian University Education. J. Educ. Soc. Res. **5**(1), 301 (2015)
16. Al-Shboul, M.: The level of e-learning integration at The University of Jordan: Challenges and opportunities. Int. Educ. Stud. **6**(4), p93 (2013)
17. Al-Shboul, M., Alsmadi, I.: Challenges of utilizing e-learning systems in public universities in Jordan. Int. J. Emerg. Technol. Learn. (iJET) **5**(2), 4–10 (2010)
18. Aljaaidi, K.S.Y.: Girls' educational crisis solving through the adoption of e-learning system: the case of Hadhramout university (2009)
19. Agarwal, R., Prasad, J.: Are individual differences germane to the acceptance of new information technologies? Decis. Sci. **30**(2), 361–391 (1999)
20. Owate, C., Akanwa, P.: Demographic variables and students use of e-learning resources in private secondary schools libraries in Rivers State of Nigeria. Rev. Eur. Stud. **10**(1), 84 (2018)
21. Taha, M.: Investigating the success of e-learning in secondary schools: The case of the Kingdom of Bahrain, Brunel University (2014)
22. Abouchedid, K., Eid, G.M.: E-learning challenges in the Arab world: revelations from a case study profile. Qual. Assur. Educ. **12**(1), 15–27 (2004)
23. Nawaz, A.: The challenges and opportunities of e-learning (2011)
24. Osika, E., Johnson, R., Butea, R.: Factors influencing faculty use of technology in online instruction: a case study. Online J. Distance Learn. Adm. **12**(1), 12 (2009)
25. Mungania, P.: Seven e-learning barriers facing employees: executive summary of dissertation. University of Louisville (Online), pp. 21–25 (2003)
26. Naveed, Q.N., et al.: A mixed method study for investigating critical success factors (CSFs) of e-learning in Saudi Arabian Universities. Methods **8**(5), 10 (2017)
27. Islam, M.A., et al.: Effect of demographic factors on e-learning effectiveness in a higher learning institution in Malaysia. Int. Educ. Stud. **4**(1), p112 (2011)
28. Al-Harbi, K.A.-S.: E-learning in the Saudi tertiary education: potential and challenges. Appl. Comput. Inf. **9**(1), 31–46 (2011)
29. Lee, Y.-K., Pituch, K.: Moderators in the adoption of e-learning: an investigation of the role of gender. In: The 2nd International Conference on Electronic Business, Taipei (2002)
30. Algahtani, A.: Evaluating the effectiveness of the e-learning experience in some universities in Saudi Arabia from Male Students' Perceptions. Durham University (2011)
31. Hussein, H.B.: Attitudes of Saudi Universities faculty members towards using learning management system (JUSUR). Turk. Online J. Educ. Technol. TOJET **10**(2), 43–53 (2011)
32. Selim, H.M.: Critical success factors for e-learning acceptance: confirmatory factor models. Comput. Educ. **49**(2), 396–413 (2007)
33. Hair, J., et al.: Multivariate Data Analysis, 6th edn. Pearson Prentice Hall, Englewood Cliffs (2006)
34. Moukali, K.H.: Factors that affect faculty attitudes toward adoption of technology-rich blended learning (2012)
35. Spotts, T.H., Bowman, M.A., Mertz, C.: Gender and use of instructional technologies: a study of university faculty. High. Educ. **34**(4), 421–436 (1997)
36. Jones, S., Fox, S.: Generations Online in 2009. Pew Internet & American Life Project, Washington, DC (2009)

Guideline for Organizing Content in Adaptive Learning System

Halina Dahlan[1], Ab Razak Che Hussin[1], and Yusuf Sahabi Ali[2](✉)

[1] Universiti Teknologi Malaysia, 81310 Johor Bahru, Malaysia
{halina, abrazak}@utm.my
[2] Ahmadu Bello University, Zaria, Nigeria
sahabiali@yahoo.com

Abstract. In the past few years, various adaptive learning systems were developed in response to a widespread desire for all encompassing educational environments. However, these learning systems were developed by educational researchers using various techniques thereby resulting in varying outcomes. This is so because there is no specified guideline that leads to the development of an efficient and effective online adaptive learning system. Therefore, the need to propose guidelines for organizing content in an online adaptive learning system that will cater for all learners regardless of their differences. Several databases and keywords were used to ascertain the lack of guidelines in organizing content in adaptive learning systems. In this study, we propose a content adaptation guidelines for different type of learners in online adaptive learning systems based on Martinez learning style model as employing the same instructional conditions to all students can be pedagogically inefficient. The guideline is developed on the adaptation mapping from information in the student model which is carried out in four stages Organizing content, Individualized content, Adaptive navigation and Control level. These guidelines will help developers as well as educators with basic steps in developing a seamless online adaptive learning system for different type of learners.

Keywords: Guidelines · Learning style · Mapping · Adaptation

1 Introduction

The emergence of the internet and information technology has changed the future of our educational settings leading researchers to develop methods, tools and environments for online based learning [1, 2]. Nowadays, online learning also known as web-based learning plays a huge role in shaping our learning process. It provide students with a high level of user control and rich materials corresponding to their learning needs [3]. Online learning system (OLS) refers to the use of computer network to conduct the process of learning in order to distribute learning course content and material to the learners [4]. Course content and materials are directly accessed by students through the internet without the need to be physically present within the four walls of a classroom. On the side of the teachers or educators, they don't have to spend too much time attending classroom to conduct the class as their burden has been

F. Saeed et al. (Eds.): IRICT 2018, AISC 843, pp. 1057–1065, 2019.
https://doi.org/10.1007/978-3-319-99007-1_98

reduced to guidance and much less supervision. The total benefits of online learning are both classroom and platform independence [5]. It offers flexible access from anywhere, anytime and allows learners to get engaged into the learning environment with little or no guidance from instructors [6]. However, online learning has its own shortcomings. Most of the course content in an OLS was organized in an arbitrary manner. Thus, making it difficult for students to get hold of the salient message that is being transmitted [7]. Based on these concerns, the purpose of this study is to propose adaptation guidelines for organizing content that can be used in adaptive learning systems (ALS) authoring tools to help teachers or educators develop an ALS with less effort accommodating different type of learners. Student's individual differences play a key role in the learning value chain including OLS [8].

Individuals have different cognitive styles that influence how they organize and process information, influencing their learning performance [8, 9]. Research on individual differences has received a sounding devotion and has been identified as the panacea to the problems bedeviling the educational field [10]. To ensure learners are responsible and engaged to their learning process, it is suggested that individual differences of each learner must be taken into consideration when preparing a learning process [11]. Thus, adaptive systems are becoming dominant in the educational settings.

Adaptation refers to making an adjustment towards learning environment in order to meet the requirement of presenting the appropriate learning content that can accommodate different learners based on their needs and preferences [4]. Adaptive learning system (ALS) is considered a new learning medium that employs online instructional strategies and hypermedia techniques [4]. This system has the capability of providing the adaptive lesson to different types of learners considering their individual differences [5]. ALS builds the adaptive learning content based on learners profile in the learner model such as student preferences, interests, goal, knowledge as their attributes that determine learners personalized features which makes them different in learning [12].

2 Theoretical Foundation

2.1 Online Learning System

Online learning has become the new choice that is employed in conducting teaching and learning in an innovative way. Through online learning, learners have direct and flexible access to the resources and information that are available in the learning platform. OLS refers to the use of internet to access learning materials; to interact with the content, instructor, and other learners. There are various terms used to represent online learning and some of which are: e-learning, computer-assisted learning, web-based learning and distance learning [13]. All of this type of online learning applies the same concept in its application where a learner uses some form of technology to access or experience learning. This form of learning has the capability to provide students with easy access to information and resources without the restriction of time and space.

Learning and teaching process can be done anytime and anywhere without overly dependent on the few available teachers or classrooms. Learners are also provided with high level of control and direct access to rich learning material corresponding to their needs, abilities in learning, and learning styles [14].

2.2 Learning Style

Learning style is a concept that followed the research from a cognitive perspective starting in the 1960's [15]. It refers to how learners are different in the way they perceive, accept, think, solve problems and learn [16]. Also, learning styles can be defined as a subset of wide ranges in individual differences that may affect the process of learning [8]. There are many approaches in defining, classifying and identifying learning styles [17]. It can be used as a preferred approach that consistently adapt in developing learners learning experience that may affect their choice in making the learning strategies to achieve their learning goal. Learning styles influence how learners go through the learning process, how teacher should teach them and how the interaction between the two of them should take place [11]. Therefore it is important for educators to determine different pedagogical procedures in approaching learners in their process of learning.

2.3 Adaptive Educational System

Adaptive Educational System (AES) is defined as a new approach in education that can make learning systems more effective by adapting the presentation of information and overall linkage structure to each individual learner preferences [18–20]. Through this assumption, each individual learner has different learning characteristic that make them different in learning and presenting a different educational setting can be more suitable for one type of learner than for others. Any form of instruction accommodating learners' individual needs can be considered adaptive, whether it is delivered face-to-face or in a technology-based format [21]. AES provide mechanisms to individualize instruction of teaching strategies (such as learning content, interface, strategies, and assessment) for learners based on their individual differences [12]. Based on the need to accommodate different types of learners, an AES with the capability to deliver online learning content adaptively to each individual learner were developed. Basically, this system can prevent information overload on the learners, discontinuous flow of learning, cognitive overload and content un-readiness [21].

3 Adaptation Components

The concept of adaptation refers to making an adjustment towards learning environment in order to meet the requirement of presenting an appropriate learning content that can accommodate different learners based on their needs and preferences [4]. There are two important components involved in the process of adaptation and they are Learners model and Adaptation mapping.

3.1 Learners Model

Learners model consist of information about learners which include general profiles, type of learner, knowledge level that are stored in the system's database [22]. In order to present an appropriate course content that fits different types of learners, the information stored in the learners' model is identified and exploited by the system into course organization, course presentation and course navigation in the OLS. There are two main sub-component of the learner model and they are: learner type and knowledge level.

Learner type can be categorized into three derived from Martinez learning style model and they are transforming learner, performing learner and conforming learner. There are also three categories of knowledge level stored in the learner model which are: Learning Goal, Learning concept and Educational Material. They are as shown in Table 1.

Table 1. Categories of Knowledge level

Knowledge Level (KL)	Description	Level of achievement
Learning Goal (LG)	Stored the learners' progress on their achievement of selected learning goal	{Beginner, Advanced, Proficient}
Learning Concept (LC)	Stored the learners' progress on their achievement on the learning concept related to selected learning goal Consist of the KL on learning outcome (LO), Prerequisite (Pr) and related topic (RT) of learning concept	{Beginner, Advanced, Proficient}
Educational Material (EM)	Stored the learners' progress on their level of performance in EM page EM page are organized in three different level of performance: Remember, Use and Find	{Beginner, Advanced, Proficient}

3.2 Adaptation Mapping Process

As one of the two adaptation components, the process of mapping between the learners' model in the system with the adaptation technologies and organizational presentation strategies is carried out in this phase. This process is conducted through four stages: Organizing content, Individualized content, Adaptive navigation and Control level as indicated by [23]. In general, the explanation on each of the four stages is described briefly below together with the adaptation technology and organizational presentation strategy used.

Organizing Content: It is the process of structuring and organizing the course content so that the presentation of domain knowledge is adapted with learners' knowledge level in the learner model. Use curriculum sequencing and adaptive navigation support as adaptation technologies and Elaboration theory (ET) together with Component Display Theory (CDT) to organize the presentation of course content.

Individualized Content: It is the process of matching the presentation of course content so that the presentation of education material knowledge modules in domain knowledge adapt with the type of learner based on their learning style (transforming, performing, conforming) that are stored in the learner model. Use CDT to match the adaptive presentation of course content by using adaptive presentation technology.

Adaptive Navigation: It is the process of allowing learners to find the optimal learning path i.e. next node to be learned.

Control Level: The process of allowing learner to tailor the system to their preferences and adapt it to their needs that change over time. The control level supports learner into several levels of adaptation.

Figure 1 shows the adaptation process which leads to the adaptation mapping on the on the right side of the figure.

Figure 1 indicates that the mapping process that is implemented when the learners' information were obtained from the learner's model which will eventually be mapped with certain adaptation technologies leading to personalized learning.

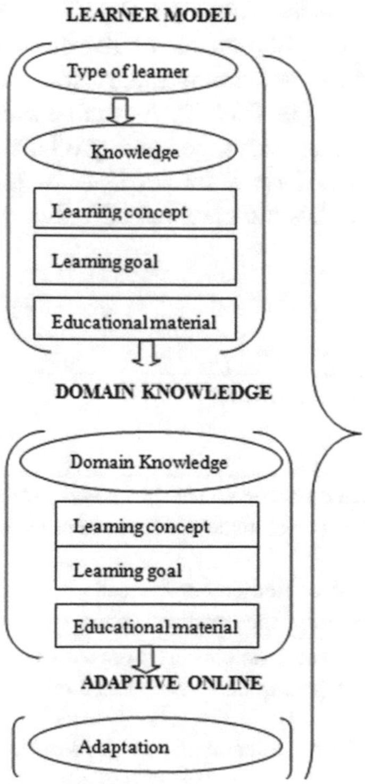

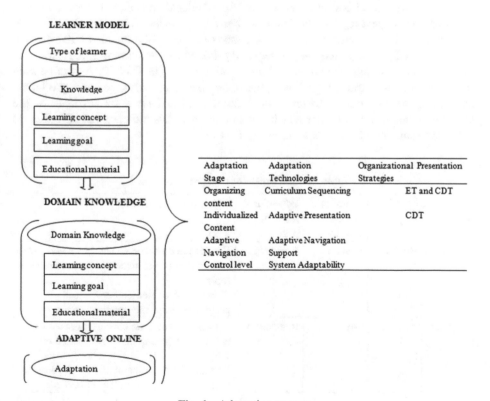

Adaptation Stage	Adaptation Technologies	Organizational Presentation Strategies
Organizing content	Curriculum Sequencing	ET and CDT
Individualized Content	Adaptive Presentation	CDT
Adaptive Navigation	Adaptive Navigation Support	
Control level	System Adaptability	

Fig. 1. Adaptation process

4 Adaptation Guidelines for Different Types of Learners in Adaptive Online Learning System

The proposed adaptation guidelines for different type of learners in online adaptive learning system designed in this study follows the adaptation process. It is built to show the sequence flow of adaptive presentation on different type of learners based on their learning style. These guidelines are significant in providing learners with the appropriate learning system that will suit their differences in the learning process. It would be used by content developers and educators in designing a befitting adaptive OLS content that suit the various types of learners. The guidelines consist of planning an adaptive OLS that specify learners needs and preferences based on their characteristics in the learners model into the development of the course content, adaptive presentation of course content and adaptive navigation.

4.1 Guideline for Organizing Content

The guidelines for organizing content depict the appropriate method to organize domain knowledge of learning concept outcome, educational material for a specific learning goal depending on the learners' knowledge level. Given the level of the learner, the adaptive system develops a pattern on how the content of the domain knowledge will be presented and subsequently altered as the learning progresses. The system adapts and map content according to the guideline in Table 2. As can be seen from Table 2, the knowledge level goes from beginner, advance, and proficient. Depending on the knowledge level, the learner is placed on a certain layer of the learning outcome concept, after which educational materials are prepared according to either the remember level, use level or the find level.

Table 2. Guidelines for organizing content

Knowledge level		Learning outcome concept	Educational material page	Curriculum sequencing
LC	EM			
Beginner	Beginner	Layer 1	Remember	Present the outcome concept of layer 1 and entire set of prerequisite concept and related topic Present EM of Remember level of performance for the outcome concept
Beginner	Advance	Layer 1	Remember, Use	Present the outcome concept of layer 1 and entire set of prerequisite concept and related topic Present EM of Remember and Use level of performance for the outcome concept

(*continued*)

Table 2. (*continued*)

Knowledge level		Learning outcome concept	Educational material page	Curriculum sequencing
LC	EM			
Beginner	Proficient	Layer 1	Remember, Use, Find	Present the outcome concept of layer 1 and entire set of prerequisite concept and related topic Present EM of Remember, Use and Find level knowledge modules of performance for the outcome concept
Advance	Beginner	Layer 1, Layer 2	Remember	Present the outcome concept of both layer 1 and 2 together with the entire set of prerequisite concept and related topic Present EM of Remember level knowledge modules of performance for the outcome concept
Advance	Advance	Layer 1, Layer 2	Remember, Use	Presented the outcome concept of both layer 1 and 2 together with the entire set of prerequisite concept and related topic Present EM of Remember and Use level of performance for the outcome concept
Advance	Proficient	Layer 1, Layer 2	Remember, Use, Find	Present the outcome concept of both layer 1 and 2 together with the entire set of prerequisite concept and related topic Present EM of Remember, Use and Find level knowledge modules of performance for the outcome concept
Proficient	Beginner	Layer 1, Layer 2, Layer 3	Remember	Present the outcome concept of both layer 1, 2 and 3 together with the entire set of prerequisite concept and related topic Present EM of Remember level of performance for the outcome concept
Proficient	Advance	Layer 1, Layer 2, Layer 3	Remember, Use	Present the outcome concept of both layer 1, 2 and 3 together with the entire set of prerequisite concept and related topic Present EM of Remember and Use level knowledge modules of performance for the outcome concept
Proficient	Proficient	Layer 1, Layer 2, Layer 3	Remember, Use, Find	Present the outcome concept of both layer 1, 2 and 3 together with the entire set of prerequisite concept and related topic Present EM of Remember, Use and Find level knowledge modules of performance for the outcome concept

5 Discussion

This study develops a guideline for developing an adaptive learning system based on Martinez learning style model. The model guides the researchers in taking care of the different styles of learning among students. This is very important as there is no one size fits all in the learning domain [5]. Therefore, the development of adaptive learning systems can now be done in an effective and efficient way as it is made to be student centered. This will also give students the confidence and encouragement to utilize this system no matter the learning style.

6 Conclusions

The call for adaptive learning system by the educational research community has been widely disseminated for several years. And rightly so, researchers have given their much precious time and resources in developing systems that caters for all learners taking their individual differences into consideration in what is termed personalization. However, these researchers employ different approaches and techniques in developing the content of these systems resulting in varying outcomes. This shed light to the need for guidelines that will lead the way for the development of an effective and efficient content for adaptive learning systems. In this paper, we proposed adaption guidelines for organizing content which take individual learning styles into consideration. The proposed content organization guideline can be used as a basic guideline for the development of content for online adaptive learning system. Therefore we conclude that, the proposed guidelines will result in the development of online adaptive systems that will motivate and also improve learner satisfaction. Additionally, in the future, we hope to develop a prototype system to validate the guidelines and also propose other guidelines for the various stages of the adaptation process.

Acknowledgement. This work is supported by the Ministry of Higher Education (MOHE) and Research Management Centre (RMC) at the Universiti Teknologi Malaysia (UTM) under the Research University Grant - Instructional Development Grant (GUP-DPP) VOT R. J130000.7728.4J244.

References

1. Tsai, C.C.: Beyond cognitive and metacognitive tools: the use of the Internet as an 'epistemological' tool for instruction. Br. J. Educ. Technol. **35**(5), 525–536 (2004)
2. Hwang, G.J.: On the development of a cooperative tutoring environment on computer networks. IEEE Trans. Syst. Man Cybern. Part C Appl. Rev. **32**(3), 272–278 (2002)
3. Lo, J.J., Wang, H.M., Yeh, S.W.: Effects of confidence scores and remedial instruction on prepositions learning in adaptive hypermedia. Comput. Educ. **42**(1), 45–63 (2004)
4. Magoulas, G.D., Papanikolaou, K., Grigoriadou, M.: Differences through system's adaptation. Br. J. Educ. Technol. **34**(4), 511–527 (2003)
5. Surjono, H.D.: The evaluation of a moodle based adaptive e-Learning system. Int. J. Inf. Educ. Technol. **4**(1), 89–92 (2014)

6. Zhang, D., Nunamaker, J.F.: Powering e-Learning in the new millennium: an overview of e-Learning and enabling technology. Inf. Syst. Front. **5**(2), 207–218 (2003)
7. Baig, F.: Comparative study of frameworks for the development of better quality adaptive hypermedia based educational systems. J. Qual. Technol. Manag. **7**(2), 63–82 (2011)
8. Graf, S., Kinshuk, K.: Providing adaptive courses in learning management systems with respect to learning styles. In: Proceedings of E-Learn: World Conference on E-Learning in Corporate, Government, Healthcare, and Higher Education, vol. 17, no. 1, pp. 2576–2583 (2007)
9. Stash, N., Cristea, A., De Bra, P.: Adaptation languages as vehicles of explicit intelligence in Adaptive Hypermedia. Int. J. Contin. Eng. Educ. Life Long Learn. **17**(4–5), 319–336 (2007)
10. Retalis, S., Paraskeva, F., Tzanavari, A., Garzotto, F.: Learning styles and instructional design as inputs for adaptive educational hypermedia material design. In: Information and Communication Technologies in Education-Fourth Hellenic Conference with International Participation (2004)
11. Dabbagh, N., Kitsantas, A.: Personal Learning Environments, social media, and self-regulated learning: a natural formula for connecting formal and informal learning. Internet High. Educ. **15**(1), 3–8 (2012)
12. Inan, F.A., Lowther, D.L.: Factors affecting technology integration in K-12 classrooms: a path model. Educ. Technol. Res. Dev. **58**(2), 137–154 (2010)
13. Sun, P.C., Tsai, R.J., Finger, G., Chen, Y.Y., Yeh, D.: What drives a successful e-Learning? An empirical investigation of the critical factors influencing learner satisfaction. Comput. Educ. **50**(4), 1183–1202 (2008)
14. Lo, J.J., Chan, Y.C., Yeh, S.W.: Designing an adaptive web-based learning system based on students' cognitive styles identified online. Comput. Educ. **58**(1), 209–222 (2012)
15. Tseng, J.C.R., Chu, H.C., Hwang, G.J., Tsai, C.C.: Development of an adaptive learning system with two sources of personalization information. Comput. Educ. **51**(2), 776–786 (2008)
16. Mampadi, F., Chen, S.Y., Ghinea, G., Chen, M.-P.: Design of adaptive hypermedia learning systems: a cognitive style approach. Comput. Educ. **56**(4), 1003–1011 (2011)
17. Brusilovsky, P.: Adaptive navigation support in educational hypermedia: the role of student knowledge level and the case for meta-adaptation. J. Comput. Inf. Technol. **6**(4), 27–38 (2003)
18. Belk, M., Papatheocharous, E., Germanakos, P., Samaras, G.: Modeling users on the World Wide Web based on cognitive factors, navigation behavior and clustering techniques. J. Syst. Softw. **86**(12), 2995–3012 (2013)
19. Brinton, C.G., Rill, R., Ha, S., Chiang, M., Smith, R., Ju, W.: Individualization for education at Scale: MIIC design and preliminary evaluation. IEEE Trans. Learn. Technol. **8**(1), 136–148 (2015)
20. Papanikolaou, K.A., Mabbott, A., Bull, S., Grigoriadou, M.: Designing learner-controlled educational interactions based on learning/cognitive style and learner behaviour. Interact. Comput. **18**(3), 356–384 (2006)
21. Akbulut, Y., Cardak, C.S.: Adaptive educational hypermedia accommodating learning styles: a content analysis of publications from 2000 to 2011. Comput. Educ. **58**(2), 835–842 (2012)
22. Tzouveli, P., Mylonas, P., Kollias, S.: An intelligent e-learning system based on learner profiling and learning resources adaptation. Comput. Educ. **51**(1), 224–238 (2008)
23. Truong, H.M.: Integrating learning styles and adaptive e-learning system: current developments, problems and opportunities. Comput. Hum. Behav. **55**, 1185–1193 (2016)

Development and Validation of Instrument for Assessing Researcher's Participation in e-Collaboration

Jamilah Mahmood(✉), Halina Mohamed Dahlan, Ab Razak Che Hussin, and Muhammad Aliif Ahmad

Faculty of Computing, Universiti Teknologi Malaysia UTM, 81310 Skudai, Johor, Malaysia
mahmoodjamilah@gmail.com, abrazakutm@gmail.com, maliif.ahmad@gmail.com, halina@utm.my

Abstract. The advancement of recent technology for conducting research helped researcher in many ways such as collaborate in real time, sharing knowledge, accessing unlimited resources of information and produce better research outcomes. However, the advantage of this technology can only be achieved if the researcher participates in e-collaboration. This paper discusses the instrument to examine researcher's participation in e-collaboration. The framework for collaborative technologies was used to develop the instrument. This study was conducted by using a questionnaire survey method. The instrument then tested by using sample of 50 researchers from five research universities in Malaysia, which are UTM, UKM, UM, USM and UPM. The respondents who took part in the survey were lecturer, postgraduate student, research assistant and research fellow who had experience in using e-collaboration tool in conducting their research activities. Smart PLS software was used to evaluate the instrument validity and reliability, and the report shows that all instrument used are acceptable. The final instrument contains of 34 items measurement scales.

Keywords: e-Collaboration · e-Research · Research collaboration
Education technology · Instrument validation

1 Introduction

E-collaboration allows team members to interact virtually to work in the research projects. E-collaboration tools advantages help researchers in many ways such as increase communication, expand the size of the group and provide a new improved methods of communication for team member to easily share and access shared information [1]. Beside that, by using e-collaboration, researchers can also produce better research outcomes and increase the productivity of their research. However, based on Mendeley analysis report, the pattern of the users group shows that not all researchers participate in e-collaboration [2, 3]. Many of researchers prefer to work independently in their research group without involving others. Some of the researchers also face problems in dealing with team members who did not commit themselves to complete the team task because of their priority on their own task. This situation will cause

F. Saeed et al. (Eds.): IRICT 2018, AISC 843, pp. 1066–1076, 2019.
https://doi.org/10.1007/978-3-319-99007-1_99

problems if team members have a different goal and opinion in completing their research project. Researchers can only derive scientific advantage from their participation in e-collaboration only if they work together with team members in achieving their common goal or shared task. Therefore, to further understand the real researcher's participation problem in e-collaboration, this paper will investigate the answer to the following questions:

1. How to develop an instrument that suitable to understand the participation of academic researchers in e-collaboration?
2. How to validate the instrument?

2 Literature Review

Influence factors of online participation in e-collaboration can give impact in providing better learning outcomes. The examples of online participation are seconds spent viewing content pages and number of written posts. Collaboration is most successful in an online environment when the user feels that they have participated effectively in the system or tool. In e-collaboration, the sense of "joint enterprise" is very important and should be fostered within team members. However, many of the researchers using e-collaboration prefer to work individually in their task without sharing much information about their work progress to others. Less interaction happens between team members because they think that they can work on their part without any help from other team member [4]. They may engage in some collaborative processes but they work individually for most of the time. The process of achieving goal is not fully shared by all team members. The participation of researchers in e-collaboration will differ depending on the degree to which team members share their goals, processes and outcome [5].

Collaboration is characterized by sharing in all of the dimensions involved; people share the processes, as well as the goals and outcomes of their work [5]. A conceptual framework for examining collaborative work with groupware technologies consist of four input factors, process variables, and outcome variables [6]. Input factors consists of individual and group characteristic, task characteristic, situational characteristic and technology characteristic [6]. While the process variable is a set of indices that reflects the patterns of activity that occur during collaborative work such as shared knowledge and amount of participation [6]. This study identifies the participation factors based on the reviewed theories and model of e-collaboration. The factors were identified using collaboration technologies input factors. The identified factors were superior influence, peer support, moral trust, self-motivation, collaboration technology experience, task interdependence, awareness, cooperation and social presence.

3 Methodology

In this study, initial instrument was tested by using data from 50 respondents, which are researchers from five research universities. Five research universities selected are UTM, UKM, UM, USM and UPM. Survey was distributed to the researchers that have prior experience in using e-collaboration tools that include lecturer, research fellow, research assistant, PhD student and master student. Smart PLS software was used to validate the instrument.

4 Instrument Development

The model in Fig. 1 describes the relationships between variables and the constructs to be measured. This model consists of 12 constructs that need to be measured. This study use multi-item scale to easure the concept of each constructs. Figure 1 depicts the illustration of the relationship of each construct along with their items. All constructs have reflective measurement model as indicated by the arrows. The indicators used to measure the constructs are listed in the Table 2.

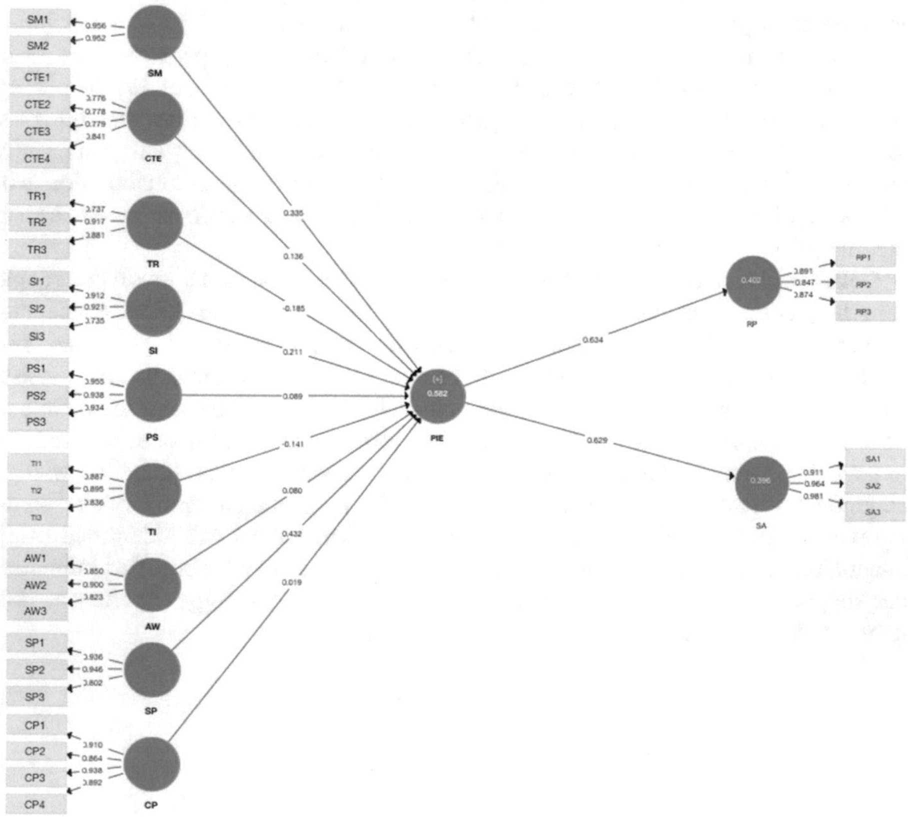

Fig. 1. Participation model for e–collaboration.

This study adapted and modified some indicators from previous studies to measure the constructs from the models. The indicators are identified based on the definition of the constructs (refer Table 1).

Table 1. Constructs definition

Constructs	Definition	Ref.
Self Motivation (SM)	Team member's ability to share or obtained knowledge based on their interest without involving pressure from others	[7, 8]
Reliance (RE)	Team member's feeling towards their responsibility that they can rely on others	[9]
Collaboration Technology Experience (CTE)	Team member's ability in using specific type of technology	[10]
Trust (TR)	Team member's action and commitments toward other and the actions in favor of a desired research outcomes	[11, 12]
Peer Support (PS)	Team member's help and support to others to share their experience and knowledge	[13, 14]
Superior Influence (SI)	Team member's belief on importance of their superior opinion that they should use e-collaboration tools	[15]
Task Interdependence (TI)	Research group needs towards support and information from others to complete their research work	[16]
Social Presence (SP)	Technology's ability to transfer non-verbal signals (e.g., gestures and facial expression) and non-word signals (e.g., voice inflection)	[17]
Awareness (AW)	Team member's awareness about what information is being shared among them and also what others can see about their behavior	[18]
Cooperation (CP)	It is the joint operation of members of the group within shared workspace for completing tasks including building, refining, manipulating shared objects. For examples to decide team member's role, sharing resources and planning activities	[19]
Participation in e-Collaboration (PIE)	A process that helps team members interacts, share knowledge and working together to achieve a common goal	[20, 21]
Research Performance (RP)	Team member's performance, which includes expressions of creativity, originality, and facts discovery	[22]
Satisfaction With Research Output (SA)	Team member's desire and expectation from the research output	[22]

Table 2. Indicator for each construct

Factor	Item	Indicator (revised item)	Ref.
Self Motivation	SM1	I like to help my research team	[21]
	SM2	I know that other members in the research team will help me so it's fair to help them	[21]
Reliance	RE1	I feel comfortable counting on research team to do their part	[9]
	RE2	I was not bothered by the need to rely on research team	[9]
	RE3	I feel comfortable trusting research team to handle their tasks	[9]
Collaboration Technology Experience	CTE1	I have a good experience in using messaging tools	[17]
	CTE2	I have a good experience in using audio conferencing	[17]
	CTE3	I have a good experience in using video conferencing	[17]
	CTE4	I have a good experience in using technologies similar to collaboration tools	[17]
Trust	TR1	Trusting my research team helps me to make a mutual understanding in order to achieve the goal of research	[22]
	TR2	I believe my research team is willing to share research information with each other	[22]
	TR3	I believe that resources and data shared by our research team are accurate	[22]
Peer Support	PS1	My research team support creative and higher order of thinking for the progress of the research	[22]
	PS2	My research team helps each other to refine research questions and research design in order to improve the quality of the research	[22]
	PS3	My research team encourages each other to share solutions to work related problems	[23]
Superior Influence	SP1	I believe that top management would like me to use e-collaboration tools to conduct research with my research team	[17]
	SP2	My supervisor suggests that I use e-collaboration tools to conduct research with my research team	[17]
	SP3	There is pressure from the organization to use e-collaboration tools to conduct research with my research team	[17]
Task Interdependencies	TI1	The results of my research work are dependent on the efforts of people from my research team	[24, 25]
	TI2	My research work often involves using knowledge or information from my research team	[24, 25]

(continued)

Table 2. (*continued*)

Factor	Item	Indicator (revised item)	Ref.
	TI3	My research work requires frequent coordination with my research team	[25]
Social Presence	SP1	Using e-collaboration tools to interact with research team creates a warm environment for communication	[17]
	SP2	Using e-collaboration tools to interact with research team creates a sociable environment for communication	[17]
	SP3	Using e-collaboration tools to interact with research team creates a personal environment for communication	[17]
Awareness	AW1	I feel that I control the availability and work progress information I am broadcasting to others	[18]
	AW2	I provide rich enough information for my research team to understand my availability and work progress status well	[18]
	AW3	I feel that people are well informed about my availability and work progress status	[18]
Cooperation	CP1	I could easily create the shared document	[26]
	CP2	I could easily refine the shared document	[26]
	CP3	I could easily manipulate the shared document	[26]
	CP4	I had access to the information needed to operate together	[26]
Participation in e-collaboration	PIE1	I am participating in e-collaboration to contribute to pool of information	[21]
	PIE2	I am participating in e-collaboration to contribute my knowledge	[21]
	PIE3	I am participating in e-collaboration to contribute my idea	[21]
Research Performance	RP1	I achieve good research results with the efforts of our research team	[22]
	RP2	I critically analyze my assigned task and perform accordingly in order to achieve good research findings	[22]
	RP3	I achieve good publication through our research results	[22]
Satisfaction with research output	SA1	I am satisfied with the research results achieved by our research team	[22]
	SA2	I am satisfied with the publication derived from our research results	[22]
	SA3	I am satisfied with the empirical data derived from our research results	[22]

5 Instrument Validation

This section consists of discussion of the instrument validation. The evaluations for the measurement model are as follows:

i. Internal consistency

Cronbach's Alpha value shows the reliability based on the inter correlations of the indicator variables. Values of 0.60 to 0.70 are acceptable [27].

ii. Convergent validity

It is the extent to which a measure correlates positively with alternative measures of the same constructs. The value of outer loading should be 0.708 or higher while outer loadings with value 0.40 and 0.70 should be considered for removal if deleting the indicator changes composite reliability value (or AVE) to increase. Indicators with value below 0.40 should be eliminated from the model [27] (Table 3).

Table 3. Construct reliability and validity

	Cronbach's Alpha	Composite reliability	Average Variance Extracted (AVE)
AW	0.821	0.893	0.736
CP	0.925	0.945	0.812
CTE	0.812	0.872	0.630
PIE	0.875	0.923	0.801
PS	0.937	0.960	0.888
RP	0.845	0.904	0.758
SA	0.948	0.967	0.907
SI	0.821	0.894	0.740
SM	0.901	0.953	0.910
SP	0.880	0.925	0.804
TI	0.852	0.906	0.762
TR	0.801	0.885	0.721

The next measure of convergent validity is AVE, and the value should be 0.50 or above. While, AVE with value 0.50 and below shows that more inaccuracy in the items [27]. According to the results for the outer loadings in Table 4, all items have loadings more that 0.70, which are all items are acceptable. For this measurement model, the AVE values are above 0.50, thus show that all constructs are valid.

Table 4. Indicators outer loading.

Items	Outer loading	Items	Outer loading
AW1	0.850	SA1	0.911
AW2	0.900	SA2	0.964
AW3	0.823	SA3	0.981
CP1	0.910	SI1	0.912
CP2	0.864	SI2	0.921
CP3	0.938	SI3	0.735
CP4	0.892	SM1	0.956
CTE1	0.776	SM2	0.952
CTE2	0.778	SP1	0.936
CTE3	0.779	SP2	0.946
CTE4	0.841	SP3	0.802
PIE1	0.925	TI1	0.887
PIE2	0.901	TI2	0.895
PIE3	0.858	TI3	0.836
PS1	0.955	TR1	0.737
PS2	0.938	TR2	0.917
PS3	0.934	TR3	0.881
RP1	0.891		
RP2	0.847		
RP3	0.874		

iii. Discriminant validity

This evaluation is to validate that the constructs are distinctive and not redundant. Cross loading and Fornell-Larcker's criterion are the measures used for discriminant validity. Table 5 illustrates the result for cross loadings. The indicator's outer loading must be higher than its loading on other constructs [27].

While in Fornell-Larcker's criterion, the square root of AVE is compared with the latent variable correlations. The value should be higher than any other constructs. In Table 6 shows that each value is higher than any other constructs. Therefore, the instrument is valid.

Table 5. Cross loadings.

	AW	CP	CTE	PIE	PS	RP	SA	SI	SM	SP	TI	TR
AW1	0.850	0.322	0.511	0.478	0.600	0.647	0.546	0.418	0.407	0.702	0.447	0.612
AW2	0.900	0.383	0.652	0.616	0.635	0.611	0.387	0.399	0.530	0.753	0.522	0.763
AW3	0.823	0.465	0.474	0.562	0.672	0.597	0.632	0.548	0.525	0.535	0.577	0.631
CP1	0.444	0.910	0.483	0.326	0.452	0.510	0.381	0.498	0.549	0.257	0.466	0.537
CP2	0.252	0.864	0.342	0.221	0.346	0.328	0.235	0.362	0.411	0.073	0.368	0.328
CP3	0.389	0.938	0.438	0.398	0.416	0.496	0.282	0.472	0.529	0.224	0.566	0.487
CP4	0.502	0.892	0.418	0.415	0.600	0.660	0.488	0.590	0.518	0.249	0.595	0.512
CTE1	0.622	0.379	0.776	0.599	0.585	0.496	0.510	0.411	0.546	0.563	0.422	0.511
CTE2	0.393	0.372	0.778	0.321	0.165	0.201	0.251	0.425	0.221	0.260	0.289	0.195
CTE3	0.398	0.302	0.779	0.399	0.236	0.193	0.217	0.378	0.193	0.397	0.285	0.293
CTE4	0.540	0.436	0.841	0.407	0.450	0.449	0.518	0.433	0.279	0.354	0.299	0.397
PIE1	0.596	0.346	0.577	0.925	0.518	0.597	0.580	0.347	0.515	0.555	0.317	0.516
PIE2	0.490	0.408	0.437	0.901	0.478	0.515	0.539	0.406	0.479	0.464	0.439	0.448
PIE3	0.647	0.310	0.523	0.858	0.420	0.585	0.567	0.379	0.501	0.716	0.317	0.547
PS1	0.729	0.474	0.489	0.526	0.955	0.562	0.522	0.389	0.570	0.526	0.532	0.739
PS2	0.716	0.483	0.513	0.503	0.938	0.547	0.479	0.487	0.542	0.473	0.571	0.650
PS3	0.647	0.510	0.398	0.458	0.934	0.464	0.492	0.435	0.532	0.457	0.557	0.621
RP1	0.693	0.466	0.421	0.676	0.545	0.891	0.545	0.330	0.641	0.670	0.498	0.675
RP2	0.496	0.566	0.365	0.433	0.436	0.847	0.546	0.457	0.557	0.321	0.549	0.486
RP3	0.651	0.405	0.391	0.494	0.457	0.874	0.616	0.382	0.485	0.450	0.575	0.485
SA1	0.598	0.403	0.511	0.555	0.624	0.609	0.911	0.468	0.399	0.402	0.400	0.425
SA2	0.542	0.368	0.432	0.594	0.416	0.618	0.964	0.468	0.391	0.383	0.396	0.313
SA3	0.579	0.372	0.488	0.645	0.482	0.630	0.981	0.462	0.404	0.415	0.352	0.322
SI1	0.523	0.548	0.426	0.391	0.540	0.447	0.564	0.912	0.287	0.161	0.474	0.393
SI2	0.387	0.527	0.574	0.393	0.319	0.382	0.425	0.921	0.246	0.134	0.398	0.271
SI3	0.471	0.309	0.318	0.295	0.326	0.278	0.236	0.735	0.025	0.431	0.499	0.318
SM1	0.583	0.607	0.407	0.546	0.556	0.634	0.439	0.295	0.956	0.420	0.548	0.669
SM2	0.509	0.468	0.414	0.518	0.555	0.611	0.355	0.142	0.952	0.540	0.437	0.740
SP1	0.732	0.184	0.429	0.676	0.495	0.577	0.432	0.200	0.515	0.936	0.335	0.661
SP2	0.746	0.244	0.484	0.628	0.490	0.532	0.442	0.248	0.475	0.946	0.442	0.661
SP3	0.576	0.226	0.562	0.386	0.390	0.439	0.195	0.278	0.321	0.802	0.390	0.512
TI1	0.548	0.518	0.402	0.442	0.594	0.613	0.439	0.388	0.581	0.395	0.887	0.536
TI2	0.450	0.451	0.342	0.243	0.430	0.479	0.287	0.495	0.310	0.307	0.895	0.344
TI3	0.561	0.516	0.348	0.290	0.457	0.465	0.264	0.529	0.376	0.391	0.836	0.469
TR1	0.517	0.601	0.312	0.443	0.648	0.540	0.367	0.249	0.730	0.347	0.444	0.737
TR2	0.761	0.409	0.424	0.538	0.604	0.590	0.328	0.386	0.634	0.668	0.434	0.917
TR3	0.704	0.356	0.464	0.451	0.569	0.513	0.241	0.317	0.518	0.725	0.493	0.881

Table 6. Fornell-Larcker's criterion.

	AW	CP	CTE	PIE	PS	RP	SA	SI	SM	SP	TI	TR
AW	0.858											
CP	0.458	0.901										
CTE	0.641	0.471	0.794									
PIE	0.650	0.395	0.575	0.895								
PS	0.742	0.518	0.498	0.528	0.942							
RP	0.718	0.543	0.454	0.634	0.559	0.871						
SA	0.601	0.399	0.500	0.629	0.528	0.649	0.953					
SI	0.530	0.548	0.519	0.421	0.463	0.435	0.489	0.860				
SM	0.573	0.565	0.430	0.558	0.582	0.653	0.417	0.231	0.954			
SP	0.772	0.238	0.529	0.652	0.516	0.582	0.420	0.259	0.501	0.897		
TI	0.604	0.573	0.425	0.397	0.586	0.613	0.400	0.523	0.518	0.426	0.873	
TR	0.785	0.531	0.473	0.566	0.713	0.647	0.368	0.379	0.738	0.690	0.536	0.849

6 Conclusion

For future study, the instrument developed can be used for survey with larger sample size. Besides, this study was conducted with the samples from five research university in Malaysia. The research can also be extend by using sample from other university in Malaysia which their researchers may have different patterns of participation in using e-collaboration tools compared to the existing sample.

Acknowledgements. This work is supported by the Research Management Centre (RMC) at Universiti Teknologi Malaysia (UTM) under Research University Grant (VOT PY/2016/06537/Q.J130000.2528.15H45), and MyPhD Scholarships (Jamilah Mahmood) from Ministry of Higher Education Malaysia (MOHE).

References

1. Luzón, M.J.: Academic weblogs as tools for e-collaboration among researchers. Encyclopedia of E-collaboration, pp. 1–6 (2008)
2. Mohammadi, E., et al.: Who reads research articles? An altmetrics analysis of Mendeley user categories. J. Assoc. Inf. Sci. Technol. **66**(9), 1832–1846 (2015)
3. MacMillan, D.: Mendeley: teaching scholarly communication and collaboration through social networking. Lib. Manag. **33**(8/9), 561–569 (2012)
4. Mahmood, J., et al.: Researcher's participation in e-collaboration. In: International Conference on Research and Innovation in Information Systems 2017, ICRIIS, pp. 1–5. IEEE (2017)
5. Jeong, H., et al.: Joint interactions in large online knowledge communities: the A3C framework. Int. J. Comput. Support. Collaborat. Learn. **12**(2), 133–151 (2017)
6. Olson, G.M. Olson, J.S.: Research on computer supported cooperative work. Handbook of Human-Computer Interaction 2, 2nd edn., pp. 1433–1456 (1997)
7. Kankanhalli, A., Tan, B.C., Wei, K.-K.: Contributing knowledge to electronic knowledge repositories: an empirical investigation. MIS Q. 113–143 (2005)
8. Wasko, M.M., Faraj, S.: Why should I share? Examining social capital and knowledge contribution in electronic networks of practice. MIS Q. 35–57 (2005)
9. Turel, O., Connelly, C.E.: Team spirit: the influence of psychological collectivism on the usage of e-collaboration tools. Group Decis. Negot. **21**(5), 703–725 (2012)
10. Carlson, J.R., Zmud, R.W.: Channel expansion theory and the experiential nature of media richness perceptions. Acad. Manag. J. **42**(2), 153–170 (1999)
11. Hardwig, J.: Epistemic dependence. J. Philos. **82**(7), 335–349 (1985)
12. Hardwig, J.: The role of trust in knowledge. J. Philos. **88**(12), 693–708 (1991)
13. Bruner, J.: Vygotsky: an historical and conceptual perspective. In: Culture, Communication, and Cognition: Vygotskian Perspectives, pp. 21–34. Cambridge University Press, London (1985)
14. Sykes, T.A., Venkatesh, V, Gosain, S.: Model of acceptance with peer support: a social network perspective to understand employees' system use. MIS Q. 371–393 (2009)
15. Venkatesh, V., et al.: User acceptance of information technology: toward a unified view. MIS Q. 425–478 (2003)
16. Fry, L.W., Slocum, J.W.: Technology, structure, and workgroup effectiveness: a test of a contingency model. Acad. Manag. J. **27**(2), 221–246 (1984)

17. Brown, S.A., Dennis, A.R., Venkatesh, V.: Predicting collaboration technology use: Integrating technology adoption and collaboration research. J. Manag. Inf. Syst. **27**(2), 9–54 (2010)
18. Szostek, A.M., et al.: Understanding the implications of social translucence for systems supporting communication at work. In: Proceedings of the 2008 ACM Conference on Computer Supported Cooperative Work (2008)
19. Shah, C.: Collaborative information seeking. J. Assoc. Inf. Sci. Technol. **65**(2), 215–236 (2014)
20. Sobel, J.: Can we trust social capital? J. Econ. Lit. **40**(1), 139–154 (2002)
21. Karna, D., Ko, I.: We-intention: moral trust, and self-motivation on accelerating knowledge sharing in social collaboration. In: 48th Hawaii International Conference on System Sciences (HICSS), pp. 264–273, IEEE (2015)
22. Karna, D., KoIllSang: Collaboration orientation, peer support and the mediating effect of use of e-collaboration on research performance and satisfaction. Asia Pac. J. Inf. Syst. **23**(4), 151–175 (2013)
23. Wang, X., Clay, P.F., Forsgren, N.: Encouraging knowledge contribution in IT support: social context and the differential effects of motivation type. J. Knowl. Manag. **19**(2), 315–333 (2015)
24. Teo, T.S., Men, B.: Knowledge portals in Chinese consulting firms: a task–technology fit perspective. Eur. J. Inf. Syst. **17**(6), 557–574 (2008)
25. Sharma, R., Yetton, P.: The contingent effects of training, technical complexity, and task interdependence on successful information systems implementation. MIS Q. 219–238 (2007)
26. Lee, M.: Usability of collaboration technologies. Doctoral Dissertation, Purdue University (2007)
27. Hair Jr, J.F., et al.: A primer on partial least squares structural equation modeling (PLS-SEM). Sage Publications (2016)

Recent Trends on Software Engineering

A Systematic Mapping Study on Microservices

Mohammad Sadegh Hamzehloui(✉), Shamsul Sahibuddin, and Khalil Salah

Advanced Informatics School, University of Technology Malaysia, 54100 Kuala Lumpur, Malaysia
{shmohammad3, skhalil2}@live.utm.my, shamsul@utm.my

Abstract. Microservices has been drawing a lot of attention in recent years. Their smaller size compared to monolithic applications makes them more maintainable, faster to deploy and hence more appealing. Together with platforms like cloud and emerging practices like DevOps, they are gaining more popularities in the software industry. However, despite their popularity very few secondary researches have been conducted to date. The goal of this study is to find out the common trends and the direction of researches on microservices. We have conducted a systematic mapping to identify and classify different categories of these researches. We have attempted to select only those relevant papers that are specifically conducted with microservices as their primary topics. We have narrowed down to 38 papers from IEEE, ACM and Scopus. These papers were published between 2016 to early 2018. These papers are classified under three main categories; infrastructure, software and deployment/management. The infrastructure and management categories are subsequently divided to several subcategories. Our study shows a rising number of evaluation papers, indicating that microservices have started to be more widely implemented in the real-world. Infrastructures related papers are more common than software related ones and cloud is the commonest platform for running microservices.

Keywords: Systematic mapping · Microservices · Literature review

1 Introduction

The computer industry has faced a lot of changes within the last decade. These changes have affected both the software development methods and the hardware requirements. One of the changes is the new software architecture known as microservices. Microservices are a new architecture in software development, in which applications are developed as some small, independent services that are working together [1].

Microservices are the result of learnt lessons from various areas of software engineering. On the software development side, they are inspired both by DDD[1] [2, 3] and SOA[2]. From the perspectives of DevOps[3] and CD[4], microservices are the solution

[1] Domain Driven Design.
[2] Service Oriented Architecture.
[3] Development and operation.
[4] Continuous deployment.

F. Saeed et al. (Eds.): IRICT 2018, AISC 843, pp. 1079–1090, 2019.
https://doi.org/10.1007/978-3-319-99007-1_100

to many of the delivery and deployment issues. Due to their smaller size, they can be delivered much faster via delivery-pipelines. This is in accordance with the concept of rapid deployment that is emphasized by DevOps [4]. As a general rule, microservices are meant to be easily replaceable, so they are designed with deployability in mind. They can be created and destroyed, at any given time with minimum impact on the application.

One of biggest advantages of using microservices is the fact that they can be developed using different technologies [5]. An application can be made of different services that are developed using different programming languages and databases. This will enable companies to choose the right tool for every task.

Unleashing the true potential of microservices requires the use of other technologies. Some of the newer technologies, like Docker and cloud contributed a lot to the idea of separation of services. Cloud services make hardware provisioning easier. Adding or dropping a server can be done merely with a few clicks. This is essential when it comes to scaling services [6]. Load-balancers, auto-scaling, and built in queuing systems are some of the major required components for scaling which are native to cloud.

Due to the advent nature of microservices, various studies have been recently conducted on various aspects of them. These studies range from software architecture and design to testing, implementation and performance benchmarking. Many of these studies are primary studies. This is of course expected given the fact that microservices are a relatively new topic. As the result, conducting a secondary study to identify the trends and focus areas of these researches is essential. This will be helpful both for the researches to identify the research gaps, and for the industries to obtain the bigger picture.

This paper is an attempt to conduct a secondary study in the form of a systematic mapping. We have conducted a systematic mapping on 38 papers relevant to microservices. The goal of this paper is to identify and classify the areas of researches that has been conducted on microservices.

We have organized this paper as the following: Sect. 2 briefly explains the microservices background. Section 3 will review the related works. Section 4 explains the research methodology. Section 5 presents data extraction the mapping. Section 6 presents the results. Section 7 shares our findings. Section 8 is the conclusion.

2 Background

Microservices are described by [7] as an architecture to implement the business logics using single-purpose services. Newman [8] however describes them as small, autonomous services that work together. Most of the microservices are sharing certain characteristics. Some of these characteristics are described by [9] as:

- Domain driven design
- Single responsibility principle
- Explicitly published interfaces
- Ability to be deployed independently

- Hidden details from outside
- Communicating via REST and HTTP

Microservices do not only change the way that applications are developed, they also ease the delivery and implementation of applications.

3 Related Works

We have found five papers that are closely related to our research. The first three papers are systematic mappings. The other two papers include a literature review and a taxonomy of microservices architecture. As we are going to discuss all of these five papers, they will not be included in the list of the papers that we have selected for our systematic mapping.

The first paper is a systematic mapping [10] conducted on 21 papers. The selected papers are all from years 2014 to 2015. This paper is one of the first systematic mappings conducted on microservices. They have used a characterization framework which uses the common terms from software engineering to classify their selected papers. They have concluded that microservices related researches are at their primitive stage. More experimental and empirical evaluations are needed. They have also mentioned about the clear link between microservices, containers and cloud. Lack of tool support for automation is another point that their research has raised.

The next paper is a mapping conducted on microservices architecture [11]. They have included 33 articles, including papers from the year 2016. They concluded that deployment is the most discussed topic, followed by cloud and monitoring. Communication followed by service-discovery and performance were the most discussed challenges in their selected papers. They also believe that security is an important topic that have not been explored enough.

The last systematic mapping was conducted on the trends and the areas of focus on microservices [12]. This is probably the most thorough study that has been conducted amongst the related works. They have included 71 papers including many from 2016. They concluded that the solutions proposals are the most common researches followed by validation researches. Evaluation researches were the least common papers. They suggested this implies that a limited number of papers are written based on the industrial or real-world examples. Another two interesting classifications presented by this paper are the quality attributes and infrastructure services categories. Performance efficiency followed by maintainability, security and functional suitability are the four most discussed quality attributes by their targeted papers. For the infrastructure services category, monitoring, system level management and service orchestration were the most discussed topics.

A literature review paper by [13] listed out the characteristics of microservices. This can be a good start to form a universal definition of microservices, as it does not currently exist. They included 37 papers, mostly from the year 2016. They mentioned that the design, performance followed by resiliency, testing techniques and cost comparison are at the top of the operational areas' list. Also, non-empirical papers outnumbers the empirical papers by 14%. And finally, Docker, Node.js, Mysql, Postgrass and Java are the top of the list of the technologies that have been used in their selected papers.

A taxonomy of microservices architecture [3] is another paper that provides a valuable insight on the categories of topics in microservices architecture. Although this paper is not directly related to our research, the breakdown of the categories helped us to gain a better understanding of microservices architectures. They have included 28 papers published up to 2016. Due to the long list of the categories, it is impossible to include them here. In summary, they listed the microservices architecture into 6 different categories and the characterization of microservices approaches into 3 main groups.

Although most of the systematic mappings in practice are using very similar approaches for their data collection, they each have their own way of analyzing and presenting their data. Different analyzing and data presentation method leads to different outcomes. On the other hand, newer mapping studies need to be conducted as more papers on microservices are being published on a yearly basis. Most of the papers included in this research are from the year 2017, although some papers from 2016 and 2018 are also included.

4 Research Method

Systematic mappings are designed to provide an overview of the current state of a certain area of research. In other words, they provide the bigger picture. And this is made possible by using classification and counting contributions for each classified section [14]. To conduct the systematic mapping on microservices in this paper, we are going to follow the steps as suggested by [15] and [14] papers.

4.1 Search and Collecting Papers

We have used ACM, IEEE and Scopus databases to search and select our papers. The keywords that we have used are: microservice, microservices (plural), micro-service, and micro-services (plural). In the first stage we have only downloaded the list of the files prior to downloading the pdf files. The combined number of articles from all of the three databases was 870 papers. After removing the duplicated papers, the number was reduced to 594.

We started to exclude the papers by reading the titles and applying our exclusion criteria. The exclusion was done in four stages. After the first stage we ended up with 283 topics. After the second stage 138. And after the third stage 85 papers. For the last stage we went through the abstracts and keywords of each paper after downloading the full pdf files. This resulted in 38 papers that were meeting our inclusion criteria. Due to the long list of the papers and the word limit, the list of the selected papers is not included. The list of the selected papers is available via the link in the footnote.[5]

[5] https://drive.google.com/open?id=1x7EVxuMuVHv7yS7l_oLpo2OTNsdW2DMq.

Since the main objective of this paper is to obtain insight on different aspects of microservices, we have only included those papers that discuss microservices as their primary topic. As the result, papers discussing the following topics were excluded:

- Comparison between microservices and other architectures/technologies.
- Migrating to microservices
- Developing applications based on microservices
- Using microservices to develop an application.

Although some of the selected papers may have similar topics to the list above, we went through their entire document and only selected those that discuss microservices as their primary topic. To maintain the quality of the research we have excluded key notes, lecture notes and abstract papers. The papers selected are either conference papers, workshops, journal or book chapters.

4.2 Research Question

Q1. What are the main focus areas in microservices' researches?

The aim of this question is to identify the current areas of interest for research on the topic of microservices. This will help us to see the direction of these researches.

Q2. Which areas require more researches?

The answer to this question can be helpful for researches in this field. At present, it is difficult to clearly see the current trends given the vast number of publications.

4.3 Inclusion, Exclusion

In order to choose the right set of papers, we have defined a set of inclusion and exclusion criteria. These criteria are designed to increase the chance of targeting the right set of papers. The list of inclusion and exclusion criteria can be seen in Table 1.

4.4 Keywords

To better understand the areas of research for each paper, the keywords from all of the selected papers were collected and analyzed. Figure 1 demonstrates the result of this analysis. To shorten the list of the words only the top 12 words are included. It is important to mention that since some papers are not providing the list of their keywords, we had to read those papers and extract some keywords based on their topics.

5 Data Extraction and Mapping

We have briefly gone through all of the selected papers. This was to gain a better understanding of those papers. Out of the 38 selected papers, 23 of them were published in 2017, 13 of them published in 2016 and 2 of them were published in 2018. Given the fact that we are conducting this research on April 2018, the lower number of published papers for 2018 was expected. However interestingly no paper from 2015 was selected, as none of them met our criteria.

Table 1. List of inclusion and exclusion criteria

Inclusion	• Studies that are discussing microservices as their primary topic • Conference, journal, workshop, book chapter
Exclusion	• Papers that are not written in English • Papers discussing migration to microservices • Papers that are using microservices as their secondary topic • Lecture notes • Key-notes • Abstract papers • Presentation-slides

Word	Length	Count	Weighted Percentage	Similar Words
microservices	13	27	9.34%	microservice, microservices
architecture	12	14	4.84%	architecture, architectures
cloud	5	12	4.15%	cloud
software	8	12	4.15%	software
service	7	8	2.77%	service, services
design	6	7	2.42%	design
model	5	6	2.08%	model, modeling, modelling, models
docker	6	6	2.08%	docker
performance	11	6	2.08%	performance, performances
oriented	8	5	1.73%	oriente, oriented
containers	10	5	1.73%	container, containers
based	5	4	1.38%	based

Fig. 1. List of the top keywords

From the selected papers 23 of them were published at IEEE, followed by ACM with 8 papers and springer with 4 papers. ECMS, ICCCA, and journal of software evolution with one paper each were at the end of the list.

The results from checking the word frequency of the selected paper's abstract was very close to the results of world cloud of the keywords.

Figure 2 shows the type of research papers. An interesting point is the equal number of evaluation papers with validation papers. Even though the number of selected papers is lower compared to researches like [12] the number of evaluation papers are higher than what they found. The lower number of validation papers can be due of our inclusion criteria.

We have classified the selected papers into three main categories. The infrastructure, the software category and the deployment/management category. Except for the software category the other two categories, have their own subcategories. Although the infrastructure and the deployment/management categories can be combined into one as they are closely related; To avoid a single group with a large number papers, we have decided to keep them as two separate categories.

Infrastructure

Table 2 demonstrates the list of papers in the infrastructure category. Out of 38 papers, 14 of them have discussed infrastructure related topics. DevOps with 8 papers has been discussed more than 7 times. Virtualization and configuration each with 4 papers are in

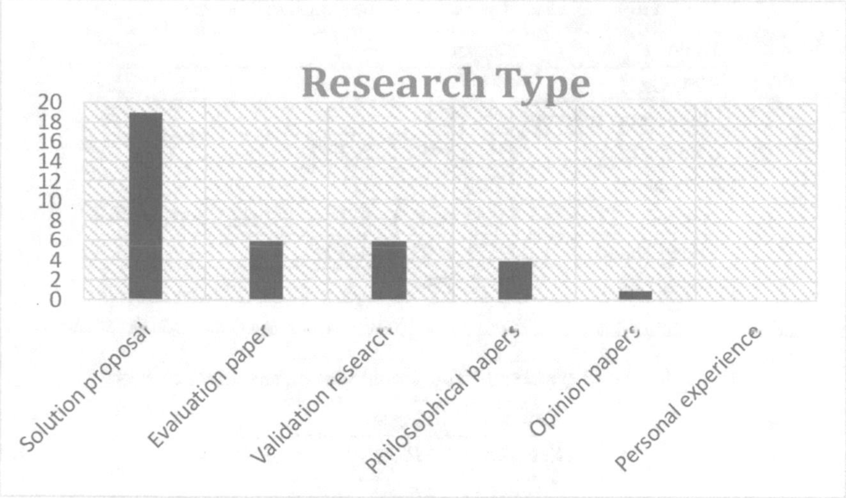

Fig. 2. Research type

the second place. Monitoring and multi-cloud solutions have been discussed 2 times each. And PaaS and CaaS with only one paper came last.

Deployment and Management

Table 3 demonstrates the list of the papers in deployment and management category. There are 9 papers in this category. Testing with 3 papers, maintenance with 1 paper, and security, testing, and deployment each with 2 papers. Paper P23, is a research conducted on the topic of high-availability on unreliable infrastructure [16]; and paper P11 discussed the topic of fault-tolerance and service discovery [17]. It is clear that maintenance is an area that requires further research.

Software

Table 4 demonstrates the list of the papers in the software category. Many of the papers have used some form of coding in their case studies. Some even went as far as developing a fully functional application. However very few papers have worked on software development aspect of microservices. Out of 38 papers only 5 were discussing a form of software related approach to microservices. From the perspective of software development many papers are comparing microservices with SOA, like paper P16 and P9. Microservices are also often being compared with DDD designs. Software development and design in microservices is another area that requires further researches.

5.1 Metrics and Measurements

Table 5 demonstrates the list of the metrics that have been used in the selected papers. Some papers have used metrics to evaluate the outcome of their research. Some other papers however, have conducted their research on enhancing some of these metrics. Out of 10 papers with metrics, performance have been discussed the most, with 5 papers using this metric. Security with 2 papers is in the second place.

Table 2. List of papers in infrastructure category

Topic	Papers
Pass, CaaS	P1
Multi-cloud solutions	P8, P29
Configuration	P29, P10, P15, P37
Automation	P29
DevOps, CD	P29, P38, P26, P18, P7, P20, P37, P3
Virtualization	P38, P26, P34, P13
Monitoring	P12, P3

Table 3. List of papers in deployment and management category

Topic	Paper
Security	P21, P17
Testing	P5, P31, P33
Cost	P30, P3
Maintenance	P6
Deployment	P6, P25
Other	P23, P11

Table 4. List of papers in software category

Approach	Paper
Guidance model for service discovery and fault tolerance.	P11
Proposing an initial UML profile for Domain-driven MSA Modeling [18]	P2
Introducing policies for microservices architecture, on size, and inter-service communication	P4
Requirement engineering with security and scalability being the priorities.	P17
An IDE for microservices.	P32
Impact of situational context on software processes for microservices	P36

Table 5. List of metrics and measurements

Metrics	Paper
Maintainability	P35
Resilience	P33
Performance	P31, P19, P28, P18, P14
Cost	P30
Workload	P25
Security	P17, P22
Scalability	P17

5.2 Microservices Architecture

Table 6 demonstrates the list of papers with the word architecture in their title. The term architecture has been used 10 times in the title of selected papers. Out of these 10 times 5 times they have been used in the context of infrastructure and cloud and 4 time in the context of software. architecture is only being used 1 time in context of deployment. Although most of the papers with architectural approaches target both software and infrastructure. We have rarely observed any paper that only targets one of these areas. As the result our classification is based on the idea that which category (software or infrastructure) has been discussed as the primary objective of the research.

Table 6. List of papers with architecture in their title

Architectural purpose	Paper
Infrastructure and cloud	P23, P25, P26, P27, P34, P24
Deployment/management	P31
Software	P2, P7, P4, P27, P24

5.3 Experiments/ Case Studies

Many papers have used some form of experiment as a way to test and prove their findings. Case studies were the most common types of experiment. Out of 38 selected papers, 22 of them used some form of experiment. Figure 3 demonstrates the diagram of the percentage of papers that have used experiment.

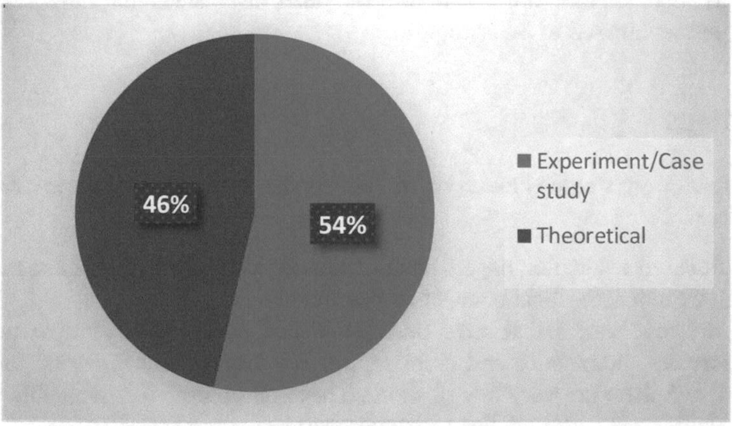

Fig. 3. The percentage of the papers that have used experiment

6 Results

6.1 Q1 (The Main Areas of Focus in Microservices' Researches)

Infrastructure is definitely at the top of this list, with DevOps and configuration being the most researched topics as the subcategory of infrastructure. Cloud together with Docker as a tool for virtualization are the most discussed platform and tool. Monitoring, automation and multi-cloud are next in this list. It is important to mention that many practices such as configuration, monitoring and automation are considered a part of DevOps practices. As the result DevOps can be considered as the most common discussed topic in the category of infrastructure.

In the deployment and management category, cost and testing are the most researched topics followed by security, maintenance and deployment.

In the software/development category there are only 5 papers. Due to the different areas of research of each paper, they cannot be classified under any specific category. But it is agreeable that at least four of these papers (P11, P2, P17, P36) have provided some level of guidelines for developing microservices.

6.2 Q2 (Areas that are Requiring More Researches)

On the infrastructure level, automation and monitoring are requiring more researches. There was only one research (P1) discussing PaaS and CaaS. We believe this is due to the fact that these topics are mainly covered in cloud related researches. On the software level there are very few researches on the topics of design and development. Microservices have mainly been compared with SOA and DDD concepts and are lacking a clear definition and guidelines.

Security, maintenance and costing are the other three areas that have been studies relatively less compared to other topics.

7 Findings

Here are few of our findings based on the papers that we have read during conducting this research:

1. The number of evaluation papers have increased. This means that microservices are moving from a theoretical concept to practical ones.
2. There are few papers that have been published, as an attempted to understand microservices' boundaries and architecture. We found these types of papers very helpful, and there are very few of them. They can be used as a valuable source to create definitions and guidelines for microservices. (exp P9, P16)
3. Few papers have worked on the topic of multi-cloud (P8, P29). These types of researches can help industries to take advantage of multiple cloud platforms at the same time. This can a great achievement.
4. One of the papers (P3) has an interesting approach to use machine learning for the purpose of resource provisioning [6] . This is a new approach that we have seen for the first time. This approach can increase the efficiency and speed of managing cloud resources.

8 Conclusion

The main goal of this paper is to provide a broad overview of the main trends and areas of focus on the topic of microservices. This can be helpful mainly for the researchers who are planning to conduct studies on this field. Given the fact that the initial researches on microservices were conducted back in 2014, they are considered relatively new. For the same reason many of the published papers on this field are still solution proposals. But the number evaluation papers are increasing which is a sign that microservices are started to be used in real-world applications.

References

1. Raj, P., Jeeva, S.: Chelladhurai. Learning Docker, Vinod Singh (2015)
2. Do, N.H., Van Do, T., Thi Tran, X., Farkas, L., Rotter, C.: A scalable routing mechanism for stateful microservices. In: Proceedings of the 2017 20th Conference on Innov. Clouds, Internet Networks, ICIN 2017, pp. 72–78 (2017). https://doi.org/10.1109/icin.2017.7899252
3. Garriga, M.: Towards a Taxonomy of Microservices Architectures. Springer. (2018). https://doi.org/10.1007/978-3-319-74781-1_15
4. Gribaudo, M., Iacono, M., Manini, D.: Performance evaluation of massively distributed microservices based applications. In: Proceedings of the 31st European Conference on Modelling and Simulation, ECMS 2017. **6**, 598–604 (2017)
5. Torkura, K.A., Sukmana, M.I.H., Meinel, C.: Integrating continuous security assessments in microservices and cloud native applications. In: Conference Utility and Cloud Computing UCC 2017, pp. 171–180 (2017). https://doi.org/10.1145/3147213.3147229
6. Alipour, H., Liu, Y.: Online machine learning for cloud resource provisioning of microservice backend systems. In: 2017 IEEE International Conference Big Data Online. 2433–2441 (2017)
7. Hunter II, T.: Advanced Microservices. Apress (2017)
8. Newman, S.: Building Microservices. Sam Newman (2015)
9. Baskaran Jambunathan, K.Y.: Microservice design for container based multi- cloud deployment microservice design for container based multi-cloud deployment. J. Adv. Res. Dyn. Control Syst. (2017)
10. Pahl, C., Jamshidi, P.: Microservices: a systematic mapping study. In: Proceedings of the 6th International Conference on Cloud Computing and Services Science, pp. 137–146 (2016). https://doi.org/10.5220/0005785501370146
11. Alshuqayran, N., Ali, N., Evans, R.: A systematic mapping study in microservice architecture. In: 2016 IEEE 9th International Conference on Service Computing and Applications A. (2016). https://doi.org/10.1109/soca.2016.15
12. Di Francesco, P., Malavolta, I., Lago, P.: Research on architecting microservices: trends, focus, and potential for industrial adoption. In: Proceedings of 2017 IEEE International Conference on Software Architecture ICSA 2017, pp. 21–30 (2017). https://doi.org/10.1109/icsa.2017.24
13. Vural, H., Koyuncu, M., Guney, S.: A systematic literature review on microservices. Int. Conf. Comput. Sci. Its Appl. **10409**, 203–217 (2017). https://doi.org/10.1007/978-3-319-62407-5

14. Petersen, K., Vakkalanka, S., Kuzniarz, L.: Guidelines for conducting systematic mapping studies in software engineering: An update. Inf. Softw. Technol. **64**, 1–18 (2015). https://doi.org/10.1016/j.infsof.2015.03.007
15. Petersen, K., Feldt, R., Mujtaba, S., Mattsson, M.: Systematic mapping studies in software engineering. In: 12th International Conference on Evaluation and Assessment in Software Engineering, vol. 17, p. 10 (2008). https://doi.org/10.1142/s0218194007003112
16. Richter, D., Konrad, M., Utecht, K., Polze, A.: Highly-available applications on unreliable infrastructure: microservice architectures in practice. In: Proceedings of the 2017 IEEE International Conference on Software Quality, Reliability & Security, QRS-C 2017, pp. 130–137 (2017). https://doi.org/10.1109/qrs-c.2017.28
17. Haselbock, S., Weinreich, R.: Decision guidance models for microservice monitoring. In: Proceedings of the 2017 IEEE International Conference on Software Architecture, ICSAW 2017 Side Track Proc, pp. 54–61 (2017). https://doi.org/10.1109/icsaw.2017.31
18. Rademacher, F., Sachweh, S., Zündorf, A.: Towards a UML profile for domain-driven design of microservice architectures. B. chapter, pp. 230–245 (2018). https://doi.org/10.1007/978-3-319-74781-1_17

A Theoretical Framework for Improving Software Project Monitoring Task of Agile Kanban Method

Hamzah Alaidaros, Mazni Omar(✉), and Rohaida Romli

Human-Centred Computing Research Lab, School of Computing,
Universiti Utara Malaysia (UUM), 06010 Sintok, Kedah, Malaysia
m7amza7@yahoo.com, {mazni,aida}@uum.edu.my

Abstract. Progress monitoring task is one of the critical steps in the software project management (SPM). Consequently, successful realization of software projects is strongly associated with used method in implementing and monitoring those projects. Over the recent years, the adoption of Agile Kanban method is being increased, however, this method still having significant challenges in progress monitoring task during the process of software development. Therefore, this paper aims to draw upon relevant theories and conducts in-depth a review in order to establish a theoretical framework to improve the progress monitoring task of Agile Kanban method. Our findings revealed that three elements need improvements, which are progress tracking (PT), limiting work-in-progress WIP (LWIP), and progress visualization (PV). In addition, the three elements are aligned with theories, which are the explicit theory of project management (TETPM), progress monitoring theory (PMT), and the program theory (TPT). These findings extend the current literature on Agile software development by providing a holistic view on how progress monitoring task should be improved.

Keywords: Progress monitoring task · Agile software development
Kanban method · Theoretical framework · Software project management

1 Introduction

Software project management (SPM) is the concept that involves knowledge, techniques, and tools, which are essential needed for managing the development process of software projects. SPM is defined as the art and science of planning and leading software projects. It comprises of a number of activities, which contains planning of project, deciding scope of software product, estimation of cost, scheduling of tasks and events, and resource management [1, 2].

In this context, the progress monitoring is a vital task in SPM, whereby it ensures that projects' plan is progressed according to budget, schedule, and quality expectations [3]. Thus, improving success rate of the software projects is important to the project management community, whereby successful projects are the goal of every project manager. In other words, SPM focuses on achieve an appropriate quality of the product, and keep under control all the other project variables, such as time and cost [1]. However, the matter of project management is that change is inevitable, since real-world changes

F. Saeed et al. (Eds.): IRICT 2018, AISC 843, pp. 1091–1099, 2019.
https://doi.org/10.1007/978-3-319-99007-1_101

during execution occur in the activities of managing such projects. Accordingly, updates to the project plan have to be made, although it is no wonder that plans are hard to maintain. Hence, providing a theory for project management has the potential to light the way towards better building of model or tool.

In this regard, Koskela and Howell [4] has confirmed that it is impossible to maintain an up-to-date plan, and therefore the plan must to be derived systematically and hopefully automatically. Likewise, Warburton and Cioffi [5] has claimed that including the dynamic nature of project management, such as the impact of changes to activities, their execution times, and their costs, is required. Besides that, they have emphasized that changing in the plans occurs during the execution of real world projects, and then the cost and schedule impacts follow. As such, the project management body of knowledge (PMBOK) guide [2] suggests that plans are input to execution and in spite of that plans are developed before execution, the key is to recognize that the entering project data is an important entity. Thus, when the project data changes, it is conceivable that related components can automatically derive an updated plan.

Currently, Agile Kanban method has being a popular method used for developing software projects in software development organizations (SDOs). This is because this method has greater consistency in managing software engineering (SE) projects [6–8]. However, the progress monitoring task of Agile Kanban method has significant lacks during the development process. This problem has negative impacts on software projects' success because the delays in project scheduling lead to late delivery [6, 7, 9]. As such, the Standish Group Chaos report on software projects showed that failed and challenged projects represented approximately two-thirds of all project outcomes, whilst only about a third of the software projects were successful [10]. Therefore, a study that investigates the lacks in progress monitoring task by developing an improved model of Agile Kanban method could remedy the situation. Hence, the research question of this study is:

How can software project monitoring task of Agile Kanban method be improved in order to ensure successful realization of software projects?

To investigate this research question, this paper aims to draw upon relevant theories and conducts in-depth a review to establish a theoretical framework to provide a strong scientific research base for this study. The theoretical framework offers scientific justification shows that research is grounded in and based on scientific theory, whereby the relevant theories and models that relate to the key concepts are presented [11].

The rest of the paper is organized as follows. In Sect. 2, we introduce the key concepts and principles of Agile and Kanban method. Next, we discuss the current problems in progress monitoring task and then present the existing models and tools that is used recently in SPM. Section 3 describes the related theories, while the theoretical framework is constructed and explained in Sect. 4. Finally, we provide the conclusion and some suggestions for the future work in Sect. 5.

2 Literature Review

2.1 Agile Kanban Method

During the past decades, several methods have been established for monitoring software development projects (SDPs) since the inception of information technology (IT). These include traditional methods, such as waterfall, which have distinct phases, and Agile methods, which are a family of lightweight software development methods that defines an iteration process for designing, constructing, and deploying different parts of software simultaneously [12]. Although there are many success stories of adopting Agile methods, there is a need to improve the existing methods to resolve the major issues in developing software projects. Subsequently, the lack of Agile progress tracking mechanism, reporting on project status, and poor communication have been identified as factors for software project failure [13].

Therefore, this study focuses on studying Agile methods generally, and Kanban method particularly. The selection is made because of this method has numerous advantages that make it performs better than other Agile methods in terms of having greater consistency in managing SE projects [6–8]. In addition, Agile Kanban method consists of best practices in monitoring software projects and the adoption of that method is continuously on the rise within SDOs as reported by Interop ITX Research Report [14] and Version One [15].

Basically, David J. Anderson, father of Kanban in software development (SD), has defined five principles for Kanban method, which are (1) limit work in progress (WIP), (2) visualize workflow, (3) measure and manage flow, (4) make process policies explicit, and (5) use models to recognize improvement opportunities [16]. This study focuses on studying only two core principles, which are limiting WIP and visualizing workflow. This is because these principles have a narrower scope with more specialized challenges that affect the projects' progress during the development process [3]. Therefore, a focus is given to overcome these challenges in order to improve the progress monitoring task of Agile Kanban method.

Currently, the issue of SDPs exceeds scheduled completion dates, and project schedule remains a challenge for SDOs. Thus, this study focuses on monitoring software project, whereby timely completion stands as one of the main measures of project success. By achieving this aim, the success rate of developing software projects would be increased and software organizations can deliver software products quickly, on the prescribed time frame, and with the highest quality and lowest cost.

2.2 Progress Monitoring Task

Progress monitoring is an essential task during any project execution, and there is no exception for Agile projects. Besides being needed to monitor a project, accurate and timeliness reporting is important for keeping the management and team updated on the project's progress. In this vein, successful implementation of software projects depends entirely on successful monitoring mechanisms, while the lack of monitoring SDPs leads to the failure of such projects [3, 17]. Recently, Agile Kanban method has lacks in progress monitoring task, therefore this method needs to be improved. To do so, this study suggests improving Kanban method through three elements and their criteria as well, whereby these elements have major effect towards improving progress monitoring task.

First element is to improve tracking mechanism by integrating Kanban method with another effective method to keep schedule of the project progresses as it is planned [3, 6]. Second element is to find out an adequate technique to determine the optimum WIP limits for each stage in Kanban board. However, determining the WIP limits is proven a major challenge that is faced by software project practitioners [7, 18]. The optimum WIP limits refer to suitable numbers for each stage that can monitor and control the team members with their tasks and ensure that project progress as planned. Third element is to visualize sufficient information and useful indicators for monitoring project progress. Nevertheless, Kanban board neither reports how much of work is left nor provides some indications of where the project ought to be [7].

To sum up, the theoretical framework of this study will embrace the above-mentioned elements, which are progress tracking, limiting WIP, and progress visualization, in order to improve progress monitoring task of Agile Kanban method.

2.3 Existing Models and Tools

This section presents the existing models and tools that is used for managing and developing software projects. Recent study [3] has reviewed various approaches that used for monitoring SDPs. It emphasized that Kanban method has lacks in progress monitoring and has a difficulty in determining the ideal number of WIP limits. Accordingly, different studies have been conducted to improve that method. For example, study has conducted to integrate Scrum with Kanban to introduce a new method called Scrumban [19], whilst other study to integrate Kanban with value stream mapping (VSM) [20]. However, these were few studies and have numerous limitations. For instance, Scrumban method has still facing challenges with progress tracking and managing WIP. Besides, the integration of Kanban with VSM was not to improve Kanban method or its progress monitoring task, instead to improve some areas that constrain and harm workflow in the value stream [3].

On the other hand, on the market today, there are several tools for managing the development process of software projects. Some are general tools such as Primavera, Gantt Project, MS Project, Jira, and so on, while others tools depend on using Kanban board, such as Leankit, KanbanTool, Kanbanery Tool, and so on. Overall, even though these tools have numerous advantages and various benefits; however, it still having disadvantages and limitations in terms of tracking progress, controlling WIP, and visualizing the crucial information and useful insights for the development process. Therefore, the proposed model for improving Agile Kanban method may overcome the aforementioned limitations of the existing commercial tools, with supporting of essential features of the previous tools.

3 Related Theories

Theories shape a discipline and guide researchers to investigate the phenomenon. Usually, theory and practice are developed concurrently, similarly to other science based fields, where theory is explicated, examined, and refined in a continuous dialog among scientific and practitioner groups. Apart of the theoretical framework, related theories should be identified and included within the key concepts and elements that have a relation with the identified problem. Although the fields of software engineering and project management are quite young when compared to other fields, this section discusses the related theories to progress monitoring task of software projects. In this study, there are three theories that are very closely related to improve the progress monitoring task in SPM. These theories are described in the following subsections.

3.1 The Explicit Theory of Project Management (TETPM)

The first theory is the explicit theory of project management (TETPM) which has various goals. TETPM can provide a prediction of future behavior, in terms of analyzing the further progress. In addition, tools for analyzing, designing, and monitoring can be developed based on TETPM [4]. Accordingly, Warburton and Cioffi [5] have proposed a guide that can be used for future improvements. The early estimates of the cost and schedule could be predicted from the existed project data. Moreover, when changes to the project are proposed, the updated plan (cost and schedule) might be derivable automatically from the new data. Furthermore, different structural assumptions could be investigated and compared to real world projects.

3.2 The Progress Monitoring Theory (PMT)

The concurrent progress monitoring is a vital task during developing software projects. Accordingly, MacGregor, Ormerod and Chronicle [21] have put forward this theory which is called the progress monitoring theory (PMT). This theory has two main features, which are maximization heuristic and progress monitoring. The first feature indicates that each decision is an attempt to make as much headway as possible towards the goal. While the second feature constantly assesses the rate of progress, and if it is deemed to be delayed or inefficient criterion failure occurred, an alternative strategy is then sought.

3.3 The Program Theory (TPT)

The third theory is the program theory (TPT), which is a systematic method for collecting, analyzing, and using information to answer questions about projects, policies and programs, particularly about their effectiveness and efficiency. TPT focuses on creating a logic model, which is a wonderful way to help visualize important aspects of monitoring and evaluation. This can help in articulating the problem, the resources, and capacity that are currently being used to manage software projects, whereby stakeholders often want to know whether the projects they are funding, implementing, voting for, receiving or objecting to are producing the intended effect [22].

Table 1 summarizes the related theories to improve progress monitoring task of Agile Kanban method in SPM.

Table 1. Summary of the related theories

Theory	Goal of Theory	Source
The Explicit Theory of Project Management (TETPM)	TETPM provides a prediction of future behaviour, whereby tools for analysing, designing, and monitoring could be developed.	[4, 5]
Progress Monitoring Theory (PMT)	PMT focuses on maximization heuristic and progress monitoring	[21]
The Program Theory (TPT)	TPT focuses on creating a logic model. Along with visualizing the important aspects of monitoring and evaluation	[22]

Overall, three related theories have been explained to establish a theoretical framework for improving the progress monitoring task of Agile Kanban method. However, these theories need to be examined and aligned with elements that have impacts on resolving the current problem. Next section discusses the study findings and proposes the theoretical framework.

4 Discussion

This study aims to construct a theoretical framework for improving software project monitoring task of Agile Kanban method. The theoretical framework grants scientific justification shows that research is grounded in and based on scientific theory, whereby the relevant theories and models that relate to the key concepts are presented.

The proposed framework is based on three theories that have been studied, and discussed by Koskela and Howell [4], Warburton and Cioffi [5], MacGregor et al. [21], and Funnell and Rogers [22]. Even though there are various theories in the area of software project management; however, those theories have been chosen due to its significant relevance with the common goals and features to the objective of study. Moreover, the theoretical framework is directed to improve three elements, which are progress tracking (PT), limiting work-in-progress WIP (LWIP), and progress visualization (PV). Significantly, there are relations between the features of the related theories along with the proposed elements to improve the progress monitoring task.

Table 2 shows the associations between the theories' features and proposed elements, whereby each element receives a check mark whenever a feature has a relation and impact on that element.

Table 2. The associations between the features of theories and the proposed elements

Theory	Features of the theory	The proposed elements		
		PT	LWIP	PV
The Explicit Theory of Project Management (TETPM)	It provides directions for analysing the further progress	√	√	√
	It helps in developing tools for analysing, designing, and monitoring	√		√
Progress Monitoring Theory (PMT)	It makes decisions to be as much headway as possible towards the goal		√	√
	It assesses the rate of progress	√		√
	It seeks an alternative strategy to avoid the occurring of delay		√	
The Program Theory (TPT)	It focuses on creating a logic model		√	
	It helps in visualizing the important aspects of monitoring and evaluation			√

From Table 2, it is clear that there are significant relations between theories and proposed elements, whereby the most of theories' features have impacts on those elements. Therefore, this indicates that these theories can be adopted to guide our research to achieve the objective of study and generate desirable results. A logical structure of the theoretical framework for this study is depicted in Fig. 1.

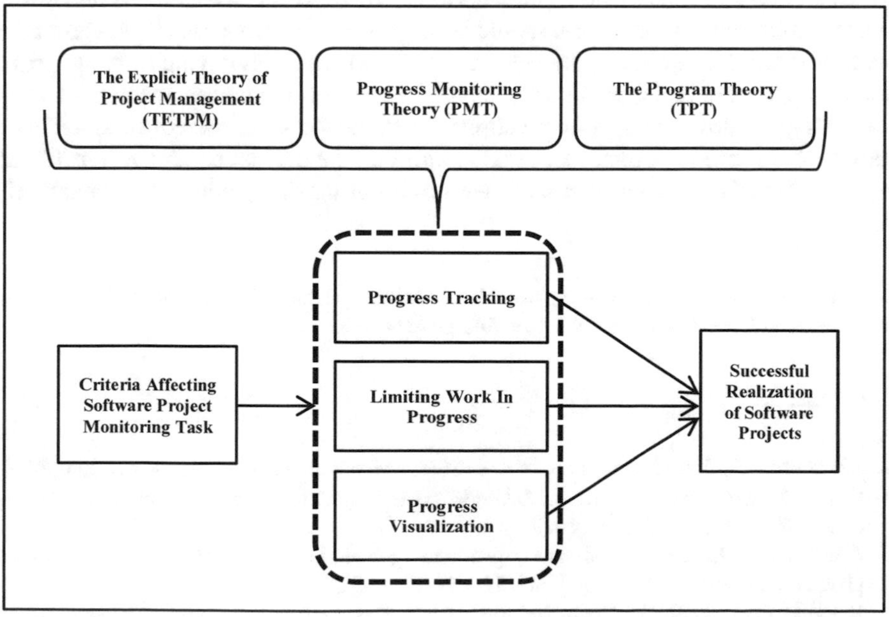

Fig. 1. Theoretical framework for improving progress monitoring task of Agile Kanban method

Figure 1 presents the theoretical framework, which is established to be a strong scientific research base, for improving software project progress monitoring task. The theoretical framework embraces three relevant theories, which are the explicit theory of project management, progress monitoring theory, and the program theory. Additionally, it involves three elements, which are progress tracking, limiting WIP, and progress visualization along with criteria that affect those elements. Generally, different criteria that affect particular elements, which are based on specific theories could be worked together to improve progress monitoring task. Thus, ensuring a successful realization of software projects according to plans' expectation would be achieved.

5 Conclusion and Future Work

This paper has conducted in-depth a review in order to establish a theoretical framework for improving the progress monitoring task of Agile Kanban method. This task is regarded as the critical step in the SPM, however, still has significant issues hinder the successful realization of software projects. Results have reported that three elements need improvements, which are progress tracking, limiting WIP, and progress visualization. In addition, it highlighted three related theories, which are the explicit theory of project management, progress monitoring theory, and the program theory. These theories have described and then aligned with the proposed elements. At the end, a theoretical framework has been constructed based on the identified problem, elements and criteria that affect progress monitoring task, along with related theories. These findings will extend the current literature on Agile software development and provide a holistic view on how progress monitoring task should be improved. Further researches might adopt this framework for developing models or tools in which of improving software project monitoring tasks of Agile or Kanban method. This is because current theoretical framework has grounded in and based on strong scientific theories. Moreover, an extended study to this framework might focus on examining the evaluation dimensions of such proposed models, which can be later measured during the qualitative or quantitative analysis.

Acknowledgment. The authors wish to thank the Universiti Utara Malaysia for funding this study under University Research Grant Scheme, S/O project code: 13853.

References

1. Villafiorita, A.: Introduction to Software Project Management. CRC Press, London (2014)
2. Project Management Institute, A.: A Guide to the Project Management Body of Knowledge (PMBOK® Guide), 6th edn. (2017)
3. Alaidaros, H., Omar, M.: Software project management approaches for monitoring work-in-progress: a review. J. Eng. Appl. Sci. **12**(15), 3851–3857 (2017)
4. Koskela, L., Howell, G.: The underlying theory of project management is obsolete. In: The PMI Research Conference, pp. 293–302, Seattle, Washington (2002)

5. Warburton, R.D.H., Cioffi, D.F.: Project management theory: deriving a project's cost and schedule for its network structure. In: Project Management Institute Research and Education Conference, Phoenix, AZ. Newtown Square, pp. 156–168 (2014)
6. Ahmad, M., Markkula, J., Oivo, M.: Insights into the perceived benefits of Kanban in software companies: practitioners' views. In: International Conference on Agile Software Development, Cham, pp. 156–168 (2016)
7. Karunanithi, K.: Metrics in Agile and Kanban Software Measurement Techniques. California State University, Fullerton (2016)
8. Lei, H., Ganjeizadeh, F., Jayachandran, P.K., Ozcan, P.: A statistical analysis of the effects of Scrum and Kanban on software development projects. Robot. Comput. Integr. Manuf. **43**, 59–67 (2017)
9. Al-Baik, O., Miller, J.: The Kanban approach, between agility and leanness: a systematic review. Empir. Softw. Eng. **20**(6), 1861–1897 (2015)
10. Standish Group: http://www.standishgroup.com/outline/. Accessed 29 Apr 2018
11. Scribbr: https://www.scribbr.com/dissertation/the-theoretical-framework-of-a-dissertation-what-and-how/. Accessed 30 Jan 2018
12. Ahimbisibwe, A., Daellenbach, U., Cavana, R.Y.: Empirical comparison of tradition al plan-based and agile methodologies: critical success factors for outsourced software development projects from vendors' perspective. J. Enterp. Inf. Manag. **30**(3), 400–453 (2017)
13. Taherdoost, H., Keshavarzsaleh, A.: A theoretical review on IT project success/failure factors and evaluating the associated risks. In: 14th International Conference on Telecommunications and Informatics, Sliema, Malta (2015)
14. Interop ITX Research Report. http://www.interop.com/. Accessed 30 Apr 2017
15. Version One Report. https://www.versionone.com/pdf/VersionOne-11th-Annual-State-of-Agile-Report.pdf. Accessed 10 May 2018
16. Anderson, D.J.: Kanban. Blue Hole Press (2010)
17. Nguyen, D.S.: Workplace factors that shape agile software development team project success. Am. Sci. Res. J. Eng. Technol. Sci. (ASRJETS), **17**(1), 323–391 (2016)
18. Tripp, J.F., Saltz, J., and Turk, D.: Thoughts on current and future research on agile and lean: ensuring relevance and rigor. In: Proceedings of the 51st Hawaii International Conference on System Sciences, USA, pp. 5465–5472 (2018)
19. Mahnic, V.: Improving software development through combination of scrum and Kanban. In: Recent Advances in Computer Engineering, Communications and Information Technology, Espanha, pp. 281–288 (2014)
20. Raju, H., Krishnegowda, Y.: Value stream mapping and pull system for improving productivity and quality in software development projects. Int. J. Recent Trends Eng. Technol. **11**(1), 24–38 (2014)
21. MacGregor, J.N., Ormerod, T.C., Chronicle, E.P.: Information processing and insight: a process model of performance on the nine-dot and related problems. J. Exp. Psychol. Learn. Mem. Cogn. **27**(1), 176–201 (2001)
22. Funnell, S.C., Rogers, P.J.: Purposeful program theory: effective use of logic models and theories of change. Jossey-Bass, San Francisco (2011)

Non-functional Ontology Requirements Specifications: Islamic Banking Domain

Ahmad Shaharudin Abdul Latiff[1(✉)], Haryani Haron[2], and Muthukkaruppan Annamalai[2]

[1] Putra Business School, 43400 UPM Serdang, Selangor, Malaysia
shaharudin@putrabs.edu.my
[2] Universiti Teknologi Mara, 45600 Shah Alam, Selangor, Malaysia

Abstract. The development of ontologies needs to meet not only the functional requirements of the ontology, but also their non-functional requirements. For ontologies, the non-functional requirements are the general requirements that the ontology should fulfil; referring to the criteria, qualities and general aspects that the ontology should satisfy. They need to be elicited during the problem exploration stage which gathers the characteristics and development criteria of the ontology. The elicited non-functional requirements need to be formally documented in the Ontology Requirement Specifications Document. Some of the non-functional requirements could be tested, while others would suffice with documented elaborations of fulfilment.

Keywords: Ontology development · Non-functional requirements Islamic banking

1 Introduction

Domain ontologies are needed to improve the dissemination of knowledge within the Islamic banking industry [1]. They are meant to attend to the prevailing grievances which had arisen from the customers' lack of understanding [2]. Besides that, the Islamic banking ontology is also required to facilitate Islamic financial engineering; particularly for the design of new Islamic banking products [3], which in turn will provide more product options to the customers. The developed ontology need to be utilised by information systems that would support the customers' decision making and the banks' product development [3]. Being software artefacts, ontologies are best to be developed using an adapted software engineering approach so that it would be more reliable, long lived and continually adapted [4]. The stages for ontology conceptualisation according to the software engineering approach [1] are as depicted in Fig. 1 below.

The Problem Exploration stage of the ontology development is the equivalent to the Software Specification stage of software engineering methods; i.e., it is a process whereby details of the services required from the ontology or software component is to be understood and well defined [1]. At this stage, the requirements of the ontology shall be derived from observation of the problems of the domain that had ignited the initial goals and purposes for the ontology development. Nevertheless, it is important that the purposes and the intended uses of the ontology to be made clear before the design and

F. Saeed et al. (Eds.): IRICT 2018, AISC 843, pp. 1100–1112, 2019.
https://doi.org/10.1007/978-3-319-99007-1_102

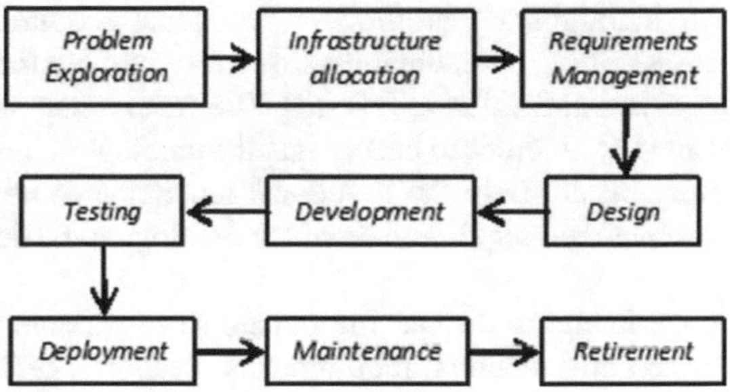

Fig. 1. Software engineering approach for ontology development [1]

development of the ontology could take place [5]. Outputs from Problem Exploration would then be deployed in the Requirements Management; which comprises of Requirements Specification and Requirements Validation. The Requirements Specification is the task that translates the information gathered during problem exploration into a formal requirements document; i.e., Software Requirement Specifications (SRS). The SRS needs to look into the functionality, external interfaces, performance, attributes and design constraints of the software component [6]. The Requirements Management includes the activities of managing the functional and also non-functional requirements of the ontology [1].

The functional requirements of the ontology are the content-specific requirements which the ontology should fulfil; i.e., it refers to the particular knowledge that needs to be represented by the ontology [7]. It includes the contents of the particular domain which comprises of the concepts and their respective relationships that are sketched into binary relations diagram or concept dictionary [8]. The functional requirements involve identification of key concepts and their relationships within the domain. It produces clear and precise text definitions of the relevant concepts and relationships [9]. On the other hand, the non-functional requirements are the general requirements that the ontology should fulfil. It refers to the criteria, qualities and general aspects that the ontology should satisfy [7]. The non-functional requirements refer to the characteristics of the domain or task for which the ontology is targeted for. In order to elicit such requirements, the intended users of the ontology need to be identified and the characteristics of the ontology need to be ascertained to clarify the purpose and scope of the ontology [1]. In the case of the Islamic banking industry, the non-functional requirements should consider several complexities of the domain that has been retarding the effective delivery of information on the Islamic banking products and processes. The peculiarities of the Islamic banking industry demand that the characteristics and development criteria of the domain to be taken into consideration in the development of its ontology [3]. Without looking into such peculiarities will render the ontology ineffective in disclosing the business models, diverse roles, distinct concepts, and also representation rules that predominates the industry [3].

Yet, it has been highlighted that the non-functional requirements of software engineering such as usability, maintainability, security and performance have been difficult to elicit, document and validate [10–13]. The same issues were also raised for ontology development [14]. In order to ensure that the ontology will be able to fulfil the needs for its development; not only the functional requirements needs to be properly adopted, but the non-functional requirements of the ontology to be developed must also be fulfilled.

This paper does not intend to discuss the functional requirements of the ontology, but focuses only on the non-functional requirements. Section 2 of the paper discusses on the non-functional requirements for the ontology and expanded into the characteristics and development criteria on an ontology for the Islamic banking domain in Subsects. 2.1 and 2.2 respectively. Section 3 reflects the domain characteristics into an Ontology Requirement Specifications Document, and Sect. 4 looks at the testing of the non-functional requirements. Section 5 concludes and offers suggestions for future works.

2 Non-functional Requirements for the Ontology

An ontology is an artefact that models knowledge and information; hence it should take into account not only the functional requirements of the ontology, but also its non-functional requirements. The non-functional requirements refer to the constraints, goals and quality attributed to the particular ontology [15]. The requirements should consider the criteria, qualities and general aspects of the ontology which in turn takes into account the characteristics of the domain or task for which the ontology is targeted for. The characteristics of the ontology refer to the distinctive nature of the concepts handled by the ontology. They are divided into general characteristics; i.e., characteristics that are also presence in other domains, and the specific characteristics that are peculiar for the particular domain of concern [3]. In the case of Islamic banks, examples of the characteristics of the ontology are the distinct concepts dealt by the domain, the rules in representing the concepts, the diverse Shariah concepts, the diverse roles played by the banks and customers, the differences of opinions among the Islamic scholars and several others. Meanwhile, the development criteria are the generally accepted quality of the resulting ontology [3]; i.e., its objective criteria expected [16]. For the Islamic banks, the criteria would fulfill the tasks of knowledge dissemination and facilitating the design of new Islamic banking products using the ontology.

2.1 Characteristics of Islamic Banking Ontology

As described in [3], the general characteristics of the ontology casted in the context of Islamic banking are as follows:

(i) The concepts of the ontology need to cater for the distinct concepts within the Islamic banking industry,
(ii) The ontology need to meet the rules that apply for the representation of the concepts in the Islamic banking system.

The specific characteristics of the ontology for Islamic banking as elaborated in [3] are as follows:

(i) The ontology need to include the business models or Shariah concepts that are being adopted by the Islamic banking product,
(ii) The ontology should possesses the concepts that reflect the diverse roles played by the banks and their customers,
(iii) The ontology should be able to portray the differences of opinion of the different juristic schools,
(iv) The ontology should the include the borrowed terms that come from Arabic terminologies.

2.2 Development Criteria for Islamic Banking Ontology

The Islamic banking ontology should also meet the generally accepted qualities or the development criteria of good ontology as elaborated in [3]; i.e., as follows:

(i) Clarity
The contents of the ontology should be made adequately clear so that the customers of the bank could make effective comparison among the different banking products.
(ii) Conciseness
Brevity of the ontology should be emphasised and redundant definitions and explanations should be avoided. This is to prevent the ontology from being counter-productive and end up confusing the customers.
(iii) Coherence; which comprise of
  (a) Individual Conceptual Integrity
  The individual conceptual integrity need to be verified to ensure that there is no contradiction in the interpretation of the concepts pertaining to entities represented by the ontology.
  (b) Collective Consistency
  The collective consistency should be verified in order to ensure that the relationships that bind the concepts reflect the dependencies between their corresponding entities.
(iv) Extensibility and Expansibility
The hierarchy in the ontology need to be ensured that it could allow multiple inheritance mechanism *via* extension and expansion.
(v) Minimal encoding bias
The ontology should be independent of certain symbol-level encoding; i.e., close to a natural textual form.
(vi) Epistemological completeness for the domain of Islamic banking:
  (a) The Shariah concepts adopted in the products
  (b) The profit rate or mark-up imposed on the product
  (c) The margin of financing
  (d) Tenure
  (e) Profit sharing ratio
  (f) Collateral

(g) Fees and charges
(h) Takaful requirement
(i) Guarantor
(j) Compensation charges
(k) Lock-in period and early settlement
(l) Rights to set-off

(vii) Competency
The ontology need to be competent to carry out the function or purpose it was designed for.

3 Reflecting the Domain Characteristics into Non-functional Requirement Specification

The non-functional requirements need to be formally documented in the Ontology Requirements Specification Document. It is done during the Requirements Specification stage, which is the activity that translates the information gathered during the stage of problem exploration into the formal requirements document. The domain characteristics and the development criteria will be reflected as the Non-Functional Requirement Specification of the document.

The Ontology Requirement Specification Document (ORSD) was actually adapted from the Software Requirement Specification Document and altered to fit the purpose of ontology development [7] (Table 1).

Table 1. Islamic banking products ontology requirements specifications document

1.	Purpose	To provide information for customers' sufficient understanding of the Islamic banking products for their decision making
2.	Scope	Focuses on the domain of Islamic banking products of consumer financings and deposits
3.	Implementation language	No specified formal representative language to remain minimal bias. Yet, the Web Ontology Language (OWL) is used for validation purposes
4.	Intended end-users	• Customers and prospective customers
		• Banks' product development staff
5.	Intended uses	• Publish Product Disclosure Sheet
		• Display product comparison
		• Display features and testing new products

(*continued*)

Table 1. (*continued*)

	Ontology requirements	
6	Non-functional requirements	Characteristics of the ontology
		General characteristics
		• Distinct concepts dealt by the domain
		• Rules of representation of concepts in the domain
		Specific characteristics
		• Shariah concept adopted
		• Diverse roles of banks and customers
		• Different opinions of diverse juristic schools
		• Borrowings of Arabic terminologies
		Development criteria
		• Clarity
		• Concise
		• Coherence
		• Extensibility and expansibility
		• Minimal encoding bias
		• Standardisation of names
		• Ontological completeness
		• Competency
	Functional requirements	The ontology should at least be able to answer product transparency requirements as sanctioned by the central bank, Bank Negara Malaysia [17]. Among the questions include:
		• Shariah concept adopted
		• Shariah source referred
		• Profit rate
		• Margin of Finance
		… and many more

4 Testing the Non-functional Requirements

Like the functional requirements, testing need to be done on the non-functional requirements too. The testing of the non-functional requirements could be done on some of the aspects of its characteristics and development criteria (Table 2).

Table 2. Testing of non-functional requirements

Non-functional requirements	Testing method
Characteristics of the ontology	
Distinct concepts dealt by the Islamic banking domain	According to Shanks, Tansley and Weber (2003), validating a conceptual model such as an ontology involves checking that it properly represents the domain that it is intended to [18]. In the case of the Islamic banking ontology, the ontology was ensured that it has included the distinct concepts of the Islamic banking industry, especially the Shariah concepts and also other distinct Islamic banking concepts such as rabbul-mal, mudharib, khiyar, ta'widh, muqasah and many more. The existence of these concepts was tested using SPARQL queries and also from graph queries made using the ontology editor The graphic result of the queries is depicted below: 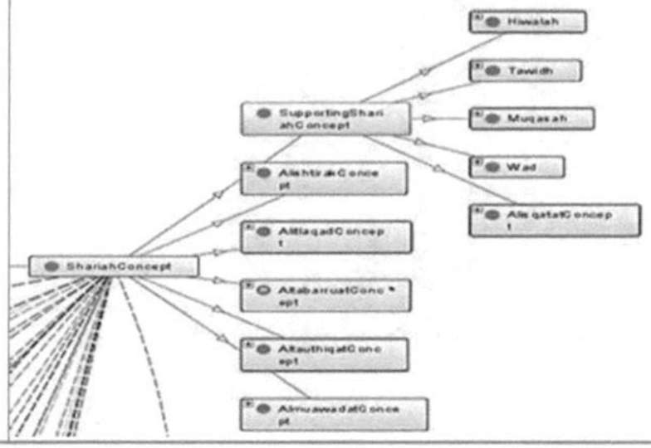

(*continued*)

Table 2. (*continued*)

<table>
<tr><th>Non-functional requirements</th><th>Testing method</th></tr>
<tr><td>Rules of representation of the concepts in the domain</td><td>As the ontology of the Islamic banking are meant to facilitate the customers' selection of the banks and also the products, but not targeted to be used for automating the banking transaction or processes; full axiomatization of the concepts and relations in the ontology may not be necessary. Nevertheless, axiomatised rules for representation of certain concepts of the Islamic banking system need to be included to attend to some concepts that need to be technically clarified
The Islamic banking ontology holds the axiomatised rules as depicted below:
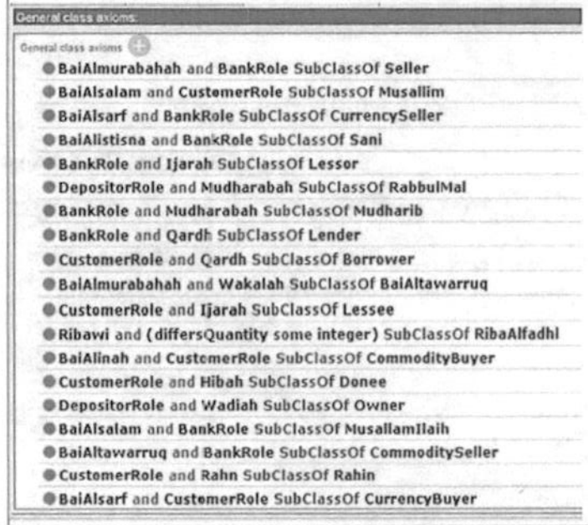
</td></tr>
<tr><td>Shariah concept adopted</td><td>The ontology need to be ensured that it includes the business models or Shariah concept commonly adopted by Islamic banking products
In the case of the Islamic banking ontology, it discloses the depicted model below when queried
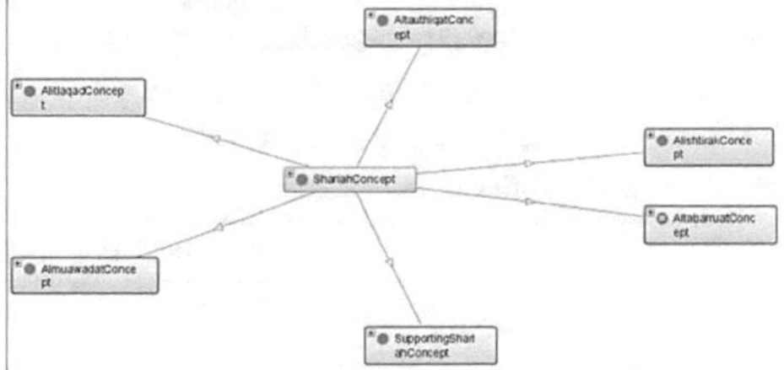
</td></tr>
</table>

(*continued*)

Table 2. (*continued*)

Non-functional requirements	Testing method
Diverse roles of banks and customers	The ontology need to be ensured that it includes the diverse roles played by the banks and their customers Visualisations made on the Islamic banking ontology using the graphic tool of the ontology editor, e.g., Protégé showed that the ontology is able to cater to the different contractual roles of the different Shariah concepts. The graphic view of the different contractual roles is depicted below: 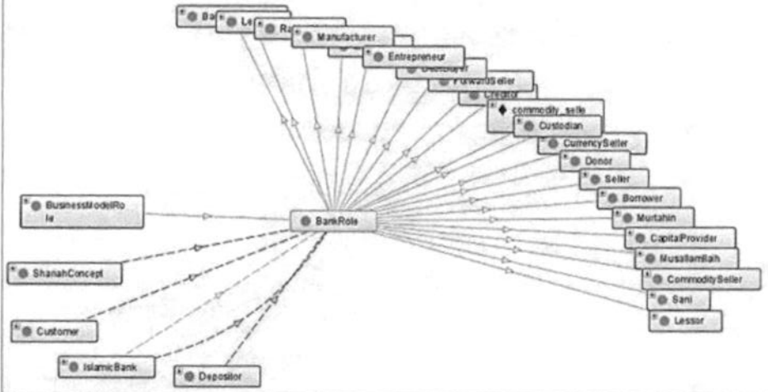
Different opinions of diverse juristic schools	The ontology must be able to portray the differences of opinion of the different mazhabs or juristic schools For the Islamic banking ontology, it was shown that query of the Bai Al-Inah on Hanafi juristic school resulted to "Forbidden" 

(*continued*)

Table 2. (*continued*)

<table>
<tr><th>Non-functional requirements</th><th>Testing method</th></tr>
<tr><td></td><td>The query of Bai Al-Inah on Shafii juristic school showed it as "Permissible"
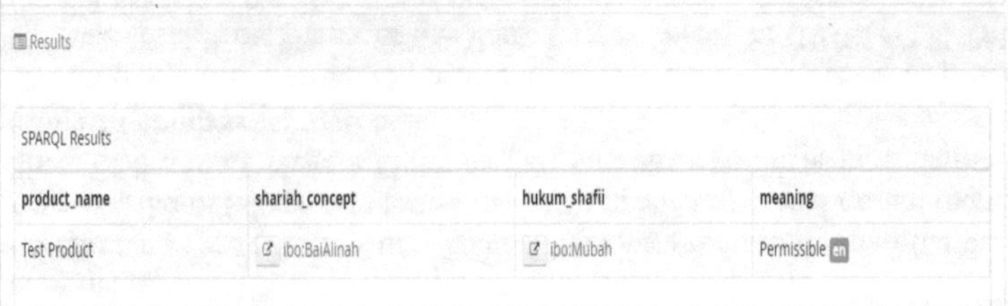
Results
SPARQL Results<table><tr><th>product_name</th><th>shariah_concept</th><th>hukum_shafii</th><th>meaning</th></tr><tr><td>Test Product</td><td>ibo:BaiAlinah</td><td>ibo:Mubah</td><td>Permissible en</td></tr></table></td></tr>
<tr><td>Borrowings of Arabic terminologies</td><td>When queried, the ontology also showed that it is be able to identify the borrowed Arabic terminologies by the industry; such as aqd, murabahah, musharakah, et cetera
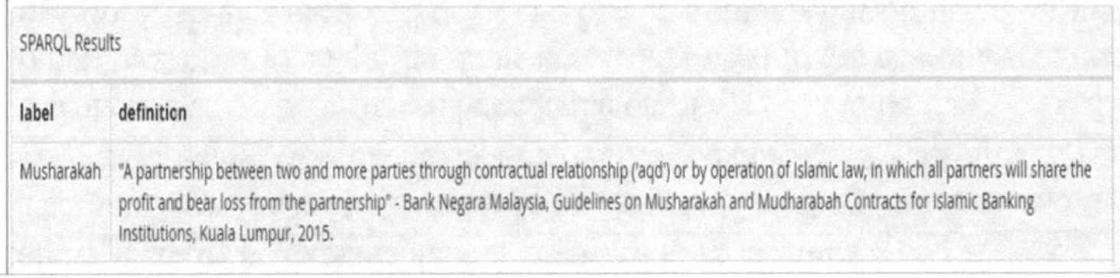
SPARQL Results<table><tr><th>label</th><th>definition</th></tr><tr><td>Musharakah</td><td>"A partnership between two and more parties through contractual relationship ('aqd') or by operation of Islamic law, in which all partners will share the profit and bear loss from the partnership" - Bank Negara Malaysia, Guidelines on Musharakah and Mudharabah Contracts for Islamic Banking Institutions, Kuala Lumpur, 2015.</td></tr></table></td></tr>
<tr><td colspan="2">Development criteria of the ontology</td></tr>
<tr><td>Clarity</td><td>The clarity of the contents was made adequate so that the customers could make effective comparison between the different products of the Islamic banks. As described by Gruber (1993), definitions should be provided in natural languages in order to provide clarity [16]</td></tr>
<tr><td>Conciseness</td><td>The ontology is brief and avoids redundant definitions and explanations that could be counter-productive and might end up confusing the customers. According to Bezerra, Costa and Okada (2009), the ontology is considered as being concise when it meets the following [19]:
- it does not store unnecessary definitions,
- there is no redundancy between terms definitions,
- redundancies are not inferred by other definitions</td></tr>
</table>

(*continued*)

Table 2. (*continued*)

Non-functional requirements	Testing method
Coherence	The Islamic banking ontology is in natural language documentation and defined coherently; individually or collectively. For its individual conceptual integrity, the definitions of the ontology were based on reliable definitions. The definitions were based on documentation by ISRA (2012) [20] and there was no contradiction in the interpretation of the concrete concept pertaining to the entities represented. The collective definitions were also made consistent and reflected the dependencies between their corresponding entities
Extensibility and expansibility	The Islamic banking ontology was ensured that the hierarchy in the ontology is diversified in order to allow multiple inheritance mechanisms. The concepts under the main class: Business were diversified in order to allow them to be extendable and expandable in the future
Minimal encoding bias	The ontology need to be ensured to be close to natural textual form and independent of certain symbol-level encoding. In the case on the Islamic banking ontology, it was built in textual natural language
Standardisation of names	The names of concepts or relations in the ontology need to be standardised whenever possible and as much as it could, it need to adopt the terms commonly used in the domain, such as Islamic banking industry. Hence, the terminologies used in the Islamic banking ontology were based on documentation by ISRA (2012) [20]
Ontological completeness	The ontology is required to serve the following purposes:- - producing information required for generating a Product Disclosure Sheet and other information needed by customers, - facilitating innovation of new Islamic banking products through the use of the ontology As such, the ontology was tested for its completeness based on the requirements of the Product Disclosure Sheet of the Bank Negara Malaysia [17] and other pertinent information on the products as listed in the designed competency questions
Competency	The ontology was also tested for its competency to at least answer or provide basic information to fill up the Product Disclosure Sheets of the Bank Negara Malaysia (BNM) [17] Thirty competency questions were drafted based on the customers' information need and the information requirement sanctioned by the BNM

5 Conclusion

Similar to the developments of other software components, the developments of ontology would require its non-functional criteria to be properly elicited, documented and tested. Documentation of the non-functional requirements of an ontology is not a trivial task. It requires proper problem exploration in order to elicit the characteristics and development criteria of the ontology. The gathered characteristics and development criteria will then be charted into the Ontology Requirement Specification Document and later tested at the completion of the development. Some of the non-functional requirements could be tested, while others would suffice with documented elaborations of fulfilment.

Research works could also be done in the future on the functional requirements for the Islamic banking products ontology.

References

1. Latiff, A.S.A., Haron, H., Annamalai, M.: Software engineering approach for domain ontology development: a case study of islamic banking product. J. Inf. Retriev. Knowl. Manag. (JIRKM) **3**(2017), 36–53 (2017)
2. Latiff, A.S.A., Haron, H., Annamalai, M.: Grievances on islamic banks: a survey. In: Islamic Perspectives Relating to Business, Arts, Culture and Communication, pp. 295–306. Springer, Singapore (2015)
3. Latiff, A.S.A., Haron, H., Annamalai, M.: Characteristics and development criteria for Islamic banking ontology. In: 2016 Third International Conference on Information Retrieval and Knowledge Management (CAMP), pp. 136–142. IEEE, August 2016
4. Annamalai, M., Rosli, M.M.: Software engineering: a pathway to enterprise-strength ontology engineering. Int. J. Digit. Content Technol. Appl. **6**(22), 98 (2012)
5. Uschold, M., Gruninger, M.: Ontologies: principles, methods and applications. Knowl. Eng. Rev. **11**(2), 93–136 (1996)
6. IEEE Standard Association. Std. 830–1998. IEEE Recommended Practice for Software Requirements Specifications. IEEE Press, New York (1998)
7. Suárez-Figueroa, M.C., Gómez-Pérez, A., Villazón-Terrazas, B.: How to write and use the ontology requirements specification document. In: OTM Confederated International Conferences on the Move to Meaningful Internet Systems, pp. 966–982. Springer, Berlin (2009)
8. Pinto, H.S., Staab, S., Tempich, C.: DILIGENT: towards a fine-grained methodology for distributed, loosely-controlled and evolving engineering of ontologies. In: European Conference on Artificial Intelligence 2004, pp. 393–397 (2004)
9. Uschold, M., King, M.: Towards a methodology for building ontologies. Edinburgh: Artificial Intelligence Applications Institute, University of Edinburgh, pp. 15–30 (1995)
10. Shahid, M., Tasneem, K.A.: Impact of avoiding non-functional requirements in software development stage. Am. J. Inf. Sci. Comput. Eng. **3**(4), 52–55 (2017)
11. Glinz, M.: On non-functional requirements. In: 15th IEEE International Requirements Engineering Conference, 2007. RE 2007, pp. 21–26. IEEE (2007)
12. Chung, L., do Prado Leite, J.C.S.: On non-functional requirements in software engineering. In: Conceptual Modeling: Foundations and Applications, pp 363–379. Springer, Berlin (2009)

13. Younas, M., Jawawi, D.N.A., Ghani, I., Kazmi, R.: Non-functional requirements elicitation guideline for agile methods. J. Telecommun. Electron. Comput. Eng. (JTEC) **9**(3–4), 137–142 (2017)
14. Li, F.L., Horkoff, J., Mylopoulos, J., Guizzardi, R.S., Guizzardi, G., Borgida, A., Liu, L.: Non-functional requirements as qualities, with a spice of ontology. In: 2014 IEEE 22nd International Requirements Engineering Conference (RE), pp. 293–302. IEEE (2014)
15. Mylopoulos, J., Chung, L., Nixon, B.: Representing and using non-functional requirements; a process-oriented approach. IEEE Trans. Softw. Eng. **18**(6), 483–497 (1992)
16. Gruber, T.R.: Toward principles for the design of ontologies used for knowledge sharing (revision; August 23, 1993). Int. J. Hum. Comput. Stud. **43**, 907–928 (1993)
17. Bank Negara Malaysia. BNM Guidelines on Product Transparency and Disclosure, Consumer and Market Conduct Department, Bank Negara Malaysia: Bank Negara Malaysia (2010)
18. Shanks, G., Tansley, E., Weber, R.: Using ontology to validate conceptual models. Commun. ACM **46**(10), 85–89 (2003)
19. Bezerra, D., Costa, A., Okada, K.: Swtoi (Software Test Ontology Integrated) and its application in Linux test. In: International Workshop on Ontology, Conceptualization and Epistemology for Information Systems, Software Engineering and Service Science, pp. 25–36 (2009)
20. International Shari'ah Research Academy for Islamic Finance: Islamic Financial System: Principles & Operations. Pearson Custom Publishing, Kuala Lumpur (2012)

A Proposed Framework Using Exploratory Testing to Improve Software Quality in SME's

Shuib Basri[1,2], Dhanapal Durai Dominic[1], Thangiah Murugan[2(✉)], and Malek Ahmad Almomani[1]

[1] Department of Computer and Information Science, Universiti Teknologi PETRONAS, 32610 Seri Iskandar, Perak, Malaysia
{shuib_basri, dhanapal_d, malek.ahmad_g02668}@utp.edu.my

[2] SQ2E Research Cluster, Institute of Autonomous System, Universiti Teknologi PETRONAS, 32610 Seri Iskandar, Perak, Malaysia
tm_gun@hotmail.com

Abstract. In the field of IT, the software industry recognizes the value of the small and medium software (SME's) enterprises in contributing valuable products and services to the economy. In SME's the quality of the software is increasingly become a concerned subject. The suitable use of testing standards was not written for the development organizations having less than 25 employees. Currently, most of the SME's do not follow any standards and do not adopt any well-structured testing methods. In this research paper, the challenges faced by the SME's towards software testing in order to produce quality products are being addressed. Various testing processes and other factors associated with software quality are analyzed and discussed in the literature review to identify the research gaps in the existing approach. In this research paper, a new framework has been proposed using Exploratory Testing that has not been used in the context of SME's which aims to improve software quality.

Keywords: Exploratory testing · Framework · SME's · Software quality

1 Introduction

Software applications are widely used in complex environment ranging from single user applications to multi-client server applications running on different platforms. Applications are written using different programming languages with variety of databases and data structure [1]. All these components are connected with different networks. When a software product is used and implemented in any applications, the developer must ensure that the product is free from errors (bug), and it should work without any interruptions under various situations. One of the major activities in software testing is to identify bugs [18]. Most of the larger organizations used more than 40% of the project development time in such activity. At present, the majority of the software companies are small and medium sized companies with 10–50 employees. These SMEs face many challenges to develop the software to a reasonable quality and acceptable standard such as project management, providing training to a new

F. Saeed et al. (Eds.): IRICT 2018, AISC 843, pp. 1113–1122, 2019.
https://doi.org/10.1007/978-3-319-99007-1_103

employee, conducting peer reviews and measurement that are quite common in small organizations [2, 3].

Currently, testing activities in SME's are done mainly without any proper structure [2]. The developer tests the module within themselves after completing certain task. When all the modules are completed, one of the developers will integrate those modules. Then a second person, probably the head of the project team will test the software after integrating all the modules. However, this is not enforced by any procedure [3]. Incidentally, there are no testing scripts made available and no testware is developed, because most of the SMEs do not apply any formal testing techniques. At the same time, the test data and testing methods are never documented.

This research work is to address the various issues surfaced in the SME's related to testing. Then to critically assess the software testing process implemented in SME's, then to propose a framework on improved testing process based on the software engineering literature, as well as the insight knowledge gained from this study.

2 Current Challenges in SME's

Majority of the software industries, across the globe, are comprises of SME's. In order to sustain and to survive in the highly competitive and dynamic marketplace, these small and medium size software companies must produce high quality software [25]. SME's face many challenges along the way to create quality software in order to maintain the survival in the market place [4]. Firstly, they do not believe in the process and methods followed by the larger software companies. Secondly, their organizational structures and processes are quite informal and it leads to a chaotic environment. Thirdly, the challenges are lacking of resources, experiences, skills, cannot afford to employ experienced software developers and the cost involved towards testing [5, 6].

Although several test case methodologies existed, all those existing testing methods are unable to detect the bug when the product is completed. There are situations where some bugs are identified by the customers themselves [7] and this is quite common when a product is developed and implemented by SME's. To focus on the improvement efforts, the test team should initiate a starting point with the awareness that different types of error may escape during the system development life cycle due to challenges in the software testing. The real fact is "Complete testing is impossible" [8].

No Structured or Standard Approach for Testing
There are no proper procedures or methods adopted for the software testing. Since the SME's main objective is to satisfy its stakeholders [3], this constraint resulted in giving too little opportunity for the developers to test the task. Besides that, last minute addition of functionality can also harm the development process. This factor causes additional obstacles which may either result in the delay of the product delivery or the product has many errors [2].

Project Delay
There is always a delay in starting the projects. The change of the requirements request by the stakeholder is one of the reasons of the delayed. The stakeholder does not know exactly what they need initially [4]. User can propose requirement changes at any stage

during System Development Lifecycle (SDLC) process. The worst part is, sometimes the stakeholders can see what can be done with the application only after the application is completed and ready to be used. SME's organizational structures and processes are quite informal and it leads to a chaotic environment which may also causes the delay of the project.

Lack of Objectives and Overviews Among the Developers
The importance of software testing is increasing due to the increasing complexity and proliferation of the software. During the development process, the overview of the project is missing within the team members since the manager will explain only certain parts of the module to each team members. This will lead to lack of awareness on the objectives and overview of the ongoing project because the managers and developers are standing too far apart from each other.

There are number of reasons on why software testing is required:-

- To identify the bugs,
- To measure quality,
- To reduce/minimize the risks,
- To ensure the product serves its purpose,
- To provide information to the stakeholders.

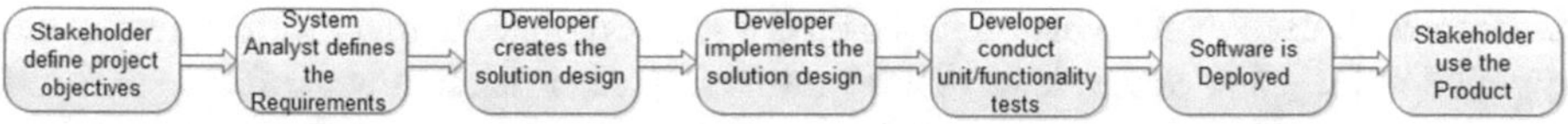

Fig. 1. Current workflow in SME's

The above figure shows the basic workflow currently used in most of the SME's. According to the Fig. 1, it is obvious that the standard testing methods or procedures are not adopted in the SME's. They even do not have any separate Quality Assurance (QA) team to do the testing. Besides that, the SME's do not have proper documentation methods to record their testing process. This would result in repetition of the same testing process for much functionality which is considered as time consuming process. In addition to that, it is not affordable to use any commercial automated testing tools to test the system. Although, automated testing is recommended, but it did mention that only organizations with stable test processes could use automation strategy for software testing in SME's [4].

3 Literature Review

Software testing is a process of finding the defect (bug) in a product to ensure that the intended product works well within its scope and boundary and also to meet the stakeholder's requirement [9]. In other words, software testing is a process of checking that the software product behaves well and provides information to the stakeholders about the quality of the product. Hence, it is essential to have good quality software.

3.1 Complexities in Testing Methods

Andrew Austin and Laurie Williams emphasized on the importance of discovering the vulnerabilities in the software [10]. The reason is security vulnerability is very expensive if it was not discovered during the development life cycle [11]. According to Gary McGraw, he wrote in his book, *Software Security: Building Security In*, claimed that "Security problem evolve, grow, and mutate, just like species on a continent. No one technique or set of rules will ever perfectly detect all security vulnerabilities" [12]. Therefore, software developers should strive to discover the vulnerabilities as soon as possible. Rodrigo Souza and Christina Chavez have done an experiment to characterize the verification of bugs found in two open sources Integrated Development Environment (IDE's) using the data available from bug the repositories. However, those bug reports are found to be inaccurate, inconsistency, misleading and ineffective [13]. It also affects the quality of the software products. It is found that the verification of the code was done through static analysis tool. This tool is not the favorable testing methods for any software projects.

Sarah Al-Azzani and Rami Bahsoon have presented a novel on the approach of using implied scenario detection for security testing with LTSA tool to test the system. In their model, they managed to detect few security threats which are closely related to one another. However, it is not clear whether this tool will detect all the security threats due to the dynamic of nature in any attacks [14]. The limitations imposed on this approach are the scalability of the LTSA tool used to detect implied scenarios. Samer Al-Zain et al. used an automation tool called Test Complete, to conduct testing for web application [15]. For any Windows applications, Web or Rich client software this tool manage to create and run automated tests that helps the test engineers to perform several types of automated tests. However, the test recorder tool shows some weaknesses when using GUI test recording, for dynamic web applications. Besides that, the tester must have the knowledge to provide the algorithm for writing robust and successful test scripts. Vesa Kettunen et al. have conducted a study, using qualitative grounded theory method, to find the differences in the testing activities between agile development methods and the traditional plan-driven approach [16]. It was found that the testing position had improved in the agile methods through early involvement of testing activities in the development life cycle and also noticed that a positive influence on end-product satisfaction. However, the study found that most of the agile methods did not give enough guidance on how testing should be aligned parallel with parallel with development work.

The benefits of using software quality metrics in SME's were explored by Haddad et al. [17] and outlined a practical framework using Goal/Question/Metric paradigm for adopting metric programs in SME's. During the process, it has been identified that project and product metrics contributes to software quality and reliability that shared a common goal. Nevertheless, the utilization of these metrics has found to be minimal or yet to be adopted by most of the SME's. The reason is, managers and the developers might have understood the high-standards for improvement on software process and metrics programs, but for various reasons they often unwilling to expand the necessary effort. Zaineb and Manarvi [18] had stated the importance of bug identification report in their research study and explored the severity level of bugs and the status associated in it. They found that technical limitations, incomplete steps reported in the bug identification report because of the poor quality of information stated in the report and insufficient knowledge of testers over the developed software are the major problems causing bug rejection. Further in their study, they concluded that rather than spending too much of time over reviews and rework to fix the bugs, it is recommended to provide necessary training to the test team before deploying the test release.

Beer and Ramler in their research study had explored the role of tester's experience towards the development of test cases in relation to regression testing and automated testing using three case studies. It was found that the tester should focus more towards the domain knowledge which is crucial in testing activities, as against the testing knowledge [22]. Kettunen et al. (2010) had also reported in their studies that tester should have the domain knowledge about the developed product and further emphasized the importance of the role of technical knowledge, especially in the context of agile development process [16]. Prakash and Gopalakrishnan emphasized on the contrasting approach in the scripted testing vs. exploratory testing and revealed that latter approach was much more helpful to find some important bugs very quickly. This clearly indicates that in the scripted testing the amount of time taken to detect a bug is longer than that in the exploratory testing [19]. By giving freedom to the tester resulted in a quick identification because the bugs were spreading across the entire application.

Juha Itkonen et al. (2007) have conducted an experiment in comparing the fault detection using test case based testing and exploratory testing [20]. In their experiment, they have found that Test Case based Testing (TCT) provides more false positive reports than exploratory testing. It was because test case based testing techniques are based on theories and it does not help to find the bugs in all the software products.

Juha Itkonen et al. (2009) had also conducted another research study to supports the hypothesis regarding various techniques and strategies employed during test execution rather than mechanically relying on test documentation [21]. Execution time techniques are found to be strongly experienced based, however, it has been applied in non-systematic fashion during test execution which is similar to test case design techniques. However, currently there are no procedure, methods or standards that exist to carry out the test execution.

As of now, a wide variety of techniques, tools and automation strategies have been developed to make testing more efficient and effective. Despite the availability of different solutions, the fundamental challenges of software testing are how to reveal the new defects in newly developed software or after major modifications in the existing software. It is still in practice that software testing largely dependent on the performance of human testers manually. Furthermore, the existing research on the role of experience in software testing raises the hypothesis that experience and domain knowledge of the tester are equally important rather than depending on scripted testing or test automation entirely. However, there are no empirical studies exist so far. Based on the above discussions, it is obvious that bugs are a significant source of the problem in developing stable and error-free software.

3.2 Why Exploratory Testing?

The two main ideas of testing are (i) based on the composition of the software, i.e. its internal structure, and (ii) based on the business or intended purpose of the software, i.e. the functional aspect of the software [23].

Based on the studies, it is evident that though many testing techniques and methods are existed, none of them can guarantee that the product is free from error, especially in the SME's when there are no standard practices adopted for testing procedure. And most of the testing is done by the developer during the development stage itself. From the literature, it is obvious that neither the testing methodology nor the tools adopted by the small software company will be able to identify the defect. Developers in the SME's write unit tests for their code because of the agile technology in the software industries.

In contexts of software development, it is crucial to ensure that the product fulfill the requirements of the users and the stakeholders with the help of professional experts and application domain experts. Hence, exploratory software testing has been proposed as an effective testing approach in this context. Incidentally, exploratory testing (ET) approaches and the traditional software testing methods differ significantly from each other. ET is not based on predesigned test cases and test documents rather it is creative and experience-based approach in which test design, execution, and learning are conducted in parallel [24]. Exploratory testing approach is a free form procedure by which the tester simultaneously explores and learns the system while testing. Nevertheless, there is still systematic structure in testing to be followed, which leads to repeatable and measurable outcomes and the results of the executed tests are immediately applied for designing further tests. This type of testing is best suited for agile development process especially in SME's industries. Thus, ET is an important complementary activity to other testing methods.

4 Research Focus

The main goal of this research is to propose a framework for Exploratory Testing in small and medium size software enterprises. There are many types of testings exists and each testing methods has its own significance and weakness. In Agile development process, the testing activities become more complicated as there will be frequent changes in the requirement from the stakeholder, which in turn the developer has to change the code and the testing strategies must repeat again. This will lead to delay in the completion of the project or it undergoes poor testing which results in the project delivered with bugs. In SME's, especially in the unstructured organization, when there is no proper documentation and guidelines for testing is exits, then chances of having bugs in the product is inevitable due the frequent changes in the requirement.

The focus of the research is towards how testing is done in SME's using Agile development process and how to apply Exploratory Testing in SME's to ensure that the product is delivered with free from errors. In Fig. 2, various testing methods and agile processes are highlighted. In the agile development process, only few type of testing methods like regression testing, unit testing and functional testing are best suitable.

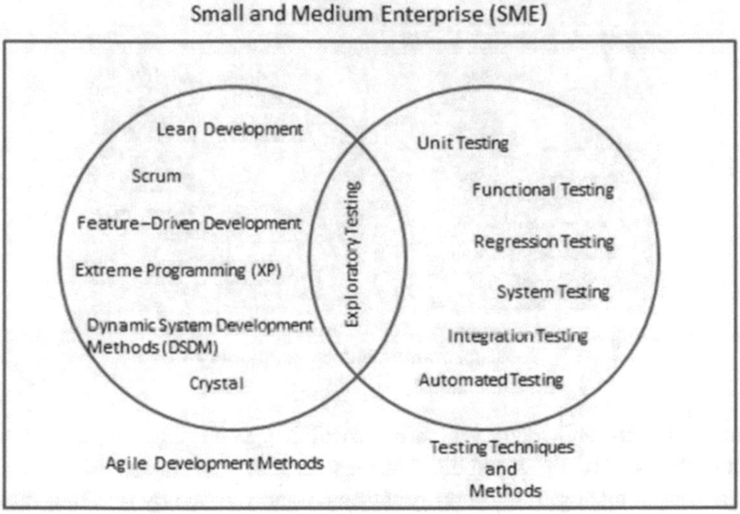

Fig. 2. Research focus

Also the testing methods used by SME's are either they use automated testing – using open source tool - or manual testing without having any test cases or test documentation. Mostly the testing is done by the developer and there may not be any separate testing (Quality Assurance) team to conduct the test. Hence, the new testing approach called Exploratory testing is proposed with the aim to produce quality software by integrating all the three domains – Testing, Agile process and SME's.

Exploratory Testing Framework

Definition: *A framework is a set of rules, goals, facts or assumptions that provide support to achieve something (or) a set of ideas, principles, agreements, or rules that provides the basis or outline for something intended to be more fully developed at a later stage.*

The Fig. 3 is the visual representation to plan and control the exploratory testing processes in general, once the conceptual framework has been evolved and it may be adopted exactly as such or with some little modifications, if successful, in SME's. This approach helps to find bugs with the simplest sequence of events, making diagnosis and bug advocacy easier.

Input combinations and constraints are the dominant factor for any testing methods. In this case, it may include data validation, verification, boundary value analysis etc., and other input combinations are also taken into consideration. The next step can be the data flow and control flow aiming to test the business logic in a prescribed manner. This helps to identify all the logical paths that are working as expected which the system aimed to serves its purpose.

Performance and Efficiency tests are aimed at where all the data goes through the system pushing to the maximum possible amount through each field and look for truncation or other forms of interruption. Various types of testing are stress testing, load

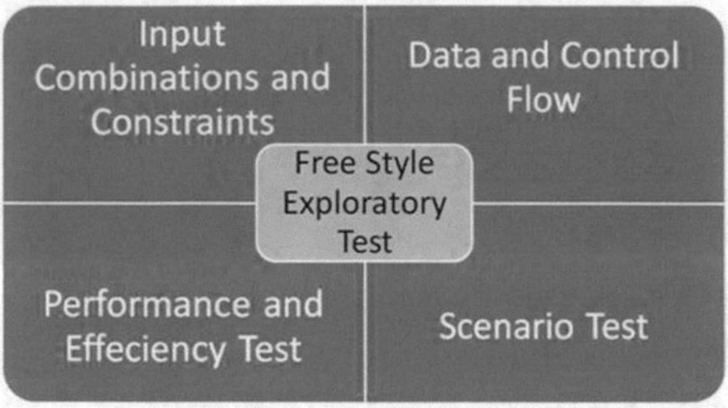

Fig. 3. A visual framework on exploratory testing

testing, sanity testing and many others are available. It all depends upon the product element and the product environment. The testers may choose any one of them.

Scenario based testing, individual functions are tested one by one that replicate the behavior and execute large number of various scenarios to simulate real user behavior over a longer period.

Free Style Exploratory test is the main aspect of ET where the testers use their intellectual skills and employ the testing procedures. The tester tends to use his/her full knowledge of the system to do things like looking for race conditions, or multiple concurrent login on the same account, or edit URLs etc.

The strength of automated testing is the fact that the tests can be run often and never get tired, which is undeniable. However, it is expected to find problems that they are expected to find, according to the test cases. Rather, people who have the domain knowledge about the application can surprisingly find new problems that automated tests are incapable of revealing it. Hence, Exploratory Testing aims at improving the quality of the software by revealing the software bugs at the early stage to be fixed before the software is released to the customers and the stakeholders. Any vulnerability in the system may come to know only after the product is in use. In such a case, ET helps to detect the vulnerability in the system which could not able to find when the system is under development stages.

The Benefits of ET Framework

- It reduces cost and save time
- Able to detect the vulnerabilities in the system
- Use the resources efficiently towards testing
- Better understanding of the domain knowledge on the product by the tester,
- To ensure customer satisfaction with reliability and efficiency in the system

5 Conclusion and Future Work

In most of the SME's, there is no tester or dedicated test team available to perform the testing activities exclusively, and therefore all the testing activities are performed by the developers. Hence exploratory testing, if used appropriately, will allow testers working on projects in SME to begin testing the system sooner, and also help them to avoid the pitfalls of intentional blindness. None of the tools or approaches can guarantee to solve all the problems and thus, exploratory testing provide clear value for the most of the projects developed in SME's when they do not have proper testing process. This research work will continue further to develop a theoretical and conceptual framework to help solve their bug identification process in a simple way to all unstructured small and medium size software company, where by the quality and standard of the system are achieved.

References

1. Popović, S., Lazić, L., Mastorakis, N.E.: Orthogonal array and virtualization as a method for configuration testing improvement. In: First IEEE Eastern European Conference on the Engineering of Computer Based Systems, pp. 148–149 (2009)
2. de l'Annee de Betrancourt, A.L.P., Lamers, P.R.: Developing a suitable structured testing approach for a small web development company (2004)
3. Tosun, A., Bener, A., Turhan, B.: Implementation of a software quality improvement project in an SME: a before and after comparison. In: Proceedings of the 35th Euromicro Conference on Software Engineering and Advanced Applications, pp. 203–209 (2009)
4. van Zyl, J.: Software testing in a small company: a case study. Technical report, University of Pretoria (2010)
5. Pino, F.J., Garcia, F., Piattini, M. (eds.) Key processes to start software process improvement in small companies. In: SAC 2009: Proceedings of the 2009 ACM Symposium on Applied Computing, pp. 509–516 (2009)
6. Von Wangenheim, C.G., Weber, S., Hauck, J.C.R., Trentin, G.: Experiences on establishing software processes in small companies. Inf. Softw. Technol. **48**(9), 890–900 (2007)
7. Moritz, E.: Case study: how analysis of customer found defects can be used by system test to improve quality ICSE (2009)
8. Kaner, C.: Fundamental challenges in software testing (2003). http://www.testingeducation.org/a/fundchal.pdf
9. Shoaib, L., Nadeem, A., Akbar, A.: An empirical evaluation of the influence of human personality on exploratory software testing (2009)
10. Austin, A., Williams, L.: One technique is not enough: a comparison of vulnerability discovery techniques (2011)
11. Boehm, B.: Software Engineering Economics. Upper Saddle River, Printice (Prentice) Hall (1984)
12. McGraw, G.: Software Security: Building Security. Pearson Education, Boston (2006)
13. Souza, R., Chavez, C.: Characterizing verification of bugs fixes in two open source IDEs, MSR (2012)
14. Al-Azzani, S., Bahsoon, R.: Using implied scenarios in security testing. In: SESS 2010, Cape Town, South Africa, May 2010

15. Al-Zain, S., Eleyan, D., Garfield, J.: Automated user interface testing for web applications and test complete. In: CUBE 2012, Pune, Maharashtra, India, 3–5 September 2012 (2012)
16. Kettunen, V., Kasurinen, J., Taipale, O., Smolander, K.: A study on agility and testing processes in software organizations. In: ISSTA 2010, Trento, Italy, 12–16 July 2010 (2010)
17. Haddad, H.M., Ross, N.C., Meredith, D.E.: A framework for instituting software metrics in small software organizations. Int. J. Softw. Eng. IJSE **5**(1) (2012)
18. Zaineb, G., Manarvi, I.A.: Identification and analysis of causes for software bug rejection with their impact over testing efficiency. Int. J. Softw. Eng. Appl. (IJSEA) **2**(4) (2011)
19. Prakash, V., Gopalakrishnan, S.: Testing efficiency exploited: scripted versus Exploratory testing (2011)
20. Itkonen, J., Mantyala, M.V., Lassenius, C.: Defect detection efficiency: test case based vs. exploratory testing. In: First International Symposium on Empirical Software Engineering and Measurement (2007)
21. Itkonen, J., Mäntylä, M.V., Lassenius, C.: How do testers do it? an exploratory study on manual testing practices. In: Third International Symposium on Empirical Software Engineering and Measurement (2009)
22. Beer, A., Ramler, R.: The role of experience in software testing in practice. In: Proceedings of Euromicro Conference on Software Engineering and Advanced Applications, pp. 258–265 (2008)
23. Meijberg, Y.: Time for testing at an intermediate Dutch SME, B.Sc. thesis, University of Twente (2008)
24. Bach, J.: Exploratory testing. In: van Veenendaal, E. (ed.) The Testing Practitioner, pp. 253–265. UTN Publishers, Den Bosch (2004)
25. Thangiah, M., Basri, S.: A Preliminary analysis of various testing techniques in Agile development – a systematic review. In: 3rd International Conference on Computer and Information Science (2016)

Author Index

F. Saeed et al. (Eds.): IRICT 2018, AISC 843, pp. 1123–1126, 2019.
https://doi.org/10.1007/978-3-319-99007-1

Printed in the United States
By Bookmasters